# Fisheries Development in India

## The Editors

**J.V.H. Dixitulu** has over 60 years of varied experiences in the fisheries sector. In this context, it may be mentioned that there was the bestowal of Life Time Achievement Award on him by the Asian Fisheries Society (Indian Branch), Delhi and also by Pillai Aquaculture Foundation, Mangalore, in recognition of his services and achievements in the fisheries sector.

Dixitulu was the architect of Fish Farmers Development Agencies (FFDA) of India. He formulated the Project on FFDA and secured the approval of the Government for the setting up of these Agencies, besides also organising the setting up of the first batch of them. He accomplished this at the time he functioned as the Deputy Commissioner (Fisheries) in the Union Ministry of Agriculture. So far, around 430 nos of these agencies have been set up in the various States of India. These Agencies now provide an integrated mechanism for an organised farm fish production. These Agencies have upgraded fish production per ha in India to a level of 2,500 kg/ha on an average from the pre –FFDA level of 600 kg/ha. In the post-independence era several fish seed and fish farms were also set up by him in Andhra Pradesh. He developed successfully the concept of relay seed farms for transporting seed over long distances for supplies to farmers in distant and inaccessible fish tanks in the State. The brackishwater fish farm set up by him at Kakinada in A.P. and the freshwater farms, set up by him at Kadiam and Balabhadrapuram, also in A.P, virtually turned out to be the centres from where the spread of freshwater fish farming took off in A.P, and blossomed to its present exalted status, over the years. The aforesaid farms at Balabhadrapuram and Kakinada were later taken over by the Central Institute of Fisheries Education, Mumbai, because these were found by the Institute to be well suited for imparting training to its students and also to fish farmers. He was also instrumental in the setting up of a Co-operative Fisheries Project for the development of Fisheries of Konaseema area in East Godavari Dist, A.P. State. He also played a pivotal role in the setting up of the South Canara Co-operative Fisheries Project for implementation in S.K. District of Karnataka State.

He earned the unique distinction of planning and introducing in Indian EEZ the first batch of 28 nos of multipurpose deep sea fishing vessels of 23m OAL range in late 1970. This development transformed the complexion of Indian sea fishing, taking it beyond the coastal waters to the offshore and deep sea zone and generating export-oriented fishing operations.

Impelled by a missionary zeal to extend his vast experiences among those engaged in the fisheries developmental activities, Dixitulu ventured into the till then uninitiated field of Indian Fisheries Journalism. He set up in 1981 the first Indian monthly English fishery journal "*Fishing Chimes*" which is now in its 32$^{nd}$ year of publication. *Fishing Chimes* has, over the years, become the most respected journalistic voice of India in the field of fisheries. His editorials in the journal on the various facets of Indian fisheries have been hailed as highly objective, constructive and thought-generating. These editorials (408 of them) are reproduced in this publication. As a recognition of this unprecedented achievement in Indian fisheries journalism, he was also conferred with an Award in 1997 by the Oriental Research Foundation, Bhopal.

**Eashwar Dev Anand** holds a First class Masters Degree in Biotechnology. He started working full time as Managing Editor of *Fishing Chimes* from January 2010. He is currently pursuing his Ph.D in Aquaculture, while taking part in the day to day activities of *Fishing Chimes.* In 2010, after joining the organisation, he started publishing a Monthly aquaculture journal with the name *Chepala Sandadi* in Telugu language. This journal now caters to the needs of the aqua farmers of Andhra Pradesh. The pattern of the contents of the journal is unique in that its objective is to serve as a medium to keep the fish farmers updated with the advances in aquafarming technologies that keep taking place from time to time and, at the same time, bringing them closer to aqua farm scientists, their field and lab work and also to the farm input suppliers. As of today, around eight farming companies have become part of this initiative of his, and are working closely with *Fishing Chimes* and *Chepala Sandadi,* thereby transferring information on the latest developments in the fisheries sector among fishery entrepreneurship.

Working in the Telugu journal has given Eashwar Dev Anand the opportunity to interact with many fish farmers of A.P on a regular basis, as a result of which he has been fine tuning the presentations in both *Fishing Chimes* and *Chepala Sandadi.* During the past two years, *Fishing Chimes* has printed and distributed around 60,000 of its copies among the farming community of Andhra Pradesh, along with a special edition on Better Management Practices for their benefit.

Eashwar Dev Anand, has been keen on bringing out this Compilation of the Editorials published in *Fishing Chimes* for the period of 30 years (1981–2011) and thereafter for the benefit of Indian fisheries development. For him, working on this compilation has been like partaking on a journey backwards through time, to understand how the industry has evolved during the period and the issues that confronted the industry from time to time. It is noteworthy that several important issues encountered during the thirty years (1981–2011) have still not been resolved in a final form. The author hopes that this compendium would serve as a further enlightening guide for the industry, for actively moving towards finding implementable solutions for the faster development of Indian fisheries.

# Fisheries Development in India

[Editorial Appraisals in *Fishing Chimes* (April 1981 to March 2011)]

*An Indispensable Desk-Side Companion for Professionals, Administrators, Entrepreneurs, Universities, Colleges, Students, Scholars, Scientists, Technologists, Financing Bodies, Voluntary agencies, etc.*

**J.V.H. Dixitulu**
(Chief Editor, *Fishing Chimes*)
&
**Eashwar Dev Anand**
(Managing Editor, *Fishing Chimes*)

**2013**
**Daya Publishing House®**
*A Division of*
**Astral International Pvt. Ltd.**
**New Delhi – 110 002**

ISBN 9789351240518



*Published by* : **Daya Publishing House®**
***A Division of***
**Astral International Pvt. Ltd.**
**– ISO 9001:2008 Certified Company –**
4760-61/23, Ansari Road, Darya Ganj
New Delhi-110 002
Ph. 011-43549197, 23278134
E-mail: info@astralint.com
Website: www.astralint.com

*Laser Typesetting* : **Classic Computer Services**
Delhi - 110 035

*Printed at* : **Salasar Imaging Systems**
Delhi - 110 035

PRINTED IN INDIA

# Acknowledgements

At the outset, the editors avail of the opportunity of thanking all contributors of articles etc., and subscribers, advertisers and the overall readership for their gracious support to *Fishing Chimes* all these years.

The Advisory Board of *Fishing Chimes* has seven scientists of global reputation. These are: Michael New, (Aquaculture without Frontiers), M.V. Gupta (World Fisheries Laureate); Mohan Joseph Modayil (Member, Indian Agriculture Scientists Recruitment Board), S.D. Tripathi (Retired Director from CIFA and CIFE of India and Chairman/ member of several Fisheries Committees), Mohan Kumaran Nair (Dean, College of Fisheries, Kochi and also Pro Vice-Chancellor, Kerala Fisheries University), I Karunasagar, presently with FAO (as a fisheries expert) and Venkatesh Salagrama (International Fisheries Expert). The Chief Editor avails of this occasion to thank them all for their support to the Journal. S. Jayaraman, presently Professor of Fisheries at the College of Fisheries, Tuticorin, and M. Sathiraju, Deputy Director of Fisheries (Retd), A.P. Fisheries Department, were devotedly helpful to the Journal as its fisheries experts, particularly in the formative period of the Journal. The editors convey their grateful thanks to them. *Fishing Chimes* has three Resident Editors. These are 1) Mr. H.S. Badapanda in Bolangir, Orissa, 2) D.B. James in Chennai, Tamil Nadu and 3) D.D. Nambudiri in Cochin, Kerala. They have been extending devoted support to the Journal and the editors thank them for their support. K. Vijayakumaran, the present Director-General of Fishery Survey of India is the Associate Editor (Hon) of the Journal. Apart from sharing his expertise with the journal in that position, his background is that, earlier, in an honourary position, he had contributed several popular fisheries articles to the Journal. H.N. Chandrasekhariah, Joint Director of Fisheries (Retd.), Karnataka Fisheries Department, V.S. Durve, Professor (Retired), College of Fisheries, Udaipur, Rajasthan, and K. Simhachalam, Retired Fisheries Development Officer, A.P. Fisheries Department, besides several others have been supporting the journal with their precious inputs. J. Santhilakshmi, the representative of *Fishing Chimes* in New York, USA, provided certain inputs and help in the finalisation of the contents of this publication for which the authors extend their gratitude. On this occasion, it is our pleasure to recall and acknowledge our association with Devashish Kar, Professor, Division of Wetlands, Fishery Science and Aquaculture, Department of Life Science, Assam (Central) University, Silchar, in the past couple of decades. Our association with him has provided to '*Fishing Chimes*' considerable information on the water bodies and fishes of North-East India.

Mr. G. Ramakrishna, Managing Director, Raamakrishna Printers Private ltd, Visakhapatnam, has been taking care of the printing of *Fishing Chimes* all these years more or less continuously, thereby facilitating distribution of the well printed monthly issues of the journal successively among its readership over all these years without any break and in a smooth and commendable manner. The Editors extend their grateful thanks to him and all his associates at the Press for their gracious support.

As an expression of their gratitude, the authors convey to Mr. Nandikala Venkata Ramana of *Fishing Chimes* their happiness for his assistance in the compilation of the contents of the publication.

**J.V.H. Dixitulu**
*Chief Editor*
*Fishing Chimes*

**Eashwar Dev Anand**
*Managing Editor*
*Fishing Chimes*

डा. एस. अय्यप्पन
सचिव एवं महानिदेशक
Dr. S. AYYAPPAN
SECRETARY & DIRECTOR GENERAL

भारत सरकार
कृषि अनुसंधान और शिक्षा विभाग एवं
भारतीय कृषि अनुसंधान परिषद
कृषि मंत्रालय, कृषि भवन, नई दिल्ली 110 114

GOVERNMENT OF INDIA
DEPARTMENT OF AGRICULTURAL RESEARCH & EDUCATION
AND
INDIAN COUNCIL OF AGRICULTURAL RESEARCH
MINISTRY OF AGRICULTURE, KRISHI BHAVAN, NEW DELHI 110 114
Tel.: 23382629; 23386711 Fax: 91-11-23384773

# Foreword

The Oriental Foundation of India, under the Chairmanship of Dr. S.N. Dwivedi, honoured Dixitulu in 1997 with an Award and citation recognizing him as the First Fisheries Journalist of India. This is also a reflection of the regard myself and many others hold towards him in various respects including his fisheries journalistic talent in this background, this is a unique occasion for me to highlight the main aspects of this Compilation of 407 fisheries–oriented editorials published in the monthly issues of *Fishing Chimes* released from April, 1961 to March, 2011.

The journal, over the years has reached the/status of a comprehensively informative monthly Fisheries Journal. The Fisheries establishments of the Union Department of Animal Husbandry, Dairying and Fisheries, MPEDA and its offices, ICAR and its fisheries research institutions, the Fisheries Universities and Colleges, besides quite a few other Universities and Colleges in the country, and several fisheries enterprises, are long standing subscribers to the journal, besides several of their scientists/technologists being contributors of articles to it. Equally as important as these are the subscriptions and contributors to the Journal from the State/Union Territory Fisheries Departments, Corporations and Co-operatives in the country. The Journal has international recognition and circulation too, with FAO being one among these.

The impressive circulation of *Fishing Chimes* stems from its comprehensive subject coverage and contents. It can be said that it serves as a medium for dissemination of scientific, technological and other fisheries developmental information gathered/received by the Journal from, various fisheries departments, fisheries scientists/technocrats, fisheries scholars, fish farmers and fisheries field executives.

I have to mention here that, as it is, there are quite a few fisheries magazines/newsletters/journals published within the country. They are published in the form of annual/half yearly or as other categories of periodicals. As one among them, *Fishing Chimes* concentrates on selective publication of contributions of fisheries technological/scientific importance, authored by those who play an active role in the development of fisheries sector of the country and some abroad, besides reports on various fisheries conferences, symposia, seminars, workshops, training programmes etc , that take place This approach has been well received by the readership, particularly fisheries professionals, including those in the industry and also fisheries P.G. and Ph.D. students of the various educational institutions.

As already mentioned, *Fishing Chimes* has come out so far with 407 editorials in its successive monthly issues up to March 2011, as contained in this publication. The editorials published in the journal (from April, 1981 to March, 2011) in a book form, are likely to benefit all those in the field of fisheries and those who are contemplating to enter the line. I am convinced that this compilation would be welcomed and well received by all concerned to whom the contents of the publication will certainly be very useful.

*Dated the 13th January, 2012*

**S. Ayyappan**

*Secretary, Department of Agricultural Research & Education and*
*Director General, Indian Council of Agricultural Research,*
*Ministry of Agriculture, Krishi Bhavan, New Delhi – 110 114*

# Preface

The inaugural number (April 1981) of *Fishing Chimes,* containing the first editorial of the journal, was released on 27 March, 1981. This was followed by the publication of 407 successive monthly editorials in its issues from April 1981 to March 2011. All these have been reproduced in this Publication. The important fisheries developmental aspects that took place in the country during the aforesaid period from time to time were evaluated in these editorials with a critical stance. Several articles of scientific, technological and of developmental importance, authored by various fisheries experts also and news items were published in the aforesaid issues, besides the editorials. All these constituted the bulk of the contents of the successive issues of the journal. An aspect of importance that may be mentioned here is that, apart from the fisheries experts referred to above, several enthusiastic, devoted and budding fisheries scientists and technologists, with knowledge and experience on various aspects of fisheries have given to *Fishing Chimes* opportunities of publishing articles and reports contributed by them. In fact, in the monthly issues of the Journal, brought out in the duration of 30 years (April 1981 to March 2011) of its publication, over 3,000 articles on subjects of fisheries scientific / technological importance were also published, besides the editorials and notes on important developments in the fisheries sector of the nation.

The editorial in the inaugural issue of the journal (April 1981) was on the topic, Financing of Fishing vessels, titled, 'Can we not follow the US System?'. This was followed successively by editorials on other thought-provoking topics in the following monthly issues, one of which was related to the setting up of a separate Union Ministry of Fisheries. All these editorials (in the issues from April 1981 to March 2011) are reproduced in full in this publication. The captions of all these are given in a separately indexed list [p i to xvii] to facilitate their location with ease.

Captions of some of the developmental aspects covered in the editorials of *Fishing Chimes* in some of the years of the first decade of publication of the journal are briefly summarised hereunder as examples of the complexion of the general run of the editorials: 1) Tough Taiwanese (May 1981 p. 86 to 87); 2) Separate Ministry for Fisheries at Centre needed; (June 1981 and July 1982 (P. 3-4); 3) Fish wealth in wrong nets, poachers pervade (June 1981 (p.7 and 9.); 4) Listen to BOBP, take to 2-boat trawling (June 1981, p.10 and 30); 5) Feasibility of Fisheries Co-operatives in Punjab (June 1981, p.11 and 13 to 15); 6) Cage Culture in Punjab and Haryana (June 1981 p.23 and Sept 1981 p.1); 7) Trawler Building in Kerala – (June 1981 p.24); 8) Constraints in Deep sea fishing development (July 1981 p.4 and 9 to 10); 9) Prospects of indigenous trawler construction, (Sept 1981 P. 33 and 53); and many others such as, 10) Hilsa thrives in Vallabh Sagar Reservoir in Gujarat, (June 1982 p.13 and 14); 11) CMFRI achieves record growth of *Penaeus indicus* in a paddy field in a farming duration of 47 days along with paddy (Sept 1982 p.17); 12) Massive fish seed production in pens set up in Tungabhadra, Karnataka, (Oct 1982, p.42 to 45); 13) Jar hatchery set up by CIFRI at Poongar farm in T.N. (Nov 1982 p. 17 to 18); 14) Floating hatcheries and nurseries go operational in Volga river in USSR – (Nov 1982 p.53); 15) Tapping underground brackishwaters in Punjab and Haryana under a central scheme (Dec 1982 p.5); 16) Cage Culture at Katli village in Punjab in river Sutlej, (Jan 1983 p. 27). In September 1998, *Fishing Chimes* wrote on introduction of Sea Cage farming in India. This work has been of course taken up by CMFRI on its own, and we can expect the realisation of this advanced system of farmed sea fish production on commercial lines very soon with good results. *Fishing Chimes* also suggested in one of its editorials the development of Pond Farming of Hilsa. This suggestion was made on the occasion of celebration of the 64$^{th}$ Foundation Day of CIFRI. In this context, there is now an Indication that CIFRI has been already working on this reform. There were quite a few of other suggestions too and *Fishing Chimes* hopes and has the confidence that the contents of this Book (publication) would be of immense interest and utility to its readership.

In the long history of Indian democracy, one would have rarely come across references to the subject of fishermen by any of the successive prime ministers of the Nation. However, *Fishing Chimes* came across atleast one rare exception. It recorded an occasion, when the then Prime Minister of India expressed his views on fishermen. These were presented in *Fishing Chimes* in the form of an editorial, published in the January 1986 issue of *Fishing Chimes.* While inaugurating a National Seminar on Integrated Development of Fisher community in New Delhi on 13 November, 1985, Mr. Rajiv Gandhi, the then Prime Minister of India, observed that there was need to demarcate fishing rights at sea. Another comment of his was related to the stranglehold of middlemen over fishermen and the need to form (fishermen's) co-operatives. It was also observed by the then Prime Minister on that occasion that an attempt would be made to improve the status of individual fishermen so as to give an industrial set up to the fisheries sector with some competition. He felt that this way both the industry and fishermen would prosper. (January 1986 issue of *Fishing Chimes*, p.5). His desire was followed up to some extent, but much more is yet to be done.

The Senior Author of this Publication is the recipient of Life Time Achievement Awards from i) Asian Fisheries Society, and ii) Pillai Aquaculture Foundation.

# Contents


***Acknowledgement*** ***v***

***Foreword*** ***vii***

***Preface*** ***ix***

| *Sl.No.* | *Category* | *Name* | *Editorial Captions* | *Vol. No.* | *Issue No.* | *Month-Year* | *Page No.* |
|---|---|---|---|---|---|---|---|
| 1 | Marine Fisheries | Financing | Financing of Fishing Vessels Can We not follow U.S. System? | 1 | 1 | Apr-81 | 1 |
| 2 | Marine Fisheries | Organisational | Upgradation of EFP into a Deep Sea Fishery Development Corporation Imperative | 1 | 2 | May-81 | 3 |
| 3 | Marine Fisheries | Charters | Chagrin over Charters | 1 | 3 | Jun-81 | 5 |
| 4 | Marine Fisheries | Organisational | Constraints in Deep Sea Fishery Development | 1 | 4 | Jul-81 | 6 |
| 5 | Marine Fisheries | Developmental Hues | Kerala Marine Fisheries Sector in a Ferment | 1 | 5 | Aug-81 | 7 |
| 6 | Marine Fisheries | Developmental Hues | Deep Sea Fishing in Blunderland | 1 | 6 | Sep-81 | 8 |
| 7 | General | Administration | Anachronistic Administration: Plight of Fisheries Sector | 1 | 7 | Oct-81 | 10 |
| 8 | | Field level including marketing | Fisherman | 1 | 8 | Nov-81 | 12 |
| 9 | General | Administration | To Get in, Get on and Get out | 1 | 9 | Dec-81 | 13 |
| 10 | General | Review | Into Nineteen Eighty Two | 1 | 10 | Jan-82 | 15 |
| 11 | General | Review | Develop India into a Fishing Nation | 1 | 11 | Feb-82 | 17 |
| 12 | General | Review | Encouraging Signs | 1 | 12 | Mar-82 | 19 |
| 13 | General | Review | Inter-State Interactions | 2 | 1 | Apr-82 | 20 |
| 14 | Inland fisheries | Culture/Farming | Fillip to Fish Farming | 2 | 2 | May-82 | 22 |
| 15 | Marine Fisheries | Developmental Hues | Don't be Fishy | 2 | 3 | Jun-82 | 25 |
| 16 | Marine Fisheries | Organisational | Trash into Cash | 2 | 4 | Jul-82 | 27 |
| 17 | Marine Fisheries | Developmental Hues | The Confrontation | 2 | 5 | Aug-82 | 28 |
| 18 | Marine Fisheries | Developmental Hues | Diversification has Diverse Problems | 2 | 6 | Sep-82 | 30 |
| 19 | Marine Fisheries | Developmental Hues | Deeps of Indian Deep Sea Fishing | 2 | 7 | Oct-82 | 32 |

| Sl.No. | Category | Name | Editorial Captions | Vol. No. | Issue No. | Month-Year | Page No. |
|---|---|---|---|---|---|---|---|
| 20 | Marine Fisheries | Developmental Hues | The Elusive Three Fifty Fishing Vessels | 2 | 8 | Nov-82 | 35 |
| 21 | Marine Fisheries | Developmental Hues | The Execution | 2 | 9 | Dec-82 | 37 |
| 22 | Marine Fisheries | Charters | Arbitrariness Stains Latest Charter Rules | 2 | 10 | Jan-83 | 38 |
| 23 | General | Fisheries Administration | Need to Revamp Fisheries Administration in the States | 2 | 11 | Feb-83 | 40 |
| 24 | Marine Fisheries | Organisational | Technocrats and Deep Sea Fishing | 2 | 12 | Mar-83 | 42 |
| 25 | Marine Fisheries | Developmental Hues | Testing through Toughness | 3 | 1 | Apr-83 | 43 |
| 26 | Marine Fisheries | Developmental Hues | The Tuna Tangle | 3 | 2 | May-83 | 44 |
| 27 | General | Review | Pull Out Sea Nets from Failures | 3 | 3 | Jun-83 | 45 |
| 28 | Marine Fisheries | Developmental Hues | In the Horns of a Dilemma | 3 | 4 | Jul-83 | 47 |
| 29 | Marine Fisheries | Developmental Hues | Ministers Miss Main Issues | 3 | 5 | Aug-83 | 49 |
| 30 | Marine Fisheries | Charters | Charter Charade | 3 | 6 | Sep-83 | 51 |
| 31 | Marine Fisheries | Developmental Hues | All Out but One | 3 | 7 | Oct-83 | 53 |
| 32 | General | Fisheries Education | Fisheries Education | 3 | 8 | Nov-83 | 54 |
| 33 | Marine Fisheries | Developmental Hues | Indigenous Deep Sea Fishing Vessels | 3 | 9 | Dec-83 | 56 |
| 34 | Marine Fisheries | Developmental Hues | Shrimps are not Scrimpy | 3 | 10 | Jan-84 | 57 |
| 35 | Marine Fisheries | Developmental Hues | Deep Sea Fishing : Salvation in Sight? | 3 | 11 | Feb-84 | 59 |
| 36 | Marine Fisheries | Developmental Hues (indigenous construction) | Silver Lining | 3 | 12 | Mar-84 | 60 |
| 37 | Marine Fisheries | Developmental Hues | Access to Foreign Fishing Vessels | 4 | 1 | Apr-84 | 62 |
| 38 | Marine Fisheries | Developmental Hues | Baseless Diatribe | 4 | 2 | May-84 | 64 |
| 39 | Marine Fisheries | Charters | Charter Chariot moves into Legal Orbit | 4 | 3 | Jun-84 | 66 |
| 40 | General | Administration | Delegation Devoid of Development Commissioner | 4 | 4 | Jul-84 | 69 |
| 41 | Marine Fisheries | Developmental Hues | Shaping of Policy | 4 | 5 | Aug-84 | 70 |
| 42 | Inland Fisheries | Aquaculture/ Farming | India Lags Behind in Shrimp Culture | 4 | 6 | Sep-84 | 72 |
| 43 | Marine Fisheries | Developmental Hues | Futility of Fishing Legislation based on Depth | 4 | 6 | Sep-84 | 73 |
| 44 | Marine Fisheries | Developmental Hues | Deep Sea Fishing Vessels in 6th Plan | 4 | 7 | Oct-84 | 75 |
| 45 | Marine Fisheries | Charters | Charter Scape | 4 | 8 | Nov-84 | 76 |
| 46 | Marine Fisheries | Developmental Hues | Anti-Outrigger Syndrome | 4 | 9 | Dec-84 | 78 |
| 47 | Marine Fisheries | Developmental Hues | Acquisition of Fishing Vessels under 100% Export- Oriented Scheme. There are Two Windows, Not one | 4 | 10 | Jan-85 | 80 |
| 48 | General | Field level including marketing | Utilisation of TV Network to Achieve Regulated Fish Supply | 4 | 11 | Feb-85 | 82 |
| 49 | Marine Fisheries | Developmental Hues | Increasing Marine Fish Production | 4 | 12 | Mar-85 | 83 |
| 50 | General | Review | Outlook under the New Government | 5 | 1 | Apr-85 | 85 |
| 51 | Marine Fisheries | Developmental Hues | Madras Fishing Harbour, an Orphan | 5 | 2 | May-85 | 86 |
| 52 | Marine Fisheries | Charters | Plus and Minus of New Charter Rules of April 1985 | 5 | 3 | Jun-85 | 87 |
| 53 | Marine Fisheries | Developmental Hues | Deep Sea Fishing Vessels: Target and Fullfilment | 5 | 4 | Jul-85 | 89 |

| Sl.No. | Category | Name | Editorial Captions | Vol. No. | Issue No. | Month-Year | Page No. |
|---|---|---|---|---|---|---|---|
| 54 | Marine Fisheries | Charters | Unending and Harsh: Detention of Chartered Vessels | 5 | 5 | Aug-85 | 90 |
| 55 | Marine Fisheries | Developmental Hues | Round the year Fishing on the West Coast | 5 | 6 | Sep-85 | 91 |
| 56 | Marine Fisheries | Organisational | Deep Sea Fishing Industry has two Windows now (With prospect of one more) | 5 | 7 | Oct-85 | 92 |
| 57 | Marine Fisheries | Financial | SDFC's Finances for Fishing Vessels: Present Hurdles Blocking Flow | 5 | 8 | Nov-85 | 94 |
| 58 | General | Field level including marketing | 'Salesmen' of Fisheries Development Needed | 5 | 9 | Dec-85 | 96 |
| 59 | Marine Fisheries | Developmental Hues | Prime Minister's Observations on Fishermen and Fishing Industry | 5 | 10 | Jan-86 | 98 |
| 60 | Marine Fisheries | Financing | Target without Adequate Funds | 5 | 11 | Feb-86 | 100 |
| 61 | Inland Fisheries | Aquaculture/ Farming | Potential of Closed Fish Farming System | 5 | 12 | Mar-86 | 102 |
| 62 | Marine Fisheries | Organisational/ Policies | Fishing Boundaries | 6 | 1 | Apr (i)-86 | 103 |
| 63 | Marine Fisheries | Organisational/ Policies | Closed Season for Shrimp | 6 | 1 | Apr (ii)-86 | 104 |
| 64 | Marine Fisheries | Financing | Fishing Vessels Financing : Setback and Solution | 6 | 2 | May (i)-86 | 106 |
| 65 | Marine Fisheries | General | Upkeep and Utilisation of Apprehen-ded Foreign Fishing Vessels | 6 | 2 | May (ii)-86 | 108 |
| 66 | Inland Fisheries | Aquacultural/ Farming | Blue Print for the Development of Fish Culture | 6 | 3 | Jun-86 | 109 |
| 67 | Marine Fisheries | Financing | Storm before Calm | 6 | 4 | Jul-86 | 110 |
| 68 | Marine Fisheries | Organisational/ Policies | Languishing Living Resources of Indian EEZ | 6 | 5 | Aug-86 | 112 |
| 69 | Marine Fisheries | Policies | Guarantees are the Need Now | 6 | 6 | Sep-86 | 114 |
| 70 | Inland Fisheries | Development Hues | Hitech Aquaculture as in the West: Implications of Introduction in India | 6 | 7 | Oct (i)-86 | 116 |
| 71 | Marine Fisheries | Developmental Hues | Erection of Barriers by Govt in the Devt of DSF, as Open Jokers (Fishery Entrepreneurs) Gape Aghast | 6 | 7 | Oct (ii) -86 | 118 |
| 72 | Marine Fisheries | Organisational/ Policies | Strategy to Achieve Diversifiction in Deep Sea Fishing | 6 | 8 | Nov (i)-86 | 120 |
| 73 | Inland Fisheries | Aquaculture/ Farming | Popularisation of Chinese Type of Hatcheries | 6 | 8 | Nov (ii)-86 | 121 |
| 74 | Marine Fisheries | Organisational/ Policies | Conflicting Strategies | 6 | 9 | Dec (i)-86 | 122 |
| 75 | Marine Fisheries | Developmental Hues | Imported Components in Indige-nously Built Deep Sea Fishing Vessels | 6 | 9 | Dec (ii)-86 | 123 |
| 76 | Marine Fisheries | Organisational/ Policies | New Deep Sea Fishing Policy: Rudderless and Without Focus | 6 | 11 | Feb-87 | 124 |
| 77 | Inland Fisheries | Organisational/ Policies | Amateurish Exercises at Enforcing Closed Fishing Season in Upper Bay of Bengal | 6 | 12 | Mar-87 | 126 |
| 78 | General | Review | On to the Seventh | 7 | 1 | Apr (i)-87 | 129 |

| Sl.No. | Category | Name | Editorial Captions | Vol. No. | Issue No. | Month-Year | Page No. |
|---|---|---|---|---|---|---|---|
| 79 | Marine Fisheries | Financing | Financing System for Deep Sea Fishing Vessels: Policy Options | 7 | 1 | Apr (ii)-87 | 130 |
| 80 | Marine Fisheries | Charters | Can Fourth Revision of Charter Policy be Avoided? | 7 | 2 | May-87 | 132 |
| 81 | Marine Fisheries | Developmental Hues | Imports under 100% Export Scheme at Standstill | 7 | 3 | Jun-87 | 134 |
| 82 | Marine Fisheries | Developmental Hues | Development of Riverine Fisheries | 7 | 4 | Jul-87 | 136 |
| 83 | Marine Fisheries | Development Hues | The Unequal Quadrangle | 7 | 5 | Aug (i)-87 | 137 |
| 84 | Inland Fisheries | Developmental Hues | A Challenge that has to be Met: *Beel* Fishery Development in Assam | 7 | 5 | Aug (ii)-87 | 138 |
| 85 | Marine Fisheries | Developmental Hues | Resources Specificity | 7 | 6 | Sep-87 | 139 |
| 86 | Marine Fisheries | Financing | No escape from Loan Guarantees | 7 | 7 | Oct (i)-87 | 140 |
| 87 | Marine Fisheries | Financing | Agricultural Licences and Market Fees on Marine Products in AP | 7 | 7 | Oct (ii)-87 | 141 |
| 88 | Marine Fisheries | Financing | Amendment to SDFC (Abolition) Act, 1986 | 7 | 8 | Nov (i)-87 | 142 |
| 89 | Inland Fisheries | Financing | Fish Culture in A.P: A Miracle at Work | 7 | 8 | Nov (ii)-87 | 144 |
| 90 | Marine Fisheries | Financing | Strategy for Development of Deep Sea Fishing | 7 | 9 | Dec (i)-87 | 146 |
| 91 | Marine Fisheries | Financing | Development of Reservoir Fisheries | 7 | 9 | Dec (ii)-87 | 148 |
| 92 | General | Administration | Separate Ministry for Fisheries Imperative | 7 | 10 | Jan (i)-88 | 149 |
| 93 | General | Administration | Effect of Drought on Fish Production: Relief Measures Must | 7 | 10 | Jan (ii)-88 | 151 |
| 94 | Marine Fisheries | Financing | Financing of Deep Sea Fishing Projects by Commercial Banks: Strategy Suggestions | 7 | 11 | Feb (i)-88 | 153 |
| 95 | Marine Fisheries | Financing | Compulsive Location of Fish Markets as Spinoff of Shrimp Famine | 7 | 11 | Feb (ii)-88 | 155 |
| 96 | Marine Fisheries | Developmental Hues | Sinking Deep Sea Fishing Industry Needs Rescue Measures | 7 | 12 | Mar-88 | 156 |
| 97 | General | Review | The Eighth Step | 8 | 1 | Apr (i)-88 | 158 |
| 98 | General | Developmental Hues | Lack of Leadership | 8 | 1 | Apr (ii)-98 | 159 |
| 99 | General | Developmental Hues | From Shrimp to Lobster | 8 | 2 | May (i)-88 | 161 |
| 100 | Inland Fisheries | Developmental Hues | Inland Fisheries: Pond Fish Production Policy | 8 | 2 | May (ii)-88 | 163 |
| 101 | Marine Fisheries | Developmental Hues | Fishing Vessels for Diversified Fishing | 8 | 3 | Jun (i)-88 | 165 |
| 102 | Maine Fisheries | Developmental Hues | Manning Requirements of Fishing Vessels | 8 | 3 | Jun (ii)-88 | 166 |
| 103 | Inland Fisheries | Aquaculture/ Farming | At Pre-paddle Stage | 8 | 4 | Jul (i)-88 | 167 |
| 104 | Marine Fisheries | Aquaculture/ Farming | Insurance of Chartered Fishing Vessels | 8 | 4 | Jul (ii)-88 | 168 |
| 105 | Inland Fisheries | Aquaculture/ Farming | Shrimpers no more | 8 | 4 | Jul (iii)-88 | 169 |
| 106 | Marine Fisheries | Developmental Hues | Need for Introduction of Tuna Long Liners | 8 | 5 | Aug-88 | 170 |
| 107 | Marine Fisheries | Charters | Snare - Ridden Charter Chariot | 8 | 6 | Sep (i)-88 | 171 |

| Sl.No. | Category | Name | Editorial Captions | Vol. No. | Issue No. | Month-Year | Page No. |
|---|---|---|---|---|---|---|---|
| 108 | Inland Fisheries | Charters | Revival of Riverine Fisheries | 8 | 6 | Sep (ii)-88 | 174 |
| 109 | Marine Fisheries | Financing | Illogical Report | 8 | 7 | Oct-88 | 176 |
| 110 | Inland Fisheries | Capture Fisheries | Major Source of Inland Fish Production: Ponds or Rivers? | 8 | 8 | Nov (i)-88 | 179 |
| 111 | Marine Fisheries | Marine Fisheries Exports | Erratic Arrival of Reefer Ships at Visakhapatnam Causing Hardship to Marine Products Exporters | 8 | 8 | Nov (ii)-88 | 181 |
| 112 | Marine Fisheries | Organisational/ Policies | Appeal to Coast Guard and Navy: Contact Fishing Vessels also in KC Channel | 8 | 9 | Dec-88 | 182 |
| 113 | General | Field level including Marketing | Purchase Tax Exemption for Sea-Foods in A.P | 8 | 10 | Jan (i)-89 | 183 |
| 114 | General Fisheries | Research | Chromosome Manipulation in Fish | 8 | 10 | Jan (ii)-89 | 185 |
| 115 | Marine Fisheries | Organisational/ Policies/Strategies | Strategies Unrelated to Basic Facts Hampering Development of Indian Deep Sea Fishing | 8 | 11 | Feb-89 | 186 |
| 116 | Marine Fisheries | Deep Sea Fishing | Strategy Aspects of Current Crisis in Deep Sea Fishing | 8 | 12 | Mar-89 | 189 |
| 117 | Marine Fisheries | Organisational/ Policies | Let us Emulate Indonesia's Successful Tuna Fishing Strategy | 9 | 1 | Apr (i)-89 | 191 |
| 118 | General | Fishing Chimes | Ascent to the Nineth Rung | 9 | 1 | Apr (ii)-89 | 193 |
| 119 | Marine Fisheries | Financing | SCICI adopts a Pragmatic Approach | 9 | 2 | May-89 | 195 |
| 120 | Inland fisheries | Aquaculture/ Farming | Problems of Fish Culturists | 9 | 3 | Jun-89 | 197 |
| 121 | Marine Fisheries | Developmental Hues | Resurrection of Deep Sea Fishing Sector | 9 | 4 | Jul (i)-89 | 198 |
| 122 | Inland Fisheries | Developmental Hues | Steps to Counter Siltation of Fish Ponds | 9 | 4 | Jul (ii)-89 | 201 |
| 123 | Marine Fisheries | Developmental Hues | Congeniality at Visakhapatnam Fishing Harbour | 9 | 5 | Aug-89 | 202 |
| 124 | General | Review | Fragmented Fisheries Set Up | 9 | 6 | Sep-89 | 204 |
| 125 | Marine Fisheries | Charters | Utilisation of Trained Indian Crew on Chartered Tuna Long Liners | 9 | 7 | Oct (i)-89 | 207 |
| 126 | Marine Fisheries | Charters | On Chartered Fishing Vessels in Indian Waters | 9 | 7 | Oct (ii)-89 | 209 |
| 127 | Marine Fisheries | Organisational/ Policy | Emerging Categories of Entrepreneurship in Indian Deep Sea Fishing Industry | 9 | 8 | Nov-89 | 210 |
| 128 | Marine Fisheries | Financing | How to Earn More from Trawlers and Start Paying to Financiers | 9 | 9 | Dec-89 | 212 |
| 129 | General | Review | 1989: Fish-O-Drome | 9 | 10 | Jan-90 | 214 |
| 130 | Marine Fisheries | Organistional/Policy | Integrated Approach Under Unitary Control: Only way to Develop Deep Sea Fishing | 9 | 11 | Feb-90 | 219 |
| 131 | General | Review | Voiceless Association | 9 | 12 | Mar-90 | 221 |
| 132 | Marine Fisheries | Developmental Hues | Another Blow to Marine Fishing Sector | 10 | 1 | Apr-90 | 223 |
| 133 | Marine Fisheries | Charters | Charter Scheme: Double Stardards | 10 | 2 | May-90 | 225 |
| 134 | Marine Fisheries | Developmental Hues | Killjoy Crisis Keeps Deep Sea Fishing Vessels Harbour-Bound at Visakhapatnam | 10 | 3 | Jun-90 | 228 |

| Sl.No. | Category | Name | Editorial Captions | Vol. No. | Issue No. | Month-Year | Page No. |
|---|---|---|---|---|---|---|---|
| 135 | Marine Fisheries | Charters | SCICI Launches the Receiver Era | 10 | 4 | Jul-90 | 232 |
| 136 | Marine Fisheries | Organisational | Near Shore Fishery Development Agencies | 10 | 5 | Aug-90 | 234 |
| 137 | Inland fisheries | Aquaculture/ Farming | Faulty World Bank's Carp Hatchery Design: Indian Experts Successfully Face the Challenge | 10 | 6 | Sep-90 | 236 |
| 138 | Marine Fisheries | Developmental Hues | Supply of Diesel Oil for Deep Sea Fishing Vessels: Invidious Attitude | 10 | 7 | Oct (i)-90 | 238 |
| 139 | Marine Fisheries | Developmental Hues | Visakhapatnam Fishing Harbour Back to Normalcy | 10 | 7 | Oct (ii)-90 | 239 |
| 140 | Marine Fisheries | Developmental Hues | Steep Increase in Diesel Price: The Heaviest Blow to Marine Fishing Industry | 10 | 7 | Oct (iii)-90 | 241 |
| 141 | General | Review | Evolving a National Fisheries Policy: Fishermen's Efforts | 10 | 8 | Nov-90 | 242 |
| 142 | Marine Fisheries | Financing | Intricate Problems Riddling Trawler Loans Scenario | 10 | 9 | Dec-90 | 244 |
| 143 | General | Review | Eventful 1990 | 10 | 10 | Jan-91 | 246 |
| 144 | Marine Fisheries | Foreign Fishing in Indian EEZ | Mysterious Quiscence | 10 | 11 | Feb-91 | 249 |
| 145 | Marine Fisheries | Developmental Hues | Processing of Fish and their Export with Australian Aid: A New Era on the Anvil | 10 | 12 | Mar-91 | 250 |
| 146 | General | Review | Chimed Through a Decade | 11 | 1 | Apr-91 | 252 |
| 147 | Marine Fisheries | Developmental Hues | What Should be Done to Improve the Economic and Social Conditions of Coastal Fishermen | 11 | 2 | May-91 | 254 |
| 148 | Marine Fisheries | Developmental Hues | The Inheritence | 11 | 3 | Jun-91 | 256 |
| 149 | Marine Fisheries | Financing | Crisis in Deep Sea Fishing Industry Unsubsided Problems of Subsidy and Working Capital | 11 | 4 | Jul-91 | 259 |
| 150 | Marine Fisheries | Organisational/ Policies/Strategies | Deep Sea Fishing Industry. Status Under the New Industry Policy | 11 | 5 | Aug-91 | 260 |
| 151 | General | Developmental/ Administration | Fisheries Developmental Strategy in the Eighth Five Year Plan | 11 | 6 | Sep-91 | 263 |
| 152 | Marine Fisheries | Developmental Hues | Trouble-Ridden ECBB Trawlers | 11 | 7 | Oct-91 | 266 |
| 153 | Marine Fisheries | Financing | The Designated Person | 11 | 8 | Nov (i)-91 | 268 |
| 154 | General | Diversification | Diversification of Fishing: Unfulfilled Promises by Government and MPEDA | 11 | 8 | Nov (ii)-91 | 270 |
| 155 | Marine Fisheries | Developmental Hues | Government's Resolve to Unshackle Deep Sea Fishing Industry | 11 | 9 | Dec (i)-91 | 271 |
| 156 | General | Developmental and Administration | Invidious attitude of Ministry of FPI | 11 | 9 | Dec (ii)-91 | 273 |
| 157 | General | Review | Editor Explains | 11 | 10&11 | Jan/Feb 1992 | 274 |
| 158 | Inland Fisheries | Aquaculture/ Farming | Extension Mechanism in FFDAs and BFDAs | 11 | 12 | Mar-92 | 275 |
| 159 | Marine Fisheries | Developmental Hues | Tuna Long Lining as a Fishing Diversification System | 12 | 1 | Apr-92 | 278 |

| Sl.No. | Category | Name | Editorial Captions | Vol. No. | Issue No. | Month-Year | Page No. |
|---|---|---|---|---|---|---|---|
| 160 | Marine Fisheries | Developmental Hues | Deep Sea Fishing Policy: Some Drawbacks | 12 | 3 | Jun-92 | 280 |
| 161 | Marine Fisheries | Developmental Hues | Development of Deep Sea Fishing: Beginnings of the Inexorable Process | 12 | 4 | Jul-92 | 281 |
| 162 | Inland Fisheries | Aquaculture/ Farming | Brackishwater Shrimp Production: 'Satellite' Farming | 12 | 5 | Aug-92 | 284 |
| 163 | Inland Fisheries | Developmental Hues | Spread of Brackishwater Shrimp Farming: Major Hurdles | 12 | 6 | Sep-92 | 286 |
| 164 | Inland Fisheries | Developmental Hues | Reflections on Future of Indian Aquaculture | 12 | 7 | Oct-92 | 288 |
| 165 | Marine Fisheries | Developmental Hues | Diversification of Marine Fishing: A Momentous Follow-Up | 12 | 8 | Nov-92 | 290 |
| 166 | Marine Fisheries | Developmental Hues | Old (Legacy of Composite Madras State) and Present Kerala Marine Fisheries Regulations | 12 | 9 | Dec-92 | 292 |
| 167 | General | Review | Fishery Fulcrum 1992 | 12 | 10 | Jan-93 | 294 |
| 168 | Marine Fisheries | Developmental Hues | Establishment of Post Harvest Shore Infrastructural Faciities | 12 | 11 | Feb-93 | 298 |
| 169 | Marine Fisheries | Developmental Hues | Marine Fishing Industry Hypnotises, Lures and Detains an Entrepreneur | 12 | 12 | Mar-93 | 301 |
| 170 | General | Review | Traversed into the Thirteenth | 13 | 1 | Apr-93 | 303 |
| 171 | Marine Fisheries | Developmental Hues | Reviving 'Deep Sea' Fishing Industry | 13 | 2 | May-93 | 305 |
| 172 | Marine Fisheries | Developmental Hues | Punishments and Rewards | 13 | 3 | Jun (i)-93 | 308 |
| 173 | Marine Fisheries | Organisation/ Policies/Strategies | Regulation of Fishing by Indian Vessels in our EEZ | 13 | 3 | Jun (ii)-93 | 309 |
| 174 | General | Field Level including Marketing | Guess the One Coming to Import our Fish | 13 | 4 | Jul-93 | 310 |
| 175 | General | Developmental and Administration | Customs Bonding and Brackishwater Farms | 13 | 5 | Aug-93 | 312 |
| 176 | Marine Fisheries | Financing | To Jog out of the Jam of Deep Sea Fishing Vessel Loans and Recoveries | 13 | 6 | Sep-93 | 314 |
| 177 | General Administration | Developmental and | Irrational Taxation on Incomes from Fish Needs Abolition | 13 | 7 | Oct-93 | 315 |
| 178 | Inland fisheries | Aquaculture/ Farming | Brood Shrimp availability for Hatcheries | 13 | 8 | Nov-93 | 317 |
| 179 | Marine Fisheries | Developmental Hues | Coastal Marine Fishery Development Agencies Needed | 13 | 9 | Dec-93 | 318 |
| 180 | General | Review | High and Low, 1993 | 13 | 10 | Jan (i)-94 | 320 |
| 181 | Inland Fisheries | Aquaculture/ Farming | Back-lash of Brackishwater Farming | 13 | 10 | Jan (ii)-94 | 323 |
| 182 | General | Developmental Hues and Administration | Re-levy of Purchase Tax on Marine Products in A.P. | 13 | 11 | Feb-94 | 325 |
| 183 | General | Developmental Hues and Administration | On Diversification of Fish Production Effort | 13 | 12 | Mar-94 | 327 |
| 184 | General | Review | Major Gaps | 14 | 1 | Apr (i)-94 | 329 |
| 185 | General | Review | Chimed into the Fourteenth Year | 14 | 1 | Apr (ii)-94 | 332 |
| 186 | Marine Fisheries | Developmental Hues | Diversification of Fishing by Shrimp Trawlers: Fund-Oriented Plan Imperative | 14 | 2 | May-94 | 333 |

| Sl.No. | Category | Name | Editorial Captions | Vol. No. | Issue No. | Month-Year | Page No. |
|---|---|---|---|---|---|---|---|
| 187 | Marine Fisheries | Developmental Hues | National Fish Workers and Deep Sea Fishing | 14 | 3 | Jun-94 | 337 |
| 188 | General | Education, Training and Manpower | Fisheries Education and Training | 14 | 4 | Jul-94 | 339 |
| 189 | Marine Fisheries | Developmental Hues | Gearing up for the Orderly Spread of Mariculture | 14 | 5 | Aug-94 | 341 |
| 190 | General | Field Level Including Marketing | Rural Fishery Development Through Voluntary Organisations | 14 | 6 | Sep-94 | 342 |
| 191 | General | Field Level Including Marketing | Transfer of Developed and Emerging Technologies | 14 | 7 | Oct-94 | 343 |
| 192 | Marine Fisheries | Developmental Hues | Pillars Cater at Their Whim Sad Maintenance and Repair Status of 'Cat' Marine Engines in Fisheries Sector | 14 | 8 | Nov-94 | 345 |
| 193 | Marine Fisheries | Organisation/ Policies/Strategies | 3-mile Non-Fishing Zone for Deep Sea Fishing Vessels Adjacent to Territorial Waters | 14 | 9 | Dec-94 | 347 |
| 194 | Marine Fisheries | Organisation/ Policies/Strategies | Indian Marine Fisheries Management | 14 | 10 | Jan-95 | 348 |
| 195 | Inland Fisheries | Aquaculture/ Farming | Angles of Aquaculture | 14 | 11 | Feb-95 | 353 |
| 196 | Inland Fisheries | Aquaculture/ Farming | Coastal and Deep Sea Fishing Policy | 14 | 12 | Mar-95 | 360 |
| 197 | Marine Fisheries | Developmental Hues | Reflect and Restore Tempo of Deep Sea Fishing Development: Appeal to Government of India to Create a Separate Fisheries Department | 15 | 1 | Apr-95 | 362 |
| 198 | General | Developmental Hues and Administration | Need for Reforms in Indian Fisheries Administration | 15 | 2 | May-95 | 365 |
| 199 | Inland Fisheries | Capture Fisheries | Reservoir Fisheries Development to be an 'Extreme Focus' Area | 15 | 3 | Jun-95 | 368 |
| 200 | Marine Fisheries | Developmental Hues | Towards an Effective Trawling System | 15 | 4 | Jul-95 | 369 |
| 201 | Marine Fisheries | Developmental Hues | Dismal Development of Deep Sea Fishing in India | 15 | 5 | Aug (i)-95 | 371 |
| 202 | General | Financing | Aquaculture and Income Tax | 15 | 5 | Aug (ii)-95 | 372 |
| 203 | Marine Fisheries | Developmental Hues | Challenge to Beep Sea Fishing in Indian EEZ | 15 | 6 | Sep-95 | 373 |
| 204 | Marine Fisheries | Foreign Fishing in Indian EEZ | The Other Side of Deep Sea Fishing by Foreign Vessels in Indian EEZ | 15 | 7 | Oct (i)-95 | 376 |
| 205 | Inland Fisheries | Aquaculture/ Farming | Diversification in Brackishwater Farming | 15 | 7 | Oct (ii)-95 | 379 |
| 206 | General | Developmental Hues and Administration | Export of Fresh, Frozen and Processed Fish & Fishery Products Order and Rules, 1995 Oppressive and Counter-productive | 15 | 8 | Nov (i)-95 | 380 |
| 207 | Marine Fisheries | Capture Fisheries | Fishing around Andamans becomes Expensive | 15 | 8 | Nov (ii)-95 | 382 |
| 208 | General | Developmental and Administration | Time for Rechristening MPEDA as FPEDA to Cover Inland Fish Exports | 15 | 9 | Dec (i)-95 | 384 |
| 209 | Inland Fisheries | Developmental Hues | New Products for Effective Shrimp Pond Management? | 15 | 9 | Dec (ii)-95 | 385 |

| Sl.No. | Category | Name | Editorial Captions | Vol. No. | Issue No. | Month-Year | Page No. |
|---|---|---|---|---|---|---|---|
| 210 | General | Review | Looking Back at 1995 | 15 | 10 | Jan-96 | 386 |
| 211 | Marine Fisheries | Organisational Policies/Strategies | Recommendations of Murari Committee: 'A Master Stroke' | 15 | 11 | Feb-96 | 388 |
| 212 | General | Education and Manpower | Training Vs Coaching | 15 | 12 | Mar-96 | 391 |
| 213 | General | Developmental Administration | Draft National Fisheries Policy | 16 | 1 | Apr-96 | 392 |
| 214 | Marine Fisheries | Developmental Hues | Turtle Exclusion Devices | 16 | 2 | May (i)-96 | 394 |
| 215 | Inland Fisheries | Aquaculture/ Farming | CIFA Poised to Revolutionise Rohu Quality in Fish Culture | 16 | 2 | May (ii)-96 | 395 |
| 216 | Marine Fisheries | Developmental Hues | Development Imperatives in Marine Fisheries Sector | 16 | 3 | Jun-96 | 396 |
| 217 | Marine Fisheries | Financing | Financing Deep Sea Fishing Industry | 16 | 4 | Jul-96 | 398 |
| 218 | General | Developmental Hues and Administration | Management of New Technologies | 16 | 5 | Aug (i)-96 | 400 |
| 219 | Marine Fisheries | Fishing in Indian EEZ | Fishing in EEZ: Soon to be a Major Bonanza to Foreign Interests? | 16 | 5 | Aug (ii)-96 | 401 |
| 220 | General | Review | Indian Fisheries Sector: Stagnation Allover | 16 | 6 | Sep-96 | 402 |
| 221 | Marine Fisheries | Organisation/ Policies/Strategies | A Legal Framework has to Precede Mariculture Development | 16 | 7 | Oct-96 | 404 |
| 222 | Inland Fisheries | Aquaculture/ Farming | Construction of Ponds in Sandy Areas: Biocrete Technology | 16 | 8 | Nov (i)-96 | 406 |
| 223 | General | Others | Deliverance from Recurring Cyclone Revages: Coastal Fishermen Deserve a Permanent Protective Mechanism | 16 | 8 | Nov (ii)-96 | 407 |
| 224 | Inland Fisheries | Aquaculture/ Farming | Control of White Spot Disease among Cultured Tiger Shrimp | 16 | 9 | Dec (i)-96 | 409 |
| 225 | General | Policies/Strategies | SC' s Order on Selective Closure of Coastal Farms | 16 | 9 | Dec (ii)-96 | 411 |
| 226 | General | Review | Fourth Indian Fisheries Forum | 16 | 10 | Jan (i)-97 | 413 |
| 227 | General | Organisation Policies/Strategies | Failure of Fisheries Management Policies: Reasons | 16 | 10 | Jan (ii)-97 | 414 |
| 228 | General | Review | An Year of Tumult, Depression and Discount | 16 | 10 | Jan (iii)-97 | 415 |
| 229 | Marine Fisheries | Organisation/ Polices/Strategies | Strategy to Augment Exports in Fisheries Sector | 16 | 11 | Feb-97 | 418 |
| 230 | Marine Fisheries | Organisation/ Polices/Strategies | Formulation of New Deep sea Fishing Policy: One Possible and Practicable Option | 16 | 12 | Mar-97 | 420 |
| 231 | General | Developmental and Management | Too many Technologies: Flow Towards Endusers Too Slow Remedy: Integrated/Auxiliary Technology Development Projects | 17 | 1 | Apr-97 | 423 |
| 232 | General | Developmental and Management | State Fisheries Corporations | 17 | 2 | May (i)-97 | 426 |
| 233 | Marine Fisheries | Developmental Hues | Fisheries Sector: The Dubious Extension of the Term "Thrust Area" to it | 17 | 2 | May (ii)-97 | 428 |

| Sl.No. | Category | Name | Editorial Captions | Vol. No. | Issue No. | Month-Year | Page No. |
|---|---|---|---|---|---|---|---|
| 234 | Inland Fisheries | Aquaculture/ Farming | Efficient Sludge Treatment: New System | 17 | 3 | Jun (i)-97 | 429 |
| 235 | Marine Fisheries | Developmental and Management | Bhavanapadu Fishing Harbour | 17 | 3 | Jun (ii)-97 | 430 |
| 236 | General | Others | American Pew Foundations' Award to Kocherry | 17 | 4 | Jul-97 | 432 |
| 237 | Marine Fisheries | Developmental Hues | Bycatch Escapement from Shrimp Trawls | 17 | 5 | Aug-97 | 434 |
| 238 | General | Field Level Including Marketing | Kerala takes the Lead: People's Involvement in Fisheries Development | 17 | 6 | Sep (i)-97 | 436 |
| 239 | General | Others | Fisheries Development: Impressive Strides in Karnataka | 17 | 6 | Sep (ii)-97 | 438 |
| 240 | Marine Fisheries | Fishing Industry | Rehabilitation of Deep Sea Fishing Industry | 17 | 6 | Sep (iii)-97 | 439 |
| 241 | General | Others | Tamil Nadu Fishery Corporation's Triumph | 17 | 7 | Oct (i)-97 | 441 |
| 242 | General | Developmental Hues | Factors Impeding Shrimp Export Tempo | 17 | 7 | Oct (ii)-97 | 442 |
| 243 | Marine Fisheries | Capture Fisheries | Mid-Sea Bunkering | 17 | 7 | Oct (iii)-97 | 443 |
| 244 | Marine Fisheries | Financing | Deep Sea Fishing: Need for Composite System of Financing | 17 | 8 | Nov-97 | 444 |
| 245 | Marine Fisheries | Financing | Rehabilitation Scheme: Missing Reliefs | 17 | 9 | Dec (i)-97 | 446 |
| 246 | Marine Fisheries | Developmental Hues | Standardisation in Marine Pearl Production | 17 | 9 | Dec (ii)-97 | 447 |
| 247 | General | Review | Looking Back at 1997 | 17 | 10 | Jan-98 | 448 |
| 248 | General | Developmental Hues | Fisheries Sector: Focal Points for Government's Attention | 17 | 11 | Feb-98 | 451 |
| 249 | General | Development and Management | Faster Fisheries Development: Consultancy Infrastructure is the Key | 17 | 12 | Mar-98 | 454 |
| 250 | General | Review | Entry into the Eighteenth | 18 | 1 | Apr-98 | 456 |
| 251 | Marine Fisheries | Organisation/ Polices/Strategies | Deep Sea Fishing Policy | 18 | 2 | May-98 | 457 |
| 252 | Inland Fisheries | Developmental Hues | The Great Indian Saline Land Stretch | 18 | 3 | Jun-98 | 459 |
| 253 | General | Others | River and Sea Ranching | 18 | 4 | Jul-98 | 460 |
| 254 | General | Education and Manpower | Flow of Expertise from Fisheries Colleges to Commercial Fisheries Sector | 18 | 5 | Aug (i)-98 | 461 |
| 255 | General | Others | The Help Small Fishermen Need | 18 | 5 | Aug (ii)-98 | 463 |
| 256 | General | Education and Manpower | Crew Crisis. Severe Shortage of Fishing Vessel Skippers | 18 | 6 | Sep (i)-98 | 464 |
| 257 | Marine Fisheries | Aquaculture/ Farming | Cage Farming in Indian Seawaters: A Developmental Imperative | 18 | 6 | Sep (ii)-98 | 465 |
| 258 | Inland Fisheries | Aquaculture/ Farming | Prevention of Whitespot Disease among Shrimps in Grow-out Ponds | 18 | 6 | Sep (iii)-98 | 467 |
| 259 | General | Development and Management | Fishing Sector's Unique Features Demand a Compatibly Structured Financing System | 18 | 7 | Oct-98 | 468 |
| 260 | General | Development and Management | Languishing Fisheries Sector: Needs Growth with Stability Supported by Technology and Creative Financing | 18 | 8 | Nov-98 | 471 |

| Sl.No. | Category | Name | Editorial Captions | Vol. No. | Issue No. | Month-Year | Page No. |
|---|---|---|---|---|---|---|---|
| 261 | General | Development and Management | Sustainable Utilisation of Trans-national Fisheries Resources of Indian Subcontinent Possible Only through a Regional Management Mechanism | 18 | 9 | Dec-98 | 474 |
| 262 | General | Review | Indian Fisheries Sector in 1998: Reflections | 18 | 10 | Jan-99 | 476 |
| 263 | Inland Fisheries | Capture Fisheries | 1999-2000: Year of Integrated Reservoir Fishery Development in Karnataka | 18 | 11 | Feb-99 | 479 |
| 264 | Marine Fisheries | Organistional/ Policies/Strategies | High Sea Bunkering System: Need for Extension to Mechanised Fishing Boats Sector | 18 | 12 | Mar-99 | 480 |
| 265 | Marine Fisheries | Developmental Hues | CMFRI Ushers in Onshore Commercial Pearl Production | 19 | 1 | Apr (i)-99 | 481 |
| 266 | Marine Fisheries | Capture Fisheries | Seasonal Closure for Capture Shrimp on Upper East Coast: Reflections | 19 | 1 | Apr (ii)-99 | 483 |
| 267 | Inland Fisheries | Capture Fisheries | Render Indian Reervoirs Sustainably Fishful | 19 | 2 | May-99 | 486 |
| 268 | Marine Fisheries | Developmental Hues | Indian Marine Fishing Sector in Regression Mode | 19 | 3 | Jun-99 | 489 |
| 269 | General | Development and Management | Augmenting Fish Exports: Relevance of Relying on Tuna and Tilapia | 19 | 4 | Jul-99 | 491 |
| 270 | Marine Fisheries | Organisational/ Policies/Strategies | Checkmating Pakistani Move | 19 | 5 | Aug-99 | 493 |
| 271 | General | Development and Management | Indian Aqua products Exports | 19 | 6 | Sep-99 | 494 |
| 272 | Marine Fisheries | Development and Management | Declining Marine Catches and Exports: Opportunities to Counter the Trend | 19 | 7 | Oct-99 | 495 |
| 273 | General | Others | Fishing Industry Victimised: Orissa Cyclone's Disastrous Affront | 19 | 8 | Nov (i)-99 | 498 |
| 273 (a) | General | Others | Farm-reared Tiger Shrimp Broodstock Production | 19 | 8 | Nov (ii)-99 | 500 |
| 274 | Marine Fisheries | Developmental Hues | Indian Tuna Tangle: Unravelling Possible only Thro' External Inputs | 19 | 9 | Dec-99 | 501 |
| 275 | General | Review | Editor's Foreward................. | 19 | 10,11 | Jan/Feb 2000 | 503 |
| 276 | Marine Fisheries | Financing | Exim Bank (with new US Creditline) Can Rescue Collapsing Indian Deep Sea Fishing Industry | 19 | 12 | Mar-00 | 505 |
| 277 | General | Development and Management | Options for Registering Cognizable Aqua Products Export Growth | 20 | 1 | Apr-00 | 507 |
| 278 | General | Development and Management | National Project on Export-Oriented Inland Aquafood Production Essential | 20 | 2 | May-00 | 509 |
| 279 | Inland Fisheries | Aquaculture/ Farming | National Fish Farmers' Day | 20 | 3 | Jun (i)-00 | 512 |
| 280 | Marine Fisheries | Capture Fisheries | Tuna Mystery of Indian EEZ | 20 | 3 | Jun (ii)-00 | 513 |
| 281 | Inland Fisheries | Aquaculture/ Farming | Tiger Shrimp Domestication | 20 | 4 | Jul-00 | 514 |

| Sl.No. | Category | Name | Editorial Captions | Vol. No. | Issue No. | Month-Year | Page No. |
|---|---|---|---|---|---|---|---|
| 282 | General | Develoment and Management | Future of Fisheries Development Rests on Technocrats of Quality and Competence | 20 | 5 | Aug-00 | 515 |
| 283 | Inland Fisheries | Aquaculture/ Farming | Tiger Shrimp Domestication: CMFRI's Progress in the Taming Game | 20 | 6 | Sep (i)-00 | 517 |
| 284 | General | Review | A Sequel on the Anvil | 20 | 6 | Sep (ii)-00 | 518 |
| 285 | Inland Fisheries | Aquaculture/ Farming | Inland Culture Fishery Sector of India: Future Scenario | 20 | 7 | Oct-00 | 519 |
| 286 | General | Others | Recommendations of Meets on Fisheries Topics: Destined Dormancy | 20 | 8 | Nov-00 | 520 |
| 287 | General | Others | Reflections on Certain Aspects of Technology Transfer | 20 | 9 | Dec-00 | 521 |
| 288 | Inland Fisheries | Aquaculture Technologies | On Nation Wide Application of Aquaculture Technologies Conducive for Export-Oriented Production | 20 | 10& 11 | Jan/Feb 2001 | 523 |
| 289 | Marine Fisheries | Developmental Hues | Prognostic Reflections on Fishery Resources on Indian EEZ | 20 | 12 | Mar-01 | 527 |
| 290 | General | Development and Administration | Forecast on Fisheries Schemes of Tenth Plan | 21 | 1 | Apr-01 | 529 |
| 291 | Marine Fisheries | Developmental Hues | Small Mechanised Boats: Diveri-fication for Tuna Needed | 21 | 2 | May-01 | 531 |
| 292 | Marine Fisheries | Developmental Hues | India's Sil/Fishing Vessel Import Scheme: Implications and Impact | 21 | 3 | Jun-01 | 533 |
| 293 | Inland Fisheries | Aquaculture/ Farming | 10th July 2001 Marks the Birth of Special Fish Farmer's Day | 21 | 4 | Jul (i)-01 | 536 |
| 294 | Inland Fisheries | Fisheries Plan Schemes | Tenth Plan Fisheries Schemes Refreshingly Forward Looking | 21 | 4 | Jul (ii)-01 | 537 |
| 295 | Inland Fisheries | Tiger Shrimp hatcheries | Tiger Shrimp Hatcheries Need Supplies of Disease-Free Brooders | 21 | 4 | Jul (iii)-01 | 539 |
| 296 | Marine Fisheries | Tuna Fisheries | Tuna Tangle | 21 | 4 | Jul (iv)-01 | 540 |
| 297 | Inland Fisheries | Shrimp Capture Resource | Shrimp Famine Stalks Upper East Coast | 21 | 5 | Aug-01 | 541 |
| 298 | General | Developmental and Administration | ....on Generation of National Fisheries Policies | 21 | 6 | Sep-01 | 544 |
| 299 | General | Developmental and Administration | On Aggregated National Fisheries Adminstration | 21 | 7 | Oct-01 | 546 |
| 300 | Inland Fisheries | Aquaculture and Administration | Trout Farming Expansion in Himachal | 21 | 8 | Nov-01 | 548 |
| 301 | Inland Fisheries | Aquaculture/ Farming | Lined Ponds | 21 | 9 | Dec-01 | 550 |
| 302 | General | Review | 2001 in Restrospect: Development, Lessons and Events | 21 | 10,11 | Jan/Feb-2002 | 551 |
| 303 | Marine Fisheries | Developmental Hues | Tuna Fishing in Indian EEZ: Training Needs | 21 | 12 | Mar-02 | 555 |
| 304 | Marine Fisheries | Developmental Hues | Time for FSI to Share its Mandate | 22 | 1 | Apr-02 | 556 |
| 305 | Inland Fisheries | Developmental Hues | ....Towards Dawn of Nation-wide Export-Oriented Aquaculture Era (of Giant Prawn, Black Tiger and Nile Tilapia) | 22 | 2 | May-02 | 558 |

| Sl.No. | Category | Name | Editorial Captions | Vol. No. | Issue No. | Month-Year | Page No. |
|---|---|---|---|---|---|---|---|
| 306 | Marine Fisheries | Organisational Policies/Strategies | Utiisation of Fisheries of Indian EEZ: Reflections on Policy Options | 22 | 3 | Jun-02 | 560 |
| 307 | Inland Fisheries | Developmental Hues | On Loktak Lake Fishery Development | 22 | 4 | Jul-02 | 563 |
| 308 | Marine Fisheries | Developmental Hues | Dilemma of Indian Deep Sea Fishing | 22 | 5 | Aug-02 | 565 |
| 309 | Inland Fisheries | Aquaculture/ Farming | Regulation of Shrimp Aquaculture | 22 | 6 | Sep-02 | 567 |
| 310 | General | Developement and Administration | Welcome Leads | 22 | 7 | Oct-02 | 569 |
| 311 | General | Review | Alarming Signals | 22 | 8 | Nov-02 | 571 |
| 312 | Marine Fisheries | Developmental Hues | Fishing in Indian EEZ: Guidelines | 22 | 9 | Dec-02 | 574 |
| 313 | Marine Fisheries | General | Shrimp Export Scene | 22 | 10,11 | Jan/ Feb (i)-2003 | 576 |
| 314 | General | Indian Fisheries Development | Indian Fisheries Sector : 2002 | 22 | 10,11 | Jan/ Feb (ii)-2003 | 578 |
| 315 | Marine Fisheries | Developmental Hues | The Challenge of India's Declining Coastal Fishing Output | 22 | 12 | Mar-03 | 582 |
| 316 | General | Marine Fisheries Industrial Situation | Status of Indian Marine Fisheries Development: An Appraisal | 23 | 1 | Apr-03 | 584 |
| 317 | Inland Fisheries | Gain! Freshwater Prawn | Giant Freshwater Prawn Broodstock: Quality Improvement | 23 | 2 | May-03 | 588 |
| 318 | General | Fisheries Administration and Association | Associations Related to Indian Fisheries Sector | 23 | 3 | Jun-03 | 589 |
| 319 | Inland Fisheries | Inland Culture Fish Exports | Fish Farmers Need Inland Culture Fish-based Expansion of Export Efforts | 23 | 4 | Jul-03 | 591 |
| 320 | Inland Fisheries | Inland Fishery Recreational Fishery | Recreational Fishing: Commercial Potential | 23 | 5 | Aug-03 | 592 |
| 321 | Marine Fisheries | Developmental Hues | India's Marine Fishing: Scenario Before and After Declaraion of EEZ | 23 | 6 | Sep-03 | 593 |
| 322 | Marine Fisheries | Developmental Hues | ....On Utilisation of Tuna Resources of Indian EEZ | 23 | 7 | Oct-03 | 596 |
| 323 | Marine Fisheries | Financing | Microfinance in Support of Coastal Fisher-women | 23 | 8 | Nov-03 | 598 |
| 324 | Inland Fisheries | Giant Freshwater Prawn | ...On production of All-male Giant Prawn Progeny | 23 | 9 | Dec-03 | 599 |
| 325 | General | Indian Fisheries Development | Balance Sheet 2003 | 23 | 10,11 | Jan Feb- 2004 | 600 |
| 326 | Inland Fisheries | Aquaculture/ Farming | Aquaculture and Aquafarming | 23 | 12 | Mar-04 | 602 |
| 327 | Inland Fisheries | Aquaculture/ Farming | ...Towards Shrimp Culture Sans Disease | 24 | 1 | Apr-04 | 603 |
| 328 | Marine Fisheries | Development Hues | Marine Fishing : Perspectives of Diversification | 24 | 2 | May-04 | 606 |
| 329 | Marine Fisheries | Development Hues | CMFRI'S Appraisal of Status of Exploited Marine Fisheries of India | 24 | 3 | Jun-04 | 609 |
| 330 | Inland Fisheries | Aquaculture/ Farming | Thoughts that Gripped on Fish Farmers Day | 24 | 4 | Jul-04 | 610 |
| 331 | General | Fisheries Administration and Association | Separate Fisheries Ministry is Urgent, in National Interest | 24 | 5 | Aug-04 | 613 |

| Sl.No. | Category | Name | Editorial Captions | Vol. No. | Issue No. | Month-Year | Page No. |
|---|---|---|---|---|---|---|---|
| 332 | Marine Fisheries | Fleet Development | Fishing Fleet Matters | 24 | 6 | Sep-04 | 615 |
| 333 | Marine Fisheries | Development Hues | Historic Commercial Tuna LL Success with Total Indian Effort–In Visakha -Kakinada Stretch of Bay of Bengal | 24 | 7 | Octr-04 | 617 |
| 334 | Inland Fisheries | Aquaculture/ Farming | On Extricating Indian Shrimp Culture Sector from its Present Travails | 24 | 8 | Nov-04 | 619 |
| 335 | Marine Fisheries | Organisational/ Policy | Marine Fishing Policy has Some New Features | 24 | 9 | Dec-04 | 621 |
| 336 | Marine Fisheries | Fleet Development | Open Sea Fishing in Indian EEZ Dominated by 'Indianised' Flag of Convenience (FOC) Vessels | 24 | 10 | Jan-2005 | 623 |
| 337 | Inland Fisheries | Cage-Based Pond Farming | Cage-Based Pond Farming | 24 | 11 | Feb-05 | 626 |
| 338 | Marine Fisheries | Development Hues | Utilisation of Tuna of Indian EEZ: Focal Look at Fishing Industry's Needs | 24 | 12 | Mar-05 | 629 |
| 339 | General | Fisheries Administration and Association | The Beginning: 'Fisheries' Secured a Place in a Union Dept's Name | 25 | 1 | Apr-05 | 631 |
| 340 | Marine Fisheries | Conservation | On Ways of Stemming Decline of Coastal Fisheries Resources | 25 | 2 | May-05 | 633 |
| 341 | Marine Fisheries | Conservation | Endangered Fish Species of India: Their Restoration to Normalcy A National Project Needed | 25 | 3 | Jun-05 | 634 |
| 342 | Marine Fisheries | Organisational/ Policy | 'Catches' in Fisheries Work | 25 | 4 | Jul-05 | 636 |
| 343 | Inland fisheries | Aquaculture/ Farming | .....On Introducing Super Tilapia (GIFT) Farming in India | 25 | 5 | Aug-05 | 638 |
| 344 | Marine Fisheries | Fleet Development | ...On Restructuring of Open Sea Indian Marine Fishing Fleet | 25 | 6 | Sep-05 | 641 |
| 345 | Inland Fisheries | Aquaculture/ Farming | Diversified Aquaculture Technology Transfer: MPEDA's Unique Initiative | 25 | 7 | Oct-05 | 644 |
| 346 | Marine Fisheries | Development Hues | On Past and On-Going Utilisation of Fishery Resources of Indian EEZ | 25 | 8 | Nov-05 | 646 |
| 347 | Inland Fisheries | Capture Fisheries | Fisheries Scenario of Kolleru Lake in A.P | 25 | 9 | Dec-05 | 648 |
| 348 | General | Review | Hind Look at 2005 | 25 | 10 | Jan-2006 | 650 |
| 349 | Inland Fisheries | Culture/Farming | Indian Freshwater Fish Farming-Problems of Major Carp Monopoly | 25 | 11 | Feb-06 | 652 |
| 350 | Extension | Field Level Including Marketing | Extension Anxieties | 25 | 12 | Mar-06 | 654 |
| 351 | General | Review | Enter Silver 'Jubilee' of Fishing Chimes | 26 | 1 | Apr-06 | 656 |
| 352 | General | Field Level Including Marketing | Nation-wide Fish Marketing Network | 26 | 2 | May-06 | 662 |
| 353 | General | Inland Fishery Recreational Fishery | Deficient Political Support: Bane of Fisheries Sector | 26 | 3 | Jun-06 | 664 |
| 354 | Marine Fisheries | Organisation/ Policies/Strategies | Need for Tuna Fishery Development Agency | 26 | 4 | Jul-06 | 666 |
| 355 | General | Indian Fisheries Development | Some of the Tasks Ahead for National Fisheries Development Board | 26 | 5 | Aug-06 | 670 |

| Sl.No. | Category | Name | Editorial Captions | Vol. No. | Issue No. | Month-Year | Page No. |
|---|---|---|---|---|---|---|---|
| 356 | Marine Fisheries | Organisational/ Policies/Strategies | Marine Fisheries Census, 2005 | 26 | 6 | Sep-06 | 672 |
| 357 | Inland Fisheries | Aquaculture/ Farming | Aquaculture in Indian Coastal Zone Eco and Farmer-friendly Legislation Introduced | 26 | 7 | Oct-06 | 674 |
| 358 | General | Field Level Included Marketing | Promotion of Organised Domestic Value-Added Fish Marketing Linked to a Network of Value-Adding Coastal Fish yards | 26 | 8 | Nov-06 | 677 |
| 359 | Marine Fisheries | Fleet Development | Promotion of Tuna Fishing in Upper Bay of Bengal of India | 26 | 9 | Dec-06 | 679 |
| 360 | Inland Fisheries | Inland Fish Exports | On Promotion of Export Market for Indian Carps | 26 | 10 | Jan-2007 | 681 |
| 361 | Inland Fisheries | Aquaculture/ Farming | Allowing Exotic Cultivable Aqua Species into India: Need for a Realistic Policy | 26 | 11 | Feb-07 | 683 |
| 362 | Inland Fisheries | Aquaculture/ Farming | Desert Fishery Development of Aqua-farming in the Saline Zone of North West India. National Fishery Board's Intervention Imperative | 26 | 12 | Mar-07 | 685 |
| 363 | Marine Fisheries | Fleet Development | Scenario, Before and After-Letters of (Fishing Vessel) Permission (LoP) System | 27 | 1 | Apr-07 | 687 |
| 364 | Marine Fisheries | Developmental hues | Tuna Exploitation in Indian EEZ: Low Key Indian Effort (As highlighted at Tuna 2006, Bangkok) | 27 | 2 | May-07 | 691 |
| 365 | Inland Fisheries | Indian Fisheries Development | Indian Freshwater Prawn Trinity: Farmed Production National Level Projectisation Imperative | 27 | 3 | Jun-07 | 694 |
| 366 | General | Field Level Included Marketing | Ignored Facet of Value-Addition | 27 | 4 | Jul-07 | 697 |
| 367 | Marine Fisheries | Organisation/ Policies/Strategies | Through Issue of LoPs to Indian Enterprises to Usher in Foreign Fishing Vessels into Indian EEZ– India Opts for a Debatable System | 27 | 5 | Aug-07 | 699 |
| 368 | Inland Fisheries | Aquaculture/ Farming | Community-Based Minor Irrigation Tank Fisheries Development in A.P, as Part of World Bank-Aided Irrigation Project | 27 | 6 | Sep-07 | 702 |
| 369 | Marine Fisheries | Financing | To Expand its Open Sea Fishing Fleet– India Needs a Related Financing Policy Starved of Investments, Open Sea Fishing by Indian Vessels in the Indian EEZ Suffers | 27 | 7 | Oct-07 | 704 |
| 370 | Inland Fisheries | Capture Fisheries | Reservoir Fisheries Development Through Partnerships | 27 | 8 | Nov-07 | 707 |
| 371 | Inland Fisheries | Research | Diamond Jubilee of CIFRI: Vintage CIFRIANS Reminisce | 27 | 9 | Dec-07 | 709 |
| 372 | General | Review | Flash Back 2007 : Year of Tuna in India | 27 | 10&11 | Jan/Feb-2008 | 712 |

| Sl.No. | Category | Name | Editorial Captions | Vol. No. | Issue No. | Month-Year | Page No. |
|---|---|---|---|---|---|---|---|
| 373 | Marine Fisheries | Fleet Development | Upgradation of Onboard Storage Facilities on Small Trawlers MPEDA's Subsidy Scheme Bestows Multiple Benefits | 28 | 12 | Mar-08 | 714 |
| 374 | General | Indian Fisheries Development | Active Political Support Needed for– Faster Fisheries Development of India | 28 | 1 | Apr-08 | 715 |
| 375 | Marine Fisheries | Sea Cage Capture | The Beginning of Sea Cage Culture: CMFRI's Successful Experiment | 28 | 2 | May-08 | 717 |
| 376 | Inland Fisheries | Aquaculture/ Farming | ...On Pangs of Vannamei Introduction in India | 28 | 3 | Jun-08 | 719 |
| 377 | General | Indian Fisheries Development | Indian Fisheries Sector Needs Fortifying Measures | 28 | 4 | Jul-08 | 721 |
| 378 | Inland fisheries | Aquaculture/ Farming | Fisheries Development in Inland Saline Waters of India NFDB has to Intervene | 28 | 5 | Aug-08 | 723 |
| 379 | Marine Fisheries | Organational/ Policies | Indian EEZ: Tuna Fishing Scenario LoP System's Benefits are One Sided | 28 | 6 | Sep-08 | 725 |
| 380 | Inland Fisheries | Aquaculture/ Farming | In this Era of Upsurge of Improved Culture Fisheries Technologies, Lentic Water Fishery Rights: Leasing Policies to be made Matchingly Production-Oriented | 28 | 7 | Oct-08 | 727 |
| 381 | Marine Fisheries | Developmental Hues | ....On Regenertion of Sea Cucumber Stocks in Indian EEZ | 28 | 8 | Nov-08 | 729 |
| 382 | General | Cage Farming | ....On Promotion of Cage Aquafarming in India | 28 | 9 | Dec-08 | 731 |
| 383 | General | Review | Fisheries Events, 2008 (Including those that took place in the last quarter of 2007) | 28 | 10 & 11 | Jan/Feb-2009 | 733 |
| 384 | Inland fisheries | Research | SPF Shrimp Broodstock Development India Far Behind | 28 | 12 | Mar-09 | 737 |
| 385 | General | Field Level Included Marketing | Empowering Marine Fishers Through Value-Addition Strategy Marine Fishers' Earnings Can Go Up Only when Identified Coastal Centres-Based Value Adding Infrastructure is Set Up | 29 | 1 | Apr-09 | 739 |
| 386 | General | Indian Fisheries Development | Impact of NFDB on Pace of Indian Fisheries Development | 29 | 2 | May-09 | 741 |
| 387 | General | Fisheries Administration and Association | Management of Fisheries of India | 29 | 3 | Jun-09 | 743 |
| 388 | General | Indian Fisheries Development | Future of Fisheries Development Rests on Technocrats of Quality and Competence | 29 | 4 | Jul-09 | 745 |
| 389 | Inland Fisheries | Giant Freshwater Prawn | Giant Freshwater Prawn : Monosex (Male) Farming | 29 | 5 | Aug-09 | 747 |
| 390 | General | Cage Farming | Role of FFDAs: Promotion of Pond-based Cage Farming | 29 | 6 | Sep-09 | 749 |
| 391 | Marine Fisheries | Fleet Development | Development of India's Own Genuine Distant Water (Deep sea) Fishing Fleet | 29 | 7 | Oct-09 | 752 |

Fishing Chimes, Editorial 1: April 1981: Vol. 1, No. 1

# Financing of Fishing Vessels: Can We Not Follow U.S. System?

US Government extends long term guarantees for duration(s) upto 20 years towards loans taken by fishing enterprises for acquisition of fishing vessels. When an advanced nation has felt the need to extend such a facility, is it not appropriate that a developing country like ours should also have a similar system?

In June 1977, the Ministry of Agriculture had issued permissions to 43 fishing companies for the import of 79 fishing vessels, valued at approximately Rs.35 crores. The Shipping Development Fund Committee granted loans to the extent of 90 per cent of the cost of vessels to several of these companies. Very few of the companies have been able to utilise the loan so far. Prices of fishing vessels have gone up in the meanwhile by 60 per cent. Why has such a situation developed, when Government are keen to develop the industry and fishing companies anxious to acquire fishing vessels?

## Reason for Non-Utilisation

The main reason for non - utilisation of loans is known to the Government. The SDFC, being a financing body, rightly insists on a guarantee either from a State Government, a Nationalised bank, or an Insurance company, valid till such time as the vessels reach an Indian port and are mortgaged in favour of SDFC. The duration of guarantee will be about a year. State Governments concerned are not interested in providing the guarantees, although the introduction of deep sea fishing vessels along their coastline would contribute to the socio-economic development of the coastal areas and would also earn additional foreign exchange. The unlikely risks during the period of guarantee, which may be failure of the shipyards to perform, loss of vessels during transhipment (although covered by insurance pledged to them) are perhaps too heavy for them, compared to the public benefits that would accrue, when the vessels arrive and operate along their coast.

## Present Picture

Commercial Banks are prepared to look at the proposal, only if the entire amount for which the guarantee is required is deposited with them. They know very well that none of the fishing companies would have that much money. If they have, they do not require a loan from SDFC.

Insurance companies say that they are forbidden by the Government to provide financial guarantees.

We now have the picture. Fishing companies cannot get guarantees. Without guarantees SDFC do not release loans. Unless loans are released, vessels will not be built. This means our deep sea fishing fleet, which is the smallest compared to any other maritime nation, with the exception of a few countries like Srilanka or Bangladesh, will have to remain stagnant, providing excellent opportunities to foreign vessels not only to continue to exploit fisheries resources of our Exclusive Economic Zone, but also to step up their efforts. Imagine countries like Thailand, South Korea, Mexico etc operating thousands of vessels, and a nation like India stagnating with a commercial fishing fleet of 58 nos.

## Short Repayment Period

The Period of repayment of SDFC loans used to be fifteen years. From June 1977, this was brought down to eight years. Why was fifteen years allowed earlier, and what was the reason for scaling it down to eight years? The reason for prescribing 15 years earlier is understandable, as it is based on the characteristics of the industry. Fishing can sometimes be an uncertain business. Incomes can fluctuate from year to year. If repayment of loans is compressed into too short a term, one may not have enough cash available during those years when income decreases.

The reason for reducing the period to 8 years is incomprehensible. Can this be the brain wave of a Bureaucrat who is not familiar with the characterestics of the fishing industry? It will be an act of wisdom by the Government to restore the earlier limit of years.

## The US Pattern

So far as guarantee is concerned, let us look at what an advanced country like USA has done. An Obligation Guarantee issued to fishermen by the National Marine Fisheries Service, a U.S. Government organisation, pledges the full faith and credit of the USA to the repayment of money borrowed by fishermen for constructing, reconstructing or reconditioning of commercial fishing vessels, to cover a repayment period upto 20 years. The fisherman is free to negotiate and obtain the sanction of loan by any financing body. Once this is settled, he gets the obligation guarantee for the asking. The NMFS guarantee is for 100 per cent (Principal and interest) of the obligation. Payment under the guarantee is in cash and will be made within 30 days from the lender's demand. There is no upper limit on the amount of any guaranteed obligation. The amount cannot, however, exceed 87.5 per cent of actual cost of constructing, reconstructing, or reconditioning a fishing vessel. All fishing vessels of not less than 3 net tons are eligible for the guarantee.

The lending agency gets certain benefits by providing loans under the obligation guarantee for fishing vessels. These guarantees are classified as Type I investment securities and may be bought or sold by banks without hesitation. These are also eligible as collateral for Federal Reserve Bank advances to the lending agency and also as lending agency's collateral for tax and Treasury loan accounts and the deposit of any other public money or funds whose deposit is under the authority of the United States. These obligations are readily assignable by the lender to secondary purchasers. In short, NMFS guaranteed obligations are risk-free, and share many of the benefits of U.S. Treasury obligations. The paper work connected with the obtaining of the guarantee and securing of the loans are minimal.

If a fisherman is unable to find a lender, the NMFS helps him in this regard.

## Are we More Advanced?

The fishing Industry in USA is far more advanced than ours. Even so, the U. S. Government, obviously recognising the need to provide support to the fishing industry, introduced the system of providing Fishing Vessel Obligations Guarantees to enterprises for acquiring fishing vessels. Such being the case, it does look anamalous that the Indian Government has not thought of providing such guarantees, particularly when the Nation is facing a grave situation in which there is practically no prospect of new deep sea fishing vessels added, in the absence of a sound arrangment for issuing guarantees for the loans to be released by the SDFC to the fishing companies. True, there may be problems cropping up in regard to a few of such guarantees issued. Can this, however, be a ground not to introduce such a system? Losing of some money on account of one or two unlikely mishaps, is preferable to facing stagnation in the industry. Further, by introducing the system of issuing obligation guarantees by the Government, pressure on SDFC funds can be reduced. The fishing companies can approach Commercial banks lor loans.

It is hoped that the Government of India would look into the subject in all its aspects and decide to introduce the system of issuing Fishing Vessel Obligation Guarantees on U.S. model, with such modifications as are necessary. Till this is done, there is no salvation to the Indian deep sea fishing industry and the fisheries resources of our Exclusive Economic Zone would continue to be a free gift to fishermen of neighbouring nations.

Fishing Chimes, Editorial 2: May 1981: Vol. 1, No. 2

# Upgradation of EFP Into A Deep Sea Fishery Development Corporation Imperative

India's Exclusive Economic Zone covers 2 million sq km. This is as much as the agricultural land in the country. The economic potential of the sea area we have has a promise of gigantic financial returns.

But all depends on how much we know about the area and where to find fishing grounds.

Exploratory fisheries project which does the job should be upgraded to meet the challenge. Its existing performance with 28 vessels operating just 126 days on an average in a year and that too, close to shore would bring in no positive help to the industry. Therefore the time is now ripe for reorganising the project into a more dynamic and purposeful body.

Soon after independence, Government of India set up the Deep Sea Fishing Station at Bombay. The objectives were to conduct exploratory surveys for locating/viable fishing grounds and to provide training in deep sea fishing for selected candidates. This organisation trained quite a few skippers and other operatives. However, with the setting up of the Central Institute of Fisheries Operatives around 1964, this objective was deleted. In regard to the exploratory work no impressive results have so far been achieved.

In 1974, the Deep Sea Fishing Station, with its few sub-stations, known as off-shore fishing stations, all along the coastline, was reorganised and renamed as 'The Exploratory Fisheries Project'. The various off-shore stations became 'Bases'. There are today 12 bases (one of these is stated to be under closure) under the Exploratory Fisheries Project (EFP), which continues to have its headquarters in Bombay. The main work of the project continues to be the same as before the reorganisation.

There are altogether 28 vessels, operated by the Project from the various bases. Of these, 23 nos. are indigenously built, 20 nos. are 17.5 m long trawlers, 1 no. is 23 m long trawler and 2 nos. are purse-seiner-cum-trawlers (built at Goa Ship-yard recently under Norweigian Aid). The others were all obtained under aid programmes very recently, two trawlers, one 40 m long and another 36 m long were obtained from Holland, two trawler-cum-purse-seiners, 35 m long each, given by Denmark and one long-liner-cum-squid-jigger, 30 m long, from Japan. So far, no perceptible benefit has flown from the Exploratory Fisheries Project to the fishing industry. Some newsletters and bulletins, containing information on results of operation of various vessels, were issued during 1979/80. The material contained in these was mostly belated. In any case, there was little in these bulletins that was of positive utility to the industry or that was new. Now these issues also seem to have been stopped.

The performance of the vessels is also disappointing. Most of the vessels operate for about 126 days in a year. Very few operate for 187 days. One vessel based at Goa is known to have operated for 220 days, on daily trips.

Surveys are mostly haphazard, for short durations and mostly in nearshore waters. Very few operations take place in depths beyond 40 fathoms.

## Working Conditions

The Exploratory Fisheries Project and its bases are run by experienced and competent officers. Yet, owing to problems relating to conditions of service of the crew members, and absence of incentives, the performance is perfunctory. Sea going personnel are also governed by the work regulations meant for those working on land. The crew members claim that they have to work for only 8 hrs a day, and that they are not expected to work on holidays. There are perhaps problems for the Government to frame a separate set of working regulations for the crew and employ them only when those regulations are accepted by them.

The Exploratory Fisheries Project now owns vessels worth around Rs. 29 crores. The annual expenditure towards running expenses, maintainance, salaries etc. is of the order of Rs. 3.5 crores. With no benefits flowing to the fishing industry, most of this expenditure is an obvious waste.

## Result-Oriented Plans Needed

The need of the hour is therefore for the Government to review the situation and devise ways and means of utilising the assets for yielding optimum results, keeping in view the urgency to progressively utilise the fisheries resources of the Exclusive Economic Zone in the shortest possible time.

The exploitation of the fisheries resources of the Exclusive Economic Zone is dependent on the existence of an effective machinery with executive responsibility to bring about the introduction of deep sea fishing vessels. At present, the Fisheries Division in the Department of Agriculture directly performs this function. By the very nature, being part of a policy making set-up, and engrossed in these matters, this Division is not attuned to implement programmes of the nature of introduction of deep sea fishing vessels. The work involves expertise and freedom of action in various directions with single minded devotion. The Fisheries Division, even if such an expertise is available with them, will be in no position to implement the programme with speed as this is not their only work.

There is thus no way other than having a separate organisation, responsible for the introduction of deep sea fishing vessels for operation in our Exclusive Economic Zone. A few thousands of Vessels have to be introduced, entailing an investment of hundreds of crores of rupees. Matters such as types, sizes, machinaries and equipment, constructional programmes and operational aspects etc. have to be looked into. One cannot imagine all these to take place on their own or from a small set-up in the Ministry of Agriculture at Delhi.

The fact that objectives of the Exploratory Fisheries Project could not be realised for over three glorious decades shows that an altogether different organisational setup is necessary. Apart from exploratory surveys and making available the results to the industry, this set-up should also be able to organise supply of suitable fishing vessels to the Industry. It appears to us that the up-grading of the Exploratory Fisheries Project into a Central Deep Sea Fishery Development Corporation is imperative, if the gigantic task ahead, to progressively introduce more and more deep sea fishing vessels for the exploitation of the fisheries resources of the Exclusive Economic Zone is to be executed.

It is suggested that the Exploratory Fisheries Project may be upgraded into a Central Deep Sea Fishery Development Corporation. This Corporation may be an independent registered company set up under the companies Act, in the Public Sector, by the Government of India.

## New Body

All the assets *i.e.*, fishing vessels, workshops etc. now owned by the Exploratory Fisheries Project may be made over to this Corporation.

The vessels may be given on time charter to various fishing companies, based on certain stipulated criteria and in accordance with all applicable guidelines and regulations now in force for chartering foreign fishing vessels. Under these guidelines charterers have to furnish all data to the Government. They have also to take a scientist on board. These arrangements will give all required data as the Exploratory Fishery Project is now striving to get. Not only this. All expenditure on running and maintenance can be avoided. Crew management problems will no longer be there. The data received can be processed by the proposed Corporation and results made available to the industry at frequent intervals and in time.

The chartering system can only be one of its functions. In addition, this organisation has to promote the introduction of deep sea fishing vessels. It can import or organise the construction of batches of suitable vessels indigenously, based on pre-determined demand, and supply them on charter or sell on hire purchase terms to fishing companies. The determination of the types of vessels, their dimensions, machinaries and equipments to be installed, the yards for construction, finalisation of drawings, manner of identifying recipient companies, conditions governing sale or charters etc. will be the function of the proposed Corporation, to precede actual supplies. After the supplies, it will keep a close watch on the progress of operations and collection of data, dues etc.

Let us not be complacent any longer. Tinkering with the problem, as it is going on now, will result in intractable complications. The utilisation of the fisheries resources of the Exclusive Economic Zone is a gigantic task requiring major effort at planning and implementation.

Government is aware of the situation and the urgency to tackle it and let us hope that a massive effort will be set in motion so as to achieve significant results of which the nation can be proud of.

Fishing Chimes, Editorial 3: June 1981: Vol. 1, No. 3

# Chagrin over Charters

The inordinate delays in the Union Department of Agriculture in regard to taking decisions on pending charter applications is intriguing the applicants. After keeping the applications pending for over two years, in February 1981, revised guidelines were formulated. It was then expected that the applications would be decided upon in the light of the new guidelines. Till the time of this writing, none of the applicants are stated to have received any communication on the subject from the Department of Agriculture.

The Directorate General of Shipping issues charter permissions for cargo vessels within seven days of receiving applications. The situation in regard to chartering of fishing vessels is such that applicants are not in a position to receive permissions even in seven months.

It is astounding that such abnormal delays are allowed to take place. The Department should obviously be aware of the effect of such delays. As it is, our reputation abroad in regard to charter permissions is the lowest. Many owners refuse even to negotiate on the ground that Indian Government does not decide on charter cases and, even if they decide, it will be years by the time decisions are taken. They say, rightly so, that they cannot keep their vessels uncommitted for long.

So far as the charterers are concerned, their agonies are undescribable. On one side, the owners threaten them of withdrawing their offer. Once an offer is withdrawn the party has again to explore and identify another owner and start negotiations. This involves visits abroad, entailing heavy expenditure. In the meanwhile costs go up, pushing up the charter hire.

The Indian fishing companies are a real wonder. They personify Indian philosophy. Their patience in pursuing and waiting for Government's clearance is something unheard of in any country of the world.

They give positive construction to all the delays and wait patiently, believing that all these may turn out to be for their own good, notwithstanding the impossible conditions governing the charters. It boggles ones imagination that Government takes so much time to decide on applications, even after imposing the strictest conditions, perhaps not obtaining in any other country.

The applying parties will reconcile themselves for the rejection of their applications, if they are not acceptable to the Government. This rejection has, however, to be done quickly, so that the owners can be informed in time. Delays frustrate both the owners and the charterers. This can be avoided by timely decisions.

It is of utmost importance for the Department to lay down a time schedule for clearing charter applications. It should not take more than 30 days either to issue a letter of intent or a rejection letter, or to call for additional material.

The fishing industry is in a mood of desperation. One company after another is withdrawing from the field, unable to withstand the rigours of Government's policies on one hand and delays in decision- making on the other. Government and the fishing industry, for a long time to come, have to function in unison, Government helping the Industry, and industry enabling Government to help them.

Government are approached for permissions mostly by nationals prepared to take all attendant risks and at the same time interested in developing the industry for securing financial benefits. Government would be doing a disservice to the nation if these are discouraged by undue delays. It is hoped that there will be realisation on the part of the concerned functionaries in the Department of Agriculture in regard to the alarming situation and they would soon see things in the proper perspective and expedite decisions on charter applications.

Fishing Chimes, Editorial 4: July 1981: Vol. 1, No. 4

# Constraints In Deep Sea Fishery Development

According to a report in Hindustan Times (June 23, 1981), Mr. S. P. Mukherji, Additional Secretary, Ministry of Agriculture stated that, for want of resources and equipment, India cannot exploit the fishery wealth in the Exclusive Economic Zone. The Government wanted the private fishing industry to help it to exploit the wealth, though it would mean little in the absence of extensive data on the fishing field.

Referring to charters, he said that the Government's scheme to permit chartering of fishing vessels from abroad has not elicited enthusiastic response. So far only 35 applications for charter had been received by the Government.

He appealed to private parties to exploit in a big way India's vast wealth of fish which could earn foreign exchange for the country and provide a supplementary protein diet for the people.

The situation, so succinctly stated by Mr. Mukherji, calls for effective measures on the part of the Government to facilitate the mobilisation of the requisite inputs by the industry. His appeal to the private industry to exploit the fisheries resources in a big way is timely. The private industry, as he knows, is keen to play its role. The Industry only wants the requisite encouragement from the Government and quick clearance of proposals.

He has stated that only 35 applications have been received by the Government for chartering fishing vessels. But these were received in a period of over two years and till the time of this writing, no company received any communication on the final decision of the Government on their applications. There can be enthusiastic response only if Government stipulates enforceable and reasonable conditions governing charters and clears the applications within a reasonable time.

There have also been enormous delays in the Government in regard to clearance of requests for changes in source of import and on increase in price of fishing vessels. It is stated that these applications have not yet been cleared. Very few applications for loans are believed to have been recommended to the SDFC. It is the ardent desire of the Industry that the Ministry clears their proposals quickly. Such a gesture would bring about a proper climate of development.

Fishing Chimes, Editorial 5: August 1981: Vol. 1, No. 5

# Kerala Marine Fisheries Sector in a Ferment

## Enfant Terrible

Ever since the promulgation of the Kerala Marine Fishing Regulation Ordinance 12 of 1980 and the subsequent enforcement of the rules thereon in G.O.M.S. 141/30, F and PD dated 29-11-1980, the fishing operations in Kerala have been in a state of nebulous ferment. The total prohibition of fishing by mechanised fishing vessels using purse seines, ring seines, pelagic trawls and midwater trawl nets within the territorial waters has to a significant extent affected the growth of the fishing industry both in terms of production and economy. The reasons attributed to the promulgation of these orders are: a) to protect the interests of persons engaged in fishing using traditional fishing crafts such as catamarans, country crafts and canoes and (b) to conserve fish and to regulate fishing in the 'specified area', the specified area being the territorial waters of the State within 22 km from the shore.

By a subsequent order, the Kerala Government prohibited totally the operation of any mechanised boat during the months of June, July and August. The provocation for this has been stated to be the need for protection of the fish population in the whole area from exploitation during their spawning period in monsoonic months. These have been 'officially' accepted as the spawning period of all the economic species of fishes in the Kerala waters. This stentorian and draconian rule was subsequently amended to exclude the mechanised boats of 32' and below operating from Neendakara base in order to facilitate the exploitation of the Karikkadi prawns during that period in those waters.

## Statements Galore

All these ordinances, regulations and rules have been accompanied by a series of utterances and statements right from the Minister of Fisheries down to the self proclaimed leaders of the smallest unions of fishermen in the State. There is hardly a week when some one does not make a statement for or against these ordinances and regulations. Agitations galore have sprung up in different parts of the State. Fishermen of non-mechanised crafts go on Satyagraha and gheraos in front of the Secretariat, District Collectorates and Fisheries Offices. In several cases Christian priests give the leadership to these. The fleet owners of the mechanised fishing boats make huge noise through statements and representations and use pressure tactics through telegrams and politicians. At least in one or two cases they used the Marine Products Export Development Authority to influence the promulgation of some amendments and changes. During this whole cacophony of comments and counter comments the experts of the CMFRI and of the Ministry of Agriculture have thought it wise to remain absolutely, totally, silent. The only statement connecting the CMFRI and the Ministry to the whole issue is the one which came from the Kerala Minister of Fisheries, who is reported to have stated that the State Government had the sanctity of support of Dr. Silas and Dr. Mammen for the amendment to allow mechanised boats of 32' and below to fish 'Karikkadi' prawns during the prohibited period from Neendakara on the ground that this species has a brief span of life and it would be a waste not to catch it when available in plenty *i.e.*, June. July and August! And the sanctity of Neendakara is obvious!

Fishing Chimes, Editorial 6: September 1981: Vol. 1, No. 6

# Deep Sea Fishing in Blunderland

If fishing companies are continuing their fishing operations it should not be mistaken that this is because of profits. If they discontinue operations, they would face several problems with the share holders and the institutions which provided financial assistance. They expect richer harvests to compensate additional expenses and losses in course of time.

The present state of deep sea fishing industry presents a depressing scenario. No solution to improve the situation is in sight.

## Present State

Let us first recount the features of the present state. The operational costs of trawlers have gone up enormously. Cost of diesel oil, the main item of expenditure, has gone up by over 160 per cent in the past four years. Maintenance and repair costs have gone up several fold. The returns from exportable component of the catches, namely shrimp, on which the economics of operation depend, have fallen. The domestic market for sale of fish is very weak. In this situation, fishing companies are reeling, not finding any rock to hold in their sea of troubles. Unable to close down, the companies are continuing their operations in deficit. Some companies, with other lines of activity, have started closing down their fishing operations. Growth of the industry in this state of affairs is impossible. Even so, there are some new companies longing to import trawlers, wanting to utilise the permissions for acquiring them. Practically no one is thinking of acquiring new trawlers, through indigenous construction, although the benefits of subsidies and 95 per cent loan facility from SDFC are offered by the Government. Government is not doing anything further to improve the climate so as to enable the fishing industry to acquire new vessels, and run existing vessels economically.

## Charters

Seven companies including two corporations have secured permissions to charter 25 fishing vessels from abroad. With the exception of the two Corporations, no others are likely to avail of the permissions. The conditions governing charter permissions, such as obligation to commence buying chartered vessels within a short period of operations, even before assessing the economics of operation and whether or not the operations are successful, need to provide a bank guarantee of Rs. 2 lakhs per trawler, etc. appear to discourage them. It is generally felt that, with the possible exception of Kerala and Andhra Corporations, no other party would utilise the charter permissions given. While the Corporations may be in a position to explain probable losses out of the activity to their Governments satisfactorily, private companies cannot take the risk of sustaining losses, and at the same time be under obligation to buy the chartered vessels and also lose Rs. 2 lakhs per trawler, owing to forfeiture of their guarantee by Government.

The officials in the fisheries section of Agriculture Dept. obviously feel that the Government's approach is realistic. From the point of view of the industry, the approach would, however, appear to be based on suspicion and conservatism. The grounds on which Government may have adopted the approach can be assessed as follows:

1. While the diesel oil costs have gone up, fishing companies are still continuing their operations. There must therefore be profit. There is no need for any substantial help.
2. Companies chartering fishing vessels will make money through sharing gross earnings with owners. Having done this, they may not purchase the vessels. The companies would thus have the benefit of making money, without the need to invest on purchase of vessels. Therefore deterrants such as furnishing of a bank guarantee of Rs. 2 lakhs per vessel, irrevocable obligation to buy vessels from owners, limitations on fishing grounds etc. are necessary.

## Fishing goes on Losses

If fishing companies are continuing their fishing operations, it should not be mistaken that this is because of profits. If they discontinue operations, they would face several problems with the share holders and the institutions which provided financial assistance. They expect richer harvests to compensate, additional expenses and losses in course of time.

In regard to economics of operation, one point mentioned is that, foreign vessels are obviously able to operate economically in our waters through poaching or through charters. Such being the case, why should not our vessels operate economically? Answer to this point lies in the fact that foreign vessels get higher prices for their fish, owing to the reason that, in their own or other countries they sell, they have networks of cold storages and marketing facilities which we do not have. Further, in several countries fuel prices are far lower to ours. In Taiwan the fuel price is stated to be less than Rs. 2/- per litre. They take their bunkering at Taiwanese ports and go back to one of their ports, after fishing for about three months, for unloading catches and taking bunker. They have the advantage of getting higher prices for fish and lower fuel prices. Further, they cover various grounds which enable richer harvests.

## Reliefs are Doubtful

Our Government is unresponsive to the representations of the industry to provide relief from the higher costs of diesel oil. There is a proposal to provide relief of Rs. 162/- per kilo litre by way of exemption of the balance of 50 per cent excise duty. This would not reduce sales tax on diesel oil by the state Governments, which is very heavy. The Maharashtra Government has exempted the fishing industry from the payment of this tax, but none of the other State Governments are willing to do so. The burden on the industry thus continues.

It is obvious that no substantial relief from the higher operational costs can be provided either by the Centre or by the States. We cannot escape from realities.

## Let Corpns Concentrate On Marketing

The coastal State fisheries corporations are now mostly engaged in the same activities as the private sector fishery enterprises. Catching of shrimp, their processing and export are their main activity. What public purpose does this serve? None. Their objective is not to leave these activities to the private sector and concentrate on marketing of fish internally, by establishing necessary infrastructural facilities. By taking up this activity in a big way, in the place of the present duplication of the activities of the private sector, the Corporations would be rendering a signal service. All fishing companies can market their products through the chain of cold stores set up by Corporations and earn better incomes, thereby making their operations economical.

## Meanwhile

Till such time all this can be done, the companies now operating deep sea fishing vessels should be allowed to charter fishing vessels, with the obligation to buy vessels in a reasonable period after the end of the charter period, subject to favourable results. Such charters should not entail guarantees of Rs. 2 lakhs per vessel for the reason that the operations are already in deficit and to find money sufficient to get these guarantees will be impossible. If they want to apply for chartering 5 vessels, they require to deposit Rs. 10 lakhs in a bank to get guarantees. One can imagine how difficult this will be, with fishing companies running in deficit.

If the companies operating fishing trawlers are exempted from the requirement of guarantee, they can probably charter a few vessels and earn some extra money which they can use to make good the deficits in their regular operations to the extent possible. The national interests do not suffer in any way by this. If a fishing company finds that it can make money from chartered vessels, they would certainly buy them. But Government should not benalilse the company concerned, if owing to delay in the release of loan by SDFC, the owners withdraw their offer to sell.

## Suggested Action

To sum up, it is suggested that, in order to get over the present crisits in deep sea fishing, Government should take the following steps:

1. Coastal State Fisheries Corporations to be entrusted with the task of providing infrastructural facilities for regulated marketing of fish in the domestic market, so that fishing companies can realise better returns and make up for the increased expenditure on account of diesel oil.
2. Companies operating fishing vessels to be permitted to charter fishing vessels, exempting them from the condition to provide bank guarantees, and to buy vessels whether or not the operations are successful. If the owners withdraw their offer to sell chartered vessels, the companies should not be penalised on that account.

Fishing Chimes, Editorial 7: October 1981: Vol. 1, No. 7

# Anachronistic Administration

## *Plight of Fisheries Sector*

The key aspects of fisheries development in the country is concentrated in the hands of Indian Administrative Service Officers whose knowledge about fisheries is a thin veneer on their mental make-up, acquired for the purpose of the job. They have neither nor the career-oriented stake in the work, nor do they have the real grasp of the ingredients of fisheries development, necessary to give shape and fillip to the programmes.

People's representatives, in the form of ministers, are the backbone of the Government. In other words, the ministers, with their eyes and ears attuned to the people, know or in a position to know the requirements of the people. Therefore, at the political or governing level, they have the responsibility to ensure effective development of fisheries, among other fields. In discharging this responsibility, they have the Secretaries to Government to help them. It is in fitness of things that Secretaries to Government in charge of fisheries are administrative officers, who alone can act as a channel between the minister concerned and the developmental apparatus. Lower to the level of Secretaries, however, it is of paramount importance to have professional fisheries personnel to deal with the subject. This is, however, not the case both at the Centre as well as several of the States.

### Lure of Foreign Trips

The Joint Secretary (Fisheries) in the Union Department of Agriculture is the key functionary in regard to national fisheries development. Of late, a keen competition among IAS officers seems to have emerged for getting this post.

This is not out of interest in the job. By virtue of his position as Joint Secretary (Fisheries), the person occupying the post gets several opportunities to go abroad. One visit to Rome every year is certain, to attend the FAO Committee on Fisheries Meeting. Besides this, there will at least be one more visit, in some connection or the other. The IAS officers holding the post (only he holds the post), even go to the extent of attending technical conferences on fisheries abroad, although he has the faintest idea of the subject to be discussed and the contribution to be made. After spending 3-5 years in the post, (The other envious IAS officers, of course, try to dislodge the incumbant in the meanwhile, and get the post themselves) and with his passport full of visas, emigration and immigration stamps of various countries, he reluctantly hand over charge to his successor. There is practically no possibility of technical officers attending important meetings on fisheries abroad, unless the joint secretary is either benign or the Additional Secretary in-charge of fisheries or the Secretary (Agriculture), for whatever reason it may be, does not approve of the nomination and suggests or makes others to suggest, often his own name, or, in rare cases, that of a technical officer.

The Technical Officers, under the system, often become by force of habit, mere yes men and in course of time lose their power of independent thinking. They organise their work in such a way as to meet the thinking of their boss. Most of the joint secretaries, particularly those coming from inland states (some not having seen the sea, let alone a fishing vessel), do not normally have any idea of problems of marine fishing. Yet, the sense of authority often impels them to indicate a decision or a view that is novel. And this often sets off a chain reaction towards the destruction of a programme.

It is not that all administrative service officers bungle fisheries administration. Some may be good. But the point is, as a principle, who should be the executive head of fisheries developmental programmes? The logical answer is that a technical person should be the head. In no country are there non-professional people in charge of fisheries development. This is the peculiarity in India alone. When things go wrong the non-professional heads tend to shift the blame conveniently on to a technical officer.

Look at the mess created and later sorted out in regard to SDFC loans and charters. All this is the handiwork of administrative service officers only. Why is the utilisation of plan funds lowest in the fisheries sector only? In the fourth plan, under centrally sponsored schemes, out of Rs. 34.00 crores, only Rs. 13 crores were utilised. In the Fifth Plan, out of Rs. 68 crores, Rs. 44 crores were utilised. In regard to Central schemes, in 1978-79, against an outlay of Rs. 59 crores, the expenditure was Rs. 20 crores. In 1979-80, against an outlay of Rs. 43 crores, the expenditure was Rs. 15 crores. In general, it is clear from the above that administrative service officers, as Joint Secretaries, could not perform well in respect of achievement of targets. Not having a stake in the specific task, and with concentration revetted perhaps on other matters, it is difficult for them to do better.

### State Fisheries Departments

Now let us take a glance at the State Fisheries Departments. The Directors of Fisheries are the key functionaries in the States to develop fisheries. The fisheries departments as well as fisheries corporations of the four maritime states on the east coast are in the hands of IAS officers. Of the four west coast states, three have technical Directors and one (Gujarat) is in the hands of an IAS officer.

The criterion followed by the State Governments in posting IAS officer seems to be one of adjusting an officer whom they are not in a position to fit in elsewhere. Any IAS officer seems to be OK to be the Director of Fisheries. In some instances an officer wants to be in State headquarters. In the absence of any other post, they are posted as the Director of Fisheries. Most of them have no long range interest in the post and manage to get another posting later. Thus, in several states we find frequent changes in the incumbant working as the Director of Fisheries. The technical officers working as Joint Directors or Additional Directors lose all hope of reaching the top and become perfunctory, devoid of any interest in the developmental work. Although they know that a particular approach favoured by an IAS Director is wrong, they just follow it without protest, thus doing damage to developmental process.

## Strategies in High Places?

IAS coterie in most of the states appears to work towards retaining as many top posts as possible in their cadre. Once an IAS officer becomes a Director of Fisheries in a State, the position remains earmarked for IAS for ever. The ostensible reason given for this is that the technical officers are inefficient and they spoil the administration. And when a Government decides by some chance to replace an IAS officer with a technical officer, the problems of the later start from the moment he takes over. The concerned IAS officials in the Government very often start scuttling his proposals and present his work to the Government in an unfavourable light, with a view to dislodging him.

The conditions in Fisheries Corporations manned by IAS officers are still worse. No corporation run by an IAS Managing Director is running on sound lines. Either they expand the organisation too fast, or they are very cautious and slow. Not having adequate knowledge in depth either of inland fishery or marine fishery activities or of processing and marketing, they get stuck up in mid stream, after launching massive programmes without reference to several connected aspects and coordination. The priorities of work are not often related to objectives. The moods and inclinations of individuals play a large part in the items of work highlighted and implemented. With frequent changes in incumbants, discontinuity in policies and emergence of new policies takes place.

## IAS Style

There are several instances of glaring decisions by IAS officers that affected the developmental programmes. An IAS Director of Fisheries in a State made provision for upper Division clerks to become Inspectors of Fisheries and Assistant Directors of Fisheries. Thus that department has now several such executives, coming from clerical ranks and without professional background. One can well visualise the effect of this. There was another IAS officer whose policy was to look into papers that were placed on his table and pass non-committal comments. There were some who concentrated on raising the revenue to the Government from fisheries by concentrating on exploitation of fisheries without development by issuing of more licenses. There were others who passed critical comments without decisions. Many IAS Directors of Fisheries have difficulty in functioning effectively at high level meetings covering technical subjects. At the same time, they do not allow technical officers to attend such meetings. At the most they take a technical officer with them for briefing. There were however, some highly capable IAS officers who worked as Directors of Fisheries. But these are only exceptions. Allied subjects such as forestry, animal husbandry, dairying etc., are left by Governments to professionals to manage. Only in the case of fisheries, the east coast states particularly, continue the folly of posting IAS officers as Directors of Fisheries.

## What is Axiomatic

Fisheries are a well recognised sector all over the world. The development of fisheries involves advanced techniques and only a professional can organise the adaption of these techniques for giving pace to developmental process, identify real problems, technical or otherwise, in the imple-mentation, and evolve corrective measures. He alone can talk to the various functionaries, fish farmers, fishermen, and fisheries experts from outside with confidence and authority. Above all, it is axiomatic that a fisheries professional is the proper person to head a fisheries department or a corporation. Any other will be a misfit, however intelligent and profound he may be. An IAS man spends one year in learning, another year to put through the schemes. Before they materialise he is transferred. He may heave a sigh for being out of the oppressive joint, which caused him mental strain in trying to do something which is not in his line, but the developmental work suffers. The new man will have a different style of functioning. He has to be educated by the technical officers, a repetitive process for them, and before they say 'Jack Robinson', the new Director will be under orders of transfer. And the trend continues.

## Professionals

It is high time that the Centre posts a professional as Joint Secretary (Fisheries). It is also imperative that all State Governments post competent professionals as Directors of Fisheries and Managing Directors of Fisheries Corporations. Besides appointing them, Government should extend just support to enable them to function effectively. If the normal tendency of administrative officers is to find fault or to make the working of technical directors difficult, there can be no progress. The present anachronistic fisheries administrative pattern, particularly in the States on east coast, Gujarat on the west coast and in Bihar, has to come to an end in the shortest possible time. Till this is done there can be no march forward in the field.

Fishing Chimes, Editorial 8: November 1981: Vol. 1, No. 8

# Fisherman

*Who* is a fisherman? Chambers 20th Century dictionary says that a fisherman is a fisher, meaning thereby, a person who fishes.

There can be commercial fishing only with fishermen. There can be any number of devices to facilitate fishing, but fishermen will have to be there to operate the nets. The fishing results are directly related to the efficiency of fishermen. It is therefore axiomatic that the fishing activity will develop well only when fishermen are developed. The problem of development of fishermen is often looked into from a narrow angle all over the country. The development of fishermen is often restricted to traditional fishing communities.

Traditional fisher communities should no doubt receive all attention. However, fishing activities are now no longer the monopoly of traditional fishing communities. Persons from other communities have now entered the field and several have become good fishermen. Therefore, whoever is a fisherman, irrespective of caste and community, deserves all facilities from the Government to upgrade his skills and rise up in his profession.

All those belonging to traditional fishing communities are not in the business of fishing at the present day. Several of these pursue avocations other than fishing. This aspect has to be borne in mind in implementing schemes providing for assistance to traditional fishermen. A long range policy is necessary in harnessing and developing the capabilities of existing fishermen on one hand and adding to the available number on the other. While a new class of fishermen, from communities other than traditional fishermen's communities is coming up, the bulk of the fisher will continue to be from fishermen communities for long. It is therefore imperative that close attention is paid to them, so that they can contribute substantially to the national fishing effort.

Traditional fishermen, by and large, have no faith in the capabilities of fisheries departmental officials. There is some justification for this, for most of the officials do not have adequate professional knowledge of the type necessary to deal with fishermen. The traditional fishermen are proud people. They think that it is impossible for others to know more than what they know of. When departmental officers visit them, they deal with them with deference and tolerance. Since they are poor, they check on only one point—whether the visiting officer can get them something free, materials or money. Their interest, mostly, does not go beyond this. Unless their attention is captured and their attitudes are attuned to upgraded and higher income yielding techniques, progress cannot be made. In order to achieve this, there has to be a long range plan.

## Social Interaction

Representatives of the Government *i.e.* fisheries officials have first to become part of the fisher community. The relationship should be so intimate that it should pave the way for planting new ideas. Fisher have a great attachment towards their traditional festivals. These should be recognised by the Government and some grants should be given for their celebration in a fitting manner. The fisheries officials should participate in these festivals. The customs and beliefs of the community have to be closely studied and understood. With this background, it will be easier to gain their attention and enlighten them. The older generation can only be enlightened. There can at last be partial success with the middle-aged and the younger generation, in planting new ideas and making them to accept improved techniques.

## Catch Them Young

More than these, action that is necessary is to catch the fisher children in their tender age, educate and train them in fisheries schools, specially set up for them. Government have to set up such schools and attract as many fisher boys as possible from the age of 8 or even less. This approach is the surest way to develop our traditional fishers. When they complete their training such boys should be able to end up as skippers or engineers or vessel-owners, with various facilities provided by the Government. With their training they will be in a position to appreciate the facilities and avail of them.

The training institutes we have at present are meant for grown-ups. Few grown-up traditional fishermen will have the qualifications to join the institutes. Hence it is necessary to have schools to provide profession-oriented education and training to fisherboys for about 12 years continuously, combining general education as in higher secondary schools with professional training and education. All boarding and lodging and educational expenses have to be borne by the Government. Such schools, need not, in fact, be exclusively meant for boys from traditional fishing communities Other boys, interested in the line, can also seek admission.

People's representatives (MPs and MLAs) from constituencies with predominent fisher population, have to champion the cause of traditional fisher in their area, Unless this is done and fisher are made into men capable of undertaking improved methods of fishing that would give them higher economic returns, the fishery sector cannot prosper.

Fishing Chimes, Editorial 9: December 1981: Vol. 1, No. 9

# To Get In, Get On and Get Out

There are several who long to get into the deep sea fishing industry. For them to be 'in' is not easy. To get in, they need advice concerning the resources available, the base of operation, type or types of vessels to be operated, manpower requirement, economics of operation, marketing facilities etc. Once the advice is received from a competent body or person, the entrepreneur has to acquire vessels. He cannot import under present regulations. He has therefore necessarily to go in for indigenous vessels. The identification of a yard which can construct and deliver vessels of the type he needs becomes a problem for him. If he solves this and accepts the offer given by the shipyard of his choice, he has to get a loan from the SDFC and the Shipyard has to get the subsidy from the Department of Heavy Industry. The subsidy part is the headache of the Shipyard.

In order to get the loan from the SDFC, he has to apply through the Union Department of Agriculture. If the present experience of entrepreneurs is any indication, how much time it will take for the application to be recommended or rejected can be anybody's guess. It may be a few months, a year or even more. This delay, which can also be because of the incomplete information given by the application, may result in increase in price and may necessitate a revised offer from the Shipyard, requiring an amended application. Consideration of this will take some more time in the Department of Agriculture. Besides this, the applicant may have problems to furnish the guarantee towards the loan of 95 per cent of the cost of the vessels, to be furnished to the SDFC. Only when this is furnished SDFC will release the loan. If he manages to get the guarantee, he may be able to see the release of the loan. For this, six months can be the minimum period on a miracular basis, and the maximum time required it is difficult to conjecture, at the present trend of attention paid to such cases. One can always expect, of course, that things would move fast and a quick yes or no emerges.

## More Vessels, More Money

Having acquired the vessels by about three years or so after conceiving the project, the entrepreneur's problems of getting 'on' start. He knows that costs of operation will be high, for he was told about it beforehand. He also knows that while shrimp component of his catch brings him sizeable returns, fish will fetch him low price. Good fishing times start alternating erratically with bad fishing periods. There will be earnings, but almost the entire amount will be offset towards annual repayment of loan and payment of interest. During bad seasons, he requires working capital which it becomes difficult to get. He struggles and somehow manages. Having got into the industry, he faces several difficulties to get on. Having got on for sometime, he learns from experience that, unless he has a viable fleet, it is difficult for him to run the enterprise economically. Safety is there in numbers, he realises. If there are six vessels, even when a couple of vessels are sick, still there will be inflow of funds.

## BUT HOW?

The acquisition of more number of vessels becomes a problem for him. He has to increase his company's equity enormously, if he has to secure loan from SDFC for more number of trawlers. He therefore thinks, why not I charter some vessels from abroad, in order to build up some capital? Then a question comes to his mind. If the operations of the chartered vessels will be successful, and if I have to buy the trawlers as per the regulations, can I raise the equity to get loan from the SDFC? He consults well-wishers and they say that the returns generated will be sufficient to form additional equity. He then makes efforts to charter vessels. The foreign fishing companies do not show much interest. Thais and Taiwanese probably think that, they are well off through poaching and do not respond well. Fishing companies of few other countries say that the conditions put forth by him, in line with the Government regulations, will be acceptable to them only with some modifications. After giving several clarifications to the Government on his proposal after convincing the foreign party, he manages to secure Government's permission. Overjoyed, he contacts the foreign fishing company to send all particulars of their vessels to be given on charter to him for securing final clearances. Then comes the information as a bomb shell that, as there has been so much time-lag, the foreign company had deployed the vessels elsewhere. He is back in square one.

He reflects over the situation. Why did EID Parry (India) Ltd., close down their fisheries wing? Why has, ITC Limited recently decided to pull out from this industry? According to their admission and what others say, the industry by itself is not uneconomic. His experience also is that it is not uneconomic. But it will be enduringly economical only with a viable fleet. Larger houses are not given loans from SDFC. They feel that they cannot raise funds from any other agency to acquire additional vessels. For this reason they closed down their operations.

## It Is Not So Easy

He feels that those who cannot expand their fleet have to get out of the industry sooner or later. But getting out is also not easy. EID Parry (India) Ltd, could get out with considerable difficulty. ITC can find several who want to buy their trawlers. But these buyers need money to pay. Although their prospective buyers are State Fisheries Corporations they also cannot raise over a crore of rupees easily. Necessarily the deal will take a long time to materialise.

One hears now and then that Union Carbide too wants to pull out, whatever be the reason. Hypothetically speaking, even if they want, it is not easy for them. Their assets are worth over Rs. 9 Crores. They are a multinational. A multinational has many conditions imposed by the Government to fulfill, to sell a property of this size. Assuming this will be forthcoming, can there be a buyer who can invest Rs. nine crores, to acquire the vessels and other assets from Union Carbide? The first question that strikes is, if the mighty Union Carbide has reasons to leave something they built up over years assiduously, are we right in taking over? Even if there is a reassuring answer, the raising of the required money comes in the way for an Indian buyer. All this is imaginary, but will prove how difficult it is to get out.

In a nut-shell the characteristics of the deep sea fishing industry are — One cannot get in easily; if one can get in he cannot get on; having understood how to get on, he cannot do so. Feeling dismayed he wants to get out but cannot; he has to wait for a pragmatic policy and strategy, devoid of the present barbs and arrows, to be formulated and implemented by the Government.

Fishing Chimes, Editorial 10: January 1982: Vol. 1, No. 10

# Into Nineteen Eighty Two

Fishing Chimes, extends warm new year greetings to its readers. We fondly wish that the new year holds a rewarding future to the Fishing Industry.

The Marine Fishing Industry was an orphan in 1981, none to help it out of the crisis of mounting operational costs and falling prices. 1981 was characterised by carrots dangled before the Industry by the Government on the one hand, and a stick brandished in the other to make sure that carrots were reached and taken by the Industry only by passing through a long suffocation tunnel. It was further characterised by the unprecedented sufferings endured by the Industry on account of the economic crisis to which it had to subject itself.

The Industry has learnt, by experience, to live on hope and perseverence, giving positive construction to various developing events in the year that were of a depressing nature and believing that they would one day lead to the good of the Industry. Darkness leads to light. Bad fishing times alternate with good fishing times.

The year was punctuated by major but unproductive measures of assistance by the Government to the Deep Sea Fishing Industry. A revised charter policy was announced, based on which some long pending applications for chartering of fishing vessels were cleared, but there was no actual arrival of any vessels. The Department of Heavy Industry announced a subsidy for indigenously constructed trawlers to the extent of 33 per cent of the cost. Permission was also given to import machinaries and equipment upto 20 per cent of the cost of an indigenously constructed vessel, without customs levies. However the concessions did not evoke any tangible response either from the Industry or from the shipyards.

A major development during 1981 was that the post of Fisheries Development Commissioner created in the Fisheries Division of the Union Department of Agriculture. Dr. P.V. Dehadrai had the distinction of being the first person to occupy that seat.

A new unit of Central Institute of Fisheries Nautical Engineering and Training was set up at Visakhapatnam. The fishing harbours at Roychowk, Vizhinjom and Tuticorin became operational. Several minor fishing harbours at places such as Jaffrabad, Karwar, Baliapatnam, Mopla Bay, Kasargode, Dhamra, Mallipatnam. Kodikkarai, and Cuddalore have also been completed, Goa Shipyard constructed and delivered fishing vessels to the Exploratory Fisheries Project and Central Institute of Fisheries Nautical and Engineering Training.

Under the Bay of Bengal Programme of FAO, new horizons were reached. New designs of surf-landing boats were tested and popularised. Pair trawling with small mechanised boats was introduced.

An Anti-Poaching Bill was passed in the Parliament, providing for heavy penalties to foreign fishing vessels caught poaching fish from Indian waters.

SDFC sanctioned new loans for the acquisition of two trawlers from Holland during the year.

In the field of research, a break-through was achieved in the hatching and rearing of eggs of giant fresh water prawn by C.I.F.R.I. at their Kakinada Unit. There was also a similar breakthrough in the hatching of eggs and rearing of larvae of tiger and white shrimp in Kerala, Tamil Nadu, Maharashtra and Andhra Pradesh.

Considerable progress was made in the country in the popularisation of carp hatcheries. The number of Fish Farmer's Development Agencies set up went up significantly.

There was a major tragedy during the year on board MFV Matsya Harini, a vessel attached to EFP, Cochin. Three persons died in the fish hold of the vessel, on account of poisonous gases that accumulated there-in, owing to putrefaction of fish. Several fishermen also died on both the coasts during cyclones and owing to mishaps.

Only one new trawler of 23m length, imported from Holland was added to the fleet during the year (in November 1981). Besides this, two wooden Thai trawlers were brought by Tata Oil Mills, but are not operational, presumably for want of clearance from the Government.

There was no addition of chartered vessels during the year. In fact, the few vessels on charter, based on earlier permissions discontinued operations, as the permissions came to a close. Poaching by Thai and Taiwanese vessels was rampant during the year. Several poaching vessels were apprehended by the Coast Guard around Andamans, near Sandheads and along Gujarat Coast.

There was a fall in export of marine products during 1980-81, 12 per cent in terms of quantity and 5.6 per cent in terms of value. There was no tangible increase in fish production.

There were enormous delays in the Government in regard to clearance of applications for chartering of fishing vessels and in respect of cases for import of fishing vessels.There were promises from the Centre that relief would be provided from the high cost of diesel oil but this did not materialise. There were also promises from the

Government of Andhra Pradesh that relief from Sales Tax on diesel oil, from purchase tax and marketing cess would be provided. But these also did not materialise. The Central Fisheries Corporation, Calcutta was closed down by the Government of India as the Corporation has failed in its functions and has incurred heavy losses.

The problem faced by the fishing companies in securing Interim Guarantees to be provided to the SDFC for obtaining release of loans continues. Government of India could not agree to provide these guarantees on behalf of owners, although such a course is justified in the interests of development of the Industry.

Several skippers left the country to avail of lucrative opportunities abroad. This drift may continue in 1982. ITC Ltd. closed-down their fishing division and have offered their trawlers for sale. Marine Products Division of Union Carbide is also stated to be functioning under considerable strain.

The outlook for 1982 can be brighter. The Minister for Agriculture, it is stated, is taking special interest to give a fillip to the Industry and it can be expected that this will be reflected in the promptness with which the officials in the Fisheries Division will handle the programme. The Second stage of the fishing harbour at Visakhapatnam will become operational during the new Year, providing relief from the present congestion. Several other harbours such as those at Madras are also likely to become operational.

Regarding increase in fishing fleet, there is no likelihood of any substantial step-up, unless the Government adopts a more pragmatic and enforceable policy. However, eight new trawlers from Holland and two from Australia are likely to be added to the fleet during the new year. Apart from this, one can expect that, in spite of various delays, construction abroad for about 12 trawlers from various Indian Companies will commence. A few chartered vessels may be introduced during the new year but the outlook for a large scale introduction is gloomy.

In regard to indigenous construction of commercial fishing vessels, there is a possibility of a beginning to be made in the year. There appears to be no prospects of any major joint ventures in the field coming up. Brackishwater farming, freshwater fish farming, prawn and seed fish production are likely to gain momentum during the year.

Future role of the newly farmed Bureau of Fish Genetic Resources at CMFRI, Cochin would bring in significant progress in data collection on important aquatic organisms, particularly of culture and commercial value. The Government of India has already sanctioned Rs. 22.67 lakhs during Sixth Plan period for the Bureau.

Two more new centres are proposed to be set up at Kakdwip Research Centre, CIFRI in West Bengal and Dhauli Research Centre, CIFRI in Orissa, for intensified work on brackishwater farming.

It is true that both the Fishing Industry and the Government have their own as well as common problems. On the premise that all problems can be solved if there is a will and the solution for any problem lies within the problem, combined efforts are necessary to resolve the problems and create the climate for faster development. Government will have to pay an aggressively benevolent role in this respect. Let us hope that 1982 holds the key to launch a new wave that leads to a swifter developmental tempo and prosperity to the industry and thousands of men depending on the industry also fill the Indian market with fish for internal consumption and the reefer spaces in cargo vessels for export.

Fishing Chimes, Editorial 11: February 1982: Vol. 1, No. 11

# Develop India into a Fishing Nation

The main occupation of the majority of the Indian population is agriculture. This is because of the vast extent of land available for agriculture. Government's policy all along has been to extend all support to promote this activity. No other profession in India has a claim to have the same importance as agriculture. India has become an agricultural nation.

Fisheries sector, which has so far had a low priority, has now to be looked at from a new angle. With the declaration of the Exclusive Economic Zone of the seas around the country to the extent of 200 miles outwards from the land mass, there is now a sea area of 2.3 million square miles added to the national jurisdiction which is almost equal to the agricultural land in the country. In other words, there is now a fishable area, available to the nation, that requires in the hands of the Government the same attention as agriculture gets. This basic fact has to be recognised by the Government. The sooner it is done, the better.

The present policy of the Government of India for the development and exploitation of our marine fisheries in the Exclusive Economic Zone, is based on limited objectives. The task of achieving a commanding position to protect and utilise the fisheries resources of the area requires a historically significant policy frame work, that would trigger off a massive and at the same time a purposeful activity.

India occupies a very unique and unrivalled position in the Indian Ocean. Piercing into the Indian Ocean in the South, and with the vast Bay of Bengal on the east, and the expansive Arabian sea on the west, great advantages can be reaped by developing India into a fishing nation.

The basic policy of the Government of India should be to shape India into a fishing nation. Once such a policy is framed, it will provide the necessary strength to draw up a major strategy to transform such a policy into action. Without such a declaration, the developmental effort will continue to be based on short range objectives, as is the case now.

Such a basic and conscious policy has various national advantages. It will provide a fillip through active state support for the introduction of a large number of deep sea fishing vessels, adequate to cover our Exclusive Economic Zone. Such an introduction will totally deter foreign fishing vessels from intruding into our area. The deployment of a large fleet can provide a massive information to the Navy about the movement of foreign vessels in the area.

## Antarctica

The recent Indian expedition to Antarctica will soon provide opportunities to us to exploit krill, available in abundance in the southern ocean. Unless India is a global fishing nation, as is the case with Norway and other countries, such a major activity will not be possible. To gain entry into the club of nations now venturing into the area, a complexion of equality is of paramount importance. And this complex can be developed only if we have a strong fishing fleet. In the 21st Century, krill will assume enormous importance as human food and we have to prepare ourselves for it.

## Strong Fishing Fleet

To be a fishing nation and thereby to build a fishing fleet of sufficient size to cover our waters of EEZ and the adjacent International waters, would mean the acquisition of the requisite confidence and ability to have a strong fishing fleet and provide adequate fish and fish products to our teeming millions and to export them to other countries. As a major fishing nation with a large fishing fleet, our exports can be stepped up enormously and we can become number one exporting nation in fish and fish products. The earnings in foreign exchange thereby will be enormous.

The development of a large-sized fishing fleet will prove to be a great support to our Navy in times of war. The large number of fishing vessels can quickly be equipped and deployed to protect our long coastline. They can also collect and provide information to the Navy on movement of foreign vessels in the area. The bringing into being of a large commercial fishing fleet will obviate the need for heavy direct investments by the Navy in adding a large number of smaller craft to protect the coast.

By virtue of its pre-eminent position in the Indian Ocean and to strengthen its economic and political power, looked at from any angle, India should be developed into a fishing nation. Smaller countries such as Thailand, South Korea, Hong Kong, Taiwan etc. have far bigger fishing fleets than us. It is no longer wise on the part of the Government to let India continue in an inferior status in the fisheries sector, in the face of the various advantages that can be cornered by having a large fishing fleet.

## Fisheries Ministry

In order to declare a policy that India will be developed into a fishing nation, several aspects have to be looked into by the Government of India. One is that a separate fisheries ministry has to be set up.

It should have a strong field organisation under it to promote the introduction of deep sea fishing vessels either in the public or private sector. Enforceable strategies and procedures to facilitate quick introduction of the vessels have to be formulated. Existing exploratory and stock assessment organisations have either to be strengthened or reformed. Exclusive yards for the construction of fishing vessels have to be set up and developed with appropriate foreign expertise from friendly nations. Training programmes have to be revamped to provide adequate man-power for operating the vessels. Ancillary industries for the production of hundreds of components required for the fishing vessels have to be developed. A well-radiating internal marketing system for fish, consisting of cold/frozen storages and ice-plants at a number of viable points have to be set up, with a proper transporting system for the flow of fish from landing points to the people has to be promoted. Effective working procedures, under statute, within the warehouses to regulate inflow, storage, auctioning and outflow of fish have to be developed.

Our sea fisheries can play, from the point of industrial development, an important beneficial role, providing food for the people, promoting employment to the weaker sections of the society, and economic emancipation. While the developmental task is gigantic, a beginning has to be made without further delay. To make the beginning, a policy frame work to build India into a fishing nation and the setting up of a Ministry for Fisheries are prerequisites.

Fishing Chimes, Editorial 12: March 1982: Vol. 1, No. 12

# Encouraging Signs

Fishing industrial circles notice a new sense of realism displayed by the higher echelons of Krishi Bhavan and the accent seems to have taken a focus towards growth and results. Overcaution and dilation seems to have yielded place in the Fisheries Division of the Agriculture Department to purposeful and enforceable action that would make the policies work. The change in the way of working seems to be the result of the infective inspiration flowing from the Agriculture Minister, Rao Birender Singh, State Minister Dr. M.S. Swaminathan, Agriculture Secretary Mukherji and Additional Secretary Acharya, the pivotal role being attributed by many to the positive dynamism of Dr. P.S. Dehadrai, Fisheries Commissioner. The fishing industry hopes that the trend will last, and will not just be a spurt.

The above comments are based on facts. Several charter applications pending for a long time have been cleared with speed, resulting in issue of letters of intent. Only one or two valid applications remained and these are also going to the screening committee. Unreasonableness of certain conditions geverning charters such as provision of bank guarantee for Rs.2 lakhs for every chartered trawler, treating a single unit of bull trawelers as two trawlers has been conceded and the conditions have been relaxed. Now only a guarantee for Rs.2 lakhs need be given for all the trawlers in an application. A pair trawling unit will be deemed as a single unit. Further, it is stated that the fisheries division has also noticed that the condition that a charterer has to confirm his decision to buy his first trawler after an operation of a chartered trawler for three months is unreasonable. One can expect that this anamolous condition would also be relaxed, as it is completely unrealistic. How can anyone judge economics of operation in three months? Unless atleast an year-round operations are completed, no assesment is possible.

There are indications that the Fisheries division may consider relaxing the conditions even further, if a fishing company chooses to charter squid-jiggers, lobster vessels and such other types that are new in our waters. There also seems to be an increasing recognition of the role larger houses can play in the development of this industry, which, unlike other industries, bristles with risks and high costs. The fact that the larger houses are getting disinterested and withdrawing from the field is affecting the complexion of the industry.

In the past one month long-pending applications concerning change in source of import have been cleared with quickness in the Fisheries Division. Like-wise several applications for loans for acquisition of fishing vessels had also been recommended to the SDFC, where they have been cleared. All the above and several other events are encouraging signs of a positive approach in the Agriculture Department in regard to fisheries matters.

Close co-operation between the fishing industry and the Fisheries Division of the Department of Agriculture seems to be progressing fast. This is a welcome sign. The credit for the development of the equation largely goes to Mr. N.P. Singh, President of the Association of Indian Fisheries Industries. The complete lifting of the veil of undue suspecion and doubt in respect of the requests or applications from the industry in Governmental circles will be the starting signal for the rapid development of the industry. It is to be hoped this will soon materialise, for, experience must have shown that undue suspicion and doubt and imposition of undue and unenforceable restrictions would retard and do damage to developmental process. The net result of this approach may only be the enhancement of the feeling of importance among the officials and the sad recognition of this by the fishing companies. If requests are just conceded there may be no realisation of the power one holds. Withold it, the other side would know.

No one wants the accumulation of this kind of importance, at a time when a fast development process has to be set in motion. Misuse of permissions and sanctions, if these are there at all, is not peculiar to the fishing field alone. These occur all over the world. Instead of anticipating these and spending enormous time to avoid the unlikely event of a misuse, it is desirable to clear all reasonably complete applications with expediency. Some of these may misfire, but that should not matter. The successful ones will form a solid gain.

One reason as to why the proposals remain pending in the Government for long seems to be that they are often incomplete. This leads to delays in processing. All officers do not think alike. Some would like to take a lighter view of omissions and other would want these to be furnished.

Lack of delegation of authority in the Fisheries Division tends to delay decisions. It is therefore necessary that greater delegation of authority is brought about in the Fisheries division. This would expedite matters. The industry also should increasingly avail of the services of consultants. This will facilitate the presentation of applications in complete form to the Government.

Let the Government and the industry hope alike and look together for the development of a prosperous fishing industry.

Fishing Chimes, Editorial 13: April 1982: Vol. 2, No. 1

# Inter-State Interactions

Heard of fish doctors? In West Bengal you find them moving with their kits during rainy season to various wet and dry bundhs to inject fish with pituitary harmone extract, so as to induce them to breed. Their time is precious and they visit places on prior engagement. Brooders are kept ready by the farmers at the appointed time. The doctor examines, gives harmone injections, issues follow-up instructions and goes away to the next place. He will visit once again to see the results and administer another dose of injection, if required.

In Karnataka there is a well developed internal fish marketing system, organised by the Karnataka Fisheries Corporation, although in a limited area. The Karnataka Fisheries Department has started a scheme under which Industrial Fishery Estates are established by Government which are given on hire-purchase to entrepreneurs. Karnataka has also achieved a break through in purse-seining with medium -sized boats.

In Andhra Pradesh, commercial seed fish raising and seed banks have been developing fast in the private sector. Freshwater prawn seed production has been standardised at the CIFRI unit in the State and Freshwater prawn farming has also been taken up there. The marginal areas of Kolleru lake in Andhra Pradesh are being increasingly converted into fish ponds. Scores of such ponds have sprung up. Brackishwater farming in a major way has made impressive strides in Kerala, Tamil Nadu and West Bengal. Marine prawn breeding and raising of their larvae to stocking stage has been achieved by Mr. K. H. Alikunhi in Kerala and also by CMFRI, CIFE and CIFRI.

Madhya Pradesh and West Bengal are proud of bundh breeding technology that they have. Madhya Pradesh, U.P. and Tamil Nadu have set up impressive records in reservoir fisheries development.

Arunachal Pradesh has developed a community fish farming system. Fish Farmers' Development Agencies at Allahabad and Jaunpur in Uttar Pradesh have achieved spectacular results. Modern carp hatcheries in Haryana and Madhya Pradesh have brought in a new trend.

Punjab and Haryana have evolved scientific systems of regulation and development of riverine fisheries. In these States where there was till recently indifference towards eating fish, a recognisable preference for fish has come about. In Orissa, new programmes of mariculture are taking shape.

In regard to organisation and implementation of fisheries development programmes, each State has adopted its own system. Several States have introduced innovative schemes to promote freshwater fish culture, brackishwater farming, small scale marine fishery development, housing and water supply for fishermen etc.

## What's Going on Over There?

What have been mentioned above are illustrative of the quiet work that is going on in various States. However, the useful work going on in a State and the techniques adopted by it are practically a sealed book to the others. Not that the State concerned does not want the knowledge to flow to the others. Nor is it that other States do not want to be benefited by the experiences gained in the State concerned. It is just that there is no established system under which inter-State interactions in the fisheries field can be fostered.

While inter-State interactions have to be fostered, there is also the need to bring about close relationship between the various central fisheries research institutions and the State fisheries departments. Several improved or new techniques evolved and conclusively tested in the field have not yet found field application in the various States. This is because of the absence of an effective machinery to handle the job. In fact, the techniques evolved tend to spread in the State in which the respective institute is located and fails to go farther. Thus, it is a case of lack of inter-state interaction:

It is accordingly imperative that a system to bring about interactions, particularly in respect of activities that will have a commercial impact, among various States on one hand and between various States and central fisheries research institutes, on the other, is brought about.

## Communication

The first step in this respect is communication. The fisheries department of a State which believes that something has been done by it, should communicate the details to other States concerned. This can be in regard to the implementation of a known scheme through improved or effective field organisation or spread of a new technology, which, in course of time will catch up in the commercial sector. Similarly the Central institutes should pass on details of their proven techniques of a new or improved nature to all the State directorates of fisheries and to commercial fisheries establishments. An updated list of existing commercial establishments and entrepreneurs interested in the activity should be maintained by the State and Central fisheries organisations concerned.

## Action

Once this is done, it is for the receiving side to investigate, evaluate and decide upon the follow-up action to be taken. If it is found worthwhile to adopt the reported system or technology, the receiving side can depute their representatives to the place concerned for a detailed field study. Another way is to secure the visit of the key person concerned in the State in which the improvement took place to visit the State on the receiving side for a field study and offering suggestions. After this, further follow-up action can follow.

Some may feel that there can be a centralised set-up, to be made responsible to handle the above-mentioned functions. They can receive details of developments from various State fisheries departments and Central Research Institutes and disseminate them, among others. This was however tried in 1960s and 1970s.

Central fisheries extension units were at one time set up in most of the States. They were closed down, probably because they were unsuccessful. Later, each of the Central Fisheries Research Institutes set up their own extension wings. Probably owing to various constraints, however, their success is not impressive. Nevertheless, their performance is far superior to the erstwhile Central Fisheries Extension units.

One way is to strengthen the extension units attached to the various Central Fisheries Research Institutes and make them really dynamic. Since any field development in fisheries work will some way or the other is connected to the results of applied research at one institute or the other, these units will be in an excellent position to maintain contacts with the State fisheries departments. If these departments do not supply information on innovative developments in their States on their own, the extension units can make them do so. On receiving the details, they can pass them on to the various concerned States and follow-up the matter to their logical conclusion.

Of course, most of the Central Fisheries Research Institutes have their newsletters. The accent and main thrust in these newsletters can be more on explaining the benefits of new techniques and guiding the State fisheries departments and entrepreneurs in popularising them in the commercial sector. If specific steps in regard to the manner in which this should be done are explained, the follow-up work will be easier. Repetitive publicity and personal contacts are essential to achieve results.

## The Catalyst

*Fishing Chimes* has no doubt a vital role to play, supplemental to the efforts of Central and State fisheries organisations, in bringing about inter-State interactions. The journal has all along been striving to perform this task. However, the support of various Governmental organisations and commercial units in providing requisite material is necessary in performing this function more effectively. Further, the journal has to be read by all functionaries in the Central and State fisheries organisations and commercial establishments, to evaluate the material presented therein and to take follow-up action. We therefore appeal to all heads of departments of Central and State fisheries organisations and all commercial fisheries establishments to permit the various functionaries upto the lowest level to subscribe to the journal.

Fishing Chimes, Editorial 14: May 1982: Vol. 2, No. 2

# Fillip to Fish Farming

Available resource statistics tell us that India has 1.6 million hectares of tanks and ponds. When properly farmed for fish, these water areas are capable of producing about 2.4 million tonnes of fish per annum valued at Rs. 1,200 crores. Against this the present level of fish production from these waters is estimated as 0.95 million tonnes valued at Rs. 475 crores. Thus only about 40 per cent of the available potential is now being tapped. It is anticipated that by the end of the Sixth Plan about 1.5 million tonnes of fish per annum will be produced, representing a level of utilisation of 63 per cent of the potential.

Besides tanks and ponds, there are areas such as *beels, mauns,* abandoned irrigation tanks, marginal areas of lakes, unused lowlying areas etc. All these have a considerable extent. When converted into ponds, they can add to freshwater fish production significantly.

There are also lowlying swampy areas along the coastline, with an estimated extent of 1.54 million hectares. One third of this area is likely to be suitable for conversion into brackishwater farms and can yield 0.255 million tonnes of shrimp valued at Rs. 1,275 crores. At least another 2.5 million tonnes of fish valued at Rs. 51 crores can also be harvested from these resources, in addition to shrimp.

## Potential

Thus intensive development of freshwater and brackishwater culture has the potential of providing the country with 2.4 million tonnes of freshwater fish, 2.5 million tonnes of brackishwater fish and 0.255 million tonnes of shrimp, all valued at Rs. 2,521 crores of which Rs. 1,275 crores will be the foreign exchange component through export of shrimp. Against this, at present, while foreign exchange earnings through cultured shrimp are negligible, internal returns to national income are a mere Rs. 475 crores.

The utilisation of all the available resources will lead to the employment of over 20 lakhs of persons.

In order to promote the progressive utilisation of the resources through intensive culture the Government of India and the State Governments had introduced certain schemes. So far as intensive freshwater culture is concerned fish farmers development agencies in 50 selected districts in seventeen States were set up by the end of the Fifth Plan. During the Sixth Plan 58 Nos. of such agencies have been set up so far under the World Bank's Inland fisheries project, in the States of West Bengal, Bihar, Orissa, Madhya Pradesh and Uttar Pradesh bringing the total number of such agencies to 108. The Government of Uttar Pradesh has sanctioned the setting up of an additional number of agencies one in each of 29 districts in the State, bringing the total of such agencies in the State to 48. With this increase, the total number of FFDAs in the country goes up to 137.

FFDA has an integrated approach to the development of intensive fish culture. Being an autonomous body, it possesses freedom of timely action, based on decisions taken by a managing committee headed by the Collector/ Deputy Commissioner of the district concerned. Candidates are recruited and trained in techniques of fish farming. Ponds are provided to them on long term lease. They are enabled to secure loans for capital investment from banks. The engineering staff assists them in carrying out improvements to ponds and also in ensuring water supply. Extension staff assists them in adopting correct farming practices. Inputs are supplied at subsidised rates. They are also assisted in marketing of catch of 15 lakh tonnes of fish at remunerative prices. All this will be done under a centrally sponsored scheme.

## Results Could have been Better

In seven years of existence of FFDA, not more than a few thousand hectares could be brought under intensive production, that too with average results not commensurate with the anticipated levels that could have been achieved through adoption of available technology. The average production level in tanks and ponds covered by FFDAs continues to be at about 600 kg/ha, which was the national average even prior to setting up of FFDAs. In other words, FFDAs have been successful only to the extent of bringing production in certain water areas from a level of 50 kg/ha to 600 kg/ha.

The net income per hectare has increased from Rs. 10.50 in 1970 to Rs. 350/- in 1978-79. However, the fish yield went up to 6000 kg/ha in a few cases and several farmers achieved production of over 1200 kg.

A World Bank Project to develop intensive fish culture in 117,000 hectares in 58 districts in the States of West Bengal, Bihar, Orissa, Madhya Pradesh and Uttar Pradesh, as already stated, has become operational from April 1980. FFDAs are the vehicles through which the scheme is being implemented. One important aspect is that Fisheries Corporations have been set up in each of the States, mainly to undertake the production and supply of fish seed to the farmers. Modern carp hatcheries are being set up in all the project States by the Corporations as part of seed farms for the purpose.

## Low Utilisation of Funds

Under the centrally sponsored scheme relating to FFDAs, the expenditure by way of subsidies incurred by the Government of India in 1978-79 was Rs. 31.73 lakhs and Rs. 0.60 lakhs in 1979-80. The Sixth Plan provision was Rs. 500 lakhs and the outlay for 1980-81 was Rs. 50.00 lakhs. Against this outlay, the expenditure was only Rs. 17.44 lakhs. The outlay for 1981-82 was only Rs. 30.00 lakhs, a considerable reduction over the outlay over 1980-81.

For the World Bank Project for Inland Fisheries Development, the outlay under the centrally sponsored scheme in the Sixth Plan was Rs. 762 lakhs. In 1980-81, against the earmarked outlay of Rs. 50.00 lakhs only Rs. 29.10 lakhs was utilised. In 1981-82 the outlay was scaled down to Rs. 35.00 lakhs.

## Dormant Schemes

Apart from FFDA projects, there is a scheme in the centrally sponsored sector in the Sixth Plan with an outlay of Rs. 800 lakhs entitled 'National Programme for Fish Seed Development'. In 1980-81 and 1981-82 practically nothing seems to have been done in implementing the scheme. The prospects of implementing the scheme in 1982-83 also appear to be bleak.

So far as brackishwater farming is concerned, the Government of India has included a scheme in the Sixth Plan for the establishment of prawn hatcheries and prawn farming in maritime States with an outlay of Rs. 500 lakhs. There has been very little perceptible progress under the scheme so far, although two years have already elapsed.

Rapid expansion of fish farming in all available water areas will provide employment to several thousands. Unproductive areas hitherto lying either waste or under utilised will start yielding economic returns. Fish in abundance will become available to the people, which will have the effect of promoting health standards. National income will go up substantially.

## Multiply FFDAs

Considering the above aspects, it is in national interest to see what further steps are desirable to give a fillip to fish farming. Looked at from any angle, the logical conclusion one will arrive at in this context is to multiply the number of FFDAs to cover all the districts in all the States, and tone up their functioning. If necessary, more than one FFDA can be established in a district, if the water resources justify. The setting up of the agencies to cover all the districts would provide a total field organisational frame work.

## Co-ordination Missing

One reason as to why several agencies have not been able to show progress seems to be the absence of an effective co-ordinating and guiding force. As the Collector/Deputy Commissioner of the district happens to be the Chairman, the Directors of Fisheries do not seem to exercise the functions of guidance and co-ordination as required. Further, tested personnel with dedication towards the type of work they have to handle are not always posted.

## Commercial Orientation

The work under the FFDAs is essentially one of a commercial nature. FFDAs perform the function of promoting the commercial activity of fish production. It is therefore necessary that the commercial orientation of the work is understood clearly.

## Corporations to play a Role

There are State-owned Fisheries Corporations in most of the States. In the same way, it is desirable to set up similar Corporations in the rest of the States. The Director of Fisheries should also be appointed as the Managing Director of the Corporation concerned. In order to help him in his work an officer of the rank of an Additional Director of Fisheries should be appointed as a Chief General Manager who will virtually run the Corporation, subject to the control of the Managing Director.

The overall function of such Corporations in Inland States would be to develop intensive fish culture in tanks and ponds inclusive of setting up of seed farms equipped with modern hatcheries. Reservoir, lake, and river fisheries development may be kept out of their purview and left to the fisheries departments, as the technologies to lead to commercially viable activities in these resources are not yet fully established. In maritime States there can either be a separate Corporation for culture fisheries activity or there can be two separate wings to a single Corporation, one to deal with marine activities and the other to deal with culture activities.

The Corporations, so set up, will be in a better position to guide the FFDAs and coordinate their activities. Financial assistance from Commercial Banks to the farmers can be organised quicker by the Corporations. As producers of fish seed, the Corporations will also have the responsibility to ensure optimal fish production, by keeping watch over training, long-term leases, stocking performance, care of the crop, harvesting and marketing etc. From these considerations, it is imperative that the FFDAs are brought under the direct control of the Corporations, to be established wherever these are not there. In other words, if FFDAs, to be set up in all districts, function under well managed Corporations, the activity is likely to spread faster with lucrative results. It should be the endeavour of the corporations to work towards the abolition of subsidies in a phased manner and also to collect from the farmers, in course of time, a small percentage of the returns from sale of fish, to cover their proportionate administrative expenses and also of the FFDAs concerned.

Development of intensive fish culture in existing ponds and tanks is one aspect. Another aspect is to form new fish farms in lowlying areas lying waste. The Corporations can invest on converting such areas into fish farms and lease them out to trained personnel on long term basis to enable them to develop fisheries in such reclaimed areas. The FFDAs in which such areas are situated can be involved in the activity from the early stage, so that, after the allotment, further attention to culture aspects, marketing, recoveries etc. can be bestowed by them as part of their work.

The Corporations can also be entrusted with the job of development of brackishwater fish farming. At the present level of experience in the field, it will be difficult for entrepreneurs to undertake the work on their own. The proposed Corporations can select suitable and viable sites for setting up farms taking into account the tidal amplitudes, soil conditions, depth contours, availability of seed etc. and undertake the construction of farms on their own progressively. The farms so constructed may be given on long-term hire-purchase lease to the applicants selected on the basis of an advertisement. Such leases can be given to a single party or to several individuals. When given to several persons, the individual ponds can be given on hire purchase, with the common supplies and services retained in the hands of the Corporation concerned. The prospective owners will pay a small fees to the Corporation concerned for the purpose.

Specific down-to-earth steps aimed at achieving results on the lines indicated above are to be taken, if the national potential for inland fish culture including brackishwater culture is to be utilised effectively in the foreseeable future.

Fishing Chimes, Editorial 15: June 1982: Vol. 2, No. 3

# Don't be Fishy

There have practically been no investments in deep sea fishing industry since 1978. This stagnation is attributed by the fishing industry to the defective policy and strategy adopted by the Government of India.

Able and deep-thinking men in the Government preside over the future of the fishing industry. There can be no second opinion as to their sincerity to promote the industry. It was they who initiated new approaches to the development of deep sea fishing. Withstanding the invincible opposition for the import of trawlers under the 1977 scheme from Department of Heavy Industry and Ministry of Shipping and Transport and neutralising it, the Department of Agriculture introduced the scheme. When the Shipping Development Fund Committee closed the loans window for fishing vessels, the Department of Agriculture wrenched it open again. It also succeeded in prevailing upon the Department of Heavy Industry to announce 33 per cent subsidy on indigenously constructed vessels. It introduced the scheme for chartering of fishing vessels. Having done all these, the department, it is obvious, must have come face to face with several hurdles in implementing the policies from within the Government and from the Industry. Otherwise, there should have been considerable progress.

## Unfair

In the past few years, after the financing of the import of Mexican trawlers, SDFC could release loans only for a few Dutch trawlers, that too, on the strength of the release of an equivalent amount as grant by the Dutch Government. In the case of these trawlers, taking advantage of the lenient attitude of SDFC and the Department of Agriculture, the concerned companies failed to pay their share (10 per cent) of the cost of the trawlers in full. This situation put the SDFC in difficulties. Earlier, under the 1973 import scheme, even after taking release of the loan for the import of two trawlers from Mexico, one company failed to bring the vessels to India, although the vessels were duly taken delivery of. This created a serious situation. The Government of Andhra Pradesh which gave the interim guarantee and the SDFC which accepted the guarantee both were placed in an embarrassing position. One or both of these authorities have to bear the loss of several lakhs of rupees.

The owning companies of the imported Mexican trawlers have to repay the loans and also pay interest to the SDFC periodically. Out of the six small companies to whom loans were granted, only one company has been paying the dues regularly. One company, as mentioned above, failed to bring the trawlers and so there has been no repayment at all. All the remaining four, with varying degrees of partial to no payments, have created a major problem to the Government as well as the rest of industry. As these companies have become defaulters, SDFC has tightened its approach.

Why have these companies not been paying their dues? It is true that shrimp catches were poor for sometime, prices also fell, and operational costs went up. These causes must have obviously reduced the inflow of returns. Yet, when one of the companies, in the same position, could pay the dues, it should have been possible, although with difficulty, for others also to pay. But it was not done and is not being done. Performance of this kind has naturally unleashed caution in Governmental circles, which has retarded the developmental process.

## SDFC Man Should Attend Board Meetings

One representative of the SDFC is there as a Director in each of these owning companies. But, it is sad to say, with the exception of one or two companies, these Directors rarely attend the meetings. This handicaps the SDFC from getting a true picture of the functioning of the companies and their financial management. Large amounts can be diverted as loans to other activities, depriving SDFC of its right to get repayments first. It is not that this happens but this can happen. If all the Government Directors have been functioning well, probably the situation would not have been such as to make the SDFC so cautious. This caution has been having a serious effect on the promotion of the industry.

## Fast Buck

Let us go a little deeper into this matter. When mechanised fishing boats were first introduced, by providing heavy subsidies and easy loans, several owners adopted a fast-buck approach. They cornered incomes and dodged repayments. When there was pressure for payment, they either handed over the boats or abandoned them. The net result was that the concerned fisheries department was saddled with used boats for reallotment to some one else and the ex-owners were happy with their money for investment elsewhere. The same trend continues even today to a considerable extent. There have also been a number of incidents of accidents resulting in total loss of boats out at sea and insurance companies suspect that these were wonten. In the case of larger vessels also, acquired with loan from SDFC, such a trend, if suspected, has to be countered. Having recovered their limited investments and having made further money over and above this, if the owners are mischievous, it is possible that they could even think of surrendering the vessels to SDFC or their bankers to whom they must have been mortgaged. Such a short sighted approach will do considerable damage to the future of the Industry.

## Tata Charters

In the case of chartered vessels, with the exception of Tata Oil Mills and New India, the Government permitted only new comers into the field to charter fishing vessels. While Tatas were prepared to acquire all the 25 Nos. of their chartered vessels, Government did not permit them to do so. Tatas managed to withhold two of the chartered vessels for operation on their own, but final permission to acquire them has either not yet come from the Government or they may have been permitted recently. None of the new companies made any effort to acquire the vessels. Consequently, Government had to tighten the charter conditions.

Both new as well as established companies are entitled to receive consideration from the Government for permission to charter fishing vessels. Yet, owing perhaps to circumstances beyond control, Government has been according these permissions mainly to new companies. A company manufacturing neck ties is one among them.

Larger houses, by virtue of their internal discipline and tradition, will be the most suitable parties to undertake and develop deep sea fishing. One major constraint they face in entering or continuing in the field is the reluctance or caution with which Government considers their requests for permissions. They are not allowed to get loans from the SDFC. So much so, one large house after the other is pulling out of the field. They want to have a stake and involvement. In contrast, smaller companies, on account of their inherent weaknesses, are unable to withstand in the field.

Government has to take a serious look and formulate a sound policy to place deep sea fishing industry on an enduring footing. The industry has to take a long range view and co-operate with the Government. Smaller and medium level companies have to realise their responsibilities and desist from being fishy, controlling the temptations of cornering the earnings to the detriment of commitments such as repayment of loans and interest. In the larger interests of development of the fishing industry, they have to stick to this code of conduct.

Fishing Chimes, Editorial 16: July 1982: Vol. 2, No. 4

# Trash Into Cash

Smaller varieties of fish such as silver bellies, lactarius, carangids etc, do not yield reasonable returns commensurate with their weight and proportionately comparable to the prices that larger varieties such as pomfrets, seer, perches, etc. command. So much so, all larger fishing vessels invariably throw them back into the sea, while sorting out quality species, mostly shrimp, for transfer into the fish hold. This practice is, to some extent, followed by mechanised boats operating trawl nets as well. About 30,000 tonnes of smaller varieties of fish is thus wasted annually. The fishes, on being pushed back into the sea, fall a pray to fishes like sharks. It is stated that there has been an increase in shark population in areas, where larger fishing vessels operate. Obviously, this must be for the reason mentioned above.

## Why Trash is Thrown

Economic compulsions make the operators to throw back the trash fish into the sea. The offers from buyers, when a bulk quantity of trash fish is exposed for sale, plummet steeply. A tonne of trash may not fetch more than Rs. 300/-. In a voyage, a larger vessel may be able to bring 20 tonnes of trash fetching Rs. 6,000/-. The work involved in sorting out and cleaning these smaller fish and storing them is so arduous that the concentration on shrimp and quality fish will have to be sacrificed, for the sake of paltry returns, if these catches are to be taken care of. At the same time, the rejection of so much of fish constitutes a huge waste.

## Financial Inducement Necessary

Only when there is financial inducement, it will be possible for the trawler operators to bring all the fish that they catch. As long as there is an absence of this inducement, the situation will remain the same.

## Problems

There is no processing set up in the country at present that can buy and process the trash fish into value added products. Fish meal plants on the west coast buy oil sardines in the glut season. They are not interested in buying fishes such as silver bellies on account of their low protein content. They can only be mostly used for human consumption.

## Possible Solution

There is much talk about converting trash fish into fish kheema etc. But no headway is made, obviously because of limited demand, There can be a solution to the problem only when a) there is an agency or agencies to buy trash fish at rates remunerative to the operators, b) there are processing units to convert the trash into value-added products and c) there is a systematic and high pressure publicity to popularise the products.

The State Fisheries Development Corporations can play a significant role in this regard. They can open trash utilisation wings at all major landing centres. A higher rate for buying small fish can be announced by them so as to induce the trawlers to bring all the fish caught for landing. They have to assure the operators that what all they bring will be purchased by them. Till such time as processing units for converting these fish into fish kheema and various other such products are brought into being either in the private sector or by the Corporations themselves, they can make arrangements for sun drying or salt-curing of such fish. Such products have a wide market and can be sold particularly in interior areas including tribal areas. The Tribal Development Corporations in the various States can assist in the marketing.

As this kind of processing and marketing goes on, the Fisheries Corporations can set up plants with deboning machines, filleting machines etc. Before releasing the products into the market a network of wholesale and retail marketing centres can be established and along with this,a relentless publicity for the products can be taken up, unmindful of temporary losses.

## Organisational Mechanism Needed

There is lot of literature on technologies available for conversion of trash fish into various varieties of value added products. All this information is of no use, till such time as an organisational mechanism under which catches of smaller fish can be pooled up, offering a remunerative price to the producers, and later made available to the processors, a category of whom has to be created, coupled with the establishment of a demand.

A purposeful beginning has to be made somewhere. Steps as indicated above are bound to lead to a major improvement in the situation. The trawler operators can then make their operations economical. Diversification of fishing can also be achieved.

Fishing Chimes, Editorial 17: August 1982: Vol. 2, No. 5

# The Confrontation

The International Conference on Deep Sea Fishing held in New Delhi from 23 to 25 June 1982 virtually started off in a mood of confrontation, the Government confronting the deep sea fishing industry. The Union Minister for Agriculture, Rao Birendra Singh, accused in his keynote address to the Conference that none of the fishing companies furnished true facts about their operations to the Government. He did not believe that the Indian deep sea fishing operations were uneconomical. When fishing enterprises of nations much smaller than India were making profits, how could the Indian companies incur losses, he questioned. The plea of the industry that the high cost of diesel oil upset the economics of operations did not appeal to him. He commented that those operating chartered vessels earned 15 per cent of the gross proceeds while providing to the Government a guarantee for only Rs. 2 lakhs. Even if this was forfeited, for non-compliance of conditions, it did not matter much to the companies.

It was unfortunate that the minister chose the occasion of the International Conference to give vent to his views. It is to be believed that the observations were based on a close study and not merely to pre-empt complaints from the industry that there existed serious delays in taking decisions on their applications on various matters to the Government.

## 'Profit' Is not a Dirty Word

The fishing industry is the vehicle on which Government has to depend for progress. Likewise, the Industry has to depend on Government for progressive and pragmatic policies in order that they may survive. The responsibility to achieve a unison between Government's policies, programmes and strategies and their implementation by the industry rests more with the Government than with the industry. While the primary motive of the Government is to encourage development and economic growth, the basic objective of any fishing company is to earn profits and there is nothing wrong in this, as long as Government's regulations are not contravened. 'Profit' is not a dirty word, as rightly pointed out by Mr. Appleyard during the conference.

## Can a Fact be Untrue?

Fishing industry is not different from other sectors in regard to the observation of codes of conduct and following Government's regulations. One allegation of the minister was that fishing companies never presented true facts about the operations. This is something that will be questioned by every fishing company. All Companies extract particulars from their books, maintained as required to meet prescribed systems and as required under Company's Act. The particulars of catches are maintained continuously in the log books by skippers. Most of the earnings are from shrimp and these are duly accounted for in the books of the companies and that of the processing plants. Always audited particulars are furnished to the Government. If Government has a doubt that any company has not presented true facts, what prevents them from verifying their records? The fact is that no company will take the risk of presenting a false picture to the Government.

The minister's statement that he did not believe that the fishing operations were uneconomical, is deplorable. There is also no logic in the observation that fishing enterprises in other countries were making profits and therefore Indian enterprises also should be earning profits. Profits or losses depend on incomes and expenses. Countries with richer fishery resources and where cost of diesel oil is low earn more and spend less.

If there are profits, the normal tendency on the part of the industry is to keep quiet. Only when losses show up the industry represents. Thai vessels which operated in Indian waters till recently are stated to have incurred heavy losses. Losses in operation of several U.K. vessels forced them to stop operations. Purse seining operations in Philippine waters are reported to be under heavy losses. In countries such as Indonesia and Taiwan profits are netted owing to cheaper oil and higher returns. Fish fetches a far higher price in those countries than in India. The conditions in India are just the opposite. It is not that there are always losses. After nearly three years of bad economics of operation, the shrimp trawling operations with larger vessels have picked up this year.

## There are No Secrets

The minister is perhaps not aware that in the fishing business, practically nothing can be kept secret, particularly regarding quantum of catches and returns. The news spreads fast. Consequently, no company can dare to furnish wrong data to the Government, for the fear that, if verified with logs, and records of purchasers truth will come out. It will therefore be a sound policy, at least in the present stage of development, to encourage and not criticise the industry unduly, and to point out and devise measures to eliminate faults and not to highlight imaginary doubts and antagonise the industry, being a sensitive field connected with catching something live from water with all attendant risks, any unjustified discouragement will have a retarding effect.

The minister's tone suggested that the private sector industry was a culprit, and that he was angry with the sector. This attitude may be the cause for the delays in the Department of Agriculture in the clearance of applications from the Industry for charters etc. If this is true, the least that can be said about this is that it was an unfair approach. It is not known whether the Department of Agriculture investigated into the furnishing of false particulars, if any, by any company or atleast written to them on the subject. So far as we know, nothing such was done.

## What's Wrong?

The minister's statement, that the companies earned 15 per cent of returns from the operations of chartered vessels and gave a guarantee for only Rs. 2 lakhs to the Government and that this is a good business, is mystifying. The companies have been following the system, exactly in accordance with the guidelines issued by the Government and the permissions accorded. Can this be wrong any way? Fishing companies are there to catch fish and earn money out of them. Does the Minister feel that fishing companies should not earn money? The logic behind the statement made by the Minister is very obscure. He perhaps feels that none of the companies may buy the vessels at the end of or during the charter as stipulated by the Government and are prepared to lose the guarantee amount of Rs. 2 lakhs for the reason that they have earned much more by way of 15 per cent of returns. If so, this line of thinking must have taken shape on wrong premises. If the operations are found to be successful, the first impulse of any entrepreneur would be to acquire the vessels, for that would be much more beneficial.

The infant fishing industry depends on the Government for its growth. To initiate and foster a confrontation with the industry will be a harmful exercise for the Government. With a positive bent of mind, the Government and the industry have to work together to achieve the ultimate objective of the development of the industry. Exaggeration of minor obstacles or faults would only cloud the main objective. The duty of the Government is to guide the industry, helping them in formulating sound proposals and in their implementation with the Government's clearances.

Fishing Chimes, Editorial 18: September 1982: Vol. 2, No. 6

# Diversification has Diverse Problems

Almost the entire fishing effort in the mechanised sector in the country is directed at shrimping through trawling. The Government as well as the industry are anxious to diversify the efforts to cover other fishing methods. Chief among these that are spoken are tuna purse-seining, tuna long-lining, and squid jigging, gill netting, lobster potting, tuna pole and line fishing, bull trawling, single vessel stern trawling with larger vessels etc.

In the recent past, bull trawlers and larger stern trawlers have been introduced through charter arrangements. Other methods through charters have not yet been introduced.

From the point of view of sheer economics, any foreign fishing company would prefer to provide trawlers on charter. Demersal and columnar fish are abundant in our waters at some location or the other, depending on the season. With their endurance, the trawlers can move from place to place searching for stocks and exploiting them. No other type of fishing can be as remunerative as trawling, be it demersal or pelagic.

Gill-netters are the next potential bet.Surface or bottom-set gill netting with extensive nets are likely to yield good results; Besides this, gill-netting operations consume less of fuel. One company, M/s Nava Bharat Ferro-Alloys, Visakhapatnam, applied to the Government for permission to charter gill-netters over a year back, but so far there has been no clearance. Once the results of gill netting under this charter are found to be satisfactory, others are likely to enter this line. One point to be noted here is that surface and bottom-set gill nets catch the same stocks as with mid-water and bottom trawls respectively. The rate of capture will however be less. While the trawls sweep the ground, in the case of gill nets, in their movements, fish get entangled in the nets, although there will be slow drifting of the nets.

Regarding squids and cuttle fish, the known stocks in our waters are well within our territorial waters. Oceanic stocks are reported but there is no tangible proof of their availability. So much so, no foreign fishing company can take the risk of providing their jiggers on charter, under the present terms and conditions of our Government.

So far as tuna fishing is concerned, we know that tuna is a open sea fish. There are extensive tuna areas within our exclusive economic zone. Foreign tuna vessels tap those resources. When such a scope is there and also when there are chances of fishing for tuna even in the farther areas of our EEZ, without fear of detection, one can understand that no foreign fishing company would like to subject itself to the various terms and conditions of charter imposed by our Government. If there are special incentives, foreign fishing companies may come forward to provide Indian fishing companies with tuna vessels on charter.

Tuna purse seining is an expensive operation, involving a high order of coordination of activities. Apart from the purse-seiner and the skiff, the foreign fishing company has to bring in a scout boat and a few light boats. If the operations are combined with the raft system (covering a ground with anchored rafts to attract tuna), the expense on rafts becomes additional.

With all these investments, running into several million US dollars, the foreign companies first want to be resonably certain about the grounds, about the type of tuna available and their sizes. In our waters, we get mostly skipjack which brings a far lesser price compared to varieties such as albacore, bluefin and yellowfin. Further, the foreign fishing companies have a belief that the average size of skipjack in our waters is on the lower side. In regard to tuna long-liners, it is stated that all major tuna fishing nations are going slow in tuna long-lining, as the operations have become unremunerative. Countries like Taiwan may like to give their long-liners on charter but on their own conditions.

When would a foreign fishing company be inclined to provide their tuna vessels on time-charter with their crew to other national enterprises? That will be when fishing in their present grounds is either uneconomical owing to depletion in stocks, or they are unable to exploit known resources in the EEZs of other nations when the fishing area is within the EEZ of that country. They also have the apprehension that they will have competition from vessels of other nations while fishing in the permitted areas, leading to confrontation. They must also be confident that the operations in the EEZ waters of other countries will be economically viable.

Fishing Companies in U.S.A. Spain, France and Japan, which are major tuna fishing nations, do not seem to have any active interest in providing their tuna vessels on charter to Indian fishing companies. Taiwan and Korea have tuna long liners, but practically none, so far as tuna purse seiners are concerned. Philippines has a sizeable fleet of tuna purse seiners. It appears that there has been an excessive fishing effort in Philippines waters. Consequently a few of the Philippine Companies started deploying their vessels in Indonesian and other waters. They may like to provide tuna purse seiners on charter to Indian fishing companies, but only on their conditions.

Thus Philippines seems to be the only source to be tried either for chartering tuna purse seiners or for entering into joint ventures, based on tuna. It is however doubtful whether the Philippine companies would agree to the standard Indian conditions for chartering of fishing vessels.

The main line of diversification of sea fishing in Indian waters undoubtedly concerns tuna. Considering the bleak picture in getting tuna vessels on charter on the basis of sharing the gross proceeds at 85 per cent to the owners and 15 per cent to the charterers, what seems necessary immediately is to open a commercial wing at the Integrated Fisheries Project, Cochin. Through this wing several purse-seiners/long liners of different capacities can be taken on charter and operated. Once the results are encouraging, the private sector will come forward to introduce such vessels. The foreign fishing companies would also develop the confidence to provide such vessels on charter, under such terms and conditions as accepted by both the sides and approved by the Government of India. The same approach can be adopted in respect of squid jiggers, gill netters and lobster-potters.

In respect of bull trawlers and stern trawlers, the results from these charter programmes are likely to be encouraging. In all such cases, Government should evolve a system under which quick clearance regarding cost of vessels and in respect of loans from SDFC can be given. If these are delayed, the chances are that the owners may divert their vessels elsewhere and withdraw their sale offer. In such an event, the Indian fleet will stagnate. It is therefore of utmost Importance that the applications for loans for the acquisition of such fishing vessels should be cleared by SDFC expeditiously.

Fishing Chimes, Editorial 19: October 1982: Vol. 2, No. 7

# Deeps of Indian Deep Sea Fishing

It would all seem mysterious. Indian press, with sudden flush of interest, has turned its search light on the deep sea fishing industry. A profusion of articles on the subject has appeared in various journals in recent months.

A write-up appeared in Eastern Economist on 11 July 1982 under the caption 'Deep Sea Fishing: Is it profitable?', followed by - spicy ones, one in India Today (31st July 1982), entitled 'Trawlers at sea', and another in Business World (August 16-29, 1982) with the catchy caption 'Indian Fishing in Deep-waters. August 1982 issue of 'Science Today' also carried the viewpoint of Dr. S.V. Gokhale on the subject with the title 'The Mania called deep-sea fishing'. There may have been many more but we have not had the opportunity of reaching them.

## Eastern Economist's

The author of the article in Eastern Economist believes that the main reason for the flight of big business from deep sea fishing is inadequate profitability, owing to rising diesel costs. The article argues that, since the Indian companies are not in a position to undertake the job with unfavourable profits, it would be desirable to licence foreign trawlers to operate in our waters. He feels that foreign trawlers must be having sizeable profits, otherwise they would not come all the way to poach often in our waters. He is against the granting of additional subsidies to make deep sea fishing operations viable for Indian enterprises.

While expressing views on the above lines, the author continues to say that deep sea fishing, on account of the huge capital costs, is certainly the game of big business. The firm stand against big business taken by the Government and its proclivity to promote State undertakings in deep sea fishing business, the author feels, will only further complicate the matter. Emphasis has been laid on the fact that EID Parry, ITC, Britannia and New India withdrew from the field, leaving only a few large houses in the arena. Pointing out the delays in clearance of the charter applications, he has emphasised the point of view expressed by the Association of Indian Fishery Industries that a time-bound programme for acceptance or rejection of the applications should be introduced. He has also emphasised the need to provide basic infrastructural facility for the development of marine fisheries.

## What 'India Today' Says

The article published in 'India Today' written by Dilip Bob also points out the progressive withdrawal of the larger houses from the deep-sea fishing field. Bringing out the several depressing features in the present state of Indian deep sea fishing industry, the article says that there should be a national fisheries policy, a separate ministry for fisheries, a clear-cut system of permitting charters and joint ventures, provision of adequate facilities such as communications, deepening of existing fishing harbours and the establishment of fisheries corporations for deep sea fishing sector.

## Business World's Observations

The write-up entitled 'Indian fishing in deep-waters' by Sita Unninair in Business World (16-29 Aug. 1982) is comprehensive, although observations therein are heavily loaded in favour of larger houses. Striking a note of alarm and panic that larger houses are progressively pulling out of deep sea fishing industry and pointing out that larger houses alone can undertake deep sea fishing, the article concludes that few small scale Indian entrepreneurs do not have that much money to throw it into the air and the Government continues to think that big is ugly. It is mentioned that, until the Government decides to do some rethinking on this basic point, the Indian deep sea fishing is likely to get into even more deeper waters.

While subscribing to the view that the flight of big business from the deep sea fishing industry is on account of inadequate profitability, the article in Eastern Economist says that the tough stand of the Government against big business and the intention to promote State undertakings in deep sea fishing business would further complicate matters. The author is against subsidising the Industry, and in the face of inadequate profitability, big business houses would not like to continue in an unprofitable field. Yet we find so much of anxiety on the part of certain larger houses, to gain a dominant place in the industry. There must be a good reason for this. In this situation what options does the Government have? Either they have to beg the big business on bended knees to continue in the field (since they alone have the money to throw into the air, as Seetha Unninayar says) or to encourage State Undertakings to enter the field in a big way. Government, obviously, as at present, have no intention of prevailing upon big business to continue in the field. The Union Minister concerned has therefore stated that he would encourage State undertakings. One has to see whether Government will be influenced to reorient their policies heavily in favour of larger houses.

## Not the Game of Big Business

The article in Eastern Economist states that even a moderate sized trawler, such as the one used for shrimps, costs nearly Rs. 2 crores. This is not correct. For about Rs. 82 lakhs a well-equipped 23m trawler can be acquired. Mazagon Dock Ltd., Bombay is willing to supply them at Rs. 57 lakhs. It is argued that, on account of heavy capital investments, deep sea fishing is the game of big business. This is certainly not true. With availability of loan to the extent of 90 per cent of the cost of imported and 95 per cent of cost of indigenous vessels, the actual investment per vessel will be within Rs. 10 lakhs and for introducing a viable fleet of two trawlers, a small/medium level entrepreneur would require a capital investment of Rs. 20 lakhs. There are many in the country, other than big business, who can invest this much. In fact there are several small/medium operators who have invested this much and are in the business. Further, through chartering of deep sea fishing vessels and later investing the incomes earned through their operation they can secure loans from SDFC and acquire fishing vessels. It is certainly not correct to think that only big business can be in it. The business is well within the scope of small/medium enterprises.

All the three authors are perturbed over the progressive withdrawal of larger houses from the deep sea fishing field. We feel that the industry can very well be developed, without the participation of larger houses. In fact, it is they who have made a mess of the whole field. While themselves not succeeding they prevented small/medium entrepreneurs to develop.

## Are Large Houses that Competent?

Seetha Unninayar's article is totally critical about the Government's role in the development of deep sea fishing. She says that observations by Rao Birendra Singh, Minister for Agriculture, which she thinks are paranoic, stem from the progressive exit of larger houses from the fishing industry. This feeling, it appears, is far from truth. Government is aware that they need not have to depend on larger houses for the development of the industry. She feels that, despite the widespread awareness that deep sea fishing is a capital intensive and a high risk business, Government is keeping the organised sector entrants into the industry on a tight leash. Although they have the capital and management capability, they are viewed with suspicion. Soft loans from SDFC are denied to them.

The authors are not aware that neither unprofitability nor absence of incentives nor of encouragement are responsible for the withdrawal. Union Carbide has a fleet of 14 vessels. How have they come to own the biggest fishing fleet in India? Has not the Government allowed them to own such a large fleet? What do they want the Government to do? Allow the acquisition of more vessels? As it is they are running their vessels on loss. Do they want the Government to encourage an enterprise, which has not been able to manage the existing vessels in a profitable manner, with all their managerial experience?

## Govt. was Good Towards Larger Houses

Government permitted New India Fisheries to charter two Japanese vessels and also permitted them later to acquire the vessels. If the company is pulling out of the field, it is only because of their internal problem and not certainly because of Government policies.

Britannia has been doing well in their fishing operations. They have not pulled out of this sector so far, as has been stated. Government permitted Tata Oil Mills to import two trawlers from Mexico and also acquire two Thai trawlers through charter. Government also permitted them earlier to operate on charter 25 vessels from a Thai Company. It is likely that Government would soon clear an application of theirs or their subsidiary for chartering fishing vessels from abroad. Is all this not an encouragement?

Konkan fisheries have not come out with any statement that they are pulling out. Government have not come in the way of their programme in any manner. If E.I.D. Parry pulled out earlier, it was owing to other reasons. Their vessels were being run on profit at the time they pulled out.

The word that went round at the time was that ITC pulled out of the field for the reason that their perspective study had shown that, in the coming years, the exports of marine products would be diminishingly attractive.

One complaint that the larger houses can have against the Government is that they are not being given loans from SDFC. Government have no where stated that granting of loans from SDFC to larger houses is barred. The Ministry of Agriculture has not been recommending their applications for reasons of their own. One reason can be that the smaller parties with lesser financial resources should get preference.

Another complaint of larger houses can be that companies such as Union Carbide and Chowgules were not given subsidies on the vessels constructed by them indigenously. There was provision for the granting of subsidy of 27 per cent of the cost of vessels or the difference between the cost of vessels and equivalent imported vessels whichever was lower. Enquiries show that the shipyards failed to provide satisfactory proof of expenditure incurred by them for the construction, to enable Government to release the subsidies.

## No Deep sea Fishing without Larger Houses?

Seetha Unninayar observes that, because the organised sector companies have been kept out of the industry, the country possesses neither suitable vessels nor the experience to undertake such extensive and ambitious operations. In other words, she means that, if the organised sector companies are encouraged with SDFC loans, with permission to sell non-exportable catches within the country and by providing other such incentives, they would have developed or utilised their expertise and undertaken extensive and ambitious operations. According to her own statement larger houses that based their operations on shrimp, which fetches the highest returns, have been pulling out, on account of non-profitability. Such being the case, is it conceivable to believe that larger houses would invest heavily on diversified fishing operations, which bring far less returns, even if they are allowed to do so? Several larger houses could not continue with success even in shrimp trawling operations. They could not develop the managerial expertise. Such being the position, it is doubtful that they can undertake extensive and ambitious operations. It is our assessment that most of the larger houses would not like to continue in the field, even when Government encourages them. It is possible they have other obscure intentions in unleashing extensive publicity about injustice to larger houses.

## DSF within Scope of Small Entrepreneurs

All the present sound and fury exhibited by larger houses seems to be to embarass the Government either to get permissions to charter fishing vessels or pull out of the industry honourably on the ground that Government is against their presence in the industry. If the charter operations are successful, they may buy the stipulated number of vessels. But then the smaller parties can also do the same. Smaller parties can probably manage their operations in a far better style than larger houses and in a far more economical manner, as the present trend shows. Preference in charters and other incentives has therefore to go to smaller companies not only from the above angle, but also from the angle of securing economic justice to those lagging behind. The present trend of publicity in favour of larger houses seems to suggest that only larger houses should be encouraged as they alone can deliver the goods. Human beings other than those in larger houses too are endowed by Almighty with resourcefulness, intelligence and wisdom. The present policy of Government in regard to charters will be a major Incentive to small parties to gain experience, build up reserves and acquire their own fleets. Small parties in the fisheries field will have only this line to concentrate. In contrast, larger houses have many. It is better they concentrate on other 'sure-fire' sectors and leave the 'Capital intensive' and 'high-risk' field to small and medium sectors, instead of using their might in prevailing upon the Government to virtually allocate the sector to them exclusively. The small men have also to live and flourish. Larger houses need not be so altruistic as to opt to enter a high risk field.

Seetha Unninayar quotes N. P. Singh as saying 'I don't see any sense in adding to the shrimp trawler fleet by about 50 per cent when the catches are falling'. There is positive proof that shrimp catches are not falling but are on the increase. The opposition to additions to shrimp fleet is obviously to preclude new entrants into the field to share the benefits. Singh says, probably with contempt, 'Small companies and fresh entrants lured by the prospect of obtaining cheap capital, thanks to SDFC loan facility, have been allowed in'. We leave it to the readers to judge the import of this. Does this give an inkling of the overbearing and haughty attitude of large houses towards small companies? Does this mean that they do not want small companies to be in the deep sea fishing industry? We are unable to fathom.

## DSF not a Mania

Dr. Gokhale in his note entitled "The mania called deep sea fishing' published in the August 1982 issue of 'Science Today' has questioned the Government's efforts to promote deep sea fishing at the cost of traditional and inshore fishing. He is of the view that the deep sea is not so rich as inshore waters. He feels that, given the necessary infrastructure facilities like transport and marketing, our fishermen can double the fish catch. He says that, while deep sea fishing will help only a handful of persons and some foreign companies, inshore fisheries can benefit a large population.

Everyone agrees that deep sea zone is not as rich as inshore waters. At the same time, it has to be conceded that deep sea zone contains a sizeable fishery, although the density is less. This resource is also a national asset and has to be exploited for national benefit. It is not correct to say that the exploitation of these resources will benefit a few persons and some foreign companies. The correct thing to say perhaps is that this activity should benefit the largest possible number of enterprises. While the units may be few, the number of persons dependant on the units will be large. Regarding inshore fisheries, there can be a positive suggestion that fishermen should be helped further to augment the catches.

Fishing Chimes, Editorial 20: November 1982: Vol. 2, No. 8

# The Elusive Three Fifty Fishing Vessels

There appears to be a curse on the Indian deep sea fishing industry. In the Fourth and Fifth Plans, it was envisaged that 200 nos. of deep sea fishing vessels in each of the plans should be introduced. At the end of the Fifth Plan the introduction went up only to the extent of about 50 nos. against 400 nos.

In the Sixth Plan, the target for the introduction was set at 350 nos. Half of the plan period is over and so far nine trawlers on ownership basis and seven pair trawlers and 6 single trawlers on charter have been introduced. If this pace of progress is any guide, one can assume that the target will remain dismally unfulfilled.

There seems to be no pointed awareness of the situation in the Union Department of Agriculture. The programme is being implemented in a perfunctory manner, bereft of the drive that the activity requires, to gain momentum.

## Slower Than a Snail

Under the scheme introduced in 1977 for the import of over 100 trawlers, so far only nine trawlers have been imported. The reason for this slow progress is the fact that officials in the Fisheries Division of the Department of Agriculture get bogged down in minor issues, loosing sight of the target. When a company applies for import from a particular source the issue of approval is delayed inordinately. Similarly, the forwarding of their loan applications is also done after a long lapse of time. In the meanwhile, the builders either increase the price or withdraw their offers. This necessitates another application for a further change in the source. Thus it becomes a vicious circle. Instead, if the Fisheries Division takes quick decisions and acts briskly the complication will be far less.

Delays in decisions have made the share holders in several companies to sell their shares. Thus changes in the controlling interests and managements came into being. These have led to the suspicion that money is being made through the companies, although it should not be a matter of real concern to the Government. The objective is to introduce deep sea fishing vessels. As long as the company is the same, changes in managements should not bother the Government, provided that those in management have the capabilities to run the vessels.

Lacking the ability to push through a programme and perhaps taking shelter under the frequent changes in managements and periodical requests for changes in source of import (for which the applicants alone are not to be blamed), the Union Department of Agriculture seems to have decided to put an end to the consideration of all pending cases under the 1977 scheme.

## Lame Excuses

The lowering down of this curtain is probably done out of desperation, but this would not give any credit to those at the helm of affairs. It is a matter of great shame that a department wedded to developmental activities and having decades of experience in implementing developmental programmes has failed in this important task. Instead of getting bogged down in petty issues, had the department found quick solutions at each stage in respect of each case, by now the nation would have had an addition of over 100 deep sea fishing vessels.

The implementation of the programme seems to have been undertaken in the recent years on the premise that the applicants are crafty men and that they should be controlled and outwitted. It was in the same way of problems being handled by district magistrates, superintendents of police and at collectorates. Such an approach is hardly the way to deal with developmental matters, which have targets to reach. Functionaries with abilities have to push aside hurdles and move towards the targets. The Union Department of Agriculture allowed themselves to be detained by hurdles. Unless there is a change in the style of functioning in the fisheries division, the outlook for development in the deep sea fisheries sector appears to be very bleak.

So much of time and energy have been expended by the Government, shipyards and the industry on indigenous construction of fishing vessels. All these bodies are anxious for the building of deep sea fishing vessels in the country. Such being the case, the lack of progress must be due to certain factors. Although these factors have been identified, very little serious effort has been made to neutralise them. This is because the shipyards have their own work and their interest in fishing vessels is only marginal. If they get orders they will book them. If they do not get, they have their other work. So far as buyers are concerned, they are wary on account of the known substandard performance of some of the vessels constructed at Indian shipyards. These two attitudes can be reconciled, not merely by announcing subsidies, but through solid work to melt them. The shipyards should secure foreign expertise to develop reliable drawings and to supervise construction. This will put some measure of confidence in the minds of buyers. On the financing aspects, if SDFC loans are made available, larger houses are likely to avail of them for construction of vessels at

Indian shipyards. So far as medium/small companies are concerned the guarantees towards SDFC loans given by the shipyards should be accepted. Concrete measures on the above lines may promote indigenous construction.

## A National Level Corporation

In our earlier issues, we had pleaded for the setting up of a national level corporation to promote deep sea fishing. Our suggestion was that such a corporation can acquire vessels through indigenous construction and give them on charter or hire-purchase to fishing companies. We understand that the Union Department of Agriculture, has a proposal to set up a National Deep Sea Fishing Corporation. One has to hope that this will materialise and the above mentioned function also would be entrusted to them. The setting up of such a corporation will provide the much needed field set-up to give a fillip to deep sea fishing.

So far as charters are concerned, it is clear that there has been some progress, although not significant. Several applications for charter permissions are pouring into Krishi Bhavan. One cannot expect clearances to come as fast as the pouring in of the applications. Yet, it will be reasonable to expect that within about three months of receiving the application, decisions would be made known. Delays in clearance, as is happening now, has the effect of inhibiting foreign owners from giving offers. It also throws an undesirable image abroad. Foreign owners can reconcile themselves with quick rejections, but not long delayed rejections or approvals.

As Mr. Chidambaram has observed in an interview (Oct. 1982 issue of FISHING CHIMES), the first need is to develop deep sea fishing. Indigenous construction will follow this automatically. Import or indigenous construction or charters, the need is to make headway. After some development, priorities can be determined.

Fishing Chimes, Editorial 21: December 1982: Vol. 2, No. 9

# The Execution

The recent trends in the processing of applications for acquisition of fishing vessels through import under the 1977 scheme from deep sea fishing enterprises in the Union Department of Agriculture would lead one to believe that the effort is to kill the proposals instead of putting life into them. It all appears to be a virtual execution. There are several instances to come to this conclusion.

## Inexplicable Attitude

One company was permitted to import two trawlers from Australia. The Department of Agriculture took over nine months to recommend their case to SDFC to grant loan, in spite of frequent requests from the company to expedite their application. In the meanwhile the Australian Shipyard withdrew their offer, as they secured a far better price from Saudi Arabia for the trawlers nearing completion. In this situation, the company had to apply for change in source of import, within the amount sanctioned. Although there was no fault on the part of the company, the request continues to be pending in the Union Department of Agriculture for over six months. The intention of the Government in delaying a decision is probably to kill the proposal.

## Confounding Hold-up

Several applications for import of trawlers from Holland were approved at a price of Rs. 67 lakhs each after a considerable time-lag. Before the parties could fulfill the formalities, the foreign shipyard increased the price to Rs. 72 lakhs. As this increase was not acceptable to the Government the shipyard agreed to conform to the earlier price. In the meanwhile the time limit given in the permission letter was over. Therefore, the parties requested for an extension of time. Although several months have passed after this, the final decision of the Government on this was not forthcoming till the time of this writing.

There are applications for grant of loans for the import of several trawlers from Poland pending for a long time in the Department of Agriculture. The department has to forward them to the SDFC for grant of loans, as permissions for import were issued earlier. For reasons not known to the buyers, the applications continue to be pending in the department.

## A Policy without Faith?

The deep sea fishing industry is alarmed at the state of affairs, keeping in mind the declared policy of the Government. If Government is not keen on development of deep sea fishing in our EEZ, it should honestly say so, so that the interested entrepreneurs can save their time and energies and divert them to other fruitful lines instead of depending on a policy in which Government have no faith for implementation.

## Is that Because of Few Votes?

From a political angle perhaps development of deep sea fishing has no potential. The men involved are few and the voting potential is limited. However, looked at from National Economic Development and earning of foreign exchange it has a great potential.

It has to be conceded that the small fishermen claim the prior attention of the Government. Their socio-economic condition has to be improved. Even so, we do not see much being done in this sector also.

## Act Fast, Please!!

The socio-economic development of fishermen involving upgradation of their professional capabilities on the one hand, and the development of deep sea fishing in our EEZ for national benefit on the other, are equally important. It is of utmost urgency that Government determines the inter-se priorities and clearly tell those involved in the sector about the same. If it is the intention to slow down on development of deep sea fishing, it should be clearly stated. On the other hand, if there is no such intention, the pending cases should be cleared once and for all, this way or that way. The dilation and the uncertainty makes the lives of entrepreneurs miserable, apart from waste of time and energy. It is also of utmost importance to take up purposeful schemes, not giving mere doles, for the socio-economic uplift of fishermen.

Fishing Chimes, Editorial 22: January 1983: Vol. 2, No. 10

# Arbitrariness Stains Latest Charter Rules

The Central Government enacted the Maritime Zones of India (Regulation of Fishing by Foreign Vessels) Act, 1981 (42 of 1981) in September 1981. Section 25(1) of the Act empowers the Central Government to make rules for carrying out the purposes of the Act by notification in the official gazette.

## Main Purpose

The main purpose of the Act is the regulation of fishing by foreign vessels. Accordingly, the Central Government, in the Department of Agriculture and Cooperation published the rules for the purpose in the notification No. G.S.R. 619(E) dated 26th August 1982 in the gazette of India (extraordinatory).

## Transgression of Purpose

The notification is comprehensive. It is a meticulous document which takes care of all conceivable aspects relating to regulation of fishing by foreign fishing vessels. There are however a few points in the rules which appear to transgress the purpose. These may prove to be of interest. Rule 8(1)(C) of the notification reads as follows:

"The charterer shall give an undertaking in the form of bank guarantee before the commencement of the charter, for an amount to be decided by the Central Government in each case to the Central Government that he shall purchase required number of vessels and put them in fishing operation in the Exclusive Economic Zone of India before the end of the stipulated period specified in the Schedule (given hereunder) II".

From a Developmental Angle Rule 8(1) (a) is perhaps justified. This is however of a compulsive nature, obviously aimed at ensuring that the Indian charterers acquire their own vessels as part of the permission given to obtain vessels on charter. The deep sea fishing enterprises who chartered foreign fishing vessels earlier did not bother to acquire their own vessels as per the stipulated conditions and they were obviously responsible for the inclusion of this harsh clause. It has however to be rioted in this connection that as the programme becomes popular the chartering companies will, in their own interest, acquire vessels of their own.

## Legal Position

The rule however sounds illegal, arbitrary and unreasonable. The Maritime Zones Act, under the provision of which the rules were framed, was enacted, as seen from the title, exclusively for the regulation of fishing by foreign vessels. Chapter II of the Act deals with this aspect only and prescribes licences for owners of foreign vessels and permits for Indians chartering foreign vessels, for operation of foreign vessels in Indian waters. The Act, even by the remotest chance or interpretation, does not deal with introduction of Indian vessels. This aspect is completely out of the purview of the Act and this cannot legally be the instrument to achieve the augmentation of Indian fleet. While the Act provides for issue of licences and permits subject to such conditions and restrictions as may be prescribed and to such additional conditions and restrictions, these additional conditions and restrictions have to be within the scope of the purpose of the Act which is to regulate fishing by foreign vessels.

## Arbitrariness

Apart from the position from a legal angle which we believe is right, the contents are clearly unreasonable. The undertaking to be given in the form of bank guarantee has to be for an amount to be decided by the Central Government in each case. According to this, this amount

**SCHEDULE II**
**Schedule of Purchase of Vessels**

| *No. of Vessels or Pair of Vessels* | *Number of Months from the Beginning of the Charter Operation when Obligatory Purpose and Fishing Operation becomes Due* | | | | |
|---|---|---|---|---|---|
| | *First Vessel or First Pair of Vessel* | *Second Vessel or Second Pair of Vessel* | *Third Vessel or Third Pair of Vessel* | *Fourth Vessel or Fourth Pair of Vessel* | *Fifth Vessel or Fifth Pair of Vessel* |
| 1 | 18 | .. | .. | .. | .. |
| 2 | 18 | 30 | .. | .. | .. |
| 3 | 18 | 24 | 33 | .. | .. |
| 4 | 18 | 24 | 33 | 42 | .. |
| 5 | 18 | 24 | 33 | 42 | 51 |

can be fixed at any level. There is no upper limit. Human weaknesses are the same at all levels. At some point of time a higher-up who does not like a particular enterprise may fix an unreasonably high figure as the guarantee amount. From this angle, the provision is clearly arbitrary. As examples, it may be mentioned that the Department of Agriculture fixed the guarantee amount at Rs. 2 lakhs for five pairs of chartered vessels in some cases, and Rs. 2 lakhs per vessel in other cases. For some others, guarantee by way of bank guarantee for Rs. 1 lakh, and a cash deposit of another Rs. 1 lakh per vessel was stipulated.

The criteria for this arbitrary fixation of amounts and mode of guarantee are not known. While the rule clearly refers to 'bank guarantee', the stipulation of cash deposit is, to say the least, autocratic and without sanction by law.

## Beyond The Maximum Penalty

Another aspect, leaving aside the basic legality of the clause itself is that, under the Act the maximum punishment for contravening any condition in the permit issued to Indian charterers, other than those relating to the area of operation or method of fishing, is a fine not exceeding rupees fifty thousand. The rule under comment contemplates the forfeiture of the guarantee amount, which may be rupees two lakhs per vessel or more (or can be less) in case the party concerned fails to introduce their own vessels as per Schedule II. When the Act contemplates a maximum punishment of rupees fifty thousand, can there be a provision relating to another penalty over and above this? To us, this appears irregular and outside the scope of the Act. We would advise those who are either operating or intend to operate vessels on charter to ponder over the above mentioned points for seeking such redressal as necessary.

Fishing Chimes, Editorial 23: February 1983: Vol. 2, No. 11

# Need to Revamp Fisheries Administration in the States

In the early post-independence period, the major role played by the provincial fisheries departments on the inland side, was to a) collect spawn/fry from rivers, grow them into fingerlings at fish farms, collect major carp fingerlings from river canal systems, and use the seed for stocking departmental ponds. Only in the northeastern States, particularly West Bengal, there was the tradition of fish culture in private ponds, In other States there was no such tradition and fish culture in selected tanks and ponds, besides a few fish farms, was the exclusive domain of fisheries departments. In the marine sector, before the advent of mechanisation of boats, the main activity was one of organising the curing of fish at fish curing yards set up at selected centres. These yards were the focal points for marine fishery administration.

Several States did not have separate fisheries departments till recently. Fisheries administration continued to be with industries, agriculture or veterinary departments in several States for long. With the Rajasthan Government conceding a separate department for fisheries recently, now all the States, perhaps with the exception of Tripura, have their own fisheries departments.

## Momentous Developments

We have seen several momentous developments in the fisheries field in the past two or three decades. On the inland side, the development of technologies to induce major carps to breed to obtain maximum output of fry/ fingerlings from spawn through operation of hatcheries and from dry bundhs, to obtain optimum fish yields from tanks and ponds through composite culture, to produce prawn fry and to undertake brackishwater farming, and of integrated farming practices, brought in a major transformation. There has been an arousal of public interest in the economic potential of the activities and situations which could not be visualised in 1950s have now materialised. Significant commercial activity in the private sector in fish seed production and fish farming in 1950s, aided by fish farmers development agencies, has come into being.

In the marine sector, the introduction of synthetic twine for net making on the one hand and mechanised boats on the other brought about a major seachange. The opening up of the export sector, particularly for shrimp, gave a major fillip to sea fishing activities. A large number of private sector units in fishing and processing of marine products have come up. A beginning has been made in the introduction of larger fishing vessels for the exploitation of deeper sea zones. The activity is gaining momentum, consequent to the declaration of the exclusive economic zone.

State Fisheries Corporations in the public sector, designed to concentrate on the commercial side of the fisheries activity, have come into being. Fisheries education and training centres have been established.

Research and extension support is now being provided by Central Institutes in marine and inland fisheries fields besides support in fisheries technology. Exploration of marine fishing grounds is being done in an intensive manner by the Exploratory Fisheries Project.

## Little Change in the Set-up

While major developments as outlined above have taken place, the pattern of the administrative setup in the States has changed very little. While there has been the formation of separate directorates of fisheries in such of the States where fisheries were being administered by other departments till recently, and also expansion in the number of posts in all the departments took place, the pattern of functioning has seen very little change. It is now high time for the State Governments to redefine the objectives and functions of the Fishery Departments.

## New Pressures

The traditional pattern of working of the fisheries departments is to take care of the departmental units. The onset of commercial activities, both in the public and private sector has put new pressures on the departments. These can no longer be borne without making changes in the pattern of functioning. An immediate problem, however, is that there is considerable confusion in regard to the role of the Fisheries Departments with the setting up of fisheries corporations and the fish farmers development agencies. More often, the Fisheries Departments incline towards the view that the corporations and the FFDA usurped their functions. The departments now feel deflated.

## At Cross Purposes

On the marine side, the construction and introduction of mechanised boats has been taken over by the corporations in most of the States. Fisheries departments have practically no say in the operations of larger vessels. On the inland fisheries sector, the departments feel that FFDAs have become a virtual threat to their initiative and authority. In most of the States, the Fisheries Departments and corporations function with distrust and suspicion toward each other. They are always on the guard to strengthen their respective positions. Working at cross purposes is not uncommon. Same is the position in respect of FFDAs and fisheries departments. The departments feel that culture fisheries sector, which is their main activity, having been taken over by FFDAs, their functioning in the districts concerned has become difficult. With the trend towards of expansion of FFDA to cover all districts, what would happen to the working of the departments?, some of the directors feel. In course of time the situation will become more serious, if immediate steps are not taken to define the respective roles.

## Overall Responsibilities of Directors

Corporations, FFDAs, and fishermen's cooperatives are instruments to achieve the objectives of fisheries development. The aim of all developmental activities is to increase fish production and create new incomes out of them, In this process, employment and economic prosperity would materialise. The overall responsibility for the planning and monitoring of the implementation of the programmes belongs the fisheries departments. The Corporations, FFDAs and Cooperatives are their creatures, created to facilitate the realisation of the objectives. While these bodies function by themselves they are there to implement the programmes of the departments but not to function as parallel departments. This philosophy and truth has not yet percolated into the minds of several of the Directors of Fisheries, Managing Directors of the Corporations or Chairmen/Chief Executives of the FFDAs. Steeped in old time traditions, most of the directors have not yet learnt the art of exercising the necessary supervision over the working of the Corporations and FFDAs. Once they exercise this inherent authority of theirs, they will be able to take credit (or discredit) for the results.

## Role of Fisheries Depts.

The role of the fisheries departments, therefore, should increasingly be to guide, coordinate, and monitor the activities of the various bodies set up for the achievement of their objectives. It should also be their function to provide necessary funds and clearances required by the bodies. They should shed ownership of fish farms, ponds etc. in favour of FFDAs progressively, keeping reservoirs and riverine fisheries under their developmental control for the time being.

The departments have also to reorganise their district fisheries setup suitably. In all the districts where there are FFDAs, the inland staff should be merged with FFDA. With such a merger. FFDAs can cover the entire area in their jurisdiction much faster. The fisheries administration in the remaining districts, not covered by FFDAs may be reorganised with the existing staff on the pattern of FFDA, with such adjustments as may be necessary to bring about faster development in pond culture, This can probably be achieved within the present allocation of funds, with marginal increases. Reservoir and riverine fisheries developmental aspects in the districts can be undertaken through a separate setup, by earmarking a few of the existing posts for the purpose. Commercial aspects relating to these can be entrusted to the Corporations.

So far as marine fisheries development is concerned, the departments should have separate wings for the purpose, both at the directorates and in the districts. The basic features and developmental patterns of the marine sector are altogether different from those for the development of inland fisheries which are akin to agriculture. Marine fisheries matters have no semblance to agriculture. The marine units should be made to implement welldrawn integrated schemes aimed at upgrading the fishing capabilities of fishermen, providing facilities for preservation, storage, transport and marketing of their catches, and directly linking the fruits of the work to improvements in the social and economic conditions of the fishermen.

The toning up of the working of the Corporations should be an important function of the departments. On the marine side, very little purpose is served if the Corporations also indulge in shrimp fishing, processing and export. This activity has already caught up well in the private sector. Their function should be to promote introduction of larger vessels, meant for fishing for varieties other than shrimp. This introduction can be for operation on their own or to help private sector enterprises. The Corporations can acquire vessels and give them on hire to private parties. They can in the alternative, acquire vessels and sell them under hire purchase terms. This line of activity will result in faster introduction of larger vessels.

## Job of Professionals

'Fisheries' are a subject with a massive content of technology. In olden days, when it was only a question of issuing licences of leasing out tanks and ponds with wild fisheries, or managing fish curing yards, it was all right to have a civil service officer to head fisheries department. In the present situation, it is imperative that technical officers with professional experience head the Fisheries Departments and Fisheries Corporations. The subject is a technical one, and it is in the nature of things that technical officers should be In-charge of Fisheries Development. The sooner the State Governments realise this, the better it will be for fisheries development.

Fishing Chimes, Editorial 24: March 1983: Vol. 2, No. 12

# Technocrats and Deep Sea Fishing

Deep sea fishing is no doubt a capital intensive industry. The requirements of heavy capital outlay, prevent small entrepreneurs to take to the activity. Financial aspect is not however the only constraint. The activity requires technical background and managerial skills. All small entrepreneurs do not have them. However, a good number of skippers, engineers and executives who have gained experience in the field, and who will, by and large, come under the category of small entrepreneurs constitute a competent nucleus to undertake deep sea fishing.

## Present Policy

The present policy of the Government is to encourage small entrepreneurs to enter the field of deep sea fishing. The fact that loans from S.D.F.C. are provided to them supports this view. While technocrats as described above, can secure loans from SDFC, it will still be a problem for them to mobilise the required share capital or their matching contribution of funds to the extent of 51 per cent of the cost of vessels to be eligible to get loan of the order of 90-95 per cent from SDFC.

## NCDC's Assistance

In the cooperative sector, the National Cooperative Development Corporation and the State Governments provide financial assistance towards share capital of fishermen's cooperatives. While deep sea fishing is not included in the schedule of assistance offered by NCDC, it should not be difficult for the Corporation to include this item as well. Once this is included, fishermen's cooperatives can get assistance from NCDC and State Governments towards share capital besides certain subsidies, for taking up deep sea fishing.

## Deep sea Fishing Coops

It will be a good step for technocrats to form into deep sea fishing cooperatives, acquire vessels and operate them. However, under the present system only traditional fishermen, mostly by caste are allowed to form into fishermen's cooperatives by the State Governments.

## Who are Fishermen?

Many of us have still the fixation in our minds that those borne into fisher families are fishermen. With the advances in fishery technologies, the concept has to be changed and the term 'fisherman' has to be redefined. Any person who is in the profession of agriculture is a farmer. He may be a Shudra, a Kshatriya or a Brahmin. Likewise, a person belonging to any community but in the field of fisheries has to be defined as a fisherman. Technocrats in the field of fisheries are professional fishermen and they should have the right to form into deep sea fishing cooperatives. It is of crucial urgency to give a serious thought to this and the Central and State Governments have to enable (through a new definition of fishermen) fishery technocrats to form into cooperative societies. The technocrats to form into a society have to consist of various categories such a skippers, engineers, engine drivers, deckhands, cooks, RT operators etc. The minimum number to constitute a society may be determined based on a single fishing unit.

## Major Breakthrough Possible

Such societies, when formed, will provide a major breakthrough in the fruitful utilisation of the talents of technocrats and in giving a fillip to deep sea fishing among entrepreneur-technocrats. The union Department of Agriculture (cooperative division) can prepare model memoranda and articles of association for such societies for circulation. The department can also popularise the idea through publicity. The financial and other items of assistance which societies require from the NCDC, State Governments, cooperative banks, SDFC, etc. can be worked out and a scheme in this respect can be framed for adoption. The cooperatives can be prevailed upon to introduce indigenously constructed deep sea fishing vessels.

Deep sea fishing cooperatives proposed as above will have the potential of becoming a success. The members being professionals and the cooperative being their own enterprise, and since profits will be shared by them all, they are likely to work hard and land sizeable find catches. Once the first one or two such societies succeed more of this nature will be set up. All this will lead to an increase in the number of deep sea fishing vessels, which is what is required now for the effective exploitation of the fisheries resources of our exclusive economic zone.

Fishing Chimes, Editorial 25: April 1983: Vol. 3, No. 1

# Testing Through Toughness

The Marine fishing industry has been confronted with an intensively toughened attitude from the union department of agriculture for the past one year. The emergence of this quality of toughness became increasingly evident from the time the International Seminar on Deep Sea Fishing was held in June 1982.

Prior to 1982, the approach of the Government towards seafishing technologist was one of benevolence and tolerance. Considerable latitude was allowed while considering representations from fisheries entrepreneurs. Applications for extension of time limits for imports and chartering of fishing vessels were considered, giving due credence to the reasons put forth for delays. As time went by, it must have become necessary for some in the Government to press their heels hard, for, this state of affairs could not obviously go on indefinitely. In any case nothing was happening. So, why not take a tough line?

Adoption of a tough line has certain advantages. It enables the grain to be sorted out from chaff. Lured by the policy of the Government to encourage small entrepreneurs, several with marginal financial capabilities made a beeline into the queue, with the hope that there will be a breakthrough on their financial front. Someone with financial means may either join them or take over their enterprise, showering benefits. Government may provide additional reliefs, seeing their predicament. It was this category possibly with these thoughts, that put a brake to the developmental process and drained the quality of mercy on the part of the Government.

The merry-go-round with various kinds of entrepreneurs had been going on and on. The revolutions have now been halted. The result is that those who are not equipped have been slowly detaching themselves from the merry-go-round. Those who are serious and are determined to be in the industry are holding on to their horses.

Entry into any commercial venture of importance would not be smooth sailing. If it is to be a smooth entry, like a hot knife into a slab of butter, the entrepreneur will tend to be callous and the full import and implications of the achievement will not sink in. The entry will have to be the result of passing through a race of hurdles. Then the requisite seriousness will be there.

Tata Oil Mills and Universal Sea Foods obtained their permissions for chartering of fishing vessels in 1978/79 without the need for any financial guarantees. The cases went through as smoothly as earth's revolution, which by itself is a celestial phenomenon of great significance but nevertheless taken for granted. Then came the system of financial guarantees from banks to be given to the Government, at first at Rs. 2 lakhs per vessel, later Rs. 2 lakhs for five vessels, and subsequently Rs. 1 lakh in cash and another Rs. 1 lakh for each vessel. These guarantees are introduced to ensure that the parties construct and introduce a number of vessels equivalent to those permitted to be chartered. This system is a device to develop India's own fishing fleet. Unlike in the past, now the applications have to be cleared by a committee and finally approved by an authority no less than the Minister for Agriculture. The evolution of this tough line is obviously intended to exclude opportunists who get in merely to earn money from the operation of chartered vessels and later lose Interest in the sector.

The tough policy no doubt creates considerable hardship to those who are genuinely interested in deep sea fishing. But this has to be endured as one cannot hope to enter the field without certain financial stake.

Toughness towards entrepreneurs on the part of the Government may be justifiable in the present situation. But the question Is how effective would it be to bring about faster development of our deep sea fishing fleet. It will take decades for the private and the present public sector organisations to introduce an adequate number of vessels to build up a sizeable Indian fishing fleet. The financial and financing problems of enterprises would retard the progress. As we had pointed out in our earlier issues, the only solution to the problem is to set up a central deep sea fishing corporation which should undertake the acquisition of deep sea fishing vessels and their operation through private and public sector enterprises in the State sector, under stipulated terms and conditions.

Compared to the tough line adopted by the Government in respect of marine fisheries sector, the encouragement given in regard to inland fisheries sector is very liberal. Entrepreneurs receive an abundant subsidy and easy loans with Governmental help for taking up aquaculture, both freshwater and brackishwater. No guarantees of the kind stipulated for entry into the marine fisheries sector are necessary there. The reason for this liberal attitude is obvious. The activity is firstly landbased. The investments called for are not as heavy as in the deep sea fishing field. Further, the type of small entrepreneurs in the inland fisheries sector are different from those in the deep sea fishing field. But a time may come when Government may tighten the liberal conditions now prevailing, in course of time, if the fish farmers fail to make repayments of their loans in time.

Fishing Chimes, Editorial 26: May 1983: Vol. 3, No. 2

# The Tuna Tangle

Tuna continues to elude us. Our lines of strategy to catch them are perhaps too weak to hook them.

For over 15 years efforts have been on to catch tuna with foreign help but without success. Kerala and A.P. Fisheries Corporations made strenuous efforts to enter into a joint venture with Sun Harbour Company In U.S.A., but proved futile. There were earlier efforts at similar collaborations with Van Camp of U.S.A. with the same fate. Negotiations in all these cases seem to have failed on one ground. The foreign parties wanted to conduct test fishing for a specified period before taking up commercial operations. The Indian view, predominantly official, was that this was not possible.

The subject came up for a renewed appraisal at the International Conference on deep sea fishing held In New Delhi in June 1982. There was an unanimity of view that our tuna resources should be exploited through collaboration arrangements with countries such as Philippines, Indonesia, and France. Later, at the 5th sea food trade fair held at Madras in February 1983, representatives from Spain showed interest in teaming up with suitable Indian enterprises in the exploitation of Indian tuna.

Shortly after the International Conference on Deep Sea Fishing held in New Delhi In June 1982, Government deputed an expert team headed by Mr. S. P. Jakhanwal, Joint Secretary (Fisheries) to Philippines and Indonesia. The fishing industry has not so far been told about the findings of the team.

How serious the Government is in regard to promotion of tuna ventures Is not clear. The Karnataka Fisheries Corporation applied for the chartering of two tuna purse seiners from Philippines almost twelve months back. Only in May 1983 Government has been able to issue a letter of Intent. A private sector fishing company had also applied for the chartering of two purse seiners from Philippines. Their application, it is learnt, was rejected by the Government. These instances tend to show that our Government must be having certain reservations in regard to promotion of tuna fishery development.

In this situation, the only ray of hope seems to be the progress made by the State Trading Corporation. Adopting a down-to-earth policy, the STC engaged a Canadian firm for the preparation of a feasibility report which is expected to be received by June 1983. The cost of the preparation of the report would be met by the Canadian International Development Organisation (CIDA). It is expected that the Canadian consultancy organisation would make recommendations in regard to the foreign company with which STC may enter into a joint venture, to undertake a 100 per cent export oriented project.

Efforts in the past for setting up a major Joint venture based on tuna failed. Let us hope STC will succeed in their efforts to enter the tuna field, where others failed. Our conjecture is that there are strong undisclosed considerations in the mind of the Government in respect of tuna Joint ventures. We wish that STC's scheme will steer clear of these considerations and rise above them. Till we know about the result we have to keep our fingers crossed.

Fishing Chimes, Editorial 27: June 1983: Vol. 3, No. 3

# Pull Out Sea Nets from Failures

'Fishing Chimes' is indeed a thought -provoking journal for the benefit of teachers, researchers, extension workers and fishing industries and also to the educated fish farmers. However, I have the following suggestions which may please be considered for inclusion in your ensuing issues, a reader of Fishing Chimes has pointed out. These are:

"1. In the 'content page' names of the authors who contribute articles may be added along with the titles for easy reference. A separate list of authors may further enlighten the readers.

2. The chapter 'Question and answers' is not only interesting but is also very informative. To this, interviews of the perspective fish farmers if added would solve many practical difficulties in fish farming.

3. As in many journals like 'Fishing News International' a new column incorporating the forthcoming Conferences, Meetings, Seminars, Symposia on fishery aspects to be held in India and elsewhere may also be started, so as to enable the Indian Fishery scientists for active participation.

4. Article may also be invited on new fish recipes from eminent food technologists/house wives for publication in 'Fishing Chimes'.

I request that my letter may please be published in your Letters to Editor column."

In the past few years, countries in the neighbourhood of India, such as Sri Lanka and Malaysia have been able to register a substantial increase in their marine fish production. In contrast, Indian marine fish production has been stagnating on the wrong side of 1.5 million tonnes. There has practically been no perceptible increase in production in the past ten years. The quantity of marine products exported also has registered a fall in 1982-83.

Fall in production in one bad year can be understood. The situation becomes alarming when there is no perceptible increase in production, in spite of concerted efforts on the part of the Government and the industry over a period of years. Someone in the higher echelons of the Government will have to take notice of the situation, analyse the causes and introduce corrective measures to augment fish production.

It is agreed by experts that there is no scope for increasing fish production from coastal waters substantially. The area is already fished to optimal level. The traditional sector, consisting of operators of non-mechanised boats and mechanised boats, conducts fishing in the zone. Whatever may be done, we cannot expect any substantial increase in fish production from this sector.

## Go Deeper

This situation leaves us no alternative other than concentrating on fishing in deeper areas. Over the past ten years what all could be achieved was the introduction of an insignificant number of 68 deep sea fishing vessels, which really exploit shrimp in nearer waters. Apart from these, permissions for chartering around 50 foreign deep sea fishing vessels were given to Indian fishing companies, of which around 25 Nos. are now operational. All these are of Taiwanese registry. One cannot be certain how long these vessels will operate, as the overzealous surveillance by coastguard is making Taiwanese owners to ponder over the situation. One need not wonder if they withdraw their vessels, for, the punishments for acts of transgression are not convincingly established, but, are too harsh leading to operational losses.The authority being on the Indian side, nothing can be done about it. They feel that it is all the same whether they poach or operate under permit, the troubles being the same. Without the baselines on the coast being established the coast guard arbitrarily determined in the case of two of the chartered vessels that operated in territorial waters, although SATNAV reading on one of the vessels impounded showed a distance of 13.7 nautical miles from the shore. By being excessively zealous the coastguard has unwittingly frightened the owners. Taiwan is the only source now for chartering deep sea fishing vessels. If this source is extinguished, as will definitely happen, the chartering scheme will have to be folded down with the present repressive rule aimed at the fishing Industry right, or wrong, from the portals of Krishi Bhavan, the charter scheme will no doubt run into serious difficulties very soon.

## Indigenous Boat Building

With the bleak scope for augmenting the fleet of deep sea fishing vessels through charters, the only way out is for India to introduce her own vessels. To say plainly, judging from the performance of the Government and the Industry, the outlook is very bleak In this respect. As matters stand, only two larger houses have a semblance of interest in acquiring deep sea fishing vessels. These are Tata Oil Mills and Konkan Fisheries, a subsidiary of Chowgules. They cannot however get loans from SDFC. Unless loans from SDFC are given to them they do not appear to have any intention to act. Regarding medium and small scale entrepreneurs, it is impossible for most of them to enter the field. The commitments in regard to 10 per cent downpayment towards the cost and meeting the debt-equity ratio stipulated by SDFC are far beyond their means. The State Fisheries Corporations can constitute an opening, but they have their own troubles. Most of them are running on losses and their Governments wouldn't allow them to create further commitments.

Considering the above situation, the Central Government has to come 'down to the sea' They have to decide whether to organise the exploitation of the marine living resources of the sea on priority. If the present dilatory approach is to continue, it only means that Government are deceiving themselves. If they are serious, the only conclusion that can be drawn by them is that deep sea fishing in India cannot be developed, in the present state, through the private sector. It can only be done through a Centralised Governmental Organisation. The simple and established fact Is that there are no entrepreneurs in the country to take up deep sea fishing on a large scale, in the present climate.

## Central Body

Deep sea marine fishery resources come in the same category of other natural resources of the sea such as oil. Government has taken up the exploitation of oil and minerals on their own, virtually as a state monopoly. It is reasonable to apply the same principle for deep sea fishing, with the relaxation that private Industry is also free to participate in the activity. Before it is too late, an appropriate solution to the problem is to be found and this has to be for the Government to set up a Central Corporation or Organisation to develop deep sea fishing under its auspices. Through World Bank's assistance, or through bilateral assistance, Government can easily secure the finances necessary and the vessels required for undertaking deep sea fishing directly or through private enterprises by hiring out the vessels acquired.

One important requirement, to be taken care of along with the introduction of deep sea fishing vessels by a central agency Is internal marketing. The Central agency shall have to establish an Internal fish marketing infrastructure, extending it to all towns and villages connected by roads. The frozen/cold stores, and retail stalls at various centres, which shall have linkage with production centres can be established either through private enterprise or directly by the central agency.

Fishing Chimes, Editorial 28: July 1983: Vol. 3, No. 4

# In the Horns of a Dilemma

The Indian charterers of foreign fishing vessels are caught In the horns of a dilemma. Within 18 months of making the chartered vessels operational, they have to introduce the first unit of their own vessels. Most of them are anxious to do this, but they have no way of achieving this. For no fault of theirs, as elaborated hereunder, their deposits are likely to be forfeited. For, it is certain, none of them will be able to introduce their own vessels within the stipulated time, however hard they may try, unless Government comes out with clearcut guidelines and directions in this respect.

## Alternatives

Imports or indigenous construction of the *pari passu* vessels are the two alternatives open to the charterers to fulfill their obligation. As per the condition stipulated in the permission for charters, vessels of over 20 m length can be introduced by the charterers. It is understood that Government are not agreeable to the import of secondhand vessels. This alternative is thus out. Charterers can afford to import new stern trawlers of 23 m OAL. International price of these is now around Rs. 80 lakhs. There is a fixation in the minds of the authorities that the cost of these vessels should not be more than Rs. 67 lakhs. Applications for importing new vessels in this size range face the inevitable risk of rejections, as has happened in the case of several other applicants. Even if approved, if past experiences are any guide, it will take over twelve months (an optimistic estimate) to secure the permission and sanction of loan by SDFC. The time limit for the introduction of the first unit being 18 months, It will be impossible for the buyers to fulfill the condition. For no fault of theirs the security deposit given by them will get forfeited.

## 'Indigenous' Matters, Confusing

Such being the situation in respect of imports, charterers can only think of indigenous construction. No situation can be more depressing and confusing than this sector of indigenous construction of deep sea fishing vessels. Firstly, no yard in India is in a position to construct and deliver deep sea fishing vessels in 18 months of placement of orders. Secondly, all the yards have only one design with them, that of 23 m long Mexican shrimp trawlers. Goa shipyard has no doubt with them designs of larger vessels meant for exploratory work but their efficacy is not yet proven. The cost of each of these vessels exceeds Rs. 3 crores and it is not within the means of any of the charterers.

## Time Limits and Prices

Charterers can only thus apply for permission to place orders for 23 m long Mexican type of vessels. This is inevitable for the reason that the buyers have to depend upon SDFC loan and subsidy. The time that will be taken for securing these sanctions, as per present trends, is likely to exceed six months, if not twelve. Within the remaining period out of the allowed 18 months, no yard will be able to construct and supply the vessels.

Apart from the above realities, choosing a suitable yard for the construction becomes a major problem. The prices quoted for these trawlers by Indigenous yards differ widely. While some yards quote around Rs. 90 lakhs, others quote around Rs. 142 lakhs per vessel.Unlike foreign yards, the prices quoted are subject to escalation with reference to 'increase in cost of materials and labour, sales tax and excise duty. The final price is anybody's guess. The date of delivery is also anybody's guess. Vessels of this category have so far been constructed In India only at Chowgule's shipyard and those constructed in the later batches were of good quality. Other yards have to experiment with the construction and how good the final product will be cannot be predicted.

The charterers may be prepared to take the risk and place orders with one of the shipyards. Even with this, they are likely to lose their deposits, as, without any doubt, they will not be able to introduce the first batch of vessels within 18 months of making their chartered vessels operations. It is indeed incomprehensible that the Ministry of Agriculture has stipulated an unachievable and unrealistic time limit. More than anyone else the officials in the ministry are aware of the situation. It is to be hoped that the objective is not to grab the deposits given by the charterers.

## Suggestions

Charterers as well as the ministry have the same objective, that deep sea fishing vessels of Indian flag have to be introduced. Introducing Indigenous ones seems to be the way obviously preferred by the Government. What is necessary then to accomplish the introduction is to lay down specific guidelines and directives. Types of vessels to be supplied, yards earmarked for their construction and the prices of the vessels may be fixed by the Government. Thereafter, the charterers may be called upon within a stipulated time to choose one of the recommended yards and the types of vessels and come up with applications for SDFC loan, together with a draft for 5 per cent down payment in favour of SDFC, contract signed with the shipyard etc. The yards will have to be directed to provide the interim guarantee towards the loan required by SDFC. The condition In respect of debt-equity ratio has also to be relaxed as no chartering company will be able to raise equity equivalent to one sixth of the loan, which will exceed over Rs. one crore for two vessels. They may be allowed to raise this in stages in a period of three years.

We feel that steps on the above lines will have to be taken by the Ministry of Agriculture. If this is not done, the situation will continue to be fluid and charterers will continue to struggle, groping their way in a slush. Such a situation would not be a creditable one to the Government either. History will only record that Government left the charterers in the lurch. To avoid this Government should come up with a specific action programme to enable charterers to comply with the conditions.

Fishing Chimes, Editorial 29: August 1983: Vol. 3, No. 5

# Ministers Miss Main Issues

Some of the Chief Ministers and Ministers in-charge of fisheries from various Maritime States met at a Conference convened by the Union Ministry of Agriculture on 27th June 1983 at Hyderabad.

The Conference seems to have served the much needed purpose of reviewing the present situation in respect of fisheries development. This is no doubt a relatively easier job compared to the formulation of a policy for stepping up marine fish production.

The Conference noted that there was a virtual stagnation in our annual marine fish production. In this context what all the Conference had done was to make a suggestion to expand deep sea fishing activities and to call upon the State Fisheries Corporations to play a more active role in the development of deep sea fishing. It was also felt that a systematic drive had to be launched for enrolling all fishermen as members of fishermen's cooperatives and a blue print prepared for organisation of fisheries cooperatives at primary, regional and State levels. A call for development of coastal aquaculture was also given.

Recommendations were also made at the Conference on the subject of welfare schemes for marine fishermen which included introduction of disaster relief schemes, accident insurance for active fishermen, group insurance for mechanised boats, development of an effective system for protection of lives and properties of fishermen, lean season relief, provision of better civil amenities, providing non-formal education to fishermen and extension of easier bank credit.

The entire fishing community ought to be grateful to the Ministers for their recommendations concerning fishermen's welfare. None of these recommendations are however new. All these have been recommended earlier either by the Centre or taken up by some of the State Governments on their own.

All maritime State Governments would like to implement these programmes but their concern will be about the financial aspects. It is not known whether this excellent and rare occasion of a gathering of the Ministers worked out any recommendation in regard to the funding aspect.

Social welfare schemes for fishermen are a necessity. Fishermen, being a part of the national population, are entitled to the benefits of the general welfare schemes launched by the various Governments. In view of their downtrodden condition of the fishermen, it is but right that fisheries departments should take special steps in this direction.

Social welfare schemes alone would not however increase the incomes of fishermen. Unless there is increase, in their incomes. their social conditions do not really improve. The schemes proposed for implementation would only tend to maintain them in the present level of poverty. The Conference does not seem to have evolved any policy for the improvement and upgradation of the professional capabilities of fishermen. Unless this is done, their incomes would never improve.

The bulk of Indian marine fish production comes from the hard work of traditional fishermen. Their area of operation is already overexploited and introduction of any new non-mechanised or mechanised units would only result in the reduction of catches per unit. Only by equipping them to reach farther waters or by introducing new techniques for exploiting hitherto untapped resources in the traditional zone would enable them to have higher income, which alone can improve their standard of living. Welfare schemes are only for the maintenance of *status quo* but they will never lead to incomes and a better standard of living. The conference does not seem to have given any serious thought to this question of upgradation of professional capabilities of fishermen.

The recommendations in regard to fishermen's cooperatives are totally ritualistic. Recommendations of this nature had been made in the past at several Conferences right from 1960s. Experience shows that co-operative activity does not suit Indian fishermen. If they join as members of co-operatives it is only to get the benefit of loans and with this their cooperation ends. They are not to be blamed for this. The activity is such that the fishermen require on the spot payments for the catches handed over by them and these have to be remunerative and more than what a fish merchant offers. The interactions among the fishermen, the fish merchant and the co-operative are influenced by the motivation for higher incomes by the fishermen and fish merchants on one hand, and lack of initiative, absence of interest and rigidity of rules on the part of co-operatives on the other. The heirarchial structure of co-operatives leaves the fishermen exasperated. In India there are only stray instances of fishermen's co-operatives functioning successfully and where they succeeded the management is in a few hands and the money goes largely to them. Efforts are needed to evolve a new pattern for restructuring fishermen, cooperatives. Readers would like to know that there is virtually no functioning set up at the Centre to promote fishermen's co-operatives. Such being the case, one can take it as a foregone conclusion that this recommendation on fishermen's co-operative is not likely to be translated into action.

It is a matter of great surprise that the Conference paid scant attention to the problem of promoting deep sea fishing. It is to be hoped that the Conference was not led away by the consideration that, since only one per cent of marine fish production comes from deep sea fishing, all attention has to go to the sectors that bring in the balance of 99 per cent.

It is to be noted that this component of 99 per cent will continue to remain so or increase marginally, whatever we may do, since the exploitable limits of inshore waters have been reached. What is therefore needed is to concentrate on the zone that is now only yielding 1 per cent of the present production. This is a priority area that requires total attention.

There are several burning questions in regard to development of deep sea fishing that are baffling and defying solution. The recommendation that public sector corporations should play a more active role in regard to development of deep sea fishing is vigue. Almost all the corporations are running on losses and that too without any concrete programmes for deep sea fishing. They are not playing any active role at present In the field except that a few of them are operating shrimp trawlers and chartered fishing vessels. The question of their playing a more active role does not therefore arise. They have to first play an active role.

The recommendations are silent on the role of private sector in deep sea fishing which is disappointing. It is not known whether the intention is to exclude private sector or to treat it, as of a second rate priority. It would have clarified matters considerably if the Conference had set in clear terms the respective roles of public and private sectors in the development of deep sea fishing.

The slow progress in the introduction of deep sea fishing vessels of Indian flag is a matter of considerable concern. The Conference does not seem to have reviewed the situation. There is no recommendation in regard to the line of action to be adopted for the introduction of deep sea fishing vessels of Indian flag.

To sum up, the Conference skirted around the crucial issues of upgrading the professional capabilities of traditional fishermen, policy to be followed for the introduction of deep sea fishing vessels of Indian flag, and relieving the country from the present stagnation in the annual marine fish production.

Fishing Chimes, Editorial 30: September 1983: Vol. 3, No. 6

# Charter Charade

The progress in the implementation of the programme for chartering of fishing vessels by Indian fishing companies has now reached a significant level. The types of chartered vessels in operation consist of bull trawlers (pair trawlers), gill netters, and stern trawlers. Of these, It is gratifying that the results of operation of pair trawlers of Taiwanese design are reassuring. These results have proved beyond doubt the existence of extensive stocks of fish and cephalopods in our exclusive economic zone beyond the territorial waters This is not to say that a major area of our exclusive economic zone has been covered by the operations. The coverage has been to the extent of a substantial belt beyond limits of our territorial waters, in the northwest and also in the southern garland below 13° latitude, on the east, west and south.

## Bull Trawling Well Suited

The main objective in permitting charter of fishing vessels is to acquire and introduce new and diversified technologies. Now it is proved that bull trawling is well suited for our waters with vessels of Taiwanese design. Urgent steps appear necessary (a) to enable Indian parties to introduce bull trawlers of their own based on Taiwanese design and (b) continue to permit Indian parties to charter such vessels till such time as the technology gets fully popular and established.

## Irrational Limit

Such being the incontrovertible situation, it is a matter of surprise that the Ministry of Agriculture, the body which has all the requisite expertise, has stipulated that the chartered vessels should return to the base port within 45 days of commencement of a voyage. This condition is perhaps stipulated to avert the likelihood of the chartered vessels making intermediate sojourn back to their foreign port for unloading the catches and continue further fishing operations on return, thereby deriving an undue advantage. If this is true, there is no substance in this line of thinking. It is impossible for these to move out of our economic zone with these to a foreign port. Even if such a thing is contemplated, no foreign port will allow the unloading from vessels which has unauthorised Indian crew on board without valid papers. The Indian counterpart crew will have no valid papers that allow them to visit foreign ports. The present restrictions on the number of days of a voyage is thus unrealistic and unreasonable and it totally defeats the purpose of fishing operations. One has to only imagine the loss in valuable fishing days and the expense if the vessels have to go back to the base port at a time when the holds do not have a economic level of catch or when they had just located promising grounds. If our ministry for agriculture is pragmatic and has the correct advice, based on expertise, a condition of this kind would not have been stipulated. This condition needs to be modified to provide for a voyage period consistent with the fuel capacity.

## Irrational Limit

The form of application for permit provides for the description of vessels, equipment and complements. This provision is justified only if the Government straightaway Issues the permit or reject the same. According to the procedure followed now by the ministry, a letter of intent is issued in the first instance. The issue of this letter of intent itself takes a very long time, often around a year. Such being the case, it would be impossible for any applicant to furnish description of vessels under nearly 27 items. As a letter of intent only is issued at first, it would be adequate for the Government to receive broad particulars in regard to the types of vessels to be chartered. Before applying for permits the companies in any case with furnish all the details.

In the permits that are issued there is provision for the inclusion of the name and address of the Master and also names of other foreign crew. The justification for including these items in the permit that will be valid for 3 years is obscure. It is common knowledge that the Master and crew on any vessel keep on changing from voyage to voyage. Such being the situation, no purpose is served by including the names of foreign crew in the permits. In any case the foreign crew lists are always separately furnished. Once the names become part of the permits, every time there is a change it would be necessary to seek amendments to the permits. Application for clearance of these changes will undoubtedly take an inordinately long time, thereby affecting the operations. It will therefore be prudent to delete the items relating to the names and addresses of the crew in the permits, and enclose a list that can be updated or replaced from time to time.

## Foreign Crew Clearance

There is a set procedure for the clearance of crew of foreign ships at various ports. This work is attended to by the immigration authorities. This procedure is also followed in the case of chartered fishing vessels. However, in addition to this, there is the requirement of the names to be cleared by the Home Ministry, based on the lists furnished by the applicants to the Ministry of Agriculture. In other words, the applicants have no hand in the particular clearance, which is attended to by the Ministry of Agriculture and Home Ministry. Knowing the situation fully well, unfortunately, the coast guard asks the applicants to produce the clearance from the Home Ministry for the foreign crew and the applicants have no means of taking up the matter with the Home Ministry, as they do not have any details of the references made by the Ministry of Agriculture to the Home Ministry. Thus, while it is unjustified to ask the applicants for this clearance, it stands to reason that this extra requirement can be given up. The immigration authorities at various ports come under the Ministry of Home Affairs and clearance from them should be adequate, as in the case of other foreign ships.

## Radio Officer without RT

In all Taiwanese pair trawlers, only one out of the two vessels has radio telephone. This is understandable because both the vessels are together and communication can be dealt with from one vessel. In spite of the fact that only one vessel has radio telephone the Ministry of Agriculture stipulates that there should be Indian counterpart Radio Officers in both the trawlers. It will be adequate if one counterpart radio officer is appointed on the trawler having the Radio Telephone.

There is a serious dearth of radio telephone operators to work on fishing vessels. All available hands have already been engaged and there is no possibility of finding any more qualified persons to be appointed as radio officers. Indian Captains appointed on the trawlers have adequate knowledge in radio telephone operation and can easily pick up expertise in the matter and also keep a watch over incoming and outgoing messages. In fact, the present trend in the shipping world is to entrust the radio telephone work to the officers in command. Our suggestion is therefore that Government may not compel chartering fishing companies to appoint Radio Officers on the chartered trawlers. Instead, appointment of the additional deckhand or secondhand will be useful.

## Declaration of Catches

The present stipulation is that catches landed by chartered vessels should be declared at the base port approved by the Government. The purpose served by such a stipulation is not clear. At the time of concluding fishing operations the vessels may be close to another port. The national interest would not suffer in any way if the vessels declare their catches to the customs authorities at the nearest port and obtain the required clearance. There will be a considerable saving of time and money if such a procedure is acceptable to the Government. In fact, in March 1982, the Ministry of Agriculture agreed to such a procedure at a meeting presided over by the Additional Secretary Incharge of fisheries. In spite of this, the fishing companies are not allowed to declare the catches at the ports nearest to the fishing vessels at the time of conclusion of fishing operations.

## Introduction of Indian Vessels

As already mentioned, the objective of the charter programme is to bring about the Introduction of diversified methods of fishing. Yet, according to the scheme, the charterers can introduce vessels of their own, not less than 20 m OAL in fulfillment of one of the conditions stipulated by the Government that the charterers should introduce vessels equivalent in number to those permitted to be chartered within a stipulated time. The condition enables the charterers to introduce any type of vessels over 20 m length. In other words, these vessels need not be designed for pair trawling. Charterers may apply for the introduction of 20 m long trawlers or gill netters or any other vessels. This freedom completely defeats the purpose of the scheme which is to introduce vessels suitable for diversified fishing. Considering this aspect, it is necessary for the Government to take steps to ensure that the charterers introduce bull trawlers only in fulfillment of their obligation. Then only the purpose of the project will be fulfilled. Otherwise the expertise gained by the counterpart Indian Personnel will become futile.

## What is to be Done

One way to achieve this introduction is to identify an Indian shipyard which may acquire designs of Taiwanese bull trawlers and undertake construction. A public sector corporation or corporations can be made to place orders for the number of vessels that will be necessary for acquisition by the various charterers. A system under which the charterers can place orders with the Corporation concerned for the trawlers on hire-purchase basis can be devised. The time taken by the Ministry of Agriculture for the issue of letters of intent and later to convert the letters of intent into permits can undoubtedly be reduced. The long gestation period has the undesirable effect of creating disputes between the owners and charterers. The delay results in deployment of the earmarked vessels to other countries by the owners, necessitating further applications for substitution. Delays in fact amount to denial of permits. A quick rejection of the proposal is far better than a delayed approval. The fishing companies can save their time and money if their proposals are rejected quickly. It may be mentioned here that countries.such as Australia, New Zealand, Pakistan and Bangladesh either issue permissions or reject the applications within a few days. Our D. G. Shipping issues charter permissions for cargo vessels within 3 or 4 days. The Ministry of Agriculture can also do like wise, if not 3 or 4 days, in 3 or 4 months. Faster decisions, approval or rejection, will alter the present image held by foreign fishing companies that our Government takes an unduly long time to decide on the proposals. Faster clearances will also pave the way for the inflow of charter offers for fishing units suited for diversified fishing. Regarding issue of permits based on letters of intent, if this is done based on a well-drawn checklist and decentralisation of authority for approval, delays can be cut down considerably.

Fishing Chimes, Editorial 31: October 1983: Vol. 3, No. 7

# All Out But One

The high tide that drove a bevy of larger houses into the vortex of deep sea fishing and into the field of export of marine products in late 1960s and 1970s, started receding in 1980s. Gnts as they are, Union Carbide, I.T.C., E.T.D. Parry, Britannia, Tatas, Konkan (Chowgule) made an enthusiastic rush into the sector. They thought that there would be encouragement forthcoming from the Government, considering that deep sea fishing is a sector involving risks. They were however too sanguine. They were disappointed that Government was not willing to extend facility of loans from the SDFC to the companies for strengthening their fishing fleets. They wanted to be in the industry in a big way. When it was clear that this was not possible and when they reached their limits of endurance, they started pulling out one after another.

The exodus started with E.I.D. Parry, first to conclude that they had no future in the industry. I.T.C., closed down their fishing operations a few years after Parry pulled out. The latest to join them are Britannia and Union Carbide. Tatas are no longer operators of an integrated setup. They closed down their processing operations long back, not wantonly, but because they were forced to close down their factory in Cochin. Although cramped and unable to expand their fishing fleet like the others who exsited, they continue their fishing operations. The main reason for this approach seems to be the tradition of Tatas who wouldn't like to retrench unless they are cornered.

Union Carbide closed down their processing plant at Visakhapatnam. They are however still in the field of fishing. Mazagon Dock-built trawlers have become a liability to them. To make up they have been operating two trawlers of Dolphin Fisheries, Bombay on charter and another four trawlers of State Fisheries Development Corporation under a management arrangement. These six vessels may soon be out of their hands, leaving the two Bender-built (U.S.) trawlers acquired by them in sixties for continuing their fishing operations as long as they can. Union Carbide has obviously lost interest in their marine fishery activities in India.

Chowgules are doggedly continuing in the field. They have their own shipyard where the vessels constituting their present fleet were constructed. They did not avail of any loan from SDFC for constructing the vessels. They thought that they would get a subsidy of 27 per cent of the cost under the rules in existence then, but they could not get this. While all others were out, Chowgules are the sole survivors among larger houses (including one multinational) with fishing as well as processing operations on their hands.

The decision of the majority of the larger houses in the field to close down their operations bristled and continues to bristle with the problem of disposal of the assets. Parry managed to sell their trawlers on outright sale basis to Nava Bharat FerroAlloys. I.T.C., has not yet succeeded in selling their trawlers. Britannia is understood to have finalised a deal at last with a fishing company. Union Carbide faces the problem of disposing of their processing plant. They would probably be anxious to sell all their trawlers and come out of the sector, but it is difficult for them to find buyers. They may therefore operate their trawlers with a skeleton staff. They have been shifting their operations from Visakhapatnam in A.P. to Roychowk in West Bengal gradually.

Larger houses no doubt wanted to play a significant role in the development of deep sea fishing industry. Sensing the lukewarm attitude of the Government and evaluating the denial of facilities that are given to others, they chose to withdraw. And tbey seem to be right in doing so. Their withdrawal is however not likely to have any perceptible impact on the present state of development of the industry. In any case the going of the industry in respect of introduction of new units of Indian flag is not good. The exit of larger houses is not going to make it worse.

It has to be admitted that the entry of larger houses into the field put the industry in focus. With their exit, it is to be hoped that the others now remaining in the field will maintain this focus.

Larger houses undoubtedly are far better equipped to venture into new areas such as tuna fishing, risking capital investments. The cost of tuna vessels being on the higher side, even with SDFC's loans, practically no small entrepreneur can mobilise the required marginal finances which by themselves will be high. Further, in a promising and yet a risky field such as tuna fishing, it would be wise on the art of the Government to give an opportunaity to larger houses to participate in the activities, within the framework of a policy, evolved after due discussions with their (larger houses) representatives or with AIFI. We believe that this approach will be in national interest.

Fishing Chimes, Editorial 32: November 1983: Vol. 3, No. 8

# Fisheries Education

Comprehensive education in the field of fisheries, covering marine and inland fisheries is now offered at various centres, chief among which are the Central Institute of Fisheries Education, Bombay, and Fisheries Colleges at Mangalore and Tuticorin. While the Central Institute of Fisheries Education, Bombay offers a postgraduate diploma course, and has facilities for research leading to M.Sc. and Ph.D of Bombay University, the fisheries colleges have courses leading to degrees as well as postgraduate degrees in fisheries. The courses include field studies to acquaint the students with what actually happens outside the college/Institute premises. At the Central Institute of Fisheries Education, and possibly at the other centres, the students are required to submit dissertations on specified topics based on intensive studies.

The students coming out of the seats of learning in fisheries will have to shoulder responsibilities of implementing developmental programmes. This requires continuous and intensive exposure to developmental activities during the course of their studies. This exposure is much more important than classroom learning.

## More of Practical Knowledge Needed

On the inland side, the students should gain practical knowledge of a retentive nature in respect of induced breeding and raising of carp spawn to fry and fingerling stages. They have also to acquire intensive knowledge in the breeding of marine and freshwater prawns and raising the spawn to fry and fingerling levels. They should imbibe adequate knowledge in respect of layout and construction aspects of freshwater and brackishwater farms inclusive of seed farms. They should secure sound knowledge about hatcheries and their operation. Techniques of reservoir and river fishery development should be acquired by them.

On the marine side there have been extensive developments relating to introduction and operation of mechanised boats and deep sea fishing vessels. Management of fishing vessels and fishing enterprises is something that can be learnt largely from field observations. Activity at the fishing harbours is a new area requiring field observations. The supporting services provided to keep the vessels in operation need to be studied in actual working. Processing and marketing of fish and fish products, particularly export activities require a close field study by the students.

Luckily, the Fisheries Colleges and the CIFE are conscious of the need for a strong field training component and field studies form a distinct segment of the course that is conducted. However, exposure of the students in a more intensive way than at present to field activities will result in a far better trained technocrats to handle developmental activities.

## Intensive Training

Organising occasional visits to field centres for the purpose of training may not fulfil the objective of such visits, particularly in respect of major activities such as those related hatchery operations, carp seed and prawn seed production, freshwater and brackishwater fish farming, fishing vessel operation, processing etc. The Institutions should have their own subcentres where intensive training in the respective fields can be provided. The Central Institute of Fisheries Education has no doubt its own brackishwater farm and major carp seed farm where its students receive training during the season. The fisheries colleges also have certain facilities, but they should also strive to have their own farms for providing intensive training. Training at farms owned by others will not be effective for the reason that often the students are considered as an intrusion and the training turns out to be perfunctory.

The students should receive their training in a climate charged with developmental atmostphere. Apart from field training, it is of paramount importance to create such an atmosphere around the students that is conducive to activating their thinking processes. They should be frequently subjected to participation in gatherings of those engaged in various fields of fisheries development. Listening to the experiences related by these persons during seminars and symposia will tone up the quality of thinking of the students and bring about precision in this respect. The students should be encouraged to ask questions, and also to talk. They should also be encouraged to present their perceptions on fisheries topics of importance.

## Quality of Leadership

The institutions should help students in developing qualities of leadership. They should be trained to express themselves precisely, properly and effectively. They should also be trained in the art of preparing proposals for initiating fresh developmental activities and also in the art of reporting. They should also be told as to how they should prepare themselves before attending meetings where they have to present their proposal for approval. All these aspects, among others, are important, if the student concerned is to make a mark.

Each of the Institutions can set up a 'Student's Forum' at which selected students in turns can be encouraged to talk on specified subjects. The manner of presentation by each of the speakers can be watched by the teachers and corrective steps necessary to improve the delivery can be brought to their notice. Organisation of lectures on specified subjects in the above manner will instill confidence in the respective students and will enable them to bring out the best in them in stages.

To sum up, concentration on class lectures alone at the fisheries centres of learning will lead to the turn out of very poorly trained candidates. Class lectures are no doubt important but play a limited role.

The requirements are a through exposure to various developmental activities by means of actual participation to the extent possible, development of qualities of leadership, creating an awareness of the dimensions of the fisheries field and stimulate the thinking processes, andprovide training on how an executive has to function if he is to succeed. The Institutes should be equipped with Centres of their own to provide sustained training in important sectors of the subject. The students should receive training in a developmental climate, interspersed with lectures on developmental subjects in 'Student's forum', field operations etc.

Fishing Chimes, Editorial 33: December 1983: Vol. 3, No. 9

# Indigenous Deep Sea Fishing Vessels

One of the Policy aspects relating to deep sea fishery development adopted by the Ministry of Agriculture is to promote indigenous construction of fishing vessels. In order to induce the shipyards to undertake the work, Government have announced a subsidy of 33 per cent on the total cost of a vessel, to be given to the shipyard concerned. While the response to the incentive is not very significant, a few companies made applications for grant of loan for the construction of fishing vessels at a couple of shipyards over two years back. The Ministry of Agriculture recommended to the SDFC that the loans may be granted. While loans have been sanctioned by the SDFC a long time back, funds to the shipyards towards the subsidy and towards the first instalment have not been released so far. The reasons for this long delay are not clear. In any case the long time lag gives the impression that the Government is not really serious about promoting indigenous construction of deep sea fishing vessels.

The Government would not have introduced the subsidy scheme just to have it on paper as a protection against likely criticism that nothing was being done for the promotion of indigenous construction. The Ministers in charge of Agriculture and of Shipping and the senior officers concerned want to promote indigenous construction. Yet, there has been no final action on the few applications which constituted the first batch, even after a lapse of two years.

## Delays

There can be a hundred problems responsible for the delay, all plausible. But all these problems would undoubtedly be the creations of the Governmental machinery, brought into the arena for someone to solve and take over responsibility for the solutions. It is impossible to imagine that any obstacle can withstand the willpower of a Government. Lack of will can be one reason for the delay. The mechanics to be followed for releasing the subsidy appear to be the reason for the delay and it is surprising that an issue of this kind is defying a solution for so long.

From one side, the industry is told that any application for loan from SDFC for vessels to be constructed indigenously would be cleared fast. On the other side, in actual practice, we find enormous delays taking place. The effect of this situation on the development of an indigenous deep sea fishing fleet can be imagined. Fearing delays, entrepreneurs tend to go slow in the matter.

The condition, stated to have been recently imposed by the Government that, out of every two vessels to be introduced by the charterers of foreign fishing vessels, one should be indigenously constructed, is an opening to give a boost to the indigenous construction. Having introduced the condition, if the Government creates delays in the approval of the proposals and eventual sanction and release of loans, the purpose would be defeated. Further, the charterers would be burdened with higher costs on account of the delays.

There has to be a realisation that all other considerations are subordinate to the main purpose of introducing indigenously constructed vessels. In order to give a boost to the activity in the initial phase, several relaxations beyond those already given such as assistance towards cost of vessels through subsidies and imports of equipment are necessary. Since larger houses are not being given loans, the only sectors in the field are the medium and small scale sectors. Of these, the medium sector has not been evincing interest and small scale entrepreneurs cannot raise adequate capital to meet the debt equity ratio stipulated by SDFC. If the activity is to make progress, this has to be relaxed and the initial ratio has to be based on the 5 per cent contribution to be made by the entrepreneurs. Within five years of operation of the vessels, the entrepreneurs may be allowed to raise the ratio to the stipulated level in annual steps. The stipulation relating to interim guarantees towards the loan also need to be relaxed by accepting the gurantees given by the shipyards and pledging of the insurance policies relating to the vessels under construction.

## Shipyards must go for Efficiency

There are only a few shipyards having real interest in building fishing vessels. Their interest should not only be directed towards securing orders and receiving payments from SDFC, but also in equipping the yards to construct good boats. The past experiences of Indian entrepreneurs with indigenously constructed vessels were very unhappy. The mistakes should be identified and rectified and repetition of faulty constructions will have to be avoided.

By proper planning, it should be possible for the Indian yards to build good vessels within the stipulated period, capable of giving excellent operational results. Once the Indian built boats gain reputation of excellence, no Indian entrepreneur would think of import of vessels.

Fishing Chimes, Editorial 34: January 1984: Vol. 3, No. 10

# Shrimps are not Scrimpy

Shrimps are not scrimpy, atleast in the Northeast of Bay of Bengal. Sandheads, an important shrimp ground in the northeast of the Bay of Bengal, are now in lime light. These rich waters are now yielding on an average around half a tonne of headless shrimp per day of fishing, with a sizeable component of whites. From what we know, since 1977, shrimp landings have always been good in the area with lean catches only in two seasons in between. However, the catches were never so good as at present, with the exception of 1978 season. Around sixty trawlers have been bringing around 68 tonnes of headless shrimp each per voyage since Nov./Dec. 1983, reminiscent of the catches by Union Carbide's Vessels in their early stages of introduction.

The intensive recruitment to the shrimp fishery is obviously because of highly successful breeding that took place in the second half of 1983 and respite in catching of juveniles in Chilka lake in 1983 due to heavy rains. The abundance is certainly not a revival but an intensification, for, annual catches in the previous years had been more or less uniform in sandheads area. The policy of the Government in restricting the introduction of shrimp trawlers is probably partly responsible for the present development, which must have given scope for part of the mature stocks to escape capture. This is, however, a debatable point. Even with 100 more shrimp boats, several mature shrimp would have escaped capture. It is not the number of mature specimens that is all that important, but the prevalence of conditions conducive to successful breeding and reduction in juveniles' capture is what is important. Successful breeding by a small percentage of shrimp stock would be enough to bring about a massive repopulation. The gloomy forecasts of depletion made by our scientists and accepted by the Government did not take into account this single factor concerning population dynamics. In the absence of reliable basic data on stocks and their composition, the tendency of scientists is to play safe by estimating stocks within a limit, based on the State of exploited resources at the time. When the exploited fishery, being an offtake from a dynamic resource, shoots up beyond the predicted limits, confusion over available stocks would arise. For example, in the case of penaeid shrimps of the upper east coast (mostly sandheads) our scientists, in 1977, estimated that the potential yield from 25 fathoms zone was 5,000 tonnes or about 3,000 tonnes of headless shrimp. The present annual catches from this depth zone far exceeds this tonnage, without any signs of depletion. It is therefore high time for scientists to shift the available data and review the situation with a view to making realistic assessment of shrimp stocks in sandheads area, taking into account the extent of likely breeding successes from year to year based on known periodicity in the past.

A section of the fishing industry, stated to be mostly those intent upon preventing new entrants into the field, supported the Government in its decision to restrict the number of shrimping boats in operation more or less at the present level. The representations from those keen on entering the field for exploitation of shrimp in the northeast bay, supported by statistics of catches that clearly showed a periodic increase in landings, fell on deaf ears. The fact that shrimp has a life of a short duration and would in any case get excluded from the population on account of natural deaths for that season, did not weigh in the mind of the Government. The policy resulted in the stagnation of shrimp catches. Had the introduction of more number of shrimp trawlers been encouraged, the nation would have maximised shrimp exports and earned more of foreign exchange.

Other countries such as Mexico and USA, with a coastline similar to ours have thousands of shrimp boats operating since several years with only one difference that the respective Governments enforce a closed season for a couple of months to allow shrimp to breed in freedom. For these crustaceans with a lifespan of little over a year, this much protection is adequate to maintain the fishery. Instead of acting in a similar manner, which is the right way, our Government controlled the introduction of shrimp boats which has actually harmed national economy. It is the greatest joke of the century that we consider that 60 medium size shrimpers are too large a number for a coastline of over 8,000 km, when other nations with much smaller coastlines operate thousands of them without any major reduction in the total output, but for seasonal variations.

We feel that the assessment that there is depletion of shrimp on the southwest coast appears to be of doubtful validity. This may be true in respect of coastal waters frequented by small boats. Medium sized trawlers never did any serious trawling for shrimp in farther waters of southwest coast as is now being done in sandheads area upto a depth of nearly 50 fathoms. Same is perhaps the case on the northwest and southeast coasts.

The need of the hour is for the Government to encourage introduction of medium sized shrimp trawlers along the southwest, northwest and southeast coasts to cover farther waters, beyond the zone of operation of small boats. There is no doubt that such an introduction, in sizeable numbers, will yield good results. The success in shrimp fishing along the northeast coast is mostly because of the large number of vessels operating and the resultant competitive spirit generated among the skippers. Government should announce schemes for the introduction of sizeable number of shrimp boats in the same way in the other sectors of the coastline (southwest, southeast and northwest) as well.

If the vessels are designed to function as combination vessels, they can switch over to other types of fishing, if, for some reason, shrimp fishing does not prove to be good.

From the point of view of likely availability of sizeable shrimp resources and stepping up of shrimp exports and foreign exchange earnings, it will be wise on the part of the Government to encourage introduction of shrimp trawlers with combination fishing facility for operation from various ports along the coastline.

Fishing Chimes, Editorial 35: February 1984: Vol. 3, No. 11

# Deep Sea Fishing : Salvation in Sight?

About eight years back, Mr. K.S. Narang, the then Secretary, Agriculture, Government of India, put a vital question to your editor. What strategy do you think the Government should adopt to bring about faster development of deep sea fishing? Mr. Narang and your editor were in Warsaw on a visit at the time and your editor had the opportunity of being asked such a question, which would never have occurred while he was in Delhi. Middle level functionaries like your editor at the time would rarely get an opportunity to meet the Secretary.

Mr. Narang has the inimitable quality of keeping his juniors at perfect ease to draw out the utmost from them. In a perfectly normal manner, I told him 'Sir, in the Indian situation, there can be only one solution to the problem and that is to set up a Central Deep Sea Fishing Corporation'. Asked to clarify, your editor told him that the industry being highly capital lintensive and involving risks, it would be difficult to attract middle and small scale operators to enter the field in a big way. While SDFC may give loans, for want of managerial skills and other predictable reasons, even those few who enter the field were not likely to create the climate for progressive expansion. The restrictions on larger houses tended to restrict their entry into the field. Therefore, which ever way one looked at the issue, there was no escape from setting up a Central Deep Sea Fishing Corporation for the promotion of the industry. Mr. Narang did agree with your editor's view at the time. Since its establishment in 1981, Fishing Chimes has been emphasising the need for a Central Deep Sea Fishing Corporation.

There are reliable reports, although unconfirmed, that the Union Ministry of Agriculture has formulated a comprehensive proposal for the setting up of a Corporation in the Central Sector for the promotion of deep sea fishing. In fact, it is understood that the proposal has been cleared by the Public Investment Board. The next step will be to obtain the approval of the cabinet for the proposal.

It is not known what exactly the functions of the proposed corporation would be. However, one longfelt need will be fulfilled with the setting up of the corporation. There will be an agency in the field to promote deep sea fishing. At present in the Ministry of Agriculture, policy as well as implementation aspects are looked into. This combination of functions has led to inevitable delays in the promotional work.

It will be a great improvement if the present Fishery Survey of India (erstwhile exploratory fisheries project) is also merged with the proposed corporation and maintained as a separate division of it. This will facilitate the availability of a more comprehensive data without trouble for analysis. When the survey and commercial functions both are with the corporation, a sense of urgency will be inculcated among the survey staff to analyse data promptly and pass on the conclusions quickly to the commercial wing.

The proposed corporation should be able to speed up the developmental process. The problems which the Government faces in issuing permissions for charters and for imports of fishing vessels and recommending loans for the acquisition of indigenous and imported vessels at present can be expected to vanish to a large extent with the setting up of a corporation as there can no longer be the suspicions and doubts which the ministry has at present towards entrepreneurs in general.

If the proposed corporation is to function on sound lines and without incurring losses, it should organise its activities in such a way that, with its help, the industry grows in the private sector. The corporation may import or construct vessels of suitable designs indigenously with their own funds or by availing of loans from SDFC. After this, they may sell them on hire-purchase or lease them out as convenient to private fishing companies under certain terms and conditions. There should be a field arrangement under which the corporation's representative will have the necessary control over sharing of the returns in such a way that the corporation's share covers the instalments due. The corporation, if it gets vessels, (on charter for diversified fishing) can also entrust their operation to private companies. The corporation can collect all its expenses besides the repayments towards the vessels, on pro rata basis from the concerned companies. It should be made obligatory for all the operators of the vessels taken from the corporation to furnish data relating to catches etc as required by the survey wing of the corporation.

Let us hope that in the new year 1984 a major breakthrough in the organisational pattern for the development of Deep Sea Fishing, through the setting up of a National Deep Sea Fishing Corporation will take place and that this corporation will undertake the onerous task with single-minded devotion. The success of the proposed corporation would no doubt depend on the competence of the Managing Director to he appointed and the freedom of action which the Government bestows on the corporation.

The comments given above have been made on the assumption that a National Deep Sea Fishing Corporation would soon materialise. If there is no such proposal and our information in this regard is not correct, then it can be taken for granted that it will take a long time for deep sea fishing industry to come up in India.

Fishing Chimes, Editorial 36: March 1984: Vol. 3, No. 12

# Silver Lining

The Ministry of Agriculture seems to have at last identified a strategy to induce fishing companies to place orders for the supply of deep sea fishing vessels on indigenous shipyards. The condition requiring the obligatory purchase of fishing vessels, equal in number to those permitted for charter is now being cleverly utilised by the Ministry of Agriculture to compel the fishing companies who have been operating chartered fishing vessels to place orders for indigenously constructed vessels. Charter programme may have its own unpalatable features; but all these will be outweighed if this leads to the introduction of indigenous vessels.

Some of the fishing enterprises involved in charter operations, by and large, appear to be unwilling to fulfill the obligation of introducing their own vessels as part of the charter programme. Most of these enterprises are those who have no long range interest in the industry. At the same time, there are several enterprises who are keen on introducing their own vessels availing of the loan facilities. This would be an opportunity for them to build up their own fleet. The advantage is that the chances of loan applications being rejected are remote as the fulfilment of the obligatory condition is not only a matter of challenge but also an important component of national targets in the field of fisheries.

Those who are unwilling to fulfil the obligation probably think that they could get out of it putting forth either unacceptable proposals or explaining the difficulties in their way. The ministry saw through the game and either cancelled or suspended certain charter permissions. It is to be seen how many of the operators coming under this category will relent and fall in line.

In regard to those who are seriously interested in introducing their own vessels, two categories of them can be discerned. One group, constituting the majority, have no faith in indigenously constructed vessels. They long for permission to import the vessels.

The strategy of the Government in this respect is however to prevail upon the parties to place orders for indigenously constructed vessels. At a meeting with the representatives of the concerned companies, the officers representing the Government are stated to have virtually told the parties that Government would only permit placing of orders for indigenously constructed vessels, although this stand is against the condition prescribed in the charter permission letters. In order to assuage the feelings of the charterers it appears that the Government's side had reluctantly agreed for the placing of orders for two vessels indigenously and for two more vessels abroad simultaneously, although the obligation is for placing orders for two vessels within 18 months and for another two vessels within 31 months. This commitment is sought to tie down the parties to indigenously constructed vessels.

It appears to us that the above strategy is sound in national interest. The strategy should not however be onesided. The concerned shipyards should undertake the work in all seriousness, shedding the perfunctory way of dealing with the subject. They should not inflate their prices and their offers should include the same norms as in the offers from foreign shipyards. The prices should be fixed and not subjected to escalation. The delivery schedule may be such that the first vessel is delivered within nine months, followed by the supply of the subsequent vessels at two nos. or at least one no. in the tenth month. Adequate attention should be paid in regard to quality of construction. If necessary, experienced foreign naval architects should be engaged to set right the shop floor drawings and for supervising the construction.

SDFC no doubt has to safeguard its interests while releasing loans for the construction of the vessels. However, the present experience is that the procedures are cumbersome and take years for their fulfillment. Unless these are reformed the object in inducing fishing companies to go in for indigenously constructed vessels will be defeated. Further, in the interests of achieving a break-through in this sector, conditions in respect of debt equity ratio and interim guarantees have to be relaxed. The risks involved in this respect are worth the while, compared to the enormous momentum they will generate in respect of indigenous construction of fishing vessels. This is a field in which Government has to shoulder the risks in national interest.

The present exercise to induce fishing companies that are operating chartered fishing vessels to place orders for supply of indigenous trawlers can at best be a short term measure. At present, the Indian shipyards have with them the designs and shopfloor drawings of only one type of commercially viable vessel *i.e.* 23.14 mts. long multipurpose trawler of modified Mexican design equipped for stern trawling for fish and outrigger trawling for shrimp. On the one hand, Government feels that increase in shrimping fleet should be discouraged and on the other it is indirectly compelling the fishing companies to go in for indigenous trawlers of the above type.

While there is ample scope for the introduction of a further no. of shrimp trawlers for operation in Indian waters, any one can see that there is no rationale behind the present policy of linking introduction of indigenous shrimp trawlers with the obligatory purchase of vessels under the charter programme.

The objective of the charter scheme is to introduce new and diversified methods of fishing. Accordingly, chartering of pair trawlers has been permitted by the Government and this method of fishing has been found to be successful in Indian waters. This should be followed up with the efforts of the Government as well as the industry directed at introducing pair trawlers. It is illogical that no effort is being made in this direction. A few fishing companies applied for the import of secondhand pair trawlers. These never fructified.

It is a glaring fact that, while the Government wants indigenous fishing vessel construction to be promoted, it has not struck them to set up an exclusive yard for the purpose. The jacks of all trades yards have neither the time nor the patience to pay the attention that fishing vessel construction demands. It appears that, just to please the Government and have the joy and satisfaction of visiting Delhi for meetings several times they keep on expressing their perfunctory interest which has no content. The prices quoted by the various yards for 23 m long trawlers, which vary abnormally, is an indication of the extent of realism they put into this work. However, if the Government is convinced that these yards can deliver the goods, then the pressing need now is for the Government to equip the various shipyards with the designs and drawings of pair trawlers of suitable dimensions. Neither the Government nor the industry can expect the Indian shipyards to acquire them on their own, as seen from the experiences so far.

Developments such as introduction of diversified methods of fishing activities normally take place once in a decade, mostly through Governmental efforts. It is not enough for the Government just to resolve to introduce diversified methods of fishing and make no investment in this regard. It is true that Government would provide subsidies towards the construction, but this forms only part two of the activity, the initial introduction of designs and drawings being the first part. For this, Government will have to make investments. It has to set up a yard or two exclusively for fishing vessel construction. Government cannot expect the shipyards to acquire drawings of vessels from abroad for the simple reason that they cannot be sure of recovering the expenses. Government can take this risk. With its might it can prevail upon fishing companies to buy fishing vessels of the new designs based on the drawings acquired from abroad.

It is high time that the Ministry of Agriculture selects certain proven designs of pair trawlers and acquire the related drawings for being supplied to the various shipyards at nominal charge. Government can apply the same formula as is now adopted for prevailing upon the fishing companies who have the obligation to introduce their own/vessels to place orders with the Indian shipyards for the acquisition of pair trawlers based on the designs and drawings secured by the Government from abroad. We feel that an approach on the part of the Government on the above lines would have meaning and content.

The success of the indigenous fishing vessels construction programme will lead to widespread economic uplift and prosperity among the entrepreneurship. Shipyards will have adequate work and they can provide livelihood opportunities to several workers. There will be increased employment on fishing 'vessels and in the ancillary activities. There will be increased exports of marine products and, at the same time, more fish will be available for domestic consumption.

In the process of coaxing the industry to go in for indigenous vessels a certain extent of harshness on the part of the Government becomes inevitable. This can cause some resentment in the industry. There is no doubt that the harshness is only a front on the part of the Government, something like a medicine a mother forces on a child for its health, and nothing beyond is meant. While the industry has to understand this, there is need for tact and toning down of the harshness on the part of the Government to the extent possible.

Fishing Chimes, Editorial 37: April 1984: Vol. 4, No. 1

# Access to Foreign Fishing Vessels

Article 61 of the International Law of the Sea Convention gave the coastal States a wide discretion in the establishment of the total allowable catch of living resources in their exclusive economic zone of 200 nautical miles. In establishing this, the coastal State concerned, besides biological factors, is entitled to take into account other pertinent factors such as economic and environmental factors. Once the allowable catch is declared, while the coastal State would be under obligation to give access to other States for the allowable catch quantum, it has also the discretion to stipulate the terms and conditions under which the access could be availed of by foreign fishing fleets.

While India has declared its exclusive economic zone, so far as we are aware, the nation is not yet a signatory to the convention under the Law of the Sea. Thus, India is not bound to declare the total allowable catch. Apart from this, the knowledge about our own fishery resources by itself is very vague. Such being the position, it is impossible for India to declare its allowable catch.

In this situation, the Maritime Zones of India (regulation of fishing by foreign vessels) Act. 1981 merely provides that no foreign vessel shall operate in Indian waters, except under a licence granted to owners of foreign fishing vessels and permit given to Indian citizens (for operation of foreign fishing vessels on charter). So far no licences have been issued to owners of foreign fishing vessels. However, several Indian fishing companies have been permitted to operate foreign fishing vessels on charter. As per the rules framed under section 25 of the Maritime Zones Act, the owners of foreign fishing vessels who are given licences have to pay specified fees per ton of marine living produce harvested by them at stipulated rates. In the case of vessels chartered by Indian parties under permits granted to them, 85 per cent of the gross proceeds are allowed to be adjusted towards the approved charter hire.

The operational conditions stipulated by the Indian Government for licensed/permitted foreign fishing vessels and the approach involved therein are oriented in the first phase more towards a) assessment of fishery resources in the EEZ, b) determination of the most suitable types of fishing vessels for exploiting available fishery resources and their size ranges, c) enabling Indian fishery entrepreneurs in developing skills at managing fishing fleets and d) enabling Indian fishing crew to receive training in diversified fishing methods on larger fishing vessels and developing endurance to stay out for longer durations out at sea on fishing cruises. Our assessment is that, after this phase, armed with reliable information on resources, knowledge of relative efficacy of different types of fishing vessels, and with a reservoir of trained fishing crew and managers, Government would move on to the second stage of evolving a comprehensive plan of utilising the fishery resources of the EEZ. It is laudable that, unlike other coastal nations such as Pakistan, Bangladesh and certain other South American countries, India has not laid emphasis on direct economic returns, the emphasis being on securing data on conditions of existence in the first instance.

No doubt the Government would, sooner or later, have sufficient material to evolve a plan for marine fisheries development on a realistic scale, in the second phase. In this connection, it has to be mentioned that the tempo of granting permission for charters, for diversified types of operations has to be intensified, say, upto the end of sixth plan or middle of seventh plan. The high the number of vessels operating, the higher will be the data available for analysis and interpretation.

Indian charterers are inclined more towards chartering of pair trawlers. They are, by and large, hesitant to apply for chartering of other types of fishing vessels such as stern trawlers, squid boats, tuna boats etc. However, there are some who have boldly applied for chartering these types of vessels. While Government have been clearing applications for chartering of pair trawlers, applications for chartering of other types of vessels remain pending with the Government for long. It is desirable that Government considers these applications and take decisions in the shortest possible time.

It is learnt that a few foreign fishing companies have also applied for licences for operating different types of fishing vessels in the Indian EEZ. These applications have also not yet received the attention of the Government. Subject to clearances from political and other angles, the issue of licenses will be in accordance with the provisions in the Maritime Zones Act to issue licences to the foreign companies. Their operational results may provide a different picture of operational efficiency and the results will add to the available data on fishery resources. Further, licences to such companies which offer to operate a mixed fleet of vessels will give a quick idea of the efficacy of the various types of operations.

The Government's announced policy is to consider applications from Indian enterprises for chartering of fishing vessels under permits and from foreign enterprises to operate vessels of their own under licences. It will be a good gesture if Government considers the applications within a time limit to be imposed upon themselves and conveys the decisions to the parties concerned. This will give relief to the parties from uncertainties and unnecessary commitments.

Foreign fishing vessels operations in Indian waters, either under licences or permits in a massive way for a short duration of 3 to 5 years are important in order to gain a substantial idea of our fishery resources, viable fishing grounds with reference to seasons, species and types of fishing. Having introduced the scheme in accordance with the provisions of the Maritime Zones Act, it is in national interest that it should be seen through to the logical end for deriving the anticipated benefit. The charter scheme has now reached the midstream stage. Government has to ensure that the activity gallops through to reach the other bank.

As mentioned earlier, fortified by the analytical picture derived from results of operation of foreign fishing vessels, in the second phase, Government can take concrete steps to promote deep sea fishing. These steps can include promotion of joint ventures on a firmer basis. As matters stand, there is not much of real interest evinced either by Indian companies or by foreign fishing firms for joint ventures, leaving stray cases apart, in spite of the various incentives announced by the Government. With the availability of basic data on resources derived from charter operations, there is no doubt that response from Indian and foreign enterprises for entering into joint ventures will receive a fillip. And joint ventures will eventually get translated into fullfledged Indian ventures, with the acquisition of technology relating to management and operation of diversified types of fishing vessels and relating to construction of such vessels in India.

Fishing Chimes, Editorial 38: May 1984: Vol. 4, No. 2

# Baseless Diatribe

The popular weekly 'Blitz' carried a news report by T.S.V. Hari entitled 'Poor Fishermen Ignored... T.N. Fisheries stinks with 'Fishy' deals', at page 7 of its issue dated March 17, 1984. Almost all the points made in the write-up are far from truth and highly misleading.

The article starts with the sentence 'The goings on within the fishing industry in Tamil Nadu really are fishy, to say the least'. None of the points made in subsequent parts of the items lend any support to the statement.

We hold no brief for the T.N. fisheries set up. Since we happen to have knowledge in depth about the various points mentioned, we feel that we should clarify the position.

It is stated that over 10,000 tonnes of marine catch was exported by Tamil Nadu Fisheries Development Corporation to many countries including Japan and U.S.A. However this export did not fetch a margin of profit since the State Government did not possess packing and freezing facilities.

The total exports of marine products from Tamil Nadu in 1983 were of the order of 6,373 tonnes. Of this, 3,924 tonnes were exported from Madras, 2384 tonnes from Tuticorin, 38 tonnes from Nagapattinam and 27 tonnes from Cuddalore. The bulk of the exports from Tuticorin consisted of fish landed by chartered fishing vessels operated by private fishing companies. The shrimp and cephaloped exports out of 6,373 tonnes exported from the State were around 3,500 tonnes. The bulk of this quantity was exported by the private sector. The exportable catches landed by vessels owned by Tamil Nadu Fisheries Development Corporation are about 150 tonnes per annum. There is therefore no substance in the allegation that during 1982-83 alone over 10,000 tonnes of marine products were exported by the Tamil Nadu Fisheries Development Corporation. The complaint should actually have been that the corporation should have exported much more. In regard to the observation that the activity did not fetch a high margin of profit to the corporation, it can only be said that the comment assumes that some margin of profit was there. This by itself should be a great achievement for the reason that the major problem now faced by the marine products processing industry is that either there are no margins and, if there is a margin that is insignificant.

It is stated that the contract for freezing and packing was given to M/s East Coast Marine Products, Visakhapatnam at Rs. 4.50/kg. If E.C.M.P. had agreed to process at this cost, it must be solely to cover the idle periods and earn something additional to cover up deficits arising out of inescapable payments to workers for idle periods. If other firms offered much lower rates and E.C.M.P. also offered to do the job at Rs. 3.50/kg on account of lower quotation from other firms, it must be for the same reason as mentioned above. Decisions on awarding contracts for a work of this type depend on several factors and not on the rate alone. It is the prerogative of the corporation to take a decision in the matter. If a processor says that he will do the work for a far lower amount than what the corporation's estimate of costs indicate, they have to think twice before awarding the contract to the lowest tenderer. The corporation's objective is to get the job done well and not to pay a low amount absurdly inconsistent with the costs involved.

The article says that, according to a scheme of the Central Government, foreign mechanised trawlers poaching in the Indian Zone should forfeit 6 per cent of their catch.Upon notification in this connection by the coastguard, Indian fishing companies, licensed to certify extent of foreign trawlers' hauls would inspect the intruder's holds.

The above observations appear to be totally baseless. The author must have been misinformed by someone. No such scheme was notified by the Central Government. There is also no system of licensing Indian fishing companies by the Coast Guard to certify the extent of foreign trawler's catches. The coast Guard has the services of the Fishery Survey of India for assessment. Forfeiture of fish caught by poaching vessels is a matter to be decided by a court of law in terms of the Maritime Zones Act.

Having made the statement as outlined above the author went on to say something much more interesting which appears to be equally baseless. To get the certification, licence, the firms have to obtain clearance from the state Government first and then approach the centre. He has stated that four firms, M/s Golden Fisheries, Delhi, M/s George Maijo, Madras, M/s Ball Exporters (He meant Bali Cold storage, Madras), and an unnamed firm (He probably meant Aparna Fisheries) could get the licences. We feel that either the author should establish with proof that such licences were given or withdraw the allegation. So far as we know no certification licence system exists, for there is no need for such a system.

It is stated that Mr. Shukla of Golden Fisheries is very close to the Union Agricultural Minister, Rao Birendra Singh. Our assessment is that Mr. Shukla is one of those who does not agree with the policies and style of functioning of the Agricultural Minister so far as they relate to deep sea fishing. Important proposals presented by Mr. Shukla on behalf of his company are stuck up in the Ministry of Agriculture. If he is close to the Agricultural Minister, he would have certainly been able to make effective representations in this respect. Rao Birendra Singh has the rare quality of not allowing anyone to be very close to him. He would certainly not allow Mr. Shukla to become close to him.

It is observed that M/s George Maijo distributes Yamaha outboard and steamer engines in India. The firm distributes Yamaha outboards and Yamaha generators but not steamer engines. George Maijo has the fundamental right to deal in outboards etc. within the confines of the rules in force and we do not see anything wrong in this. It is alleged that the Managing Director of George Maijo is also one of the Directors of MPEDA. MPEDA has little to do with distribution of outboard motors by private companies. Even if it has something to do with this, it does not mean that he cannot hold an office on the Board of MPEDA.

The bringing in of the names of Bali Cold Storage, and the concern owned by Mr. Srinivasan and others does not seem to have any relevance. We are unable to understand what 'NOC' of 'licence' of theirs was cleared by the State Government/Commerce Ministry.

The author says that the deals that the State Government hands down to poor fishermen stink. So far as we are aware, the Government of Tamil Nadu is in the forefront in providing various reliefs to fishermen. The Government provided fishermen with houses, roads etc., besides undertaking several other ameliorative measures. The Government of Tamil Nadu is one of the few State Governments in the country that has adopted progressive and benevolent schemes for the uplift of fishermen.

Fishing Chimes, Editorial 39: June 1984: Vol. 4, No. 3

# Charter Chariot moves into Legal Orbit

The Indian charterers of foreign fishing vessels have been nurturing a feeling, perhaps not wholly justified, that the Government's attitude towards providing helpful solutions to problems that keep on cropping up in the course of implementation of the charter programme, are inflexible and inconsiderate. The barrier between the wavelength at which the Government has been administering the scheme and the wave length at which the charterers have been operating, has been steadily growing. While Government is cautious and suspicious and therefore rigorous, the charterers feel that Government is being unduly harsh and strict, and delaying decisions on their applications relating to charter matters. Their requests are often rejected, that too after a long delay. One day, this widening gap has to lead to the need for a court to judge who is night and who is wrong and this development did take place all on a sudden.

Two of the charterers *viz.*, M/s. Bottleglass Private Limited, Delhi, and M/s. Tropical Shipping Co. Ltd., Visakhapatnam filed writ petitions in Delhi High Court in April 1984 seeking redressal on the orders of the Ministry of Agriculture cancelling the permits issued for the operation of their chartered trawlers and invoking the bank guarantees given by them to the Government. These orders are stated to have been issued for the reason that the companies failed to comply with some of the charter conditions.

Some of these vessels were engaged in fishing operations at the time of the orders and as per the orders they should stop fishing and report to the base port in India within a week or so. This order put the charterers in a spot. If the vessels came to the port they would be impounded and it would be anybody's guess when they would be allowed to go. The owners would press for the return of the vessels and the charterers would not be able to do anything about it. The companies seem to have met the senior officers in the agriculture ministry, presented their point of view and sought the lifting of the cancellation. They seem to have pointed out that no opportunity was given to them before the cancellation to explain their standpoint and no enquiry was conducted as required under Maritime Zones Act. They pleaded that they did apply for permission to acquire vessels of their own as required under one of the conditions governing their charter permits, but the officials did not obviously relent.

The reason for the stand seems to be that, as we understand, the Government gave them sufficient time to fulfil the conditions but the companies were coming up with some problem or the other, resulting in dilation. We are not however certain to what extent this position is correct. Thereupon the companies seem to have decided to seek legal remedy. They filed writ petitions in Delhi High Court seeking stay of the cancellation and guarantee invocation orders, pending final hearing. The court granted stay orders and the cases are expected to come up for hearing.

The two cases mentioned above relate to operational aspects of chartered vessels. Apart from these, it is understood that two companies, *viz.*, M/s. Fish Products Ltd., Kakinada and M/s. Southern Sea Foods, Madras, have filed writ petitions in Delhi High Court questioning the action of the Ministry of Agriculture in rejecting their applications. They have been in the field of marine products processing and export for the past two decades and they are eligible to receive letters of intent. It may be mentioned here that, in fairness, these two companies are well qualified to receive letters of intent, better than necktie manufacturers, and other nonfishery establishments who have been issued with letters of intent earlier. The necktie company, a partnership firm at the time, was given a letter of intent, when rules clearly stated that only limited companies are eligible to apply. Of course, much later in the Maritime Zone Act it was provided that even individuals can apply for chartering of fishing vessels.

No doubt those who are not in the fisheries field also need to be encouraged to take up fisheries line, such as those in the manufacture of neckties. After all, there has to be a beginning and those who are in the industry cannot say that they alone should be the vehicles for further development. At the same time those already in the field should not be excluded, if they fulfill all conditions. Southern Sea Foods and Fish Products feel that they were unceremoniously excluded. The reason for this feeling is that while others not in field and whose applications were similar to theirs were given letters of intent, their applications were rejected. It is however possible that those companies did not fulfill certain criteria which must have weighed on the mind of the Government while rejecting the applications.

Before rejection, all aspects of various applications will no doubt be looked into by the Government and it is of course the Government's prerogative to reject. However, grouses build up in the minds of applicants and feeling of injustice dominate when they get the impression that they are discriminated against. Let it be hoped that the decision of the court in respect of the writ petitions of Fish Products and Southern Sea Foods may lead to the laying down of sounder guidelines.

Aggrieved parties are always free to go to court. However, if the administration is organised in a manner that gives the impression to the parties that nothing unjust would be done, the scope for litigation becomes minimal. Right from the stage of submitting their application, the parties happen to be in the dark, under the present system. They do not even know whether their application was registered or not. This is because the ministry neither acknowledges an application nor indicates a likely date by which the decision would be communicated. It is necessary that the ministry introduces a basic reform in this regard.

The applicants also do not know what types of vessels and from what countries the Government would consider for charter permission. The ministry does not educate the industry in this aspect. From decisions in the recent past, an impression that vessels of Thai and Taiwanese registry would not be cleared by the Government for charter permission, has taken roots in the minds of those interested in charters. Government, however, has never told the public about this.

It is worth mentioning here that, while vessels of Taiwanese registry may have been excluded, Taiwanese vessels registered in Panama continue to be in the green list. In the recent past several permissions involving Taiwanese vessels with Panama registration were issued. It is understood that Government is now seriously considering to exclude Panama registered Taiwanese vessels also in considering charter applications. It this is done, that will be another way of singing the swan song of the policy. This is because no other country, except Thailand and Taiwan, is willing to give vessels on charter. We understand that Taiwanese owners themselves are now gradually withdrawing their existing vessels on charter and diverting them to Australia and New Zealand where getting permissions is quicker. The logic in not allowing Taiwanese vessels of Taiwanese registry one can understand but allowing Taiwanese vessels of Panama registry is not clear. The vessels are the same, the crew are also the same, same Taiwanese. The only difference is that the owners get them registered under Panama flag, to get over a technicality brought in artificially. Before registration under Panama flag, the vessels are 'Tweedledum' and after registration they are 'Tweedledee'. It is very difficult to understand this approach of the Government in this respect. Very obscure indeed.

If the Government are against operation of foreign fishing vessels of any particular nationality or flag, this should be indicated in clear terms to the public, so that they may explore possibilities of chartering vessels from other sources. If there are political implications, there must be a system under which the applicants can be told whether vessels registered in a particular country would be considered or not. In other words, there should be an indication beforehand that the application will not be rejected on that ground. If they are not acceptable the parties should accordingly be advised.

The applicants pay a fees of Rs. 500/with each application. Atleast in consideration of this, they are entitled to hear about the decision of the Government on their application within, say, a maximum of 90 days. As matters stand for months together applicants do not know what has happened to their applications. At the moment, over 50 applications are understood to be pending in the ministry for nearly six months.

It appears that the system for approvals of charter cases has been made highly cumbersome by the Government. Clearance of charter cases is the prerogative of officialdom at the appropriate level, as this will be done on the basis of a policy laid down by the Government and the decisions are mostly technical in nature. At present, an Interministerial committee considers the applications and makes recommendations. Based on this, the cases are put up to the cabinet minister for approval (so we understand). The recommendations of the interministerial committee do not seem to have any sanctity for the reason that the cabinet minister often approves only some of the recommended ones and rejects or defers the others. This approach is of course again his prerogative. The minister is performing the final act of approval in respect of this executive or administrative function, taking it away from the officers to whom it rightly belongs. This is no doubt a good system to the extent that the officers are protected in case of any legal offshoot arising thereafter. We however do not feel that the minister, whose job it is to approve of policies and strategies of development, should burden himself with the task of according approvals to charter applications which can be cleared by authorised officials, as long as the applications fulfil the norms. For cargo vessels charters, officials in the office of DirectorGeneral of Shipping issue permissions and licences. The applications do not even go to the Ministry of Shipping and Transport.

If the D.G. Shipping takes as much time as the ministry of agriculture takes in clearing fishing vessel charter cases, by now our merchant shipping would have come to a close. In the case of fishing vessels, we feel that the activity would soon come to a close, for reasons of long delays in clearing not only the applications and issuing letters of intent, but also in respect of delays in issuing permits, orders on substitution of vessels etc. Whether such a development will be good or bad for the nation, it is difficult to say. It is true that cargo vessels stand on a different footing to fishing vessels. At the same time both the types are engaged in commercial operations. If a cargo vessel is detained for a day they suffer heavy losses. In the same way, a fishing vessel of 38 m OAL loses about Rs. 10,000 by way of unproductive expenses per day. In addition, fishing days are lost. Our Government should either run the scheme with minimum delays or fold it up. Rejections are reconciled with. But these can be done fast. It is not good for the nation to run a scheme in a dilatory manner as is happening now.

Out of 80 pairs of foreign trawlers stated to have been permitted to operate in Indian EEZ on charter basis, at the moment around 12 pairs are in operation. 80 chartered pairs are there only for record. The chartered vessels often sustain heavy losses when they are stranded at Indian ports for want of some formal clearance or the other. Such vessels rarely come back, unwilling to compound their losses.

It is very odd that the D.G. Shipping, who has to issue voyage licences to chartered cargo vessels, has no say in the granting of charter permits. One shudders to think what would happen, if after an ordeal of over six months, a party gets a permit from the ministry of agriculture and later D.G. Shipping refuses licence. There should be a system under which issue of licences would be automatic, once the permit is there. For this D.G. Shipping has to be involved in the process leading to the issue of permits.

It is necessary for the Government at top level to review the progress and take a decision whether to continue the charter scheme or not. If it is to be continued, as is being done by other countries such as Australia, New Zealand, Canada etc., then ways and means of implementing the scheme with the required pace consistent with the characteristics of the activity, the losses that will result in case of delays, both financial as well as in terms of reputation, should be evolved. These can be on the same lines as for cargo vessels, since their charters are cleared under an excellent system by the D.G. Shipping.

It is desirable that Government deputes an officer to countries such as Australia and New Zealand to make a study of the manner in which they are administering their charter schemes. Such a study would help in streamlining procedures.

Fishing Chimes, Editorial 40: July 1984: Vol. 4, No. 4

# Delegation Devoid of Development Commissioner

The World Fisheries Conference was held under the auspices of FAO in Rome from 27 June to 6 July, 1984. Nearly 150 countries are stated to have sponsored their delegations to participate in the Conference. Conferences of this type are organised once in a while and this conference has a momentous significance to all the countries of the World, more so to the developing and underdeveloped countries. The occasion was a forum to put forth problems that came in the way of increasing fish production and to propose solutions for international acceptance. This was a unique platform for experts engaged in fisheries developmental activities to exchange views and to listen to experiences in other countries, so as to evolve views and strategies for the development of their own fisheries. The coming together of representatives from various countries could well lead to possibilities of establishing joint ventures.

The Indian delegation to the Conference was appropriately headed by Rao Birendra Singh, Agriculture Minister. The rest of the team consisted of a) Mr. P.S. Kohli, Additional Secretary (Trade and Coop), Mr. S.P. Jakhanwal, Joint Secretary (Fisheries and Trade), Mr. K.C. Saha, Director (International Cooperation) and Dr. S.N. Dwivedi, Director, Central Institute of Fisheries Education. The first three delegates in the team are administrators, of whom Mr. Jakhanwal deals with the fisheries subject. Dr. Dwivedi, a technocrat, deals with fisheries educational matters.

Mr. Jakhanwal and Dr. Dwivedi have sound knowledge, a national perspective and an overview of the Indian fisheries situation and the Indian needs of fisheries development. It is, however, to be appreciated that it is Dr. P.V. Dehadrai, Fisheries Development Commissioner who is required to have the necessary exposure at the Conference by way of meeting his counterparts from other countries, exchanging views with them, and picking up useful ideas related to developmental aspects, besides taking part in the proceedings of the Conference in an effective manner. The reason for this is that he is the person in overall charge of fisheries development of the country, from technical angle in particular. It is felt that, for reasons known to Government only, he was not made part of the delegation. It is packed with administrators and with marginal representation from technocrats. The weakness in the composition of the delegation is so transparent.

The Conference was meant to cover not only matters related to increasing fish production, but also exports of marine products. There was no representative either from the Ministry of Commerce or Marine Products Export Development Authority in the delegation. It is felt that this too was a serious omission.

Irrespective of the composition, there is a general appreciation of the speech delivered by Rao Birendra Singh, Minister for Agriculture, at the Conference. It touched on all vital problems confronting fisheries sector. His proposal to declare 1985 as the International Year of aquaculture was timely and reflected his keen perception of the role aquaculture can play in reducing the gap between fish requirements by the turn of the century and the present level of annual world fish production. It is heartening that, taking the cue, the Conference recommended that 1985 should be declared as 'International year of fisheries'.

Fishing Chimes, Editorial 41: August 1984: Vol. 4, No. 5

# Shaping of Policy

The marine fisheries resources of our exclusive economic zone are estimated at 4 to 5 million tonnes. Its present annual marine fish production is of the order of 1.5 million tonnes, consisting of demersal, columnar and surface swimming fish. Most of this production comes from inshore waters upto a distance of about 15 miles from the coast and upto a depth of about 20 fathoms. This zone is now saturated with fishing effort, (barring a very few unexploited pockets), by mechanised and non-mechanised craft operated by small fishermen. About 60 larger vessels fish in waters beyond 20 fathoms depth, mostly for shrimp.

The nature of resources in the zone extending from 20 to 40 fathoms depth are known but there is no estimate of the extent of resources in this area, out of the unexploited balance of estimated resources (2.5 to 3.5 million tonnes) in the exclusive economic zone. What exactly is the resource strength in the rest of the economic zone upto 200 nautical miles distance from the coast is not known. It is reported that experimental fishing conducted for limited durations indicated the existence of demersal fish stocks at depths beyond 40 fathoms, in isolated pockets consisting of non-traditional species, not having a market at present, such as bulls eye, Indian drift fish, black ruff, deep sea lobster, deep sea prawns, threadfin bream and Carangids. Besides these, conventional varieties such as perches are also located. The average rate of catch per hour of all these varieties is very low in the exploratory hauls. Besides these, the resource potential would presumably consist of pelagic and midwater stocks.

The situation thus is that larger vessels to cover farther areas are required to exploit, a) demersal and pelagic stocks (including columnar fish) in the region from 20 to 40 fathoms, b) demersal stocks such as perches and deep sea shrimp and lobsters in areas having depth of over 40 fathoms and in newly located grounds hitherto unexploited and c) pelagic (and columnar) stocks in the areas beyond.

Dealing with b) above, exploitation of these resources will necessarily be experimental and can be undertaken to start with only by a Governmental organisation such as the proposed National Fisheries Corporation. Without establishing the feasibility, private enterprises would not be able to risk capital investment to acquire expensive vessels for the purpose. There is no likelihood of foreign fishing vessels being available on charter for the purpose. The experiences of unremunerative results of polish vessels chartered by the Ministry of Agriculture and M/s Kelbex Co, which operated for sometime in deeper waters, are still fresh in the minds of entrepreneurs. Further, vessels with capability to trawl in deeper waters are scarce and mostly East European countries have them. For these countries economics of operation are not as important as providing fish to their people.

In regard to (c) above, while the existence of substantial stocks of open sea fish such as tuna and tuna-like fishes is asserted by our scientists and experimental fishing has yielded some results, the extent of availability and the economic viability in the matter of exploitation of the non-traditional species is not yet established. Thus, fishery of this zone can only be exploited with the objective of establishing commercial feasibility, by the proposed National Fisheries Corporation. There is no doubt of some interest in the private sector to introduce tuna long liners on charter but Government has not yet approved of such applications which were filed over six months back.

It is now reasonably established that operation of larger vessels for demersal stocks in the 20-50 fathoms zone will be economically viable, on the northwest coast and over the Wadge Bank. This is the result of the operation of chartered pair trawlers in the past two years. Results of purse-seining along Kerala and Karnataka Coasts have also established favourable economics of operation of purse-seiners along Kerala, Karnataka and Goa Coasts. The 20-50 fathom zone happens to be virtually virgin and hence it is an economically feasible fishable area immediately outside the zone of the presently fished areas. The Taiwanese pair trawlers are currently exploiting these waters, particularly on the northwest coast and at the Wadge Bank, and the annual output is estimated at 15,000 tonnes. In the situation when the disposition of fish stocks, particularly demersal, in the areas beyond 50 fathoms is not fully known and the economic feasibility not yet established, the logical endeavour of any enterprise is to concentrate on fishing effort in 20-50 fathom zone. As and when adequate economically viable exploitation of resources in this area is reached, the time would be ripe to extend the activity further.

The dimensions of the area to be covered by the fishing activity are vast. Heavy capital investments, involving serious risks, are needed to extend the activity all over the EEZ (although in stages), backed by feasibility studies, capability to introduce sophisticated vessels, trained men, harbour facilities, other shore infrastructural needs, and marketing facilities. Even with the availability of money for all these, it will take a long time to achieve end results. It is therefore, desirable to plan initially for the exploitation of fisheries in the 20-50 fathom zone and implement the same with single-minded devotion. The entire area can never be covered at one stretch. We can only progress from stage to stage. There can no doubt be efforts directed at covering the zone beyond 50 fathoms as and when enterprises interested in this regard make approaches, but the broad policy has to be to cover the 20-50 fathom area intensively in the next phase.

Fishing in the 20-50 fathom zone calls for introduction of vessels having an endurance of at least 30 days for economically viable results, and of a superior design over the medium sized vessels, about 60 nos. of which are now in operation. The introduction of these vessels in larger numbers is bound to attract resistance and agitation from traditional fishermen on the main plea that these vessels catch the fish that enter inshore areas from farther waters and therefore their fishing results will be affected. Such resistance and agitation will probably be led mostly by political groups interested in the votes of traditional fishermen, as is happening now in respect of conflicts between non-mechanised and mechanised boat operators in several coastal States.

Development of sea fishing in areas beyond the operational range of traditional fishermen does not affect the catches of traditional fishermen in any significant manner. Even if it affects to any extent, development of fishing in farther areas of the sea cannot be withheld on this account. If this is withheld there can be no development. Either we keep our fishing effort at the present level in order to protect traditional fishermen on the ostensible ground that any expansion would impinge on their interests or take up formulation of a bold policy to introduce a sizeable number of larger vessels to exploit fisheries resources in the 20-50 fathom zone.

Introduction of new units for fishing in inshore areas shall have to be stopped, as these waters in fact are being over-exploited. At the best, assistance can be extended for replacement of old and unserviceable units. Efforts have to be directed at enabling traditional fishermen to go in for larger vessels. This requires a programme to provide training to youngsters drawn from traditional fisher families and to assist them in acquiring their own vessels through trained groups or co-operatives. Upgradation of skills and capabilities of traditional fishermen is imperative in order to bring them in tune with the inexorable winds of change in technology. If this is not done and Government clings to the policy of according unqualified protection to traditional fishermen, the policy would not pay. Even the present protection which is in terms of not allowing mechanised boats to operate in nearshore areas would be of no avail as the operations of non-mechanised boats by themselves may become non-remunerative by and by. In course of time, non-mechanised boats would disappear leaving the related fishermen groping for livelihood. Analogy lies in the developments in land transportation. Hand-drawn rikshaws had to yield to cycle-drawn rikshaws and cycle-drawn rikshaws are now slowly yielding place to auto-rikshaws. In the same way, fishermen being backward, and owing to technology aspects, it is necessary to make a conscious effort to upgrade their capabilities so that they can go in for better boats.

Priority in granting loans by SDFC for acquisition of larger boats no doubt could go to trained fisher groups consisting of skippers etc. or their co-operatives or companies set up by them. Next preference would have to go to fishery enterprises consisting of other trained men followed by fishery enterprises with managerial talent and in a position to engage the required trained personnel. As matters stand, however, considering the enormous scope, all the above categories can be encouraged, taking care that applications for loans from the first and second categories mentioned above are invariably cleared once these are in order. In order to encourage these categories certain relaxations in respect of debt-equity ratio and guarantee towards loans granted by SDFC would become necessary.

Fishing Chimes, Editorial 42: September 1984: Vol. 4, No. 6

# India Lags Behind in Shrimp Culture

Indian Government is against increase in the national shrimping fleet. The authorities feel that the present fleet of about 68 shrimp trawlers, which alone are capable of catching offshore shrimp, are far more than the number warranted to exploit the resources. Around 16,000 mechanised boats are also now engaged in catching shrimp from coastal waters. The Government's reasoning must be that the offshore shrimp comes inshore and serve as the brooder stock for the coastal waters where mechanised boat operators conduct fishing. If the number of trawlers goes up, a greater quantity of shrimp may be captured offshore, thereby affecting the fishing for shrimp in coastal waters.

No person with knowledge about life span of shrimp and their biology would subscribe to the above line of reasoning. At the same time, the wisdom of the Government in restricting the shrimping fleet has to be viewed with respect.

Number of mechanised boats cannot be increased further as there is a saturation in this respect already. There is no scope for increasing the number of trawlers also. In this situation, the questions to be answered are: Are we to reconcile with the present level of shrimp landings forever or, are we to take other steps to increase shrimp production?

We cannot obviously remain static in respect of capture shrimp production. Therefore the other available step *i.e.*, production of culture shrimp has to be taken up.

It is not that the nation is not aware of the need for the development of production of culture shrimp. All concerned are conscious of this. In fact considerable work has been done in regard to production of shrimp fry and the culture practices at the farms for the growth of shrimps in farm ponds.

The cause for concern now is that no concerted measures have been set in motion to develop culture shrimp fisheries in a big way, to utilise the 1.6 million hectares of brackishwater areas which are, by and large, remaining fallow. Our Government as well as those in the industry are thinking about the programme in a very small way, whereas the dimensions of the development work needed are big. We have to think and plan in a big way in this sector, not only to catch up with the other nations who have already progressed far, but also for boosting up exports and foreign exchange earnings, for providing employment and contributing to national wealth.

The Central Government has a plan which envisages assistance to Coastal State Governments for setting up of commercial farms of about 100 ha. each. At the most this programme may cover 800 ha. Beyond this there is practically no organised effort to develop shrimp farming. Our farmed shrimp production, mostly from West Bengal and Kerala, runs to a few hundred tonnes.

Compared to the situation in India, let us see what is happening in a few other nations. In Taiwan, with far less of resources than ourselves, production from brackishwater farming per annum is 86,750 tonnes of which 10,000 tonnes are shrimp. In terms of per ha/ production per annum, the production, inclusive of shrimp, from brackishwater farms is 24 tonnes. Indonesia has developed shrimp farming in 31,000 ha so far. The Indonesian Government has introduced a scheme to bring another one lakh hectares under shrimp farming. Srilanka has launched a major coastal aquafarming project with a loan of US $ 17.27 million from Asian Development Bank.

Farming of shrimps in brackishwater has grown into an important industry in Philippines. In 1968, export of shrimps from this country was only 168 tonnes. By 1983, this had increased to 4,400 tonnes, most which came from farm ponds. There are about 200,000 hectares of ponds in that country used for polyfarming of milk fish, tilapia and shrimp. One firm claims having achieved a production level of 12 tonnes/ha/yr by applying Taiwanese techniques of intensive feeding and paddle wheel aeration.

Under its national aquaculture programme, the Ministry of Fisheries of Mexico plans to build some 50,000 acres of shrimp ponds. Some of these ponds have already been built on the Pacific coast. The Ministry also plans to invest US $ 12 million in oyster production over the next ten years.

The Spanish Government has plans to produce around 600,000 tonnes of shrimps, mussel and finfish from brackishwaters all along Spanish coastline. Ecuador produces 18,000 tonnes of shrimp annually from culture ponds. The country has plans to bring 12,500 hectares under aquafarming in three years. In this connection they have invited Taiwanese experts to Ecuador. Panama has brought extensive brackishwater areas under aqua-farming.

The initiative taken by the various other countries mentioned above deserve to be emulated by us in national interest. Let not the coming generation blame us for apathy and lethargy in the development of our brackishwater resources suitable for development of shrimp farming. Much depends on to what extent our Government applies its mind to this crucial sector and how big and wide the vision in this respect will be.

Fishing Chimes, Editorial 43: September 1984: Vol. 4, No. 6

# Futility of Fishing Legislation Based on Depth

Notification No. 30035/71/82Fy (Tl) dated 23rd April 1984 of the Ministry of Agriculture (Department of Agriculture and Cooperation) contains an amendment to the Maritime Zones of India (Regulation of Fishing by foreign Vessels) Rules, 1982 in regard to limitations to areas of fishing in respect of foreign fishing vessels covered by licences and permits. The new rule [No. 8 (1) Q] lays down that such vessels shall fish beyond territorial waters or in the high seas of depth more than forty fathoms whichever is farther from shore.

As per the Maritime Zones of India Act 'fishing' means catching, taking, killing, attracting or pursuing fish by any method and includes the processing, preserving, transferring, receiving and transporting fish.

The new rule is based both on horizontal and depth scales. The horizontal scale is related to territorial waters which extend upto 12 nautical miles from the appropriate base line. This base line has not yet been determined. This moves further inland during high tide and recedes during low tide. Thus there is no means of accurately assessing the distance of a vessel from the base line, which by itself has not been worked out.

Regarding the depth limitation, while the charts issued by the hydrographic office give the depth lines in metres and the echosounders also give the depth in metres, the new rule speaks in terms of fathoms. Although calculation based on fathoms has been given up long back, the Ministry of Agriculture obviously has not yet given up this system.

The definition of sea depth is not clear either from the Act or from the rules notified thereunder. An echosounder displays the depth from the lower end of the transducer at the bottom of a ship to the seabed or any major obstruction between it and the seabed. If a rock outcrop is there at the bottom, the depth becomes less till such time as the vessel passes over it. The height of the transducer has to be added to the reading shown by the echogram to arrive at the correct depth. Further, depth need not necessarily connote the vertical distance upto the bottom. It can also be a vertical measurement from the surface to any required level. Without a definition of depth, it will be difficult to undertake corrections of this nature.

The charts do not show the 40 fathom line, which by itself, as in the case of other fathom lines, will not be in straight line. It therefore becomes an arduous ask for the captain to steer his vessel always to be beyond 40 fathoms. His concentration and manipulation of his vessel's sojourn so as to be away from 40 fathom depth totally affects his fishing strategy and goes against the objective of the voyage.

The depth over which a vessel is navigating is not necessarily the depth of fishing. This basic fact is totally overlooked when the location of fishing activity of chartered vessels is checked. Nor are the rules specific in this regard. Let us visualise a vessel conducting trawling. The vessel is, say, over a depth of 39 fathoms (whatever is the definition of depth). This depth is certainly not the fishing depth. The trawl net which the vessel is dragging will be at a distance of, say, about 200 to 250 m away from the vessel where the depth will be different. According to the depth over which the net will be passing during the drag, the length of towing rope will be adjusted by the captain. In other words, at a given time, the net may be at a depth of 42 fathoms and the vessel may be over a depth of 39 fathoms. The depth at which the vessel is found cannot be construed as the depth of fishing.

The objective of the rule seems to be to prevent capture of fish at depths below 40 fathoms by chartered vessels so as to reserve them for being caught by coastal fishermen. In this context the location of the net which catches fish is important and not the location of the vessel. It may also be mentioned here that there is no easy means of checking the depth at which the trawl net in movement is actually located. The sea bed is not uniform and the net may cross over varying depths. Further, there can be a situation where the vessel is situated over an area having a depth of 42 fathoms but the net can be at a depth of 39 fathoms (provided it can be established). Would the location of the vessel or the location of the net be the criterion to judge an offence in such a situation?

The problem of judging the depth of fishing as per the notification is not restricted to trawling alone. Consider purse-seining at a point where the depth is 100 fathoms. A purse-seine net may go upto a depth of 38 fathoms from the surface. In other words, fishing is done with the net upto a depth of 38 fathoms from the surface. Technically, although the overall depth at the place down to the sea bed is 100 fathoms, since fishing is done upto a depth of 38 fathoms, an offence is committed under the rules. This is because fishing has to be done in the high seas of depth more than forty fathoms. The rule does not say that, in the high seas of depth of more than 40 fathoms fishing can be done within 40 fathoms from the surface.

A similar intriguing situation will arise in the case of surface long lining, surface gillnetting and pole line fishing. At all these types, fishing will be done within 40 fathoms from the surface and farther to the outer limits of territorial waters, from the shore. The term 'depth' does not mean depth upto the bottom. Because of this, fishing within forty fathoms from the surface, although the depth at the place is much more than 40 fathoms, also becomes a violation of the rule.

The coast guard will undoubtedly be facing serious problems in enforcing any regulation based on depth. Such a regulation is also a serious hardship to the operators, particularly in respect of trawling. First it has to be established what the expression in the notification 'shall fish beyond territorial waters........' specifically connotes. If this means that the trawler should always be at a depth of 40 fathoms and beyond, the charterers will be free to adjust the tow in such a way that the net is beyond 40 fathoms and the vessel at an appropriate point in relation to it. It may also be noted that, in actual operation, the vessel and the net pass over various depths, since sea bed is not uniform. If at a given moment, a vessel is found at a depth of 38 fathoms while towing a net, the operators would be contravening the regulation, for no fault of theirs. Again, wind, current and other physical forces push a vessel in different directions and this may cause a drift into shallower waters. This would also constitute a violation of the rule.

Fish enter into the net in the process of fishing. Therefore, the location of the net is the criterion to judge the fishing depth. Where the vessels are fishing can be assessed only with reference to this. The depth at which a vessel is located is not the fishing depth, seen from any angle. Unfortunately the fishing depth cannot easily be established precisely. Sooner or later, therefore, the rule has to be modified to provide only for a horizontal scale of limitation. It is a curious and ambiguous combination to have both horizontal and depth scales.

Fishing Chimes, Editorial 44: October 1984: Vol. 4, No. 7

# Deep Sea Fishing Vessels in 6$^{th}$ Plan

Mr. S.A. Dorai Sebastian, M.P. posed questions of crucial importance in regard to additions to the Indian deep sea fishing fleet in the 6th Plan in the Lok Sabha on 23rd July 1984. One of the questions was whether it was a fact that only 11 Indian flag vessels had so far been added to the fleet of deep sea fishing vessels against a target of 350 vessels in the 6th Plan which had now been scaled down to 200 vessels. In answer, while denying that the position mentioned in the question was not correct, the Minister of State in the Ministry of Agriculture Mr. Yogendra Makwana said that a target of additional 300 deep sea fishing vessels was fixed, during the 6th Plan period. This target was subsequently reduced to 200 nos. The fact therefore remains that the target was scaled down to 200 nos.

The minister also said that 16 Indian flag vessels had so far been acquired during the 6th Plan period by the Indian entrepreneurs including three deep sea fishing vessels "which arrived a few days back". At the time of putting the question, excluding the newly added vessels, there were 13 nos. of Indian Flag vessels as pointed out by the M.P. The position thus is that, against a depressed target of 200 nos. of vessels to be introduced, only 16 nos. were added. This works out to an achievement of 8 per cent.

Analysing the target of 200 nos. to be introduced, the minister stated that the target figure would consist of 100 nos. of chartered vessels to be introduced, besides 40 nos. of imported vessels, 30 nos. of indigenous vessels and another 30 nos. to be introduced against the condition of obligatory purchase imposed on the charterers of fishing vessels. The fallacy of this break-up would at once be noticed by even a layman, not acquainted with the subject of deep sea fishing. Chartered vessels can never be construed as an addition to Indian fishing fleet. Therefore 83 nos. of chartered vessels that operated in the Indian exclusive economic zone by March 1984 cannot be counted against the target. There were no additions in the 6th plan so far against the subtargets allocated for introduction of indigenous vessels and obligatory purchases against chartered vessels. Consequently, the virtual addition of deep sea fishing vessels to the Indian fleet is only 16 nos. against the target of 200 nos.

Even assuming that chartered vessels could be counted against the target of 200 nos., the total number introduced comes to only 99 nos. It may be noted in this connection that all the 83 chartered vessels would never be in operation in Indian waters at any given time. Not more than 1/3 of these vessels will be in Indian EEZ, the others being away at the home port. At the moment of writing only two pairs of these chartered vessels are operational in Indian EEZ.

The reply given by the minister as outlined in the first Para, on analysis, is a clear admission, although indirectly, that the target in respect of addition of deep sea fishing vessels to the Indian fishing fleet could not be fulfilled. In other words, the ministry of agriculture failed in this respect. The reasons for this are several, One is that the ministry or agriculture takes normally a long time to decide on the applications for import of fishing vessels and for recommendation of loans to the SDFC for the indigenous construction of vessels. The SDFC also takes its own time to clear loan applications, although the programme by itself is of immense importance in the context of increasing fish production, stepping up of exports, earning of foreign exchange for the country and providing employment.

Hitherto the Ministry of Agriculture as well as the Ministry of Shipping and Transport had been complaining that the Indian shipyards were not making efforts to secure orders for the construction and supply of fishing vessels and that the Indian fishing companies were also complacent about placing orders. The situation had changed considerably in the past one year and several Indian fishing companies placed orders with indigenous shipyards.

Now three Indian yards hold orders for the construction of over 25 vessels. The Indian fishing companies applied for loans from the SDFC long time back. However, except in the case of 5 vessels the other applications continue to be under consideration, although over 3 months had elapsed. If the approach of the ministry of agriculture/ministry of shipping and transport in dealing with the subject is so perfunctory, one cannot expect the realisation of the target. No wonder there is a major gap between the target and the performance.

Fishing Chimes, Editorial 45:November 1984: Vol. 4, No. 8

# Charter Scape

The Maritime Zones Act came into being for the sole purpose of regulating fishing by foreign vessels in our exclusive economic zone. The Act provides for this regulation through the devices of issuing licences to Indian enterprises for operating foreign fishing vessels on charter and by giving permits to foreign owners to operate their vessels themselves.

While several licences have been issued, no permits have been issued so far. In regard to licences, there are three participants in their implementation, *viz.*, the Government, foreign owners, and the Indian licensees. All these three passed through various experiences in the past three years in the course of implementation of the programme.

On account of the newness of the programme, the activities thereunder virtually took place in the form of a test to all the three actors concerned. Neither the Government nor the charterers had any previous experience in the sector of fishing vessel charters. The owners, while they may have had experience in other areas, lacked experience in conducting operations in Indian waters and under Indian regulations.

The results of the operations, whatever be the reasons, in our assessment, proved to be disappointing to the Government. To the majority of the charterers and owners they proved to be not only disappointing but also economically disastrous.

The novelty of the activity and limitations on the part of the Government to visualise and assess the likely course of events as a result of the regulations and approaches adopted seem to be largely responsible for the virtual failure of the licensed charter operations. Shaken by the burden of a programme to which they are not accustomed to, the ministry endeavoured to channelise the scheme on the right course but it could not fortify the work with correct and timely decisions on various problems that have been coming up in the course of implementation. The ministry had to encounter representations against charters, mostly emanating from those who failed to get letters of intent and also from the members of parliament. Apart from these, in the process of application of the charter rules, several decisions, probably sometimes arbitrary, had to be taken. In consequence, a couple of charterers were forced to contest some of the decisions in a court of law and the court's verdicts were against the Government. The charterers get access to the officials in the ministry with great difficulty. Even when they get access the meetings mostly end in futility without any encouraging responses. One charterer has observed that the attitude of the officials as one of treating the charterers as criminals, although they are only engaged in implementing the permissions issued by the Government themselves.

The charterers came into the field only in response to the scheme notified by the Government. Owing to rigours impinging on them from the Government on one side, and complaints and non-cooperation from the foreign owners of the vessels on the other, the charters were caught in a pincer embrace. Having got into the suffocating chamber of charters, investing heavily towards deposits and guarantees, towards international travel and national travel mostly to Delhi, they could neither get on nor get out. Delays in official clearances of chartered vessels on first arrival at Indian ports are common and frowned upon by owners. Mandatory clearances and the time taken for this are also resented by them. Some of the vessels are apprehended by the Coast Guard for alleged violations and the vessels are detained for months together. The Charterers are not told what for the vessels were apprehended. The charterers are forced to make conjectures and are placed in a disadvantageous stance in explaining the position to the owners.

The foreign owners' problems are basically economic. Their aim is that their vessels should spend the least possible number of days at port and maximise the number of fishing days. Most often these aims are not fulfilled and therefore they are dissatisfied.

Implementation of charter scheme requires considerable experience and background knowledge. The Director-General of Shipping implements the charters of cargo vessels as a matter of routine and chartered cargo vessels are rarely held up. One reason for the current problems in the implementation of fishing vessel charters is probably the lack of understanding and experience on the part of both the Ministry of Agriculture and the charterers.

The Ministry of Agriculture has recently stipulated that chartered vessels should operate beyond a depth of 40 fathoms or outside territorial waters whichever is farther. In September1984 issue of *Fishing Chimes* as effort was made to explain how unenforceable such a condition is. Any arrest by the Coast Guard on this basis will often be difficult to prove. Apprehension on this basis will lead to withdrawal of chartered vessels by the owners. In fact, this process is already in motion. Some of the chartered vessels have set the precedent of unlawfully leaving Indian waters after completing fishing operations without enabling the charterers to declare the catches to the customs. This tendency appears to have materialised because of the fear that the Coast Guard might apprehend them, as there can always be some lapse, however small, to point out. The owners probably feel that the entire Indian official machinery is against them.

Apart from the style of implementation of the scheme by the Government and the behaviour of the charterers and owners, it appears that the scheme lacks a sound basis. In other countries, operation of foreign vessels in their EEZ, is related to specific surplus quotas and the grounds where such surplus fishery is available. They determine the quantity of fish to be allowed to be exploited by foreign vessels and the areas from where they should be exploited. The number of vessels to be permitted is also determined. In our case we have no estimate of available quantities, let alone the surplus quantity.

Even concluding that the scheme has failed in the matter of fisheries exploitation in the EEZ within the rules, one would expect a breakthrough in the matter of introduction of their own vessels by the charterers as envisaged under the scheme. Firstly Government have not encouraged the introduction of the same type vessels as permitted to be chartered. An encouragement of this kind would have led to diversification of fishing effort. One can, however, reconcile with permissions to acquire any type of vessel of over 20 m OAL as provided for in the scheme. Several of the charterers have applied for the acquisition of 23-25 long vessels and for loans from SDFC for the acquisition. In the past two years not a single acquisition has materialised. In this respect also the failure continues. The conclusion that cannot be escaped, therefore, is that the charter scheme has failed.

The authorities have now to do their best in this bad situation. Expeditious clearance of permissions for acquisition of vessels and sanction of loans by SDFC is something that is in their hands to be accomplished. This will atleast connote a partial success of the objectives of the scheme.

In the light of the experiences, it is now of utmost urgency for the Government to give a new shape to the rules governing operation of foreign fishing vessels in Indian waters. If the ministry is not in a position to determine the surplus fish available for exploitation and the number of foreign vessels, type-wise, to be allowed to operate, it is desirable that the implementation of the provisions in Maritime Zones Act for the operation of foreign fishing vessels may be kept in abeyance. If these can be determined and Government can evolve a system in which punishments to boats violating the rules can be administered quickly, it is advisable to continue the charter programme under a suitably revised set of rules.

Fishing Chimes, Editorial 46: December 1984: Vol. 4, No. 9

# Anti-Outrigger Syndrome

An anti-outrigger syndrome seems to have taken roots at all levels in the Fisheries Division of Ministry of Agriculture. This irrational fixation in the minds of those in authority appears to have arisen from a vector concept that outriggers and shrimp fishing go together. And, since the misplaced and baseless notion of discouraging shrimp trawling is now one of the cardinal policy elements of the Ministry, the syndrome's sway continues unabated.

Outriggers are only a device to cover a wider fishing area. They are widely used for catching larger quantities of fish. It is all a question of the design of the nets used. Whatever type of nets are used, a certain quantity of shrimp is bound to make entry into the nets, although the bulk of the catch will be fish. Even in nets specially designed for shrimp, over 90 per cent of the catches will be fish. In Holland outriggers are used for catching flat fish. In Nigeria, outriggers are used for catching ground fish of all types. Larger stern trawls with head rope length of even 140 ft can be operated from the outriggers of vessels of 23-25m in length, giving fish catches as good as those taken in an equivalent size bull trawl. In other words, a single vessel will give almost the same results as from two boats operating a single bull trawl. It is a wrong premise that shrimp can be caught in nets operated from outriggers only. Shrimp can be caught in substantial quantities in shrimp trawls operated from the stern. Thus, if outriggers are discouraged, operators can still catch shrimp by using stern trawls.

Government perhaps wants shrimp trawling to be discouraged for fear of depletion of shrimp. This fear is unfounded. The levels of shrimp population are directly related more to success in breeding and less to the offtake. What is needed is protection to brood shrimp during breeding season which extends for a few months. For this reason, shrimp nations like Mexico enforce a closed season for the short duration of breeding period. They do not discourage introduction of outrigger vessels. Shrimp, like fish, is a dynamic resource. Its lifespan is short. Both these features are in favour of uncontrolled fishing for shrimp, except during the breeding period. Depending on success in breeding, the quantity caught in a given year by all the boats remains the same. If the number of boats concentrating on the resource are more, the catches per boat will come down. This should not be mistaken for depletion. At the stage when a larger number of boats are introduced, the margin of profits per boat dwindles. This is the point when further introduction of boats will stop automatically, as no entrepreneur would like to run his boats on loss. The shrimp catches in respect of small mechanised boats, per unit, have come down. By and large, this is mistaken for depletion. The fact, however, is that a larger number of boats are sharing the same resource. Consequently, further introduction of mechanised boats has slowed down.

Larger trawlers with outriggers, mostly of 23m in length, are engaged in shrimp fishing along the north-east coast. These vessels are operated at depths upto 55 fathoms, and mostly beyond 30 fathoms. Results in the past few years show that the shrimp catches are on the increase. Not only this, the fish catches brought by these trawlers are also on the increase.

One of the objectives of the Government is to step up exports of marine products. Discouraging introduction of outriggers, particularly where there are no signs of decrease in catch per unit effort, will be an unwise policy. With increased number of units, each of the outrigger vessels operating off northeast of Bay of Bengal is landing around 9 tonnes of shrimp, with heads removed, in a voyage of 21 days, compared to 34 tonnes in a voyage a few years back. There is no doubt that increase in the number of boats operating does not as yet have any effect on unit catches, which in fact is on the upswing.

Nets operated from outriggers of trawlers do not bring 100 per cent shrimp catch to warrant restrictions on grounds of overfishing for shrimp. Proportion of shrimp caught by these nets is 5-10 per cent, if not less. Outriggers are only a means to augment the area of coverage for fishing, utilising the engine power to the maximum. There is no plausible reason to discourage introduction of outrigger vessels, whether they operate shrimp trawl or fish trawls. There need be no bar on these trawlers to catch shrimp, as their catches have not yet reached the maximum sustainable level in the areas where the vessels now operate. Whatever type of trawl net is operated, there is bound to be some shrimp catch and this is needed. Without some percentage of shrimp catch, the operations will turn out to be uneconomical.

Demand for fish in southeast Asian countries is now becoming apparent and there is an awareness of possibilities of exporting bulk quantities of fish to foreign countries, among Indian entrepreneurs. Likewise, foreign firms are also evincing keen interest to import quality fish in bulk from India. There is no doubt India would soon be in the forefront in the matter of fish exports. For this, however, introduction of vessels of sizes that are economically optimal through operation of the largest possible nets has to be encouraged. Vessels of 23-25 m length with outriggers are the most advantageous for the purpose, since single trawl nets with head rope length as much as 140 ft can be operated with the help of outriggers. Operation of such nets will yield catches equivalent to the operation of single trawl nets by two trawlers, such as in bull trawling. Such trawl nets are now-a-days becoming popular and are referred to as pseudo-bull trawls.

Fishing Chimes, Editorial 47: January 1985: Vol. 4, No. 10

# Acquisition of Fishing Vessels under 100% Export-Oriented Scheme:

# There are Two Windows, Not one

Most of fishery entrepreneurs have the impression that when clearances are given under the 100 per cent exportoriented scheme by the Ministry of Industry, such clearances would be in the nature of a 'Single Window' clearance. This impression ought to be correct but the evidence is against this. A recent development shows that the Ministry of Agriculture has also to clear the vessels permitted for import under 100 per cent scheme separately. Otherwise, the Mercantile Marine Department refuses to register the vessels.

This position came to light when M/s Dev Fisheries (Pvt.) Ltd. Bangalore imported one second hand tuna vessel in accordance with a letter of intent issued by the Ministry of Industry under 100 per cent export scheme. After the arrival of the vessel at Visakhapatnam, the company approached the M.M.D. for the registration of the vessel. M.M.D. expressed inability to register the vessel, as they had received a communication from the Ministry of Agriculture stating that the vessel should not be registered until they gave clearance from them. The company then approached Ministry of Agriculture with the request that M.M.D. may be advised suitably. The response was that the company should apply separately for the required clearance. On receipt of this, the matter would be placed before the Fishing Vessel Acquisition Committee. As and when the Committee cleared the matter and the recommendation approved by the concerned authority, permission would be accorded. The company pointed out that they had supplied all the required particulars long back to the Ministry of Agriculture on their own, although there were no instructions to follow such a procedure. His contention was reported to have been checked and it was found that the company had been keeping the Ministry informed of the developments with the required material. If it was needed that the company should formally apply for clearance from the Ministry of Agriculture, the company should have been so told by Agriculture or Industry Ministry in the beginning itself. But this was not done. For no fault of theirs, the vessel is detained.

The company had accordingly made a fresh application which was considered at a meeting of the Fishing Vessel Acquisition Committee which met a considerable time after the application was made. The Committee was kind enough to clear the import with certain stipulations that a) within six months the company should place orders for an indigenous vessel, b) it should provide a bank guarantee for Rs. 5 lakhs and c) Japanese experts should not be engaged for more than six months. More than four months after the arrival of the vessel were over and there is no indication at what point of time the approval of the authority concerned would be available.

M/s Southern Seacraft Ltd., Madras faces exactly the same situation as Dev Fisheries. The Company was allowed to import two 22.80 m long vessels under the 100 per cent export-oriented scheme. One of these reached Visakhapatnam recently but was stuck up as M.M.D. required clearance from the Ministry of Agriculture. As an interim measure, however, M.M.D. accorded temporary registration and the vessel sailed on 23rd Dec. 1984.

The permission for import was given by the Government of India in the Ministry of Industry/ Commerce. As per the Merchant Shipping Act, registration of ships has to be done by the surveyor concerned at ports declared as ports of registry by the Central Government. The reported refusal by the surveyor at Visakhapatnam to register the vessels permitted to be imported by the Central Government, on the ground that he requires clearance from the Ministry of Agriculture is intriguing. Obviously he does not recognise Ministry of Industry/Commerce as part of the Central Government. It is surprising that the Ministry of Industry has not reacted as it should at this affront and insult.

So far as we know, Ministry of Industry circulates copies of applications received under the 100 per cent export-oriented scheme among all concerned ministries for their comments within a stipulated time. Representatives of ministries concerned are invited to attend the meeting which considers these applications. Representative of the Ministry of Agriculture must have been invited to attend the meetings along with others and it has to be believed a representative from that Ministry attended the meeting. A procedure of this kind automatically means that Ministry of Agriculture has either subscribed to the decision to permit the imports, or, if it had not subscribed to the decision, their views must have not been accepted by the majority of the participants. In this situation, the decision of the Central Government to permit imports being based on the recommendation of the concerned committee, the M.M.D. has no ground to say that it requires clearance from the Ministry of Agriculture.

The gravity of the deplorable situation is that the Ministry of Agriculture, instead of saying that separate clearance from them for registration of the vessels is not required for the reason that the Ministry of Industry/Commerce on behalf of the Central Government issued the letters of intent, seems to have taken the view that, irrespective of the letters of intent, a separate clearance by it is necessary. The authorities concerned in the Ministry of Agriculture probably feel that permission for import by another Ministry is something different from the permission they have to give based on the recommendations of their Fishing Vessel Acquisition Committee. On this point, it may be mentioned that the ministries represented on the Fishing Vessel Acquisition Committee in the Ministry of Agriculture are also represented in the Committee that considers applications under 100 per cent export-oriented scheme.

It is quite possible that the second window clearance from the Ministry of Agriculture is also necessary in national interest. It is not for the industry to dispute this. What can be disputed is the fact that in between these inter-ministerial gaps, the applicants are pushed down and put to great financial losses and mental agony. The Ministry of Industry/Commerce should have mentioned in the letters of intent that these were issued with the concurrence of Ministry of Agriculture. If such a concurrence is not there, they should have introduced another condition that the letters are subject to further clearance by the Ministry of Agriculture, which should be obtained by the applicants. For want of clear instructions and directives, the applicants are victimised.

The vessel of Dev Fisheries (Pvt) Ltd. remained at Visakhapatnam port for over 120 days, and recently left for Mangalore. The company spent enormous amounts to bring the vessel to India. They continue to incur expenditure on port charges, maintenance and on the two Japanese experts who came with the vessel. The management thought that the Government, which is wedded to the achievement of fast progress in deep sea fishing, would help them in making the vessel operational in the shortest possible time. But the management is too sanguine. The nodal ministry that wants to promote deep sea fishing has caused avoidable delay in the clearance. In the light of this approach we have to watch what will happen to the imported vessel of Southern Seacraft. Unless an element of wisdom dawns on the Ministry of Agriculture and MMD or on someone superior to the Ministry of Agriculture and MMD, and tell them not to create artificial hurdles, the permanent registration of the vessel may take a longer time. (It is learnt that the D.G. Shipping has issued orders for the temporary registration of the vessels. Permanent registration will be considered only when the Ministry of Agriculture gives clearance).

Ministry of Industry should have the guts to tell the M.M.D. to permanently register vessels imported under 100 per cent export-oriented scheme, if the vessels are in order. If their guts are not strong enough to do so, they should delete fishing industry from the list of items eligible for clearance under the 100 per cent export-oriented scheme. It is not fair on their part to issue letters of intent and later leave the applicants in the lurch when the vessels are actually imported.

The events show that, so far as import of deep sea fishing vessels under the 100 per cent export-oriented scheme is concerned, there are two delivery windows, not one as publicised. Department of Industry should see that the second window is closed. If they cannot do so, they should close their own window, so far as fishing vessels are concerned.

Fishing Chimes, Editorial 48: February 1985: Vol. 4, No. 11

# Utilisation of TV Network to Achieve Regulated Fish Supply

The need for developing a regulated fish marketing system in the country has long been established. This is because this system will provide, by and large, fair returns to the fish producers. The consumers would also be able to buy fish at reasonable prices. A regulated fish supply will facilitate the linkage of sale prices with production costs, as in other sectors, and will remove the anomaly of the present system of settlement of prices based on quantity of landings and distance to the markets, and in some cases on availability of preservation facilities.

A regulated fish marketing network necessarily involves storage of fish in cold or frozen storages/packing of fish in ice. The bulk of fish consumers, with the exception of those in major centres such as Calcutta, Bombay and Delhi, prefer to buy fresh fish. Iced fish or fish coming out of cold/frozen storages are considered by them to be unworthy of consumption. They prefer a tainted fish to preserved ones. If the consumer is educated on the superiority of preserved fish over unpreserved fish, from the point of view of hygiene, change in outlook is bound to emerge. The deleterious effects of consuming unpreserved fish in areas away from landing points have to be continuously dinned into the ears of consumers.

This consumer education will lead to a public reaction. There will be demands for a set-up under which people can buy well preserved fish. Popular Governments will then have no alternative other than creating a network to provide people with the commodity, hygienically preserved. As long as demand does not come from the people, regulated fish marketing system is not likely to emerge. Without demand, investments to bring about the system will never be made.

All these years no progress could be made in educating the public on the benefits in consuming preserved fish. It is true that audio, visual and other publicity media have all along been available to popularise the advantages. However, these media have not been made much use of. Even if these are used, it is doubtful whether the idea would have registered on the minds of the people, with the desired impact.

Television has now spread to the four corners of the country. There are over 120 TV Stations. The number of TV viewers is on the increase. Vivid shots, displaying the contrasts between tainted and well preserved fish and the ill-effects of consuming tainted fish, will have a telling effect on the viewers. The process of deterioration, how preservation at low temperatures stems this deterioration, how fish from landing points can be taken to various cold storages located at selected distances, how fish can be preserved at the cold storages and regulated supplies can be organised therefrom can be brought to the notice of the viewers. Enterprising persons can be invited to participate in schemes aimed at establishing a fish marketing network to provide well preserved hygienic fish to the people.

The Central and State Governments should evolve a plan for getting fish marketing networks and simultaneous popularisation of preserved fish through television. As the demand builds up entrepreneurs should be encouraged to invest on fish transport vans, on cold storages and on retail stores at selected centres from which fish supplies can be radiated to places around.

The advantages of introducing a marketing net work are that a) people will get hygienic fish as in advanced countries. This will promote their health which is good for the nation, b) people can buy fish at rational price levels and sudden ups and downs in fish prices will disappear. c) producers can get fair returns, linked to investments and economics of operations. A marketing system can come into being out of popular demand. Popular demand will arise when there is awareness. Awareness will come when, through T.V, people are enabled to see and feel the need. The Central and State Governments should buy TV time on a regular and daily basis to popularise the system, supplemented by exhortation through Radio and other means of mass communication.

Fishing Chimes, Editorial 49: March 1985: Vol. 4, No.12

# Increasing Marine Fish Production

Government wants to increase our marine fish production. Government also wants to step up our marine products exports. To match with this, there is an unprecedented response from entrepreneurs to participate in this endeavour, as evidenced by the registration of a large number of fishing companies recently all over the country.

These companies were established over the past few years mostly to operate deep sea fishing vessels on charter and eventually to acquire their own vessels under the provisions of the charter scheme. For want of clearance of their applications for charter or for acquisition of vessels or for other reasons such as non-availability of vessels on charter, most of these companies are dormant. The entrepreneurship of all these companies can be harnessed by the Government under a realistic programme, implemented in an understanding manner. The nation cannot afford to ignore and waste so much of manpower, enthusiastically awaiting Government's support.

It is axiomatic that the nation's fish production cannot be stepped up without the introduction of new units to exploit farther marine areas. The nearshore waters are more or less fully exploited and there is very little scope to increase the number of craft to intensify fishing in these waters to augment production (of course, no increase in the number of crafts to exploit coastal waters has also been taking place in the past few years). The only way to increase production is therefore to introduce a good number of larger vessels to exploit new areas. In this task the nation has failed miserably. A large number of applications seeking permissions for imports and loans for acquiring indigenous or imported vessels have piled up in the Ministry of Agriculture, although a few have been cleared recently. Entrepreneurs sometimes come across large piles of these applications in the corridors of Krishi Bhavan, obviously kept there for want of space.

It is not an easy matter for the Government to clear such a large number of applications. They require scrutiny and processing. At the same time the long time taken between presentation of applications and their clearance creates problems. Apart from loss of time in increasing our fishing fleet, the delay leads to increase in prices, and gives time to other shipyards to influence buyers to change their options. The delays cause frustration among the applicants, multiplies their working expenses and saps their energy and enthusiasm. In the Ministry of Industry much more complicated applications for setting up industries and taking up joint ventures are cleared under a time schedule. Surely, a similar system can be adopted for clearing applications for acquisition of fishing vessels. The introduction of such a system is all the more necessary, as, more the applications are delayed, the greater will be the invisible losses. Augmentation of fish production gets delayed. In the protracted period, some of the fish in the sea either die or are poached. All this is a national loss. As long as we do not increase our fishing fleet, there will be no increase in fish production. There is no point in wailing about lack of increase in production, either in inventing obscure causes for this or in covering up the situation through evasive writeups. There is no magic to make more fish to get into any single net.

The scrutiny of prices and specifications is probably the major obstacle in the clearance of applications. It is understandable that those concerned perhaps either have reservations or fear in recommending either a price or a specification as in a reasonable order, for, it is difficult to dwelve deep into these. Precisely for this reason, perhaps, the Ministry of Agriculture, in spite of having all the decision-making powers, has involved Ministry of Shipping and Transport in the exercise. All troubles and delays seem to have started from the time this arrangement came into being. The objective of Ministry of Agriculture is to increase fish production. In total contrast, the exclusive objective of the Ministry of Shipping and Transport is to promote indigenous construction of fishing vessels and discourage their imports. They are not directly concerned whether fish production goes up or not. Their objective is laudable, but only when implemented on a long term basis. Our indigenous capability is limited to construction of 23 m long shrimp trawlers. It takes a long time for the few indigenous yards who are interested in the line to acquire capability for building vessels based on other designs. Recognising this fact, the Ministry of Agriculture should have been applying their own judgement in according clearances. But this has not been so. The tight link-up such as the one with Shipping and Transport Ministry perhaps comes as a boon to the indecisiveness, as this would protect them in the event of any problems cropping up later on.

The augmentation of Indian deep sea fishing fleet is a mammoth task. In accomplishing this, there is no doubt there will be several undesirable offshoots. Whichever way the scheme is implemented, these cannot be avoided. Any effort of this kind bristles with problems and unhappy offshoots and, in the interests of development these have to be taken in the stride, for, these do happen but in a good cause. Such being the case, a wiser course will be go to ahead with speed taking all reasonable precautions with a positive mind and achieve the augmentation of the fleet. By being critical at every turn of implementing the scheme, not only do we fail to achieve the target, but would also experience even a larger number of offshoots that are more problematic.

The need of the day is to forge ahead, inculcating a sense of discipline among the fishing companies and converting their emergies and enthusiasm into positive results. We therefore urge upon the Ministry of Agriculture to develop a pragmatic strategy that would ensure faster introduction of deep sea fishing vessels.

Fishing Chimes, Editorial 50: April 1985: Vol. 5, No. 1

# Outlook under the New Government

There is general consensus that the performance of the fisheries sector in the Sixth Plan was far from satisfactory in the marine sector, particularly in the deep sea fishing field. Marine fish production stagnated at around 1.5 million tonnes per annum. There was no perceptible increase in the introduction of deep sea fishing vessels. Only about 16 nos., of vessels of Indian flag were introduced during the Sixth Plan period against a depressed target of 250 nos.. The introduction of deep sea fishing vessels is totally controlled by the central Government (Ministry of Agriculture) and therefore the nation has the right to blame it for the failure. It may be true that hundred and one causes are responsible for this. The Industry may not have played their part properly. Other Ministries may not have co-operated. But the responsibility for the failure will be attributed by the nation to the Ministry of Agriculture.

Inland fisheries are a State subject. However, the Centre plays an important role in assisting the States in the promotion of the activity, by way of setting up Fish Farmers' Development Agencies and organising projects in the sector with international assistance. The number of FFDAs in the country increased to 147 Nos. in the sixth plan which is a commendable achievement. Fish seed production exceeded the target, thanks to the massive efforts in West Bengal. Inland Fisheries Projects with World Bank's assistance in the States of West Bengal, U.P., M.P., Orissa and Bihar made some headway, although the progress has not been altogether satisfactory. In the Seventh Plan, under the new Government, it is to be hoped that the second phase of the World Bank's project for the development of fisheries in small waters will become a reality and will contribute substantially for increasing fish production and uplifting the lot of small fish farmers. A scheme under the World Food Programme for the development of *Maans* in Bihar and *Bheels* in Assam is taking shape and we are confident that the new Government would ensure that it materialises and is implemented with success.

The general atmosphere arid attitudes in Krishi Bhavan counts a lot in building up confidence among entrepreneurs and developing the right equation between the Government and the Industry. As a refreshing contrast, from what the fishery entrepreneurs feel, the new Government has brought in a change in the general approach at Krishi Bhavan which is one of easy accessibility of the entrepreneurs to the authorities to represent their problems, and a realistic approach to deal with them. This sea change appears to be a major development and it should go a long way in translating the objectives of the Government into action in the matter of development of fisheries, with the willing interest and cooperation from the entrepreneurs. There are several examples of quick decisions on matters of importance to the fishing industry and to individual companies after the new Government took over. If this is an indication of the benefits of the new approach, the fishing industry can look forward to a bright future. The nation can also look forward to a quick growth of its fishing fleet and increase in marine fish production. The fish farmers can look forward to new avenues of fish production and for setting up of a larger number of hatcheries so as to augment seed production.

The tide seems to be picking up speed in favour of fisheries development of the country in the seventh plan, judged from the progressive mood of the Government and its officers. Entrepreneurs should make the best use of it, honestly and sincerely, thereby bringing prosperity and name for themselves and for the nation. There is no doubt that, all encouragement will be forthcoming from the benign Government.

Fishing Chimes, Editorial 51: May 1985: Vol. 5, No. 2

# Madras Fishing Harbour, an Orphan

Two years have passed after completion of the construction of Phase I fishing harbour at Madras. For want of various facilities, no commercial fishing company has been able to operate their vessels from this harbour, which is meant for providing berthing facilities for deep sea fishing vessels. The Director of Fisheries, Madras was expected to take care of the management of the harbour. However, it is learnt that he could not take over the work as he had no funds. The Port Trust had also no funds for running the harbour on its own, it is stated. So much so, the harbour is like an orphan. No maintenance is being done. There is no management whatsoever.

There are practically no shore facilities. There are neither bunkering nor water supply facilities. There is no security arrangement. The vessels of the Fishery Survey of India have been operating from this harbour on compulsion. Every time the vessels have to go to main harbour for water, wasting one day.

The fishing harbour has now become more or less a scrapyard. One old ship was dismantled here. The keel and some plates which were the remains in the harbour area were washed away around December 1984 and these are now lodged right in the middle of the main channel inside the harbour. This has become a major navigational hazard. No vessels can enter the harbour after 5 p.m.

Because of lack of control, the local goondas are terrorising lawful users such as FSI and CIFNET staff. Steel materials are stolen. Even mooring ropes are not spared. Consequently the vessels drift. They also enter the fish hold and steal fish. They lift baskets of fish and jump into the waters and it is found very difficult to catch them. It is learnt that two staff members of CIFNET were beaten by goondas recently. The victims sustained serious head injuries. This incident took place around. 28th February 1985 and Police filed cases. Some of the mechanised boat operators, who utilise the harbour facilities, berth their vessels in the harbour but are stated to be not paying any charges to the authorities. CIFNET and FSI are the only parties paying berthing charges to the Port Trust at present.

For want of attention, thus, a major fishing harbour constructed at a huge cost, is virtually remaining unutilised. Located strategically to enable vessels to undertake fishing in Southern Bay of Bengal with potential resources, it is sad that such a situation is allowed to continue. It is hoped that the authorities concerned would pay particular attention to the matter and make this important fishing port operational.

Fishing Chimes, Editorial 52: June 1985: Vol. 5, No. 3

# Plus and Minus of New Charter Rules of April 1985

The Maritime Zones of India (Regulation of Fishing by Foreign Vessels) Rules were notified for the first time on 26th August 1982. These rules provided for the operation of chartered foreign fishing vessels in the areas of Indian EEZ beyond a distance of 12 nautical miles. These rules were subsequently amended on 23rd April 1984, according to which foreign chartered fishing vessels had to operate beyond a distance of 12 nautical miles or a depth of 40 fathoms whichever is farther from the Coastline.

The amended rules notified on 23rd April 1984 evoked a spontaneous protest and opposition from the charterers of fishing vessels as well as the owners. Representations were made to the Government to restore *status quo ante*. Several discussions were also held between representatives of the fishing industry and the Government on the subject. It was pointed out from the Government's side that the President of the Association of Indian Fishery Industries had given his willingness for the amendment notified in April 1984. But for this, Government would have held further consultations with the industry. The contention on behalf of the association however was that the President never gave his willingness.

Whatever this may be, a volley of representations from the industry were given cognisance by the Government, and obviously after very careful review of the position, a new notification containing further amendments to the rules was issued on 4th April, 1985. According to this, in general, fishing on the west coast by the chartered fishing vessels would have to be conducted beyond 24 nautical miles from the Coastline However, off Maharashtra and Gujarat coastline and off a part of Kerala and Tamil Nadu coastline, certain restricted areas are delineated by certain coordinates. These restricted areas are out of bounds for the chartered fishing vessels.

According to the new notification on the East Coast, in the Southern Section *i.e.* from the southern tip upto Nizampatnam (150N Lat.) fishing could be conducted by the chartered vessels beyond 12 nautical miles from the shore. From Nizampatnam to Paradeep Port, however, fishing can be done only beyond 24 nautical miles. In addition to this restriction, from north of Chilka upto Bangladesh boundary, five coordinates were given, fishing within which was not open for the chartered foreign fishing vessels.

Believing that the fishing industry would be interested in knowing the implications of the new set of rules promulgated in April 1985, we have made an effort to make an assessment of the area which is now available for fishing by chartered vessels with reference to the earlier rules of April 1984. Judged from the standpoint of the notification of April 1984, our estimate is that a net additional area of 21,280 sq.km towards the coast has become available for operation by the foreign chartered fishing vessels on the west coast. So far as the east coast is concerned, it is computed that there is a reduction in the availability of fishable area towards the coast, of the order of 10,138 sq. km. Thus the net additional area that has become available towards the coast, for the operation of chartered fishing vessels as a result of the notification of April 1985 comes to 11,142 sq.km. This represents (from an academic angle) an average additional area of 1.63 km in width all along the coastline.

The additions to the fishable area on the west coast are noticeable at eight points along the coastline, measuring in total around 7,000 sq. nautical miles which works out to about 25,000 sq. km. The coordinates of these points where additions are noticed are furnished hereunder:

- ☆ 7030′N77°20′E to 7°30′N77°40′E
- ☆ 8040′N76°E to 10°40′N75°40′E
- ☆ 8042′N75°42′E to 12°40′N74°30′E
- ☆ 12042′N74°32′E to 15°30′N73°20′E
- ☆ 15°32′N73°22′E to 18°N72°40′E
- ☆ 19020′N71°40′E to 19°52′N71°05′E
- ☆ 20025′N70°25′E to 22°05′N68°32′E
- ☆ 22°40′N68°E to 23°N67°20′E

The reduction in areas on the west coast has taken place at five points as under:

- ☆ 7°45′N77°05′N to 7°30′N77°20′E
- ☆ 7°45′N77°52′N to 7′45′N77045′E
- ☆ 18°33′N72°E to 18°N72°31′E
- ☆ 19°20′N71°40′E to 19°02′N72°E
- ☆ 20024′N70°25′E to 19°50′N71°04′E

The area reduced by these points is of the order of 1,100 sq. nautical miles or 3,720 sq.km. Thus, on the west coast, the additional fishable area being 25,000 sq.km. and the reduction is of the order of 3720 sq. km., the net additional area made available for chartered foreign fishing vessel operations by the Government is estimated at 21,280 sq.km.

In regard to the east coast, it appears that there is no change in the operational area on account of the rules notified in April 1985, upto Nizampatnam from the southern tip. Northeast wards, from Nizampatnam to Chilka lake, because of the increase in width of prohibited area for fishing to 24 nautical miles, and also because of certain coordinates given by the Government further northeast (from Chilka to Bangladesh border) there is a reduction in fishing area. The coordinates on the east coast referred to above (15°30'N-80°40,E to 20°42'N-880E) exclude the area from the coast to these points. The total reduction in the area is estimated to be of the order of 13,517 sq.km.

In regard to addition to fishable zone on the east coast it is seen that it is of the order of 3,379 sq.km., lying between the coordinates 20042'N-88°E to 210i6'N-84°14'E in sandheads area.

To sum up, it is to be concluded that the notification of 4th April 1985 has provided some additional fishing areas to the Indian charterers. Although there is no demonstrably welcoming reaction from the fishing industry in respect of the new notification, from the fact that there has been some increase for the coming back of the chartered vessels from their home ports abroad, it is to be construed that the relief has had some effect on the trend of chartered vessels operation which came almost to a standstill from May 1984.

It is believed that the Coast Guard would no longer have any difficulty in patrolling and apprehending erring vessels as the depth factor is no longer there. There is, however, still a lacuna in the rules in respect of the point, whether the location of the fishing vessels or fishing net is the criterion to judge a violation of regulations.

The coordinates have been fixed in such a way that they are almost straight and contiguous with the demarcation of the rest of the prohibited areas for fishing by chartered vessels and this is a welcome feature. The coordinates really do not represent additional restricted areas. From the fact that there have been very few transgressions of the rules after April 1985, although the time is too short to judge, it is felt that there will be an orderly implementation of the programme relating to the operation of the chartered foreign fishing vessels from now on yielding the expected benefits by way of training to Indian nationals onboard chartered vessels in locating fishing grounds, estimation of abundance of availability of various species in the grounds located, augmentation of exports and inflow of additional foreign exchange.

Fishing Chimes, Editorial 53: July 1985: Vol. 5, No. 4

# Deep Sea Fishing Vessels: Target and Fulfillment

Government proposes to introduce a deep sea fishing fleet consisting of about 300 vessels, from a cumulative angle, by the end of the Seventh Plan. The present fleet and the clearances given so far may take the fleet to over 150 nos. by the end of the seventh plan. Thus, the task before the Government and the Industry is to organise the work in such a way that the cumulative target of 300 nos. is achieved, by ensuring the introduction of the vessels for which permissions have already been given and by organising issue of clearances and further follow up action for the remaining.

In planning the introduction of deep sea fishing vessels, it will be pertinent to keep a few aspects in view, particularly the lessons learnt in the past. In the shipping sector, the strength of merchant naval fleet and the targets to be achieved in this respect are referred to in terms of G.R.T. This system is adopted obviously for the reason that it is the capacity and not the number of vessels that count. The introduction of a large number of vessels with lower tonnage may not add up to the targeted cargo-carrying capacity.

In the fisheries sector, the number of deep sea fishing vessels to be introduced can have a meaning only when considered from the angle of NRT. Net rated tonnage is practically equivalent to the fish hold capacity. If the intention is that each of the vessels on an average should bring in 500 tonnes of fish/shrimp in a year of 12 voyages, the net capacity of the fish hold has to be 45 tonnes, or in terms of NRT around 90 tonnes. Taking the existing number of active vessels as 60, the additional 240 nos. (or 21,600 NRT) to be introduced may bring in 1,20.000 tonnes of extra annual fish production. Considering financial, technological and managerial constraints, it appears that this target is reasonable.

Past experiences show that clearances by the Government for imports and sanction of loans by Shipping Development Fund Committee for indigenous and imported vessels to the required level shall have to be given by the second year of the Seventh Plan and this part of the activity will have to come to a stop by that time if the target is to be achieved by the end of the seventh plan. In other words, it is necessary to look at the work in two parts, one part dealing with sanctions relating to introductions in seventh plan and the other part devoted to the fulfilment of the contracts, delivery of vessels and making them operational by the end of the seventh plan period. Opportunity can be given to the interested enterprises to apply for loans till middle of 1986. Any application received after December 1986 or any sanction issued in or after April 1987 should be related to Eighth Plan only.

Government may have to draw up and adhere to a strict time table, upto the stage of sanctions, to cover the first part. The countdown should start from the time the FVAC clears an application for import and the loan committee clears it for releasing the loan.

There are quite a large number of Indian and foreign shipyards interested in building fishing vessels for Indian fishing companies. Because of this, there is a severe competition among all shipyards aiming at bulk orders in order to bring down the price per unit. The larger the number of vessels a yard undertakes to build, the longer will be the time of delivery of the total lot. An Indian shipyard, generally, may take about 14 months to deliver the first vessel and take about two months for delivering each of the subsequent vessels. In other words, if a yard has orders for 12 vessels, which is probably the minimal workable number to earn reasonable remunerative margins, the yard may take 36 months to deliver all these vessels, if everything goes on well.

So far as foreign built vessels are concerned, assuming 12 vessels are to be delivered by each one of the yards, each of them may be able to deliver the first vessel in 9 months and one vessel every subsequent month. In other words 12 vessels may be delivered in 20 months. This may not however take place in actual practice as release and transfer of funds in foreign exchange may take further time, particularly because of fluctuations in foreign exchange rates.

It has been mentioned that not more than two years should be earmarked for the phase culminating with permissions for imports and sanction of loans by SDFC. Experience shows that it takes not less than 12 months for all clearances for import/indigenous construction to be obtained. With a faster tempo this duration can at best be brought down to six months. Thus, upto March 1987, three rounds of clearances can be given, covering 150 vessels or more and this should be adequate to meet the balance of the target of 300 vessels. It is desirable to give clearances for 20-30 per cent more number of vessels than is actually targeted, so as to take care of dropsout.

The introduction of a system of following up the progress of each of the applications through certain select indicators depicted on a chart that is made up-to-date through extension of lines from stage to stage, at the seniormost level once in two months, appears to be necessary, in order to monitor the progress and take corrective measures from time to time.

Fishing Chimes, Editorial 54: August 1985: Vol. 5, No. 5

# Unending and Harsh: Detention of Chartered Vessels

Four pairs of chartered Taiwanese fishing vessels were apprehended by the Coast Guard on July 19, 1984 while these vessels were allegedly engaged in fishing in an area having a depth far less than 40 fathoms, off Quilon on Kerala coast. Under the regulations in force the vessels should have operated 12 nautical miles away from the shore or beyond a depth of 40 fathoms whichever was farther. The Coast Guard's case was stated to be that they were well within 40 fathoms depth.

The vessels were taken by the Coast Guard to Bombay soon after the arrest. Thereafter, the Coast Guard filed cases in the Court of the Chief Presidency Magistrate, Bombay, against the charterers and Captains of the vessels. The cases have not yet come up for final hearing before the Magistrate till the time of this writing. In other words, over twelve months have elapsed after the apprehension.

The contention of the charterers/owners/masters is that the vessels did not violate the regulations. According to the Coast Guard, however, the vessels violated the rules. This can well be believed, although one cannot be certain, for, fishermen, not only Taiwanese, but all over the world, have the temptation to go where fish are. If the vessels transgressed our national regulations willfully, penalties as justified under law are inescapable. If there is no wilful transgression, and the intrusion was beyond the control of the Captains, because of current, wind etc., sympathetic consideration appears called for. As matters stand, there appears to be a total lull in regard to the consideration of the matter. All the eight vessels, with around 120 crew members have been under detention for twelve months at Bombay harbour. We do not know whether under law, vessels with crew members can be kept under detention indefinitely.

Undoubtedly the authorities are aware of the hardships. Families of most of the crew members in Taiwan are reported to be virtually starving, with the breadwinners away from the country. For want of income some of the womenfolk of the detained crew members are reported to have been forced into prostitution. The detention of the vessels and the men must have had a telling effect on the other Taiwanese vessels operating in Indian waters.

It will not be in the best traditions of our country to ignore human suffering for such a long time. Either punishments commensurate with the alleged crime, if proved, have to be awarded in accordance with the law of the land, or, the vessels and the crew should be set free.

The crew members follow the directions of the Captain while aboard. They are under his command. If there was any lapse committed by a fishing vessel, it will be that of the Captain and not the crew. In the present case, the crew cannot even be charged with abetting, as they would not know the exact position of the vessel at a given time. Only the captain knows this.

The rules governing operation of fishing vessels, framed under the Maritime Zones Act are, if we may say so, are still in need of refinement from technical, law and order and punishment angles. For example, it is not yet clear from the rules whether the position of the vessel or the net is the criterion to judge the distance from the coast. Our coastline, like any other coastline, is not straight. It has several convolutions, indentations, etc. A base line to measure distances out on to the sea is to be delineated. Our experts are reported to be still on the job. In the absence of a fixed baseline, to judge the distance at which a fishing vessel is operating will be empirical and certainly not precise. The punishments prescribed for licensed vessels as well as poaching foreign vessels for transgressions are unfortunately similar. This will no doubt encourage poaching. The rules undoubtedly require refining and there may be occasions to amend the rules several times, based on the experiences gained by the Government in enforcing the rules from time to time. The arrest of the four pairs of vessels referred to above and the handling of the subject must have given considerable insight to the Government in regard to such situations. Obviously there was something amiss somewhere; otherwise there does not appear to be any reason why the vessels should be under detention for so long without trial.

We appeal to the Minister for Agriculture and Prime Minister of India to look into the subject personally with a view to ending the hardship to so many human beings, living under dire conditions on board their vessels at Bombay harbour, with practically no means to have food and survive. Charterers in India probably provide money for their food but how long can this go on? All the crew members are now emaciated. Their families are languishing in their country. With the breadwinner of the family away, it does not require lot of thinking to imagine their ordeals. It can be decided quickly, whether the vessels committed a serious crime or not, in the present state of evolution of the rules. If punishment is called for, the Court concerned should be urged to decide fast. Otherwise, the vessels and the men should be released. There has to be an end to this melancholic drama. The crew members at Bombay harbour have become mentally rattled, loosing virtually their sanity. Sickness haunts them with no one to care. The nation cannot be a spectator to this kind of situation.

Fishing Chimes, Editorial 55: September 1985: Vol. 5, No. 6

# Round the Year Fishing on the West Coast

Sea fishing on the westcoast of India by Indian fishing vessels comes to a complete halt during the southwest monsoon period, say, for the period from June to August, except in certain waters very close to the Coast. The fury of the winds will be so severe that it is impossible for the smaller craft to operate during the period.

Taiwanese chartered vessels, mostly pair trawlers, operate invariably on the west coast, far away from the coast, at a distance beyond 24 nautical miles. They operate along Kerala, Karnataka, Goa, Maharshtra, and Gujarat coasts all round the year. The fury of the monsoon from June to August does not affect their operations. The reason for this is that the vessels are of over 350 g.r.t. and over 35m in overall length. We have to learn a lesson from this in order to exploit the rich fisheries resources in distant waters of the Arabian Sea. Only when the introduction of vessels of over 35m OAL. is promoted, the aberrant concentration of all vessels of over 20m length at Visakhapatnam can be broken. Taiwanese skippers are reported to have made favourable observations concerning existence of stocks of white and tiger shrimps off Mangalore beyond a depth of 40 fathoms. It is also stated that rich stocks of squid and cuttle fish occur off Goa coast in deeper waters.

There is no doubt that fishery resources off the west coast beyond the traditionally fished areas, are scarcely touched by Indian flag vessels. What Taiwanese have been able to do, we should also be able to do. Only the will to do this is needed.

Our traditional approach involving several considerations such as indigenous angle, large or small companies, debt-equity ratio etc. come in the way of evolving a programme for introducing larger vessels of 35 m OAL and over. One way of achieving the introduction is for a Government Corporation to import the detailed shop floor drawings of Taiwanese 35m long vessels through a third country like Singapore and make them available to an Indian shipyard, to construct and supply one or more pairs of vessels to that design to the corporation concerned. The proposed Rashtriya Matsya Nigam is best suited to undertake the job. Several Indian personnel have been trained on Taiwanese vessels and these can be employed to operate the vessels. If necessary, for a short duration, a master fisherman and a few crew members from Taiwan can be employed.

A few Taiwanese companies are reported to be interested in selling their trawlers, not more than five years old, subject to the condition that the entire cost of the vessels will be repaid in fourteen half yearly instalments with a reasonable rate of interest on declining balances. The companies concerned will themselves buy the catches at prevailing international prices. If the Indian owners do not have sufficient confidence in the Indian trained personnel to operate the vessels successfully, required Taiwanese crew can be engaged for a short duration.

It is a great anomaly that the nation has no programme for introducing vessels of the type specifically operated under charter scheme to test their suitability and, having operated such vessels and having been convinced of their viability, no measures are contemplated to introduce such vessels. Instead of adopting the negative approach of criticising the chartered vessel operations, it is desirable to have a positive approach, taking into account the beneficial demonstrations that emerged out of the programme and pressurise the Government for taking steps to introduce such vessels.

Fishing Chimes, Editorial 56: October 1985: Vol. 5, No. 7

# Deep Sea Fishing Industry has two Windows now

## (With prospect of one more)

The deep sea fishing industry has been urging upon the Government for quite some time past to set apart only one window for conveying decisions of the Government in respect of various applications from the fishing industry. This the Government could not accomplish so far.

While the deep sea fishing industry is no doubt at the receiving end, it should also have a single window to represent its point of view to the Government, in the same way as it wants a single window delivery of its dicisions by the Government.

Till 6th September 1985 there was only a single window for the deep sea fishing industry. On this day, a few fishing enterprises formed a separate association under the name 'All India Trawler Operators Association' (AITOA). Its headquarters is at Visakhapatnam. The main objective of this association is basically to prevail upon the Government not to grant permissions for chartering of foreign fishing vessels as these are felt to be not in national interest. This Association is also against joint ventures with foreign fishing companies. It wants that national deep sea fishing fleet should be built up, without resorting to chartered or joint venture operations. In other words, this new Association wants the Government to extend all support to deep sea fishing enterprises but only for the acquisition of deep sea trawlers, either through domestic construction or imports.

The approach of the new Association is thus clearcut. One needs joint ventures or charters only when a fishing method other than singleboat trawling is to be introduced. The newly formed Association, as its name indicates, seeks to champion the cause of trawler operators. Presumably, other types of vessels such as purse-seiners, longliners, gillnetters, pole and line vessels, lobster potters etc are not of interest to this Association. Our conclusion as outlined above is presumed to be correct, for, the articles of this Association talk of trawlers only. If our presumption is correct, the approach to have an Association to concentrate on one type of vessels *viz.*, trawlers, which happen to be the predominant type of vessels now in operation can be said to be right.

It is quite possible that, in course of time, there will be Associations of companies dealing with purse seiners, longliners etc based on this precedent.

No one can dispute the rights of like-minded companies to form an Association. From this angle none can find fault with the formation of a trawleroperators Association. One question in this context is why should any one or more companies think of forming an Association of this kind, when the Association of Indian Fishery Industries (A.I.F.I.) exists and is stated to be taking care of the problems of trawler operators and when trawlers happen to constitute the bulk of the Indian fishing fleet. Obviously there is some motivation for the formation of a new Association. One likely motivation is that the A.I.F.I.'s principal office is in Delhi and is unable to take care of the problems of the operators all of whom function from Visakhapatnam. AITOA has its headquaters in Visakhapatnam for greater ease of co-ordination. On this aspect, however, AIFI has a point that it has a regional committee at Visakhapatnam. If this argument fails, the other reason can be that the members and the elected office bearers of A.I.F.I have no voice in its the running and that the interests of AIFI members are thereby suffering. A check on this point lends evidence that this may be the motivation for the formation of another Association.

Both AIFI and AITOA are private limited companies. Both these companies will be making representations to the Government on some subject or the other from time to time. These representations may not be alike, and so the Government may have to consider two points of view. It is possible that Government may not take either of the points of view seriously and decide in its own way. Neither of the associations can find fault with the Government, since, from the Government's side, it can always be replied that another association has something else to say. The ultimate result is that the voice of the Industry gets weakened. It is to be hoped that inter-association rivalries will not surface but the possibilities are there. Neither of the Associations can any longer harp on 'one window theme'. There are now two windows on the Industry's side and the confusion this will create will have a chain reaction that will rebound.

There is a talk that another Association under the name 'Trawler owners welfare association' has been formed. The members obviously feel that welfare of the trawler owners is important. They no doubt feel that if the owners are not doing well, development suffers. The formation of this Association is probably a good step, provided the activities are restricted to their welfare. If the objectives of the Association go beyond their 'welfare' and are in conflict with those of AIFI and AITOA, then there will be a virtual anarchy. Dealing with two Associations may provide amusement to the concerned officers in the Government. To deal with three associations, there will no longer be any amusement to them. They have a serious job to do and they may encounter difficulties in discharging their functions with three Associations in the ring, and the Associations will become a laughing stock.

Fishermen are basically individualistic. They, particularly (fisherwomen), create lot of din in fish markets. These inherent characteristics are related to the nature of the industry. We have seen the promoters of several fishing companies working together, but only upto a particular point. After the threshold stage, each entrepreneur tries to keep his essential promotional work confidential, thereby displaying individualistic tendency.

Deep sea fishing sector is so small, hardly around 70 Indian vessels now operate in Indian waters. To secure the interest of these seventy and to take care of future development of the industry there are two Associations, with the prospect of another being set up or already set up. The proliferation of the Associations in this situation only confirms the general characteristic of the industry – to be individualistic. One need not wonder if a few more Associations spring up. But we appeal to the fishing companies not to think of forming any more Associations and avoid confirmation as a laughing stock.

Before concluding, we are constrained to comment that there is a considerable extent of immaturity of action atleast among some of the fishery enterprises. It is to be hoped that an improved order in respect of positive unity among deep sea fishery enterprises would emerge in the near future.

Fishing Chimes, Editorial 57: November 1985: Vol. 5, No. 8

# SDFC's Finances for Fishing Vessels: Present Hurdles Blocking Flow

There are reliable indications that the Shipping Development Fund Committee would be wound up with effect from first January 1986. It is stated that, while no further loans for the acquisition of ships and fishing vessels will be granted from this date, the Committee would continue to shoulder the responsibility of realising loans already sanctioned and to be sanctioned until 31st December 1985 and of recovery of instalments of loans already released, and discharge all the responsibilities cast on it under loan agreements, tripartite agreements etc.

The Shipping Development Fund Committee has been playing a commendable role in financing the acquisition of deep sea fishing vessels. If the Indian deep sea fishing fleet is now around 70vessel strong, with a further substantial number in the offing, the credit goes largely to the Shipping Development Fund Committee. It is unfortunate that the fishing industry faces the likelihood of the closure of this financing window. The link between the fishing industry and SDFC would continue for atleast another fifteen years, which is the duration for the repayment of the loans and the latest batch of loans released will take this much time for repayment.

The organisation for the granting of loans for the acquisition of deep sea fishing vessels from the SDFC is the single boldest step of this century taken by the Government for the development of deep sea fishing industry. While this facility, created with so much of imagination is being unimaginatively closed down, in continuing to deal with the release of loans already sanctioned and facilitating delivery of fishing vessels in the pipeline to the companies, SDFC will have to be enabled by the Government for adopting a more pragmatic approach.

The entrepreneurs entering the fishing industry are weak financially. Considering this, apart from granting loans at a concessional rate of interest, Government have recently taken a decision to add the interest due from the loanee companies to the loan sanctioned, to be recovered in six half-yearly instalments, after the vessels become operational. But these concessions are not adequate. A financially weaker sector such as the deep sea fishing industry, otherwise possessing technical skills, needs a number of props. If Government provides a few props and expect the weaker entities, which the fishing companies are, to mobilise the rest, it would never happen. Governent should either provide all needed props or just not implement a programme of financial assistance. Having implemented the programme for some years, Government has probably decided to withdraw the facility with a view to establishing an alternate facility. Be this as it may, it is of paramount importance for the Government to provide a few other props that are needed in the case of loans already sanctioned by the SDFC. In other words, it is not wise to leave things halfway. Having done something, the mission has to be carried on to its logical end. The props needed relate to i) debt - equity ratio and ii) mortgaging.

So far as debt-equity ratio is concerned one has to admit that the present stipulation of 6:1 undoubtedly constitutes a highly diluted ratio. In fact, sound financing norms require a far tighter ratio. In other words, if financing norms are the criterion, Government would not have fixed 6:1 ratio but stipulated one at par with other sectors at the level of 2:1. The reason for fixing a diluted ratio is to facilitate introduction of fishing vessels by fishing companies known for their financial weaknesses. Let us assume a vessel of 23 m length to be built at an Indian yard costs Rs.70 lakhs less subsidy. A fishing company has to invest Rs.3.5 lakhs, and secure a loan of Rs.66.50 lakhs from the SDFC. To get this loan, however, the contribution of Rs.3.5 lakhs by the company, sufficient for acquiring the vessel, is not adequate as it does not satisfy the debt-equity ratio of 6:1. For fulfilling this ratio, the company should have an equity of Rs.11.10 lakhs. Many of the entrepreneurs cannot raise this much equity. Further the stipulation of 6:1 debt-equity ratio is not consistent with the requirement of 5 *per cent* contribution. The beneficial effect of this provision is neutralised by the debt-equity ratio and several companies get stuck on this. Any developmental measures of this type should either be total or there should be no effort in this direction. Developmental concessions of a half-way kind cause greater harm than good.

In order to give the needed relief from this condition of debt-equity ratio, Government should permit SDFC to stipulate a revised condition under which the company concerned should raise the debt-equity ratio to 6:1 within 2 years of commencement of operations in specified stages, while raising adequate equity equivalent to little over 5 per cent of the cost of a vessel less subsidy.

The second major irritant is the value of the mortgage to be given to the SDFC by the fishing company, before taking delivery of a vessel. Under the present rules, security equivalent to 120 per cent of the loan value should be provided in the case of both imported and indigenous vessels. An imported vessel costs say, Rs.80 lakhs. 10 per cent of this, coming to Rs.8 lakhs, is contributed by the party. The loan component is thus Rs. 72 lakhs. 120 per cent of this comes to Rs.86.40 lakhs. The cost of the vessel being Rs.80 lakhs, the party has to provide additional security of Rs.6.40 lakhs in addition to mortgaging of the vessel. The stipulation of 20 per cent security is no doubt reasonable and sound from the angle of financial discipline. But few entrepreneurs will be in a position to find a way to provide the additional security, a need that arises at the time he is totally sapped and eagerly looks forward to get hold of the vessel, to commence fishing operations and start earning for repaying the loans incurred. There is no justification for asking for security beyond the cost of the vessel.

In the case of indigenous vessels, the position is still worse, while it should actually be advantageous. Let us say, the cost of an indigenous vessel is Rs.96 lakhs. Of this, the shipyard gets 33 per cent as subsidy, of the order of Rs.30.58 lakhs, the final cost for the buyer being Rs.65.42 lakhs. He contributes 5 per cent of this, coming to Rs.3.271 lakhs and the rest of the amount coming to Rs.62.149 lakhs is taken as a loan. If 120 per cent of this, which is about Rs.75 lakhs is taken as the security amount for mortgage, the entrepreneur should not have any problem as the total value including subsidy is Rs.96 lakhs. However, the definition of 'price of vessel', so far as SDFC is concerned, is stated to exclude the subsidy amount. This is something that goes to the yard direct from the Government and is not part of the price paid by the entrepreneur. One can probably feel that from a financial interpretation this definition is correct. However, from a common sense angle, the definition has no substance. Government gives subsidy only for the reason the cost is high. The value of the vessel, reckoned from the angle of the value of materials and labour that went into her construction is Rs.96 lakhs and certainly not Rs.65.42 lakhs. It therefore stands to reason that SDFC should take into account the total cost of the vessel including subsidy into account and not the cost less subsidy.

Considering the above aspects, Government should also favourably consider the removal of the unfair condition concerning additional security over and above the cost of the vessel.

Fishing Chimes, Editorial 58: December 1985: Vol. 5, No. 9

# 'Salesmen' of Fisheries Development Needed

Development of inland fisheries, and marine fisheries within the territorial waters fall within the sphere of the state Governments concerned. Development of marine fisheries beyond our territorial waters is the responsibility of the Government of India.

The availability of a sizeable potential for the development of fisheries in our waters, both inland and marine, and the need for the utilisation of this potential is well recognised.

Technologies adequate enough for the development of inland freshwater and brackishwater culture have been established in the country. Likewise, in the marine sector also, technologies in respect of various types of fishing, particularly with large vessels, capable of exploiting the fishery resources in waters beyond the traditional zone, are becoming increasingly apparent both to the Government and marine fishing enterprises.

The utilisation of the national fisheries potential is dependent not merely upon availability of finance and technology but also on willingness on the part of financing agencies to part with money and the availability of suitable entrepreneurs and trained men to absorb technology and utilise the money for the purpose for which it is meant. There is a pre-operational and an operational phase in both marine and inland fisheries development activities. There is no doubt that there are several aspects to be taken care of concerning these. In the inland culture sector, entrepreneurs should be able to secure water resources for culture operations either through fresh excavation or on lease. Inputs such as fish seed are needed for which again hatcheries have to be set up. A great relieving factor in inland fish culture development is that there is less of dependence on imports.

In the marine sector the need is for total concentration on the development of deep sea fishing effort in the area beyond the presently fished traditional zone. Like the inland fisheries sector, this activity has the pre-operational and operational phases. The pre-operational phase, mostly relating to construction and acquisition of fishing vessels, calls for experience. There is a heavy dependence on imports of machinery and equipment and drawings in this connection. There should be entrepreneurs prepared to buy vessels from the shipyards without Governmental coercion and the entrepreneurs should have technological and financial support. We have the Government at one end to lay down policies, strategies, accord various permissions, and recommend granting of loans etc. At the other end are there, a large number of enterprises to participate in the programme. In between the efforts of these two, there are so many gaps and weaknesses. The entrepreneurs have to wend their way through a maze of problems to achieve the objectives and it is impossible for a Government, without a field machinery, to assist the entrepreneurs in solving the problems.

In order to fill this missing link a field organisation consisting of a number of trained salesmen' is needed. Appropriately headed and structured, the trained 'salesmen' of the organisation, vibrant and active, should be able to perform 'miracles' in the inland and marine sectors. Through them Government should be able to come to know the conditions, developments, and needs in the field on the one hand, and the entrepreneurs should be able to have the comfort of having an agency close to them, which can understand, evaluate and appreciate their problems and assist in getting over them, on the other. The 'salesmen' have to be professionals with a thorough knowledge and experience in respect of the subject they are handling. Otherwise no entrepreneur will trust them. And if trust is not established lot of damage will be done.

Extension efforts in the fisheries sector have been a virtual failure in India. If some development has taken place, it is in spite of the extension organisations. Instead of extension workers, therefore, we should have a band of dedicated fishery technology/technical salesmen. Besides their normal emoluments, Government should pay them extra amounts as incentives for the jobs done successfully. They should also be allowed to receive payments for the services rendered from the enterprises on a suitable scale to be laid down. Ability to sell and introduce new technologies, and to help in the setting up of new production units shall be the criteria in selecting men for the jobs It shall be their function to meet their 'clients' often and help them in solving their problems and not vice versa. In the same way as extension workers, the few fisheries consultants that there are, are also a failure in bringing about fisheries development, for, their job mostly ends in preparing project reports. Several of them have no field experience and the preparation of their reports is based on published material. While there can be consultants to perform functions of a limited nature, such as preparation of project reports, there is a definite need for supplemental men to assist the entrepreneurs in translating the projects into realities. This job very few of the presently available consultants can perform. A band of professional men under the proposed organisation can fulfill this function.

The 'salesmen', or by whatever designation they may be called (field operatives for example) would have to be trained men. Special courses have to be designed for them and conducted at an institution such as the Central Institute of Fisheries Education, Bombay. These men should have to specialise in the various branches of field activities of fisheries development.

On the Inland side, the functions of the present field staff in the various State fisheries departments handling inland fisheries schemes should be redefined to meet the needs of salesmanship of technology and to take over work of this kind and their performance should be judged with reference to certain selected indicators.

There is an inescapable and imperative need to have a field fisheries developmental set up in the central sector for promoting deep sea fishing and allied activities and in the State sector for promoting inland fish culture and allied activities.

The present state of affairs is mostly because of the conspicuous absence of such a set up. We hope that the Ministry of Agriculture would give due attention to this aspect and fill up the gap.

Fishing Chimes, Editorial 59: January 1986: Vol. 5, No. 10

# Prime Minister's Observations on Fishermen and Fishing Industry

The Prime Minister inaugurated a National Seminar on Integrated Development of Fishermen Community in New Delhi on 13 November 1985. It appears that this event happened to be the first opportunity for the Prime Minister to acquaint himself with the problems of fishermen *and* the fishing industry and to express his views publicly on the subject.

## Fishing Rights

In his inaugural speech the Prime Minister made a few specific observations. One was that there was need to demarcate fishing rights at sea as on land. The Australian Government is also trying to do the same within its marine territorial waters but has not been able to make any headway. Traditionally, all over the world, nationals of any country have the inherent right to conduct fishing freely within their nation's marine territorial waters, subject to whatever conservation measures the State imposes. Same is the case in India, except that the Government has no legal powers at present to impose any restrictions on Indian nationals. However, since fishermen normally conduct fishing in waters adjacent to the coast on which their villages stand, they get a feeling of having specific fishing rights in the respective areas. Where others enter this area and undertake fishing, clashes occur. Such situations are inter-regional in nature. In distinction to this, there are also intra-regional conflicts. Non-mechanised boat operators resent fishing by mechanised boat operators in the traditional zone of fishing, which belongs to fishermen operating mechanised boats as well, on the ground that mechanised boats catch more of fish to their detriment.

The Prime Minister's observation in respect of bestowing fishing rights at sea must obviously have been made to avoid these conflicts. Rights on ownership on land is related to individuals mostly single or a family group. Conferring of such rights over sea waters is not the same as on land, while this is possible in backwaters and estuaries for stake nets as is in vogue in some parts of the country. Firstly, limits of holdings of each cannot be delineated easily in practice on water. Even when delineated, in situations of disputes, it will become very difficult to establish the encroachment. This problem can probably be got over by conferring rights of fishing on all fishermen in a village in specifically defined waters adjacent to that village. But again, this may lead to disputes between villages and establishing encroachments in such cases also will be difficult.

It appears to us that the subject matter is not amenable for regulation through conferring of property rights. A feasible solution can only be found in the long run through a developmental process, not one of conferring benefits by way of doles but one of enabling them to earn more through augmentation of fishing skills and means of fishing. Younger generations in fisher families are gradually moving towards owning mechanised boats. Owing to this and other social pressures the intensity of the problem is bound to get minimised over a period of time. Government should encourage the upgradation of skills on a family to family basis.

Fisher children should be weaned from the traditional way of thinking at an young age and put through a process of training over a period of 8-10 years to equip them to become Captains, engineers and skilled deck hands. Training centres for the purpose should be set up. If this is done real progress can be seen in 8-10 years. The present efforts, as we feel, are a waste.

Fishermen go for fishing to such spots where fish is there. Artificial legal restrictions on fishing over vast open waters of the sea, which are not enforceable, will only increase litigations and lead to wastage of the hard earned money of fishermen. Therefore we suggest that Central and State Governments should work on two lines a) bring about upgradation in the working life of each of the fisher families by enabling them to own mechanised boats and b)providing training to the younger lot in the operations as suggested in the previous paragraph.

## Exploitation by Middlemen. Is it True?

Another comment made by the Prime Minister related to the stranglehold of middlemen over the fishermen, and the need to form fishermen's co-operatives. This appears to us a trite comment and not in line with the new outlook of the Prime Minister. Perhaps he has not yet taken fishermen and fishing industry seriously. Officials usually use this arrow as a defence to cover up inabilities to introduce a nationwide integrated system of providing credit facilities to fishermen to own improved boats and nets and providing preservation, processing and marketing facilities. !n fact, the conditions of fishermen would be worse, if the so-called middlemen are not there in this situation. With no other facility, fishermen will be in a vaccum and plummet down. Fishermen prefer to have dealings with these so-called middlemen instead of co-operatives for the reason that co-operatives rarely provide credit on time and rarely pay in time for the catches sold to them. Solution to the problem of extending credit to fishermen, purchase of fishermen's catches will lie only in having a nation-wide fish marketing grid as a part of an integrated system covering various stages from production to marketing. This is a subject which cannot be tackled piece-meal and that too in certain pockets only. The Central Fisheries Corporation failed for this reason only. The so-called middlemen did a far better job than the co-operatives and they continue to survive. The irony is that the producers were happier with the services rendered by middlemen and this continues to be so. Government dabbles with the situation in such a way that fishermen secretly make fun of the approach, and either try to take advantage if they can get some money or ignore the overtures. They are the ones who are actually in the profession and their satisfaction is obviously important. Very few of the fishermen are unhappy with the men who help them in getting credit and buying their catches. In fact, if one makes a deep study, fishermen are much more astute than the middlemen. Wisdom therefore lies in either not disturbing the traditional pattern or in introducing an integrated programme to be operated all over the country. The politicians and officials, so to say, try to be more catholic than the pope himself, without the requisite fervour. Fishermen do not want a wasteful effort on the part of a co-operative that disturbs their working life. And politicians and officials disturb them by saying that they are exploited by middlemen which, by and large, fishermen do not experience, and it is our feeling that, when the good samaritans meet fishermen, they (fishermen) tell several stories of bondage, only to get rid of the men as quickly as possible. Even assuming that fishermen are exploited by the so-called middlemen, the solution does not lie in the direction of co-operatives, but a nation-wide well knit marketing system supported by financial assistance to fishermen. It is the politician who tries to exploit them not the men who help them in providing credit and assist in marketing.

The Prime Minister also observed that an attempt would be made to improve the status of individual fishermen and to give to the industrial set-up in the fishing sector some healthy competition. He felt that this way both the industry and fishermen would prosper. But what we feel is that fishermen can prosper only when the fishing industry prospers. Raising the status of individual fishermen does not give any competition to the fishing industry. On the other hand, it helps in the growth of the industry. Trained fishermen, all said and done, form the back bone of the fishing industry and also part of it. It is not good to create competition between the men who are part of the industry and the industry as a whole. The need is to absorb active fishermen into the industrial set-up and make them a part of it and encourage them to have their own enterprises.

Any person who has the ability to fish or knows the art of fishing or allied activities is a fisherman. There is an erroneous thinking that only those who are fishermen by caste are fishermen. This line of thinking is dangerous and Government should not encourage this, for, all trained and competent skippers, second hands, engineers of fishing vessels and engine drivers of fishing vessels would at once become non-fishermen. All these are far better fishermen than traditional fishermen.

The social and economical conditions of fishermen may improve if they are classified as a 'Scheduled Tribe' as recommended at the seminar, for, they will receive educational and other benefits. But this will not help in any way in promoting the national fishing effort. The recommendations, if implemented, will wean away the fishermen from fishing and they would make a beeline to take up other jobs. The move does not help in the promotion of fishing but the development of a particular caste. From other angles the step may have merit but not from the point of view of development of exploited fishery wealth. The fishing villages need connecting roads to the nearest main roadheads, sanitation, electricity, power etc. The provision of these is the responsibility of the State Governments concerned and provision of these should not be confused with the development of fishing effort.

## Traditional Fishermen are Not the Only Fishermen

Let us not get further confused between caste-based fisherman and a fisherman, to whichever community he may belong, with skills to run larger and more efficient fishing vessels and operate diversified fishing gear to catch and land larger quantities of fish to feed the teeming millions of our soil. The objective should be to upgrade the skills of traditional fishermen and assist them in acquiring improved craft and tackle so that they can come closer to those who already have superior skills at fishing. Let us not work towards bringing a rift between the traditional and non-traditional fishermen. We should work towards bridging the gap and make their activities complementary to each other.

Fishing Chimes, Editorial 60: February 1986: Vol. 5, No. 11

# Target without Adequate Funds

A repetitive but unfulfilled feature in the successive plans from the fourth plan onwards had been the fixation of target to reach a level of introduction of 350 deep sea fishing vessels, by the end of the respective plans. Various factors conspired against the realisation of the target and by the end of the sixth plan the number of vessels introduced remained at a level of about 80 nos., inclusive of the non-functional ones. What had been happening all along was that the funds earmarked under the successive plan schemes for the purpose were either surrendered or revised to lower levels of provisions because of poor performance in the utilisation of funds.

The Finance Ministry and the Planning Commission must have got accustomed to the above pattern and the past perfunctory way of implementing the programme. So much so, one can guess, the Planning Commission earmarked an outlay of Rs.50 crores for being given as loans by SDFC to fishing companies for the acquisition of deep sea fishing vessels for the entire seventh plan, being under the spell of the past impression that even this amount would not be utilised.

The number of vessels to be introduced during the seventh plan, to reach the level of 350 nos. will be well over 250 nos. Of this, the provision of Rs.50 crores will take care of around 60 vessels, leaving a gap of 180 nos. to be financed. In other words, the provision needs to be increased by about Rs.180 crores to Rs.230 crores.

For the Government of India to allocate a mere Rs.230 crores for financing the introduction of deep sea fishing vessels, which are vital for increasing fish production ought not to be a difficult job. It is a compulsive task as well. Firstly, this investment will be earned back by way of foreign exchange in about three years from the time of operation of the vessels concerned. Secondly, the investment, being in the way of loan, will come back to the Government in 15 annual instalments with interest. Added to these are certain other important national benefits. Besides contributing to increase in exports and foreign exchange earnings, the presence of 350 nos. of larger fishing vessels will act as an indirect protection to the national interests relating to our exclusive economic zone. If the captains of the vessels are trained, they can undertake surviellance and report presence of foreign fishing vessels and other vessels indulging in anti-national activities to the coast guard.One can agree that the Planning Commission is justified in providing an outlay of Rs.50 crores considered from the angle of past performance. The Planning Commission obviously did not take into account the change in the past style of functioning in the Union Department of Agriculture in respect of increasing the deep sea fishing fleet. There is now application of mind in the department in a positive direction. The present administrative set up in the department seems to feel that, if there is a target, it should be fulfilled. In fact, it is stated, the top officers feel that vessels upto a level of 500 nos. shall have to be introduced to exploit available resources to a perceptible level, depending on the performance of the industry.

As matters stand, the same target *is* being brought forward from plan to plan. While there had been stagnation till now in the realisation of the target, now an enormous demand for the acquisition of fishing vessels has built up and this has lent support to activities relating to the realisation of the plan objectives. In other words, there are receptive implementing agencies (fishing companies), a progressive administrative set-up, a financing agency (SDFC) and also popular backing. Politicians who know the pulse of the people are lending support. Because of the conjunction of these aspects, several applications for imports of vessels and loans for acquiring imported and indigenous vessels have been cleared and for the first time in the history of this sector, the entire outlay for the 7th plan (Rs.50 crores) now stands fully committed. As matters stand, ground has been prepared for raising the number of vessels to a level of over 180 nos. This performance calls for increasing the outlay for granting loans for the acquisition of deep sea fishing vessels to an extent of Rs.230 crores. Unlike the shipping sector, repayments from fishing companies currently operating fishing vessels, acquired with SDFC loans, are stated to be well over 85 per cent. This is an encouraging feature which should convince the Planning Commission and the finance ministry to provide the required funds.

Earlier, in order to explain away the poor level of introduction of fishing vessels of Indian flag, the Ministry of Agriculture resorted to a kind of break-up, giving the numbers of vessels to be imported, indigenously constructed, to be operated under charter, and under joint ventures. Those operated under charter cannot be deemed to form part of the Indian fleet and will not count towards the target. Those under joint venture also would need financial assistance for acquisition. It is therefore necessary to raise the outlay for providing loans for the acquisition of deep sea fishing vessels from Rs.50 crores to 230 crores. This will be sufficient to lend financial assistance for the introduction of vessels to reach the level of 350 nos.

If there are problems in raising the outlay, to be utilised for granting loans either through SDFC or any other media, supplementary alternatives have to be thought of. One is that fishing companies may be freely permitted to import vessels by raising foreign loans. This will however mean retardation of indigenous construction. There are two problems connected with raising of foreign loans. One is that the margin money to be paid is 20 per cent of the cost of a vessel. This will be too big a burden for most of the Indian companies. Another problem is that the foreign banks want a guarantee for the loan amount from an Indian Bank or SDFC.

To get over these problems, the Government and SDFC can consider granting a loan to the extent of 90 per cent of 20 per cent of the cost of the vessels, and the SDFC may extend a guarantee for the loan amount of 80 per cent of the cost which the companies can raise from foreign banks.

So far as the guarantee which may be given by the SDFC in respect of the loan component to the foreign bank is concerned, the regulations governing the same will be fulfilled by Indian companies, subject to such relaxations as may be necessary, which may have to be granted. We understand the problems connected with financing for the acquisition of deep sea fishing vessels are receiving the top-most attention of the higher authorities in the Ministry of Agriculture. There is optimism in fisheries industrial circles that a lasting and workable solution to the problem will soon be found and implemented.

Fishing Chimes, Editorial 61: March 1986: Vol. 5, No. 12

# Potential of Closed Fish Farming System

Closed water circulation systems for fish farming are gaining ground in the past few years. The advantages of a closed system are several. This system cuts down the quantity of water needed, in terms of weight of fish produced per unit space, production of over 100 tonnes/ha can be obtained from closed systems annually, compared to around 4 tonnes that can be achieved on an average through intensive farming in open systems. And this *is* possible with the utilisation of a far less quantity of water and stocking with around 1.5 lakhs of fish seed per ha companed to about 10,000 nos. ha in open systems. Heavy feeding has to be however resorted to for achieving the high rate of production in a closed system, which may push up costs, but the returns will be equally high, going by the present day price trends. The remnants of heavy feeding and the resulting biological wastes cause no damage because of the water circulation which removes all wastes. Mazharuddin Javeed has contributed a popular article on the subject which is published in this issue. This should prove to be of interest to the readers. We expect that soon Javeed or another will write on the economics of closed farming systems.

The closed system of fish farming has an industrial complexion and is well suited for the progressive entrepreneurship in the country, on the look out for avenues of this kind. Open systems of farming, with all their advantages have also several drawbacks. Pilferage is the most disadvantageous of all. All the hardwork put in by a farmer goes to a naught, because of pilferages which are common all over the country. The other drawback is the vulnerability of the fishes towards diseases. In a closed system, both these drawbacks can be controlled. Further, the fear of pollution will be minimal in a closed system. Inputs and outputs can be monitored accurately in a closed system.

Closed circulation system is no doubt a supplemental means to augment fish production. This is because our available open lentic water, large in number and with a vast combined extent of over 7 lakhs ha, have to be brought under fish production and cannot be allowed to remain fallow. This technology cannot easily be adopted by all fish farmers, because it calls for greater investments, compared to open water systems, which will continue to be the mainstay and the produce from these may well lead to a competition as prices of fish grown in an open system can possibly be cheaper. It may also turn out, to the amazement of all, that the fish grown in closed systems will be cheaper because of the higher rate of production which may counteract the higher costs.

Promotion of closed systems deserves attention in the hands of the Government. The Central Inland Fisheries Research Institute can take up a pilot project on a commercial scale and demonstrate the results. The economics of operation, if favourable, will undoubtedly attract the attention of progressive fish farmers and other investors. They would then think of having integrated farms consisting of a hatchery, and closed systems for producing seed and producing fish. It is quite possible that the Central Inland Fisheries Research Institute has already done considerable work in this direction. If so, it will be useful if the results are made known to the entrepreneurs. After the pilot work, the Institute could make available project outlines to those interested in the activity so that they may approach commercial banks for financial support.

The Central Institute of Fisheries Education is reported to have already made considerable headway in pond aeration technology. Through this technology the Institute is reported have achieved very high rates of fish production. It is also stated that the Institute has taken up experiments on closed systems of farming through aeration and removal of metabolites. It is hoped that this Institute would initiate work on closed system of farming on a commercial scale. More important than this is to organise training courses in this technology once it is proved to be economically viable, for the benefit of private entrepreneurship.

If closed circulation system of farming is proven to be viable and the technology is standardised, it can be said that a new vista in inland fish farming has been opened up. The beneficial effects of this will be far-reaching. There will be a multiplying effect which will lead to increase in useful employment and in fish production. Incomes will go up which will contribute to national prosperity.

We hope that the Union Department of Agriculture, and the Indian Council of Agricultural Research will give the importance that this programme deserves and encourage their subordinate organisations to give the needed priority to this work.

Fishing Chimes, Editorial 62: April(i) 1986: Vol. 6, No. 1

# Fishing Boundaries

By tradition and under the Law of Sea, marine fishing in India is an activity that does not require a licence. Certain categories of fishing vessels however require registration under the Merchant Shipping Act and rules framed under MPEDA Act. Indian fishing vessels are also subject to such restrictions as may be imposed in the State fishery enactments. There is no central enactment at present to regulate sea fishing.

There are fishery enactments in several States such as Tamil Nadu, Kerala, Maharashtra and Goa. These basically delineate areas of operations for non-mechanised and larger fishing vessels. The enactments do not place any restriction on fishing vessels from other States fishing in the waters of the particular State. This is so, for the reason that under Article 301 of the Constitution of India no fetters can be placed on inter-state flow of business and commerce. Under this Article, trade, commerce and intercourse throughout the territory of India shall be free. Article 302, however, empowers parliament to impose such restrictions on freedom of trade, commerce or intercourse between one State and another in public interest. At the same time as per Article 303, neither Parliament nor the legislature of the State shall have power to make any law giving or authorising the giving of any preference to one state over another. When an Article requiring the issue of licences to fishermen outside the State of Gujarat was made in the Gujarat Fisheries Bill, this was not agreed to by the Union Ministry of Law on the ground that it offends the provision in the Constitution of India referred to above.. Obviously, for the same reason no restriction on fishing by fishermen from other States is placed in any of the State Fishery Legislations.

In the above background it is surprising that, in the Orissa Assembly, on 6th March, totally ignorant of the Constitutional position, the Minister of State for Fisheries, Ms.Frido Topno said that three Andhra trawlers at Balasore and few others at Puri were released, after the owners furnished undertaking for not conducting fishing in Orissa waters again. On the oppositions charge of releasing the trawlers and persons involved, the Minister maintained that this was done as it was their first offence. For the concerned boats this may have been the 'first offence'. The position however is that Andhra boats have been fishing off Orissa coast, mostly on invitation from fish merchants of Orissa right from 1960s.

The fact of the matter is that the boats did not commit any offence whatsoever. They are fully entitled to conduct fishing in Orissa waters. The statement of the Minister of State for Fisheries, Orissa is therefore totally incorrect and misleading. At present vessels from Tamil Nadu fish in Andhra and Kerala waters. Karnataka vessels fish in Kerala waters. Gujarat and Maharashtra vessels fish in each others waters. What the Orissa Fisheries Minister said was therefore not correct both in letter and practice. If the visiting vessels commit a crime it is a different matter.

It is now a matter of utmost urgency that the Governments of Andhra Pradesh and Orissa examine the matter with reference to provisions in the Constitution and lend such protection as is necessary to Andhra fishermen who have to move to the unexploited or sparsely exploited waters of neighbouring States for livelihood.

The problem is not restricted to Andhra Pradesh and Orissa alone. This is a common Inter-State problem. Therefore, it will be helpful to the various State Governments, if the Ministry of Agriculture clarifies the position to the various maritime States.

Fishing Chimes, Editorial 63: April(ii) 1986: Vol. 6, No. 1

# Closed Season for Shrimp

The majority of trawler owners on the one hand and skippers, engineers and engine drivers of fishing vessels on the other developed a sudden interest in voluntarily instituting a closed season for catching of shrimps from the waters of the upper east coast. The Deep Sea Technocrats Association, headed by Capt. N. Venkaeswarlu is given credit for intensifying follow-up action on this idea. It has to be said that the acceptance of this idea, that too on a voluntary basis, is a major achievement and is a tribute to the foresightedness of the owners as well as the technocrat operators. However, whether this voluntary effort would prove to be a success has to be seen. We say this because, a few vessels taken delivery of by three companies were given time for a period of three months by the SDFC to make up the mortgage value to a level of 120 per cent of the loan amount The concerned companies cannot fulfill this obligation if they stop fishing for 60 days. They may therefore continue fishing. An alternative however is for them to ask SDFC to give more time because of this situation.

It appears that this awakening burgeoned, for some strange reason, after the incident of shooting at the trawler YSF 101 of Yamuna Sea Foods by the Coast Guard recently. This incident seems to have brought about unity among the floating staff as well as the owners and this development has now been made use of for a good cause which is to achieve a closed season.

Governments either at the Centre or in the majority of the States have no legal power to declare a closed sea fishing session. In this situation the trawler owners and the floating staff who have adopted a code of conduct to protect brood shrimp deserve to be congratulated. Both the sides have agreed to have a closed season from 15 April to 15 June 1986.

What has been done is undoubtedly a good interim measure. We should however know whether there is need for such a closed season, and if so, for what duration and during which part of the year. A very careful thinking with reference to available data has to precede any enforcing of a closed season on an enduring basis.

In favour of the need to have a closed season are certain facts such as the indiscriminate exploitation of a large number of juveniles during certain months of the year in the sea and also in the Chilka lake, particularly in the channel connecting the lake with the sea, and the catching of brood shrimps, again in certain months of the year. The other side of the picture is that shrimps happen to be an annual crop. A large number of eggs are released by each female shrimp and this crustacean dies a natural death after a life of around 12 months. If the shrimp is not caught within these 12 months they are bound to be lost.

Biologists tell us that brood shrimps should not be caught for the reason that they have the potential to increase shrimp population. Biologists are also against catching of juveniles as they have the potential to grow. It can also be said that those in mid-age also are not to be caught as they would grow into brood shrimp. The question that arises from this situation is at what stage should shrimp be caught? The answer can be that shrimp should not be caught at any of its stages. This is however not a practical approach. The sensible way is to exclude juveniles from the harvest and catch all others in quantities that would give maximum sustainable catch. The index of this is to check the total landings and size parameters from time to time. If there is a decline in the total catches or a consistent fall in the lengths and weights of shrimp caught, there is a case for resorting to stoppage of fishing for juveniles during the months in which these are available in large quantities.

So far as we could see, no one has been able to establish any decline in shrimp landings in the upper-east coast. There are surmises that the sizes at which shrimp are caught during certain months of the year are coming down. This is certainly true, particularly, in late May and June.

The members of the fisheries associations held a meeting with the scientists of Central Marine Fisheries Research Institute's Centre at Waltair recently. The scientists are reported to have told the members that shrimps breed almost throughout the year in some area or the other, with a periodicity of duration of 2 months. Being an annual crop, they felt that there would be no need for rigorous conservation measures. Their view is stated to be that fishing for shrimp could be stopped for any two months in a year. The mesh sizes of the codends could also be increased to a certain extent so that juveniles could escape. They also advise that catching shrimp in shallow waters should be stopped.

Having taken heed of the recommendations of the scientists, it is learnt that the owners and floating staff have expressed preference to declaration of closed season for shrimp for any two months instead of an arbitrary declaration of closed season by the industry itself. Let the Government advise and the industry follow the same, on a voluntary basis, without the need for legal regulations to be enforced by the Government. Mechanised fishing boats could also adopt the same norm of observing a closed season from 15 April to 15 June for the time being and later follow such advice as may be given by the Ministry of Agriculture.

The closed season will be a great opportunity for foreign fishing vessels to take advantage of the lull. During this period, Coast Guard should intensify patrolling.

Fishing Chimes, Editorial 64: May(i) 1986: Vol. 6, No. 2

# Fishing Vessels Financing: Setback and Solution

Fishes are exposed to the danger of being caught at any time, although such a danger is much less for fishes in Indian waters. The deep sea fishing industry is exposed in the same way to repetitive dangers that materialise in respect of choice of indigenous or imported fishing vessels to be acquired and of financing arrangements thereof. There can be smooth running in respect of this choice for sometime, but one can expect a stumbling block round the corner impeding progess.

A lot of confusion and conflicts exist among the various agencies concerned. By 1984, a system emerged under which, for every one vessel allowed to be imported, the buyer had to place orders for one indigenous vessel. In early 1986, the fishing companies found that the few indigenous shipyards interested in constructing fishing vessels had signed contracts for such a large number of vessels each that, barring one or two yards, others were not in a position to deliver the vessels for quite sometime (in any case) before the time limit for claiming investment allowance comes to a close (April 1987). Further, the shipyards are equipped to build only one type of vessels. The situation was that, based on 1:1 formula, even if loans were available, the likelihood of reaching a level of introduction of 350 vessels by the end of the Seventh Plan was remote.

Sensing the situation, the Government seems to have shown an inclination to increase the proportion of imported vessels i.e for every four or five vessels imported, buyers have to place orders for one indigenous vessel. Atleast this was the understanding of the indigenous shipyards and the Shipyards Association of India strongly protested against the policy. This must have made the Ministry of Agriculture to seek the approval of the highest possible body to their proposal. It is quite possible that by the time these notes are printed, Government would have taken a decision that would protect the interests of the Indian shipyards, without sacrificing the need to increase the strength of the Indian deep sea fishing fleet to 350 nos., atleast by the end of the seventh plan. Till a policy decision is taken, one of the causes for the present slowdown in clearances will persist.

A reason for the Government to think in terms of liberalising imports of fishing vessels seems to be the heavy drain on funds by way of subsidy to Indian shipyards. Subsidy amounts could be saved if imports are liberalised, but there will be obvious repurcussions. One is the slowing down in the flow of orders to indigenous yards in the long run (for the time being all the yards seem to be busy). This is not in national interest. However, if bulk imports are organised, the foreign suppliers can be made to provide shopfloor drawings free of cost and Government can supply them to indigenous yards. Construction can pick up if financing arrangements are there and both builders and buyers agree on whatever Government's policy is there at the time concerning subsidies.

Another point relates to outflow of foreign exchange, if imports are intensified. As the President of SAI observed, if Indian yards are allowed to import whatever is required freely off the shelf, they can build vessels of excellent quality at prices comparable to those of imported vessels. If the yards have spare capacity and have the drawings, this approach is preferable to import of fully built vessels. Whichever system is followed, the money spent on imports would be earned back by the buyers in foreign exchange in a short duration of five years or even less. Therefore, the best approach seems to be that the Indian shipyards deliver the first batch of vessels now under order and let the Government and the industry see how good the vessels are. If the vessels are good, this gives confidence and orders for the further construction of vessels at indigenous yards, who should equip themselves with drawings of various types of vessels, would flow in automatically.

It is wrong to think that Indian fishing companies have any fascination for imported vessels. Imports are disadvantageous to Indian fishing companies in vital respects. One is that they have to pay 10 per cent of the cost of vessels (provided SDFC would continue to be the financing agency) as margin money, whereas they have to pay only 5 per cent of the cost as margin money for indigenous vessels. Further, the rate of interest on loan for imported vessels is more. Moreover, if the dollar gets stronger, the Indian party has not only to invest more in terms of rupees but has also to involve itself in various complicated procedures. All factors point out to the beneficial nature of indigenous construction, provided that the shipyards and the fishing companies play their part rightly.

So far as financing of deep sea fishing vessels is concerned, certain problems started showing up in this established financing system followed by SDFC. At a time when the Ministry of Agriculture, the fishing industry, the shipyards and the fishing companies have found level, procedures took shape and matters started moving smoothly, SDFC's loaning system was frozen. No plausible reason could be ascribed to this by the industry, as recoveries by the SDFC from the fishing companies are stated to be of the order of 85 per cent. Because of this unforeseen development, several companies who received letters of sanction of loan from the SDFC could secure sanctions but those awaiting receipt of sanction letters are in a state of considerable exasperation and worry. The least the Government could do is to release funds for those to whom sanction letters are issued, and the issue of sanction letters for the remaining are expedited, to be followed up by firm financing arrangements for the acquisition of deep sea fishing vessels by the industry.

There has to be some agency to provide loans to the fishing industry. SDFC had been doing well in this respect. Unless Government is unable to provide funds to the SDFC, there is no reason why the system should be suddenly discontinued. Whether it is SDFC or someone else, money will have to be found to be lent. So, why not allow SDFC to continue with the job? There may be complex issues that are not visible to the myopics in the industry, but they may be of vital importance to the Government which led to the decision to divest SDFC of the function. (There are however indications that the arrangement may be extended for sometime, which is indeed a silver lining), If money is the problem, Government can without much difficulty negotiate foreign credit on long term repayment basis with such of the countries which have either a good stock of used vessels (but not old) or are in a position to construct, and deliver them. This loan arrangement can be tied up with the SDFC, who will recover the amounts in instalments from the loanee fishing companies and pay back to the foreign countries concerned as per mutual terms and conditions as agreed to. We are told that enormous opportunities exist for this line of action.

NABARD has funds. Government can give a directive to NABARD to support schemes for the introduction of deep sea fishing vessels, following the same terms and conditions as imposed by the SDFC. In fact, NABARD can be a second window for the purpose.

Finally, our appeal to the Government is to announce firm arrangements for financing introduction of deep sea fishing vessels with reference to various aspects of imports and indigenous construction and implement the same without major changes during the entire seventh plan period.

Fishing Chimes, Editorial 65: May(ii) 1986: Vol. 6, No. 2

# Upkeep and Utilisation of Apprehended Foreign Fishing Vessels

Several fishing vessels which were found fishing within the EEZ had been apprehended by our Coast Guard off East as well as the West coasts of India and taken to the various ports for further action. The Coast Guard officials filed cases in the concerned courts. Legal proceedings which followed took a considerable time, atleast two years. In the beginning all such fishing vessels were being kept under the custody of the Coast Guard. As their number grew, however, they experienced difficulties because of shortage of manpower and certified officials to man them. Consequently, they requested the Ministry of Agriculture who are concerned with fishing to take custody of such confiscated vessels even before the proceedings in the court were completed. The Ministry of Agriculture, in turn, directed the Fishery Survey of India to take custody of such vessels from the Coast Guard without perhaps realising the limitations in respect of availability of certificated hands. It is learnt that neither the Coast Guard nor FSI has the man power to keep such vessels in their custody. Moreover, the work involved in manning the vessels requires certificated hands and also funds to meet expenditure towards port dues, fresh water, fuel etc.

The sequel in regard to the seized vessels was therefore to keep them completely idle at anchor without putting them to use. To cite an example, the four pairs of trawlers (8 nos) which were apprehended in July 1984 off Quilon were brought to Bombay port by the Coast Guard authorities and till now (April 1986) the legal proceedings are not yet over. Retention of vessels at anchor for so long, particularly the old vessels, make them practically unserviceable and as such may have to be disposed of as scrap. In this connection, it may be mentioned that many of these vessels can be put to gainful use. The following measures are suggested in this connection which deserve consideration by Coast Guard/Ministry of Agriculture.

1. First of all, the Ministry of Agriculture should organise a separate wing to take custody of such trawlers with certificated hands and see that the different machineries and equipments are kept in trim condition for which sufficient budgetary provision is to be made so that all port charges, apart from cost of fuel, fresh water, minor repairs, etc can be paid.

2. This wing should also have a legal expert, who should take up the matter with the courts in such a way that legal proceedings are expedited.

3. In order to expedite the legal proceedings it would be appropriate for the Coast Guard and Ministry of Agriculture to take up the matter with the Ministry of Law to have a separate Tribunal to deal with such cases for quicker disposal.

Needless to say that the quicker disposal of legal proceedings will enable the Government to reduce expenditure for the upkeep of such vessels which can then be disposed off to any of the interested fishing corporations or companies who are trying to get vessels either on charter or under the scheme for the import of fishing vessels. If such vessels are properly kept, there is no doubt that many buyers would be interested to have such vessels because of their proven design, fishing efficiency and endurance. Most of these vessels have more or less all their machineries and equipmerits manufactured from one and the same source. This would help in getting spares through some authorised agencies.

This approach would help to increase the number of Indian fishing vessels to a certain extent. This will lead to a greater coverage of the Indian EEZ. A thought in this direction will be worthwhile.

Fishing Chimes, Editorial 66: June 1986: Vol. 6, No. 3

# Blue Print for the Development of Fish Culture

Fish culture has been making progress in the country. This is taking place at a faster pace in the sector of freshwater fish culture, than in the sphere of brackishwater farming.

The setting up of Fish Farmers Development Agencies has no doubt provided a mechanism to popularise fish culture through the creation of a class of fish farmers and coordinating the functions of various concerned Governmental agencies and Commerical Banks. More than this, the agencies are in a position to tell exactly the extent of water area developed by them for fish culture and the quantity of fish produced.

One aspect is that in the same districts where the agencies exist, fish culture goes on outside the purview of the agencies too. This is no doubt something good. However, not all the details of this activity are known either to the agency concerned or to the unit of the State Fisheries Department having jurisdiction over the area, The effect of this situation is that the authorities do not have an adequate and clear idea of the total fish production achieved from culture activities. The removal of this gap is of vital importance in the larger context of development of inland fist culture in the country. Beside this, there should be a clear idea of the areas that can one day or the other be converted into fish ponds, b) there should be a system under which Information will come to the Government as and when agricultural or other lands are utilised for pond excavation and eventual culture of fish in such ponds so excavated is taken up. Once an enforceable system is found to achieve all the above aspects the State Governments can follow them. The result will then facilitate a precise understanding of the developmental stages from time to time, so necessary for planning further lines of development.

A prelude for the development of an enforceable system is to have an annual census covering all villages and urban areas in all the States to collect all required information on water areas brought under culture, with reference to categories such as those owned, taken on lease etc., and the names and addresses of the farmers concerned. Similar information can also be collected in respect of fish seed farms and hatcheries.

The first annual census of this kind should concentrate on collection of information on vacant areas suitable for eventual utilisation for fish culture and the subsequent annual exercises. It can be checked up to what extent these areas are utilised for fish culture and the balance that is available for similer utilisation in the coming years. Census of the type mentioned above by deploying all available departmental staff for a duration of 10 to 15 days to undertake the work in such areas allotted to them will be very useful.

At present practically no State fisheries department in the country is in a position to state the exact area available for fish culture, area brought under fish production, number of seed farms and hatcheries set up and their capacities etc. This is undoubtedly an embarrassing and somewhat shameful situation and this can be altered for beneficial use by following the line of action suggested above.

While various incentives are offered for the development of brackishwater farming, initiatives in this respect are left to the entrepreneurs. What is needed is a systematic approach in this respect by the various coastal State Governments and Union Territories. This approach should consist of: (*a*) preparation of an inventory of available area for the development of brackishwater farming, (*b*) dividing the same into viable plots, (*c*) calling for applications from entrepreneurs for allotment of plots with reference to sound policy guidelines, (*d*) introducing supportive measures relating to extension management, provision of finance, etc. for the effective implementation of the scheme. The work should be organised in such a way that the Government concerned should know where it stands in respect of this development at a given time.

It is high time that the State Governments give up the perfunctory and piecemeal way of dealing with the developmental aspects in culture fisheries sector and evolve a blue print under which the activity should be developed, within the precincts of well formulated guidelines, in specified stages. Such a line of action will contribute to national income, foreign exchange earnings, production of additional fish food for supply to the people and provide new means of employment.

Fishing Chimes, Editorial 67: July 1986: Vol. 6, No. 4

# Storm before Calm

Storm signals indicating the impending destabilisation of the well established arrangements for the financing of deep sea fishing vessels first emanated from 'Kailash' and 'Krishi Bhavan' in November, 1985. There was a warning pause soon after, saying that the Shipping Development Fund Committee would continue to extend financial assistance until December, 1985. This unexplained part of the extension for a short duration was a real riddle to the deep sea fishing industry, It may be true that Government never gave a categorical assurance that funds by way of loans and subsidies for acquiring deep sea fishing vessels would definitely be forthcoming from the SDFC. Yet there was an unwritten commitment that financial support for the purpose from SDFC would be there. Without such a promise, Government would not certainly set a target level of introduction of 350 nos. of deep sea fishing vessels to be reached by the end of the Seventh Plan. No capital intensive target of this kind, particularly with the exclusion of larger houses from the financing facility of SDFC, could be set, with middle and small scale entrepreneurs as the media for promoting the activity. There would have to be a firm system of financing.

The entrepreneurs waited with bated breath to know the decision of the Government in regard to the financing arrangement for the acquisition of deep sea fishing vessels after December, 1985 Then came the news that the arrangement was extended upto the end of March, 1986. Several sanctions were issued by the SDFC upto the end of this month. Funds were released to the extent available and later the activity came to a grinding halt. Thereafter came the news that Government decided to close down SDFC.

The commercial activities of fishing vessels have little in common with those of cargo ships. Fishing vessels catch fish and sell. Cargo ships carry cargo. Yet, by definition, fishing vessels are ships. Several shipowners who took loans from SDFC became defaulters, owing over Rs.500 crores to SDFC as overdues, according to newspaper reports. Unable to reconcile itself with such a heavy pendency, the Finance Ministry seems to have taken the decision to suspend or close down further lending by SDFC. Fishing vessels being ships, this decision became applicable to them as well, in spite of the fact that recoveries of loans given to owners of fishing vessels are stated to be of the order of 85 per cent.

A couple of years back SDFC did not want to handle the subject of giving loans, for the acquisition of fishing vessels. This led to a gap. The Ministry of Agriculture explored possibilities of having an alternate arrangement but finally came to an agreement with the SDFC to revive the earlier system. If the ministry had made an alternate arrangement at that time itself instead of going back to the SDFC, the present impasse, which is mainly because of the defaults of shipping companies, would not have become applicable to fishing vessels.

Granting of loans by Governmental bodies for the acquisition of fishing vessels is an accepted system all over the world. The Indian Government has a commitment to raise the national deep sea fishing fleet to 350 nos. by the end of the seventh plan. None of the entrepreneurs have the means to invest heavy amounts on their own to acquire fishing vessels. Further, there are pending commitments with the SDFC in various stages for granting loans based on their issued sanctions or approvals. It is learnt that the ministry considers that its first responsibility is to ensure that all these commitments are fulfilled.

Here it has to be mentioned that the Planning Commission seems to have erred in approving a level of 350 nos. of deep sea fishing vessels to be achieved by the end of the seventh plan, while at the same time making a provision of Rs.40 crores only for being given as loans, knowing fully well that middle and small scale entrepreneurs alone have been allowed to come forward to take to deep sea fishing and financial assistance to them is inescapable. This was probably done based on the past performance of the Ministry of Agriculture in issuing approvals and apparently the later change in the conservative set-up in the Ministry was not taken into account. Apart from this, it is known that, unless the national fleet increases, the ingress of foreign fleets cannot be controlled. The national importance of the sector is well recognised. Despite this, it is a matter of great surprise that targets and the financial provisions have not been related. Without a provision of at least Rs.250 crores for being given as loans, there is no way of stepping up the national Deep sea fishing fleet's strength to 350 nos.

We are now in the midst of a storm. Despite this, we are certain that the storm will soon lead to a calm and a proper mechanism for providing loans for the acquisition of fishing vessels will be evolved. It is a matter of great satisfaction to the deep sea fishing industry that the Union Department of Agriculture and the concerned Central Government Departments are actively engaged in solving the problem. News reports say that ICICI has been identified as the agency for providing loans for the acquisition of ships. Since fishing vessels are also ships it is felt that the facility will cover these also. Almost every day, ft is learnt, a meeting is held in the Government to resolve the issue with various apex financing institutions and selected commercial banks participating in some of the meetings, and the industry can be confident that what all is possible will be done by the Government. Those at the helm of affairs are experts in the field and it can be believed that no stone will be left unturned and no avenue will be left unexplored to find a solution.

Several foreign banks are willing to extend loans for the acquisition of fishing vessels by Indian fishing companies, provided that a Prime Indian Bank or the Government give a guarantee. An Indian Bank would agree to give a guarantee only when a collateral security is provided in addition to a substantial fixed deposit. The guarantee amount being high, about Rs.1 crore per vessel, the Ministry of Agriculture alone can help in this matter. They can prevail upon the Department of Banking to issue a directive to the nationalised banks to provide the needed guarantees on condition of hypothecation of the vessels after their arrival in India. This is what Is now being done in respect of SDFC loans.

After the operations commence, it could be a condition that the fishing companies keep the concerned banks informed of the voyage details so that the banks can keep a watch over.the earnings, out of which a part can be taken as fixed deposit, and another part towards repayment of loan, from voyage to voyage. The banks can also appoint an officer of theirs to be stationed at the concerned company's premises to keep a close watch over the operations and make recoveries. The salary and other expenditure on the officer, while met by the bank, could be recovered from the company. This system will ensure recoveries on time.

The pressure on the Government to organise funds for being given as loans can be reduced considerably by introducing this device of extending non-funding guarantees, and by not prescribing any limits on the foreign loans which will be repaid in instalments out of foreign exchange earnings through exports. EXIM Bank can provide the requisite back-to-back guarantee to the commercial bank concerned which will extend the needed guarantee to the foreign bank. It is possible that the Ministry of Agriculture is already looking into an alternative of this type and it is hoped that the storm that is now raging will blow over very soon, and an enduring and progressive system will emerge.

Fishing Chimes, Editorial 68: August 1986: Vol. 6, No. 5

# Languishing Living Resources of Indian EEZ

The living resources of the Indian Exclusive Economic Zone continue to languish. Both the Indian Government and the Indian deep sea fishing industry are keen on exploiting these uptapped fishery resources. There was some confidence until the close of 1985, that the progressive introduction of deep sea fishing vessels would reach a level of 350 nos. by the end of the Seventh Plan would be accomplished, in spite of the fact that convoluted and time consuming procedures were evolved for clearing the applications for issuing letters of intent for import and for providing loan assistance from the SDFC. The final link *i.e.*, the system of loaning by the SDFC, having been disrupted, the entire process came to a grinding halt in 1986. All applications for acquiring deep sea fishing vessels now gravitate towards 100 per cent export scheme. These applicants expect that they can secure 100 per cent deferred payment loan for financing the vessels from foreign sources supported by a guarantee from an Indian Bank and with the needed clearance from the Department of Economic Affairs. While the present situation is discouraging, it however keeps three groups happy. One of these consists of successful poachers. The second group comprises some of those who are now operating deep sea vessels and who would be happy if there are no new entrants. The third group consists of the fishes themselves which are allowed to live longer and are either doomed to die a natural death or preyed upon.

The present state of affairs, all artificially propped up, are not something that a Government can be proud of, particularly when the Government announces often in the parliament that the commercial deep sea fleet will be stepped up to 350 nos. by the end of the seventh plan. Such announcements obviously warrant the backing of an efficient working system but this is lacking. Yet assertions on this point continue to be made.

One has to concede that bad situations that come in the way of developmental activities do develop. SDFC was kept in suspended animation because of heavy overdues from shipping companies and the fishing industry was put to suffering along with it. Added to this was the position that there was no available budgetary provision for providing loans for financing the acquisition of deep sea fishing vessels. For adding over 250 vessels to the fleet (to reach a level of 350), a provision of Rs.40 crores was made for the entire Seventh Plan. This had already been utilised, with another three years of the plan to go by. Obviously the planning commission did not take the EEZ, its unexploited living resources, and the imperative need to exploit them seriously. If the intention is to go slow, because of financial constraint, a target of adding just 40 nos. of deep sea fishing vessels to the present fleet could have been made. Keeping the target high arid making a ridiculously low provision is something that denigrates the capabilities, the seriousness and the solemnity of a Government.

The industry felt that under the leadership of Mr.Dhillon and Mr.Makwana the deep sea fishing industry would make giant strides. But this is not to be. Work on setting right the situation and keeping it back on rail has been going on since December 1985 with no solution in sight. We do not think that the matter is so intractable that it defies a solution. The inordinate delay in evolving remedial measures is contributing to the slippage of valuable time.

If money is the problem, and Government could not allocate funds owing to constraints, one way is to allow fishing companies to borrow 100 per cent of the cost of the vessels proposed for import by them from foreign banks and extend Governmental help in providing guarantees to the foreign banks concerned, taking the needed safeguards with the exception of the dreaded collateral and sizeable margin money.

Another effective way is to set up a Central Fishing Vessel Import Organisation. The fishing companies holding letters of intent for importing vessels can be asked to approach this organisation to arrange foreign loans and supporting guarantees. This organisation could have a field set-up to supervise the activities of the fishing vessels concerned and to ensure recoveries. The cost of the engaging personnel employed may be recovered on a pro-rata basis from the fishing companies concerned.

Government can also allow purchase of second hand vessels on hire-purchase basis, with an arrangement for the payment of the entire cost to the owners out of earnings from sale of fish through a marketing tie-up and in specified instalments. If this is allowed, hundreds of vessels can be introduced, provided the MMD exempts these proven vessels from having certain standards as prescribed by them.

So far as indigenous construction is concerned Government should encourage Indian shipyards to have collaborations with foreign yards with arrangements for supply of different designs of vessels and transfer of technology. Imports of machineries, equipments, steel plates etc., should be allowed but under credit terms, to be repaid by the buyers out of earnings, to the foreign banks concerned extending the credit. There should be an arrangement for securing 100 per cent foreign credit by the buyers supported by the required guarantee to the foreign bank from a Governmental organisation/ Nationalised bank. Through skillful negotiations and package deals with Foreign Governments and banks, the present problem can undoubtedly be solved, in case Government is unable to extend and organise loan facilities to the Indian fishing companies out of its own funds.

There has been an inordinate timelag. Further delay, apart from enriching poachers' pockets, will result in economic ruin of the budding entrepreneurs who have entered the arena with high hopes, based on Government's declarations on the subject.

Fishing Chimes, Editorial 69: September 1986: Vol. 6, No. 6

# Guarantees are the Need Now

Readers are aware of the swift shift in the complexion of the activities in respect of introduction of deep sea fishing vessels.

The tempo concerning the introduction of deep sea fishing vessels gained momentum because of the charter scheme in which a condition was imposed by the Government that the charterers should introduce vessels of their own, equal in numbers to the chartered vessels, both imported and indigenous, following 1:1 ratio. The Shipping Development Fund Committee gave a real thrust to the activity, by providing loans for the acquisition. There was optimism in the fishing industry as well as in Government circles that the longheld dream of introducing a large number of deep sea fishing vessels to facilitate the exploitation of the rich fishery resources of our EEZ, estimated at 4.5 million tonnes, would soon materialise if the trend had continued.

Circumstances always conspire to create problems and bottlenecks, particularly when a way is cleared for achieving a certain target. The Finance Ministry decided to stop granting of any further loans by the SDFC for the acquisition of ships. Fishing vessels being also ships by definition, the facility of granting of loans for fishing vessels also was withdrawn. The only saving grace was that in all cases where loans had been approved or sanctioned by the Shipping Development Fund Committee, the ban would not be applicable.

This is something the main question of alternative arrangements for the financing of deep sea fishing vessels is stated to have been also solved. Government is reported to have decided to set up a Shipping Credit and Investment Corporation of India. The proposed corporation will be controlled by the Ministry of Finance. Its management is likely to be undertaken by the Industrial Credit and Investment Corporation of India. A Working Group was set up to work out the details for the implementation of the decision. It is conjectured that SDFC would be closed down by 15 Oct 1986 and all the balance of work with it pertaining to the existing loans would be transferred to SCICI. In the meanwhile Rs.25 crores are stated to have been earmarked for being given to SDFC to take care of the current commitments. However, at the time of this writing these funds had not yet reached SDFC.

The consideration of fresh applications for the granting of loans for the acquisition of deep sea fishing vessels by the Shipping Development Fund Committee came to a stop towards the end of 1985. The reaction of the fishing enterprises to this situation has been to gravitate towards the 100 per cent export oriented scheme. There has thus been a major shift in the disposition of applications for permissions to import vessels. Most of these are now being made under 100 per cent export-oriented scheme. The main reason for this is the reported norm that foreign financial loans would be allowed by the department of economic affairs to meet the entire cost of the imported vessels.

Several entrepreneurs are reported to have contacted foreign banks and these foreign banks are reported to be willing to extend loans, repayable at a reasonable rate of interest in about 8 annual instalments. The hitch in the transaction is that the foreign banks require a guarantee from an Indian bank for the repayment of the loan amount in the instalments as agreed to, together with the interest on declining balances.

Each of the fishing vessels costs around Rs.1 crore in foreign exchange. The banks normally require securities to cover the guarantees, besides a cash deposit to cover a part of the commitment, although nonfunding in nature. If a fishing company that is given a guarantee does not pay the instalments in time, the bank will have to pay the amount and deal with the nonpayment separately with the owners. In order to cover this contingency a guaranteeing bank requires a security.

The banks have a fear that the fishing companies may not pay the instalments in time. They may even completely avoid payments. This fear they have developed based on the behaviour of the owners of certain small mechanised boats. The operational configuration of the small mechanised boats is totally different from the deep sea fishing vessels which are much larger in size and are registered under the Merchant Shipping Act. Their operations are governed by a set of stringent regulations which impose a serious discipline on the owners and the concerned crew. This makes things difficult for the owners to play hide and seek with the bank. Further, since the vessels will be imported under the 100 per cent export-oriented scheme, letters of credit for the exports would have to be opened by importers through the bank of the owners which will obviously happen to be the guaranteeing bank. This will enable the bank concerned to deduct proportionate amounts towards instalments from the remittances specialised features of the fishing industry would require a specialised mechanism to monitor the recovery of the loans. In fact it should be possible for the Government to set up a separate organisation for issuing guarantees towards the loans obtained by fishing companies for the acquisition of fishing vessels from foreign banks. This organisation could build up a share capital base by taking shares from the various commercial banks and from the public. This guaranteeing organisation should be recognised by the Government of India. If the guarantees issued by this body are not acceptable to a foreign bank, it can give a guarantee to an Indian prime bank which can provide the guarantee needed to the foreign bank. The proposed SCICI may be structured only to provide loans. The separate organisation suggested above should be entrusted with the task of providing guarantees, to financing bodies and even to SCICI for the loans granted by it.

The proposed body should have a separate wing with expertise to take over defaulting fishing vessels and operate them on its own. As matters stand, the real weakness from which the Indian financing institutions now suffer is the absence of such a set-up. This could be solved in the above mentioned manner. This system coupled with the flow of funds as commissions for guarantees etc. will make the operations reasonably safe for the proposed organisations.

If a separate body cannot be set up for providing guarantees, SCICI should have a wing for taking over and operating defaulting vessels.

## Norwegian Example

In fact, in Norway there is a body owned through sharecapital contributions by several banks to perform the guaranteeing function. This organisation is known as FISKERIKREDITT. This has been recognised by the Government of Norway and it provides long term guarantees on loans issued by the Fisheries Bank to fishing companies. Apart from the banks, several companies and individuals have invested in this company. This bank takes only first charge over the vessels. If the fishing company fails to repay any of the instalments on time it takes over the vessels and operates them on its own. The body has the needed expertise in this respect. This safety device is stated to be the secret of the success of this body. It is learnt that this guaranteeing body also lends money upto 50 per cent of the cost and provides guarantees for the balances, which may be obtained by the owners from commercial banks. A similar body based on participation by the public and participation by commercial banks and various companies, if set up in India, would certainly be able to take care of the guarantees to commercial banks that provide financial assistance for the acquisition of deep sea fishing vessels.

Fishing Chimes, Editorial 70: October(i) 1986: Vol. 6, No. 7

# Hitech Aquaculture as in the West: Implications of Introduction in India

It is common knowledge that aquaculture, in some form or the other, is one of the earliest of human avocations. It is being practised in China since the past 3000 years. The first text book in fish culture is reported to have been written in China by Fanly in 475 B.C. There are indications that fish culture is much more ancient, as revealed by the earthen ponds located along the margins of an area which was once a lake in Poland. The age of these structures is estimated at 7,000 years.

## Aquaculture in the East

Aquaculture in SouthEast and Far East Asian countries developed on certain lines in the past two decades. The accent has been on the development of intensive systems of aquaculture in tanks and ponds. In Japan and several other East Asian countries integrated farming combining fish culture with poultry, cattle, and pig farming has become popular. In Taiwan, Indonesia, Philippines etc. brackishwater farming, mostly in respect of shrimp, has been developed based on high technology. Coupled with the development of hitech culture systems, mostly in tanks and ponds, production of fish seed in hatcheries both in respect of fish as well as prawn, have also reached a high stage of development, particularly in Taiwan. Certain countries like Japan have taken to sea ranching through the medium of cages. Small sized cages for fish culture in canals etc., have also become popular, mostly as a family enterprise, in countries such as Thailand and Japan.

In India, while high technology in respect of fish culture in tanks and ponds is now gaining ground, the production per ha of water area continues to remain at levels far less than one tonne per ha, compared to far higher levels of production in similar waters achieved in several other Asian countries, where the per ha production on an average is reported to exceed two tonnes. The extensive nature of cultivable resources in India, next only to China, and the gradual spread of improved technology have given the advantage to India of being the second in Inland fish production among all the countries of the world. This is no doubt commendable, although some may say this is at the best a consolation.

One aspect in the Indian inland fish culture sector is that the annual fish production from culture resources cannot be related to the volume of annual fish seed production for reasons not clearly known.

There has been a lopsided development in the country in respect of fish and fish seed production in certain States such as West Bengal. In this State, there is an abundance of production of fish seed, while in other States the production is inadequate to meet the stocking needs. The progress in respect of setting up of fish hatcheries in the various States is also very slow, although a begining has been made in the States of Uttar Pradesh, Madhya Pradesh, Bihar, West Bengal and Orissa under the World Bank's programme.

## Cage Culture in the West

In several West European countries an unprecedented revolution in inland fish production through the utilisation of cage culture systems has taken place. The fish produced mostly consist of trout and salmon.

Trout can be cultured only in waters in certain parts of the country. These are located in Kashmir, Himachal, Uttar Pradesh, Arunachal, Nagaland and Tamil Nadu. Salmon does not occur in our waters. It cannot also thrive in Indian waters. While these are the constraints, the same principles of producing trout and salmon in cage systems in the west can be adopted by promoting cage systems in our waters, with the needed adjustments, for the production of larger quantities of fish.

In West European countries rows of cages are fixed permanently in such water areas as permitted by the respective Governments. With cages on either side, there are floating pathways in between, over which those working at the cage farms can walk or even take mechanically propelled trolleys. The cage systems can consist of several cages, but most of them have around 20 cages varying in diameter from 15 to 40 metres. The depth of the cage goes up to 50 feet. The feeding of fishes and the reading of the various physico-chemical parameters are done through a computerised system. The cages are intensively stocked with advanced fingerlings/yearlings. The feeding is done at very short intervals automatically through conical plastic jars poised above the water sheet of each of the cages. One can have the most adoring and translucent sight of plumpy and well grown fishes meandering about in bursting health in the cage waters at various levels. The rate of production from each of the cages is reported to be atleast ten times more than what is obtained per hectare from earthen ponds under intensive system of culture.

## Ship Model Cages

Recently 'ship model' cage systems have been introduced in a few West European countries. Old ships are anchored at a selected place in a water area. In these ships the deck and the sides of the vessels are suitably opened up by cutting the plates. Ballast is provided all around and on this, platforms for movement are improvised. Two or more cages are inserted in the open hollows created by the cutting out of the deck plating. Advanced fingerlings/yearlings are farmed in these cages.

There is no doubt that the hightech cage culture systems as followed in Western Europe and Japan are highly efficient. However, the investments involved in this activity are high. The farmers in these countries are able to invest heavily in these systems for the reason that, in spite of the heavy investments involved they are able to get profit yielding returns. Each kg of trout/salmon fetches equivalent of over Rs.80/-

## Hitech Cages Need to be Introduced in India

In countries like India, similar cage systems can be introduced with such modifications as are needed. When such systems are put to use, it is likely that in the initial period of introduction, the operations may or may not prove to be encouraging, One factor that will contribute to the success of the introduction is that unlike trout and salmon, Indian major carps grow very fast. This will give the advantage of a thicker turnover, yielding better results. Another aspect is the systems that are in vogue in the west will have to be suitably modified to suit Indian conditions. One point in this connection is that the depth of the cage will have to be suitably reduced, as a depth of 50 feet in our inland waters is not very common. In sea ranching, however, the depth can be 30 feet. Cage materials used in the west are very tough and the possibilities of the meshes giving way are very remote, most of the cages are fabricated out of knotless netting. It is not yet known whether these materials will withstand crab bites. This has to be tested. It may become necessary to provide some kind of chafing gear or a metal grill around the cages.

To sum up, there is an imperative need to undertake experiments on cage culture in a serious manner in India, with proper planning going behind it. CIFRI can probably undertake this job, with technical assistance from a country like Norway in the west and Japan in the east. If the experimentation yields good results, the way will be paved for spreading fish culture activities in large sheets of water in a faster manner.

Fishing Chimes, Editorial 71: October(ii) 1986: Vol. 6, No. 7

# Erection of Barriers by Govt in the Devt of DSF, as Open Jokers (Fishery Entrepreneurs) Gape Aghast

The Government of India has undone the ongoing working systems in vogue for the development of the deep sea fishing industry, only to re-do the same for no apparent advantage. This backswimming exercise is something unique in the history of marine fishery development in the country. Total disruption in the financing arrangement for the acquisition of deep sea fishing vessels, although stated to be temporary, was carried out in a purposelessly relentless manner. Indigenous deep sea fishing vessels building industry which showed signs of revival in the past few years, has been led on to a path of slow death. The provisions in the Maritime Zones Act concerning chartering of foreign fishing vessels have been made non-operational, a rare instance oh the part of the Government to keep the regulations under the Act, framed for national benefit, in a deplorable state of dormancy.

## Hanging Halfway off the Hook

A well established system for financing the acquisition of deep sea fishing vessels has been closed with relentlessness, only to introduce another system, which will take atleast 180 days front the date of closure of SDFC (Oct. 25, 1986 ?) to find its feet on the ground.

## SCICI to Come Up

A new body, a consortium of ICICI, IDBI and IFCI, who are hardcore financiers, is expected to be established to manage the proposed Shipping Credit and Investment Corporation of India, which will function under the Ministry of Finance. The anatomy of this proposed body, its vascular system and the stretch of its hands are not yet known. One expects that the Government could share some of these details with the industry, so as not to keep its constituents guessing and wilting, but no such sharing has taken place so far. Would it not be reassuring to the fishing industry if the Government tells the industry the reasons for the undoing and redoing the financing arrangements? Out of sheer disgust, in the present situation, it is possible that the entrepreneurship may loose their present interest in deep sea fishing. This may or may not be good for the country. What pains one is that something that has been assiduously built over years is being demolished, only to build a new edifice. This transformation and the entailing delays do not obviously cause any serious concern to the Government, and this is to the delight of those that are now entrenched in the deep sea fishing industry and the poachers who will have the fishing resources ail for themselves for an unspecified period. The shutting off the present financing system has left the deep sea fishing industry hanging half way off the hook.

## Plight of Indigenous Fishing Vessel Building Yards

The Minister for Agriculture, according to reports, expressed disappointment on 24 September 1986 at a press conference in Delhi at the slow progress in indigenous fishing vessel construction. He would not have expressed this disappointment if he had been informed of the ordeals to which the indigenous yards are subjected to. Apart from enormous delays in examining proposals for recommending the granting of loans to fishing companies by the SDFC from the angle of designs and specifications, prices and in the actual sanctioning and release of loans, the yards have been made to wait for extraordinarily long periods for getting licences to import various machinaries and equipments. The yards were told that they could import machinaries and equipments equivalent to 20 per cent of the cost of a vessel without payment of duty. This was no help to them as they could very well import these items without payment of duty as these are for fitment in sea going fishing vessels which come under the classification of ships, and imported machinaries and equipments for ships are exempted from payment of customs duty. With this ostensibly helpful concession, the authorities took a longer time than usual for granting import licences. The Minister has only to ask for a blow-by-blow account of the successive dates on which each of the stages were crossed from the Government's side. It is a wonder that the indigenous yards could manage to deliver five trawlers in a duration of about four years, in spite of the hurdles.

## Faults of Indigenous Yards

What we have said above is not by way of defending the indigenous shipyards in any way. Some of the yards were given successive instalments but they did not take interest in progressing with the building work. It is any one's guess what was done with all the money released by the SDFC during the intervening periods of reported inactivity. In any case the closure of the financing window, even for a short duration, has caused major damage to the basic structure for the development of deep sea fishing and a great loss of time, which could have, been utilised for hastening the introduction of vessels.

## The Death of the Charter Scheme

The third 'kill' was the charter scheme, which came into being under the provisions enshrined in the Maritime Zones Act. The Minister for Agriculture is reported to have stated on 24 September, 1986 in Delhi that Government no longer favoured charters. At the same time, at the meeting of the Central Board of Fisheries at Trivandrum on 26 September, 1986, the Minister is reported to have said that Indian deep sea fishing fleet would be built up through acquisition, charter and joint ventures. While it is difficult to know what exactly is the Government's policy in this respect, from the trends and from what the Minister said in Delhi, it is clear that Government is unable to favour charters. One reason for this seems to be that the activity is not conducive to our way of administration. In other words, Government probably finds it difficult to run the scheme further, for reasons best known to it. Transgressions cannot be taken as the reason for his approach. Wherever there is law there are law breakers. There are several other sectors which Government administers where regulations are transgressed, atleast now and then. Government is not winding up those sectors for this reason. In this background, the reasons for disfavouring charters are inexplicable. Developed countries such as Australia, New Zealand in the east and several countries in the west allow fishing by foreign vessels in their waters. What they could administer well, could also be administered well by us.

We would not like to deal at length on the merits and demerits of the charter scheme. Suffice it to say that it is probably an uncharitible allegation from the Government side at the press conference on 24 September 1986 that the chartered vessels transfer fish catches illicitly on high seas. We do not know whether Government has any concrete evidence of such transfers. Such activities are operationally hazardous. We have not heard of any such activity by larger fishing vessels so far. All said and done, the chartered vessels increased our fish exports and foreign exchange earnings. The operational results of these vessels told us for the first time about our own fishery wealth. Money in foreign exchange has been earned out of fish caught by the chartered vessels. If these chartered vessels were not operational, the fish they had been catching would have been lost for ever through natural death and poaching.

## Joint Ventures

The present thinking of the Government seems to be that joint ventures between Indian and foreign fishing companies should play a major role in deep sea fisheries development. If joint ventures were to be easy, under the existing regulations itself many joint ventures would have come up in India by now. We do not expect any major changes in the new policy that is expected to be announced soon, in respect of conditions governing joint ventures. There are several intractable problems that are bound to come in the way of implementation of joint ventures. These are related to equity participation by the foreign party, and acquisition of vessels by the joint venture company either through charter or purchase. Our Government seems to be of the view that the chartered vessels should fly Indian flag. This may not be possible as chartered vessels will be of foreign registry. Only when sold to an Indian company, the vessels can be registered in India, thereby giving the entitlement to them to fly Indian flag. Most of the foreign vessels suitable for operation in Indian waters do not satisfy the regulations of our M.M.D. and these will not therefore be eligible for registration in India unless exempted in certain respects by the D.G. Shipping. Apart from this, the formulation of a system for the payment of the cost of the vessels to the foreign partner who will supply the vessels by the joint venture company will bristle with several hurdles.

## Open Jokers

In this gloomy picture, the entrepreneurs are now like open jokers, exposed to ignominy in the overall industrial set-up of the country, victimised from all sides and made vulnerable for being used as a pawn by the Government in a manner it feels the entrepreneurs can be fitted in. The situation may make some of the entrepreneurs more spiritual in outlook and leave the line. They may feel that fishery wealth in our seas has been remaining sparsely exploited for centuries and we do not lose any more, if the same state of affairs continue for some more time. Several in authority feel that with around 90 larger vessels operating in our waters, our shrimp and fish resources are overexploited. May the Lord bless them. At the same time let us hope that no one will put an idea in the mind of the Government that fish should be declared as part of 'wild life'.

Government tells us of its seriousness in raising the national deep sea fishing fleet to a level of 300-500 nos. by the end of the seventh plan. At the same time the strategy adopted by the Government for the purpose is one of losing valuable time for long spells, through creation of needless problems and working hard thereafter on solving them. It is hoped that those in authority will realise the harm that is being done to this important sector of national endeavour through frequent policy and strategy changes and the consequential delays, with the result that the living resources of our EEZ continue to be an open invitation for foreign poachers. Further, avenues for increasing fish production and fish exports and the earnings of foreign exchange thereby are being jammed. It has to be noted that by investing around Rs.One crore in foreign exchange per vessel, not only this amount is recovered in foreign exchange in two years through exports, but also, whatever is earned thereafter will become a substantial addition to the national foreign exchange earnings. We are confident that the authorities concerned are already looking at the situation in the right perspective and would remove all the hurdles in the way of promoting Indian deep sea fishing in the fastest possible time.

Fishing Chimes, Editorial 72: November(i) 1986: Vol. 6, No. 8

# Strategy to Achieve Diversification in Deep Sea Fishing

The deep sea fishing industry is very keen that its constituent companies should diversify their fishing activities by diluting the present concentration on catching shrimp. The real problem faced by them in this respect is related to the economics of operation.

Diversification of fishing effort, as the Government is aware, is related to the availability of more of fishery resources of commercial value and the existence of markets for the sale of the diversified catches.

As matters stand, shrimp stocks on the upper east coast are the only known commercially viable resources for trawling by larger vessels, beyond the area of operation of small mechanised fishing boats. For this reason the few large sized fishing vessels of Indian flag are all fishing in the area with Visakhapatnam as the base.

The industry is aware that there are now proven resources of tuna on the east as well as the west coast, available for a period of about 7 months (Nov-May). The entrepreneurs can certainly take the risk of diversifying into catching tuna during this period. The problem that will however remain in this regard is that the vessels will have to be idle for the remaining 5 months. In the same way, vessels can concentrate on catching squid and cuttlefish for about four months a year in certain areas where the availability of these stocks is not however as yet fully proven. In this approach too, the same problem as mentioned above *i.e.* keeping the vessels idle for a certain duration will become inevitable. It is these aspects that keep the deep sea fishing entrepreneurs in a dilemma in regard to diversification of fishing effort.

It will be unrealistic to expect all entrepreneurs to introduce vessels equipped exclusively for the catching of varieties other than shrimp. It is also not good for the nation in the sense that this may lead to total ruin of several fishery enterprises. The transformation has to materialise in stages and a suitable strategy in this regard has to be formulated.

In this connection, it is imperative for the Government to adopt a strategy that would allow the introduction of combination vessels which can undertake all types of trawling and also equipped for one new type of fishing out of methods such as tuna long lining, pole and line operation, purse seining, squid jigging etc. Once the entrepreneurs gain experience in the additional fields of fishing and start getting encouraging returns, they will gradually reduce their emphasis on shrimping. In any case dilution of concentration on shrimping, which will result from the above mentioned approach will by itself be a good gain from the standpoint of the present policy of the Government to discourage shrimping.

Considering the above aspects, it is strongly recommended that the strategy of encouraging entrepreneurs to take to the introduction of combination vessels with facilities for two types of fishing may be adopted. One of the systems has invariably to be trawling (various types). The other system will be for catching of tuna, squid/cuttle fish etc. This will be a realistic approach and will be a far better one, than adopting the strategy of clearing vessels described as nonshrimpers but would actually undertake shrimping.

In the end, it has to be mentioned that the way to diversification in deep sea fishing lies through shrimp, keeping shrimp catching system as one of the methods on a combination vessel. Any other route taken to achieve diversification will not only result in loss of time but also in disappointing results.

## Delay in the Announcement of New Deep Sea Fishing Policy

The Deep Sea Fishing Industry has been gathering from the press reports that the Government would soon be announcing a new deep sea fishing policy.

Pending this announcement, Government have withheld the issue of new permissions for chartering of fishing vessels and joint ventures. Because of the disruption in financial arrangements, there has also been a major setback to the introduction of new deep sea fishing vessels.

The intention of the Government to introduce a new policy has been on the cards for the past 3 to 4 months. So far this has not been announced. This situation created a great vacuum in the progress towards the development of deep sea fishing industry. It will be agreed that the harm that has already been done to the industry in the past few years will intensify further if the announcement of the new policy is further delayed. It is accordingly suggested that Government may finalise the new policy in the shortest possible time and announce the same for being followed by the fishing industry.

Fishing Chimes, Editorial 73: November(ii) 1986: Vol. 6, No. 8

# Popularisation of Chinese Type of Hatcheries

We all applaud and are happy at the impressive achievements in the field of major carp and exotic carp spawn/fry production in West Bengal. Representatives from the various States gather periodically at National Fish Seed Congresses held in Calcutta and the single major theme of the congresses has been to praise the State Department of Fisheries, West Bengal and the fish farmers of West Bengal for their achievements. The routine has been that the representatives attending the congresses are taken to a farm, where spawn is produced. They all feel well impressed after seeing the demonstrations. After this and participation in the final session in Calcutta, the delegates go back to their States. With this the story ends. There is very little follow up after this.

If the present imbalance in the country in respect of fish seed production is to be corrected, the fish farmers all over the country are to be motivated to produce larger, or additional quantities of fish seed by setting up mini-bundhs, and Chinese type of hatcheries, both permanent as well as the portable ones. This is necessary to meet the present demand and also to create new demands.

The Ministry of Agriculture alone can induce this motivation. It is not that the Directors of Fisheries in the various States or the State Governments cannot do this. Not all the State Governments, unfortunately, give the needed weight to the proposals of the Directors of Fisheries, mostly because of the additional expenditure involved.

Fish farmers can be motivated to set up spawning pools, overhead tanks, mini-bundhs and circular hatcheries, only through subsidies. And granting of subsidies means money. Without this motivation, fish farmers will not be anxious to increase spawn/fry production/set up new hatcheries for two reasons. One is that they can get spawn from West Bengal, grow to fry stage and sell. This is easier than making fresh investments. Another is that they can be sure of offtake if they do not increase their seed production. Further, the West Bengal farmers would like to sell their spawn to other States, rather than witnessing a situation in which this demand dries up. They therefore offer spawn at competitive prices as an inducement. This indirectly counteracts the interest among farmers in other States in setting up hatcheries of their own.

In the above situation, the Ministry of Agriculture will have to expand the scope of Fish Farmers' Development Agencies by providing attractive subsidies to private fish farmers for the modernisation of facilities at the existing seed farms for production of spawn through the setting up of hatchery complexes, and also encouraging the new entrants to take up fish farming activity to set up seed farms with hatchery complexes.

The success of the hatcheries set up under the World Bank Project in the States of West Bengal, Orissa, Bihar, U.P. and M.P. is not yet visible. Once success in this respect is established farmers can be assisted to set up similar hatchery farms, although of smaller extents. The option whether to set tip World Bank type of jar hatcheries, or the latest West Bengal type of hatchery complexes involving spawning pools, mini-bundhs and circular hatcheries could be left to the farmers. The need, however, is to encourage selfsufficiency of seed in each of the States.

Coupled with this, there should be a relentless drive for bringing more and more of tanks and ponds under fish culture, particularly in West Bengal, so as to create an outlet for the additional quantities of fish seed produced. Cage culture can also be popularised so as to achieve quick results. In this context, work has to be done in regard to standardisation of cage materials, their construction, dimensional aspects etc and entrepreneurs should be encouraged to take to the manufacture of cages. CIFRI will have to undertake extension work to teach the farmers the techniques of setting up of cages, and about stocking and feeding schedules, harvesting methods etc. Administrative regulations concerning allotment of water areas for setting up cages have also to be worked out.

It is time that those in authority in the Ministry of Agriculture and in the State Governments think of the various aspects of giving fillip to a wellconceived programme for the purpose of augmenting fish seed production in the various States to a level where they can be self-sufficient from time to time, in step with the growth of the fish culture sector.

Fishing Chimes, Editorial 74: December(i) 1986: Vol. 6, No. 9

# Conflicting Strategies

The Union Department of Agriculture has taken a stand that further permissions for the import of deep sea fishing vessels equipped with outriggers should not be given. The reason for this seems to rest on the assumption that outriggers are meant for catching of shrimp and the ministry does not want any addition of fishing effort for shrimp by larger fishing vessels. There can be an increase of any number of mechanised boats for catching shrimp. But there should be no additional effort through larger fishing vessels for catching shrimp. For some strange reason, restriction on outriggers is forced on the buyers not realising the fact that outriggers are not designed for operating nets for catching shrimp but to increase the sweep area. Outriggers are operated in countries such as Holland for catching fishes such as flat fish. It is also not realised that only a certain percentage of catches either from lateral (outrigger) trawls or bottom single stern trawls, not more than 5 per cent of the total catch, will be shrimp. Above all, there is no justification to think in terms of controlling the deep sea fishing fleet *vis-a-vis* shrimp landings, when in the areas where these vessels now operate, there has been an increase in the annual shrimp landings from year to year, with no reduction in sight, notwithstanding the increase in the number of vessels added to the fleet.

While the position in regard to this subject is as outlined above the fishing industry will have to respect the decision of the Government in the matter. Strange as it may seem, while the Union Department of Agriculture has adopted this approach to restrict permission for the introduction of deep sea fishing vessels equipped with outriggers, the State Minister for Agriculture, Mr. Makwana is reported to have told the Lok Sabha on 24 November that there has been an increase in the shrimp landings along Andhra Pradesh coast from 8,887 tonnes in 1984 - 85 to 10,506 tonnes in 1985 - 86. Giving reply to a question from Mr. Balagoud, MP, the Union Minister of State for Agriculture Mr. Yogendra Makwana further said that, according to the information from the State Government, there was no decline in shrimp catch during 1986-87 as well.

Replying to another question as to whether the Government had monitored the extent of the fall in shrimp landings from July 1986 as compared to the corresponding period last year, the Minister replied that no fall had been indicated. In the light of the above clarification one really wonders at the rationale behind the approach of the Union Department of Agriculture in restricting permissions for the introduction of deep sea fishing vessels with outriggers. In fact there is an inherent contradiction between what the Minister said and the policy adopted by the department, The story does not end with this contradiction. The State Minister for Commerce clearly mentioned at the previous Parliament Session that there was no fall in Indian shrimp catches. The Ministry of Commerce has been exhorting the fishing industry to step up exports of marine products. This will be possible only through a substantial increase in the export of shrimp and other exportable species.

We see here also a major contradiction between the approaches of the Ministry of Commerce and the Ministry of Agriculture. This confuses the fishing industry further. It will be desirable that the Government as a whole tell the industry in which direction it has to function, so far as catching of shrimp and other exportable species is concerned.

It may be mentioned here that, according to the available reports, the subcentre of the Central Marine Fisheries Research Institute located at Visakhapatnam has conducted a detailed study of the shrimp situation along the upper east coast recently. The results of the study are understood to have revealed that there is no fall in the shrimp catches, either in terms of the gross production or in terms of per unit catch. On the other hand the production is on the increase. Another conclusion that emerged, as understood, is that there is no reduction in the size composition of the catch also. If Government disregards this scientific conclusion and persists in the stand that no more permissions for fishing vessels equipped with outriggers should be given, the only conclusion that can be drawn is that the decision is not related to a scientific conclusion. The fact that shrimp has a maximum life of about 12 months and its bionomics are totally different from fin fishes and that the shrimp populations are usually restored to normal levels year after year because of the annual cycle, should never have made the Government to think in terms of imposing the restriction, particularly when hardly 86 deep sea fishing vessels are operational along the entire coast line. We do hope that the Ministry of Agriculture will review the position and adopt a rational approach on the subject.

Fishing Chimes, Editorial 75: December(ii) 1986: Vol 6, No. 9

# Imported Components in Indigenously Built Deep Sea Fishing Vessels

The deep sea fishing vessels have to be equipped with machineries and equipments that will work non-stop all through a fishing voyage. The breakdown of any of the machineries and equipments in a fishing vessel will create several problems. Firstly the fishing operations would come to a stop. Secondly the catches stored till the time of failure of machinery may soon get spoiled. These two offshoots will have an adverse effect on the earnings. Apart from these, the captain has to be frantically on the look out for help for towing the vessel back to port.

All the above contingencies can be eliminated to a large extent by equipping the vessels with tested and proven machineries and equipments. Such tested and proven ones, however are available from other countries and therefore will have to be imported.

The installation of indigenous machineries and equipments on the fishing vessels constructed by the Indian shipyards will no doubt provide encouragement to the indigenous machinery manufacturing industries. At the same time, if these machineries are not proven the owners will be at a serious disadvantage. As matters stand, there are very few machineries and equipments for installation on deep sea fishing vessels that are proven to be of adequate quality and meet the requirements. Some machineries may be of good quality, but they occupy a lot of space in the engine room. It is not good for fishing vessels where space is very valuable.

For the above reasons most of the buyers prefer to instal imported machineries and equipments that are proven beyond doubt. In this context, a condition is imposed by the Government that machineries and equipments costing not more than 20 per cent of the total value of a fishing vessel can be imported without payment of duty. This restriction is not liked by the fishing companies. Government permits import of fishing vessels as such with full exemption from the payment of customs duty. Such being the case there is no justification whatsoever to fetter the hands of the fishing companies as well as the Indian shipyards, preventing them to import all the needed machineries and equipments. Common sense should tell any one that it is always desirable to permit import of machineries and equipments for indigenous fishing vessels rather than import of fully built vessels without payment of customs duty.

In any case, this restriction that allows not more than 20 per cent of machineries and equipments for import in the installation of indigenous constructed vessels is a major constraint in the development of Indian deep see fishing industry and Indian deep sea fishing vessel construction industry. The restriction retards the indigenous construction of deep sea fishing vessels. It also has a very serious effect on fish production and export of marine products on which the Government lays so much of emphasis.

It will be wise on the part of the authorities concerned to review the situation and allow the Indian shipyards to Import all the machineries and equipments that are required for installation in deep sea fishing vessels as per the owner's requirements till such time as the deep sea fishing industry develops and the exports improve to the expected level.

Fishing Chimes, Editorial 76: February 1987: Vol. 6, No. 11

# New Deep Sea Fishing Policy: Rudderless and Without Focus

The Ministry of Agriculture brought to the notice of the public in the year 1987, eve a new policy in respect of the development of deep sea fishing industry. This new policy is aimed at encouraging Indian enterprises to invest in deep sea fishing and thereby step up fish supplies for internal consumption as well as exports. The policy covers the following aspects:

A. Charter of Foreign Fishing Vessels.

B. Liberalisation of *Pari passu* condition for the import of new vessels.

C. Import of second hand fishing vessels.

D. Permissible size and types of deep sea fishing vessels.

E. Joint ventures in deep sea fishing with foreign collaboration.

The earlier charter policy, which was discontinued about a year back, came under serious criticism by certain Members of Parliament. Apart from this, the Agriculture Ministry also felt that the charter programme was benefiting the foreign interests and that they were taking away all the fish caught in our waters. Mr.Rao Birendra Singh, the then Minister for Agriculture went to the extent of saying that entrepreneurs, who were getting 15 per cent of the gross proceeds out of sale of fish caught by chartered vessels, had been making easy money. The general consensus was that the charter programme should either be discontinued or tightened further.

In the above background it was expected that the new policy would tighten the charter terms and conditions further. Instead, it is seen that the new scheme has been designed to liberalise the terms governing charters further. A fishing company can now, when permitted under the new regulations, operate chartered vessels upto 10 numbers for a period of one year, earn a significant amount and withdraw from the scheme without fulfilling the obligation of introducing its own vessels, equivalent in number to those that are chartered. It may also avoid entering into a joint venture as required under the scheme, as another available alternative. Such a company's loss for adopting the above mentioned lines of action would be that it would lose the security deposit of Rs.3 lakhs. Crafty entrepreneurs would not mind foregoing the security deposit. The reason for this is that by operating the maximum permissible limit of 10 vessels for one year or even half of this number, they can well earn at least about 20 lakhs at Rs.2 lakhs per vessel. With the accrual of earning of this order an entrepreneur would not hesitate to forego the security deposit. He would no doubt put forth some explanation or the other for his inability to introduce the vessels as required or to enter into a joint venture. If these are accepted, he will be lucky as he will get back his security deposit.

Otherwise, he will not mind losing the deposit. In order to enable the Indian chartering company to get a higher share, the sharing of gross incomes between the Indian and foreign sides has been stepped up to 20:80 against the earlier ratio of 15:75. This is a greater incentive to forget about the security deposit, which is now set at Rs.3 lakhs per company. While the wording in respect of the security amount is ambiguous, it appears that this will be Rs.3 lakhs per company and not per vessel. It is not known whether this has to be given in cash or as a bank guarantee, as the notification is silent on this point. Whatever this be, a company which can succeed in chartering the optimum number of 10 vessels would make around Rs.20 lakhs. Assuming that the security deposit will not be returned in case of failure to fulfil the other obligation at the end of the first year of operation as required, the fishing company concerned will stand to lose only Rs.3 lakhs which will be all right for it.

The notification being thus an inviting one for unscrupulous companies, who intend to make a quick buck it is extremely doubtful whether the new charter policy will yield the anticipated result of introduction of equal number of vessels as permitted to be chartered by the company concerned either directly or under a joint venture programme.

The liberalisation of *pari passu* condition in respect of import of new vessels is not likely to tilt the balance in a more beneficial manner than the present position. Those who are keen on importing fishing vessels will continue to apply for permission in this respect, and such of those who have understood the advantages of having indigenous vessels will go in for such vessels. The replacement of the stipulation that for every one vessel imported one indigenous vessel should be introduced in the new liberalised policy, with the new stipulation that for every two imported vessels, one indigenous vessel can be added is not likely to improve the situation in any significant manner.

According to the new notification, introduction of specialised and resource-specific vessels, such as tuna long liners, purse-seiners, pole and line vessels, squid jiggers, gill netters and other vessels for exploitation of non-shrimp resources only will be allowed. The Department of Agriculture will monitor the size and type of deep sea fishing fleet based on resources sustenance from time to time. This means that even when the fishing companies apply for the import ot resource-specific vessels other than those for exploitation of nonshrimp resources an entrepreneur can never know whether his application would be accepted or not. An application for import of resource-specific vessels such as long liners etc., takes generally a long time for clearance by the Government, which extends to 12 months or even more. As these delays which are likely to intensify because of the monitoring there are bound to be problems with such of the shipyards which offer to supply the vessels at fixed prices, in respect of maintaining the same beyond a time limit.

The provision in the notification concerning joint ventures in deep sea fishing does not say anything new except that the intending parties should apply to the Ministry of Commerce or the Marine Products Export Development Authority in this respect. While it appears from the notification that the intending parties need not get into touch with the Ministry of Agriculture in respect of joint ventures, the likelihood is that there will be the inevitable need in the form of the technical clearances from the Ministry of Agriculture in respect of the vessels proposed to be imported under the joint ventures which will be part of the monitoring of the level of resources exploitation which that Ministry has to do. This would mean the passage of the application through the well-known Fishing Vessel Acquisition Committee in the Ministry of Agriculture and approval of the recommendation by this Committee by the Agriculture Minister. All this process entails considerable timelag, because the Ministry of Commerce/Industry will not clear any of the proposals unless the recommendation from the Department of Agriculture is there.

In order to achieve the target of reaching a level of introduction of 350 - 500 numbers of deep sea fishing vessels by the end of the Seventh Plan, there can be only one pragmatic move on the part of the Government. This is to announce a package of imports and indigenous construction of a certain number of the stated specialised and resource-specific vessels such as tuna long liners, purse-seiners, pole and line vessels, squid jiggers, gill netters and other vessels for exploitation of non-shrimp resources. (The list of the types of vessels in the notification does not include trawlers. Obviously it is a taboo to think of trawlers, to talk of trawlers, to hear of trawlers, or write about trawlers). Such a package deal will have to be coordinated by the Government, by selecting the sources of supply of vessels the number of vessels to be obtained from each of the sources, and link these to the number of various types of vessels for which applications will be received in response to a notification to be issued by the Government for acquiring deep sea fishing vessels. Thereafter a Governmental agency specifically created for this job will have to tie up the various components of the programme such as the suppliers, the buyers, the financial arrangements etc. The free-for-all system as is adopted now will not lead to the expected results in a developing country like India. Results may be achieved under this free-for-all system in a period of 13 years but not by the end of the Seventh Plan period. The entire programme for the development of deep sea fishing is rudderless and without focus. The provisions in the new notification are too diffused, tempting entrepreneurs to stray into the various lines, only to land up at the far end of blind lanes. Till the Government provides a rudder to the programme and imparts a focus to it there will be no major improvement in the present situation.

## The Other Side of SDFC

The Shipping Development Fund Committee put a stop to the consideration of fresh loan applications for the acquisition of deep sea fishing vessels from about October 1986. For the same reason, several sanctions of new loans, although approved by SDFC, were not released. The various applicants spent enormous sums in promoting their companies, securing offers for supply of trawlers and obtaining clearances before applying for loans from SDFC. All these were greatly disappointed at the decision of the S.D.F.C. obviously at the instance of the Government, not to release the loans that were approved. A Governmental agency withholding loans sanctioned for a developmental purpose is something unprecedented. Apart from this type of action by a Governmental agency it is not in keeping with the status of a Government. Sanctions are solemn promises and we do not expect a Governmental agency to go back on its word.

All the affected companies had to keep mum, mainly for the reason that the policy parameters of the Government are much more paramount than their problems, which really stemmed from the Agriculture Ministry's declared developmental policy to increase the deep sea fishing fleet of the country to a level of 330 to 300 nos. The unjustified action of the Agriculture Ministry in stopping the release of the loans, understood to be at the instance of the Financial Ministry, is in severe contradiction to the declared developmental policy in respect of deep sea fishing.

In the above situation it was all the more surprising to learn that, on 7th January, 1987, long after the release of new loans was stopped, sanction of a fresh loan for the import of one deep sea fishing vessel from M/s.De Hoop Shipyard of Holland by Machael Marine Pvt. Ltd was accorded. Readers can clearly see the invidious attitude of the SDFC in stopping the sanctioned loans on the one hand and sanctioning of fresh loans on the other, at the time it was due for closure The affected parties holding of sanctions that will lapse by March 1987 seem to have no way of securing redress.

Fishing Chimes, Editorial 77: March 1987: Vol. 6, No. 12

# Amateurish Exercises at Enforcing Closed Fishing Season in Upper Bay of Bengal

The All India Deep Sea Technocrats Association, which consists of deep sea fishing operatives as its members is reported to have recently met at Visakhapatnam and virtually called upon its members not to operate shrimp trawlers north of Visakhapatnam from 1st April to 30th June, 1987. This step was taken to protect shrimp stocks.

This is a good approach, from the point of view of operatives as well as owners, looked at from a particular angle. From April to June shrimp fishing will be dull and uneconomic. Such of the owners who send their vessels out during this period will suffer serious losses because of poor fishing. So far as the operatives are concerned, there is of course the unavoidable disadvantage of not getting incentives, because of the likely poor landings. They will atleast have the satisfaction of having their salaries, although their readiness to go but for fishing may not be availed of by all the owners. This 'satisfaction' will no doubt turn out to be the dissatisfaction of some of the owners, mostly for the reason that they have to pay the salaries of the floating staff who are forced to be idle at shore and which they greatly dislike, being sailors. In a way this is good, as, after nine months of hardwork with or without incentives, they are entitled to a duration of rest, although on their mere salaries.

It is possible that the owners feel that it is their prerogative to tell their employees whether to go out for fishing or not and that there is no justification for the employees to declare a 'closed season' for three months on their own, stop going out for fishing, and at the same time want to be on the pay roll of the company concerned. The owners may feel that they cannot be dictated in this regard by their employees.

The interest evinced by the floating staff to lend protection to the shrimp stocks in the Upper Bay of Bengal is to be lauded. What could have been done by the owners has been sought to be done in a forcible way by the floating staff.

What perplexes one however is the nature of thinking that may have led the operatives to arrive at that specific duration 'April to June'. This is a time when shrimps almost totally disappear in the Upper Bay. It is possible that the view taken by the operatives is that the stoppage of fishing during this period saves companies from losses. This way of reasoning may not altogether be altruistic, for, as already mentioned, the crew will have the much needed rest and they do not miss anything, as they will not get incentives in any case during this period and their salaries are assured.

There can be no two views on the need for a closed season in respect of catching of shrimp. Although an annual crop, the various types of shrimp that constitute the crop need protection at some stage or the other. This can be related to a) breeding season, b) times of abundance of Juveniles at sea as well as in adjoining backwaters, estuaries, brackishwater lagoons, c) mesh size, d) fleet strength, e) present extent of availability of protection, natural or otherwise, etc. The subject has thus multiple angles from which it has to be sifted before a cogent line of action is evolved. The complexity of the matter has led the Government to appoint a Committee to study the subject and make recommendations. The issue for consideration now is: Before the Committee's report is received by the Government, is it correct for any single body with limitations of knowledge and restricted interests to arrive at arbitrary conclusions in respect of having a closed season and the duration of such a closed season?

On the Atlantic Coast of Mexico no closed season for shrimp is observed, the reason being that the prevalence of heavy weather conditions for a substantial number of days in a year provides a natural protection from over-exploitation. Do similar conditions prevail in the northern Bay of Bengal, and, if so, what should be the approach? Has any of our scientists made a study from this angle? Why is this blessed duration of 'April to June' thought of, when the waters are more or less barren of shrimp during this period in most of the years? Why not decide upon clearly known durations of breeding peaks or Juvenile concentrations, as determined by our scientists, in order to declare closed season or seasons through mutual consent?

There may be several alternatives that can be worked out on a scientific basis, one out of which can probably be adopted.

First it has to be seen whether there is rationale behind the 'April-June' flash thinking. If there is no rationale and if there are vested interests playing a part in adopting this approach, there is need for a reconsideration of the subject.

There are dangers involved in adopting any approach that is not sound. Such an approach may change the complexion of the fisheries of the area in an unpredictable way and this may prove to be much more disadvantageous. Further, the economic elements involved in the adoption of arbitrarily framed closed seasons may lead to undesirable conflicts between managements and floating staff, which may create embarassing situations for the Government to resolve. If the managements force the crew to take the vessels out during the months in which the floating staff do not like to go out, the outcome can be either removal of the crew from service by the managements, or the crew resorting to non-cooperation. Such developments can be avoided through a rational approach to the problem. It appears that the Associations concerned should approach the Government to make a detailed study of count ranges of exported shrimp by the various processers in the area and determine the peak periods of export of very high and very low counts. These periods may provide a guideline for enforcing a closed season, since mature shrimp/juveniles need protection.

## Leasing of Fishery Rights in Andhra Pradesh

The fishermen community in Andhra Pradesh is not happy with the present system adopted by the State Government in the disposal of fishery rights over inland waters in the State, which, by and large, is based on open auction system. The community wants that leases of fishery rights should be given to fishermen's co-operatives on a long term basis. The State Government has taken cognisance of the feelings of the community and accordingly set up an expert committee to make recommendations.

The leasing system in respect of fishing rights in the state has to be truly integrated with the imperatives of fishery development in the water areas of the State, the fishery rights of which are mostly vested in the Government.

While formulating new policy guidelines in this respect, the distinction between a 'fisherman' and a 'fish farmer' has to be borne in mind by the Government. While a fisherman knows the art of fishing in the main, a fish farmer is conversant with the art of fish farming basically. The fishermen community is, by and large, handicapped by the fact that most of the men in the community are fishermen and not fish farmers. At the same time, the community has a claim over getting priority in regard to fishery leases. The gap between the two positions mentioned above has to be bridged by the State Government. This task can *be* accomplished by a total reform, which can be achieved only over a period of time that need not be long.

The first step is to encourage suitable fishermen to form into new fish farmers' cooperative societies in distinction to the existing fishermen's cooperative societies whose role should be restricted to riverine fishing. The activities of fish farmers' societies should be related to development of fish of ponds/tanks and reservoirs.

The jurisdiction of each of the newly formed societies should be restricted to a viable area, consistent with the membership, which may be at the rate of ten persons per hectare of water area in the precincts of a village or villages and subject to a minimum of ten.

Before registering the societies, the concerned authorities should ensure that no middlemen creep into the set up. To start with, one third of the membership of each of the societies should be given practical training in fish seed production and fish culture for the needed duration at one of the fish farms in the district concerned and each of the trained candidates that come upto the mark should be awarded with certificates of competence in fish farming by the Farm Manager concerned or the Director of Fisheries. Once there are a minimum of three trained candidates per hectare in a society, fishery leases of water areas in the jurisdiction of that particular society may be awarded to that society on a fixed annual rental, which should be equal to one third or one fourth of the estimated returns from the anticipated annual fish production based on development. The lessee society should furnish to the concerned authorities details of stocking with seed, and the details of harvest, as prescribed. The Fisheries Corporation/Department should guide the members of the society from time to time in regard to the culture operations, besides providing extension support. Long term lease for a period not less than ten years should be given to the societies to enable them to secure loans for the initial stage of operations from Cooperative Banks. The leases are to be extended from year to year within the ten year period, besides revising the annual rentals on a rational basis. Only when there is a gross transgression of the terms and conditions cancelation of leases should be resorted to.

In the case of the reservoirs or large irrigation tanks which need development of capture as well as culture systems, the societies should be entrusted with the task of producing their own fish seed (by leasing out along with reservoir fishery rights), rights over any adjacent fish seed farms. If such farms are not in existence close to the leased out water areas, Government should take steps to establish such farms for being leased out. production of seed in cages also could be encouraged. An integrated approach on these lines will facilitate the development and maintenance of a maximum sustainable catch. The leasing system can, however, be on the same lines as in the case of tank and pond fisheries.

Fish farmers' societies need not necessarily be formed with traditional fishers/farmers as their members. Rural youth, having aptitude towards fish farming can also be encouraged to form into such co-operatives.

Any leasing system not governed by an inbuilt mechanism that will ensure development of fisheries in the concerned waters will be anachronistic and out of tune with the aspirations of the people and national interests, which call for optimal utilisation of water areas. These are national assets, as they can provide employment, provide larger quantities of fish to the people and generate higher incomes to the members of societies and various others who can make out a living by participating in the various links of the activity.

## Integrated Approach in Implementing New Charter Policy

The objective of the new charter policy is to achieve ultimately the introduction of resource-specific fishing vessels, through the mechanism of enabling the charterers to operate similar vessels on charter in order to train their personnel in the concerned new technology and also to enable them to secure the needed addition of funds for the acquisition of their own vessels.

The achievement of the above mentioned.objective is very closely linked with the availability of an enforceable and implementable financing system, for the acquisition of vessels. In the absence of such a system, at the end of one year of operations of chartered vessels, the charterers would forego the deposit of Rs.3 lakhs given to the Government and give up the idea of acquiring their own vessels.

Considering the above position, it is necessary to bring about an integration between chartered fishing vessel operations and acquisition of own vessels by the charterers in the beginning itself and not at the end of one year of operation.

Along with the application seeking permission to operate chartered vessels, it should be stipulated by the Government that the charterers should also submit clearcut proposals for acquiring their own similar resource specific vessels, before issuing letter of intent. Government should consider stipulating the condition that the chartered vessels should be made operational in Indian waters only after providing specific proof to the Government in regard to firm financing arrangements, not in principle, but on ready-for-release basis, in respect of.loans or deferred payment guarantees. By the time the charter operations are completed, trained men will be available for placement on the newly acquired similar resource-specific vessels, which would have been acquired by this time.

It will be a very good approach for the Government to give priority to such of the companies which already have permissions 'to acquire resource specific vessels. These could be encouraged, to charter similar vessels, so as to ensure training of personnel for the operations of the new vessels as soon as they are acquired. The letters of intent for chartering issued to this proposed priority group should also include the condition that the chartered vessels should be made operational in Indian waters only when proof of firm financing arrangements as mentioned above is produced. The companies will be able to produce such a proof only when the Government ensures that SCICI extends the needed financial help by way of loans or deferred payments guarantees. This aspect is the most crucial element in the entire programme. If the financing arrangements are not effectively evolved in consonance with the present characteristics of the fishing industry, the programme is destined to fail in the achievement of its objectives. If charter permissions are accorded without integrating these with the introduction of vessels of Indian Flag, the result will be that at the end of one year of operations all the new entrants will get out of the industry unceremoniously, leaving behind their deposit of Rs.3 lakhs with the Government. Those who are already in the industry may continue with their present operations, leaving to the Government the deposit of Rs.3 lakhs which will become forfeited money.

Fishing Chimes, Editorial 78: April(i) 1987: Vol. 7, No. 1

# On to the Seventh

'Fishing Chimes' has completed six years of service to the fisheries field by March, 1987. The seventh year of its existence has now commenced.

The combined effect of the patronage of our subscribers and advertisers and our own unstinted and unceasing efforts to keep the journal going, with such material that we believed the readers would need, has enabled us to run the journal in a way that caters to the specific needs of the fisheries sector which, although small at present in the national canvas, is nevertheless important.

When your Editor embarked on the publication of 'Fishing Chimes', there was a wide and deep scepticism in regard to the need for such a journal and its future. Very few of those consulted before the commencement of the publication could visualise and comprehend the need, which was lurking to be identified. Your Editor was finally convinced about the need, and started the journal, subjecting the hypothesis to test by registering the journal with what appeared to be a banal label 'Fishing Chimes'.

The fact of the matter was that your editor applied for the allotment of the name 'Fishing Times', but the Registrar of Newspapers could not approve of this probably because there were already several 'Times' around. A search was then made for a word that came the closest in rhyme to 'Times' and the word 'Chimes' popped up as an instinctive choice. This choice made even a professional like Mr.Peter Hjul, Editor of Fishing News International, London curious, prompting him to ask your editor for the rationale behind the name. Your editor was overwhelmed at this attention and explained the history behind. In reply to enquiries from several other baffled fishery conscious men also, the import of the selection was conveyed.

According to the Chamber's Twentieth Century Dictionary, 'Chimes' means "A set of bells tuned in a scale; the ringing of such bells in succession; A definite sequence of bell like notes sounded as by a clock'. Now, dear readers, the explanation may look a bit far-fetched but the intention behind was to have a name that reflects the objective of reporting the developments in the fisheries field from time to time in the same way as the ringing of bells in succession. In fact, some fishes (like Gourami) are trained to ring a bell when hungry. The other way round, we now ring bells through 'Fishing Chimes' to alert the readers about what is happening in the fisheries sector in and out.

Our assessment is that readers of 'Fishing Chimes' have atleast come to the stage of liking the journal, and looking forward to receiving it month after month.

For us this is a joyous occasion (and hope it is the same with the readers too) as we have been able to reach 'Chimes' to the readers continuously without break, month after month, although a little belated some times. So much so, at this happy moment, it will be inappropriate for us to chronicle our woes, in order to catch the attention of the subscribers and advertisers, for making a case for further support to make the publication viable. So, we desist from this, for we have also several encouraging aspects to relate.

One is that our readership has gone up, from about 14 nos. in 1981 to around 1,000 nos. now. From some little known corner point in the country we keep receiving enquiries for subscription rates etc., almost in every few days from interested persons. These make us feel proud. In the earlier years of publication, there was not much of response from those in charge of fisheries development, technologists and scientists to contribute material for publication. The situation has changed over the years and now we receive several contributions on fisheries developmental aspects from professionals, and other knowledgeable men who have obssessive attachment to the subject of fisheries. There is now abundant evidence that the publication of these contributions have been helping in bringing about dissemination of knowledge and spread of new technologies implemented in one part of the country to other parts. We are indeed indebted to the various authors, subscribers, our correspondents and representatives for all the support that they have been extending.

From a financial angle the main support to 'Chimes' comes from the advertisers. The amount that they pay us, which is based on our low tariff, keeps us going, although under strain. In the early years we were taking the sums towards advertisements with a sense of guilt, because we were not getting any feedback about their impact on the readers. Now we do not have that guilt any more. The advertisements are noticed and acted upon substantially by several of the readers.

It is a different story with subscriptions. At any point of time over 70 per cent of the annual subscription amounts remain unpaid. Not for very long, of course. The dues are cleared sooner or later, but simultaneously some other subscribers move forward to restore the deficit back to the earlier level. Thus the cycle runs and our journal survives.

It is the patronage from you all that is important. This encourages and impels us to move further forward towards excellence and hold readers' interest and their patronage.

Fishing Chimes, Editorial 79: April(ii) 1987: Vol. 7, No. 1

# Financing System for Deep Sea Fishing Vessels: Policy Options

Dr.N.K.Thingalaya, Deputy General Manager, Syndicate Bank, Manipal is a seasoned banker having wide experience, among others, in dealing with problems connected with extending financial support for the acquisition of fishing vessels. For the first time, your Editor has had the opportunity of listening to his talk at the National Seminar on Export Strategy for Indian Marine Fisheries, held in New Delhi under the auspices of M.Visvesvaraya Industrial Research and Development Centre (World Trade Centre). Bombay, with the co-sponsorship of MPEDA, recently. In his talk he dealt with the various problems that are peculiar to the financing of the fisheries sector. He said that, from his long experience, his assessment was that the characteristics of the deep sea fishing industry are such that the conventional type of banks will have difficulty in extending financial assistance to the sector. He felt that the only alternative to meet the financing needs of the deep sea fishing industry seemed to him to be to establish a separate fisheries bank. The same view was expressed earlier by several fisheries professionals, particularly Mr.R.K.Verma, past President, the Association of Indian Fishery Industries. This line of thinking has now acquired credence, now that a seasoned senior banker has come to the same inevitable conclusion. It is time, although late, for the Government, to examine this highly crucial suggestion. If this is followed up, the deep sea fishing industry would surely develop. Otherwise it would stagnate.

One might say that Government had already acted by closing down the Shipping Development Fund Committee and facilitated the setting up of a new private sector financing company (Shipping Credit and Investment Company of India), charged with the responsibility of providing financial assistance and deferred payment guarantees for the acquisition of deep sea fishing vessels, for the setting up of processing plants etc. The main function of this organisation is to provide financial assistance to shipping companies.

Compared to the needs of the shipping companies, before whom fishing companies are dwarfs, the fishery enterprises are most likely to be accorded secondary priority in the granting of loans etc. Further, the characteristics of deep sea fishing industry and other ancillary industries being different from those of the shipping industry, it would be very difficult for the new company to think of having a different set of norms for fishing companies, although it proposes to impart a considerable extent of flexibility in considering the various applications from fishing industry for financial assistance.

In countries such as Norway, separate fisheries banks have been set up, obviously because of the distinctive characteristics of the sector and the need to have a separate banking set up for the purpose. The approach can in no way be different in a developing country like India.

No one, even the Government, can ignore the basic characteristics of the deep sea fishing industry, based on which only the financing pattern for the introduction of deep sea fishing vessels has to be evolved. Chief among these is the feature that it is only the medium/small enterprises who do not have much of money, but are nevertheless are those that have the infusion and flair to enter the field and succeed to come forward to enter the line. These media cannot fulfil the existing norms of commercial banks and financing institutions in respect of debt-equity ratio, collateral etc., nor are the banks and financing institutions have the time, expertise, and inclination to enter the sector. In contrast to the above situation, the most glaring characteristic is that practically no large company is interested in investing in the deep sea fishing industry. All those who were interested earlier entered the field ceremoniously and later came out of it unceremoniously. The policy option before the Government, therefore, is to recognise the factual position and set up a financing body that will exclusively handle matters pertaining to financing of fishery industries with such norms as are realistic and enforceable. If a separate fisheries bank cannot be established, the other option is to set up a separate wing in the Shipping Credit and Investment Company of India to deal with the financing of the fishing industry. While the norms of lending funds and granting of deferred payment guarantees for the industry can be worked out by a team of experts drawn from the Government, fishing industry and the financing institutions, to start with, and the proven norms followed by SDFC till recently could be adopted by this wing. In order to protect the interests of the financing body, its fisheries wing should have an executive branch that should always be in readiness to take over and run such of the ponds and tanks covered by intensive fish culture, incorporating composite stocking technology. Levels of fish production exceeding 4 tonnes/ha have been obtained not only at research centres but also in the private sector. The keys to increase fish production from culture sources, besides proper training to fish farmers, are the availability of water for filling up ponds, manures/fertilisers, feed and last but not the least, fish seed. Renovation may also become necessary in the case of ponds and tanks that have become derelict.

Of all the items mentioned above, the main limiting factor is the availability of fish seed. If seed is available close at hand, a rural community will develop the desire to utilise the seed for the stocking of the tanks and ponds within the jurisdiction of the community. Repairs to ponds, application of fertilisers/manures, and organising water supply are not that very difficult as securing fish seed.

The bulk of the fish seed production in the country now takes place in West Bengal. Most of the States depend on this source to meet their needs of fish seed. This practice has enchained the enterprises/fish culturists in the various States to the suppliers of fish seed in West Bengal. It is the duty of the various State Governments and the Government of India to encourage entrepreneurs to produce their own fish seed. No doubt, the Government of India has set up modern hatcheries for producing seed in the States of West Bengal, Madhya Pradesh, Orissa, Uttar Pradesh and Bihar under a World Bank Scheme. It has also introduced a national level scheme for the setting up of fish seed farms. These steps have not however changed the basic complexion of the seed production and distribution pattern in the country and West Bengal continues to reign supreme in this sector. There is no reason why this system should continue any longer, particularly after the introduction of the Chinese type of hatcheries. At a very little cost every village or a group of villages having a viable pond area could be encouraged to have an arragement for the production of fish seed to meet their needs. What all that is necessary in this respect is for the State Governments or the Central Government to introduce a scheme under which a) candidates sponsored by each of the village communities having viable water areas will be trained in the breeding of major carps and with operation of Chinese type of hatcheries. These candidates may be drawn from a village fish, farmers' Cooperative Society exclusively set up for producing fish seed and undertaking pond culture in the given area, b) an agency which can provide the needed equipmerit to the societies for breeding of major carps and for setting up of Chinese type of hatcheries to the societies may be established. c) as the outlay on these items is not substantial, the Government concerned could arrange loans for the purpose to the societies from the NCDC with a component of subsidy, d) in order to minimise the capital requirements, the old system of breeding major carps in hapas could be revived. The available ponds can be made use of for the purpose and Chinese type of hatcheries can be set up close by. The spawn produced can be grown to fry stage in cloth hapas set up in the ponds, or one or two ponds can be set apart for stocking the early fry for seed production. Out of the seed produced, the requirements of the village ponds can be met first, and any surplus can be sold to others in need.

The training programme coupled with the installation of Chinese type of hatcheries in rural areas under a properly drawn-up scheme would revolutionise village economy. Apart from providing employment, fallow tanks and ponds are national assets which can be put to effective economically productive use. The returns earned will make villages prosperous.

Fishing Chimes, Editorial 80: May 1987: Vol. 7, No. 2

# Can Fourth Revision of Charter Policy be Avoided?

Ministry of Agriculture has gained considerable experience in the formulation of policy in respect of permitting of operations of foreign fishing vessels on charter. This experience has been accumulating since 1977. In early 1980s, Government suspended the later scheme, but later came up with a revised version was in operation till 1986. After this, there was again another suspension, followed by a re-revised scheme, notified in January 1987. This was further followed by another notification published in newspapers on 9 April, 1987.

According to this notification of 9 April, applications for chartering of fishing vessels should be received in the Ministry of Agriculture on or before 8th May, 1987. Those received after this date will not be entertained. Further, the total number of vessels that will be permitted for charter will be fifty, consisting of resource-specific vessels such as purse-seiners, longlining vessels, squid jiggers and stern trawlers, for pelagic and midwater trawling etc. Pair trawling will not be permitted.

The notification also lays down priorities. First preference in the allotment goes to public sector undertakings, followed by fishermen's co-operative societies/groups of fishermen, groups of technocrats, small/medium companies and finally the larger houses.

Those interested have been advised to write to the Deputy Secretary (Fisheries) for details of the scheme. Enquiries at the time of this writing gave the impression that the various terms and conditions were being drafted but were not ready for being made available to the interested parties well before 8 May, 1987. Companies and representatives of companies having offices in Delhi will have an advantage in that they can get hold of the guidelines soon after issue and act on them.

The Ministry deserves congratulations on evolving a specific approach, concerning type of vessels and number of vessels permitted to be chartered in all. Realistic or not, the charter scheme is open only for certain types of vessels, and these have to be obtained only from countries with whom India has diplomatic relations.

Major interest in providing fishing vessels on charter to Indian enterprises lies with Taiwanese fishing companies. As Taiwan has no diplomatic relations with India, Taiwanese fishing companies have taken the system of registering their vessels under Panamanian flag. It is not known whether such vessels would continue to be acceptable to the Government for operation in Indian waters, at the time of this writing.

One basic objective of the charter programme is to enable the Indian enterprises to introduce vessels similar to those operated by them on charter through persons placed as understudies on the charter vessels for receiving training. According to the scheme, before the end of one year of operations by the chartered vessels, the Indian companies concerned should be able to take effective steps for introducing their own vessels or enter into joint ventures with foreign fishing companies for acquiring and operating vessels flying Indian flag.

It is practically impossible for any Indian company to complete all formalities connected with the acquisition of fishing vessels under the charter scheme within a period of one year. The main bottleneck that they will face in the present situation concerns financial assistance for the acquisition of fishing vessels which has become gloomy after the closure of the Shipping Development Fund Committee.

Practically no company which will operate chartered vessels under this scheme can show proof of having made all financing arrangements leading to commencement of construction of their own vessels. While it is the ardent desire of the entire deep sea fishing industry that the newly established SCICI would be able to extend the needed financing facilities including provision of deferred payment guarantees, it is very doubtful whether this august financing body would be able to condescend and evolve this policy in tune with the conditions of existence prevailing in the deep sea fishing industry. Very few deep sea fishing companies would be able to subscribe to a debt-equity ratio better than 6:1. The fishing companies are bound to have difficulties in this respect and therefore because of the problems of financing acquisition of vessels, the ultimate objective of charter scheme is not likely to be realised. This would only mean the forfeiture of the deposits made by the companies and virtually no addition to Indian flag vessels under this scheme. It is also not known whether SCICI, which is a private company, can be entrusted with funds from the Consolidated Fund of India for purpose of providing loans. Without this SCICI will have problems of cash flow and functioning effectively.

Foreign owners agree to give their vessels for commercial operations only when they are sure of availability of adequate fisheries resources. For this reason, chartered vessels operations should not be considered as something experimental to lead to the acquisition of vessels. This means that linkage between the chartered vessels operation and acquisition of their own vessels by the Indian companies concerned will have to be brought about in the beginning itself and not at the end of the one year of operation. This will, however, be possible only when the financing arrangements are fully settled before the commencement of the operation of the chartered vessels, whose main purpose will then be to provide training to Indian counterpart personnel to enable them to operate the newly acquired vessels without leaving a vaccum.

We feel that, along with the application for permission to obtain foreign fishing vessels on charter, Government should stipulate that contracts signed for acquiring similar resource specific vessels should be submitted simultaneously by the companies concerned with proof of having made all financing arrangements, not in principle, but on a ready-for-release basis.

Such of the companies which have already been given permission to acquire resource-specific vessels should actually be given preference and be encouraged to operate on charter similar vessels, provided financial arrangements as mentioned above are made for the acquisition. This will enable the companies to have trained men in readiness by the time their own vessels arrive. Another suggestion is that, while letters of intent can be issued to operate chartered vessels, these should be made effective only when the first down payment for the acquisition of own vessels of the related companies is released to the shipyard concerned, on the basis of deferred payment guarantee or letter of credit, which will have to come from an Indian financing institution.

The adoption of above strategy alone would ensure the fulfilment of objectives of the charter scheme. And the above strategy cannot be implemented as there is no effective financing system in existence at present. There is thus a vicious circle. The need is, therefore, first evolve a suitable and an effective system for the financing of deep sea fishing vessels, taking into account the characteristics of the entrepreneurship in the industry. It appears that what is being done is placing the cart before the horse. Any steps that are taken without establishing a feasible financing pattern will not lead to anticipated results.

It is learnt that some suggestions have been made to the Government, one of which was that out of 20 per cent of the gross earnings from chartered vessels that are allowed as the share of the chartering Indian companies a major part should be deposited with the SCICI so that the money could be utilised for the acquisition of obligatory vessels by the Company. There was a further suggestion that Government should prescribe a minimum deposit in this respect so that the companies may not be evasive in this respect. It is difficult to say how far these suggestions will be practicable, unless there is a field agency to keep a watch over earnings and compliance with the various conditions. The SCICI will have a right to insist on such payments only if it agrees beforehand to extend financial assistance in some form or the other. A point to be kept in mind here is that of the total accumulations of these deposits are not likely to be sizeable, and there will not be improvement in the level of deposits to an extent that will satisfy the SCICI fully. Further, penal provisions for non-compliance will only mean that the suggestion will not be a sound one to ensure deposits. In any case, this will be an uncertain way to motivate the entrepreneurs for acquiring their own vessels and an admission on the part of the company concerned in regard to their inability to make effective financing arrangements as has been done by the several other companies. Government would have to face the calculated risks involved in the development of deep sea fishing industry and in this light make effective arrangement for the development of deep sea fishing on charter or other wise, mainly linked up with the financing needs which should be first tackled. Otherwise, Government would find a need for the fourth time to revise the charter policy.

Fishing Chimes, Editorial 81: June 1987: Vol. 7, No. 3

# Imports under 100% Export Scheme at Standstill

Clearances have been given by the Ministry of industry for the import of over 100 deep sea fishing vessels under the 100 per cent Export Oriented Scheme. Of these, around 6 numbers have so far been imported. Apart from these, there are a large number of applications pending with the Ministry of Industry for the issue of letters of intent. All these applications have been cleared by the Department of Agriculture, but are now waiting for clearances from the Department of Industries.

There has been a major shift in the trend of applications for the import of fishing vessels in the recent past, from the normal programme of the Department of Agriculture to the 100 per cent Export Oriented Scheme implemented by the Ministry of Commerce/Industry. The attraction that led to this shift was that foreign loan to the extent of 100 per cent of the cost of the vessels would be allowed. The Department of Economic Affairs did extend this facility in the case of some companies. However, it is learnt that, the Department of Economic Affairs has of late decided to issue such permissions only to the extent of 80 per cent of the cost of the vessels. The reason for this is not known.

The fishing companies are already reeling under the impact of the closure of the Shipping Development Fund Committee by the Government which put a stop to an on-going system of providing financial assistance for the acquisition of fishing vessels. It will obviously take some time for the entrepreneurs to understand the style of functioning of the newly formed Shipping Credit and investment Company of India. This new financing body, a subsidiary of ICICI accustomed to financing large houses and big companies has to adjust itself to the needs of fishing sector. The fishing companies who have received sanctions of loans from SCICI are experiencing the problem of high debt : equity ratio, and a few other conditions stipulated by the SCICI.

On one side the Government intends to promote introduction of deep sea fishing vessels to bring the national fleet to a level of 500 numbers at the end of the Seventh Plan. On the other side, the Government has chosen to introduce as many blocks as possible to prevent the fishing companies from acquiring the vessels.

The blocks are a) delays in release of letters of intent b) closing down an on-going financing system and opening a new one that stipulates conditions that the industry is unable to fulfill c) reducing the scope for the availment of foreign loan from 100 to 80 percent and d) leaving all fishing companies who were granted loans by the erstwhile SDFC for the acquisition of two or more vessels each, without any alternative arrangement for fulfilling the approvals.

There are no strong reasons for disconnecting the targets and the necessary steps needed to fulfil them. Government will have to be purposeful and make sure that a proper climate is established for achieving the targets. If Government has problems in promoting the industry, the best way is to tell the industry to wait till such times to come instead of keeping the constituent companies in suspense and subject them to lot of expenses.

## Ignominy of Govt. not Standing up to Commitments

It was decided in early 1986 that SDFC would be wound up and in its place a specialised financing agency would be created to take over its functions. The industry had apprehended at that time, which had now proved to be correct, that the implementation of the project for introduction of deep sea fishing vessels to raise the fleet strength to a level of 500 numbers would not be possible if the SDFC was to be wound up.

The applications for the acquisition of indigenously built and imported vessels, which had been approved by the SDFC but sanction letters not issued, are now lying in the archives of the office of erstwhile SDFC, now taken over by the Department of Economic Affairs (Banking Division) of the Ministry of Finance.

The Ministry of Agriculture imposed a condition on such of the companies which had operated foreign fishing vessels on charter, to acquire vessels of an equal number as permitted to be chartered. These companies gave interest free financial deposits and also bank guarantees to the Government as securities as stipulated by the Government, for the fulfilment of the commitment.

The companies applied for loans from the SDFC when it was alive. The applications were recommended by the Agricultural Ministry to the erstwhile SDFC which had also approved them. The issue of sanctions by the erstwhile SDFC was held up only for the reason that they had to obtain formal clearance from the Ministry of Shipping and Transport.

The loans in each case exceeded Rs. One crore. At this stage, the SDFC was wound up throwing the fishery companies into a total jeopardy for no fault of theirs. There is a moral responsibility on Agriculture, Finance and Surface Transport Ministries to accept the situation and find a way of extending the financial assistance that was promised, in the shortest possible time. This matter is of utmost urgency, as prices of vessels are going up. If the Government is unable to provide the loans as promised the least that it can do is to inform the companies about this and absolve them from the obligation to introduce the vessels, refund the guarantee amounts, and also release the companies from the bank guarantee obligation.

Fishing Chimes, Editorial 82: July 1987: Vol. 7, No. 4

# Development of Riverine Fisheries

A symposium on the Impact of Current Land Use Pattern and Water Resource Development of Riverine Fisheries, held at the Central Inland Capture Fisheries Research Institute, Barrackapore, in the last week of April 1987, highlighted the declining trend in the fisheries of our major: rivers. Emphasising the need to restore the fishery wealth of the rivers, the symposium made several recommendations. One important recommendation was to prevent the pollution of the rivers which has a disastrous effect on the riverine fisheries. Another suggestion was to introduce measures to counteract the effects of the construction of dams and other barriers across the rivers. In this way the migration of fish into and out of rivers for the purpose of breeding/feeding is not affected. Yet another main recommendation was to organise the production of fish seed for the stocking of major rivers all along their river courses at selected points, with a view to resuscitating the fishery wealth of the rivers.

The recommendations made at the symposium deserve to be followed up with speed in national interest. The nation certainly wants the prevention of the extinction of the important species of our riverine fishes which are part of our national heritage.

The problem of the present decline in riverine fisheries could not be tackled in isolation. The steps that are recommended have to be multidisciplinary and integrated in nature. The work relating to the tackling of the problem has to be worked out in great detail and made applicable to the various river basins, which such variations as are necessary to suit the local conditions before implementing them.

The Ministry of Agriculture, with the help of ICAR, will have to draw up a national plan in this context, in the shortest possible time, with the help of the various experts and implement these plans over a period of time that is not too long. If this is not done, very soon the nation has to witness the extinction of some of the species. The decline of mahseer fisheries in Himalayan waters was pointed out by Mr. C.B. Joshi of the Bhimtal Research Centre of the Central Inland Capture Fisheries Research Institute, Bhimtal. Hilsa fisheries of the Ganga, Cauveri, Krishna, and Godavari have been declining from year to year at a fast pace.

Dr. A.G.K. Menon of the Zoological Survey of India, Madras has listed 25 endangered fish species. These Include *Notopterus chitala,* seven species of mahseer, five *Schizothoracid* species, and last but not the least, *Thynnichthys sand khol* of the Godavari, and Krishna systems, which represents one of the rare examples of discontinuous distribution.

There is an urgent need to stem the process towards extinction of fish species of the country. And this can be.achieved only through a major programme of producing large quantities of the seed of these species and stocking them in the waters, while also adopting whatever regulations are needed, and implementing what are already there.

Fishing Chimes, Editorial 83: August(i) 1987: Vol. 7, No. 5

# The Unequal Quadrangle

There are four participants in the activities at the fishing harbour, Visakhapatnam. These are a) the owners of fishing vessels, b) the certificated officers who are in-charge of the vessel operations; c) the deckhands who conduct the deck activities and d) the Port Trust which is in charge of the infrastructural facilities.

These categories form the four sides of the quadrangle. Only when these four arms are properly balanced the operations run smoothly. Any imbalance in the quadrangular equation automatically results in disruption of the fishing activities.

The administration of the facilities at the fishing harbour by the Port Trust has all along been, by and large, smooth and responsive to the need of the operators. Lack of proper understanding among the other three constituents sometimes used to result in difficult situations but these had always been resolved amicably.

From April 1987, however, the situation became somewhat different. The Deep Sea Fishing Technocrats Association, rightly or wrongly decided not to take the trawlers out for fishing from 1st April to 30 June, on the premise that shrimp required protection, during these three months which also happen to be a time when shrimp availability comes down considerably. The owners did not react to this move on the part of the technocrats during the period from April to June 1987. At the same time, they never told their vessel officers to take the vessels out, with the exception of a few. In all these cases, the vessels were taken to areas south of Visakhapatnam, in accordance with the decision of the Technocrats Association, In effect, the decision of the technocrats prevailed without any commitment to this decision on the part of the owners.

An idle period of three months at port became a boon to the deckhands and their leadership. They worked very hard during these three months to emerge as a strong arm of the quadrangle. They formed into an union, although with an alien leadership. The Union, Vizag Trawler Operators Union, has been affiliated to CITU. Under the leadership of CITU, very systematically, their movement progressed. And all of a sudden, in the last week of June, the Charter of Demands, which was served by the Union on all the owners, became a document of great attention, particularly when the deck-hands said that they would not work on the trawlers unless there was a negotiation on the basis of their demands.

The owners as well as the trawler officers were in a fix. The trawler officers found that their command over the deckhands was getting weakened. Some of them felt that the long berthing of the trawlers at port, because of the closed season, gave an opportunity to the deckhands to become a major force and work up to a position in which they could dictate terms not only to the owners but also to the officers. Thus, an inequality has got established itself in the quadrangle, between the owners and the deckhands on the one hand, and to some extent, between the officers and the deckhands on the other.

The owners did not have any alternative other than negotiating with the Trawler Workers Union. Fishing vessels come under the Merchant Shipping Act and are not covered by the Industrial Disputes Act. There was some reluctance on the part of the owners to negotiate because of this legal hurdle. Notwithstanding this, the owners set up an Action Committee which, it is stated, successfully negotiated certain terms and conditions, utilising the good offices of the police officials concerned. As a result of these negotiations, there was peace and the vessels sailed out towards the middle of July.

How long the negotiated terms and conditions will hold good can be a matter of conjecture. The nature of work on board a trawler is such that, if a deckhand is not found fit, his services will have to be dispensed with. Such actions might be construed as a repraisal by the owners or by the trawler officers, although such a line of thinking may not have any basis. With such cases multiplying, for which there is every likelihood, the negotiated equation may get again disturbed either for good or for bad. It is certain that the leadership of the Workers Union would understand soon that the characteristics of the work requirements on board the fishing vessels are quite different from the standards applicable to land-based industries. Otherwise, the system of engaging seamen on contract for specified periods would not have come about in the merchant navy. In the light of such an understanding the leadership may decide not to do anything further. Looked at from another angle, the leadership can insist on the continuance of the workers, based on the reports given by their own members.

The situation will then be that the captains cannot have such deckhands who cannot work on board. The insistence by the Union to continue to employ them will create problems to the owners and the fishing activities can come to a standstill. It is hoped that such difficult situations will not arise and the labour leaders will gradually develop an understanding of the characteristics of the fishing industry. In the coming one or two years there will definitely be further anxious situations at the fishing harbours. When these are understood and properly solved, the accumulated experience will lead to viable solutions that will keep the quadrangle well balanced.

Fishing Chimes, Editorial 84: August(ii) 1987: Vol. 7, No. 5

# A Challenge that has to be Met: *Beel* Fishery Development in Assam

*Beels* in Assam are extensive waterspreads connected to rivers such as Brahmaputra, and their tributaries. These are a part of geography of Assam. It is estimated that there are about 760 *beels* with an area of over 100,000 acres. In terms of potential, these water areas are capable of producing an annual fish crop of 70,000 tonnes. The present production is estimated at about 10,000 tonnes.

Only some of the *beels* in Assam come under the registered fisheries list maintained by the State Revenue Department and the registered fisheries alone are settled under lease terms by the Government.

Several efforts have been made by the State Government to reclaim *beels* and bring them under fish culture. A project under World Food Programme has been taken up for the development of fisheries of Assam *beels* through the State Fisheries Development Corporation, but this has not taken off so far. The progress achieved in *beel* fishery development has not been perceptible. One reason for this situation seems to be that the rights over development of fisheries of these *beels* have not been transferred to the Corporation. On account of this, the utilisation of these water areas in an effective manner for augmenting fish production in the State has suffered. So as to meet the heavy demand for fish in the State, it is essential that the State Government reviews the present position concerning the development of *beel* fisheries and institute necessary urgent steps for their development.

A Workshop on development of *beel* fisheries in Assam was held at Guwahati in April, 1987 under the auspices of the Assam Agricultural University. At this workshop, several recommendations were made. However, no organisational framework for the development of *beel* fisheries was recommended by the workshop. The workshop, nevertheless, wanted that the economic viability of the development of *beels* into productive water bodies should be ensured by the State Fisheries Development Corporation before taking loans for their development. It was further recommended that a phased programme of the development of *beels* of Assam should be formulated by the Directorate of Fisheries. Another recommendation said that the Directorate of Fisheries may undertake detailed investigation and model development of one *beel*. At the same time, another recommendation was made stating that, for making finances available for the development of *beel* fisheries, advantage should be taken of loans from the country's nationalised banks. This recommendation is not in consonance with the other recommendations suggesting that the State Fisheries Development Corporation should make studies on the economic viability before taking loans for development of *beel* fisheries and that a phased programme of *beels* should be formulated by the Director of Fisheries.

One of the main functions of the State Fisheries Development Corporation is to develop *beel* fisheries. A phased programme for the development of *beel* fisheries of Assam would have to be carried out by the fisheries Development Corporation. This phased programme would, of course, be cleared beforehand by the Directorate of Fisheries.

It is surprising that it was felt at the workshop that economic feasibility of development of the *beel* fisheries should now be undertaken. The data gathered over the past several years must have thrown specific light in this respect. Otherwise, it would not have been recommended by the workshop that the development of *beel* fisheries should be taken up with financial assistance from banks.

The need of the hour, keeping in view the high demand of fish and the soaring prices of fish in Assam, is to make a beginning in the development of a few of the *beels* that are considered to be relatively more amenable for taking up the development work immediately. Based on the experience gained in the first phase, further phases involving the development of fisheries in other *beels* could be taken up. The inland Fisheries Technical Committee of the Central Board of Fisheries (way back in 1973) and the National Commission on Agriculture in 1976 made recommendations in regard to the development of *beel* fisheries, suggesting specific lines of action. It is advised that the State Fisheries Development Corporation take into account these recommendations, which are still valid, in evolving a programme for the development of *beel* fisheries in Assam.

Fishing Chimes, Editorial 85: September 1987: Vol. 7, No. 6

# Resources Specificity

One of the achievements of the Union Department of Agriculture in the 1980s is the introduction of the concept for permitting the introduction of Resource- specific fishing vessels. Shrimpers are Resource- specific fishing vessels, but these do not fall within the purview of this new classification. On the other hand the department believes that multipurpose fishing vessels are shrimpers. According to a recent notification, permissions for the introduction of multipurpose fishing vessels will not be issued. This statement being unqualified, it means that facilities for two types of fishing even through both these are not for shrimping will not be permitted. Further, there is no opening any more for introducing vessels such as trawlers that will have capability of catching various types of fishes. This means the main fish stocks will die a natural death, thereby contributing to losses in Indian fishing.

Introduction of Resource-specific vessels, as they are called, entails several considerations. One is that fishery resources should be available throughout most part of the year. The fishes harvested should fetch reasonably good prices. This obviously means that they should have a ready market. In other words, Resource-specific vessels should also be economy specific.

Scombroids such as tuna and Cephalopods such as squid, cuttlefish and octopus, Crustaceans such as lobsters are the specific resources one can think of while deciding on acquisition of Resource-specific vessels. The fishery of all these types is seasonal, around six months in a year, in the Indian waters, as per the present knowledge. The prime questions that will arise in this context are: what are the vessels to do during the period of non-availability of the specific resources? Should the owners keep the vessels idle incurring expenses without incomes? Would the operations for part of the year fetch adequate returns to meet commitments in respect of repayments of loans? and, Would it be correct to leave vessels which are assets unused for a long duration?

The Minister of State for Agriculture observed at the Annual General Meeting of the Association of Indian Fishery Industries held in New Delhi that, if tuna disappeared in Indian EEZ for a part of the year, the Tuna vessel operators should show the 'initiative of following them wherever they would be within or outside Indian EEZ. The same advice obviously holds good for the other fisheries resources of a specific nature.

The Minister's advice is a 'great one', but only in respect of resource-specific vessels that have endurance to stay out for six months, or more. There will then be an adequate time to loiter on high sea hoping to find the specific resource. Larger vessel owners will no doubt be able to either absorb or bear the extra unproductive costs, if the catches are poor. This line of argument however is not sound from the angle of economics. A owner cannot lose money just to subscribe to a whimsical policy of the Government that cannot withstand any test to justify it. In the case of medium-sized Resource Specific vessels, aimless fishing efforts during off seasons will prove to be a great economic burden.

Enterprises taking up operations of medium-sized resource-specific vessels will flounder. Entrepreneurs are aware of this position. Therefore, the short point is that introduction of medium or small-sized resource- specific vessels cannot prove to be economically viable or economy-specific. Financing institutions do not lend money for the acquisition of vessels whose operations will be uneconomic. Therefore, the idea of promoting resource-specific vessels is not likely to gain momentum.

Promotion of introduction of resource-specific vessels equipped with arrangements for harvesting two types of resources, other than shrimp, (if the Government, out of its obsession, is averse to its capture in spite of the short 12months life of shrimp), may well prove to be an economically viable measure. While the Government can retain its excessive love for safeguarding shrimp stocks whose total annual catch has so far not fallen, in a statistical sense, there is no case for applying the same affection towards ground and pelagic fish which can be harvested by bottom and midwater trawl nets respectively. Arrangements for hauling these resources as a secondary one on resource -specific vessels could be encouraged.

Vessels having arrangements for undertaking more than one type of resource-specific fishing also are multipurpose vessels. There appears to be a fixation in the minds of certain ignorant persons that the description as multipurpose vessels is always a substerfuge or a cover-up for shrimp trawlers. This, as any knowledge person would say, is a baseless way of thinking and has to be expelled. Multipurpose vessels need not necessarily be shrimporiented.

If the harping is on introduction of resource-specific vessels capable of hunting for one type of resource, India's plans for the development of deep sea fishing will not materialise in any significant measure. The only way to promote the activity is to allow multipurpose vessels, resource-specific or otherwise, coupled with financing arrangements, infrastructural facilities, marketing arrangements etc. If Government so desires, it can be stipulated that there should be no special rigging for catching shrimp.

Fishing Chimes, Editorial 86: October(i) 1987: Vol. 7, No. 7

# No escape from Loan Guarantees

The progressive introduction of deep sea fishing vessels for exploiting the deep sea fishery wealth of Indian seas is inexorably linked to the provision of guarantees towards loans for acquiring deep sea fishing vessels by an independent organisation such as the Marine Products Export Development Authority to the enterprises concerned. The present stagnation in the introduction of deep sea fishing vessels can only be broken by the intervention of the highest authority in the country who should issue orders that guarantees towards loans should be made available to deep sea fishing enterprises and these provided by a designated Governmental body.

In the United States there is an organisation known as the National Marine Fisheries Service set up by the U.S. Government, whose main function is to provide guarantees of this nature. When an advanced country such as the USA has found it necessary to have an organisation to provide guarantees, it is all the more necessary that a developing country such as India should have a similar system. It is unfortunate that the Government of India does not have the will to take the risk to establish a system for providing guarantees to deep sea fishing entrepreneurs. Compared to the benefits that would accrue to the nation from such a system the risk that will be taken by the Government is negligible.

The US National Marine Fisheries Service (NMFS) issues an obligation guarantee to any financing institution/bank on behalf of a fishing entrepreneur, guaranteeing to his lender that the loan provided will be repaid. This guarantees the repayment of money borrowed by an enterprise for constructing, reconstructing or re-conditioning commercial fishing vessels. The purpose of the obligation guarantee issued by the NMFS is to enable the fishery enterprises to get loans to which finance or refinance of an adequate portion of the cost of major capital investments at reasonable interest rates, and for lengths of time commensurate with enterprise's ability to repay.

The obligation guarantee is available from NMFS for financing or refinancing the acquisition of fishing vessels or reconditioning them upto 87 per cent of the cost.

The NMFS guarantee obligation reduces the lender's risk of loss to zero. The lender can be certain that if the borrower will not repay, the US Government will pay him the amount. The additional feature of this system is that NMFS guarantee can be used as collateral, as approved by U.S. Federal Reserve Bank, 'by the lending bank. These guarantees can also be assigned by the financing institution to secondary purchasers.

Under the US system a lender's paperwork is extremely minimal. There are only two simple documents which will be prepared by the NMFS. These are (a) a one page obligation (a promissory note) and (b) a one half-page guarantee.

The NMFS guarantee is given for 100 per cent of the obligation including interest payments under the guarantee are made in cash and will be made within 30 days from the date of lender's demand.

If an entrepreneur is unable to find a lender, NMFS will make all approaches to find one.

The Indian deep sea fishing industry is now at crossroads. The present drift can only be continued by the Government to the detriment of national interest. The development of deep sea fishing is a major programme. This cannot be achieved by the Government without taking any risk. The previous Union Finance Minister Mr. V.P. Singh did the greatest of disservice by closing down the facilities available to the industry from the Shipping Development Fund Committee, without creating a viable alternative for providing ready facilities. It is now the duty of the then Finance Minister Mr. N.D. Tiwari to set right the situation by taking a clear, imaginative and bold decision to provide a simple and enforceable system of extending financial assistance for the deep sea fishing industry which is to play a vital part for increasing fish production, for stepping up of exports and for earning foreign exchange and acting as a second line of sea defence.

Fishing Chimes, Editorial 87: October(ii) 1987: Vol. 7, No. 7

# Agricultural Licences and Market Fees on Marine Products in AP

The A.P. Government is perhaps the only coastal State where the anachronistic, illogical and ill-conceived system of levying market fees on fish and fish products exists. The machinery of the A.P. Government has subjected the various fishing and marine products processing units in the State to pay licence fee and market fee on the marine products handled by them under Sec. 7(1) and Sec. 12(1) of Andhra Pradesh (A.P. and L,S.) Market Act, 1966 and Amendment Act, 1987. On a representation made by the industry the Commissioner for Development of Marketing and Director of Marketing, Andhra Pradesh, Hyderabad stayed the implementation of the order on 10 December, 1980. This ban has been lifted on 1 June, 1987 leaving the doors open for the officials of the directorate to resume collection of the licence fees and market fees.

The Act referred to above is related to agriculture and live stock. The fish and prawn caught from the sea are not agriculture produce by any stretch of imagination. These are caught from the sea and there is no culture process involved in this. It is not therefore clear as to how authorities have interpreted the catches from wild stocks at sea as a product from culture. If catches of fish from the wild are agricultural products, then forest produce, oil and gas etc are also agricultural products and merit inclusion as taxable items under the Act.

Deep sea fishing is a central subject under the Constitution. Most of the sea fish and shrimp catches are from deep seas. Assuming for a moment that these catches are agricultural products, it is difficult to support or justify the State Government's right to legislate on products that come out of an activity under the central list of the Constitution. This being the position, the inclusion of fish and fish products in the schedule under group IV in the AP. Act should only refer to fish and prawns produced in fish ponds, as a result of fish culture, an activity akin to agriculture. The Market Councils being the creatures that come under the Andhra Pradesh Market Act, they are not competent to hold that any of the provisions of the Act are legally applicable to marine capture produce too. This competency vests in higher judicial authorities.

Article 286(1)(b) of the Constitution prohibits a State law to impose tax on a sale or purchase which takes place in the course of the import of the goods into or export of the goods out of the territory of India. To give effect to the above constitutional provision the constitution amendment Act 1976 (Act 103 of 1976) inserted the following sub-section under section 5 of the CS.T. Act.

"Notwithstanding anything contained in Subsection (1), the last sale or purchase of any goods preceding the sale or purchase occasioning the export of those goods out of the territory of India shall also be deemed to be in the course of such export, if such last sale or purchase took place and was for the purpose of complying with the agreement or order for or in relation to such export." This amended Subsection became necessary to effectuate the interpretation given to the words "in the course of occurring in Article 286(1)(b) of the Constitution".

The purchase of shrimps by the processors is "in the course of exports". This is because of the fact that soon after the purchase, the shrimps are sent to a processing factory of the company purchasing shrimp where the material is processed for export. Therefore, the purchases immediately preceding the processing are the last purchases "in the course of exports" and such sales/ purchases are not taxable under the Constitution of India.

It may be noted that under the AP. Act it is only when an area has been declared as a market area and it is functional, that the authorities enforcing the provisions of the Act will become eligible to collect the taxes/fees under the Act. Firstly an area is to be declared by the State Government as a market area. A Committee has to be then constituted and thereafter the Committee must actually function. Before embarking upon the collection of fees the Committee has to establish markets in accordance with the directions issued by the Government. This is mandatory. In other words, there must be directions from the State Government to establish marine products markets which have to be followed up by the Committee. Then only the Committee can exercise the power to collect fees. It is not known as to why the authorities of the State Government are keen on enforcing the provision of the A.P. Market Act on a developing activity which does not really come under the purview of the Act. It is suggested that the higher authorities of the State Government may take a close look at the matter and advise the Commissioner for Development of Marketing and Director of Marketing, Andhra Pradesh to desist from the collection of licence fees under section 7(1) and section 12(1) of the Act. The Commissioner of Fisheries would no doubt be able to advise the Government on this subject, the intricacies of which will not be apparent to the uninitiated *i.e.* those who do not know that deep sea fish catches are not agricultural produce.

The deep sea fishing industry is advised to bring the unreasonableness of the levy to the notice of the authorities concerned. If the obvious position does not register in the minds of the authorities concerned, the industry would not have any alternative other than approaching a competent court of law for redressal.

Fishing Chimes, Editorial 88: November(i) 1987: Vol. 7, No. 8

# Amendment to SDFC (Abolition) Act, 1986

In exercise of the powers conferred by Subsection (2) of Section (1) of the Shipping Development Fund Committee (Abolition) Act, 1986 (66 of 1986) the Central Government appointed 3rd April, 1987 as the date on which the Act came into force *i.e.*, the date on which SDFC was closed down.

In another notification, in exercise of the powers conferred under the Shipping Development Fund Committee (Abolition) Act, 1986, section 16 and subsection(1), the Central Government delegated to the Shipping Credit and investment Company of India Limited all its powers and functions under Chapter III of the said Act with effect from 3rd April, 1987. All powers under the remaning chapters vest in the Central Government at present.

The main Act was notified very recently *i.e.*, on 24th December, 1986 by the Ministry of Law and Justice (Legislative Department) in the Gazette of India, Extraordinary, Part II Section (1) dated 26 December, 1986. The Act provides for delegation of powers under Chapter III only to a 'Designated Person'. This is now sought to be amended, by the Central Government to acquire powers to delegate its functions under all the Chapters of Act to a Designated Person.

The intention seems to be to transfer to the Shipping Credit and Investment Company of India all past contractual obligations of the erstwhile Shipping Development Fund Committee, which have now devolved on the Central Government under the said Act. It appears that the proposed amendment is only a device on the part of the Government to avoid direct accountability. The Government is certainly capable of shouldering the responsibilities connected with extension of financial assistance for the development of deep sea fishing fleet on its own. A professional banking institution is not expected to, and is also not equipped to undertake developmental financing. SCICI cannot be expected to shoulder developmental obligations of the Central Government pertaining to deep sea fishing. Why should the obligations be transferred to SCICI? It is not correct in principle for a Government to entrust responsibilities of this kind to a firm. Government should shoulder the contractual responsibilities and not subject the fishing companies coming under these obligations to the rigidities of professional bankers whose approaches are not attuned to developmental banking.

From April 1987, the month in which it was entrusted with certain functions under the Act, the Shipping Credit and Investment Company of India has released only one or two supplemental loans for the acquisition of fishing vessels. It has been insisting on conditions that cannot be complied with by the entrepreneurship, such as depositing of 25 to 50 per cent of the cost of the vessels proposed to be acquired with it in the first instance. With each of the vessels costing not less than a crore of rupees, no company can have so much of liquid cash to deposit with SCICI. Because of this realism only, the erstwhile Shipping Development Fund Committee which existed from 1958 had adopted a debt-equity ratio of 6:1 for loans and 10:1 for guarantees. In the case of fishing vessels for which SDFC gave loans the recoveries had been of the order of 85 per cent of the dues, which, by any standard, are a high rate of recovery. SCICI can of course say that they can lend only to enterprises that can conform to their rules. But this does not promote development of deep sea fishing.

The total stoppage of financing for the acquisition of deep sea fisting vessels, caused because of the abolition of SDFC and entrustment of the financing function to a professional financing body (SCICI) has led to the stagnation in the development of our deep sea fishing fleet. The Central Government (believed to be because of the Ex-Finance Minister Mr.V.P.Singh, who spearheaded the SDFC [Abolition Act] is responsible for this situation. An ongoing system was disturbed and a new practically non-working system was introduced.

The 7th plan envisages the raising of the deep sea fishing fleet level to 500 nos. and it now stands at about 110 nos., achieved in two decades. Over 3 million tonnes of marine fishery wealth remains unexploited and the progress towards its exploitation has come to a halt because of what we believe to be an unwise step taken by the former Finance Minister Mr.V.P.Singh to close down the SDFC. The result is that exports of marine products are stagnating. Marine products exports can be promoted only by enlarging the deep sea fishing fleet but certainly not by closing down a financing system for the acquisition of deep sea fishing vessels for no valid reason. If owners of some of the cargo ships and fishing vessels could not make repayments of loans taken and the dues accumulated, further lines of credit to these defaulters could probably have been stopped. In the case of fishing industry, recoveries were reasonably good and therefore the SDFC could have been allowed to continue to exercise its function. In order to stepping up funds for lending, and in regard to a Secretariat, MPEDA is a body with expertise on the subject and is vitally interested in the development of deep sea fishing fleet with a view to boosting up exports. Instead, of considering this excellent alternative, these powers were delegated to a newly created company *i.e.*, the Shipping Credit and Divestment Company of India which will need time to build up back-ground knowledge in regard to deep sea fishing. So much so, since April, 1987, this body could not make any progress in this vital sector, a positive proof of its unsuitability for the purpose. There is no justification for entrusting functions of a developmental nature which are the prerogative of the Government, to a financing firm. The abolition of SDFC has thus virtually created a disaster in regard to additions to deep sea fishing fleet and exports of marine products.

In the end, it is felt that the passage of the Bill (already introduced in the monsoon session) to amend the Shipping Development Fund Committee (Abolition) Act, 1986 to provide for delegation of powers by the Central Government under all Chapters of the Act to a 'Designated Person', which will be the Shipping Credit and Investment Company of India, is not in developmental interest. In fact, the repeal of the Shipping Development Fund Committee (Abolition) Act so far as fishing vessels are concerned will be a correct step to undo the wrong that was done.

In any case, there is no reason why the Act should be amended so soon after it is placed on the statute book. Government should have thought of the need in the beginning itself. In case the Act cannot be repealed the powers under the Act should be vested in the Marine Products Export Development Authority and, if the Act is to be amended, it should be to provide for this purpose only, since, only this autonomous body has the needed developmental expertise. By having a suitably equipped financing wing for the purpose, MPEDA can promote deep sea fishing actively. The amendment bill will come up for discussion in the parliament session that would commence on 6 November, 1987. The passage or otherwise of this bill will set the tone for the future development of deep sea fishing in the country.

Fishing Chimes, Editorial 89: November(ii) 1987: Vol. 7, No. 8

# Fish Culture in A.P: A Miracle at Work

Twenty five fully laden truck loads of pond-grown major carps wend their way every day to Calcutta from East Godavari, West Godavari and Krishna districts of Andhra Pradesh. Ten years back, only a few tins containing magur and singhi used to be despatched by train daily to Calcutta from these districts. Now, Distinct offices, exclusively to meet the demand of trucks for transporting fish, have come up at places such as Akividu, Bhimavaram and Kaikalur.

From practically nothing, in s period of ten years, there has been an economic growth involving production of over 40,000 tonnes of pond grown freshwater fish and marketing the same in a distant city. The turnover exceeds Rs.100 crores. Thousands of men eke out living out of the activity now.

The spread of the avocation in the private sector has been so unrelenting and overpowering that in an Andhra village by name Potumarru, all paddy fields of the village, 500 acres in at extent, have been converted into fish ponds. There is no paddy cultivation in this village now. Fertile paddy fields in deltaic areas are being gradually converted into fish ponds. It is a common site in most of the coastal districts and to some extent in Telangana and Rayalaseema districts of A.P. to see fish ponds all over.

There is no precise estimate of the extent of pond area brought under fish culture. It is unfortunate that no serious effort has been made to conduct a census and to evaluate the level upto which the latest technology has taken roots, the precise extent of additional water area brought under culture, quantum of production etc. The State Fisheries Department would no doubt take needed steps to collect relevant data.

There is no doubt the initial spurt came as a result of the efforts by the A.P. State Fisheries Department and the substation of Central Institute of Fisheries Education, Bombay at Balabhadrapuram in A.P. After this, the entrepreneurship took over and there was no looking back. The spread of the activity was so fast, the official machinery could not catch up and the initiative went into the hands of the resourceful farmers. They imbibed the technology and adopted it to meet their specific conditions of existence.

It is believed that fish production in the private ponds that have come up in the recent past varies from one to ten tonnes per hectare. A survey may well reveal that over 25,000 hectares additionally must have been brought under culture, enabling under its wake tens of thousands of men to become either economically prosperous or have a satisfying means of living.

The revolution in fish production is only half of the story. In the sector of freshwater fish seed production also there has been a major breakthrough in the State, unbelievable in terms of proportion. In East Godavari district alone one can count atleast 33 seed farms, and how many of which have Chinese type of hatcheries is not clearly known. Some of them atleast do have them. In Krishna district it is stated that there are over 30 Chinese type of hatcheries, producing over 40 crores of major carp spawn of which atleast 30 crores are believed to be of Catla.

Your editor has had an occasion to meet a few of the farmers owning fish farms as well as fish seed farms. They speak in a truly professional language. The depth of their knowledge in respect of preparation of ponds, stocking schedules, care of crop, control of diseases, harvesting patterns etc is amazing.

As a parallel to the freshwater sector, a revolution in brackishwater shrimp farming and brackishwater shrimp seed production is now brewing up in A.P. Several farmers all along the coastline of AP. have already taken up brackishwater farming. They are now frantic about feed and seed. Most of them have acquired good knowledge about prawn seed production and prawn culture. Some of them already visited Taiwan, Philippines and other countries and studied various aspects of this industry. Professionals in the field working in Government institutions will be amazed at the depth of knowledge of these farmers.

It is stated that over 2,000 hectares have already been brought under prawn farming. Several thousands of brackishwater areas have been purchased by enterprising farmers for setting up brackishwater farms. They are totally confident about what they are doing. The developments in the brackishwater sector are virtually due to the unrelenting promotional work undertaken by the Subcentre of the Marine Products Export Development Authority at Machilipatnam to promote the activity. The farmers are guided and educated so well in the technical and economic aspects of brackishwater shrimp farming and brackishwater shrimp seed production that, with unbelievable devotion, several farmers in coastal A.P. are now totally dedicated to it and the activity has now become virtually a religious practice. In about five years, without the slightest shadow of doubt, one can predict that the coastal fringe of Andhra Pradesh will have a flourishing shrimp production industry, coming close in comparison to the production levels in countries such as Ecuador. And credit for this, mainly goes to MPEDA.

Inland States, with the exception of those in the northeast will benefit greatly if they can send teams of fish farmers to A.P. to study the progress in inland fish farming and fish seed production. Likewise farmers in coastal States may gain from the infective enthusiasm of Andhra brackishwater farmers in the development of this activity and also in observing what has already been accomplished by them.

This is the time when the Central and State Governments can take advantage of the upwelling enthusiasm of the farmers in order to strengthen it and channelise it. Subsidies are not really that important. A few firm and practical steps to introduce mini-shrimp hatcheries and minifeed production units will go a long way in hastening the pace of the activity. As already mentioned, the A.P. State Fisheries Department may undertake a systematic and comprehensive survey of the fish farms and fish seed farms that have come up in the private sector, in the State.

Fishing Chimes, Editorial 90: December(i) 1987: Vol. 7, No. 9

# Strategy for Development of Deep Sea Fishing

Strategy and structure are the hardware of any developmental activity. Indian deep sea fishing, being a developing activity, is no exception to this axiom.

The introduction of a proper strategy and structure for the development of deep sea fishing in Indian waters continues to evade an effective solution. The process of evolving one has been going on since the third Five Year Plan without any success.

## Past History

There have been four schemes introduced by the Government so far for the development of deep sea fishing (in 1968, 1973, 1977 and 1981). Of these, in terms of timelag and achievement, the ratings were low in respect of 1968 and 1977 schemes. Same is the case in respect of the 1981 scheme. The scheme introduced in 1973, however, yielded comparatively superior results, in a duration of four to five years. 28 out of 30 targeted vessels were introduced. Logically, the strategy adopted under the 1973 scheme deserves to be revived with the anticipation of repetitive good results under the programme. In this connection, it has to be mentioned that the strategy aspects of the 1973 scheme had been good until the stage of introduction of the vessels. All aspects, including financing, were co-ordinated by the Government.

In the post-introduction stage there were several defaults in repayments on the part of the fishing companies, although the scale of this came down over a period of time. Under the other schemes there was no coordination by the Government. The selection of vessels, prices, source of supply etc., were left to the entrepreneurs while applying for clearances, provided the vessels are over 20 m in length.

Loans from the erstwhile SDFC were made available under the schemes introduced in 1973, 1977 and 1981 until March 1986, to the extent of 90 per cent of the cost in respect of imported vessels and 95 per cent in respect of indigenous vessels. The rate of interest charged was 4.5 per cent p.a. with a repayment period of 15 years and a moratorium of one year. The debt-equity ratio stipulated was 6:1. Loans in foreign exchange from foreign banks were allowed to the extent of 100 per cent of the cost in respect of imported vessels under EOU scheme. All these financing facilities did not result in the expected increase in the deep sea fishing fleet, which now remains at about 110 nos.

The policy till recently had been to consider applications for the introduction of all types of vessels of over 20 m length. The response from the entrepreneurship had however been in favour of operating trawlers for the main purpose of catching shrimp. The reason for this bias was the feeling that only with some shrimp catches, operations can become economical. Government has now modified its policy to the extent that permissions for the introduction of multipurpose vessels will not be given. This is on the misplaced assumption that multipurpose vessels are synonymous with shrimping vessels. Even a lay man can see that the argument is weak for the reason that a multipurpose vessel need not have any arrangement for the purpose of catching shrimp and can have arrangements for catching other varieties.

## Financing Drawbacks

No financing body can offer terms that are more liberal than those adopted by the erstwhile SDFC. Yet, the actual availability of funds from the erstwhile SDFC and utilisation of the facility was poor, considering the fact that assistance could only be extended to raise the fleet strength to about 110 nos. against the 7th plan target of 500 nos. There must surely be something seriously wrong somewhere for this bleak situation. Fear of financial risks because of fluctuating fishing results, crew problems, efforts needed in following procedures, limited availability of expertise in the field, absence of shore facilities etc., might have contributed to the disappointing performance.

The performance of the erstwhile SDFC and that of the entrepreneurship could have been improved by the authorities concerned by taking certain steps, taking advantage of the improving repayments position, which was stated to be of the order of 85 per cent of the dues, at the time of the closure of the SDFC. Instead of toning up the situation, Government abolished SDFC with effect from April, 1987. The work was entrusted to a new financing body, namely Shipping Credit and Investment Company of India. Even with offer of extremely soft loans by the erstwhile SDFC, much progress could not be made in developing the deep sea fleet to the needed level. The terms and conditions governing sanction of loans by the SCICI for acquiring deep sea fishing vessels not being within the reach of the entrepreneurship, the situation became much worse. On top of this came the revised policy of the Government that only resource-specific vessels will be permitted to be introduced. Entrepreneurs feel that operation of resource-specific vessels will be un-economical. The problems in respect of securing finance and uncertain economics of resource-specific vessels which can operate only for part of a year have retarded the activity.

## Need for Change in Strategy

This situation calls for a change in strategy consistent with the characteristics of the deep sea fishing entrepreneurship which consists mostly of those hailing from small and medium levels, so as to bring about a faster realisation of the 7th plan target of introducing 500 deep sea fishing vessels. A change in strategy is needed as the revised strategy adopted for financing acquisition of deep sea fishing vessels by 'Shipping Credit and Investment Company of India' virtually turned out to be a damp squib. Its function is to provide financial assistance to entrepreneurs for the acquisition of cargo ships and fishing vessels. Ships carry goods. Fishing vessels catch and store fish. Patterns of investments and returns and expertise in respect of these two categories are vastly different. Yet the clubbing of the two for financing by SCICI, as in the case of SDFC, continues to be an anamoly.

So far as financing pattern is concerned, being professional bankers, SCICI rightly insists on a debt-equity ratio of 2 : 1 to 3.5 : 1, according to merits of each of the cases. Apart from this, SCICI's norms require that one has to deposit with it in cash at least of 25 to 35 per cent of the cost of a vessel for which loan is sought. This is clearly beyond the capability of most of the entrepreneurs having an inclination to enter the field. The net effect of the closure of SDFC and the assumption of the function of financing deep sea fishing vessels by SCICI has been that, since April, 1987 there has practically been no financing for the acquisition of deep sea fishing vessels.

Financing bottleneck is one aspect. The other aspect is the irrational stipulation that multipurpose vessels will not be permitted to be introduced. Only resource-specific vessels for a single purpose will be allowed. Most of the fishery resources are available seasonally and are not amenable for commercial exploitation with profit, if they are oriented for catching a single resource only. This means that vessels acquired through heavy investments have to be kept idle for a period of 3 to 6 months in a year due to non-availability of the same single resource for which a vessel is equipped.

## Conclusion

The main conclusion that can be drawn from the present situation is that something radically different has to be done in order to achieve progress in the sector. In other words, several variables which are interrelated but are left loose have to be related to each other and coordinated in a way that loose pieces are made to fit into a mosaic. While evolving a new strategy the following inescapable conclusions have to be borne in mind.

i) Substantial financing, either direct or indirect (in foreign exchange or in rupees, through guarantees or other-wise) is necessary. ii) Co-ordination by the Government in respect of selection of shipyards, vessels, prices, companies to operate etc., which are the most important of the variables, next to financing arrangements, is imperative. iii) Introduction of a substantial no. of vessels for operation in specified areas to catch specific resources, more than one, is essential to ensure successful commercial operations. Operational areas and resources to exploit are the other important variables.

## Compulsive Role of Government

All the above variables virtually indicate the need for financing by Government or its agencies, and coordination by the Government. Under earlier SDFC norms 90 to 95 per cent of the cost of vessels was coming from the Government. Such being the case, Government can as well venture into this business of deep sea fishing investing 100 per cent of the cost, adopting a suitable strategy. There is no point in lending upto 90 to 95 per cent of the cost and looking at others (entrepreneurs) to do the job. The additional 10 to 5 per cent can as well be invested by the Government and total control over the sector can be assumed by the Government. This will resolve the present problem faced by entrepreneurs in securing loans, providing guarantees, collaterals etc. What can probably be done by the Government is to acquire vessels first and lease them out later under certain terms and conditions to selected entrepreneurs or agencies on payment in advance quarterly/half yearly lease amounts as fixed. The vessels acquired shall have to be equipped for the exploitation of two types of resources so as to achieve commercially viable operations.

## Strategy Needed

The above approach calls for a fresh strategy and structure. The strategy should be for the Government to enter into bilateral arrangements with willing countries such as Holland, Japan etc., for the supply of a specified number of resource-specific vessels, with facility for two types of fishing, by their shipyards under Government to Government credit, repayable in a specified number of years. These vessels may be deployed for operation by public sector undertakings.

Fishing Chimes, Editorial 91: December(ii) 1987: Vol. 7, No. 9

# Development of Reservoir Fisheries

The development of reservoir fisheries is closely linked with an assessment of their seed stocking needs and actual stocking from year to year with advanced fingerlings of fast growing fish species. The assessment of the stocking requirements is connected with the present status of fisheries of a given reservoir, the existing fishing effort and the type of fishery to be promoted, consistent with the conditions of existence and the types of fishing that are suitable for the waters.

Determination of the rate of stocking of fingerlings per annum in a reservoir in a precise manner is a difficult process. Pending conclusive studies to arrive at the annual seed requirements, for a particular reservoir, it should be possible for the State Fisheries Departments to cordon off a part of the reservoir concerned in a suitable location and rear major carp fry of a suitable type therein till such time as they reach advanced fingerling size. The present availability of major carp fry during April and May, well before the onset of floods in most parts of the country is an advantage as the effect of floods can be avoided during the rearing period. Thereafter the seed could be released into the reservoir concerned. This step may lead to an increase in the fish production results from our reservoirs.

It may not be technically feasible sometimes to have a carved nursery area as mentioned above because of the prevailing conditions at some of the reservoirs. In all such cases floating cages made out of synthetic webbing of appropriate size could be utilised for raising the needed number of fingerlings for stocking a reservoir. In the alternative, it should be possible to earmark specified fish seed farms close to the target reservoir, grow spawn/fry to advanced fingerling size in them and utilise the same for stocking of a nearby reservoir. Pending detailed studies of stocking requirements ad hoc measures on the lines mentioned above may go a long way for stepping up fish production from our reservoirs.

Fishing Chimes, Editorial 92: January(i) 1988: Vol. 7, No. 10

# Separate Ministry for Fisheries Imperative

The Ministry of Agriculture, as a nodal Ministry and the ICAR as the main agency for evolving new technologies and extending them to the field and for spreading fisheries education, have accomplished something very significant in the past two or three decades in the matter of development of fisheries of the country. The notable achievements, among others, are: (a) setting up of fish farmers development agencies, (b) introduction of technology of induced breeding of fishes and setting up of carp hatcheries, (c) introduction of the technology of composite fish culture, (d)introduction of mechanised and motorised fishing boats and medium sized fishing trawlers for sea fishing, (e)setting up of a large number of minor fishing harbours and several major fishing harbours, and (f)promotion of fisheries education and training through setting up of various institutions. The technology of integrated farming is at threshold stage. Likewise the stage is set for augmenting the deep sea fishing fleet by introduction of fishing vessels of various types as well as combination vessels for undertaking diversified fishing operations with a view to utilising available fishery resources of the Indian exclusive economic zone. But for the various developmental programmes promoted by the Ministry of Agriculture and the Ministry of Commerce and MPEDA, the nation would not have been able to augment the annual exports of marine products to a level of Rs.460 crores by 1986-87.

The achievements represent the starting point for a massive expansion of the programmes ensuring quality in production. The sad part of the good story is that the pace of fisheries development has come to a virtually standstill, particularly from the Sixth Plan onwards. The reason for this is not lack of expertise or capabilities on the part of the Ministry. They are available in abundance. The real reason for the stagnation lies in the fact that the Ministry of Agriculture is a heavily burdened creature, called upon to take care of several subjects, besides agriculture which is its main charge. These subjects are supposed to be allied to agriculture but are not so. Marine fishing comes no where near agriculture. The daily burden on the senior officers in the Ministry, most of whom are bureaucrats, is so high, that it is physically and mentally impossible for them to pay the requisite attention in depth to the subject of fisheries which is an area with great potential for development and calls for detailed application of mind and the closest attention. Exclusive attention to the subject of fisheries will lead to generation of employment, reclamation of fallow areas under fish production, introduction of new technologies relating to mariculture, which can augment fish production significantly, contributing towards boosting up of exports and earning additional foreign exchange. If there is a separate minister to take care of fisheries development aided by a set of devoted officers, there will be a sea change from the perfunctory attention now paid to the subject to an exclusive, vibrant, innovative and forward looking approach.

The main problems faced by the Department of Agriculture, in respect of co-ordination of roles of various other Ministries connected with fisheries would become minimal, once there is a separate Minister and a separate developmental and administrative set up to deal with the subject of fisheries.

At present, the Ministry of Agriculture has not been able to bestow individual attention towards even important issues such as exploitation of fishery resources of our exclusive economic zone, provision of needed infrastructural facilities at various fishing harbours and several other aspects. A separate Ministry for fisheries at the Centre will pave the way for major improvements in the sector.

The job ahead in respect of development of fisheries is of immense proportions, not yet adequately apprehended at political and administrative levels. A task involving the achievement of an additional annual production of 2.5 million tonnes of sea fish, over 2 million tonnes of freshwater fish and over one lakh tonnes of shrimp from brackishwater certainly calls for specialised attention. This is an activity that has potential to add over Rs.300 crores to national foreign exchange earnings through exports. It has also the potential to add Rs.100 crores per annum to the national income. Programmes for utilising this potential have to be carefully drawn up and implemented assiduously and it is not correct to allow them to languish as is unwittingly happening now.

Under the allocation of business rules, no doubt, the subject of fisheries is the responsibility of the Ministry of Agriculture. The Ministry however does not enjoy the powers to take decisions on important fisheries developmental matters in the same way as it can do in respect of agricultural matters. It is dependent on several other ministries to give shape to its decision.

In the same way as there is no separate Ministry for Fisheries at the Centre, none of the State Governments, with the exception of Governments of West Bengal and Assam, have separate departments for fisheries at secretarial level. There is an urgent need for all other State Governments to set up separate departments of fisheries at the secretariat level and appoint the directors of fisheries as *ex-officio* Secretaries to the Governments. This will ensure faster decisions and quicker implementation of the various developmental projects. In several States there are ministers to deal with fisheries in addition to one or two more subjects. This system is a major improvement on the conditions in existence with the Government of India. The present system can be further improved by having ministers for fisheries exclusively, as in the case of States such as Assam.

Most of the ills now plaguing the development of deep sea fishing and other sectors of fisheries will get resolved to a large extent, once a separate ministry for fisheries is set up at the centre. This ministry should however encompass the present functions exercised by the Union Department of Agriculture, and Union Department of Ocean Development in respect of fisheries, besides those exercised in the Ministry of Commerce and Ministry of Surface Transport. The development of fisheries in the State sector could also be improved to a great extent, once separate departments of fisheries are set up at the Secretariat levels, and separate ministers for fisheries are appointed in the States.

Fishing Chimes, Editorial 93: January(ii) 1988: Vol. 7, No. 10

# Effect of Drought on Fish Production: Relief Measures Must

1987 was a very bad year for fish production. The annual statistics of fish production for the year, as and when published would establish the fall in fish production which would entirely be because of severe drought conditions that gripped almost all parts of the country.

The inland as well as marine fisheries sectors have suffered this year because of drought conditions. For want of rains a large number of tanks and ponds remained unproductive and empty. Some areas did receive water supply, but very late and in small quantities. This situation did not permit stocking of the tanks and ponds with fish seed, which, in any case, could not be produced in adequate quantities for the reason that maturation of fish suffered greatly in the absence of rains. At very few places, however, it had been possible to produce major carp seed and also seed of exotic carps under controlled conditions but this production was negligible.

In the case of reservoirs where natural production of major carps normally takes place, production was very poor. Same was the case in all major rivers of the country.

There is no doubt that fish farmers all over the country suffered severe losses. It is not known how many farmers are covered under fish crop insurance. It is hoped that those who are covered under fish crop insurance have taken care to preserve all needed evidence to prove the effects of drought conditions on fish seed production and also fish production.

In the case of marine fisheries, all along the coastal belt and also in the offshore areas arid deep sea zones the fishery has suffered considerably during the year. The reason for this is that none of the major rivers received the normal floods and in time. The reduced flow of flood waters into the sea has had the effect of altering the hydrographical conditions and nutritional status of the seas. These not being congenial for fish growth and reproduction of fish, many fishermen have experienced a major reduction in the fish catches. Those operating larger fishing boats have also experienced the same, causing a major set back to the economics of operations. The frequent cyclonic conditions over the Bay of Bengal have paralysed fishing operations from June to December at least six times, upsetting the fishing operations. The few days during which fishing could be conducted resulted in lower catches, attributed to poor reproduction and growth in fishes. Similar bad years were experienced in the recent past on two occasions.

Most of the fishing vessels, from small mechanised boats to larger ones were acquired by the enterprises with financial assistance either from commercial banks, State Financing Institutions or the erstwhile Shipping Development Fund Committee/Shipping Credit and Investment Company of India Limited. The capabilities of fishing companies to make repayments depend mostly on the earnings out of the sale of fish catches landed by their vessels. During years when the fishing seasons are good and earnings are average or above average, repayments towards instalments of loans taken would be facilitated. If any enterprise fails to make repayments during a good fishing season, the lending institutions will have no option other than instituting coercive measures for collecting the amounts due.

During bad fishing years, when it is clearly established that the fishery had failed in one or more regions of the coastline, it will be incumbent upon financing institutions to consider providing relief measures to the concerned enterprises. This could take the form of postponement of collection of instalments or rescheduling of loans. These relief measures will have to be exactly on the same lines as is adopted in the case of agricultural farmers.

The level of success or failure of fisheries can be estimated by the authorities concerned on an annual basis with reference to the selected parameters. A scale to measure the volume of fishery can be evolved so that, in drought conditions, the scale level of "fish famine" can be arrived at. When the fishery comes down to lower levels in the famine zone it will be necessary for the financing institutions, based on the recommendations of the appropriate authorities, to provide the needed reliefs. The declaration of the status of fishery and the stage at which it falls on the scale in any particular year should be the responsibility of a specified Government organisation which should determine this based on the recommendations of a Standing Committee to be set up by the Government. This Committee should have representatives drawn from the Ministry of Agriculture, the Fishery Survey of India, the Central Marine Fisheries Research Institute, the Department of Ocean Development, State Fisheries Departments, fishing industry, financing institutions etc.

It is felt that the Ministry of Agriculture should set up this Committee for this purpose in the shortest possible time. This committee should first evolve the various creteria to be taken into account for judging and giving rating to the status of fisheries during any particular year. Based on these criteria the Committee should be able to give rating of fishery levels. The Committee should also make recommendations to the Government concerning the level at which the fishing industry should be eligible to acquire reliefs by way of postponement of collection of loan instalments or re-scheduling instalments.

The fishing industry, for want of experience, has been enduring the ordeals arising out of fish famine in bad years, without seeking reliefs, as is common in the agricultural sector. The enterprises have been dumbly putting up with the hardships flowing out of bad fishing years. This has resulted in accumulation of dues to financing institutions. The industry should now wake up to the situation and seek the help of the Government in developing a machinery to judge the level of fishery output every year and to provide the needed relief when the fishery level drops down to an uneconomic level.

Fishing Chimes, Editorial 94: February(i) 1988: Vol. 7, No. 11

# Financing of Deep Sea Fishing Projects by Commercial Banks: Strategy Suggestions

Several Commercial Banks have the basic interest to extend financial assistance for the promotion of deep sea fishing projects. Efforts of these Banks, barring a few which participated in joint financing along with other banks, have not yet been harnessed into a sizeable commercial lending activity.

The reasons for this are often stated to be heavy overdues, and lack of adequate expertise on the part of the fishery enterprises. Apart from this, commercial banks, for want of adequate knowledge concerning deep sea fishing industry are not equipped to monitor the operational aspects of the deep sea fishing vessels.

Some of the banks also feel that financing of deep sea fishing projects is the responsibility of the Shipping Credit and Investment Company of India.

The SCICI is a financing institution set up to extend financial assistance to projects connected with the acquisition and operation of cargo ships and deep sea fishing vessels. The existence of this company does not in any way prevent the commercial banks from extending financial assistance to deep sea fishing projects, unless they are dissuaded from taking up this.

Commercial Banks are eminently suited for extending financial assistance for deep sea fishing projects. Firstly, unlike SCICI, all major commercial banks have branches at all important port towns. This single factor is a plus point, from the angle of monitoring and supervising the operations of fishing companies to whom loans will be given.

Banks now reject applications for assistance for taking up deep sea fishing projects invariably on considerations such as lack of professional knowledge, risks involved in investments and several other reasons. In order to get over these problems and undertake developmental financing for promoting deep sea fishing projects, banks have to adopt a specialised approach. By following the normal banking norms and practices, it will be difficult for any commercial bank to extend financial assistance for promoting these projects. As matters stand, the initiative for the selection of fishing vessels to be acquired, negotiations of their prices and all other connected matters will have to be done by the entrepreneurs. Comprehensive project reports have to be prepared by them and submitted to their bankers. Thereupon, the bank concerned applies its mind to the proposal.

The Indian Exclusive Economic Zone has an estimated unexploited fish resource of the order of 2.8 million tonnes. The Government of India is very anxious to utilise this resource not only for the purpose of providing additional proteinaceous food for the national population but also for augmenting exports of marine products to other countries for earning additional foreign exchange. In other words, the development of deep sea fishing in our EEZ is one of the national priorities and it is the responsibility of Commercial Banks to participate in promoting this programme.

It will be a wrong way of reasoning to say that, because SCICI is there, Commercial Banks need not extend financial assistance to the deep sea fishing industry. The presence of major branches of Commercial Banks in all port towns gives an edge to them over the SCICI in participating in such a programme from the point of view of monitoring etc.

Commercial Banks should not leave everything to prospective entrepreneurs to come up with suitable proposals. When bargained for bulk supplies of fishing vessels, shipyards are likely to bring down prices. If an entrepreneur asks for two vessels, a shipyard may quote, say, a rate of Rs.1crore per vessel. Instead, if a coordinating body asks for the supply of 20 vessels, there is a likelihood of reduction in' price to the extent of about 10 to 15 per cent. In other words, scheme formulation should come from the banks and not from the entrepreneurs to start with. This major change in approach will alter the pace of introduction of vessels significantly.

Banks have to notify schemes on the above lines, after doing all the homework, inviting applications from entrepreneurs. Applications may be accepted on first-cum-first served basis, subject to fulfillment of various conditions. Prior clearance to such schemes from the Ministry of Agriculture will be needed with reference to available fish resources, ports from where vessels can be operated and such other details. Another advantage is that the Commercial Banks can secure the shop floor drawings of the vessels from the shipyards as part of the deal free of cost.

The additional advantage of the system is that the time factor involved in clearing applications would be reduced.

In a system of the kind referred to above, it will be necessary to have special wings at the branches of the concerned banks at port towns concerned, equipped with professional personnel to monitor the operations and collect instalment amounts due periodically. Returns can be prescribed by banks concerned to enable the fishing companies to furnish particulars such as date of departure of vessels, expected duration of voyage, grounds expected to be covered etc. On return from a voyage, particulars of catches and earnings will have to be furnished. Banks will be able to make proportionate deductions out of the earnings deposited with them by the owners.

It is understood that Canara Bank is exploring possibilities of formulating new guidelines in the field of financing of deep sea fishing projects. It is likely that other progressive banks like Punjab National Bank, State Bank of India, Syndicate Bank would also evince similar interest. It is hoped that our suggestions would receive their close attention.

Fishing Chimes, Editorial 95: February(ii) 1988: Vol. 7, No. 11

# Compulsive Location of Fish Markets as Spinoff of Shrimp Famine

The hydrographical and physico-chemical conditions of the north-west Bay of Bengal adjoining Indian coast seem to have altered in 1987-88 in a manner that is adverse to the thriving of shrimp and fish populations. The reason for this is believed to be that the flow of river waters into the Bay is greatly retarded because of poor precipitation in the catchment areas of the various rivers. The direct result of this is the degradation of the nutritional status of the waters. Another aspect is that there has been a total change in the current pattern as experienced by the captains operating vessels in the area. According to them, there is a practical cessation of currents in the north west Bay.

The major failure in shrimp fishery of the north-west Bay during the year has forced the operators to strive at catching more of fish, so as to improve the economics of operations atleast to the extent of meeting operational costs.

Although the availability of fish stocks in the Bay is also stated to have been adversely affected during the year, this does not seem to be as serious as that of the failure of shrimp fishery. In fact, fish landings by deep sea fishing vessels have increased considerably at Visakhapatnam. This has led to the need for organising marketing for the fish landed. This has become a challenging task to the owners and fish merchants. A few fish merchants operating from the base seem to have risen to the occasion and have created a few channels of supply to various places such as Madras, Calcutta, Hyderabad, Bhilai, Rourkela etc. For the time being despatches are being made in big baskets with fish packaged in ice. If further attention is paid to the creation of channels of supplies to the interior places from production points, the nation will have a domes-tic fish distribution system. This can be made into a well organised one, but dedicated efforts are needed for this. It may be necessary to have whole sale cold storages, main and subsidiary, along pre-determined routes. From these places fish can be radiated to retail stalls suitably equipped for keeping fish in fresh condition.

Another development at Visakhapatnam is that large quantities of low value fish are being either sundried or salt-cured by fisherwomen. The womenfolk sell the cured fish to merchants who take the material to various places. Large quantities are also exported to States such as Madhya Pradesh, Assam etc, mostly for sale in the tribal areas. All the links in this activity are capable of being developed further for an increased utilisation of low value fish landings for securing higher incomes.

Now that it is established that there are ways of marketing large quantities of fish, the Union Ministry of Agriculture can organise a study by a group of experts for suggesting a suitable and practicable scheme in this respect. With the implementation of such a scheme as approved by the Government, there will be a major change in the economics of operations of fishing vessels. Apart from this, a large number of people in far flung areas will gain access to buying fish. The various links in the marketing system will provide employment to several persons.

Fishing Chimes, Editorial 96: March 1988: Vol. 7, No.12

# Sinking Deep Sea Fishing Industry Needs Rescue Measures

The deep sea fishing industry is sinking. The functioning of its main engine (which is the financing system) is impaired because of the super status of the engineer (SCICI) manning it. Its rudder, which is the types of vessels to be permitted for introduction by the Government is also malfunctioning. The third cause for the advanced stage of collapse of the industry is the declining entrepreneurship interest in the activity.

The switching over of the function of extending financial assistence for the acquisition of deep sea fishing vessels from the erstwhile SDFC to the SCICI has virtually served the purpose of testing the real capabilities of the existing and emerging entrepreneurship to make investments based on the generally accepted debt-equity ratio applied by financing bodies while lending funds to conventional industries.

There would have been cause for rejoicing if the debt-equity ratio adopted by SCICI, condescendingly and reluctantly diluted by it from 2:1 to 3:1, has attracted entrepreneurs. But the dilution has attracted very few. This dilution should normally have proved to be a great attraction to the larger companies to take advantage of, coming as it is from a high status financing body, whose functionaries are all rightly accustomed to and believe in stringent financing norms. But, alas, this is not to be, although one company with non-resident Indian participation and a few others availed of the facilities extended by SCICI. With the abolition of SDFC and the advent of SCICI, entrepreneurship interest has waned. This is apparently because of the fact that the class of entrepreneurship interested in participating in the programmes of the Government for the development of the deep sea fishing industry could not stretch their financial resources any further from a level that meets with a debt-equity ratio of 6:1 as per erstwhile SDFC's norms, to 3:1 as required by SCICI. The deep sea fishing industry is now sinking fast for want of expansion. And for expansion entrepreneurship is important.

With the creeping in of disinterest among entrepreneurs, chiefly because of the unavailability of finance any more on concessional terms, the plummeting process to which the industry has been subjected to, probably unwittingly, continues unabated. The financial capabilities of the entrepreneurship interested in deep sea fishing to meet a higher debt-equity ratio as required by SCICI have now been conclusively tested. The results are by and large negative. Therefore, if the industry has to progress, it has to be rescued by the Government through reversion back to norms as adopted by erstwhile SDFC.

The setback in financing arrangements is, however, no doubt the main problem that confronts the deep sea fishing industry. Added to this is the unimaginative policy of the Government in respect of the types of vessels to be permitted for introduction. According to the present policy, only resource-specific vessels will be allowed to be acquired. In formulating this policy, no thought seems to have been given to a inadequacy or utter lack of information on fishing grounds, extent of availability, markets and other aspects in respect of several economically important fishery resources.

Regarding unconventional deep sea resources located but not exploited because of unfavourable economics of operation and utilisation of catches, it will be beyond the means of fishing enterprises to concentrate on this activity without sizeable support and encouragement from the Government. The seasonal nature of availability of most of the resources prevents entrepreneurs from investing on resource-specific vessels. One reason is that, in general, seasonal operations will not be economical. Second one is that it will be difficult to keep vessels idle for part of the year, bearing all overheads.

Resource-specific operations are mostly meant for global fishing. The large purse-seiner recently acquired by Indus Sea Foods is suited for global fishing. Tuna fishing in Indian waters being seasonal, resource-specific vessels of this kind can move to other oceans, according to seasons of availability. Vessels that have limited endurance and can fish within shelf areas in a range of less than 200 n. miles have to necessarily restrict their operations to national waters. Such being the position, it would have been much more prudent on the part of the Government not to have complicated matters by introducing the term "resource-specific vessels", not in vogue any where in the world. Government should not have also banned the introduction of 'multipurpose' and combination vessels, equating them with trawlers concentrating on shrimp catching. It may be true that vessels introduced in the name of 'multipurpose' or 'combination' vessels concentrate on catching shrimp. But this is no reason to ban vessels of this kind, which will be far more economically viable than non-global resource-specific vessels. Shrimp are mostly caught in bottom trawl nets. Whichever type of bottom trawl net one uses, the entry of shrimp into it cannot be prevented, particularly after the codend is partly full. In any type of bottom operated net, no more than 10 per cent catch will be shrimp. Anti-shrimp obsession should be controlled by realities and it must have limitations.

The number of applications for permission for introducing new vessels is understood to have come down after the term 'resource-specific vessels' entered the scene. Progress in the development of deep sea fishing will continue to be retarded until the usage of this ugly phrase and introduction of such vessels become minimal. Government will have to evolve a policy which enables introduction of deep sea fishing vessels that can be operated economically with reference to the present knowledge of fishery resources, their level of exploitation, seasonal occurrence, markets etc, and also the availability of shore facilities and scope for disposal of catches.

The most important of the causes that have led to the sinking phenomenon is the precipitously waning interest among entrepreneurs. This is accelerated by the virtual stoppage of fishing operations at Visakhapatnam, the only base along the entire Indian coastline from where deep sea fishing vessels operate. Entrepreneurs are scared of taking up this line of activity seeing the situation. Without enthusiastic entrepreneurship, development of deep sea fishing industry, as in the case of any other industry, cannot be achieved. The foremost task on the part of the Government is therefore to decide whether the development of deep sea fishing in Indian waters is to be followed up seriously or not. If the decision is affirmative, then enthusiasm among entrepreneurs interested in the field is to be restored to the earlier level first, then maintained and intensified.

The sinking deep sea fishing industry needs, in the main, the above mentioned rescue measures, if the country is to be benefited from the available fishery resources of its EEZ by way of additional fish production and increased exports and foreign exchange earnings, and to ward off incursions by foreign fishing vessels.

Fishing Chimes, Editorial 97: April(i) 1988: Vol. 8, No. 1

# The Eighth Step

*Fishing Chimes* has entered into its eighth year of life in April 1988. With this, seven years of fisheries journalism fortified by the patronage extended primarily by its readers, advertisers and contributors behind, *Fishing Chimes* now looks forward to a more intensively active future that would give greater opportunities to continue to serve all those connected with the fisheries sector. The progress of the journal towards maturity gives us great pride and we believe that readers of the journal also share the pride with us. The journal could survive and keep its flag flying in spite of the steep rise in printing costs and running expenses inadequately offset by marginally increasing returns.

During the past seven years those of us engaged in bringing out *Fishing Chimes* month after month could do so without break, although sometimes with inescapable delays. We believe that our closeness to the readers has been helping us in identifying what the readers want. The editorial staff have learnt to think in terms of identifying the specific information readers of various categories would need and have been striving to provide the same.

During these seven years, our set-up has been able to collect lot of literature and a wide spectrum of information on the sector. Efforts are now being made to acquire a computer so that analysed information on various aspects of fisheries could be stored for retrieval as and when needed by us or by any of the subscribers. These days we receive quite a number of enquiries seeking information on various aspects of fisheries and we do our best to answer them.

Readers will be glad to know that, apart from themselves, the journal has been referred to in favourable terms in several international journals of standing. The journal is now widely known in international circles. Several foreigners connected with the fisheries field visit our office to gather information on aspects of interest to them. The representatives of the journal were afforded opportunities to attend international fairs such as Nor-Fishing and also International Fisheries Conferences, organised by FAO. Within the country, the journal receives invitations to cover the proceedings of several seminars and symposia. All these developments are a great source of encouragement to us.

The circulation of the journal has improved, marginally though. While there is considerable improvement in the financial position of.the journal, the extent of increase is not yet commensurate with the galloping production costs. Our printers are doing their best to improve the quality of printing and we are confident that during 1988 the subscribers, advertisers and authors of various articles published in the journal will find a perceptible improvement.

*Fishing Chimes* continues to provide free advisory service to its subscribers as well as those who are new to the field. The advice given by us proved to be of great use to several. These contacts have helped us in increasing the strength of our subscribers.

It is our sincere belief that the journal will continue to be informative to the readers and will continue to serve the purpose of bridging the communication gap among the fisheries workers in the Government sector at various levels on the one hand and between the Governmental organisations and the commercial sector on the other. We also believe that the journal has been helping the various commercial fisheries organisations in providing information on various developments in the fisheries field.

We renew our pledge to the industry to strive hard and do our utmost for the faster development of the fishing industry through publication of material that could be of real use to those concerned with the fisheries development, be in the Governmental or in private sector.

Fishing Chimes, Editorial 98: April(ii) 1988: Vol. 8, No. 1

# Lack of Leadership

The critical problem that confronts the deep sea fishing industry is the lack of active leadership. The degree of leadership exercised both in the Governmental as well as in the commercial sectors is woefully inadequate to bring about any significant improvement in our deep sea fishing industry. Unless there are improvements and changes, it is quite certain that the industry will continue to be stagnant.

The dispersal of deep sea fishing effort is directly related to the availability of infrastructural facilities for the operation of the vessels at various identified ports. It is known to all those connected with the industry that these are available at only one fishing harbour (Visakhapatnam) in the country. The imperativeness of providing these facilities at various ports to achieve dispersal in fishing effort has not evoked the needed attention among the Governmental authorities concerned. Several representations from the industry to the Government, have not made any real dent. If there is real leadership among the deep sea fishing entereprises, the situation would have been different. Considering the present day standards of manoeuvres and agitations, obviously the efforts of the Associations relating to the industry have not been up to the mark. Considering the fact that the authorities concerned should also make efforts to lead the industry forward in national interest, the present state of affairs clearly indicates that the needed quality of the leadership is particularly lacking in the Governmental circles as well as in the private sector.

Leadership develops when the objectives are clear and the strategy to achieve the objectives is well formulated. Apart from the absence of infrastructural facilities, no one connected with the industry is clear in his mind about the target stocks to be exploited and in what manner this has to be organised. The approach adopted at present is to say what should not be done. The Government's guidelines on shrimping by larger vessels are one in this direction. On the positive side, the Government wants resource-specific vessels to be introduced, without in any way taking a look at the economics of operations. Any good leader connected with fisheries development should be able to clearly explain to his followers the various details of new programmes and how exactly they should set about, with the assurance of good returns. There is no leader in the country who can say something, on the subject authoritatively. There is thus a lack of leadership. The result of this is that the nation does not really know what it is missing because of the non-exploitation of the marine fishery wealth in the farther regions of our exclusive economic zone.

In this confused situation, the Shipping Credit and Investment Company of India is called upon to provide loans to deep sea fishing entrepreneurs for deep sea fishing projects involving, among others, the acquisition of resource specific deep sea fishing vessels. Appraisals apart, as matters stand, no one can be certain about the results of a deep sea fishing project which is divested of shrimp or tuna fishing, although the fond hope is that these ought to be successful. With the economics of the operations of resource-specific vessels not having been established, it is hardly surprising that the SCICI wants a margin money of not less than 30 per cent of the cost of a project and very few companies naturally come forward to actually avail of this facility. At the rate the loans sanctioned by SCICI are being actually availed of, it would take several decades for the fishery resources of our exclusive economic zone to be exploited by the Indian enterprises. It is really for the national leadership to take a close look at this in order to solve this serious problem and find solutions. However, our national leadership has no time to look at the fishery sector.

It appears that the representative bodies of the deep sea fishing industry should tell the Minister of Agriculture very clearly what they require and in what time frame. The details of the problems have to be personally explained to the Agriculture Minister and he should be requested to take spot decisions on the further line of action concerning the various problems that are riddling the deep sea fishing industry. If these decisions are not implemented within a reasonable time, the wisest thing that the enterprises could do is to stop making efforts to increase their fleet strength. Those who have received sanctions for financial assistance from SCICI or other financing bodies and who are unable to pool up the needed finance towards margin money should honestly say so and withdraw from the field, instead of wasting their time and money on projects that are elusive and which can never materialise. Those who intend to enter into the field afresh should look into the various aspects and if they are not in a position to pool up the needed margin money they should desist from making applications for financial assistance and wasting their energy, time and money on pre-project activities. Instead, they should concentrate on other sectors where they feel they can make progress.

## Rays of Hope

A welcome development that took place recently is that the two major institutions that provide financial assistance to the fishing industry (SCICI and NABARD) have inducted representatives from the industrial sector; to be on their respective Boards.

Mr.N.P.Singh has been taken as a Director on the Board of the Shipping Credit and Investment Company of India. This progressive step has revived the sagging hopes of the deep sea fishing industry in regard to securing the needed financial assistance on reasonable terms from the SCICI.

N.P.Singh has a very intimate knowledge about the various aspects of the deep sea fishing industry. It was he who brought the deep sea fishing industry into lime light during his tenure as the President of the Association of Indian Fishery Industries. His skills at presenting facts, suggestions and correct solutions in a persuasive way are unique. With this ability, coupled with his knowledge concerning the industry, it is quite certain that he would tender correct advice to the Board on policy decisions pertaining to extending financial assistance for deep sea fishing projects. A long time after the closure of the SDFC, this is the first time that the deep sea fishing industry has a refreshing news. : Another ray of hope : to the fishing industry is the reported decision of NABARD to provide financial assistance for deep sea fishing industry as well. There has been a lacuna in the composition of Board of Directors of NABARD till recently. Apart from a representative from the department of agriculture on its Board there has been none who has a close knowledge of the problems and prospects of the fishing industry relating to both inland and marine sectors. No less important than the introduction of Mr.N.P.Singh on the Board of SCICI, is the inclusion of Mr.V.S.Prasad, a noted fishery industrialist as a Director on the Board of National Bank for Agriculture and Rural Development. Mr.Prasad virtually grew up as part of the deep sea fishing industry in India. He has also gained considerable experience in the fish culture sector. With his innate management skills coupled with accumulated field experience he is one of the few competent fishery industrialists in the country. His presence on the Board of NABARD will enable this august body to understand the problems faced by the fishing industry very clearly, so that, on that basis, they can take pragmatic decisions to promote developmental activities.

There are also two other rays of hope. One is the lead taken by Mr.N.S.H. Prasad, President of the Association of Indian Fishery Industries, in being instrumental in transferring a sizeable number of deep sea fishing trawlers from Visakhapatnam to Cochin to undertake fishing for deep sea lobsters. If this experiment proves to be successful, it is quite possible that a large number of vessels now concentrating at Visakhapatnam will move seasonally to Cochin. This will facilitate the exploitation of the reportedly rich deep sea lobster grounds on the south west coast and also part of the south east coast.

Another important ray of hope is the pioneering work being done by Mr.Pratap Shergill, President of the All Indian Trawler Operators Association, for shifting a sizeable number of the deep sea fishing vessels now at Visakhapatnam for operation from Roychowk. He has been working very hard of late in organising the needed facilities for the operation of trawlers from Roychowk. The workshop and the processing plant set up by the State Fisheries Development Corporation some time back are now being restored for active use by the entrepreneurs expected to shift their trawlers to Roychowk. The clear advantage of this shifting is that the trawlers operating from this harbour will be able to reach the fishing grounds much quicker than those stationed at Visakhapatnam. Mr.Shergill has rightly understood that facilities at any fishing harbour can improve only with the stationing of some vessels first and making efforts simultaneously to organise all the needed facilities. The entire industry will be grateful to him, once he succeeds in his efforts.

Fishing Chimes, Editorial 99: May(i) 1988: Vol. 8, No. 2

# From Shrimp to Lobster

A perceptible swing towards exploitation of deep sea lobster resources on the Southwest Coast by a good number of vessels, hitherto engaged in shrimping operations from Visakhapatnam, took place recently with Cochin as the base. This sudden development may look miraculous, since, deep sea fishing companies were being reluctant to shift their base of operations from Visakhapatnam on the premise that suitable facilities were not available for operations at other ports. The present shift has obviously happened because of economic compulsions. Facilities at Cochin Fishing Harbour, considered to be practically absent for the operation of larger vessels, has suddenly been found to be manageable. Draft is not adequate but captains are able to negotiate and manage to berth their vessels at the harbour, with a couple of inches of water between the bed and bottom of the vessels. The spirit of enterpreneurship, and financial strength, enabled a few of the companies to take the risk of making efforts to exploit the much talked about stocks of deep sea lobsters adjacent to the south-west coast. More compelling was the firm policy of the Government which has led to this sea change in the scenario of deep sea fishing operations.

Under its policy formulation, Government have been exhorting the deep sea fishing industry to diversify their activities from shrimping to other types of fishing, particularly, longlining, purse-seining and pole and line fishing for tuna and fishing for squid and cuttle fish. As one of the steps in this direction, Government put a stop to according permissions for the introduction of vessels fitted with outriggers, which are largely used for catching shrimp. While this step has stemmed the increase in the fleet of shrimpers of 20m OAL and above, the industry could not simultaneously develop the needed boldness to switch over to a)stern trawling with the existing vessels for catching finfish and b) to catching other species. The sporadic introductions of a few tuna fishing vessels did not.lead to any significant diversion from the accent on shrimp harvesting.

The motivating factors for the migration of about 12 trawlers which are equipped for both outrigger and stern trawling from Visakhapatnam to Cochin seems to be: a) the congestion at the Visakhapatnam harbour b) frequent allegations of play of tactics on the part of floating staff to keep vessels detained at port during times of low catches which cannot be true, c) agitation by deckhands for redressal of their grievances which are genuine and d) seasonal failures in shrimp fishing off north-east coast.

There were no shrimp fishing operations along the upper east coast from January to July 1987. This had placed most of the fishing companies in the red. There was hope that in late 1987 and early 1988 the losses could be made up. However, as it turned out, poor returns from operations even during the main fishing season in 1987-88 pushed the owners into a deeper red. Nonetheless, fishing operations on the upper east coast continued until December, 1987. From January 1988 there was a slump in the occurrence of shrimp along the upper east coast. Consequently, almost all the vessels stopped fishing operations and berthed their vessels at the Visakhapatnam fishing harbour. The floating staff adopted an ambivalent attitude signifying their readiness to sail if owners wanted to but at the same time they were also on the alert to put forth their problems in taking vessels for fishing in the northern bay. This was a Catch 22 situation.

A major reduction in cash flow in the past one year has made a few enterprising fishing companies to shift their base of operations to Cochin from Visakhapatnam to try their luck at exploiting deep sea lobster resources. These vessels, 12 nos., commenced fishing operations at Cochin from February/March onwards.

Lobster trawling differs considerably from shrimp trawling by outriggers. In the later case, one net is operated from each of the outrigger booms. In the case of lobster a single lobster trawl net is operated from the stem of the vessel.

The design of the lobster net is also different from that of the outrigger trawl net. The design of otter boards is altogether different in the case of lobster trawls. There are a few captains who have had previous experience in lobster fishing, which they gained mostly on Government vessels in which they worked before. It is these captains and those whose services could be obtained from Central Government institutions who are now conducting lobster fishing with Cochin as the base.

The owners could succeed in securing the cooperation of the various officers in the Central Government institutions such as the Integrated Fisheries Project, Central Marine Fisheries Research Institute, Central Institute of Fisheries Nautical and Engineering Training and the Fishery Survey of India. The Directors of these institutions and also others working at these institutions having experience in lobster fishing rose to the occasion and have been assisting the industry considerably. The Ministry of Agriculture has granted all needed permissions to enable these institutions to extend all help to fishing companies which have come over to Cochin to undertake deep sea lobster fishing. The industry is indeed grateful to the Government for providing the needed support which is extended for the realisation of the objective of diversification. The strategy of the Government to introduce the deep sea fishing enterpreneurs to take to other methods of fishing has thus ultimately suceeded. Once taken up and found lucrative, normally, there will be no looking back. What Government has to think of and plan is as to how to regulate and control the activity so as to avoid repetition of crowding of vessels as has happened on the north east Bay of Bengal.

According to reports, the 12 trawlers which started operations from Cochin could catch substantial quantities of deep sea lobsters in the few voyages completed till now. Apart from deep sea lobsters, some of the trawlers could also catch impressive quantities of deep sea prawns and also squids and cuttle fish. While the catches are stated to be at a fairly satisfactory level, the operators faced the problem of processing and exporting of catches. This has been solved, thanks to the help extended by MPEDA.

Processing of lobster tails is something new to the Indian deep sea fishing entrepreneurs. They joined in groups and took on hire three processing plants in and around Cochin. At these plants the processing operations are stated to be in progress. While engaged in this task, the entrepreneurs understood the various problems involved in processing. Firstly, the shell is very tough and there are several spiny projections on the sides. All these had to be pruned by using scissors. Unlike shrimp tails, it was found that handling of lobster tails took a considerable time. While, in a day, in the case of shrimp, about 2 tonnes could be handled by about 30 processing workers, the same number of persons were able to handle only about 300kg of lobster tails. So much so, the entrepreneurs had to increase the number of processing girls for quicker processing of the lobster tails.

It has to be mentioned in this connection that Mr.N. S.H. Prasad of Srinivasa Sea Foods played a pioneering part in securing co-operation from Governmental agencies and foreign buyers. But for his pioneering efforts, this important development would have taken a longer time to materialise. Mr. Uday Kumar of Seaman Fisheries was, in fact, the first to enter the lobster field in 1987 season but this didnot lead to a wider interest at that time. The various Central Fisheries Institutions rose to the occasion, and behaving in a way more catholic than the Pope himself, helped the entrepreneurship in parting with various aspects of lobster fishing technology with a missionary zeal and with swiftness. They have strengthened the confidence of the industry in the capabilities of the institutions.

Another helpful factor was the dilemma of the floating staff of the trawlers in regard to conducting fishing operations from April to June in the Northern Bay of Bengal for the reason that during these months shrimp fishing would be dull and it would not be worthwhile to spend time in catching small quantities of shrimp which would not In any case lead to economic results. Therefore, in a way, the captains and other officers led by Capt.N. Venkateswarlu did a great service to the deep sea fishing industry in forcing at least some of the owners to shift their activities to the south west coast and south east coast. Capt. Venkateswarlu has thus, indirectly though, did yeoman service to the industry.

Another reason is the dawning realisation on the part of the owners that they could no longer build their future around shrimping activities alone. Sooner or later they would have to adjust their activities to catching other stocks too. The beginning of lobster fishing on the south-west coast is a refreshing development, which has blazed a new trail. There is no doubt that during the next off-season for shrimp on the north-east coast in the sand-heads area, a large number of trawlers would definitely move over for fishing along south-west and south east coast for lobsters, deep sea prawn, and squid and cuttle fish.

The considerations that govern fishing for shrimp are different from those relating to lobsters. In the case of shrimp, this being an annual crop, fishing can be done continuously and there is hope of revival from year to year because the life span of shrimp, by and large, is 400 days and shrimp breeds several times in a year. In the case of deep sea lobster the biological aspects are not fully known. It is quite likely that the deep sea lobsters have a life span of over 4 years. In such a case the fishery activity has to be regulated very carefully. If there is over-fishing, the entire ground will be denuded of lobsters and it would take another 3 to 4 years for the fishery to revive. The Central Marine Fisheries Research Institute would no doubt conduct further research into the biology of the deep sea lobsters and advise the industry in regard to the manner in which the fishing activities are to be regulated. For example, it is quite possible that the conclusion of CMFRI would be that a certain number of vessels alone should be deployed for exploiting these resources.

The industry rnisunderstood the Government for insisting that the accent on shrimping should be diluted. Looking back, it has to be admitted that this is for the good of the industry itself. Any likely decline in per unit earnings from shrimp catches can be counteracted through the exploitation of the virgin deep sea lobster resource along south-west and south-east coasts in a judicious manner.

It is to be hoped that, in the same way as a beginning has been made in the diversification of fishing operations from shrimp to lobster, there will soon be a fillip to move to other areas of fishing, such as tuna purse-seining, tuna longlining, pole and line fishing, squid jigging etc.

Fishing Chimes, Editorial 100: May(ii) 1988: Vol. 8, No. 2

# Inland Fisheries: Pond Fish Production Policy

That policies need to be evolved and implemented in such a way that all available natural resources are exploited in an optimisingly sustained manner is axiomatic. More specifically in a predominantly agrarian economy like ours, harvesting of natural resources assumes paramount importance. Under the head agriculture, fisheries, horticulture, sericulture and others are clubbed in our planning system and administered by the Union Agriculture Ministry. This has led to a situation in which subjects other than agriculture under the common head receive step-motherly treatment. The fisheries sector, like the other sectors under the major head of Aquaculture, does not receive the attention it deserves.

Fisheries are neglected in our water management. All water resources are national assets and are primarily used for irrigation purposes besides power generation. Besides these two purposes, there are several other needs such as those for fish production and no integrated water management policy which would ensure equitable distribution of water among the different sectors has so far been evolved. India is blessed with rich inland fisheries wealth. Its river systems measure 27,000 km. Out of 7.53 lakh ha. of ponds and tanks, only 1.5 lakh ha. are.presently under farming. Fish prodution from these water resources during 1986-87 had been provisionally estimated at 1.2 million tonnes which is a six-fold increase over 2.18 lakh tonnes estimated during 1951-52. Our present annual fish production is far less compared to what has been achieved in China.

For the past few years fish production in China has been reported to be rising at a remarkable rate of a million tonnes a year. China's fish and shellfish production target of 9 million tonnes seems likely to be achieved well in advance, as evidenced from further periodic boost of harvest in that country from its fish culture sector. According to available information, China's fish production exceeded 7 million metric tonnes in 1985 and a yearly rise of around 4 lakh tonnes up to 1990 is considered attainable. Preliminary reports for 1986 indicated that the total harvest rose by 13 per cent over that of the achievement in 1985 and was estimated at around 8 million metric tonnes. The most impressive achievement is one of 30 per cent increase in fresh water fish culture and most of this was due to a jump in the farm output. But there are also big increases in farming of marine prawn, clams, abalones and other shellfish. According to a report, Chinese farmers dug 1,16,000 ha. of new ponds and improved 31,200 ha of existing ponds during winter and spring in 1986. Some 8 lakh ha of paddy fields were also used for raising fish.

Recently, China was reported to be embarking on an ambitious plan of doubling its present fish, shellfish and seaweed production to more than 18 million metric tonnes by the end of the century. It was emphasised that substantial development of freshwater farming throughout the country would be crucial to the achievement of the target.

A Shanghai fish farmer, for example, was quoted as saying that the pace of freshwater fishery development was so fast and intense that the whole area smelt of fish. It was claimed that the whole fish output per m (about one-fifteenth of a ha.) has gone up from 500 to 700 kg, a record yield for China. This explains why China is the leading fish producer in Asia. Two reasons were attributed to this phenomenal success, the first one being continued improvements in aquaculture technologies both in increasing the yield from existing ponds and in opening up new farms. This has been stimulated by increasingly flexible policies providing for rewards for outstanding farming and allowing fish prices to respond to market demand.

The second and more Important reason is that China had long been implementing an appropriate and integrated water management policy which ensures equitable distribution of water among the needy sectors. An FAO mission which travelled extensively in China recorded the phenomenal success China had achieved in the freshwater fish culture sector as early as 1974 and attributed this to its water management policy. The view expressed was that this policy was one of the main catalytic factors for the success achieved. Comparing this with our situation, no doubt, we are far behind. Therefore there is need for us to study and adopt the Chinese water management policy with appropriate changes for the effective utilisation of our water resources. We had already borrowed their hatchery technology which had proved its utility. In fact, a massive inland fisheries development project with World Bank assistance for the proportion of fishes raised through the Chinese hatchery technology is now being implemented in some of The Indian States. The adoption of water management policy may well give a further fillip to increase in fish production, besides agricultural development. A team of national scientists should be sponsored to visit China and study its water management policy in conjunction with the factors which continue to cause remarkable success in fresh water fish culture in that country. Based on the recommendations of the team, an integrated water utilisation policy has to be evolved, with major emphasis on achieving optimal fish production from the waters. The policy to be so evolved should also cover training of man power to implement the programmes drawn up under the policy which should be available for field application.

Requirements of water for the purposes of irrigation need to be Integrated with the requirements of fish culture in reservoirs and for ensuring adequate supply of water to ponds and tanks for fish culture purposes. The policy should also cover controlling of pollution In the river stretches so that fish and other aquatic as well as non-aquatic biota are not affected. Introduction of fish farming systems like cage farming systems should also be integrated with irrigation projects. It should also provide adequate numbers of appropriate fish passes for fishes to safeguard their migration to their spawning grounds. Provisions should also be made for protecting the breeding grounds of the fishes.

Fishing Chimes, Editorial 101: June(i) 1988: Vol. 8, No. 3

# Fishing Vessels for Diversified Fishing

There are varied and extensive fisheries resources in our Exclusive Economic Zone. The fisheries of the inshore area of this zone are being more or less optimally exploited, with the exception of cephalopod resources which still remain to be adequately exploited. This situation calls for the introduction of fishing vessels specially equipped for catching cephalopods, particularly along northwest and southeast coast, where sizeable stocks of cephalopods exist, according to available broad knowledge of fisheries resources in the EEZ.

It is stated that on the northwest as well as southwest coasts also good stocks of cephalopods exist. Considering the above position, it is of utmost importance to introduce squid/cuttlefish catching vessels of about 16 M length all along our coastline, particularly along north-west and south-east coasts. Vessels of this length are adequate to fish in inshore waters.

Several 16 m long vessels equipped for trawling have already been operating on the northeast coast, mainly engaged in catching shrimps. Government can evolve a scheme under which these vessels can be made suitable for cephalopod fishing. The concerned Central Fisheries Institution (IFP or FSI) can take up a project to work out the manner in which these vessels can be equipped for this new type of fishing. On an experimental basis, squid/ cuttlefish catching devices can be imported and installed on one or two commercial vessels of 16 M length. Once it is seen that the operations from these vessels are satisfactory, other 16 M long vessels can be progressively converted for this diversified fishing activity in the Inshore zone.

In regard to the exploitation of our offshore and deep sea fisheries areas of our Exclusive Economic Zone, which are sparsely exploited, It has to be recognised that vessels for diversified fishing should be introduced in these areas too. This introduction has however become difficult owing to (a) lack of a clear picture of economic feasibility (b)financial weaknesses of entrepreneurs c)lack of technical expertise and (d)lack of detailed shop floor drawings to enable Indian shipyards to undertake construction of diversified types of fishing vessels.

All Indian shipyards have drawings of shrimp trawlers. No efforts appear to have been made so far by any of the Indian shipyards or any of the Governmental agencies to secure drawings of vessels other than those meant for shrimping. If drawings for categories of vessels such as longliners, pole and line vessels, purse-seiners etc are available with Indian shipyards, they would be able to offer competitive quotations for these vessels, lower than those of similar imported vessels, because of the availability of subsidy. The primary need is therefore to have a strategy under which the Indian shipyards would be able to secure the detailed shop floor drawings of various types of fishing vessels meant for diversified fishing. Once, these drawings are available, the need to approach foreign shipyards for the supply of such vessels will get minimised. With foreign expertise where needed, Indian shipyards can bid and supply various types of vessels other than shrimpers as per the requirements of entrepreneurs.

The basic step needed for the Introduction of vessels for catching squid, cuttlefish, octopus etc and also of vessels such as tuna longliners, is to create an organisation which can acquire these drawings from competent agencies abroad or within the country and supply the.same to the Indian shipyards on payment of stipulated charges.

Once the shop floor drawings are available and Indian yards are in readiness to build vessels based on these, entrepreneurs will have to be motivated and assisted to contend with other problems such as financing Institutions such as SCICI which are already in readiness to provide financial assistance. It is true, that, as matters stand, the margin money requirements insisted upon by the SCICI are on the high side. This insistance, it is stated, is for the good of the enterprises themselves. There will certainly be several entrepreneurs who would be in a position to conform to the norms of SCICI. These enterprises can avail of the offers made by Indian shipyards for acquiring vessels for diversified fishing by securing financial help from SCICI.

It is certain that unless tangible steps are taken for equipping Indian shipyards with detailed drawings of various types of vessels, progress in regard to diversification will be very slow.

Fishing Chimes, Editorial 102: June(ii) 1988: Vol. 8, No. 3

# Manning Requirements of Fishing Vessels

The Amendment to Merchant Shipping (Amendment) Act, 1987 (No.13 of 1987) was notified by the Union Ministry of Law and Justice on 22 May, 1987. As per this amendment, every Indian fishing vessel, when going to sea from any port or place to India, shall hve with it: a) if the vessel is of 24 m or more in length and Is operating beyond the contiguous zone, a certificated Skipper Grade I and a certified Mate of a fishing vessel; b) if the vessel is of 24 m or more in length and is operating within the contiguous zone, a certificated Skipper Grade II and a certificated Mate of a fishing vessel; c)if the vessel is of less than 24 m in length and is operating beyond the contiguous zone, a certificated Skipper Grade II and a certificated Mate of a fishing vessel; d)if the vessel is less than 24m in length and is operating within the contiguous zone, a certificated Skipper Grade II, e) if the vessel has a propulsion power of 750 KW or more, at least one engineer of a fishing vessel, who shall be designated as Chief Engineer and an Engine Driver of a fishing vessel; f) if the vessel has a propulsion power of 350 KW or more but less than 750 KW, at least one Engineer of the fishing vessel who shall be designated as Chief Engineer, and g) if the vessel has a propulsion power of less than 350 KW, at least one Engine Driver of a fishing vessel who shall be designated as Engineer-in-charge.

The above mentioned provisions will come into force from such a date the Central Government may by notification in the official gazette and different dates may be notified for different provisions of the Act.

As per the rules in force now, an engineer of a fishing vessel to be designated as a chief engineer, is required for manning vessels fitted with an engine of 50 NHP (about 350 HP) and over. According to the amendment Act, vessels fitted with engines of 50 HP and over but upto 350 KW (466 HP) an Engine Driver of a fishing vessel will be adequate.

The majority of deep sea fishing vessels owned by Indian fishing companies have horse power not exceeding 4 20 HP. Only about 20 vessels have engines of over 420 HP. The requirements of engineers of fishing vessels thus becoming limited, engineers of the fishing Vessels becoming surplus is inevitable for sometime to come. Till the next generation of vessels are introduced some of the engineers of fishing vessels who may become surplus may have to accept jobs on lower salaries.

Engines of vessels manned by engineers of fishing vessels will undoubtedly be far better maintained. Owners therefore may show preference for employing engineers of fishing vessels whenever available. The adequacy of having engine drivers of fishing vessels to work on vessels fitted with engines upto 450 HP would however eliminate the need for M.M.D. to provide dispensations which is a welcome feature.

Prior to the amended Act there was only one category of Skipper. The amended Act provides for Skipper Grade I and Skipper Grade II Vessels of 20M OAL and above operating beyond the contiguous zone (beyond 24 nm from the coast) would now have to be under the command of Skipper Grade I. There are no Skippers of Grade I at present. Till such time as this category of officers is available, M.M.D. will have to grant dispensations liberally in the case of vessel of 24 M OAL and above desirous of operating beyond 24 nm from the coast.

As already mentioned, the new provisions will come into effect from a date or dates to be notified by the Central Government. The date or dates have not yet been notified. Further, the examination rules and syllabi also have not yet been notified. Unless these are notified, candidates will not be able to appear for the examinations concerned. In all likelihood, until middle of 1989 the provisions in the amended Act are not likely to become effective.

There is thus a breathing time. The syllabi for the various courses can be made available by the D.G. of Shipping, pending their formal publication, to the CIFNET so that the institute can take needed steps to train candidates to sit for the relevant examinations based on the new syllabi. CIFNET can also organise compressed courses for candidates who wish to sit for Skipper Grade I examination.

Now that it is clear that engine drivers of fishing vessels can sooner or later become Engineers-in-charge of fishing vessels fitted with engines having horse power upto 450, M.M.D. may consider granting dispensations to engine drivers of fishing vessels till such time as the appointed date is notified by the Central Government.

Fishing Chimes, Editorial 103: July(i) 1988: Vol. 8, No. 4

# At Pre-Paddle Stage

Most of us believe India has made a spectacular breakthrough in inland fish culture, achieving substantially high rates of farmed fish production per ha. In actual fact, compared to countries such as Taiwan, China and U.S.A, India is, so to say, at 'pre-paddle' stage. Our levels of fish production per hectare are far lower than Taiwan and several other countries.

It is no doubt true that technology for achieving high levels of fish/shrimp production from inland confined water areas has been introduced in our country. But this introduction did not cover the ways and means of keeping the pond waters conducive to the healthy growth of fish. In the case of shrimp, in addition to this, perfected feed technology has not yet reached the farmers, Heavy stocking and heavy feeding that is resorted to often results in diseases and heavy mortalities. This has led several countries such as Taiwan to introduce aerating paddles set up in required numbers at various points in a pond. The paddles are located within a pond, partly under water. The unit has a submerged electric motor that drives the paddle wheel. The paddle wheels rotate so fast that a strong aerating effect is set in motion which replenishes oxygen content. While paddle wheel technology has reduced the incidence of disease and at the same time increased fish production, the technology for keeping pond waters pollution-free is imperfect for the reason that the aeration generated by the paddle wheels covers the top and columnar layers, leaving the bottom water layers scantily aerated Because of this, aeration systems, which cover bottom waters as well have been standardised and patented by several companies in the West. The application of the water aeration system which covers all layers is reported to have resulted in the near-total combat of diseases, besides promoting growth. The availability of substantial quantities of dissolved oxygen because of aeration, coupled with copious feeding, is resulting in increased fish/shrimp production.

In India, Dr.S.N.Dwivedi, former Director, Central Institute of Fisheries Education and Additional Secretary in the Department of Ocean Development was one of the prominent advocates of the beneficial role of aeration in increasing fish/shrimp production. He used to exhort fish farmers to aerate their ponds to banish the evils of disease and step up growth. The advice does not seem to have been adopted by most of the farmers. The fault, however, is not that of the farmers. There have been no field demonstrations and there has been no extension effort. The result is that farmers tend to put up with heavy mortalities and consequential losses. While several countries moved forward and adopted the deeper water aeration technology, India stands detained at pre-paddle stage. No fish pond in India seems to have even one paddle wheel unit. Deep water aeration systems are still a far cry.

The State Fisheries Departments and the Central Inland Fishery Extension Agencies have not yet succeeded in educating fish farmers on the importance of aeration of their ponds. This will be possible only through demonstration and organising the ready availability and setting up of aerating units, in fish ponds. The Central/ State Governments should introduce a scheme under which the farmers are enabled to acquire suitable aeration equipment, with provision of subsidies to start with. Commercial banks will have to come forward to provide financial assistance for acquiring and installing the units. Power supply should be ensured for operating the aerating units. There can also be a system of encouraging enterprising persons to own these equipments and providing them on hire to the various pond owners for the required duration.

Urgent efforts should be made in this direction by concerned authorities and the farmers. Otherwise the field of inland fish culture would soon be a field of losses and the industry will collapse at the same speed with which it expanded.

# Insurance of Chartered Fishing Vessels

The Union Department of Agriculture recently issued letters of intent in favour of several fishing companies for chartering of foreign fishing vessels, mostly tuna longliners. One of the conditions in the letter of intent relates to insurance of the vessels. It says that the vessels and crew shall be insured by the owner against all risks, Acts of God, insurrections etc with an approved insurance company for the entire period of the agreement.

Insurance of chartered vessels by the owners will scarcely provide any relief to the Indian charterer or the Indian crew on the chartered vessels. The owners will insure their vessels with a foreign insurance company. The benefits out of this will accrue to the owners in their country. Securing any share out of this by the charterer from the owners is extremely difficult. Even when there is provision for certain benefits out of the insurance of these vessels by a foreign insurance company to accrue to the charterers, it will take years to actually reap these benefits. Further, similar problem will be faced by the Government in collecting customs cess and the value of the catches in foreign exchange.

In the case of foreign merchant ships taken on charter by the Indian Shipping Companies, it is learnt that there is a policy known as "charterer's liability policy" which is obtained by the charterers from an Indian Insurance Company. This protects the interests of the charterers of foreign ships. Stipulation of a similar condition in the case of chartered fishing vessels will serve the purpose for which provision for insurance is included in the letters of intent.

The charterers have certain interests, limited in nature, in the operation of chartered vessels. Besides technology transfer, these include (a) securing 20 per cent of the value of the catch (b) getting compensation for the loss of life of the Indian crew members, (c) ensuring return of the deposits made by the charterer with the Government reimbursement of the benefits that he is likely to lose if the chartered vessel, (d) for some reason leaves Indian Exclusive Economic Zone and will not come back, and (e) earning back promotional expenses. A policy taken by the charterer, to meet the above needs and also covering the customs cess likely to be lost by the Government in the event of the chartered vessels leaving the Indian EEZ on their own, will be far more useful then asking the owners to insure the vessels.

It appears that the condition pertaining to insurance should be restricted to the above or other similar aspects of interest to the charterer and the Indian Government and the insurance to be taken by the charterer with an Indian insurance company.

Owner's insurance will be mostly related to total loss. If there is a total loss, the compensation paid by the insurance company concerned would clearly go to the owner. In a policy to be taken by the owner, it will be very difficult to incorporate any condition that would protect the interests of the charterers as there are no standard policy systems to provide for this type of payment. In view of this position, it is desirable that Union Department of Agriculture modifies the relevant condition in the letters of intent to provide for the charterer's liability policy to be taken from an Indian Insurance Company. A policy of this kind to suit chartered fishing vessel operations by Indian fishing enterprises can be standardised by the General Insurance Corporation at the instance of the Ministry of Agriculture.

Fishing Chimes, Editorial 105: July(iii) 1988: Vol. 8, No. 4

# Shrimpers no more

Something very disappointing to those in the Government and also in the fishing industry who have branded the deep sea fishing vessels now in operation from Visakhapatnam as shrimpers (in the garb of multi-purpose vessels) has happened. Readers are aware that companies have been accused of acquiring these vessels describing them as multi-purpose vessels. The Ministry of Agriculture was also accused by several of allowing import of such vessels in the name of multi-purpose vessels, knowing fully well that these were in fact shrimping vessels. The accusation was that, while providing outriggers stating that these were meant for stability, stern gallows were also provided on these vessels only to hoodwink the Government, but never to be used.

Over 12 multipurpose vessels undertook single net bottom trawling for lobsters making full use of stern trawling facility available on the vessels from January to May 1988. The results of these operations are stated to be very encouraging. Seeing this demonstration, all vessels now operating from Visakhapatnam having the needed winch capacity and driving power are planning to shift to West coast for fishing in the southern Arabian Sea for lobsters, around January 1989. This development amply proves that the vessels acquired by the fishing companies are not mere shrimpers but are real multipurpose vessels. The reason why vessels were operating only shrimp trawls from the outriggers until recently is that they could not have any opportunity of conducting single net trawling from the stern with profit till now.

This development should rectify the feeling of critics in the Government as well as in the industry that shrimp trawlers were being imported in the garb of multi-purpose vessels. The development would also no doubt take all the wind out of the sail of the companies who filed a writ petition in the High Court of Andhra Pradesh contending that all vessels except those of the petitioners are meant to be non-shrimping vessels but are not actually so. The fact is that almost all deep sea fishing vessels of the Indian fishing fleet are equipped both for lateral trawling as well as stem trawling.

Fishing Chimes, Editorial 106: August 1988: Vol. 8, No. 5

# Need for Introduction of Tuna Long Liners

Hitherto interest on the part of Indian entrepreneurs has been focussed on securing financial help for the introduction of multi-purpose deep sea fishing vessels, mainly aimed at catching shrimp and fish through lateral trawling with outriggers. These trawlers, which became popular as multi-purpose vessels, are also equipped for stern trawling. With the failure of shrimp fishing on the east coast in 1987, some of these have been deployed for trawling for deep sea lobsters. Having studied the situation, the emerging set of entrepreneurs seem to have come to the conclusion that acquisition of any more multi-purpose vessels would prove to be disastrous for the reason that the addition of new units for fishing for shrimps may not be adequately paying. Trawling for deep sea lobsters is no doubt possible from these trawlers for five months in a year but it is considered a cumbersome activity. The result of this line of thinking is that practically no applications for the acquisition of deep sea fishing vessels now reach the authorities concerned for the needed clearances. This appears to be causing alarm to SCICI, a dedicated body to provide financial assistance for the acquisition of deep sea fishing vessels.

In this situation, unless there is a movement for the introduction of economically viable tuna long liners and other vessels for diversified fishing, there will be a stagnation in deep sea fishing. Fortunately, it has been adequately proved that there are abundant resources of tuna all along the coastline and around the islands. Experts as well as some of the entrepreneurs are convinced that tuna long lining is the best method for harvesting these resources. However, the availability of tuna in Indian waters being seasonal, one major consideration is: What should these vessels do during that part of the year when tuna is not available? The general inclination on the part of entrepreneurs in this regard is that these vessels should also be fitted either for stern trawling (demersal and pelagic) or squid-jigging so that the non-tuna period can be covered by these income-yielding activities.

The next question relates to the overall length of the tuna long liner-cum-stern trawlers/squid jiggers to be introduced. Being a new commercial activity, by and large, there can be no two opinions that, to start with, the introduction of vessels of a specified length fitted with an engine of a matching horse-power that would cost the lowest without sacrificing efficiency and would give optimal economic returns commensurate with the investments, is needed. In Taiwan, a large number of 16 m long tuna long liners are in operation with success. Vessels of similar size would probably be the most suitable for catching tuna around Andaman and Nicobar Islands and Lakshadweep Islands. For operations adjacent to the mainland, which have to be undertaken some distance therefrom, 20-25 m size range might well prove to be successful. In other words, it is of paramount importance to make a beginning with vessels of small and medium size ranges (16-20 m and 23-25 m OAL) and gradually move up towards introduction of larger vessels, based on experience gained in the operation of these vessels. This approach would also be in tune with and in consonance with the financial standing of the class of entrepreneurs that is now coming forward to take to deep sea fishing.

Certain public sector undertakings are interested in arranging financial guarantees etc., to fishing companies for acquiring tuna vessels with a tie up for canalising exports through them with a view to augmenting their export performance. One such organisation is reported to have.been advised by a Committee of Government officials that the vessels to be supported by them should have atleast 30m overall length. It is not known on what basis such an advice had been tendered. This advice cannot stand any test, as it is difficult to explain as to how tuna long liners in the range of 16-22 m OAL operate off Taiwanese and US (Florida) coasts and from several other places. It is true that Taiwanese long liners of around 36 nm OAL have been operating successfully from our waters. These are, however, global and long-distance fishing vessels, equipped to stay out for long durations at sea. Indian entrepreneurs do not have to go in for global tuna fishing vessels at this stage. They need vessels which can stay out for about 30 days out at sea, as tuna grounds are within 50 mile range all along the coastline. Overinvestments at this stage are clearly uncalled for.

The Ministry of Agriculture has to play an important rote in advising the financing bodies concerning minimum size range of tuna long liners to be considered for extending financial assistance, keeping in view the applications for the import of tuna long liners and long-liner-cum-trawlers/jiggers earlier cleared by them. The Committee referred to in the preceding paragraph cannot be wiser than the Government which cleared applications mentioned above on the recommendations of a high-power Committee consisting of experts.

Fishing Chimes, Editorial 107: September(i) 1988: Vol. 8, No. 6

# Snare – Ridden Charter Chariot

The declaration of Exclusive Economic Zones by various maritime nations has had the effect of placing restrictions on the fishing activities of global fishing nations and also on nations traditionally fishing in certain areas which have now come under the declared EEZs of their neighbouring countries. So far as Indian EEZ is concerned, having experienced the futility of trespassing into this Zone, Taiwanese and Thai fishing vessels could find no alternative other than subjecting themselves to the system of operating their vessels in Indian EEZ under charter permits issued by the Government of India. While Thai vessels could operate for a few years and had to later withdraw for the reason that the Indian Government discontinued the issuance of permissions for their operation on charter. In the case of Taiwan, a few pair trawlers and long Liners of theirs, mostly provided by Singapore companies, still continue to operate in Indian EEZ. The present mood of Taiwanese owners is, however, stated to be to fish in areas other than the Indian EEZ, after the few vessels now operating under charter complete their term. Besides Thai and Taiwanese vessels, a couple of Japanese vessels operated on charter in Indian waters for about three years.

In this background, the Union Ministry of Agriculture notified a fresh scheme in June 1987 under which procedures for chartering of foreign deep sea fishing vessels were laid down. The *inter-se* priorities for considering the applications received in response to the notification, category-wise, were also laid down. While response from the entrepreneurs was unfortunately luke-warm, an encouraging feature was that the release of letters of intent to several of the applicants was swifter than before. Out of several letters of intent issued, however, only one is stated to have materialised, so far. This Company was given a permit to operate one Japanese longliner. No progress in respect of the operations is in evidence.

Considered from the angle of the progress achieved under the scheme,so far, it is unlikely that a tangible number of Indian enterprises with letters of intent will be able to obtain and operate foreign vessels on charter for the reason that the major supplier *i.e.* Taiwanese owners, have reportedly chosen not to send their vessels for operation in Indian waters. Indian Government is also against permitting any fishing vessels of Taiwanese registry to operate in Indian waters. There is thus no disappointment on either side.

Undeterred, the Union Ministry of Agriculture announced yet another scheme, on the same lines of the previous one, in July, 1988. This had also not triggered off a responsive chord any better than the earlier notification, thus stifling in a cruel way all hopes of the Government and the industry in reaching the targeted level of introduction of 500 nos. of deep sea fishing vessels in The Indian EEZ by the end of the Seventh plan at least through this system, as the last resort. The picture concerning introduction of deep sea fishing vessels of their own by Indian enterprises is already bleak, with no signs of improvement. The flow of applications into the precincts of Krishi Bhavan from fishing companies to acquire fishing vessels has plummeted to the lowest level. So much so, the SCICI, the financing body, is in a situation under which, if is unable to play adequately its role of extending financial assistance for the acquisition of vessels.

The charter scheme of the Government of India is primarily aimed at training of Indian personnel in new fishing technologies of locating fishing grounds that are not yet known to us and for earning foreign exchange through export of the catches. An incidental objective appears to be to utilise the charter scheme as a tributary for reaching the Seventh Plan targetted level of introducing 500 nos. of deep sea fishing vessels.

Indonesia, another developing country like ours, issues permits for the operation of a specified number of foreign fishing vessels in its EEZ for a specified period in favour of foreign or national applicants, on payment of stipulated license fees. In contrast, Indian Government does not give licenses to any foreign company directly for operating their vessels in Indian EEZ, although rules provide for this. (Indian Government magnanimously enables Indian enterprises to earn a fixed percentage of the gross income from fish sales as their share). The permitted foreign vessels catch fish in the specified zones of Indonesian waters and go back home after providing whatever particulars that Government requires. Training of personnel, location of grounds and such other aspects which form an important part of the objectives of the Indian scheme have no place of prominence in Indonesia.

The objective of developed nations in allowing other nations to fish in their EEZ is different and is connected with the sharing of surplus stocks. These nations make estimates of fishery resources in their EEZ from time to time and arrive at the quantities exploited by the national fleet and the surplus that will be available for allocating quotas to other countries. Because of this difference, the monitoring of the activities of chartered/licensed fishing vessels differs considerably in respect of the developed nations from that of developing nations. Once a permitted foreign vessel harvests to the extent of the quota allotted, she has to leave the EEZ of the developed nation concerned. The quantity of fish caught by the foreign vessels is calculated based on daily information conveyed by the Captains of the foreign vessels on a day-to-day basis.

The poor response to the recent two notifications issued by the Government inviting applications for according permissions for the chartering of foreign fishing vessels is in sharp contrast to the situation that prevailed in late 1970s and early 1980s.

Taiwan is the only nation that had been showing predominant interest in fishing in our EEZ till recently. Their preference had all along been to conduct pair trawling, although some interest was there for operating tuna long liners.

Under the new schemes of 1987 and 1988, Government excluded pair trawlers from their purview. The reason for this is stated to be that, pair trawling being an effective fishing method, our seas would soon be denuded of fishery wealth and the harvest would be enjoyed by Taiwanese people. The exclusion of pair trawlers in the Government's notification has led Taiwanese to divert these vessels mostly to Indonesia. Tuna long liners are no doubt owned by several Taiwanese enterprises. However, they prefer deploying their vessels in the EEZ of nations bordering the Atlantic and the Pacific and in the international waters of these oceans and of the Indian ocean.

The very little interest that Taiwanese owners continued to have now stands extinguished. Certain circumstances (which are difficult to evaluate) seem to have impelled them to move out of our EEZ without intimation to the concerned authorities. Seven Taiwanese chartered vessels seem to have left Indian EEZ in the past few months with probably no intention of coming back. The position seems to be that the Taiwanese owners of fishing vessels have decided not to send any vessel for fishing in Indian waters. This decision, fortunate or unfortunate, has stemmed the progress in the implementation of the charter programme under the two new schemes. The second scheme is reported to have attracted very few applications. Next to Taiwan and Thailand, Japan is the only country that gave vessels under charter for a long term for operation in Indian waters.

Despite lack of interest, when Indian fishing companies well known to them approach and ask for offers, some Japanese companies oblige. It is possible that under 1988 scheme one or two Japanese companies offered but these may not make any dent. U.S.A. is reported to have built up a sizeable tuna long lining fleet, and if Indian owners approach the concerned companies in U.S.A., there is some possibility of getting offers, provided Indian terms and conditions governing charters are attractive to them.

The present state of affairs, however, is that the Indian companies interested in the chartering of foreign fishing vessels are now dependent heavily on Taiwanese vessels, mainly long liners and squid jigging vessels. Long liners could operate for a period of about 7 months at the most in Indian waters. The incomes realised may not be attractive to the charterers as the operations will have to be limited to one year in the initial phase. The deposit of Rs. one lakh per vessel which would have to be made by these companies with the Government cannot be earned back in this short duration. At the same time, the time schedule prescribed by the Government for the acquisition of the vessels by a charter under the scheme is so tight that it will be impossible for any one to acquire fishing vessels of his own as per the terms and conditions either directly or through joint venture. It has been the experience so far that very few foreign countries come forward to enter into joint venture in the field of fisheries with Indian entrepreneurs.

Another constraint is the condition imposed by the Government that, if vessels of Taiwanese Registry are to be chartered, the Indian charterer should give a letter to the concerned authority of the Taiwanese vessels stating therein that as long as the vessels operate in Indian EEZ they will be under the jurisdiction of Government of India. The concerned authority should give a reply accepting the contents of the letter mentioned above. Photo copies of these two letters should be filed with the Government of India before the concerned vessels sail out for fishing. Having prescribed these conditions, it is learnt from some of the charterers that it was understood by them that Government will not now accept such letters. In other words, Government will not permit chartering of vessels of Taiwanese Registry. This means that Taiwanese owners will be forced to register their vessels in Panama, involving considerable additional expenditure. This situation is also a deterrent to the Taiwanese owners to send their vessels on charter to Indian waters, and if Taiwan is not in the picture there will be practically no chartered vessels in Indian waters.

Foreign fishing companies often complain of enormous delays on the part of Indian authorities in getting clearances to their charter offers, little realising the many angles from which the offers have to be screened. The result is that by the time the letters of intent are issued, the owners deploy vessels offered on charter for operations in other oceans. The Ministry of Agriculture uses typed forms for issuing permits and the foreign companies suspect these may not be genuine.

Another problem faced by the foreign companies is the time-lag involved in securing GOI's clearance to their crew. The lists have to be furnished to the Government atleast one month in advance *i.e.*, long before the vessels are in readiness to sail from their Home Port.

The problems involved in the assembling of crew all over the world are such that the final list of the names will be clear only at the time of departure of the vessels from the Home Port. This necessitates the furnishing of a revised crew list for a fresh clearance and this means the detention of a shattered vessel at the Indian port concerned for about 15 days during the clearance. The expense involved and the fishing days lost because of this cause concern to the owners and charterless.

For some reason or the other, several Taiwanese vessels often face a situation that exposes them to measures of confiscation. This gives rise to a psychological fear and this requires analysis for formulating remedial measures. This fear also is partly responsible for the reluctance on the part of Taiwanese owners to send vessels on charter to India. As has already been mentioned. If Taiwanese vessels are not available for charter, possibility of getting vessels on charter from other countries appears to be extremely limited.

Yet another deterring factor is the conditions stipulated by the Government concerning insurance of chartered vessels. The Indian owners feel that this does not cover the interests of Indian side. The Indian charterers are interested in the coverage of 20 per cent of the gross earnings which they have to get under the scheme and the earning back of the amounts deposited by them with the Government, the operational expenses, etc. The insurance of the vessels as stipulated by the Government will only benefit the owners and not the charterers. Owners will get the insurance amount in their country in foreign exchange and there is no way in which any part of the amount could be secured by the Indian charterers. What is required is a specially insurance cover available to the Indian charterers from Indian insurance companies. Another factor that weighs heavily on the minds of the foreign owners is the condition that they should make an initial deposit of 20 per cent of the estimated gross returns from one voyage. This condition is prescribed to safeguard the interests of the charterers in case the vessels leave the Indian EEZ with fish catches without declaring them to the customs. The above mentioned deposit has to be made before the issuance of the permits. The owners feel that it would not be possible for them to make this deposit, when they do not know whether they would get a permit or not. However, it would perhaps be proper to insist on a deposit of this kind to be made before the chartered vessels are permitted to leave the prescribed Indian port on the first voyage.

Another aspect is the fear on the part of the Indian owners concerning their ability to fulfill the obligation imposed under the scheme for the introduction of their own vessels within a short period of one or two years. Deep sea fishing vessels are expensive and the problems involved in selecting the vessels to be purchased and in securing loans take a considerable time. The proposals given by the charterers in most cases may not be acceptable to the Government, leading to the confiscation of the deposits. The objective of the Government being to ensure the introduction of an additional number of vessels and not the confiscation of deposits, the conditions in this regard would have to be made flexible so as to attract Indian entrepreneurs to go in for the chartering of fishing vessels. It is well known that there is very little scope for developing joint ventures by the Indian fishing companies with foreign fishing companies. Difficulties in availing of this alternative make it all the more necessary to restructure conditions relating to introduction of wholly owned Indian vessels. Ministry of Agriculture has personal experience of the long time it takes to acquire vessels for their own organisations.

These and several other shares have made the charter scheme a juggarnaut, impeding its progression. Since the Government is serious about the successful implementation of the scheme, it is high time that it is reviewed afresh and suitably revised avoiding all the conditions that come in the way of making progress under the schemes.

Fishing Chimes, Editorial 108: September(ii) 1988: Vol. 8, No. 6

# Revival of Riverine Fisheries

Dr.A.V. Natarajan, one of the top Indian inland fisheries experts of international standing, participated in September 1986 in the International Large River Symposium held in Toronto, Canada. In the background of international approaches to production and management of fishery resources of large rivers, he has brought out (in a report entitled "International Approaches to Production and Management of Fishery Resources of Large Rivers and their significance to India", published by I.C.A.R. in 1988) the significance of these approaches in the Indian context and presented a plan for the management of fishery resources of Indian rivers. According to the report, nearly half of the world's commercial catch of inland fish of 5 million tonnes comes from running waters and floodplains, while about a million tonnes of sport fish are taken from the rivers of North America alone.

Rivers are national heritages. They are the home of a large number of fish species. Concentration on the development of culture fishery sector has led to neglect of riverine fish production and development. The result of this is an increasing tempo in the seasonal massacre of brood fishes, and an increase in pollution levels and consequential mortality of fishes. The waning interest in the development of riverine fisheries is responsible for the low-key resistence to the setting up of factories (that release toxic pollutants) adjacent to river banks. This situation and the neglect to which fisheries of running water systems have long been subjected to, calls for specific attention towards this sector.

The multiple uses for which the river systems had long been subjected to have led to introduction and accumulation of inimical pollutants into the aquatic realm. The consequent adverse effects have been impairing the health of these realms as well as the dependent biota. Being a common property, they provide unrestricted open access resulting in conflicts among users and concomitant overfishing. This has brought about changes in hydrological regimes, degradation of habitats and decline or disappearance of many local fish stocks in several river systems of the world.

Refreshingly different from the general run of reports on the subject, Dr.Natarajan's report presents a profile of the impacts of river and river basin developments in the country on the fishery resources of the Himalayan and Peninsular river systems. Taking into account the global perceptions, perspectives and practices of river fisheries management, he has outlined a plan that aims at mitigation of the habitat degradation. Rehabilitation of the dwindling river fisheries, augmentation of fish production and protecting genetic diversity of the valuable river fish resources of the country.

The report tells us that the Brahmaputra river has the largest river island in the world. The low-gradient Brahmaputra is described as one having "vast sheets of flooded water in depressions adjoining river banks and the formations as characteristic for its oxbows".

The author feels that fish habitat models employed for evaluating the habitants of ichthyofauna had overlooked floodplains, resulting in overestimating of depth and under estimating of available percentage of fish habitat. It is a valid point as the floodplains form a large part of channel area and 70 per cent of fishery in many rivers.

It depresses one to note that 'the fish pass devices installed in tropics for local species have met with mixed results; fish pass at a barrage at R.Parana was successful, that at K.Niger a failure, and others in India not successful. Obviously, studies are needed to ascertain the reasons for the failure and to evolve designs of passes to meet Indian riverine conditions.

On the need to adopt a coordinated approach for effective control and management of river fishery resources, he says: There are various Federal Act enactments to control river abuses in India. We have the (Prevention and Control of Pollution) Act 1971, 1977 Wildlife Protection Act, 1972 Environmental Protection Act, Insecticide Act etc. The Indian Fisheries Act was the earliest which was enacted at the turn of the last century and was framed to protect fish stocks from destructive fishing methods and to regulate size and mesh of fishing gears, to enforce closed season etc. The States framed their own enactments within the frame of Central Acts, emphasis and focus differing, however, according to their perceptions of the problems and the priority the fishery enjoyed in their overall development plans. Basically, perceptions and emphasis in various States vary even among those that come within the same river basin. The author emphasised the need for a co-ordinated approach among the States within a river basin for effective control and management of river fishery resources. This is especially paramount for River Ganga basin, according to Dr.Natarajan.

Dr. Natarajan has proposed a research programme comprising initiation of studies on all aspects of river morphology, crucial to assess the event of loss of spawning, nursery, rearing and grow-out habitat of fishes and studies on productivity, biological characterisation by sub basins and stream orders to assess structural changes in biological communities from river modifications and undertaking elaborate studies on genetic structuring for stock differentiation within species at bio-chemical level through starch-gel electrophoresis for protein profile and to determine allele frequencies for elected enzyme systems. Such studies would help in selective stocking, stock enrichment and maintaining genetic diversity.

Furthermore, preparation of river inventories, monitoring of fishing effort and catch, experimental fishing, stock assessment with methods discussed in the report and studies on energy flow in the rivers are other components of the research programme outlined by him.

The research programme discussed by him is of considerable relevance. It remains to be seen how it can be translated into action to develop our riverine fisheries which is vital, considering the astounding length of our river systems (27,000 km) as well as the people including fishermen depending on it to eke out a living.

Fishing Chimes, Editorial 109: October 1988: Vol. 8, No. 7

# Illogical Report

Readers may be aware that there have been concerted efforts by a small section of the deep sea fishing industry in the past few years to convince the Ministry of Agriculture to take up a stand against further introduction of trawlers with outriggers. Succumbing to this relentless activity, the Ministry, while taking an illogical and enforced stand against further introduction of trawlers with outriggers, was allowing for sometime introduction of multi-purpose or combination vessels, which included outriggers. In the specifications of most of these multi-purpose/combination vessels, outriggers were being described as stabilisers. Fully aware of this euphemistic expression, the Ministry issued permissions for the import/indigenous construction of several such vessels. This approach was construed by some as a way of circumventing the irritating and repetitive approaches from the lobby against the introduction of vessels with outriggers. The lobby, incidentally, consists of companies which operate vessels with outriggers and claim proprietory rights for fishing with outriggers along the north-east coast.

Government gave clarifications in the Parliament that it was not against introduction of trawlers with outriggers but at the some time would not encourage introduction of such vessels. In order to settle this pestilential issue, Government set up two or three committees one after another to make recommendations on the subject, the latest one being the Working Group on multi-purpose fishing vessels appointed by the Ministry of Agriculture on 29th October, 1987. This Group came out with a frontal attack on further introduction of vessels fitted with- outriggers, backed up with arguments that are spineless and demolishable.

As per the terms of reference, the Working Group was to indicate the maximum sustainable yield (MSY) of penaeid shrimp in the Indian EEZ and the number of shrimp trawlers/outrigger trawlers including multi-purpose fishing vessels with facilities for shrimping that it could sustain. In their recommendations the Working Group neither made any reference to the maximum sustainable yield of penaeid shrimp in the Indian Exclusive Economic Zone nor has it determined the number of shrimp trawlers/outrigger trawlers of 20m OAL and above that can be operated in the Indian EEZ. On the other hand, the Group made its recommendations directed exclusively at the north-east coast, forgetting the rest of the Indian coastline. The exports of shrimps from the three ports on the north-east coast for the period 1982-83 to 1987-88 are given hereunder:

**Table 1: Port-wise Exports of Frozen Shrimps North-east Coast**

Tonnes

| *Year* | *Visakhapatnam* | *Paradeep* | *Calcutta* | *Total* |
|---|---|---|---|---|
| 1982-83 | 4,170 | 2,372 | 4,510 | 11,052 |
| 1983-84 | 3,922 | 2,276 | 4,419 | 10,617 |
| 1984-85 | 4,955 | 2,968 | 3,763 | 11,686 |
| 1985-86 | 5,0B4 | 2,217 | 3,333 | 10,634 |
| 1986-87 | 4,811 | 2,140 | 4,440 | 11, 411 |
| 1987-88 | 3,465 | 1,075 | 4,707 | 9,247 |

Commonsense tells us, on going through the above particulars, that the maximum sustainable yield of shrimp from the north-east bay should be atleast more than 11,000 tonnes per annum (headless),allowing for processing loss and the reduction in exports because of unusual and unfavourable hygrographic conditions of the bay. According to available statistics on shrimp landings by deep sea fishing vessels at Visakhapatnam, the year 1984 saw maximum landings of 2381.37 tonnes which had always been less than 1900 tonnes during the period 1980 to 1987, except 1984. (Table 2) Almost all these trawlers operated in the sand-heads area in the north-east.

**Table 2: Shrimp Landings by Deep Sea Trawlers at Visakhapatnam**

| *Year* | *Quantity (mt)* | *Growth Rate (per cent)* |
|---|---|---|
| 1980 | 736.63 | — |
| 1981 | 1,649.15 | +123.88 |
| 1982 | 1,714.99 | +3.99 |
| 1983 | 1,638.34 | –4.43 |
| 1984 | 2,381.37 | +45.30 |
| 1985 | 1,418.78 | –40.42 |
| 1986 | 1,860.87 | +31.16 |
| 1987 | 1,049.53 | –43.60 |

Such being the case, it is astounding that the learned Working Group estimated the maximum sustainable yield of shrimp from the resources exploited by the larger trawlers on the north-east coast as 3,140 tonnes (headless). Stating that in 1986-87 hundred large trawlers landed 3,077 tonnes (headless) from this area, it concluded that shrimp resources were being exploited at the near optimum level by these 100 trawlers. The conclusion thus appears to be incorrect, it is also somewhat odd to notice a group of scientists making estimates of MSY in terms of headless weight and that too based on export figures. It is very difficult to predict what this Working Group and the Government that set up this Group will say when the exports in 1988-89 from the north-east coast would exceed 10,000 tonnes again. This is distinctly possible as indicated by the trend of shrimp catches from the north-east bay in 1988-89 season. If the argument of the Working Group is that 3,077 tonnes reportedly were caught by large trawlers with outriggers and therefore further increase in numbers of these should be stopped, one has also to consider that the balance comes from mechanised and non-mechanised boats. It will therefore stand to reason to suggest control of fishing activities by these smaller craft than by the larger vessels for the reason that it is the former that are over-exploiting the coastal resources. The larger trawlers with outriggers fish areas far beyond coastal zone.

The Working Group was probably under the impression that larger trawlers with outriggers fish in the coastal zone cheek by jowl with small fishing craft. The scientists are unfortunately not aware that brown shrimp is available in good quantities in the sand-head areas in deeper waters. The larger trawlers presently catch this variety upto a depth of 60 fathoms, far away from the range of operation of smaller craft. It is quite possible that this variety could be caught in still deeper waters as well with higher winch capacity. Captains sware that they do catch whites and tigers also in waters beyond the coastal zone, although not in such large quantities. Unfortunately, our scientists refuse to believe that whites and tigers occur beyond the coastal zone. What is sought to be brought out here is that bulk of the shrimp catches from the larger trawlers comes from deeper waters and by doing this they are rendering a distinct service. Instead of appreciating the effort, it is very strange that scientists have made a deliberate attempt to malign an effort that is being made by these larger trawlers with outriggers for catching shrimp in areas which are not accessible to smaller fishing vessels.

A rose is a rose whether you call it by that name or by a different name. Vessels with outriggers are outrigger vessels whether you call them this way or as multipurpose vessels, since they have arrangements for stern trawling as well. The Working Group obviously feels that description of outrigger vessels as multi-purpose vessels is to circumvent the reluctance of the Government to permit introduction of vessels fitted with outriggers, overlooking the fact that these vessels are also fitted with gallows for stern trawling as well. Several trawlers with outriggers (mini-trawlers of 16M length) remove outriggers for part of the year and concentrate on stern trawling. The larger vessels undertake trawling from the stern for catching deep sea lobsters, deep sea prawn and fish. While it may be true that this is a recently commenced activity the fact still remains that these vessels are capable of conducting stern trawling with success.

India is the only nation which has irrational views on outriggers. Vessels with outriggers are operated in several countries, notably in Mexico and USA and none of their Governments had ever imposed a ban on operation of vessels with outriggers. This distinction has obviously been cornered by the Government of India. The Government of India has not understood that outriggers merely increase the area swept by nets and that the shrimp component of the catches increases marginally in these nets, rarely over 6 per cent of the total catch.

Working Groups of this kind mislead the Government and divert its attention from the main purpose of intensifying marine fishing effort. We do hope that the Government will not be swayed by the illogical and seemingly biased recommendations made by these Working Groups. We describe the reports as "biased" for the reason that a strong lobby appears to be at work to prevail upon the Government to allow proprietory rights over fishing for the first batch of Mexican and Dutch-built trawlers with outriggers, introduced in late 1970s and early 1980s. This lobby's aim is to corner for itself rights of fishing for shrimp along the entire north-east coast. We strongly urge that the Government should not fall into this trap, which is being opened wider by the recommendations of the Working Group.

The Working Group has recommended that this strength of trawlers with outriggers operating along the north-east coast should be frozen at the present level. This is certainly not a well considered recommendation for the reason that it has not as yet been established that the presumed over-fishing is on account of the outrigger trawlers and not by the smaller craft. Having said that shrimp are a coastal resource and knowing that smaller crafts fish in the coastal zone, there is no justification to accuse trawlers with outriggers which always fish in the zone beyond the area of operations of smaller vessels that overfish shrimp.

We feel that restriction in catching shrimp in northern Bay of Bengal or in any other area is called for only when the unit catch plummets to an uneconomical level. Instead of checking on this and making recommendations in the light of the findings, it is unfortunate that the Working Group that described the MSY of shrimp from the north-east bay as little as over 3,000 tonnes, when the exports themselves averaged 10,000 tonnes per annum. It is surprising that the Group has not said anything about the effects of shrimp fishing by the multiplying smaller fishing crafts along the north-east coast. Instead, it has painted the outrigger trawlers which fish beyond the coastal zone as the culprits.

**Table 3: Total Shrimp Production In India**

| *Year* | *Tonnes* |
|---|---|
| 1982 | 2,09,678 |
| 1983 | 1,92,897 |
| 1984 | 2,03,186 |
| 1985 | 2,32,489 |
| 1986 | 2,14,695 |

The Working Group has observed that an all-time peak catch of 2,20,750 tonnes of shrimp was recorded during 1975. According to the Group, after this, there had been a. declining trend followed by a revival in 1986 with a catch level of 2,21,616 tonnes. To the best of our knowledge these figures are not correct. In recent years peak shrimp production from Indian waters was reached in 1985. The catches during the year were 2,32,489 tonnes and not 2,21,616 tonnes in 1986 (PRIME, Vol. No.16 dated 22-4-1988 published by the MPEDA). The particulars of shrimp production from 1982 to 1986 are shown in Table 3.

The mentioned particulars of Table 3 relate to both penaeid as well as non-penaeid shrimps. There is no reason for the Working Group to give the total shrimp landings in its report, particularly as the terms of reference are related exclusively to penaeid shrimps. The reasons for not bringing out the trends in penaeid shrimp landings are not known. These are shown in Table 4.

**Table 4: Penaeid Shrimp Production In India**

| Year | Tonnes |
|---|---|
| 1982 | 1,08,529 |
| 1983 | 1,05,482 |
| 1984 | 1,15,447 |
| 1985 | 1,08,721 |

Fishing Chimes, Editorial 110: November(i) 1988: Vol. 8, No. 8

# Major Source of Inland Fish Production: Ponds or Rivers?

(While engaged in the collection of material in connection with the preparation of an outline on development of domestic fish marketing in the country, our editorial staff referred to the 8 volumes on the study made by a team experts of the Indian Institute of Management, Ahmedabad, on Inland Fish Marketing in India in early 1980s. In this process, certain thoughts emerged which are presented hereunder).

The Indian Institute of Management, Ahmedabad, has been making a major contribution to the fisheries development of the country through critical studies of various sectors of the fisheries field. One such study was Inland Fish Marketing in India.

The results of the study were published by the Institute in 1985 in eight volumes. These have become important source of reference to planners, technocrats and administrators. Some of the observations and conclusions of this study would be of considerable interest to the readers. An effort is therefore made here to present these in some detail.

The team of experts who conducted the study estimated the National inland fish production at 1.687 million tonnes (Pages 17, 68, 75 and 86, Vol.I) as against the officially published figure of 0.909 million tonnes in early 1980s. It can be conceded that official statistics sometimes go wrong, but it is hard to explain a variation that is of the order of 0.728 million tonnes. Obviously, there is some error somewhere.

By early 1980s inland fish farming in tanks and ponds had already attained an upward swing. In fact, a production level of about one tonne of fish/ha/yr in the main north eastern States (West Bengal, Orissa and Bihar) had already been reached. In States such as Andhra Pradesh, vast areas had been brought under inland fish culture yielding an average production of over one tonne/ha/yr. Such being the case, while one can reconcile with a marginal increase in the official estimates, it would be difficult to believe that the production from tanks and ponds had actually been as low as 0.181 million tonnes per annum in early 1980s as estimated by the IIM team.

In fact, the annual fish production was more than this in West Bengal alone in early 1980s. There are several fisheries scientists who are of the view that the national inland fish production is grossly underestimated by the Governmental agencies. It accordingly appears that the national inland fish production per annum would have been not less than 0.909 million tonnes in early 1980s and not 0.18 million tonnes per annum. It appears that the IIM team came to the conclusion that the majority of inland fish supply in early 1980s came from riverine fishing and that the same trend was likely to continue for some time. They also believed that riverine fish production played an important role in inland fish production. Accordingly, they came to the conclusion that major attention would have to be paid to increase the production from the riverine fisheries. There is no doubt in our mind that this conclusion of the IIM team is incorrect and that a major portion of the inland fish supply must have come from farmed fishery resources in early 1980s as it is now, but in a more conspicuous way.

The above mentioned conclusion arrived at by the IIM study stands on a slippery ground, judged from the fact that, according to their own statement, they could record only 205 tonnes of fish catch from 14 river centres annually. Scientists of the erstwhile CIFRI, who had the competence to work out the production estimates, came to the conclusion that in the case of rivers, particularly the Ganga, fish production at the best was about a tonne for every one km. distance. On this basis, a production of 29,000 t of fish had been estimated, as the length of all Indian rivers put together is 29,000 km. The maximum production that can be expected from the rivers, with all the present day impact of pollution, would therefore, in all likelihood, be about 29,000 tonnes.

Accordingly, it is difficult to give credence to the conclusion of IIM team that a major portion of inland fish supply in early 1980s came from riverine fishing. It appears that the IIM team must have had problems in working out a realistic estimate of the available extent of ponds and tanks in the country. The team's estimate of a total extent of 156,000 ha of ponds under fish farming for the entire country, however, can be considered reasonable, if the Government, had not arrived separately the possible achievement of 2.18 lakh tonnes annually by the end of 1988-89.

According to a statement made in the IIM's study, the average annual fish production from ponds in West Bengal was 1340.25 kg/ha/yr. The team also estimated that the total area covered by freshwater fish farming in West Bengal was 82,560 ha. This shows that the annual fish production from West Bengal in early 1980s was 0.11 million tonnes per annum. In this background it is not clear as to how the total production from tanks and ponds for the entire country was estimated at 0.10 million tonnes. (Vol.3, P.17).

There also appears to be a technical error in the working out of the production from tanks and ponds in various parts of the country. As the area of a pond increases, the production per ha tends to go down, unless effective developmental and management measures are taken. In the study, the production from ponds was estimated based on certain selected ponds which varied in extent from 16.1 ha to 30.6 ha. In Kurnool district, the average pond area was 16.1 ha, Nalgonda 20.5 ha, Chandrapur 17.8 ha and Ajmer 30.6 ha (Vol.3, Table 3.5). Further, according to the estimate, the national average of maximum depth of fish ponds was 4.7 m and average minimum depth was 1.7 m. These observations appear to be unsound. Another point is that the study mentioned that all ponds in Gujarat, Madhya Pradesh, Orissa, Uttar Pradesh and West Bengal are perennial, and that only 24 per cent are perennial in Bihar. (Vol.3, Page 109). This observation also does not reflect the actual conditions that exist in those States.

One omission in the study of IIM seems to be that the results of the studies conducted for over a period of 15 years under the All India Co-ordinated Project of ICAR on Composite Fish Culture and its field impact, was not taken into account by the team. According to published material relating to these studies, composite fish culture gives higher production than culture of Indian major carps alone. In fact, composite fish culture, based on the technology developed under the All India Project, had already becoming popular in the country among fish farmers in early 1980s. The team seems to have come to the conclusion that the overall growth rate of fish under composite fish culture was lower which does not appear to be correct.

The study made a major contribution in respect of riverine fish marketing aspects. One recommendation made was that marketing centres for riverine fish should be located at places where the volume of business would be about 200 tonnes of fish per day. Even in the city of Calcutta, which happens to be the biggest freshwater fish marketing centre of the country, the daily arrivals in the city's markets are stated to be far less than 200 tonnes per day. According to the study itself, 'the total landings at 14 riverine centres was about 205 tonnes per annum. Such being the ease, it is obvious that the market development strategy for the riverine fish recommended by the study rests on contradictory grounds.

It will be useful for all those connected with development of inland fish culture in India to ponder over the points made above and react for the benefit of entrepreneurs, planners, technocrats and administrators. We feel that CIFE, Mumbai, a Deemed Fisheries University, CIFA and various other fisheries institutions in the country should react and express their views to set the record straight.

Fishing Chimes, Editorial 111: November(ii) 1988: Vol. 8, No. 8

# Erratic Arrival of Reefer Ships at Visakhapatnam Causing Hardship to Marine Products Exporters

Exporters of marine products at Visakhapatnam have been experiencing severe hardship because of the postponement or delays n the arrival of reefer cargo vessels or in lifting marine products consignments for export to Japan and other countries from Visakhapatnam. The reason for this situation, stated to have been deliberately brought about by the shipping companies, is not clear. Some say that because of inadequate quantities of cargo, vessels are not calling. This cannot be true, because most of the processors hold heavy stocks for export. For want of facility most of them are forced to send their cargo by road to Madras, incurring heavy and avoidable expenditure. It is the bad luck of the local processors at Visakhapatnam that there are none to help them in their predicament. It is stated that MPEDA could not help them much in this respect. The Vice-Chairman of MPEDA who has promised all help to the processing industry at Visakhapatnam also seems to have not done anything to alleviate the suffering of the local industry.

The community of marine products exporters expect that the MPEDA would understand the gravity of the situation and do its utmost to ensure regular arrival of cargo vessels to lift the stocks of marine products from Visakhapatnam.

One result of this situation is that, the long gaps in the arrival of reefer vessels at Visakhapatnam are resulting in reduction in the quantity of marine products exported from this port. The credit for these diverted exports will go to Madras port. This diversion will thus present a distorted picture of the total quantity exported from Visakhapatnam. This will also, in a wrong way, strengthen the protagonists of the theory that, shrimp landings from the north-east Bay of Bengal have been coming down. The persons engaged in the study of shrimp catches largely go by the actual exports and they may or may not take into account diversions of cargo from one centre to another, because of artificially created situations resulting in delayed and erratic arrivals of reefer ships. Some wonder whether all this is being done deliberately to suppress the growth of marine products exports from Visakhapatnam.

Fishing Chimes, Editorial 112: December 1988: Vol. 8, No. 9

# Appeal to Coast Guard and Navy: Contact Fishing Vessels also in KC Channel

There are about 145 commercial deep sea fishing vessels of Indian flag. Most of these vessels are equipped only with SSB Radio Telephone sets. They do not have VHF sets. At the various meetings which the Association of Indian Fishery, Industries has had with the Coast Guard officials, it was pointed out by the later that all the deep sea fishing vessels should be equipped with V.H.F. sets to facilitate easy communication by the Coast Guard with the fishing vessels.

There are several constraints faced by the deep sea fishing enterprises in regard to equipping their vessels with Radio Telephone sets. One is that these sets are not easily available for purchase. The various electronic companies in India manufacturing these equipments, take about 6 months to supply, that too at an exorbitant price. The import of these sets is prohibited. The Association of Indian fishery Industries has requested the Coast Guard authorities to take up the matter with the Chief Controller of Imports and Exports and secure permission for importing the sets.

The economics of deep sea fishing operations are not that good to allow additional investments on purchase of VHF sets by the firms. Considering the fact that the installation of VHF sets on deep sea fishing vessels is not a mandatory requirement, the coast Guard would have to reconsider the matter and allow the deep sea fishing vessels to continue to have SSB Radio Telephone sets only.

One reason for the Coast Guard desiring that the deep sea fishing vessels should be equipped with VHF sets is that Coast Guard's vessels do not have SSB Radio Telephone units. It is felt that this is not an adequate reason to insist on fishery enterprises to equip their vessels with VHF sets. It would strike any one that it would be easier for a few coast guard vessels to have SSB Radio Telephone sets installed on their vessels instead of asking a.large number of deep sea fishing vessels to install VHF sets. Another aspect is that the international distress channel on 2182 kc is common to both the SSB sets and also the VHF sets. All vessels are expected to keep this channel open always. The deep sea fishing vessels can be asked to be on watch on this frequency at all times, and whenever the Coast Guard vessels come across a fishing vessel and want to talk with her captain, contact could be established on this particular channel, if there is no response on other channels. Considering all these aspects, we expect that the coast guard would not insist that deep sea fishing vessels should equip themselves with VHF sets as well, since this would be too much of a hardship.

Till such time as a decision is taken in the matter, it is desirable that the Association of Indian Fishery Industries takes up the matter with the Coast Guard and Navy so as to ensure that whenever the Coast Guard or Navy has to contact a fishing vessel, they always switch on to the 2182 kc channel, which is common for SSB and VHF sets, being an emergency channel. Like-wise, the fishing vessels should be on watch on this particular channel, specially when they see a vessel which may well turn out to be a naval or Coast Guard vessel. This way the problem that is now faced can be got over. In the meanwhile, the Coast Guard can certainly equip their vessels with SSB Radio Telephone sets.

Readers are aware that on the night of 16 November, 1983 an unfortunate incident took place off Trincomalee Port in Sri Lanka. An Indian fishing vessel Shri Shreyas was on an innocent passage towards Madras. At that time, it so happened that, according to a report, an Indian naval vessel patrolling the area tried to contact the fishing vessel on VHF. The vessel Shreyas could not reply because she did not have a V.H.F. set. The Captain of Shreyas said that he was trying to contact the Naval vessel on 2182 kc channel but there was no response. This led to firing at the fishing vessel by the Navy. Fortunately there were no casualities, although the vessel sustained several bullet holes. This was an unfortunate incident. Such situations can be avoided if the Association of Indian Fishery Industries requests the Navy and also the Coast Guard always to contact the fishing vessels on 2182 kc channel, if they do not happen to have a SSB. Radio Telephone set, while at close quarters.

Fishing Chimes, Editorial 113: January(i) 1989: Vol. 8, No. 10

# Purchase Tax Exemption for Sea-Foods (Shrimps) in A.P.

Fishing operations have come to a standstill at major fishing centres in Andhra Pradesh, consequent to the levy of purchase tax on sea foods (shrimps) for export with retrospective effect. Over 1450 mechanised fishing boats, about 100 mini-trawlers, over 80 sona boats, nearly 145 large trawlers and thousands of non-mechanised boats are forced to suspend fishing operations with effect from I5 January, 1989, at a time when catches of exportable shrimps species are encouraging. The loss in terms of earnings of fishermen per day.because of this is estimated at Rs.5 lakhs in respect of mechanised fishing boats alone. This apart, several others depending on ancillary activities have lost their earnings.

Sales tax is a familiar impost because we pay this whenever we buy something. Few would have, however, heard of purchase tax, which is levied on certain types of purchases, one of which is shrimps. Where these purchases are for export, the purchasers are eligible for exemption from payment of this tax from the State Government concerned, subject to fulfillment of certain conditions.

Most of the maritime State Governments tried their hand at levying this tax on shrimps exporters on some ground or the other, but had to withdraw the levy either owing to public opinion being against the levy or because of judgements of legal bodies. As matters stand, it is only the Government of Andhra Pradesh, having earlier granted exemptions, has suddenly issued notices to various exporters that they should pay the tax with retrospective effect from 1982 onwards. According to tax officials, the purchases were not in the course of export, as required under rules, although it had always been physically true that what all were purchased were being exported.

There are practically no internal sales for the exported varieties of sea foods for the reason that the selling prices are not within the reach of even the most affluent in the country.

The exporters, according to tax authorities, could not produce before them confirmed export orders for shrimps, clearly stating the quantities to be exported, variety-wise, and the prices thereof and relate the same to the actual exports. According to the authorities, most of the exporters purchase shrimps, process, and keep stocks anticipating orders for export. The exporters, however, say that they do receive purchase orders. And on the basis of the purchase orders only they buy material for processing. It is however true that both the sellers and buyers have serious difficulties in committing themselves to any specified price well in advance for the reason that the international market for sea foods fluctuates widely. If the prices go up before shipment, it will be an advantage to the buyer but not so to the exporter. In order to avoid such situations and to have a fair trade practice, it appears that both the sides make their decisions a few days before shipments with reference to the stocks available and prevailing international rates. It is pointed out that the authorities concerned are least bothered about these practical aspects.

Government officials are proverbially 'blind' to realities. They go by the literal interpretation of the rules. Shrimps are to be caught from wild waters. It is difficult to predict the varieties and the quantities in which these will be caught. Unless different varieties are purchased, stored and processed from time to time, it is impossible for the processors to be in the business of marine products exports. In a land-based industry, one can accept on order for the export of materials, based on production capacities. In sea food exports, such a streamilined work is impossible. One has to depend mostly on what nature gives and react to export demands on this basis. The importers know this, but the Government officials are not aware of this. The present modalities of the marine products export system, *i.e.*, determination of prices shortly before shipment, based on prevailing international prices, have evolved over a period of time and are accepted by the various State Governments, including the Andhra Pradesn State Government, but for the present revisional approach and various Central Government Organisations.

The effect of the present restoration of purchase tax levy was the starting point of a tax storm. The exporters stopped procurement of sea foods. This had led to stoppage of fishing operations, mostly from Visakhapatnam and Kakinada harbours. The effect of this is particularly felt by the artisanal fishermen, who have, for the time being, lost their means of living. Exporters are reported to be planning to shift their operations to other States. Owners of larger trawlers, unable to suspend fishing operations for long, because of the stance of the exporters, are thinking of shifting their bases to centres in other States, for the reason that eventually exporters could depress the purchasing prices so as to meet payments towards purchase tax. If this trend is allowed to continue the State of Andhra Pradesh may soon have the distinction of a) allowing a fall in earnings of fishermen, b) suffering a reduction in exports and foreign exchange earnings and c) generating substantial unemployment, and d) sustaining revenue losses in sales tax collections if a large number of vessels migrate to other States and take supplies of diesel oil at ports in these States. In a month, oil worth about Rs.one crore is purchased by deep sea fishing vessels at Visakhapatnam. The loss in revenue on account of sales tax on this will be over Rs.17 lakhs per month.

Several are of the view that the intent of the rules providing for exemption of purchases of shrimps for export is more relevant than the literal interpretation of the rules. So long as there is no proof of local sales of the material purchased for export, the interpretation of the rules will have to be in favour of the exporters, it is felt. The floating staff and owners of all types of fishing vessels operating from major fishing centres in Andhra Pradesh extended full support to the exporters of sea foods. The owners of large and medium sized vessels and exporters as well were in particular sympathy with small mechanised boat operators who were the worst sufferers.

A laudable quality of the present Chief Minister of Andhra Pradesh is that he knows the pulse of the people, more so of that of the weaker sections of the society. When the artisanal fishermen made a fervent appeal to him for waiving the purchase tax which has been causing undue hardship to them, he comprehended the situation with great sagacity. He placed the matter before the State cabinet, which has at once decided to waive the purchase tax, taking into account most of the aspects delineated in the preceding paragraphs. Democracy survives because of men of mettle of the stature of the State Chief Minister who always comes to the rescue of the poor and down trodden. The entire fishing industry and the fishermen in particular heaved a sigh of relief at the decision of the State Government to provide the much needed relief.

Fishing Chimes, Editorial 114: January(ii) 1989: Vol. 8, No. 10

# Chromosome Manipulation in Fish

The prime mover of Green Revolution has been the judicious application of carefully chosen genetic and husbandry methods. This had brought out the need for an altogether different scenario in the production of high yielding and disease resistant fish and their aqua foods as well as cash crops. Such a revolution has not yet taken place in the case of fisheries sector. This is, it is said, partly because of the much shorter history of aquaculture which for most species is less than a century, and partly because rearing aquatic organisms, which frequently have complicated life-cycles, is intrinsically more difficult than rearing land plants and animals. All farmed animals, with the possible exception of carps, are genetically indistinguishable from the wild populations from which they were captured but there is considerable scope for genetic manipulations which will significantly improve the productivity of aquaculture.

Most of the commercially important fishes invest about 20 per cent of their assimilated energy on reproduction and if this energy could be channelised for somatic growth, then the growth of the fish gets enhanced. This is obtained by producing sterile or monosex fish employing several methods. Adoption of endocrinological techniques for either suppressing gonadal development or producing monosex population (which grow faster than the other sex) has not yet yielded total success because it demands great skill. Increasing consumer resistance to such fish, though not based on any fact, and the attitude of several Governments to ban use of the required hormones, have also limited the scope of this technique.

Sex determination and differentiation in fish sets in sometime after the fertilised egg starts developing. Sex might not have been determined in early juveniles of fish. Hence, it is possible to manipulate naturally the sex of the fish, before it sets in, by modifying the chromosome set. Temperature, hydrostatic pressure or chemical shock treatments are the common ways adopted for this purpose.

Dr. Pandian, the only one fisheries scientist having the honour of receiving Bhatnagar Award, and his collaborator Mr.Varadaraj, both attached to the School of Biological Sciences, Madurai - Kamaraj University, have for the first time in the country produced haploid, triploid, hybrid triploid gynogens and supermales in *Tilapia*. Of these, viable haploids have been produced for the first time by them.

Viability of haploids is not high. Triploids grow faster than normal diploid fish. Triploid hybrids often have higher survival rates, fewer developmental abnormalities and higher growth rates than corresponding diplold hybrids. Moreover, diploid hybrids, obtained through hybridisation are sometimes fertile but triploid hybrids could be sterile. Tetraploid fish has not been reported to be viable beyond fry stage. Supermale or superfemale fish can grow at a very high rate.

What has been ably accomplished by Pandian and his team is the development of a new technique of producing a fast growing sterile fish, something like broiler chicken. In fact, Pandian calls these fish as broiler fish. Tilapia was used as a test animal as it has several advantages for genetic studies. This technique needs to be worked out for other commercially important species like major carps and others and standardised for transfer to the field for adoption.

Pandian and his team, who deserve to be congratulated on their achievement, have to now simplify and standardise the technique to facilitate its adoption in the field. That would transform the aquaculture scenario of the country to fortify it for chromosome manipulation in fish.

It is unfortunate that although India is credited to be having the second largest pool of scientific talents, there are hardly a few scientists working on fish genetics, especially genetic engineering in fish. The National Bureau of Fish Genetic Resources, Lucknow was formed to initiate work to fill up this gap. It is hoped that it would chalk out a national programme on the subject and implement the same.

Fishing Chimes, Editorial 115: February 1989: Vol. 8, No. 11

# Strategies Unrelated to Basic Facts Hampering Development of Indian Deep Sea Fishing

The Deep Sea Fishing Industry of India is expanding at a leisurely gait. The reasons for this are relatable to the basic conditions of existence which are not compatible with the developmental strategies being implemented.

The basic aspects from which strategies for the development of deep sea fishing have to spring are:

## i) Knowledge of Fisheries Resources

Our proven knowledge of viable fisheries resources in the EEZ, beyond the zone of operation of small and medium fishing vessels covers to some extent shrimps, deep sea lobsters, and tuna among finfish. Knowledge on certain fishing grounds where there will be availability of fishes such as *Priacanthus, Decapterus, Nemipterus* and other deep water forms is also available.

## ii) Economics of Fisheries Projects

Major Indian fisheries of commercial value are seasonal. It has not yet been established conclusively that any form of deep sea fishing in Indian waters, aimed at seasonal exploitation of any single species, will be viable. The economics of operation of vessels, which have of late been fishing for a part of the year for shrimp on the north-east coast and for the rest of the year on the southwest coast for deep sea lobsters, however may prove to be viable.

## iii) Infrastructure

Infrastructural facilities at various fishing harbours, except one or two, could not so far be developed to the extent needed, mainly because of uncertainty in the utilisation of facilities and justification for investments. Lack of these facilities, coupled with slow pace of introduction of vessels, hampers development.

## iv) Trained Manpower

Trained manpower to man vessels other than trawlers is, by and large, in short supply.

The deep sea fishing industry being capital intensive, only those with ability to invest at least 20 per cent of the capital needs, will be able to secure loans for investment in the acquisition of deep sea fishing vessels etc. Larger houses and Government corporations are the two main categories, who can take risks in investments of this kind. These categories, by and large, face constraints to come forward to introduce deep sea fishing vessels.

As at present, small and medium scale entrepreneurs have no chance of taking up this line, unless they can mobilise adequate share capital. Problems are encountered in this regard because of shyness of investors to buy shares in fishing companies.

## vi) Joint Ventures

Fishing vessels can be acquired by Indian companies under joint ventures with foreign fishing companies. There is a general reluctance on the part of foreign companies opting for such joint ventures in India and for providing vessels to the joint venture companies, unless full payment or substantial payment is made towards their cost. Very few Indian enterprises are in a position to make needed investments towards margin money to secure funds from financing institutions for the purpose.

## vii) Chartered Vessels

Operation of chartered vessels is expected to lead to introduction of vessels for diversified fishing. The returns from the operations not being sufficient to invest towards downpayments or to meet the debt-equity ratio, the likelihood of the realisation of the stipulation that, equal number of vessels as permitted for charter should be acquired within one year of commencement of chartered operations, is bleak.

## viii) Indigenous Fishing Vessel Construction

Indian shipyards have capability for the construction of deep sea fishing vessels. They do not, however, have drawings for the construction of vessels for diversified fishing. This has accentuated dependence for supplies of shop floor drawings on foreign yards.

## ix) Financing System

The institutional facilities in India through the Shipping Credit and Investment Company of India (SCICI) for providing financial assistance for the acquisition of deep sea fishing vessels by fishing enterprises or for the setting up of integrated deep sea fishery projects are probably the best available among developing countries. Yet, anomalously though, the pace of utilisation of credit facilities offered is not upto expectations. The main reason for this seems to be the financially weak entrepreneurship. All those who are interested in securing loans for developing a deep sea fishing enterprise face enormous difficulties in pooling up share capital to meet the debt-equity ratio, which has been diluted to a considerable extent by the SCICI, ignoring the traditional norms, only in the interests of providing a fillip to the industry. Even when few of the enterprises, mostly private limited companies, succeed in raising the needed level of share capital to serve as margin money, it appears that the managements concerned are subjected to tremendous pressure from the share holders for returns with interest equivalent to their share holding, within the first few years of commencement of operations. This not being possible, certain compelling situations appear to arise, leading to arrears in the payment of loan instalments etc.

### x) Lack of Direction

Entrepreneurship interested in commercial deep sea fishing is in wilderness. Enterprises applying for permissions to acquire vessels of their choice are unmindful of the consequences. There is no overall framework available giving particulars of the types of vessels with certain specifications, determined after taking into account the fishery resources position and other parameters, to enable enterprises to formulate their acquisition programme within this frame work. It is virtually a 'hit or miss' process that is in vogue.

All the aspects brought out above are interrelated. Homogeneous promotion of all these alone can bring about development of sustainable fishing in our EEZ. And this will be possible only when well determined substantial additions to the national fleet are made. Such additions are possible only when the hands of the Government are forced to tackle all the points mentioned above.

The presently available entrepreneurship does not have, among others, the financial strength for the development of deep sea fishing. The absence of financial stamina on the part of the category of entrepreneurship that comes forward to take up deep sea fishing is a clear pointer to the need for the adoption of a strategy that will counteract this weakness. Keeping all the above aspects in view, a new strategy to replace the present system, (that has the effect of stagnating the developmental process) has to be evolved. The ingredients of the new strategy may be as follows:

1. Considering the fact that there are very few enterprises in the country capable of fulfilling the norms prescribed by the financing institutions, there is no way other than the setting up of a Government-owned, professionally-oriented autonomous commercial organisation capable of fulfilling the norms of financing bodies for the acquisition of deep sea fishing vessels. If fishing in EEZ is to be developed, the writing on the wall, needs to be recognised. In the oil and natural gas sector, Government has set up a Commission. Fisheries are no less a natural resource than oil, and fisheries resources being living and oil non-living. Applying the same principle, the least Government can do is to set up a professionally-oriented, autonomous commercial organisation for the development of the fisheries of our EEZ.
2. Taking into account, the existing strength of fishing fleet of various types, and the present levels of exploitation of various commercially viable species, the suggested organisation shall have to chalk out a programme for the introduction of deep sea fishing vessels, delineating the types and their broad specifications and numbers in which these are to be introduced and the total financial outlay, main bases of operation, facilities to be provided there at, and requirements of trained personnel of various categories. Such a programme can be revised from time to time, to include joint ventures in the field, with foreign fisheries enterprises, which could be tackled from strength, the proposed body would have to emerge as a professional Governmental Organisation.
3. Once a programme of this kind is formulated, under a guarantee from the Government, the proposed new organisation should negotiate with foreign yards for the supply of half of the total number of vessels, of each category to be introduced. Finance may be arranged by this body by way of credit or grant from selected foreign countries reputed for fishing vessels construction, under the Indian Government's financial guarantee.
4. One condition for such supplies by foreign shipyards will have to be that, the seller will supply free of cost, all shop floor drawings. These can be made over into a Design Bank, for supply to various Indian shipyards. Orders for the remaining half of the planned number of vessels can be placed by the Government owned commercial organisation with the Indian yards who can construct them utilising the drawings obtained from the Design Bank. After this, in stages, total indigenisation can be accomplished.
5. The proposed organisation can think in terms of adopting one of the following operational systems, as appropriate.

   i) Operate them on its own from the various selected bases following commercial practices, and meet loan repayment commitments out of the earnings, or ii) Lease out these vessels to selected parties against appropriate guarantees taken from them for the payment of annual lease amounts in advance, or iii) Entrust the vessels on hire-purchase basis to State Fisheries Corporations or private sector companies, against payment of hire amounts in advance, or iv) Set up joint sector firms to which the vessels may be entrusted for operation, either on lease or hire-purchase basis.
6. Domestic and export marketing linkages may be established suitably, either by a)entering into arrangements with existing exporting and domestic marketing enterprises, or b) developing a mechanism for undertaking the work on its own.

In justification of the above suggestions, it may be mentioned here that whichever way the matter is looked at, bulk of the investment will have to come mostly from Governmental sources, but with a small component coming from the fishing companies. This virtually means that the loanees accept a position under which they will go by the dictates of the financing body, the only satisfaction being that they own the mortgaged vessels. With the maximum chunk of investment coming from it, the financing body can as well have a single Government-owned professional body as the loanee, with the loan guaranteed by the Government. Government has to consider favourably the need for providing these guarantees to a professional organisation to be set up by it, in the interests of development. Without this sacrifice, Government will have serious difficulties in promoting the activity. The autonomous professional body can have a ramifying set up to guide the lessees or hire purchasers to monitor their activities and recover periodical payments with alertness. This system will solve the present problems of conforming to the debt-equity ratio etc, by fishing enterprises and communication gaps will be reduced. Then the pace of investments in this sector by financing bodies will pick up.

Fishing Chimes, Editorial 116: March 1989: Vol. 8, No. 12

# Strategy Aspects of Current Crisis in Deep Sea Fishing

India has a commercial fleet of about 160 fishing vessels of over 22 M overall length. Justified or not, the Union Government as well as the fishing industry refer to them as deep sea fishing trawlers. In addition to these, there is a fleet of over 40 nos. of trawlers popularly known as 'mini-trawlers', and also a couple of tuna long liners. Until a few months back, the fishing industry was proud of having at least one company owning a large tuna purse-seiner. This pride is no longer there as the vessel has been sold away to a foreign company, presumably because of unfavourable economics of operation.

All the trawlers mentioned above were introduced between 1978-88 for catching penaeid shrimps, principally along the northeast coast. With the fleet gaining strength from time to time, the managements as well as the skippers of the trawlers had to increase the duration of voyages, firstly to reach and cross the break even level and secondly to enable the crew to get incentives. Increase in duration of voyages led to a reduction in the number of voyages.

The net result was a steep fall in trawler-wise annual shrimp landings and earnings therefrom, while the overall harvest remained the same. Most of the companies ran into deficits as a consequence.

Mr.Daulatsinhji Jadeja, M.P., one of the most knowledgeable of Indian politicians in matters relating to fisheries brought to the notice of the Government through a question in Lok Sabha on 10th March 1988, the crisis that had engulfed the deep sea fishing industry, for want of facilities from the Government. Mr.Eduardo Faleiro, Minister of State in the Ministry of Finance, replying to the question, denied that the fishing companies were facing crisis for want of facilities from the Government. He attributed the crisis to lack of proper management.

Mr. Jadeja must have raised the question of providing facilities for the deep sea fishing industry by the Government with a view to bringing the matter into sharp focus, in the context of diversification of deep sea fishing effort, particularly by some of the shrimp trawlers to exploit deep sea lobster resources on the southwest coast and the consequential problems faced by them. It is some consolation that the minister conceded the existence of a crisis in the deep sea fishing industry, although he felt that the crisis was the result of poor management.

What exactly the minister had in his mind when he made a reference to 'poor management' is not known. Whatever this may be, there must have been bad management somewhere. Otherwise the industry will not be in the state of mess in which it is now.

Concentration on the exploitation of a single species *i.e.*, shrimp, by almost the entire commercial Indian fishing fleet has been denounced by all, including those who were responsible for it *i.e.*, the Government. The situation was mostly the making of the Government. In other words, it was mismanagement of fleet development. Mr.Jadeja must have had this in view while talking about facilities for the development of deep sea fishing. Analysing his question, and the casual answer given by the Minister, it would not be incorrect to ask several questions such as: a) Has the Government provided needed facilities at any major fishing port other than Visakhapatnam? b)Could the Government tell the fishing enterprises, with conviction, anything tangible about the types of vessels they could acquire, linking these with adequate availability of one or more resources of value other than shrimp? c)When it is noticed by the Government that shrimping is no longer economical for larger vessels, has any facility or encouragement been provided for equipping them with other types of fishing and for internal or export marketing of finfish or other marine living species? d)Knowing fully well that major commercial fisheries of India are seasonal, has any facility or help been extended to fishery enterprises to equip their shrimp trawlers for tuna long lining, squid jigging or stern trawling? Many more such questions can be asked to counter what the minister said in the Parliament but these will not annul his statement that there was poor management, which, was by none other than the Government itself.

The saving grace however was the help and advice rendered by the central Governmental organisations such as the Integrated Fisheries Project, the Fishery Survey of India and the Central Marine Fisheries Research Institute, based on which several of the owners of shrimp trawlers switched over to the strategy of fishing for deep sea lobsters off southwest coast, during the offseason for shrimp on the northeast coast. Most of the trawlers that switched over to fishing for lobsters have chosen Cochin as the base. Although Cochin is a major fishing centre, practically no facility has been provided by the authorities for the berthing of the vessels, for undertaking maintenance and repair works, bunkering, unloading of catches etc. Can we fault Mr.Jadeja when he asks about facilities that are provided for deep sea fishing vessels, for example for those operating from Cochin? Is this situation and the inability of owners of deep sea fishing vessels to operate from any of the ports other than Visakhapatnam, the result of poor management on the part of fishing companies? Certainly not. It is the result of poor planning in the introduction of deep sea fishing vessels and lack of development of major fishing harbours, notwithstanding diversion of fishing effort of several trawlers for catching deep sea lobsters off southwest coast.

It appears that neither the Governmental agencies nor the fishing enterprises have adequate experience in working out the manner in which our deep sea fishing effort should be developed. Almost all the larger fishing vessels introduced so far are shrimpers. All important commercial fisheries in Indian waters being seasonal, there can be no difference of opinion that the nation requires multipurpose vessels, with more than one fishing facility on deck.

These facilities should enable each of the systems to be operated independently without hindering the other.

A beginning in this direction has been made by several owners of shrimpers. They have modified the deck facilities in such a way that, with the same set of winches, outrigger trawling as well as bottom stern trawling can be conducted. While the stern trawling facility is utilised for catching deep sea lobsters from September to May on the southwest coast, outrigger trawling facility is availed off for harvesting shrimps on the North-east coast.

Most of the presently operated deep sea fishing vessels in Indian waters are designed for fishing upto a maximum depth of 200 metres. The exceptions are probably the four South Korean built vessels owned by M/s.Shrimp India and M/s. Akama Marines and those built by Bharati Shipyard, Bombay. The winches of nearly sixty vessels that are now fishing for lobsters on the Southwest coast are constrained to haul back bottom stern trawl nets of nearly 38 m head rope length laden with lobster catch from a depth of over 350 metres, the length of wire rope used being around 1,000 metres. Unable to withstand the strain, winch systems of most of the trawlers have been failing. Although rectifications have been undertaken, these are not of an enduring nature. In the case of some of the vessels, hydraulic motors and pumps also could not withstand the heavy load. It may be mentioned here that none of the shipyards which built the vessels can be faulted. Winches installed on these vessels were not designed for operation beyond a depth of 200 metres.

From shrimp trawling in shallower waters to the present development to the stage of trawling for lobsters in deeper waters is a major step forward indeed. This step has been taken by the companies regardless of the risks involved and therefore it is no surprise in its wake several operational as well as marketing problems came up. No Governmental agency ever warned the entrepreneurs about the risks. Now that a bold step aimed at diversification in fishing effort has been taken, there is a responsibility on the part of Governmental agencies to identify the operational and other problems precisely, work out remedial measures and help the enterprises with schemes of financial help to strengthen or modify the winches, through replacement of gear boxes, hydraulic pumps and motors, with those of higher capacity. And Mr.Jadeja obviously wanted to know details of whatever was accomplished by the Government in this direction.

Unlike shrimps which have a life span of about 12months, deep sea lobsters have a longer span of life. Our scientists have not yet been able to work out the actual span. It is however believed that the span will be about four years. This means close monitoring of the effort is needed. September 1988 to June 1989 will be the third successive exploited fishing season for lobsters on the southwest coast. In this short time of three seasons, it is seen that the sizes of lobsters harvested have come down. There has also been a reduction in the average quantity of lobsters landed per vessel. Export market for the lobsters is stated to be not all that encouraging. In this context, another facility which Mr.Jadeja would certainly like the Government/MPEDA to extend is location of lucrative export markets.

Deep sea lobster and tuna fishing seasons more or less run parallel to each other. If lobster fishery fails, fishing vessels should be able to switch over to tuna long lining, this being the most effective method for catching tuna from Indian waters. Trawlers of about 18 m OAL and above, in relation to the specific overall length, can be equipped for tuna long lining with the latest American monofil method for operating 30 to 60 km of monofil long line of 3.5 diameter with about 600 to 1,200 branch lines, without causing any operational hindrance to switch over to trawling when required. Several of us still think in terms of Japanese twine method of long lining, a very cumbersome and expensive system, requiring larger vessels and enormous space for storage of main line, branch lines etc. The American system obviates the need for a line hauler. Instead, a compact net reel performs the function.

The reel is the place where the line is stored. Compact Discs take care of storage of branch lines.

Government can certainly help the present trawler owners in restructuring the winch systems for trawling for lobsters and equipping the vessels for long lining. These measures will go a long way in achieving economic stability of the presently operated trawlers and reviving the hopes of the Government/financing bodies in recovering their loans back. Otherwise, the outlook seems to be that there will be continued stagnation in repayments to financing bodies, and the market value of the trawlers will come down, making matters much more difficult. Furthermore, the present trend of declining interest on the part of entrepreneurship to invest in deep sea fishing will peter down further, if there is continuing complacency. This will certainly harm the nation's interest in the exploitation of the fishery resources of our EEZ.

Regarding future programme for augmenting the national deep sea fishing fleet, a realistic approach that aims at introducing multi-purpose vessels that can catch diverse fishery resources throughout the year in our EEZ, through a suitable organisational mechanism would have to be formulated and followed. It is hoped that the recently constituted National Fishery Development Board will apply its mind to this most urgent task, taking into account the need for augmenting fish supplies for the domestic market, for boosting up exports, and, above all, to deter incursions of foreign fishing vessels into our waters.

Fishing Chimes, Editorial 117: April(i) 1989: Vol. 9, No. 1

# Let us Emulate Indonesia's Successful Tuna Fishing Strategy

Indian waters hold sizeable stocks of tuna, as proven by the good tuna fishing results achieved by chartered long liners and long liners of CIFNET and FSI, over the past few years. The waters around Andamans are particularly tuna-rich.

There will be an immense advantage in introducing tuna long liners of about 16m overall length for tuna fishing operations around Andaman Islands. Taiwan and Indonesia are now operating vessels of this category for catching tuna, with great success, although the boats have only iceholds. We have to emulate them and adopt a similar strategy that will give a boost to small/medium scale tuna fishing operations, which will be in accordance with the avowed social objectives of the Government. Tuna fishing in Indian waters is no doubt seasonal but the vessels can operate for tuna for at least six months in a year, based on the present knowledge of tuna grounds. With location of new grounds the duration may go up.

There is a strong but misplaced opinion, deep rooted and obsessive, that vessels so small can never undertake tuna long lining, as space to keep the main line, branch lines etc on board will be inadequate. If one goes through the report at page 6 of the February 1989 Issue of Fishing News International, this impression will be dispelled. A private fishing firm of Indonesia, P.T.Minasanega Pertiwi has chartered about 42 wooden tuna long liners (Iceboats) from Taiwan to fish in the EEZ of Indonesia to the South of Java Island. Although this company started its tuna operations only in September 1988, it has already reached an export level of the order of 30 tonnes a day. Since November 1988, the company has chartered a weekly cargo flight from Jakarta to Singapore. The fresh tuna is exported from Singapore later the same evening aboard planes operated by Nippon Cargo Air to Tokyo for the following morning's market. The company now plans to expand its charter fleet to about 75 vessels.

The vessels being operated by the Indonesian company on charter are no doubt ice boats, Yet, introduction of similar vessels, for operation around Andamans or other suitable locations will be a good strategy as Andaman Islands are very close to Singapore and it will not take more than a couple of days at the most for the vessels to go to Singapore to unload the catches and come back to Andamans waters to resume long lining operations. Alternatively, an appropriate organisation can buy the catches, store in a reefer vessel or at a shore frozen storage and transport them in a reefer ship to Singapore from where the tuna can be flown to Tokyo. There can also be an arrangement to lift tuna from Port Blair to Singapore by chartered cargo air craft. The returns from Sashimi tuna in Tokyo being so high the transport expenses are worth the while.

A similar strategy as outlined above can be adopted for catching tuna off the Southern Bay of Bengal and off Southwest coast with arrangement for pooling up the catches and transporting the same to the Western ports of Thailand or to Singapore. Surely, a viable and highly attractive scheme could be worked out to introduce smaller tuna long lining vessels for profitable operations. In fact, India should launch a scheme under which a few 16 to18 M long tuna long liners are taken by the Government on time charter from Taiwan directly or through Singapore, Hong Kong or Indonesia and entrust them to CIFNET for the purpose of providing training on them to Indian fishermen for a short duration. Once training is over, a large number of such liners can be taken on time charter without any Taiwanese crew and entrusted to Indian fishermen for operation under the auspices of an Andaman's Co-operative Organisation or a Corporation or through private sector companies. Entrepreneurs as well as small fishermen will be greatly benefitted by this approach. The constraint of foreign fishermen being on the vessels gets eliminated once Indian crew are inducted for fishing around Andamans. Once there is confirmation about the suitability of the vessels, they can be constructed at Indian yards, with needed additions such as refrigeration.

*Fishing Chimes* has been advocating for the past few years and also during discussions with various fisheries experts in the Government that tuna long lining should be encouraged through the introduction of medium sized vessels (about 23 m OAL) for undertaking stern trawling and tuna long lining according to seasonal needs. Trawlers of 23m length range with freezer hold at -35°C and with tuna long lining arrangement are the best, from the point of view of operational economics, as such vessels can operate throughout the year (trawling for part of the year and long lining for the rest of the year). This was not acceptable to the authorities on the ground that a long liner should be at least 30m long, otherwise there will not be adequate space for storing the main line, branch lines, floats etc. This view was taken in spite of being told of the fact that long liners of this range have been successfully operating 60 km long longlines along Florida coast of USA. They were also told that by replacing the cumbersome line hauler of Japanese type by a net reel, and heavy main long line (as in the traditional Japanese style) with monofilament long line of 3.5 to 4.0 mm dia and using branch lines of 3.0 mm dia monofilament, the Americans economised on storage space needs and what a 30 m long vessel with heavy investments could do, they have been able to accomplish with a 22m long vessel, with far lower investments.

It will be of strategic importance to (a) encourage those presently operating 23- 28m range trawlers in the country to add long lining equipment on board without in any way affecting the trawling arrangements or operations. During times when conditions are not conducive for trawling for shrimp or lobsters or finfish other than tuna, the vessels can switch over to the catching of tuna for about six months by using the long lining system. By this method the vessels in this size range can be operated beneficially with improved operational economics and (b)encourage introduction of new long liners with or without trawling arrangement in smaller size ranges from 16 to 18m length onwards. Included in this issue is an informative paper on the subject by Mr. Conrad Birkhoff, a German Naval Architect, who has an intimate knowledge of Indian fishing situation.

For implementing a programme of the kind envisaged in this write-up, Government will have to provide incentives by way of subsidies and loans without 'fuss'. It is estimated that the total cost of such conversions per vessel would be around US$ 60,000 and there are foreign firms who can help Indian yards in accomplishing jobs of this kind.

For formulating and implementing a programme of this kind the first requirement is that the officers concerned should shed the idea that only vessels of 30m OAL and above can be used for tuna long lining. Even if they do not want to shed this idea they can at least take pains to study the subject and examine the reasons why a country like Indonesia is using long liners of about 40 to 50 tonnes GRT (16-18 m OAL) and U.S.A. about 100 t GRT liners (22 m OAL and above) for catching tuna and how they have increased prosperity of the fishing sector in the respective countries.

Initiative in this respect has to be exercised by the Chairman, MPEDA. Driving force from that level is needed, in view of the available high potential for harvesting and exporting of tuna from Andaman waters by smaller boats once a week at the rate of about 750 kg per boat of 16 -18m and from there to Sashimi market of Tokyo by air. This is an activity which has to be treated as a project of great importance because of its potential for augmenting foreign exchange earnings. MPEDA may therefore consider appointing a special officer to organise the work through private or public sector organisations.

Fishing Chimes, Editorial 118: April(ii) 1989: Vol. 9, No. 1

# Ascent to the Nineth Rung

*Fishing Chimes*, in active association with its distinguished readership, has reached yet another landmark in its sojourn. Having succeeded in completing eight years of service to its readership, keeping them informed of various developments in the fisheries field, the journal has embarked on its nineth annual journey, keeping up all along the distinction of coming out month after month, without any break, although with certain inescapable delays now and then. The delays have been because of certain constraints which it continuously strive to get over.

One important function the journal has assigned to itself is to make efforts to provide an appraisal of the trends in the fisheries field and their impact on the developmental processes and to bring into focus the corrections that are honestly considered necessary in regard to approaches, policies and strategies. We have been striving hard to fulfill these functions in respect of both marine and inland sectors. Readers are aware of our suggestions concerning the direction in which diversification of deep sea fishing effort could be brought about. We have been taking special interest in proposing new approaches for achieving the augmentation of our deep sea fishing fleet. On the Inland sector too we have endeavoured to stimulate a debate on basic aspects relating to culture and capture fisheries, apart from advocating approaches for promoting intensive fish culture, and developing our riverine fisheries which are the heritage of the nation.

Popularisation of various aspects of new and emerging technologies relating to freshwater and brackishwater farming and marine fishing have been our obsessions. Papers written by scientists and technologists of professional experience and repute on these matters published in the journal have proved to be of great value and use, particularly to the needy readers.

It is a matter of great pride to us that our journal has become a medium of reference to various fisheries administrators, scientists and technologists. Our pride has become all the more accentuated by the fact that most of the students undergoing fishery courses at the various colleges gather the latest information on various aspects of fisheries from *Fishing Chimes* for enriching their knowledge, particularly while preparing for their examinations. The journal has also established itself as a medium of reference to trace the history of various fisheries developmental activities. In fact in vintage establishments such as Fishery Survey of India, the issues are preserved in a bound form for reference purpose.

It has been our longing to improve the get up of the journal and also to set up a computer for storing information relating to various vital aspects of fisheries for easy retrieval when needed, not only for our internal use, but also for the use of others. Owing to financial constraints these objectives could not be achieved so far. We are however confident that, sooner or later, we will be able to convey the glad tidings of having achieved these goals to the readership.

It is also our earnest desire to set up our own phototype setting arrangements, coupled with a baby offset printing system. These would cost money, but we are hoping to find the needed funds. Once we succeed in this, which we hope would be sooner than later, we will be able to present a well designed journal with expressive captions, telling photographs, and contents printed in a style that would not strain the eyes.

It is a matter of immense satisfaction to us that, although the fisheries sector in the country has its own limitations, we have been able to expand the circulation of the journal from mere 14 numbers in April 1981, to over 1000 nos.now. Neutralising this increase are the unfortunate and detestable factors such as the spiralling increases in the cost of paper, and cost of printing and publication. Added to these are the limitations concerning cash flow. A large number of subscribers have difficulties in the payment of their subscriptions on time, or within a reasonable period of these becoming due. With only around 20 per cent of subcription amounts being received in a month and the limited extent of income from advertisements, we continue to be in the process of evolving the technology of survival, particularly with the help of our bankers, the State Bank of Hyderabad, Dwarakanagar Branch, Visakhapatnam. Our account having been with them for the past several years the successive managers at the bank have developed some faith in us and without any security allow us the needed accommodation which keeps us going and makes us more spiritual in outlook. We have to thank the Bank profusely for the help extended, apart from conveying our grateful thanks to the advertisers and subscribers, from whom we sometimes receive demand drafts/cheques at the most unexpected moments thereby providing a great relief and an avenue to tide over tight financial tangles.

We continue to receive a number of enquiries from our subscribers as well as others for advice on several matters relating to various aspects of fisheries. We endeavour to guide them to the best of our ability and we are glad to say that several of them have been kind enough to acknowledge our help. The journal, besides circulation within the country, has also made inroads into several fisheries outfits outside India. A couple of foreign journals keep quoting important newsitems published in *Fishing Chimes*.

It is believed that *Fishing Chimes* has reduced the communication gap among the various State Fisheries Departments, among fishery workers, scientists, technologists, industrial establishments and others concerned. The journal has been of particular utility to the commercial sector, which is a matter of satisfaction to us.

On this solemn occasion of entering into the nineth year of publication, *Fishing Chimes* humbly renews its pledge to the fishery sector to contribute its utmost for the: promotion of the fisheries development of the country and play its own modest role in baling out the industry, particularly the deep sea fishing industry from its present plight to an era of prosperity and plenty; and we do hope that, in this transition, they would bring about a quantum jump in foreign exchange earnings through exports of marine products.

Fishing Chimes, Editorial 119: May 1989: Vol. 9, No. 2

# SCICI adopts a Pragmatic Approach

According to available reports, the Shipping Credit and Investment Company of India has decided to set up a deep sea fishing company of its own. We consider this decision as realistic and pragmatic, in the background of the present anamolous situation of entrepreneureal and Governmental apathy amidst plenty. In other words there are fisheries resources in plenty with few to reap the bounty.

Since its inception, SCICI has been treading a progressive path under the inimitable leadership of its present chairman. SCICI has virtually reversed the traditional systems of interaction between a financier and a client. At a time when the clientele was inactive, which normally should not bother a financing body, the chairman and his officers condescended by way of convening meetings of representatives of various fishing companies and interested entrepreneurs at various places and explained to them in detail the flexible nature of their financing system. In fact, SCICI has been more catholic than the Pope himself. Yet, the response to the inviting stance is apparently poor, considering the fact that the sanctions issued by the SCICI in 1988-89 covered a mere 26 vessels costing about Rs.35 crores and there is no demand from the industry for a more encouraging response.

Who is to be blamed for this situation? The inert Government? the passive clientele? or the ready-to-lend SCICI, on case to case basis? or the complex inter-relationships and equations among all concerned, that has been impeding a forward thrust?

Whatever be the reasons, it is clear that the SCICI has taken the poor response from the industry as a major affront and a challenge and probably as an insult. The SCICI, although a chip of the old block (ICICI) is still in its burgeoning phase and appears to be sensitive to discouraaging situations. Consequently, its approach seems to be: if the entrepreneurship has hesitancy, let it be removed. One way is to enter the line by itself, an unothodox approach; and show to the interested parties, what can really be achieved. This kind of demonstration is bound to have a salutory effect on the various entrepreneurs who want to take up this line.

It is admirable that SCICI is free from shrimp mania. In 1990s the shrimp export market is bound to crash. Most of the sea fishing effort would have to be therefore finfish-oriented; chief among these fin fishes is the tuna which has a good export market to Japan in Sashimi form and for canning to Thailand. Several other species of finfishes command a viable market in several south-east Asian countries. No less important are the squids, cuttlefishes and lobsters.

Establishing companies of their own is not the normal business of financing institutions in India. If SCICI is thinking of setting up a deep sea fishing company, reasons for this must have been compelling. The intention, we are certain, is not to impart any disadvantage to the ongoing industrial entrepreneurship. On the other hand, it must be only to show the way and to dispel the fears that are nurtured by the entrepreneurship in taking up sizeable deep sea fishing projects. In fact, the setting up of an enterprise by the SCICI could be a boon to the ongoing industry. The SCICI's proposed subsidiary can advance funds to the existing fleet to make all needed alterations, for tuna fishing by gill nets and/or longliners, or stern trawling. All the harvested fish can be purchased by SCICI's another ship and out of payments due, the SCICI's company can make reasonable deductions towards repayments due to SCICI by the company concerned.

Fish purchased from vessels not belonging to SCICI's subsidiary cannot always be transferred to the proposed mothership. In fact, there will be a sizeable collection of such catches which require shore storage space. The new company to be set up by the SCICI can establish frozen storage at -55°C close to major fishing ports. Fish purchased by this company from the various fishing companies, when not transferred to the mothership, can be stored at these plants for export or domestic sales as the case may be. The deep sea fishing company of SCICI or SCICI on their own can provide the working capital needs of the companies and recover them out of the earnings, based on a well designed mechanism of advances and recoveries. Adopting an approach of the kind referred to above, and linking it to a marketing tie-up with a foreign company/importer, major changes for the better could be brought about. Once the activities of the proposed company are set in motion and as experiences are gained, various new beneficial ways of functioning may well unfold.

Once a large fishing company under the auspices of the SCICI is set up, a stronger state of leadership in the field is bound to emerge. Once there is a strong leader, there will be a full going. And the emergence of such an encouraging situation will alter the quality of functioning of the sector and will bring about a greater measure of organisation, in terms of efficiency of operations, accountability, financial discipline etc. SCICI's subsidiary can also acquire vessels and give them on lease to fishing companies. With a large fishing company of its own, SCICI can anytime take over vessels of companies defaulting in the payment of instalments without adequate reason, and run them, because of the availability of an organisation.

Where would SCICI like to locate its new deep sea fishing company? We feel that Madras should prove to be the best choice because of the nearness of tuna and other finfish grounds off the southeast as well as southwest coast and also the Wadge Bank. There can be a subsidiary base at Porbandar or any other suitable base along Gujarat coast to facilitate the exploitation of the rich potential of the Great Kori Bank.

Fishing Chimes, Editorial 120: June 1989: Vol. 9, No. 3

# Problems of Fish Culturists

Indian fish farmers are a persevering lot, compared to their counterparts in agricultural sector. Agriculturists have political patronage. Their problems are looked into by politicians and administrators in authority with lightening promptitude. If this is not done, those in power have to face their fury, with the resultant loss in their support. Agriculturists therefore are able to get financial credit at lower rates of interest and their problems of crop husbandry are carefully looked into and solutions offered. They are helped in the marketing of their produce at the highest possible rates. Electricity is supplied to them at low tariff. Water is also supplied to them at nominal rates.

Agriculturists are land farmers. Fish culturists are also land farmers with the difference that their culture activities take place in water that gets stored or is stored in depressions over land. While land crops take nutrients directly from soil, water crop *i.e.,* fish crop takes the soil nutrients through the water medium. Water areas that agriculturists cannot make use of, are brought under fish production by fish farmers. For rendering this service to the nation, fish farmers should actually be bestowed with far more benefits than the land farmers get. Instead, they are forced to absorb indifference and apathy. Neither the Central Government nor the State Governments bother much about them.

It is no doubt true that certain subsidies are granted to fish farmers coming under the purview of Fish Farmers Development Agencies. This is however of very little solace. Firstly this help alone is not adequate. Secondly, there are a large number of fish farmers who are not covered by Fish Farmers Development Agencies.

What the fish farmers need most is the supply of inputs including financial credit at concessional rates at par with Agriculturists. What they also require is an avenue for marketing their produce at remunerative prices.

In most of the States fish farmers have to buy seed and feed at enormously high rates. Water and electricity is supplied to them at rates vastly and uneconomically higher than those applicable to agriculturists. No attention is paid to the problems of fish/shrimp diseases faced by them. In regard to fishery leasing policies, a committee appointed by the Union Ministry of Agriculture in 1970s conducted a thorough study of the situations obtaining in various States and made several recommendations, implementation of which would go a long way in promoting production and thereby helping the fish farming community. Very little has been done by the State Governments to implement the recommendations which were commended by the Ministry of Agriculture.

Water and electricity tariffs related to fish culture vary considerably from State to State. So do prices vary in respect of seed and feed supply. The extension programmes relating to fish culture are also weak in most of the States.

The proclaimed aim of the Central and State Governments is to promote freshwater and brackishwater fish culture. Such being the case, it is of utmost expediency for the Ministry of Agriculture to appoint a committee of officials and fish farmers to review the present state of support being given to fish farmers in various States (covering electricity and water tariffs, fallow water/land lease durations and lease amounts for promoting of fish culture etc.), to recommend measures to be implemented by the State Governments so as to enable fish farmers, a) to secure, in the main, supplies of electricity, water supply etc at tariffs at par with agriculturists b) to get seed and feed supplies at subsidised rates, c) to receive marketing help for their produce at assured and remunerative prices, d) to strengthen extension support to the farmers and e) suggesting measures for leasing of public tank and pond resources to fish farmers for undertaking fish culture.

Fishing Chimes, Editorial 121: July(i) 1989: Vol. 9, No. 4

# Resurrection of Deep Sea Fishing Sector

The Indian deep sea fishing sector is now in a state of suffocation. The position is comparable to the "Black hole of Calcutta". In the same way as a large number of persons were shut up in a limited space by the British in Calcutta, over 160 deep sea fishing vessels and over 70 minitrawlers are now huddled together at Visakhapatnam fishing harbour, The vessels conduct fishing, mostly for shrimp, for a duration of about six months in the upper Bay of Bengal (mostly sandhead area). The operations of these vessels are characterised by fluctuations in per-unit catches and incomes and increase in per unit operational costs. Payments of heavy incentives to the floating staff during three to four peak fishing voyages, in total disregard of the overall annual income has become an established practice. During the offseason, most of the vessels remain berthed at the harbour, with the owners meeting all commitments concerning salaries and wages of crew, port charges, drydocking charges etc. mostly out of unsecured loans. Lack of cash flow during offseason forces fishing companies to obtain unsecured loans. The economics of operations of these vessels, probably with the exception of a few which conduct fishing for deep sea lobsters with Cochin/Tuticorin as the base during the offseason for shrimp, are thus in a state of jeopardy.

In this situation, it is now commonly felt that diversification of fishing and fanning out of vessels for operations from ports other than Visakhapatnam is of crucial importance. A beginning in this direction has been made by a few companies by shifting their base to Madras, but fishing continues to be done in Sandhead area. Equally important is the need for restriction of future fleet strengths of different types of vessels at various selected ports either by deployment of present fleet at Visakhapatnam to other ports or by additions to the national fleet strength. The truth learnt by the hard way is that increase in the number of vessels beyond a carefully worked out limit at any one port not only reduces catch per unit effort and returns but will also lead to agitations from crew because of their numerical strength, and results in payment of higher wages, often disproportionate to incomes of companies concerned.

The wide publicity given in the newspapers on the periodic confrontations between the owners and the floating staff results in the reduction of whatever little enthusiasm is there on the part of prospective fishery entrepreneurs. So much so, as at present, there are few coming forward to take up this line. The existing companies have practically given up ideas of expansion. The industry being thus in a state of suffocation, an emergency has arisen for the Government and the industry to take a purposeful and concentrated look at the alarming situation which, if not effectively dealt with forthwith, may well wipe off India from the map of world's deep sea fishing.

The genesis of the present situation can be traced to lack of adequate knowledge and commercial experience in deep sea fishing in the country at the time India entered this sector. In the early stages there was very little knowledge of fisheries resources of the present day EEZ for the experts at the time to advise the Government in regard to the manner in which the development of the industry should be planned and developed in stages. Fishing for shrimp along the upper east coast was equated with development of deep sea fishing, and when the shrimping fleet developed relentlessly upto a particular point the concept of diversification has burgeoned. There is now a realisation that, if a well thought-out diversification plan is not drawn up and implemented, greater damage will be done to the larger cause of utilising the deep sea fishing resources of our EEZ.

As days went by, and when the trawler fleet strength increased and when a major reduction in the number of fishing days per vessel and fall in catch per unit is noticed, there was remorse at the thoughtless introduction of tradition of high salaries, irrational payment of incentives etc. There was also resentment at the Government's merciless action of withdrawal of exemption of excise duty on diesel oil, hike in charges levied at fishing ports etc.

Thus cornered, as an immediate step to balance their budgets, the owners thought of introducing a system of sharing the gross proceeds with the floating staff with the hope that this would ensure rational payments to crew from voyage to voyage, leaving sufficient funds to meet their own overheads and also repayments to the financing institutions. With this proposal not being acceptable to the floating staff, the discomfiture of the trawler owners mounted up. The floating staff are unable to accept the sharing system as they feel that this system would decrease their incomes.

The question of introduction of new vessels by the present category of entrepreneurship can virtually be ruled out, at feast for some time to come, because of the financial plight of the majority of the deep sea fishing companies that they see, whatever be the reasons. The immediate step needed now therefore is to find out ways and means of dispersing the existing fleet at Visakhapatnam to various other potential ports along the coastline by mounting special schemes.

The Ministry of Food Processing Industries and other concerned Central Ministries have now to initiate effective steps in this regard on warfooting, offering attractive incentives to the companies to spread out their vessels to various other selected ports all along the coastline where the needed infrastructural facilities have to be provided by the Government. This approach may stem the trend of mounting losses to fishing companies. Otherwise, it may well be a matter of time that the existing fishing companies will be forced to give up deep sea fishing as a profession, in stages, owing to frequent operational losses, although 1989 may prove to be a pacifying year for shrimp catching vessels.

The fanning out of the existing vessels to various ports involves considerable financial risks to the fishing companies. At the same time, this may well open up new opportunities for profitable fishing. As already stated, attractive incentives, if offered by the Government, would be a major motivating factor for the owners to equip their vessels for another type of fishing and pull out of the jinxed Visakhapatnam fishing harbour and move over to another suitable port.

Offering a suggestion for remedial action in this connection appears to be in order. The Ministry of Food Processing may consider entrusting to the Fishery Survey of India a new line up activity, an activity of working jointly with selected fishing companies for undertaking commercial surveys all along the coastline from specified ports, based on presently available resource data in respect of tunas, cephalopods, lobsters etc. FSI may select a specified number of vessels from the present commercial fleet and render advice for converting the same to undertake an additional type of fishing and for deploying them from various selected ports and operate them under its guidance. Owners can be told to contribute their vessels for this commercially oriented exploratory work. The companies may be advised by FSI to modify them suitably for undertaking a viable additional type of fishing to enable them to organise the needed modifications. Government may provide 100 per cent subsidy to owners in this regard. Operational expenses may be borne by the FSI initially but recover them later from owners out of earnings. A scheme on these lines, setting out the respective roles, can be worked out and implemented. If successful, this approach will pave the way for extending diversified deep sea fishing activity to various suitable bases all along the coastline.

Coupled with this is the need to provide infrastructure facilities at each of the selected harbours. This can probably be done by encouraging the private sector to set up workshops and outlets for supply of deck and engine side requirements. This scheme should also provide for measures to equip a predetermined number of existing vessels (over and above those already converted for experimental work) for diversified fishing systems, in addition to the present trawling arrangements, based on the results of commercial surveys by FSI and private enterprises as mentioned above. Government of India should offer, at this stage, at least 50 per cent subsidy towards the cost of additional equipments and the rest of the amount could be provided as loan by SCICI under a Government guarantee. These guarantees will be a very small sacrifice on the part of the Government in the interests of development of the industry and to secure the needed enthusiasm and support from the entrepreneurship.

The proposed additional equipments can include those relating to squid jigging, tuna longlining, tuna purse-seining, bull trawling etc. A break up in this regard, harbour-wise, will have to be worked out by the FSI in consultation with CMFRI, taking into account the resources position.

Simultaneously, the Government would have to bring into being a centralised marketing agency with branches at each of the selected fishing ports to buy fish catches on payment of reasonable but remunerative prices and organise internal sales and also undertake exports.

Along side the introduction of diversification system, it is necessary to provide supporting action by way of training facilities at CIFNET to candidates sponsored by the concerned fishing companies. The nucleus crew can be provided from the Central Fishery Institutes such as FSI to begin with.

Readers are aware that SCICI has decided to set up a mammoth deep sea fishing company. An alternative way of stemming the deteriorating situation is for the Government to prevail upon the SCICI to set up this proposed new company as quickly as possible. SCICI is an advocate of the system of mergers of companies so as to make operations viable. SCICI can implement this concept by inviting all willing existing companies to merge with the company proposed to be set up by them, under such terms and conditions as are decided. This would solve the problem of pending repayments of principal and interest from various fishing companies, whose vessels, upon merger, would no doubt be well managed from financial angle by the proposed new company of SCICI. With a vast fishing fleet at its command, SCICI's proposed new company can set up their own facilities *i.e.*, workshops, frozen storages, marketing network etc. at various ports, and can gain a near total control over Indian deep sea fishing and marketing of catches.

It has been suggested in the preceding paragraphs that the existing vessels may be provided with additional fitments for diversified fishing. It will be easier for SCICI to prevail upon the Government to provide 50 per cent subsidy towards the cost of the addition of systems for diversified fishing on the existing vessels. It may not be a problem for the proposed new company to finance the balance of money required for the additional equipments for diversified fishing.

For the industry to survive, the Government of India will have to consider granting of remission of interest on loans given for the acquisition of vessels, because of the past bad fishing years and also approve rescheduling of the repayments of principal amount, allowing the meter in respect of interest to start afresh. Other-wise the problem of repayments is bound to intensify further. The company to be newly formed by SCICI can plead with the

Government to provide remission of interest to the amalgamating companies. Once most of the existing fishing companies are merged with the proposed new company of SCICI, remission of interest could be secured, and 50 per cent subsidy on additional equipments for diversified fishing could be obtained. With this, a new shape and framework for the Indian deep sea fishing industry will emerge.

All fishing companies now face problems of crew management. For a large company of the size contemplated by SCICI, these problems and also those of crisis management will get accentuated further. In this context, a basic statutory framework is of utmost importance. Lack of legal frame-work for the development and regulation of deep sea fishing activities by national enterprises has already created chaotic conditions. It is hoped that the Ministry of Food Processing Industry would take the needed steps in this regard.

Added to all this is the lack of any enforceable statutory regulatory mechanism to control relations between the trawler owners and the floating staff. The service conditions of the crew in deep sea fishing vessels come within the purview of Merchant Shipping Act. There being no machinery and no regulations laid down to sort out various service problems of officers and workers on trawlers and their relationship with the owners, virtually a pandemonium prevails often. On such occasions, the owners, the officers and the workers take law into their own hands as they like. Only when they get tired of conducting prolonged negotiations and only when they realise that they are losing fishing days they come to some kind of agreement and resume fishing operations.

The Ministry of Surface Transport/Director General of Shipping has to take cognisance of this situation and evolve and implement effective measures to regulate the relationship between the owners and crew of fishing vessels registered under the Merchant Shipping Act.

## Our Final Appeal to the Government

Save the collapsing Indian deep sea fishing Industry, in national interest. Let not foreign fishing vessels take away the fishery wealth of our seas, for the mere reason we are not equipped to fish adequately in our own EEZ.

Fishing Chimes, Editorial 122: July(ii) 1989: Vol. 9, No. 4

# Steps to Counter Siltation of Fish Ponds

All the State Governments and the Central Government have been making laudable efforts to step up inland fish production, particularly from tanks and ponds. According to statistics presented, inland fish production has been going up year after year, the latest as at the end of 1988-89 being around one million tonnes. The statistical validity of the estimates has however not reached a stage at which one can believe in good faith that the estimates projected represent the correct position.

It is common knowledge that in almost all States such as Tamil Nadu, A.P., Rajasthan, Uttar Pradesh etc., tanks, ponds and reservoirs have been getting silted up at a fast rate. With the volume of water then thus coming down, in spite of the density of stocking with fish seed, production from farming sector must have been coming down, which is perhaps not reflected in the State-wise statistics.

Let us examine one example. Cultivable waters in Tamil Nadu are no different from those in most other States. In this State, it is seen that the inland fish production which was of the order of one lakh tonnes in 1987-88, came down to 82,000 tonnes in 1988-89. There was not much of difference in the monsoon conditions between the two years. There is also no reduction in the quantity of seed stocked in the various waters in the State. Such being the case, there is no explanation as to how the production has plummeted. Either there is some grave error in the collection of production statistics or the quantity of seed stocked has been boosted up. We do not think either of these two would have happened, for the reason that the persons engaged in the collection of statistics or stocking the seed, must have been following the same pattern year after year and there is no special motivation for deviating from the beaten track. Further, the general tendency is always to show an increase in the production rather than bringing it down. That there has been a fall in inland fish production in Tamil Nadu therefore clearly shows that the reduction in water area is probably the main reason for the fall in production. The officers of the department have to be congratulated for truthfully bringing out the level of production instead of making efforts to cover this up, which is no problem at all.

When the situation in Tamil Nadu is as explained above, one can presume that things may not be much different in other States. It is therefore necessary that, with central assistance the various State Governments undertake a study of the siltation problem from the point of view of fish culture. If siltation is the main culprit in the falling levels of inland fish production, it is necessary to evolve schemes for deepening these water areas, by pooling up funds for the purpose from various interested departments such as panchayats, fisheries, irrigation etc.

Fishing Chimes, Editorial 123: August 1989: Vol. 9, No. 5

# Congeniality at Visakhapatnam Fishing Harbour

The Visakhapatnam fishing harbour would not have survived all through the past ten years but for the balanced equation and relationships of friendship and comradarie among those handling the various components of work at the harbour, whose integration resulting in the successful operation of fishing vessels from the harbour, which continues to function with vibrancy.

The owners of trawlers of various categories (larger ones, mini-trawlers, mechanised boats etc.) have their own associations. The officers who man the trawlers have an association of their own. This Association is known as 'All India Deep Sea Fishing Technocrats Association'. (The members however do not qualify to be known as technocrats). The term 'technocrats' is derived from the word 'technocracy', which means 'a Government by technical experts'. In the same way as the term 'bureaucrat' is associated with administrative officers in a Government, the term 'technocrat' is also connected with 'technical experts' in a Government. The members of the Association are in fact 'fishing vessel officers; who are not technocrats. According to New Webster's Dictionary, technocracy means 'a theory of Government, prominent in 1932, advocating control and management of industrial resources and reorganisation of society by technologists and engineers'. The same dictionary defines a technician as 'one highly trained in the technicalities of a subject, profession or occupation'. Thus the trawler officers can aptly be referred to as technicians and not technocrats).

There is a union consisting of all trawler workers who form a critical segment of the fishing activity that is undertaken from the harbour. The owners of the various workshops have also set up an association of their own. The workers who undertake the unloading of catches from the trawlers have also a union of their own. Last but not the least are the exporters who have a strong association on whom the survival of all operations depend to a large extent.

In regard to the Governmental Sector, the Port Trust and the Mercantile Marine Department, apart from the State Fisheries Department, have a paramount share in the working of the fishing harbour. Basing on the adage that the best governance and administration comes out of the least possible interference, these organisations are benovelently close to the human and economic activities that take place in the harbour area, while at the same time maintaining a calculated distance.

The Hindustan Petroleum Corporation and the Indian Oil Corporation are two other organisations which play a major part in the successful operations at the harbour. The running of the trawlers totally depends on the promptness with which these Corporations supply diesel oil and lubricants to the trawlers. It is entirely to their credit that these organisations and the various Governmental agencies referred to above do a splendid job. Further there are the officials concerned in the police department who keep a close eye on law and order in the harbour area and do their best to prevent clashes and thefts within the premises. They also make efforts at pre-empting clashes by timely interventions.

With so many agencies involved, one can visualise what would happen if proper equations are not established among these. If the owners do not pay adequately the officers and workers can paralyse the operations. Such situations do take place now and then but are averted because of settlements with mutual understanding and cooperation. The C.I.T.U. which is a great body but sometimes disliked by some, deserves to be complimented so far as the fisheries sector is concerned, for their deep understanding of the characteristics of the deep sea fishing activity, its fluctuating fortunes, effects of cyclones and bad weather on the operations, ups and downs in the catches, their prices etc. Unless they have no alternative other than asking for something more, they do not come up with demands. And when they come up with demands they discuss in a very realistic way and settle matters in a manner that does not deeply hurt both the sides in the end.

No one knows better about the characteristics of the industry than the trawler officers. They rarely raise issues that would spoil fishing operations. When they do, it will always be because of some injustice they strongly believe is being done to them. Realising the plight of the owners during the past two years the trawler officers association lent support to them. It was only the owners who wanted the officers to accept a system of sharing gross returns between floating staff and the owners with a view to achieving a rationalised system but this was not acceptable to the officers. This honest expression of their views deserves to be appreciated. Belying the prediction of a certain section of the officers and also the owners, the present President of the All India Deep Sea Fishing Technocrats Association proved to be a good leader, as it turned out. It is only hoped that he will not go by any unsound advice given by such of the office bearers who do not take a holistic view of industry but look at things with a cockroach vision and from a selfish angle.

It is to be conceded that the proprietors of the various workshops and small suppliers of materials are placed in a critical state because of non-payments for a long duration by the fishing companies for the services rendered. The owners have to accept blame for two reasons. One is that, when they entrust a job to a workshop they should have a clear understanding of the charges that would be levied. After the work is over, the owners feel that they have been overcharged. This can be one of the reasons for delayed payments. Once the trawler owners establish a system of prompt payments, the proprietors of workshops would also automatically develop a system of raising bills that would take care of their services in a reasonable way, providing for justifiable profits. It is desirable that those representing the managements of the fishing companies and the workshops sit together and evolve a schedule of payments in three categories for the various types of works normally undertaken, with reference to the nature of work of each involved. The owners will have to accept higher charges where the works have to be done on a war footing. However, for jobs done in due course with a sense of urgency but without a specified time limit, the charges would have to be moderate. Another category can be the type of jobs that can be done in a period of 7 days or so. Once such a system is developed the misunderstanding regarding payments to workshops may, diminish and the efficiency of fishing operations may improve. Another aspect is the reported tendency on the part of some trawler officers to show favouritism towards certain workshops, although this is not a common occurence. It is for the fleet managers or owners of trawlers to look into the merits of various situations and take appropriate steps.

The most disciplined among all sections concerned perhaps are the unloading parties. They do their job well. They seek occasional increases on payments which they receive in terms of number of tonnes unloaded and this is conceded to the extent possible. The unloading parties have however one weakness. This is to collect a fairly good quantity of fish as their share for having unloaded fish, for which they do not receive any direct payment. Owners sometimes exhibit a grudge that they take away a substantial quantity of fish. The payments that they receive are from the processors and are related to unloading of shrimps.

Now a word has to be said about the Exporters Association which adopts a sophisticated approach in the fixation of prices in regard to purchase of shrimp landed by mechanised boats and Sona boats. A Committee presided over by the Deputy Director of Fisheries of the State Fisheries Department fixes up prices for shrimps of various categories from time to time and these are adhered to. So far as the catches from the larger and mini-trawlers are concerned, the owners always try to secure the highest rate possible from the processors. The converse is not fully true in the sense that the processors do not always try to pay as less as possible, because of the competition among processors to secure as much material as possible. The export rates are known to the trawler owners and they are in a position to judge to what extent they can bargain. In fairness to exporters, one has to say that, in ultimate analysis they veer round to payments moving closer to the demands of the owners.

The Port Trust is always in touch with the fluctuating fortunes of the trawler operations. Where the situation demands, the administration extends concessions by providing additional time for payments which by itself is a great relief to the operators. When matters relating to sanitation, renovation of old electric wiring system etc have been brought to the notice of the authorities concerned, they have taken very prompt rectificatory action. The owners of workshops and also of trawlers however still nurture grievences in matters such as lack of adequate number of power points for taking connection for work inside the trawlers, supply of potable water, unjustified allocation of prime space in the harbour to the ONGC etc. and cutting out a few oil supply outlets etc. The industry continues to have the help of the Chairman of the Port Trust who pays special attention to all harbour grievances with a view to removing them in the shortest possible time.

The fishing industry has to pay a compliment to the surveyor-in-charge of the Mercantile Marine Department, Visakhapatnam and his colleagues and staff for the promptness with which the problems faced by the various fishing companies are attended to. For them it is just performance of their duty but some of their prompt actions prove to be major reliefs to the operators. Not only is there a free access to the surveyor for representations by the fishing companies which is as it should be, the surveyor-in-charge extends the needed advice with reference to the details of the subject matter, whenever approached by the representatives of fishing companies.

To sum up, it is our hope that the existing equation among the various constituents concerned will improve further and the Visakhapatnam fishing harbour will continue to be model for the other major fishing harbours in the country, which are now coming up one after another.

Fishing Chimes, Editorial 124: September 1989: Vol. 9, No. 6

# Fragmented Fisheries Set Up

India now has a fragmented fisheries administrative set up at the national level. All matters pertaining to inland fisheries with a national jurisdiction and application are dealt with in the Ministry of Agriculture. The subject of deep sea fishing, to the extent of granting permissions for acquisition and charter of fishing vessels is dealt with in the Ministry of Food Processing. Activities which are an integral part of deep sea fishing such as development of fishing harbours, training of personnel for deep sea fishing, introduction of new fishery products etc continue to be with the Ministry of Agriculture. Joint ventures in deep sea fishing are handled by the Marine Products Export Development Authority, apart from promotion of exports of marine products.

The situation as brought out above contradicts the oft-repeated adage that a well-coordinated set up alone can lead to an effective promotion of developmental activity relating to subjects such as fisheries. Defying this principle and bereft of coordination in respect of the various components connected with deep sea fishing, the Ministry of Food Processing could manage to achieve impressive results in the introduction of chartered foreign fishing vessels, holding the burden of a fragmented and dismembered chunk of the integrated subject of 'Deep Sea Fishing', transferred to it from the Ministry of Agriculture.

As a result of the efforts of the new ministry, now over 40 nos. of chartered long liners are under operation, reportedly yielding good results. This gives hope that tuna long lining with Indian owned vessels win soon materialise as a follow-up of the demonstration by the chartered long liners. Apart from this achievement, the ministry has succeeded in lifting the ban on the operation of chartered pair trawlers imposed by the Ministry of Agriculture earlier. The credit for these developments goes to the Minister, his Secretary and other Officers. These achievements should set us to imagine what progress can be achieved if an effective coordination or centralisation of decision making is established in the Ministry of Food Processing in respect of all aspects connected with deep sea fishing, particularly in respect of fishing harbours, training and adequate support by way of technical personnel. In other words, a well coordinated set up will lead to wise and rational investments in deep sea fishing industry and bail it out of the present problem of stagnating investments and inadequate returns as seen from the present situation of high operational costs, dwindling export prices and concentration of operation of deep sea fishing fleet only in one area, *i.e.*, along the upper east coast for about six months, with the fleet tied up at port in idleness for remaining part of the year, owing to unviable fishing operations.

Sensing that it is not wise to encourage the development of national deep sea fishing fleet any longer on the present lines, the Ministry of Food Processing has done well in demonstrating, through allowing operation of chartered tuna vessels, the availability of sizeable stocks of tuna in the Indian EEZ. The results of this demonstration have already led several entrepreneurs to take steps for securing their own tuna vessels. Further, some of the entrepreneurs are contemplating to equip their trawlers with tuna long lining equipment as well to facilitate fishing operations throughout the year. Not resting on these achievements the Ministry of Food Processing has now decided to promote the operation of pair trawlers on charter so as to prove once again to the entrepreneurship the effect of this type of fishing for catching finfish, cephalopods etc. The stocks of these animals are there, not to be retained as in a showcase or to lure the foreign fishing vessels to exploit them, but for utilisation to our advantage.

Worried about reaching export target of marine products, the Ministry of Commerce and MPEDA are struggling hard to increase the national deep sea fishing fleet, for the obvious reason that exports can go up only when the fleet strength is increased. They know that the increase in chartered fishing vessels is only a palliative and is not a solution of enduring value. They know that a big chunk of the earnings from the chartered fishing vessels which is in foreign exchange is offset towards charter hire, and only 20 per cent of the earnings accrue to the foreign exchange earnings of the nation. They therefore desire to have a speedier method of development of national deep sea fishing fleet spread over all along the coast line. The fragmentation of deep sea fishing set up at the Centre, having freed the Commerce Ministry and MPEDA from the grip of the Agriculture Ministry and, seeing that the department of food processing is concentrating almost wholly on the subject of charter of fishing vessels, they have succeeded in snatching initiative in the matter of formulation of strategies for the development of deep sea fishing fleet, a subject held tightly by the agriculture ministry till recently, in spite of express provisions in the MPEDA Act that this subject comes under its purview.

One has to admit that the commerce ministry/ MPEDA have been doing well in the matter of formulation of strategies for the development of deep sea fishing on new lines. With lightening speed, they constituted several groups of experts to make recommendations for the promotion of deep sea fishing and to step up exports of marine products. The groups have made recommendations and the Ministry of Commerce has already initiated steps to implement them. These recommendations are timely in the background of the present declining trend in the development of deep sea fishing. Availability of fishery resources in our EEZ at an economic level throughout the year to make the operation of "resource-specific vessels" economically viable could not be established and the policy of the Agriculture Ministry to encourage introduction of resource-specific vessels alone has failed miserably. Response for joint ventures from foreign fishing companies is disappointing, although, MPEDA, the nodal agency to promote this activity is working hard on this subject. The development of fishing harbours has been very tardy. In the absence of infrastructure facilities at various ports, overconcentration of fleet at only one fishing harbour *i.e.*, Visakhapatnam has materialised.

Those who are allowed to operate foreign fishing vessels on charter create various problems and most of them do not fulfil the conditions of obligatory purchase of vessels. Although the Government does not want Taiwanese vessels with Taiwanese flag to be chartered for operation in Indian waters, the charterers manage to bring Taiwanese vessels with Taiwanese crew, although with Panama flag, which makes only a technical difference. Proper statistical information on landings is not often made available to the Government by the chartered as well as Indian owned vessels. Several companies often do not clear the instalments of loans taken for acquiring fishing vessels and for setting up processing plants, although they are not to be wholly blamed for the situation. All these factors have led to the retardation in the introduction of deep sea fishing vessels of national flag and the Ministry of Commerce/MPEDA are trying their utmost to create a climate in which the national fishing fleet will grow.

The Ministries of Food Processing and Commerce now face the serious problem of reversing the present trend in the development of deep sea fishing, which has taken roots over the past few years, with an unworkable financing system and the misconceived policy of encouraging introduction of resource-specific vessels in our EEZ which is characterised by seasonal fisheries that requires the application of different methods of fishing according to types of fishery available in a particular season.

Development of fishing harbours, training of personnel for manning fishing vessels, and introduction of new fish products are an integral part of promotion of deep sea fishing. Provision of financial help for acquiring fishing vessels is another important aspect that has to be an integral part of the system. While financing part vests with SCICI, who seems to be making a bid for the closure of the Fishing Vessel Acquisition Committee in the Ministry of Food Processing, all the remaining aspects listed above continue to be under the control of the Ministry, of Agriculture. The Fishery Survey of India is the only body that has been brought under the control of Ministry of Food Processing. With export endeavour vested in the Ministry of Commerce and most of the other links vested with the other organisations, one can imagine the hurdles that come in the way of development of deep sea fishing, particularly in the development of national deep sea fishing fleet.

Certain other aspects can also be pointed out here, although they may sound frivolous.

Obviously because of space considerations, the Minister and the officers and staff of the Ministry of Food Processing have their places of work in three separate buildings in Delhi. The Minister, his personal staff, Secretary, Deputy Secretary, etc are located at Transport Bhavan. The Joint Secretary, a few Deputy Commissioners and Assistant Commissioners have their place of work in Krishi Bhavan. The Under Secretary and a few others function from Indian Oil Bhavan. Movement of files from one Bhavan to the other, takes time, even when all possible steps are taken to minimise delays. This situation is obviously unavoidable at present and no doubt the Ministry is anxious to find a place where all officers and staff can be together. This search takes time and cannot be held against the Government, considering the problems of securing accommodation in Delhi.

As already mentioned, one subject of priority rating handled in the Ministry of Food Processing at present relates to granting of letters of intent and permits for chartering of fishing vessels. The fragmented set up has led to this concentration. Distances between the three Bhavans and the frequent need to keep messengers running with files from one to the other comes in the way of taking quick decisions and some charterers say that it takes an enormous time to receive communications from the Ministry conveying crucial decisions or orders. Nevertheless, the work being done gives hope that the target of introduction of 500 nos. of deep sea fishing vessels can be realised through introduction of chartered vessels, if not out of those of our own national flag. Those at the helm of affairs seem to have realised that the presence of 500 nos. of vessels at any given time, chartered or of national flags is of the essence. This gives two advantages straightaway, it must have been comprehended. One is that, operation of chartered vessels augment export earnings. The other is that no investments need be made for acquiring vessels. The only weak point is that the bulk of the chartered vessels come from one country *i.e.*, Taiwan, and practically no fisheries enterprise of any other country is interested in giving vessels on charter to Indian enterprises. The fishing companies that charter Taiwanese or other fishing vessels take abundant risks in various respects, but seem to be happy because they stand to get 20 percent of gross proceeds and, even if they do not fulfil the obligations such as introduction of fishing vessels of their own, they will still not lose, if everything goes on smoothly, particularly because, under the latest scheme, permits will be given for operating chartered vessels, including pair trawlers, for two years at a stretch. The charterers will no doubt endeavour to fulfil the obligations imposed under the scheme. Even if they fail, they will not lose their investment which is a reassuring feature of the scheme. The charter scheme protects the developmental and commercial interests of the charterers, which is a laudable feature indeed.

The charter scheme should definitely lead to something good for the promotion of the deep sea fishing industry of the country eventually. This would however depend largely on the will of charterers to introduce vessels of their own and the helpful support extended by the Government and financing bodies in the matter of provision of infrastructural facilities at selected ports, introduction of suitable vessels, training of crew etc. One of the main functions of the Food Processing Ministry being the promotion of processing of fish products for domestic consumption and export, it is to be hoped that a major thrust in this respect, in tandem with the Ministry of Commerce/MPEDA would be imparted. If major components of marine fisheries sector such as development of fishing harbours, training schemes and applied processing activities such as those undertaken at the Integrated Fisheries Project are brought under the purview of the Food Processing Ministry, with all its officers housed in one building, there can certainly be a breakthrough in the development of integrated development of deep sea fishing activities.

The exclusion of the subject of deep sea fishing from the Agriculture Ministry, has enabled it to concentrate on Inland fish culture development. This is evidenced by the recent sanction of 100 more FFDAs.

There has been a persistent demand from the fishing industry that a separate Ministry of Fisheries should be constituted at the centre. It is unfortunate indeed that, instead of taking this step, the Government indulged in an irrational fragmentation of the fisheries set up, leaving various sectors of the industry to develop or stagnate without direction.

Fishing Chimes, Editorial 125: October(i) 1989: Vol. 9, No. 7

# Utilisation of Trained Indian Crew on Chartered Tuna Long Liners

There are quite a substantial number of chartered tuna long liners in operation in Indian waters. As per the regulations governing these chartered operations the charterers have to place on board of each of the vessels a' certain number of Indian crew for receiving training in tuna long lining from the foreign captains, master fishermen, engineers and others. Normally, one skipper/ fishing secondhand, one engineer / engine driver of fishing vessel, one radio telephone operator and around four deckhands are placed for training on each of the chartered vessels. The purpose of this provision is to enable Indian personnel to receive training in the operation of tuna long liners of Indian flag as and when introduced by the Indian charterers in terms of one of the conditions contained in charter permits.

India will have to develop its own tuna long lining fleet sooner or later. For manning this fleet the personnel now being trained on foreign chartered tuna long liners will form the nucleus. Close attention need to be paid at this stage itself as to how to ensure that the trained candidates will be of effective utility for the operation of Indian tuna long liners as and when introduced.

In this context, we may have to examine whether what is now being done is conducive to the realisation of the purpose for which Indian crew are placed on the chartered tuna long liners. Most of the charterers engage the counterpart Indian officers and crew on the chartered vessels on voyage to voyage basis. Those who may work for a voyage on a chartered Long Liner may later go back to work on a trawler. There is no system under which vital aspects such as continuity of training, number of voyages that are needed for the Indian crew to pick up the technology, the extent to which men placed on foreign tuna long liners are able to acquire the technology, and several other aspects of this kind would be taken care of all through, so as to avoid problems in securing suitable Indian personnel for the operation of Indian owned long liners which will come into operation sooner or later.

As matters stand, the entire training system is haphazard and to what result it would lead to in the coming years it is very difficult to comprehend.

The chartered tuna long lining operations offer an excellent opportunity to provide practical training on board to Indian personnel. This opportunity has to be afforded in a proper and systematic manner. There should be a central point such as the CIFNET where there should be effective coordination of all the aspects of this training programme on chartered long liners. A pool of trained candidates in long lining will have to be created under this training programme and only those awarded with a certificate after having completed such a training as per requirements (after a field test) should be allowed to join the future tuna long lining fleet.

The present system of asking the charterers to appoint captains/fishing secondhands, engineers/engine drivers of fishing vessels etc as counterparts to the foreign crew, to say the least, is defective and counter-productive. Preference for placement on chartered tuna long liners should be given to candidates who have completed institutional training at CIFNET. Undertakings should be taken from those candidates that they will work on Indian owned tuna long liners, subject to such terms and conditions as are approved by the Ministry of Food Processing Industries. CIFNET should determine the number of sea days the candidates who are placed on the tuna Long liners, should put in to get their respective tickets. Candidates emerging out of such training would then be specialists in tuna long lining. What is happening now is that those having experience in operating trawl nets are placed on chartered tuna long liners. Most of these do not have interest to work on long liners in the future. The training therefore goes waste atleast in the case of the bulk of the candidates. Therefore, we feel that the first reform to be introduced by the Ministry of Food Processing Industries is to stop this system of placing all those who are already possessing tickets issued by the Mercantile Marine Department for training on chartered vessels.

Such of the candidates coming out of CIFNET, who are willing to gain experience in tuna long lining should be given an orientation in regard to the manner in which they have to function on the chartered tuna long liners. The tendency on the part of the captains, master fishermen and also engineers of the foreign chartered vessels is to put up a show of providing training but not allowing the candidates to learn much. The absorption of technology is therefore largely dependent on the prior knowledge of the candidate on the subject and also on the knowledge as to how to acquire the technology without excessive dependence on the foreign crew. The candidates have to be acquinted about various aspects on which they have to concentrate and pick up knowledge. For example, the main aspect they have to learn is as to how the foreign master fishermen locate tuna and how they arrive at the depth of their occurrence. They should also understand as to how the foreign crew poise line at the correct depth. Most of the trainees who work on chartered vessels are not aware of critical aspects such as those mentioned above, concerning which they have to pick up knowledge. If there is an orientation course the candidates would be able to pick up the technology in a systematic manner.

Such of those candidates alone, to whom the orientation course is imparted should be allowed to be employed on the chartered tuna long liners. The number of days of training required on chartered vessels should be determined by CIFNET, A system under which they write down what all they learn daily in a diary should be introduced and the diary should be shown to the CIFNET authorities on conclusion of the training programme. After evaluation of their performance CIFNET should issue certificates to the candidates that they are fit for placement in Indian chartered tuna long liners in a specified capacity provided that they obtain the relevant ticket from the Mercantile Marine Department.

The Indian tuna long liner owners should follow the system of approaching the CIFNET for obtaining the services of qualified candidates for employment on their vessels. By adopting this system of centralisation, the available manpower resource can be properly regulated.

We suggest that the Ministry of Food Processing Industries should work out a scheme on the above lines in consultation with the Ministry of Agriculture and CIFNET and take the needed steps for implementing the same with immediate effect. The sooner the present haphazard and "free for all" system is given up and CIFNET transferred to the control of Ministry of Food Processing Industries, the better it will be for the emerging tuna long lining industry. In the meanwhile, until a system is evolved, we suggest that Government should allow as an interim measure only CIFNET trainees to be appointed by charterers as counterpart crew in the place of the present irrational and counterproductive system of appointment of ticket holders who have no long term interest, by and large, in making a career on long liners, as their expertise is wholly in respect of trawling only. As must have been noted by the Government, trawler men on Taiwanese vessels do not work as long lining men. Crew who work on long liners are always a separate lot. This kind of specialisation exists in all countries and India ought not to adopt an irrational and counter-productive approach of allowing placement of trawler men for training on long liners. What would happen to trawler operations, if trawler crew move over for the operation of long liners? To avoid likely repercussions, it will be a good and sound strategy to develop a cadre of operatives for long lining, instead of creating confusion and patting companies to a dilemma of knowing whether a particular person is a good trawlerman or a long liner man.

We do hope that the suggestions given above would receive the close attention of the Government.

Fishing Chimes, Editorial 126: October(ii) 1989: Vol. 9, No. 7

# On Chartered Fishing Vessels in Indian Waters

There was a time when almost every one in the fishing industry knew how many chartered vessels were exactly in operation Indian EEZ. Same is not the case now.

To the extent information could be gathered, at present, there are six pairs of bull trawlers in operation, two pairs by Ganga Kaveri, one pair by Young Fisheries, one by Coastal Trawlers and two numbers by Matsyika.

All these are operating based on an extension granted to permits issued under an earlier scheme. The permits of these companies contain a stipulation for granting extension for a further period of two years. It is learnt that M/s.B.R.S. Fisheries also has been given, an extension for operating two pairs for a duration of four months, probably as a prelude to further long term extension. A few other companies are also understood to have received such extensions (Leo Sea Foods, Blue Chrome, Golden Ahar etc.). Some of these companies fulfilled their obligations for the introduction of their own vessels in full and some are in the process or have taken steps to fulfil them.

In regard to long liners, under the earlier schemes, according to available information, two numbers each of long liners are being operated by Young Fisheries, Coastal Trawlers, Artina Fisheries, Cherita Fisheries, Triven Exports, Pisces Exports, Omini, Manpra Sea Foods, and N.G. Marines. There are thus 6 sets of pair trawlers and 18 long liners in operation under the old schemes.

Under the latest scheme, two numbers each of long liners have come or will soon be coming under operation by the following companies: Surya Seafoods, Kelamaval, Smart Marines, Pink City Fisheries, Khetan Fisheries, Sheetal Seafoods, Inter Trade Fisheries, Tropic Trading Company, Chandur Seafoods, A.V. Fisheries, Seasonics, Buoyancy Fisheries, Creative Export and Imports, Mavidi Marines, Ganga Fisheries, M/s. Matsygandh, Newage and Seagull. The A.P. Fisheries Corporation is reported to be operating three numbers of long liners with the prospect of two more being added. A.P. Fisheries Corporation seems to have secured a carrier vessel also on charter which is stated to be awaiting clearance. The All India Schedule Caste Co-operative Society, Karolbagh, Delhi, is reported to be operating five numbers. This Society is reported to have already been granted extension as a special case for the operation of the vessels for a further period of one year. Whether they have fulfilled the obligation under the scheme within such a short period is not known. The number of long liners introduced/shortly to be introduced comes to 44. Inclusive of 18 nos. under the old scheme, it can be taken that 62 nos. are in operation or will soon be in operation, as soon as some of them (about 18 nos.) complete all formalities.

M/s.Target Marines is the only company which ventured into the introduction of three stern trawlers on charter from Italy. Of these, one is stated to be already operational.

Readers are aware that towards the end of August, 1989 the Ministry of Food Processing Industries issued yet another notification inviting fresh applications for chartering of fishing vessels. The difference between this notification and the previous ones is that chartering of pair trawlers would also be permitted. The deposits to be made with the Government have also been raised, to Rs.1.5 lakhs per vessel, subject to a minimum of Rs.5 lakhs. In response to the latest notification it is learnt that 46 companies applied for permission to charter, mostly pair trawlers, by the end of September, 1989. The screening committee was expected to meet on 20th and 25th October for considering the applications.

Several of the remaining companies who were given permits for operation of pair trawlers under the 1983 scheme and who had completed three years of operation are understood to have applied for extension of validity of permits for a further period of two years. In the permits issued to these companies, as in the case of the four companies referred to above and who are operating six pairs of bull trawlers and a few others, there is a provision that these are extendable for a further period of two years. However, although the requests of these companies are similar, Government could not make up their mind concerning the extension so far. Several of these companies are reported to have again applied to Government for consideration of their requests favourably, on the ground that the counterpart crew employed by them on chartered bull trawlers could not absorb the technology and also because the companies could not generate adequate funds to fulfill the obligation to introduce their own vessels. There are 11 such companies who are awaiting extension of permits from the Government for a further period of two years for operation of two numbers of pair trawlers each. These are a) Varuna Marine Products, b)Tropical Shipping, c)Golden Protein, d)Four Season Fisheries, e) Shrimp India, f) Akama Marines, g) Srinivasa Enterprises, h) 0ceanic Shipping, i) G.P. Marines. The total number of pairs involved under these proposals come to 22 numbers.

Fishing Chimes, Editorial 127: November 1989: Vol. 9, No. 8

# Emerging Categories of Entrepreneurship in Indian Deep Sea Fishing Industry

The introduction of shrimp trawlers in early 1970s by Union Carbide and later by several other companies was the starting point for the generation of interest among Indian entrepreneurship to enter the line of 'deep sea fishing'. The declaration of the Exclusive Economic Zone by the Government in 1981 has led to diversification of this interest for operating foreign fishing vessels on charter, of schemes for the chartering of deep sea fishing vessels of various kinds by the Government, based on powers derived from the provision in the Maritime Zone of India Act, 1981.

Broadly stated, the present position is that three categories of entrepreneurship in the field are discernible. One category consists of those who have or want to have vessels of their own. They have no belief or inclination towards the operation of chartered foreign fishing vessels, for various reasons. The other category consists of those who are devoted to the operation of chartered fishing vessels only, with problems in getting into the industry by acquiring their own vessels. These relate to getting clearances for operation of such vessels.

The third category consists of enterprises who want to operate vessels on charter and at the same time acquire vessels of their own in accordance with the terms and conditions of the charter schemes.

All the three categories have their own sufferings. In regard to the first category, the entrepreneurs face problems in getting into the industry by acquiring their own vessels. These relate to getting clearances from the Government, and securing loans from financing institutions. After passing through this phase, they have to contend with tough situations such as delayed release of loans and also cope up with the delays caused by the builders in delivering the vessels. Because of these delays, the interest over the loans accumulates but without any returns, and this forms a backlog by the time the operations commence. After this, they realise, somewhat late, that the returns are not as anticipated at the time of project formulation. They steadily notice that they are in the total grip of the crew and their unions. While they own the vessels (mortgage to financing institutions not being taken into account) as per the documents, they realise that the ownership is virtually in the hands of the crew and that the vessels conduct fishing at the will of the crew and not based on the directions of the owners. Apart from increase in operational expenses, and demand for higher wages etc by the crew members, they find that they have to reconcile themselves with fluctuations in prices which show mostly a downward trend. They are under perpetual dread of what the officers and crew of the vessels will do next that will adversely affect the operations. They notice that all kind of unsuspected requirements relating to maintenance and running of machinaries come up, even before the completion of one year of their operations. They find that the various machineries soon become due for overhauling, or for some other repairs. They remain in the fear of receiving some communication from the captain or engineer of their vessel while out at sea conducting fishing that something disastrous happened, and something should be done to get over the same urgently. With no money in hand, they moon around their bankers who keep on reminding them about the heavy overdraft. They also notice that there are very few lucky operators who run their vessels smoothly without any operational and financial constraints, and not receiving some kind of communication or the other from the financing institutions inviting attention towards the payments due. They get the deceptive impression that they are working for the benefit of the financiers and not for themselves.

They envy their brothers who are in the business of chartered fishing operations, who do not face the kind of operational problems and financial difficulties that they face. At the same time they have the fear that they may run into difficulties with the foreign owners and with the coast guard and the Government in regard to the observence of various regulations relating to the operation of chartered fishing vessels, if they enter that sector.

The second category has to utilise the opportunity given to them for operating foreign fishing vessels on charter as a means of training their own men for operating similar vessels on conclusion of the duration of charter, by introducing similar vessels of their own. The Indian counterpart crew on the chartered vessels mostly keep on changing from voyage to voyage and this does not enable them to develop a cadre for utilisation on their own vessels, if and when introduced. Those in the non-chartering sector think that the charterers are comfortable financially because of the provision made by the Government that 20 per cent of gross returns would accrue to the charterers. The charterers feel that the reverse is the situation. With mounting financial commitments such as payments of salaries to the Indian counterpart crew, customs duty, port charges etc, the likely savings will not be sufficient to meet the margin money requirements to acquire their own vessels. This leads them to forego the various deposits kept by them with the Government. Some enterprises, however, do their best to acquire their own vessels. While some succeed and get back their deposits, others tend to work towards getting back their deposits through some device or the other and often prevail upon the Government to allow them to charter some more vessels or secure extensions for the operations of the vessels already permitted for operation on charter. Some feel that they will avoid the exercise of having their own vessels, preferring to lose the deposits. For them this is a far better solution than acquiring their own vessels. The acquisition of technology in respect of securing charter permissions and operating chartered vessels over a period of time enables most of the ex-charterers to encourage their friends to set up new companies in which they may or may not have substantial interest and advise them as to how to secure permissions for chartered vessels. In fact, it should be possible for those who have specialised in the operation of chartered fishing vessels to be in the field on a continuing basis without giving any chance whatsoever to the Government to know that it is the same charterer who continues to get vessels on charter under the banner of different companies.

The third category are those who strive to fulfil the obligations imposed under the charter schemes in respect of acquisition of their own vessels. They try to make up what they have lose in the operations of their own vessels by whatever they gain they get in the operation of chartered vessels. Operators of this kind, having gained experience in both the sectors are often in a position to enter into joint ventures with foreign fishing companies and increase their own fishing fleet. They are in a position to do this by virtue of their foreign contacts. Some of the entrepreneurs falling under the second category will also be in a similar position to develop joint ventures but they do not make efforts in this direction mainly for the reason that they have no commitment towards the fishing industry. Most of those coming under this category do not have basic interest in the field of deep sea fishing and their main objective is to treat the activity exclusively as a profit earning venture, with no commitment towards the profession.

Apart from the three categories of entrepreneurs referred to in the above paragraphs, there is perhaps another category as well. This particular category consists of those who have gained some experience in the operation of chartered fishing vessels earlier or those who have picked up contacts with the owners of foreign fishing vessels. The role played by this category of entrepreneurship is to develop a consultancy system under which they provide a package of services to those interested in chartering of fishing vessels on payment of a certain fees. This category may also have the luxury of operating chartered vessels of their own, besides the consultancy work. They lend considerable support to the foreign vessel owners and they are often in a position to tell them about the number of vessels each category that would be required for operation by their own clients and to what extent the foreign owners have to block vessels either of their own or of other sister companies in their own countries or in any other country for supply to prospective charterers.

Although over the past 10 years or more, several vessels were permitted by the Government on charter so far, not a single vessel of the same type as the one operated on charter could be introduced by any of the charterers. It is not that all Indian charterers are against introducing vessels of the same type as chartered by them. In fact, many have interest in doing so. The reason for this appears to be that almost all the chartered vessels are of Taiwanese construction and all documents, with the exception of the one relating to registration, are of Taiwanese origin. For the reason that India has no diplomatic relations with Taiwan, a subterfuge or deceptive approach of registering these vessels under Panama or other flag recognised by India is adopted. The acquisition of these vessels is however not permitted by the Government. The Indian entrepreneurs are also unable to acquire new vessels of similar type constructed in Taiwan as this will not also be acceptable to the Government for various reasons. Countries such as Australia and New Zealand who also do not recognise Taiwan allow the operations of Taiwanese vessels on charter flying Taiwanese flag. The reasons why our Government cannot also allow Taiwanese vessels to operate flying their own flag, discarding the devious and transparent approach of accepting the same vessels with Panama registry are not known.

Used trawlers/long liners of Taiwanese construction should be allowed to be acquired either on outright payment basis or on the basis of payment of the cost in instalments. Acquisition of new trawlers of Taiwanese construction whose cost is often far lower than those built in other countries is also possible. If the Indian charterers are allowed to acquire secondhand Taiwanese vessels or newly constructed Taiwanese vessels, be they long liners or other types of vessels, it appears the development of Indian fishing fleet would be faster. In addition to allowing acquisitions as indicated above, Government should also permit the employment of Taiwanese and Philippine crew on vessels so acquired for sometime with a certain component of the Indian crew. This will enable the Indian crew to take over the operations completely over a period of time.

It is heartening that all the categories mentioned above do co-exist to a large extent. However, it is often seen that those who are unable to secure charter permits or those who did not apply for charter permits within the time stipulated, make representations to the Government from time to time seeking the scrapping of the charter scheme, putting forth several ostensible reasons. However, the same companies, sooner or later veer round to the point that they were wrong in not taking up chartered vessels operations and at the earliest opportunity join the relevent category, sacrificing their earlier objectives.

Fishing Chimes, Editorial 128: December 1989: Vol. 9, No. 9

# How to Earn More From Trawlers and Start Paying to Financiers

Majority of skippers take their trawlers out with the fond hope and resolve to bring the highest possible commercially viable level of catches, be they shrimps, lobsters, cuttlefishes, squids or fin fishes. While some come back with proud faces signifying good catches, others wear countenances that depict sadness and philosophy, which at once tells that their catches are poor. The later group have several reasons in rationalising their bad luck, notwithstanding the fact that they may have also fished in the same grounds as the successful set of skippers. While some trawlers merrily steam into the harbour with, say, eight tonnes of shrimps, some others which probably fished in the same area amble into the harbour with, say, 2.0 tonnes of shrimp. What could be the reason for such major variations?' The known ones are, old machinaries, old nets, inefficient winches, non-cooperative crew etc. There is however a little known but a major reason that is responsible for the variations.

The trawl net is extremely sensitive to speed, rigging, (mouth opening, and doorsetting on which the angle of attack depends etc.), floating, bottom conditions, and so on. The Indian skippers have no way at present to know whether their trawl is moving as they want. If they have a way to know the headline height, any bad aspects in rigging, foul shooting, deficiencies in door sizes and weights, wrong dragging speed etc, and also the means to set them right, the fishing results would be dramatically different. Irrespective of the dragging power of their vessels, several captains just copy designs adopted on other vessels. This hit and miss method appears to be responsible for the poor landings of some of the trawlers.

The problems referred to above have been resolved by trawler owners in several countries by installing net monitoring equipments which are not really expensive. It is a matter of real surprise that neither the Indian Government, the financing institutions nor the owners have ever thought of this aspect. For doublerig shrimping, for example, what all are additionally required in the wheel house are a cabinet with a display screen and a hydrophone, with a distance sensor, height sensor, catch sensor and temperature sensor attached to the nets. The distance sensor is the primary unit for double rigged operations. On one of the wings of such of the nets one sensor can be attached. Alternatively sensors can be fitted on the doors. These sensors will enable the captain to rapidly find out correctness of the size of doors and angle of attack and whether size can be reduced to save fuel.

A height sensor will enable the captain to maintain the highest possible headline height, if he is pursuing higher schooling shrimps.

Catch sensors will help in knowing the state of codend, whether it is full up with shrimp or fish. This knowledge saves many wasteful tows. Shrimp migrates extensively and lies in narrow strips. This behaviour is related to temperature and other factors. By using appropriate sensors, a captain can match information obtained from catch sensor placed on try net and actual catch data and draw valuable conclusions. The temperature sensor can be fitted easily on try net doors that can support it. Alternatively, the sensors can be attached to main net doors.

Sensors are a great means to adjust designs of trawl nets, after watching deficiencies on the screen in the wheelhouse, for optimising profits, from single boat bottom trawling, pair bottom trawling, single boat pelagic trawling or doublerig shrimp trawling.

At this critical stage in the development of the Indian 'deep sea fishing industry', with practically no Authority to care for its needs, it is high time the owners associations or the Governmental agency or financing bodies concerned apply their mind to this crucial aspect of improving operational economics of presently operated trawlers, characterised by escalating costs, plummeting incomes, and expanding idle periods at port with the number of fishing days rarely crossing 150 mark. Several makes of net monitoring units that suit specific purposes, such as Scanmar, Netimer, Netset, Net mar etc are available in the world market. Being a developmental measure, Government should share the investmental risks involved and help the industry in the selection and installation of the most suited net monitoring units, in the presently operated trawlers.

We have been advising the Government and fishing companies with great perseverance often and on to take steps to make Indian trawler fleet operational throughout the year so as to augment incomes. How can the trawler fleet, mostly dependent on shrimp fishing for six months and surviving on the income realised in this short period for the whole year, (paying salaries and other allowance to the crew who stay idle at port for six months, meeting maintenance expenses etc without incomes), endure for long and enable financing institutions get back their money, unless the vessels are equipped to operate the year round? All concerned are aware that all major Indian fisheries are seasonal and without every vessel being equipped for undertaking atleast two types of fishing, a major part of the industry will collapse. What is urgently required now is, therefore, to say again, a developmental strategy with Government's support to equip the existing trawlers for tuna long lining or another type of viable fishing. The strategy should involve a major element of subsidy and a granting of loans for the balance required on such terms and conditions that they do not lead to 'tears' in the eyes of the owners.

From what we hear two companies *i.e.* M/s. Nav Bharat Ferro Alloys Ltd., Visakhapatnam and M/s. Pron Magnate (Pvt) Ltd, Visakhapatnam are the only firms making repayments uptodate relating to loan and interest to SCICI. There may however be a few others too in that category. One hears of Sapphire Fisheries, Visakhapatnam as one of those paying its dues on time but the fact is the majority of the fishing companies are defaulters. All companies cannot have the same style of management as the few successful ones, because all managers are not of comparable capabilities. For the Government or the financing bodies to help the industry to earn more and pay back dues on time, it is much more important for them to change their style of functioning and monitoring, much more than the fishing companies. SCICI appointing Chartered Accountants to check accounts or to supervise the working of the companies may not improve the position. The situation should not be brought to a stage when the few successful companies would have reason to think, "why should we work hard and pay to financing bodies which are functioning on a different footing?". To keep the successful ones to continue to be successful and to motivate other relevant organisations to achieve better results, Government or any other relevant organisation can introduce a system of recognising merit, through awards for those who clear dues to financing institutions in time, once a year. Presentation of a shield or a memento will be grossly inadequate to accord adequate recognition to such of the companies that achieve distinction. The awards will have to be substantial, and should be presented by the Chairman of SCICI or the Minister concerned.

SCICI must be wondering as to why a couple of companies pay their dues on time and others not. It is possible that they have already initiated a study in this respect, to identify differences in management policies of the successful ones and the unsucessful ones. The publication of the results of such a study would be of great use to the industry.

Fishing Chimes, Editorial 129: January 1990: Vol. 9, No. 10

# 1989: Fish-O-Drome

Fisheries development during the year 1989 passed through situations of despair as well as moments of achievement. The situations of despair stemmed from the fact that the administrative set up at the national level became fragmented, leaving in its wake certain inaction and retardation of progress in several respects.

Looked at from a positive angle, promises for the development of the sector could be discerned from the gravid events that developed during the year, notwithstanding the situations of despair, which may well turn out to be the harbingers of major developments both in the marine as well as the inland fisheries sectors. This is our crystal ball prediction for the decade into which we have now entered and which would eventually launch us into the 21st century. The pointer to the prediction is the building up of developmental activity in the private sector, notwithstanding the organisational setbacks created by the Government, and in spite of Governmental apathy.

## National Fishery Development Board

The National Fishery Development Board has been set up by the Government of India during the year. It was expected that the Board would be tendering advice from time to time to the Government in regard to the lines on which the fisheries of the country should be developed.

## Eighth Indian Sea Food Fair

Another event that took place was the eighth Indian seafood fair at Madras from 10-12 February, 1989. The fair was a resounding success.

## Veterans Honoured

The Indian Fisheries Association had established for the, first time a tradition of felicitating outstanding fisheries scientists of the country, Dr. D.Bal, Dr. C.V. Kulkarni, Dr. V.G. Jhingran, Dr. T.V.R. Pillay, Dr. S.Z. Qasim and Mr. K.H. Alikunhi, all renowned Indian Scientists, were honoured by the Association.

## Fish Production

The world fish catch reached a level of 97 million tonnes in 1987-88, registering a growth of 2 million tonnes over the 1986-87 level. So far as Indian fish production is concerned, the total annual production at the end of 1988-89 is stated to be 3.15 million tonnes of which 1.35 million tonnes were inland produce. So far as exports are concerned, a major breakthrough had been achieved in that the value of exports crossed Rs.600 crore mark.

## Massive increase in FFDAs

One notable factor that may herald a major development process in the inland fisheries sector is the courageous step taken by the Ministry of Agriculture for the development of freshwater fish culture in an intensive manner through a massive increase in the number of Fish Farmers Development Agencies in the country, covering almost all the districts in the various States. From 5 numbers at the end of the 5th Five Year Plan, 313 agencies had been set up covering 20 States and one Union Territory, by the end of 1989. The largest number of these agencies had been set up in Uttar Pradesh (48 nos.) followed by Bihar (36 nos.), Madhya Pradesh (24 nos.), Andhra Pradesh (22 nos), Maharashtra (18 nos.), Karnataka (17 nos.), West Bengal (16 nos.), Assam (15 nos.), Orissa, Rajasthan and Tamil Nadu (13 nos. each), Punjab, Gujarat and Haryana (12 nos. each), Kerala (10 nos.), Manipur (7 nos.), Tripura (3 nos.), Himachal Pradesh, Jammu and Kashmir and Nagaland (2 nos. each), and Mizoram,. Pondicherry and Arunachal Pradesh (1 nos. each).

## Fish Farming

The year had seen an unprecedented growth in fish seed production in various States (West Bengal being the foremost) and also in the setting up of a large number of new freshwater fish farms/ponds. The activities had come up significantly in Andhra Pradesh, mostly owing to the efforts of the farmers themselves but with help received from the Andhra Pradesh Agricultural University, College of Fisheries, Mangalore, Central Institute of Fisheries Education, State Fisheries Departments, and CIFA. Progress in this sector had also received a major fillip in the various other States such as Tamil Nadu, mostly because of the guidance from the State Fisheries Departments and the initiative exercised by the farmers themselves.

In regard to freshwater fish culture, while there are no clear cut statistics available about the area covered by this activity, it is estimated that around 14 lakh hectares in all have been brought under fish culture of which 2 lakh hectares are stated to have been brought under culture by 2.5 lakh numbers of trained fish farmers, coming under the fold of fish farmers development, agencies. Average level of annual fish production achieved by ponds covered by FFDAs is estimated to be 1560 kq/ha/yr, benefiting 2.5 lakhs numbers of trained fish farmers.

## OVAPRIM

The miracle drug "OVAPRIM" of Canadian

manufacture had been successfully tested and introduced for the induced breeding of major carps and exotic carps by the College of Fisheries, Mangalore.

The Government of M.P. had banned the import of fish seed from West Bengal on the ground that the seed may bring in Ulcerative Disease Syndrome among cultured fishes in the State. A depressing feature of the year was the attack of Ulcerative Disease Syndrome in various eastern States.

## Brackishwater Farming

The Marine Products Export Development Authority had transformed itself into a dynamic "AVATAAR", performing miracles in the brackishwater farming sector, The relentless crusade launched by it under the leadership of Mr.T.K.A.Nair, the then Chairman, MPEDA and spear-headed ' by Dr.M.Sakthivel, Director, MPEDA, for the development of brackishwater farming is unparalleled in the recent history of fisheries development in India. Hundreds of farmers had been motivated all along the coastline of the country, to take up brackishwater shrimp culture. A major infrastructural facility had been provided by the Authority by setting up two modern prawn hatcheries on the East Coast for the production of tiger prawn seed in millions under the leadership of Dr. M.P.Haran, by MPEDA. The success of these hatcheries, one located near Visakhapatnam in A.P, and another at Gopalpur in Orissa, set up with foreign expertise, had surprised skeptics both among the farmers as well as scientists and technocrats. The seed produced at these hatcheries is avidly procured by the farmers. Prawn seed production centre at Mangamaripeta near Visakhapatnam was declared open the then Union Minister of State for Commerce Mr. Priya Ranjan Das Munsi on 31st September, 1989. A prawn culture project for implementation in Orissa in collaboration with Hindustan Lever Ltd., had been formulated.

## Land Lease Policy

The development of brackishwater farming is largely dependent on the policies of coastal State Governments to provide brackishwater lands, mostly owned by them, to interested entrepreneurs, on long term lease. There has been practically little progress in this respect during the year.

## Deep Sea Fishing

An additional effort initiated by the MPEDA, again under the leadership of Mr. T.K.A. Nair, the then Chairman, MPEDA and innovatively directed and followed by Mr. Ousep Attokaren, Director, MPEDA, related to development of deep sea fishing. The unprecedented work done had led to the stabilisation of deep sea lobster fishing activity on the South West Coast and paved the way for the setting up of important joint ventures and introduction of new methods of fishing such as tuna long lining and purse-seining. Having realised the unfavourable economics of operation of the presently operated shrimp trawlers, the MPEDA has taken up the challenging task of finding out ways and means of equipping these vessels with an additional type of fishing, and equipping these vessels for operational round the year, so as to upgrade the operational economics. MPEDA had also introduced a number of schemes to assist fishing companies in the acquisition of deep sea fishing vessels and processing companies in the diversification of marine fishery products for export through provision of several incentives.

## Joint Ventures

MPEDA had been making strenuous efforts to bring about joint ventures in long lining/purse-seining for tuna, facing severe odds.

While there had been an improvement in sea shrimp landings during the shrimping season along the upper east coast, the fall in export prices had led to declining trends in economics of operation of these vessels.

## Irrational Fisheries Administrative Setup at the Centre

The most depressing aspect that took place during the year was the irrational fragmentation of the fisheries administrative setup at the Centre. If such a fragmentation had not taken place, the progress in the fisheries sector would have reached new heights. The subject of introduction of deep sea fishing vessels, new or used, or chartered, had been entrusted to the newly created Ministry of Food Processing industries, along with control over the Fishery Survey of India. For some inexplicable reason, control over the Central Institute of Fisheries Nautical Engineering Training, Central Institute of Coastal Engineering for Fisheries, Bangalore, and the Integrated Fisheries Project, Cochin, whose activities are of vital importance for the development of deep sea fishing along had been left with the Ministry of Agriculture, there had been no evidence of co-ordination between the steps taken for the introduction of deep sea fishing vessels and the training programmes etc which formed an integrated part of the activity. The work concerning joint ventures in fisheries was entrusted to the MPEDA, while clearances for the vessels to be introduced had to be given by the Ministry of Food Processing.

## Deep Sea Fishing Vessels

The Ministry of Food Processing Industries had no doubt succeeded in introducing about 26 new fishing vessels (although cleared much earlier to the formation of the ministry) and around 70 numbers of chartered vessels most of which are long liners. The number of commercial fishing vessels of over 23 m OAL at the end of the year is stated to be around 165 nos. During the year, in the indigenous sector, Bharati Shipyard reached a level of delivery of ten vessels, with eight more in the pipeline. The yard had also acquired technology for equipping existing trawlers with Florida type tuna long lining system. N.N. Shipbuilders had delivered one vessel of Bender design, and Hooghly Dock and Goa Shipyard two vessels each. Chougule Shipyard is reported to have delivered one vessel during the year. Sesa Goa delivered one mini-trawler. In the import sector, two nos. of 27 m long vessels were imported from Korea by Surya Sea Foods. In addition to this, a substantial number of medium class vessels in 14-18m range had been added to the national fishing fleet.

## Crew Training for Longlining

The Ministry of Food Processing Industries had introduced a system of training of Indian crew on chartered tuna long liners and on trawlers on an enduring basis so as to make them available for the operation of any such new vessels of Indian flag to be introduced by Indian fishing companies in course of time, which however seems to be a remote possibility. The unrealistic nature of the charter scheme had, by and large, prevented the introduction of vessels of Indian flag as per the terms and conditions of the scheme. Several inevitable irregularities are believed to have crept in because of the several conditions that are unenforceable.

## No Reliefs to DSF Industry

The Centre could not help the deep sea fishing industry by way of reliefs from the mounting operational expenses, the major component of which is the excessive cost of diesel oil on which subsidy equivalent to excise duty and sales tax levied was requested by the owners. The owners also asked for help by way of subsidy to the extent of 50 per cent of the cost for the installation of an additional system for diversified fishing besides trawling, to make the vessels operational around this year. This had not however evoked any reaction from the Government.

## Crew Problems: Danger Ahead

It was expected that during 1989 there would be a break-through in the evolution of a system to regulate the service conditions of deep sea fishing vessel crew.

This did not however take place. On the other hand major setbacks to the functioning of the industry took place, on account of crew problems. In June-July 1989 the owners had to lose over 30 fishing days owing to demands of higher wages by deckhands, in the face of mounting costs. The owners had to concede several of the demands. Towards the close of the year, the trawler officers came up with demands that are considered realistic by them but are felt as totally bereft of reason by the owners. These demands are bound to create a major havoc for the reason that, even with utmost economy, a minimum loss of Rs.6 lakhs per trawler is bound to emerge, if the demands are conceded by the owners. This situation may force the owners to close down deep sea fishing operations towards the middle of 1990, notwithstanding their commitments to financiers etc.

## SCICI's Help

1989 had seen a major reorientation in the understanding of the characteristics and the present state of deep sea fishing industry by SCICI, the only financing institution in the country to finance deep sea fishing projects.

Having noticed heavy accumulation of dues pertaining to loans granted by it and the erstwhile SDFC, the institution had been making valient efforts to develop a rapport with the various companies constituting the fishing industry with a view to helping them to come out of the jam into which they had forced themselves in. It held several meetings with the representatives of fishing companies and concerned ministries, and organisations. There is no doubt that in 1990, SCICI will introduce some bold measures to bring a further measure of discipline in its financial dealings with the deep sea fishing industry, notwithstanding the likely stoppage of fishing operations by a good number of companies, taking the view that the demands of trawler officers as unwarranted and 'irresponsible'. A section of the deep sea fishing industry Is likely to think that the officers will have the distinction of emerging as the 'killers' of the industry. Others may feel that the present earnings by a skipper annually would hardly be sufficient even for his living expenses and that it is a poor excuse to say that the owners save pretty little and even this is not sufficient to clear the dues to the SCICI.

SCICI is now engaged more in stepping up measures to recover pending dues from trawler owners rather than providing fresh loans for which it is stated that there are very few sound applicants, SCICI's hopes that larger houses will take up deep sea fishing projects with its help appear to have been shattered. SCICI, however, did extend financial help for the acquisition of a limited number of vessels, of which the notable one was the help extended to M/s.Sumura Maritime for the acquisition of two tuna long liners from Japan. The help given to M/s. Indus Fisheries Ltd., for the acquisition of a purse seiner was of no avail as the company sold the vessel soon after the acquisition and paid back the loan with interest, after conducting a couple of voyages. Some entrepreneurs seem to have obtained letters of sanction of loans from SCICI but are not availing of the same being content with payment of commitment charges to SCICI. Of what help such a line of action is to the entrepreneurs is not clear.

The SCICI had also announced its plans to set up its own deep sea fishing company involving the operation of a mother ship and several satellite vessels. It appears to be engaged in working out various details in this regard,

## Resources Survey

The Fishery Survey of India, CMFRI (through operations of Sagar Sampada) and CIFNET located several new marine fishing grounds in the deep sea zone of our EEZ, consisting of threadfin breams, bulls eyes, Indian drift fish, scad etc. An Atlas of tunas, bill fishes and sharks in Indian EEZ was published by the Fishery Survey of India. Similar publications on tunas and a chart depicting the various commercially important fishes had also been published by the MPEDA.

## Pearl Production

The CMFRI conducted outstanding work on the culture of holothurians and standardised technology in this regard. Based on the totally indigenous technology developed by CMFRI at Tuticorin, artificial breeding of pearl oysters with a view to producing oysters for the production of pearls had been taken up by the Institute at

Mandapam, near Rameswaram. It was expected that, once the work at this unit moves into large scale production of oysters, India would soon become self sufficient in the production of pearls, apart from reaching eventually a stage of exporting pearls. It may be mentioned here that CMFRI was the first to locate blacklip pearl oysters in the seas of Andaman and Nicobar Islands, According to Dr. James, then Director of CMFRI, India has a big market for pearls and CMFRI had the technology to produce the same, Two Japanese built long liners had been added to the fleet of the Fishery Survey of India. While one was conducting exploratory work on the East Coast with Madras as the base, the other was operating with Goa as the base. It was for the first time a systematic work to locate tuna fishing grounds in the Indian EEZ had been initiated by FSI, although some knowledge on the subject was already available.

## Fishing Craft Development in Kerala

An agreement was signed by the Government of India with FAO for taking up a project on fishing craft development in Kerala. A cold storage plant at Cochin owned by MPEDA had been entrusted for management to the sea food exporters association. The construction of a dry dock for deep sea fishing vessels within the premises of Visakhapatnam fishing harbour was in progress. The Calcutta high court lifted the ban imposed by the Andamans administration in respect of fishing for sea cucumbers in Andaman waters.

An achievement during the year was the establishment of an export market for deep sea lobsters. Owing to the pioneering efforts of private sector companies such as Seamen Fisheries, and Srinivasa Enterprises, market for lobster tails in countries such as USA had been established. M/s. Shrimp India and Yeduguri Seafoods emerged as the pioneers in respect of export of frozen whole deep sea lobsters and shrimp to Italy and Spain.

## Freshwater Prawn Culture

A prawn seed bank had been opened at Kolisalem, Balasore district in Orissa, a major addition to the existing ones in the country. Considerable progress had also been achieved in this State in the culture of fresh water shrimp *Macrobrachium malcolmsonii*. The State Bank of India, Sambalpur is reported to have extended considerable financial assistance for the development of this activity. Similar progress in the culture of fresh water prawn had been achieved in Tamil Nadu and a few other States.

## Reservoir Fisheries

A significant beginning had been made in the introduction of cage culture for the production of major carp fry in Tawa reservoir in Madhya Pradesh. Experiments on cage culture had also been taken up in U.P., Tamil Nadu etc as a sequel to similar successful work done earlier in Tungabhadra reservoir.

Owing to the management measures introduced by the State Fisheries Department, a remarkable recovery in the fisheries of the reservoirs of Himachal Pradesh had taken place during the year. Another major event in this State that deserves mention is the substantial progress achieved in the setting up of a Trout Fishery Project at Katrain in this State which is expected to become operational by March, 1990. The Central Inland Capture Fishery Research Institute had conducted outstanding work in respect of fisheries of two major reservoirs in Rajasthan. The findings are expected to bring about revolutionary increase in fish production from these reservoirs. The institute had also completed an outstanding work in respect of Ganga river project. The project spelled out the steps needed to preserve the fishery heritage of the river. The Institute had also standardised the technology for the culture of giant African snails which have an export market.

CIFA made major contributions to the monoculture and multi-species culture of freshwater fish and in the evolution of new fish strains through hybridisation. Likewise, significant advances had been registered by CIBA in respect of shrimp seed production and brackishwater farming. The National Institute of Fish Genetics at Allahabad acted as a watch dog to prevent introduction of exotic species that would harm the indigenous fishery wealth.

## Fishermen's March

A cross section of fishermen of the country staged a marathon march under the auspices of the National Fisheries Forum in April, 1989. This march emanated from two points, one from Maharashtra and another from West Bengal. The marchers converged at Kanyakumari and highlighted at a meeting the threat to environmental and ecological balance because of indiscriminate fishing by larger fishing vessels.

## State Bank's Overseas Branch

A matter of relief for the trawler operators, marine products processors and exporters in Visakhapatnam had been the setting up of an Oversea Branch by the State Bank of India in Visakhapatnam. It is stated that this was the 7th branch of the kind set up by the Bank. Another reassuring feature for the exporters of marine products from Visakhapatnam was the decision of APL (American President Lines) vessels to call at Visakhapatnam port once a month, to lift marine products cargo to various destinations, mostly to Japan, in containers.

The Governments of Karnataka and Andhra Pradesh had granted exemption from payment of purchase tax on shrimp meant for export. This had proved to be a great relief to the exporters.

## Lalbahadur Shastri and Udyog Patra Awards

Three prestigious awards were presented to three outstanding personalities in the fisheries sector. One of this, Lal Bahadhur Shastri Award was given at a well attended function in Delhi to Dr. V.R.P. Sinha, then Director, Central Institute of Fisheries Education (Deemed University), Bombay by the former Prime Minister, Mr.Rajiv Gandhi. In the Private Sector, Udyog Patra Award was presented by the Vice President of India to

Mr.N.S.H.Prasad, Managing Director of M/s. Srinivas Enterprises, Visakhapatnam. Lal Bahadur Shastri Award for outstanding performance in fish farming was bagged by Mr.Diwakar Reddy, a fish farmer hailing from A.P. M/s. Seamen Fisheries, Madras received an award from the hands of Mr.Priya Ranjan Dasmunsi at a function held at Visakhapatnam on 31st May, 1989. The award was given for the pioneering work done in the development of commercial deep sea lobster fishing and export.

## Conflicts

There were serious conflicts between traditional fishermen and mechanised fishing boat operators in Kerala in respect of fishing areas during the shrimp season. The Tamil Nadu fishermen who traditionally fish along the southern coast of A.P. had to face hostilities in the hands of Andhra fishermen of Nellore Coast. Likewise, minor conflicts between Andhra and Orissa fishermen took place in Orissa waters. Further, fishermen of Tamil Nadu had to face harassments in the hands of Sri Lankan navy in Indo-Sri Lankan waters. On the inland side agitation in respect of leasing rights took place in U.P., M.P., and A.P., between fishermen and State Fisheries Departments concerned.

## Coast Guard

The coast guard of India apprehended nearly 40 foreign fishing vessels in various locations in our EEZ. These vessels consisted mostly of Taiwanese and Burmese origin. Several Indian fishermen who allegedly encroached into the waters of Pakistan were reported to be languishing in the jails of that country. Captains of three Taiwanese trawlers were convicted by the Bombay High Court by levying of fines and also ordering imprisonment for one month.

## Trawler Officers in CITU

A significant development during the year had been that the All India Deep Sea Fishing Technocrats Association, consisting of officers who man fishing vessels joined CITU, a trade union which is related to workers' interests. With this, both the trawler workers union as well as the officers union had come under the purview of CITU.

## Conclusion

In conclusion, we have to say the following: The Government of India may consider formulating a National Fishery Development Policy, which may cover among others, i)The setting up a separate Ministry/Department of Fisheries at Central level, and the setting up of a field organisation for the promotion of deep sea fishing that would take care of introduction of vessels for diversified fishing and providing capabilities to presently operated deep sea fishing vessels for conducting round-the-year fishing so as to make the operations economical. Considering the crisis that now grips the deep sea fishing industry, a policy decision on the future course of development of this sector either in public sector or private sector, or both have to be spelt out with the needed details. The policy in regard to the following has to be formed: i) To lease Governmental lands for the setting up of fish farms in the private sector/or through public sector agencies. ii) To promote polyculture involving *Gracilaria* in brackishwater ponds so as to make the operations economically viable. iii) To identify major centres for setting up of fishing harbours and provide infrastructural facilities at fishing harbours. iv) To introduce a viable fish preservation, storage and marketing system and promote diversification of processed fish products and their export as well as internal marketing. v) To integrate training programmes with introduction of deep sea fishing vessels. vi) To replace the system of chartering of vessels in stages so as to divert the value of fish which is being taken away by foreign owners, for the development of fleet of Indian flag with substantial subsidies and loans on easy terms; Similar assistance should be extended for incorporating additional systems of fishing in existing trawlers to make their operations viable. vii) To provide subsidies on diesel oil consumed by fishing vessels and on salaries and on wages now being demanded by the trawler crew which ranges from Rs.8,500 per month for captain and at slightly lower level to others, over and about various allowances, so as to avoid additional losses estimated to be at least Rs.6 lakhs per year per vessel and, viii)To evolve and introduce service conditions to govern fishing vessel crew.

Fishing Chimes, Editorial 130: February 1990: Vol. 9, No. 11

# Integrated Approach Under Unitary Control: Only way to Develop Deep Sea Fishing

Fishes are the oldest among vertebrates. The first incarnation of the Lord in Hindu Pantheon is the fish. In this form, with facundity that no other vertebrate has, God seems to have transformed himself into a highly protienacious digestible food for several other aquatic animals and certain terrestrial beings, particularly humans. He must have wanted his divinity to enter into the bodies of other animals including *Homo sapiens,* while Himself remaining in tact, 'As I multiply and offer myself and you take me unto yourself without disturbing the core of my existence, so shall you be blessed with health and wealth with no curse coming out of me. Be warned, you humans, fail to catch my replicas, you stand cursed'. This is not a biblical or a koranic or a vedic edict, but an inspired 'divine chime' of Fishing Chimes. Let us listen to the lines of the Lord that have come out of trancy peals of sounds of '*Fishing Chimes*'.

The foremost job is therefore to catch more of fish from our EEZ which is again a gift of God, whatever way we look at it, as an incarnation that has restrained from multiplying Himself beyond 4.5 million tonnes level of total fish stocks as our fishery scientists assert, or as a Diety he has imposed upon Himself a sort of family control system and exists at a much lower tonnage level, as Mr. Engvall probably deduces.

Our marine and inland fish production is depicted as going up. On the inland side we can believe reports of increases in production, because of advances in inland fish culture technologies. Can we have the same level of belief in respect of marine fisheries, with the exception of 1989 season in which there was a bumper harvest of mackerels on the west coast? With no significant increase in fishing effort, can sea fish production go up to any significant level, unless there are sudden hydrobiological changes? According to our statistics, with effort remaining more or less constant, marine fish production has been going up steadily at about 8 per cent per annum in the past few years although the additional contribution is insignificant in reality. The contribution from chartered vessels, may be around 10,000 tonnes, all of which goes to other countries by way of export and only 20 per cent of the invoiced value accrues to the nation as foreign exchange. The coastal zone is already overexploited. In this situation, with no increase in effort or efficiency of existing tackle or diversification of effort or coverage of farther areas by vessels of Indian flag, not taking into account possible poached Quantities, it is difficult to believe there has been an increase in marine fish production of the country. Many outside the Government as well as inside will certainly be sceptical about accepting the situation as realistic.

The remedy to the situation *i.e.* lack of increase in marine fish production is for the Government as well as the entrepreneurship to take risks, for, fishing business is fraught with risks, some seasons turning out to be good ones, others at an average level or bad. Anyone who wants to be sure of assured economic level of returns always, should be aware of this situation. Some years there may be surpluses, and losses in others, depending on the disposition of the fishery.

In making a reference to risks, we mean only calculated and surmountable risks. The risks are related to conditions of existence. Fish being mobile, their movements are influenced by the prevailing environmental conditions. Fishing effort and post-harvest activities are to be organised in a manner as to make the best possible use of the exploitable fisheries, based on current knowledge and information. Government understands this position, but it will be difficult for bankers to appreciate factors of this kind and this is also understandable. Developmental banking by a Governmental agency, such as the erstwhile SDFC is therefore well suited for the promotion of this sector.

In order to rectify the present state of affairs, an integrated developmental framework consisting of various components of the activity for phased and regulated implementation should be drawn up. Whatever be the odds and problems, the project should be implemented with periodic appraisals and with no going back or development of cold feet. The implementation should go on until it reaches the logical end. The main component of the framework will have to be a field Agency. This should be set up with full authority to implement the programme. It should be manned by technically knowledgeable and competent personnel. It shall provide the needed mechanism for the developmental process. The main purpose of the Agency, irrespective of costs involved, is to arrive at, once and for all, the maximum that can be done to exploit the known or unknown fisheries resources of our EEZ. The Agency should have the following wings.

1. Fishery Survey of India (the present one) to be brought under its fold; and
2. An infrastructure development enterprise has to be set up and brought under the purview of the Agency. The Integrated Fisheries Project would

have to be merged with this. Its function will be to identify specified selected port locations and provide major harbours, through port trusts or otherwise, with all needed infrastructure facilities. These can be organised through public or private sector as the Agency deems fit. These will include all post. Harvest facilities needed upto marketing stage. Any existing organisation/organisations dealing with this work (such as the fishing harbour unit at Bangalore) should be amalgamated with this enterprise;

3. A fishing fleet development wing is needed. Its job is to evolve schemes to upgrade the capabilities of the presently operated vessels for round-the-year fishing based on the advice of the F.S.I, and C.M.F.R.I., and deploy the same for operation from various ports. It will also implement a programme for introducing new vessels built indigenously or imported. This wing will acquire designs and drawings of various types of vessels capable of fishing almost round-the-year from abroad or internal sources through a separate unit set up for the purpose. This unit should be in a position to provide detailed drawings on nominal payment to various Indian Shipyards. One of its jobs is to determine the number of vessels to be introduced and how best to introduce them. It can acquire the vessels and operate them on its own or give them on lease to fishery enterprises, subject to well framed terms and conditions. Securing foreign expertiser formation of joint ventures, and chartering of foreign vessels should come within the purview of this wing;
4. There should be a financing section under the field agency which should be provided with the needed funds to finance harbour construction, provision of infrastructure facilities and acquisition of vessels. The financing should be in the form of investments and not lending. All returns should accrue to this wing, whether as lease amounts or as income from fleet development wings, by way of direct operations;
5. The present CIFNET should be brought under the purview of the proposed Agency. The training centre should be suitably revamped to provide to the candidates the types of training needed;
6. There should be a marketing cell, which should take care of domestic marketing as well as export marketing. This cell should build up a domestic fish marketing net work which should have its origin from production centres and move inland along various routes to the consumer centres. Not only fish but various fish products can be disseminated for marketing under the network. Once a domestic marketing system develops, it will pave the way for exports. Foreign buyers, watching the hygienic way of preservation of fish and well planned marketing within will come forward to import; and
7. The main Agency will coordinate the activities of all its limbs as outlined above.

Loss or gain, the Agency should cover the entire EEZ with the operation of its own or sponsored vessels for a period of ten years. Thereafter it should recommend to the Government the lines on which the fisheries of our EEZ should be developed under its auspices. The expenditure involved will prove to be far more economical than the present expenditure on deep sea fishing most of which is probably going down the drain.

The Agency should be responsible only to one master in the Government. It may be the Ministry of Food Processing or Ministry of Commerce or Ministry of Agriculture.

Lot of money is involved in the development of deep sea fishing now. Further, the expenditure is haphazard and losses are heavy. Losses are no doubt inevitable in the developmental phase. Without passing through this phase, however, we have straight away adopted a commercial way of functioning, leading to a heightening of losses. When losses are inevitable, let us experience the losses in a systematic way and purposefully to learn something and to establish certain facts and prepare a sound basis for development of deep sea fishing.

Fishing Chimes, Editorial 131: March 1990: Vol. 9, No. 12

# Voiceless Association

Fishes have no voice. The few that have voice merely croak. In the same way, the Sea Food Exporters Association of India and the Association of Indian Fishery Industries, to name the two main Associations, now and then croak about their problems, with no succour coming forth from the Government.

The Sea Food Exporters Association of India is the main organisation in the country that is supposed to champion the cause of the seafood exporters who provide to the nation over Rs.600 crores of foreign exchange through marine products exports.

The Marine Products Export Development Authority, a statutory body set up by the Government to promote exports of marine products, no doubt, introduced several innovative measures to improve the quality of marine products exported. It has also initiated steps to promote brackishwater shrimp farming to increase the quantum of shrimp exports and succeeded to a significant extent in this respect. In the matter of diversification of marine fishing efforts, because of inherent constraints in the functional responsibilities, MPEDA could not however make much progress. The basic structure for stepping up of production of marine fish suitable for exports continues to be stagnant for want of a proper policy, strategy and a dynamic outlook on the part of the authorities concerned. With very little increase in efforts, therefore, it may look baffling that the value of Indian Marine Products Exports has been going up from year to year. The apparent reason for this perhaps is the increase in per unit value of exports and because of the fact that total earnings from export of catches by chartered vessels are taken into account, while only 20 per cent of this actually reaches Indian Banks. The increase in export earnings because of the recent addition of cultured shrimp exports continue to be significant. Till such time the Ministry of Food Processing Industries and MPEDA are in a position to enable fishery enterprises to build up fishing fleets suitable for catching varieties other than the traditional ones presently exported and lend full support through provision of infrastructural facilities by way of provision of harbours at key points with integrated facilities and find markets for their exports, notwithstanding the risk of losses which are to be subsidised for sometime, and providing subsidy on diesel oil, there can be no progress in the marine products export front. The MPEDA can achieve pretty little to step up marine products exports, unless it has the strong backing of the Government, and lends support to deep sea fishing enterprises on the above mentioned aspects. The deep sea fishing activities having reached a position where there is a virtual stalemate, the only way open now is to undertake solid ground work to build up a strong fishing fleet equipped for two types of fishing with the needed infrastructure. Any number of meetings were held on the subject by the Government and all aspects of the subject are on record with the Government. What is lacking is translation of the decisions into action.

In this background, neither of the above mentioned associations have representation on the 30 Member Board of MPEDA, the only autonomous body that is trying to do something to develop deep sea fishing and processing and export sector. Why has the Government not given representation to these two important bodies on the Board of MPEDA? We feel that this is so because it feels that these Associations have no vocal cords, they have no spines and they are sterile. No other conclusion can probably be drawn from a study of the nominees to the MPEDA's Board from the seafood export industry and trade. With the exception of M/s. V.B.C. Exports Ltd., Visakhapatnam, none of the other nominees have their own means of production. Their job is to buy shrimp from the open market, process and export. An activity of this kind can be undertaken by any trading company. If a large house is able to achieve a higher level of exports, this is because of their capital strength but not because of their ability to produce any exportable raw material on their own. In other words, they do not play any role in augmenting production of exportable marine species. Their presence on the MPEDA's Board will not thus in any way help increasing exportable marine produce unless they have the backward linkage of means of production. A just way of constituting the Board is to eliminate or keep to the minimum the category of procurer-processor-exporters and increase the number of producer-processor-exporters. Further, the representatives of Seafood Exporters Association of India and the Association of Indian Fishery Industries or any other associations representing these two activities should be taken as members on the Board. Unless they are made part of the Board and enabled to function effectively on the Board by making suggestions for increasing production of exportable marine produce and diversification of exports, the Board will be far less effective than the previous Boards some of which were by themselves mostly ineffective.

The Government should evolve certain criteria for the selection of nominees from the marine fishery industry to serve on MPEDA's Board, taking into account performance in respect of production and exports. Lowest priority should be given to traders as they do not play any significant role in increasing basic production. Those who produce the highest quantity of exportable varieties and succeed in exporting and bringing good returns in foreign exchange should get preference to be represented on the Board. Next in the order of preference should be major producers of marine exportable species, who do not however export on their own at present. Their representation will enable the Government to know the reasons why major producers do not take any part in export activity. Based on this, MPEDA should be able to find ways and means of encouraging them to take up exports.

Another aspect we feel the Government has to bear in mind is the need to provide region-wise representation to the industry on the Board. According to the list, there are two nominees from Kerala, and one each from Tamil Nadu, Maharashtra, Goa, West Bengal, Gujarat and Andhra Pradesh. Orissa and Karnataka are totally ignored. What are the crimes committed by the exporters of these two States to merit their elimination?

Government also will have to take into account performance of the various units at the time of selection, based on certain selected criteria such as quality rating, diversification of exported products, popularity of the products abroad, markets explored or covered, modernisation of plants etc. In fact, all those connected with the industry and interested in being represented on the Board should be asked to furnish particulars under specified heads. These returns can be screened and a Committee set up by the Government can select the best of the lot to serve on the Board within the framework of a set of guidelines. There can be a stipulation that no one can serve on the Board for more than two terms.

We are unable to understand why there should be so many from the Government sector represented on the Board. The Board consisting of 30 members is too unwieldy. We do not find any justification for having political representation on MPEDA's Board. One particular M.P. from Rajya Sabha has been a recurrent inclusion on the Board and two representatives of the Lok Sabha may soon be nominated. It appears to us that one experienced, representative from the Parliament should be adequate to voice the views of MPEDA in the Parliament. More of them can of course be there as they may be the fittest to serve on MPEDA's periodic delegations visiting other countries. We however feel that instead of having these on the Board, representatives of the major associations pertaining to the industry could be nominated. They will be able to contribute substantially to the deliberations of the Board and assist the same in taking meaningful decisions that would go a long way in the development of the marine products export activity.

Fishing Chimes, Editorial 132: April 1990: Vol. 10, No. 1

# Another Blow to Marine Fishing Sector

The Indian marine fishing sector consists of around 20,365 boats. A majority of these, coming to 20,000 numbers, are small mechanised fishing craft operated by artisanal fishermen all along the coastline. Of the remaining 365 numbers, around 35 numbers are the largest of all in the size range of 25 to 28 metres imported from Australia and Korea, and built by Bharati and N. N. Shipbuilders in India, 130 numbers are in the range of 22 to 23.5 metre OAL, built in Mexico, Poland, Australia and India. Apart from these there are around 100 numbers of mini-trawlers of about 50 ft length and about 100 numbers of Sona boats of about 46ft OAL.

In the past few years fishing operations for all the above mentioned categories of vessels have become highly uneconomical for reasons of high cost of diesel oil, increase in cost of spare parts and repairs, hike in wages bills etc, The marine fishing industry has been doing its utmost to somehow come out of the adverse conditions into which they have been gravitated. The owners have made several representations to the Government for assistance towards diversification of fishing effort to enable the fishing vessels to operate throughout the year, when only there is a possibility of profitable operations. At present for about four months in a year there are practically no fishing operations because of off-seasonal and bad weather conditions.

At a time when there are indications that Government has understood the plight in which the fishing industry is in, and it has taken up examination of the representations of the various sectors of the industry in order to extend the needed assistance for diversification of fishing effort, which alone would enable the vessels to operate throughout the year, there came as a bolt from the blue a hike of 54 paise per litre of diesel oil in the union budget of 1990-91. The industry, which is already indebted to the financing institutions to a significant extent because of non-payment of instalments of loans and interest, has now to bear an additional expenditure of nearly Rs.50 crores annually. Those in the Government connected with marine fishery development are certainly aware that the various segments of marine fishing industry would not be able to bear this additional cost. It is ironic that Government hikes up prices but does very little to provide infrastructure facilities at various harbours, the only exception being Visakhapatnam.

There is no doubt one has to admit that the increase in the cost of diesel oil and other items such as steel which also affects the marine fishing industry considerably has been done on a national basis and the fishing industry has become an unwitting victim. The impost has come as a major blow to the already ailing industry. Increase in the diesel oil cost, leaving aside other indirect increases such as cost of petrol, steel items etc., is unbearable.

It is ironical that the industry is placed in a position of bearing the additional burden as detailed above, in spite of the fact that representations, often and on, are being made by the industry in the past few years to provide some relief on the cost of diesel oil by way of subsidy or excise duty exemption, as the cost of diesel oil happens to be the major input, coming to nearly 60 per cent of the operational costs of vessels. This shows that Government has utter disregard towards the problems of the marine fishing sector, which, unfortunately, does not have the understanding and support from any of the political parties. The financial state of the industry is such that most of the companies are unable to contribute funds even to mount a delegation to make representations to the authorities concerned.

It is a matter of great shame that the authorities concerned at the top who are handling the subject are a great failure in the matter of development of this vital and crucial industry. If they had taken some measure of interest in the subject, the position would have been different. While the industry is prospering in all neighbouring countries, in India the industry is languishing.

The Shipping Credit and Investment Company of India has been very tolerent and has shown a great sense of understanding in regard to the problems faced by the marine fishing sector, particularly the problems relating to the operations of deep sea fishing vessels. The commercial banks have also adopted the same approach in regard to the various other types of fishing vessels financed by them. The additional costs that have to be borne now by the owners of fishing vessels will undoubtedly create further problems in regard to repayments of loans and interest to the financing bodies. Even if the financiers want to take over the vessels and attempt to do something with those vessels in order to get back the money, it is quite possible that they will be able to find few enterprises coming forward. If the present operators are unable to earn adequate funds, there is not much scope for other enterprises to do so.

There are several countries where the cost of diesel oil is considerably lower, compared to the costs prevailing in India. The earnings from fish landings are also higher in several countries compared to the Indian market conditions. Considering this, it might prove to be a feasible solution, if SCICI or the other financing bodies permit the

**Budget 1990-91: Increase in Diesel Oil Cost: Quantification of Burden on Marine Fishing Industries**

| *Sl.No.* | *Type of Vessels* | *ASI, Korean and Bharati Trawlers* | *Mexican, Dutch, Indian 'K' Shipyard etc. Vessels* | *Mini Trawlers* | *Sona Boats* | *Mechanised Boats* | *Total* |
|---|---|---|---|---|---|---|---|
| 1. | Number of Vessels | 35 | 130 | 100 | 100 | 20,000 | 20,365 |
| 2. | Consumption of Diesel Oil per voyage.KL. | 65 | 50 | 1 | 8 | 0.2 | 133.2 |
| 3. | Increase in cost of Oil per voyage/vessel/Rs | 540 | 540 | 540 | 540 | 540 | - |
| 4. | Increase in cost of Oil per voyage/vessel/Rs. | 35,100 | 27,000 | 5,400 | 4,320 | 108 | |
| 5. | No. of voyages per annum per vessel | 9 | 9 | 15 | 15 | 200 | |
| 6. | Total increase in cost of diesel oil per vessel/annum/Rs.lakhs | 3.16 | 2.43 | 0.81 | 0.65 | 0.22 | |
| 7. | Total burden on each type of fleet per annum in Rs./lakhs | 110.6 | 315.90 | 81.0 | 65.00 | 4,400.00 | 4,972.50 or 49.725 crores |

owners to operate their vessels in the waters of other countries, if they are able to enter into suitable agreements with enterprises in those countries. According to reports, Namibia is interested in entering into joint ventures in the field of fishing with other countries. India has a very good political relationship with the present Namibian Government. This can probably be taken advantage of and a bilateral agreement with that country and such other countries could be worked out by the Indian fishing companies particularly those having larger vessels which could be moved for operations in those waters. Under this system, while Indian entrepreneurs would stand to earn considerable amounts of foreign exchange by way of their share of earning, out of the joint venture operations, there will be very little dent on total direct foreign exchange incomes of the country out of marine products exports. It is estimated that out of nearly Rs.620 crores of annual foreign exchange earnings by India from export of marine products, the component of foreign exchange earnings from the operations of deep sea fishing vessels hardly works out to about Rs.25 crores. Much more than this can be brought into the country from joint venture operations in other countries by Indian enterprises. It is desirable that additions to the deep sea fishing fleet, for that matter, additions to any fishing fleet in the country could be stopped for some time till the prevailing conditions turn to be favourable for Indian operators in their own waters. The owners of vessels that are now under construction at various Indian shipyards would have to face the same problems as the present ones in operation, once they also become operational. They have therefore to work towards sending these vessels for conducting fishing operations to any of the neighbouring or distant countries, under mutually acceptable terms and conditions.

Another alternative seems to be that the Government of India should prevail upon the Japanese, United States and West European countries to advise importers of marine products from India in those countries to take into account the increase in operational costs and readjust their import rates suitably so as to cover the additional costs involved. Because of the friendly relations which the Indian Government has with all importing countries, it may be possible to convince the Governments concerned to assist India in the manner mentioned.

Another alternative is for the Government to set up a Central Fisheries Export Marketing Organisation to purchase all acceptable varieties of shrimp and fish at remunerative prices, worked out on the basis of the actual operational costs and the capital investments and export the material under their banner to other countries through the mechanism of setting up of their own import organisations in those countries. Such organisations should be able to secure higher prices in the importing countries which will enable them to offer realistic prices to the Indian exporting enterprises.

Fishing Chimes, Editorial 133: May 1990: Vol. 10, No. 2

# Charter Scheme: Double Standards

Several schemes for permitting Indian fishing companies to operate foreign fishing vessels on charter in the Indian Exclusive Economic Zone (EEZ) were notified by the Government from time to time in the past few years. The objective of all these schemes is to introduce new deep sea fishing technologies through provision of training to Indian personnel, employed by the national fishing companies concerned, on chartered foreign vessels concerned, through the foreign crew and to ensure the introduction of similar vessels by the Indian enterprises for operation in the national EEZ.

Conditions governing the issue of permits to Indian companies for operating foreign fishing vessels on charter have undergone various changes from time to time, particularly in respect of the type of vessels to be permitted for operation, the areas where the vessels can conduct fishing and the amounts to be deposited by the Indian parties as security for the fulfilment of the main condition in the letters of intent for the introduction of their own vessels.

The schemes were implemented in the earlier stages, particularly when Mr.Rao Birendra Singh was the Minister of Agriculture, and Mr.S.P.Jakhanwal was the Joint Secretary (Fisheries) in the Ministry of Agriculture, with great alertness and promptitude, ensuring all needed efforts to see that the Indian charterers took the needed steps for introducing their own vessels. As time passed by, under the successive ministers, secretaries and joint secretaries in charge of fisheries, a perfunctory attitude towards the objective of the scheme crept in. The result was that, after Mr.Rao Birendra Singh relinquished his ministership, the successive ministers incharge of the scheme seem to have lost sight of the objectives. So much so, the implementation of the scheme acquired business overtones, with the promotional objective becoming subsidiary. The scheme is treated as a money yielding one for the charterers, from the stage of application until the permitted operational period is over.

A mental revolution had taken place among the charterers. Instead of thinking in terms of promoting the activity as laid down in the scheme, the quest of the majority of the charterers has been to find ways and means of circumventing the conditions laid down by the Government in the scheme according to which within one year of the issue of permissions, the companies would have to show proof of having taken steps to introduce similar vessels as permitted for chartered operation, either through joint ventures or in a direct way, in a period of one year of receiving permits. Quick calculations seem to have shown that, if they are unable to fulfill the obligation and if Government does not accept the 'pseudo' steps taken by them in regard to entering into joint ventures or acquiring similar vessels, the worst that would happen to them is that they would lose their deposits, which would probably be almost equal to the income derived, if not marginally more, out of 20 per cent of the gross proceeds which they would get from the foreign owners. Government have also helped the owners and protected their interest by stipulating that certain amounts should be deposited by the foreigners in favour of the Indian charterers concerned as advance towards payment of 20 per cent of gross returns from the foreign owners to the charterers from time to time. This has also helped the Indian charterers to organise funds for making deposits with the Government.

While the objectives of the scheme are laudable, the provisions, relating to payment of 20 per cent of the gross proceeds by the owners to the charterers, the system of allowing disponent ownership, the manner in which vessels of Taiwanese flag are being made to be brought into our EEZ under Panama flag, for the mere reason that we have no diplomatic relations with Taiwan, if all these and other aspects are gone into deeply, there may come out several issues that can be questioned. Countries such as Australia, New Zealand, U.S.A., etc have no diplomatic relations with Taiwan, but Taiwanese vessels are allowed to operate in the EEZs of their countries. Taiwanese cargo vessels call at Indian ports and Indian vessels call at Taiwanese ports. Unless repurcussions Vis-a-vis Mainland China are visualised, there is no rationale in not allowing chartering of vessels of Taiwanese flag. The bulk of chartered vessels now in operation are *de facto* Taiwanese vessels, whatever be the coverup we have (Panama registration). Mainland China can still raise the issue as the vessels are basically of Taiwanese origin and Taiwan allows dual registration. However Mainland China does not seem to be concerned about Taiwanese vessels operating in Indian waters.

There appears to be highly flexible norms that can be swung to justify any specific case in the issue of Letters of Intent/Permits to the various Indian parties for the chartering of fishing vessels. Those similarly situated as the others who could get Letters of Intent, are dismayed at their elimination, cause for which is not told to them. There can be no doubt that double standards have been adopted in the matter of issue or extension of Letters of Intent and in the follow-up action concerning various provisions in the permits issued.

One important aspect of double standards relates to the granting of extensions to permits as provided in the Letters of Intent issued to chartering companies around 1983 and thereafter. According to one of the conditions, those who fulfilled the obligation of introduction of vessels in the same numbers as permitted for operation under charter would be given extension for operating chartered vessels for a further period of two years. There are several companies who did introduce an equal number of vessels as operated under charter, thanks to the strict implementation of the scheme in the past. Most of these companies applied for extensions but few of them could get them so far. In contrast, those who did not introduce any vessels of their own were given extension of permits. Reasons for this are known only to the Government.

Only an enquiry will show as to why such double standards were adopted. The right way of dealing with the subject should have been to first grant permissions to such of those who fulfilled the obligation of introducing vessels of their own. This has not been done, by and large.

We suggest that the Prime Minister and the Minister concerned may order a thorough enquiry into the working of the scneme from time to time with a view to setting right the faults, and to evolve a promotion-oriented chartering scheme that would ensure the development of the fisheries of our EEZ. The tendencies on the part of Indian charterers to avoid fulfilment of obligation and move from scheme to scheme in the garb of new companies (if they make applications in the name of the operating companies they would not be eligible for issue of new Letters of Intent, as they have not fulfilled the obligations), will have to be curbed by the Government through some means or the other, if there will be any new scheme.

No country in the world has adopted the system of allowing their own nationals to receive a particular percentage of the gross earnings, and remaining altruistic by not taking any fees from the foreign owners, unlike other countries. It is only the Indian Government which provided a means of earning to the Indian charterers (20 per cent share of gross earnings from the foreign owners) with devices to ensure these payments. Our Government is perhaps wiser and more benovelent towards Indian chartering fishing companies. It is necessary for the Government to review this and see that foreign vessels are allowed to operate under the licensing scheme as provided in the rules framed under provisions in the Maritime Zones Act. Under this system the foreign owners pay particular fees to the Government based on harvested catches, and there is no question of any Indian charterer coming into picture. If the intention of the Government is to ensure transfer of technology, there can be no doubt that under the present system of chartering, this can never take place. Transfer of technology can only take place by the engagement of foreign or Indian experts by the Government to train men sponsored by fishing companies on board training vessels specially meant for the purpose, with an integrated arrangement for the interested Indian companies to acquire vessels for the utilisation of their nominees trained under foreign or Indian experts, on the new vessels as and when introduced. This kind of system was adopted under the erstwhile Deep Sea Fishing Station successfully.

In conclusion, we strongly appeal to the Prime Minister of India, the Minister for Food Processing Industries and the Commerce Minister, to order a review of the implementation of charter scheme with a view to identifying their weaknesses, and in that light to introduce a fresh scheme that would serve the purpose of development of deep sea fishing in our EEZ.

## Charter Scheme: Information Barrier

The progress in the implementation of, the fishing vessel charter scheme which involves the operation of foreign owned fishing vessels in the Indian Exclusive Economic Zone is of considerable national importance. However, very little information concerning this comes out of the Governmental sources, the only exception being occasional answers given to a few questions on the subject put in the parliament. The contents of these answers are as good as not having any answers in the sense that they do not often say much.

The fishing industry and the Fishery Research Organisations such as the Central Marine Fisheries Research Institute have to know the position in regard to the progress of this scheme from time to time. The only Governmental agencies who have information on the latest position are the Coast Guard and the Fishery Survey of India and probably the MPEDA. The last mentioned organisation had to resort to a warning to the charterers that, if they do not file the returns concerning the results of operations relating to the previous voyage/voyages, they will not get clearance concerning prices for the catches landed in the latest voyage. The information gap that exists within the Governmental bodies and between the Governmental agencies and the public is something that needs eradication.

It is not that the Government alone has to evaluate the success or failure of the charter scheme. The fishing industry should also have the opportunity to evaluate and react. The ingredients of the scheme are not in consonance with the objectives and therefore are unrealistic but of a short term utility (although the scheme is stated to have been drawn up with long term objective of building up the Indian Deep Sea Fishing Fleet).

At least once in a month, it should be possible for the Government to tell to the people the number and type of chartered vessels of various types in operation in Indian EEZ, against the number of permits and letters of intents issued, and the stage in the fulfilment of the various terms and conditions imposed upon the charterers who are now operating chartered vessels. The information on the above aspects will reveal the solidity or hollowness of the scheme. While it can be expected that the Government will have no alternative than to give out the details sooner or later, probably in the parliament, it is known that no chartering Indian company has taken any tangible steps (to be distinguished from perfunctory steps known to both the sides as meant to get extension of charter permits for one more year) for the fulfilment of the obligations. If what is stated above is taken into account, it is high time that the scheme is reviewed and recast suitably, for, the nation needs the operation of various types of foreign vessels in Indian EEZ to provide information on the potential and the types of fishing to be introduced. The attachment of Indian personnel on each of the foreign fishing vessels is good only to keep a watch over the operations. This system will not however be of any real use if the intention is to develop a cadre of trained Indian personnel in the types of fishing undertaken by the chartered vessels, No single person, by and large, continues to work on a specified vessel and the foreign counterparts who have to teach them keep changing every time. Further, there is no linkage between the trained men and their future appointment on the vessels supposed to be introduced by the Indian charterers. This linkage has to be brought about in the early stages of commencement of operations.

It can be anybody's guess to know how many chartered vessels are now actually operating in the Indian EEZ. To the extent information can be gathered, which can only to be taken as indicative, there are five pair trawlers in operation, one pair each belonging to coastal trawlers, young fisheries, GangaKaveri, Leo Seafoods, Kanchanaganga and, BRS Fisheries(?). In addition, two pair trawlers of Matsyika are also reported to be in operation. One vessel of Ganga-Kaveri and one of Leo Seafoods are stated to have been apprehended by the Pakistani Authorities. It is quite likely that these have been released by the Pakistani Authorities by now. Apart from these, it is stated that there are 13 chartered Taiwanese long liners and one Italian stem trawler now in operation. Most of the long liners are likely to go back to Taiwan, as the results of long line operations at this time of the year are stated to be uneconomical. It is likely that these will return back around September, if the permits are kept valid till then and thereafter.

It is stated that Taiwanese find long lining operations inconvenient due to various constraints in India, most important of which, it is stated, are procedural delays, relating to clearance of foreign crew and clearance of vessels as such. Transhipment of catches by reefer vessels other than those flying Indian Flag is not being allowed. This is acting as a deterrent for the Taiwanese to continue operations. It is not really known as to why the Government.should not allow foreign vessel owners to transport the catches purchased by them through foreign cargo vessels of their own choice. It is stated that even Japanese reefer vessels are not allowed for the transhipment. If transhipment by foreign cargo vessels is not allowed the owners have to tranship the catches by the chartered vessels to Taiwan or to any other country. This would take away lot of time, reducing the number of fishing days. In several cases the validity of the letters of intent obtained by the Indian companies expired before the vessels could come back after completing transhipment of catches. Apart from allowing transhipment by foreign flag vessels, the system should be that the date of initial reporting of the vessels at an Indian Port should be the date of commencement of validity of the permits issued. This will give maximum number of fishing days to the chartered fishing vessels.

The Association of Indian Fishery Industries should take up all issues connected with the chartered fishing vessel operations with the Government and it should ensure that all irritants are removed so as to make the scheme really effective for the eventual expansion of the Indian Deep Sea Fishing Fleet. The Association should also prevail upon the Government to make public all vital information concerning all the chartered vessels once in a month through the Fishery Survey of India, omitting any information that may have security implications. Unless this system is introduced, there will be a major gap that will have a serious retarding effect on deep sea fishery developmental activities.

Fishing Chimes, Editorial 134: June1990: Vol. 10, No. 3

# Killjoy Crisis Keeps Deep Sea Fishing Vessels Harbour-Bound at Visakhapatnam

An unproductive but probably justifiable tussle between the owners of deep sea fishing vessels based at Visakhapatnam on the one hand and the deep sea fishing technocrats and the trawler workers on the other has kept over 120 deep sea fishing vessels harbour-bound. For the owners the tussle constitutes a last ditch battle. Having started this for their survival and to meet the commitments to the financing institutions, they should stick to their justified stand relentlessly, whatever be the consequences, till there is an equitable settlement. Most of the owners are of course determined to operate their vessels only when the floating staff agree to a rationalised system of sharing of the net annual returns, 40 per cent to go to them and the rest to the owners. In addition, the owners would pay a fixed salary, daily sea allowance, mess allowance etc. Such a system, they feel, would optimise the number of fishing days, thereby improving the incomes of the owners as well as the floating staff. The owners would then have adequate cash flow to make repayments towards loans taken.

So far as the floating staff are concerned, they are averse to any sharing system, even when linked to assured salaries and allowances. Initially they demanded very high salaries and allowances. As it is, a captain gets a salary over Rs.48,000 per annum, and a deckhand gets over Rs.7,000, with various gradations in between. These do not include the very high incentives they receive.

The owners, seeing the attitude, became desperate and not made any effort to discuss the matter with the floating staff based on their terms. Instead, they proposed the sharing system mentioned above. They are willing to discuss the details within the contours of the scheme. The matter was nonetheless discussed at a meeting convened by the Government in Delhi in which representatives of both the sides (owners and floating staff) participated, but with practically no progress. The floating staff organised rallys, dharnas, met various local officers, but all these have not led to any solution. The floating staff even went to the extent of conducting a dharna, raising slogans, scribbling graffiti, and tampering with records, vehicles etc at the office of the President of the Association of Indian Fishery Industries. At the commencement of one of the rallys, there were clashes between the floating staff and the traditional fishermen, resulting in injuries to some. The mechanised boat operators suspended fishing operations for a day or two as a protest.

This stalemate between the owners and the floating staff of the larger vessels, which continues at the time of this writing, is in a way a boon to mechanised boat, sona boat and mini trawler operators who have recommenced fishing operations, as the off-season has come to a close. The mechanised fishing boats whose owners were allowed to continue fishing during the ban period also on a day-to-day basis, were landing 2 to 3 kg. of shrimp per day prior to the recent cyclone. After the cyclone, their shrimp catches went upto a level of 50 kg/day. Further, one of the training vessels of CIFNET which conducted trawling south of Visakhapatnam could catch about one tonne of shrimps, mostly of 21/25 count, in two days. There are thus indications of prospects of good fishing during this year. Both the owners and the floating staff want to avail of this, but are unable to do so. If the owners resile from their present stand, they will not be able to build up the tempo again. The solidarity among the owners has become total, to the dismay of the floating staff. Some of the owners who sent their vessels out relented and called them back, giving a great boost to the cause for which the owners are fighting. The floating staff are understood to have proposed negotiations, seeing the solidarity.

To recall the past history, on 20 June, 1989 the representatives of the All India Deep Sea Fishing Technocrats Association and the representatives of the trawler owners came to an understanding that the salary and incentive systems applicable to the floating staff should continue without any modifications. The floating staff had not agreed to the sharing system. They had however agreed that the various aspects including the development of fisheries could be discussed between the two sides in March 1990.

At a time when the owners were in a state of shock and dismay over the increase in diesel oil prices, the withdrawal of concession on excise duty on diesel oil by the Government, the increase in cost of materials and expenditure on repairs of trawlers, a salvo was fired by the technocrats well in advance demanding a major increase in salaries, payment of sea allowance, higher rates of mess allowance and various other benefits. It is true that the technocrats said that the matter could be discussed so as to reach an amicable settlement over their demands. They particularly pointed out the increase in cost of living and 'high earnings' by the Owners.

The presentation of charter of demands of this kind has been a routine annual or biennial feature. This particular charter of demand, however, converted most of the owners into men of steel. On one side, owing to escalating prices almost all the companies have been in the red, unable to make repayments to the financing institutions, pay salaries to the crew members, keep up insurance on vessels, and maintain the vessels well. The technocrats are fully aware of the problems faced by the owners, although, probably for practical reasons, they kept harping on the same tune that the vessel were acquired by the owners taking almost the entire cost of the loan from the Government, and that they are not deliberately making repayments to the Government, so as to divert the profits. They continue to overlook the realities of economics of operation, which had resulted in the owners ending up in losses, or not leaving any profit or margin or very little margin to the owners.

In this situation, the presentation of a demand for higher payments to the floating staff acted as a major challenge and affront to the owners.

Accepting the counsel of the sobre elements among the owners, they announced a share-catch system as mentioned above for the acceptance of the floating staff.

The Associations representing the floating staff felt that, while theoretically, the basic salary and the sea allowance paid would almost be equal to the present salary levels, if the owners detain the vessels at shore for long, their incomes would come down. Apart from this, they felt that the offer of payment of 40 per cent of net profits would never take any beneficial shape. The owners would not show any profits and therefore there was no question of actually receiving any incentives under the scheme. The owners have expressed surprise at this interpretation.

A section of the owners pointed out that the floating staff might show a large number of sea days, without putting their heart and soul into the operations, thereby leading to a reduction in catches and the consequential lower returns. The floating staff feel that they will never sink to such low depths.

The trawler officers and deckhands were retrenched around February 1990 by most of the fishing companies for the reason that the vessels were remaining idle with no incomes. Owing to the irrational method of payment of incentives to the crew existing prior to the retrenchment, whenever there was some increase in income in a particular voyage, a substantial part of it went out as payments towards incentives to the crew, resulting in shortages of funds to send the vessels out after some of the voyages that resulted in poor catches. Further, in the earlier stages of development of the industry, the owners, in a short-sighted manner, introduced certain pernicious systems of treating fish catches, shark fins, cuttle fish etc on a footing different from shrimps. They were leaving all these items to the crew to be disposed of by them for enjoying the returns exclusively. Later, systems of payment of around 50 percent of returns from sale of fishes to the crew developed. Shark fins and also squid and cuttle fish which bring in good returns continue to be apportioned to the crew for their benefit, in the case of most of the companies. These systems have resulted in a major drain of funds to the fishing companies. For this reason they were forced to retrench the crew for want of cash flow, the owners say.

Considering these aspects and the irrational system of payment of incentives over the returns from sale of shrimp, the owners, taking all aspects into account, evolved a salary linked share-catch-system which they felt would be equitable and beneficial to both the sides. Broadly stated, under this system, the floating staff would be given the following salaries:

| *Sl.No.* | *Designation* | *Salary per month (₹)* | *Sea Allowance per Day (₹)* | *Sea Messing Allowance per Day (₹)* |
|---|---|---|---|---|
| 1. | Skipper | 1600/- | 140/- | 30/- |
| 2. | Engineer | 1400/- | 135/- | 30/- |
| 3. | Bosun | 850/- | 80/- | 30/- |
| 4. | II Engineer | 650/- | 65/- | 30/- |
| 5. | Cook | 400/- | 23/- | 25/- |
| 6. | Oilman | 400/- | 23/- | 25/- |
| 7. | Sr. Deck Hand | 400/- | 23/- | 25/- |
| 8. | Jr. Deck Hand | 300/- | 13/- | 25/- |

They will not however be entitled to Sea Allowance at home port, base port or any port of shelter. They are however entitled to Sea Messing Allowance while at sea and also at any port of shelter but not home port, or base port. They would be given productivity bonus (40 per cent of net returns). This would however be paid at the end of the fishing season or at the end of the financial year. The floating staff would be eligible for other benefits which include Group Accident Insurance, Medical Insurance, Leave Travel Allowance, Provision of Uniforms and all statutory benefits for which they are entitled.

The retrenchment of crew by the owners acted as a major irritant to the floating staff and this is understandable. With no monthly income, there is no means for them to meet their expenses. So much so, they approached the High Court for a direction to the owners to reinstate them. What exactly the orders of the Court were on this are not known. However, according to the owners, the Court seems to have directed the owners to consider the claims of the retrenched staff for employment on the trawlers as and when recruitment action is taken. The floating staff however say that there was a clear directive from the court to the companies to reemploy them. Whatever be the position, several fishing companies issued advertisements calling for applications from qualified persons for appointments under various categories. They have prescribed a form in which applications have to be made. These are to be obtained from the company concerned on payment of Rs.5. However, those who earlier worked in the company concerned, can obtain application forms free of cost.

In the terms and conditions, given as enclosure to the application form, the payment pattern as already outlined, was given. In addition, certain rules governing discipline and other general conditions were included. These are as follows:

1. The date and time of sailing of the trawlers will be decided by the Management at its sole discretion

and all floating staff shall be required to be present on board the sailing trawlers at least 2 hours before the scheduled sailing time. Failure to do so will be construed as desertion and will invite summary dismissal.

2. The Management shall also exercise its right to terminate the services of any of its floating staff in the event of acts of indiscipline or misdemeanours on the part of such persons a few of which are listed below:

   i)Insubordination; ii)Leaving the trawlers at any port of shelter without the prior permission of the Management; iii) Drunkenness onboard the trawlers; iv)Damage to or theft of Company's property; v)Habitual late coming; vi)Poor performance; vii)Any careless action that might endanger the lives of other floating staff at sea and or the Company's property.

3. The Management reserves the right to terminate the services of any of its floating staff upon issuing one month's notice to such employee or employees or upon payment of one month's pay in lieu thereof. In the event of dismissal of an employee on disciplinary grounds, one month's notice pay is not payable. In case the floating staff desires to leave the services of the Company, he may do so by giving one month's written notice to the management.

4. The decision of the Management to conduct any type of fishing activity anywhere in the Indian EEZ shall be final and binding on the floating staff.

5. The Management reserves the right to recruit extra hands for any post for eventualities like leave reserve or for works connected with the trawlers, and to rotate such staff on different voyages as per its own discretion.

6. The Management reserves the right to place the services of any of its floating staff on shore duty or at the Company's office for any operational reasons at any time and to lend their service to other friendly company's trawlers or place them on any of its own vessels elsewhere in India, including the Company's chartered trawlers.

7. The Management reserves the right to change its base or port of operations at any given time.

8. The Management reserves the right to engage on board of its trawler as many floating staff as it desires, but not more than the maximum number stipulated by MMD.

9. The Manning requirement of the trawlers shall be governed by the provisions of the MS Act. Selected candidates shall be required to sign an Agreement with the Company before appointment.

The Unions concerned are totally against this approach adopted by the owners. They wanted the managements to take back the retrenched crew without preconditions and they should be paid all arrears of salaries, wages etc. This is not acceptable to the owners.

The Unions said that their members would be willing to continue to work as per old terms and conditions and they would conduct fishing for about 270 fishing days. This did not receive the acceptance of the owners.

In the meanwhile, the A.P. Deep Sea Fishing Technocrats Association came up with a proposal which they discussed with the Trawler Owners Association. According to this proposal, the Skipper should be paid per month Rs.5,200 to Rs.6,000, Engineer Rs.4,900 to Rs.5,700, Second Hand Rs.3,300 to Rs.4,000, Engine Driver Rs.2,500 to Rs. 3,500, and Senior Deck Hand/Oil Men/ Cook Rs.1,200 to Rs.1,800 and Junior Deck Hand Rs.900 to Rs.1,500. These emoluments are inclusive of Sea Allowance and Mess Allowance. However, incentives would have to be given based on the tonnage of shrimp landed. Incentives would also have to be given on fish landings, the value of which should be calculated on mutually agreed prices per kg. They have guaranteed 225 to 250 fishing days, provided the managements take steps towards the diversification of fishing methods to be adopted during April to June. There was a talk that the All India Deep Sea Fishing Technocrats Association utilised the A.P. Deep Sea Fishing Technocrats Association, as a ploy, to discuss the proposal, so as to get an idea about the extent to which the owners would agree to meet the demands. This talk was probably a conjecture based on the fact that most of the technocrats in the A.P. Association are also members of the All India Association.

In the present situation the floating staff seems to be the direct losers as their incomes have probably totally dried up. The owners are also seriously affected both directly as well as indirectly. Their only consolation is that they can prevent the incurring of further losses by keeping the vessels tied up at port rather than acceding to the 'greedy' proposals of the floating staff. It is true that the crew have to lead a very hard life out at sea. This is not however a justification to ask for salaries far higher than what a Director of Fisheries gets. The owners feel that the floating staff on their own opted for a sea life and they should not have any complaint about the 'hardships' they 'face' out at sea. In fact many of the crew members are stated to be happier out at sea than at shore, the owners feel.

The major effect of the present situation seems to be that the deep sea fishing industry would not be able to play any significant role in contributing foreign exchange by way of export earnings. According to reports, the present contribution of deep sea fishing industry towards the marine products export effort is less than 10 per cent of the total export earnings. The Ministry of Commerce wants that this sector should contribute Rs.129 crores per annum by way of foreign exchange, against the present level of around Rs.40 crores. There is practically no interest on the part of the new entrepreneurship to enter the field of deep sea fishing, apart from the fact that the fleet strength has been gradually dwindling. Such being the case, there is no way the deep sea fishing sector can contribute Rs.129 crores to the marine products export effort by way of foreign exchange earnings. It will be a great achievement if the present level of contribution can be maintained. Looking at the present trend of employer-employee relations it is quite possible that there will be a drop in the contribution by the deep sea fishing sector in respect of exports.

Government also do not seem to be really serious about lending help in resolving the dispute between the owners and the floating staff. (The local officers may try to do something only when there is threat to 'law and order' situation). One reason for this seems to be that the feeling in MPEDA and the Ministry of Commerce is that by promoting prawn culture the export achievements can be far higher than through concentration on the development of deep sea fishing sector. There can be no other reason for the apathy being shown towards deep sea fishing sector by all central ministries concerned.

It is really not understandable as to why the floating staff are not willing to accept the productivity linked share catching system suggested by the owners. If they are really serious about working hard and bringing good catches by increasing the number of fishing days, it appears that they stand to gain by accepting the scheme. Firstly their present monthly incomes would in no way be affected. Secondly, because of the likely higher output, they would certainly get something substantially extra by way of 40 percent share of net profits. If they are not confident about this, they could ask for the laying down of certain norms concerning administrative expenses etc and also about a minimum bonus guarantee linked to their salaries or 40 percent of net profits which ever is higher. On the owner's side, they should not agree to pay earlier dues and re-employ all those who were retrenched but only those who were adjudged as loyal and devoted to the work. The managements would have to develop a benevolent attitude towards the crew, in spite of what all had happened. The floating staff can promise a minimum of 250 fishing days of fishing and they could ask for compensation if vessels are detained at shore daring inter-voyage periods for a period exceeding, say, seven days, as that would affect the total number of fishing days guaranteed by them.

Lastly, we have to admire the sagacity and pragmatism of CITU, to whom the All India Deep Sea Fishing Technocrats Association and the Trawler Workers Union are affiliated. Sensing the situation and the plight of their members, it has boldly and spontaneously proposed the continuance of the old system with an additional offer of putting in 270 fishing days, although knowing that the owners may have problems in sending out vessels for fishing to give a total number of 270 fishing days. This needs good management, money support and efficient work to bring down the number of days of stay at shore to the minimum. One can only hope that the owners would examine to what extent this proposal can be accepted.

We feel that, out of the two systems, however, the share-catch system will ultimately prove to be beneficial to both the sides. The continuance of the old system can at best be an inescapable alternative, provided the owners would still have the freedom to recruit their crew afresh based on their advertisements. If there is no agreement, it is suggested that the Government may appoint a panel consisting of representatives of the Fishery Survey of India, SCICI, Mercantile Marine Department, Port Trust and the CMFRI to work out a system taking all aspects into account. This should be binding on both the sides.

Fishing Chimes, Editorial 135: July 1990: Vol. 10, No. 4

# SCICI Launches the Receiver Era

The programme for the financing of deep sea fishing vessels was initiated by the erstwhile Shipping Development Fund Committee around 1976. The functions of this body which was wound up, had subsequently been entrusted to the Shipping Credit and Investment Company of India as per the provisions in the SDFC (abolition) Act, 1986 from September 1988.

The agreements entered into between the erstwhile SDFC and the beneficiary fishing companies allowed the appointment of receivers if the Committee was satisfied that the borrowing company failed to make the payment of instalments of principal/interest in time or commits any other default. However, the Committee never invoked this right in respect of fishing vessels during its life time, although there were several defaulting fishing companies in respect of payment of dues, at the time of its closure.

For the first time in the history of deep sea fishing industry of the country, the SCICI, after several warnings and prolonged correspondence with the beneficiary companies, recently embarked on the unpleasent task of appointment of a Receiver in respect of some of the defaulting companies. On one fateful morning in early March, deep sea fishing companies at Visakhapatnam passed through a traumatic experience of resting their eyes on the notifications of SCICI, in Deccan Chronicle, appointing a Receiver for the trawlers and other assets of Marshall Seafoods, M/s. Uni Marine, and Marine Fisheries. After this there was a gap, a lull before a gathering storm. The executives of fishing companies started looking into the advertisement columns of Deccan Chronicle in which the notification pertaining to the above mentioned three companies appeared. As expected, one more advertisement appeared in Deccan Chronicle appointing a Receiver for the trawler and other assets owned by Satya Sai Marines. A few days after this, another notification appeared about the appointment of Receiver for the trawler owned by M/s. West Coast Marines and later another one of the same kind in respect of the company owning the vessel 'Bhavani'. In the case of all the companies, the Receiver is one and the same.

None could point a finger at SCICI for the appointment of a Receiver to take over the trawlers and other assets of the various notified companies. The appointment was done after giving due notice and also allowing sufficient time for replies, which were perhaps either not given or the attitude of the company concerned was found unsatisfactory. Several defaulting owners now feel that their vessels may be the next to be put in charge of a Receiver. Those who have the shadow of the Receiver over them now, however, are still in the dark as to what would ultimately be done by the Receiver, and in what manner he would deal with the assets. It is quite likely the SCICI had already informed the affected fishing companies about what the Receiver would do. In the alternative, the Receiver might have already done something. For the time being, the dilemma of the 'owners' is about their responsibilities and functions from now onwards. If the Receiver is going to take over the vessels and operate them in a manner as decided by him, without the participation of the owners, then the owners would not like to incur any expenditure about the upkeep and operations of the vessels. Whatever money they would spend on these from now on would become fruitless for them, if the receiver is going to take over the vessels and operate them in his own way. They, however, have the lingering hope that the Receiver would allow the owners to operate the trawlers under his hawk's eye and ensure repayments to the extent possible. 'They are also probably conscious that the Receiver might give the vessels on lease to competent agencies or persons. Sale may not be possible, as the parties concerned would have to be given loans by SCICI to pay the sale prices which will go back to SCICI, only to restart the repayments story afresh, albiet from a new party. It is now of utmost expediency that SCICI or the Receiver clearly tells the affected fishing companies about their commitments and responsibilities consequential to the appointment of the Receiver. Let it be presumed that the Receiver sells the trawlers and makes arrangements with the new buyers for the payment of a certain amount which would fall short of the amount due to the SCICI by the owners. In such an event and even after the additional guarantee for 20 per cent of the loan amount is also invoked, if some amount is due from the owner, it will be fair to apprise the owners about the kind of approach to which they will be subjected to by SCICI. It is likely that the owners will feel, once the vessels hypothecated to the erstwhile SDFC are taken over and guarantee for 20 per cent is invoked, they would not have to pay anything further. However, if through sale of vessels, SCICI's Receiver realises a higher amount than is due, it will be fair to pay the excess to the owners.

Normally, the owning companies ought to have the inclination to retain the trawlers with them explaining their position and problems to SCICI. It was obviously their fault in not responding to the notices isued by the SCICI regarding payment of pending dues. Had they explained their problems SCICI would have taken a sympathetic attitude.

A lot of thinking must have gone behind the appointment of one Receiver for the vessels of several companies by SCICI. Whichever way the Receiver would act, he may have problems to take over and operate the vessels. But then in an intricate matter of this kind, problems will be there and the function of the Receiver is to face them and solve them. If the Receiver wants to sell the vessels or give them on lease to others, what happens in consequence will be very complicated although the results may ultimately turn out to be favourable. If the intention of the Receiver is to keep a close watch over the operations of the owners without taking over the vessels but ensuring repayment due to SCICI, proper monitoring and liaison with the buyers of the harvested material will become necessary. He has to keep watch over the disposal of catches of a good number of vessels which he may sooner or later take over. It is quite possible that some may feel like unloading their catches at ports other than Visakhapatnam and reach Visakhapatnam with a slender catch so as to outwit the Receiver.

Now that SCICI has shown that it means business, it appears to us that it should revert back to its original idea of having its own Company, with the present Receiver as the Chief. This company can be entrusted with all the vessels taken over by the Receiver and operate them with the help of professionals. The possible alternative of entrusting the vessels to technocrats may not really work since, in the matter of operations they are no better than the present owners. In fact, there is a likelihood of some of the technocrats unwittingly neglecting aspects relating to the upkeep of the vessels and this would be very dangerous. They are likely to bestow attention exclusively, on operational aspects, and unknowingly many other matters relating to the administration of the working of the vessels, personnel management and various other aspects such as external relations may suffer.

In conclusion, we have to say that several in the industry have not expressed any concern over the action of the SCICI in appointing the Receiver. It is nevertheless of utmost expediency for the SCICI to institute a system for the administration of the operation of these vessels for recovering their own dues and let it be known to the industry. We feel the most feasible alternative is for the SCICI to set up a fishing company of its own and entrust all the vessels taken over to this company for operation, under the Managing Directorship of the Receiver or a professional person or body in the live.

Fishing Chimes, Editorial 136: August 1990: Vol. 10, No. 5

# Near Shore Fishery Development Agencies

Management of near shore marine living resources is one of the most neglected aspect in our marine fisheries management system. The activity is in the hands of traditional fishermen, mostly engaged in various methods of artisanal fishing, from isolated coastal fishing villages. The traditional sector contributes significantly to the nation's fish production. Even so, practically no serious efforts are made to organise in an effective manner the nearshore fishing activities and to upgrade capabilities of the fishermen who live close to the coast.

The fishermen face several handicaps such as lack of proper housing, drinking water supply, medical facilities, communications, educational facilities etc., apart from the absence of various infrastructural facilities and an organisational mechanism to channelise their energies towards beneficial fishing for their benefit. Further, there is no effort to replace their boats and nets which outlived their life and no longer repairable.

The present state of affairs is often explained away by saying that the nearshore areas are already overexploited and therefore there is no scope for increasing the fishing effort in the zone. The fact that the near shore waters are exploited for harvesting only certain marine species with specific types of nets and that too only in certain zones of the near shore area is overlooked and no efforts are made to increase the fishing effort by introducing integrated facilities needed for adequate harvesting of fishery wealth in the zone and undertaking postharvest jobs.

In order to bring about a major reform in this sector along our coastline, the establishment of an organisational set-up is of crucial urgency. This should provide integrated covering infrastructure facilities at shore for repairs and maintenance of vessels, pooling up of catches, transportation to central cold/frozen storages and from there to markets, and banking facilities to meet working capital needs. Facilities for auctioning of catches in a regulated manner, and immediate payment of the highest bid amount to the fishermen then and there, should be introduced.

The organisational setup should be such that it covers a viable group of villages located as close to each other in a stretch extending 20 or 25 km. Such stretches should be first identified and these could be covered in a phased manner by the activities under an organisation in the form of 'Near Shore Marine Fishery Development Agency (NSMFDA), which should provide the needed mechanism for implementing the developmental programme. These NSMFDAs should be headed by a professional Fishery Promotional Officer with the needed staff support. He should be allowed a time of one year (pre-project phase) to assess the fishery potential of the area with reference to the present extent of harvested fishery, fishing effort etc. During this period he will also assess the species that are available in significant quantities but are not being exploited and economically utilised. For example, information could be collected on the scope for the exploitation of underfished species such as squids and cuttle fish and other molluscan shell fish, *gracilaria* etc. Scope for pen culture, culture of *gracilaria* etc, can be explored. Apart from the present activities and the catches, he should be able to work out the various infrastructural facilities required for the exploitation of these species and their economic utilisation, with reference to the types and numbers of fishing boats and nets needed, the communication facilities required, and educational, health, training, water, housing and other facilities needed.

Another important function of the Promotional Officer will be to assess the number of working fisherpopulation within the jurisdiction of the Agency limits, of men and women who are already gainfully employed and those who also need gainful employment. For the unemployed section of the population, employment schedules should be drawn up and these would have to be implemented by the Promotional Officer. These schedules can be related to long liners for tuna, tuna-like fishes and sharks, gill netting for coastal tuna, sharks and finfishes, and fishing methods for columnar species and setting up of pens in protected areas, if available, introduction of squid jigging, culture of *gracilaria*, molluscan shell culture etc.

It shall be the responsibility of the Fishery Promotional Officer to ensure the setting up of integrated organisational facilities for the smooth flow of work under the project. Each Agency may have a Management Committee headed by the Revenue Divisional Officer having jurisdiction over the area. The other members of the Agency may be drawn from the various concerned departments, such as the fisheries (Vice-Chairman), roads and buildings, health, education etc. Wherever necessary the Inspector of Police of the area can also be co-opted on the Committee. The Fishery Promotional Officer can be the Member-Secretary and the Convenor.

As part of the integrated scheme and to fortify the organisational mechanism, it is necessary to provide certain common facilities inclusive of cold/frozen storages, fish curing yards, processing plants, fish meal plants, apart from communication facilities and marketing set-up. These should be set up at appropriate points within the reach of members of each Agency concerned and these should be manned separately, but with the needed integration and administrative arrangements.

So far as the marketing aspect is concerned, the Agencies concerned will have to ensure the delivery of the produce to the marketing set-up created within an Agency and link the same to the Integrated Marketing System which would cover the activities of a viable number of agencies, as mentioned above. The system should be such that the fishermen receive payments for their produce sold then and there and there should be no purchases on credit basis.

A field training programme for the upgradation of the skills of fishermen and shore based skills of fisher women covering aspects of fish preservation storage and marketing as part may be promoted with an appropriate coverage that spreads to all coastal villages. Within the broad outlines mentioned above, various detailed innovations can be made such as arrangements for the collection of fish out at sea from traditional boats by large mechanised fishing vessels, under a suitable system of pooling them up at a central point, be at a fish meal plant or a frozen storage. It may also be possible to leave the marketing aspect entirely to fishermen to sell their catch to whomsoever they want, but within the system. Once the needed facilities are provided at shore, the marketing system in a suitable form would develop and this system would give freedom to the sellers and buyers in the dealings as per market conditions, with provision for leaving a Commission to the Agency.

In regard to the setting up of cages there should be a national policy. Such of the areas found to be suitable for the setting up of cages and those to be identified later should be covered by a legislation to be introduced by the various coastal States (based on a model legislation to be circulated by the centre) for giving such areas on lease over a certain period to Fishermen's Co-operatives, Government corporations, fishing companies, individuals etc. The introduction of a legislation for this purpose can be preceded by an ordinance for the commencement of the activity without much delay. It will be desirable to involve a foreign agency with expertise to set-up cages in suitable areas, linked with training to traditional fishermen drawn from selected agencies. To start with, a certain number of cage units can be set up with imported technology as needed. Once the system takes roots, all future efforts can be through indigenous inputs.

Fishing Chimes, Editorial 137: September 1990: Vol. 10, No. 6

# Faulty World Bank's Carp Hatchery Design: Indian Experts Successfully Face the Challenge

Major carp hatchery farms were set up in the past few years with World Bank's financial and technical expertise, in the States of West Bengal, Bihar, Uttar Pradesh, Madhya Pradesh and Orissa. The layout of the ponds for raising the hatchlings into fry stage and the feeder channel, as part of the hatchery farms, is very impressive although some may feel this as somewhat elaborate and probably overcapitalised. In regard to the hatchery part of the farms, the operations had unfortunately shown beyond doubt that the design was faulty. So much so, the hatchery complexes set up with World Bank's aid were left in a forlorn state, although at a few places, the spawning ponds are being used, with some modifications, as hatching ponds.

There is a fixation in the minds of the Government, financing bodies and entrepreneurs themselves that Western Experts have superior expertise to impart to the Indian technologists, particularly in respect of fish seed production. Mesmerised by this mental attitude, the Indian experts seem to have succumbed to the dictations of the foreign experts deputed by the World Bank, in respect of the hatchery design.

The experts must have given the design in good faith. This is however no consolation, as most of the financial help of the World Bank, channelised through NABARD to the various State Fish Seed Corporations turned out to be a wasteful investment, with the obligation on the part of the corporations remaining in tact. The Indian technologists who fortunately have adequate expertise in respect of designing and setting up of seed farms and hatcheries had to face this grim but challenging situation. They have done a memorable job by way of setting up of additional workable hatchery units, of modified Chinese design. This was a way out and the result is that the hatchery farms set up with World Bank's aid, with these additions are now slowly improving their performance. The reformatory job accomplished is a tribute to Indian expertise.

All concerned had a part in the finalisation of the faulty World Bank's hatchery design. The fault mostly revolves round the unsuitable hatching baskets and ill-conceived disposition of water inlets into spawning pools as well as into the hatching pools.

Each hatchery complex, conforming to World Bank's design, is understood to have cost around Rs.36 lakhs, inclusive of the farm which consists of 18 numbers of ponds. The total water area is 10 ha. The ponds are dispositioned on either side of a median feeder canal. Each pond has an area of 0.55 ha. Out of these ponds a few are used for stocking brood carps. The remaining ponds are used for raising seed. Apart from the ponds, the World Bank's layout includes two nos. of spawning pools of 3 m dia/1.8 m depth. There are also 5 numbers of hatching pools, each with a capacity to hold 10 hatching baskets. Each basket can take a little over one lakh of eggs.

The spawning pools were found to be suitable to the extent of achieving spawning only. In the subsequent operations, however, a serious problem in regard to collection of eggs released was encountered. The reason was that the bed of the egg collection chamber was at a level 60 per cent lower than the bed of the spawning pool and this gradient was found to be detrimental to the lives of the eggs. The limited diameter of the spawning pools also became a constraint to undertake large scale breeding operations. Further, the disposition of the inlet pipes within the spawning pools was such that the faulty generation of current of differing velocities from the various inlet pipes did not merge to form a harmonious current of the needed velocity conducive to carp breeding.

In order to get over these deficiencies and also similar dificiency in the hatching pools, as typified at the additional constructions at Danapur farm near Patna, two numbers of spawning pools of 10m dia with a depth of 1.8m were constructed. Three numbers of inlet pipes were provided at a height of 15 cm from the bottom and two more inlets were provided at a slightly higher level to meet the requirements of the circulation of eggs after spawning. Overhead showering system had been incorporated. 225 numbers each of brood female and male fishes could be released into each of the spawning pool at a time. The fertilised eggs debouche at a gentle gradient into a collection chamber from where they are transferred into hatching pools. It is stated that three crores of fertilised eggs could be produced because of the setting up of the new spawning pools at Danapur farm. These pools have been improvised at Danapur by modifying the spawning pools built under the World Bank Scheme suitably. Each of these pools can hold 75 lakh numbers of fertilised eggs. The ten hatching chambers built under the World Bank's Scheme (L15xb5xd6 feet), however, are in a state of total disuse.

The disuse of the hatching chambers is attributable to the faulty design of the immersion type of funnel basket for incubation, made of 0.2 cm mesh mosquito net clothing, which gives rise to lateral dissipation of the water that flows into the basket from the lower extremity. The cone of the basket is made of canvas cloth, while the upper cylindrical portion (75 cm high and 50 cm dia) has bands of bolting cloth with a middle screen of 0.2 cm mesh mosquito netting cloth. Water flows (through a simple stopcock on the cistern wall) into the basket through a shower with convex face and no doubt there is uniform jetting out of water without any chance of low oxygen pocket. The upward flow of water towards the top of the basket from the bottom of the basket however is such that the fertilised eggs tend to get attached to the sides of the basket and get injured. Because of this, the retrieval of larvae under this system is stated to be rarely more than 30 per cent. However, as in the case of the spawning pool of WB's design the 60 per cent difference in the level of the bed of hatching pool and the spawn collection chamber have resulted in physical damage to the fertilised eggs.

As another example, it may be mentioned that, at the World Bank's hatchery farm at Sitamarhi (Bihar) the spawning pools as well as hatching pools have been in disuse. In their place a new set of pools for both spawning as well as hatching were designed, based on scientific principles. It is stated that the results have been extremely satisfactory. At this farm the diameter of the spawning pools (2 nos.) has been reduced to 6 m taking into account the focal climatic conditions. The results have been found to be good. It is stated that realisation from spawning and hatching pools at Danapur and Sitamarrhi farms were extremely good, because of the position of the inlets, 6 nos. of one inch dia capped with duckback openings. The inlet pipes have been given curvatures based on engineering calculations and the arrangement is such that the flow of water is clockwise and harmonious, and this design is based on Ferrel's Law, it is stated. There are o to 8 of such inlets dispositioned at an angle of 40° and a height of 15 inches from the bottom.

According to Ferrel's Law it is stated that lower pressure is created to the left of the direction of the current. So much so, as the eggs move to the left of the current, they do not stick to the circular screen at the centre of the hatching pool. In contrast to this, in the World Bank's design, it is stated, the flow of water through the hatching baskets is such that the eggs tend to stick to the corners and create congestion, preventing water exchange, thereby injuring the eggs, as already stated.

In the World Bank's system the depth of water in the hatching pools is maintained at 1.3 m. Under the new system introduced at Danapur and Sitamarrhi hatcheries in Bihar, the depth maintained is 90 cm. The reduction in depth, it is stated, facilitates vertical movements of the hatchlings after 48 hours, which is not the case in the World Bank's design.

In all the other four States, where World Bank's hatchery farms have been set up, modified Chinese type of hatcheries have been added. There is thus now hope that the Corporations concerned would be able to earn adequate money through sale of seed produced at the farms and pay back loans extended by NABARD through the commercial banks concerned, without burdening the State Governments unduly, who would have provided guarantees to NABARD/Commercial Banks concerned.

The sum up, the moral to be learnt from the episode is that the technology brought in by the foreign experts, particularly those from the West is not necessarily always reliable. With proper encouragement and trust and financial backing, the Indian experts would be able to accomplish a far better job, particularly in the field of inland fish culture. This is also true in the case of the brackishwater farming sector.

Fishing Chimes, Editorial 138: October(i) 1990: Vol. 10, No. 7

# Supply of Diesel Oil for Deep Sea Fishing Vessels: Invidious Attitude

Until recently deep sea fishing vessels fitted with engines of 150 HP and above were eligible for exemption from the payment of excise duty on the diesel oil consumed by them to the extent of 50 per cent of it straight away. The remaining half of the excise duty used to be released based on export performance. Last year, all on a sudden, the Union Government withdrew this concession, thereby increasing the financial burden on the operators of deep sea fishing vessels. In fact, the economics of operation which were already adverse had been made far worse.

Having withdrawn the concession very much needed by the deep sea fishing companies, Government have recently announced a new scheme for providing total exemption from the payment of excise duty on diesel oil consumed by mechanised fishing vessels upto an overall length of 20 m. The main stipulation to become eligible to this concession is that proof should be shown that at least 25 per cent of the harvested catches are exported. Administrative arrangements for Implementing the scheme are stated to be under formulation.

This concession will help mechanised boat operators in improving the economics of their operations. What is intriguing, however, is the rationale behind granting this concession to smaller boats and denying the same to the larger vessels whose operational economics are under far greater stress than the smaller boats. This invidious attitude of the Government would give the impression to those engaged in deep sea fishing Industry that the Government Is not in a mood to come to their rescue or provide a prop to the ailing deep sea fishing Industry, in spite of the fact that development of this sector is the need of the hour. Huge stocks of fish varieties, although of low value at present, apart from high value varieties such as tuna, lobster etc., are awaiting exploitation by larger vessels, and it Is unfortunate that the deep sea fishing industry is placed in a state of abject discouragement and financial crisis through denial of the concession of giving exemption on excise duty as in the case of smaller vessels. Unless this facility is extended to the sector along with several other much needed incentives, our Exclusive Economic Zone would either remain unexplolted or would be harvested by foreign fishing vessels who can have an easy access to the resources for the reason that they lie in the outer Zone of our EEZ, not amenable for easy detection.

It is of utmost urgency for the Association of Indian fishery Industries to take up this matter in a forceful manner with the Union Government and ensure that the benefits given to smaller vessels are also extended to larger fishing vessels.

Fishing Chimes, Editorial 139: October(ii) 1990: Vol. 10, No. 7

# Visakhapatnam Fishing Harbour Back to Normalcy

Undeterred by the aggressive and provocative stance taken by the CITU affiliated Unions of floating staff (Vizag Trawler Workers Union and All India Deep Sea Fishing Technocrats Association) in respect of their salaries, incentives etc., the representatives of 18 deep sea fishing companies, who happen to be members of the Association of Indian Fishery Industries, took a firm stand not to be bullied by the Unions. They stood firm and refused to deviate from their resolve to run their trawlers on the principle of sharing the net profits between the owners and floating staff based on a specified ratio. They also did not yield ground on the demand of the floating staff for the payment of the salaries for about six to seven months during which period their services stood retrenched and they did not do any work for the companies. On the other hand they are stated to have acted in unison, against the interests of all the employers concerned and also the Industry as a whole. The floating staff also went to the extent of creating unprecedented violence and disturbance within the harbour premises. Even police officers became victims of the violence.

There have been prolonged negotiations between the CITU-led Unions of the floating staff and the members of the Association of Indian Fishery Industries. There were also efforts on the part of the district administration to resolve the dispute but the net result was that there was only a partial but, in a way, a major success achieved by the fishing companies. The CITU, on behalf of Its affiliated Unions of floating staff, had accepted the principle of sharing of the net profits as mutually agreed to. They, however, had demanded that salaries of the floating staff during the period of strike, which was about six months, should be paid, although their services stood terminated. This was not acceptable to the other side.

Getting weary of the prolonged negotiations and the consequential effects on their means of living, several members of the floating staff appear to have ignored CITU and took up jobs on one fishing trawler or the other. Some took up jobs in Bombay and in Nigerian fishing companies. The actual terms and conditions under which they had rejoined the Vizag-based trawlers are not clear but obviously they differed from case to case. It is possible that in some cases the companies had refused to make payments towards past salaries. A few others seem to have agreed to make partial payments, in various forms. All those who have joined are stated to have been treated as new appointees.

There had been strikes by the floating staff before at Visakhapatnam Fishing Harbour, and they had always won. It is for the first time that they had agreed to a basic change In the operational terms, thanks to the "Active eighteen", who leaving their vessels in the lurch and losing prime fishing time and prepared to lose this further whatever be the consequences, stuck to their stand which enabled them to secure acceptance of a basic change. This envisages payments to the crew by the owners linked to a minimum number of guaranteed fishing days (performance), acceptance of a share of net profits and a subsistence level salary plus sea allowance for the number of sea days spent and also Mess Allowance for the number of days they are actually out at sea and at shelter ports during bad weather conditions etc. In addition, they would also receive an advance payment towards the proposed share of annual net profits In the form of Incentives calculated at about 13 per cent over and above the breakeven levels, which are fixed separately for each category of vessels. Under the new system, the floating staff will in any case get nothing less than what they were receiving before. Over and above this, they have distinct chance of earning higher amounts as their share of net profits because of the likely Increase In number of fishing days put in as mutually agreed to. The owners also will be able to earn higher amounts.

Some of the members of the Unions of the floating staff who have been responsible for triggering off violence at the harbour are understood to have attracted stern measures from the police to keep them under control. Some of their violent actlvities, it is learnt, left no alternative to the police than to think In terms of prosecuting some of them who were trouble shooters. It is to be hoped that floating staff will not resort to violent methods any longer and stick to the mutually agreed formula of sharing net profits, optimising fishing days and doing their best to promote good relations with the managements for common good. It should be the policy of the managements to do their best to keep their floating staff happy in all respects, by sticking to the various terms and conditions, mutually agreed to.

One aspect which we would like to mention is that It was expected that, since the floating staff were no longer dogmatic about their balance of demands, the managements concerned would send their trawlers out of the harbour to undertake fishing operations. This was not however the case. Out of a fleet of over 135 vessels, It Is stated only around 55 nos. of Vizag based ones are now conducting fishing. Apparently the companies concerned are not either serious about sending their vessels out or they have cash flow problems in this regard or they have problems of employing suitable floating staff. The continued detention of vessels at the harbour means that crores of rupees of investment is just lying Idle, without yielding returns. Some have however already instituted the needed steps to make their vessels operational and there is a likelihood of total restoration of normalcy in their activities very soon.

It is the floating staff who have to work hard out at sea to secure good harvests. We congratulate such of the officers and men who had decided to get back to work instead of wasting their time any further at shore, as this will be counterproductive. Their pragmatism is to be appreciated. Likewise, the owners who braved to send their vessels out for fishing, particularly during the days of turmoil deserve appreciation for their pragmatism (without losing self respect) and realisation that vessels are meant to be out at sea for fishing and not for being moored at port indefinetely and sustaining losses indefinetely.

Fishing Chimes, Editorial 140: October(iii) 1990: Vol. 10, No. 7

# Steep Increase in Diesel Price: The Heaviest Blow to Marine Fishing Industry

The Deep Sea Fishing Industry which has been reeling under several adverse conditions such as unrealistic demands by floating staff for increasing their earnings and increase in running costs including the hike in diesel oil prices in March, 1990, has received yet another killer blow in the form of the second increase in diesel oil price by 25 per cent. This increase has pushed up the cost of oil per kl from about Rs.3,900 (March '90 prices) to Rs.5,432, with effect from 14 October midnight. This increase means an additional expenditure of about Rs.1,500 per kg. per vessel. For 25-27 m vessels the additional expense per voyage will be Rs.75,000, for 23 m vessels about Rs.60,000, for mini trawlers and sona boats about Rs.11,000 and mechanised boats about Rs.700.

Sea fishing, particularly fishing by larger trawlers has now become an ill-fated activity. Although the fishing vessels are responsible for a significant part of the national foreign exchange earnings through marine products exports, the industry has become virtually singled out by the Government for imparting blows one after another.

It is certain that the sea fishing industry will collapse eventually under the impact of the steep rise in operational costs which are not restricted to the increase in cost of diesel oil alone, but also because of the increase in costs of various supporting and ancillary activities depending on fuel energy. Overheads will also Increase because of higher transportation costs etc. One need not be surprised if the operations of sea fishing vessels come to a standstill soon, posing problems of stoppage in earnings by the trawler owners and inability to repay to the financiers. It is unfortunate that the price hike has come at a time when deep sea fishing is being given a special thrust in order to boost the country's sea food exports. The encouraging trend in marine products exports during the current year may get reversed because of the rise in prices.

It is learnt that several large, medium and small fishing vessel operators have already made representations in the matter to Mr.C.T. Sukumaran, Chairman, MPEDA. Trawler owners operating from Visakhapatnam also sent S.O.S. representations to MPEDA, Ministry of Commerce and other ministries concerned. Acting on the representations, MPEDA is understood to have sent a representation to the Government urging it to exempt the fisheries sector from the price hike by providing subsidies.

It is now a matter of utmost urgency for the Government to provide a package of reliefs to the fishing vessel owners, consisting of waiver of all existing dues to the Government, SCICI, reducing future interest rates, abolltion of penal interest system and providing cash compensatory support, In some form or the other to enable the operators to balance their budgets. A Committee under the Ministry of Commerce/Department of Economic Affairs (Banking Division) would have to be set up immediately to determine the needed reliefs for the approval of the Government. If this is not done, the Government should be prepared to expect a reduction in foreign exchange earnings from marine products exports. Even shrimp culture fishery sector will be effected for the reason that the shrimp farms are pump fed and therefore the hike in cost of diesel oil will upset the economics of shrimp culture farms too.

Because of good sea shrimp season in 1990 several were expecting to come out of their financial problems to some extent. At such a hope-laden moment, the increase in diesel oil price has neutralised the expectations. SCICI has also been expecting that there would be some repayments from fishing companies in 1990-91. Now It is doubtful whether these hopes will be realised. The distress of SCICI, so far as fishing vessels are concerned, will now become steeper. Its position will be far worse than the operators for the reason that they will have a serious problem in dealing with the trawler owners who are now in deep distress.

Apart from the reliefs suggested, the Government can also take up programmes of introducing methods of fishing which will be fuel efficient and do not consume as much fuel as in the case *e.g.* trawling.

The Association of Indian Fishery Industries and MPEDA having already taken up the question of reliefs with the Government, we are confident that both these organisations, particularly the AIFI, would induce the Government to put life back in the sea fishing industry to allow the industry to grow and thrive, thereby stepping up marine products exports and consequential foreign exchange earnings.

Fishing Chimes, Editorial 141: November 1990: Vol. 10, No. 8

# Evolving a National Fisheries Policy: Fishermen's Efforts

The National Fishermen's Forum, New Delhi has accomplished an outstanding exercise in evolving an approach to the National Fisheries policy for the fuller and sustainable development of fisheries during the 8th five year plan. A dedicated set of scientists and those deeply involved in the welfare of traditional fishermen took part in a seminar on the subject held in New Delhi on October 12-13, 1990.

One important aspect that has been brought out while delineating the approach is the need to temper 'Production-Oriented schemes, with emphasis on development of the traditional fishermen who depend on fishing and the sustainbility of the fishery resources for their livelihood. It is to the credit of organisers of the seminar, particularly the Task Force headed by the Veteran, Prof. P.C. George, a seasoned professional in the field of fisheries, apart from being a champion of the cause of traditional as well as artisanal fishermen, to have brought out sharply that production- oriented development of fisheries so far pursued had led to contradictions causing increased wideremployment and unemployment. It has also been aptly brought out that uncontrolled fishing activities in marine as well as inland waters often resulted in overinvestments in the sector. This has led to over-exploitation of the fishery resources. Another important aspect which has been sharply brought out at the seminar is that the rivers, lakes and the marine waters are being treated as waste disposal dumps of the nation, with utter disregard towards the ecological balance and the future survival of the human population of the nation.

We feel that the Planning Commission should keep in view all the above aspects in finalising fisheries development plans relating to the Eighth Five Year Plan. It has to be realised that, without human development and lack of emphasis on this aspect no fisheries scheme can acquire production orientation.

The seminar has expressed its serious disappointment that not much has been achieved in respect of the socio-economic uplift of artisanal as well as traditional fishermen, eventhough it was an important objective of the Seventh Plan itself. The plan schemes have not been able to help to any significant extent in bringing an upgradation in the living conditions of traditional fishermen. The absence of any congnisible transformation in the quality of clear objectives in this respect had led to major conflicts among fishermen themselves.

The fishermen operating traditional boats were placed at a disadvantage in the socio-economic sphere compared to those who operate mechanised fishing boats. In other words, a community that was once one, became fragmented, one group ranged against the other. This has led to serious situations:

If the State Governments had adopted sound near-shore management policies in regard upgradation of skills of fishermen and their socio-economic interests, things would have been different. What the State Governments should have really done is to bring all the fishermen under one major scheme, for the upgradation of their skills, by helping them with financial and technical inputs and enabling them to pursue their avocation with greater effectiveness. Instead of doing this, the misplaced policies of the State governemnts enabled a part of the community to corner the benefits provided by the Government and the financial assistance extended by the commercial banks, to the disadvantage of others.

The Central Government provided heavy financial help to a few entreprenuers for the development of deep sea fishing without the simultaneous development of infrastructural facilities at various ports. This approach proved to be a damp squib as this led only to a negligible contribution of less than one per cent of the total value of annual marine products exports of the country.

If a different and a well-conceived approach had been adopted, it would have led to the development of the deep sea fishing as well as coastal fishing, in a balanced way. Our Government has never taken into consideration the developmental policies adopted by neighbouring countries such as Srilanka, which have helped small fishermen to undertake offshore fishing, even to a distance of about 200miles from the coast, with small mechanised boats of about 38 feet length. These boats mostly conduct fishing for tuna which is a foreign exchange earner. The vessels stay out for a fairly long duration. Had such a project been adopted, there would have been a major spurt in the activity in the small scale sector benefitting both the small fishermen and also simultaneously increasing fish production particularly tuna, and helped in the earning of additional foreign exchange.

So far as development of Inland fisheries are concerned, one major reason for the failure in the functioning of a majority of Fish Farmers Development Agencies in achieving fish production levels as expected has been pointed out at the seminar. The wrong choice of fish farmers was stated to be responsible for the shortfalls in fish production through FFDAs. The seminar advocated the introduction of a price support scheme and producer-controlled marketing facilities. These steps, if they were taken, would have provided the needed incentive to the farmers to increase production. The main reasons for the short fall in production in the FFDA sector seems to be a) lack of reconciliation between the interests of panchayats and fish farmers which can only be brought about by the State Governments concerned through policy adjustments, b) conflicts between the fisheries department and the FFDAs concerned, c) absence of needed technical support and d) non-availability of timely financial help.

In the sector of brackishwater aquaculture the seminar deplored the attitudes of State govememnts in allocating brackishwater lands to big entrepreneurs, instead of encouraging the traditional fishermen and small-scale fish farmers in the sector. While in the schemes connected with the activity of allotment of lands for brckishwater farming, it has been proclaimed that preference would be given to the traditional fishermen and their cooperatives. In actual fact, the brackishwater farming sector is gradually moving into the hands of medium and large scale entrepreneurs. In fact, atleast in the State of Andhra Pradesh, schemes for the development of brackishwater farming for the benefit of the weaker sections of the society appear to have been implemented in such a way that the farms set up under the programme would never become operational. It should not be difficult for the State Government to draw up plans in such a way that community brackishwater farms can be developed for the benefit of the weaker sections of the society, with the needed infrastructural, organisational and financial support from the Government.

The seminar has aptly pointed out that the present State-level fishery legislations and the manner in which they are implemented by the State Governments reflect the fact that the authorities consider the subject as something connected with law and order problem. There can be no argument against the view expressed at the seminar that what is required is a broad national policy fortified by a regulatory legislation for the scientific management of all national aquatic living resources. One has also to agree that regulations need to be viewed as an integral part of sustained development of fisheries and there should be a monitoring of the activities of the entrepreneurs in the sector. It has also been rightly pointed out that there is need to register all crafts and gears and introduce a licencing system in this regard. The State Governments as well as the central Governments have no doubt done some work in this regard. Nothing tangible had however happened.

The Seminar has made various recommendations concerning the development of marine fisheries, inland fisheries, processing of marine products and marketing of fishes. Recommendations concerning welfare measures and reorientation of fisheries administration have also been made. We have reproduced these recommendations separately in this issue itself. A good part of these recommendations are highly relevant and timely and one has to hope that the Central and State Governments take the recommendations seriously, get them evaluated, and adopt them as the basis for the reformulation of the plans to be implemented during the Eighth Five Year Plan.

Fishing Chimes, Editorial 142: December 1990: Vol. 10, No. 9

# Intricate Problems Riddling Trawler Loans Scenario

The erstwhile Shipping Development Fund Committee granted loans to the tune of around Rs. 80 crores for the acquisition of over 120 trawlers untill its closure in 1987. The release of the loans was governed by loan agreements entered into between the SDFC and the fishing company concerned. The enactment of the SDFC's Abolition Act brought to a close the equation between the SDFC and fishing industry. Invoking the provisions in the Act, the Government designated the Shipping Credit and Investment Company of India as the "Designated Person" for the recovery of the loans and interest thereof due to the erstwhile SDFC as per the terms and conditions contained in the Loan Agreements. The SCICI also emerged as the only body for providing financial assistance to fishing companies. In the past two years, SCICI has enlarged the list of activities for which it will be providing loans. These now include not only lending finances to fishing companies for the acquisition of fishing vessels but also for the setting up of integrated deep see fishing projects covering processing plants, additions to existing vessels, marketing units etc, for the establishment of brackishwater farms, hatcheries and so on. As the "Designated Person", a responsibility has devolved on SCICI to recover overdues from 59 out of 85 companies to whom the erstwhile SDFC granted loans. Ending March 1990, the overdues were Rs. 21.64 crores from the 59 companies. The remaining 26 cases are related to old cases and also the new companies to whom loans were sanctioned by the SDFC a year or two before its closure. Out of the 26 companies, it appears that Pron Magnate, Matsyka Exports, T.N. Fisheries Corporation, and may be a few others had been paying instalments relating to both principal and interest without leaving any dues. SCICI called upon all the defaulting companies to clear the dues. For various "Valid" reasons, the bulk of the companies could not pay. What seems to be intriguing SCICI was the ability of a few companies to pay and inability of the others to do so. Those who could not pay point out that those who had paid had pumped in funds from other sources, and not out of incomes generated by these fishing companies. This can be partly true as Pron Magnate has no other income. So is the case with Tamil Nadu Fisheries Corporation, who must have been paying out of their earnings.

When the defaulting companies were unable to pay the dues, the SCICI agreed to consider proposals from the companies for the reschedulement of the pending loans and interest, with provision for a moratorium period. After obtaining clearance from the Department of the Economic Affairs (Banking division), the financing body communicated the approved reschedulement schemes to such of the companies concerned, with the request that they may signify their willingness to the approved scheme. Almost all the companies, concerned are understood to have not given their concurrence, purportedly for the reason that SCICI included two conditions not acceptable to them. One is that personal guarantees have to be given by all the Directors of the Companies concerned to the SCICI. The companies have also to agree for setting aside 30 per cent of the gross proceeds in a "No-lien Account". The amounts accumulating in this account maintained by the concerned bankers of the companies, would be offset towards the repayment of loans.

We have to mention here that, looked at from the angle of SCICI, they have to come up to the expectations of the Government which wants the Company to collect the pending amounts due from the various loan companies. The Company has been working restlessly to find a way of collecting the pending amounts through dialogues with the companies concerned. They have also organised a Top Management Programme recently to educate the companies in respect of financial management by the top brass of the Companies concerned.

According to quarters close to SCICI, the financing body is perplexed over some of the present characterstics of the industry. They have to unduly suspect that, the equity contribution of the Companies being very low, the first priority of the share holders is to somehow get back amounts equivalent to their investments. Their expectation, it is felt, is to get this through some means or the other. In other words, the possible suspicion seems to be that, funds are diverted so as to reach the safety point, so far as the share holders part of the investaments are concerned. It is further surprised that they are not much concerned about the major amounts invested by the Government through the erstwhile SDFC to enable them to acquire the trawlers. If what is stated above is true, one has to say that, while they are not certainly justified in having this line of thinking, it is possible that because of their distress and the goading from the Banking Division of the Department of Economic Affairs, SCICI is forced to think on those lines. Not only this, they also seem to feel that, having acquired the alleged habit of diversion of funds, it continues to linger, whether a fishing season is good or bad. They seem to be almost now certain that, whatever reliefs are offered, the bulk of the owners, barring a few, would continue to avoid payments of amounts due to the erstwhile SDFC. SCICI is understood to have come to the conclusion that the companies are mentally prepared to hand over the trawlers to them, as was done by Ms. Marshall Sea Foods and Uni Marine. The financing institution seem to feel that the owners want to have the vessels already taken over by the Receiver back, only to make some more money and divert the sums for other purposes. If true, what a wrong line of thinking! According to our enquiries, the owners are very serious to operate the vessels in a very efficient manner and pay back the amounts due to the erstwhiule SDFC as quickly as possible. SCICI, undoubtedly, realises that the trawlers taken over by the Receiver are lying idle. The lack of any returns from these trawlers would mean keeping heavy investments in animation without any returns. This situation is certainly deplorable. However, SCICI must be having its own schemes to make the vessels taken over operational. They must also be aware, to make these vessels operational, lot of money has to be invested.

Matters seem to have reached a stage of SCICI taking over trawlers of 11 more companies through the Receiver appointed by the SCICI. The Association of Indian Fishery Industries seem to be certain of this. The Association is anxious that the SCICI should hand over all the vessels taken over back to the owners.

Several fishing companies who feel that they may be the targets for the take over of their vessels next, are understood to have taken steps to prevent SCICI from taking them over on certain legal grounds. While it is very difficult to say what will be the outcome, because of the interplay of several factors involved.

It is stated that SCICI is unable to understand one aspect. Why is the Association of Indian Fisheries Industries not exhorting its members to strive to pay the money, particularly when it is known that the companies have earned substantially during the 1990 season? Instead of doing this, why is the Association mounting the campaign against the SCICI in order to secure reliefs that it will certainly consider as unrealistic? Why are the companies having objections to set apart 30 per cent of the gross sales in 'no-lien' account, when there is an assurance that in times of genuine need, the companies will be allowed to withdraw quite a part of the funds available from this account? It is said that SCICI does not see any effort on the part of the companies to make repayments. The exceptions are a few companies which have been making repayments regularly. When these could make repayments, why not others? This perplexes SCICI. So far as the owners are concerned, they have a strong feeling that they are being harassed by the SCICI. After three bad seasons, in 1990, they could undertake a few voyages, which no doubt have brought attractive returns. Balances out of these returns, after meeting operational expenses, may look as if they disappeared but this is because of the clearance of several unsecured loans, which they had to incur all along in the past to keep the trawlers in a running state. It is easy for them to hand over the trawlers back to SCICI as was done by some, but they feel that this would not solve the problem. The vessels have to be operated by someone. Otherwise the investments will be wasted. It will be difficult for the SCICI to operate the vessels on their own. They may have to probably give the vessels on lease or organise outright sales. In this situation, the owners want the SCICI to have a reappraisal of the situation and formulate a suitable and acceptable reschedulement schemes, without including conditions such as 'No lien' Account, and furnishing of personnal guarantees by the Directors concerned.

It can be argued that it is the prerogative of the financier to recover the loan amounts, particularly in situations when there are heavy overdues. In a democracy, however, it is also the prerogative of the owners to secure the needy redressel from the Government for the postponement of the collection of dues for a specified period.

There are clear indications that the owners would agree for reschedulement as long as there is no insistence on the two conditions referred to above. There are also clear indications that SCICI also will not agree to the deletion of these and thereby weaken the reschedulement plan. SCICI seem to be apprehensive that the owners will revert back to their alleged old methods, as unwarrantedly suspected by it.

The problem is thus very intricate. It can only be solved by an Authority higher in status to that of the SCICI, keeping in view the larger objective of development of deep sea fishing in the country. Otherwise, if the present situation continues, the collapse of the industry would become imminent.

It appears to us that, since the Government has announced certain reliefs on the cost of diesel oil, the industry could be given an adequate moratorium without the two repelling conditions referred to above, to enable the owners to rehabilitate themselves. The conditions in respect of personal guarantees and maintainence of "No-lien" account can be readjusted suitably as mutually acceptable. The other alternative seems to be for the SCICI to convert a substantial part of the loan into equity of their own and allow the companies to operate the vessels, under the supervision of management experts appointed by it. The proposed high equity share of SCICI in the companies may neutralise the unwarranted suspicions regarding the working of the companies. In any case, the problem being complex, urgent attention should be paid at the highest level in the Government to find a solution so that there may not be any hurdles in resuming fishing in the coming season.

Fishing Chimes, Editorial 143: January 1991: Vol. 10, No. 10

# Eventful 1990

1990 has to be deemed as an eventful year for the fisheries sector. The events of the year were of various hues. In the Inland Culture Fisheries field there have been revolutonary developments along the path of progress, both in the freshwater as well as estuarine sectors. In the marine sector there have been turbulances, although limited to the larger vessels, with operational and financial tangles. The commercial fleet strength of larger vessels and mini trawlers came down for some reason or the other. Some of the larger vessels attracted the iron grip of SCICI and had temporarily become non- operational. Some others stopped operations for financial and other reasons. Near-shore operations were encouraging, except along Kerala coast, where the confrontation between traditional fishermen and mechanised boat operators continued.

## Marine Products Exports: Unprecedented Growth

Fish production in the marine sector was estimated at 2.0 million tonnes and it is believed that inland fish production must have crossed 1.6 million million mark, during the year. The marine products exports had registered a growth of 39.62 per cent in terms of value and 25.90 per cent in quantity, by the end of December 1990. The exports touched Rs.612 crores compared to Rs.438 crores during the corresponding period last year. The quantity exported was 90,522 tonnes compared to 71,875 tonnes in the previous year. This was an unprecedented growth and the major growth took place in Tamil Nadu and Andhra Pradesh. The growth in A.P. was because of increase in shrimp landings off Orissa and West Bengal Coast. The boatside prices of shrimp reached an all time high towards the close of 1990. These developments are certainly eventful.

## Balkanisation of Administrative Setup

The state of balakanisation in the fisheries administrative setup at the Centre continued during the year. Inland Fisheries, Fisheries training, setting up of harbours and infrastructure facilities and control over integrated Fisheries Project remained with the Ministry of Agriculture. The subjects of according permits for chartering of foreign fishing vessels, permitting acquisition of fishing vessels, and exploratory fishery surveys (including inland fishery statistics) remained with the Ministry of Food processing. The Ministry of Commerce/MPEDA continued to concentrate on marine products exports, apart from playing a major role in the development of export-oriented deep sea fishing activity and also marine products exports. MPEDA moved into its own building complex in Cochin during the year.

## Tamil Nadu and Andhra Pradesh

A major cyclone hit Andhra coast rendering thousands of fishermen homeless. So also was the case along Tamil Nadu coast. Fishermen lost their boats and nets. The State Governments provided relief to them. In A.P., relief was provided out of funds received from the World Bank, under the cyclone relief project.

## FFDAs Proved their Worth

Fish Farmers Development Agencies, around or over 313 nos. in various States became established in their own way. Although they could not succeed in enabling the farmers in their fold to perform better than those in non-FFDA sector, yet they succeeded in introducing an integrated mechanism for the development of freshwater fish culture, achieving production of over one to one and half tonnes of fish per ha or even more. It is stated that 0.17 million ha of tanks and ponds have been brought under farming under the auspices of FFDAs. An achievement indeed.

## CIFA's Contribution

CIFA succeeded in developing technology for the production of pearls from freshwater mussels. Its scientists made major contributions to the technology of Indian and exotic carp seed production, integrated and composite farming on an intensive scale.

Brackishwater Fish Farmers Development Agencies have come up in almost all coastal States. Their activities have been gaining momentum. Now there is no looking back and the multiplication of FFDAs and coming up of BFFDAs are events to reckon with.

## Chinese Type of Hatcheries Blooming

The most spectacular development in the freshwater culture sector is the sprouting up of Chinese type of hatcheries in almost all the States, in the private sector. More important than this was the acquisition of expertise by the farmers in the operation of the hatcheries. Most of the States became almost self-sufficient in freshwater fish seed production and supply in 1990.

Indian technologists succeeded in making innovative changes to the hatchery design of World Bank model in states such as M.P. and Bihar.

## *Beel* Fishery Development

Bihar and Assam were poised to take up *beel* fishery development schemes with international assistance.

## Reservoir and Riverine Fisheries

Revolutionary developments took place in the riverine and reservoir fisheries development sector. CICFRI was wedded to the task of developing these fisheries and making the waters pollution-free. Collection of carp spawn from rivers was no longer an activity that received encouragement. Production of carp seed through hypophysation for stocking not only tanks and ponds but reservoirs and rivers as well, was popularised as a general practice.

## Cage Culture

Cage culture in reservoirs had gained momentum in the States of Tamil Nadu, Karnataka and Madhya Pradesh, among others. Foundation had been laid in these States to develop cage farming of major and exotic carps in a big way.

## Blue Revolution in Punjab

A major development is the blue revolution taking place in Punjab in a big way, amidst all the turmoil in the State. Several entrepreneurs took up fish farming in a big way.

## OVAPRIM

The Indian Branch of the Asian Fisheries Society had put the progressive Indian fish cultural activities on the Asian map. This society took the initiative of popularising "Ovaprim", a drug that would induce breeding in carps with far greater ease than through other media.

## Big-Head Carp

Big head carp had made its entry into the list of exotic carps introduced into our inland waters. Its potential to control water hyacinth continued to be experimented upon in Madhya Pradesh. Tamil Nadu Government was stated to have banned its introduction in that State.

## Commercial Linkages in Seed Production

In certain parts of States such as Andhra Pradesh production of carp spawn, rearing spawn into fry, and fry into fingerlings, became separate commercial but interlinked activities.

## Freshwater Prawn Culture

Culture of *Macrobrachium rosenbergii* and *M.malcomsonii* and production of their seed became popular in States such as Tamil Nadu.

## Developments in North-Eastern States

Of great eventful importance was the development of fish culture in a significant manner in north-eastern States, particularly Arunachal Pradesh, Manipur and Nagaland. The work done was pioneering in nature. This apart, the development of trout culture in Arunachal, Himachal and in Jammu and Kashmir were milestones indeed.

## Mahseer in Maharashtra

In Maharashtra, FFDAs continued to perform with considerable success in the co-operative sector. The development of Mahseer seed production and culture of Mahseer by Dr. Kulkarni at Lonavla, under the banner of Tata group was indeed a major advance. Several States received Mahseer seed supplies from Lonavla.

## CICFRI's Contribution

CICFRI continued to play a major part in the development of riverine fisheries, striving to make the rivers pollution-free and for the restoration of stocks of dwindling rivering fishery wealth, once our heritage to be boasted about, but unfortunately now subjected to the ravages of pollution. CICFRI has done commendable work on fish stock assessment in inland waters. It had also developed methods of fish conservation. Work on genetic upgradation of food fishes, continued with encouraging results at the National Institute of Fish Genetics Resources Allahabad.

## Revolution in Brackishwater Farming

A major revolution had been ushered in by MPEDA/CIBA in the field of brackishwater farming. The relentless work done by Mr.T.K.A. Nair, Ex-Chairman of MPEDA, Mr. C.T. Sukumaran, its present Chairman and Dr. M. Sakthivel in collaboration with Dr. Algarswami of CIBA brought about a major and a sweeping upsurge of activity in brackishwater farming all along our coast line. Over 60,000 ha of brackishwater area had been brought under farming by the end of 1990. Private farmers in States such as West Bengal, Orissa, Andhra Pradesh, Tamil Nadu and several other States became the proud possessors of brackishwater farms. Had the State Governments allotted saline land to the farmers who applied for them, the country would have been able to produce several thousands of tonnes of extra farmed shrimp and added a new dimension to our exports of marine products and in the process provided additional employment to several, as in the freshwater sector.

## Farmed Shrimp Production Level

Farmed shrimp production per hectare had ranged in general in the year between 3,000 to 4,000 kg per crop. MPEDA had so far identified another 10,500 ha for being brought under farming and over 3,354 farmers were trained in the activity by the end 1990. There was a breakthrough in tiger shrimp farming in Karnataka. An eventful achievement indeed.

## Devoted Work by MPEDA

One must mention here the special fillip given to brackishwater farming during the year by the Ministry of Commerce and MPEDA. Joint ventures in the activity including feed production were encouraged. Farmers were allowed to import various farming requisites under OGL at a concessional customs tariff. Introduction of aerating devices and feed manufacture formulations had been encouraged. A number of training courses and seminars and workshops were conducted by MPEDA on the subject in association with other interested organisations. Plans had been drawn up to double the shrimp seed production capacities of the hatcheries at Visakhapatnam and Gopalpur. Encouragement was given to entrepreneurs to set up prawn seed banks and prawn hatcheries through offer of subsidies. So was the case in respect of construction of brackishwater farms. It could be said that the services rendered by MPEDA for the promotion of brackishwater farming would remain in golden letters in the history of brackishwater culture fishery development in India.

## Incentives to Processors

The marine products processors and MPEDA had done equally well in the matter of developing infrastructure for the promotion of exports. The Ministry of Commerce/MPEDA introduced several incentives, and concessions for modernisation of processing plants, installation of IQF plants and other value adding machineries.

## BOBP's Beachlanding Craft

BOBP's beachlanding craft has become increasingly popular in A.P. and other Coastal States on the east. BOBP was engaged in finding ways and means of value-addition to 'trash' fish landed at Visakhapatnam.

In respect of development of deep sea fishing, the Ministry of Commerce/MPEDA did a commendable job. They were instrumental in the setting up of an Empowered Committee, which would act as a single window for giving clearance to all types of applications relating to deep sea fishery development. Entrepreneurs would be allowed to apply for test fishing, taking vessels on lease and pruchasing of vessels, even 15 years old, entering into joint ventures, importing of fish for value- addition and export etc. Above all, financial help to an extent of Rs. 16,007 per kl of oil consumed by deep sea fishing vessels engaged in the production and export of fish was announced. Similar cash reimbursement scheme equivalent to the excise duty payable on diesel oil was announced in the case of vessels less than 20 m in OAL. A promise was made that clearance for foreign crew to be engaged in vessels to operate in Indian waters would be given within three days of application. MPEDA also offered subsidies for modifications to existing trawlers to catch varieties other than shrimp, suitable either for value-addition or export as such. There were several other concessions, given in 1990. The year was truely eventful for the fisheries sector.

Fishing Chimes, Editorial 144: February 1991: Vol. 10, No. 11

# Mysterious Quiescence

Reports of relentless, unabated, and unchallenged fishing by over one hundred Thai trawlers well within our territorial waters and close to our coast in the upper Bay of Bengal, north of Paradeep and adjacent to Sandheads indicate the low rating the Thai Fishing Industry gives to our alertness in safeguarding fishery wealth of our territorial waters, not to speak of our EEZ. Stray fishing of this kind in our exclusive economic zone by one or two foreign fishing vessels is something that can be understood. The brazen violation of the law of land by a teeming trawler fleet of another nation should make any Government to take notice of the violation and sternly deal with the situation. According to reports from skippers of Indian trawlers fishing in the area, Indian trawlers operating in the zone are in a minority, less than 30 nos., compared to over 100 nos. of Thai trawlers. The fishing activity by the foreign trawlers has been going on unchecked, according to reliable reports from our skippers, since October, 1990. This Quiescence on the part of the Indian authorities appears mysterious.

The Thai trawlers must have taken away by now over Rs.50 crores worth of shrimp and fish, with no sign of the activity diminishing. Strangely, according to reports, while the Thai trawlers are fishing with utter abandon so close to our coast, our own trawlers are conducting their fishing operations in the area with a feeling of constraint and restriction. The reason in that the Thai Vessels are stated to be fishing in such a formation that the Indian trawlers have become aliens in their own waters and the Thai trawlers well at home. Reports speak of motherships from Thailand coming periodically to lift the catches and to provide supplies and services, all in our waters. In short, the entire operation by the Thai trawlers seems to have been well planned and executed. How such a thing could happen, when we have a coast guard with three bases along the upper east coast and a strong navy in the form of Eastern Naval Command is, to say the least, preplexing. While the Thai fishing fleet has encroached into our waters with inpunity, they could not do so in the neighbouring waters of Bangladesh and Burma, in any case not on this scale.

Several representations were made by the industry to the various authorities concerned. From the fact that the activity goes on shows that very little is done to check the encroachment. Coast guard did send a vessel on a few stray rounds with practically no success, although on one occasion they apprehended two Thai vessels and one more on another occasion. These apprehensions have to be seen in the background of over 100 Thai trawlers fishing in the area. One of the apprehended vessels is reported to have sunk. The crew members, around 44 nos., continue to be under remand for over three months in Visakhapatnam jail, with practically no visible interest shown by Thai embassy in India or our Ministry of External Affairs, to deal with the situation.

The strategy followed by the Thai fleet seems to be to move out of the area before a patrol vessel of coast guard reaches the place where the sin is committed and come back the moment the vessel is out of sight. This explains the chasm between the wailing of the fishing industry about the fishing onslaught by Thai vessels and their total absence when a coast guard vessel reaches the area. How do the Thai captains know that a coast guard vessel is coming to apprehend them? Obviously, they have their own methods.

Because of the presence of the Thai fleet, a steep fall in the catches of Indian trawlers is reported, in the past few months. For this season, a large number of Indian trawlers have preferred to remain at port. Thai trawlers are understood have had the temerity to enter Paradeep Port for certain supplies and resume fishing later. How far this information is true is not known.

Apart from the ingress into our fishing waters, the other issue is one of our territorial integrity and national honour. The nation cannot allow our laws to be broken by foreign fishing vessels as a matter of course and for such a long duration. We cannot be helpless spectators to this, unless, of course, our Government have permitted them to fish and we do not know about this.

What is now required is continuous partolling in the area by the coast guard over a period of time and keeping a watch over the trawlers while they are on their way to sandheads area. The movement of the fleet can be checked from air and ground action can be planned on that basis. In any case, some regard should be shown by the Government towards the responsible and genuine complaints from the fishing industry and the fishery wealth in the sand heads area should not be allowed to be harvested by trawlers of another nation. Our Government should lodge a protest with the Thai embassy in India and demand the discontinuance of this brazen practice.

Fishing Chimes, Editorial 145: March 1991: Vol. 10, No. 12

# Processing of Fish and Their Export with Australian Aid: A New Era on the Anvil

At last there are visible signs of plugging of a major lacuna in the Indian marine products export system, thanks to the proposed offer of Australian Aid for the setting up of a comprehensive fish processing facility at Visakhapatnam. Once the project receives the needed approvals and becomes operational, the stage will be set for organising similar activity, with the needed modifications, at other needy fishing ports.

The MPEDA is greatly concerned over the matter of diversification of marine products exports, considering the availability of large quantities of finfish all along the coast line in the sea. MPEDA's priority is the promotion of exports of finfish and finfish products in an organised manner. Keeping this in view, the Agency has succeeded in attracting the attention of the Australian fisheries authorities to spare their expertise and also extend financial aid for the development of this latent vital part of Indian Marine Products exports.

Accordingly, a team of Australian fisheries experts visited India in March 1991. The team consisted of Ms. Ray J. Winger, Professor of Food Technology, Massey University, New Zealand, Jon R.Cook, Director, Sloane Cook and King Pty Ltd, Crows Nest, Austrialia, Dr.John Sumner of M and S Food Consultants Pty Ltd, Albert Park, Victoria, Austrialia, and Mr.Don Calde Cotte, Refrigeration Engineer and Mrs. Bedi, Field Representative, Community Aid Abroad, Pune.

The team visited Visakhapatnam, and was there from 4 March to 19th March, 1991. Ms.Margret Rayner, First Secretary (Development Assistance) Austrialian High Commission, New Delhi accompanied the team and was at Visakhapatnam for some days and organised meetings in consultation with the Association of Indian Fisheries Industries and MPEDA. Mr. S.M. Shukla, President, and Dr.C. Babu Rao, Vice President, of A.I.F.I. respectively, Dr. K.A.R. Prasad, President, Sea Food Exporters Association, Visakhapatnam Zone, and Mr.M.V.Krishna Rao, Secretary, A.I.F.I. played an active role in working out the requirements of the industry and appraising the Australian team about the actual needs. Dr. Haran, Director, TASPARC and Mr.Dey, Assistant Director, MPEDA, performed the crucial task of organising discus-sions between the representatives of the industry, and the members of the team. MPEDA further or-ganised a very important meeting between the team and Mr.P.V.R.K.Prasad, Chairman, Visakhapatnam Port Trust. At this meeting, the Chairman explained in a very lucid manner the present status of the operations at the Visakhapatnam fishing harbour, the progress of the third phase of the harbour extension work now going on and the anxiety of the Port Trust to bring about a fish processing and a frozen storage facility for the fish landings at the Visakhapatnam harbour within the port limits. He said that the Port Trust would be glad to provide the needed land for the setting up of the infrastructural facility.

After a series of meetings which were held between the team, the members of the AIFI and the MPEDA a consensus has emerged. The MPEDA would be the nodal and implementing agency and the Australian Aid, when available, would be passed on to MPEDA, for implementing the project. There was also a general agreement on the need for implementing a Project for Shrimp Resources Management in the Upper Bay of Bengal.

It is learnt that the team has formulated their recommendations, the main ingredients of which are as follows:

1. A fish storage, and processing complex needs 'to be set up. This can be located in the 10 acre land, near the new flyover bridge, adjacent to the Visakhapatnam Port. The Port has agreed to provide this land subject to completion of various formalities. A plan for the setting up of the various units has been drawn up by the team and this received the acceptance of the representatives of AIFI and also MPEDA.
2. a) A fish processing unit, having a blast freezer and a spiral freezer and other types of processing facilities for value addition has to be set up. The total capacity of the various units will be 40 t/day. The final products will be utilised for export and also for internal consumption; b) There will have to be a frozen storage of 500 t capacity; c) There is also need to have a shrimp feed mill with a capacity of 4,000 mt per annum.
3. The Australian authorities may provide the services of a gear technologist for the introduction of a high opening trawlnet to facilitate catching of more of fish and less of shrimp.
4. A Project Director from the Indian side may be appointed. The services of a counterpart Director from Australian side may be provided.
5. A cold storage may be set up in Hyderabad with a capacity of 30 tonnes/day. A national fish

distribution grid for marketing will be developed in stages.

6. The total financial outlay of the project is estimated at Rs.7 crores, which may be provided as aid by the Australian Government.
7. The MPEDA may be nominated as the implementation Authority, with an arrangement for the routing of funds to the MPEDA in cash or kind.

It is expected that well before the end of 1991, the project will receive the approval of both Governments and project work will commence soon after.

The recommendations outlined above, once implemented, will transform the contours of the marine fisheries developmental activities in India. Over a period of time, the accent on fishing activities will shift from shrimp to finfish, because of the likelihood of progressive development of infrastructural facilities at various fishing ports. The relationship with Australia will facilitate exports of finfish products to Australia. As this catches up, other foreign markets will unfold. The Australian assistance by way of making available the services of a gear technologist and a processing technologist will go a long way in the introduction of advanced fish harvesting systems and processing technologies. The activities will result in increased foreign exchange earnings to the nation and also a fresh lease of prosperity to the sea fishing and processing industry. The activity will give rise to employment opportunities on a large scale.

This Indo-Australian project, which we ardently hope will definitely materialise, will be one of importance and will prove to be a milestone in the history of exploitation and utilisation of finfish resources of our EEZ. The names of MPEDA, the AIFI, the Visakhapatnam Port Trust and above all, of the Australian team, the Australian High Commission in India and the Governments of India and Australia would live in the everlasting memory of the sea fishing and marine products exports industry, as the ones who were responsible for the ushering in of a new era of prosperity.

Fishing Chimes, Editorial 146: April 1991: Vol. 11, No. 1

# Chimed Through a Decade

Laden with writings on weighty matters of great import for the development of Indian Fisheries Industries, and journeying through various 'weather' conditions month after month over a period of ten years (120 months), *Fishing Chimes* has triumphantly sailed into its eleventh year of its publication in April, 1991,in its usual readiness, in spite of odds, to move further, and forward.

The passage through the decennial barrier was an event of joy that has awarded to us the excitement of achievement and also a renewed sense of service. And we are agog to tell our subscribers, contributors of articles/papers, advertisers and our readers about what they have enabled us to achieve and also to avail of this occasion to rededicate ourselves to the cause of rendering service to all those engaged in fishery and fishery related activities and also to those who want to enter the arena. The first number of the eleventh volume of the journal (April 1991 issue) is now in your hands. As you will no doubt observe, it has been brought out as a special issue to mark the completion of a decennium of devoted service to the fisheries industries on the one hand and to herald the beginning of the next phase of service on the other.

Our subscribers, advertisers and other professional fisheries persons, who support us by contributing articles/papers on fisheries subjects of topical interest are the three stanchions to which we are moored and our journal subsists and grows on the strength imparted by these. Scientists, technologists, farmers, technocrats and administrators who contribute articles and papers form the main base that gives shape to our journal month after month.

We believe all our subscribers, advertisers and contributors of articles/papers read or atleast scan through the contents of the successive issues of the journal. This esteemed readership has created and continues to create a secondary line of readership, which may or may not be substantial though. Some of these secondary readers and a few others who come to know about the journal keep becoming our subscribers, thereby giving us a chance to expand our subscription base. We want to avail of this auspicious occasion to thank all these categories of readership.

The senior readers of the journal would re-call that one of the important objectives of the journal has been to bridge the communication gap among the fishery fraternity. Knowledge of the developmental works going on and innovations being made at other places and the results thereof and the new policies and strategies adopted by the various Governments as published in the journal, set the readers thinking and this enables them to sift the information, and adopt in their area the best out of it that suits them. Availability of information on the progress made in the required fisheries field of endeavour activates the cerebral cells and the thinking process gains momentum.

Of late, there has been a major upsurge in the number of seminars, symposia, conferences, training courses, both in the culture and capture sectors, in the various States under the auspices of the Central and State Organisations. Innovations made and experiences gained in their fields of work by the various scientists and technologists are presented at these gatherings. *Fishing Chimes* has been playing the crucial role of publishing summaries of the proceedings of these gatherings and the recommendations made thereat. We have the humble conviction that this practice has, to some extent atleast, contributed towards the upgradation of the relevant activities. Our 'Comments' on topics of critical value in every issue of ours are known to have imparted the quality of vibrancy needed for hastening decision making processes.

It has been our endeavour to encourage junior scientists, technologists and entrepreneurs to contribute articles embodying their findings on problems, and their experiences. The reason for this, as can be imagined, is that it is the younger set of dedicated persons will come to occupy positions of higher responsibilities sooner or later, and the inculcation of the habit of reducing the details of the work done and results achieved by them together with their suggestions into the form of papers at this stage of their career, will help to bring out the best in them, in the future, from time to time. Several youngsters are grateful to 'Fishing Chimes' for the encouragement given to them through publication of their contributions, and we reciprocate this gesture.

Several seasoned professionals of course send their papers to us and we publish them with pride. A majority of these contributions are momentous in nature, and the contents, in several cases, have made a major impact on the course of fisheries developmental activities in the country.

Our periodical 'Chimes', according to some of those engaged in fisheries development of the country, have brought about a measure of qualitative as well as quantitative change in inland as well as marine fisheries development. Over the past ten years, the thinking in respect of fisheries developmental matters has changed perceptably and also in a remarkable way. It is hard to imagine that there has come about such a sea change in the thinking of entrepreneurs all of whom firmly believed till recently that harvesting of shrimp is the only way to get profitable returns. The location of deep sea lobster grounds on the southwest, followed by, similar but more extensive grounds east of middle Andamans, have led to a major effort in diversification of fishing. Similar efforts at diversification, based on results of operation of chartered Taiwanese longliners are in the pipeline. Fishing for squid, cuttle fish and sharks is gaining momentum. Our trawlers may, any season from now, switch over in a big way to stern trawling to catch deep sea prawns and finfish. A beginning is already discernible.

The decade has also seen major policy changes in respect of fisheries development, highlighted by various incentives provided by MPEDA on the one hand, and professional banking norms and approaches instituted by SCICI and commercial banks in providing loans on the other. The setting up of a separate Union Department of Fisheries is probably round the corner, which will be a major development of the decade. Production from freshwater and brackishwater culture has reached new heights of achievement. Almost all the districts in the country now have Fish Farmers Development Agencies and Brackishwater Farmers Agencies have been coming up in several coastal States.

MPEDA has emerged as a major force in stepping up marine capture fishery output and in organising brackishwater farming for shrimps, from the angle of stepping up of exports. Our worthy farmers, scientists, technologists, entrepreneurs and Government Agencies played a major and magnificant role in stepping up marine animal production and exports. MPEDA has proved the adage 'power are taken and are not given', in furthering a good cause.

We have had the opportunity of chronicling or recording the developmental '*Chimings*' from time to time, apart from providing constructive comments on the developing situations, sometimes laced with fresh ideas.Was it possible at any time that our writings influenced Government's polices and strategies or the operational planning of fishing companies? A difficult question (although our own, but some readers may also like to ask this) which is partly answered by the reactions of several in authority who did tell your editor that the observations and suggestions in the journal helped the authorities in the evolution of policies and strategies for the development of fisheries of the country. Whatever be the position, we at *Fishing Chimes* are happy that we have kept on record notes on most of the developmental activities during the years after 1981 and that in several offices and libraries both private and public sector, the successive issues are being carefully preserved in a bound form. As my good friend Pradip Tagore says, the successive issues of *Fishing Chimes* now constitute reference material to all dealing with fisheries and more so for students at Fisheries Colleges.

On this joyous occasion it will be out of tune to say anything about our problems, although we have a right to share them with our readers. All have problems. No one is free from them and cemeteries are the only places without problems to those residing inside. We only want to say that it should have been possible for us to present the journal to our readers, with more of news, with an improved and a more inviting get up, and free from mistakes, and on time. We do hope that pernicious factors that have been haunting us, one among which is our inability to employ a larger number of persons to take care of the various links that govern the production of the journal, should soon loosen their grip over us.

Once this stranglehold disappears, we should come up to a level of diversification and co-ordination that will enable the production of a better journal with improved features and get up, hopefully well before diversification of sea fishing and exports of marine products makes further progress, brackishwater farming attains semi-intensive culture status all along the coast line in a big way, and freshwater fish farming reaches an intensive and integrated production stature in a way that causes envy to our neighbouring nations.

In the end, our dear readers, allow us to thank you all for your support, cooperation and encouragement, but for which we would not have been able to maintain continuity in the publication of the journal.

Fishing Chimes, Editorial 147: May 1991: Vol. 11, No. 2

# What Should be Done to Improve the Economic and Social Conditions of Coastal Fishermen

Fisher communities have been an isolated lot from times immemorial. Some say that soldiers of invading armies, left behind to fend for themselves, settled down close to the sea shores and took to fishing as a profession,as the mainstream of the society did not accept them. The growth of urban surroundings have not also succeeded in any perceptible way in absorbing them into the mainstream of civilisation. They continue to live a closed life of their own. No fisher boy could virtually succeed in coming up to the level of a captain or an engineer on a fishing vessel of over 20 m length. Practically no fisherman owns a large fishing vessel.

Fishermen are the sons of the sea. Their life is intimately entwined with fishing activities. It is sad that, during the 44 years after independence what all they could get from the successive Central Govenments was some sops and no more. The situation is just the same as the British allowing the bulk of Indians to rise to the level of clerks and no more.

Our constitution allows anyone to take up fishing as a business activity. This is fair. At the same time, what all is possible should be done to bring up the communities traditionally dependent on fishing, to a place of pride. The development of the activities of traditional fishermen is of crucial importance to increase fish production and exports. More than this, a well-developed, enlightened and skilled fishing community along our coastline will be a great asset to the nation. The fishermen living along the coastline can be trained to keep a watch over movements of foreign ships and foreigners landing on our coasts. In fact, there were several instances of fishermen informing the Government about suspicious movements of foreign vessels Our Members of Parliament, particularly those elected from coastal constituencies should prevail upon the new Government to adopt short term and long term measures to alleviate the lot of traditional fishermen.

## Short Term Measures

Coastal fishing villages lying along a coastal strip of about 25 km should be covered by an organisational mechanism that can bring about integrated development of fishing activities as well as the socio-economic life of fishermen of the zone. The organisation can be called as 'Near Shore Fishery Development Agency'. It has to be registered under the Societies Act. As a non-official organisation, it should be run by a Committee of officials and non-officials. Government should provide the needed funds for providing training to fishermen, laying roads, providing power supply, setting up cold/frozen storages, providing fish transportation facility to carry processed products to the port for exports and also fish for sale in internal markets. There should be a price fixation system and payments to fishermen for the catches sold by them should be paid by the Agency or its Agent then and there. A housing scheme should be implemented by the Agency, by organising a tie-up wiih the Housing Board or a Housing Financing Agency. Recoveries towards loans for houses can be made by the Agency in instalments at the time of payment of value of fish, prawn etc purchased. Schools for fisher children have to be strengthened or set up afresh.

The Management Board of the 'Agency' can be headed by the Revenue Divisional Officer or Sub-Collector having jurisdiction over the area, as the Chairman. This is necessary for securing land and other requirements of the Agency. One or more of representatives of fisheries, education, health, roads, electricity, transport and other connected departments have to be taken as members on the Board in addition to one representative from each of the villages covered. There has to be a professional person (with the needed staff), to function as the Chief Executive of the Agency.

The work of the Agency will encompsass the following facets : a) Development of an organisational mechanism to achieve coordinated development of the activities of the proposed Agency, b) provision of infrastructural facilities with funds granted by the concerned Government departments and on land given by the Government department concerned, c) employment of staff to run the facilities with subsidies given by the Government and d) organisation of a strong wing for supplying to fishermen improved nets and boats with loan and subsidy facility. This wing should develop strong linkages with lead banks and help the fishermen in the preparation of project reports for securing loans and ensuring their clearance and release of loans for the acquisition of the needed units e) setting up of a wing for technology transfer, f) organisation of a marketing set up, and g) maintenance of liaison with the various external agencies concerned and good relations with the fisher community.

## Long term measures

There will be real transformation in the socio-economic life of fishermen only through a long term strategy. This strategy has to revolve round the fisher child.

However much we may try for any number of years to introduce improved fishing systems etc among coastal fishermen, there will be no real progress, because of the strong roots of tradition. The reason is that the traditional way of work will always be in the back of the mind of fishermen and they will be averse to absorbing new systems. This is probably justified because, traditional fishermen, led by their isolation and monopoly over coastal fishing, have the justifiable pride that they are the best and they meekly listen to what development people tell them with benevolent patience and tolerance and stop at that. Grown-ups among fishermen community do not bend to accept outside advice, the only exceptions in recent times being the adoption of synthetic twine for making nets, FRP boats and beachlanding craft and installation of outboard/inboard motors for mechanisation of boats. The result of this attitude is that the community continues to be practically static.

## Reform

Reform is required to get over the situation. Fisherchildren have to be caught young. about the age of seven well before the traditional attitudes take roots in their minds, to the exclusion of new ideas. These boys should be recruited and trained in all the latest aspects of sea fishing from larger fishing vessels of various types. The training should be over a period of 2 years, by which time the youngsters would be fit to command deckside work and engine room work as they choose. In other words, the training should equip them to appear and pass the concerned examinations. Once they settle down in their new jobs, they will form the nucleus for the development of the community. Apart from taking up jobs, the candidates can also set up enterprises, acquire their own vessels and run them on commercial lines. The first batch of trained candidates will come out after 12 years of education and training, to be followed by other batches year after year. These trained boys will be able to change the complexion of the community. The organisation to undertake this job will have to be a voluntary one. A name that can be suggested for this is 'Society for the Uplift of Rural Fishermen. The acronym can be 'SURF'. There can be one society for each of the Coastal States to run training centres for fisherboys. Foreign organisations such as NORAD, DANIDA etc would be willing in all likelihood to participate in such projects. They could provide to the proposed society training vessels and teaching equipments and models. They could be prevailed upon to bear the salaries and other expenses relating to foreign experts. From the Indian side, investments on all civil works can be contributed with help from the Government. Salaries of the project director who will be from the Indian side and of the Indian instructors and other staff can be borne by the Indian society, with grants given by the Indian Government.

The Members of Parliament and the authorities concerned should give deep thought to this strategy for bringing about the uplift of our fishermen and making them prosperous.

Fishing Chimes, Editorial 148: June 1991: Vol. 11, No. 3

# The Inheritence

The Minister of State in-charge of Food processing who is also in charge of development of deep sea fishing, has inherited a 'disorganised' and ramified structure for the development of deep sea fishing. One positive aspect of the inheritance, however, is the setting up of the Unit for clearance of applications for the outright acquisition of deep sea fishing vessels new or second hand by purchase or on lease. The Unit referred to above is known as "The Secretariat for Approval of Fishing Enterprises", the acronym being "SAFE".

While the Food Processing Ministry has very little to do with joint ventures in deep sea fishing, integrated or otherwise, (which is taken care of by MPEDA) it plays a part in the matter of according clearances for the acquisition of fishing vessels by the prospective joint ventures enterprises.

The Ministry of Commerce and MPEDA gave as a gift to the Ministry of Food Processing a set of detailed guidelines to be provided to the fishery enterprises to apply for permissions to acquire deep sea fishing vessels in order to secure financial assistance from SCICI. While a few entrepreneurs could secure these permissions and applied for loans to SCICI and secured sanctions, not much of a head-way could be made as yet under the new single window system ie., 'SAFE'.

The guidelines referred to above were well drawn up but it has to be seen how they will work in practice. One omission seems to be the lack of a procedure for release of cost of second hand vessels (not connected with joint ventures) to enable the companies to pay the same to the owners who demand this payment within a few days of striking the deal.

We may sound pessimistic when we say that there must not have been much of progress in respect of any of the links of the activities referred to above. Firstly, it is difficult to secure offers from foreign fishing companies, particularly by progressive entrepreneurs who constitute the bulk of those who are interested in entering the sector but have no adequate financial backing. The foreign fishing companies do not reciprocate with the same enthusiasm as the Indian parties. The vessels to be contributed by the foreign party for operating under the Indian joint venture company concerned would have, in most of cases, values more than 40 per cent of the equity that can be allotted to them. When the vessels are taken over by the joint venture company for offsetting the amount due towards equity by the foreign party, the Indian parties will invariably have problems of securing funds towards cost of balance of the amount by raising additional equity to the extent of 60 per cent of cost of a vessel so as to secure loan as required from SCICI, unless permission for leasing of vessels can be obtained or deferred payments arrangement is organised, under a guarantee which again is a problem. In the case of leases, the lease amounts being always on high side, the problem of payments towards instalments of lease amounts would loom large from time to time.

Presuming that the permission for acquisition of fishing vessels could be obtained smoothly and quickly, the problem of financing will continue to be a major stumbling block. Very few companies can meet the conditions stipulated by SCICI (although these may be reasonable). This means there will be practically no further addition in the near future to our deep sea fishing fleet. This situation is an open invitation to foreign fishing interests to clandestinely or deliberately fish in our Exclusive Economic Zone without let or hindrance as had happened in 1990-91 season when the Thai Fleet was on the rampage not only in our EEZ but well within out territorial waters. This reflects clearly the results of our inaction and consequential losses of fish wealth and export earnings. The Navy loses the advantages of an expanded fishing fleet as a second line watchman of our seas which would mean a constriction of the media which can provide crucial information on movement of foreign vessels in Indian waters. Further, we would have to continue to bear the ignominy of being a third rate marine fishing power, shamefully inconsistent with the size of the nation and the extent of its EEZ (2.1 million sq. km) and our fishery resources, which far outweigh, compared to the status of all the countries around us and several of the countries far way, which are smaller than us but have larger fleets with far greater fish catching power. This anamolous situation does merit the immediate attention of the Ministry for Food Processing Industry, atleast to protect our national honour.

Another aspect that calls for an immediate review is the haphazard manner in which the development of deep sea fishing industry in channelised. There is no field organisation to take care of this great task, except MPEDA, which fortunately has taken over the job to a considerable extent, although its main function is to promote exports of marine products. No doubt, the MPEDA Act provides for development deep sea fishing by the MPEDA. However, from the very beginning untill recently this responsibility was being discharged by the Department of Agriculture, but now taken over by the Ministry of Food Processing. There is now an urgent need for a field organisation to plan the manner in which our deep sea fishing fleet has to be expanded, with reference to resources, area-wise, and numbers of each type of vessels that can be operated from each of the bases and monitor the activity.

This suggested organisation has to take into account the financing problems for the acquisition of deep sea fishing vessels and the manner in which these can be got over. Several foreign shipyards are on the look out for orders for building fishing vessels and the foreign Governments concerned are anxious to help their idle shipyards to secure orders and execute them based on long term financing arrangements, if necessary, which may take the form of loans or grants. Whatever is the case, it should be possible to enter into bilateral arrangements with friendly foreign countries by way of package deals' for the supply of a certain number of vessels of specified types within a time frame. Within this framework, on the Indian side, applications can be called for from genuine and interested fishing companies to acquire these vessels coming under the packages, which should have Government's guarantee and support. Finances (loans/grants) can be channelised through a financing organisation. The advantage of this system, which could cover not only the supply of deep sea fishing vessels but also processing equipment etc., is that there will be no immediate outflow of foreign exchange. On the other hand, there will be foreign exchange earnings out of exports from vogage to vogage, which can be utilised for the needed periodical repayments.

One is tempted to mention here that, the problem of financing acquisition of deep sea fishing vessels has become very acute, because of the decision taken by the Finance Ministry sometime back, when Mr. V.P. Singh was Finance Minister, to close down the erstwhile Shipping Development Fund Committee which was in existence for over 25 years. This committee was closed down for the ostensible reason that the shipping companies (and also fishing companies) were not making repayments on time and the arrears were accumulating. In order to set right this situation the Shipping Credit and Investment Company of India, (SCICI) was created and this body was entrusted with the task of recovering the pending dues and also for providing loans for fresh acquisitions of vessels. So far as we are aware, this new body has been working very hard to recover the pending loans and also to give loans for the acquisition of new vessels. The progress in this regard, however, has not been perceptible. In any cause there is only a very marginal addition of deep sea fishing vessels to the existing fleet after the SCICI took over financing functions and the very few sanctions given, particularly for acquisition of tuna long liners, remain mostly unutilised. Over and above this, a barrier of strains have come into existence between SCICI and fishing companies, with very few fishing companies having a good word to say about SCICI. SCICI's business like attitude is in great contrast with the approach of the erstwhile SDFC and there is no wonder a chasm has come about, in spite of a number of interactions between the two sides, explaining respective positions to each other.

There is no doubt the present situation in this vital sector is very depressing, in spite of its vast potential and ability to alter the complexion of the Indian economy substantially for the better. This is the only sector which can bring in returns in foreign exchange, equivalent to the investments made, in the first two or three years, if not in the first year itself. It is unfortunate that there is no realisation of the potential of this sector at political level. Had there been this realisation, the Prime Minister (the earlier and also the present one) would have heeded to repeated representations from the Association of Indian Fishery Industries to create a separate Ministry of Fisheries but this was not to be, although this should certainly have been possible with a Cabinet strength of 58 Ministers as at present. Once there is a separate Ministry for Fisheries, there will be an effective application of mind and the consequential followup action. For example, the present fragmented setup (as explained here under) would come to an end and there would be cohesion which is so very necessary for the development of the activity. Can we imagine the tagging of Fishery Survey of India and the function of looking after acquisition of vessels to the Ministry of Food Processing Industries on the one hand and attaching related functions of development of infrastructural facilities such as fishing harbours and training of personnel for high sea fishing to the Ministry of Agriculture? Arrangements of this kind are not expected to be made by modern Governments. Our Government is probably primitive in not paying attention to anamolies of this kind that hamper progress.

One good decision of the earlier Government (as understood) was to close down the system of according fresh permissions for chartering of foreign fishing vessels by Indian fishing companies. One hears about a spate of irregularities in the granting of extensions to old and eligible charter permissions. While some who had fulfilled the conditions governing the operation of vessels permitted under the charter scheme and who are eligible for extensions were not accorded extensions that are due to them, those who never fulfilled any of the conditions in reality or otherwise could secure extensions. There must have been valid reasons for these decisions but these are not obvious to several in the industry. By saying this we do not mean that there are many skeletons in the cupboards in the Ministry of Food Processing Industries. It is possible that whatever was done was totally above board. Nevertheless, a checkup would be worth the while. In conclusion we would like to emphaise the following aspects :

1. A Ministry of Fisheries should be formed at the centre.
2. There should be a field organisation to promote deep sea fishing industry.
3. There should be bilateral agreements, between India and competent friendly countries for the supply of deep sea fishing vessels of the types we need and in accordance with a well drawn out programme. The supplies should be available at reasonable costs to selected and deserving applicants through

the new organisation suggested to be created, on easy repayment terms. Training programmes could be an integral part of the agreements.

Since payments have to be made for second hand vessels to be acquired by Indian parties within a short duration of striking of the deal, a separate procedure in this regard should be formulated and implemented, and a radical reform in the financing for the acquisition of deep sea fishing vessels will have to be brought about.

Fishing Chimes, Editorial 149: July 1991: Vol. 11, No. 4

# Crisis in Deep Sea Fishing Industry: Unsubsided Problems of Subsidy and Working Captial

The operational aspects of deep sea fishing vessels continue to be in grave straits. For want of working capital most of the deep sea fishing vessels continue to be portbound. The Commercial Banks have shown their unwillingness to provide working capital even towards bunkering of fishing vessels and this has worsened the situation. While several ostensibly valid reasons have been advanced by the banks (instead of advancing money), scant attention is paid by the authorities concerned to the problem, particularly when the need is to keep the vessels operational, in order to ensure their contribution towards marine products exports. A specific suggestion hasbeen made that if commercial banks have no funds to spare or are not in a position to help, Government may provide a guarantee to the banks for the amounts advanced as per a pre-approval given, for disbursement to the fishing companies concerned. If there is credit squeeze which indirectly means that the banks have no money to lend, it was suggested that MPEDA may provide funds to the banks, taking the needed assurances and commitments from the owners and the AIFI in regard to repayments. It is a matter of utmost regret that an important issue like this is allowed to drift.

The authorities are aware that, the fishing companies would be able to make repayments towards the pending dues in respect of loan taken for the acquisition of the vessels only when the vessels are in a position to operate. While a few are able to operate securing private finance, at exorbitant interest, most of the others are forced to keep the vessels idle, just for want of working capital facility of Rs.5,00,000 per vessel. Surely, something can be done by the authorities concerned, particularly the MPEDA, an organisation directly interested in the production of marine living species for export. The Ministry of Food Processing Industry and Ministry of Commerce have to step in at this moment of crisis to help the industry. An effective intervention will lend proof to the intention of the Government to do all its best to promote marine products exports. The amount involved will be less than Rs.5 crores, and can be recovered in a period of three months or even less. We do hope that the present complacency will be shed and the authorities will find a way out, on priority. In regard to granting of subsidy of the order of Rs. 1,600 per kilolitre, or 25 per cent of value of exports, whichever is less, the scheme is now restricted to producer-exporters. Unless proof of having exported the catches is shown, the subsidy will not be granted.

There are several producers of exportable species such as shrimps. They sell their catches to exporters for the reason that they have no processing plants of their own. They cannot also have their own plants for the reason that the order of their catches do not warrant the setting up of their own plants. There is thus no lapse on their part. It will be invidious if the Government and MPEDA ignore all these producers, who actually need the help. This category provides material in substantial quantities for processing and export and there is no justification to deny the benefit to them. They are already burdened with heavy expenses to keep their vessels operational.

It is conceded that the scheme is linked to export performance. This does not mean that the producers who do not export on their own are non-performers. They do provide the material for processing and export. Considering this established fact, the approach of the Government should be to find a way out, rather than denying the much needed facility of working capital.

It is known to the Government that transactions in respect of sale of shrimps etc. by the producers are recorded in the books of the producers as well as the processors/exporters. Such being the case, it should be an adequate procedure to ask the producers to furnish certificates issued by the exporters in respect of quantity purchased by them from a particular producer and the quantity exported. This should be adequate proof of contribution of the producers towards export effort. Another way is to evolve a scheme under which the producers can entrust the processing and export function to established processors/exporters. The exports can be made on behalf of the owners by the exporters, and both their names can find place in the export documents. The producer can give a waiver in favour of the exporter in regard to REP licence benefits etc. One indirect advantage of enabling the producers to come within the purview of the subsidy scheme is that the possible malpractice of diversion of part of the catches for generating unaccounted money can be checked. The producer will be forced to account for all his catches in order to get the subsidy.

There were announcements sometime back from the Ministry of Commerce that all subsidies were withdrawn. It is believed that this particular subsidy is not withdrawn, as no mention of this is made in the budget for 1991-92. Believing this to be the true position, we strongly appeal to the Government to set right the anomalous situation pointed out above. The Association of Indian Fishery Industries could probably put in a little more effort and explain the position to the Government in a persuasive manner, although repetitive, so as to achieve the extension of subsidy scheme to producers as well.

Fishing Chimes, Editorial 150: August 1991: Vol. 11, No. 5

# Deep Sea Fishing Industry: Status under the New Industrial Policy

The new guidelines simplifying procedures to apply for approval of joint ventures between Indian and foreign fishing companies for fishing in Indian EEZ with provision for such Indian companies to take permissions for the operation of foreign fishing vessels on lease or for acquisition of fishing vessels and for test fishing were introduced in the first quarter of 1991.

## Test Fishing

The Indian and foreign companies can undertake test fishing subject to entering into an agreement with MPEDA in respect of aspects such as the area of operation, the targeted species, provision of data to MPEDA, FSI etc. Test fishing will be permitted in respect of trawling for deep sea lobster, deep sea shrimp, long lining etc., but pair trawling will not be allowed. There are several other conditions such as those governing payments to Government, employment of foreign crew etc.

## Leasing

The guidelines also provide for obtaining permissions for leasing of foreign deep sea fishing vessels by Indian companies, for a period not exceeding 15 years. Several conditions have been stipulated to govern the leasing of fishing vessels. There is also provision for acquisition of vessels at the end of the lease period provided the vessels are new. Several other conditions governing the leasing system have been stipulated. There is also provision to apply for taking up deep sea fishing activity under 100 per cent EOU scheme, apart from provision to apply for acquisition of vessels, new or second hand.

The four methods described above (joint ventures, 100 per cent EOU's, test fishing and leasing) are governed by the provision of Maritime Zone Act and the rules framed thereunder.

## Composite Application Form

A composite type of application form has been prescribed. This covers a) import of capital goods; b) joint ventures; c) leasing of vessels and d) test fishing. This application would be adequate to obtain permissions for payments in foreign exchange towards various items, engaging of foreign crew, transportation of catches to foreign ports etc.

So far as the applications under 100 per cent EOU scheme and joint ventures are concerned, these have to be submitted to the Entreprenurial Assistance Unit, Secretariat for Industrial Approvals, Udyog Bhavan, New Delhi through MPEDA. In the case of all other schemes, the applications are to be submitted to the Secretariat for Approval of Fishery Enterprises (SAFE).The stipulated fees will have to be paid by way of challan for each of the activities along with the applications. (Details are given in our March, 1991 issue).

After obtaining the needed permission, entrepreneurs are expected to take steps for securing the needed finances for implementing the project concerned.

In this background, the deep sea fishing industry has to examine the effect of the new industrial policy announced by the Government recently which contains several liberalised provisions.

## New Industrial Policy: Extent of Application

The basic principle behind the new industrial policy is to provide automatic approvals for such of the projects for which the needed foreign exchange, not exceeding 51 per cent of the equity, can be mobilised from abroad. Since the present policy in respect of deep sea fishing industry has been largely framed under the provisions of the Maritime Zones Act, (particularly in respect of joint ventures, leases etc.,) the extent to which the new relaxations would apply needs clarifications from the Government, taking into account the extent of availability of deep sea fishery resources, the number of vessels of each type that can be introduced, are-awise, and the foreign (companies) countries with which financial and technical collaborations can be taken up by enterprises.

The relaxations that have been introduced in the new industrial policy imply that the provisions under the Maritime Zones Act and the rules framed thereunder should normally have minimal application, in all cases where the Indian fishing companies take up deep sea fishing projects, availing of assistance, technical or financial or both from foreign sources or otherwise.

Any conflicts between the new policy and MZA rules have to be removed considering the fact that the genera! thrust is that an enterprise has to find its own source of foreign exchange for developing a project, out of foreign investments or own earnings in foreign exchange. If there is a conflict, no enterprise can make progress, although foreign investments upto 51 per cent of the equity is allowed, apart from acceptable arrangements for securing foreign exchange to finance the acquisition of fishing vessels etc, under the general provisions.

## Foreign Investment Approvals

The provisions in respect of foreign investment approvals appear to be not conducive to the development of deep sea fishing industry. Deep sea fishing vessels are expensive, new ones of even 30 m OAL costing over Rs. 8 Crores. The latest regulation is that the foreign equity in high priority industries (Deep sea fishing is stated to be in high priority list) can be upto 51 per cent. However, approvals will be available if the foreign equity covers the foreign exchange requirement for import of capital goods *i.e.*, fishing vessels. 51 per cent foreign equity will not be sufficient to buy a new vessel. At the best, it will be adequate to get second hand vessels. The policy says that the Government will not allow import of second hand capital goods under the new provisions. The relaxation in terms of foreign equity will be helpful to the industry only if second hand vessels are allowed for import. Another way which the Government may adopt is to authorise SCICI to provide guarantees for loans arranged by the companies to meet requirements over and above 51 per cent foreign equity. It may be mentioned here that permission will be issued by RBI for availing of 51 per cent foreign equity.

51 per cent foreign equity holding will be allowed in trading companies engaged in export activities. Presumably 100 per cent EOUs coming under this category will have to file application for release of foreign exchange to the RBI.

According to the notification, the application for foreign equity investment should clearly state the description of the article to be manufactured. The proposal should also contain all information regarding import of capital goods for the project. Deep sea fishing vessels harvest exportable animals. These may or may not be processed on board the vessels as such.

According to the policy, if new vessels can be purchased within 51 per cent foreign equity, there will be no need for indigenous clearance. This means that a situation of this kind would not require any application for clearances to SAFE or Secretariat for Industrial Approvals. The position needs confirmation.

The policy envisages the import of components, raw materials and intermediate goods and payment of know how fees and royalties, in accordance with the provision under the general policy applicable to other domestic units. In other words clearance by RBI would be required. The catch would often require shifting to a shore plant for processing. This characterstic of marine products production will have to be brought within the meaning of the policy as manufacture may take place at a place, outside the vessels, both old and new of which will have to be allowed to be imported under the provisions of the scheme. For the operation of these vessels only Indian crew will be allowed.

According to the policy, dividends will have to be paid in foreign exchange to foreign investors, out of foreign exchange earnings. This is a reasonable provision.The present system of applying to SAFE or to other agencies, in the light of the new industrial policy, may have to be sooner or later be replaced by a system of making a declaration to the Government about the details of the project being taken up. This means that any formal permissions for these will be issued automatically. It will then be for the concerned parties to obtain permission for release / utilisation of foreign exchange from the Reserve Bank of India, particularly for engaging foreign technicians and crew.

## Non - Licenseable Industry

Our present comments are based on the new procedures lanuched by the then Industries Minister Prof. P.J. Kurien at the Entrepreneurial Assistance Unit counter of the Department of Industrial Development. According to this, the industries, which have been exempted from licensing of which deep sea fishing is one, (Deep sea fishing is a non licenseable industry) can put up a new unit immediately after they submit the Memorandum of Information. According to the Minister, even the substantial expansion of projects can be undertaken after the Memorandum of Information has been filed with the Secretariat for Industrial Approvals, or in the case of fishing industry with SAFE, presumably. Forms are now readily available at the counter concerned at Udyog Bhavan, New Delhi, it is stated.

## Imports Without Restraint

It may be mentioned that the Government is also establishing an automatic approval system for import of Capital Goods for all proposals which fall within certain parameters. Capital Goods imports will be allowed under this system if they are fully covered by foreign equity or if they do not exceed 50 per cent of the value of plant and equipment subject to a ceiling of Rs. 3 crores. Acquisition of deep sea fishing vessels being capital intensive, entrepreneurs can import them without restrictions, if we go by what the minister has said.

## EXIM Scrips

So far as EXIM scrips are concerned, it is stated that value added products including marine products would be eligible for EXIM scrips to the extent of 40 per cent of fob value.

## Foreign Exchange Accounts in India

According to reports, the Government has also decided to allow established exporters to open foreign currency accounts in approved banks and allow exporters to raise external credit for export related imports from such accounts and credit export proceeds to such accounts. Government is also expected to enlarge the list of items eligible for import under OGL It is further expected that fishing vessels and other fishing requisities would find a place in this enlarged list. It is also mentioned that secondhand ships (which include fishing vessels) can be imported under OGL against EXIM scrips. One feature of the new policy is that EOU companies will be allowed to sell 25 per cent of their production in the domestic market, if their use of indigenous raw material is more than 30 per cent.

## Quick Clearance

All proposals with automatic approval parameters, according to reports, will be cleared by the Government within two weeks. Other proposals would be cleared within 45 days.

## Frame - Work Needed

Considering the distinctive characterstics of the fisheries sector which are related to fish stocks availability from time to time and the number of vessels of various types that can be introduced or operated from time to time, while applying all these relaxations under the new industrial policy to the fisheries sector, intervention of the Government is of utmost importance to enable the entrepreneurs to set up deep sea fishery enterprises within a certain framework and a feeling of freedom. Government could determine once in three years the number of vessels of various types in terms of OAL or tonnage, that can be introduced, and the bases from where they can be operated. In the case of certain types of fishes *viz.*, tuna, squid, cuttle fish etc., there must be no restrictions for quite sometime to come. So can be the case with deep sea shrimps and deep sea finfishes. In the case of fishery resources that need regulated fishing, the number of vessels that can be introduced can be specified areawise and the introduction can be regulated within these numbers. We feel that barring these restrictions, no impediments need be placed in respect of introduction of deep sea fishing vessels. Submission of Memorandum of Information should be applicable in this regard and also for the setting up of shore processing plants and other facilities. The new policy provides for engagement of foreign technicians and also for development of technical and marketing collaborations. Payment of royalties for marketing of fish products abroad is also allowed under the new policy. It would be heartening to the deep sea fishing industry if the Government clarifies and confirms that these provisions are equally applicable to the deep sea fishing industry.

## Charter Scheme's Failure

The scheme for allowing the Indian fishery enterprises to operate foreign fishing vessels on charter has been in vogue for the past several years. The objective of this scheme is to achieve transfer of technology and the introduction of similar vessels as operated on charter. These objectives have not been realised at all so far. Further, under this system, there is no increase in the Indian fishing fleet. The foreign exchange earnings are also restricted to 20 per cent of the export returns. Therefore, in the place of the charter scheme, it would be advantageous to the nation if a lease cum purchase scheme is introduced, the lease extending to a duration of not more than 15 years as already provided for in the present scheme. This system will have the effect of increasing the Indian deep sea fishing fleet. Unlike the charter scheme under which vessels just operate for some time and go back practically leaving no long duration production benefit to the nation, a lease cum purchase system as already stated, would be a great improvement. This will enlarge the strength of the Indian fishing fleet.

## The Need

Another reform that is of paramount importance is that no questions should be asked if a company can import vessels and operate them with foreign technology and men, irrespective of the equity aspects and their proportion, Indian or foreign. The same principle should be applicable for manufacture of fish and prawn feed, setting up of modem farms with cages or other ways for fish production. We do hope that Government would look into the various aspects and appraise the Indian fishing industry concerning the extent of application of the new industrial policy to the fishing sector, through an announcement *vis-a-vis* the existing guidelines. Unless these reforms are introduced, it is doubtful whether the policy will be of much help in increasing the strength of the Indian deep sea fishing fleet, upon which depends the exploitation of the untapped deep sea resources such as lobsters, deep sea shrimps, and deep sea fishes.

Fishing Chimes, Editorial 151:September 1991: Vol. 11, No. 6

# Fisheries Developmental Strategy in the Eighth Five Year Plan

India has a large number of experts belonging to various fisheries disciplines. There are also a number of bureaucrats who have experience in the administration of fisheries sector. These two assets, coupled with the availability of vast extents of marine, freshwater and estuarine resources should have taken the nation forward in the matter of development of fisheries. In actual practice, however, for want of specialised attention and coordinated thinking, the sector has been lagging behind. Only in the matter of exports of marine products there has been good progress.

Some seemingly clear cut policies have been formulated by the Ministries of Commerce and Food Processing in respect of development of marine fisheries sector. The most important of these are the recent concessions in regard to foreign equity participation, allowing larger houses to take part in the fisheries sector, allowing the employment of foreign technicians and crew in fishery enterprises, and granting of export scrips.

Prior to announcements of these relaxations, some policy guidelines were announced by the Ministry of Commerce in respect of joint ventures, 100 per cent EOUs, test fishing and leasing of fishing vessels. A system of single window clearance through an organisation known as "Secretariat for Approval of Fishery Enterprises" (SAFE) was created in the Ministry of Food Processing which would process all applications relating to fishery projects with the exception of 100 per cent Export Oriented Units and joint ventures. In the case of 100 per cent EOUs the applications have to go to the Entrepreneurial Assistance Unit and in the case of Joint Ventures they have to go to the MPEDA, who would send them to the Secretariat for Industrial Approvals. MPEDA has also taken up a scheme to provide a subsidy of Rs. 1600 per kl of diesel oil used by larger vessels. This is linked to export performance. MPEDA also grants a subsidy not exceeding Rs. 2.5 lakhs for equipping vessels for diversified fishing or processing.

The impact of the policy formulations in respect of joint ventures, acquisition of new vessels, taking vessels on lease, test fishing etc. were carefully formulated keeping in view the development of deep sea fishing and exports of marine products in view. The problems of the entrepreneurship have however been one of finance and the recent relaxations can be availed of only by those having connections with NRI's or other investors in foreign exchange.

There are very few of this category in the marine fishing industry, in general. On the processing side proposals for assistance from Australia, Norway, World Bank etc have been initiated.

On the inland side no specific policy formulations were made with the exception of augmenting the number of FFDAs, luring World Bank to assist in the setting up of brackishwater farms, reservoir fisheries development etc. Nevertheless the nation has achieved considerable progress in respect of inland fish culture in the private sector in almost all the inland States, the average production coming to nearly 2,000 kg/ha.

The upsurge in activity, particularly in brackishwater farming, is largely due to the general awareness created among the entrepreneurship by the Government through an infective chain reaction. Even a pessimist has to admit that considerable alround progress has been achieved with production per hectare coming to about 2.5 tonnes per two crops, with a few exceptions. There are reports of freshwater fish production per hectare crossing 11 tonnes and brackishwater prawn production crossing 5 tonnes per annum in some cases. This is good progress indeed.

Coming back to marine sector, the coastal waters of our EEZ are overfished. The measures called for in regard to this sector are therefore one of regulation and not of intensification. Any new strategy for the development of coastal waters has to take into account this aspect and the manner in which fishing in farther seawaters has to be developed with needed linkages with financing, infrastructural facilities, trained manpower, marketing facilities etc. All the best brains in the country having knowledge of deep sea fishing field could not so far formulate any implementable strategy for this sector. The deep sea fishing fleet (so called) remains stagnant and most of the vessels constituting the fleet conduct profitable fishing for part of the year and unprofitable fishing or detention of the vessels at port for the remaining part. The fact that any Indian fishery is seasonal is overlooked by both the entrepreneurs and also the Governmental authorities concerned.

The exploitation of deep sea fishery resources (including open sea fish such as tuna) is obviously of utmost urgency for augmenting production and stepping up exports. The problems involved in this have already been stated.

The strategy guidelines that need the earnest consideration of the authorities in this respect are as follows:

1. The deep sea fishing fleet has to be augmented for exports of products from deep sea fishing to go up. Vessels of 30m OAL and above equipped for bottom stern trawling at depths beyond 200m for the exploitation of known resources such as deep sea prawns, deep sea lobsters, deep sea fishes etc and also to catch sub surface swimming fish such as tuna and cephalopods such as squid and cuttle fish have to be introduced with reference to areas and seasons of availability,. The numbers that can be introduced should be determined by the Government who should allow entrepreneurs to introduce vessels of the determined types and numbers.
2. Considering the fact that indigenous capabilities at designing and construction of such vessels is on the lower side and entrepreneurs face financial problems both in regard to indigenous acquisition or imports, the only way out is for the Government to create a separate Field Organisation to deal with these problems successfully on the following lines.
   (*a*) After determining the number of vessels of each type to be introduced, the proposed Organisation should manage to enter into bilateral agreements with suitable foreign countries for providing long term credit facilities to the proposed organisation to provide funds to such of the entrepreneurs who can come within the fold of the bilateral scheme. Such a strategy will enable payments out of foreign exchange earnings to the Foreign Governments concerned over a period of time while securing the vessels of the type required well in advance.
   (*b*) The proposed organisation should be able to establish bilateral links as mentioned above which would include infrastructural facilities, introduction of vessels, induction of foreign personnel, organising training to Indian personnel, establishment of repairing workshops, marketing tie-ups etc. It should be in a position to advise entrepreneurs on the type of vessels to be acquired, mode of acquisition, prices, source of acquisition etc. It should acquire and keep with it drawings of different types of vessels, and also maintain a list of companies from whom various items of equipment can be obtained.
3. The proposed organisation must have two tiers. One should be the Ministry of Fisheries to be constituted as a separate entity headed by a Minister having knowledge of the fisheries field and a Secretary with technical and administrative background on the subject. This Ministry should establish an integrated marine fishery development company with share holding coming from fishery entrepreneurs, Ministry of Fisheries, MPEDA and others such as Port Trusts, from whom substantial share capital should be forthcoming. This company should be able to obtain fishing vessels under bilateral arrangements for different programmes. These would include outright purchase, leasing, test fishing, joint ventures and so on.

   This company should have a separate wing to take over all vessels of defaulting companies and operate them on its own. The company should be equipped for this purpose. Any single default in repayment without justification should be an adequate reason for taking over the vessels obtained by the company and supplied by it to an Indian entrepreneur.

   The spread of the deep sea fishing vessels should be such as to cover the entire EEZ along the Indian Coastline so as to exploit resources such as tuna, deep sea shrimps and lobsters and others (as already stated).
4. In view of the acute shortage in foreign exchange, where companies are in a position to obtain vessels on hire purchase basis and to be registered under Indian flag, they should be allowed to do so (within the precincts of the national scheme) and the companies should be allowed to make repayments out of the foreign exchange earnings. The initial working capital required in foreign exchange should be allowed to be put in by the foreign agency. There should be provision for granting guarantees by SCICI, where needed.
5. The existing fishing fleet is mostly engaged in trawling for shrimp. Season for catching shrimp is for a duration of about six months. Unless the vessels are equipped to conduct economically advantageous fishing in the remaining months their operational economics will continue to suffer. Government and the Association of Indian Fisheries Industries should therefore encourage owners of all vessels which are amenable for installation of additional equipment for catching tuna, sharks etc., to do so. This will facilitate round the year operation of fishing vessels. A fleet of mini trawlers can be sent to Andamans during tuna season to catch tuna with a mothership arrangement to carry the catches in ice to Singapore which can be reached from Andamans in 3 or 4 days. Tuna can be preserved in ice for the duration of 5 days of fishing and 4 days of journey.
6. When a deep sea fishing policy, broadly on the above lines is developed foreign exchange earnings from this activity are bound to increase and the operational economics of the vessels are bound to improve.

So far as inland fisheries development is concerned the strategies needed to implement appear to be as follows.

1. Now that almost all districts in the country are covered by FFDAs there has to be an integration of the staff working in FFDAs and normal culture fishery development schemes, with defined fresh targets of production to be achieved. The objectives of FFDAs should be readjusted so as to cover all inland farming fishery areas. The role of FFDAs should be progressively changed to enable the

organisations to lend financial and technological support to the fish farmers. No impediments should be placed in the way of agriculturists who intend to convert their lands into fish ponds.

2. The number of brackishwater farming agencies should be stepped up progressively. Their main function should be to arrange selection of pond areas and supply of inputs, apart from providing technological support.

3. Developmental activity pertaining to reservoirs, lakes and riverine fisheries should come under a separate set up to be coordinated by the Director of Fisheries Concerned. There should be a separate set of officers at various levels to deal with these fisheries.

   It is frequently noticed that there is a major gap in the transfer of technology from the research institutions to the field. In this context, it should be the policy of the Government that those who conduct research and produce results should not be disturbed to propagate the same, unless absolutely essential. Extension officers concerned from all over the country should be trained by the research workers in the new technologies. The extension officers should then spread the technology in the field utilising known effective methods.

4. The Central Institute of Fisheries Technology, Integrated Fisheries Project, and MPEDA should improve upon their present strategies of formulation of new products for export and domestic consumption and popularise them through suitable marketing agencies.

5. The financing problems of the existing trawler owners would have to be resolved in a pragmatic manner. It is true that several owners have not been paying the dues to the SCICI which is the 'designated person' of the Government to make recoveries on behalf of the Government, as provided for in the SDFC Abolition Act. SCICI has taken its job seriously. After using several persuasive methods to get back the loan repayments, it had experimented with the taking over of the vessels of several companies into its custody through the appointment of a receiver. This has led to a spate of litigations. The action taken by SCICI did not improve the situation in any way. On the other hand, the detention of a good number of vessels led to fall in exportable catches. In view of these, SCICI has started taking steps to return the vessels to the owners. They have also offered rehabilitation packages to the various defaulting companies. This rehabilitation package consists of down payment of 10 per cent of arrears and the balance of arrears to be paid within a few months. The SCICI set up their branch office at Visakhapatnam to recover the dues. Their view seems to be that the presence of their representative at the base of operation would improve collection, particularly as the 1991 - 92 shrimp season seems to be promising both in terms of fishing and price realisation.

One point that has to be borne in mind by the Government is that deep sea fishing is a developmental activity. The arrears of loans that have accumulated are not that substantial when compared to other loan utilising sectors such as agriculture. The objective of the Government in the present situation should be to promote the activity and not to frighten prospective entrepreneurs.

Steps should no doubt be taken to recover the pending arrears but these cannot be on the lines of the present rehabilitation packages offered by SCICI which are harsh, from the viewpoint that the commercial banks have stopped providing the needed cash credit limits towards working capital and the owners borrowed money from private financiers at exorbitant rates of interest. The owners would have to first pay to the private financiers so as to get out of their stranglehold. This means that they would need a longer time to pay back the arrears towards the loans earlier given by the erstwhile SDFC.

Keeping all the aspects in view, we are of the opinion that the policy of the Government should be to advise SCICI to a) give a moratorium of two years for the payment of pending dues b) collect interest at the rate of41/2 per cent as was in vogue earlier and after capitalising the same and making it part of the loan amount to be collected in suitable instalments from the third year onwards, and c) Any refinements that are required within the precincts of the above mentioned proposal could be made.

Something done on above lines would pave the way for the promotion of deep sea fishing industry not only by relieving the present owners of their anxiety and tension but also by attracting new entrepreneurship. The new entrepreneur who would come forward can be brought within the new policy of the Government that envisages diversification of fishing effort through acquisition of new vessels, leasing of used vessels, test fishing, joint ventures etc.

Fishing Chimes, Editorial 152: October 1991: Vol. 11, No. 7

# Trouble-Ridden ECBB Trawlers

M/S East Coast Boat Builders and Engineers, Kakinada, A.P. managed to secure orders from 17 Indian Fishing companies for the construction and supply of 24 fishing trawlers in a period of one year, commencing from 1985. Of these, 4 trawlers were supplied after a lapse of six years, two numbers to M/S. Golden Ahar and two numbers to Varuna Group. The yard could not supply the remaining trawlers to the other buyers. It is not that they do not want to build and supply. Their problem is one of economics of construction.

Shortly after the signing of the tripartite agreements for the supply of trawlers between the parties concerned, there was a steep rise in the value of Dutch Guilder. This placed the yard in a difficult situation as the contracts were based on fixed price. The problems now are three-fold. One is that the cost of the trawlers would have to be enhanced and the buyers would have to agree for the same. This is apparently not acceptable to the buyers, but for some marginal adjustments. Another problem is that the yard requires lot of additional funds to complete the construction of balance of trawlers. These funds have to come from some organisation to the buyers who would have to pass on the same to the shipyard. Several high level meetings were held under the auspices of the Government on who should provide the needed additional funds for completing the trawlers. And the Government is in no mood to do this.

The buyers have recently mounted an agitation against the shipyard for the supply of the trawlers. The yard laid off a substantial no. of workers for obvious reasons. It is quite clear that whatever the buyers will do, they will not get the trawlers, unless there is some way of pumping in more funds.

The tripartite agreement signed between the shipyard, buyers and the erstwhile SDFC contained provision for the incomplete trawlers to be taken over by the buyers in case they were not supplied as per contract conditions. The buyers can entrust the balance of construction work to another builder and the cost of such construction has to be borne by ECBB. The enforcement of this provision is found to be not possible.

In these circumstances the shipyard found a solution to the problem which will enable it to pay back the loans and subsidies taken from the Government with interest and also the down payment made by the parties together with interest. The shipyard has managed to export two trawlers at a fairly good price, stated to be without infringing any of the regulations. The talk is that the shipyard may export another 4 trawlers very soon.

In the above background one has to understand that the propositions of all concerned in the tangle are justifiable. At the same time there is a deliberate avoidance of finding a solution.

From 1985 to 1991 the fishing industry has passed through several phases. Several Government committees set up on the question of fishing for coastal shrimp came to the conclusion that there had been over fishing of shrimp and that there must be regulation of shrimping fleet strength. It appears to us that this factor was not taken into account by the three parties concerned in arriving at a solution. When there is a strong feeling among the scientists that the shrimping fleet should be kept under control, the Government has to ponder whether they should allow the supply of balance of shrimp trawlers by the shipyard to its clients. If this supply really takes place, shrimping pressure along the upper East Coast would go up further, bringing down per unit catches. The shrimping vessels operating at present for a duration of six months in a year have some signs of survival now because of the additional incomes accruing, consequential to devaluation of the rupee. In this situation, it may appear to the Government that it will not be in the interests of the buyers and the shrimp resources position to add a further number of 18 shrimp trawlers to the shrimp fishing fleet. This is however a very unpalatable observation and no buyer would like this statement.

Even the longest river winds its way safely to debouch into the sea. In the same way this presently intractable problem must also be resolved and this can be done only by the Government. The intention of the shipyard should not be to trouble the buyers. In fact, the shipyard seems to be having no such intention. The buyers have also suffered heavily. In spite of this, they do not want the shipyard to suffer. What all they want is the trawlers to be delivered. In fact, so much of interest has accumulated over the past 6 years that their capital investment, even with the devaluation, it is very doubtful whether they can make enough money to pay back the loan to the SCICI. We all know the approach of SCICI towards recoveries and the buyers have to be after the Government for relief and redress. In the process of these agonising activities, it will be difficult for them to concentrate on fishing. We do not like to be mistaken by any of the buyers for touching upon these sensitive issues. The only pragmatic solution that comes to our mind is that the buyers and ECBB should discuss the matter thoroughly. ECBB should agree to pay back the down payment made by each of the buyers at a substantial rate of interest plus some premium amount for the inordinate delay and the ordeals through which the buyers had to pass. What ECBB would do with their trawlers must be their own problem. If they want to export, before doing so, they should pay money as mentioned above or through a system mutually agreed to. Exports may bring in some extra money to the shipyard. A reasonable share of this can be passed on to the buyers to compensate for the long delay etc. mentioned above. There can be other solutions but the one mentioned above appears to be the most pragmatic. ECBB can rebuild its reputation by diversifying their building activities through joint ventures and other means.

Fishing Chimes, Editorial 153: November(i) 1991: Vol. 11, No. 8

# "The Designated Person"

For the first time, those in the fisheries sector have seen the term "Designated Person" in the rules published under provisions of SDFC Abolition Act. With the closure of SDFC, the Government in the banking division of the department of economic affairs took over all the functions of the erstwhile Shipping Development Fund Committee. The Banking Division cannot obviously perform the function of a financier. Accordingly, a provision in the rules framed under the Act gave the Government powers to appoint a "Designated Person" to perform the functions of erstwhile SDFC. As a follow up action to this provision, Government appointed the SCICI as the "Designated Person". From the time this appointment took place, problems started confronting the Banking Division, the fishing companies and the SCICI itself. Appeals for reliefs from the fishing companies started pouring in forcing SCICI, a professional financing agency, to take a stiff but at the same time a flexible stand to enable the companies to perform the function of harvesting marine animals and processing and exporting the same. The export activities relating to marine products gave to the nation in 1990-91 a foreign exchange of Rs. 900 crores. Each of the fishing trawlers, although owners of the bulk of them are in arrears in respect of repayments of loans and interests, have been virtually contributing about Rs. 50 lakhs per vessel annually in foreign exchange.

The supplication of the fishing industry has been that they were not having adequate surpluses to repay instalments with interest on the loan to the SCICI, the Designated Person. In contrast, the later, taking into account the levels of shrimp catches and earnings thereof by the owners were convinced that the enterprises had been making enough money sufficient for making repayments. On one side, it is claimed that the empowered committee is in favour of granting a moratorium of 2 years for the repayment of loans, apart from other concessions such as dilution of debt-equity ratio etc. Certain sources in SCICI say that no such decision was taken by the SCICI. All the confusion arose because of a press statement in the Economic Times recently stating that a moratorium of 2 years was favoured by the Empowered Committee Presided over by the Finance Secretary.

As repayments were not forthcoming, the SCICI took an experimental step of confiscating a few vessels of the defaulters. Efforts were made to auction these vessels but there was no success. SCICI was forced to return some to the vessels taken over. Some of those whose vessels were taken over did not want to get them back. The lesson learnt by SCICI after confiscation seems to be that, they have found themselves between the devil and the deep sea and that the only beneficiary was the receiver. If they had not taken over the vessels the fishing companies will get the obvious impression that SCICI had some constraints in taking over the vessels. Looking back, it is possible one gets the feeling that for reasons of lack of an organisational structure to auction such apprehended vessels or for running the vessels on its own, the experiment failed. All through its life, SDFC rarely ventured to confiscate vessels although confiscation is provided for in the agreements, This was probably because they were aware of the implications.

It looks to us that the ultimate result will be that the enterprises still holding the vessels would continue to operate them and SCICI will continue to be in a helpless position to make substantial recoveries.

We should say that SCICI, not really comprehending the consequences, has accepted to be the Designated Person. In our view they should have not agreed to act in this capacity, considering the peculiar characteristics of the situation. Their goodness has not helped them. On the other hand the organisation has become unpopular among fishing companies, although without justification. The banking division of the Department of Economic Affairs also probably feels that SCICI is being unduly harsh towards fishing companies, going by what the fishing companies say. Because of the interplay of circumstances ail concerned are bearing a share of the problem.

The result is that over 60 large fishing vessels have remained idle in the 1990-91 fishing season, forcing the nation to lose over Rs.30 crores of earnings in foreign exchange. Some situation must have been responsible for this. The strategy of handling the owners is obviously the foremost. In any case loan recoveries were slow in coming from the bulk of the companies. For this reason, should the nation have to lose so much of foreign exchange? The amount to be recovered is stated to be so small that it does not call for very harsh measures to counteract loss in foreign exchange, by keeping so many vessels idle. There has obviously been a failure on the part of someone in formulating a strategy to keep all the vessels operative during 1991-92 season and for no fault of theirs, SCICI bears the responsibility for the non-recoveries.

In the present situation we feel that the selection of SCICI by the Government as the Designated Person to look after post-SDFC affairs has proved to be a dismal failure.

SCICI could have told the Government that they would not like to be the 'Designated Person', so far as recovery of loans given by the erstwhile SDFC is concerned. Then the Government would have selected another alternate organisation to perform this function. All said and done, the extent of arrears are not so huge for the Government to get alarmed. It is only the notes written by some of the officials on the files exaggerating the situation makes the minister concerned to give assent to drastic measures which are not really warranted.

To our mind the activities built up under the auspices of erstwhile SDFC should not be mixed up with the post SDFC activities. Just because a company owes money lent by the earlier SDFC, this matter should not come in the way of the development of deep sea fishing activity further. The reason is that only when the fleet is expanded, marine products exports will go up. They do not materialise from thin air. Therefore, the accent should be to enable entrepreneurs to acquire additional vessels of suitable type instead of dampening the enthusiasm of the new generation of entrepreneurs who hear several stories from fishing enterprises about the attitude of SCICI towards recoveries of loans. Alarmed by this they prefer to take up some other line. The important thing at this juncture is therefore to create a climate that enables new entrepreneurship to go in for vessels with Government's assistance. If this is to be achieved it is preferable for SCICI to tell Government plainly that they are not interested in being the Designated Person of the Government under the rules framed under SDFC Abolition Act.

Fishing Chimes, Editorial 154: November(ii) 1991: Vol. 11, No. 8

# Diversification of Fishing: Unfulfilled Promises by Government and MPEDA

It will be recalled that the Ministry of Food Processing, which was very keen on diversification of fishing effort so as to boost up exports of marine products, announced a subsidy scheme around December 1990. Accordingly 30 per cent of the cost of diversification or Rs. 2.5 lakhs whichever is lower will be given to the concerned entrepreneur as an incentive. The entrepreneurs themselves are very keen to diversify their fishing operations not only for augmenting exports but also to make their vessels operational throughout the year. Accordingly, several owners equipped their vessels for diversified fishing for deep sea lobsters etc. They submitted prescribed application forms, duly filled up, to the MPEDA. Although a long time has elapsed, we understand that till date no applicant has received the promised financial assistance.

The aggrieved owners have made several representations to the Ministry of Food Processing and also to MPEDA which is nominated as the disbursing agency. So far there are no indications of any steps taken for releasing the amounts. What we hear is that the MPEDA is helpless in the matter for the reason that the Ministery of Food Processing has not provided to the MPEDA some clarification relating to disbursements sought by MPEDA. The fishing companies which have installed equipments for diversification are now restive. They feel that they have been let down very badly by the Ministry of Food Processing and MPEDA, in spite of the fact that they have installed equipments for diversification of fishing effort and have also commenced diversified operations.

Very few commercial banks are lending additional funds towards working capital or as medium term loan to fishing companies. Because of this, substantial amounts invested by these companies for diversification are now locked up. Owing to this situation the fishing companies are anxiously awaiting the disbursement of subsidies due to them. The reasons for the long delay are not clear. The authorities concerned have to look into the matter and ensure that the funds are released as quickly as possible. Further, the volume of documentation needed could also be reduced or simplified.

## Import of Fishery Requisites

Certain items of fisheries equipment, not available indigenously, if allowed to be imported free of duty or with a nominal duty levy, will go a long way in stepping up marine products exports. Authorities in the Ministry of Commerce, Ministry of Food Processing Industries and MPEDA are no doubt aware of this. It is learnt that MPEDA has been making serious efforts to help the entrepreneurs in securing the concessions, not only for their benefit but also for augmenting export production.

The items required so far as culture fisheries are concerned are prawn and fish feed and aerators. Aquaculture production will go up considerably and will bring a value addition of over 50 per cent to the harvested produce if the import is freely allowed.

In regard to the processing industry, import of IQF, blast freezing and fish processing equipment of various types including tuna canning will boost up exports in a big way. Added to these is the need to permit the import of packing materials of quality as acceptable to the importers.

For augmenting marine fishery production itself, the deep sea fishing industry needs to import equipments for diversified fishing such as long lining equipment and various types of fishing nets of new designs. The industry also requires certain electronic equipments to improve fish catches, some of which are: the net monitoring equipment, satellite navigators, radio telephones, VHP sets, weather display units, fascimile machines etc. Government should allow all these things to be imported by the industry freely without burdening the industry by import levies. As a special case Government has to release the needed foreign exchange for these items instead of asking the companies to depend on exim scrip for the purpose or clearance from Department of electronics.

Government has been invidious in regard to issue of exim scrips to fishing companies when compared to other industrial units. It is well known that there are traditional as well as value added products for exports. It is unfortunate that exim scrips are given to both the types of products at the same level, which has no justification.

The traditional exports consist of headless block frozen shrimps. The value of exim scrips for these could be the same as given at present. In respect of other items which have an element of value addition (items such as IQF products, blast frozen products, head-on or headless) the value of exim scrips should be at least 20 per cent more than those given for traditional products irrespective of type of freezing and importing countries. This particular aspect has been brought to the notice of MPEDA by the AIFI. So far nothing has been done in this respect. We feel that MPEDA should take up the matter strongly with the concerned authorities. Without this reform the fishery enterprises are not likely to take adequate interest in the production of value-added items.

Fishing Chimes, Editorial 155: December(i) 1991: Vol. 11, No. 9

# Government's Resolve to Unshackle Deep Sea Fishing Industry

All has not been well with the Indian deep sea fishing industry. The rise in operational costs of fishing vessels forced almost all fishing companies to continue to be in the defaulters list of SCICI.

Several efforts are being made by the Government and the SCICI to rejuvenate the industry which is a foreign exchange earning sector. The results, however, have been very discouraging from what we hear, with the exception of two companies (Fisheries Division of Navabharat Ferro Alloys Limited, and Prawn Magnate Limited, both located at Visakhapatnam), all others who took loans are in arrears to the financing institutions. How these two companies alone have been able to manage repayments of loan and interest up to date and others unable to do so is something about which some study may be called for. We do not mean that the two companies mentioned above are doing something special and others are lacking in managerial talent. In spite of their best efforts all remaining companies big or small, ran into arrears, low or high. One has to presume that the situation is mostly because of genuine reasons. Responsible companies running into arrears must be on account of serious problems.

In order to educate the nascent industry, the SCICI conducted a Top Management Programme through the IIM, Ahmedabad from 19 to 24 November, 1990 at Visakhapatnam. SCICI also formulated rehabilitation packages for most of the companies. Although a few accepted these packages most of the companies could not adopt them for fear of not having adequate cash flow. Most of the vessels are engaged in seasonal fishing and the rising costs and the imperativeness to spread earnings of six months to meet expenses for, over 12 months must have deterred them from accepting the packages. To have round the year fishing and earnings they have to equip the vessels for diversified fishing and they are unable to raise money for this. The AIFI represented to the SCICI and the Government about the plight of its members and sought certain reliefs, expressing their eagerness to contribute to the foreign exchange earnings of the nation. It is now common knowledge that the persuasive as well as practical actions of SCICI failed to induce the trouble ridden companies to pay the dues. The financing body took over the vessels of defaulting fishing companies and made efforts to auction them. The episode ended in a failure, whatever be the reasons. The rehabilitation packages offered so far also have met with the same fate. Added to the problems of operators is the reluctance of the commercial banks to extend or augment working capital support, despite several entreaties from the operators that they cannot operate deep sea fishing vessels without adequate working capital availability.

The fishing companies feel that the amounts payable by them are so small compared to other sectors that they deserve sympathy, encouragement and support from the Government. An approach of this kind would have led the companies to work hard, stepped up exports of marine products and brought in additional foreign exchange earnings to the nation. The harsh steps initiated led to the non-operation of over 60 trawlers and partial operation of most of the remaining vessels. It is now proved that whether the Government adopts harsh or other steps, the result would be the same. It would therefore probably be wise for the Government and the SCICI to think in terms of promoting and encouraging the industry, keeping aside for some time the concentration on recoveries of arrears. The normal way of prevailing upon the companies to clear the arrears, could however continue. The AIFI could be made increasingly responsible for the recoveries, since it is this body which makes representations to the Government and SCICI on the one hand, and is in a position to prevail upon its members to pay the dues on the other.

Being a popular Government, the representations of AIFI have atlast registered themselves on the minds of authorities and they have started considering the matter seriously. According to press reports the Empowered Committee set up by the Government under the Chairmanship of Finance Secretary to deal with matters of this kind advised the SCICI to take into consideration the representations made by AIFI for a two year moratorium period, provision of adequate working capital and give to the committee a revised rehabilitation package. It is hoped that this directive may yield results. The subject matter, in the context of over all development of deep sea fishing came up for discussion in the Rajya Sabha on 12th December, 1991. The Minister for Food Processing, Mr.Giridhar Gomango reacted positively, in his main answer to a question on the subject put by a member and also in his answers to the supplementaries. This attitude was very well received by the industry and restored their faith in the Government that it (Government) will stand by the industry, keeping in view the nascent state of the industry and its potential for earning foreign exchange.

In the national context, it appears to us that some sane person in authority in the Government would have to take a strong view that for the sake of overdues amounting to a few crores this entire potential sector cannot be paralysed and the activities cannot be jeopardised. When each of the deep sea vessels is contributing about half a crore of rupees in foreign exchange (in all over Rs.90 crores annually), it would be a correct policy to strengthen the sector despite the arrears. All said and done, the Government, which is mighty and all powerful, would not have any real problem in recovering the arrears. The reliance of the industry is on the benign nature of the popular Government, understanding of the problems of the industry in the right perspective and to provide the needed reliefs as requested for by the AIFI, chief among which are a) a moratorium for the repayments for a period of 2 years, b) release of adequate working capital, and c) certain concessions with regard to rates of interest. With these reliefs, a proper equation will get established between the Government and the industry, clearing the atmosphere and paving the path towards recoveries of pending dues.

Fishing Chimes, Editorial 156: December(ii) 1991: Vol. 11, No. 9

# Invidious Attitude of Ministry of FPI

The Ministry of Food Processing Industries has been given, among other things, the responsibility of stepping up the deep sea fishing fleet of the nation and accordingly the Ministry deals with clearances for the introduction of deep sea fishing vessels. This Ministry assists an Empowered Committee set up for the purpose through a Secretariat entitled "Secretariat for Approval of Fishery Enterprises (SAFE)" in taking decisions on various applications. To the extent the industry knows, pretty little seems to have been accomplished by this Ministry so far as the deep sea fisheries field is concerned.

The recent decisions of the Ministry in respect of permissions for operation of chartered fishing vessels would make anyone to raise his eyebrows. The word in circulation is that the ministry has decided to discontinue granting of new fishing vessel charter permissions. How this can be done is not known, when rules framed under MZI Act give a procedure for applying for such permissions. A provision of this kind automatically means that chartering of fishing vessels in all deserving cases will be allowed subject to various terms and conditions. This provision does not make any sense if the charter scheme is discontinued. Only by withdrawing or amending the rules, the charter scheme can be discontinued.

The letters of intent and permits issued to some contained a solemn assurance that an extension to the permits will be given for a further period of 2 years subject to satisfactory performance. In spite of this, there are quite a few instances, where, ignoring satisfactory performance, a few companies who fulfilled all the conditions and applied for extension have not been granted the needed extensions so far, although the applications were made long time back. On the other hand, some of those who have not fulfilled the conditions are understood to have been given extensions. We do not intend to mention the names of the functionaries, who are responsible for the situation.

The Ministry, so to say, has reduced the charter fishing sector into a major business proposition. The activity, ostensibly, is supposed to bring in new technology and provide training to Indian personnel. The Ministry succeeded, not in the fulfilment of these objectives, but in the non fulfilment of these main objectives. There are many in the industry who believe that the charter scheme has become a racket, the benefits of which are cornered by some, who became professionals in the line. In other words, these persons have developed a technology for continuously having foreign fishing vessels on charter. In this process one may get the feeling some persons in the ministry are giving scope for these unscrupulous elements to derive financial advantages.

The industry welcomes those making money through application of technological and investmental inputs, coupled with a non-crafty management approach. Unfortunately, the rules framed under Maritime Zones of India Act allows the Government to permit charters on an almost continuing basis through interpretations as suitable to it. It is high time the matter is reviewed by the Minister and Secretary of Ministry of FPI and set right the faults, omissions and comissions, if any, and make changes in the personnel in charge of the work for long, as overstayal in a post tends to increase chances of developing vested interests.

There are probably several companies who have been thriving well under the charter scheme for the past over 6 years through some device or the other, while some companies who have fulfilled the obligations and are eligible for extension of permits are unable to get the extensions as at the time of this writing. If there are valid reasons it is the duty on the part of the Government to clarify the reasons thereof and clean the stables of the charter scheme so that a healthy beginning can be made. The scheme as such is excellent and will be beneficial to the country.

Fishing Chimes, Editorial 157: Jan/Feb 1992: Vol. 11, No. 10 and 11

# Editor Explains

Almost all fisheries professionals forming part of Government fisheries departments, fisheries educational, research and training institutions, and those in various commercial fisheries enterprises feel the need from time to time to have information in respect of one or more of activities coming under the fisheries sector. Same is the case with entrepreneurs who intend to get into the fisheries line. Often, it so happens that great inconvenience is experienced by professionals and others in locating the needed information, which will often be required at short notice. The discomfiture is felt all the more, at the time of drafting articles, either to be contributed for being read at seminars, symposia, workshops etc. or for preparing project reports, or for evaluating the conditions of existence to set up a new fisheries project.

Keeping all the above aspects in view, your Editor has decided to bring out a Fisheries Diary, to contain statistical and other useful information. This was to be published as our January 1992 issue. We wrote to various authorities concerned in November 1991 for favouring us with certain information, knowing fully well that all concerned, being busy functionaries would either have to take some time to respond to our request or would not be able to send the information for some reason or the other. We started becoming panicky as time was passing by and the end of our compilation process was not in sight. We then reviewed the position. The options before us were either to go ahead with this adventure, although time taking, in view of the usefulness of such a compilation, or bring out a routine issue as we have been doing for the past over 10 years.

After a detailed deliberation, we came to the conclusion that we should bring out this compilation not as a diary (not being the correct caption) but as a Year Book, notwithstanding the inconvenience that will be felt by our valued subscribers, advertisers and contributors. It was felt that, in a good cause, some sacrifice has to be made by all concerned and the year book should be brought out.

As the readers would agree, the commencement of the publication of the first ever Year Book by us is important. When the first one is published, the subsequent year books annually would become somewhat easier. It took quite some time for us to realise that this is a challenging task and this made us much more determined to bring it out. We passed through exciting and also exasperating moments in the process.

We have to confess that we have not been able to acknowledge the sources of material included in this Year Book individually for the reason that we picked up the particulars from various sources in a somewhat haphazard manner. Nevertheless, with great gratitude, we hasten to thank MPEDA, Cochin and its various regional and subregional offices, and the various Central Fisheries Institutions;more particularly the CMFRI, Cochin and its various regional centres, CIFT, Cochin and its regional centres, CIFNET, Cochin and its regional units, CIFE, Bombay and its attached training centres, EIA, Madras and its various units, CICFRI, Barrackpore and its various regional centres, CIBA, Madras, NBFGR, Allahabad, NRCCF, Haldwani, U.P., Union Department of Agriculture, (Fisheries Division), Union Ministry of Food Processisng (Fisheries Division) and Directors of Fisheries of various States, more particularly Gujarat, Maharashtra, Orissa, Karnataka, Bihar, Madhya Pradesh, Punjab, Meghalaya and Mizoram and the Union Territories of Lakshadweep and Pondicherry.

We are somewhat unhappy that the Directorates of Fisheries of several major States, such as West Bengal, Andhra Pradesh, Kerala, Tamil Nadu, Haryana, Rajasthan, Uttar Pradesh, Assam etc. could not help us by providing the basic data we requested them to help us with. We do hope that in the future we will be favoured with the needed information which is no way of a confidential nature.

We do not claim that the Year Book is complete. As we gain experience we are confident that we would be able to improve upon the year book from year to year. Valuable comments and constructive criticism from our readers will give us strength in this respect.

Considering the fact that the issue has become voluminous, and taking into account the long time taken for the release of the issue, we have spread out the material in Jan and Feb 1992 issues. (Vol.11, No.10 and 11). In a compilation of this type there is a possibility of some mistakes, omissions and commissions made inadvertantly. We will be grateful if these are kindly pointed out to us for needed corrections in the next Year Book.

H.S. Srikishen, our Executive Editor, has virtually engrossed himself in pooling up data, and putting the same together. His assistance has been very invaluable to your Editor.

Fishing Chimes, Editorial 158: March 1992: Vol. 11, No. 12

# Extension Mechanism in FFDAs and BFDAs

Several technologies aimed at either increasing fish production or simplifying the production work have been evolved over the past several years by the Central Fisheries Institutions, particularly CIFA, CIBA, CIFE and certain Agricultural Universities. The process continues. The practices which are amenable for extension in the field, such as composite fish culture, integrated farming the latest in induced breeding, hatchery operations etc. have become extremely popular all over the country. To be specific, several techniques of achieving increased production of fish seed as well as table fish have been worked out by our scientists. Farmers in various States have also evolved improved practices on their own for augmenting production. By trial and error methods, the farmers themselves are able to develop artificial feed formulae of their own to the best of their abilities and also disease control measures, often taking tips from experienced scientists in research organisations. Several farmers in the country are known to have achieved high rates of fish and shrimp production, exceeding 8 tonnes per ha/annum. Likewise, percentage of realisation of fish fry has crossed 90 per cent level in the case of most of the farmers who are engaged in this work.

One sad aspect is that, barring a few instances, production levels in tanks and farms coming under the purview of Farmers Development Agencies, both freshwater and brackishwater, continue to be lower, compared to water areas brought under production by other farmers. This is happening in spite of the fact the agencies have an organisational mechanism to promote the activity and also have extension support.

## Weakest Link

It appears to us that extension aspect continues to be the weakest link in the FFDA and BFDA set-up, barring a few exceptions as in Orissa and Maharashtra. By not strengthening and making the extension function far more effective than what it is, the Governments concerned are undoubtedly neglecting this vital link. The forward and backward linkages of extension function have not been adequately taken care of. The result is that the per ha production which ought to have been the highest under these agencies tends to be far lower. It is true that the production per ha. in FFDAs, on an average, has reached nearly two tonnes per ha/annum but it will be agreed that there is still a long way to travel. We concede that achieving a level of nearly two tonnes per ha. on an average, is something to be proud of compared to production level of 600kg/ha/annum prior to the setting up of FFDAs.

## Need : Concerted Extension Effort

There can be no two opinions on the fact that the extension effort in the area of the agencies is too thinly spread. It should be possible to ensure very sharp concentration on the production work being done by each of the farmers, infusing into it the latest technologies that are flowing in swift waves. The introduction of OVAPRIM for induced breeding work with no chance of failure is one instance. Technologies which were once considered to be the best are now treated as obsolete. The foremost job is therefore to ensure that the extension officers update their knowledge at frequent intervals. The research institutions should introduce a system of sending monthly newsletters containing everything useful to the Extension Officers attached to FFDAs and BFDAs. Apart from this, the periodicity of training courses aimed at updating their knowledge should be increased and should be conducted at selected places in each of the FFDAs or in groups of FFDAs in every State. Duration should be so fixed that nothing is done in a hurry and whatever is told is absorbed by the E.Os. There should be no rushing through the courses. The extension officers should be encouraged to boldly bring out their problems and seek solutions. There should be sessions exclusively devoted to tackling all likely or unlikely situations that may arise in the farm work of a farmer, through simulated exercises. The chasm between the farmer and the Extension Officers has to be totally eliminated. In other words, the officers should function like another farmer, first among equals, and mix with farmers freely. A congenial climate has to be created. Once the backward linkages are developed on well drawn lines, the extension officers will have sufficient strength and technological capability to take care of forward linkages.

It will not be correct to entrust a vast extent of water area and several farmers to one extension officer to handle. This work should be developed in such a way that there will be a sharp focus on the farmer and his work. The extension officer should devote special time in the case of inexperienced farmers. The functions of the extension officers would no doubt include organisational aspects such as timely supply of inputs and adhering to the calendar of operations which a farmer has to follow. When the farmers are knowledgeable and become technologically competent, production increases.

Application of extension strategy as outlined above provides increased quantities of fish to the people. Increased production of shrimps from brackishwater as well as prawns from freshwater would generate higher earnings of foreign exchange.

The Ministry of Agriculture is no doubt doing all it can to strengthen the extension set up. The need however is to step up the equation with the State fisheries departments and the chairmen of the agencies to pay special attention to the extension function, on which depends the prosperity of the farmers and the flourishing of the agencies.

We have no precise estimate of the extent of land in the country suitable and can be allotted for brackishwater farming. How much of land of this category has already been allotted to various categories of entrepreneurs does not seem to have been compiled accurately as yet. However, according to the available estimates 11,00,900 ha of land is available in various coastal States, of which a very small extent has been actually made available to the entrepreneurs for development.

All the coastal State Governments have proclaimed their developmental interest and anxiety to allot these lands to eligible entrepreneurs. This is understandable. These lands are now lying practically fallow, yielding practically no income. It is therefore appropriate that the popular Governments in various coastal States are now committed to bringing these areas under productive use. The reason for the commitment is to provide means of livelihood to a significant section of the population who will put the land to productive use, and also pave the way for augmenting foreign exchange earnings through export of shrimps grown at the farms, to be set up over these lands.

## Slow Progress

The objectives are thus laudable. What is not laudable is the snail-like progress. In some States, such as A.P. there can be satisfaction that a part of available lands are allotted. But there has been no further progress therafter, by and large. It is felt that this situation stems from the fact that the subject goes beyond mere allotment. There are several aspects which will have to be taken care of by the Government concerned for realising the objectives.

## Pre-Allotment Phase

Let us look at the pre-allottment phase. During this phase, the fisheries departmental officers conduct a broad survey with as many details as possible and identify the lands suitable for the purpose. Although part of the same Government, since the ownership of selected lands vests with the Revenue Department, the problem of physical demarcation of boundaries of sites delineated on paper for allottment to entrepreneurs comes in. Hoping that this job would be done, the State fisheries departments (ex:A.P.) allot the lands. And in order to provide a positive report to the Government, the allottees are made to sign the needed agreements. The lessees, however, have no idea of the boundaries. For that matter, neither the State fisheries departments nor the revenue departments have a tangible idea of this. This situation naturally calls for a comprehensive survey and demarcation of plots. This is a laborious process, which nevertheless would have to be accomplished. Without this, lease registrations are not possible. This is one aspect which has been impeding progress. To get over this, surveys will have to be done in the pre-allotment phase itself. The extent of the job, the heavy work involved in accomplishing it needs planning and co-ordination. These calls for a separate cell. It will be a grievous mistake to ignore this aspect.

There is a much more important aspect from a technical angle which has to be fulfilled by the State fisheries department concerned. This relates to the pro vision of infrastructural facilities. Without these, the entrepreneurs will have diverse difficulties in setting up the farm and making them operational. The entrepreneurs will have to depend on the Government concerned for provision of power supply and communication facilities. It will also be necessary to see that the main sources of natural brackishwater supply are without blocks. In several cases the river mouths are silted up. Unless these blocks are removed by the State, the string of farms to be located along an estuary will fail to receive water supply. And this is enough to make the farms non-starters. In the case of power supply as well as road communication, unless Governments take initiative the entire developmental process will slow down which is what is happening now. It is our considered view that, prior to issuing notifications calling for allotment of lands for setting up brackishwater farms, Government will have to completes these exercises.

Another point relates to the need of trained personnel. To the extent we know of, no State Government has taken any organised steps to create a corpus of trained managers and other technical personnel for running brackishwater fish farms. We feel that this is the responsibility of the State, being a common problem. Dependence on other organisations in respect of training programmes has made the State Governments complacent. They could atleast ensure that there is a good measure of co-ordination between the central organisations such as CIBA and MPEDA and the State Departments themselves to achieve this objective. As matters stand, there seems to be no such coordination and the entire burden of training rests with MPEDA and BFDAs. Very little use is made of the organisational mechanism created as part of Brackishwater Farmers Development Agencies. A plethora of seminars, workshop and short term training courses are organised mostly by MPEDA at random, and without linking these exercises to a main framework, embodying the essential training needs.

## Allotment Phase

Now let us take a look at the allotment phase. On one side it is clear that the nation has expertise in setting up brackishwater farms and in successfully undertaking shrimp farming activities. TASPARC of MPEDA is stated to have achieved a production of over 8 tonnes of shrimps per ha in two crops in a year. So is the case in experiments conducted in Orissa with OSSPARC's participation. A few private organisations are also stated to have achieved a high level of production. This may be a plus point but what has been achieved by a Government organisation is of irrefutable value. It lends proof to a statement that, there should be no need to bring in foreign technology as proven technology is available within the country. Likewise, there are atleast two shrimp hatcheries in the country operating successfully. Technology relating to these operations is available. It should be the function of the concerned Government agency to popularise this technology, instead of stipulating a condition that a tie-up with a foreign agency to set up hatcheries is essential, irrespective of the consideration whether a piece of land allotted by the Government for setting up a farm can get water having the needed salinity and quality. Further, the training imparted by TASPARC in hatchery operation and management from time to time to successive batches seems to be no consideration to delete this untenable condition of foreign collaboration. Those who are lucky to have land with all facilities can certainly be allowed but this cannot be a universal condition. Further, to ask all the prospective allottees to set up hatcheries bor-ders on the ridiculous. Government cannot ignore technological aspects and at the same time tell an allottee having a land adjacent to an estuary to set up a hatchery there, for producing penaeid shrimp seed, when the salinity of these waters is much lower than what is required, apart from deterrants such as pollution. In fact, it should be the responsibility of the Government either to set up on their own or encourage suitable enterprises to set up hatcheries at selected locations congenial for the purpose. They should be helped in having the needed land.

Another condition that is stipulated at the allotment stage seems to be that each of the allotees must agree to set up their own feed mills. This, would lead to irrational sprouting of feed mills with odd capacities, creating a chaos in the future prawn/shrimp feed market. Hatcheries and feed mills will have to be dealt with as part of a separate plan integrated with the planning for setting up farms, but limited to the likely requirements of the farms to come up.

## Post-Allotment Phase

The post allottment time has to be totally devoted by the entrepreneurs to farm development work and farming activities. Financing agencies should be prevailed upon to extend the needed assistance, but results will emerge only when the State Government and other agencies stand solidly behind the farmers. If this is not done, there are bound to be cost overruns and a great time lag between allotments and reaching of the production phase to materialise. The allotees will be driven from pillar to post. They will be subjected to considerable harassment and their developmental mood will get eroded leading to desperation. All this can be avoided by organising the entire work in a systematic way, most of the aspects being take care of at the pre-allotment stage itself.

For the present, we refrain from commenting on the duplication of work and lack of co-ordination among the various promotional agencies concerned. Suffice it to say that these aspects also have the effect of impeding progress in this vital sector, which has a vast potential, of which readers are fully aware. The benefits, when work is organised and executed well, are three fold. One is bringing the fallow areas under production. Another is employment potential. The third one is the ability of the sector to earn for the nation considerable amount in the form of foreign exchange.

Fishing Chimes, Editorial 159: April 1992: Vol. 12, No. 1

# Tuna Long Lining as a Fishing Diversification System

The economics of the presently operated larger fishing trawlers, as is being often brought out, are moving from bad to worse year to year. These trawlers whose economics depend upon catching shrimp from the upper east coast are now able to perform around four voyages a year. A few of these trawlers cover up the remaining period to the extent possible by deploying the vessels for catching deep sea lobsters. This activity, however, has intermittent success.

Looked at from any angle, any vessel that operates for about 120-150 days in a year and is confined to the harbour for the remaining period, is bound to sustain visible or invisible losses. No wonder financing institutions are chagrined at this situation. More than the financing institutions, particularly the Ministries of Commerce, Food Processing Industries, Agriculture, Finance, and MPEDA have been facing a baffling situation that has been defying a solution. Apart from aspects such as inadequate utilisation of the capacities of the vessels, non-recovery of loans given for the acquisition of the vessels and interest thereof, policy and strategy matters relating to deep sea fishing have been going awry, with no solution in sight. The larger issue of harvesting the available potential of our EEZ stands relegated. Without all presently operated vessels becoming fully operational and without increasing the deep sea fishing fleet, there cannot be any substantial increase in our marine products exports.

There is some consolation that, with the shrimp culture sector coming up, exports could be stepped up, whether or not the deep sea fishing sector develops. If one goes by the present pace of progress of culture shrimp sector, from a national angle and the frequent slump in shrimp export prices, the future of marine products exports can be clearly assessed. Unless capture fisheries are further developed, the future will look bleak. Slow rate of growth, if nor stagnation, now faces us straight in our face. There is no future to marine products export sector without diversification of production and exports, as is being told to us by the authorities concerned.

The AIFI, under the leadership of its former President, Mr.S.M. Shukla visualised the impending peril and requested the Govt. of India to take the help of FAO to find a solution. This was indeed one of the few practical suggestions given by AIFI to the Government. With surprising quickness, Government approached the FAO. Thereupon FAO mounted a one-member mission (Capt.Marcel Guidicelli) to study the subject.

*Fishing Chimes* has been advocating for the past over four years that one solution is to incorporate the Florida style of tuna long lining system on the presently operated trawlers. The availability of adequate tuna resources during the months in which shrimp catches become slack has been amply demonstrated by the long liners of FSI, CIFNET and the operational results of a large no. of Taiwanese tuna long liners which visit Indian waters year after year. The repetitive long liners operations by Taiwanese chartered long liners clearly prove the potential. If they are not making money, they would stop coming, but this is not the case.

Capt. Marcel, after making a thorough study, came to the conclusion that the presently operated trawlers must be equipped with diversified fishing systems, chief among which is tuna long lining. He has preferred the American System, taking into account the simplicity of the system and ease of operations. He has also suggested, the adoption of trap and pot operation systems, chiefly for lobsters during the non-shrimping season. He has also advocated deep sea stern trawling, pelagic, as well as demersal, in such of the vessels which are suited for the purpose. Deep sea trawling can be done during the period when the traditional shrimp varieties can only be harvested in very small numbers.

The responsibility before the trawler owners now, if they want their vessels to become economically viable, is to instal tuna long lining system on their trawlers, before prices go up further. This installation would in no way come in the path of the normal arrangements for shrimp trawling. The Ministry of Food Processing (through MPEDA) has already announced a grant of 25 per cent of the cost or Rs. 2.5 lakhs whichever is less for the installation of a diversified fishing system on a trawler.

There is a duty cast on AIFI to initiate and see through a movement of the kind envisaged above, that all its members equip their trawlers for diversification of fishing effort. The Association must also prevail upon the authorities concerned for extending help to solve two problems which the owners are likely to face.

## Problems and Solutions

One major problem is the availability of finance. There is no doubt that SCICI would provide loans for the installation of the new system on the presently operated trawlers. One factor which may weigh in its mind is that the trawlers must be made economically viable and this is the only way to achieve this. They may, however, say no to such of the applicants who have either not paid the pending dues or have not accepted the rehabilitation package. On this point, the Association must make a strong representation to all concerned and, of course, to SCICI to treat all applications for loans for the new installation as a separate category of loan. A lenient view needs to be taken and loans granted irrespective of past performance which, in any case, is accounted for by the diminishing shrimping days.

The second aspect relates to customs duty. Tuna long lining equipment forms a permanent fitment on the trawlers. Under existing rules all permanent fitments on fishing vessels (as in other ships) do not attract customs duty. In the same way as the wire rope wound around winch drums is a permanent fitment, the long line and branch lines wound around the respective reels or tubs are permanent fitments. Same is the case in respect of the entire system. Customs authorities at various points, particularly Visakhapatnam, Cochin, Bombay, Madras etc., should be told not to levy customs duty on long lining equipment. In order to avoid doubts on the part of the assessing authorities at the time of arrival of the equipment, the Association should ensure that no duty is levied and orders to that effect are passed on. All this is necessary to facilitate swift installation and commencement of operations which will mainly have the effect of augmenting exports.

Fishing Chimes, Editorial 160: June 1992: Vol. 12, No. 3

# Deep Sea Fishing Policy: Some Drawbacks

Of late, there have been a spate of publicity releases concerning relaxations made by the Government for the development of deep sea fishing. These are based on the premise that exploitation of deep of sea fishery resources holds one of the major means of augmenting marine products exports, the other being brackishwater farming and/or possibly mariculture.

With a view to achieving higher marine fish production and exports of marine products, Government have formulated, amongst others, guidelines in respect of joint ventures in deep sea fishing. It has been mentioned that all benefits given to 100 per cent export-oriented units would be applicable to 100 per cent EOU fisheries units as well. Guidelines for a liberalised system of test fishing in Indian EEZ have also been announced. It is stated that leasing of foreign deep sea fishing vessels would be encouraged and broad guidelines in this regard have been announced. The liberalised procedures for applying to the Government for clearances under the various schemes have been published.

In response to the new policies announced by the Government, quite a few applications have been received by the Government and, of these, several have been cleared. These cover joint ventures, test fishing and leases. Of these, there are distinct signs of a few, but not all projects, materialising.

Broadly stated, such of the applicants having a close personal and business relationship with the counterpart foreign companies concerned and also with such of those who have a reputation abroad, known for their financial strength and business capabilities have a good chance of forging relationships in the fisheries field with known or introduced foreign enterprises.

The two categories of activities mentioned above, although dominant, would form, unfortunately, a very small segment of the Indian deep sea fishery entrepreneurship. In regard to the remaining categories, it appears to us that they may not be able to achieve a breakthrough. According to reliable sources, there is no doubt that there are a large number of used but good deep sea fishing vessels in countries such as Japan and Spain, The owners of such vessels seem to prefer selling their vessels based on outright payment and the prices are negotiable. Since full payment may not be possible, they may agree to the furnishing of a deferred payment guarantee from a prime bank acceptable to them and their bankers, coupled with a down payment. Another aspect is that there is a deeply ingrained suspicion or impression in the minds of foreign deep sea fishing companies that clearances from Indian Government for such deals would either not be forthcoming at all or may take a very long time, causing serious problems to them. Many seem to entertain the view that the liberalisation in regard to clearances etc. announced by the Indian Government does not have much of an effect on the bureaucratic procedures which they believe are mainly responsible for the delays. They feel that automatic clearances announced by the Indian Government would also entail lengthy bureaucratic procedures. Considering this situation, it is of utmost urgency that these deep rooted impressions are removed, by whatever device the Government considers the best, if the country is to make cognisable progress in the development of integrated deep sea fishing in the Indian EEZ as the main fishing area. At least, as a strategy, if Government clears a few applications within 15 days of their receipt, a favourable publicity wave may begin rolling.

One has to agree that it will not be an easy task for the Indian fishery enterprises and the Indian Government to erase the misgivings nurtured by foreign fishery enterprises. Another step in the direction of removing the misgivings therefore is to involve the foreign Government concerned and develop a basic frame-work, acceptable to both the countries, which would facilitate the enterprises of both the countries to cooperate in this mutually beneficial field.

Another step needed is to identify certain Indian banks and entrust the role of providing the needed deferred payment guarantees or other financial instruments to foreign banks concerned. The Foreign Governments concerned can be requested to identify and nominate a few prime banks and such foreign banks should be prevailed upon by the concerned Foreign Governments to take part in the endeavour which will be beneficial to both the sides.

Fishing Chimes, Editorial 161: July 1992: Vol. 12, No. 4

# Development of Deep Sea Fishing: Beginnings of the Inexorable Process

Indian Marine Fisheries Development progressed from an under-developed state that existed from 1950s to a developing status in 1990s. Starting with mechanisation of existing fishing craft, and introduction of synthetic twine (patiently inducing fishermen to adapt the reforms), the process moved on resulting in the introduction of several new designs of small boats and establishment of boatbuilding yards, where construction of boats commenced. Effective training courses were introduced. These innovative steps launched by the Government led to a total coverage of the entire coastline with vibrant fishing activities. Having achieved this part, the Government's move had been to extend the coverage to farther waters of the seas. By this time shrimps shot into prominence as an exportable commodity. The step taken consisted of introduction of a few medium-sized vessels, some suitable for catching shrimp and finfish from demersal layers and some designed for catching shrimp alone. The next step on the developmental ladder was to introduce a further number of medium-sized vessels for catching finfish and shrimp, the presumptive thinking being that vessels so equipped for two types of trawling take care of seasonal fishing for shrimp and finfish. The positive signals from the exploratory fishery work of the Fishery Survey of India and the operational results of chartered vessels having clearly proved the existence of heavy stocks of finfish and crustaceans in both mid-water and demersal zones, impelled the Government to formulate a new policy and strategy support to achieve the progressive exploitation of the virgin stocks referred to above, in the columnar and bottom zones of the deep sea, forming part of our exclusive economic zone. The Fishery Survey of India and the Taiwanese chartered vessels (trawlers as well as tuna long liners) have established, as already mentioned, the existence of sizeable stocks of finfish and cephalopods in the mid-water layers, and crustaceans including lobsters, and finfish in the demersal layer. Alongside, as part of the overall effort, a greater understanding of the economics of operations, financing requirements of technology support, suitable vessel designs to cope up with seasonal fisheries of India has emerged. The awareness and process of moving towards establishment of professionalism continues. New aspects learnt by the enterprises from the SCICI, the accumulating experiences gained by the enterprises, and above all the responsive support extended by the Food Processing Ministry, Ministry of Commerce and MPEDA to the Industry has been making a major contribution to the evolutionary process.

In this background, the inexorable process towards the utilisation of the deep sea fishery resources, demersal as well as pelagic, to the farther zones of our EEZ from the coastline has begun. The foundation which tottered for sometime because of the decision of the Government to abolish the erstwhile Shipping Development Fund Committee, withstanding the ordeal and the curses of entrepreneurs who assessed that this abolition meant the lifting of the ban on larger houses to obtain loans from the financing institutions for either acquiring fishing vessels or taking up Integrated Marine Fishery Projects.

By now the problem became two-pronged. One was to stem the deteriorating economics of the presently operated medium-sized vessels (true or partly true) which are heavily depending on shrimp trawling, only for the part of the year. The other was to achieve the introduction of larger vessels through join ventures, financial, technological, or both, or through leases, acquisitions etc.

The liberalisation policy of the Government of India has given a fillip and vigour to private entrepreneurship to take up projects to exploit fishery resources of our EEZ's deep sea section.

Readers are aware that, at the instance of Government of India, recently the FAO had fielded Capt. Marcel to study the problems impeding progress of deep sea fishing activities in India. The Captain's recommendation was that about 85 vessels out of the total fleet of around 176 could be revitalised and made economically viable, through additional investments, for the installation of equipments for undertaking another type of fishing during non-shrimping season. His formulation seems to be that such a step would pave the way for improving the economics of operations of these vessels.

From his report, it is surmised that, unless the presently operated vessels are refitted to undertake round-the-year operations, the owners will have serious problems of repayment of loans.

In this background, a view seems to have been taken at a higher level that a seven member Mission needs to be set up to work out the approach and detailed plans for the exploitation of deep sea fishery resources. The report of the FAO contains a recommendation that a Rs.15crore project, largely consisting of technical assistance and the needed consultancy services, needs to be taken up. Many in the industry have welcomed this recommendation but they feel that the estimate of financial requirements is on the lower side. The report of the FAO detailing the manner in which the presently operated fleet can be made viable is with the Government of India and the World Bank. It is stated that the World Bank has shown interest in extending assistance for revitalising the activity.

At an inter-ministerial meeting held in New Delhi recently the report of the FAO is understood to have been considered. The consensus is stated to be that a project report on the basis of the recommendations of the FAO should be worked out by the Food Processing Ministry in consultation with the Association of the Indian Fishery Industries. It is learnt that the Association had prepared an outline project and given the same to the Food Processing Ministry.

It appears that a recommendation has been made in this project that a Project Coordination Body to take care of the initiation and implementation of the project including periodical reviews, should be set up. Representatives of all concerned organisations are proposed to be taken as the members of Project Co-ordination Body.

The main provisions in the outline project, stated to have been prepared by AIFI, runs on more or less the same

| *Name of the Enterprise* | *No. and Types of Vessels to be Operated* | *Nature of Project* | *Projected Investment (Rs. in Crores)* | *Foreign Investment (in thousand dollars) and Equity percentage (in brackets)* | *Base of Operation* |
|---|---|---|---|---|---|
| Fishing Falcon Ltd, Hyderabad | Two tuna long liners | 100 per cent EOU Acquisition, Joint venture | 29.50 | 860 (40) | Madras |
| Oceanic Merchandise Ltd., Hyderabad | Three stern trawlers | Joint Venture,100 per cent. EOU Leasing | 17.47 | 167 (40) | Visakhapatnam |
| Target Marine Ltd., New Delhi | Two stern trawlers | Joint venture 100 per cent EOU Acquisition | 6.30 trawlers | 40 (40) | Goa |
| Leo Seafoods Ltd., New Delhi | Four stern trawlers and one factory trawler | Joint Venture by vessel acquisition | 15.94 | 56 (40) | Goa |
| Shivganga Fisheries Ltd. New Delhi | One tuna purse seiner | Joint Venture 100 per cent EOU Acquisition | 140 | (42.4) | Cochin/ Madras CM |
| Trading Co. Calcutta | Two stern trawlers and one tuna long liner | Joint venture 100 per cent EOU | 15.75 | 1452 (49) | Gopalpur/Ltd., Paradeep |
| Greaves Cotton Bombay | One unit of tuna purse seiner consisting of three vessels (catcher and two scout boats)- fish aggregating device | Joint Venture100 per cent EOU Test fishing | 5.40 | 800 (48) | Madras Ltd., |
| Leela Sea Foods Pvt. Ltd., Visakhapatnam | Four stern trawlers | Joint Venture100 per cent EOU Acquisition | 3.73 | 14 (19.35) | VSP |
| INKO Fisheries Hyderabad. | Two stern trawlers | Joint venture100 per cent EOU Acquisition | 11.25 | 167 (51) | Madras/ Cochin/Mangalore |
| Buouancy | Two stern trawlers | Joint venture100 per cent EOU Acquisition | 18.75 | — | Goa |
| Sea Joy Fisheries Pvt. Ltd. New Delhi | One stern trawler | Joint venture100 per cent EOU Acquisition | 3.97 | — | Goa |
| Sovin Sea foods Pvt., Ltd New Delhi | One factory freezer Trawler | Joint venture100 per cent EOU Acquisition | 47.05 | 116 (70) | Goa |
| Chaika Exports Pvt., NewDelhi | One stern trawler | Joint venture100 per cent EOUTesting Fishing | 10.53 | 43 (50) | Madras Ltd. |
| Indian Fisheries Ltd., New Delhi | 48 mini-lines and 12 multi-liners and two supply vessels | Joint venture100 per cent EOU Leasing/Acquisition | 598 | 80 (40) | To be determined |
| Indamar Fisheries Pvt. Ltd., New Delhi | Three stern trawlers | Joint venture100 per cent EOU Acquisition | 2.90 | 40 (40) | Madras |
| Oriental High Sea | One factory trawler | Joint venture100 per cent Calcutta EOU Acquisition | 14 | 230 (40) | Vizag Fisheries Ltd., Madras/Goa/Mangalore |
| K.S.K. Fisheries Ltd., Calcutta | One unit of tuna Purse seiner consist of one catcher, three skiff coats and two carriers | Testing Fishing | 18.90 | 920 (40) | Madras |

The thrust and direction of the developmental strategy now seem to be clear. Once the proposed National Fishery Development Board with adequate powers comes up and the Board adopts an effective system of channelising the activities on sound technical lines, there is no doubt that India, having such a long coastline and being one of the few nations having an Extensive Exclusive Economic Zone, will become one of the frontline nations, in the marine fisheries field.

lines of the FAO's report. The additional fishing methods recommended by Capt. Marcel, and also by the AIFI in its outline project, and one out of which is to be added to the presently operated trawlers are: (a) oceanic pelagic long lining, (b) demersal long lining, (c) lobster potting, trapping and deep sea trawling.

As matters stand, several technical and financial parameters of the project are still to be precisely worked out, apart from the technical personnel required to organise and push through the programme. According to available information, there were suggestions at the inter-ministerial meeting referred to above that the FAO should first constitute a versatile Mission consisting of an experienced Fishing Vessel Naval Architect, a Master Fisherman and a Resource Person, apart from the needed Indian counterparts. In regard to funds, these are expected to be provided by the World Bank. How far this Project would materialise is not yet known. Some more meetings between the representatives of the Government of India, AIFI, FAO, and the World Bank may have to take place for the project details to emerge in a precise manner.

According to information furnished in the Rajya Sabha, the Union Government has cleared as many as 16 applications for joint ventures in deep sea fishing under 100 per cent EOU and one application for test fishing. The total investment envisaged under these 17 projects is stated to be nearly Rs. 845crores. The most ambitious of all is the cleared application of M/s. Indian Fisheries Limited, New Delhi. It intends to introduce 62 vessels with an outlay of Rs. 598 crores. The foreign investment in the project will be of the order of US $800 thousand.

Fishing Chimes, Editorial 162: August 1992: Vol. 12, No. 5

# Brackishwater Shrimp Production: 'Satellite' Farming

Brackishwater shrimp farming has gained unprecedented popularity all along the coastline. Several entrepreneurs who have taken a leap forward as part of this momentum, now face the problems of securing quality seed and feed. In order to move towards solving the problems the Marine Products Export Development Authority has taken a major lead in the production of tiger shrimp seed. A few large companies have also set up small hatcheries for producing shrimp seed, both white and tiger. However, while the demand exceeds over one thousand million nos. (one billion) of seed per annum, the production is within 150 million numbers annually. In regard to shrimp feed, the dependence is, by and large, on imports from foreign countries, mostly Taiwan. No doubt there are several indigenous manufacturers of shrimp feed. None of them could so far succeed in producing feed that comes close to the quality of imported feed. There are, however, one or two exceptions. Amalgam Group, Cochin, has set up a feed mill near Cochin to produce feed in collaboration with a Japanese firm, M/s Higashimaru. The feed bears the same brand name as that of the Company. The feed is expected to be released for sale in the market, as soon as the trials, which are now on, are concluded.

An entrepreneur belonging to Nellore in Andhra Pradesh claims that the feed manufactured by him (Standard Feed) gives a good conversion ratio. According to this entrepreneur the shrimp harvested as part of semi-intensive farming by TASPARC at a farm near Nellore gave a production of over six tonnes per hectare. Extensive trials using this feed are probably required.

In regard to technology, the concerned Indian Research Organisations and several Indian consultants claim to have the needed technology and this appears to be true, although quite a few foreign technologists have been engaged by certain Indian enterprises. In this situation, the provision of these three inputs (seed, feed and technology) to cater to the needs of the various entrepreneurs, particularly the small farmers, has become a matter of paramount urgency. So far as shrimp seed production is concerned, the main organised public sources of supply are the two hatcheries set up by MPEDA through bodies registered under the Registration of Societies Act, one located at Visakhapatnam (TASPARC) in Andhra Pradesh and another at Gopalpur in Orissa (OSSPARC) with a total annual seed production capacity of about 120 million numbers. The demand from the farmers on the East coast alone exceeds 800 million numbers. Thus there is a formidable gap which has to be covered. This calls for herculean effort. The activity not only entails investments but also the availability of technology. Besides seed production technology, the other important part is the feed production technology. While much cannot be said about possibilities of manufacturing quality feed by individual small farmers, major enterprises in the field of shrimp production will have the capability of setting up feed mills of their own, basically to meet their own requirements. The surplus feed available can be provided by them to other needy farmers. If it would take time for the affluent and competent enterprises to set up feed mills, they would have no way other than importing the feed.

So far as production of shrimp seed is concerned, there are various constraints, not of finance alone. It might be possible, in the long run, for the small farmers to have their own hatcheries, adopting a system evolved by our scientists to meet the basic requirements of water, of needed salinity etc. However, as matters stand, there is no solution in sight for this. It would be difficult for small farmers to set up their own shrimp hatcheries. They have to therefore depend on large companies for securing seed. Water Base Ltd. of Thapar group has already progressed well in this regard.

Small farmers may have difficulty in acquainting themselves with the needed technology, which has to come from institutions such as the Central Institute of Brackishwater Aquaculture, MPEDA and state level organisations. One cannot visualise farmers from all over the country making a bee-line to CIBA located in Madras or to other units in the line for gaining knowledge in respect of technological inputs, from time to time. CIBA, as the main organisation, must have plans of setting up their sub-stations at various centres around which there is a cognisible brackishwater farming activity. This will however take a long time.

The efforts at neutralising the constraints dealt with above have led to the formulation of a solution which revolves round the subject of 'Satellite Farming'. The general belief is that this concept is the brain-child of the Marine Products Export Development Authority. MPEDA apparently has taken upon itself the tasks of developing brackishwater shrimp farming in the country, because of its interest in promoting exports of shrimps. The Authority is forced to play an aggressive role for the promotion of the activity because of the mandate it has from the Government in respect of stepping up exports of marine products. The Authority probably nurtures a feeling that the Coastal State Governments and the Union Ministry of Agriculture ought to play a more predominant role. These bodies however do play a major role in matters such as land allotment, although there has been no cognisible progress in this sector so far, but for frequent statements by concerned ministers and officials on the progress made which do not tally with the actual field situation nor in the matter of providing infrastructural facilities. Their main occupation at present seems to be to make the entrepreneurs to moon around them and tell them about sweet nothings. Probably finding the role of the Coastal State Governments, barring a few of them, not equal to the exigencies of the situation, and realising that further delay can be harmful to national interests, MPEDA continues the work. The pioneering work of MPEDA on the subject is laudable. MPEDA is taking care of the problems of small farmers in securing seed, feed and technology inputs. One can say that MPEDA is being more catholic than the Pope himself, so far as the shrimp seed/feed production is concerned.

With the 'Satellite System' hopefully coming into being, no doubt several practical problems may emerge and hopefully these would also be neutralised through appropriate solutions. In this context, one basic point that can be raised is "Would small farmers become Satellites to the Large Farmers"? It may be mentioned here that the intention is not to make the farmers satellites to the larger enterprises. The concept is believed to be based on the premise that the relationship between the larger enterprises and the small farmers should be of benefit to each other, without in any way materially upsetting the interests of both. If the proposed basis for introducing 'Satellite Farming' is considered to be in order, probably the expressive phrase that would reflect the real concept can be adopted.

The Water Base Limited belonging to Thapar Group will soon be introducing the system of 'Franchisee Shrimp Farming'. This system seems to be more or less on the same lines as that of 'Satellite Farming' now suggested, with the difference that in the Franchisee System the relationship between the farmer and the company will be formalised and spelt out clearly. Under this system of water Base, the target farmers (Franchisees) will be assisted in various ways. They will be given training. They will be assisted in the preparation of project reports and in securing financial assistance from Commercial Banks. They will be assisted also by way of supply of seed, feed, and other raw materials and will be provided also with technical support. The company, of course, will ultimately purchase the farmer's produce. It appears that the phrase "Franchisee Farming System" is much more expressive than "Satellite Farming", provided the franchise becomes acceptable.

No system can be foolproof. Human weaknesses can come in the way of working of the system. For this reason the WaterBase Limited has created a Division for promoting the System.

There are possibilities of both the small farmers as well as the large enterprises having advantage of the system. Apart from supply of feed and seed by the larger enterprises to the small farmers, the programme envisages the supply of the harvested produce to the larger enterprises by the small farmers at prices as mutually acceptable, or some times, as dictated.

There are possibilities of small farmers deriving certain benefits, such as money advances etc, from the larger enterprises. One has to hope that these advances would be wisely utilised and repaid, and that small farmers would not divert their produce to others, who will always be around to corner part of the produce offering inducements. However, if any of the farmers fall a prey to such manouevres by unscrupulous elements, this will no doubt be reprehensible. Further, another aspects is that there can be disputes arising concerning several aspects of feed and seed supplies. These can arise although it will be in the interests of the larger farmers to supply feed and seed of the highest quality to the small farmers at reasonable rates.

So far as the larger enterprises are concerned, taking advantage of the benefits to be extended to them under the system, they may step up the prices of feed and seed. They can penalise the small farmers for any default on their part (small farmers) by denying supplies, thereby leaving the small farmers in the lurch. They can also refuse to buy harvested produce, if farmers do not agree to sell the same at a particular market price. In situations such as incidence of damages to their farms, requiring repairs etc, small farmers may approach the company concerned for help. It is to be hoped that once good relations are established, the company giving the franchise will provide the needed help.

It may be true that in countries such as Thailand, the 'Satellite System' functions smoothly. The system however cannot be free from problems of the kind dealt with above. It would be therefore advisable for a small team consisting of an official, preferably from MPEDA, in company with a couple of competent small farmers and representatives of large enterprises to visit, say, Thailand and study the subject in great detail. This will enable the formulation of suitable system devoid of deficiencies.

Fishing Chimes, Editorial 163: September 1992: Vol. 12, No. 6

# Spread of Brackishwater Shrimp Farming: Major Hurdles

Several hectares of saline low-lying areas have been brought under production in various coastal states of the country. The MPEDA has made an unique contribution for the development of the farms by providing financial incentives and technical and technological inputs. In contrast, the contributions of the State Governments are far less unique. Most of the farms were set up in privately owned lands, mostly acquired by the entrepreneurs. There are, however, very few instances where Government lands have been made available to the entrepreneurs for the purpose.

Coastal State Governments are the largest single owners of saline lands along the coastline. The State Governments are development-oriented or at least expected to be. Such being the case, there must be some deep rooted cause inhibiting the work of allotment of lands by the various coastal Governments to the entrepreneurs. Otherwise, any Government infused with a developmental outlook and with all their inherent might and with the men and material at their command, should not normally have much of a problem in accomplishing the task of allotment of lands, on lease or otherwise, to progressive entrepreneurs or others.

By now we are accustomed to the routine. Right from the ministers up to the lowest functionaries, in the manner of a parrot, keep repeating that 60 per cent of the land will be allotted to the cooperatives and small enterprises, and that the lowest priority will go to the large companies. They even go to the extent of saying that the weaker sections complain that lands are being given to large companies to the detriment of the interests of weaker sections and that this is not true at all. In actual fact, with the exception of Tatas in Orissa and Hindustan Lever in West Bengal, no one else among large houses seem to have received any perfect allotment from the State Governments.

In Andhra Pradesh, a few allotments were made a few years back and the allottees were "forcibly" made to sign transfer documents related to allotments on lease. The allotments so made, are totally ineffectual, like marriages without knowing the bride and clearance for inter-course. Neither the Government nor the allottees know where the lands are. No physical transfer has taken place, mostly because the boundaries are not known. This is because there seems to be no driving force to accomplish the task. Sooner or later, no doubt all this will be done when there are development-oriented persons in positions of authority, at some positions right from the Minister concerned or the Chief Minister, upto to the level of Assistant Directors. For the time being one will be justified to conclude that there must be a mystery behind the dithering.

These are days of scams and allegations of various kinds. It is advisable that Governments act fast, allot lands on lease without fear or favour, unless the Governments want to deal with another serious problem to the existing list, if any.

Referring again to the situation in Andhra Pradesh, one would wonder, not having achieved anything tangible so far in regard to the first batch of allotments referred to above, another scheme was notified by the State Government for the allotment of some more saline lands on lease, with a third to follow (we understand). The various applicants were interviewed quite sometime back. So far none of the applicants could get any allotment letters. It is a great surprise that on a developmental scheme of this kind, capable of bringing in foreign exchange, neither at political level nor administrative level there are any visible signs of any decision making, except that all the big shots in the Government keep repeating the same clichés at various meetings, as already referred to. There is no way other than describing the situation as mysterious. Is the Government contemplating to demarcate the sites, allot, and make matters easy for the allottees to take over the allotted lands without problems or, is the Government thinking in terms of providing concessional tariffs on electric power utilised by the farms or other infrastructural facilities, before actual allotment ?

Nothing concerning any of these has ever come out which only shows that the delays are not because of developmental thinking. If this indecision continues for long, matters will be left open for the public to draw various conclusions, most of which will be probably unjustified.

It is axiomatic that one of the important functions of the Government is to bring to fruition developmental items of this kind, more so because of the export potential. Now there is no evidence so far that this responsibility is being discharged, so far as brackishwater farming is concerned, leaving aside the pocketful of World Bank sponsored projects concerning which also progress seems to be slow in several States. So much so, enterprises have resorted to acquiring private lands, which they have been developing with the help of the benign MPEDA and a few commercial banks and the SCICI. There is a strengthening feeling that, in terms of expenditure and incomes, taking into account the saving of time (which is also money, by way of not waiting inordinately long for allotment of lands on lease by the Government), the setting up of farms in privately acquired lands will cost about the same as setting up of farms on lands taken on lease from State Governments. There is a further advantage. Setting up farms on private lands will not be subject to several conditions which Governments are normally prone to impose, unless Governments want to inconvenience such prospective farmers. Time is coming closer for the tide to change, and interest to take Government lands on lease may wane and will be aspired for only as a last resort. This implies that Governments will have to work hard to put the lands into use for brackishwater farming, if there is going to be much more delay in the matter of allotments etc and very few coming forward. Of course, it will be a different matter if the Governments are not responsive and are least bothered about the utilisation of these lands.

Let Governments comprehend the writing on the wall and pre-empt the events that are likely to surface and not give room to misplaced conclusions such as lack of real developmental interest on the part of the Governments and to indicate that there can be some obscure reasons for the present state of affairs.

Fishing Chimes, Editorial 164:October 1992: Vol. 12, No. 7

# Reflections on Future of Indian Aquaculture

Dr. S.D.Tripathi, in one of his presentations at a seminar, said in a telling way that "India is a carp country". Most of these carps, whose original home, by and large, is the rivers of the north such as the Ganges, have now spread all over the country, through transplantations into a few other rivers and also simultaneously through the spread of major carp culture practices.

Major carp farming practices of a traditional nature were at one time largely confined to the north-eastern States of India, mainly West Bengal, Bihar and Orissa. Technological developments and extension work have taken major carp culture to the various regions of the country, with the exception of high altitude areas, where exotic mirror carp, trout, tench etc. have been established. A few exotic species, mainly common carp, grass carp, silver carp, etc. were also introduced. They have now established themselves all over the country and their seed is produced and utilised for farming, with the exception of some high altitude areas. Indian scientists and technologists, evolved several new technologies amalgamating the traditional systems therewith.

The combined efforts of the scientists and the farmers have brought fish farming into sharp focus as an economically viable venture. We now see all over the country several prosperous fish farmers and their prosperity, as the readers know, is the result of the relentless toil of fishery scientists, technologists and the farmers themselves.

The days of collection of major carp spawn and fry from rivers now largely remain as part of history, although collections of this kind are still undertaken here and there. Hatcheries have become the order of the day for the production of spawn. The nursery practices for raising spawn to the fry stage have become well defined. The activity has become amenable for commercialisation at various intermediate stages. This has unburdened the production system from pressures of work load and made the accomplishment of the job convenient in its entirety.

Types of hatcheries in use underwent various changes. Several versions of hatcheries, mostly based on the Chinese model have become popular. Artificial fish breeding systems also underwent several refinements, from the stage of injecting hypophysed fish pituitary gland to the stage of availability of bottled injectable liquid (OVAPRIM).

Various farming systems for achieving higher yields such as composite fish farming, integrated farming etc. have been introduced for achieving higher yields. Artificial feeding systems have come into vogue. Work on hybridisation involving genetic improvement with a view to achieving faster growth improved flesh quality etc. has been making steady progress. Innovative work on many other aspects is also in progress.

While much has been done and considerable research activity is in progress, one has to admit that all these are by way of extensions or improvements to the culture practices already achieved. The need now seems to be to open up new lines of culture activity for popularisation, without disturbing the environment. Cage culture systems, indoor culture systems and running water culture systems seem to be the openings to give a further boost to aquaculture, work on which CIFA is stated to be engaged upon. However, much more has to be done to give a further fillip to these activities.

Readers are aware that, in almost all countries of the world cage culture has an important place, either in the form of mariculture or freshwater or brackishwater culture. Sophisticated hi-tech systems have become very common. In several States of India cage culture has been tried on an experimental basis but without much of an impact. Modem cage culture systems consist of batteries of cages around which there are platforms improvised to facilitate movements of workers for feeding, harvesting, inspection etc. Automatic feeding systems are made part of the set up. The designs are such that easy transportation between and around cages becomes possible, because of heavy load capacity of working platforms and stable pathways. One can work on cage platforms even in rough weather. The design facilitates the cages to adjust themselves to wave movements. Cage systems are computerised. All technical data relating to the cages are stored and retrieved when required. Feeding is made automatic and can be done at regular intervals.

It is of utmost urgency that cage farming systems are introduced in India in a big way, with such modifications as are necessary with reference to existing foreign designs. A good beginning has to be made in this respect. The Government of India should sanction a pilot project, if necessary, under a bilateral assistance programme. Integrated cage culture centres, one in a suitable freshwater area and another in a protected bay can be set up. After the needed experiments at the respective centres, training schools could be attached to the centres to provide training to a selected no. of candidates from time to time. Policy decision has to be taken for the allotment of water areas to eligible candidates for undertaking cage farming with such conditions as are necessary to govern the allotments. The candidates must be assisted by means of suitable financing programmes to enable them to implement projects. Work on similar lines can be initiated for popularising running water culture practices and indoor culture systems. Entrepreneurs can be permitted to develop culture systems, particularly for river cat fish which can be exported in live condition to U.S.A. where the fish is popular. Joint ventures or technical collaborations in this respect can be promoted.

Fishing Chimes, Editorial 165: November 1992: Vol. 12, No. 8

# Diversification of Marine Fishing: A Momentous Follow-Up

The Ministry of Food Processing Industries has initiated with promptitude follow-up action in the desirable direction on the report of Capt. M. Giudicelli, Fisheries Consultant of FAO on the subject of economic up gradation of the presently operated shrimp trawlers along the upper east coast. The Ministry has set up a small Group of experts headed by the Joint Secretary in the Ministry, Mrs. Promilla Essar, to identify three deep sea fishing vessels either belonging to private sector companies (all these are shrimp trawlers) or owned by the Fishery Survey of India to carry out modification and demonstrate the efficacy of the operations with such modified vessels. The Group has also to consider and make recommendations in respect of the investments part, whether the entire cost of additional installations can be given as subsidy or a part thereof, in case private sector vessels are selected. It is indicated that the selection of the vessels of private companies may be done with reference to the present status of repayment of loans.

The main recommendation of the FAO expert is to add monofilament long-line system. This system originated in U.S.A. None of our shipyards have experience in the installation of monofilament long line system. The yard having some knowledge of this is the Bharati Shipyard, Bombay. Mr. Kumar of this yard has made a detailed study. The working out of the specifications and the preparation of drawings has a great lot to do with the existing machineries and their capacities and the additional ones to be installed. An Indian shipyard should no doubt be preferred. However, it would be desirable for the chosen Indian shipyard to team up with a foreign naval architect of experience, so as to make sure that everything is done properly and the possibilities of preparation of defective specifications and drawings, although unlikely, are eliminated. There were one or two past instances of defective designing, whatever be the reason. If the demonstrations fail in spite of this precaution, what should be done further will no doubt be thought of. Another aspect is that there is only one vessel in India having monofilament longline system. What we hear is that, foreign crew are essential for a fairly long duration to train Indian crew. In this light, it may be necessary for the Group to consider employing a tuna master fisherman and atleast three other crew members on each of the vessels selected for carrying out modifications. Unless this is done, the results, if they turn out to be unsatisfactory, will keep all concerned looking askance at the whole exercise. Capt. Guidicelli can probably be consulted in this regard.

CIFNET will soon be taking up a short term training programme to train tuna men. But this training will be in the context of the Japanese system which differs in several respects from the American system. Further, the Indian deep sea fishing vessels (Shrimpers) may not be suitable for incorporating the Japanese system. Hence there is no escape from incorporating the American system.

Squid jigging and trapping system can be incorporated as alternatives, as suggested by Capt. Marcel. However, it is doubtful whether any of these systems can be carried out for a six month period (non-shrimping period, in general). Stern trawling was tried for lobsters by several of the shrimpers having stern trawling arrangement too, but without much of long term success. Stern trawling for finfish may probably work.

The dimensions of this project need to be appreciated. A separate cell for the purpose may have to be set up at the Fishery Survey of India, in view of the technical content of the work, apart from earmarking an officer in the Ministry of Food Processing to look after this work from this stage itself. Several technical issues have to be looked into and followed up. It is suggested that the Group that has been set up for the purpose may look into this aspect as well.

One or two trawlers of Fishery Survey of India which are not yet equipped for longlining can probably be selected for the installation of American style of tuna long lining. This will have to be supplemental to the demonstration conversion of private sector vessels but cannot be considered as a substitute.

The considerations that weighed in favour of not including representatives of Fishery Survey of India and of Central Institute of Fisheries Nautical Engineering and Training in the Group that has been set up are not known. F. S. I. deals with fishing, whether it is survey or other wise. CIFNET deals with training. The representatives of these two bodies, in our view, should have found a place in the Group. Taking advice from them as members outside the group will be vastly different from taking advice from them as members of the Group. MPEDA also deserves a place on the Group or atleast deserves to be consulted. There is no provision in the order setting up the Group to consult others, other than FSI, a Naval Architect, Mr. C.S. George, and Capt. Giudicelli.

It is unfortunate the scope of the report of Capt. Giudicelli is restricted to 'deep sea fishing vessels'. It may be mentioned here that operations of mini trawlers have become uneconomic (There has been some improvement in the recent shrimp season, but the outlook is gloomy thereafter) and 'Sona' boats are on their 'Way' of joining mini-trawlers in the same plight. Monofilament long line system (30 km) can well be incorporated on both the types of vessels which have a OAL of about 50ft. Fishing News International, a reputed globally circulated fisheries journal published quite a few accounts and pictures of vessels of about the same OAL, with long line systems installed. The system can catch tuna, tuna-like fishes and sharks, depending on the size of hooks used. While these vessels are not classifed as deep sea fishing vessels, they still deserve attention, as they do fish outside territorial waters, in actual fact. In Sri Lanka vessels smaller than these (38ft) fish outside Sri Lanka's territorial waters with long lines, staying out for over ten days. NABARD can be consulted in regard to providing R&D finance for experimental conversion of one each of these vessels for undertaking long lining.

The intention of the Ministry of Food Processing Industries seems to be to keep the Group as small as possible. If this be true, there is considerable merit in this. Large Groups tend to complicate issues and prolong the deliberations. Yet the Group must have strong representation from the technical side. Mr. R.K. Verma is no doubt there representing AIFI and is a competent entrepreneur.

We feel that a representative of FSI and a fishing vessel Naval Architect must have an official position in the Group. Outside consultations with them may not guide the deliberations that well. It is not known as to why no provision is made to co-opt other experts in the Group as and when necessary. This is often provided while setting up Committees and Groups. Notwithstanding the foregoing observations, we feel that the Ministry of Food Processing Industries has done an excellent follow-up job.

Fishing Chimes, Editorial 166: December 1992: Vol. 12, No. 9

# Old (Legacy of Composite Madras State) and Present Kerala Marine Fisheries Regulations

Historically, Kerala fishermen have the misfortune of being brought under unjustified fisheries regulations, considered essential at the time of introduction, but later proven to be wrong.

In the composite Madras State, the North Kerala coast (possibly, South Kerala coast also) faced a major decline in sardine and mackerel fisheries in 1940s. Dr. Sundar Raj and Dr. Devanesan made a study of this phenomenon. Later a Special Cell in the Fisheries Department of the Composite Madras State was created with Dr. Devanesan and Mr.Chidambaram in-charge of the Cell. Dr. G.N. Mitra is stated to have been associated with the cell for sometime. After prolonged studies, the special cell came to the conclusion that oil sardines had a life of 14 years. Pointing out that the young ones were heavily harvested by small meshed nets such as *mathichalavala,* the decline in the annual (seasonal) fishery from 22 lakh tonnes to 9 tonnes was attributed to overfishing, mainly because of the usage of this net. Legislation was then introduced in respect of mesh size etc but the legislation could not be enforced successfully. As it happened, the sudden disappearence of sardine fisheries in certain years was later attributed mainly to upwelling and other changes in hydrobiological conditions. It was felt that there was no other way to explain as to how, around 1948, sardine fisheries revived dramatically, almost coming to the predecline fever. In any case it was proved that the regulation was misplaced and there was no perception at that time that the Indian stocks are part of the fishery of the region as a whole. When the conditions were congenial, the fishes just returned back. The story of sardines is applicable to mackerels as well.

The present rules to regulate marine fishing along Kerala coast seem to have its defeciencies, if not the same as in the case of sardines. Prior to trawling becoming popular, there was no emphasis on shrimp fishing in the sea. Shrimp fishing was restricted to backwaters. The only exception was that during the monsoon season, (*Chakra* period) fishermen used to operate cast nets to catch shrimps.

It has to be admitted that overfishing of shrimps takes place not out at sea but in backwaters too, to where young ones of shrimps migrate for feeding. It is this destruction that has to be stopped. In the sea, almost all the shrimps caught are adults. Shrimp is an annual sea crop consisting mostly of adults. In contrast, the shrimps caught in backwaters are juveniles. A major part of the backwaters of Kerala is not only polluted but it also abounds in stake nets for catching shrimps. (The same situation exists in several other Coastal States). The point sought to be made is that what needs regulation is catching of shrimps in the backwaters, if banning is not possible. From a scientific angle, banning will be the ideal solution. Fishing in the sea for shrimps would not require any harsh measures for controlling, as the available stocks would be adults mostly, having a life span of a little over 12 months.

By restricting the number of vessels operating during the shrimp season it should be possible to ensure the retention of adequate stocks for breeding and for leading to the next progeny. The restriction can be enforced without causing undue social distress. The present fleet of boats engaged in trawling can be divided into three sections. Boats falling under each section may be allowed to fish for two days. Six days can be covered this way, and one day can be a fishing holiday. Total banning of shrimp fishing does not seem rational, as the shrimp will die a natural death, if not exploited during the season. It is impossible to catch all the available shrimp. In other words, some will always remain behind, and they form the breeding stock. Even from a small shallow pond, it is impossible to harvest all fishes. Such being the case, harvesting the entire sea shrimp stocks also is not possible. This leads to the conclusion that there can be no fear of overfishing but only fear of reduction in unit catches. Battle for survival of shrimps should be fought in the backwaters and not in the sea. Shrimps of various kinds breed in the sea. The young ones ascend into the backwaters of Cochin in October-November. The only exception is Karikkadi *(Parapenaeopsis* spp*).* In earlier days the ascent used to be so heavy that the devotees at Vaikom temple used to have the impression that the shrimps were coming to worship the Lord. All this is gone now.

In Kerala, on the fish front, it appears that what is required is the banning of *Nonnavu* fishing which is carried out with a kind of cloth netting. In this fishing system, young ones of various types of fishes are massacared, particularly in the Vizhinjam area.

In this background, we have to view and evaluate the impulse of the successive state Governments of Kerala to appoint committees, one after another, to resolve the dispute between the two "Categories" of fishermen over fishing for shrimp in the territorial water during the ban season. The purpose may be either a dilatory one or a political or a socio-religious move but the position is that these committees have not helped in finding a solution. All believed in a banning system, although only during the brief *Chakara* season. The eighth committee has now been formed. Most of the previous committees helped the Government not in solving the problem but in strengthening the barrier between traditional fishermen operating non-mechanised boats (some using outboard motors) and traditional fishermen operating mechanised boats, who have now formed into groups, one ranged against the other. There is no realisation that the problem can be solved only by making efforts to bring the groups together. Political and religious interventions and support without having a proper understanding of the basic issue in proper perspective seem to have led to the deterioration of the situation. Obviously the situation must be so bad that it required so many successive committees, with no solution in sight and the Government continuing to be in a state of indecision. First Dr. Mammen tried his hand to solve the problem followed by several Committees headed by various experts. The Eighth Committee, the latest in the series, is the one set up under the Chairmanship of Dr. P.S.B.R.James, then Director, CMFRI. The other members of the Committee are stated to be Dr. Gopakumar, Director, CIFT, Dr. M. Sakthivel, Director, MPEDA, Dr. C.P. Verghese, Director, CIFNET, the State Director of Fisheries and Mr. John Kurein. It would be a miracle if this Committee could come up with an implementable solution, if it follows the same old lines of thinking. Among the earlier Committees, the one which had come closest to a solution was the Kalawar Committee. Being an experienced scientist and administrator, under the leadership of Kalawar, that particular Committee laid emphasis on restricting the number of boats to be allowed to undertake trawling operations during the monsoon season. Both the Groups consist of small scale fishermen. Non-mechanised boat operators include also those using outboards can be considered as small scale fishermen. Mechanised boat operators also fall under the category of small scale fishermen. Let it be hoped that the eighth committee (James Committee) would be able to work out a scientifically sound and an enforceable solution or formula, keeping our suggestion to introduce a phased system of regulating fishing during the shrimp season and lift the ban on monsoon trawling altogether.

Fishing Chimes, Editorial 167: January 1993: Vol. 12, No. 10

# Fishery Fulcrum 1992

The Indian Fishery Fulcrum could not receive in 1992 an adequately sound upward swing. This was because of forces of stagnation at work. The year turned out to be an appraisal period for the authorities concerned and the industry, with a view to identifying ways and means of imparting the needed momentum, particularly to deep sea fishing. Another consequential reason for the stagnation was stated to be the financial problems faced by the fishing companies, already under loan obligation to SCICI. The companies could not develop a proper equation and converge for talks at the same wave length with the financing body and the Government. One unprecedented event that took place during the year and in the history of operation of larger trawlers was the taking over by the SCICI of the vessels of chronic defaulters in repayments of instalments. SCICI succeeded in auctioning four of them. The auctions resulted only in recovery of throw-away prices.

Activities in the small scale marine fishery sector continued to be as in the previous year. Inter-state marine fishery disputes were very much in the news mostly on the east coast between Orissa and Andhra fishermen on one hand and between Tamil Nadu and Andhra fishermen on the other. The seasonal eruption of confrontation between traditional fishermen and mechanised boat operators recurred as before in Kerala State. The fishing disputes between Indian and Sri Lankan fishermen were on a far lower key than before.

Mechanised and non-mechanised boat operators in Gujarat, Kerala, Andhra Pradesh etc were greatly benefited because of satellite data interpretations conveyed on a regular basis to the various fishery organisations and companies by the National Remote Sensing Agency. Small mechanised boat operators developed the system of "Voyage fishing" extending over around seven days for catching chiefly shrimps and finfish. This system had improved the economics of operation of the boats. Small and medium-sized mechanised fishing boats developed confidence in moving farther out at sea for fishing.

Walkie-talkie system was introduced for the benefit of small fishermen at various coastal centres and fishermen accepted this innovation which was becoming popular. Fibreglass boats and beach landing craft became increasingly popular in the small scale sector.

There had been a perceptible improvement in the quality of life of fishermen because of several facilities provided for their catches to reach markets, and an increase in incomes thereby. Wider areas are now covered by an improved network of fish marketing but a lot remains to be accomplished. The long standing request of fishery enterprises for provision of adequate infrastructural facilities at various fishing harbours remains unfulfilled.

Fishermen are able to absorb new technologies and adopt the same in a far better manner than before.

FISHCOPFED, New Delhi implemented well the Accident Insurance Scheme for fishermen in collaboration with the insurance company concerned, thereby providing substantial relief to fisher families.

There were several accidents leading to deaths of fishermen, mostly out at sea, during the year, but these appeared to have come down compared to the previous year.

Bay of Bengal Programme of FAO conducted several important studies for the benefit of small scale fishermen. These were in respect of feeds for artisanal shrimp culture, manual fish hauling devices, introduction of economically viable fishing craft, socio-economic conditions of fishermen as they exist and the changes taking place, and on several other topics.

There were reports of poaching mostly by Thai fishermen in the Indian EEZ, particularly in upper part of the upper east coast. The chartered fishing vessel operations came down considerably and may cease altogether in the coming few years.

Cyclones were much more frequent all along the coastline which affected fishing operations. There were quite a few cases of Burmese fishermen drifting in their boats to coastal centres in Tamil Nadu and Andhra Pradesh.

In the case of larger trawlers, the number that effectively conducted fishing operations from Visakhapatnam and Madras bases, came down to around one hundred. The shrimp and fish landings fluctuated considerably, although the average shrimp landings per voyage remained around five tonnes, with marginal variations. An increase in brown shrimp landings in most of the voyages was conspicuous in the later part of the shrimp season, on the upper east coast. Because of the lower value of browns, the total earnings, in most cases, was stated to be just adequate to send the vessels back on the next voyage.

The new policies of the Government for increasing deep sea fishing fleet through joint ventures with foreign fishing companies, allowing Indian companies to take foreign fishing vessels on lease, and allowing the companies to undertake test fishing evoked fairly good response from several enterpreneurs, although the results seen in 1992 could have been much more impressive. Government no doubt cleared several proposals for joint ventures, leasing and test fishing (24 nos) as 100 per cent EOU projects. This was a very good begining which signified the opening of a new chapter in Indian deep sea fishing sector. Five factory vessels in the OAL range from about 60 to 110 m are stated to be operating by the end of 1992 with varying results. Considerable interest among entrepreneurs in acquiring factory vessels for exploiting deep sea resources took shape in 1992.

1992 saw the virtual non-existence of a financing body for the acquisition of fishing vessels. SCICI which was extending financial assistance until recently started showing symptoms of disinterest in this line of lending, ostensibly owing to the risk factors stated to be inherent in this sector. The stipulation of 1:1 debt-equity ratio for providing loans virtually left the industry in the lurch.

SCICI did extend financial support for a few integrated brackishwater farming projects. However, it is learnt that, towards the end of 1992, SCICI is stated to be not receptive towards brackishwater projects as well. The last one to have been approved by them and piloted for a public issue appeared to be that of Rank Aqua Estates Ltd.

Several entrepreneurs thought that they would be able to get vessels from countries which were constituents of the former USSR under joint ventures but, with the exception of a few companies, not much of progress could be noticed. The Russian States merely were willing to hand over the vessels to the Indian joint venture enterprises, subject to the condition that all repair costs as well as operational costs would have to be borne by the Indian side. Net profits to be shared on 50:50 basis. This detracted several Indian entrepreneurs as the cost of repairs was found to be very high. There were several others who had tried to develop joint ventures with Spanish fishing companies without any tangible result.

1992 marked a turning point in the perceptions of the Government as well as the industry concerning exploitation of deep sea fishery resources, both pelagic and demersal. There was a near unanimity that without diversification of fishing, the operations of existing trawlers operations could not be made viable. Foundation for introducing a larger number of tuna lengliners had been strengthened during the year. New companies such as Fishing Falcon, and Bay liners, are expected to introduce Japanese tuna longliners for operation in the Indian EEZ soon. This development is a sequel to successful long lining operations by chartered Taiwanese vessels, although there were wide variations in the catches. Out of the companies that introduced tuna vessels earlier, M/s. Lewis and Lewis is stated to have more or less stabilised their operations. The long liner of Sumura Maritime is stated to be on its way to stabilising operations. 1992 has the unique distinction as the year in which a 110 m long factory trawler flying Indian flag, bearing the name "High Sea Angel", was introduced by Oriental High Sea Fisheries Ltd. Visakhapatnam.

Considering the increase in price of diesel oil, Government introduced a scheme for providing a subsidy of Rs. 1.60 per litre of diesel oil consumed by deep sea fishing vessels. The subsidy would be limited to 10 per cent of the value of exports and the subsidy was available only to exporters. A diesel subsidy scheme was also in force for small mechanised boats.

On the processing side, adoption of diversified processing systems gained in strength. Several processing plants added IQF units and other machinaries for undertaking diversified processing.

The Export Inspection Agency modified inspection procedures to a large extent to enable the exporters to step up exports of marine products with greater ease.

The shrimp export prices fluctuated widely during the year, affecting the incomes of the producers as well as the exporters.

There had been a major improvement in the movement of reefer vessels, mostly belonging to Sea Land and APL for lifting frozen sea foods from the various parts all along the coastline.

The subject of Fisheries continued to be dealt with by the central Government, without much of coordination, mainly between the Ministry of Agriculture and Ministry of Food Processing Industries. While the subject of deep sea fishing and the control over the Fishery Survey of India continued to be with the Ministry Food Processing Industries, CIFNET, IFP and CICEF continued to be under the administrative control of the Ministry of Agriculture.

Quite a few fishing companies came out with public issues during the year. These were M/s. Alsa Marine Harvests, Fishing Falcon, Bay Liners and Water Base of Thapars. These companies came out triumphantly with over-subscriptions to their issues. Financing bodies such as SCICI emerged as successful manager for public issues of this kind. There were quite a few others in the pipeline in 1992. One of these was Rank Aqua Estates.

Matsyafed of Kerala, and Tamil Nadu and Gujarat Fisheries Corporations are understood to have operated their fishing fleet during the year with considerable success, although the operations suffered from the same drawbacks that caused anxiety to the private sector operators. Their success is attributed to their ability to secure adequate working capital which was stated to be not available to private sector companies. One aspect which makes the SCICI to point out often was that public sector corporations were not facing unfavourable economics of operation of their larger trawlers. When such was the case it was unable to appreciate the arguments put forth by the private companies in support of their inability to repay the loans.

On the issue of repayments to SCICI towards the loans given by the erstwhile Shipping Development Fund Committee, the Association of Indian Fishery Industries seemed to have taken the stand that the offer of rehabilitation for the sick units by the SCICI was fraught with difficulties. The Association had put forth before the Government their problems in operating larger shrimp trawlers. Government approached FAO who deputed Capt. Marcel of FAO for a study. In his report, he said that the operations of the vessels would continue to be uneconomical until such time as diversification in fishing through installation of one or more additional systems of fishing, consistent with the design of vessels, is undertaken and the results demonstrated. This report set in motion a major change in thinking and is expected to alter the complexion of the industry in the future. According to the Association the completion of this job would take some time. After a successful demonstration, companies would take time to equip their vessels for diversified fishing in accordance with the demonstrated techniques. It was accordingly requested by the Association that their members may be given time for repayments until such time the operations stabilised after the proposed demonstrations. The Union Ministry of Food Processing Industries set up a Group to study the subject and make recommendations. These were not forthcoming by the end of 1992.

From the foregoing, the year 1992 can be deemed as the harbinger of a new system and policy formulation In respect of exploitation of the deep sea fishery resources of our EEZ.

For the first time, the year 1992 witnessed shortage of skippers, mates and engineers of fishing vessels. The Central Institute of Fisheries Nautical Engineering Training drew up plans for meeting this challenge. Short term training programmes has been formulated. The Fishery Survey of India has also taken up a scheme for imparting training to 120 candidates in various kinds of fishing systems on their vessels.

The long delays in providing Government lands on lease for the development of brackishwater shrimp farming by the various coastal States prevented the achievement of the anticipated progress in this Sector. Development of brackishwater culture in Chilka area became a subject of severe controversy between environmentalists and Tata Aquatics, a joint sector company floated by the Government of Orissa and the Tata Group.

On the freshwater culture fishery sector, several FFDAs in the country fell short of the anticipated achievements, although the average rate of production on a national basis went up considerably, going upto nearly 2.1t per hectare.

Not much of improvement is seen on the reservoir fishery development sector except probably in Himachal Pradesh and in a few other States such as Tamil Nadu and Bihar.

The revolution in brackishwater shrimp farming received a fillip in 1992, in spite of the fact that there was practically no effective allotment and handing over of Government lands on lease to entrepreneurs. A large no. of private brackishwater shrimp farms sprung up. The extent of increase in this activity was so high that a good part of our shrimp exports in 1992 consisted of farmed shrimp. Another encouraging factor was that several brackishwater shrimp farmers imbibed a scientific outlook. Many of them adopted modern tehnologies in preference to rough and ready methods. Several private shrimp hatcheries came up. Quite a few entrepreneurs also set up shrimp feed mills. C. P. Group of Thailand is poised to make a major entry into the Indian shrimp feed market. They appointed a few sole selling agents for their feed in different coastal States to sell their products under different brand names. During 1992 several Large Houses and Companies also made an entry into the field of brackishwater farming with varying successes. Their representatives visited leading shrimp farming countries such as Thailand, Taiwan and China to study the systems.

The Gujarat Fisheries Aquatic Science Research Institute, Okha, made considerable progress in the culture of Artemia, a major feed item for shrimp larvae. The work done at the institute generated the hope that dependance on other countries for Artemia could be reduced to a large extent in the near future.

A major World-Bank-Assisted Project for the development of brackishwater farming, small reservoir fishery development, freshwater prawn culture was taken up in the States of West Bengal, Orissa, Andhra Pradesh, Tamil Nadu U.P and Bihar. The proposal to set up a National Marine Fishery Development Board had been receiving the attention of the Government for quite sometime and it is expected that in 1993 this Board would be set up with wide ranging powers. The likelihood of this Board being entrusted with the function of financing fishing vessels also is visualised. A proposal for the setting up of a National Fishing Harbour Authority had also been under the consideration of the Government. The ultimate decision in this regard is not yet known.

On the freshwater culture fishery sector the introduction of OVAPRIM by Glaxo Laboratories had ushered in a revolution in the breeding of carps with almost 100 per cent success.

There had been a silent revolution on a large scale in the various States in the setting up of Chinese type of hatcheries. Scientific fish culture under various systems, such as composite culture, polyculture and integrated farming took roots and several progressive farmers became professionally competent in freshwater fish culture activities. The per ha fish production went up considerably because of scientific application of manures/fertilisers, and adoption of sound feeding and aeration systems. An awareness among fish farmers to take preventive steps to avoid diseases among farmed fish was noticeable coupled with a demand for fish pathologists. Technology for producing freshwater pearls was perfected and efforts were on to transfer the technology to farmers in rural areas, preferably among enterprising womenfolk.

UDS surfaced in several States and farmers took several steps to tackle the disease but not with much of success. The general situation was the disease vanished in a few months of attack.

A significant achievement of the year was the emergence of a large number of fish farmers in States such as Haryana, Punjab and Rajasthan which were backward in this respect until recently. Considerable progress in this sector had also been noticed in the north-eastern States including the hill States. U.P, Madhya Pradesh, Bihar, Maharashtra and all southern States also improved their performance in freshwater fish production. The Trout project with Norwegian (NORAD) assistance, set up in Himachal Pradesh, made tangible progress.

Several freshwater prawn hatcheries were set up in States such as Tamil Nadu, Andhra Pradesh, Orissa, West Bengal, Gujarat, and others. The practice of farming freshwater prawns along with major carps is understood to have made cognisible progress.

A new college of fisheries was started in Andhra Pradesh at Eppuru near Nellore. Construction work on new buildings to cater to the increasing needs of CIFE, Mumbai, consequent to its declaration as deemed university was expected to commence soon. The Forum of Fisheries Professionals was inaugurated by Mr. Giridhar Gomango, the then Minister of Food Processing Industries, in September, 1992.

Fishing Chimes, Editorial 168: February 1993: Vol. 12, No. 11

# Establishment of Post Harvest Shore Infrastructural Facilities

Edgar Allen Poe said in one of his books that, "It is the most obvious that is often overlooked. This observation, although related to detection of clues relating to crimes, is aptly applicable to our marine fishery industry too, be it in relation to the small scale or large scale sector.

Value addition to marine products can be imparted only when there are adequate facilities for storage and processing of harvested marine or freshwater produce at lower temperatures, at all main centres of fish landing. This requirement should not be confused with medium to small scale storage and processing facilities that already exist in the private sector at several centres along the coastline or at certain interior centres. These are designed to cope up with the peak seasonal landings that the owners of the plants have to handle. When some storage space is free during lean seasons, because of reduced catch flow, it is mistakenly concluded often that there is ample surplus space. The same illusion haunts in respect of processing facilities. Capacities of plants are always designed keeping peak conditions in view.

The fishery of the entire coastal strip or the territorial waters which are part of our Exclusive Economic Zone is stated to have been fully exploited, if not over-exploited. On this premise, a breakthrough in increasing marine fish production can be achieved only by the fishing effort moving beyond our territorial waters and upto the edge of our exclusive economic zone. This will be possible only when we know the potential and we have trained personnel to conduct high sea fishing, in the operation and maintenance of shore plants and similar plants in factory vessels, and in external and internal market network.

We have certainly moved forward in all these aspects to some extent. But we are so much stuck up in the early phase of these things, we are neither in a blissfully primitive state nor in an advanced stage. Unless we get out of this 'Thrisenk' stage, it appears that no progress can take place.

On the resources front, the Fishery Survey of India told us that the estimated potential of our EEZ is 3.9 million tonnes of marine fish inclusive of cephalopods, crustaceans etc., category and depth-wise. Some evidence that these estimates are promising has been lent by the results of operation of an Italian vessel, and several pair trawlers besides tuna long liners and, very recently, by a factory vessel of Japanese origin but registered as an Indian vessel. These instances show that the future of the marine fishery industry lies in the deep sea zone, pelagic or demersal. The dreams about this future can be realised, it appears, by introducing large-sized deep sea fishing vessels, in numbers as required and by setting up post - harvest shore. Infrastructure facilities on a large scale. When these facilities are made available meaningful introducton of deep sea fishing vessels can take place. Most of the joint ventures in deep sea fishing cleared by the Government so far are *avataars* of the earlier charter scheme, remnants of which are still there, thanks to the ingenuity of our entrepreneurs. Ultimately, the joint ventures cleared so far may lead to the same end, if not in all cases, but in respect of most of them. The ultimate result will be the accrual of about 20 per cent of the gross earnings to the Indian side, with more or less the same involvement as the charter scheme, but for all official purposes having the features of joint ventures. The aim of several of the emerging joint ventures is to secure access to foreign vessels to fish in Indian waters. There will be a good number of genuine ones, though. In fact, all of them may prove to be genuine. One well motivated company is 'Oriental High Sea Foods' which is operating courageously an acquired foreign vessel, flying Indian flag. Quite a few out of the vessels likely to be deployed under joint ventures will be factory vessels. These and other joint venture vessels will take the catches or the products to a foreign port, by themselves. Very few will export such catches on cargo ships. All this means only one thing. The nation gains or loses foreign exchange (depending on the way we look at the activity) in the same way as in the case of the charter scheme. In all these cases, barring processed catches, there is practically no value addition. In the case of factory vessels certainly there will be value addition, but this cannot be the same kind of real value addition imparted at shore plants. The observations made in the preceding paragraphs, if judged as the correct position, bring us to the conclusion that the nation must establish common storage and processing facilities at all main ports for general good. These common facilities, set up at each of the main centres, must have several independent sections, each consisting of a frozen storage facility, and an adjacent space, to be leased out or hired out, along with the storage unit to entrepreneurs. The common or general facility will have to be under the control of a co-ordinating organisation. All expenses connected with the running of frozen storage unit have to be met by the entrepreneurs themselves. The entrepreneur concerned can also make use of the spare space as he wants, subject to the approval of the co-ordinating organisation. He can have a plate freezer or an IQF unit, with a filleting plant or a fish meal plant, a bone and meat separator or a surimi-plant, as he decides, depending on the nature of operations of his vessel/vessels and consistent with the availability of leased space adjacent to the frozen storage and as a part of its space. The products will then have real value addition.

Provision of storage facilities at pre-determined central points in the interior will help domestic marketing of freshwater or marine fish considerably, while opening up avenues for exports eventually, because of the facility and possible improvements or developments in the future. These facilities will also help in building up a network of frozen or cold storages to facilitate domestic marketing of freshwater, estuarine and marine fish and other aquatic species.

Those in authority have been talking of shore/interior based infrastructure for storage of fish and allied catches such as cephalopods, crustaceans etc along with processing facilities, for the past over three decades. All these averments remained as sound and fury signifying only pious intentions (and may be pretentions) with little useful infrastructure coming up that will be of utility for deep sea fishing operations and the common man. Apparently our Government does not take the need for infrastructure seriously, but at the same time cry hoarse for stepping up exports of marine products, little realising that, without augmented and specialised shore infrastructure facilities for handling deep sea catches, (to be landed by deep sea fishing fleet to be introduced), exports cannot go up. The confidence in certain quarters that increase in foreign exchange earnings can come through the development of culture shrimp fishery sector may prove to be 'illusory' soon. There are various inexorable forces arrayed against shrimp culture fishery development. As is known, most of the brackishwater lands are with the State Governments. For all practical purposes the attitude towards allotment of these lands to entrepreneurs on the part of the coastal State Governments, is mostly perfunctory. Heart of hearts, it appears, the Governments do not want to lease out these lands, although there may be one or two exceptions. A. P. Fisheries Department is understood to have prevailed upon the allotees of lands on lease to sign documents to the affect that they have taken over the allotted lands. This amounts to a subterfuge as no physical transfer has taken place. Neither the Government nor the allottees know the boundaries.

Apart from this, villagers object to entry of the allottees to take over the lands that are in their village limits on the ground that they would not allow outsiders to make use of the lands allotted. Further, environmental factors are heavily against conversion of most of the coastal lands into fish farms. The problems faced in the development of shrimp farm in Chilka area by the Orissa State Government in collaboration with Tatas is a glaring example of opposition from villagers and environmentalists. With the expansion of shrimp farming sector poised to slow down sooner or later, expansion in marine products export efford is bound to get retarded. The only outlet for expansion of exports will be the progressive but judicious exploitation of the fisheries resources of the deep seas. And this can be done, not by mere introduction of vessels, through joint ventures or otherwise, but by establishing the needed structural facilities at selected harbour points.

The setting up of shore facilities, in anticipation of the certainity of increase in deep sea fishing effort, is a major and crucial task. This cannot wait until deep sea fishing vessels are actually introduced. Only when an entrepreneur knows that there are shore facilities, he will work hard to introduce deep sea fishing vessels, with encouragement from the Government. He would not like to introduce the vessels first, and look for shore facilities to come up later. Entrepreneurs by themselves will have serious problems in setting up shore facilities of their own. These have to be part of common facilities to be provided by a central organisation.

The National Federation of Fishermen's co-operatives (FISHCOPFED) with its headquarters at Delhi has been performing a commendable job in helping fishermen in more than one way. The Federation has now international recognition for which the credit goes largely to Mr. S. Chandra, its then Managing Director. He was responsible for the widely appreciated and welcomed insurance scheme aimed at helping fishermen and their families in distress. The Federation has introduced a fish marketing scheme, although on a limited scale.

Being a National level apex body, this Federation is eminently suited to set up shore-based as well as interior-based infrastructural facilities. By establishing these facilities and running them, the Federation will be rendering yeomen service to the cause of fishery development and marine products and freshwater fish exports. It will be an extremely narrow view to think that it should exclusively take care of fishermen's co-operatives. As a body of fishermen and as a body connected with fish, it cannot isolate itself from the larger canvas of the fishery sector, of which fishermen's co-operatives form a part. In fact, this isolation that has been erroneously cultivated, has been the bane of the fisheries co-operative sector. As the Apex National Co-operative, it has to perform its legitimate job of helping not only the coastal and inland fishermen, but should also work for the development of all ramifications of the industry. Lop-sided developmental effort will ultimately prove to be disastrous. Whatever will be done in the co-operative sector should become part of the total national effort. It should not fall a prey to the motivated propaganda that coastal fishermen alone need attention, to the exclusion of all others. The area frequented by coastal fishermen already stands over-exploited and the Federation has a duty to show to the increasing fisher population an improved way to eke out a living. Their skills have to be upgraded and they have to be equipped to deploy themselves to fish in distant waters. For augmenting their incomes out of their present activities and to organise facilities for their development, shore or interior based infastructural facilities are needed. It will be also a sound promotional proposition to help other entrepreneurs who can be welcomed to join the co-operative field. There can be no exploitation nor encroachment of the coastal fishing zones now frequented by fishermen, by other entrepreneurs as the coastal zones are already overfished by 'Small' fishermen and through a code of conduct, operators of larger vessels do not fish in coastal waters, by and large. Deep sea creatures, rarely move into the coastal zone, nor would they generally influence coastal fisheries, contrary to what is normally believed. Such being the case, with the interests of small fishermen not being affected, by and large, FISHCOPFED would be performing a national duty by taking up a scheme for setting up shore/interior infrastructural facilities. Such facilities would enable small fishermen to store their catches and release them to the market, based on supply and demand, preferably through their co-operatives or otherwise. Such a strategy will yield higher returns. Co-operatives need not be the monopoly of small fishermen any longer, particularly as the fisheries in their area of operation are over-exploited and they have to spread out. Large scale fishery operators can also become Members of Fishery Co-operatives. The Apex body can earn money by taking the viable business proposition of setting up shore/interior infrastructure facilities and share the proceeds with members of the co-operatives, big or small. This would be additional income to them. They can participate progressively in deep sea fishing operations and eventually reach the stage of owning such enterprises through co-operatives or otherwise.

FISHCOPFED is the best suited organisation to bridge the existing missing gap in respect of infrastructure facilities. NCDC and the Government or the World Bank or a bilateral organisation such as NORAD or DANIDA would certainly come forward to extend the needed finance and organisational support and the Ministry of Agriculture and MPEDA can provide the needed expertise, if properly approached.

Port-wise, the basic task is to estimate the number of deep sea fishing vessels that are likely to be introduced. For this estimation, the permissions already granted for the introduction of vessels by the Government have to be taken into account, although it is difficult to say how many joint ventures would materialise. To this number, the vessels likely to be added over a period of time, consistent with the resources, have to be estimated. This estimation can be used to work out the likely extent of landings. On this basis a certain number of integrated frozen storage units can be set up at each of the bases. All such units located at a given harbour, have to be part of a complex and having access from one to another from an inter-connecting adjacent platform or varandah of sufficient height and width so that transport vans can come very close, flush with the platform for loading and unloading operations. The frozen storage of each unit shall have adequate covered extra space where, the hirers of the unit concerned can set up processing machineries of their choice with the needed facilities for work and also for the workers. Residential blocks for the workers can be constructed close-by to the storage units. Organisations such as co-operatives of fishermen, fishing companies or processors can take the units on hire from FISHCOPFED, which will fix up hire charges for specified periods. Processing equipments can be taken back by the hirers, when they leave the premises after the lease period, and after carrying out whatever repair work that may become necessary. State Governments can be prevailed upon to provide the needed land, power and water connections for the complex to be set up.

The general plan mentioned above can be one of the several alternatives that can be thought of and defined. The Centralised Operational and Management aspects have to be taken into account in working out any project.

FISHCOPFED should take the initiative, interpreting its role in a broader perspective and not limiting the scope of their work to co-operatives of small scale fishermen. The Federation can certainly take business ventures of this kind. Several imaginary problems would present themselves in deciding upon undertaking a mammoth national level job of this kind. These have to be got over, and with a missionary zeal the task has to be accomplished without giving scope to anyone to think that FISHCOPFED could not rise to the occasion.

Fishing Chimes, Editorial 169: March 1993: Vol. 12, No. 12

# Marine Fishing Industry Hypnotises, Lures and Detains an Entrepreneur

Barring a few exceptions, all entrepreneurs in the marine fishing line say, may be ostensibly, that they are losing money, not just from 1986, but before that as well. Before 1986, the position was bad. From 1986, frequent decline in shrimp catches, drop in incomes, and increase in operational expenses, reported to be taking place, made the position worse. It may be true there has been a substantial increase in incomes from 1986 because of weakening of rupee and higher dollar rates for shrimps etc., but this increase was not sufficient to offset the escalation in operational costs, according to the enterprises. These unfortunate developments jeopardised the industry at micro as well as macro level, the owners of fishing vessels wail.

There are several, specially the financing bodies, who feel that the fishing companies certainly do make money but they do not pay back loans. They may not be earning as much as they were getting before, but certainly earnings are adequate to discharge their loan commitments to financing bodies and others, the financiers feel. Their riddle is simple. Any one not making money in any business would like to pull out and try to move into another line. They are unable to understand fishing companies continuing to be in fishing business in spite of the proclaimed losses. Well, Marshall sea foods and Unimarine did pull out, after over ten years of operation of their vessels. This act of the two companies is against the general law that it is difficult to 'get into the fishing industry; Once one gets in he cannot get on; once he learns to get on, he cannot get out'. We said this in early 1980s when also the industry was in the throes of a crisis. The crisis now faced is probably proportionately more alarming than at that time. But situations of this kind are a common feature in the industry with varying intensities. The reason for fishing companies continuing in fishing business seems to be the lingering hope that they can soon recoupe tosses and earn profits. Who knows when jackpot catches would flood them. Bumper catches may be round the corner! May be this will happen next season! This kind of equivocation appears to be the burden of their optimism but turn of events of this kind had not been happening in adequate measure, observe these optimists. For them, in ultimate analysis, probably fishing is a gamble; and this hypnotises, lures and detains entrepreneurs, unless they are of the mettle of Marshall and Unimarine at one end of the scale or of the mettle of Navbharat, Nekkanti and a few others who pay instalments more or less regularly to SCICl at the other end of the scale.

Is the industry really in a crisis ? SCICI, and commercial banks may not agree and this is understandable. They lent their money or they have been asked by the Government who lent the money to the enterprises to collect the dues and it is their pious job to do so. All loanee owners virtually say that shrimp fishing has been so bad over the years and, because of this, they are not in a position to meet the loan repayment commitments. SCICI, among others, could not appreciate the stand and probably thought that only when the noose is tightened, repayments would start forthcoming. The noose did have some effect but they probably feel it is not that effective. For, the noose wouldn't make much of a difference to the owners in their present ordeal. One factor is that the trawlers stand hypothecated to the lenders. When payments are not made and even rehabilitation schemes also do not work, the ultimate that will happen is that trawlers will be taken over and later auctioned by SCICI as was done in the case of some of the trawlers. And probably they are prepared for it.

The owners of the trawlers must have evaluated the aftermath of SCICI taking over the trawlers on behalf of the Government who gave the loans through the erstwhile SDFC. Firstly, very few owners are understood to be interested in getting back their trawlers. This disinterest has passed on to the payment relating to insurance, port dues, maintenance, watch and ward, receiver fees etc to the SCICI/Government. These payments work out to be a fairly large amount, if the result of the auctions referred to are any guide, the amounts realised may not be adequate to meet the expenses involved. The final outcome of the exercise may turn out to be counter-productive. Firstly, interested new enterprises, after seeing all this, will think twice before opting for this line. Let us presume that the loanee owners do make money but give the impression that they are incurring losses. Such a presumption would not hold water since impartial agencies like FAO and DANIDA, after a detailed study, confirm the unfavourable economics of operations of larger trawlers, and even go to the extent of saying that at zero level of investments the trawlers just make a marginal profit. This corroboration is hair-raising. So far as the remedy is concerned, we have capt. Marcel Guidicelli of FAO telling us that the economics would certainly improve if the vessels are made to operate throughout the year and not just for six months, that too for catching coastal shrimp during the season of availability. The systems for incorporation on the shrimp trawlers suggested by him during non-shrimp season cover tuna long lining, use of traps for catching lobsters etc.

The enterprises now engaged in shrimp trawling with larger vessels have come to the conclusion that their days of shrimp trawling are over, particularly because of their heavy indebtedness and overdues, the drop in per unit catches and rising operational costs. This (likely) way of thinking probably explains their apparent apathy. Looked at from more than one angle, there appears to be no way for them to repay the loans. In this situation, Government/ SCICI will have to swallow the bitter pill and make consequential arrangements, which probably can take one of the following forms (The first step would have to be, of course to identify vessels that are now in a bad and irrepairable state and separate them for auction or other means of disposal). (a) to (e) given hereunder.

(a) Negotiate with the public sector corporations and hand over to each of them specified number of larger shrimp trawlers out of those found to be operable based on mutually acceptable terms.

    Financial assistance for diversification may be offered to the corporations.

(b) Take over the trawlers and give such of them found to be operable on lease or on negotiated rates to companies with good credentials for specified durations, stipulating suitable terms and conditions.

(c) Give them in lots to such of the companies having integrated facilities and whose repayment performance is relatively encouraging, on lease terms under mutually acceptable terms and conditions.

(d) Provide further financial help to the present owners for the installation of an additional system for diversification of fishing, to enable the owners to augment the incomes. Alternatively, the suggested beneficieries or allottees under a), b) and c) may be provided with the needed help.

(e) Leave things as they are so that atleast there can be some exports. Lenders can continue their strategy for recoveries, and loanees can continue their efforts to repay.

Strategy to auction the vessels will prove to be a very bad solution. Against this, it can be argued: In any case, the outlook is that there is no way of getting back the loans from the bulk of the loanees. Therefore, trawlers can be auctioned and the chapter can be closed. The implication of such as argument is that development of the industry will receive a set-back. Financing bodies may care less but Government cannot take this attitude. The present situation is unfortunate. But on this account, national requirements cannot suffer. Government no doubt invested lot of money in the industry. But the nation also enjoyed the benefits by way of foreign exchange earnings, deemed ones or otherwise.

The alternatives mentioned above may not be acceptable. Whatever be the position, there is no escape from searching for the most feasible alternative.

To be fair, Government has all along been taking more or less correct decisions and implementing correct strategies. At the time the fleet strength of larger trawlers was around 80 nos. Government did declare that no more of these trawlers would be allowed to be imported, as such an approach would have flooded the sea with these trawlers and consequential drop in per unit catches. But then the stipulation of the Government under the old charter scheme that for every one vessel operated on charter by an enterprise, one indigenous vessel has to be acquired, led to proposals for the acquisition of indigenous shrimp trawlers. Unfortunately, the availability of the shopfloor drawings of fishing vessels in India was limited to trawlers of 23 m class for quite some time and later to 25-28 m trawlers. Applications were therefore made by the charterers and also others for the acquisition of vessels of these ranges. Since only shrimp trawling was popular at that time all trawlers came to be equipped for that purpose.

Fortunately, and largely because of the circumustances, no one wants to go in for new shrimp trawlers now, nor would Government allow their further introduction. A beginning has been made and the nation is one the anvil of reaching farther and deeper parts of our EEZ for exploiting the fisheries of the area. In conclusion, it may be said that, at this developmental stage, Government may identify all older and uneconomical trawlers first and phase them out, treating the amounts due from the loanee owners as compensation given to them for agreeing to phase out their trawlers. A Committee can be appointed to identify such trawlers. The remaining trawlers can be equipped for diversified fishing and used for shrimping for six months and deep sea stern bottom/ mid-water/trawling and long lining for the remaining period by the beneficiaries, according to the final system to be evolved for healthy operation of trawlers, keeping in view the alternatives mentioned above.

Fishing Chimes, Editorial 170: April 1993: Vol. 13, No. 1

# Traversed into the Thirteenth

*Fishing Chimes* greets all connected with the fisheries field, on this occasion of its completion of 12 years of service to the sector.

Our subscribers are our Sultans. Readers are our Rajahs. Contributors of articles/papers to the journal are our kings. Advertisers are our archdukes. All of them are our patrons and are an integral part of the life of the journal. Fishing Chimes traveled with them all through the twelve years of its existence and continues to march on. Our gratitude goes to all of them.

The thirteenth volume of the journal has sprouted now, as signified by this issue (Vol. 13. No. 1). All of us here at *Fishing Chimes* no doubt work hard so as to bring the journal out issue after issue, but this is expected of us. What makes us salute the sultans. Rajahs, Kings and the Archdukes is the overwhelming support we get from them.

Over the past twelve years, bonds between us and the readership have grown stronger. The contributors of articles/papers have imparted a recognition and status to the journal. The journal has earned a place on the tables of Chiefs of fishing companies, Directors of Fisheries and Professors of Fisheries Colleges, Administrators, Technocrats, up-and-coming entrepreneurs and several others of distinction who are habituated to the journal and read it regularly. Most of the students of fisheries colleges in the country have acquired the habit of taking notes from *Fishing Chimes* as a means of enriching their knowledge and as an aid to perform well at examinations. The rewards we are given by way of spontaneous compliments spur us on to realise our responsibilities and upgrade our performance, to be of far superior service to the readership.

We believe that our journal provides material to those in authority on matters of strategy and policy formulation, and in respect of fisheries development.

By providing informative material on developments in the field in various States, the journal has been rendering service to the readership to the best of its ability. Now, FFDAs, BFDAs, and various fisheries centres in a particular State come to know of events in other States, thereby having the benefit of appriasal of work outside their State, in implementing their own programmes.

The deficiencies in the World Bank's freshwater fish hatchery project and the compelling conditions under which the Corporations concerned had to add to the farms the Chinese type of hatcheries in a modified form were brought out in *Fishing Chimes*. Many aspects concerning FFDAs, BFDAs, reservoir fisheries development, matters concerning trout fishery development in Himachal, deep sea fishery aspects, achievements of BOBP in small scale marine fisheries sector and in brackishwater farming, training programmes and scores of innovative happenings found place in the successive issues of *Fishing Chimes*.

The Union Department of Agriculture, the Ministry of Food Processing Industries, and the various State Fisheries Departments have been striving hard to develop our fisheries. Our Central Fisheries Research Institutes have been making remarkable contributions to the development of fisheries of the country. Events of importance that emerge as a result of their efforts that come within our knowledge and the comments of *Fishing Chimes* thereon are reported in the journal.

During the past twelve years there have been sweeping changes in the fisheries scenario of the country. Private sector has made a bold beeline to participate in the fisheries developmental endeavour of the Governments. Several new companies, large and small, and countless number of small farmers and entrepreneurs took to brackishwater farming, and freshwater prawn culture, in an integrated manner and also in isolation. In fact, the upsurge of fisheries activity is so phenomenal that it is feared to be of a magnitude that may upset environmental balance. On the marine side, a crisis in the 'deep sea' fishing sector has developed, spewing problems around. Diversification in deep sea fishing has been identified as a major solution to diffuse the crisis. All known developments connected with this subject along with appraisals have been presented in the journal.

In respect of brackishwater farming and in the matter of exports of marine products the contributions made by MPEDA are unparalleled. MPEDA has left no stone unturned for lending support to deep sea fishing and in implementing ameleoratory schemes of Ministry of Food Processing Industries. This Ministry made a major contribution to the deep sea fishing sector by granting certain subsidies and in giving fillip to joint ventures, leases, test fishing etc. in our EEZ. The Union Department of Agriculture and the State Fisheries Departments have been encouraging freshwater fish culture, specially through FFDAs and brackishwater farming through BFDAs and also through World Bank Projects. Fishing Chimes has been striving hard to keep abreast of the developments and report about them in the journal. Over the past twelve years one important role of *Fishing Chimes* that has emerged is to cover, in a large measure, proceedings of seminars, workshops, conferences etc on various fisheries subjects and present informative reports thereon in the journal.

On this occasion, we will be failing in our duty if we do not make a mention of our shortcomings too. Although the monthly issues of *Fishing Chimes* come out without gaps, time-lags of varying durations in the release of the issues take place, for reasons beyond our control. In this connection, we solemnly and humbly express our unflinching determination to progressively eliminate the time -lags, bringing under control the factors that have been leading to delays. The all-round co-operation, support and encouragement we get, gives us great strength and galvanises us to go ahead with a steady gallop and grace. We seek the blessings and best wishes of all of you for the good health, progressive prosperity and long life of the journal.

Fishing Chimes, Editorial 171: May 1993: Vol. 13, No. 2

# Reviving 'Deep Sea' Fishing Industry

The anouncement made by the Minister for Food Processing Industries, Mr. Tarun Gagoi, that auctioning of seized larger fishing vessels would be stopped came as a relieving news to most of the desperate larger fishing vessel owning companies. Some doubts are expressed in a section of the industry that the announcement of the minister is related only for stopping auctions of seized vessels. It does not cover seizing of the vessels. Whatever this be, SCICI already has the bitter experience of investing heavily on payment of port dues, insurance amounts, watch and ward, receiver fees etc in respect of seized vessels. One can believe that it would not like to seize some more vessels and spend further on the above mentioned items. The other problem is the question of thefts. There are unconfirmed reports that lot of materials from the seized trawlers have been stolen. It has to be hoped that this is not true. Nevertheless the temptation of pilfering things from seized vessels will be irresistable to some because of opportunities to get hold of accessible items on board, and get away. From these angles, if the report that Government has issued instructions for stopping auctions alone, there will be despair in the industry. A point of view is expressed that the Ministry of Food Processing has very little to do with seizing and auctioning or otherwise of the vessels, as the subject matter is the concern of the Ministry of Finance, Department of Economic Affairs (Banking Division). One cannot rely much on this argument for the reason that the decision is stated to have been announced by a minister who is bound by the combined responsibility principle on which the Government functions.

In this background and to achieve the best in this bad situation, it would be wise on the part of the SCICI to release the seized vessels, to enable the owners to repair them in order to make them ready for the impending fishing season. There is no doubt that, by holding on to the seized vessels SCICI will be increasing its problems, which may have to be avoided. One aspect is that there will be reduction, although small, in the total marine products exports and foreign exchange earnings. Whether exports, however small, can be sacrificed on the ground that the owners have to clear arrears to the Government is an issue which will no doubt be weighed carefully, keeping in view the severe shock already administered to the owners. The owners may be now in a chastised mood and may strive harder to clear the dues to the extent they can, if they have an opportunity to get back their vessels and resume operations once they can manage working capital. If the owners are not in a position to clear the dues there is not much that the Government can do. Some of them may just keep quiet prepared to lose the vessels as they are hypothicated to the Government. It can also be said that as the vessels are under seizure, owners could not have any means of earning in order to pay towards the dues. Looked at from any angle, it is apparent the SCICI would have to discontinue the policy of seizures and auctioning as the results of such a policy have proved to be counter-productive.

Several say that SCICI issued advice to commercial banks not to enhance the working capital limits of fishing companies and that this is one strong reason the defaulting owners put forward to explain their inability to earn and pay.

All concerned have led the industry into a vicious circle. If the companies pay, SCICI will release the vessels. Only when working capital is forthcoming, the owners can operate the vessels, earn money and make payments towards the dues. And only when the dues are paid, SCICI can release the vessels.

Another important angle is that the nation is fast losing credibility in international marine fishery circles. The policy of the Government is to encourage exploitation of deep sea fishery resources in our Exclusive Economic Zone, with technical and financial support from foreign fishing companies. The news that the Indian deep sea fishing industry is in doldrums and is on the brink of disaster as evidenced by the seizures and auctions has spread far and wide. It is stated there is considerable reluctance on the part of capable fishing companies abroad to have tie-ups with the Indian fishing companies. Unless urgent steps are taken by the Government to counter this wave and give a proper orientation to the news now spreading all over the world, by clarifying that the deep sea fishing industry in Indian waters is good, very little improvement in this sector can be expected. With the exception of a few fishing companies in countries of former USSR and a few other such countries, there is a general reluctance to come forward to enter into joint ventures with Indian companies. Countries like Thailand and Taiwan which have a very good idea of our waters can be an exception for the reason that these are the main countries whose interest in collaborations is restricted to such of the Indian companies about whom they have previous knowledge (having operated their vessels on charter etc.). In the above background, it is imperative that the Government takes a pragmatic decision to resolve the present crisis.

The SCICI and some of the Officers in the Government feel that the fishing companies concerned are making money but are not making payments. But problems of this kind will have to be resolved by the authorities in a manner that the international reputation is not besmirched, and the industry is put on an even keel.

Experiences show that seizures and auctions would not help in any way to solve the problem. A section of the industry may be taking advantage of the situation. To prevent undesirable developments, wisdom lies in returning back the seized vessels to the respective companies, treating the investments made by SCICI towards port charges, insurance etc. as additional loan to the respective companies.

SCICI feels, rightly so, that most of the loanee companies, by and large, do not pay at all. They offered a rehabilitation schedule and the response to this is disappointing. On the other side, a strong feeling exists among the loanee companies that the SCICI is harassing them. This situation is unfortunate. Government alone can find a solution. In fact, it is the function of the Government to find a solution. Some owners can be crafty, but it will not be correct to think that all owners are crafty. There must be a real problem which has led to the present situation.

It appears that Government should restore normalcy in the industry and keep the 'harassed' companies at ease. Similarly, Government should not burden SCICI with an intricate subject like this which involves developmental aspects of great import and defence considerations as well. Government should consider relieving SCICI of this difficult function, particularly when this financing body no longer deals exclusively with fishing vessels and ships. There can be no two opinions on the misconceived seizing and auctioning system. Many believe that a strong scrap dealers lobby is active to buy the vessels for a song.

Apparently, taking into account the report of FAO on the present situation and representations from the industry, Government has set up a high power technical committee recently to identify the problems and recommend measures to restore normalcy in the larger vessel fishing sector.

Fleet strength has to be no doubt reduced as recommended by the FAO expert but selling the vessels in an auction for ridiculously low amounts does not make sense. Instead, leaving the vessels with the owners, and exhorting them, for what it is worth, to pay the dues, with the consideration that atleast there will be some extra export earnings would be a better proposition in this difficult situation. Scrap merchants already purchased from SCICI two expensive vessels which are worth over Rs. 30 lakhs each for a token sum of around Rs.3 lakhs each. Government can surely prefer to allow the vessels to be operated by the present owners, instead of earning just Rs. 3 lakhs by sale of each of such vessels. This amount may not be sufficient even to meet the expenses on port charges, insurance, receiver's fees etc. The officials of the Food Processing Ministry could have advised the Food Processing Minister to visit Visakhapatnam and make a personal appeal to the owners to strive hard to operate the vessels effectively and to pay the dues, pointing out the gesture of the Government (Provided this will be forthcoming) in returning the vessels. Done this way, there can be a distinct possibility of a good response. The Minister could appeal to the companies to see that the national honour in the international fishing sector is protected. A pliable and accomodative repayment schedule could be announced by the Minister. An alternative solution put forth by a section of the industry is that SCICI may take possession of such of the vessels under heavy default and give the same under a management contract either to the owners concerned or to another body for operating them and paying the net profits to SCICI for offsetting towards the loan amounts.

It is true that, with the exception of a few companies such as Navabharat Ferro Alloys and Pron Magnate, practically all others are in varying degrees of default. We have had an occasion to talk to persons representing the management of these two companies and they swear that they pay the instalments and interest out of earnings from their vessels only. Those representing T.N. Corporation (another good pay master) also says the same. While this situation raised various questions, one has to agree that so many owners could not have been managing their affairs wrongly and are being crafty so as to avoid repayments. The explanation for non -payments seems to lie in the cost of the vessels and loans taken. Older Vessels costed far less than those acquired after 1985 and the loan instalments were lower accordingly.

It has to be remembered on the part of the Government that deep sea fishing industry contributes per annum by way of foreign exchange much more than the initial investment on the vessels, in foreign exchange. No other sector of the economy can do this. As repayments are not being made by a substantial section of the industry, it will be prudent to go into the causes, identify real problems and set right the situation. The erstwhile SDFC never seized or auctioned any vessels which they financed. It was a Government body, presided over by the Secretary, Shipping and Transport. The officers knew that any resort to seizing and auctioning would make the problem much more intricate and complex. Government in the Ministry of Shipping and Transport were only keeping in view the larger interests of the nation which is that our larger sea fishing fleet should increase and that certain unavoidable sacrifices would have to be made.

History tells us that in all new and innovative schemes there would be success only through sacrifices. Government spent money on subsidies under several new developmental schemes and when the activity took roots and when there was no further need to give subsidies, they were withdrawn. We do not say that Government should just write off loans. Government should get back its money. In order to realise the dues, harassing and bullying the companies and allegedly telling commercial banks not to augment working capital limits of the companies, can lead only to situations like this. The vessels stand hypothecated to Government and can be taken over at any time, although they may be ageing. But then, a persevering and benovelently firm policy would be far better than selling working vessels to scrap dealers. One cannot make omelletes without breaking eggs; And a

Government cannot develop deep sea fishing without financial disadvantages to the exchequer. The introduction of nylon twine, mechanised boats, outboard motors and many others could take roots only because of well thought out policies. It will not be correct to say that Government's money was lost. In fact, the Government and the nation had gained. The new activity attracted the attention of a large number of entrepreneurs who took so much of trouble to get the vessels. For this participation in the endeavour of the Government, the industry has to be rewarded by a sympathetic and understanding attitude, particularly at this time when future of the industry is in peril, by formulating a pragmatic line of action.

In any case, auctions are not a solution to the problem. Owners should not be given the opportunity to formulate a strategy to get back the seized vessels at a very cheap price in an auction, without the botheration of paying the accumulated dues and to prevent the scrap merchants from buying the vessels in an auction. An owner can manage to set up a representative of his to participate in the auction and acquire the vessel at a low price, thereby getting rid of the accumulated dues. We say this only to prove that the present approach will be counterproductive. Selling working vessels at a low price is worse than leaving the vessels with the owners and helping them to improve the operations so that they may have cash-flow for repayments.

As we had pointed on several occasions earlier, the need is to enable the vessels to operate round the year. As Capt. Guidicelli of FAO also recommended, diversification of fishing activities can alone improve the situation. If Government can succeed in pulling out around 40 vessels to undertake types of fishing other than shrimping for part of the peak season atleast, with whatever incentives that are necessary, considerable progress can be made. It is good news that Government set up a technical committee to go into the question in totality and make recommendations.

We strongly advocate that (a) all the seized vessels be returned to the owners (b) make arrangements for providing the needed additional working capita! (c) the Minister of Food Processing Industries may personally intervene, announce a new package of measures for rehabilitation of the industry, pointing out the expectations of the Government to place India in the upper slots of global deep sea fishing operations, (d) set up as quickly as possible the National Fishery Development Board and also entrust to it the function of financing the industry and nominate this body as the designated person in the place of SCICI for the reason that SCICI is far too busy with numerous programmes not relating to ships and fishing vessels, and last but not the least, (e) enter into bilateral enabling agreements with selected foreign nations with a broad frame-work of cooperation for bringing about joint ventures between India and the countries concerned. We may point out here that emergent steps have to be taken so that this industry of crucial importance to the nation survives, and repercussions such as providing an encouraging situation for other nations to intrude into the fishing areas of our EEZ and a humiliating situation in which the Indian deep sea fishing industry would be forced to sink are eliminated.

Fishing Chimes, Editorial 172: June(i) 1993: Vol. 13, No. 3

# Punishments and Rewards

The 'semi-deep sea fishing fleet' of India consists of around 175 vessels several of which are non-operational. Most of the vessels were financed by the erstwhile Shipping Development Fund Committee and the remaining few by the SCICI Ltd. A majority of the companies owning the vessels are known to be under heavy default in respect of the loans and interest thereof. Several of the companies are stated to have not repaid any part of the loans taken by them. There are however a few companies who have been maintaining a good track record of repayments.

The SCICI Ltd, happens to be the 'designated person' nominated by the Government of India to exercise the functions of the erstwhile SDFC, which was its own. SCICI Ltd appears to have come to its wit's end in respect of recovery of loans and consequently resorted to offering rehabilitation packages on one side for some and taking possession of such of the loan vessels which are under heavy default. A beginning was made for auctioning some of these vessels. Initial failures in this exercise were faced. There was success in auctioning four vessels subsequently but to the advantage of the buyers, as can be expected. The effect of this inescapable measure was that a) the owners concerned were punished by depriving them of their means of income and b) a warning signal was sent across to the others on what would happen, if repayments were not made, c) the defaulters faced problem of securing needed working capital and d) the image of the industry was damaged through loss of credibility and e) No bank or private credit as needed was forthcoming to the owning companies. One can say that if there are deliberate defaulters they are well punished. And along with them, few others also are forced to face problems of working capital etc.

The non-defaulting companies have the satisfaction of good performance. The defaulting companies have a point that the few non-defaulting companies could pay the instalments for reasons of lower quantum of loans. However, as there are several others similarly situated with a lower quantum of loans but could not repay, the argument is not considered forceful, although there could have been extraordinary operational and managerial reasons for the inability to pay.

Satisfaction of having performed well is a supreme emotion and no reward can surpass this. However, some of the few who did well have other feelings. Should they not have come down to the performance level of those who defaulted, and avoided repayments, and be at par with the others and enjoy the benefits such companies enjoy? What have they got now for their good performance, except the vessels that outlived their life, as is the case with others, from whom, in any case, very little money towards the loan can be collected by the Government or SCICI now ?

The non-defaulting companies, in the same way as the defaulting companies, took loans. The difference is that the former paid back the loan instalments. They did what was expected of them. Yet, considered from the angle of the unimpressive performance of the majority they probably deserve demonstrative recognition, so that it will have the effect of toning up the developmental activity and taking the industry forward.

Reduction of shrimping effort and taking up other types of fishing has been the new strategy that is being advocated by the Government. Some developmental expenditure on the part of the Government is inevitable to achieve this transformation. Those who have been found to be making repayments in a satisfactory manner, full or substantial are good media to serve as demonstrators of diversified methods of fishing. They can be given a substantial subsidy to be treated as margin money and helped in securing loans for equipping their vessels for diversified fishing, apart from carrying out all needed repairs. Such an action on the part of the Government will not only promote the diversification programme but will also act as a reward. It will also motivate others to perform well so as to become eligible for similar assistance.

Fishing Chimes, Editorial 173: June (ii) 1993: Vol. 13, No. 3

# Regulation of Fishing by Indian Vessels in our EEZ

The Indian Fisheries Act, an Act providing for certain matters relating to Fisheries of India came into force in February 1897 during British colonial rule. That was the time when focus on fisheries regulation was mainly on inland fisheries. So much so, this Act did not contain any provision in respect of Marine Fisheries Regulation. This is understandable. Those were the days when sea fisheries were more or less considered inexhaustible. The level of sea fishing effort at the time was not alarming. So much so, there was no motivation for introducing any regulatory measures to control sea fishing at the time.

There has been a sea change in the situation, with the increase in fishing effort in our seas particularly in respect of shrimps. Depletions and reductions in catch per unit effort of shrimps became discernible since 1970s. This situation led the Government of India to suggest to the maritime State Governments a model Marine Fisheries Regulation Act and also a set of model rules to be promulgated, deriving powers from the provisions of the said Act.

Under our Constitution, State Governments have control over the territorial waters contiguous with the coastline of the respective States. Accordingly, Marine Fisheries Regulation Acts and rules framed there under came into being in States such as Kerala, Orissa, Kamataka, Tamil Nadu etc.

So far as the seas beyond territorial waters (considered as deep sea fishing zone) are concerned, as per the provision in the constitution, the jurisdiction over this activity is vested in the Central Government. In these waters Indian nationals can fish unhindered, but restrictions are imposed on fishing by foreign fishing vessels in the area. Under the Maritime Zones of India (Regulation of Fishing by foreign vessels) Act 1981, and the rules framed thereunder, no foreigner can fish in the area (Exclusive Economic Zone) without a valid licence or permit issued by the Government of India.

The Indian fishing fleet having capability to fish in our EEZ, beyond the territorial waters, is increasing gradually. So much so, the restrictions imposed on the foreign vessels to fish in our EEZ will have to become applicable to the Indian fishing fleets sooner or later, although in a modified form. There is thus some urgency to frame legislation to regulate fishing in our EEZ by the Indian fishing fleet too. Among species that may call for regulation in the near future are the tunas, tuna-like fishes, squid and cuttle fish and several others. There may not be an immediate, need to impose fishing restrictions on Indian fishing vessels in our EEZ, but there has to be provision stipulating certain restrictions when called for.

There is no legal frame-work at present to introduce restrictions on Indian fleet as already mentioned above. The application of the Maritime Zones of India (Regulation of Fishing by foreign vessels) Act, 1981 is to regulate fishing by foreign vessels only. The Marine Fishery Regulation Acts and the Fisheries Acts of various Maritime State Governments have no application beyond the territorial waters. The question that arises now is therefore: What should be the provision for regulations, and what should be the mechanism to regulate fishing by Indian vessels in waters beyond our territorial waters as and when it becomes necessary.

Our Exclusive Economic Zone is a open area for fishing by Indian vessels. While Government can control fishing by foreign vessels as per the Act of 1981, it cannot do so in respect of Indian vessels. It is learnt that a new Indian Fisheries Bill has been drafted with a view to replacing the Act of 1897. The Bill will obviously cover all the present day needs of regulating inland as well as marine fisheries, the later restricted to fisheries of territorial waters but not beyond. This bill cannot obviously include any area beyond our territorial waters for regulation, for the reason that the Exclusive Economic Zone minus the territorial waters is with the Government of India only for purposes of exclusive economic use and regulation and there are no other powers vested in it over the area. The jurisdiction over the continental shelf is also with the Central Government, which would bring in some overlapping jurisdiction as part of the shelf is in the territorial waters.

At the time the Maritime Zones of India Act, 1981 was introduced, the question of regulating fishing by Indian vessels did not arise. The situation was that the Indian Industry did not have the capability to operate its own vessels in the area and had to depend totally on foreign fishing vessels to fish in the area and had to obtain them on charter, joint ventures and other means along with the crew for operating in the area. It was this perception that led to having rules to regulate fishing by foreign vessels. Now that the Indian industry has made considerable progress and several Indian fishing vesselsare now operating in the deep sea Zone, fishing vessels of Indian registry, (which cannot be treated as foreign vessels,) although of foreign construction). There is need to amend the Maritime Zones of India (Regulation of Fishing by foreign Vessels) Act, 1981 to provide for making rules by the Government of India to regulate fishing in the EEZ by vessels of Indian Registry beyond territorial waters, in the shortest possible time.

Fishing Chimes, Editorial 174: July 1993: Vol. 13, No. 4

# Guess the One Coming to Import our Fish

Our marine products exports appear to have reached a plateau. There was a fear that there may be a drop in the quantity exported in 1992-93. Deservedly the final figures have shown a rise in quantity and value of overall marine products exported from 1,72,000 tonnes in 1991-92 valued at Rs. 1,375 Crores to 2,08,602 tonnes in 1992-93, valued at Rs. 1,767 Crores. The substantial contribution made by the farmed shrimp sector has undoubtedly stemmed stagnation, if not a fall in marine products exports. MPEDA's imaginative programme to promote brackishwater shrimp farming and the strategy of the Union Department of Agriculture and certain States of to set up Brackishwater Fish farmers Development Agencies for stepping up shrimp production, in the main, have neutralised the evil effects of indescriminate trawling for shrimps along the upper east coast. In order to normalise the situation, Capt. Guidicelli, an expert fielded by the FAO, suggested a substantial reduction in the shrimp fleet along the upper east coast, through diversification of fishing effort etc. The work of a high power technical committee set up recently to consider the situation arising out of mounting overdues on loans given to fishing companies *vis-a-vis* recommendations of Capt. Guidicelli, has commenced but there is no momentum as yet. On top of this, the representative nominated by the Department of Economic Affairs to serve on the committee has chosen to resign and no substitute is appointed. In this situation it will take a long time for the Committee to formulate recommendations acceptable to the Government, All this means that the vessels seized by the SCICI will continue to be non-functional along with a sizeable number of other vessels whose owners are stated be not having the capacity to mobilise working capital. These will be stranded atleast for the current season. Several smaller vessels (mini-trawlers, Sona boats and others) also are stated to be in the same boat. The net result is that, so far as larger trawlers are concerned, out of a total of over 175 vessels (or around 110nos. of operable vessels) only around 55 nos. are now engaged in shrimping. Thus, the conspiracy of circumstances has reduced the shrimping fleet in a manner that might benefit about half of the 'trawler owners' and the remaining 'left in the lurch'. While this is an undesirable situation, atleast the dream of Capt. Guidicelli is achieved. There is a saying 'God protects the working women'. In the same way, the soothsaying or oracle that can be coined is 'God protects the port-bound trawlers'. How God will give this protection is not apparent as yet. Probably Government will extend a hand of forgiveness to them to relent on the Government's way of functioning that created all this mess, which undoubtedly exists.

One suggestion in the present plight and the deceptive comfort or satisfaction offered by increase in exports of marine products is that there should be a shift from over concentration of activities to step up shrimp exports to fish exports. This strategy is acceptable to all concerned. A handful have started exporting fish but it will take quite sometime for an enticing or overpowering interest to take hold. One reason for this is the lack of infrastructure facilities. We hear from scientists, technocrats, administrators and several others that there is plenty of fish available and that a trawler can perform ten-day voyage trips and coming back to port with holds fully laden with fish, but the only drawback in this connection is the absence of infrastructure. The main component of this that is missing is the non-availability of integrated storage complexes consisting of processing, and storage facilities at needed temperature levels.

In this context only we have to guess who will provide the storage facilities and import our fish. Norwegians, EEC, and several others who showed interest, made a study, only to lose interest later. Australia is the only country that has retained a sustaining interest. They want to organise the work on commercial lines. Their study is stated to have indicated availability of adequate raw material, particularly at Visakhapatnam Port, where the local port trust authorities have agreed to provide a convenient site and extend all facilities at their command. Australian Government, based on a few studies conducted by teams sponsored by it, is stated to have cleared a project to set up an integrated fish processing and storage complex at Visakhapatnam and the ball is now in the court of the Government of India.

The latest on the subject known to us is that the Australian Government fielded Mr. Robert Cardover, a marine fisheries specialist to visit centres such as Cochin, Madras, Visakhapatnam, Bombay etc., and collect information to provide answers to certain questions that the Government has. In the wake of his visit, another mission from Australia, depending on developments, would be sent, it is learnt.

So far as Robert is concerned, he finds distinct possibilities for setting up infrastructure for exports of finfish from Visakhapatnam. He is aware that the Government of Australia and Government of India are keen to co-operate and introduce a system which links marine fish production with marketing, *viz.*, mostly exports to Australia. The system would include the setting up of a processing complex and a fish collection system.

Robert thinks that funds for the project will be forthcoming from both the Governments. No investments are contemplated to introduce new fishing vessels. There is already a surplus and owners can deploy them for catching fin fish. He is hopeful that by the end of the current shrimp season the project will take shape. The processing plant proposed will be fully equipped and facilities, to be organised, would be available to proceessors and exporters. There will be no total dependence on exporters and the Project Authority will have its own system to procure finfish etc and export them, if gaps in co-operation surface in actual working.

Robert says that he was asked to go to India in Aug-Sept 1993 again to visit Madras, Cochin, Visakhapatnam etc. He feels that action will be at Visakhapatnam. During his proposed next visit he will be in a better position to judge whether integrated marketing system should be run with individual exporters as the basis. He has his own doubts whether such a system would turn out to be haphazard and move into the hands of non-technical persons.

Of one thing Robert is certain is that, before end of the calendar year 1993 there will be action. Relationship and understanding between the two Governments will produce ultimate decisions, he feels. Whatever be the decisions, the real secret is in management, Robert is sure. His personal view is that there is good scope to run the project on commercial lines.

One aspect mentioned by him is that the marketing activity need not be restricted to marine fish alone. There is abundant demand for carps in East European countries and this can be fulfilled on commercial lines. Australian Government is also interested in importing marine fish from India to Australia as part of the proposed project. Australians do not consume much of freshwater fish but what will happen when the project gains in age and carp culture comes up in a big way, he is unable to guess.

Robert says that in Australia very largest shrimp farms have been in operation since 1960. There are some good signs of business relationships, by way of joint ventures, linked to marketing, that would open up between India and Australia. There are not many carp farms in Australia and he does not see much of a future in carp sector in Australia. There are no serious carp producers in Australia.

*His warning:* If Indian entrepreneurs come to Australia and try to do farming for prawns or carps for exports, they will lose. Indonesia tried to produce carps and prawns and export them because of liberalisation, but failed. Same thing would happen to others as well. They would not have a chance to export also. To find a market to absorb the production will be a Herculean task. Therefore, the only alternative is to produce them in India and export. After reading all this, you can guess who is likely to import fish in a big way from India.

Fishing Chimes, Editorial 175: August 1993: Vol. 13, No. 5

# Customs Bonding and Brackishwater Farms

The letters of Intent issued by the Government to 100 per cent export, oriented integrated brackishwater farms contain a primary provision for allowing exemption from payment of customs duties on imports meant for the project and also excise duties on indigenous items. It says that all operations including processes incidental and ancillary for the production of exportable goods should be carried out only in an area that is declared as warehousing station and as a bonded area under the Customs Act, 1962.

Accordingly, the entire sprawling integrated brackishwater farm of Water Base Ltd. of Thapar Group near Nellore in Andhra Pradesh, a 100 per cent EOU unit, is stated to have been allowed by the customs authorities as a bonded area. This approval is in accordance with the provision in the 100 per cent EOU letter of intent. Water Base offers feed and seed for supply to farmers, probably either under the Franchise system which is integrated to 100 per cent export or out of percentage of production allowed for domestic sale. Prawns, which are totally utilised for export, are grown from stage to stage in farm ponds and therefore the farm has been bonded. At the hatchery, prawn seed is produced and this seed is stocked in the nursery ponds of the farm. The prawn production (manufacturing) activity in the integrated farming is conducted employing a series of processes entailing two farming rounds of about six months each in a year, in the bonded area.

In contrast, with the bonding of the total integrated farm area of Water Base Ltd, other similar integrated farms are reported to have been found by the customs as not amenable for bonding, although they are also based on approved integrated production system. Their projects also include the same interdependent components.

This invidious attitude of the customs has dismayed the other 100 per cent export-oriented shrimp producing integrated brackishwater farming companies. The unrealistic, field-offensive and counter-productive action of the customs is deplorable, as it offends the national policy of augmenting exports. It is learnt that orders for partial bonding system came form the Board of Central Excise and Customs. Such being the case, even when one or two components have not been set up, customs and excise authorities will have to bond such of the components that have been set up. The provocation for the illogical thinking that, excepting the farm owned by Water Base Ltd, none of the others deserve bonding in Nellore Dist. of Andhra Pradesh, seems to be misplaced.

The effect of this approach is that the concerned companies cannot import essential items such as aerators, shrimp feed and water purification units which have to be set up near inlets and outlets of the farms and several other essential items as well, without paying customs duty. This position goes against the very basis of 100 per cent export-oriented units. If these companies have to pay duty, they can as well be in the shrimp production job without recourse to registration as 100 per cent export-oriented units. Companies were formed mainly to secure exemption from payment of customs and excise duties, as such an exemption would reduce cost of production, thereby making their products competitive in external markets. What customs authorities have done, as can be seen, is something that is harsh and harassing and without justification. It defeats the very objectives of 100 per cent EOU scheme and goes against the unambiguous provisions in the letters of intent and the customs and Excise Tariff Acts. The relevant notifications under the Act (13-81-Cus dated 9.2.1981 and 123/81 cus dated 2.6.1981) as amended clearly specify that all capital goods, raw materials, consumables, spares and packing materials bought in connection with all operations (Processes) in the approved undertaking shall be exempted from levy of these duties.

One has to concede that until the processing plant is also added to the farm complex, the project will not be ready to undertake exports. All farm components do not take shape at one time. Ponds will come up first. To secure raw material, production part has to be stablised in the first instance. In this connection, aerators and feed will have to be imported under the 100 per cent export scheme, pending completion of the entire complex. These and other imports, duty free, should have to be allowed, with a direction, should it be necessary, not to operate the integrated complex until all the components are in operational condition and the entire area is bonded. There is no other way Customs can withhold bonding merely on the ground the area is vast. It is not the fault of entrepreneur if the area is vast. Export production of shrimps needs the area.

Customs authorities will have to realise the anomaly, pitfalls and after-effects of their action in not bonding the entire farm area of the 100 per cent EOU companies, with the exception of Water Base Ltd who seem to have set up the processing plant as well, but were allowed imports of aerators etc long before the processing plant was set up. It is not known whether the plant has become operational as yet. In the case of other 100 per cent EOU prawns units, it appears that the customs authorities want to keep under bond only the processing plant where the final product takes shape as an export commodity in contrast to the bonding status accorded to Water Base Ltd. This reluctance on bonding some of the components of integrated farm units would mean that a large part of the manufacturing area, of which shrimp production is a part, would be out of bonding, thereby creating situations in which the shrimps produced may not all be utilised for exports by the company concerned, but may be sold to other processors for ready cash flow. Apart from this there is the possibility of seed from hatcheries getting diverted, thereby reducing inflow of raw materials (shrimps) to the processing plant. Shrimps can be brought into the area from outside also. All these would mean encouraging 'fraudulent' activities.

Because of denial of duty-free imports and non-availability of excise-free machinaries within the country, the investments will go up, thereby rendering the shrimp production work and subsequent processing activities uneconomical. The customs duties being substantial at 25 per cent on cost of aerators and 15 per cent on cost of feed, along with the excise duties, the anticipated incomes by way of exports, would get neutralised, if not leading to losses.

Denial of the explicit concessions allowed under 100 per cent EOU scheme for 100 per cent export-oriented shrimp farming projects by the authorities, would be against the scheme itself. The conclusion that vast areas cannot be bonded has no substance. Vast factories are bonded without any problems. Further, to say that all the components, particularly the processing plant, have not yet been set up for bonding the entire farm has also no substance. Feed, aerators and machinaries would have to be imported by the companies beforehand for starting the integrated operations. In fact, Water Base Ltd. was allowed, as already stated, to import all these items without duties, long before their processing plant was set up. If there are reservations that shrimps produced, before setting up processing plant, may be sold to others, the line of action should be to take a guarantee for an amount equivalent to duty not paid. This would be a deterrant to the misuse of the facility provided.

In fact, to guard against such eventualities as brought out, when the construction process is on, bonding of the entire area is essential. Customs and excise authorities should not expose themselves to the criticism that they gave facility for bonding their entire farm to Water Base Ltd. and denied the same to others. Now that a problem seems to exist, Government should lay down a clear-cut system for bonding of 100 per cent export oriented brackishwater farms, considering the fact that this activity has certain special characteristics.

Fishing Chimes, Editorial 176: September 1993: Vol. 13, No. 6

# To Jog Out of the Jam of Deep Sea Fishing Vessel Loans and Recoveries

The meeting of the Technical Committee of Experts to make recommendation to vitalise the deep sea fishing industry and find a way to collect the pending loan instalments from the vessel owners who were financed by SDFC/SCICI was held on 31.8.1993 at Visakhapatnam where the Industry is concentrated. The Committee also met several representatives of owning companies and gathered a wide range of suggestions. The main suggestion is that the capital loans on each of the vessels should be brought down to the present market price that emerged in a recent auction held by SCICI. Another suggestion seems to be to allow the vessels to operate unhindered and accept 10 per cent of the net returns towards repayments, if there will be net returns. Some seem to have emphasised on the need for working capital loans and for granting of loans for equipping the vessels for diversified fishing.

*Fishing Chimes* offered certain suggestions in respect of operational aspects which have direct relationship with unit catches and earnings, in its August, 1993 issue. On the capital restructuring and repayment aspects, the basic point to be kept in mind is that in situations of this kind it has to be ensured that operational aspects should be divested from undue pressures for repayments. SCICI or the Government should not be deemed to be in a state of confrontation with vessel operators. There is now a realisation that vessels should be operated well and loan dues should be extinguished. These perceptions have to be fostered further, and keeping developmental aspects in view, loan recovery aspects should be pursued.

The only way out can be to sell vessels to foreign fishing companies or entrusting defaulting vessels to the few well run Indian companies or to a company to be set up by the Government/SCICI to operate the defaulting vessels. There appears to be no other way of recovering the dues for which, in any case, most of the owners say they have no money for the purpose. However, none of these remedies are likely to work. Foreign companies offer very low amounts for purchasing the vessels. This is as good as reducing the capital cost and collecting the same in instalments from owners.

Companies which are now running their vessels successfully would not like to take a greater burden. A fleet of four or six vessels is considered optimal for a company for efficient operations. Forming public sector companies is now out of date. Government's present policy is to off-load shares of public sector corporations.

In this situation, a sense of recognition and realisation that a realistic developmental perspective, taking lessons from past mistakes (allowing introduction of a very large number of shrimp trawlers by import in the garb of multipurpose vessels ignoring scientific advice etc.), is of paramount importance for evolving a pragmatic approach. Past should be totally forgotten, storing the experiences for future use. We sugggest the following approach.

## SDFC-Financed Vessels

The debt-equity ratio is 6:1 for the vessels financed by the Government through erstwhile SDFC. This ratio may be changed by converting the entire balance of debt into equity, leaving the equity of the present share holders as such. The management structure may be reorganised or strenthened as needed. The present share-holders may be given the option to utilise the net earnings to buy back the new shares taken by Government or SCICI by conversion of balance of debt. If this fails, the Government/SCICI owned shares may be sold to others. If this line of action fails, SCICI would know the last resort.

## SCICI-Financed Vessels

The loan given to the companies by SCICI may be around 70 per cent of the total cost of the vessels. This loan amount may be converted into equity shares of SCICI by broadening the equity base. The management may be left to the present owners for a specified period, while strengthening the management structure as needed.

The companies concerned may be given additional finance as considered inescapable for strengthening the operations of the vessels. The companies concerned may be encouraged to buy back the equity shares acquired by the SCICI in a phased manner, allowing them to utilise the net earnings. SCICI may also sell their shares in the company to NRIs or foreign fishing companies, in case they find the present share holders not responsive or the management is unable to generate profits. The last resort is known to SCICI.

Fishing Chimes, Editorial 177: October 1993: Vol. 13, No. 7

# Irrational Taxation on Incomes from Fish Needs Abolition

Income derived from Agricultural products are exempt from income tax. However, Incomes from sale of fish, particularly farmed fish, is subject to income tax. This disparity has the effect of upsetting the economics of fish production. Government do not appear to have made any effort to examine whether this taxation has any basis. In fact, the scientific considerations, for levying income tax on earnings from fish sales, notwithstanding the general policy of the Government exempting agricultural produce from levy of income-tax, require an urgent appraisal. This apart, the levy of income tax on value of fish catches and fish products despite several clarifications by the Government and Reserve Bank that fisheries are part of agriculture and that they fall under the definition of 'Agriculture', requires abolition in order to remove this irrational anamoly, particularly in the context of the policy of the Government to promote exports.

It is known to all that Agriculture is a land-based culture. In the same way, 'Fish Culture' is also a land-based culture, although this simple fact is not widely realised. Plants grown on land derive their nutrients through their roots from the soil in which they grow. Fish culture is no different, looked at from a scientific angle.

In the same way as agricultural/horticultural plants (or trees) derive nutrition from the land underneath, fishes grown in tanks, ponds etc. primarily derive nutrition from the bottom soil. Being swimming animals, fishes are grown in water that absorbs nutrients from the soil in the same way as land plants. The soil nutrients are extracted by the water medium in tanks, ponds etc, that stands on the bottom soil. Utilising these nutrients, minute microscopic plant organisms sprout in the water in tanks, ponds etc., which are referred to collectively as 'phytoplankton'. The phytoplankton forms the food of microscopic animal organisms, referred to as 'Zooplantkon'. Farmed fishes feed on these two kinds of organisms. The subtle difference, which has biological validity, is that, while agricultural plants take nutrients directly by extracting the same from the soil along with moisture, fishes take the nutrients in a transformed state from the water medium, from mere depressions on land that are not suitable for agriculture but suitable for fish culture. Looked at from any angle, they do not call for distinguishing them from food production angle from Agriculture. The levy of income tax on earnings of fish farmers from sale of fish is therefore irrational as there is no basis for the differentiation.

In agriculture, manures/fertilisers are used. In the same way, in fish farming too, manures/fertilisers are used. This similarity adds strength to the point that taxing the earnings of fish farmers is unjustified. Further, in quite a natural way, fish growers are also referred to as 'farmers' as in the field of Agriculture. If the activity is not similar they would not be referred to as farmers.

It can be argued that, while cereals and pulses are essential food items, fish are not. Incomes from fish have to be therefore taxed. If fish are not essential food items, oranges, coconuts etc. which are produced on land are also not essential food items but incomes from these are not taxed.

Another aspect is that heavy electricity charges are collected from fish farmers, whereas their counterparts engaged in 'agriculture' are charged at concessional rates. It is unjust that these concessions are not extended to fish culturists. It will be just and fair to do so.

Apart from the imposition of income tax on fish farmers without any basis and for no reason whatsoever, in the State sector sales tax under the name of purchase tax is levied, particularly in Coastal States. Exporters who purchase exportable marine products from the producers are unwittingly harassed for collecting the tax. Under an ill-framed and ill-thought out rule, if purchases are made and processed without a specific export order on hand, purchase tax has to be paid by the buyers who are invariably the exporters.

The rationale behind such an obnoxious rule is totally obscure, particularly when the proclaimed policy of the Government is to promote exports. An exporter has to buy the available produce fit for export and convert into a suitably processed form as and when it is available. For want of an immediate order on hand he cannot desist from purchasing the material. The nature of the activity is such that as and when exportable stocks are available they are to be purchased and stocks built up to sizeable level, adequate for a shipment. An exporter offers the commodity to foreign importers. Upon receipt of a firm order he exports. Any one can see that the building up of stocks for export should be done only after receiving an export order, does not stand upto any rational test. Some sensible States withdrew this levy of purchase tax. But some still continue this practice, causing misery to exporters.

At present, while procurement prices are moving up, the export prices are sliding down. And the levy of purchase tax converts any small profit that the exporters may expect to get, onto a loss. This tax is clearly a measure that discourages exports and makes the capital investments unproductive. The sooner a uniform national policy is evolved to abolish this tax the better it will be for promoting exports.

Another aspect is the 'market cess' imposed in some of the States such as Andhra Pradesh. This is imposed ostensibly for maintenance of market yards. How absurd this concept is, so far as fish are concerned, can be gauged from the following facts:

1. Fish are landed at fishing harbours or at scattered landing points. For handling the produce, the harbour authorities collect a certain cess.
2. There are no market yards at fish landing centres and therefore there is no cause for asking for payment of any separate cess, particularly when no separate facilities of the kind required are provided.
3. It will be an unnecessary and unproductive excercise to the fishermen to take the fish to the yard (if it is there) and take them back again to the market engaging extra labour merely to pay an unjustified cess, and
4. Fish will get spoiled in the process resulting in losses to fishermen.

The sooner this irrational and impracticable system is removed wherever it exists, the better it will be for the socio-economic fabric of the fisheries sector.

The latest blow to the sector is the innovation of 'Turnover Tax' payable by all on their annual turnover of sales, as introduced in Andhra Pradesh. It is very difficult to reconcile with this concept of 'turn-over' tax particularly by those in the fisheries sector. Exporters pay cess on every consignment they export. They also pay a fees for pre-inspection of products meant for export. The common fisherman toils hard to catch fish and sell them. What he earns may not be enough to meet even his daily needs. Whether his 'turn-over' is subject to this new tax is not known. As it is, no sales tax is collected on fresh fish. In order to avoid confusion, it will be generous on the part of the Government to issue a clarification that fish and fish products, sold or exported in fresh or frozen or frozen processed condition are not subject to turn-over tax. This is the least that can be done for propping up this largely export-oriented sector.

Governments will do well by desisting from imposing any taxes on the fishing industry which is struggling for survival. The industry which is passing through severe ordeals deserves this consideration.

Fishing Chimes, Editorial 178: November 1993: Vol. 13, No. 8

# Brood Shrimp Availability for Hatcheries

World-wide efforts to grow marine shrimps in farm ponds to advanced maturity stage for hatchery purposes have not yielded any satisfactory results so far. As early as 1964, Fujinaga made efforts to grow shrimps in confined waters to a breeding stage but he could not succeed. Similar efforts were made in Taiwan but without success. SEAFDEC in Philippines has also been making efforts for quite some time to grow shrimps to a size of maturity. Success continues to evade this organisation, too. The Central Marine Fisheries Research Institute, Cochin, experimented in various ways to achieve a breakthrough. However, finding that there were no positive results, it is understood that the experiments have been stopped for the time being.

It can be said that farming of marine shrimp to maturity and to a breeding condition in confined waters is at the same stage as that of major carps in inland confined waters in late 1940s. In 1950s a breakthrough was achieved in inducing major carps to breed in confined waters, through the development of induced breeding technology. Until this technology was developed, farmers were totally dependent on natural carp seed collections from rivers, in various stages. Spawn used to be collected on a large scale from rivers and grown to advanced stages of fry/early fingerlings for stocking in tanks and ponds and also in reservoirs, until recently. With the standardisation of induced breeding technology at the Cuttack Centre of the Central Inland Fisheries Research Institute in late 1950s and the subsequent advent of inducing agents such as those based on hypophysed fish pituitary, HCG and Ovaprim, and the setting up of Chinese type of modified circular hatcheries all over the country, seed problem in the inland freshwater sector has been, more or less, resolved.

Now, in the inland freshwater fish production sector, there is a major reduction in natural collections of spawn. This has facilitated the revival of carp fisheries in our major rivers. Farm-grown major carps are now induced to breed utilising the developed technology. Spawn so produced is grown to advanced fry/fingerling stage and these are used for stocking in tanks, ponds and reservoirs.

As in the case of carps (until 1950s), shrimps do not attain maturity in confined waters. Our Scientists have to identify the reasons for this through concerted efforts. It is understood that shrimps of less than 100g are generally not suitable for hatchery operation as oogenesis does not start in ovaries. Only in specimens of over 100g oogenesis takes place and these alone are fit for use in hatchery operations. In other words, specimens of more than 100g in weight harvested from the sea alone could be made to release eggs. Eye-stock ablation is done to help in the further growth of ovaries and for release of eggs. The same brooder releases eggs at intervals three or four times but the number of eggs released and the size of the eggs released would become smaller.

It is known that the breeding grounds of shrimps are at a depth of 24m and over in the sea. Considering this, if suitable structures to replicate water pressure condition at 24m and over are constructed and are used for experimental work of breeding shrimp in waters of same composition there can be some hope of success. Besides water pressure, about which scientists know already, there may be some other factors such as effect of salinity, water currents and their velocity etc on breeding results. It may be desirable to study the conditions of existence at the breeding grounds to facilitate replication of those conditions. Thapar Water Base Ltd. is understood to have constructed one or two structures with depths not normally required. If this is true the purpose is not known. It is quite possible that these are intended for experimentation to grow shrimps to maturity.

In any case, it is of considerable urgency for CMFRI to renew experiments on producing mature shrimps with the determination to achieve success. Until a breakthrough is achieved the problem now being experienced in securing brood shrimps in adequate numbers will continue, resulting in further decrease in sea shrimp stocks.

As is happening now, the removal of large quantity of natural shrimp seed for stocking brackishwater farms and exploitation of brood shrimp to meet the requirement of hatcheries has been having the effect of reduction in natural stocks. The problem was not so acute in the early stages of setting up of the premier shrimp hatcheries near Visakhapatnam and Gopalpur. With more and more hatcheries coming up and a large number of brood shrimps being diverted to hatcheries and wild shrimp seed being exploited more and more, the situation is assuming alarming proportions. Fishermen and fishing vessel owners are already up in arms agitating against the present practice of collection of brood shrimp and young ones from the sea. This agitation is likely to intensify and reach alarming proportions. In this context research efforts have to be restored by CMFRI by mounting a major project for achieving the needed success, with the help of the Department of Biotechnology, if necessary.

Fishing Chimes, Editorial 179: December 1993: Vol. 13, No. 9

# Coastal Marine Fishery Development Agencies Needed

The lot of coastal fishermen has been improved in the past several years to a significant extent by the Central and State Governments. On the social side, roads to several fishing villages have been laid and subsidised houses have been built, drinking water supply is provided under various Government schemes. Insurance cover to fishermen for accident and death by way of subsidised premia has been introduced. At all amenable centres fishing with mechanised boats has been popularised and outboard motors with State subsidy are being supplied. Some progress has also been made in respect of fish transport and marketing facilities. Fishermen's co-operatives in several zones play a significant role in supplies and services to fishermen. Saving schemes have made beneficial inroads in several States.

While all these measures have resulted in some improvement in the life style of fishermen, much remains to be accomplished, for, our coastline is very long dotted with thousands of fishing villages. Haphazard and fragmented developmental measures now under implementation do not make a visible overall impact. Only when the needs of coastal fishermen are met under an integrated organisational and well co-ordinated mechanism, systematic and measurable development would be possible.

The suggestion to introduce an integrated organisational mechanism for upgrading the socio-economic conditions of the coastal fishermen is not a new one. In several States particularly, in the south-east and south-west coasts, during British days several fish curing yards were set up at various centres. These centres used to be the focal points for collection of information on the fishing conditions along coastline. The officials in charge of the Fish Curing Yards, known as Petty yard officers, used to gather all available information on fishing conditions and communicate to the Inspectors of Fisheries. The primary function of the yards, however, was to provide facilities for the fishermen for curing their catches. The complexion and the Infrastructural facilities available have no doubt, changed considerably over all these years.

Several of these yards still exist in some of the States. These yards have to be now upgraded and more such yards have to be set up all along the coastline, one or more in each of the coastal districts. The structure and functions of the yards have, however, to be enlarged to meet the developmental needs of the coastal fishers of the present day.

The coastal fishers now require integrated facilities and an organisational mechanism to channalise them for the benefit of the fishers. This can be achieved by enlarging the concept behind the setting up of Fish Curing Yards over 50 years back, in view of the latest developments in terms of technologies etc. In other words, Coastal Marine Fisheries Development Agencies have to be brought into being absorbing fish curing yards wherever they exist to serve the coastal fishers, and such Agencies have to be set up all along the coastline, according to the needs, one or more in each coastal district. The Agencies should be provided with funds, originating from the Central or State Governments concerned and channelised suitably for setting up landing centres, as minor harbours or jetties; if such scope does not exist, crafts (Wooden or FRP catamarans or canoes/navaas) capable of being beached, have to be introduced, so as to meet the requirements of all the fishers. Outboard motors for propelling these crafts may be supplied to them, through the Agencies proposed, to reach the grounds faster so as not to lose fishing time. It now happens that from several coastal centres several of these crafts are still sail-propelled by wind power.

The fishermen have also to be supplied with suitable nets and containers and ice for bringing back the fish in a well preserved state. The suggested Coastal Marine Fishery Development Agencies can disseminate the information received from the National Remote Sensing Agency concerning the location of fishable stocks for the benefit of fishermen. The Agencies have to be equipped with the needed communication facilities to receive messages from N.R.S.A. for the purpose and also to the fishers at the grounds, if they are already out there. Further, the crafts have to be provided with walkie-talkies to enable them to communicate with each other out at sea and also with the Agency whenever necessary. All centralised facilities for preservation of catches at shore before releasing them to the market as required, would need to be set up. The Agencies can also act as conduits for the transfer of emerging technologies. All the services can effectively reach the fishers only through an Integrated and Co-ordinated set up, *i.e.*, Coastal Marine Fishery Development Agencies set up, by and large, on the lines of Fish Farmers Development Agencies. The Management pattern may differ considerably and this can be evolved suitably.

Once these Agencies are set up, they will become a suitable medium for channelising the efforts of the Government for the socio-economic development of the coastal fishers effectively. The Central and State

Governments may have to seriously consider the imperatives of setting up and making good use of such Agencies for more rapid development of fishers so as to upgrade the quality of their life, and to enable them to improve their fish catches and secure better returns. The development of coastal fishing villages would largely depend on a tangible step of this kind. Such Agencies would facilitate the implementation of various innovative schemes of the Government for the uplift of the coastal fishermen.

Fishing Chimes, Editorial 180: January(i) 1994: Vol. 13, No. 10

# High and Low, 1993

Upsurge in farmed shrimp production, successful tapping of capital market by promoters of several integrated shrimp culture and tuna fishing projects and what turned out to be a stable start in high sea fishing, marked the year 1993. On the debit side one can see a cessation in operations of several so-called deep sea fishing vessels, a plummeting in capture shrimp production and auctioning of several hypothecated fishing vessels by SCICI, and steps by Visakhapatnam Port Trust to auction vessels in order to collect port dues.

In a section of near-shore marine fishery capture fishing sector, however, there have been isolated and aberrant developments. Mini-trawlers and other smaller vessels suffered bad innings. Although addition to Indian built fishing vessels has come to a standstill, four Chougule built shrimp trawlers stagnating at the yard for over four years were added to the fleet. The Ministry of Food Processing Industries had been a brave spectator to the stagnating Indian fishing vessel tonnages. Ways and means of deploying the present shrimping fleet so as to reduce shrimping pressure along the upper-east coast continued to be explored. Based on representations of the trawler owners, a study of result of shrimp trawler operations was undertaken by an FAO expert at the instance of the Government. The expert recommended experiments on three types of diversified fishing on three selected trawlers. This is being followed up. Government also set up a technical committee to go into the problems of deep sea fishing. A committee for formulating a new National Fisheries Policy has been set up.

Several fishing companies entered into joint ventures with foreign companies for undertaking deep sea fishing in Indian EEZ. These covered introduction of vessels with or without onboard processing facilities for harvesting mid-water and demersal fishes through various systems. Indian vessels for open sea fishing for tuna through longlining were introduced, but operations were partly successful. Capital market responded favourably for integrated tuna longlining projects. Marine fish production however remained almost stagnant. Inland fish production is believed to have gone up, particularly in respect of farmed shrimp.

Several new aquatic products processing plants came up. Apart from new processing technologies, diversification in procesed products for exports came about. Export of fresh and frozen freshwater and marine fish has gained further momentum. Export of aquarium fishes has become attractive. M/s. Alsa Marine and Harvests started a fully owned subsidiary in Belgium for marketing its processed products in Europe. Alsa is also planning to set up a subsidiary in Dubai. Kings Group and Amalgam Group of Kerala had expanded their activities, mostly on shrimp culture and processing side. Like-wise, Rank Aqua, Magunta and few others had come up well in regard to Integrated shrimp culture activities.

Marine Products Exports to EC countries went up. New outlets for exports of marine products such as main land China have materialised.

A task force on Krill project development has been set up by the Department of Ocean Development under the Chairmanship of Prof. M.G.K. Menon.

Inter-State fishing disputes, particularly between Orissa and Andhra Pradesh, Tamil Nadu and Andhra Pradesh erupted several times in the year. Furthermore, intra-State disputes between non-mechanised and mechanised fishing boat-owners of Kerala recurred. Mr. Biju Patnaik, then Chief Minister of Orissa inaugurated the fishing harbour at Gopalpur in Orissa.

There has been an upgradation of the Fishermen's Saving's cum relief scheme in Pondicherry. The scheme now allows a contribution of Rs.45 each per month for a period of 8 months (March to October) in respect of Pondicherry/Karaikal/Yanam and from October-May (8 months) in respect of Mahe region. Fisher Housing Schemes have made progress in several States.

The National Federation of Fishermen's Cooperatives Ltd (FISHCOPFED) set up a data bank. The organisation is also implementing fishermen's accident insurance scheme. MPEDA has announced a scheme to assist mechanised fishing vessels below 20 m OAL for diversified and multi-day fishing in the EEZ. A fishing vessel of CIFT's design, 15.24m long, was commissioned in Cochin in April 1993.

MMTC has entered into the shrimp export sector in a big way. The Corporation became part of a couple of joint venture agreements with shrimp based companies involving equity participation and exports.

Government of India has decided to phase out the scheme that allows chartering of foreign fishing vessels. The scheme has been closed virtually by end of 1993, by which time only a few chartered vessels were operational. Government came forward to grant extensions to those who applied for them before Dec 1993 but how many such extensions were given is not known.

The Fishery Survey of India added a few Japanese survey vessels to its fleet. The Integrated Fisheries Project, Cochin recently received as a gift two Japanese vessels. The Bhaba Atomic Research Centre developed a colourless, odourless protein powder from low value sharks.

The Society of Fisheries Technologists (India) Cochin has instituted a biennial award christened as SOFTI award. This award carries a medallion and a cash of Rs.5,000.

MATSYAFED in Kerala set up a very large fishing net making factory near Cochin, called M/s Covema Filaments Ltd. This commenced export of mono-filament longline.

The concerned Parliamentary Committee recommended to the Governement the closing down of the Ministry of Food Processing Industries, of which deep sea fishing is a part. Efforts were on to rescue the Ministry. It is stated that the Ministry of Agriculture from whom the subject of 'Deep Sea Fishing' was 'snatched' away, is making a strong bid to retrieve the lost empire. One has to concede pretty little that is worth while that has been done in the field of deep sea fishing after the subject was taken away from the Ministry of Agriculture, except largely unfulfilled promises from foreign companies to invest.

Quite a few poaching Thai trawlers were apprehended by the coastguard. Fishing disputes between India and Sri Lanka persisted.

One outstanding development during the year was the emergence of oyster culture, mussel culture and Holothurian culture as commercially viable mariculture activities. Artemia cyst production has also become pronounced in the country reducing dependence on imported cysts.

Concessions on customs duties/excise duties/ inspection of exporting marine products have been accounced by the Government. Production of shrimps from brackishwater farms under semi-intensive farming conditions increased to over 6 t/ha/crop as achieved by some of the farmers. Excavation of brackishwater farms along the coastline particularly in forest areas and in a few other areas is stated to be leading to pollution. Several shrimp hatcheries have come up and atleast two reputed feed mills, that of Higashimaru and of Thapar Waterbase have come up. Others like 'standard' feed came up but there has been no cognisable impact.

Chilka shrimp project established by Tatas and the Government of Orissa more or less has aborted. However, the shrimp hatchery set up by Tatas near Puri is now ready to commence production. It is stated that 28 shrimp hatcheries are in the offing all along the coastline. TASPARC and OSSPARC expanded their seed production capacity. A shrimp hatchery has been set up at Kumta in Karnataka by BFDA, Kumta with CIBA technology. Another hatchery with OSSPARC technology came up in Goa and one more at Chandrabaga in Orissa, also with OSSPARC technology is in the offing.

MPEDA announced several subsidy schemes aimed at shrimp farmers. One is to provide 25 per cent of cost of setting up farms or Rs.30,000 per ha upto a maximum of Rs.1.5 lakhs per individual/company/firm limited to developing 10 ha of new area. Another is the scheme to grant a subsidy of 25 per cent on the cost of seed and feed upto a maximum of Rs.450 and Rs.3,000 per ha respectively to each beneficiary/unit for a maximum area of 50 ha at a time.

The Regional Centre (Prawn Farming) of the MPEDA, Bhubaneswar, Orissa provided training to fishermen on wild shrimp seed collection. Several beneficial improvements have been made in several States in respect of accident insurance scheme. Integrated fish farming is becoming increasingly popular. CIFAX was found to be a successful curative for UDS, by CIBA. Mr Dayanand Modi, Dangalpura, Madhupur, Deoghar District, Bihar is stated to have achieved a breakthrough in the cure of EUS through application of Chacop and plantomycin. Azolla has been found to form a good organic manure by CIBA and it has standardised technology for producing the weed and converting into fertiliser. Freshwater pearl culture at rural level was also introduced by CIBA. Glaxo Labs introduced Ovaprim for induced breeding of Carps with success.

Research work on fish genetics has been stepped up in several universities and at the National Fish Genetics Research Institute.

Punjab and Haryana States have made enormous progress in raising farmed freshwater fish. Haryana achieved good results in producing brackishwater shrimps and fishes. Haryana has underground saline water. Gobindsagar reservoir in Himachal Pradesh has come to the fore-front by producing 96.9 kg of fish per ha/ annum. Himachal also acheived outstanding progress in cultured trout production under an Indo-Norwegian project.

Indian scientists could produce seed of Hilsa in confined waters, although on an experimental scale. Hilsa ranching is now considered possible. Experiments conducted in Orissa have proved the commercial viability of multiple breeding of major carps. Orissa University of Agriculture Technology successfully demonstrated integrated fish farming.

Lipton India set up a new modern fish feed plant near Vijayawada in A.P.

Mahaseer is stated to have staged a remarkable recovery in Bhandardara reservoir in Maharashtra, according to Valsangkar, Chief Executive Officer, FFDA, Satara. There has been an increase in FFDAs and BFDAs during the year.

To conclude, the farmed shrimp sector was buoyant in 1993 with good indications of stepping up of exports of shrimps. The point of reversal from the present stage of growth of farmed shrimp production, with the hidden but bursting dangers of pollution is somewhat far away, with some more time still available to keep the production environment clean.

*The following were some of the awards given to distinguished scientists:*

1. Prof. H.P.C. Shetty received Asian Fisheries Award from the Asian Fisheries Forum held in Singapore in October 1992,
2. Mr. Nisheeth Bhat, a progressive fish farmer, Haryana received the first prize for achieving highest fish production per ha in Amabala district, Haryana in 1991-92
3. "Young Manager of the year" Award was given to Mr. K. Srinivasrao, Executive Director, Rank Aqua Estates, Hyderabad.
4. Award was given to Dr. V.V. Sugunan, Sr. Scientist CICFRI for his outstanding achievements.
5. A.P. Best Farmer Award was given to Mr. Ch. Hariprasadarao, Shrimp Farmer of Prakasam district in A.P.

Fishing Chimes, Editorial 181: January(ii) 1994: Vol. 13, No. 10

# Back-lash of Brackishwater Farming

The speed at which brackishwater shrimp farming is progressing along the Indian coastline has added a new dimension to farmed shrimp fishery developmental activities. This has led to a spurt in the exports of shrimps from the country. The shrimp farmers and the processors deserve to be complimented on this. But for the special drive imparted, incentives given, and technology transfer organised by MPEDA, the developmental process would not have peaked to the present upward trend. Through these special measures MPEDA has rendered a signal service to the nation, in boosting up shrimp exports, with support from Central Institute of Brackishwater Aquaculture and various Coastal Fisheries Departments. The leadership developed by MPEDA in this regard is unparalleled.

MPEDA is fully conscious of the side-effects of uncontrolled and non-monitored development of brackishwater shrimp culture. The authorities at MPEDA are also well aware of the disastrous effects of intensive shrimp farming on environment, as had taken place in Taiwan and Ecuador. Having studied this aspect, Dr. Sakthivel who was the Chairman, MPEDA until recently, and who is presently the Director MPEDA has been cautioning farmers not to be lured by the pseudo-attraction thrown by intensive shrimp farming system but to stop at adoption of semi-intensive method in order to avoid a disaster of the kind that took place in Taiwan and Ecuador. MPEDA has been simultaneously stipulating special measures to avoid pollution within the brackishwater farms and also in the nearby water resources from where water is drawn for filling up of the farm ponds and also for periodical discharge as is required. Considerable harm has already been done to mangrove vegetation because of conversion of mangrove lands into brackishwater farms. Several farmers have already been experiencing the effects of pollution which have been leading to mortalities of shrimp crops, although sporadic at present.

A large number of brackishwater farmers utilise sea water or brackishwater from estuaries or backwaters or other such sources for filling up their farm ponds. They also exchange water periodically. As it happens, around the duration of a low tide giving way to high tide, the water that is let out from one farm, in several cases polluted, is taken in by an adjacent farm. As several farms depend on the same water source both for taking in supplies and also for discharging the waters from the farms (often polluted), deterioration in the quality of near-shore waters takes place. The polluted waters lead to various degrees of mortality of marine animals and plants in the sea. The water discharged from farms carry deteriorating left-over feed materials, faceal matters and also bottom detritus. This gets partly redistributed among the various farms, and the rest debouches into the sea. During the next high tide these polluted waters, most of which come back from the sea during high tide are again used for filling up farm ponds. In other words, the pollutants forming part of the discharges are buffeted from farm to farm and from the near shore of the sea to the farms.

Apart from these aspects, it is seen that the groundwater within the reach of food crops grown on lands adjacent to brackishwater farms are affected slowly by the salinity; The agricultural farmers complain that their lands are gradually becoming saline because of adjacent brackishwater ponds. Consequently the yield from land crops, it is said, is coming down or land itself is becoming unfit for such crops in some cases. They are left with no option other than selling the land to shrimp farmers or the owners themselves undertaking shrimp farming themselves.

This alarming situation seems to be connected with the inordinate time taken by Coastal State Governments in alloting brackishwater lands owned by them after taking the above mentioned factors into account, and controlling sale of privately owned lands that may be vulnerable to the effects of inroads to brackishwater farm sector. It is possible the Coastal State Governments and the centre are studying carefully the implications of the invasion of brackishwater farm conditions into lands adjacent to productive paddy fields. Several State Governments, particularly Tamil Nadu and Andhra Pradesh are understood to have ordered a micro-survey of all lands belonging to them and this is no doubt a wise step, provided the survey covers not only boundaries but also all other connected aspects.

In other words, there is now an urgent need to evolve a suitable policy for allotment of these saline affected lands and also those belonging to other owners, at national level, subject to such adjustments as may become necessary with reference to local conditions.

No one can advocate against spread of brackish water shrimp farming as such. The activity adds to aqua production and our export earnings. This should not, however, mean that all coastal lands need to be converted into brackishwater farms. Such of the areas which are saline and are unfit for raising paddy or other land crops and are uneconomic for raising land crops, have to be earmarked for conversion into brackishwater farms, leaving a buffer area between the farms and the paddy fields through a suitable device so as to safeguard against the spread of salinity.

One undesirable effect of setting up of brackishwater farms is on capture marine fisheries. Apart from polluting the sea, there has been an increase in the capture of shrimp seed stocks of economic variety, such as Tiger and White shrimps, from the sea, leading to depletion in capture shrimp fisheries, thereby upsetting the ecological balance and national wealth of shrimp available from the sea.

The total requirement of shrimp seed for stocking farms on the east coast alone is very vast. Ending 1991-92, Orissa has 7,417 ha, Andhra Pradesh 8,100 ha and Tamil Nadu 480 ha under shrimp farming totalling 15,997 ha. Taking into account a round figure of 10,000 ha at conservative stocking rates of one lakh seed in two crops in these three States, the seed requirements in a year are of the order of 10,000 lakhs. Of this requirement, hatcheries on the east coast can supply at the most 2000 lakh nos. This means about 8000 lakh nos. net (not counting mortality) are removed from the sea for stocking the farms. This much of seed, left to grow in the sea into adults would have given a production of 24,000 tonnes presuming an average growth of 30g per individual seed. Such a heavy removal from the sea reduces marine capture productivity drastically. No wonder, therefore, a gradual depletion in shrimp capture fisheries is experienced.

If the seed is allowed to grow in its natural environment, there will be several chances for the seed to become adults and grow to maturity and breed. Because of intensive collections of seed, the chances of their growth into maturity are substantially extinguished. Shrimps raised in farms do not attain maturity.

The operations of vessels depending on capture shrimp fisheries are now reported to have become uneconomical because of intensive removal of shrimp seed from the sea. Consequently vessels are now forced to conduct intensive fishing in order to catch shrimp in adequate numbers so as to improve the economics of operations. This is affecting the population further. From this situation it is quite clear that several more hatcheries have to be set up to meet the entire requirement of shrimp seed for culture so that natural seed collections from the sea can be stopped.

Brood shrimps are now collected from the sea for hatchery operations. It is true that in any case these would have been utilised for export. Instead, a good part of the catch is being used for hatchery operations. Looked at from another angle, because of heavy exploitation of wild seed, the extent of stocks of brood shrimp in the sea has also been coming down. Therefore, there is an urgent need for the concerned Research Institute to pay special attention for growing shrimp to maturity in farm ponds so that such shrimp can be used for breeding in hatcheries instead of depending on brood shrimps captured from the sea. This is not an easy Job. World over experts failed to grow shrimps to maturity in confined waters. This situation should accentuate the determination to achieve a breakthrough in this regard and the ICAR and the Department of Biotechnology would have to mount a special scheme for the purpose.

The Nation has the advantage of learning lessons from countries such as Taiwan and Ecuador, in developing shrimp culture on proper lines, free from greed. In these countries brackishwater farming has virtually collapsed. Noticing this, these Governments have imposed strict regulations in respect of brackishwater farming. Before the situation worsens in India, it is imperative for the Government of India and the Coastal State Governments, with the help of experts, to deliberate on the problem through a suitable committee constituted for the purpose to evolve policies and strategies for preventing or bringing down exploitation of seed from the sea and stepping up research to grow shrimps to maturity in farm ponds. Steps are also needed to prevent salinisation of lands adjacent to brackishwater farms.

Fishing Chimes, Editorial 182: February 1994: Vol. 13, No. 11

# Re-levy of Purchase Tax on Marine Products in A.P.

## '*Reddy gaaru* has come; Start the show from the begining: A Telugu saying

There are over 40 marine products exporters in Andhra Pradesh. These have been struggling hard to survive during the past several years to make their operations viable and contribute to the nation's exports and foreign exchange earnings. The State Commercial Tax Department identified seafood exporters as one of their target groups for spreading their tax net. Collection of money being their main concern, they found out a misplaced and illogical way of interpreting a rule in the Central Act which specifically says that exported marine products do not attract sales or purchase tax. The concerned officers demanded the production of export orders based on which shrimp or other marine products were purchased by the processors. It was not practicable and easy for the processors to comply with this requirement.

This 'Tweedledom and Tweedledee' situation came up in late 1980s also. At that time, the State Government, based on appeals from the industry to exempt them from purchase tax as per the practice in the neighbouring States, the State Government examined the subject thread-bare and ordered the stoppage of collection of the tax. The points that went in favour of the decision were :

1. Marine animals are harvested whenever they are available. These have to be pooled up, processed, packed and kept ready while simultaneously making efforts to secure export orders. It is possible that a processor may receive export orders in advance when he has no stocks but based on which he could be able to procure stocks but such orders are exceptions. In other words, it will not be possible always to produce export orders to prove that stocks are held against those orders.
2. A processor will be able to try for orders when he has stocks. Without stocks he cannot compete in the international market for securing export orders.
3. A processor has to procure material when available. He cannot get material when he wants. If he does not procure when available, he cannot fullfil orders. He will lose orders.
4. The fear of taxation is having the effect of the material getting diverted to States such as Tamil Nadu and Kerala, where there is no imposition of purchase tax. Because of this, the export performance gets reduced.
5. Shrimps, the main item of export have practically no local consumption because of the high price. Like-wise, cuttle fish and squid, which are exported have no local market at all. Taxing such exports does not make sense.

This situation clearly implies that, whether backed by export orders or not, they will have to be exported.

It is a matter of immense surprise that the Government, which was convinced already and lifted levy of purchase tax on exported marine products in 1989, has again revived the practice only to prove the adage: '*Reddy gaaru has come: start it all over again:* There is no doubt, the then Chief Minister, Mr. Vijayabhaskara Reddy is not obviously aware of what is going on or the subject has not been explained to him well. Your Editor as well as several of those in the fishing industry are aware of the rational and developmental approach of Mr. Vijayabhaskhara Reddy. It was Mr. Reddy, in late 1970s, when he was a Minister in Andhra Cabinet, extended *sue motto* support to the subject pertaining to granting of a financial guarantee in favour of various fishing companies who were then importing trawlers from Mexico. But for his benevolence and intervention, deep sea fishing would not have made an entry into Indian waters. A great man of that stature, always wedded to the cause of development, cannot be expected to behave differently in the matter of an unjustified taxation, that too without adequate reason. It does not need a 'Vijayabhaskara Reddy' to understand the artificially created problem, just for the sake of netting a few lakhs of rupees towards purchase tax. He cannot be oblivious to the fact that the opening of the old chapter would unleash a spate of problems. Producers, most of whom are common fishers belonging to the weaker sections of the society, will receive reduced returns for their efforts because of the tax. This can be stemmed only by discontinuing the re-extension of the pernicious system of taxing exportable marine products with purchase tax. Even the dreaded income tax people have cause to exempt exported marine products from the tax. Such products cannot be subject to sales or purchase tax just because Government needs money. Such a tax is totally irrational.

If the matter goes to the notice of Mr. Vijayabhaskara Reddy, one can be sure he would certainly set right the situation. Taxing exportable marine products! What an abominable idea! That too when Mr. Vijayabhaskara Reddy, the champion of fisheries development is the Chief Minister! This can never materialise. If it materialises, Government will face a coastal revolution that would need lot of repression to quell and lot of rethinking to gain lost ground. Prevention of trouble is far better than allowing it to brew and later trying to quell the same. In the meanwhile, producers were planning to stop producing, and processors planning to stop processing and export.

Fishing Chimes, Editorial 183: March 1994: Vol. 13, No. 12

# On Diversification of Fish Production Effort

Longevity of fishery enterprises, by and large, is dependent on periodical technological upgradation, expansion or intensification of fishing effort, taking care of backward and forward linkages, and diversification of fishing effort aimed at judiciously harvesting known but, by and large, unutilised resources. These would have to be an integral part of the periodical thrusts. Enterprises that stagnate for long are not likely to survive. Let us look back at periodic developments on the commercial side in the inland as well as marine fisheries sectors. Those who incorporated beneficial technological developments that have been taking place from time to time, have improved their earnings. Further, as is known, product demands keep changing, leading to changes in economics of operations and this aspect also is kept in view by progressive producers and processors.

Diversification in the inland freshwater culture sector will possibly take place in the near future in the form of mono-sex farming, running water farming or systems relating to other types of farming hitherto not tried, but have now become important. If a research institute standardises systems for culturing, say murrels, or climbing perch or improved systems of culturing pearl spot, eel, mullets or any other fish or any edible aquatic animals, it would be a valuable contribution towards diversification. Those who are now engrossed in brackishwater farming but with facility for pumping in or taking in sea water may like to diversify their activities for culture of mud crab, edible oysters, holothurians etc.,

Take the specific case of marine capture fishing for shrimps off upper east coast. In this zone the same stock of shrimp is shared by a large number of vessels, thereby bringing down catch per unit. In this situation there is no escape from reducing the strength of this fleet, big or small. This can be achieved by equipping a substantial number of these vessels for another type of fishing and diverting them for this newly added type of fishing for a suitable duration in a year or all through the year. This strategy will gain ground only when the economics of operations are not affected.

The above mentioned diversification strategy has to cope up with two dimensions. One is that shrimping is done only for about six or seven months and most of the vessels are idle during the rest of the year. To get over the problem some have taken to stern trawling for lobster, caphalopods and deep sea shrimp but not often with good results. Should diversification of fishing cover the entire year or for only part of an year when fishing for shrimp in the Upper Bay of Bengal is poor? or would the shrimping durations need regulation on a rotation basis organising access to the area by a determined number of vessels of the fleet for such durations as can be worked out, and the other vessels concentrating on some other source with a suitable fishing system added? There can also be other alternatives. From September for a few months shrimping will be promising on the north-west coast in areas beyond the fishing zone of traditional boats and the vessels can probably fish in this zone. Vessels can also be equipped for catching items such as cephalopods and tuna.

Another way of diversification can stem from the successful deep sea (200m depth) demersal trawling results achieved by mechanised fishing vessels operated by the Indo-Danish Project at Tadri in Karnataka. This system may be well suited for emulation primarily by Sona or such other boats, smaller mechanised boats and also by larger fishing vessels. Marine Products Export Development Authority thinks that any future increase in exports would have to come from deep sea fishing as farmed shrimp fishing has its own limitations. The Chairman of MPEDA would no doubt consult his experts as to how to plan and organise to obtain marine produce from deep seas. Owners of larger fishing vessels feel in general that they are in the lurch at present and are unable to move ahead.

One encouraging factor is that in regard to fishing and processing effort there is a general consensus on the need for diversification among technocrats, administrators and scientists. The manner in which this has to be done and the technical, procedural, financial and organisational aspects of it do not seem to have been worked out as yet. One point has to be borne in mind in this context. A monitoring system having legal force has to be introduced. At present, provision for the legal enforcement of a monitoring system for Indian fishing vessels in the EEZ does not seem to exist. Government may think of promulgating an ordinance to meet this need. Most of the current problems seem to stem from the absence of a monitoring system to regulate activities of Indian vessels in the EEZ.

There are portending signals emanating from those few who are engaged in deep sea fishing. They appear to feel rudderless and are exploring possibilities of moving out their vessels for operation in the waters of foreign countries with needed permissions, based on their present perceptions in respect of fish stocks in our EEZ. There are a few who seem to have found through actual fishing that deep sea shrimp stocks as per the particulars given by the authorities are not really at or near the locations mentioned. The stocks may have drifted to other locations. Further, they say that deep sea lobster stocks that cannot withstand fishing by half-a-dozen medium sized vessels for one season are not worth depending upon. There is no known or tested fishery in our EEZ which can sustain economic operations for a few years even by a limited number of vessels, as per present indications. Bull-trawling alone seems to have given good results, although the duration of fishing in a voyage is long. Those engaged in tuna long lining operations feel that the fishery, apart from being seasonal, does not always offer adequate fishable stocks.

Views of this kind are unmistakably gaining ground. This is not a good trend, particularly when our Government tells us, based on the excellent work done by Fishery Survey of India and Central Marine Fisheries Research Institute that our EEZ has excellent fishery potential capable of giving a maximum sustainable yield of 3.9 million tonnes including the coastal fisheries. One main reason for the negative conclusions seems to be the pressures from the financing institutions for repayments of loans. Their current earnings are poor and no working capital loans are forthcoming. No one is in a position to break the vicious circle. Without working capital, companies cannot earn. And, if they do not earn they cannot repay. It is not that the industry alone faces the challenge but the Government is also now face to face with it. The Chairman of the Marine Products Export Development Authority thinks that the only way in the long run to increase our exports is through deep sea fishing. As a follow-up of this, some bold and pragmatic steps are needed on the part of the Food Processing Ministry and the MPEDA to achieve an enduring breakthrough. A solution to the problem may lie in introducing mid-water trawling and stepping up long-lining efforts and FSI may have to give special importance to investigate and arrive at feasible diversification systems for augmenting marine harvest.

Fishing Chimes, Editorial 184: April(i) 1994: Vol. 14, No. 1

# Major Gaps

The nation has done well in respect of the various lines of endeavour in fisheries sector. We have been doing fairly well in inland fish seed production and in inland fish farming. Brackishwater shrimp farming has attained new dimensions. Monofarming of sterile Tilapia and Catfish culture, although having restrictive dimensions, have now a place in our fish culture systems. There has been progress, although limited, in freshwater prawn culture, freshwater pearl production and development of supporting culture systems such as those of *Azolla, Spirulina*, and several diatom species. *Artemia* which was not prominent two decades back now has a place in the production systems. Aeration methods have come to stay. There are many more developments that have come to be adopted by the farmers, a great tribute both to our research establishments and developmental bodies to have brought them into the arena and the farmers who have adopted them imperceptibly. Fish Farmers Development Agencies and Brackishwater Farmers Development Agencies have come to stay as an organised mechanism for the development of culture fisheries. There has been a significant improvement in the availability of technically trained personnel but a gap still persistsr although on a reduced scale. Reservoir fisheries have shown considerable improvement in States such as Himachal.

The developments in Trout culture in Himachal with Norwegian aid has led to remarkable results. Although much remains to be done, one has to congratulate the Union Ministry of Agriculture of which ICAR is a part and the Central Inland Capture Fisheries Research Institute, Central Institute of Freshwater Aquaculture, the Central Institute of Brackishwater Aquacunure, and the Marine Products Export Development Authority for having achieved so much. The achievement of MPEDA in catapulting exports of marine products to Rs.2000 crore level is phenomenal. One has to concede that fishery development in India is well administered and well co-ordinated. No wonder India has a leadership status in this sector.

India has a long history of reservoir fishery development. The expertise that India has in this line is immense. Yet no impressive progress could be achieved so far in raising average annual production from reservoirs per hectare to a significant level. No doubt projects with World Bank's assistance have been taken up in certain States such as Karnataka but there has been no overall impact. Notwithstanding these development programmes, there is an utmost urgency to fully utilise the expertise available at CICFRI. There is a wealth of information available on the fisheries status of almost every reservoir in the country. In order to proceed further, concrete step-by-step plan of action would have to be worked out and implemented. In order to achieve this, the nation needs an organisational mechanism at reservoir level (large reservoirs to start with) and the needed monitoring system at the State level. CICFRI should be entrusted with a predominant role for vitalising technological inputs and manpower structure and it should be the main vibrating thread running through the system to be evolved. Reservoir Fisheries Division of CICFRI has to be reorganised or strengthened substantially to enable it to undertake programming, co-ordinating, monitoring and over-seeing reservoir management and development functions all over the country. Our total inland fishery development work would have to be considered as a major failure, until such time as our reservoir fisheries are well developed and brought up to a graceful balance between capture and culture fisheries. Millions of ha of reservoirs producing a mere 10 kg/ha/annum, on an average, diminishes the achievements in the culture fishery sector. With the availability of technology in abundance in the country, the continued neglect of reservoir fishery development by State Governments has to be construed as a major gap and this gap has to be filled up.

Another major failure is the complacent attitude or neglect in imparting a spread effect to farming of freshwater prawns. Although technology is available, both for producing seed or collecting seed from nature and also for culturing them, not much is being done to provide fillip to this segment of activity, although farming of giant freshwater prawn and production of giant freshwater prawn seed has made several strides. A few hatcheries have been set up and culture technology has been standardised to a considerable extent. The giant prawn has considerable economic importance. It has export potential. There is certainly a gap here and this has to be removed.

We have to go a long way in introducing cage farming in inland lentic waters and also in reservoirs when the river concerned is in floods. Considerable attention and application is needed for removing this gap.

We have done pretty little to promote closed systems of fish farming. CIFA no doubt has done some work but the work needs intensification. In the private sector Hindustan Lever has no doubt made a begining. Mono-sex farming now confined to Vorion Distilleries can be popularised by CIFA. This gap has to be bridged.

Practical application of results of genetic manipulation for popularising, producing and farming of improved strains has to reach the farmers.

There is an enormous gap in the mariculture sector. In fact, we have not yet made a beginning. As mariculture has to be undertaken mostly in coastal waters, developments in this sector are to be linked to a legislation to provide proprietory rights to private enterprises. We do not have any such base. This has to be created. Only when this base is there, progress in mariculture can be achieved through application of available technology.

Brackishwater farming is spreading all along the coastline in a haphazard manner. The need for introducing regulatory measures so as to ensure eco-friendly development restricted severely to saline soils over a period of time and to eliminate incidental conversion of non-saline soils into saline soils is often emphasised. Compulsory treatment of in-coming and out-going waters to eliminate or filter pollutants would have to form part of the regulatory measures. MPEDA has no doubt initiated steps in this direction but results are not yet apparent. The removal of these gaps in the on-going system are of vital importance, looked at from any sensible angle.

In the small-scale sea fishing sector there is a major organisational gap. This gap has to be removed. When this is done, Government can get a total picture of the situation all along the coastline. When a total picture at macro and micro levels is available, something meaningful and useful for equipping fishers with improved means of production, training them as required and improving their socio-economic conditions, can be achieved. The real problems of fishers will be known, paving the way for their integrated development by setting up developmental units all along the coastline. This will go a long way in bringing about organised development.

Developments in the deep sea fishing sector are taking place in such a way that they have a form but very little content that would establish the deep sea fishing industry in the country on an enduring basis. No real training gets imparted to Indian hands from the foreign companies who now operate their vessels in Indian waters under some system or the other. An analysis shows that the permissions accorded by the Government for bringing in foreign vessels either under charters, joint ventures, leases or test fishing largely serve as a means for the foreign companies to gain access into our waters. The Indian counterparts are able to learn very little. We have heard comments that in most cases, it would be difficult to reconcile/perceive the actual state of affairs and what the invoices, returns say. Those at the helm of affairs in the Government such as the Fishery Survey of India or at the Ministry of Food Processing will no doubt have a true evaluation of the situation.

Our evaluation is that, irrespective of the nature of collaboration, larger vessels of global fishing nations with enormous endurance move from one fishing area to other, (often from waters of one nation to other) based on fishing conditions. The operations, therefore, will not be confined to the EEZ of any one country. If the joint ventures are true ones, not based on mere gaining access into our waters for part of the year, with whole time commitment and with a proper climate for providing training, there can be real progress, provided there are adequate resources. The subject is complicated and so far no suitable strategy could be found for the exploitation of the fishery resources of our distant and deep sea waters, particularly in the mid water zone where the bulk of fisheries of the deep sea zone are believed to inhabit.

The utilisation of the deep sea waters nearer the coast and those that are distant consists of both culture and capture fisheries. Offshore cage culture systems linked to land based mariculture aimed at fingerling or yearling production will facilitate realisation of substantial harvests particularly for export. Relevant technology as well as cage systems can be imported. Considerable hard work is involved in achieving this, linked to site/leasing systems.

In regard to deep sea capture fishery development, Government as well as entrepreneurs would have to contend with a number of problems. The present development process which has progressed to some extent has two main characterestics. One is that the very few foreign companies that have entered the Indian scene have a commitment. Their objective, as already mentioned, is to gain access. The other is that their vessels, particularly from former Russian States, are languishing in their countries and they want to put them to use. The enterprises on the Indian side have commitment but what naturally top most in their mind will be to earn money but they have problems in gaining real control over the foreign vessels. The factors in the way are investments, technical expertise, fish stocks, processing facilities for value addition and marketing matters. The gap that the nation has in the deep sea and distant water fishing sector is not easy to bridge, but this has to be achieved if we are to acquire our own capabilities.

Simultaneous to the steps already under way, it appears that we should have a well conceived and upgraded training system. This should be mostly on board vessels and very little at the Institute. There should be a fleet of four or five vessels of 80 m OAL and over with modern electronics and modern gears, value-adding processing systems and adequate accomodation for trainees. The vessels should have facilities to undertake atleast one main fishing system. Vessels should be capable of being out for a long duration. Those on board should be prepared to stay out at sea for long durations. Once batch after batch of captains, mates, engineers, deck-hands, engine room hands are trained for the needed duration to withstand the ruggedness of the job involved, there will be our own men for employment on commercial high sea fishing vessels, first as apprentices and later as regular employees on real Indian Vessels. Our own fleet has to be thus developed alongside the present strategy in the hard way, to replace the present system, the weaknesses of which are known to several.

Regarding shore establishments and the personnel to man them for producing processed products with value addition, our processors have the resourcefulness to enter into joint ventures with importers/foreign processing companies. There are several capable Indian entrepreneurs who can establish integrated deep sea and distant water fishing projects. Through a well formulated scheme the energies of interested entrepreneurs should be channelised to develop a good infrastructure.

The entrepreneurs (represented by their associations) and the Government would have to take on the challenge that faces them, without restricting it to the present system of operations under joint ventures, leases and test fishing, which, at the best provides awareness of the field. Government should launch an integrated project to impart training to Indian candidates on really large vessels for long durations, organising investments to companies with commitment and intending to acquire large vessels with needed endurance to undertake long distance fishing inside and outside our EEZ, setting up infrastructure having shore plants with facilities for processing involving value addition and linked to marketing systems etc. Above all, fishing harbours to take in large vessels of 80 m OAL and above would have to be established in phases. The first such harbour with infrastructure has to come up as part of the upgraded training institute. Providing training on medium-sized vessels for short spells will not lead to needed results in deep sea and distant water fishing.

Fishing Chimes, Editorial 185: April(ii) 1994: Vol. 14, No. 1

# Chimed into the Fourteenth Year

*Fishing Chimes* has chimed its way into the fourteenth year of its life in April, 1994, having completed 156 chimes. This issue represents its 157th monthly chime. The journal jogged along all these years, with marching chimes, reinforced the patronage, extended by the subscribers, advertisers and contributors, who are the fountainheads of knowledge, over the years. Sound bonds between the journal on one hand and the subscribers, advertisers and various authors on the other have come to be established. The quality of readership improved and the readership response imparted prideful strength to *Fishing Chimes*. It gratefully acknowledges the support lent by the readership.

Since its inception in 1981, *Fishing Chimes* recorded several significant events in the development of Fisheries of India, apart from strides made in the sector in other countries. The recent history of development of fisheries, particularly in India, breathes in the various volumes of *Fishing Chimes*. As a record, it is a major source of reference for those who look for material on most of the aspects of Indian fisheries development since 1981. The various articles/papers published in it demonstrate the capabilities of the general bulge of the fisheries workers who have been toiling hard on some problem or the other relating to the subject incessantly and relentlessly.

There are several working in the Central Fisheries Research Organisations, State Fisheries Units and a large number of fishery enterprises who are on the constant quest to find solutions to various problems having a bearing on the developmental aspects of fisheries. Journals devoted to the scientific, technical and technological aspects of fisheries are few in the country; their periodicity of publication is protracted and this does not give an opportunity to the enthusiastic workers to relate their experiences, achievements, and views to the others in the field. Further, the few journals which one can say are conservative and non-compromising in respect of traditional norms for publication, are not accesible to the young and aspiring workers for publishing the accounts of their work. *Fishing Chimes* has by now established itself as a medium to provide expression to this category of up and coming fisheries scientists, technologists, technocrats, entrepreneurs and others.

Seminars, Symposia, Workshops and Conferences on fisheries are held in some part of the country or the other almost round the year. Your editor with his camera is a common sight at most of these gatherings. *Fishing Chimes* has been in the forefront in providing a faithful summary of the papers presented, views expressed and recommendations made at these gatherings. There is no doubt these reports have been of considerable help to the readers, particularly those who form part of the links to help the Governments concerned in the formulation of policies and strategies for the developments of fisheries of India.

The journal has been chronicling the giant strides made in freshwater fish culture and brackishwater shrimp culture sectors including the development of hatchery systems, the breakthrough achieved in mariculture and the achievements of various voluntary agencies in the socio-economic sector, to upgrade the quality of life of fisherfolk through improvements in their professional calibre, so as to bring about a sea-change in their lives. The journal has seen the growth of Fish Farmers Development Agencies and Brackishwater Fish Farmers Development Agencies and the process of their gaining strength as organisational mechanisms to develop the culture fisheries sector. All these aspects are recorded.

The journal has been recording the periodical growth of marine capture fisheries sector, which has matured, although in a small way, from a predominantly coastal fisheries stance to one of deep sea and distant water fishing. *Fishing Chimes* has also had the privilege of recording from time to time the phenomenal developments of our marine products processing sector and the significant growth in the exports of marine products besides the great beginning in the export of marine as well as freshwater fish. The journal has the pride of recording for the posterity the major role played by the Ministry of Commerce (Marine Products Export Development Authority) and the Union Ministry of Agriculture and the State Fisheries Departments in bringing about respectively a major spurt in the development of integrated brackishwater shrimp farming and in the development of integrated freshwater fish farming, apart from the contributions of the Ministry of Food Processing Industries in leading the country to become a distant water nation.

*Fishing Chimes* has progressed into the present stage through devotion to the cause of development of fisheries of India in all its aspects and its abiding responsibility towards the readership. The result is the reward of acceptance of the journal in a large measure by the fisheries sector of India.

Fishing Chimes, Editorial 186: May 1994: Vol. 14, No. 2

# Diversification of Fishing by Shrimp Trawlers: Fund-oriented Plan Imperative

Following the introduction of small mechanised fishing boats in 1950s, Government took an important decision about 30 years back (around 1963) to introduce larger sized vessels. In early 1970s, 10 vessels imported from Mexico, U.S.A. and Iceland were introduced followed by the introduction of around seven vessels, constructed at Mazagaon Docks, by Union Carbide. Thereafter, in 1978, 28 vessels out of 30 permitted were allowed to be imported by various companies from Mexico with financial assistance from SDFC and a Mexican Bank. Later, vessels were allowed to be imported from Holland, Australia, Korea, U.S.A. etc. Several vessels constructed at Indian shipyards were also introduced raising the fleet strength to over 170 vessels all of which are in the 23m OAL range.

Loan assistance was given for acquiring vessels from abroad to the extent of 90 per cent of the cost and upto 95 per cent of the cost for indigenously built vessels through the erstwhile Shipping Development Fund Committee, from around 1977 untill enactment of the SDFC (abolition) Act around 1987 and taking over of the functions of financing acquisition of fishing vessels by SCICI. All the vessels introduced were shrimp trawlers fitted with outriggers. The companies maintained a stipulated debt-equity ratio of 6:1 for all the loans given by SDFC. This was done in the interests of development and to move towards the goal of acquiring deep sea fishing capability. The vessels introduced in 1978 could exploit the shrimp fisheries off Orissa and West Bengal in an impressive manner.

The story of introduction of shrimp trawlers would have had a good ending had Government given weight in time to the recommendation made by an expert committee set up by the Ministry of Agriculture under the Chairmanship of Dr. James in middle 1980s. The committee recommended an upper limit of around 120 shrimp trawlers. Despite this, the Government continued to allow import and also indigenous construction of shrimp trawlers.

The addition of shrimp trawlers beyond the number recommended by the expert committee was mainly because of the scheme for chartering of fishing vessels. Under this scheme, the charterers were under compulsion to add an equal number of vessels as they operated under the charter scheme. If they were asked to introduce vessels similar to those as under operation on charter, there would have been a totally different set of developments. Firstly, the swelling of the shrimping fleet would have been stemmed. The strength of the shrimping fleet went up because the condition stipulated was vessels over 20m OAL should be introduced by the charterers in fulfillment of the *pari passu* condition. Because of the demonstrated profitability of shrimp trawling charterers made a beeline for acquiring shrimp trawlers only. There were others as well who managed to import or acquire indigenously built shrimp trawlers.

Government did act on the recommendations of the expert committee referred to above. An amendment was made that trawlers with out-riggers used for shrimp trawling would not be allowed to be acquired. However this was not of much avail. The companies managed to have outriggers described as stabilisers in the specifications. In spite of the transperancy of this euphemistic expression, the Government accorded permission for further acquisition of shrimp trawlers was with these renamed ones. This was the main reason for the increase in the shrimp fishing fleet.

The effect of the increase was the reduction in the catch per unit effort (CPUE) but not the total fishery, by and large. The total shrimp landings on the Upper East Coast continued to be more or less constant but for statistically valid variations. The fishery normally dips once in three or five years for biological reasons, although coastal shrimps are an annual crop.

The fall in catch per unit effort (CPUE) of the larger shrimp trawlers is attributable to other developments. Lured by the successful operations of the shrimp trawlers during the first few years of their introduction, several entrepreneurs, acquired a class of trawlers which came to be called as "mini-trawlers". These are 16m long each and their strength grew to a level of over 100 numbers. Their operations brought in good returns. The happiness of the owners of these mini-trawlers, however, was short lived. While mini-trawlers also had out-riggers, a new class of low-cost stern trawlers for shrimping without outriggers (Sona Boats) were introduced initially by a few entrepreneurs. As the landings from these vessels were found to be encouraging and there was some visible prosperity, many others acquired similar vessels, taking the number of such boats to over 600. The introduction of these two categories of trawlers (mini-trawlers and sona boats) led to a decline in the per unit catches of the larger trawlers as well as others for the reason that the two new categories of vessels also learnt to operate their boats in the same area (Sand heads), where hitherto the larger vessels alone used to operate.

It may be mentioned here that defaults in repayments of loans taken for acquiring vessels came into being in stages. For some reason or other, when the mini trawlers came into being and started their operations, repayments by the larger vessels slowed down. When Sona boats joined the race, repayments by larger vessels as well as mini trawlers suffered a set-back. Sona boats did well for some time. However, they also faced the problems of repayment to banks because of the diminishing CPUE. The point to be noted here is that the authorities concerned completely ignored the trends in the shrimp fishing efforts in the Upper Bay of Bengal. The portending signals thrown by unregulated growth of various categories of vessels were not apparently well evaluated and acted upon. From real prosperity at one stage the owners of all categories of vessels drifted into a sense of false prosperity mostly by postponing payments to the financiers. Finally they landed in a position of declaring that the operations were uneconomical. The situation acquired such characteristics 'that experts had also to agree that the operations were uneconomical and the vessels could be run with some profit only if there was no burden of repayment towards investment loans.

At this stage, when the enterprises are not in a position to repay the loans taken by them, and based on the representations of the Association of Indian Fishery Industries, the Government appointed a technical committee to make recommendations, mostly on rehabilitating the industry.

We intend to digress here to discuss what deep sea fishing is: There is lack of clarity in our thinking as to what constitutes deep sea fishing, and distant water fishing. In contrast, most of us know what coastal fishing is. An effort to recapitulate and consider the specific import of these terms and to reclassify the existing fleet of larger vessels according to their capabilities would be purposeful and helpful.

## Coastal Shrimp Trawlers

The 23 m long Mexican built trawlers and others in the 23-28m range come under this category. These are basically designed for trawling for shrimp, making use of outriggers. Two nets and sometimes four are operated laterally from the outriggers one or one set of two on the starboard and another one or one set of two on the port side. These vessels are also equipped for operating a single trawl net from the stern but with severe limitations imposed by the capacities of hydraulic system, the trawl winch, the main engine and the gensets. As equipped, these vessels cannot operate nets of both the types beyond 50 fathoms range. These and other vessels of this kind built in India have to be excluded from the definition of deep sea fishing vessels. They are basically coastal shrimp trawlers.

## Glorified Deep Sea Fishing Vessels

All the other vessels constituting our larger fishing vessel fleet (23 to 28m OAL) are also basically equipped for lateral (out-rigger) shrimp trawling, besides stern-trawling as in the case of the Mexican trawlers. They can be distinguished from the Mexican trawlers (of Gulf design) by a few vital aspects. They have higher engine horse power. They have generators of higher power output. The fish holds are better equipped in terms of space and preserving capacity. The trawl winches have higher capacity and are driven by a hydraulic system of far greater capacity than the Mexican trawlers and their like. These variations have given them the capability, although unwittingly, to fish up to a depth of 200m and beyond. Here, let it be remembered that the design of all vessels of this category is basically for shrimp trawling. Even so, because of the features referred to above, imported trawlers built in Australia, South Korea, U.S.A. and trawlers indigenously constructed at Bharati Shipyard, Alcock Ashdown etc are seasonally deployed by some of the owners for catching lobsters, fin fish and deep sea shrimp, on the continental shelf of the south-west coast, in Australia, and in a few countries which purchased trawlers from Australia as India did, the trawlers are used exclusively for shrimping. These vessels in India come under the category of glorified deep sea fishing vessels.

Two small wooden mechanised boats 43/48 ft with trawl winch that can take 600m of wire rope were successfully tested for deep sea fishing by the Indo-Danish Project, Tadri in Karnataka. The boats are fitted with GPS (Global Positioning System). It is not the overall length but the capacities of the equipment on the boat that makes a boat a deep sea fishing vessel. GPS (Global Positioning System) enables the fishermen to take the boat to grounds found to be promising on earlier trips. Having seen the results, around 200 vessels are now under construction in Tadri area. The constraints in small boats used for deep sea fishing are the lower fuel capacity and fish storage capacity. There can be a problem of safety during cyclonic conditions. These constraints make them multi-day or short duration boats.

Deep sea fishing vessels would have certain features designed for deep sea fishing. The hydraulic and power-generating systems, main engine H.P. and trawl winch system, would be of a higher capacity. The facilities on deck for handling of fish would be of a far improved nature. The electronic aids will be far more sophisticated and would invariably include Global Positioning System (GPS). The vessels would also have modern communication systems, apart from radio telephone (items such as a fax). Deep sea fishing trawlers conduct not only bottom deep sea trawling but also mid-water trawling.

We do not have, to the extent known to us, any true deep sea trawler of Indian registry, having adequate endurance to stay out for more than forty days. However, there are a few deep sea fishing trawlers of foreign registry operated by Indian companies, now in operation under joint ventures, leases, or test fishing systems having higher endurance. Some of them are probably registered as Indian vessels.

## Other Deep Sea Fishing Vessels

Longliners, gill-netters, jiggers and mid water trawlers and other deep sea vessels meant for specialised fishing conduct fishing in deep sea zone but in the surface and columnar layers. Indian entrepreneurs now operate six tuna long liners in the pelagic zone with varying degrees of performance.

## Distant Water Fishing

There are a few distant water fishing nations. These are countries such as Japan, Korea, Taiwan, former U.S.S.R. France etc. These vessels have the capability of staying out at sea for a long duration. These vessels return to their home ports mostly when the holds become full with fish. The duration of stay away from their home port may extend for a very long time, sometimes for one year. They manage to obtain supplies and services from certain foreign ports under some arrangement previously entered into. Distant water fishing vessels engage themselves in fishing alone or they may conduct fishing and also undertake on-board processing or only on-board processing by being a medium of collection of fish from a number of satellite vessels under some arrangement. There are quite a few such distant water fishing vessels belonging to various countries such as N. Korea, S. Korea, Japan, former States of U.S.S.R. etc., operating in waters of other nations or in open waters outside Exclusive Economic Zones. The operation of vessels in EEZs of other nations can be under joint ventures, leases/test fishing, license etc.

It would obviously take considerable time in India for the Government and industry to develop a real deep sea trawling and distant water fishing fleet. As matters stand we are now at a stage of having the help of foreign deep sea and distant water fishing vessels to gain experience. Our EEZ is so vast that we may have to eventually graduate into a distant water fishing nation to harvest fishery resources within and outside our EEZ as well.

The Technical Committee appointed by the Government is understood to have recommended reduction in shrimp trawling fleet and a rehabilitation package for the repayment of pending loans connected with the present fleet. We have heard of several other recommendations relating to policy measures, infrastructure, finance, marketing, legal aspects, training and research.

## Reduction in Shrimping Fleet

The Technical Committee appointed by the Government told the obvious to the Government again that the shrimping fleet should be reduced by equipping the majority of the shrimpers for diversified fishing. Government would now need a recommendation on how to achieve this. The Technical Committee appears to have been silent in its recommendations in this respect. It has been mentioned, as we understand, subsidy may be given for diversification on case to case basis. This recommendation does not take the matter any farther.

## Finance

The Committee seems to have recommended granting of subsidy to those who want to equip the shrimp trawlers for diversified fishing. While this will no doubt be an incentive, the companies concerned will still face the problem of investment for which credit facility and technical expertise is required. In view of the present jaded reputation of the bulk of the companies, there appears to be very little possibility for the companies to secure credit, with or without diversification. Almost all the companies face the problem of working capital. As no bank is stated to be coming forward to provide either block capital or working capital, there is no likelihood of any improvement in this sick industry in the near future, in spite of the recommendation of the Committee which mainly emphasises diversification aspect but has not dealt with the financing and technological support part.

In our view, Government has to take the blame for the present tragic situation. Government lent several crores of rupees to the industry with the prime objective of paving the way towards utilisation of our deep sea fishing resources. As part of the lending programme, Government had not introduced any well-knit scheme to ensure recoveries. This could have been in the form of a special cell as part of Fishery Survey of India to monitor and follow up closely the operations of the vessels. There could have been a separate agreement with owners to enable the financing agency to deal with the processors to whom the catches are handed over and to have the owners authorise the processors to pay a determined part of the amount direct to the financing agency.

Government feels that they have been let down badly by the industry in the matter of repayments. There is a surmise that incomes from the trawlers are diverted, taking advantage of the concessional terms under which the loans were granted. A debt-equity ratio of 6:1 was adopted and the vessels were hypothecated to the lending agency which would help in recoveries. The argument in favour of the alleged diversion of funds is that the owners would have earned long back their meagre share of investment and it would matter very little for them even when the trawlers are confiscated for non-payment of dues. The companies can carry on fishing without paying the arrears until such time the vessels are taken over. The problem of the lending agency is that they have no means of running the confiscated vessels. They have no option other than disposing of the vessels and be satisfied with the sale proceeds. If the financing agency takes over the vessels, in spite of the above mentioned drawback, it is burdened with the cost of insurance, repairs, watch and ward and port charges.

The owners rebut the views as outlined above and brand them as slanderous. They reiterate their commitment and say that they have been losing money, particularly from 1987. They find fault with the banking structure for denying working capital. By this denial the industry has been deliberately strangulated as if by conspiracy. Their surprise is that the financier, who really knows that without working capital the vessels cannot operate, has not done much to help them in this regard and, when the vessels do not operate there is no inflow.

In this situation, the reported recommendation of the Technical Committee on diversification has entered the scene. Whichever way the recommendation for diversification is looked at, the conclusion has to be that this programme has to be supported by an additional investment lending component with subsidy, working capital loan and technical and technological support, with provision for working capital support even without diversification. Coupled with these, Government or the financing agency must enter into a fresh or a supplemental tripartite agreement in which the buyer of the shrimp, as nominated by the owner from time to time be a party, through an enabling provision in the agreement. This supplemental agreement has to be totally different and has to be designed to make it obligatory on the part of the owners or the processors, under authorisation from the owners, to pay a certain part of the dues to the lending agency. Further, the lending agency of the Government must develop a list of processing plants whose managements are willing to come under the system. These recognised plants alone can take material from the owners and the selling activity should be restricted to these processing plants. The list can be enlarged or reduced from time to time. This system can be suitably modified to suit companies having their own processing plants.

To state briefly, a very close and active interaction between the three agencies concerned has to be brought about. This will improve the situation in respect of repayments. As and when the catches are good and surplus amounts realised, over and above the financial needs for sending the vessels back on a voyage, the estimated excess can be determined by the proposed cell in FSI mentioned above for organising repayments towards the loans taken and the processors to pay the amount direct to the financing agency under intimation to the owners and Government.

Sick fishing companies, like any other sick companies, would need pumping in of additional funds. If this need is not met, most of the vessels would continue to be sick, which is not either in the interests of the Government or the industry. Government cannot afford to keep valuable assets idle, just for the reason that repayments towards the loan given are not forthcoming. Only an elephant can pull a trapped elephant from a trench. Government will have to adopt a new strategy for nursing the vessels/companies back to health by integrating their activities with a closely circuited monitoring system, as already mentioned. The additional expenditure involved for setting up a separate multi-functional monitoring cell at FSI will provide a close grip on the situation to the Government. The recoveries are likely to improve and the entire activity may get systematised. The expenditure involved is worth the while.

## Rehabilitation Package

According to unconfirmed information available, the Committee offered three alternative packages to the ailing industry. Broadly stated, one of the alternatives is to provide a waiver of penal and other interest payable, freezing of 50 per cent of the pending loan amounts on the condition that the remaining is payable in instalments over the remainder of the computed life of the vessels. Another alternative is that the owners could pay 30 per cent of the pending loan and take over total ownership of the vessel. Waiver of penal interest etc. is applicable. The third alternative is that the owners are free to surrender their vessels to the financing agency. All the loan and interest dues would have to be waived.

Alternative two, if our information is correct, is no doubt attractive. However, owners of sick vessels, unless they have some means of mobilising funds may not be able to opt for this alternative. For those who want to continue in the line and unable to comply with the condition of one time payment of 30 per cent of the pending dues (second alternative), the first alternative is obviously the best, provided Government helps them to make the vessels operational. Alternative three has the veneer of vindictiveness. Surrendering the vessels has several implictions on the negative side as can be visualised. Instead of forcing the owners to surrender the vessels, Government/financing agency can help by bringing them under the umbrella of an integrated revival package *i.e.,* pumping in additional funds for equipping vessels for incorporating a diversified fishing system, providing working capital and setting up of an effective monitoring system and an improved repayment collection system.

Faults are on both the sides. Starving the industry further is not in the national interest. The additional funds that have to be pumped in will not be a major burden to the exchequer. This is a crucial phase for the industry. If it sinks, the implications will be very serious. The unconfirmed but reported alternatives, unless fortified by additional investments from the Government by way of loan or subsidy, will not lead to any viable improvement in the situation. Into the alternatives stated to have been recommended by the Technical Committee, Government should put flesh and blood, as already indicated, through a separate ancillary scheme for good results. This is the time for the Government to plan for introducing real deep sea and distant water vessels, provided FSI confirms again availability of resources. It would be difficult to say how realistic the data on catches harvested by vessels operated under joint ventures etc. Some uncharitably and probably incorrectly say that the financial arrangements are an inversion of the charter system, unofficially followed but fitted in. This conjecture, if valid, may mean availability of adequate resources. Otherwise foreign vessel owners may decide not to continue operations. In any case, at this juncture, efforts at implementing the recommendations of the Technical Committee as they are, without fortification will be futile and may not take the matter any further.

Fishing Chimes, Editorial 187: June 1994: Vol. 14, No. 3

# National Fish Workers and Deep Sea Fishing

The National Fish workers Forum [NFF], it is stated, stands for the betterment of small fishermen and for conservation of our fishery resources. Small fishermen need leadership to voice their interests and the Forum apparently is striving to provide this leadership. Fishermen are a vulnerable group. They can be both led and misled. Central and Coastal State Governments have done so much to improve their lot. In any case Central and State Governments do not seem to have done anything less for the uplift of fishermen compared to other weaker sections of the society. Fishing harbours, and landing jetties are provided. Fisheries Training Centres have been set up. Massive assistance had been given for mechanisation of traditional boats, supply of synthetic twine, construction of houses etc., over the years. Several roads have been formed to connect fishing villages with towns nearby.

Fish workers lack adequate leadership out of their own ranks. So much so, the gap was filled up by service-oriented outsiders. Representatives of fish workers, participated in a national seminar on 'Deep sea fishing: Fisher people's perspective' held in Bombay on 28-29 September, 1993 conducted by the NFF. Groups from various States asserted their traditional capabilities in deep sea fishing at the seminar. The seminar urged the centre to evolve a new policy approach for deep sea fishing which should ensure expansion of the area of operations of the smaller fishermen farther for harvesting the resources in deeper waters. The seminar expressed great concern at the 'present policy of investment support to the big business houses and other merchant capitalists', neglecting the enterprising working fishermen who had got much skill and knowledge in deep sea fishing. A suggestion was made to confer suitable legal rights and reserve exclusive fishing zones for small scale fishermen at least upto the contiguous zone. The seminar wanted the Centre to ensure adequate measures for proper resources management to protect our fishery wealth at sustainable level avoiding overfishing and depletion.

The stand taken by NFF is laudable. Like other weaker sections, fishermen need special attention from the central and the State Governments. The flaw appears to be that fisher leaders have a fixation in their mind that fishermen by caste alone are fishermen. In the same way as those from fisher communities drift from their traditional calling to other professions, those by caste who are non-fishermen also have a right to participate in fishing or other fishery-related industries. No discriminatory demand will be sustainable. Such of those who are accomplished and belong to fisher communities, have always the right as Indian citizens to make their way into any of the fisheries industries. But this approach should not deny opportunites to others.

In a democracy like ours no area can be reserved for the use of any exclusive class of persons. All nationals, irrespective of caste or creed have the right to fish all along the coastline, subject to laws, rules and regulations in force. Even the Marine Fishing Regulation Acts promulgated by the various States show no discrimination in this respect. In other words, all Indian Nationals have the right to fish or organise fishing in Indian waters and this right devolves under the provisions of the constitution, provided the regulations governing such fishing are complied with.

A new policy approach, as pointed out, is of crucial importance. How it should be and in what way it should be different from the present policy has not been brought out clearly at the seminar. The present position is that no big business house is keen to enter deep sea fishing or is in it. Merchant capitalists, as mentioned, have been allowed to enter into joint ventures with foreign fishing companies or take deep sea fishing vessels on lease which cannot be prevented under the law of the land.

To the extent we know of, there is no investment support from the Government to large houses in deep sea fishing sector. Their past ventures were a flop and they closed down their operations long back. Further, several foreign fishing vessels operating under joint venture, lease or test fishing terms have been noticed to be pulling out whatever be the reasons. This development is certainly no solace. We have to reach a stage of achieving capabilities of introducing our own deep sea fishing vessels and operating them successfully, linked to a processing, preserving and marketing system. As voiced by the Forum, small fishermen and fish workers should be enabled to play their legitimate role in the developmental process. In this context, the leadership of the Forum has to come up with a concrete plan as to how any perceptible improvement can be achieved. Firstly, vessels are needed. Secondly, trained men as per provisions in the Merchant Shipping Act are required. Shore facilities and marketing infrastructure at several places where they are not available now are needed. The Forum has to do its home work in this respect. Once they start addressing themselves to this task they will understand the depth of the problem. One ray of hope is that the Indo-Danish Project at Tadri in Karnataka has succeeded in equipping small coastal trawlers of 40 and 43 ft for deep sea fishing for a short duration of 7 or 8 days at a time.

The ire of the Forum is now directed at joint ventures. Obviously the Forum feels that we will be in a far better position, if the joint ventures are not there, whether we are now in a position to fill up the vacuum or not, should the foreign vessels are sent away from our waters. Government, fish workers and the fishing industry have to ponder over the issue.

Fish workers are a part of an integrated whole of the fishing activity in coastal or deep sea zones. Their interests cannot be looked at in isolation. Fish workers and/or small fishermen alone cannot accomplish deep sea fishing.

Reservation of exclusive fishing rights for the national fishermen upto the contiguous zone is another point raised. This is a just demand but one would see that this provision is already there in the rules framed under the Maritime Zones Act. Along most of the coastline foreign vessels are not allowed to fish upto 24 n.m. from the coastline and totally not allowed upto 12 n.m. from the coast. And 24 n.m. from the shore covers territorial waters (upto 12 n.m.) and the contiguous zone which has an extent of another 12 n.m. from the edge of the territorial waters. Further, Centre has notified several zones where fishing by foreign vessels is totally prohibited.

Mechanised boat operators are small fishermen. Traditional boat operators are also small fishermen. Economic disparities, if any, are incidental. It is most unfortunate to read from reports from time to time that, in Kerala, the most enlightened of all Indian States, an unbridgeable spasm has been created and fomented between traditional boat operators and mechanised boat operators. Who created and fomented the chasm between traditional boat operators and mechanised boat operators it is difficult to state with assertion. The fact however remains that the chasm continues and the Forum should strive to remove this.

The most relevant resolution of the Forum is the demand that Centre should ensure adequate measures for proper resources management to protect our fishery wealth at sustainable level avoiding overfishing and depletion. Another appropriate resolution that probably could have been passed is to demand from the Government, institution of measures on top priority to equip fishermen to undertake deep sea fishing on a large scale in augmentation of their traditional skills on Tadri model. The Forum would certainly be right in putting forth a demand to cancel all licenses issued under joint ventures provided the workers are in readiness to operate deep sea fishing vessels and take the place of present licensees. Fishes move from place to place and do not wait and be alive untill our traditional fishermen come up to the standard of operating deep sea fishing vessels. Indian skippers, including those drawn from fisher communities would be able to take command of large deep sea fishing vessels under joint ventures, only after receiving considerable training under foreign skippers. Commanding a large deep sea fishing vessel, as is known, is vastly different from commanding a mechanised fishing boat. The foremost task of the Forum is therefore to ensure creation of a band of competent skippers and workers to man deep sea fishing vessels.

The operation of foreign fishing vessels in our waters helps the Indian small scale or large scale fishing industries in two ways. One is that our men get trained. The other is that, we can know the hitherto unknown resources potential through Indian counterparts, if not from the foreign hands. Indian crew are placed on these vessels as counterparts for receiving training.

There is no evidence that Governments in any of the coastal States or for that matter the governement at the centre are apathetic to the demands of fish workers. Thus, we feel that there is no compelling case for giving a strike call. The objective of the proposed strike is to achieve cancellation of licenses issued to foreign fishing vessels. It is to be hoped that the proposed rally on July 20 by fish workers would be successful from the angle that it would demonstrate the solidarity of the workers, who are not really aware of the ingredients of "apathy" stated to have been shown by the Governments. The proposed All-India strike on November 23 would, in the end, prove to be the most unwise step. Instead of asking for cancellation of licenses issued to foreign fishing vessels, the purpose of the strike should have been to prevail upon the Government to take immediate practical steps to upgrade the capabilities of fishermen to a level at which they can acquire and operate large deep sea fishing vessels far better than the merchant capitalists allowed to enter into joint ventures and also to evolve a scheme under which they will be enabled to acquire such vessels for operation with the provision of the needed infrastructural facilities.

Indian fishing vessels operate as per Indian regulations. It is not a crime for Indian nationals to invest in such projects under Indian laws, rules and regulations. Trained workers can also join together and acquire such vessels seeking Government help. The catches under joint ventures are the property of Indian Nationals.

It will be wise for the forum to prevail upon the Government to introduce and implement schemes that would impart capabilities at deep sea fishing to small fishermen and fish workers and also provide capital and other inputs to them. In their present mood, Government would certainly take up schemes to intensify the on-going efforts in that direction by bringing them under the fold of an integrated scheme. Fish workers would stand to lose if they really adopt the strike path. In fact, the forum has not defined the term 'Fish workers', as they understand and have not examined whether they are equipped to take to deep sea fishing. CITU tried to bring officers and deckhands on fishing vessels under trade unionism at Visakhapatnam few years back and failed. The result was the officers and deckhands suffered. Let not such a thing happen again. Government will be receptive to developmental ideas, but not to suggestions that would prove to be counter productive. By stopping joint ventures, we not only lose chances of improving our awareness and training opportunities (Indian Crew, as already stated, are posted on every foreign fishing vessel in certain numbers), we also deny to ourselves fresh information on fishing grounds, harvesting systems and additional information on catch pattern in our waters. The suggestion to cancel licenses under joint ventures to foreign fishing vessels will sound negative, until such time as we have absorbed technologies and are in a position to introduce deep sea fishing vessels and operate them on our own.

Fishing Chimes, Editorial 188: July 1994: Vol. 14, No. 4

# Fisheries Education and Training

Indian Marine Fisheries Sector until independence had been totally one of traditional fishing. Same was the case in most of the countries until world war II. In various countries having fishing tradition including India the traditions and culture of fishermen had been playing the role of transmitting knowledge of fishing grounds and fishing skills from generation to generation. In contrast to the problems such as overfishing unleashed by modern methods of fishing, lack of fishing capabilities of a higher order happened to be the main reason for stock preservation prior to World War II. None of the present day measures such as Marine Fishing Regulation Acts, issuing permissions for operating foreign fishing vessels on lease or under joint ventures were thought of then.

Designs of fishing vessels, fishing systems, shore infrastructural facilities, processing technologies and export as well as domestic marketing patterns in the country have changed enormously over the years. The transformation would be astonishing to the generation prior to 1947.

India did recognise after independence, although in stages and at a tardy pace, the need for fisheries education and training. While several fisheries colleges to impart fisheries education, several state level fisheries training institutes to impart training to fishermen, and the Central Institute of Fisheries Nautical and Engineering Training to impart skills in high sea fishing have been set up, the fact remains that on the training side much more has to be done to cope up with the emerging and future requirements.

Let us look at the first global developments in respect of fisheries education and training. Erstwhile USSR and Japan were among the earliest to institute steps in 1950s to create a new category of skilled workers and officers for manning their increasing global fleets. In this endeavour several technical schools and fisheries colleges were set up in those countries. The emphasis was however on producing sea going officers and men with fishing and navigational skills. Processing and aquaculture sectors were neglected. Needed attention was not paid to resource management, environmental aspects, socio-economic developments and management practices. The concentration was on stepping up fish production.

The activities of these countries, later joined by Taiwan and others, started affecting the interests of other countries. The result was the development of educational and training programmes in other countries, including those with maritime tradition. The programmes were oriented mostly towards navigational aspects, with marginal emphasis in providing training if gear technology, location of fish concentrations, post-harvest technology and fishing vessel management etc. Much later, initiatives in respect of providing training in these aspects were taken mostly in developing countries at that time, including India. International agencies provided support to a wide range of training programmes. In short, specially designed courses, revolving more around practical training on board, and less of class room lectures, started gradually yielding place to courses mostly concentrating on navigational aspects, to the neglect of others.

The State Level Fisheries Training Institutes in India produced and continue to produce trained personnel for manning small mechanised fishing boats. The Central Institute of Fisheries Nautical and Engineering Training at Cochin and at its centres at Madras and Visakhapatnam provide to candidates the training needed to appear for the qualifying examinations conducted by the Mercantile Marine Department for securing the tickets needed for the respective categories.

The marine fisheries sector now looks ahead to prepare itself for greater efficiency in operations. Major changes have been taking place round the world aimed at making operations of smaller vessels much more effective with addition of electronic equipments, introduction of efficient and environment-friendly nets, communication equipment etc. Deck and fish hold designs and connected arrangements are being upgraded. Winches and hydraulic systems of greater capacities to equip the vessels to fish in farther and deeper waters are being installed. The present need is that the training programmes at the institutes have to be revamped to meet these requirements, particularly those relating to electronic equipments such as fish finders and equipments such as GPS navigators (Global Positioning System). GPS helps the operators to reach anyone on the known grounds with precision and ease.

In the case of larger vessels too, training programmes in the country would have to be upgraded to meet the looming challenges and to come up to the standards of advanced fishing nations. Training in gear technology needs considerable strengthening to cope up with the new designs introduced for various kinds of fishing nets, traps and lines. With a 200 n.m. wide Exclusive Economic Zone, our officers and men have to acclimatise themselves to stay out for longer durations for distant water fishing. The Institute and its centres have to be provided with well equipped large training vessels in sufficient numbers so that intensive onboard training to the candidates for longer duration can be provided. Such a strengthening will go a long way to create a category of well seasoned and tough vessel officers. The syllabi should give greater or equal importance to fishing and ancillary matters such as electronics. The strength of training vessels should be such that the extent of on-board training needed for the various categories of positions can be provided, without much of a need to formally work on commercial vessels to put in the needed seatime. The period spent by the candidates for theoretical lessons would need to be kept at the minimum.

Fleet Management Course needs special attention. CIFNET as well as CIFE (Central Institute of Fisheries Education) conduct these courses for a short duration once a year or occasionally. Well trained and efficient Fleet Managers are the key for successful fleet operations and the fleet managers that we now have are those who came up on their own through experience. A well designed course of adequate duration backed by a comprehensive syllabus must form the basis for producing capable fleet managers one each of whom is required by every fishing company. Fleet managers are now in short supply. CIFNET will have to start a regular fleet manager's course of a longer duration so as to cover the entire gamut of functions of fleet managers and to fill up this lacuna.

On the fish culture front there is an acute shortage of farm managers, farm operatives and technicians, hatchery managers and hatchery technicians. Whomever we have now ate those who have had opportunity to gain experience in the connected functions. This unorganised way of allowing manpower to develop will not be enduring. The need is to have regular courses for these categories, which are very crucial. Untrained and inadequately trained hands can do lot of harm. It is surprising that Government has not done much in this respect. With over 70,000 ha under shrimp culture and requirement of over 70,000 million seed necessitating a large number of hatcheries, the complacency is baffling.

Prior to setting up the premier fisheries educational and training institutes in the country, Government of India set up high level committees headed by eminent fisheries experts to make recommendations to have men educated and trained in the fisheries field. In the same way, it is of utmost expediency for the Union Department of Agriculture to set up a committee to go. into the question in a comprehensive manner and make recommendations for the steps to be taken to organise effective training courses for producing men of calibre for the above mentioned crucial categories. Government is aware that a large number of farms and hatcheries are coming up. The present situation is that, owing to shortage of trained personnel, a large number of brackishwater farms and shrimp hatcheries in particular are run in an ineffective manner, giving rise to mortalities and environmental hazards.

Mariculture has immense potential, particularly cage culture, in our waters. Very little has been done to develop this sector. Government would have to institute measures to promote mariculture by introducing integrated maricIture projects. The main component of these projects would of course be the creation of trained man power. Union Department of Agriculture has to take initiative in this respect to advise MPEDA and CMFRI to work out schemes and plans to develop trained men for the promotion of this activity.

Fishing Chimes, Editorial 189: August 1994: Vol. 14, No. 5

# Gearing up for the Orderly Spread of Mariculture

Time is not far away for the mushrooming of mariculture units. Now that standardised technologies for culture of edible oysters, pearl oysters, mussels, sea weeds and several other marine species such as crabs are available, entrepreneurs are in various stages of readiness to take to mariculture.

We are aware of the sprouting of brackishwater shrimp farms in a haphazard manner all along the coastline jostling with each other and some endangering agricultural lands by imparting salinity to them. Freshwater wells at several places have been salinised. Fishermen are forced to pace over long distances skirting the farms to reach their traditional coastal centres to set sail for fishing. Polluted waters from the farms are released into creeks or the sea. Such released waters are often taken into other farms causing diseases/mortalities to the farmed aqua-animals. All these are happening because of the sudden growth of brackishwater farms, without leaving any time to the Government to evolve norms and a machinery, to regulate and monitor the growth of the activity.

The Nation should not be trapped into such a bad situation, although may be different, in the sector of mariculture which will soon emerge as another major activity with export potential This looming problem has to be examined by the scientists and the Government from social, economic, environmental, and ecological angles. More than these, the prospective development has to be examined from the angle of granting leases of specified coastal and offshore zones to entrepreneurs for a reasonably long duration. In the absence of such a regulation, as is there in several other countries with monitoring arrangements, the growth of mariculture will lead the nation into having unhealthy water areas bordering the country. The kind of species which can be allowed to be cultured have also to be listed out for the benefit of the entrepreneurs. As cage/pen culture systems would lend attraction to several entrepreneurs a fresh set of rules under the Maritime Zones Act may have to be framed with Indian nationals as the beneficiaries under joint ventures or otherwise, in the case of waters beyond the territorial waters and by the States in respect of areas within territorial waters. Cages and pens and coast-based mariculture units will have to be set up in such a manner that capture fishing activities in the sea and passage of vessels are not hampered.

Seminars on the subject to be started right from now will throw light on the various angles of the problem and the recommendations made at the seminars are likely to form a crucial basis to give shape to the kind of regulations to be framed and enforced. The identification of areas suitable for setting up mariculture units can be undertaken through a survey which should take into account the various angles of the activity and the likely good and bad offshoots. A code of measures to be taken by the entrepreneurs during and after setting up a culture unit would have to be work out. The training programmes to be instituted right from now to equip the entrepreneurs to undertake the work on judicious lines have to be planned well in advance. Let there be a thorough preparatory work done so that the central and state Governments can be in readiness so as to create a proper base and not give rise to difficult situations later.

Fishing Chimes, Editorial 190: September 1994: Vol. 14, No. 6

# Rural Fishery Development Through Voluntary Organisations

The role of voluntary organisations can play in the development of both inland and marine fisheries needs to be expanded and fully utilised. There are quite a few voluntary organisations engaged in this endeavour in several parts of the country. There are distinct indications that their work has been yielding cognizable results in terms of socio-economic development of fisherfolk. Their work in several cases has been able to bring about community action for the upgrading of professional skills of fishermen and thereby securing improved fish landings and incomes therefrom through a more conscious and purposeful way of marketing. Examples are the work done by the Bay of Bengal Progamme, Madras, Society for Rural Industrialisation, Ranchi, Confederation for the Coastal Poor Development Action Network India, and so on.

The various Central Fisheries Research Organisations developed technologies amenable for adoption by the rural poor. The adoption of these technologies in the rural sector has however to reach dimensions of a distinct level. Induced breeding of major carps, technology for which was developed as early as 1960s, has become part of the rural scene but in a limited way and rural farmers still depend on outside supplies of seed. Composite fish culture and integrated fish farming have to make deep inroads into the rural sector. Reservoir and *beel* fishery development has not been progressing, in spite of availably of technologies. There are immense possibilities of enriching our riverine fisheries. Environmental and ecological conditions in fishery areas need protection and upgradation. Technologies for the production of freshwater and marine pearls have been developed in such a manner that they can be popularised at rural level. There are several aqua-based items that can be produced in rural areas.

The socio-economic conditions of inland and marine fishermen await to be upgraded mainly through equipping them for adopting improved ways of culture and capture fishing systems and practicing viable post-harvest technologies. Through voluntary action the dreaded degradation of environmental and ecological conditions in rivers and such other running waters, and in the coastal zones can be countered.

The primary task of developing fishery and fishery-based technologies including those suitable for adoption in rural sector, by and large, now rests with the Government The responsibility for extending the technologies too largely rests with the Government and the organisations such as Fish Farmers and Brackishwater Fish Farmers Development Agencies.

FFDAs and BFDAs have demonstrated that such agencies are capable of promoting an interface between the farmers and the Government. The farmers have come to rely on the agencies and have come closer to them. This result proves that agencies of this kind can play a vital role in popularising developmental programmes aimed at achieving socio-economic prosperity of fishermen and fish farmers and also those engaged in aqua product-based ancillary industries.

Women play a significant role particularly in the post-harvest part of fishery development. There are outstanding examples of successes achieved by fisherwomen's cooperatives directly or with the help of women's voluntary organisations. Fisher families need considerable support and help when the womenfolk are out to attend to post-harvest or net fabrication/repair jobs. The support has to come from voluntary organisations in terms of taking care of the fisher children, their health and education.

Fishery development aimed at socio-economic upgradation of fisher families/fish farmers will gain momentum, once the organisational base absorbs voluntary organisations which can have a wider and deeper reach. In this context Government would have to work out a plan that will systematically induct voluntary organisations in to the fisheries orbit and lend support to them in various ways. The design should be that the fisheries related voluntary organisations will function within a set of guidelines framed by the Government. Their performance and progress achieved from time to time should be subjected to a review so as to facilitate corrective measures. Government can extend financial and other support to these and also to reviewing organisations, to be identified, which should be promoted taking care to ensure that there is no overlapping areas of operations.

Fishery related voluntary organisations will be able to function effectively if they are headed by those having experience in fisheries development and having knowledge of the socio-economic conditions of fishermen and in the fisheries work in rural areas. Retired fisheries officials from Government departments who have interest and aptitude in the line of work will be very well suited to run or play a key part in such organisations.

To sum up, it is desirable that Government should evolve and declare a policy to promote socio-economic development of fishermen in fishing villages and in rural areas having culture/capture fishery potential.

Fishing Chimes, Editorial 191: October 1994: Vol. 14, No. 7

# Transfer of Developed and Emerging Technologies

The climate for transfer of developed and emerging technologies to entrepreneurs in the fisheries sector needs to be improved in a great measure and the interface between the research institutes that standardise the technologies and the entrepreneurs who are on the look out for putting into practice commercially viable technologies would have to be strengthened, apart from providing the needed linkages and integrations.

Absence of infrastructure with an organised and effective system for the transfer of technologies continues to bo a weak link in the fishery developmental work. It is true that the need is known and extension mechanisms are at work but their impact requires toning up. Krishi Vigyan Kendras and the Trainers Training Centres set up by ICAR do not seem to have brought about the targeted transformation.

The strengthening of extension technology has to be two fold. One is to organise and fortify a sound system of educating entrepreneurs including farmer entrepreneurs through a field-oriented infrastructure based on developed technologies. The other is to impart effective work-oriented field training to various executives/operatives on pre-determined lines with a combination of class-room lessons with video-aids and an actual run-through field demonstrations as a follow-up including use of gadgets for checking various parameters and recording the same. First-hand knowledge and experience imparts strength to entrepreneurs and equips them to ensure effective work and results at their farms or other units. Further, executives who receive effective work-oriented field training would be able to conduct operations with efficiency and success.

It is known that, in the past, there have been major breakthroughs in the inland and marine sectors that materialised over a period of time. Notable among these were the induced fish breeding technology, carp seed production technology, composite fish culture, brackishwater shrimp seed production and shrimp culture technologies on the inland front. Introduction and popularisation of synthetic twines for fabricating fishing nets of various new designs both for marine and inland operations had been distinct epoch-making achievements. No less important were the achievements in respect of introduction of mechanised fishing boats, high sea fishing vessels and processing technologies such as block-freezing and IQF. All the breakthroughs were the result of the emergence of a strong interface between the developmental agencies concerned and the target groups.

Let us take a look at the present state of affairs. Technology for freshwater prawn seed production and culture has been developed and standardised. The latest results in this respect came not only from the Central Institute of Freshwater Aquaculture but from Central Institute of Fisheries Education and other centres as well. Technologies for mud crab culture, edible and pearl oyster culture, holothurian culture, mussel and clam culture, production of marine and freshwater culture pearls, land snail culture, frog farming etc., have been developed. On the marine side, technology of high sea fishing has been absorbed by Indian captains and other crew members from chartered foreign fishing vessels and vessels operating under joint ventures. In order to improve and intensify the trend, the Central Institute of Fisheries and Nautical Engineering Training would have to broad-base the infrastructure available to meet the emerging needs. Quite a few value-added processing technologies have been introduced on their own by private sector enterprises and the Central Institute of Fisheries Technology and the Central Food Technological Research Institute possesses the needed value-added technologies for dissemination. MPEDA, in association with these institutes must develop an infrastructure for the transmission of value-added technologies to the processors and their technicians. On the mariculture sector, technologies are made to lie in the show cases of the institute concerned. Melas, training courses etc. have not been able to make a congnisible dent. And this happens when a multitude of entrepreneurs are looking for new avenues of commercially viable endeavours.

The strategy to be adopted for transference of technologies and providing the needed services for bringing about integration with the various linkages would have to be flexible from subject to subject. So far as provision of services is concerned, the experiences gained by the MPEDA and the FFDAs/BWFFDAs would provide guidelines for the same. All the emerging production technologies being export-oriented, MPEDA would have to take upon itself the responsibility in respect of provision of services to entrepreneurs in projectisation, organising financial support etc. and co-ordinating the same with the work relating to transfer of technologies.

Further, for every major technology developed, there should be set up commercially-oriented Centres of Technology at carefully selected places, with all needed infrastructural facilities. Careful planning is needed for setting up such centres in respect of cage culture, both in respect of inland and sea cage culture. Same is the need in respect of processing technologies. At these places field training course for entrepreneurs should be conducted. These courses should extend to cover one cycle of operations, from the stage of preparation of producing units and all the way through until a crop is produced and marketed. These centres of technology must be under the control of the Research Institute concerned with a scientist having specific experience in the line in charge and with needed supporting staff. Integration of all connected matters such as projectisation and organising investment finance etc., should commence from the training stage itself. Such of the entrepreneurs truly interested should be admitted to undergo training programmes conducted at the centres through the device of a well drawn agreement, with the needed safety clauses to protect the interests of both the sides. Official nominees of State Governments should be admitted for training only on the basis of a specific undertaking that the trained officers would not be transferred to unrelated units for at least a period of five years, a reasonable period for ensuring that the entrepreneurs trained by them take up the activity and practice it. Recruitment for practical training at any of the centres should be based on this criterion and only such candidates who are willing to undergo training for one cycle of operations at their cost but with the needed facilities for stay in the farm campus itself should be admitted for training. The selected candidates should be able to provide some evidence of availability of sites for setting up farms and also of their capability to meet margin money requirements and financial security requirements, if any. Banks should be prevailed upon by the Government/MPEDA to waive collateral security requirements other than the farm site and the harvested crops.

The setting up of Centres of the kind suggested above are all the more imperative in the present situation when transfer of developed technologies for culture systems such as those relating to freshwater prawns, running water and recirculatory culture systems, cage culture, culture of mud crabs, oysters, holothurians etc. have not yet gained any cognisible momentum. And this happens to be the situation when the positive pulse of the entrepreneurs and their favourable inclinations are known. The pulse beat and the aptitudes of entrepreneurs towards commercial culture of above mentioned aquatic species are an asset by itself. They should not be frozen by inaction on the part of the authorities concerned but harnessed for economic benefit to the entrepreneurs and the nation.

The same approach as outlined above is applicable to diversified marine fishing effort. New and viable diversified fishing systems would not catch up fast, unless well equipped infrastructural facilities are created, and suitable fishing vessels, are added. Batches of candidates (entrepreneur-skippers, skippers in service and those interested to come up but having basic training) should be recruited for advanced training in long lining on high seas for tuna, purse-seining, squid jigging, cuttle fish catching, lobster and crab potting, deep sea trawling for shrimps, lobsters etc. They should be given training on-board under competent skippers and master fishermen. Same is the need to provide training in on-board processing and in the fabrication of new types of gears. In the case of gears, training can be imparted at CIFNET, where suitable flume tanks for testing would have to be set up for testing new designs to suit our conditions.

It is axiomatic that implementation of emerging culture systems would largely depend on training of executives to a near-professional level. This can be achieved by organising sufficiently long field training courses based on well-drawn syllabi that would be suitable for the concerned category of executives and the training system, as suggested in the case of entrepreneureal education and training.

These should be totally field-oriented and supported by preceding class room lessons. After the training, the candidates should be enabled to undergo apprenticeship for a determined duration at the farms or in functional marine fishing units/companies as the case may be. A combination of enlightened entrepreneurs and well trained executives will pave the way for taking the new technologies for field adoption and towards diversification and spread of various forms of production effort for national benefit.

Fishing Chimes, Editorial 192: November 1994: Vol. 14, No. 8

# Pillars Cater at Their Whim

## Sad Maintenance and Repair Status of 'Cat' Marine Engines in Fisheries Sector

Marine Diesel Engines of Caterpillar Make entered the Marine Commercial Fisheries Sector of India in early 1970s under extraordinary circumstances. Neither the Caterpillar tractor Inc., Peoria, the manufacturing company in U.S.A. nor its regional office in Hongkong (Caterpillar Far East Ltd. Sun Hung Kai Road, off Hongkong harbour) had not earned the gratitude of the then budding Indian deep sea fishing industry on their own effort. The project framers opted for the fitment of Caterpillar Marine Diesel Engines on their own and for Caterpillar Tractor Inc. it was a cakewalk. The industry opted for cat engine because of its well established reputation, not only for supplying good engines but also for providing excellent servicing support. One can say this was not only a windfall to Caterpillar Inc. but also a great tribute to the reliability and sturdiness of the marine engines of this company. The reputation of cat marine engines was greatly enhanced in India, mainly because of the excellent servicing support provided by Larsen and Toubro at that time Those days we used to Call Cat marine engines 'stupid'. Crank the engine and it does not stop. This is a tribute enshrined in this uncomplimentary sounding phrase. Once you crank a Cat engine, it does not stop chugging, until of course the engine is turned off.

It was the greatest fortune of the owners of the first - and subsequent batches of Deep Sea Fishing vessels, all of which were fitted with Cat engines. M/s. Larsen and Toubro was appointed by Caterpillar Tractors Inc in that period as their representative in India. This company has a distinct culture of its own, the central part of which was customer service orientation in a overall sense. The manner in which this company organised its service for the repairs and maintainance of the engines during those days was unparalleled. Their work was based on a well-conceived and an orderly system which enabled the company to become aware of the individual functional aspects of the various engines from time to time. Based on this, with a mere approach from any of the operators the maintenance and servicing staff of the company located at Visakhapatnam used to swing into action and used to work without break until the problem was solved. The effect of this was that the owners could undertake fishing operations without interruptions for the maximum number of days in an year. The fishing companies used to be extremely happy at the services rendered by M/s. Larsen and Toubro for the mere reason that they optimized the number of fishing days and thereby shrimp landings which enabled them to earn good profits.

Alas! Gone are those good days. Fortunes of the fishing companies appear to have changed from the time the Caterpillar Tractor Inc replaced Larsen and Tourbo as their representatives. Within a short period, the operators could see the contrast in the services available. Your editor has not heard from any of the owners of fishing vessels fitted with cat marine diesel engines that they continue to get the same kind of services as before. In fact, many turn nostalgic recollecting the past. Unable to get servicing support, several companies are forced to entrust their engines to 'Quacks' professing to have expertise with Cats. Cat engines are no longer their pets. With more days of sickness than health, the owners operate their vessels not with any sense of assurance but with constant dread any time anything can happen. Even those days of rigorous import controls, L and T used to manage supply of spares and repairs. With all the present day liberalisation about which our Government and ourselves boast of, owners have no sound infrastructure support for receiving supplies of Cat spares. Is this not an amazing situation ? There is no doubt that M/s. Caterpillar Tractor Inc., Peoria, Illinois, U.S.A. has to bear the moral responsibility for all the damage done to the Indian Marine Deep Sea Fishing Industry. The reduction in the number of fishing days and the consequential fall in shrimp catches are intimately connected with the fall in standards of repairs and maintenance of Cat marine engines for which the Indian deep sea fishing industry, in a moment of weakness that arose owing to advice of experts at that time, although sound, opted for Cat engines.

As late as October 1994, the present representatives of Caterpillar engines in India told the deep sea fishing industry that they were working out a plan under which they would know the requirements of all the owners of the fishing vessels fitted with Cat engines. Imagine, a highly responsible company recognising such a need after handling the subject as the representatives of Caterpillar Tractor Company for over 8 years. The company openly admitted at the meeting held in October 1994 that there were faults in their set up but had promised to do their utmost in the future to see that no problems arose. The representatives of the company could not give any effective answers to most of the questions raised by the participants at the meeting referred to above concerning servicing etc.

They expressed their inability to maintain stocks of spare parts for supply to various companies as and when needed. Fishing vessels are designed to be out at sea for a long duration subject to fuel availability, to conduct fishing operations and not for remaining harbour bound awaiting receipt of spare parts or servicing support from a company which has the legitimate responsibility to fulfil the duty of supplying spares for replacing worn out ones etc. Their primary responsibility is undoubtedly to undertake periodical repairs and maintenance of the engines. What we hear from engineers and owners of the vessels is that the support from the Indian representatives is far below expectations. They have been victimised both in terms of operating the vessels and also losing sizeable shrimp harvests. Catches are being continuously affected because of the short-comings. A thorough audit of all the happenings *vis--a-vis* Cat marine engines installed in fishing vessels after field enquiries, reviews of views of the owners, by M/s. Caterpillar Tractor Company Inc., Peoria IL 61629, U.S.A. or Caterpillar, Far East Ltd., Sun Hung Kai Road, 26/F30 Harbour Road, Hong Kong appears essential.

*Fishing Chimes* is of the view that such an audit or check can be conducted by the Principals and whatever remedial measures are needed, should be taken by them, so that atleast in future, the operators would not continue to face the problems as they are now facing and their shrimp harvests do not suffer.

Fishing Chimes, Editorial 193: December 1994: Vol. 14, No. 9

# 3-Mile Non-Fishing Zone for Deep Sea Fishing Vessels Adjacent to Territorial Waters

Readers are aware that the Expert Committee set up under the Chairmanship of Dr. D. Sudarsan, Director-General of Fishery Survey of India by the Ministry of Food Processing Industry submitted its report. The main finding of the Committee was that fishing operations in the territorial waters had very little relation with the operations of larger fishing vessels which conducted open sea fishing very far away from the coastline. There is however, a possibility of larger vessels entering into the territorial waters to exploit the fisheries of the area which occur in higher concentration in terms of varieties, weights, prices, etc.

The Committee seem to have examined the issue in depth and made a recommendation that a 3-mile sea corridor may be created between the edge of territorial waters and the open sea so as to ensure that the deep sea fishing vessels or other non-coastal fishing vessels would not transgress and move towards the coast.

According to a press report, the Government of India had accepted the recommendations including the 'sea corridor' mentioned above which will be created along the North-West Coast, considered to be a highly vulnerable area. It is not yet clear whether this corridor will extend to the rest of the coastline. The ire of the fishermen is directed towards foreign fi shing vessels operating under joint ventures. The coastal fishermen feel that these vessels will deplete the coastal fishery wealth rapidly.

The Committee is understood to have examined this issue and recommended that the joint venture vessels should operate only beyond 15-nautical miles from the coast. In other words, they should operate only beyond the 3-mile corridor. The committee was assisted by representatives of Fishermen, State Fisheries Departments and the Coast Guard in arriving at the recommendations.

The committee has also recommended regulation of all fishing effort. It is of the view that manifold increase in the small fishing boats in coastal area is resulting in heavy fishery exploitation far beyond maximum sustainable limit and that this will ultimately harm the interests of traditional fishermen. It has also recommended that such of those coastal State Governments which have not enacted suitable legislations to regulate coastal fishing should do so immediately.

In a move to protect interests of the local fishermen, the Central Government has decided to constitute a standing coordination committee to ensure close coordination between the Centre and State Governments. The committee will suggest measures and policy initiatives to promote and develop fishing activities. The Secretary of the Ministry of Food Processing Industries will chair the committee. Representatives from the State Government will be the members of the committee. It will also have representatives from the Ministries of Agriculture, Surface Transport and Coast Guard.

It may be pointed out that as on date only 2.6 million tonnes of fishery resources are being exploited from the Indian Exclusive Economic Zone (EEZ) area as against the estimated potential of 3.9 million tonnes. The Government is committed to allow operation of not more than 200 joint ventures vessels in the EEZ area during the eighth plan period. At present only 19 such vessels are operating. The Government had given approval to 184 joint venture fishing vessels. After a review it has already cancelled permission to 57 such ventures having found out that their implementation was not satisfactory.

We feel that the recommendations of Sudarsan Committee are timely and appropriate. The decision to set up a standing coordinating committee to keep a watch over the implementation of the recommendations of Sudarsan Committee and ensure coordinated development of fishing activities will bring about an orderly development of marine fishing effort.

Fishing Chimes, Editorial 194: January 1995: Vol. 14, No. 10

# Indian Marine Fisheries Management

India can be proud of several aspects of its marine fisheries development. At the same time, the nation has reasons for being unhappy over the complacency in the utilisation and management of the marine resources.

Presently the marine fisheries resources are exploited by 170,000 nos. of traditional craft, over 60,000 nos. of smaller motorised and mechanised fishing craft a fleet of medium range fishing vessels, effectively around 150 nos. in 22-28 m range, and about 400 nos. in the 15 to 17m range. In addition, there are eight tuna longliners in a higher range of over 35 m OAL There have been also several foreign fishing vessels operating in our EEZ by Indian companies under joint ventures, charters, leases and test fishing permits but the number has come down to a few vessels now. Government has since stopped issuing fresh permissions for charters and joint ventures and probably under other systems.

So far as the main aspect that gives us pride is that we have a well established fisheries research and fishery exploration infrastructure. We have the Central Marine Fisheries Research Institute (CMFRI) with its presence all along the coastline. For the exploration of the fishery wealth on a continuing basis the nation has the Fishery Survey of India (FSI) capable of undertaking exploratory surveys in the entire Indian Exclusive Economic Zone. Another aspect is that the country has a major man-power base for fishery research and exploration. CMFRI and FSI have fairly well developed infrastructural facilities.

Technological support for the industry comes from the Central Institute of Fisheries Technology (CIFT) which too has its presence at several places along the coast and is well equipped. Then there is the Integrated Fisheries Project, which demonstrates integrated operations with an emphasis on processing and production of value-added products. In addition, there is the Central Institute of Coastal Engineering and Fishery (CICEF) that takes care of investigations and drawing up of designs for the setting up fishing harbours. The nation has a well equipped training infrastructure in high sea fishing, the Central Institute of Fisheries Nautical Engineering and Training (CIFNET) on the west coast with two additional units along the east coast.

The Coastal State Governments develop, regulate and administer marine fisheries of our territorial waters. In addition, the States take care of the socio-economic development of fishermen, implement schemes for training of fishermen, provide housing, organise road connections, provide fish transporting facilities from fishing villages etc.

The Government of India is, in the main, in charge of fishery matters beyond territorial waters, particularly deep sea fishing. The Centre also takes care of exports of marine products, organising the activity and helping a large number of processors through the Marine Product Export Development Authority (MPEDA) which provides financial help to processors in upgrading their processing technologies. This organisation also evinces keen interest and has positive involvement in fishing activities, particularly in compensating for the high cost of diesel oil and promoting diversification in deep sea fishing. Ministry of Food Processing Industries also provides subsidy of 33.30 per cent towards the cost of indiginously constructed vessels. However, the scheme is dormant as no entrepreneur is coming forward now to build vessels at Indian yards.

So far as coastal fishing is concerned, the activity is well organised, boats operating from various centres along the coastline, having harbours or landing centres and the operations are generally trouble free on all fair weather days. The activity however suffers from inter-state and intrastate conflicts which the State Governments do their best to control, regulate, resolve or eliminate. The activity is such that conflicts of the kind that surface now and then are inevitable, for the reason that fishermen struggle for their existence and they tend to move to places where fish are available. Most of the Coastal State Governments introduced Marine Fishery Regulation Acts. While these Acts do not discriminate against fishing by units from other States, the main complaint happens to be that issue of licenses to fishing units from other States is evaded.

So far as the conditions of existence at the Central set-up is concerned, in respect of development of fishing in the zone beyond territorial waters, there is an utmost urgency for major reforms. It is a matter of great concern that, in spite of the availability of a strong research base at the field level as mentioned in the preceding paragraphs, Government could not take the deep sea fishing industry much farther than the introduction of shrimp trawlers, (most of which are not really deep sea trawlers) and a few tuna longliners. Not having really touched deep sea stocks in any significant manner through vessels of Indian flag, foreign deep sea fishing vessels had to be allowed by the Government to operate under the provisions of the Maritime Zones Act for specified durations. One benefit that can be reckoned is the beginning made in the diversification of fishing by some of the trawlers introduced, for catching deep sea lobsters some years back and this year for catching squid, cuttlefish and deep sea shrimp. The anticipated technology transfer to the participating Indian crew either on chartered vessels or leased vessels or on vessels engaged in experimental fishing or the vessels in operation under joint ventures has not, by and large, materialised. Nor any introduction of the same type of foreign vessels operated by the charterers or lessees of foreign vessels has taken place.

It may be mentioned that the subject of deep sea fishing is administered by the Ministry of Food Processing Industries. The fisheries wing of this ministry needs additional expert support to strengthen the set-up from technical and technological angles in order to formulate well-conceived deep sea fishery development programmes.

What appears to be of considerable urgency is to transfer CIFNET, IFP and CICEF (deleting the function of designing brackishwater farms and other aspects connected therewith which can be entrusted to any one of the other existing organisations) to the control of the Ministry of Food Processing Industries so as to ensure proper co-ordination.

Mr. S.K. Amin, a veteran leader with decades of experience in marine fisheries development says "Lack of expertise at the Centre and in the State Governments was the cause for confusion and agitation in the fishing industry". Mr. Amin is the President of the All Karnataka Fisheries Federation.

Mr. Amin further says that much of the agitation against foreign fishing vessels fishing on the high seas is ill-advised as around 80 per cent of the catch by local fishermen is from coastal waters. Pointing out that he is a fisherman himself, he observes that he has no quarrel with deep sea fishing activity by 'multinationals'. In using the term 'multinationals' he means foreign companies, as no multi-nationals in India have taken to this activity. Apprarently the agitation is for 20 per cent of the fishery wealth in Indian waters and our fishermen are not in a position to exploit these resources. The question that arises then is: Are we to allow the fishes to die a natural death or utilise the situation by allowing foreign vessels to get an idea of the resources and utilise the vessels for providing training to our fishermen as the best way out in this situation. If the foreign vessels are not permitted to fish in our EEZ, they will, in all likelihood, resort to poaching which would necessitate intensive efforts for patrolling for which it is not clear how well our Coast Guard is equipped. They do not have even the ability to prevent Thai fishing vessels fishing close to Orissa coast right now in spite of several representations.

The agitation launched by the National Fisheries Forum acquires justification only from the angle of non-introduction of Indian-owned deep sea fishing vessels, barring a few tuna longliners. The foreign owners gain access to fish in our EEZ, through payment of 20 per cent of the gross proceeds relating to each of the voyages to the Indian charterers, as per one of the conditions stipulated in the permits issued to the Indian charterers or lessees. The interest of the foreign owners is mostly to secure access to fish in our EEZ. Once this materialises, by and large, they appear to neglect the training part and without the cooperation of the foreign captain the Indian crew will not be able to pick up the work. When there is no tangible transfer of technology the purpose of the scheme is not fulfilled. In other respects, such as gaining knowledge of the fishing grounds and types of fishes that are available and can be fished, and getting habituated for long duration voyages, the charter and other schemes relating to operation of foreign vessels in Indian waters have undoubtedly helped the Indian industry. Further, for the first time we have come to know about the availability of yellowfin tuna, squid and cuttlefish stocks in viable quantities, although seasonally, in our waters, confirming the conjectures concerning their stocks prevalent during the days before the advent of the charter scheme.

The comptroller and Auditor General of India, in his report No. 1 of 1994, says that 20 per cent of the gross proceeds from the operations of chartered vessels credited to the charterers account by the owners was trivial. The total catch during 1987-91 was estimated to exceed two lakh tonnes, valued at Rs. 430 crores, but the voyage reports given during the period 1987-89 had disclosed a value of less than Rs.9 crores received by the charterers. CAG had drawn the inference from this that the exports must have been grossly under-invoiced. He also expressed doubts about the integrity of the Indian Companies concerned.

In this connection, it can be mentioned that the under-invoicing and the integrity aspect on the part of the Indian companies is probably relatable to adjusting the invoices to tally with the quantum of payment made by the owners, but one cannot be sure.

The general impression in the fishery industrial circles is that the Central Government, although interested, in marine products exports, has only a perfunctory interest in the utilisation of the marine fishery wealth of the deep seas even upto a sustainable level.

Developmental activities spring from needs of the people and attitude of the politicians either in power or in the opposition. In the case of marine fisheries, both the groups have not demonstrated so far their will to develop the sector, but for some sporadic initiatives.

The reasons for this complacency can be that the fisheries sector constitutes a minor segment from a political point of view. A politician knows the pulse of his voters and fisheries sector does not constitute a major attraction or vote bank for political parties.

No other plausible reason can be identified for fragmenting the body of the sector by the Government and distributing its limbs to more than one Ministry for administration, that too without any coordinating mechanism. It is, of course, true and transparent, that coordination of such fragmented administration will be extremely difficult. The resultant complacency springs from the fact that the time and mental energy that goes into coordinating the implementation of the activities under various ministries leads to frustration and the results often turn out to be a damp squib. If the efforts are channelised through a single ministry or department the work will flow smoothly and speedily. This aspect has not made any dent on the top brass in the Government.

Whoever is the strongest and has the initiative among the concerned ministries or department snatches the work, provided others do not come in the way. Thus the developmental process progresses in fits and starts, and often limps.

Added to this is the anomaly of the location and functions of the Fisheries Development Commissioner. He is supposed to be the Development Commissioner for the entire sector. He is however attached to the Union Department of Agriculture. His functions are limited to some of the fragmented fisheries subjects that happen to be under the control of the Union Department of Agriculture. He has no role in deep sea fishing matters which are vastly inter-related with shore infrastructure facilities such as harbours, and training programmes which are no doubt handled by the Agriculture Ministry but not effectively co-ordinated with the main subject of deep sea fishing, which is with the Ministry of Food Processing Industries. No wonder mistakes take place in the location and determination of capacities of harbours, and in strengthening CIFNET for providing training and producing needed personnel well equipped for manning deep sea fishing vessels. Coordination with CMFRI and FSI in determining harbour location with reference to areas of concentration of resources is deficient. Designing of suitable nets for deep sea fishing has to be done by CIFT and there is a conspicuous absence of any application to this subject because of lack of coordination.

The Ministry of Food Processing Industries is given the portfolio of developing deep sea fishing as mentioned above. The Fishery Survey of India alone is attached to it. But the pillars for the integrated development are in the Ministry of Agriculture. The Institute for training manpower for deep sea fishing is with Agriculture Ministry. The subject of setting up of fishing harbours from where deep sea fishing vessels are to operate is also with the same Ministry. Integrated Fisheries Project is also controlled by Agriculture Ministry. In all likelihood, owing to lack of inter-ministerial understanding or for some other reason, formulation of an enduring deep sea fishing policy continues to be in a formative stage. The Central Board of Fisheries meets once in two years but these meetings turn out to be a formality with no impact. Very little that is tangible has come out of the meetings of the Board.

Once can visualise the dithering in the Government from the fact that no decision has been taken on the recommendations of the Murari Committee set up in April 1994 mainly to have recommendations on providing reliefs from the loan burden and accumulated arrears of instalments due from companies given loans by the erstwhile SDFC and now taken care of by SCICI and on diversification of fishing effort from the shrimp syndrome. The report was submitted by the Committee long back. At least the worthy part of the recommendations relating to the diversification of effort could have been accepted and follow-up measures instituted. It will be recalled that in the report submitted by Mr. Giudicelli, an FAO Expert, it was strongly recommended that diversification of fishing by the existing shrimp trawlers might be brought about as there were far too many of them for exploiting the limited shrimp stocks available in the upper Bay of Bengal and that this was the reason for the fall in unit catches and returns. Murari Committee also strongly recommended financial help from the Government to the owners to achieve this diversification but nothing has been done by the Government so far, leaving the position of shrimp fishing in the area at non-sustainable level, according to owners of the vessels.

Murari Committee also has made a recommendation consisting of three alternatives that may be given to loanee trawler owners (who borrowed money from erstwhile SDFC and now administered by SCICI). Government has not yet accepted the three alternatives. One ray of hope now is that the recommendations of the Murari Committee, excluding the recommendation that such of the companies who have difficulties in operating the shrimp trawlers, if they so desire, can hand over the trawlers to SCICI, are understood to have been accepted by the Ministry of Food Processing Industries and the Finance Secretary covering option A (50 per cent loan to be frozen and rest to be paid in instalments with some concessions in respect of interest and option B to allow such of the owners opting for deversification a period of moratorium and a few other concessions.

It may be true that the Ministry of Food Processing Industries is unhappy at the general attitude of the industry in not making repayments towards loans taken, even during seasons when shrimp fishing results, as reported to them, are viable. But Government knows that it will not be correct on their part to match their wits with the industry. Government is far above such situations which arise out of weaknesses in the schemes which get built in, in view of Government's anxiety to push through projects in an implementable form without impediments. The industry has to evaluate these aspects and co-operate. The need for such an evaluation on their part in the larger interests of development is well understood by the Government and this is the main reason for the emergence of Murari Committee.

The lone voice that proclaims that our exports can go up only through development of deep sea fishing is that of the Chairman, MPEDA. It is the Ministry of Food Processing Industries which has to organise introduction of new deep sea fishing vessels or diversfying the operations of the existing fleet, and follow up the matter. The Ministry apparently faces probems in accomplishing these.

The present deep sea fishing policy of the Government is under attack by fishermen's organisations and they are up-in-arms against the charter scheme, temporarily in abeyance, and against allowing operation of foreign vessels under joint ventures (or on lease or for test fishing in Indian waters). They feel that these activities are having an adverse effect on employment position of fishermen and on Indian coastal fish stocks. Government has been able to counter these grievances effectively, by saying that along north-west coast foreign fishing vessels will not be allowed to operate upto 15 n.m from the coastline. In other areas, fishing by whatever numbers of foreign fishing vessels that still have permitted time can obviously fish under the existing rules, framed under the Maritime Zones Act. According to these rules, foreign vessels cannot fish upto 24 nm. along some parts of the coast and upto 12 nm. (territorial waters) in other parts of the coastline. There are also areas declared out of bounds even for foreign vessels having permits. The decision to introduce 15 nm limit was in pursuance of the recommendations of Sudarsan Committee which suggested creation of a three-mile corridor between territorial and deep sea zones so that the joint venture and other foreign vessels would be allowed to operate only beyond 15 nautical miles from NW coastline. This Committee also suggested monitoring of operations of mechanised boats in coastal areas so as to promote the interests of small fishermen. Government has also announced that it will not allow any more joint ventures in deep sea fishing. Government is also considering the setting up of a National Fisheries Development Board. Further, the Minister for Food Processing Industries announced that a Review Committee would be set up to make recommendations to protect the interests of small or traditional fishermen and also for the development of deep sea fishing, keeping in view the demands of fishermen's organisations.

While the views of fishermen's organisations are made clear, it is surprising that the deep sea fishing industry, which is largely represented by the Association of Indian Fisheries Industries, has not come out with their views on the controversy. Government also does not seem to have elicited their views. The attack mounted by the fishermen's organisations is not just aimed at Government alone, but at the industry as well. The AIFI should have responded by eliciting the views of its members and others concerned through a Seminar and conveyed the recommendations of the seminar to the Government so as to facilitate consideration of the subject by the proposed Review Committee.

Fishermen apparently feel the impact of operations of foreign fishing vessels for the reason that these vessels encroach often into territorial waters for fishing and also for the reason that they are denied employment opportunities, because of employment of foreign crew on foreign vessels which are supposed to be under the control of the Indian Companies concerned.

The Government is aware that no superstructure without foundation will withstand the forces around. The superstructure built for deep sea fishing through joint ventures etc has started collapsing because of forces around and as it is not enduring. A sound and well thought out policy in this regard with long term and short term objectives is the need of the hour. Sustainable development of deep sea fishing will have to be the objective. In order to realise this, considerable planning effort is needed and it is hoped that the proposed Review Committee will lay the foundation for this.

As a long term plan, selected fisher boys, based on their fisher family background, traditional knowledge, physique etc should be caught young and trained, at an academy to be set up, for a sufficiently long and needed duration, at Government's expense. Others also can be recruited depending on their suitability. They have to receive training and education for a duration as required to qualify themselves to become deck-hands, skippers or engineers. Governmental support can be extended to them to acquire deep sea fishing vessels as necessary with reference to resources on their own and become entrepreneurs in the field on their own or secure exployment in fishing companies operating larger vessels. Such a programme can be evolved by the Government with reference to reliable information on resources provided by CMFRI and FSI and manpower requirements worked out on the basis of number of vessels to be introduced. Various types and number vessels needed for maintaining the resources at sustainable level based on the quantities that can be harvested annually as advised by CMFRI from time to time can be computed. The proposed Academy can be set up as an adjunct to CIFNET.

The present courses relating to high sea fishing conducted at CIFNET would have to be continued with the courses upgraded suitably and the Institute equipped with adequate number of larger training vessels (and other needed facilities) which should be used for providing on-board training for a far longer duration than at present. The training vessels will have to be equipped for various types of high sea fishing. The gear technicians course conducted at the institute should be upgraded in such a way that the candidates can design and fabricate larger nets of various types and designs.

In the meanwhile, as a short term measure, Indian entrepreneurs with aptitude, knowledge and experience in the line can be helped to acquire certain number of desirable types of vessels from other nations without any involvement of foreign fisheries interests. Trained Indian hands as available can be entrusted with the operations. (Foreign crew could probably be permitted for employment when there is no alternative). A crash training course for this purpose for those already having tickets can be mounted to meet the immediate needs. It is needless to say that marketing arrangements with foreign buyers would have to continue.

Government should set up an exclusive field organisation to promote deep sea fishing in the EEZ or expand the functions of FSI to implement deep sea fishing development programmes. This organisation or FSI as decided should function as a deep sea fishing development authority. At present this function is discharged by the Ministry of Food Processing Industries, whose function, in the main, is to lay down policies. It will be difficult for the ministry both to lay down policies and implement them. A separate agency is required for the purpose. As and when the National Fisheries Development Board is set up the Agency or FSI can come under its fold.

There appears to be no substance in the argument of fishermen's organisations (National Fisheries Forum etc) that deep sea fishing will have an adverse impact on coastal fishing. The complexion of fisheries in the coastal waters has very little to do with the fishery in deeper waters, either pelagic, columnar or demersal. However, there is considerable force in their view that the stepping in of foreign fishermen will deprive jobs to our skilled fishermen, however unsuitable they are at the present level of their skills. At the same time it has to be understood that fishing activity cannot be the monopoly of traditional fisher communities alone. It should be the monopoly of fishermen by definition and not by caste. Others are also equally entitled to take to fishing as a profession. Times have changed. Many from fisher communities are now in other professions and they are of course entitled to be in other professions. Further, restricting the sector to traditional fishing communities will contravene the provisions in the constitution. Moreover, the fishery wealth will be cornered by foreign vessels, if the present systems are allowed to continue. The strategy can be to allow entrepreneurs to buy suitable vessels from abroad and entrust them to Indian personnel trained under crash courses or to those already having the experience to undertake operations. As matters stand, the step taken by the Government to stop further operation of foreign vessels under joint ventures should substantially meet the demands of fishermen's organisations. In any case a plan of introducing vessels of Indian flag would have to be drawn up and implemented, although this may take time, keeping the employment of Indian personnel on the vessels, resource utilisation, and the sustainability of resources in view.

Finally, the larger issue of development of fisheries of Indian EEZ up to a sustainable level can be addressed to effectively only when all the units connected with deep sea fishing (CIFNET, CICEF, IFP) are transferred and brought under the control of the Ministry of Food Processing Industries, and when a field organisation is developed to implement and monitor deep sea fishing development activities, and the Ministry has a Development Commissioner of its own to handle the subject and assist the Ministry effectively.

Fishing Chimes, Editorial 195: February 1995: Vol. 14, No. 11

# Angles of Aquaculture

All of us are familiar with the term 'policy', and its connotation, which is wide. The term sounds obtuse but at the same time it has a powerful meaning. How does a policy emerge? It emerges from forethought or predictive perceptions arising out of accumulated knowledge. Any formulated policy will have a goal, an objective to fulfill, or an anticipation of something specific to materialise as a result of the implementation of the policy.

A policy obviously needs lot of supportive action to arrive at the anticipated goal. It requires a plan of action and a strategy to implement it and there should be an organisational mechanism for the purpose.

The policies of the Central and the State Governments in respect of aquaculture development have a positive orientation with a developmental face. The policies could however been supported by far better plans of action than at present, prescribing measures for a more orderly development of the activity. This disorder, particularly in the brackishwater farming sector, led to haphazard location of farms, for the reason that, in the main, coastal State Governments had been slow, by and large in alloting Government lands for the development of brackishwater farms. As a consequence, no regulations have so far been stipulated for the pollution-free and eco-friendly development of the activity, and the private sector, unable to wait indefinitely for the allotment of Government lands, preferred acquisition and use of private coastal lands for setting up the farms, which, for want of guidance and regulation were located in a disorderly and eco.-unfriendly manner. This led to problems of environmental hazards, pollution etc. After noticing the adverse effects, the coastal State Governments and their Pollution Control Boards started working on evolving rules for regulating the activity in the private sector by prescribing norms, long after considerable damage had been done. Among all the States, Tamil Nadu and probably Andhra Pradesh have made some progress for introducing legislation or framing regulations for the sector. The readily available accounts concerning the travails of Taiwan, Ecuador, Thailand etc. were ignored or by-passed and while the policy evolved by itself through compulsions of the Law of the Situation, the formulation of plan of action and strategy to achieve orderly development failed miserably. None can fault bacteria, protozoans, virus, and others for taking hold over shrimp seed and shrimp under farming. Nor can the farmers be faulted as they were left directionless. The lapse was on the planners and the executives.

## Fish Production Regime

Most of us are conditioned to think of culture and capture fisheries as two separate activities, overlooking the fact that culture and capture are closely related aspects of fish production regime and policies have to be adjusted keeping this aspect in view.

In respect of revival of our river and reservoir fisheries and sea shrimp stocks we have to bear in mind that "as we get, so we shall give". Once this aspect forms a major part of aquaculture policy, a major hurdle in the maintenance of environmental and ecological balance, will be removed. If necessary, needed legislative provision may be made for achieving the proposed policy objective.

## Resurrecting Carp Heritage

We do so much of farming of major carps, but we pay very little attention for stocking our rivers and reservoirs with their yearlings. The natural carp fisheries of our rivers and reservoirs are on the decline. This trend can be reversed by introducing a system, if necessary under law, to make it incumbent on fish farms raising seed, to earmark a specified percentage of seed produced, for stocking in rivers and. reservoirs. There has to be a movement for this in order to achieve restoration of ecological balance in the rivers and reservoirs.

## Vanishing Carps

Biologists and environmentalists are alarmed at the decreasing trend in the population of some important carp species such as *Labeo fimbriatus, Labeo calbasu* and several other species. These fish are seen in harvested catches only now and then. Special steps are necessary to revive these declining fisheries.

## Segments of Aquaculture

Indian Fisheries Act defines fish as including all aquatic animals and plants. Therefore culture systems cover not only culture of fish but also crustaceans, molluscs etc and also aquatic weeds.

Aquaculture, as we all know, has three segments; 1) Culture of aquatic animals in Freshwater (Freshwater Aquaculture). 2) Culture of aquatic animals in brackishwater farms (Brackishwater aquaculture), and 3) Culture of aquatic animals in marine waters (Mariculture).

Policy and planning in respect of development of fresh-water aquaculture has taken roots. These aspects in respect of brackishwater aquaculture are in a formative stage. Policy and planning for the development of mariculture does not seem to have been as yet evolved either by the coastal State Governments or by the Central Government.

Aquaculture policy, as can be perceived at National and State levels, can be construed as aiming at bringing all inland lentic resources under Aqua-production, preferably in the private sector, in an eco-friendly manner, maintaining pollution-free environmental balance, and in such a way that the activity produces sustainable quantities of aquacrops for domestic consumption as well as for export. The strategy to translate the policy into ground reality is primarily and partly through the organisational mechanism provided by FFDAs and BFDAs, and also through entrepreneureal initiative which is mostly outside the purview of the organisational devices (FFDA and BFDA) created. The Policy is also to maintain an infrastructure for training of candidates to become farmers, managers and operatives. The system also provides for granting subsidies to a limited extent so as to serve as incentives to farmers.

The State Fisheries Departments, which are primarily responsible for inland fishery development under the constitution, come within the umbrella of the policy outlined above, for which co-ordinating mechanisms at District, State and Central levels need strengthening.

For implementing aquaculture schemes, planning efforts, coupled with a sound strategy is an obvious means. The components or parameters of planning and strategy for promoting freshwater aquaculture had been worked out by the authorities visualising various problems that may surface during implementation. The exercise of updating or modifying the strategy no doubt goes on, but strategies need adjustments from time to time based on local, regional and national needs based on developing situations. Regarding brackishwater aquaculture we are in mid-way of strategy formulation. No policy or strategy is apparent in respect of mariculture.

## Extension Infrastructure Needed

The policy framework and the strategy thereof and the present extension infrastructure is not strong enough to promote transfer of technologies. Policy formulation will have relevance when guidance and help from Government is well directed to push through a programme. Leaving peripheral programmes aside, although they are important, State Governments, supported by the Centre will have to announce and push through policies with planning effort and a mechanism for implementation in respect of emerging schemes proven to be viable such as freshwater prawn culture, catfish culture, culture of freshwater mussels for production of freshwater pearls, running water fish culture, integrated fish farming, cage culture etc., in respect of freshwater aquaculture, and various 'components such as shrimp culture, crab culture and finfish culture in respect of brackishwater farms, and molluscan culture, cage culture of marine fishes etc., in regard to mariculture. In other words, in all areas where technologies have been standardised, the strategy to popularise them has to be made an inherent part of the policies for follow-up. We have not as yet seen any policies and strategies formulated to popularise them with or without incentives.

The problem of an unbridged gap between the development of technologies of economic importance and their transfer to entrepreneurship has not yet been satisfactorily resolved. The persistence of the problem is the result of the weaknesses in planning and efficiency in effectively implementing formulated strategies. The establishment of a coherent and effective infrastructure to meet the extension needs continues to evade us. We can expect unimpeded transfer of technologies, only when an extension infrastructure is established. The importance of extension is widely recognised and extolled but attention to the subject is, by and large, perfunctory, so far as fisheries are concerned.

## Extension Infrastructure

Extension infrastructure can be defined as consisting of specialised units, managed by scientists and technologists and equipped with all facilities that can be termed as ultimate and capable of providing on-farm/on-hand training to selected candidates for atleast one round of culture operations, inclusive of hatchery operations. In other words, the system of taking candidates to farms for short duration visits to observe the activities and the system of short-term training programmes should be replaced by participatory training at integrated farms culminating in rigorous tests at units specially meant for the purpose to test the level of acquisition of technologies. Candidates found deficient should be disqualified. In fact, selection of candidates to participate in the training programme should be based on an entrance examination.

There are infrastructural facilities for extension, set up by ICAR, but these are not tailored fully to meet requirements of fisheries development. This lacuna calls for a separate infrastructure to provide a full range of training to candidates.

## Freshwater Aquaculture

Policy and planning at State or Central level in respect of spread of freshwater aquaculture can, by and large, be considered as yielding results. Freshwater fish culture, as part of inland fisheries development, is the responsibility of the State Governments and, the initiative exhibited by them is relatively good but the dependence is more on initiatives taken by the Centre. Composite culture and integrated farming technology have gained wide-spread acceptance and adoption of these is relatable to successful policy formulation, and planning/strategy for technology transfer in a consistent manner. The urge to acquire technology has taken hold and is now transparent among farmers. Seed production technologies as formulated by our research institutes, mostly because of compulsions imposed by seed requirements, have spread all over the country, with regional variations. Eco-Hatcheries/Mini-Hatcheries and modified Chinese type of hatcheries became popular among farmers. In a way freshwater aquaculture has become self-propelling but for a few gaps, which relate to spread of freshwater prawn culture technology etc.

## Running Water Aquaculture

This system merits inclusion as an important part of Aquaculture Policy. This facilitates stepping up of our fish production both for domestic consumption and also for export marketing. Further, there is need to link production from our Aquaculture systems other than shrimp (which is already in the export basket) to the export market systems.

## Freshwater Prawn Culture

There is an increasing interest among the fish farmers to diversify their activities from shrimp culture. The two main species among the freshwater prawn, *Macrobrachium rosenbergii* and *M.malcolmsonii* have already gained popularity as species suitable for aquaculture. Some farmers have already diversified into farming of these species, but the farming systems have not yet attained the dimensions they deserve. The reasons for this tardy progress is that the farmers are unable to have access to the technology in normal course. Our Aquaculture policy must have a prominent place in respect of popularising and ensuring the spread of freshwater prawn culture. We are all aware that the freshwater prawns have a good export market too.

## Integrated farming and Polyculture

There are other benefits that flowed from the 'freshwater aquaculture development policy. The visible ones out of these are the increasing popularity of polyculture, and integrated farming, but the strategy should be such as to encourage agricultural farmers to take to integrated farming, covering duckery, piggery, dairy, poultry etc and also paddy-cum-fish farming on an extensive scale. Integrated farming is becoming increasingly popular, particularly in Orissa, Tamil Nadu, Punjab, Haryana etc and through an extension of infrastructure this beneficial activity could be further popularised.

## Frog Farming

We have totally stopped thinking about frog farming bacause of what appears to be an unrealistic policy formulation. No doubt frogs are in the list of wild animals in the Wild Life Act needing conservation. Although most of the riverine fishes are not in the endangered wild life list, with the exception of a few aquatic mammals such as manatee, existence of most of them such as major carps is ensured whenever necessary through replenishment stocking with farm-raised yearlings. In the same way, we can enrich wild stocks of frogs through culture as is being done in countries such as Thailand. Cultured frogs can be used partly for export in live condition. While it is to be conceded that the policy in prohibiting catching of wild frogs is well conceived, there seems to be no justification in not promoting frog culture. By not allowing frog culture, malpractices by way of diversion of live frogs caught from the wild, through neighbouring countries have come into vogue. This apart, through promoting frog culture, wild stocks can be supplemented and at the same time frogs can be exported. Frog culture in no way interferes with protection to wild fish stocks.

## Snail Farming

Government of India has to formulate a specific policy in respect of capture, culture and processing of edible snails for export. The giant African snail, *Achatina fulica,* hitherto considered as a pest in the agricultural farms has acquired importance as an exportable commodity to several western countries. CIFRI has developed a technology for rearing these snails to marketable size at a very high rate of survival. They can be raised on kitchen refuse. The technology is simple and easily adaptable. No policy has been formulated as yet to popularise culture of this species. A strategy has also to be evolved in this regard, as the commodity has the potential to earn foreign exchange.

## Freshwater Pearl Production

CIFA has achieved a breakthrough in the culture of freshwater mussels and production of freshwater pearls. The technology developed continues to be confined to the Institute. This technology is amenable for introduction in the rural sector. Intensified initiative for developing a mechanism for introducing the activity primarily in the rural areas is required.

## Reservoir and Coldwater Fishery Development

The policy in respect of reservoir fishery development which covers both culture and capture fishery development, aimed at achieving integrated operations and a maximum sustainable level of fishery yield, has not yet generated any significant results. The strategy as envisaged is to undertake stocking of yearlings in determined numbers untill such time a maximum sustainable yield is reached. Policy in respect of cold water fish culture, and fishery development in respect of trout, in Himachal, U.P and J&K (with Norwegian assistance in Himachal and EEC assistance in J&K), is yielding encouraging results, but for setbacks due to floods now and then. Nevertheless, transfer of technology took place, although the departments experience constraints in passing on the technology to entrepreneurs. Here also extension infrastructure is lacking.

## Brackishwater Farming

The awareness in respect of policy formulation and evolution of strategy for giving the targeted physical shape to the policy components of promoting brackishwater farming is transparent at Central as well as at the State level, more so at Central level. While the policy in broad terms has mostly evolved by itself with support lent by MPEDA, the spread and ramification of the activities under the policy needs closer attention of the Centre and coastal State Governments. There has been an excessive concentration on farming of tiger shrimps. Apart from white shrimp, there are 13 varieties of cultivable shrimps. Policy initiatives to bring about diversification in shrimp culture are needed. Further, there is need to mount a research project for bringing shrimps to full maturity under controlled conditions. A policy formulation in this regard for taking up the work on a wider scale is needed. Otherwise, continuance of dependence on wild stocks will lead to depletions of brood shrimps, necessitating either the placement of wild shrimp in the category of endangered species or taking up of a regular programme of stocking with farm-grown shrimp. MPEDA and CIBA are the most enlightened organisations in the country on the various policy and technological facets of brackishwater shrimp farming having a firm feel of programmes that should enter the policy canvas.

The policy in respect of promotion of brackishwater shrimp farming had a very simple beginning in late 1980s. Although the debacle in Taiwan, Ecuador, Thailand etc., surfaced by then, the policy makers did not take this aspect into account seriously, and ignoring these, made a beginning by announcing bald Government-owned coastal land allotment policies, based mostly on social considerations and distance for setting up the farms from the coastline, ignoring infrastructural needs and environmental considerations. Environmental considerations entered the policy framework much later. The implementation of the policy turned out to be protracted and it has not reached a decisive phase as yet. Centre circulated guidelines in this respect among various State Governments. Politicians, particularly those in power, most of whom are agriculturists, have a great attachment to land. The concerned politicians in power such as fisheries ministers, chief ministers, revenue ministers and others realised the importance of the coastal waste lands while dealing with allotment to various applicants. Allotments presented serious problems of various kinds, political, social and technical. An interim short term but valid step the State Governments could identify was to order a survey of the coastal zone so as to identify areas suitable for allotment. The Remote Sensing Agencies came to their rescue in this respect. They knew such a survey and consequential allotments will not only take time but will also highlight the problem of removal of encroachments, determination of zones etc., before finalising allotments.

The land allotment policy was designed, only for allotment of lands owned by State Governments without taking into account, by and large, infrastructural needs, as already mentioned. Because of delays, they soon faced a situation in which entrepreneurs lost interest in Government lands, with the exception of a few States, much to the disappointment of several State Governments. Entrepreneurs chose to buy private lands along the coastline and set up farms and the damage was done, by way of haphazard farm locations coupled with neglect of environmental aspects and possibilities of pollution. Brackishwater farms sprung up in a haphazard manner along the coastline creating problems of pollution, drinking water supply to villages around farms, salinisation of lands adjacent to agricultural lands, constraints in respect of access to fishermen to reach the coastline and so on. In other words, the State Governments were caught unawares, in spite of the availability of a fund of experience in respect of all these aspects from countries such as Thailand.

In this setting, the need for a comprehensive policy in respect of promotion of coastal or brackishwater aquaculture on eco-friendly lines suddenly came to the forefront in various coastal States of India. The policy and strategy aspects were considered in the light of the present situation at a Technical Meet convened by NABARD recently and the following points deserving incorporation in the policy/strategy relating to the subject emerged:

1. Sustainable development of shrimp farming alone should be allowed to be taken up;
2. Precaution should be taken to prevent entry of water from brackishwater farms into the irrigation canal systems;
3. Infrastructure for monitoring the development from the stage of project initiation until harvesting should be developed and an integrated monitoring system should be introduced;
4. In areas where there is poor tidal water exchange, brackishwater farms should not be located;
5. Benefits of Settlement Tanks may be studied;
6. The system of zonation of farm areas should be adopted preferably through satellite imageries;
7. Fishery estates may be set up;
8. Indigenous feed production should be promoted;
9. Brood shrimp collection from the sea should not be banned;
10. Contract farming system should be promoted;
11. Power supply, transport and marketing systems should be improved;
12. Processors should be prevailed upon to gear up for meeting the standards laid down by the importing countries;
13. Courses for training manpower for hatchery and farm management should be started so as to counteract the acute shortage of trained personnel in these disciplines;
14. Administration of antibiotics for control of diseases should be discouraged. Infrastructure for disease control and monitoring should be set up;
15. A package of practices for shrimp culture should be developed and it should be ensured that farmers adopt them;
16. A master plan for developing shrimp farms is necessary;
17. Guidelines formulated by the Pollution Control Boards should be rigorously followed;
18. Safety of shrimp crops from poaching should be ensured;
19. Implementation of research projects for producing mature shrimps under controlled conditions should be pursued on priority;
20. Non-saline lands should not be converted into brackishwater farms;

21. Survey of mangrove zones may be undertaken;
22. Land lease policies may be implemented speedily by State Governments;
23. Quality standards relating to the various stages of culture may be evolved and introduced;
24. Value-addition of processed products to be promoted;
25. Deep sea resources should be increasingly exploited;
26. Culture of viable aquatic animals other than shrimps and conventional fishes, may be promoted;
27. Shrimp farms should be allowed to be set up only on the waste coastal lands not suitable for agriculture;
28. Collection of wild shrimp seed should be banned;
29. A well designed system of brood shrimp management from the collection stage onwards should be followed;
30. Buffer zones should be set up between brackishwater farms and the adjacent villages/ agricultural lands;
31. Concentration on tiger shrimp farming alone is not desirable and there may be diversification into the farming of white shrimps and other non-conventional varieties of shrimps suitable for the purpose;
32. Farms should not be set up on mangrove lands,
33. A system of registration of farms should be introduced and new farms should be allowed to be set up only after obtaining permission from the authorised officers or organisations;
34. Utmost care has to be taken in the siting of the farms, their design, and in respect of pond preparation;
35. The number of shrimp hatcheries in existence now are inadequate to meet the needs and therefore more number of hatcheries should be allowed to be set up. Entrepreneurship should be encouraged to set up more number of hatcheries;
36. Feed quality should be ensured and feed management should be organised on scientific lines;
37. Water quality should be properly taken care of;
38. Effluents should be released only after proper treatment as per the procedure laid down;
39. Adoption of aeration system at all farms is desirable;
40. Farm designs should be such that there would not be any seepage;
41. Tubewells/Borewells would not provide the needed quantity of water to fill the farm ponds within the needed time limit and therefore these cannot be depended upon. Further, they will deplete the ground water resources. Therefore setting up of these should not be encouraged;
42. Farm operations should be mechanised, and;
43. Traditional farms may be developed for Integrated farming.

The most unfortunate and surprising aspect is that, having full knowledge of the fatal consequences of improperly designed and managed shrimp farms, authorities concerned let the situation drift with a complacency that is difficult to imagine, to a stage at which it would be difficult to retrieve the industry from the perils of collapse, by way of upsurge of disease attacks mostly to shrimp under farming. Let it be hoped that the Governments on the west coast would organise the activity on sounder lines than at most parts of east coast (Where effects of pollution, attack of viral and bacterial diseases, particularly in A.P. and Tamil Nadu became conspicuous), even at the risk of investing more of time in proper planning.

The present situation in respect of coastal aquaculture gives a clear indication of the revised policy approach needed. Irrespective of ownership, coastal lands, Government owned or privately owned, would have to be surveyed, and zonation of all suitable lands would have to be completed first. An organisational mechanism by way of setting up Fishery Estates/Strengthening BFDAs should be firmly established, supported by master-plans and infrastructure as envisaged therein. The basic survey work, which should cover aspects such as eco-friendliness, amenability to control pollution and keeping environment in tact in the area, infrastructure needs and feasibility of providing training to suitable farmers, farm managers, technicians etc preferably in the zone itself, state of water salinity and its possible encroachment into adjacent agriculture lands etc. will have to be undertaken. A blue print of development for various zones so prepared should serve as the main document for according clearances by the authorities concerned. In the case of farms already set up they may be brought under one or the other demarcated zones and suitably integrated. The State Governments should have district level boards to consider applications and accord clearances on a realistic basis without fear or favour. The Boards should be headed by knowledgeable senior officers either drawn from the State concerned or from outside (CIBA, MPEDA). Clearance should be given only when undertakings are given to fulfill all the conditions stipulated so as to avoid similar situations as have materialised in Andhra Pradesh and Tamil Nadu. A system of registration of farms, to be set up with prior permission of designated authority, should be introduced in each of the coastal States.

The policy should be to compile comprehensive data of the zones and document the same. Brackishwater aquaculture in diverse forms can be developed and the supporting structures such as hatcheries, and a regulated system of supplies and services in the private sector of keeping stocks of quality and pretested feed, seed, water testing kits, pharmaceuticals, equipments, zeolite etc for supply to farmers in each of the zones can be maintained as required. Proper treatment of influent and effluent water from farm through filtration devices should be ensured. No farm should be allowed to be set up either in private or public land without registration with the designated authority and a system of monitoring the activity should be developed. The Government of India should introduce an enabling legislation under the provisions of which State Governments should be able to introduce legislations as suitable for the State concerned and frame rules as required as per powers given.

Above all, diagnostic laboratories in the proximity of farm zones with mobile units will have to be set up, manned by trained biologists and pathologists. One element of the policy or the strategy should be to provide for the training of farm staff or farmers to maintain farm records. No farm without water filtration system and equipped with water and soil analysis kits etc., and trained men to analyse samples should be registered. Government should appoint monitoring staff to periodically study the state of affairs at each of the farms with reference to records maintained.

## Progressive Arresting of Collection of Wild Brood Shrimp for Hatcheries

Wild brood shrimps are being captured from the sea in increasing nos. This accelerated activity is having serious depleting effect on capture shrimp fisheries. It appears necessary in common interest that Government should introduce a mandatory system of earmarking a specified percentage of shrimp seed produced at hatcheries for sea ranching. One of the Government organisations has to be made responsible for this work and equipped to receive the seed from the hatcheries and transport to certain pre-determined points for release.

Apart from this, research on raising tiger shrimp to full maturity in farms would have to be stepped up. According to information, this work appears to have been taken up by Aquastride Bio-tech Ltd, Madras under the leadership of Dr. K.O. Isaac, Managing Director of the company at the R&D farm of the company at Shertalai in Kerala and achieved success in the work. The Central Institute of Brackishwater Aquaculture is also understood to have taken up experimental work in this regard. Dr. S.N. Dwivedi is also understood have initiated experiments on the subject in cages set up in the sea. It is of utmost importance to coordinate efforts on this important aspect under the leadership of Central Institute of Brackishwater Aquaculture and ensure that the technology as and when perfected is transferred to various farms.

## Transfer of Shrimp Farming Technology

It is heartening that the national workshop on "Transfer of technology for sustainable shrimp farming" held at Madras held on 9-10 Jan 1995 came up with several recommendations in order to promote sustainable shrimp aquaculture without upsetting our environmental and ecological heritage. A report on these recommendations has been included in this issue for the benefit of our readership, particularly shrimp farmers.

## Finfish Culture

While the present policy covers shrimp culture, very little attention is paid towards promotion of brackishwater finfish culture. This needs urgent attention. Very little is known to the farmers on culture of finfish in brackishwater farms. Further, there is no dissemination of information on availability of seed, and technologies for hatchery production of brackishwater finfish seed have not been developed in the country as yet. Our scientists have to work on developing hatchery technology for producing seed of sea bass, groupers etc for brackishwater finfish culture.

## Crab Culture

There is sizeable scope for the development of mudcrab farming in brackishwater farms. The fillip given to the activity is marginal at present. Areas suitable for mud crab farming will have to be identified and setting up farms for farming of mud crabs needs to be encouraged. Untill such time as hatchery technology for producing mud crab seed is developed, a prudent system of collecting and utilising megalopa of crabs for culture or water crabs from the wild for fattening should be formulated for the benefit of the farmers. In fact, parts of several brackishwater farms meant for shrimp farming can be modified for mud crab farming. Farmers can be educated and trained to undertake mud crab culture.

## Mariculture

Let us take a look at the policy aspects of mariculture. Now there is no specific policy formulated by the authorities for the pursuit of modern mariculture. What all we have now is state control over exploitation of pearl oyster beds off parts of Tamil Nadu and probably Gujarat coast and auctioning of molluscan shell beds along the coastlines of the some of the States. There is no element of farming in these traditional systems.

The State and Central Governments accord the lowest priority to popularisation of molluscan seed production and farming technologies (raft culture, longline culture, rack and string method, stake culture, bottom culture etc.) relating to oysters, mussels, clams, gastropds, holothurians, and of sea weeds etc., developed by the Central Marine Fisheries Research Institute. Government/ ICAR might be contemplating steps to popularise the techniques but there is no transparency concerning the extension of the technology developed for the benefit of the entrepreneurship, with the exception of one oyster mela held at Tuticorin sometime back.

The establishment of units for farming of marine animals and sea weeds is coastal water oriented, be they coves or bays or such intrusions of sea water into land depressions or behind the surf zone. The development of culture of marine animals such as fish or molluscs is inter-linked with granting of leases of demarcated zones to entrepreneurs by the Government concerned. Similar demarcation is also needed for allowing persons to set up cage systems in sea. If an arbitrary or unplanned beginning is made, the activity is bound to enter into a far more complicated situation than the present state of brackishwater farming development. At this stage when the mariculture sector has all the potential for development into a major activity, Central Government should enact an enabling legislation which would allow coastal State Governments to legislate on leasing of demarcated areas on a long term basis in territorial waters and integrate the same with technology transfer, culture and financial support. These apart, norms regarding species for cage culture would have to be prescribed. Lessons have to be learnt from the undesirable effects of delays and complacency in dealing with matters relating to development of brackishwater farming, for promoting mariculture on a sound basis. The suggested legislation should also cover powers to Central Government to demarcate and permit the setting up of off- shore cage systems beyond territorial waters and in our EEZ. Foreign collaboration in respect of cage farming out at sea would be required for quite some time and this should be permitted until such time Indian entrepreneurs pick up the technology. Import of cage systems should be allowed as long as this is necessary as cage systems are not manufactured in India now.

Fishing Chimes, Editorial 196: March 1995: Vol. 14, No. 12

# Coastal and Deep Sea Fishing Policy

Policies and strategies in respect of coastal and 'deep sea fishing' are reviewed periodically and readjusted according to prevailing conditions from time to time by all maritime nations. Nations tend to be careful about managing their fisheries resources in a sustainable way. At the same time some of the nations deploy their vessels for fishing in the waters of other maritime nations, known for their undeveloped or developing status and the amenability of Governments of such nations to allow foreign vessels to fish in their waters. This state of affairs is particularly true in respect of countries developed from fisheries point of view such as Taiwan and Thailand, to quote two aggressive nations. Even developed countries such as Australia, Canada, USA etc are not free from the onslaught of Taiwanese Vessels. No wonder India too came under the spell of Taiwanese, Thai vessels and vessels from other countries such as former States of USSR, Korea, Spain etc. The position in respect of Indonesia is far worse. That country has reopened their waters for trawling by an increasing number of foreign vessels.

India made a good beginning with sound policies of coastal as well as offshore and deep sea fishing. As the activities progressed, there have been no doubt certain revisions which are not however adequate mainly from the angle of monitoring and control, particularly in respect of coastal fishing. In the case of offshore and deep sea fishing there is a conspicuous absence of long term planning so as to ensure development of its own deep sea fishing fleet, dependence being solely on allowing operation of foreign fishing vessels and, as it turned out, resulting in an ineffective system of training of Indian personnel on foreign vessels which is one of the objectives. The factors not visualised in this context are i) the aim of owners of foreign vessels is to gain access to fish in our waters, and ii) the foreign crew have no interest in imparting training to Indian counterpart personnel as this is not in their interest and iii) no simultaneous programme has been taken up to strengthen the Central Institute of Fisheries Nautical Engineering Training so as to train candidates to prepare them to operate larger deep sea fishing vessels. On the coastal fisheries side, the need has been to extend their area of operations of the vessels. A breakthrough in this respect has since come from the efforts of the Indo-Danish Project in Karnataka, who succeeded in demonstrating trawling in farther waters, particularly through use of GPS. Several fishing units adopted the system with profitable results. The pressure on coastal fishing, however, continues unabated, bringing down per unit catches and crossing limits of sustainable levels.

So far as deep sea fishing is concerned, thanks to the agitation launched by fishermen's organisations, Government has taken the much needed step, although through compulsions, of freezing for the time being the system of issuing further permissions for joint ventures with foreign companies, taking foreign vessels on lease by Indian enterprises, etc. It will be wise to give up the systems once and for all, as it is seen that this policy has not helped the nation in any way, except that the availability of resources is established. Malpractices in a system of this kind are inevitable because of the inherent weaknesses in the policy. It attracted several enterprises to avail of the scheme, more for making quick money, rather than developing an enduring stake in the Industry. It was convenient under the system to have an understanding for fixed payments in lieu of 20 per cent payment on gross proceeds to be received, through the needed adjustments, although all expenses are supposed to be borne by the Indian side and within the system of payment of lease amount or share of profits. The problem of weighment of catches and determining the exact quantity harvested had come in handy to both the sides and also to Inspection agencies to virtually ignore this aspect.

There is considerable force in the stand of the National Fisheries Forum and Nation Fisheries Workers Action Committee that, in the above situation, operation of foreign owned fishing vessels in our EEZ should be totally given up. But this will leave a gap which needs to be filled up by Indian - owned deep sea fishing vessels. The Forum is silent in this respect. We cannot afford to leave deep sea resources unexploited.

The coastal fisheries and deep sea fisheries differ totally in their complexion. Real deep sea fishing, not impinging on coastal resources, will not affect exploitation of coastal fisheries. Introduction of large sized deep sea fishing vessels with such draught limitations that they cannot come close to the coast is needed to ensure exploitation of deep sea stocks only. Such vessels should be Indian-owned, manned by Indian personnel. Captains and engineers of needed calibre can be drawn from merchant navy and ex-navy personnel for such time as is necessary. In addition to these categories, there should be Indian master fishermen and crew to undertake fishing. Senior Indian skippers can be trained in locating and operating various types of gears at CIFNET under intensive short term courses, while organising long term courses for others. Should it be necessary, for a short duration, foreign master fishermen can be engaged. Indian skippers can receive training under captains drawn from merchant navy and ex-naval personnel to acquire the needed expertise in navigation.

In replacement of the present system, used larger fishing vessels from abroad can be allowed to be purchased by Indian companies, with the officers and men recruited for operating the vessels as outlined above. Indian personnel for operating electronic equipment can be trained and employed. There may however be risks involved in the recruitment of personnel which will have to be overcome by proper planning and borne untill such time a fleet develops.

What kind of deep sea fishing fleet should be developed? Undisputed ones are those of longliners, both for pelagic and demersal long-lining. Even by any stretch of imagination they cannot affect coastal fisheries. Mid-water trawling is another system that can be adopted in deeper zone. This too will not affect coastal fisheries as long as mid-water fishing is conducted in deeper zones. Bottom trawling in deep waters cannot also affect coastal stocks as the stocks will be mostly of deep water shrimp which have no relation with coastal shrimp whatsoever. Purse seining in deep sea zone and far away from coastal zone would not affect coastal fisheries. Further, squid jigging in deeper waters for oceanic squids may be allowed to be undertaken.

Gill netting in the deep sea zone close to the coastal zone may have an effect on coastal stocks. This may not be allowed in the deep sea zone. Deep sea trawling, either mid-water or demersal, may be allowed beyond a depth of 50 fathoms, so that it may not have any effect, even by a remote chance, on coastal stocks. There should be, however, a sound surviellance system.

Coastal fishermen should be increasingly sucked into deep sea fishing activities. This is certainly possible through short term and long term strategies. Short term policy should be to recruit practising fishermen on coastal trawlers and impart training to them so that they can function effectively on deep sea fishing vessels. Long term strategy should be to catch fisher boys young, give education and training to them over a long period on the lines of army academies totally at State expense. The candidates should be selected through testing of traditional skills and aptitudes. Once they come out of the institutes they can shape themselves into deckhands, officers and entrepreneurs. This course of action will end the present conflicts.

Days have changed. Youngsters from traditional fisher communities are able to perceive and assert their right to have an increasingly prominent place in the deep sea fisheries set up. They are entitled, to have all help from the Government in this respect.

The present problem is the sequel to a liberal approach on the part of the Government and also a set of unscrupulous and greedy entrepreneurs with no stake in the industry, but only bent upon making hay while the sun shines. It is a normal human weakness to avail of opportunities unwittingly given to them by the Government. The Association of Indian Fisheries Industries, in spite of the fact that several of its members were lured by the opportunities of availing of Government's policy of permitting joint ventures etc., without any seriously governing responsibilities, have extended support to the fishermen's organisations in their agitation. It has however totally overlooked outlining its views on development of deep sea fishing, the main subject on which its foundations are laid and their membership is oriented towards the objective of development of deep sea fishing.

Fishing Chimes, Editorial 197: April 1995: Vol. 15, No. 1

# Reflect and Restore Tempo of Deep Sea Fishing Development

## Appeal to Government of India to Create a Separate Fisheries Department

A Sizeable deep sea fishing fleet can prove to be the sentinel of our EEZ. Such a fleet will have the opportunities of supplementary observation of foreign vessel movements in our EEZ beyond territorial water. Apart from exploitation of the untapped deep sea fisheries resources for national benefit, deep sea fishing can play a crucial role by way of providing proteinacious food to our population and also by contributing to exports. The development of the activity will be a deterrent to poaching by foreign fishing vessels. Development of the activity will enable us to lift our head as one of the deep sea fishing nations of the world. At present we have a very small deep sea fishing fleet of our own.

The Government of India recently created six new departments in various Union Ministries. These do not include a separate department for fisheries development, although there has been a need and a persistent demand for a seperate department for a co-ordinated development of the fisheries sector and to replace the present fragmented and un-coordinated set-up. The worst affected is the deep sea fishery development because of its separation from Agriculture Ministry without its limbs (Harbours, training etc). The subject of deep sea fishing comes under the union list in the Indian Constitution. The coastal waters are overfished and deep sea resources are barely touched. Fishing in territorial waters comes under the jurisdiction of coastal State Governments who are not in a position to divert the over fishing effort towards deeper waters, with a recent isolated exception in Karnataka.

Deep sea fishing which until recently was under the Union Department of Agriculture had been tagged on to the Food Processing Ministry. This Ministry is seen by several to be totally unequal to develop a proper policy for the exploitation of deep sea fishery resources, and reconciling and co-ordinating the activity with coastal fishery development and regulation. Unfortunately the initiative to co-ordinate coastal fishing with deep sea fishing does not rest with Food Processing Ministry and it can only be handled by the Union Agriculture Ministry which handles coastal fishing. Food processing Ministry is thus not in a position to bridge the gap between coastal fishing and deep sea fishing through a process of upgradation of the skills of coastal fishermen. It could not also develop deep sea fishing, handicapped by the position that organising the needed infrastructural facilities such as fishing harbours, the needed training facilities etc. continue to be with the Agriculture Ministry.

The result is that the Ministry of Food Processing industries continues the tradition of Ministry of Agriculture of super-imposing deep sea fishing through joint ventures, leasing of vessels from abroad etc., instead of simultaneously starting from the base and laying a proper foundation for the development of deep sea fishing. It was hoped that charters and its various *Avataars* such as joint ventures and leases would facilitate transfer of technology, leading to introduction of deep sea vessels of Indian flag eventually, as the weakness and the opportunist interests of entrepreneurs had not yet surfaced menacingly by then. This policy has proved to be a failure, as it turned out. A new class of entrepreneurs sprung up, allured by the possibilities of making easy money, availing of the opening that does not entail capital investment and which enabled them to contact foreign companies, avidly looking for access to the grossly unutilised Indian deep sea resources, and obtain offers from them for providing vessels on charter, lease, test fishing or under joint ventures. No development in respect of introduction of Indian owned vessels and development of needed training programmes and development of infrastructural facilities to the level needed took place. Indian counterpart personnel placed on foreign vessels more in a perfunctory way by the new class of entrepreneurs could not pick up the anticipated transfer of expertise for utilisation. What is talked about all over is that the system boiled down to two parameters a) providing access to foreign fishing vessels to fish in our EEZ, with some of the vessels stated to be intruding into the coastal zone ex: Thai vessels intruding into territorial waters off Orissa coast; and b) entrepreneurs receiving a fixed amount, as is talked about, from the owners on a voyage to voyage basis, through some device or the other. Whether these indelible impressions of observers are true or not can be established only through a detailed investigation involving scrutiny of invoices and export documents and collecting evidence probably by the Second Murari Committee (as referred to hereunder).

While the complexion of coastal fisheries and of the deep sea fisheries are totally different, at the time of initiation and progressing stages of charter scheme, the imperativeness of a scheme to upgrade skills of coastal fishermen to eventually enable them to enter deep sea fisheries sector was not thought of, either by Agriculture/ Food Processing Ministries or by the fisher leaders themselves. The agitation by the fishermen's organisations, a belated one and designed more to embarass Food

Processing Ministry and teach a lesson to neo-entrepreneurs, started all on a sudden without warning, probably unable to bear the sight of a few securing benefits from charter and other such schemes. The Agriculture ministry is totally silent on the subject. The other aspect is that very few Indian-owned deep sea fishing vessels have been introduced by a few dedicated companies directly (Companies such as Fishing Falcon, Bay Liners, Sumura Maritime, Lewis and Lewis and Oriental High Sea Foods). These few Indian companies suffer from several problems, compared to the other allegedly pseudo-deep sea fishing companies that mushroomed in recent years on a totally different footing. Their standing and commitment are questioned by several in the fisheries sector, particularly by the coastal fishing interests. That only 2 per cent of total exploited fisheries comes from deep sea fishing is seen as a specious argument to assuage the feelings of coastal fishermen. Because of the super-imposed development policy, India's own deep sea fishing has not come into being with the few exceptions mentioned above. The use of this argument of 2 per cent share does appear to those in the field as a subterfuge to cover up a wrong policy, fragmented developmental set up and absence of achievement. It is unfortunate that Government has a deep sea fishing policy over the years that accounts for only 2 per cent of the fish production attributable to deep sea fishing. What all has been accomplished under the fragmented set-up is the creation of a category of entrepreneurs attuned to having fast financial returns as long as the policy lasts or lasted. The generally deplorable progress in deep sea fishing is visible to one and all.

The only breakthrough in deep sea fishing arising as an extension of coastal fishing sector and with a foundation, has been achieved recently by the Indo-Danish Project at Tadri in Karnataka and the Ministry of Agriculture, following the convincing demonstration made by the project for successfully conducting deep sea fishing by suitably equiping the largest of smaller vessels in the coastal sector. Around a thousand medium sized vessels, equipped for deep sea fishing, are stated to have come into being along the Karnataka, Kerala and Goa coasts. This is a development achieved from grass roots in contrast to developmental effort through super-imposed devices such as allowing foreign vessels to be taken on lease or permitting joint ventures etc. which is devoid of a foundation.

The other achievement, that too by the Ministry of Agriculture, when it was handling deep sea fishing, was in respect of introducing around 190 medium class vessels which turned out to be shrimpers, much against the intention of having them function as multi-purpose vessels equipped for stern trawling as well as lateral trawling. The first Committee set up by the Ministry of Food Processing Ministry in 1993 under the Chairmanship of Mr. P. Murari had not led to any tangible solution towards diversification of fishing by these vessels. Recommendations of this Committee to provide a rehabilitation package for repayment of loans taken by the owners of these vessels does not help promotion of deep sea fishing. The only field-oriented recommendation of this Committee is to extend financial assistance to the owners of these presently shrimping vessels to diversify their fishing effort so as to wean them from shrimping in the shrimp over-exploited upper east coast. No follow-up action is visible on this recommendation. In the wake of the ineffective recommendations made by the Committee, again another Committee referred to in one of the following paragraphs, under the same Mr. P. Murari has been set up, probably to buy time, as is believed in fisheries circles, leading to further stagnation of the activity, to recommend policy measures for development of deep sea fishing vis-a-vis problems of coastal fishermen. It has to be seen what this Committee would do. Let us hope the Committee was not set up to counteract the agitation launched by the National Fishworkers Union. Otherwise Government would have selected a known expert in deep sea fishing to head the Committee. There is no dearth of such experts. Dr. G.N. Mitra, Dr. C.V. Kulkarni, Dr. Silas, and the recently retired Dr. Sudarsan and several others are there. Mr. Murari is a seasoned administrator but not a seasoned deep sea fishing expert.

No doubt the Ministry has lot of work relating to deep sea fishing to deal with. However it faces formidable problems to proceed, now that the entire edifice built up in a super-imposed way through joint ventures and other similar ones is on the threshold of being dismantled (issue of permissions for these activities has already been frozen), and in this process, it of course, stands to lose only part of 2 per cent of national fish production, and with no programmes in sight to increase the proportion of catches through increase in Indian-owned deep sea fishing vessels, with the exception of around 1,000 vessels launched in the small scale deep sea sector, emulating Tadri results.

The Food Processing Ministry, several say, has clearly failed in the job entrusted to it by the Government. If the Government is wise, it should send back the subject of deep sea fishing to the Ministry of Agriculture and a separate Department of Fisheries should be established there The achievement of the food processing ministry, according to the general impressions in the fisheries sector, is limited to marginal benefits to the nation, through introduction of a few Indian-owned vessels. The detachment of the subject of deep sea fishing from Food Processing Ministry helps this ministry in getting rid of the fifth wheel, and in enabling it to utilise the time spent on deep sea fishing for doing even better in food processing sector. The deep sea fishery sector does not fit into the subject of Food Processing in general.

The Fishermen's organisation (National Fish workers Union) has an agenda consisting of a single programme of achieving the elimination of fishing by foreign fishing vessels. It does not talk of poaching by several foreign vessels. It does not offer any suggestion either for upgradation of skills of fishermen or an alternative for the exploitation of our deep sea resources such as deep sea shrimps, cuttle fishes, squids and columnar finfishes far away from the territorial waters. This approach indicates that the aim is only to prevent the new class of entrepreneurs to make easy money from joint ventures etc.

and to force the Government to pull out the subject of deep sea fishing from the purview of Food Processing Ministry. It is possible that there is a political force behind this strategy but one cannot be sure. The programme of the organisation is negative, although there is a point that foreign vessels should not fish in our waters on an indifinite basis through permissions issued from time to time. Its agitation however, lacks strength as there are no suggestions to utilise deep sea fishery resources, including columnar and demersal.

It appears the only benefit that may accrue to the nation out of the agitation launched by the National Fish Workers Union is to drive home the point that the present set up at the Centre for development of deep sea fishing should change. Hitherto the Food Processing Ministry, by and large, has been surviving, on issuing permissions for operation of foreign vessels in Indian EEZ. Now that this regime is no longer there, having been suspended temporarily, it is difficult to imagine the worthwhile task it proposes to accomplish, except waiting for the recommendations of the second Committee, (which is again headed by Mr. P. Murari), at a time when the only developmental recommendation made by the first Committee headed by him has not yet been cleared by the Government and no follow-up measures to translate those recommendations into action seem to have been formulated. Mr. Murari should strongly protest and refuse to head the new Committee, unless the fate of his first Committee's recommendations are known. The Ministry of Food Processing appears to be virtually presiding over the liquidation of deep sea fishing in Indian waters unintentionally. The subject continuing in the wrong ministry may spell the doom of the industry, the beginings of which are evident.

We have to look at another problem. The stoppage of permissions for operating foreign deep sea fishing vessels would unleash the poaching activities and the nation is ill-equipped to control poaching, in spite of having a coast guard. When it could not control poaching by Thai vessels close to the coast of Orissa at a depth as low as 20 m (so the shrimp trawler operators in the area complain frequently) we cannot expect control of high sea fish poaching by the coast guard. Even presuming that these are Indian-chartered foreign vessels holding permits, they are not expected to fish within territorial water.

One alternative to the present policy to develop deep sea fishing in Indian EEZ seems to be to help entrepreneurs to buy suitable second hand foreign deep sea fishing vessels, to start with, by providing needed financial assistance, which by itself is a problem considering the dismal repayment performance by enterprises who availed of loans from the erstwhile SDFC and present SCICI. This has to be somehow solved. It has to be mentioned here that loans given by the Government through financing bodies should be deemed as developmental or promotional expenditure and the problem of recoveries should be considered as secondary to the main purpose. We cannot make omelettes without breaking eggs. Vessels purchased should be supported by trained crew from CIFNET for operating larger vessels under an emergency scheme, with the barest employment of foreign crew as long as necessary. There should be a specially created field organisation as part of FSI to check the quantum of catches, and keep a watch over returns, so as to ensure that the owners make repayments out of earnings which aught to be sizeable. Another alternative is for the Government to negotiate with one or more nations for the supply of selected types of deep sea fishing vessels on a bi-lateral basis finance for which may be sought under a World Bank Project or ADB Project and with Indian experts organising the operations. Government may give such vessels on lease for operation in Indian EEZ or outside EEZ, under stipulated conditions to selected Indian entrepreneurs. With projectisation and having an adequate Indian supervisory and operational personnel, the project can be run successfully. Training programmes can be inbuilt into the project. No Ministry is equipped to deal with field work relating to this subject directly and there has to be a field organisation and FSI can be strengthened for the purpose.

The fact that the Association of Fishery Industries has lent support to the agitation launched by the National Fisherworkers Union shows that the Association is also disillusioned. The Association should have told the Government in what manner it wants deep sea fishing to be developed. It appears that there is no equation between the Association and the Food Processing Ministry. If there is, one representative from the Association would have been co-opted as a member of the second Committee, a Committee stated to be consisting of bureaucrats exclusively, who, by and large, would have no depth of knowledge in deep sea fishing, and will not be able to develop a sound deep sea fishing policy for the country by themselves.

Marine fishery development is a composite whole. Upgradation of skills of coastal fishermen in various steps to move from the coast to deep sea, as a long term policy is of paramount Importance, which has been spelt out by us earlier, is necessary in this regard. In the meanwhile, a sound short term policy linked to a specialised training at CIFNET and financing arrangements for acquiring deep sea fishing vessels by Indian enterprises, has to be set in motion. The effort should have backward and forward linkages. The details of this concept were spelt out by us earlier.

The problems of deep sea fishing sector started intensifying when the subject was taken out of the fold of Ministry of Agriculture. They will get minimised only when Government sends back the subject to that ministry and entrusts it to a newly created department of fisheries in that Ministry. The subject does not fit into the overall subject of Food Processing.

Fishing Chimes, Editorial 198: May 1995: Vol. 15, No. 2

# Need for Reforms in Indian Fisheries Administration

Major achievements in the fisheries sector took place in the past. In the inland fisheries sector these are mostly related to culture fishery development in lentic waters such as tanks and ponds. The credit for the development is attributable primarily to the erstwhile Central Inland Fisheries Research Institute, from the point of view that the scientists at the Institute developed various technologies and standardised them and the work is being continued by CIFA. Farmers and various State Fisheries officials shared and continue to share the credit. Extension of the technologies was undertaken by the erstwhile Central Extension set-up. Subsequently, in the wake of the need for an organisational mechanism in this regard, the job has been entrusted to the Fish Farmers Development Agencies, set up under a centrally sponsored scheme. Brackishwater Farmers Development Agencies have also been set up at State as well as Central Governments' initiative on the same premise.

In the marine sector, coastal fishery development owed its origin to the erstwhile Indo-Norwegian Project. Mechanisation of traditional boats received a fillip based on the successful results of the work done by the Project. Later, the Central Institute of Fisheries Technology introduced several designs of mechanised fishing boats. The introduction of synthetic twine in 1950s has transformed coastal fishing effort, coupled with mechanisation of boats, into a viable socio-economic activity. CIFT has also introduced new designs of nets. It has also a unit for research and demonstration of fishing techniques in reservoirs. At a particular stage there was a buzzing activity with the help of FAO experts in the development of reservoir fishing technology.

Minor fishing ports, followed by major fishing harbours, have been established under Centrally sponsored and central schemes to provide improved berthing and landing facilities.

In middle 1950s, pioneering developments took place in the initiation of export of marine products, which have led to unprecedented progressive augmention in the exports by 1994-95 and the trend continues. Infrastructure for preservation, storage and processing has come into being. MPEDA is encouraging entrepreneurs in the setting up of machinaries for producing value added products.

Trawling for shrimps, because of their export value has become an established practice. Medium-sized shrimp trawlers entered the scene as a supplement to the trawling efforts of small mechanised boats.

Infrastructure for providing fisheries education and training has been established. Several Research Institutes, all of which are now under ICAR, have been set up and these provide technologies for fisheries development.

In the field of deep sea fishing, coming under the central sector, Fishery Survey of India, earlier known as Deep Sea Fishing Station, an organisation to continuously assess the fishable stocks, mostly in off-shore and deep sea areas, has been set up. The FSI has located several new grounds in deep sea zone of our EEZ.

In order to demonstrate integrated fishing operations in the marine sector, an organisation (Integrated Fisheries Project) has been established. Further, to lend support to development of fisheries harbours and coastal fishery development with engineering aspect as the main component, a Central Institute of Coastal Engineering for Fishery has been established.

The momentum provided by all these Institutes and the Marine Products Export Development Authority led to the development of a strong private sector which now forms the base for commercial activities in the fisheries field. The momentum provided by the pioneering work in the past decades however now shows symptoms of stagnation.

The State Fisheries Departments constitute the main agencies for the implementation of the various development programmes, which have now assumed a different complexion. Expansion of areas to be brought under freshwater and brackishwater culture is one aspect. A more important one is the injection of refinements in technologies developed by the Central Fisheries Research Institutes for increasing per unit production which is in the process of achieving further cognisable momentum. However, reservoir fishery development continues to be a neglected sector.

On the marine side, upgrading the skills of coastal fishermen to enable them to extend their operations farther, with a view to bringing them closer to deep sea fishing activities on the one hand, and to reduce excessive fishing effort in the coastal zone on the other is of utmost urgency. Further, the coastal State Fisheries Departments have to tackle the challenges posed by the present state of development of brackishwater farming, without being content with the work being done by MPEDA and CIBA in this sector. The work basically belongs to them and MPEDA's intiative is largely because of their responsibility for export development. Efforts of MPEDA are to be substantially supplemented by the coastal States from the point of view of ensuring eco-friendly development of the activity and taking care of the social and professional interests of coastal villages.

Compared to past performance, developments of significance in the recent years are largely limited to haphazard shrimp culture fishery development, the outstanding demonstration and consequent spread of deep sea fishing with the aid of GPS by medium sized trawlers on the south west coast, wholly attributable to the Indo-Danish Project, Tadri, Karnataka, and stepping up of exports of marine products by entrepreneurs helped by MPEDA. It is for the discerning among those who formulate fisheries policies to ponder over the present stagnation. There are no serious efforts at diversification of brackishwater farming activities, particularly in the face of threat of diseases engulfing farmed shrimp. Although some progress is visible in respect of crab farming technologies, in respect of marine finfish farming no progress is visible.

These would have to be perfected and extended for the benefit of the farmers. Freshwater prawn farming for which technology is available has not gained the needed momentum. Technologies in respect of mariculture, regarding molluscs developed by CMFRI have not yet been extended for field application. The measures needed to achieve these need to be formulated and implemented.

On the inland freshwater fish culture sphere, the intensive culture technologies developed at CIFA have not yet come out of the Institute's gates in a significant way. Likewise, reservoir fishery development technologies continue to await adoption by the State Governments, for want of an organisational and developmental mechanism.

These and many other tasks indicate that the administrative set up in the State Governments need urgent reforms. The Central Fisheries Research Institutes are no doubt well geared up to develop technologies and refine them but they face problems mostly relating to extension and financial aspects. Unfortunately there is no developed extension infrastructure.

The Research Institutes are believed to be starved of funds. Presuming this is not true and they have adequate funds, it is learnt that they cannot spend according to requirements because of constraints imposed by unreasonable rules and controls applied by the Administrative Officers and Financial Officers of the institutes who literally stick to their interpretation of rules and make the life of the Directors and Scientists miserable. The only gain, if it can be called a gain, is the promotion of a philosophical bent of mind among Directors and Scientists of the institutes. But this is of no use as it blocks real work. No wonder we hear of anything worthwhile coming out freshwater/brackishwater, and marine research centres only now and then. In respect of role of Central Government, one can say to the credit of Union Departments of Agriculture and Commerce they have a positive attitude, devoid of obstructional tactics, to take the developmental process forward. The departments strive to maintain good liaison with the State Fisheries Departments but responses or reactions from the State fisheries set-up but these are believed to be inadequate. The Central set-up has also to positively influence the policies of the State Governments with greater vigour in respect of inland or marine fisheries development for better results. So far as deep sea fishing is concerned, in the absence of a co-ordinating mechanism between coastal fisheries and deep sea fisheries developmental work, and lacunae in the developmental strategies, the benefits are mostly cornered by foreign fishing interests with marginal financial benefits going to Indian counterparts. In national interest and for proper co-ordination, the subject of deep sea fishing should go back to the Union Department of Agriculture, preferably to be brought under the umbrella of a separate Department of Fisheries, before greater damage is done.

It is axiomatic that any Director of Fisheries in the State sector must have an in-depth knowledge of the complexion of fisheries resources of the State and a professional background to enable him to formulate and implement developmental strategies. His officers would follow his directions wholeheartedly and with devotion only when they are sure that the Director knows his job. A Director may have any number of technical officers, but their energies can be channelised purposefully only when the chief has a broad and deep perspective, knowledge and aptitude towards the subject. State Governments should locate officers of this kind, who will be available primarily from ICAR institutes. Most of the officers in ICAR are recruited based on knowledge and merit and through a competitive examination. They will be in a position to liaison with ICAR institutes and the Central Fisheries set up with efficiency, because of their background and the nature of their knowledge and work equips them to direct and undertake developmental work well. The techno-administrative system at top level in respect of fisheries development in the State sector would need to be reviewed at national level and a viable and progressive system designed for application of technologies and achieving results is to be introduced. Seniority in the department could be the criterion for the top slot only where capabilities and background of the official concerned are of a high order. The appointment of officers from administrative service when it is inescapable, should be related to the suitability of the officer concerned for the job. There are instances in the past, although stray, where an administrative officer heading a Tuna Mission to a foreign country and taken to a tuna frozen storage, pointing out at tuna asked what the fish was; and another officer used to refer 'spawn' as 'sprawn'. Notwithstanding these, in the past there have been devoted administrative service officers whose contributions are still fondly remembered. There are even today several competent and devoted administrative service officers but the general approach on the part of the State Governments continues to be one of appointing an officer for whom a place has to found. Such officers often aim at moving out at the earliest opportunity. Such appointments would do more harm than good, as has happened in several States.

The general tendency on the part of the Central and State Governments is to accord the lowest priority in respect of fisheries developmental matters. Further, not much is being done by the Centre and the States for the upgradation of the professional and economic life of fishermen. In this situation an inescapble alternative for the State Governments is to select a suitable outstanding technocrat from an ICAR institute or a technocrat with known proven record connected with the main line of development work to be pursued, to head their fisheries departments. Further all fisheries matters dealt with at central level should be brought under the Union Department of Agriculture for effective co-ordination, as already pointed out.

Fishing Chimes, Editorial 199: June 1995: Vol. 15, No. 3

# Reservoir Fisheries Development to be an 'Extreme Focus' Area

Reservoir fisheries developmental work, more in the nature of study of conditions of existence, is being carried out for the past over four decades, both by the CIFRI (present CICFRI) and the various State Fisheries Departments. Massive data in respect of most of the reservoirs have been collected and analysed. Actual fishery development work has been taken up in the light of data available in respect of several reservoirs. But fishery development work in respect of a large number of reservoirs remains to be taken up. Integrated efforts at development of reservoir fisheries are lacking.

It is stated that there are 2365 reservoirs in the country (large 58: medium 152: small 2155) having an area of nearly two million ha of surface area. According to Mr. V.R. Desai per ha average production per annum is 22.84 kg. The total annual production on this basis comes to 45,680 t. By raising the production to an average level of 50 kg/ha, production can be stepped up to an average level of around one lakh tonnes per annum, provided there are determined efforts coupled with facilities.

There are a few major instances of increase in annual fish production in reservoirs. One is Bhavanisagar in Tamil Nadu where production is stated to have gone up from 25 to 80 kg/ha/annum. In Sathanur reservoir, also in Tamil Nadu, production as high as 83 to 189 kg per ha/annum was reported at one stage. In Gobindsagar in Himachal, production has been worked out to have gone up from 25 to 75 kg/ha. Tungabhadra reservoir in Karnataka is stated to have produced 111 kg/ha in 1980-81 but the production fell drastically thereafter. The average annual yield per ha from Gujarat reservoirs has been given as 48 kg/ha. Girna reservoir in Maharashtra is stated to have given 40 kg/ha per annum at one stage. Jaisamund lake in Rajasthan yielded 50 kg/ha per annum in 1980s. In Madhya Pradesh average production from reservoirs per ha in 1992-93 was computed at 17 kg/ha/annum.

Seed stocking programmes in reservoirs have been taken up in several States. While the activity led to increase in production in some States, in several others no improvement was noticed. While stocking with Tilapia in some of the reservoirs of Tamil Nadu led to good results (but for a short duration), in Jaisamund lake in Rajasthan, the entry of Tilapia resulted in a diminishing fishery. The introduction of common carp in Loktak lake in Manipur is reported to have adversely affected its fishery wealth. So is the case in the water of Kashmir valley. Introduction of Tilapia in Powai lake in Maharashtra is stated to have resulted in a phenomenal reduction in the average weights of major carps in two decades. In Gobindsagar, the entry of silver carp no doubt increased production but has unleashed an imbalance in the fishery composition. Its predominance had a disastrous effect on the native fisheries. Further, silver carp led to problems of preservation and marketing. There are alarming developments in several reservoirs where many prominent fish species are facing extinction. In our May, 1995 issue Sreenivasan brought out the details. This disastrous phenomenon will have to be stemmed. The Central and State Governments and ICAR no doubt take the situation seriously.

Considering the extent and potential, reservoir fisheries development has to be declared as an 'extreme focus' area, as has been done in the case of marine products exports. Apart from an increase in internal demand, an export market for major carps is fast developing and this scope can be well utilised by developing reservoir fisheries, in an integrated manner co-ordinating seed production and stocking programmes, growth aspects, fishing systems, and preservation and storage of catches for marketing including export. Such an integrated system of development is particularly called for in all States particularly in H.P., from where trout too can be developed for export.

Development of reservoir fisheries will turn out to be a very rewarding task, once an organisational mechanism for the development of various categories of reservoirs is worked out and implemented. The developmental plans have to be moulded to take care of a) increasing production, b) stemming the on-going process of depletion of native species, c) allowing stocking of exotic species only, after a careful study, d) Introducing an organisational mechanism which will not only ensure development coupled with collection of data on continuing basis, but also involvement of local population in the activity.

Fishing Chimes, Editorial 200: July 1995: Vol. 15, No. 4

# Towards an Effective Trawling System

INFOFISH has done a great service to traditional trawlermen by publishing an outstanding paper in its Mar-Apr '95 issue on an effective way of trawling, divesting the present system, although modern, of its shortcomings. This paper was authored by M.I. Shenker, associated with development of special products and technologies, Cape Town, South Africa. Shenker has patented 'Variable Thrust Vector Devices' (VTVDs) for use in the place of Conventional trawl doors (otter boards). He calls the trawling system that uses VTVDs as 'Active Trawling System'.

What are VTVDs? Shenker says that they provide control over the force that lifts the trawlnet, independently of the trawl speed and also by changing the angle at which the lift force acts, to control both speed forces and the vertical forces of each device, at any time during trawling operations. The depth and lateral position of the trawl net can thus be adjusted without changing the trawl speed or the length of the warps, as is being done in the present modern trawling system.

Shenker tells that the present day otter boards of trawls, referred to by him as 'passive gears', just spread the trawl horizontally and provide downward force to keep the net in contact with the seabed at a set depth. Once the trawl is shot, one cannot change the mass of the otter board, nor the relationship between lift force and speed.

In the 'Active Trawl' that uses patented VTVDs, several factors are controlled. Instead of going in a straight path the net can be maneuvered to move in variable paths so as to tap resources located in areas lateral to the trawl path by moving into them, fish there and come out again to resume fishing along the main path. The depth and lateral position of the trawl can be adjusted without changing the trawl speed or length of warps.

The real shift in the technology is concentration on control of the warps over the gear end of the trawl, instead of the conventional system of concentrating at the trawl winch end.

Shenker explains the differences between the present system and his Active Trawl Gear system in Tables 1 and 2 reproduced on this page.

He says that VTVDs provide continuous self compensation for the effects of undersea currents and heavy seas. Further the active trawl system has a bottom-contour mode in which the VTVDs maintain a light contact with the bottom, and the net does not dig into soft muddy bottoms and act like ploughs as otter boards do. It eliminates most of the shortcomings of otter boards such as stretch of chains and bending or wear of warp attachment brackets. It also eliminates the possible imbalance in the performance of the otter boards in the conventional system.

**Table 1: Bottom Trawl**

| *Conventional Gear* | *Active Trawl Gear* |
|---|---|
| Drags on sea bottom | Sea bottom drag or controlled height above sea bottom |
| Fixed trawl grounds based on seabed conditions *i.e.*, roughness, softness, obstacles, slope | Previously "untrawlable" grounds can be exploited as trawl can easily be raised, lowered, moved sideways or angled |
| Deep water trawling requires heavy otter boards | Light variable thrust vector devices can be used to pull the trawl down to great depths |
| Restricted control limits hunting capability | Precise and fast control over trawl position improves hunting capability |
| Trawl speed governed by otterboard characteristics. High power required | Trawl speed can be changed any time during trawling from low to high reduced power requirement. |

**Table 2: Mid water Trawl**

| *Conventional Gear* | *Active Trawl Gear* |
|---|---|
| Trawl path determined by trawler course | Trawl can be moved to either side of trawl path |
| Depth proportional to trawl speed and trawl warp length | Depth is independent of trawler speed and warp length |
| Delayed response to change in trawl position. Large net required | Reduced response time to change in trawl position. Improved hunting capability and higher trawl speed allows a smaller net to be used. |

*Source*: From M. L Shenker's paper on Active Trawl System: a revolution In trawling technology, published in INFOF1SH, No. 2/95 Mar/Apr.)

Unlike the conventional system, the Active Trawl System needs no mechanical adjustment of warps and doors, as it constantly monitors and controls the magnitude and direction of lift force of each device. Shanker says that it is possible to have VTVDs of a different size on each warp and still have a stable trawl.

VTVDs have to be matched. For this, device size, net size and type, bridle length, warp diamater and length, trawler type and available power to control the system through the computer keyboard are to be known, to facilitate initial installation. A programmed test trawl is required for checking the system and maintaining the above mentioned parameters constantly. Shenker observes that a major advantage of the system is that the same VTVDs can be used for small or large trawl nets, bottom or mid-water without any mechanical change.

The active trawl system optimises lift capabilities, tones up drag, maintains trawl speed with minimal variations, and minimises need to change warp length. All these features reduce fuel consumption, saves time and reduces wear and tear on the engine and winch, it is stated.

These Variable Thrust Vector Devices are no bigger or heavier than the otter boards and are controlled by a computer keyboard which will obviously be in the wheelhouse. As the device is patented, Shenker probably could not give details of it. He, however, says that the Active Trawl System is a new generation trawl system, which uses patented Variable Thrust Vector Devices. It offers precise and responsive two-dimensional control of the position of the trawl and also new standards. It has the potential of exploring, developing and exploiting new trawl grounds.

It is planned to publicly display the system at FISH AFRICA '95, to be held in Cape Town from 30 Nov to 2 Dec 1995. It will be a major help to the trawling industry in India, if the Government sponsors trawling experts from the Fishery Survery of India and from the private sector to witness the demonstration at FISH AFRICA '95.

(*Source*: INFOFISH, No. 2/95: March-April with due acknowledgement)

Fishing Chimes, Editorial 201: August(i) 1995: Vol. 15, No. 5

# Dismal Development of Deep Sea Fishing in India

The Fishery Survey of India is supposed to be the beaconlight for the deep sea fishing industry of India. This premier organisation has however just 12 survey vessels purchased between 1978 and 1979. These vessels conduct demersal trawling, midwater trawling and tuna longlining. In 1992-93 demersal trawling was conducted in 57,000 sq. km against a low target set for this survey activity in 1,12,000sq.km. Tuna longlining was conducted in 66,000 sq.km, against a target of 139,000 sq.km. No purse seining was done. The performance in 1993-94 was no better. Against 1,12,000 sq. km to be surveyed by demersal trawling 40,000 sq. km were surveyed, and 24,000 sq. km were covered by survey for tuna through longlining against a target of 92,000 sq.km. In 1994-95 (upto January '95) 45,000 sq.km were covered by demersal trawling against a target of 101,000 sq. km. So far as tuna longlining is concerned, 47,000 sq. km were covered against a target of 81,000 sq. km. No midwater trawling was done from 1992-93 to 1994-95. No purse seining was also done during the period.

The Indian EEZ has an extent of 2.02 million sq. km. Of this, barely 0.142 million sq.km (1.42 lakh sq. km) were covered by demersal trawling surveys so far. 0.137 million sq. km were surveyed by longlining to locate tuna grounds. Of this area, there is no overlapping in the areas surveyed during the three years 1992-93 to 1994-95 as per the reports but some overlapping must have been there.

In this situation, the FSI is hardly in a position to confidently guide and promote the deep sea fishing industry. Firstly, its survey fleet is too small and that too the vessels constituting this small fleet have become old. The ministry of food processing industries, which presides over development of deep sea fishing could not do well both in respect of toning up the performance of the existing fleet and in adding new vessels for intensification and diversification of surveys. This ministry could not develop an indigenous deep sea fishing fleet worthy of the extensive Exclusive Economic Zone the nation has. Instead of concentrating on strengthening of survey base and developing an indigenous deep sea fishing fleet, it has on hand schemes such as providing grants-in-aid for Indian built deep sea fishing vessels (in the addition of which there has been no progress whatsoever in the past several years), and for processing ventures, for providing interest subsidy on loans for acquisition of deep sea fishing vessels, and for providing grant-in-aid for diversified fishing which never took off (Except granting subsidies for activities which cannot probably be construed as any real diversification), for providing communication facilities to coastguard which in a devious way only can be justified, and for schemes for setting up of infrastructural facilities for preservation and processing of fish which form part of "major" fishing harbours (No information on any such facilities provided is available), a subject that in any case comes under the purview of the Union department of Agriculture. No efforts have been made to render assistance to the existing shrimp fleet for addition of systems of diversification such as longlining, in spite of recommendations from FAO experts, and a Committee set up for the purpose, emphasising the need for reducing pressure on shrimping and secondly to make the operations economical.

So far as utilisation of funds is concerned, against an outlay of Rs. 63 crores in the VIII th plan for deep sea fishing, the ministry of food processing utilised a mere Rs. 0.15 crores in 1992-93, Rs. 0.06 crores in 1993-94, and had a revised estimate of Rs. 0. 68 crores for 1994 - 95 on capital account. In 1995-96 it proposes to utilise Rs. 0. 50 crores. The bulk of the allocation is utilised or earmarked for revenue expenditure. The unutilised funds could have been channelised for assisting the concerned fishing companies for the purpose. It is true that most of the companies are defaulters in the repayment of loans taken for acquiring the vessels, but it ought not to come in the way of diversification and a way out could have been found, in national interest.

Much can be said on the slack progress in respect of development of deep sea fishing. There has been an all round failure looked at from any angle. Achievements can be measured by the level of transfer of technology, introduction of deep sea fishing vessels that are really our own and flying Indian flag and not based on alleged devious arrangements, and increase in catches from deep sea fishing, particularly from columnar zone and in equipping the existing shrimp trawlers for diversified fishing. The ministry could not accomplish the task of equipping shrimp trawlers for diversified fishing systems such as tuna long lining, squid-jigging etc. Introduction of deep sea fishing vessels of our own and adding new vessels to FSI's fleet could not be expected, when diversification, which is relatively achievable, could not be accomplished. The industry is unable to find any achievement of which the food processing ministry can be proud of in respect of development of deep sea fishing. The sooner the policies and strategies for development of deep sea fishing are overhauled, the better it will be for the nation. In this situation, the first job to be done in this respect is probably to shift the subject from Food Processing Ministry to the Ministry of Agriculture with which the subject can merge homogeneously.

Fishing Chimes, Editorial 202: August(ii) 1995: Vol. 15, No. 5

# Aquaculture and Income Tax

It is now a settled matter that exemption from Income tax is applicable only to Agriculture under the Income Tax Act, as the Provisions under the Act are very clear about this. Government has no intention to extend the relevant provisions in the Act to cover aquaculture. This appears to be a great injustice to aquaculturists.

The issue has to be looked at from a scientific angle. This has not been done so far. Aquaculture is undertaken in depressions on land where water either accumulates or is let in. Land is the main source for the water in ponds and tanks to derive fertility. Water just happens to be a medium overlying the land. The soil nutrients are absorbed by land plants directly in agriculture. In aquaculture the underlying land imparts fertility to the water. Utilising the dissolved nutrients derived from land underneath, phytoplankton (small microscopic plants) develops. These microscopic plants are consumed by small microscopic animals (zooplankton). Fishes and other aquatic plants consume both phytoplankton and zooplankton as their food. In other words, land is the basis for aquaculture. Aquaculturists till land and apply organic and inorganic fertilisers and undertake disinfection as in the case of agriculture but before turning in water. This clearly shows that this activity is akin to agricultural operations. Instead of taking a direct land crop, they take what would appear to be a water crop but is essentially dependent on land.

Further, it is well known that several aquaculturists undertake paddy-cum-fish culture and trapa-cum-fish culture. Trapa fruits so cultured with fish are obviously akin to agriculture. So is paddy-cum-fish culture.

If your editor's memory serves him right, the Privy Council decided on an appeal from a land lord or *Zamindar* from the former composite Bengal that fish ponds come under the definition of land, as they are mere depressions on land and therefore land revenue would have to be paid. You editor's memory is that this judgement is an appendix to the report of the first Royal Commission on Agriculture. Even if it is not an appendix, the judgement should be available in legal records.

This historic judgement proves that culture in fish farms comes under the definition of Agriculture. It is not known what the provisions were in respect of levy of Income tax during those days. Whatever was the case, in the present context, when agriculture does not attract income tax, fish culture too should not attract levy of income tax.

This being an intensively legal issue, the Associations of fish farmers should revive the matter based on science and the old judgement of the Privy Council, which was the highest court during those days. Any eminent lawyer should be able to look at the issue from a factual basis and make out a realistic case for exempting aquaculture as in the case of agriculture.

Fishing Chimes, Editorial 203: September 1995: Vol. 15, No. 6

# Challenge to Deep Sea Fishing in Indian EEZ

The Ministry of Food Processing Industries now faces the basic issue that directly concerns the imperativeness or otherwise of promoting deep sea fishing in Indian EEZ. Under the existing laws, rules and regulations, the Ministry is in-charge of the implementation of the deep sea fishing policy of the Government. This policy envisages the operation of deep sea fishing vessels of foreign or Indian flag in the Indian EEZ through joint ventures, leasing of vessels, or those owned by the Indian enterprises themselves. Permissions in respect of joint ventures etc., are issued by the Ministry for specified durations. Earlier, licenses were being issued for the operation of chartered fishing vessels. This system was subsequently withdrawn.

In this background, organisations of fishermen, particularly the National Fisheries Action Committee Against Joint venture has spearheaded a movement demanding that all rules introduced by the Government of India under the Maritime Zones Act, which enable foreign fishing vessels to operate in Indian waters should be withdrawn forthwith. The permissions issued so far should also be withdrawn is another demand so as to make the Indian EEZ free from the operations by foreign fishing vessels.

The leadership of the Committee has embarked on this confrontation mostly on the ground that the foreign fishing vessels are taking away the fishery wealth that belongs to the nation and the Indian companies that secure these vessels under one of the permitted systems are earning easy money because of one of the approved conditions that 20 per cent of the gross value of the total catches from the fishing vessels would accrue to the Indian counterparts. At the same time the nation has not been able to derive any tangible benefits by way of acquisition of technologies relating to various types of deep sea fishing from the foreign operators. The subject has been raised by the Committee and others with the Ministry of Food Processing Industries and also in the Parliament.

The Indian fishermen and the politicians, by and large, are not well informed about matters relating to the development of the fisheries of the country and its relationship with the socio-economic fabric of the fishing community. This deficiency led several politicians to the display of persistent agitational confrontation with the Government. When the issue was raised in the Parliament, the Minister for Food Processing Industries made an effort to explain the poor state of utilisation of deep sea fisheries resources (2 per cent of the total catch) and the various measures taken by the Government for the development of the sector which falls under the central list of subjects as listed in the Constitution of India.

The Ministry, having taken cognisance of the views expressed in the parliament, has expanded an existing Committee to recommend measures for the formulation of a deep sea fishing policy under the continuing chairmanship of Mr. P. Murari. This Committee now has 41 members of whom 16 are members of parliament. This Committee has already held preliminary meetings and set up five sub-committees to undertake field enquiries based on the results of which it is expected to consider the main issue in its forthcoming meetings.

An incisive analysis of the observations by fisheries professionals and the Government on the subject would reveal the underlying causes for the agitation. One is that the present leadership of the fishermen, which has taken upon itself the task of protecting the interests of fishermen, would soon have to address the basic problems of fishermen, which is its primary function. They seem to give the impression that, in order to avoid the wrath of fishermen they may have to face, owing to their inability to be of real service to them, they have chosen the tactical option of diverting the fishermen's attention to misinterpreted injustices being done to them by the Government of India by permitting foreign fishing vessels to operate in Indian EEZ through joint ventures etc. The various coastal Governments with aid from the centre have done an enormous service to fishermen over years starting with the introduction of nylon twine, new designs of nets, introduction of mechanisation, providing training, construction of housing colonies, providing credit through co-operatives etc. Human memory is short and past is fast forgotten. The current developments, whatever be the reasons thereof tend to sway the fishermen, but the discernible and the non-dogmatic elements in the leadership would have to atleast make an assessment. Although joint ventures etc are allowed to be formed under law, Government are so responsive, that irrespective of the merits of the agitation, they kept in abeyance the implementation of the scheme. While there is no doubt that operations of foreign vessels in Indian EEZ by Indian companies through joint ventures, leases etc. approved under the law of land have not led to the anticipated technology transfer and training of Indian personnel but only provided financial benefits to the concerned Indian companies without their direct investments in the operations, to say that the exploitation of the deep sea zone, (the fishery wealth of which has little relationship with coastal fisheries), has an adverse impact on the inshore fishery wealth is a travesty of the truth.

The coastal fishery wealth is over-exploited by the traditional sector. Scientists categorically say that small sized quality fishes such as seer and pomfrets are over-exploited in the coastal zone. A major component of the catches taken from the coastal waters consist of small sized fishes (Juveniles) mainly because of indiscriminate reduction in mesh size of nets. Fishing done during spawning seasons is affecting reproduction and growth of fish. The stepping up of traditional fishing activity is so high that the annual fish landings in the Gulf of Mannar increased from 1,224 t in 1950 to 33,000t now. The catches show that the stocks of tuna, seer, sciaenids, milkfish and sea-bass in the Gulf of Mannar have alarmingly declined. Sea turtles and dugongs, have been reduced to a residual population. Average size of seer fish has decreased from 645 mm in 1984 to 535 mm now. Seer fish landings declined from 900t in 1984 to 545t now. Coastal tuna catches have also decreased to 700 t per annum from 1300 t in 1984. Juvenile fishes are being caught indiscriminately for the past over 10 years. Many coastal marine fishes are now almost extinct. There is a conspicuous absence of fully grown seer fish in Tuticorin market.

There has been a sharp decline in sardine landings along Kerala coast. In 1968, the catches of sardines were a little over 2.47 lakh t. Last year the catches were a mere 1554 t Catches of Dara, Rawas etc. have fallen drastically from the inshore waters of Maharashtra. All these are the findings of scientists.

No scientist or fisheries professional has attributed this state of affairs to the operations of foreign high sea fishing vessels, most of which concentrate on catching open sea tuna such as yellowfin and cuttle fish and squid which occur mostly in waters far away from the coastal zone.

The present state of coastal fishery depletion is thus directly related to the transparent overfishing by the traditional sector, which does not get any guidance from its leadership to conduct their activities in a manner that will keep the fishery at a sustainable level.

Having failed in this task, the leadership, for reasons of (their own but as widely talked about, seem to have chosen the irrational and inappropriate alternative of brain-washing the fishermen into believing that the operations of foreign fishing vessels are harming their interests, which is not clearly the case.

In saying this, no brief is sought to be held wholly in favour of operations of foreign fishing vessels, although their operations are allowed in accordance with the rules framed under the Maritime Zones Act. The non-accrual of any tangible benefits to the nation, barring a negligible net inflow of foreign exchange, the general impressions of malpratices that are believed to go with this, lack of progress and the total non-realisation of the cherished objectives are the main results of this scheme, so laboriously worked out by distinguished professionals. At the same time, it is correct to say that our deep sea fishery resources should not be allowed to remain unexploited. Objectives of acquiring capabilities in the operation of deep sea fishing vessels, achieving technology transfer, building up of trained man-power, acquiring capabilities for constructing deep sea fishing vessels and, lastly, to totally become non-dependent on foreign operations, have not yet been realised. But then what has been realised is one of enabling foreign vessels to gain access to fish in our EEZ, making it possible for Indian companies to convert the scheme into a money-spinning source, without the realisation of the concommitant objectives and obligations aimed at gaining a place for India as a deep sea fishing nation. There are however a few Indian companies operating in the deep sea zone successfully.

The Ministry of Food Processing Industries does not seem to have undertaken any serious review of the progress in respect of the achievement of the objectives of permitting foreign vessels to fish in our deep sea zone. It is generally seen that the Government is content with issuing licenses and permits to Indian companies to bring in foreign fishing vessels operated by foreign crew and with a component of Indian crew who have not been able to wrest any opportunity from the foreign crew to pick up the technologies. There in no evidence of any serious effort on the part of the Government to reorient the scheme to set right the situation. Further, it is unfortunate that Government has not applied their mind to bring about an integration between coastal and deep sea fishing effort and to upgrade the capabilities of coastal fishermen to undertake deep sea fishing which is possible only when both the subjects are dealt with by one department of the Government. The only break-through in the upgradation of skills of coastal fishermen is seen in the outstanding results obtained by the Indo-Danish project at Tadri in Karnataka.

This complacency has provided fuel to the leadership of fishermen to mount up agitations which do not seem to be not in national interest in the present form. There is no maritime nation in the world which discourages deep sea fishing. There are several countries which allow foreign fishing vessels to fish in their respective EEZs. But the manner of issuing permissions is different from ours. The system is based on an assessment of available resources, their own fleet strength and the catches it can land, the balance of estimated resources that may remain unexploited and a calculation of number of vessels and the kinds of vessels of foreign origin that can be permitted to fish in their waters based on quotas. Once they complete their permitted quota of catches the vessels have to go back.

So far as India is concerned, the system of issue of permissions and licenses is irrational. The activity is ostensibly related to acquisition of technology, which has not taken place so far. The question of allocation of quotas never arose as the resources are mostly unexploited. A few Indian-owned vessels do exploit the resources, but the quantity is so small that it has not been significant. The stated quantity of 2 per cent of total catches forming deep sea catches are mainly the catches by foreign vessels operated in the name of Indian companies. The Government has also never seriously applied their mind to work out a purposeful, practicable and implementable integrated scheme to eventually develop an Indian deep sea fishing fleet, or have the needed Indian man-power for managing and operating the vessels, providing the needed post-harvest infrastructure etc. No long range project has been launched to train the traditional fishermen to equip them to operate deep sea fishing vessels. Government will have to review the situation and bring the administration of all components relating to deep sea fishing under one ministry. Until then there does not seem to be a way towards progress.

The columnar resources should be central to our deep sea fishing. Mid-water trawling for columnar fish and long lining for tuna and such other fishes, apart from the needed deep sea trawling effort consistant with the available resources would have to be promoted on a sustainable basis. It will no doubt be ridiculous to think of abandoning deep sea fishing as demanded and allowing the fishes to die a natural death.

Government alone are to be blamed for the mounting agitation spearheaded by the leadership of fishermen. It has provided the opportunity and climate for the situation. The leadership, which is believed to be in a predicament as to how to hold their members, in the face of their failure to solve the problems of over-exploitation of coastal waters (which cannot be achieved by seeking non-exploitation of deep sea resources even by Indian deep sea fishing vessels), lack of growth, reduction in financing avenues etc, seem to have found a diversionary way-out, from the manner of development of deep sea fishing by the Government itself which is impregnated with omissions and commissions.

Offence is the best form of defence. The present situation has obviously provided an opportunity to the vocal and bullying section of fishermen's leadership to follow this dictum. The climate is such that this section may be enabled to monopolise and harangue the representatives from the Government's side, in the newly constituted Committee under the chairmanship of Mr. P.Murari with their monologue, as they have all the grist. The Government may have incontrovertible, firm and scientific developmental ground and is in a position to put forth the same before the committee. Yet, it has to be seen as to how well the representatives will face the onslaught. They may have a tough time to counter the view points of the leaders of fishermen on the Committee. The representatives of the Government may be impeded by their inability to stem the vehemance of presentations of fishermen's leadership, unless the team has persons who can equally vocally and forcefully present their case, which one can be certain, they have.

The Association of Indian Fishery Industries and the Deep Sea Fishing Association, whose members are on the Committee would no doubt bring out the imperative need to formulate a cogent and integrated programme for the development of deep sea fishing in the EEZ, which is barely touched, on right lines.

It appears that there are rays of hope of formulating cogent and meaningful recommendations because of the presence of 16 Members of Parliament on the Committee. These are men of experience, who can certainly be unbiassed as national interest is paramount and close to their heart. They should certainly be able to assist the Chairman by presenting their rational views, no doubt based on the various facts relating to the problem.

Fishing Chimes, Editorial 204: October(i) 1995: Vol. 15, No. 7

# The Other Side of Deep Sea Fishing by Foreign Vessels in Indian EEZ

It is possible that the Murari Committe may come up with a recommendation that Deep Sea Fishing by Indian vessels alone should be allowed. Hopefully, it may suggest in this connection a way out for securing finance for the acquisition of Deep Sea Fishing vessels, effective arrangement for which is not there now. Assuming that it recommends the setting up of a separate fisheries development bank, it would not make any difference, unless it also recommends an adjunct as a fully equipped operational wing of the Bank with professional skippers, engineers and a force of deckhands, ready to take over the operation of loanee vessels whose owners fail to make repayments on time.

Otherwise, this bank will also be in the same straights as the erstwhile SDFC, and the present SCICI (so far as financing of fishing vessels is concerned) and a few commercial banks have been.

A lot has been said against the regulations introduced by the Government under the provisions of the Maritime Zones Act, which empowers the Government of India to permit operations of foreign fishing vessels in Indian EEZ, by way of charters, leasing, test fishing and joint ventures by Indian fishing companies. These rules must have been made not merely for acquiring technology relating to Deep Sea Fishing but because of the expected problems of finance which had become much more acute in the recent past.

There is substance in what is being said against operation of Deep Sea Fishing Vessels of foreign origin by Indian companies under various systems mentioned above. This feeling has transformed itself into an abnormal agitation on the part of the traditional fishermen's organisations, which has forced the Government to control the same through referral of the issue to a high level committee headed by Mr. Murari. There is criticism that the systems of taking foreign vessels on leases etc. are enabling the Indian Companies to make easy money through securing of 20 per cent of gross proceeds out of sale of the catches without achieving the technology transfer to the crew and acquisition of similar vessels as permitted to be operated.

The concerned Indian Companies cannot be faulted that uncharitably as charged. One reason for this is that they have only took permissions under the schemes announced by the Government itself and there is nothing wrong to this extent. So far as technology transfer is concerned, the companies concerned did post counterpart personnel on board the foreign vessels as stipulated. If these counterparts could not pick up technology the owners could not be blamed, unless they had not posted the right type of men for the job. One fact that is transparent is that the foreign crew did not facilitate the Indian counterparts to pick up their technology. In fact they are deliberately not given an opportunity to learn. According to what is understood, the foreign officers on board the vessels keep the Indian crew comfortable but at the same time manage to ensure that they do not get an opportunity to learn. At present the Indian Companies have been helpless in inducing or convincing the foreign crew to enable the Indian crew to imbibe the new technologies of fishing operations. The Indian Companies may be making "easy money"' but they could not be faulted for this. The rules framed by the Government clearly say that 20 per cent of gross proceeds should accrue to the Indian Companies concerned. Further, the allegations of manipulations in the invoices may or may not be true. It is for the authorities concerned to check up these to their satisfaction. It is possible that manipulations become necessary owing to the play of the law of the situation. The Indian Companies could have probably utilised atleast part of the 20 per cent of gross proceeds given to them (by the owners for it is they who market) for acquiring their own vessels by saving atleast part of these amounts. The inability of the Indian Companies to do so is not clear. Probably the money went into several items of expenditure that they cannot talk about and also towards the funds spent earlier for the arrangements to secure foreign vessels on lease etc.

In respect of non-acquisition of vessels by the Indian Companies under the stipulated *paripassu* clause forming part of the permissions issued by the Government, this can be very well understood, if they have no savings out of accumulated incomes earned by them out of 20 per cent of returns of gross proceeds, and if no financing institution is willing to extend financial assistance for the acquisition of the vessels for reasons of sickness of the industry.

The present representation of the Association of Indian Fisheries Industries is to allow Indian Companies to introduce and operate Indian owned Deep Sea Fishing vessels in the Indian EEZ or outside the zone. This is a rational approach and is well justified. Probably the Association implies that their members can enter into joint ventures with foreign companies, with Indian vessels as the base. The only problem is the investment pant. They have no agency to provide them with the needed finance. Unless this question is answered there can be no development of Deep Sea Fishing in Indian waters, barring a few vessels introduced by five companies who could succeed in organising finance for the acquisition of the vessels through public issue or otherwise.

No financing institution is willing to lend money for reasons of alleged lapses on the part of several Companies now operating shrimp trawlers (of 23-27m OAL), in not making repayments of loans taken by them for the acquisition of the vessels, on concessional terms mostly from the erstwhile SDFC.

This performance on the part of the Companies may look dismal, but only when viewed superficially. Once we look deeper, it will be seen that the operational and other costs have gone up so high, and operations are limited to part of a year, thereby leading to negative results. An added unfavourable factor is that the companies pay salaries and mess money to the crew for the entire year although they operate the vessels only for a part of the year (around six months). Probably for these reasons the companies concerned do not practically have any savings to repay the loans taken. With the commercial banks refusing to increase the cash credit limits to meet the escalating costs, the owners are forced to borrow money at heavy interest from money lenders and also take supplies of spare parts etc. on credit at very exhorbitant rates, and also pay for the services at far higher rates, because of the credit factor. When the vessels are moored for part of the year at the harbour lot of additional burden is imposed on them by way of port charges.

Deep Sea Fishing vessels are designed for operation throughout the year, (leaving such periods ridden with cyclonic weather, time taken for dry docking etc.). Mooring such vessels at port naturally brings down incomes substantially and expenditure also goes up on account of the watch and ward arrangements, payment of port charges etc.

Recommendations for rehabilitating shrimpers have been to equip for diversified fishing so as to extend their operations over the entire year but these have not been implemented. While the reasons for this are not known, to a common observer it appears that this is because of disinterest on the part of the Government as well as the owners. An additional and important reason can be that no financial help for this diversification is forthcoming, although MPEDA offers subsidies. Government could have taken steps for having the drawings and designs prepared at Government cost and provide the same to the various operators to enable them to explore possibilities of securing financial assistance for incorporating the systems on-board their vessels. This way at least the position would have been clear as to how much investment would be required for the diversification programme. It might be quite possible that under bilateral or other foreign assistance programmes funds could be obtained for achieving this reform. It is probable that Government never moved in this direction for the reason that most of the Companies could not pay the pending instalments of loans taken. However, on this ground, developmental process cannot be detained but unfortunately it is so. Deep Sea Fishing Sector is not the only one which defaulted in making repayments. Several other sectors are in a similar position. Yet, Government continues to lend support to these sectors. The same could have been done to the deep sea fishing sector. In any case the non-intervention of the Government to find solution to the problem is deplorable. It is true that a few larger vessels of 25m OAL and over moved over to the west coast for deep sea trawling but this development is an exception as those few vessels alone can undertake fishing in deeper waters beyond the present range. Thus, this development has not solved the general problem. The owners of 23m OAL or vessels even smaller than this OAL could have been helped to equip their vessels with systems such as long lining or for deepwater trawling based on the Tadri Model.

Many really believe that Indian Deep sea fishing vessels were introduced in good numbers compatible with sustainability and therefore they make money. The line of argument for this view is that foreign fishing vessels holding permission to operate in Indian waters seem to be harvesting in an economically viable manner.

It may be mentioned here that there is absolutely no truth in the allegation that multi-national and larger houses are involved in joint ventures with Indian Companies. Multi-nationals such as Union Carbide learnt through experience that Deep Sea fishing is not their cup of tea. The same kind of lesson was learnt by other companies such' as Britannia, ITC, EID Parry etc. The set-up and culture of large companies do not fit into the type of organisational system required for implementing deep sea fishing schemes. It would be wrong to construe fishing companies of Taiwan, Korea etc as multinational companies. These are just distant water fishing companies and have no characteristics of multi nationals, although they operate in the waters of various countries.

In this background the policy for the development of deep sea fishing in Indian waters would have to be evolved. A few indicators in this direction are attempted hereunder.

1. Set up a Fisheries Development Bank and organise the transfer of all the accounts of the presently operated Companies to it Allow the loanee vessels to operate unhindered and without any harassment The proposed bank may be provided with an effective operational wing equipped not only to monitor the operations and collecting the loan amounts due to the extent possible but the development work should also be designed in a way that the system could provide finance for acquisition of new/second hand Deep Sea Fishing vessels from various sources and also for diversification of operations of existing vessels that are at the brink of sickness.
2. Set up Marine Fishery Development Agencies charged with the responsibility of introducing Tadri model vessels but suitably modified to have refrigerated holds and higher fuel holding capacity to stay out longer while simultaneously providing training to the enlightened and educated section of traditional fishermen to operate such vessels. Modifications such as provision of refrigerated holds are suggested for the reason that vessels of Tadri model or similar in size will have to come

back to port now after fishing for about a week only. At present they do not have refrigerated holds and have limited tank capacities for fuel, water etc.

3. Introduce a long range scheme for setting up of specialised training centres for fisherchildren over a period of time. The children should be caught young at an age of around nine and they have to be trained till they gain adequate knowledge and experience to become skippers/engineers and acquire ability to have their own vessels with financial support forthcoming through concerned Marine Fishery Development Agency, with effective monitoring and loan recovery arrangements.
4. In the meanwhile, as a short term measure allow Indian Fishing Companies to acquire new/ secondhand deep sea fishing vessels from abroad on such terms as can be approved (joint ventures, deferred payments etc.) and simultaneously create a cadre of Indian crew through crash on-board training programmes at CIFNET (which could be equipped with suitable vessels for the training purpose) for a sufficient duration to coincide with the arrival of acquired vessels, on-board of which one foreign master fishermen and one foreign engineer may be allowed to function for such durations as necessary, to provide training through demonstration of operations. These vessels should of course be of Indian ownership and Indian flag. This short term measure can also be taken with bilateral financial help for supply of vessels from friendly nations having the needed expertise.

In the absence of support from Indian financing bodies to enable Indian enterprises to acquire deep sea fishing vessels, it appears that the four routes of development as mentioned above would need serious consideration by the authorities concerned. Any one out of these suggestions would enable the small scale sector to upgrade their capabilities to exploit the Deep Sea Fishing Resources, pelagic or demersal. The term "pelagic" covers mid-water or columnar fish resources. The programme to upgrade the small scale sector to cope up with the requirements of deep sea fishing would necessarily have to be a long range programme. We do not see any other routes to ensure the effective exploitation of our Deep Sea Fishing resources in the long run. There has to be a compromise in respect of one aspect or the other. We cannot make omelettes without breaking eggs and we cannot leave the Deep Sea Fishery Resources totally unexploited. If left unexploited the situation will provide temptations to foreign vessels to poach and this poaching is not easily amenable for detection. Further, any complacency will either facilitate foreign interests to take away fish that is legitimately ours or we leave them to die a natural death.

Fishing Chimes, Editorial 205: October(ii) 1995: Vol. 15, No. 7

# Diversification in Brackishwater Farming

The brackishwater farming sector continues to be in a crisis. Some of the farmers have started stocking the farm ponds with additions of sedimentation and reservoir ponds to the farms, and moving towards a recirculation system, open as well as closed. The industry may come out of the crisis because of these innovations, but crises may be repetitive which no one wants. Many brackishwater farmers are on the look out for alternative systems other than brackishwater shrimp farming. They are in a lurch in this respect. They run from pillar to post for advice on alternative farming systems but they do not get any firm advice. This situation should be set right. The job rests with the Central Institute of Brackishwater Aquaculture and partly with the Central Institute of Freshwater Aquaculture in respect of freshwater prawn culture, wherever possible. They should come out with clear alternatives, if any, for the farmers to take up tiger shrimp culture and disseminate their advice.

Commercially viable finfish seed resources which can be utilised for farming in brackishwater farms does not appear to have been well located in Indian waters. This scarcity can be solved for the time being and untill adequate seed resources are located, by allowing import of seed of sea bass, groupers, breams etc., from other countries linked to a a quarantine system.

The species no doubt occur in our waters but location of seed sources has not yet been done systematically and development of hatchery systems for the species will take time. A policy has to be evolved in this respect. CIBA will no doubt play an active role in this direction.

There are several brackishwater farms having access to freshwater supply. Several farmers having this facility want to utilise their farms for freshwater prawn culture. Their wail is that they have no access to any extension agency to advise them in respect of details. CIFA should develop an extension wing for the purpose with the needed ramifications. This would be a great help to the farmers.

Fishing Chimes, Editorial 206: November(i) 1995: Vol. 15, No. 8

# "Export of Fresh, Frozen and Processed Fish and Fishery Products Order and Rules, 1995"

## Oppressive and Counter-productive

The Ministry of Commerce has published under 21st August, 1995 rules imposing monitoring fees on exports of marine products. It is learnt that the representatives of Industry/Association concerned were not consulted before finalising the rules.

According to the Joint Director (EIA), Mr. Srinivasan, EIA need not consult the Industry before notifying orders and rules. In the same way as Income-tax, EIA have imposed the following fees on F.O.B. value of material which is exported.

| *Unit Export Turn Over* | *Monitoring Fees* |
|---|---|
| Under Rs. 10 crores per annum | 0.2% of F.O.B. value of export |
| Rs. 10 crores and above per annum | 0.15% with a minimum of Rs. 2.00 lakhs and maximum of Rs. 5.00 lakhs per annum |

A back dated letter (31-8-1995) from E.I.A, Madras is stated to have been received by the exporters to pay the fees every fortnight with the following details.

i) Name of Processor ii) Name of Exporter iii) Countries of Exports iv) Shipping work, Invoice No. and Date. V) F.O.B. Value.

On the basis of the above statement received, fees may be realised. In the year 1991, Mr. Chidambaram, Commerce Minister, and Mr.A.Didar Singh, Commerce Joint secretary, removed E.I.A. inspection stating corruption and inspection by unqualified officers were hindering exports and that no importer was happy with the methods of Inspection and refused to accept their certificate for quality.

In fact, almost all importers clearly mentioned in their purchase order that EIA, certificate was not required and that buyers inspection was final. With such a type of reputation for E.I.A. the reintroduction of collection of fees without inspection, quality control and no liability of any claim demands made by importers is improper.

EIA wants 0.2 per cent fees on F.O.B. value for just monitoring exports. In 1993, E.I.A. announced a scheme of voluntary retirement of their staff. Accordingly, most of the experienced, qualified staff took voluntary retirement and settled in private sector. Now the personnel left over who could not get any employment anywhere else continue in E.I.A. The exporters allege that unqualified, corrupt and inefficient officers are now trying with their union force to come back to E.I.A.

Presently the exporters are paying following taxes in A. P. 1 per cent as Cess on F. O. B. value to G.O. I.

9.7 per cent as Sales Tax to State.

10 per cent as Excise Duty to Centre.

1 per cent as Marketing Cess on total purchases (now under stay order)

6 per cent as Purchase Tax (kept in abeyance).

Government of India used to give value-based advance licence for 5 per cent of F.O.B. value to import packing material. Shockingly, without any announcement the authorities concerned brought this down to 0.26 per cent, it is stated. The new rules are clearly oppressive and counterproductive. With the exporters paying so much by way of taxes and cess, it has become extremely difficult to exporters to compete with other exporting countries, who are exporting at much cheaper prices to Japan.

From the last six months and over, Japan stopped buying Brown Shrimp. The Japanese importers also started imposing restrictions on sizes of Tiger and White shrimp. This has made exports by Indian exporters extremely unworkable.

The economy of Japan, and weakening of its Currency completely upset Indian Exports of Marine Products and all the Companies have run into heavy losses, particularly as the shrimp catch composition has changed (70 per cent browns, 25 per cent whites and 5 per cent tigers) to the great disadvantage of exporters.

Several countries increased their efforts to dethrone India and increase their marine product production and exports, taking advantage of the situation. These countries are: Equador, Philippines, Indonesia, Thailand, Taiwan, Bangladesh, Pakistan, China, and Sri Lanka.

They stepped up farming of brackishwater Shrimp and started exporting them at much lower prices to Japan compared to Indian export prices of shrimps.

Under above circumstances the Indian exporters have no alternative other than closing down their factories.

It is high time the Central Government and Coastal State Governments start giving incentives by way of waiving all taxes and duties and make export of Indian Marine Products competitive with other countries so as to earn valuable foreign exchange. This will also help fishermen, boat-owners, farmers and exporters. Motivation of this kind alone will promote exports of marine products. The exporters strongly feel that the Government of India should immediately withdraw monitoring of exports by EIA and atleast stop collection of the newly levied fees.

Fishing Chimes, Editorial 207: November(ii) 1995: Vol 15, No. 8

# Fishing around Andamans becomes Expensive

The waters around Andaman and Nicobar Islands are known to have rich fishery resources. Fishing vessels from countries such as Thailand, Taiwan, Korea etc., have often been found conducting fishing in the Exclusive Economic Zone around these islands.

A few Indian Fishing vessels also have extended their fishing operations to this zone since the past few years.

At this time when the vessels of Indian fishing companies have gained courage and familiarity with the fishing grounds around Andaman and Nicobar islands, the Administration of the islands slapped a royalty system on catches landed on the islands which would no doubt act as a disincentive to Indian fishing vessels to expand the operations in the area. The notification in this regard indirectly helps the foreign fishing vessels to conduct unauthorised fishing in the EEZ around the islands and take away the catches directly to their home ports. It will be difficult for the Indian Coast Guard or Indian Navy to keep a constant and continuous vigil over the clandestine operations.

The worst affected category would of course be the Indian fishing companies. They will prefer to return to a port in Andamans for declaring the catches to the customs authorities and to make necessary arrangments for transportation to a foreign country. Being Indian vessels it would not be possible for them to take the catches to foreign ports, in the same way as foreign vessels can do. They can of course go to one of the ports on the mainland but that would be inconvenient and expensive.

The royalty rates imposed range from Rs.200 per kg for shark fins, Rs. 257 per kg. for shrimps/prawns, crabs, lobsters, Rs. 20 per kg for fish caught in deep sea fishing vessels and Rs. 10 per kg. for cultured prawns, shrimps, lobsters etc. The royalty payable for shark flesh is Rs. 5 per kg. By any standard the royalty rates notified are not only uncalled for but also very much on the high side. The profit margins earned by the operators would be far less than the rates of royalty notified.

The Andaman and Nicobar Administration has not spared small fishermen a/so. A levy of Rs. 15 per kg. for fish caught within 12nm from the coast by local fishing vessels, is not only on the high side but it also imposes a compulsive and burden-some levy on the fishermen. The small fishermen belonging to the islands have no other places to go for their fishing operations, for they can only continue their operations in their traditional grounds around the islands and it is doubtful whether they will have so much of margin as Rs. 15 per kg. However, while the levy is limited to the catches within 12 nm for local fishing vessels, it is not known what kind of system the administration wants to introduce to determine whether the catches came from the area within 12 nm or beyond. It is not known whether the small fishermen are restricted from fishing in areas beyond 12nm. The system of payment of royalty introduced would certainly lead to considerable confusion and problems in the enforcement of the system, besides being burdensome, counter productive and anti-development in nature.

There can be several ways of avoiding declaration or underdeclaration of the catches and it will be impossible for the licensing officers or others to ensure proper declarations. Further, local fishing vessels may catch animals other than fish such as cuttle fish, squid etc. and there can be dispute that these varieties are outside the purview of the term "Fish". It is possible that in the relevant Act all aquatic animals are brought under the definition of "Fish". Even then there will be scope for disputes.

The system of levying royalty on marine fish or other catches is not adopted by any of the coastal states. It is somewhat obscure as to why the Andaman and Nicobar administration has felt the necessity to levy royalties, particularly at this time when the emphasis is on encouraging the exploitation of the fishery wealth and utilising the same for export through the setting up of the needed infrastructure. Exports are not subject to such levies under present rules. The action of the administration does not appear to be in line with the developmental policy of the Government of India. The Union Department of Agriculture/Ministry of Food Processing Industries/ Ministry of Commerce do not seem to be aware of the introduction of the new system, before the issue of the notification by the A&N administration. If the notification relating to the levy of royalties as mentioned above was issued without the approval of the Government of India, it is a matter of considerable urgency for the Central Government to review the position and set right the situation. These reviews are all the more necessary as the levies would prove to be counterproductive and totally against the present policy of the Government of India in respect of development of fisheries. In case the levies are imposed with the unlikely approval of the Central Government then the policies of the Central Government will be in conflict with the notification issued by Andaman and Nicobar administration which totally stands apart when compared to the regulations in force in the various maritime States.

The concerned organisations are no doubt aware of the situation and would take the needed steps to ensure uniformity in the regulations and prevention of unrealistic and totally unreasonable and uneconomic levies slapped by the Andaman and Nicobar administration. The Government of India will have to look into the matter with the attention it deserves and ensure that the notification is withdrawn by the Administration.

Fishing Chimes, Editorial 208: December(i) 1995: Vol. 15, No. 9

# Time for Rechristening MPEDA as FPEDA to Cover Inland Fish Exports

The Marine Product Export Development Authority now presides over not only on exports of marine products, but also on inland water products. The export basket now includes not only farmed marine shrimp but also freshwater prawns which are inland water products. Freshwater fishes which too are inland water products are being increasingly exported. This change in the composition of exports calls for consideration to change the name of the Authority as Fish Products Export Development Authority. In the Indian Fisheries Act the definition of 'Fish' includes not only fish but also all aquatic animals and plants.

A revision of the name is called for, not only because it reflects the fast changing picture of fish products exports, but also because it would pave the way for providing an impetus for the exports of fast increasing inland fish production all over the country. A modification of the name will inculcate interest among the inland fish producers, aided by the Authority and with organisational support by the Authority, to participate in the national endeavour to step up fish products export. The suggested change in the name is all the more necessary in the present situation of possibilities of decline in the exports in the sector in 1995-96 and probably thereafter.

Setting up of frozen and cold storages/IQF facilities at selected points in the inland zone, coupled with promotion of container transport facilities to the nearest major ports would become necessary to enable the inland fish producing/pooling enterprises to participate in the activity. The Authority has to step in to evolve schemes and putting them in motion to give a fillip to the materialisation of the activity. Work in this direction is all the more necessary as inland fish such as major carps, cat fishes, Tilapia etc. now find an increasing demand and acceptance in the Middle-East, South-East Asian countries, in Europe, and even in USA.

Fishing Chimes, Editorial 209: December(ii) 1995: Vol. 15, No. 9

# New Products for Effective Shrimp Pond Management?

M/s WOCKHARDT AQUAVET of Wockhardt Ltd. Dr. Annie Besant Road, Bombay-400 018 is the supplier of 'WOLMID', stated to be a universal pond water disinfectant. According to the company, the application of this product releases bio-degradable chlorine, ensures rapid destruction of micro-organisms, releases active ingredients as per demand and improves yield. The product has no bacterial resistance. It is stated that WOLMID constitutes a solution for white spot disease problem.

K.R. INTERNATIONAL, Madras, claims that their product marketed under the name "NS Series, Super SPO" is effective in the treatment of shrimp farm waters. It is stated that the product is a scientifically blended concentration of selected, adapted and cultured bacterial formulation plus enzymes and special buffers fermented with cereal and mineral substrate for use in shrimp farms. According to them the product removes $H_2S$, $NH_3$, and $NO_2$ from shrimp ponds. It is effective in the breakdown of organic and faecal wastes. It is further claimed that it reduces bottom sludge and increases the availability of nutrients for stable growth. The suppliers also say that the product stimulates growth rate of shrimps, prevents diseases, increases survival rates (and minimises pond bottom acidity). The product, it is stated, has the ability of preventing drastic biochemical changes that will upset the metabolism of shrimps, of increasing FCR and controlling suspended solids because of high organic loading. The company no doubt would be in a position to support their claims by bringing to the notice of the farmers the results actually obtained, to reinforce the features attributed to their product.

Another product for effective shrimp pond management much publicised is '*Epicin*'. It appears as if Environmental Networks, Indonesia has created history in the management of waters of shrimp ponds, by using '*Epicin*', a product, stated to be mirobial based and is environmentally sound. '*Epicin*' is a product developed and introduced by Epicore Networks Inc, Canada with an office in Jakarta, Indonesia.

According to the publicity given, *Epicin* led to achievements of outstanding results in pond water management with impressive results. In two trials at farms where there was a production crash in seven and four year old farms respectively, declining production is stated to have been arrested. Levels of ammonia are stated to have been maintained at less than 0.7 mg/l throughout production cycle and in any case at lesser levels compared to untreated ponds. Levels of other metabolites/nutrients came down. Production levels in Epicin treated ponds increased to an average of 120 per cent of that in virgin ponds and to nearly three times of similarly aged, but untreated ponds. The increase in production is attributed largely to increased growth rate and food consumption of shrimp reared in Epicin-treated ponds, leading to increase in profits, it is claimed. It is further claimed that the product reduces concentrations of not only ammonia, but also nitrite, nitrogen and phosphate.

Another beneficial aspect attributed is that it improves pond water quality and controls pollutants, promotes higher crop yields, reduces grow-out time and extends productive pond life and pond down-time between crop cycles.

The efficacy of the products has been successfully demonstrated in farms of Equador and Mexico as well, it is averred. It may possibly have a role in hatcheries and freshwater fish farms. If the claims are true, the product deserves applicational efforts by the farmers with promotional input by the agencies concerned.

M&M Suppliers of Richmond, Canada, have introduced three products for effective water management. The Company says that the product *Mirror Clear"*, clears the water safely to bind Common Contaminants. Their *Envi-restorer'* a customised bacteria - enzyme combination maintains proper pH and controls organic build-up, it is publicised. They have yet another product *Bountiful Harvest Survival Formula'* which stimulates disease resistance and enhances survivability of shrimp.

Fishing Chimes, Editorial 210: January 1996: Vol. 15, No. 10

# Looking Back at 1995

Entrepreneurs in the fisheries sector shudder at the straights into which the year 1995 transported them uncermoniously. True, the year was bad, but the badness had the effect of chastening the entrepreneurs. The year built up an awareness of proper water management among aqua farmers at a meteoric pace, and about sustainable fishing in a substantial measure among wild fish hunters and on the need for honesty among those rendering services and supplies. So far as positive aspects are concerned, the best one can say is that freshwater culture fishery sector has made impressive progress in various land-locked States as well, particularly in Punjab and Haryana. Integrated farming systems, freshwater prawn seed production and culture had entered the scene in a significant way. Several Fish Farmers Development Agencies had achieved a significant level of functioning. Freshwater pearl production was stated to have made a real breakthrough at Tata Electric Companies at Lonavla in Maharashtra.

Although it could be said that there was a general stagnation in reservoir fishery development, in some of the reservoirs, located in Madhya Pradesh, Tamil Nadu, Karnataka, Kerala, Rajasthan, U.P. and in a few other States, there had been a trend towards progress with certain gradations. Several State Fisheries Departments and the Union Ministry of Agriculture had now veered round to the view that reservoir fishery development was one bastion that they had not yet been able to conquer in any remarkable manner. Efforts in most of the States had been perfunctory in this sector untill now but the situation is bound to change for the better in 1996 and thereafter, owing to compulsions imposed by the Law of the Situation. Reservoirs, *beels* and lakes now constitute the main resource for fisheries development. One bright point that can be listed is the successful effort at reviving the population of several species of mahseer at Lonavla lakes in Maharashtra by Tata Electric Companies under the leadership of late Dr. C. V. Kulkarni and Mr. Ogale.

The worst that had happened was in the brackishwater shrimp culture sector. Continued incidence of white spot disease among farmed shrimp threw the farmers into a state of anxiety, disarray and concern and imposed on them severe losses. Lessons were learnt and farmers started following improved systems of pond preparation and water management as a follow-up to 'crop holiday' adopted by them. Several converted some of the farm ponds into reservoir/sedimentation ponds and this had improved the situation. Most of the farmers reduced stocking rates. Several shrimp hatcheries became operational but seed prices came down.

Further spread of brackishwater shrimp culture ran into serious problems during the year, culminating in a public interest litigation in the supreme court which had spilled over into 1996. This has happened at a time when several farmers set up sedimentation/reservoir ponds at their farms, adopted improved water management techniques and are moving towards semi-closed and closed recircuation systems in order to fully protect the environment. The court appears to be now engaged in sifting the evidence presented by the petitioners and respondents to determine the alleged extent of harm coastal aquaculture is actually inflicting and the stated benefits it is bestowing to the coastal population. It is widely known that coastal aquaculture helps the nation by way of increase in production, foreign exchange earnings, and a remarkable contribution to economy, and providing employment in rural areas and to trained technical personnel.

Mud crab culture continued to hold the attention of the farmers in 1995 too, as one of the alternatives to shrimp farming. Dr. D.E. Babu of Andhra University and scientists of CMFRI had demonstrated technologies of culture of mud crabs. Dr. Babu had also reported significant progress towards setting up hatcheries for producing mud crab juveniles.

Total disorder came into being in the marine sector. The real and self-styled leadership of coastal fishermen was up in arms and continue to be so against deep sea fishing, particularly by foreign vessels, and totally against the provisions in the Maritime Zones Act, which hitherto had withstood the test of time. The Committee set up in this regard under the Chairmanship of Mr. P. Murari embarked upon resolving the matter and the efforts have spilled over into 1996, with the chairman doing his best to reconcile the different points of view, several of which are believed to be not in the interests of the nation, seen in global context, and to channelise the deliberations on proper lines and to ensure that the final recommendations are produced, preferably around end of January 1996, but are not likely to be framed before the ensuing elections, as the issues are intricate and are already rendered complicated.

Fishing Companies which raised money through pub-lic issues had fared badly in 1995 also. Exceptions were Alsa Marine, Rank Aqua and one or two other companies, whose names alone appear in the investor guides in economic dailies.

There is one bright spot in the entire scenario. The Indo-Danish Project at Tadri in Karnataka succeeded in equipping a vessel in 40 ft OAL range for deep sea fishing upto 120 m depth and probably beyond. The resoundingly triumphant operations motivated a large number of vessels to follow the example.

There had been a retardation of exports of marine products in the last three quarters of 1995, leading to the prediction that the exports in 1995-1996 may not be beyond Rs.2500 crores and in any case beyond Rs. 3000 crores. The debacle in the cultured shrimp sector, consequent drop in shrimp export prices and falling demand for brown shrimp are stated to be the reasons for the decline in export earnings.

CMFRI developed technologies of proven value for setting up of Artificial Fish Aggregating Devices. The Institute's efforts for propagating marine pearl culture and edible oyster culture had received a fresh fillip during the year. Farmers' meets and training programmes were organised at various centres of CMFRI in the country. CIFT's technologies for producing chitosan and chitin and for production of surgical sutures from fish guts received international acclaim.

Fishing Chimes, Editorial 211: February 1996: Vol. 15, No. 11

# Recommendations of Murari Committee: 'A Master Stroke'

The Committee set up by the Government of India in February, 1995 to review the deep sea fishing policy is reported to have submitted its report to the Government. The Committee, consisting of Members of Parliament, Senior Officials, Representatives of Fishermen's Organisations and Associations of Deep Sea Fishing Industries, if one goes by reports, has made far reaching recommendations. Of these, it is learnt that there are two valuable recommendations of great import which may alter the course of fisheries development of the country. These are i) A Ministry of Fisheries at the Centre may be set up and ii) A National Fisheries Authority may be established. Some of the other recommendations may invite mixed reactions. One of these seeks cancellation of all permits/ licenses issued to foreign fishing vessels. This recommendation is applicable to all categories of vessels, those relating to chartering, leasing, test fishing and vessels operating under joint ventures as well, be they of Indian flag or otherwise. Another recommendation is that all vessels upto 20 m length must have an earmarked area in the EEZ on the west coast extending upto 100 n.m. from the coastline or upto a depth of 150m whichever is greater and in regard to the east coast 50 nm distance from the coast or 100 m depth, whichever is farther to the reserved area.In this area no vessel of over 50 m length should be allowed to fish. The third important recommendation is to strengthen shore infrastructural facilities.

It appears that there are no tangible recommendations in respect of utilising the fishery potential of deep sea zone of EEZ, although the reported main purpose in setting up the Committee is to review present deep sea fishing policy and make recommendations on development of deep sea fishing in that light. The Committee is reported to have expressed the bald view that in future, in areas open to vessels of over 20m length, all operating joint venture vessels in the EEZ should have atleast a 51 per cent Indian stake in both equity and debt. No other measure is recommended. It is not known whether the intention of the Committee is to relegate development of deep sea fishing to insignificance.

Further, the measures needed to enable traditional and mechanised fishing sector to extend their operations upto a distance of 50 n.m. or 100 nm from the east and west coast respectively seem to have been covered in the recommendations. It is reported to have said that Government may take active steps to make available finances for upgradation of technological skills and equipment used by fishermen. It would no doubt be seen that fishermen require a training support and a mechanism for providing finance to them to acquire fishing units without serious constraints. So far as infrastructure is concerned, the need for stregthening the same is widely recognised and the recommendation has reiterated the obvious.

The recommendations of the Committee, so far as the operations of Indian vessels of over 20 m length are concerned, appear to be biased to the extent that the rights now being traditionally enjoyed by such vessels under national laws and international conventions are virtually denied, ignoring the realities that sea fishing is non-licenseable and the Indian deep sea fishing vessels of over 20 m length were introduced under specific approvals given by the Government, with freedom to conduct fishing in the Indian Exclusive Economic Zone, barring the present zone of operations of traditional boats, which coincides with the national jurisdiction, fishing in which is regulated by the Marine Fishing Regulation Acts introduced by the coastal state Governments. The Maritime Zones of India Act exclusively deals with regulation of fishing by foreign fishing vessels and it has no provision for regulation of fishing by Indian vessels as defined in the Act and this is also ignored. In fact, there is no Indian law at present for enforcing such a regulation and no such law can probably be there in future in the area beyond territorial waters.

The difference between the National jurisdiction and the jurisdiction over E.E.Z. is well known. In the territorial waters, which constitute the national jurisdiction, but part of E.E.Z, there is no obligation whatsoever to permit foreign fishing vessels to operate. In the E.E.Z (excluding territorial waters) the concerned nation has the obligation under international law to grant permissions for fishing by foreign fishing vessels through quotas or otherwise as provided in the Maritime Zones Act of India and in accordance with the international conventions to which India is a party. This obligation, however, may be fully or partly exercised or not exercised at all according to the state of development and traditional rights of other nations. However, neighbouring nations have a right to ask for quotas under international conventions. and we have no data to determine quotas, which is an impediment. Exclusive Economic Zone does not constitute 'National Jurisdiction'. It is carved only for the economic benefit of the nation.

It appears that Government would have to protect the inherent fishing rights and economic interests of deep sea fishing vessels of over 20 m length to fish in Indian

E.E.Z., excluding territorial waters. Traditional and mechanised sectors would always be free to fish within or outside the territorial waters. The stakeholders concerned would have to be assisted by the State and the Central Governments to upgrade their skills through a long term coaching, training and education programmes to enable them to extend their operations progressively towards the outer edge of EEZ, with diversified fishing activities and with provision of the needed infrastructure for them (and others as well),-which should include not only fishing harbours and landing centres, but also processing plants capable of manufacturing not only traditional products but also value-added items, besides a chain of domestic market outlets linked to fishing harbours and other landing centres. It is however laudable that the Committee is reported to have recommended that Governnment should provide duty concession alongside concessional finance for navigational and fishing equipment, which, it is hoped will include fishing vessels as well.

This apart, Government would have also to consider providing upgraded coaching, training and educational facilities for various categories of deep sea fishing operatives. The need for these coaching and training requirements appear to have been scantily touched by the Committee.

The imperative need for development of deep sea fishing industry is covered inadequately in the recommendations, overlooking the fact that in the context of globalisation of economic activities, the development of deep sea fishing, an identified thrust area, is inescapable. This exclusion will remain a conspicuous and harmful one, in the context of international conventions on the subject, including the latest one on High Sea Straddling Stocks, if not rectified. Unless this aspect is covered there will be international and national repercussions to the disadvantage of the nation. There has to be a balanced development of fishing in the entire EEZ. There is, however, rationale in extending fishing activity of vessels upto 20 m length, away from the over-exploited coastal zone, provided it is aimed at diversified fishing.

There seems to be a recommendation that the Fishery Survey of India would have to be technically upgraded for locating fishing grounds for all types of fishes. This recommendations would be welcomed by all in the marine fisheries sector.

The recommendations have not spelt out in which of the areas bilateral and multilateral assistance has to be availed of in the development of the much neglected Deep sea Fishing in Indian waters, which cannot be developed in isolation, particularly when technology in respect of vessel design and construction, introduction of electronic gadgets such as GPS plotters, fish finders etc., and the connected operational technology has to come from outside. At the present stage of knowledge and experience of the nation, without this support, albeit for a limited duration, there cannot be much of progress. It is of critical importance to cover these aspects. Otherwise, illegal fishing by foreign vessels would continue.

No measures are recommended to equip the existing Indian ship building yards for the construction of fishing vessels of larger sizes of various types. This is a clear omission and would have to be looked into by the Government.

The recommendations seem to ignore the need for bringing into being stable and workable arrangements for providing financial assistance to fishing companies on an enduring basis for acquiring deep sea fishing vessels. This would no doubt be looked into by the Government. The fact the financing system adopted by Indian finanncial institutions inexperienced in financing fisheries projects and the consequential situation, should not come in the way of working towards a stable system.

An earlier Committee set up by the Government under the Chairmanship of Mr. Murari recommended a financial package to the sick deep sea fishing industry for its rehabilitation. This aspect does not seem to have been looked into by the present Committee.

The earlier Committee, referred to above, recommended the provision of financial assistance for equipping the existing shrimp trawlers with additional systems of fishing such as tuna long lining. There is no reiterating recommendation on this crucial requirement.

The present Committee also has not said anything on the implications of operations of foreign fishing vessels with or without permission and on the economics of operations of Indian deep sea fishing vessels. This Committee expressed only a perfunctory desire to improve the fishing capabilities of traditional and small mechanised sectors to enable them to achieve appropriate technological upgradation without any mention of means of achieving this.

As per the provisions in section 3 and section 7 of the Territorial waters, continental shelf, EEZ and other Maritime Zones Act 1976 (80 of 1976), national jurisdiction extends over territorial waters, control over which is vested in State Governments. Centre has jurisdiction over the continental shelf (only the sea bottom), apart from deep sea fishing which too is a Central subject. So far as the regulations governing operations of fishing vessels are concerned, Indian vessels are free to conduct fishing in the EEZ without any legal restraint, subject to Marine Fishing Regulations Acts, introduced by coastal State Governments. Legal restrictions and regulatory measures enshrined in the Maritime Zones Act of India are strictly applicable only to foreign vessels and these do not cover Indian vessels, as per the provisions in the Act.

The Committee appears to have overlooked the fact that the suggested legally untenable introduction of legislation with regard to area, and depth are neither reasonable nor workable. If this is done, it will only introduce restrictions over harmonious development of fishing activities and will unleash social tensions. Restrictions on fishing by Indian vessels beyond Indian territorial waters will be against international conventions to which India is a party.

Indian owned vessels operating under joint ventures in Indian EEZ cannot be prevented from operating in the EEZ, under the provisions of Maritime Zones of India Act. It is the unfettered previlege of Indian owned vessels of over 20 m OAL to undertake fishing in the Indian EEZ, with the exception of territorial waters, (without in any way being constrained by 100 nm and 150 nm limits recommended on east and west coast respectively), fishing in which is governed by the provisions of the Marine Fishing Regulation Acts introduced by the various Coastal State Governments as per the guidelines circulated by the Indian Government, as already mentioned.

The reported recommendation that the indigenous deep sea fishing fleet of over 20 m length may be given time to fish in Indian EEZ for 3 years, is invidious and contrary to the existing legislation and contravenes natural justice and the provisions in the Indian Constitution which provides for equal opportunities under Law to Indian citizens and does not allow impediments to flow into interstate and intrastate commerce.

The recommendations, as understood to have been made, to say that vessels of over 20m length shall not fish upto 100metres depth or upto a distance of 100 nm on the east coast and 150 m depth and upto 100 nm distance on the west coast appears to be clearly unlawful, unrealistic and in any case may not be amenable for implementation.

Indian vessels over 20m are fully entitled to fish in EEZ around Andaman and Nicobar Islands and Lakshadweep Islands, without restriction barring the territorial waters around the islands, unless prevented by law.

It is widely and authentically known that the Indian Ocean is the only zone among all oceans with heavy unexploited stocks of Yellowfin tuna. Leading experts confirmed this at length at the recently held international Tuna Conference at Manila. India has the most strategic and enviable position among Indian Ocean countries for exploiting this resource. The Committee should have made a comprehensive policy recommendation on this important subject, which has the potential for contributing substantially to the national economy. The Committee is also silent on the policy options relating to exploitation of Antarctic krill.

In conclusion, it can nevertheless be said that the Committee has given to the marine fisheries sector in a camouflaged fish plate an oceanful of hope in respect of its future. The Committee has indirectly neutralised in a probably artful way through its 'time-serving' recommendations, if they can be so referred to, by recommending the setting up of a separate Union Ministry of Fisheries and also a National Fisheries Authority. These bodies (as and when set up) on the one hand, and the Union Ministry of Law and the Legal Division in the Ministry of External Affairs on the other, would notice in the course of examination the various flaws in the recommendtions *vis-a-vis* provisions in the Constitution and existing national and international laws governing fishing in EEZ. In other words, the two excellent recommendations (as identified above), beneficial to the nation, would eventually neutralise the retarding effects of various other recommendations several of which appear to be short-sighted. This balancing act incorporated in the recommendations appears to be a concealed 'Master-stroke', which can serve as a safety valve in the formulation of future policies by the suggested Union Ministry of Fisheries/National Fisheries Development Authority.

Fishing Chimes, Editorial 212: March 1996: Vol. 15, No. 12

# Training *Vs* Coaching

A large number of fishery executives and farmers are imparted training in various systems of fishery development, mostly at Research and Training Institutes and at field units run by the various Governments. Training through formal courses is given for a fairly long duration, while short duration training programmes for various categories of personnel and commercial field operatives are conducted, mostly by the research institutes both as part of formal courses and also as ad hoc courses at other Governmental units now and then.

Recruitment of candidates for formal training under various courses is done through advertisements. Most of the candidates, who have acquaintance with the subject and some background knowledge, join the courses.

So far as short duration training courses are concerned, these are mostly related to in-service candidates to acquaint them with new technologies, with the objective of enabling them to pass the newly learnt technologies on to end users in the private sector. However, the duration of these training courses are for very short period and often do not cover a full cycle of operations. Further, the candidates so trained, are not often posted, after the training, in a job that will enable them to utilise their new knowledge.

Training courses are also conducted sporadically for the direct benefit of end-users but these are for short durations and are often inadequate.

The remedy is to have a few well equipped Centres of Excellence, where, mostly the end users of the developed technologies are enabled to experience for themselves for an adequate duration the entire cycle of operations in such a way that they can be confident and certain that they have picked up the technology concerned.

In fact, what the candidates, whatever be the category, need, is coaching and not training. What is the essential difference between training and coaching? Coaching has greater depth and completeness and is individually oriented and imparted at a well equipped centre designed for the purpose and with facility of comfortable stay during the coaching period at the Centre itself.

Coaching aims at providing specialised guidance to enable the participants to function effectively at their places of work, after the coaching is over. Coaching differs from training in that the later takes place in an institutional and formal environment through structured programmes to be completed in a time frame. In contrast, coaching is characterised by an intensely individualised and in an informal environment and this permeates through the entire duration of intensive they at the workplace which will have to be a Centre specially set up for such coaching. The functionaries (Coaches) to undertake the coaching would have to obviously receive training or coaching at the concerned research institutes in an intensive manner before the they commence intensive coaching work. The job of a trainer and a coach differs in that while the trainer will tend to teach more towards enabling the candidates to observe and listen, a coach will do much more. He will present himself as an inspiring example of what one has to do to emerge as a person laying standards. He can be compared with 'a Coach' in performance at coaching candidates at games. He will be an all-in-one combining the jobs of providing inspiration and of serving as a task-master. He will be an action-oriented person for the candidates to emulate. A coach should be a picture of confidence and he has to be always in command of the situation. He would have to be intimately one among the candidates to be coached and at the same time one above all of them.

The emerging technologies in the marine fisheries sector such as those related to long lining, jigging, use of electronic equipments etc need coaching on board, but not of the traditional pattern of training on shore at the institutes most of the time. Likewise, coaching has to replace the concept of training in respect of new seed production and aquaculture technologies with special accent on innovative links such as application of inputs and bottom upkeep etc., and in various components of integrated farming in respect of culture fisheries and purposeful use and operation of equipments in marine fishing. The same principle is applicable in respect of processing etc.

Fishing Chimes, Editorial 213: April 1996: Vol. 16, No. 1

# Draft National Fisheries Policy

A draft National Fisheries Policy was placed by the Ministry of Agriculture before the 20th Meeting of the Central Board of Fisheries held in Calcutta on 3rd November, 1995.

The contours of the policy enunciated in the document would no doubt be well received by several. At the same time many would consider the document as an exercise in the nature of continuation of the present policies. It will be difficult to perceive therefrom much that is tangibly new.

It recounts what the Ministry has been telling us often in the past. It is probable that, having known that no tangible breakthrough could be made in sectors such as reservoir fishery development and in the upgradation and regulated development of small scale marine fisheries sector, the Ministry appears to have confined itself to repeating the old formulations. This is understandable as the effect of the policies relating to these sectors of endeavour have not made the anticipated impact. The document no doubt reiterates that fisheries sector continues to be a thrust area. Ostensibly, this sounds well. The expression of intention to treat the fisheries sector as a "thrust" area has by now become routinely repetitive and perfunctory in nature. In the formulation of programmes to place the policy on ground, not much of thrust effort is seen. It is to be hoped that the prevailing impression that the fisheries sector would continue to be a low priority area in actual practice, as is seen by general state of development, would be erased in the Ninth Plan.

One lacuna experienced and frequently pointed out is the progressive and relentless evolution of an uncoordinated development of marine fisheries sector, which is largely due to what one would believe to be an indifferent and perfunctory attitude of the Government towards it. This attitude probably stems from the fact that the fishermen and those engaged in the fishing industry have no political backing and politicians do not have much of a reliance on the vote banks of fishermen and others dependant on the industry.

Efforts to provide a thrust to the fisheries sector would be visible only when the policy declares that suitable mechanisms to provide the thrust would be established so as to give a fillip to the present slow growth of the important schemes of the sector. The draft does not speak of any new policy guidelines to bring about an integrated and coordinated development of the fisheries of the nation. Recommendations galore have been made by several responsible and competent bodies to the Government to constitute a separate Ministry/Union Department of Fisheries,but, for some unknown reason or owing to complacency these appear to have been disregarded. This means that the subjects of small scale marine fisheries sector and fishing harbour development, the high sea or deep sea marine fisheries training programmes would continue to be with the Ministry of Agriculture along with centres such as the Integrated Fisheries Project. The subject of development of deep sea fishing, which is integrally connected with the above mentioned subjects would continue to be with the Ministry of Food Processing Industries. Likewise the efforts to upgrade the capabilities of small scale fishermen to extend their area of operations would continue to suffer because of conflicts and the uncoordinated control over the subject between the Ministry of Agriculture and Food Processing Industries. The future of the fishermen would no doubt continue to be directionless and stagnate unless a well conceived scheme to be administered by one Ministry/Department is set in motion. Then only it would become possible to provide the needed infrastructural facilities and organisational structure to bring about the recommended upgradation of the capabilities of small scale fishermen. Such an upgradation would solve most of the problems now besetting the marine fisheries sector.

Further, the Document does not say anything about the policy to be adopted for providing financial assistance to marine fisheries projects, taking into account the prevailing problems. Now practically no credit flows towards the sector.

The policy does not seek to rectify the present fisheries administrative system in the States upon which the fisheries development of the country largely depends. It is widely known that most of the State sector fisheries departments are headed by officers drawn from the Indian Administrative Service. This policy by itself can probably be justified, notwithstanding the general position that most of these officers have scanty knowledge of fisheries and their development aspects.

This system is anachronistic and not in tune and harmony with the developmental needs. In fact, the services of senior fisheries scientists belonging to the Indian Agricultural Research Service, who form the cream of fishery expertise in the country should be utilised where necessary for directing the activities of the State fisheries departments. As Directors of Fisheries such Officers would be able to perform remarkably well.

Coming back to the Indian Administrative Service, there is no doubt that there are some from the service who take real interest in the development of fisheries. However, by the time they begin to develop a grasp over the subject, they get transferred. The fast turn over of Directors of Fisheries in Tamil Nadu is an example. In the case of some others, postings themselves are seen as stop-gap arrangements. Those who know that they will be in the post for a short duration would be hesitant to take bold decisions and their interest towards the work will have its own limitations. There is a common belief, which may not be correct, that some of the officers from the Indian Administrative Service accept posting as Director of fisheries, nurturing an ambition of going abroad. (Fisheries are one sector where chances of officers to go abroad for study tours etc., are highly rated). The drawbacks of frequent changes in the incumbant holding the post of Director of Fisheries give raise to deviation in approaches and attitudes towards fisheries developmental work. Although the postings of Directors of Fisheries are a State subject, policy guidelines on the subject can be framed and recommended by the Centre. The policy document is silent on this vital aspect on which development process depends. In the same way as the draft policy formulation covers traditional coastal fishing which is a State subject, the system to be followed by the states in regard to posting of Directors of Fisheries, which is also a State subject can come within the draft National Fisheries Policy. Effective interaction between the Central Ministries and State Directors of Fisheries would be possible only when the Directors have depth of knowledge in the subject and experience to back it.

It will no doubt be in the interests of the nation to exploit the renewable resources of the Indian EEZ as observed in the draft document. But this observation is more in the nature of a statement. It does not delineate the policy aspects. This appears to be for the reason that the Ministry of Agriculture and the Ministry of Food Processing Industries prefer to leave this matter vague for reasons probably best known to them. It is very odd that the document does not cover the policy to be adopted in this regard, particularly when the sector is now having conflicts among sections of fishermen and a situation that is impeding the development of marine fisheries beyond territorial waters, which is already stagnating, is looming large. The inbuilt void in regard to this matter in the policy will have to be set right.

The general policy of the Government is to encourage joint ventures. However, so far as marine fisheries are concerned, particularly in regard to deep sea fishing, there appears to be dissidence in respect of application of this policy. What exactly is the present shape of the policy, particularly in regard to deep, sea fishing? The document does not specify this. Without a clear-cut policy the fisheries of our EEZ cannot be developed. The future policy approach regarding bilateral or multilateral relations with other fishing nations is also not covered in the policy.

The policy guidelines do not say much about strengthening the training infrastructure needed either for deep sea fishing or for development of inland fisheries. The observations in the document are too general even for a policy document and are accordingly totally inadequate.

The policy document does not say much on development of cage and pen culture linked to leasing policy in respect of coastal waters for mariculture development which is of crucial importance for the production to go up. A legal frame work is needed for the purpose and the policy is silent in this regard.

Maritime Zones of India Act deals exclusively with regulation and operation of foreign fishing vessels in Indian EEZ. What is the policy of the Government in respect of operation of Indian vessels in Indian EEZ outside territorial waters? Does it have powers in this regard? If so, what is the policy? We draw a blank on this aspect from the draft policy document.

The role of women in fisheries and promoting the same particularly in respect of fish marketing and support in processing plants is not highlighted in the policy. There is also no reference to the policy in providing training to local women to acquire capabilities for working in processing plants in the area. The policy in regard to fisherwomen would have to find a place in the document, so as to ensure homogeneous development of talent in all the coastal states.

It is to be hoped that the Government would convert the draft into a purposeful and a really thrust-oriented policy which would enable the derivation of specific activity-oriented programmes in a homogeneous manner. This would promote fishery industries and the socio-economic lot of fishermen in a significant way.

Fishing Chimes, Editorial 214: May(i) 1996: Vol. 16, No. 2

# Turtle Exclusion Devices

The intrusion of US policies in respect of exports of marine products from India is widening. Until recently the intrusion has been in respect of quality, spearheaded by USFDA. This is justified and cannot be questioned. However, now it has reached the basic level, affecting the capture of shrimps. If turtles are also caught along with shrimp, USA says, it will not accept exports of such shrimps into USA. Exports will be allowed only when the U.S. authorities are satisfied that trawl nets used for catching shrimp are fitted with Turtle Exclusion Devices (TEDs). If they find that TEDs are not fitted to trawls, they will not allow entry of shrimps caught in such trawls into their country.

U.S.A. seem to think that they alone are concerned on the subject. Under provisions of Indian Wildlife Act, Indian Government too is concerned about protecting turtles. Further, Indian fishermen, who are predominently Hindus, avoid taking in turtles on the deck of their boats. Even when caught they put them back into the sea. Thus Indian fishermen avoid catching turtles. Even when caught they put them back into the sea. This is the Way of their fishing life. For Hindus, turtles represent the second incarnation of Lord Vishnu. They have an innate feeling of sin to catch or kill turtles. In fact, what fishermen do is to ceremoniously release turtles back. No doubt, fish too represent to the Hindus the first incarnation of Lord Vishnu, but it is believed that the Lord endowed this phase of his evolutionary existence with enormous fecundity for his recurring existence, which is also necessary to maintain the food chain in the sea and to provide food for humans. The veneration towards turtles stems from an innate urge to ensure that this form of His life, which exists in far lesser numbers, continues.

The socio-religious aspects mentioned above may be construed as irrelevant based on various other grounds. While these aspects can be brushed aside as irrational and unscientific and are not wholly convincing, one cannot ignore the fact that there is no scientific basis established for extending the TED policy for application in Indian waters. There has been no study of the extent of turtle population in Indian waters, the extension of depletion and whether it is lower or more than natural mortality. The system should not be imposed arbitrarily. Such an action would give the impression that a deliberate effort is on to bring down shrimp exports from India to U.S.A.

It may be mentioned here that the strengthening of socio-religious beliefs will be a far more effective means of safeguarding turtle stocks, than introduction of TEDs. Need for conservation of turtles in a country like India can be achieved much faster through the plank of religion, which, by itself discourages overexploitation of not only turtles but other animals as well. Religious beliefs have a stronger hold on Indians much more than regulations. Where there is a law, there are law-breakers. This is the case all over the world. But where there is a religious dictum, there will be very few who will disregard it. The TED regulation can be circumvented and the violation cannot be easily detected. Further, the Government of India is equally concerned, in the same way as several other national Governments, to lend protection to turtles. The really dangerous phase in the life of a turtle is its land-ward migration for laying eggs. The Indian Government deploys Coast Guard and naval personnel to prevent any unlikely incidents of removal of eggs by ignorant persons.

This situation calls for a reiterating appeal to fishermen not to catch turtles and, if caught, to release back the captured turtles as has been the age old practice. This will be very well received by them. A step of this kind coupled with stricter enforcement of provisions of the Wild life Act should be reassuring to US authorities.

The present stipulation is that all trawls should have TEDs attached to them. This will be an economically harsh proposition. TEDs not only exclude turtles but also larger fish in significant numbers. This will totally upset the economics of operations, as catching shrimp alone in trawl nets will not be remunerative. Further, new problems, revolving round objections on what are declared as cultured shrimp are actually capture shrimp caught in trawls without TEDs may arise. Elaborate checking and identification procedures and certification would eventually become necessary and all this would be an expensive and unnecessary exercise.

USA imposed similar restrictions on purse-seiners in regard to incidental catches of dolphins and very harsh penalties are imposed if the regulations are violated. It is learnt that the US Government has now realised the inaptness of the ban, after subjecting the fishermen to so much of hardship, as it is later seen that the capture of dolphins by purse-seiners is not more than 7 per cent of the total weight of catch which is equal to or less than the natural mortality. The imposition of the stipulation that trawl nets should be fitted with TEDs may, in all likelihood, turn out to be a similar 'wrong' step.

Fishing Chimes, Editorial 215: May(ii) 1996: Vol. 16, No. 2

# CIFA Poised to Revolutionise Rohu Quality in Fish Culture

The first revolution in fish culture was ushered in by the erstwhile CIFRI in late 1950s through development and dissemination of technology for induced breeding of major carps. This revolution transformed the complexion of inland fish culture, which was later developed into composite and polyculture systems. After a gap of nearly forty years, the Central Institute of Freshwater Aquaculture is now on the anvil of releasing upgraded strains of seed of Rohu, raised from selective breeding of Rohu. The dedicated work accomplished by the Institute will live in the history of development of inland fish culture in India as a land mark, in the same way as the induced breeding work accomplished in late 1950s became indelible. The work on Rohu is done under a Indian Norwegian Project and in collaboration with ICLARM. The scientists are now monitoring with great satisfaction the improved rate of growth of the new strain of Rohu seed. By the end of 1996, the scientists expect to reap a rich harvest of the improved stain.

Norwegians brought about a major upgradation in cultured Salmon through genetic selection. In accomplishing the Rohu miracle, CIFA teamed up with Norwegian experts under a Indian-Norwegian Project supported by NORAD AKVAFORSK, (Institute of Aquaculture Research, Norway) is the Norwegian counterpart Organisation. This project became operational in 1992 and the first phase came to a close in March 1996. The experiences gained by the scientists of CIFA in implementing the project would be one of crucial advantage in that these experiences will now be of immense value in bringing about 'genetic' upgradation of other commercially important fishes. It should now be possible for CIFA to broadbase and decentralise their efforts by taking up similar work *vis-a-vis* other fishes of commercial value at several centres.

The design of the work undertaken is well conceived and implemented. Dr. S.D. Tripathi, Director of CIFA until the advanced stage of implementation of the project, and a team of scientists under the leadership of Dr. P.V.G.K. Reddy have been responsible from the Indian side for the outstanding results. From the Norwegian side Dr. Bjane Gjarde, Dr. Morten Rye and Dr. Morris put forth utmost effort to give shape and lend support to bring the first phase of the project to fruition. The need for genetic improvement in respect of Rohu arose from the fact that the growth performance of the fish under polyculture is poor in general compared to other species in the culture system.

The work consisted of selecting from the wild rohu stocks collected from various rivers and transported them in live condition to CIFA farm. Care was taken to select only healthy stock, taking into account six different parameters including two relating to diseases. The selection was also done after taking into account detailed studies made on the subject and after finalising methodology, largely in association with Norwegian counterparts and Dr. Eknath of ICLARM, Philippines. Similar work was done by ICLARM, Philippines under the 'Genetic Improvement in Farmed Tilapia (GIFT) Project.

The starting point for genetic improvement, in the present state of available technology is to procure stock of the fish for experimentation from its natural habitat. Accordingly, Rohu seed from the Ganga, the Gomati, the Yamuna, the Sutlej and the Brahmaputra were procured by the project. These were reared separately and stocked in grow-out ponds after marking them, and grown upto brooder size. Suitable mating designs wore drawn up, in a way as to avoid inbreeding. Family selection method by producing fullsib and halfsib families or group was followed. Another mating design followed was diallele design which is a form of mating design suitable particularly for the estimation of both additive and non-additive genetic effects. In this, each strain is crossed with itself and with remaining other strains to produce pure-bred and cross-bred groups.

Fully matured brood fish were generally selected for induced spawning. Fertilised eggs of each fullsib family were kept separately until they were fully swollen and thereafter transferred to outdoor hapas separately.

Before stocking in grow-out ponds each of the fingerlings were tagged with 'Passive Integrated Transponder' (PIT) tags as a marking device for individual family selection. Estimated heritabilities were recorded. A substantive additive genetic variance for growth rate when reared in monoculture was noticed. In polyculture this variance was small, attributed to food competitions. The overall results were found to be remarkable. The institute expects conclusive growth results of various categories of seed produced from selective breeding by end 1996.

The work on selective breeding of fish undertaken by CIFA, is the first ever attempt on the subject in India, Rohu fish being the first one chosen. The final results, which are expected to be very promising, going by the present trends, would undoubtedly pave the way for further work not only on Rohu but on other economic species, leading to increase in quality fish production which has immense potential in various productive directions.

Fishing Chimes, Editorial 216: June 1996: Vol. 16, No. 3

# Developmental Imperatives in Marine Fisheries Sector

Erection of superstructure on a weak foundation has been the hallmark of our marine fishery developmental work.

Whatever the Government may do for marine fisheries development, not much can be achieved on a sustaining basis without having a strong foundation. There is no doubt there is a foundation but it is weak and is allowed to be weak. A multitude of traditional fishermen remain in the same stage where they were. Such a huge human resource, which should be an asset, has been scantily developed.

Education and Training constitute the spring-board for development of fisheries. The inescapable imperativeness of imparting education, coaching and training to fishermen in upgraded systems of fishing, which should start, to be effective, from an impressionable age of their life, still remains unregistered on the minds of the authorities concerned. Fisherboys have to be caught young and inducted into the fisheries education system to enable them to acquire the needed attitudinal changes, awareness, build-up and of knowledge over a period of time to emerge as national assets in off-shore and high sea fishing. Many do not realise that moulding the attitudes takes time and there is no alternative to the hard way of achieving it. Close to half-a-century has passed since independence and all along efforts have been on to improve the lot of fishermen. But the achievements are few and these have not touched the core of fisher population.

Schemes to build housing colonies for fisher families is a good social measure but such schemes alone do not upgrade skills of fishermen. While the bulk of fishermen remains at the traditional level, a few of them who could break through the shackles of tradition are stuck up at mechanised fishing boat level. Is there no way for them to reach higher levels of endeavour? There is. The fishers have to be caught young, around the age of nine, inducted into well organised and well equipped fisheries schools, and given education, training and coaching for a duration of, say, ten years. Such reformed candidates in thousands would constitute the foundation to marine fisheries development. An enduring superstructure can be built only on such a foundation. Government should devise a scheme to lay this foundation by setting up a chain of fisheries schools all along the coastline with the needed infrastructure, support and linkages.

The State Fisheries Departments may be entrusted with the task of setting up of these schools; The curricula and syllabus for the various classes spread over ten years can no doubt be drawn up by chosen experts. The organisational aspects, manner of financial support etc. would require the attention of a National Committee.

Along side the setting up of fisheries schools, an attendant reform needed is to set up a chain of Integrated Coastal Marine Fisheries Development Agencies. All such Agencies in course of time will be able to upgrade the professional life of coastal fishermen. With the help of the Agencies and the fisheries schools in the area, various opportunities of benefit to fisher families will unfold themselves. The varied interactions between fisher families, officials of the Coastal Marine Fisheries Development Agencies and Fisheries Schools, under the umbrella of the organisational mechanism developed, will equip the fishermen with improved means of fishing, processing and marketing and skills thereof. One feature of utmost importance that will emerge in the process is the upgradation of skills of fishermen to extend their area of fishing far beyond the traditional zones.

Readers are aware that over 170 medium-sized trawlers equipped for stern trawling and lateral trawling have been introduced on the east coast in stages, 1978 onwards. Now the number of such trawlers operational are around 90, of which over 12 nos. are operating on the west coast. Most of the owners have not been able to pay the instalments towards loans taken, for some reason or the other. Most of the trawlers are now over fifteen years old. Some of them have become non-operational. The SCICI, on whom the burden of recoveries of loans has fallen, is in a Catch-22 situation. If the trawlers are not taken over for defaulting in payment, the owners will be given an undeserved advantage. If the trawlers are taken over, payments towards insurance and port dues will become a burden to SCICI. If sold, the returns will be of scrap value.

In this background, Government has to decide on the future of these trawlers, taking into account the beneficial aspects of the project and bearing in mind, the undesirable experiences in loan recovery management as lessons. The beneficial aspects are : a) The activity generates trained manpower, employment and incomes; b) It contributes to foreign exchange earnings much more than what was released by the Government for the acquisition of the trawlers; c) The problems involved in the management of such trawlers and the non-suitability of large houses to take part in the activity is highlighted; d) The shore facilities needed for providing safe anchorage, landing and berthing facilities, dry docking and repairing facilities to keep the trawlers continuously operational would become apparent; d) Need for diversification of fishing effort would come out sharply; and e) The problems of financing trawler projects would come to the fore. There may be many more resultant aspects.

The other aspects are: a) Recoveries of loans will be very difficult, unless there is a field organisation closely in touch with the industry, b) Flow of working capital towards running expenses has to be ensured, c) MMD's regulations concerning dry-docking of trawlers need to be modified, consistent with safety so as to ensure continuing operations, d) Traditional fishermen, for want of education and training, had to content themselves by working as deckhands, engine room hands etc., on trawlers; and so on. It is to the credit of Murari Committee to have recommended certain formulae for solving the problem of recovery of loans. One such formula is out-right payment of a value, based on certain criteria. Government would be acting wisely, if they approve of this formula, so as to enable the companies to opt for this or one of the other formulae for one time settlement.

This way Government will get back at least some money through the financing body. It is now time for the Government to launch a second scheme for introducing various types of vessels as estimated and also for equipping the existing trawlers for diversified fishing such as tuna mono-fil longlining. While a large number of countries have taken advantage of adopting this kind of long lining system, Indian Government has not done much in this respect, although it is widely known that Indian Ocean has the largest known resources of tuna, particularly Yellowfin tuna. The second scheme has to be designed in the light of experiences gained in the implementation of the first scheme. Bad experiences will no doubt be taken by the Government as lessons and not a provocation to impede further development. Any inertia in not taking up a second scheme, in whatever form it may be, will have a disastrous effect on the development of fisheries in our EEZ.

Finally, we have to refer again to the often repeated need for the Government to set up a separate Union Ministry or Department of Fisheries, preferably under the Ministry of Agriculture. The sooner the system of having the subjects of training of marine fisheries operatives, setting up of harbours, integrated fishery operations under one ministry (Ministry of Agriculture) and Fishery Surveys and introduction of new fishing vessels under another ministry (Ministry of Food Processing Industries), which calls for close co-ordination and which is found to be difficult, is given up the better it will be. It is to be hoped the new Government would bring the administration of all fisheries subjects at central level under a single umbrella. This will quicken development process. Fishery enterprises and State Governments would then have one place to approach.

Fishing Chimes, Editorial 217: July 1996: Vol. 16, No. 4

# Financing Deep Sea Fishing Industry

India has, as is known, over 2 million sq.m. of Exclusive Economic Zone. It is estimated that the Zone has around 4 million t of fishery potential, of which around 2.7 mt are currently exploited. The deep sea zone, it is estimated, has about 1.2mt of fish stocks. Around 20 deep sea fishing trawlers are currently engaged in trawling mainly for squids in the deep sea zone. (We do not consider the present shrimping fleet of 23 m OAL as deep sea trawlers). In addition there is one tuna longliner operating from the West Coast There are five more tuna long liners, but, according to reports, these are portbound at present In addition it is stated that around 20 foreign deep sea fishing vessels are operating in the Exclusive Economic Zone. The correct position is not, however, known. In this situation, in normal course, commercial banking structure ought to have come forward to extend financial support for introducing a larger number of deep sea fishing vessels. However, chagrined by the present depressing state of, repayments towards loans taken, the commercial banks and other financing institutions continue to be reluctant in the matter of extending financial support to the deep sea fishing industry. There is also not much of enthusism on the part of entrepreneurs to enter the field of deep sea fishing. Another reason for the reluctance of the financiers to extend support is that they are not equipped to deal with the fishing vessels hypothecated to them, and as and when taken over because of defaults in payments. Such vessels would become an additional burden to them as was experienced by the Shipping Credit and Investment Company of India. If matters are allowed to continue to drift as at present, India will be out of global picture so far as commercial deep sea fishing is concerned. This is not good for India which has been playing a major role in international fisheries affairs. There is a notion that the incursion of foreign fishing vessels into our EEZ will be intensified, availing of the advantage of the insignificance of Indian deep sea fleet strength. Government will, therefore, have to review the situation and find a sound solution.

As already stated, there has been a general reluctance on the part of commercial banks and other financing institutions to provide financial assistance to the Indian deep sea fishing industry. Until a few years back financing bodies were well disposed towards extending financial assistance to the deep sea fishing sector. The trend towards a negative attitude in regard to providing the needed funds to the industry set in around 1986. The reason for this reluctance is that most of the beneficiaries who availed of loans have not been making repayments. The reasons put forth by the beneficiaries to explain their inability to make repayments can no doubt be genuine, but the banks feel that unless they receive repayments towards loans given together with interest their lending is counter-productive and their funds get blocked.

In a recent writeup on the subject published in "Professional Fisherman" Mr. Warwick Kneale, Senior Executive, Agribusiness Commonwealth Development Bank, Sydney, Australia, made a serious effort to tell the industry as to what banks look for while lending money to the fishing industries. He conceded that growth and quality control in the fishing industry can contribute heavily to the banking industry's future. Australian banks, he says, are willing to provide funds for small or medium sized fishery entreprises, where the above mentioned aspects are upto the mark. He has rightly pointed out that, as the provision of finance is in the nature of meeting medium and long term requirements, it is essential that the banks are fully satisfied with the system of control and management of the industrial enterprises concerned. This will pave the way for release of funds.

In the Australian situation, the extension of financing facilities to fishing companies is related to fishing quotas for their fleet and also for foreign fleets fishing in their waters and also licencing. In the case of India, unfortunately, there is no system of fixing catch quotas probably for two reasons. One is that Government is not adequately equipped to fix quotas for Indian vessels. Secondly, there is no legal provision, as matters stand, for fixing such quotas for Indian vessels. Regarding licences, these are necessary in India only for the acquisition of vessels over 20m OAL and these have to be obtained from the Ministry of Food processing industries. Licences are not however required for fishing by Indian fishing vessels. Fishing is free for Indian enterprises, subject to provisions in the Marine Fishing Regulation Acts in force in various coastal states. There is no legal provision under Indian Laws for issue of such licences for fishing, although such a provision exists for issuing fishing permits for foreign vessels under Maritime Zones Act, beyond territorial waters on the east coast and 24 n.m from the coast line on the west coast (broadly speaking).

Government of India is reported to have recently banned deep sea fishing off Kerala coast beyond territorial waters for 90 days from July 15,1996. But it is not known under what legal provision this ban has been promulgated.

No one can dispute the contention that the commercial banks would have to lend money to fishing enterprises to enable them to pursue their activities. At the same time a tangible hold on the concerned company's fishing activities, connected financial transactions and the vessel itself may have to be there with the Bank concerned so that there can be a way to recover the money lent, in case of default. Hypothecation of the vessels itself has been found to be not of much utility for the banks for the simple reason that the bankers are not equipped to handle the vessels when taken over in case of default. It is very difficult for them to operate the confiscated vessels on their own. This forces the Banks to sell the confiscated vessels to get back the money lent. As has been the experience of several financing institutions whenever such vessels are auctioned, the general reaction is for them to strive to get the vessels at the possible price which will often be far less than the amount to be recovered by the concerned financing body. Owing to this main constraint commercial banks seem to be desisting from extending financial assistance to the existing or new enterprises.

The Australian system of financing revolves around the hypothecation of the licences and quota allocation orders issued by the competant authority for undertaking fishing. With the fishing licence and quota allocation order hypothecated to them, the Australian banks are in a position to handover these licences to another enterpreneur interested in taking over the activity under mutually acceptable terms and conditions. In the Indian situation there is no such system of issuing fishing and quota licences. It may therefore be necessary to introduce a similar system so as to protect the interests of banking sector, while lending funds to the concerned for operating fishing vessels.

Warwick Kneale says that although in recent years Government regulations (in Australia) have been tightened considerably, yet some new fishing companies have benefited greatly both in terms of quantity of fish landings and by way of protection to price received for individual products. He continues to say that the regulations enhanced the value of the licences which in turn have enabled some, though not all financial institutions, to accept the unregistered charges for the facilities provided. It is further pointed out that, as the commodities prices on international basis are in a way outside the control of the industry they are also certainly outside the control of the banks.

While accepting fishing licences and quota permits as security, Australian banks insist on other securities too as a safety factor, preferring real estate. In addition, in the case of fishing licences a form of charge as a "prescribed interest" is maintained in a register by the authorities. This prevents any licencing transfer without the lending banks permission. This form of security is useful in controlling the enterprises, particularly where the licence has significant value. The other main security is mortgage of fishing vessel, registered under Australian shipping registry which acts in much the same way as mortgage over real estate. Warwick Kneale has explained as to why only a limited number of banks in the field provide loan facilities to the fishing industry in big amounts. This is because of the uniqueness and the historical uncertainty of the industry. In recent years this uncertainty has been mitigated to some extent due to Government regulations and quota systems enforced.

Warwick Kneale has laid great emphasis on industry research as one of the most important areas from lender's point of view. The results of such research, as compiled by bankers, will enable them to decide on providing lending support. He says that what is considered as satisfactory progress in one area need not be the same in another zone. Bankers must have local knowledge which is of paramount importance. It has been pointed out that lending to fishing industry is in many ways not different to any other primary industry, except that it requires substantial knowledge and expertise on the part of lenders, attained by many years of experience.

The main areas of risk which have been identified by the Commonwealth Development Bank, Sydney, through its experience, include fluctuations of marketing demand and supply for the commodity; boat construction cost over-runs, financial structure of entrepreneurs entering the industry under-capitalised, unseasonal weather conditions, poor management causing losses and non-acceptance of insurance claims due to technical requirements etc.

The banker says that several of the areas mentioned above are difficult to control from the lender's perspective. The problems can be mitigated at least partially by excercising close control over progressive payments or through employment of credible surveyors. Control, whether by Government or by self-regulations can only enhance the industry's and the lending body's preparedness to involve themselves. The observations of Warwick Kneale as outlined above can be of some interest to the Indian lending bodies and the concerned Ministry in the Government of India may have to think of introducing a legislation that will enable the Government to frame rules for issuing fishing licences and prescribing quotas to Indian enterprises. The documents concerning these items when pledged to the lending Banks, may bring about a totally revised financing environment for the fishing industry in the country.

Fishing Chimes, Editorial 218: August(i) 1996: Vol. 16, No. 5

# Management of New Technologies

Our Fisheries scientists are the men behind technological developments that take place in the fisheries sector from time to time. The technological developments form the grist for the formulation, refinement and strengthening of the various fisheries developmental schemes. It has to be explicitly realised that without the fisheries scientists the sector will be mostly restricted to social development schemes.

The fisheries scientists devote their time to bring out hidden truths that have the potential to add new dimensions or to refine the existing practices to be of superior support for augmenting fish production. The scientists convert the results obtained by them into technologies which get released to the field for testing viabilities.We tend to take the technological developments in the stride, not being really conscious of the hardwork put in by men behind the development of technologies and those who prepare them for field application.

There are several additions to the technologies in the past several years. On the Inland fish culture sector, several technologies that will fortify the present practices have been developed at CIFA but they have not penetrated adequately to field level. Technologies relating to running water culture systems, carp seed production through multiple spawning, seed production and culture of air breathing fishes, azolla's role in aquaculture, single cell protein as feed supplement, water recirculation systems, and culture of freshwater mussels and using them for producing freshwater pearls and several other technologies have been developed. Various ways for development of reservoir fisheries based on refinements of technologies made by CICFRI and several state Governments keep accumulating. Ways of using abandoned mines, burrow pits, and jute retting tanks for fish culture or seed production are formulated.

So much of knowledge and experience has come to be compiled in respect of culture of freshwater prawns on the one hand and marine shrimps on the other, besides production of seed of these crustaceans.

The nation is now close to the frontiers of mud crab seed production and organised mud crab culture. Evolution of technologies for lobster culture is not far behind.

The scientists of CMFRI have done a splendid job in standardising mariculture technologies relating to marine pearl oysters, marine pearl production, edible oysters, mussels, clams, holothurians etc. The technologies are being extended by CMFRI to the field through workshops, farmers meets etc. The marine capture fishery sector is the single sector where the entrepreneurs themselves tend to add new technologies for adoption.

There must be a convenient and free flowing system for extending the various technologies without over burdening the scientists to perform the job of extension as well. We really do not know how effective the present extension network is, and how effective the Krishi Vigyan Kendras set up by ICAR are in respect of spreading technologies to the field.

We have twelve fisheries colleges in the country. These are the points from where newly developed technologies should move to the field along with the candidates emerging out of the colleges and acting as the carriers of the new technologies to the field. There must be a specific provision in the syllabus of the fisheries graduate and postgraduate courses under the heading "New Technologies". The Research Institutes should make available full details of the technologies developed by them to all the fisheries colleges, advising them to include the technologies in the syllabus, with provision to enable the candidates to gain field knowledge alongside.

The candidates who leave the colleges and get employed in whatever capacity and in whichever fisheries organisation, will be in a position to pass on the knowledge on related technologies to the field and work towards its percolation to the field. With a large number of candidates coming out of fisheries colleges, their service ought to automatically become available for extending new technologies to the field. The successive batches of candidates coming out of fisheries colleges should be the primary vehicles for the spread of new technologies to the field.

Fishing Chimes, Editorial 219: August(ii) 1996: Vol. 16, No.5

# Fishing in Indian EEZ: Soon to be a Major Bonanza to Foreign Interests?

What are the implications of the Government's anticipated formal decision to ban fishing by foreign fishing vessels and to cancel all fishing permits/licenses/ joint venture permissions given to foreign fishing vessels? Does the Maritime Zones of India Act provide for banning of fishing by foreign vessels? or only for mere regulation of fishing by foreign vessels? Does banning come within the definition of regulation? Is banning in consonance of International fishery law? Would banning of foreign fishing in Indian EEZ invite international protest? What will be the new equation that may emerge ? The following other questions may be visualised by several.

What is the real genesis of the much publicised fishermen's agitation to ban fishing in our EEZ by foreign fishing vessels? Who can really be behind the agitation, judged from its features and contours? The fishermen? or their leaders? or foreign fishing interests? What can be the real motive and can this be financial in nature and does this throw any unlikely doubts of a scam?

Can there be any truth that the traditional fishermen, asked about their views, say that they are not really affected? Again, can there be any truth in a view expressed privately in industrial circles that the foreign interests are keen on the imposition of the ban so that they can freely poach, their experience being that, with or without the ban they can operate in Indian EEZ with impunity and with the same results? Do they feel that with a ban they can stand to gain by avoiding payment of 20 per cent of gross proceeds (or a fixed sum) to Indian counterparts and also feel that with or without the ban their problems and solutions thereof are the same? They have probably only to contend with the traditional sector only in addition? Do they think that with a ban the balance of advantage will accrue to them, rather than without? In the background of these questions, can the foreigners be behind the agitation with support possibly from the close and tight invisible circle of Indian interests? And as a result of the expected ban:

1. Will our EEZ become an open ground for foreign fishing vessels to poach? Particularly by Thai fishing vessels in the upper east coast, as has been continuously happening unhindered and as frequently reported by Indian vessels operating in the area for the past several years?
2. Is our Government equipped adequately and has the determination to control poaching by foreign vessels?
3. If not, will the foreign fishing companies be benefitted having the best of both the worlds? No official payments of any kind and good fishing?
4. What may be the new patern that would emerge? Will the presently interested Indian companies operating foreign vessels find a way to indirectly enjoy the benefits as at present ?
5. Will the foreign owners evolve a system of deliberately allowing Indian authorities to apprehend earmarked junk vessels to give the needed satisfaction of having done their job?
6. Will the present agitators be tempted in unimaginable or imaginable forms by the foreign countries, particularly Thailand and Taiwan who know all about Indian situation and Indian fishing waters, to continue their operations in Indian EEZ?
7. Will the damage be less or more once the lease, permit and other systems are removed?
8. Will the corruption, if it is there now, vanish, come down or increase?
9. What would be the plan of action of the Government to develop deep sea fishing in Indian EEZ, in the light of the likely fresh developments? Treat the fish in the zone as wildlife not to be touched or will some plan of securing economic benefits will be introduced?

Many more questions of this kind may arise. Let us hope that Government would evolve an effective system to tackle all the above mentioned and other aspects. Will the fisher leaders start a fresh movement to meet challenges of the offshoots of the change in policy, if their objectives are not fulfilled?

Fishing Chimes, Editorial 220: September 1996: Vol. 16, No. 6

# Indian Fisheries Sector: Stagnation Allover

There is a conspicuous and spreading stagnation in the fisheries sector of India noticeable for the first time after independence. The situation is attributable to the rapidly declining political will, and entrepreneural and banking sector's interest towards the sector.

The main constraint stemming development is the apathy and complacence of the politicians in power towards the sector. The developments in the sector are presently limited in a large measure to the management of crises whenever they erupt. Developmental needs in relation to potentials and resources are by and large neglected. Global integration of the fisheries sector, as in the case of other sectors, on which progress largely depends, is left uncared for. Recent liberalisation measures introduced have little impact on the fisheries sector. Indigenous technological achievements with potential to contribute to the economy of the nation are left in the lurch. There are no indications whatsoever to show that the Government has evaluated the present appaling situation of total stoppage of flow of finance to the sector, and is making efforts at devising a policy to correct the situation with effective measures. Starved of financial support the existing fisheries industries are moving in the reverse direction. The interest on the part of the present and prospective entrepreneurs is waning. The absence of fillip, drive and encouragement from the Government has obviously led to this situation, and this is a reflection of the Government's apathy. Clearly, fisheries sector is not only not a priority sector as envisaged, but is not even a normal sector deserving Government's attention.

The deep sea fishing sector has been reduced to a mess. The decline of activities in the sector started from the time the subject was unrealistically and irrationally chopped off from the Union Department of Agriculture and tagged on the Ministry of Food Processing Industries. This Ministry is now presiding over the capping of the industry with no addition to the deep sea fishing fleet of Indian flag. The indigenous fishing vessel construction industry has been rendered defunct. Exports have come down in 1995-96.

Government have failed to advise the commercial and apex banks on the kind of organisational mechanism they should have in respect of financing deep sea fishing industry and its adjuncts and for recovering loans given to the industry, keeping in view the industry's unique features so different from others. Banks should have been advised to finance batches of resource-specific deep sea fishing vessels at a time, based on appraisals coupled with an effective system of monitoring the operations, and with linkages for sales of catches to approved processing plants and an inbuilt system of recovery of a share of sale proceeds towards loans. Collection from the enterprises concerned, a proportionate amount of the salaries of the monitoring personnel of the financing bodies as determined could have been stipulated in a suitable manner and in instalments along with loan recoveries so as to instill a sense of greater responsibility and a belonging to the system by them. Evolving a system on these lines may have helped in diminishing the scope for evasion of payments. Further a system of submitting monthly returns by the enterprises concerned to the lending Banks would have led to better results alerting Banks when recoveries are going out of hand. Further the system would have mitigated the possible tendency on the part of enterprises to evade payments.

Just because there are meagre recoveries of loans given, it is not correct to detain a developmental activity. A Committee, consisting of experienced professionals set up by the Government will certainly be able to recommend appropriate measures for ushering in a healthy financing system for the sector. No Government worth its salt can neglect the development of fisheries in the EEZ under its charge. Such a neglect will prove to be of benefit to foreign fishing interests. Under charter and other similar schemes 20 per cent of gross proceeds are paid to Indian permit holders by foreign fishing companies concerned. With the anticipated withdrawal of charter and other schemes, foreign interests will expect to gain, arrogating to themselves the right of poaching in a manner that will evade detection, thereby saving the present outflow of 20 per cent out of gross incomes, to the concerned Indian enterprises.

Another gap noticeable is the sealing of the fate of ailing shrimp trawlers. Addition of tuna long lining or other diversification systems to these trawlers was recommended by several experts to improve their performance and incomes. Government has not done anything to follow-up the recommendations, but is sitting tight on the recommendations.

According to reports, the Ministry of Food Processing Industries is working on a new deep sea fishing policy. It is to be hoped that it will prove to be an enduring one, facilitating the utilisation of the fisheries resources of our EEZ in a sustainable way and that it will come out soon, so as to end the present stagnating situation. As matters stand, the conspicuous positive factors in this situation are the reported initiatives taken in the private sector by the Amalgam group under the leadership of Mr. Abraham

J. Tharakan to develop the much neglected domestic fish marketing system by creating a national network coupled with processed marine products manufacture with American and Japanese collaboration; and by Mr. Elias Sait of Alsa to pursue intensification of production of value added products at their plants in Madras and Visakhapatnam with foreign technical collaboration and improving export marketing of diversified products further.

There has been no introduction of a viable system of upgrading the capabilities of Coastal fishermen and channelising their efforts towards securing economic benefits to the nation and to themselves. A well-conceived integrated and coordinated effort is necessary to provide a fillip in this regard. State and Central Governments are apparently apathetic and complacent in this respect too. The lead to remove this lacuna has to come from the Union Ministry of Agriculture.

Our reservoir fisheries are the least developed. Several seminars/workshops made useful and purposeful recommendations, amenable for implementation. What the Union Ministry of Agriculture and the State Governments are planning to do to develop the sector is not known.

Annual fish production from freshwater culture fishery sector too is stagnating. The organisational mechanism available through FFDAs is not being fully tapped to step up fish production from these resources and integrating the same with preservation and marketing infrastructure. The initiative taken by farmers in States such as Haryana and Punjab in adopting culture fish production has not been well availed of by the concerned Governments to lend the needed support to the progressive arid other farmers. Technology for the production of freshwater pearls is reported to have been standardised by CIFA, but very little has been done to take the same to the commercial sector.

In the brackishwater farming sector, neglect of advance planning, in spite of the availability of information on the experiences of countries like Taiwan, Thailand etc., much before the activity was started in India, has led to chaos in the sector and to losses of shrimp crops besides the reverse interest among farmers. There are no visible initiatives at helping the farmers in taking up recirculation systems and other measures to avoid diseases and promoting growth, used successfully in neighbouring Thailand. Further, shrimp hatchery owners are brought to a position of selling seed compulsively at uneconomic prices on account of unprecedented increase in the number of hatcheries all along the coastline, with the authorities being just onlookers. What steps are contemplated to improve the situation are not clear.

Mariculture is full of promise. Reports speak of development of technologies for the culture of various marine shellfish and finfish species by CMFRI. The lack of tangible progress in imparting a spread affect to the technologies linked to formulation of legislation and policies for leasing out identified suitable coastal and offshore waters to entrepreneurs for cage culture or other systems of culture is deplorable.

Introduction of a system of allotting seawaters for culture activities such as cage and pen culture through a legal framework is of paramount urgency, so as to avoid in the future chaos of the kind that erupted in the sector of coastal land aquaculture.

What is apparently now needed is an incisive study of the situation in the various fisheries sectors of endeavour and an intervention to monitor, regulate and develop the activities on sound and viable lines.

Fishing Chimes, Editorial 221: October 1996: Vol. 16, No. 7

# A Legal Framework has to Precede Mariculture Development

CMFRI has developed technologies for culture of pearl oysters, edible oysters, clams, mussels and marine finfish etc. The Institute is striving to extend the technologies to the field. Soon entrepreneural interest in these economically viable technologies will gain momentum. As matters stand, the interest in the application of the developed technologies is oriented towards coastal waters. Soon the activity would spread to offshore waters and protected bays in the form of cage culture. A beginning in offshore cage culture has already been made in Indian waters. A cage culture project with shore support for finfish culture is in progress in Andaman waters.

The Government has to pave the way for the development of Coastal water mariculture as well as offshore water mariculture, which are distinguishable from the presently developing coastal land based shrimp/fish/crab culture activity, commonly referred to as 'Coastal aquaculture'. Without preparatory measures taken at this stage itself, there is a likelihood of chaos, disputes, conflicts and legal battles emerging, detaining the development process as time goes by Public have no right to utilise territorial waters for culture, unless given on lease to individuals or organisations for culture under law by the Government concerned. Unlike terrestrial waters, the legal utilisation of seawater plots for culture by those other than the Government of India is possible only when the plots are given on lease as aforesaid. No ownership rights can be bestowed on individuals, enterprises or organisations as per the present national or international laws. As granting of leases by the designated authority is necessary for utilising the areas for culture, an enabling legislation for the purpose may have to be introduced by the Government of India to empower the state Governments to further legislate and frame rules, for the reason that territorial waters also form part of EEZ, and so the area and the floor underlying the shelf comes under the control of the Centre under the present law. Utilising the powers vested under the suggested legislation State Governments will be able to frame rules for granting leases in territorial water. The subject would have to be obviously examined in detail by the Government of India.

The granting of leases of sea water plots for culture bristles with several issues that need close attention. Either the Government or the applicant has to identify the plots, with coordinates. It will be ideal for the Government to undertake this job, taking into account the distances needed from one plot to the other with reference to the type of culture envisaged, such as culture systems for molluscs, cage culture for finfish etc., and possibilities of allotment of adjacent coastal land for setting up shore infrastructure facilities such as hatcheries and nurseries for seed production for culture in cages. The other aspects that would need consideration are the environmental, ecological, biological and navigational aspects. A comprehensive guidebook containing information on the Authority to be addressed, the form of application which will have provision for providing information on technological and financial support, the applicant's own experience, the fees payable, the number of years for which lease can be given and such other details, the design of the cage systems or rope or other culture systems to be adopted, the species to be cultured etc., may have to be furnished to the Designated Authority, by the prospective applicants. The format of permit to be issued may have to be standardised and included in the guide book. Provision for periodical inspections, submission of periodical reports by the lessees may have to be there in the system.

Unlike landbased aquaculture systems, it may be necessary for the Authority concerned in respect of sea-based cage culture to issue notifications in the Gazette to ascertain any objections on the applications submitted before granting leases for the plots to be alloted. If necessary, hearings on mariculture lease applications may have to be conducted by the competent Authority to ascertain any serious environmental or other objections.

The distances between two cage systems permitted, having a minimum of 2,000ft., may have to be stipulated. A time limit may have to be stipulated for the applicants to set up the cages or other culture systems. There may have to be a stipulation for marking the corners of the boundaries of cage systems conspicuously and for erecting sign boards with adequately large lettering. Lease rentals may need reviews from time to time.

The offshore areas may fall within territorial waters or beyond this zone. In the later case, the Central Government alone may have to issue the permissions to set up cage systems under appropriate legal provisions, taking into account well drawn guidelines and the international conventions and practices and the shore infrastructural support needed including vessels for transportation of materials and men from shore to cage locations and vice versa.

The International Law does not seem to recognise the right of property or right of ownership by any person in areas of continental shelf, or high seas of EEZ. According to the International Law, it is stated that even Nations do not possess these rights, the sovereignty of a Coastal Nation over the territorial water is excercised subject to some specific conditions. The most important rule is to honour the right of innocent passage of foreign vessels.

The International Law does not provide for legal possibilities for a Coastal Nation to grant a person or a company or an organisation the right of property over the teritorial waters or a part of the sea. Leases, according to experts, can however be granted. In other words, the International Law does not seem to impair the right of a Coastal Nation to establish the specific ownership in any part of the territorial waters, provided that such regimes are consistent with International law. Beyond the limits of national jurisdiction, the rights of property and ownership are legally solved only in respect of marine mineral resources. However, this principle is considered by some specialists as applicable to marine living resources too, although the concept is not confirmed by International Law, beyond conceding the rights of utilisation of resources by the nation and the nationals and of conceding the estimated surplus fishery to other nations through quotas. Nevertheless, the setting up of cage systems in the EEZ has to be construed as legal under International Law as several countries have been setting up cage systems or molluscan culture systems in their EEZs.

At this early stage of introducing mariculture, it is of utmost expediency to consider the need for legislative measures to ensure the smooth development of the activity which has enormous potential to contribute to employment and the economic wellbeing of a significant section of society and to add to food production. Above all, the Nation can proudly lift its head by becoming a part of the developed nations who have advanced so much in mariculture, particularly, cage culture.

Fishing Chimes, Editorial 222: November(i) 1996: Vol. 16, No. 8

# Construction of Ponds in Sandy Areas: Biocrete Technology

Vast tracts of sandy areas remain unproductive, particularly along the coastline. Productive utilisation of these areas has been evading a sound solution, while some sandy tracts along the Tuticorin and Nellore coasts are being utilised for shrimp culture, with inferior results.

It is heartening to note that a Dutch Company developed a system christened as *Biocrete System* (Registered) and applied it with success in waste sandy tracts in the southern part of West Java. The technology was first made known to the industry during the Third Fisheries Conference in Singapore in November 1992. It took over six years to develop the technology. Experiments in ponds, constructed based on *Biocrete* technology were continuously monitored by specialists of the Faculty of Fisheries of the Bogor Institute of Agriculture and the Dutch Company PT Triasta Citarate. The technology is now considered as a proven success.

The salient aspects of the Technology were described by K.H. Stroethoff and APHM Hovers of PT Triasta Citarate in a paper published in *INFOFISH International* (Number 5/96). Although the technology was developed for shrimp farming in sandy areas, it appears to be equally applicable for freshwater fish/prawn farming in such areas.

The technology takes its inspiration from ferrocement technology. Instead of iron, non-corrosive natural materials such as bamboo frames and natural fibres are used for reinforcement. The system enables the use of locally available construction materials of natural origin as far as possible and it is found to be economically feasible. In essence, *Biocrete* is a mixture of cement, sand and natural fibres including bamboo frames used for paving sides. It appears that the bottom area can also be similarly sealed where necessary.

The ponds constructed in Java using *Biocrete* technology are rectangular in shape with a maximum area of 3,000 sqm., provided with a central drainage outlet through which water exchange can be undertaken and accumulated sludge drained periodically. A 2.5 cm thick inlay of *Biocrete,* comprising bamboo frames and thin open bamboo sheets, is constructed to reinforce the bunds and central drainage tower.

It is however stated that the permeability can be totally stopped by spreading a membrane to seal off the bottom. Over the membrane a layer of sand is to be placed. This layer will become solid after the pond is filled with water.

This system will facilitate cleaning of the pond bottom during the last 45 days of the shrimp production cycle. This cleaning will tone up the quality of shrimp under culture as tail/gill rot does not occur.

Under the *Biocrete* system the outer faces of the bunds can also be protected with a membrane to form a secondary seal.

One distinct advantage of the system is that the sealing of the bottom and sides of the ponds prevents salinisation of adjacent areas, thereby removing conflicts between adjacent agriculturists and shrimp farmers. Further, ground freshwater reserves also will not get salinised. The operations being sludge-free with periodical cleaning, the need to use antibiotics will get minimised. Feed wastage also will come down for the reason that waste feed will not sink in the sludge at the bottom. pH will improve and will not go up beyond desirable limit, as the system is amenable to ensuring a good and balanced plankton bloom. It is stated that because of improved environmental conditions there will be higher survival rates of stocked material. Further, there will be no need under the system to dry the pond bottom. This means that time lag between harvest and the following stocking with seed will get reduced. The authors clarify that the investment per unit of production from ponds constructed based on *Biocrete* technology will be 30 per cent less than the conventional clay ponds with comparable production.

We strongly endorse that MPEDA may examine the various aspects *of Biocrete* system and encourage farmers to adopt the same with such modifications as needed, as the system appears to be eminently suitable for the utilisation of sandy areas not suitable for agriculture for aquaculture and the system is environmentally safe and ecofriendly.

Fishing Chimes, Editorial 223: November(ii) 1996: Vol. 16, No. 8

# Deliverance from Recurring Cyclone Revages: Coastal Fishermen Deserve a Permanent Protective Mechanism

Nature repeats the same scene of cyclonic havoc several times a year and year after year. It creates a low pressure area, as else where either in the Arabian sea or the Bay of Bengal mostly during monsoon seasons, transforms it into a cyclone of various intensities, often severe, imparts a rage to it. and propels it landwards. Causing severe turbulance along its path it gushes on mercilessly towards the coast, sometimes unleashing tidal waves and impinging initially on the coastline in its path and in the process venting its laden fury and cruelty on the fishermen living along the coast and who are the children of the sea and who depend on it for their livelihood, and eventually destroying their dwellings boats and nets. Nature enacted all this on 6 Nov, 1996 along Central Andhra Pradesh Coast, covering principally East and West Godavari districts.

Cyclones, as already stated, are a recurring occurrence both during southwest and northeast monsoons and the coastal belt bears the first brunt of the devastation that follows. The state Governments publicise warnings of impending cyclones. In practice, these warnings do not reach all fishermen on time. Further, some of the boats may be out at sea when the warnings are received and they are then at the mercy of the sea. Fishermen have their own traditional system of sensing an impending cyclonic weather and relying more on this, they tend to ignore warnings particularly when they do not tally with their own assessments: The system of conveying warning through T. V., Radio, Newspapers and through official machinery have often been ineffective. What is being done now largely amounts to post cyclone reflexes. Enduring precyclone steps to protect the lives and properties of fishermen are, by and large, not either thought of or taken. In this situation, for saving the lives and properties of coastal fishermen, a well conceived permanent mechanism has to be drawn up by the Union Ministry of Agriculture in unison with the Meteorological Department and entrust the same for implementation to the coastal state Governments. This may cost money but it will be far less compared to loss of precious lives and properties and the huge amounts the Governments spend to provide relief after the devastating event.

Protection to the fishermen and their families when a tidal wave generated by a cyclone submerges the village is one aspect that requires attention. Permanent and adequate accommodation for the fisher families will have to be provided at an elevated place in the village or a nearby place so that the families can go there once they notice the signs of an approaching cyclone. This infrastructure can be utilised as training centres or other schools, for making fishing nets, and for community activities during non-cyclone period.

The other aspect is to relay the news of impending storms on time. This can be achieved by equipping all mechanised boats with compasses and VHF sets./SSB transreceivers at state cost or with a high subsidy. Shore stations have to be set up at various points to pass on detailed storm warnings to the boats out at sea indicating the path of the storm, the direction in which the boats at sea should move so as to be away from the eye of the storm. Training centres should be set up at Central points to train all active fishermen on how to use the VHF/SSB sets. A way of establishing similar facility in nonmechanised boats with a power source to activise the units may have to be examined. In the alternative patrol boats will have to be introduced and deployed in areas where nonmechanised boats are conducting fishing to spread the message and guide them. Shore centres manned on a continuing basis to disseminate the information, have to be set up. Establishment of infrastructural facilities on above lines will mitigate the hardships to fishermen. It may not be possible to organise this reform overnight but this should certainly by possible over a period of time. Several fishermen may be uneducated; yet their robust common sense and their ability to absorb skills would certainly counteract the impediment.

We had pleaded earlier in this column that coastal Fishery Development Agencies should be set up at selected points all along the coastline, each headed by a Chief Executive Officer. The Ministry of Agriculture had also informed us that a scheme for setting up such Agencies in the 9th Plan would be considered. It is not known whether a decision has been taken in this regard. If not taken, it is of paramount urgency to set up such Agencies to shoulder this responsibility in addition to other objectives of the proposed Agencies, such as upgradation of skills of fishermen, and also of fisher children through imparting training in aspects of upgraded fishing methods on a long term basis in order to attune them to improved technologies. Fishermen being the nation's sentinels along the coast their role has to be strengthened by providing them with better means of fish production and utilisation and imparting awareness. In conclusion, we have to say that the present general complacency and waking up to extend aid after the damage is done, is to be replaced by an effective system of protection on a long term basis. Fishermen may not be 'green' as agriculturists are, who may claim the major part of assistance extended by the State, but they are 'blue' with blue blood in their veins. Being an untouched class of the society, living in isolation from the mainstream of the society they deserve a well conceived system of protection from ravages of cyclones, apart from a project for upgradation of their skills and provision of sturdy and operable boats and nets of quality for betterment of their lives.

Fishing Chimes, Editorial 224: December(i) 1996: Vol. 16, No. 9

# Control of White Spot Disease among Cultured Tiger Shrimp

***Does Latex of a Swallow Wart (Calotropis gigantia), and also Neem Cake with Methylene blue offer a lasting solution to the problem***

Of late, reassuringly positive results in the control of white spot disease (SEMBV) among tiger shrimps under culture through application of latex of *Calotropis gigantea*, or neem cake with Methylene blue are being reported. The reports on treatment with latex have come from R. P. Raman, Scientist at the Aquaculture Research and Training Centre, Kakinada in A.P., and from Maruti D. Yaligar of Marine Products Export Development Authority, stationed at Karwar in Karnataka, together with Laxminarayana Pai, Aquaculture Consultant, Coondapur, also in Karnataka, and on Neem cake treatment with methylene blue from Madhumita Mukherjee, Deputy Director, Microbiology and Parasitological Research Centre, Captain Bhery, Calcutta, West Bengal.

Raman starts the account of his work by explaining the manner of collection of the latex from the Swallow Wart, *Calotropis gigantea*. He says the latex is collected through incisions or slicing of any part of the plant preferably branches or stem of older plants. The latex that trickles out is collected in a beaker/bottle. A wide mouthed pipette is used for the collection. 2,050 ml of latex is taken from each grown-up plant without destroying it.

Following laboratory experiments, Raman says he conducted two sets of field experiments, one being spray treatment with the latex of the plant diluted with water before it coagulated and another with bath treatment with water-diluted latex of the same plant.

In spray treatment, four litres of latex of *Calotropis* plant was mixed thoroughly with four litres of brackishwater, well before the coagulation stage of the latex. The mixture was sprayed all over the surface of the culture pond, in such a way that the rate of application did not exceed 20 ppm. The treatment was continued on a weekly basis for five consecutive weeks.

Raman says that the results were dramatic. Mortality, which was there in the beginning stopped in 24 hours of first spraying. Moulting, which stopped after the disease outbreak, revived in the second week of the treatment. With subsequent moultings white spots progressively diminished. In four weeks the spots totally disappeared.

In the bath treatment the affected shrimps were transferred into plastic pools of 1,000 litre capacity each. 20 g of the latex diluted with brackishwater in a bucket of 16 litres capacity was added in each of such pools before transferring the afflicted shrimps harvested from the affected pond. The shrimp were left in the pools for 30 minutes for treatment. They were later transferred into a freshly prepared pond of 0.02 ha. The same treatment was continued for five weeks. The pond was stocked in successive batches with 2,000 nos. of PL.

The treated shrimp soon started accepting feed. Moulting began in five to seven days of stocking the treated shrimps in the freshly prepared pond. By 14-16 days there was a significant reduction in white spots, with complete disappearance by the 30th day.

Further, according to reports, work on treating white spot affected shrimp with latex of *Calotropis gigantea* mixed in feed was conducted by Maruti D. Yaligar and Laxminarayana Pai.

The Scientists selected for their experiment (a) pond of 0.27 ha at Adarsh Aqua Farms at Hasargatta in North Canara district in Karnataka, stocked with 50,000 nos. of PL tiger shrimp. Before stocking, immersion treatment was given to PL with *Vibrio* vaccine and gluconC. After netting out the shrimp in the third week of stocking, they were subjected to the same treatment again. During the first two months of culture, the survival rate and growth were good. On the 61st day it was noticed that the feed consumption came down. Examination of the specimens revealed that they were positive to SEMBV (white spot disease). Iodine, Chlorine and antibiotic treatments were given from 62nd to 74th day which resulted in slowing down of mortality, but SEMBV persisted. From 75th day normal rate of feed mixed with 20 ml of latex of *Calotropis gigantea* per kg was given 3 to 6 times a day for fifteen days.

On the 4th day of the latex treatment feed consumption improved. Mortality stopped. Moulting at regular intervals took place. White spots were nevertheless seen externally on the moulted shrimps but the number of spots came down progressively. There was 20 per cent mortality during the third month of culture. After the harvest, samples out of 100 shrimps were taken randomly for examination. 34 shrimps out of the sample were seen to have developed clear white spots. They were nevertheless healthy and the guts were full with feed. The shrimp were in a good condition.

The results of work done at Kakinada and Hasargatta may have to be considered encouraging. Yet they may prompt the raising of a few questions such as stocking density, treatment system etc. Notwithstanding these, there can probably be no doubt that *Calotropis* latex treatment has the potential of preventing and controlling the white spot disease. Unfortunately, the scientists concerned have not come out with full details. For example, pictures of shrimps or their PL before and after treatment and pictures showing the actual treatment have not been released by any of the authors. Efforts to secure them have not yet borne fruit. Yet, it appears that there is substance in the claims of cure.

This is a subject of adequate importance that merits the attention of the Central Institute of Brackishwater Aquaculture and other concerned research organisations to pursue further research to evaluate and follow-up the results reported with a view to standardising the technology of the treatment and developing supporting measures such as raising *Calotropis* plants in sufficient numbers in the farm premises itself to meet the requirements of the farm. If there is confirmation of the leads given, a viable herbicidal alternative to the application of antibiotics would emerge.

## Neem Oil Cake Solution and Methylene Blue Treatment

Madhumita Mukherjee, Scientist, (Deputy Director) heading the Fisheries Research Laboratory at Captain Bhery, Calcutta, West Bengal has reported a breakthrough through use of Neem oil cake solution in water and Methylene blue in the control of the white spot disease in tiger shrimp under culture. She collected 150 specimens from different affected shrimp farms in Midnapore and North and South Paragana districts of West Bengal. The specimens were having, as she says, white spots on carapace. Having noticed infection in them with *Vibrio parahaemolyticus* and also SEMBV, she reports undertaking detailed remedial treatment in 177 farms with an area of 2200 acres that have been affected with white spot disease.

She applied neem oil cake soaked in water for 18 hours and later mixed it in water taken from the ponds, broadcast the same on pond surface at the rate of 34 kg/ha/metre depth in a diluted form once in a week. Methylene blue at 90g/ha/metre depth at fortnightly intervals was also broadcast on the surface mixed with the neem oil cake solution. After application, water exchange was suspended for four days. An additional measure taken was to mix stable Vitamin C at one gram per kg of feed. The feed so fortified was fed to the shrimps at fortnightly intervals during the occasions when neem cake liquid and Methylene blue was applied. On other days vitamin C was not mixed in the feed given, Madhumita says.

Response was noticed, according to her, within two days of application of Methylene blue together with diluted neem oil cake solution. However, the application did not improve the normal frequency of moulting, although the exuvia showed white spots. She further says that shrimps affected with white spots are susceptible to fungal attack as a secondary infection. The neem oil cake solution was found to act as an antifungal agent.

According to Madhumita's observations, Methylene blue is the main agent that becomes active in the presence of neem oil cake solution and serves to cure white spot disease.

The scientist says, the response to the treatment as outlined above led to positive results of the order of 95 per cent.

It appears that it will be in order for apex research institutes to follow-up these results and come up with a technology based on more broadbased experiments in various coastal states. Raman, Maruti, Laxminarayana and Madhumita deserve appreciation and recognition for their contributions and for having led the way to tackle the pernicious problem of incidence of white spot disease among cultured tiger shrimps. No doubt due notice of this will be taken and further work to standardise the technologies will be undertaken by the concerned institutes, based on the leads provided.

Fishing Chimes, Editorial 225: December(ii) 1996: Vol. 16, No. 9

# SC's Order on Selective Closure of Coastal Farms

The judgement of the two member Bench of the Supreme Court delivered on the public interest petition filed by Mr. S. Jagannathan delineates where shrimp farmers stand now. An outstanding feature of the judgement is that small farmers, having their farms 500 m away from the coastline, reckoned from the high tide mark, are not affected by the judgement. Further, farms located away from the banks of creeks etc., do not seem to be affected by the judgement. These categories constitute more or less 80 per cent of the farms along the coastline. The judgement, as per news reports., permits extensive and improved extensive shrimp culture by traditional farmers. Out of the remaining 20 per cent also, several will be away from the 500m HTL mark, if the inlet channels are not taken into account. Whether these channels will be considered as part of a farm is a matter of legal interpretation. Most of the Hatcheries set up so far conform to exemption granted under the Coastal Regulation Zone of the Environmental Protection Act (1986).

The Bench also directed that an Authority may be set up to regulate the activities and this should be welcome. One provision in the judgement is that the farms within 500 m high tide line should be destroyed. Another provision is that wages of those employed at the farms should be paid as compensation equivalent to their salaries of six years. It will be recalled in this connection that these farms were set up with clearances taken earlier from authorities concerned.

There is a view that semi-intensive and intensive shrimp farming, mostly carried out in sea based farms will be affected by the judgement. This is no doubt true. In the present situation and in the spirit of the judgement, it will not be appropriate to carry on these practices any further in any of the farms for known reasons related to disease, environment and ecology. Notwithstanding this aspect, there is a strong case for the Aquaculture Foundation of India and other bodies/farmers to go in appeal for review of the judgement by a full bench of the Supreme Court. In this connection, the following points appear relevant.

1. Water recirculation systems in farms that are ecofriendly have now been found to counter environmental degradation and are therefore becoming a common practice. These are developed specifically to counteract ecological and environmental imbalance. Once water is taken in, no part of it is let out but recirculated within the farm. Further, at the inlets and outlets, systems are developed to ensure absorption of harmful gases such as ammonia and treating the sludge before releasing. Practically no harmful substances reach the water source once the adoption of the systems is made compulsory. This aspect may have to be emphasised before the full bench. Further use of antibiotics has been mostly given up in preference to herbicidal treatments.
2. Almost all farms, large or small, have adopted far reduced stocking rates. Standard for this may have to be stipulated under law. This will prove to be a major feature to protect against formation of excessive sludge imbued with rejected or unconsumed feed which, in any case will receive neutralising treatment before being let out.
3. The hatchery managers regularly monitor various water parameters to ensure proper water management.
4. Proven technology (*Biocrete* technology) to prevent salinasation in adjacent agricultural lands or wells has been developed which can be stipulated for adoption where required.
5. Closure of some of the farms cannot be considered as an anti-social and anti-economic measure although there has been so much of positive and eco-friendly technological advancements. Further, the farms can be used for growing freshwater prawns, crabs, fish, oysters etc., if our scientists recommend this. The culture of these is, however, known to be eco-friendly. Closure of farms may throw thousands of workers drawn from fishing villages who now earn a lucrative living from the farms out of work, but this has to be considered as in the interests of environmental protection.
6. There appears to be reason for some to believe that the present situation arose because of apathy on the part of the coastal State Governments in being passive spectators to the haphazard setting up of the farm. Without a thorough and comprehensive enquiry into this aspect, it does not appear to be clear to several whether a drastic order to close down and destroy the units, which were permitted to be established earlier by the authorities concerned themselves, including Pollution Control Board, would not be a hardship to the owners. Several comments on the validity of NEERI's report on

which the Supreme Court Bench seems to have depended were raised and it is not known how convincingly these were answered. The lapse appears to be mostly on the defendants side. We feel that they could not defend the cause well, particularly when a strong case to prove that their activities border on eco-friendliness exists and they are willing to follow the guidelines already formulated and as amended from time to time. There is however a general feeling that the judgement is a well balanced one which took into account various aspects of the case placed before it.

Fishing Chimes, Editorial 226: January(i) 1997: Vol. 16, No. 10

# Fourth Indian Fisheries Forum

It is reassuring that the Fourth Indian Fisheries Forum held at Cochin from 24–28 November, 1996 attracted the participation of a large number of fisheries scientists, students and technocrats. As many as seventeen sessions on various topics were held. The election of Dr. M. Devaraj as the Chairman of the Indian Fisheries Forum, reflects the wisdom of the members and the approach to continue the tradition of having a scientist of proven track record as head of the forum.

The recommendations of the deliberations would no doubt br released very soon. These will be of great benefit to the various Governments to strengthen their fisheries policies But, as understood, the trend of the reported deliberations do not appear to reflect any optimism in this regard.

What causes concern, however, is the reported common view among several participants that the Forum failed to come up with concrete suggestions or proposals in respect of improvements to fisheries education system and fisheries research and development options. At this juncture when there are declining fish production levels, perfunctoriness in research activities, that too taking place in areas around research centres and the cadaverousness in developmental activities being the order of the day, all those who feel disturbed at the present cheerless situation, would expect well considered recommendations from the Forum. Further, the senior level participants, who are expected to say with vigour what is needed to be done to upgrade the capabilities of traditional or non-traditional fishermen with a view to improving their earning, did not seem to concentrate on this vital area at all. It is not clear what had been said on development of deep sea fishing in our EEZ and in regard to development of reservoir fisheries and mariculture. The Central Board of Fisheries which met in Bangalore recently seem to have done well in surprising contrast some of its earlier meetings.

Scientists from the School of Marine Sciences, Cochin, seem to have brought out a new facet in the marine fishery resource scenario, which is a major contribution. According to them, it was found that productivity in the benthic zone of the sea is four times more than that at the pelagic strata. This revelation coming, as it were, from these scientists would necessitate an upward revision in the marine resources estimates. One feels sad that this crucial revelation was not at all discussed at the Forum. This tantamounts to neglect of this issue of far-reaching implications in national marine fisheries developmental planning.

Ashtamudi lake and its connected backwaters are of vital importance to Kerala State. A couple of scientists seem to have observed at the Forum that the lake is biologically barren and is productive only for a few months. The validity of these observations ought to have tempted several scientists working on various fisheries aspects of the lake to raise questions arising out of such a statement, but none are reported to have been raised. **A situation of this kind at an important Forum in which several experienced scientists participated does not seem to have given much of value addition to the proceedings of the Forum.**

Some of the participants are stated to have noticed improper conflicts among some of the scientists and in between those representing institutes. There was a transparent tendency on the part of scientists (egoistic?) to project themselves and their institutions relegating facts and findings to the rear seat. The emergence of Dr. M. Devaraj, a dedicated scientist, as the Chairman of the Forum would no doubt usher in a new era of endeavour to raise India to a pedestal of eminence in scientific and technological achievements.

It is sad that the Forum could not derive any advantage out of the presence of Dr. Y.S. Yadava, Fisheries Development Commissioner, by projecting fresh policy options and projects for consideration of the Government for inclusion in the ninth plan fisheries schemes; and this happening when Yadava specifically told the gathering that a National Policy on Fisheries was on the anvil.

Fishing Chimes, Editorial 227: January(ii) 1997: Vol. 16, No. 10

# Failure of Fisheries Management Policies: Reasons

In a Write-up entitled 'Management in trouble', Menakhem BenYami gives in the December 1996 issue of World Fishing, a gist of the criticism on the subject contained in the book Crisis in World Fisheries authored by J. Russ Mc Goodwin. The criticism is three pronged: a) the assessment part of the management process is lacking, for the process is based on mathematical models which do not take into account either natural fluctuations (changes in abundance that are not due to fishing) of the fish stocks, or the human component of the fishery system; b) the fact that the theory "the tragedy of the commons", popular among fishery economists and managers, is too simplistic to fit into most fishery situations, and c) Regulation and implementation parts of the management process are lacking. The main reason for this is that they are not understood well and the harvesting sector is not attuned and integrally involved in these aspects. Further, very few social scientists are employed and consulted in the framing of management policies. Mc Goodwin says that the social science of fisheries management are relatively new and, so far, it is an uncoordinated field of study. He observes that social and culture values are important in taking fisheries management decisions. The points mentioned above merit an appraisal by the ICAR and the CMFRI for the reason that formulation of marine fisheries management techniques from time to time springs from the results of fish population studies which are conducted by CMFRI The observation of Mc Goodwin in regard to the importance of integration of social and culture aspects in fish population studies is worth taking notice, the reason being that social and culture background of fishers will have a major impact on the successful implementation of management plans. In fact, it is advisable to include social science of fisheries management as a part of the syllabus of graduate and postgraduate fisheries courses.

Fishing Chimes, Editorial 228: January(iii) 1997: Vol. 16, No. 10

# An Year of Yumult, Depression and Disconent

1996 was a frustrating fishery year; It was interspersed by agitations which upset fishing operations in the Indian EEZ, and also caused set backs in fishing in Indian territorial waters. Rare and recurring cyclones ravaged the coast, the brunt of which was mostly borne by the fishermen who are the setinels of the sea, leaving in their wake deaths of thousands of them, loss of their boats and nets and dwelling houses. Hundreds of boats remained port bound for months together, resulting in retardation in fish landings and consequential hardship among fishermen owing to disastrous fall in their earnings and of those dependent on fishbased activities. Fishermen were pushed into folds of hunger during the periods of disasters.

Fall in capture fish production during the year is not restricted to sea fishing alone. River fish production too continued to come down. Fish production from reservoirs showed no signs of overall improvement in per ha production, although in reservoirs such as Gobind sagar, production went up mostly because of the establishment of silver carp as a fishery but without a marketing infrastructure. Fishery in Mettur Reservoir has registered a survival. Organised development work in certain reservoirs in Madhya Pradesh had brought about an improvement in their fisheries, leading to socio-economic benefits to fishermen. On the other hand, further depletions in hill stream fisheries were recorded in the year in Himachal, U.P. and J&K.

Freshwater fish culture sector, in contrast, continued to perform well as in the past year, with additional areas brought under production, particularly in Haryana, Punjab, U.P., M.P., Bihar and some other States. A significant development in the freshwater sector was the upsurge in interest in freshwater prawn culture in all inland zones, particularly in the Vidharbha area of Maharashtra, in Madhya Pradesh in Rajasthan, in U.P, Punjab, Bihar and Haryana, notwithstanding seed constraints. *Clarias gariepinus* and its hybrids had entered the culture sector. Pioneering experimental work in Madhya Pradesh had paved the way for waste land development for fish production.

The major set back in shrimp culture in 1994 owing to attack of white spot disease that led to fall in production had been reversed with the disease coming under control and production going up, only to be nullified by the effect of cyclones in some of the areas. On shrimp hatcheries in Andhra Pradesh, the Government levied 1 per cent market cess on the seed purchases for exports, while not providing any facilities for marketing. The result was a total strike at all levels, by fishermen, processors and exporters. Processors and exporters had no way other than passing on the burden to producers who were not in a position to take on the levy. The stalemate continued for long, but later the Cess was reduced to 0.5 per cent by the Government.

In order to reduce the financial strain experienced by exporters of marine products, the Ministry of Commerce reduced the cess payable to EIA from 0.5 per cent to 0.3 per cent. The Ministry of Commerce has also waived the import duty on imported aqua feeds, although imported artemia cysts still attracted duty. The burden of heavy sales tax in States such as A.P. continued.

Projects taken up with assistance from World Bank were in disarray in almost all the States, particularly in West Bengal, Orissa and A.P. However, surpris-ingly enough, significant progress in the implementation of foreign aided Cold Water Fishery Projects in J and K took place, despite the unsettled conditions in the state.

On the research front, significant progress in the genetic improvement of Indian major carps with Norwegian assistance took place. FFDA of Tanjavur district in Tamil Nadu demonstrated with success paddy-cum-fish culture with impressive results. Cage culture was introduced in Bakra canal Irrigation system. The other discernible research results are mostly related to management of shrimp culture on aspects such as combating or controlling diseases, particularly white spot disease afflicting tiger shrimps, water recirculation and aeration, sediment removal, pollution control etc. Several advances have also been made in shrimp hatchery management. Considerable work has been done on optimising production from fish ponds.

Tata Electric Companies achieved a major breakthrough in the breeding and conservation of various species of Mahseer in their reservoirs supported by the on land research facilities. They also achieved significant success in the production of freshwater pearls, based on technology developed by Central Institute of Freshwater Aquaculture, Bhubaneswar.

The Central Marine Fisheries Research Institute took the initiative of organising entrepreneur farmer meetings at their various regional centres to disseminate the technologies of mariculture relating to marine pearl oyster culture and pearl production, culture of various shellfish such as mussels and clams and holothurians. The Institute also took up a new project on tissue culture. The institute could also achieve experimental success in producing seed of prime cultivable sea fishes. CMFRI and a voluntary organisation in Orissa, Project Swarajya, developed artificial fish habitats. A new Surimi plant had been set up at Visakhapatnam by S.K. Big Star Co.

Implementation of fishermen's housing projects gained momentum in Tamil Nadu, Karnataka, Kerala and in some other States. There had been notable contributions made by the Central Institute of Fisheries Education, College of Fisheries, Mangalore and College of Fisheries, Panangad, and CIFA in respect of culture of freshwater prawn and its seed production. The faculty members of College of Fisheries, Panangad, Cochin conducted extensive investigational work on the fisheries of Vembanad lake, particularly in regard to production of giant freshwater prawn.

The foundation stone for the new complex of the Central Institute of Fisheries Education Mumbai was laid by the Union Agriculture Minister. MPEDA participated in several International Fisheries Conferences and Fairs, which had imparted publicity and popularisation of Indian products abroad.

A notable achievement was the Antarctica Fishery Exploratory Mission under the leadership of Mr. M.K.R. Nair, Director of Integrated Fisheries Project. Utilising the Krill catches during the expedition, the project could develop various krill products.

There was no progress in deep sea fishing industry. There have been no additions to the Indian deep sea fishing fleet. There were however some efforts at diversification with some of the vessels switching over to cephalopod fishing on the west coast, braving several problems. There were several cases of transgression of licensed foreign fishing vessels into prohibited areas and apprehension of all detected intruding vessels by coastguard. Srilankan Navy apprehended several Indian vessels but released some of them. Pakistan apprehended several Indian vessels, stated to have strayed into their waters. Several Indian fishermen continue to languish in Pakistani Jails.

Several seminars/workshops were held during the year, the major one of which was that of the Fourth Indian Fisheries Forum held in Cochin from 24–28 November, 1996. 21st Meeting of the Central Board of Fisheries was held in Bangalore in early November 1996. The Union Department of Biotechnology, ICAR, and Union Department of Oceanography sanctioned several projects for solving problems be setting fisheries development.

Some of the other events of the year are listed hereunder.

1. Calcutta High Court stayed introduction of exotic species into the country
2. Murari Committee submitted its second report in Feb'96 but no tangible decisions were taken by the Government on recommendations such as rehabilitation measures to revive deep sea fishing industry, promotion of joint ventures, setting up Fishery Authority of India etc.
3. FAO published World Fish catch 1994.
4. Seaweed culture in a shrimp farm was taken up in A. P. with the help of CMFRI.
5. CMFRI standardised technology for clam production.
6. Malpe II stage harbour in Karnataka sanctioned.
7. NCDC assisted Reservoir Fisheries development Project taken up in Karnataka.
8. Biotechnology laboratory started at College of Fisheries, Managalore.
9. NCDC had accorded sanction for an Integrated Marine Fisheries Development Project in Karnataka.
10. Training Programme in intensive fish seed production was conducted for women students at Government Benazeer College, Bhopal in June 1996.
11. Commercial prawn culture was inaugurated in Madhya Pradesh on 2nd July 1996 under Rajiv Gandhi Fisheries Development Mission.
12. ICAR and ICLARM entered into a scientific and technical cooperation agreement.
13. Training programme on Breeding and Hatchery Management of Giant Freshwater Prawn was held at Kakinada by the Prawn Breeding Unit, Kakinada of CIFE from 23.8.96 to 31.8.96.
14. Amalgam Group announced setting up of nationwide cold chain network.
15. EUS struck parts of Kerala again.
16. Training in Magur culture was held at Nagaon in Assam on 24 and 25 Aug1996.
17. Cage and Pen culture was taken up in Indira Gandhi canal area in Rajasthan.
18. Indigenous production technology for production of Limulus Amoebocyte Lysate was developed by Odissi Research Laboratory, Bhubaneswar.
19. *Mystus oar*, giant catfish, found in Chambai river in Rajasthan.
20. ICAR awards to Fisheries Scientists announced.
21. NGRI developed water salinity sensing device.
22. Cyclones and tidal waves devastated Central and South-east coasts of India.
23. New office bearers of Seafood Exporters Association of India, and Association of Indian Fisheries Industries were elected.
24. Aquaculture Exhibition conducted at Darbhanga in Bihar for 27 to 29 Sept., 1996.
25. New fishing harbour at Puthiappa on Malabar coast was commissioned.

## Some New Appointments

☆ Dr. M.R. Sinha took over as Director CIFRI.

☆ Dr. S.A.H. Abidi took over as Director, Central Institute of Fisheries Education (Now Retired)

☆ Dr. S. Ayyappan took over as Director, Central Institute of Freshwater Aquaculture, Bhubaneswar (Now DG, ICAR, New Delhi).

## Seminars/Workshops/Inaugurations

1. National Workshop on Aquaculture Nutrition was held in Bombay by the Indian Aquaculture Society at CIFE Bombay, on 29–30 Dec., 1995.
2. Orissa shrimp Farmers Association's Sunamuhee congress was held at Sunamuhee in January 1996.
3. A workshop on development of Fisheries in reservoirs of M.P, in the co-operative sector was held in Bhopal on 22nd and 23rd Jan., 1996.
4. Multipurpose project on scientific fish farming established by HIFFCO Farmings Ltd., inaugurated at Shitalpur Brindabhan villages in East Champaran District in Bihar on 28 Jan., 1996.
5. Seminar on Small Scale Deep Sea Fisheries Development in Karnataka was held on 31st Jan 1996 at the College of Fisheries, Mangalore.
6. National Symposium on Integrated Coastal zone Management was held at University of Berhampur, Orissa on 25 and 26 Feb., 1996.
7. National Seminar on Diseases in Aquaculture was held at Kakinada (A.P.) on 15th March, 1996.
8. A workshop on 'Marine Fishery Resources on the Upper East Coast' was held at Visakhapatnam on 8th March by Fishery Survey of India.
9. Workshop on 'Peninsular Aquac-ulture conducted at Bangalore on 21 March, 1996.
10. International Workshop on 'Molecular Methods for Rapid detection of Seafood Associated pathogens in Seafoods" conducted in April 1996 at College of Fisheries, Mangalore.
11. Seminar on Prawn farming was held in May 1996 in Bhopal under the auspices of Rajiv Gandhi Mission.
12. Pearl Beach Hatcheries Pvt. Ltd, Visakhapatnam took up marine pearl oyster culture with the help of CMFRL.
13. Workshop on seaweed and pearl oyster culture conducted by CMFRI at Visakhapatnam on 25th May, 1996.
14. National Seminar on Fisheries Education was held by CIFE in Mumbai on 23 and 24 May, 1996.
15. Farmers Meet on clam culture held at Visakhapatnam by CMFRI on 24 July, 1996.
16. Seminar-cum-Workshop on Aquaculture Industry held at Bhubaneswar on 22nd and 23rd Aug., 1996.
17. Symposium on 'Fish Genetics and Biodiversity conservation for sustainable production' held in Lucknow by National Bureau of Fish Genetic Resources' on 26 and 27 Sept., 1996.
18. National Workshop on Fish and Prawn Disease Epizootics and Quarantine Adoption in India held at CIFRI, Barrackpore on 9th October, 1996.
19. A Workshop on TEDs was held at Paradeep from 11th November, 1996 to 14th November, 1996, organised by Department of Fisheries, Orissa and Project Swarajya, Cuttack.
20. Seminar on Probiotics held at Nellore A.P. on Nov. 15, 1996 by Prism Corporation.
21. Fourth Indian Fisheries Forum held in Cochin in November, 1996 from 24–28.

Values of shares of Fishing companies continue to fall in the stock market. fered economic losses owing to fall in seed prices, caused by higher levels of production, and with demand remaining constant or diminishing. The judgement of the Supreme Court towards the fag end of the year ordering demolition of shrimp farms within 500 m of HTL of the sea had caused gloom among farmers and hatchery owners.

The farmers in Chilka and Pulicat lake areas stand totally in gloom because of the order that all farms situated upto 1000 m distance from the lake would have to be destroyed. The owners of farms damaged during cyclones were rendered helpless, with no financial means to renovate their farms. The banking sector had continued their total apathy in respect of extending financial assistance to the sector. Insurance cover to aquacrops had been stopped.

The fall in capture shrimp catches for a period of over 90 days on the east coast had led to severe losses to producers, processors and exporters. On top of these, the fall in international prices of processed shrimps had subjected the exporters to abnormal losses. The constraints imposed by U. S. insisting on use of TEDs in shrimp trawls had contributed to reduction in shrimp and fish catches. The outlook was that earnings from marine products exports might fall. In Kerala, there was stoppage of fishing, processing and exports, because of the stipulation by the Government of Kerala that all exporters should pay 1 per cent of the turn over value to the Kerala fishermen's Welfare Fund. Although the Government had given an assurance that this imposition would be reviewed and as a result the fishing operations and exports may be recommenced, what would ultimately happen was not clear.

Fishing Chimes, Editorial 229: February 1997: Vol. 16, No. 11

# Strategy to Augment Exports in Fisheries Sector

Marine species such as shrimps, fish, crabs, squid and cuttlefishes, be they from capture or culture, are the main base on which Indian exports in the fisheries sector depend. Around 40 per cent of shrimp exports come from culture sector. Attacks of virus disease on cultured shrimps and seasonal fall in captured shrimps and fish have led to a drop in shrimp as well as fish and other catches. These have resulted in a dip in exports in the fisheries sector.

Of late, freshwater prawns (*Macrobrachium rosenbergii and M. malcolmsonii*) whose culture is gaining momentum in a few pockets of the country have been marginally contributing to exports in the fisheries sector, in addition to freshwater fishes. In the inland areas of coastal states and in the inland states of the country, it is estimated that there are 28.55 lakh/ha of tanks and ponds, 20.50 lakh/ha of reservoirs and 7.88 lakh/ha of heels derelict waters. By formulating a workable strategy to bring as much of these areas as possible under freshwater prawn production through monoculture or polyculture and implementing the same through an effective field mechanism a large quantity of freshwater prawns can be produced for augmenting exports. At a modest estimate of production of 200 kg of these prawns per ha (although, over one tonne can be produced per ha) with a target of five lakhs of ha to be covered under the programme (at a rate of one lakh ha per annum) an additional production level of one lakh tonnes in a period of, say, five years, at the rate of 20,000t per annum can be achieved on a cumulative basis, despite the constraint of nonconducive conditions in the northern states in winter months to culture freshwater prawns. Export earnings from 20,000 t, (at US $ 5 per kg calculated on a conservative basis), can be US $ 100 million or Rs. 360 crores. When five lakh ha are developed for freshwater prawn culture, the export earnings per annum can be US $ 5 00 million or Rs. 1800 crores. And this is, not including possible augmentation of freshwater fish exports.

The proposition no doubt bristles with seemingly insurmountable problems and one is apt to brush this aside as wishful thinking. Developmental problems considered much more intractable were got over in the past not only in India but in other countries.

The concept can be translated into practice through the following steps.

1. Identification of strategic places close to coastal belt where brackishwater and freshwater is available and also in States like Rajasthan, Haryana and U.P. where underground saline water is available for setting up and main freshwater prawn hatcheries.
2. Preparation of plans and estimates and financial outlays needed for setting up the hatcheries with seed storage ponds based on capacities decided upon.
3. Specially designing suitable transport or carrier vans for carrying the seed to the interior inland areas of coastal and inland states for storage in feeder or relay seed banks for further distribution among farmers for stocking in their ponds.
4. Locating feeder or relay seed banks/farms as necessary at predetermined distances preferably in areas central to places where there is concentration of ponds around, and along routes carefully worked out leading to Inland States such as Madhya Pradesh, Bihar, Assam etc., as one category, U.P., Haryana, Punjab, Rajasthan etc., as another category, besides coverage of interior parts of coastal states.
5. The net work of seed banks so set up can be used for storage of seed received from the main seed or a nearby relay bank. It can be so organised that the seed received from the main bank or feeder banks or relay banks through the carrier vans can be conveniently supplied to the farmers around. The relay system as envisaged supported by the transport system as mentioned above cuts down distances of one time transportation to the ponds, thereby reducing mortalities on the way which take place during long distance transport from the hatcheries to the pond sites.
6. The capacities of the chain of hatcheries closer to the coast or in Rajasthan, U.P and Haryana have to be designed in such a way that all of them put together will produce adequate quantities to meet the needs of a minimum of five lakh ha of water area.
7. Ice plant, cold and frozen storages and processing plants have to be set up with quality control laboratory support coupled with inspection facility at selected centres in the interior in such a manner that farmers can bring their produce to these plants for sale or the processors can organise a system of collection from the ponds at harvest time. This facility will enable storage, processing and packing of the material for export. There has to be a supporting container supply system equipped for transportation by road or rail with needed

temperature maintenance facility and ability to carry the containers to the nearest port of export.

It is felt and believed that the proposed system although extensive in nature, is within the realm of feasibility. The system, however, calls for the setting up of a separate organisation coupled with an implementable mechanism, to be handled by professionals. Considering the coordinating effort needed, it is desirable that MPEDA or the Union Department of Agriculture invests on the infrastructure and lease out the facilities to private sector enterprises, coupled with a monitoring system. The Centre can consider appointing a Study Group to make recommendations on the system proposed.

The production potential of the tanks and ponds in the areas away from the coast to produce freshwater prawns and exportable varieties of fish is immense. The effectuation of the envisaged concept is no doubt a stupendous and arduous task. However, once the proposition is debated and an acceptable plan of action is evolved and implemented, the benefits, that will accrue to producers, processors, farmers and to the nation by way of additional foreign exchange earning will be vast. Further, the declining trend in fish products exports can not only be arrested and restored to the past level, but can also be augmented substantially. The main aspect is that the entire nation can be brought under fish products export endeavour.

It has to be mentioned here that either preparing synthetic brackishwater of around 12 ppt as required at the hatcheries or transporting saline from the coast to interior for use at hatcheries to be set up or already set up at a few places have been, by and large, found to be not practicable and economically disadvantageous.

Dedicated and devoted work on the lines of the proposed concept, particularly at this time when a) sea shrimp catches have reached a plateau, b) Culture shrimp production is facing several constraints and c) fish product exports are displaying a declining trend, will usher in an era of prosperity to the fish farming community. In this context, an all-out effort to augment production of freshwater prawns and fish, solving the logistic problems of supply of seed, processing of the products and their packing and storage at interior places and taking the products to ports for export, is of utmost expediency.

Fishing Chimes, Editorial 230: March 1997: Vol. 16, No. 12

# Formulation of New Deep sea Fishing Policy: One Possible and Practicable Option

In the present situation, when Indian deep sea fishing is at the crossroads, option before the Government appears to be to introduce around 16 m long vessels equipped with machineries and equipments of higher capacities to fish in farther and deeper waters and to equip coastal fishermen to operate such vessels through appropriate training and certification.

There were two major schemes implemented by the Government of India for the promotion of deep sea fishing in Indian waters in the past over two decades. One of these related to introduction of deep sea fishing vessels in the range of 23-27m OAL and the other related to granting permissions to Indian enterprises for operating foreign fishing vessels on charter/lease and under joint ventures and also for test fishing.

Regarding the results of the scheme for the introduction of deep sea fishing vessels there have been two negative and several positive results. The negative ones are: i) there are heavy overdues from the enterprises which took loans for the acquisition of vessels from the erstwhile Shipping Development Fund Committee and later from the Shipping Credit and Investment Company of India, and ii) there has been a total stoppage of flow of finance for the acquisition of fishing vessels.

On the positive side there are several points to enumerate. Some of these are i) Development of infrastructure by way of major fishing harbours coupled with shore facilities for providing various supplies and services for repairs and maintenance of vessels and supplies of spare parts, ii) formation of a cadre of competent trained personnel for manning vessels, iii) significant contribution to exports of marine products, iv) Imparting capabilities to Indian shipyards to build larger fishing vessels, v) location of new fishing grounds not known earlier, vi) Inclusion of a factory vessel to the Indian fishing fleet, vii) ushering in the era of diversified fishing effort and above all, viii) securing a place to India among the galaxy of deep sea fishing nations.

The negative aspects enumerated above are attributable to the apathy of the Government and the financing institutions.

Having given the money it was presumed by the financing bodies that the repayments would come automatically. No well-designed machinery was set in motion to ensure recoveries and to develop a system under which the buyers, mostly processors, would be obliged to pay to the financing institutions a part of amount they would be paying to the loanee fishing vessel owners, through tripartite agreements.

Deep sea fishing is a new sector to the country. The birth of any new sector will involve a sacrifice and pangs. It is to be deemed that this sacrifice was made by the Government through financing institutions in a good cause *i.e.*, the development of deep sea fishing industry. Government may have lost around Rs. 150 crores but with this much outlay Government could revolutionise fishing effort in farther waters of the EEZ and in setting up fishing harbours, training men for manning the vessels and in other connected spheres as enumerated above. The achievements are in the nature of a breakthrough and call for an expression of satisfaction by the Government. Instead, in Governmental circles there appears to be a feeling of disenchantment arising out of non-recoveries of loans. Since late 1970s the deep sea fishing sector contributed export earnings in foreign exchange several times more than the loans the Government could not recover. Over years, diversification in fishing effort for deep sea lobsters, deep sea shrimps and cephalopods, all export-oriented, has come about mostly because of the Government's initiative. Instead of being proud of what it has done, the Governmental circles display dissatisfaction where it is not really called for. Government broke the egg to usher in the omelette of deep sea fishing, although with unpleasant reactions from the fisher leaders.

The schemes pertaining to chartering and leasing of foreign fishing vessels, obtaining foreign fishing vessels for test fishing, and permitting joint ventures for deep sea fishing are the wisest contributions of the Government to the nation. Some companies are no doubt benefited, but this has to be expected. After all, companies work for financial gains. What has been gained or not gained or the depredation of the resources believed to have been caused because of the operations of foreign vessels would have to be treated now as a closed chapter, as there is no other option an unavoidable development to be.

Now that the availability of resources such as cephalopods has been established and Indian enterprises have been able to exploit the resources, Government has taken the pragmatic decision at this stage to continue the schemes. The only thorn in the flesh is the question of joint ventures between Indian companies and foreign enterprises, which are stated to have legal implications. Some doubt the genuineness of these joint ventures. They believe that these are an inverted system of charter arrangement. Government has now agreed to cancel all charters and leases in force. Normally the Indian companies concerned should have no objection to these cancellations as by now the Indian counterpart crew must have picked up the needed expertise. Abnormally, however, joint ventures being allegedly inverted charter arrangements with the vessels flying Indian flags, the problem of paying the balance of the price of the vessels by the Indian companies to foreign suppliers, if any, would arise. The problem nevertheless should be amenable to settlement, considering the alleged involved arrangement in arriving at joint venture arrangements. The foreigners who 'sold' the vessels to the Indian companies can be asked to take back the vessels and the joint ventures can thus be closed. The foreign partners will understand the logical end.

Looking back, the demands of the National Fisheries Action Committee Against the Joint Ventures calling upon the Government to forthwith stop the operation of foreign fishing vessels in Indian waters appear to be justified. The Government have been referring ostensibly to some repercussions or the other all along probably only to prolong the process of implementing the contents of their own statement in parliament that all licenses issued would be cancelled. It is now a comforting news that the Minister for Food Processing Industries had recently reaffirmed this position, reacting to the harbour blockades organised by the NFACAJV at Mumbai, Kandla and Visakhapatnam. The NFACAJV has however no full faith in the re-assertion of the Minister. To neutralise this, it will be wise on the part of the Government to categorically act on this reassertion. Dilation in the matter would only throw doubts on the Government's real intentions. These can be interpreted as lending protection either to Indian or foreign interests involved.

The NFACAJV deserves full credit for their determined agitation to rid the Indian waters from the insurgence of such of the foreign vessels, which are being allowed to continue to operate, despite assurances to cancel their permits, whether current or otherwise. The foreign vessels may have proved the existence of unexploited resources, but several feel that the operations proved to be more harmful to national interests than imparting full benefits as anticipated. Apart from alleged malpractices in the invoicing of exported products, the progressive depletion of resources will take a long time to regenerate. The scheme to permit operation of foreign fishing vessels introduced with good intentions will have to be therefore construed as having now turned out to be harmful to national interests, with the resources covered by foreign interests, with the benefits flowing to them in a large measure, and a small part, with marginal investments, flowing into the coffers of Indian entrepreneurs, who, as it turned out, became mere middlemen. The small fishermen suffered because of the alleged intrusion of the foreign vessels into the coastal waters, often for exploiting of the resources available in concentration, and which are the rightful stocks for capture by small scale fishermen.

Thanks to the agitation, legal fishing by foreign vessels will no longer be there. Poaching may continue and it is to be hoped that Government will tighten measures to prevent the menace of poaching.

With our seas to be soon rid of the menace of foreign vessels, legally (which some say will be welcomed by foreign interests, as they can revert back to poaching avoiding apprehension and also saving the money they now pay to their Indian partners) the problem will be one of having a sound deep sea fishing policy, oriented around operation of Indian vessels in the deep sea zone. Organising Indian-owned fishing fleet to replace the foreign vessels bristles with several problems. Larger houses have scant interest in entering into this line as their management culture is not compared with fishing activities. Even if they have interest, it will not be in consonance with the interests of the small fishermen. Small and medium level Indian enterprises will not be able to step in mainly owing to problems of investments. The commercial banking structure is averse to extending credit facilities to deep sea fishing sector. At the same time, the EEZ cannot be left unexploited. Such a situation will not only openly invite clandestine fishing by foreign vessels, but also to natural mortalities of whatever fish stocks that now remain.

The strategy to fill the vacuum is linked to the evolution of a sound deep sea fishing policy revolving round the upgradation of the capabilities of small scale fishermen, who have the inherent skills but do not have adequate support to develop themselves and to acquire vessels having the capability to operate in farther and deeper waters. There has to be therefore a scheme, which can be albeit implemented on a long term basis.

At present small scale fishermen operate vessels upto around 16m OAL,. These vessels, however, are now ill-equipped to operate in deeper waters. The first step under the new policy is therefore to redesign such vessels, providing them with engines of higher horse power, installing winches of greater capacity and bollard pull, providing them with refrigerated holds, installing generators of higher capacity, upgrading hydraulics, providing nets of improved design and equipping them with modem electronics. Available space will be adequate for improvements. So strengthened, even 16 m OAL vessels can perform well same way as larger vessels. Foreign vessels are large in size as they have to operate as distant water vessels. We do not need such large vessels as the operational distances are nearer. 16 m long vessels may not have high endurance as the foreign vessels have but this is not necessary, as they can return to port as often as necessary, once the holds are full with fish.

The second step relates to finance. Small fishermen will not be able to invest. Government should institute a programme for building such vessels of upgraded design and give them on lease to trained fishermen under predetermined terms and conditions. An Organisation to select the parties, entrusting the vessels to them and to closely monitor the operations of the vessels will have to be set up. This Organisation must have a well-knit system that covers the operators and the identified buyers of the products to whom the operators will sell the catches, who would have to be made to commit themselves to pay a reasonable part of the sale proceeds to the organisation towards the lease amount, which will be allowed to accumulate into equated instalments to be recovered towards the cost of the vessels. Once the cost of the vessels is fully recovered by way of lease amounts their actual ownership can be transferred to the operating owners. There has to be an arrangement for the Organisation to extend working capital facilities, also recoverable out of sale proceeds, voyage after voyage.

The third, and a simultaneous step to be taken is to organise training programmes for the upgradation of the set up. Another step needed is to provide shore fish storage facilities at identified ports at State cost to enable the fishermen to store their catches and if necessary to process and export their catches, in case the buyers do not concede payment of just prices or to hold the catches until such reasonable time needed to secure better returns.

It will be desirable to seek bilateral assistance from an Organisation in a developed country to develop and implement a project of this nature, should it be necessary.

The implementation of such a project will promote the capabilities of our coastal fishermen, and equip them to exploit the resources of our Exclusive Economic Zone, without the demeaning need to depend on foreign vessel operations to exploit the resources.

*Fishing Chimes* has advocated several times in the past to set up Marine Fishery Development Agencies along the coastline to serve as Organisations to achieve the abovementioned objectives. Such agencies, at field level can be set up for the implementation of the activities in an integrated manner. These Agencies can be brought under the fold of the National Level Organisation, as suggested above and which can be in overall charge of the implementation of the suggested new deep sea fishing policy.

Government should stop thinking in terms of developing deep sea fishing in our EEZ starting from the top. It has to start from the base with the development of the capabilities of small scale operatives. Then only there will be ah enduring development, although it may take time.

Fishing Chimes, Editorial 231: April 1997: Vol. 17, No. 1

# Too Many Technologies: Flow Towards Endusers Too Slow Remedy

## *Integrated/Auxiliary Technology Development projects*

The various National Fisheries Research Institutes have the distinction of evolving several viable, proven and innovative technologies. These proved their potential to augment national fish production substantially, when effectively adopted by the end-users. The scientists at the institutes are talented; Like their seniors, the recent entrants are men of proven worth, selected as the cream out of those who come out triumphantly through successive competitive examinations. Under the direction of their seniors, they evolve technologies for increasing fish production and utilisation of the produce for domestic consumption as well as far exports, based on scientific results.

A few epoch-making technologies such as induced breeding of carps, quality seed production, composite culture and polyculture evolved in the past decades, have taken roots, in spite of the inherent weaknesses in the prevailing extension systems. The sheer sway caused by the economic worth of the developed technologies, enabled the gravitation of the technologies to the target groups, in spite of the weaknesses in the extension mechanisms available to the institutes.

Scores of tenable technologies have been evolved in recent years too. However, according to those who have experience of the limitations that the institutes have, most of these could not as yet reach the end users effectively owing to a few problems in the extension systems. They point out that one problem is that the efforts at extension are dissipative, perfunctorly spread. The technologies first move into the hands of extension wings of the Institutes wherever they exist. From there they mostly move on into the laps of field functionaries of the departments, some of whom are handicapped for want of financial and organisational support from their superiors. Several of them are also unjustify dubbed to be deficient, owing to lack of adequate indepth knowledge. Consequently, it is believed that they are unable to effectively pass on the technologies to the farmers or endusers most of whom are adept at evaluating the capabilities of the men earmarked for the purpose. This situation calls for direct contacts between scientists and endusers, besides providing training to officials of State Fisheries Departments, which will be of great help. The endusers being men of robust common sense and having a sound base of field experience, are equipped to judge the capabilities and competence of the roving men charged with the task of extension. The introduction of majority of the technologies so assiduously developed by the scientists largely fail to click at field level, consequent to the weakness in the strategy of extension.

Yet another weakness, as already hinted, is the absence of an organised interface between the scientists and endusers. There are alleged instances where the endusers visit the research institutes to secure first hand information on certain specific technologies but return back sometime with disappointment, unable to establish validity but one aspect emphasised is that the visits of the endusers to the institutes tend to provide only a surface exposure to the field work shown to them at the units of Institutes concerned. This makes them either curious to learn more in depth or leave the matter severely alone. At this point, disappointment starts and sometimes some of the visitors start nurturing doubts about the efficacy of the technologies imagining various reasons, which of course is an undeserved way of reacting.

Developing technologies is an arduous and challenging job. As is known, the job that entails a greater challenge as the part relating to field extension of the technologies developed, which calls for a comprehensive strategy, and which, when implemented in the field, may not prove to be effective. No doubt the institutes are more conscious about this than anyone else but the real test is to know to what extent the technologies penetrate into the field.

The transmission of technologies in the initial phase must be first hand and take place in the field, outside the permises of the institute concerned. The brand of scientists who develop the technologies have no alternative other than moving to selected field locations of the endusers and be with them as many times and as long as necessary till the technologies are picked up by them, but situated as they are, scientists will not be able to spare the needed time for the purpose. There are, however, several Workshops and Seminars conducted by the institutes to convey lot of information, but endusers who attend such events are often heard to comment on their ineffectiveness to educated them.

The main Job of a scientist is to work out technologies for application or for other purposes with results of scientific work as the base. It will be too much to ask them to extend the technologies to the field to the detriment of their concentration on scientific work, If they divert their time to extension work. Their scientific pursuits will suffer, unless they are allowed to stop lab work, until they achieve the spread of a technology developed by them in the field.

In this situation, one strategy that can be adopted for transferring proven technologies to the field is for the ICAR to set up specialised commercially-oriented technology transmission units under special projects to serve as centres to transmit technologies with professionals at the helm. These units must have facilities, in the same way as CIFE's field units have for the stay of a manageable number of endusers for one cycle of operations. These endusers have to be fully involved in the operations in such a way that they can pick up the technologies in full, and they will come to know what all they need to adopt the technology concerned in their units. Conditioning or acclimatising the endusers for the use of technologies, is only one aspect. This has to be followed up religiously by frequent visits by the scientists to the farms or units of the endusers to monitor the progress and offer advice/suggestiom for adopting corrective measures where needed. In this context, the setting up of as many field- oriented auxiliary units as necessary, technology wise; or group of technologies-wise or associated with specific systems. Institutes by themselves are seen to be not inherently equipped, to attend to, as already stated, extension aspects. It is for this reason they have to concentrate on their own scientific work, which is their primary responsibility. On this premise only, it has been suggested above that there must be special extension field projects. These can be Centres of excellence for transfer of technologies to selected farmers/entrepreneurs.

## Farmers-Entrepreneurs-Scientists Meets

seminars and workshops as conducted now by the institutes are no doubt traditional or modified traditional devices to promote awareness among entrepreneurs and endusers and motivate them, but the gap between the outcome of these gatherings and field application of the results thereof tends to remain largely uncovered. We are conscious that the institutes are fully aware of the organisational shortcomings which they have to endure and await to derive satisfaction that the technologies they have evolved get established in the field. There are too many technologies; new or improved ones, and aiming at transferring them all to the field but without adequate field infrastructural and organisational support is not possible. The absence of the support would dilute the focus aspect of the efforts.

It is appropriate here to list out some of the technologies developed whose extension to the field has yet to gain momentum. CMFRI has evolved standardised technologies relating to mariculture of several economically important aquatic animals, prominent among which are pearl oysters, mussels, clams, and holothurians etc. The scientists at the Institute also worked on mud crab fattening and shrimp hatchery technology. A standardised technology for raising pearl oysters and pearls therefrom form a land- based pond system, besides technologies of production of seaweeds and products of value therefrom such as agar agar, alginic acid, carrageenan etc., have been made known. Sea ranching techniques have also been developed. CIFT has to its credit the development of technologies for the production of Chitin, Chitosan, gut sutures etc., besides several marine products processing technologies, vessel and net designs etc, CICFRI has accomplished so much in the field of development of sustainable fishery systems in rivers, reservoirs, *beels*, and such other areas, besides land snail culture, cage culture, hilsa seed raising, hilsa ranching etc. CIFA has several newly evolved technologies to its credit. These include super-intensive fish culture, Azolla culture for biofertilisation, sewage treatment system through aquaculture, technology for culture of freshwater mussels and production of freshwater pearls from mussels, freshwater prawn culture and freshwater prawn seed production, genetic upgradation of major carps, fish culture in waste waters, cage culture, production of a specialised nutritive feed CIFACA for major carps at an economical price, multiple breeding of carps, gamete cryopreservation, breeding and hatchery management of cat fishes and their culture, production of sterile triploid grass carps, utilisation of biogas slurry for fertilisation, formulation of an antidote, CIFAX, for control of EUS, etc. CIBA has made significant contributions, one of which was feed formulation for shrimps under culture, technology for crab fattening, shrimp hatchery technology etc., The Institute also formulated guidelines for brackishwater shrimp and fish culture, covering pond management etc.

There are no doubt initiatives taken to extend the technologies for adoption by the endusers but with a low percentage of successes owing to organisational and strategy constraints. By any reckoning, the development of so many technologies, as partly listed above is an unparalleled achievement by the Indian fisheries research institutes. All these are no doubt important to serve as tools to boost up production but they need extension effort. The available organisational dimensions and extension mechanisms are so inadequate that they do not permit the scientists to concentrate on the extension aspects as well.

As already stated, the system of establishing auxiliary units, specially meant for field extension work, possibly with support from Union Department of Biotechnology, which extends assistance for such endeavours, has to be developed. The DBT should not content itself with its present role of ensuring the development of technologies. The assistance should go farther, so as to ensure that the technologies reach the end users, which is the logical end. Integration between development of technologies and their extension through Centres of Excellence should be presented as a package before DBT issues sanctions for technology development projects. This approach is particularly necessary for all new projects to be sanctioned by DBT. For projects already sanctioned, the integration of extension component should be included. A system for sharing the additional capital costs for providing field infrastructure between ICAR/other organisations and DBT can be evolved. One thing that appears clear is that without the innovation of providing exclusive field infrastructure to transfer technology to enduser, enduring development will largely prove to be futile. Technologies are not developed to be kept in the glass cases of the Institutes in the form of publications and papers presented at seminars etc. Further, the scientists at the auxiliary extension units or Centres of Excellence should not expect the trained end-users to meet them frequently at their Centres, but the scientists should be meeting the trained end-users often at their units, so as to monitor the activities atleast for a reasonable duration in the end-user's domain. Considering the arduous nature of the work, ICAR/DBT wilt have to choose not more than a few technologies of priority at a time for imparting training and to ensure flow of the technologies to the field. These field centres will then develop into Centres of Excellence and serve as focal points for providing spread affect. Once the job of disseminating selected technologies on priority basis is achieved, a similar work on extending other technologies, as per determined priorities can be taken up subsequently in stages. Taking too many of the technologies on the plate of extension will dilute the efforts and the results.

The need for field oriented extension mechanism is far greater in respect of aquaculture of various animals, particularly, crabs, freshwater prawns, pearl oysters and mussels and its their components such as hatcheries, seed raising, pond management, disease control etc.

Unfortunately there are no well established and professionally managed farmers, Associations at various levels, primary, secondary and tertiary, barring a few here and there. There are cooperative bodies at various levels but most of them are not professionally managed from technological angle. They have not played any significantly effective role in taking the technologies to the field but by strengthening these bodies this might become possible. The need is therefore for such organisations to emerge not only to voice the needs of the farmers but also to act as a medium to liaise with the scientists and organise and ensure the transfer of technologies to endusers and the visits of the scientists to the users in rotation to help them in monitoring the implementation of the technologies. Very encouraging field results have already been obtained in respect of intensive fish culture through fattening in certain States without much of extension effort, literature or film shows, but based on the sheer force of economic advantages of the technologies and help of knowledgeable technologists, and with the farmers improving upon the technologists on their own. This only proves that field is central for the convergence of all related support for securing results and to serve as the focal points for further spread of technologies.

Fishing Chimes, Editorial 232: May(i) 1997: Vol. 17, No. 2

# State Fisheries Corporations

Several State Governments set up fisheries corporations, with the main objectives of I) undertaking commercially-oriented integrated fishery development activities, to show the way to the private sector the manner of utilising the available fisheries potential in a sustainable manner for additional fish production and earning profits, and 2) to promote socio-economic development of traditional fishermen. (A few State Governments such as Bihar and Orissa also set up Fish seed corporations for implementing a World Bank Project for constructing and operating fish hatchery farms, finance for which was channelised through NABARD. The project was implemented in the States of Bihar, Orissa, W.B., U.P and M.P. The seed corporations have since been wound up).

Both the above mentioned objectives turned out to be weak in content as the private sector had already made significant commercial strides both in marine and inland fisheries sectors by the time of formation of the various Corporations by the respective Governments. There were, however, certain aspects left for the Corporations to demonstrate to the private sector at the time of their formation, but some of them were not adequately equipped to demonstrate them. This situation resulted in a parallel development of commercial activities that had already developed in the private sector.

Several Corporations that undertook marine fishing operations with larger trawlers converged at Visakhapatnam. Thus the purpose of demonstrating the activity to the entrepreneurs in the respective States was basically defeated. From the Commercial angle also, the purpose was not served with several Corporations in the coastal States incurring losses, the A.P. Corporation being the foremost in loss making. The A.P. Fisheries Corporation, with the dawning of some wisdom on the State Government, has been recently wound up, although belated, (in the footsteps of West Bengal, Orissa etc.). The A. P. Corporation tried to justify its existence all along on the ground that it was offering higher rates to the small fishermen for the purchase of their produce for processing at the Corporation's plants. Maharashtra, which had also closed down its Corporation, has since revived it a few years back, the provocation for which is not clearly known.

The Tamil Nadu and Kerala Fisheries Corporations are exceptions to the general trend although in different ways. Tamil Nadu Corporation's trawlers too are based at Visakhapatnam. (Those of the other Corporations *i.e.*, Gujarat, Kerala, West Bengal were also based at Visakhapatnam. While the vessels of Gujarat Corporation stopped operations, those of West Bengal Corporation were sold away. Kerala Corporation has recently shifted its trawlers from Visakhapatnam to Cochin. The management of the trawler operations of Tamil Nadu's Corporation with Visakhapatnam as the base has been consistently good and profitable, by and large, in the past nine years. This is because of supervisory, operational and managerial skills of the personnel in charge. One consequential development is that several entrepreneurs from Tamil Nadu acquired their own trawlers and put them into operation with Visakhapatnam as the base. Later, some of them shifted their base to Madras. A net making plant was also set up by the T.N. Industrial Development Corporation which proved to be beneficial to the fishing enterprises. The Fisheries Corporation has set a good example in the development of some of the reservoirs in the State, establishing an increase in per ha production in several of them. CICFRI has played a crucial role in demonstrating the developmental techniques. The Corporation has also developed a marketing system for the benefit of the fishermen.

The Kerala Corporation operated its trawlers with Visakhapatnam as the base successfully for several years, because of the managerial and operational skills of the personnel, until they were shifted to Cochin.

In contrast, the A.P. Corporation failed in the operations of their trawlers. Although symptoms of failure were visible at one stage, yet the Corporation built a 25m long wooden trawler at its yard at Kakinada. What was talked about when the trawler was launched was that the design was faulty and the trawler was listing and for these main reasons, she could do no better than the others in the Corporation's fleet. The Corporation could not run its processing plants successfully for long, so much so it leased them out to private companies for sometime but now closed.

One achievement of the Corporation is, however, to build several mechanised boats at its yard at Kakinada and supply them to groups of fishermen/co-operatives. The building activity motivated a few boat builders to set up their own yards which have been flourishing, although the building activity at the Corporation's yard languished. It is also to the credit of the Corporation that it constructed several fiberglass boats based on BOBP design and distributed them on subsidy basis among fishermen, with outboard motors. The Corporation ventured into domestic marketing of fish besides exports of shrimps but failed in both, in the long run.

As already stated, the West Bengal Corporation sold its trawlers and closed down operations. The Gujarat Corporation has suspended the operation of its trawlers and seems to be in the process of disposing of the trawlers which are now lying at Visakhapatnam. It, however, continues its other activities such as reservoir fisheries development and marketing in Gujarat.

A decision that was different from other States was taken by the Kerala Government. It transformed the Corporation imperceptibly into a Cooperative Federation with the acronym 'MatsyaFed', and imparted a wider role to it. The new body set up fishing net making plants, which proved to be of direct benefit to the fishermen. MatsyaFed took up purchase and marketing offish harvested by the fishermen with an integrated arrangement for payments towards purchases linked to savings by the members. This system improved their lot.

The Organisation has been involved in housing programmes for fishermen and other socio-economic activities. However, after the shifting of the base of the trawlers to Cochin, MatsyaFed, as the owner of the trawlers, could not prove that running of the trawlers is one of its strong points.

The Karnataka Fisheries Corporation continues to function with accent on marketing. It is stated that the Corporation is moving towards profitable operations.

On the inland sector, M.P. Fisheries Corporation and U.P. Matsya Vikas Nigam are reported to be functioning reasonably well. Both the Corporations are stated to be closely engaged in the development of fisheries in the reservoirs entrusted to them. M.P. is implementing a World Bank- aided Reservoir Project and, U.P a similar project for *Beel* Fishery Development. They have adopted scientific development strategies with well designed integrated marketing systems for the benefit of the fishermen. The Corporation in Assam is entrusted with implementation of a World Bank aided *beel* Development Project.

Over a couple of decades back a Central Fisheries Corporation was set up and it did function for some years. The aim of the Corporation was to procure fish from various neighbouring States and sell them in Calcutta markets, in competition with private merchants. As it turned out, with the sale prices at the Corporation's outlets being lower, the merchants used to purchase all the fish through a number of buyers drafted by them, so as to impart the complexion of purchase in retail. The fish so procured were then pooled up and deployed for sale through their own outlets. The Corporation could not withstand the situation, and eventually the Corporation had to be closed down.

To sum up, in the inland States, with the exception of M.P. Fisheries Corporation and UP. Matsya Vikas Nigam, and possibly Assam, then appears to be no other inland Fisheries Corporation in the country that has been performing reasonably well. It is believed that the Assam Corporation is also doing a good job in implementing the *Beel* Fishery Development Project. In the marine (and inland) sector Tamil Nadu Fisheries Development Corporation and Kerala Fisheries Corporation (Now part of MatsyaFed) seem to be performing well, while Karnataka Corporation has been showing progress in the marketing line. The moral to be drawn seems to be that Governments should keep away from involving themselves directly in commercial fisheries activities.

State Corporations can engage themselves in developmental-cum-commercial activities attuned towards diversified fishing, processing, and marketing systems and these to be done in a such a way that they set an example for the private sector to emulate. Corporations would need to desist from functioning in competition with the private sector unless the situation so warrants.

Fishing Chimes, Editorial 233: May(ii) 1997: Vol. 17, No. 2

# Fisheries Sector: The Dubious Extension of the Term "Thrust Area" to it

The subject of 'Fisheries' is ostensibly one of the developmental "Thrust" areas, as proclaimed by the Central Government. In actual terms, by and large, we do not see any signs of the sector being actually treated as a 'Thrust' area. All evidence is mostly to the contrary, looked at from several angles, So much so, fish farmers, fishermen and entrepreneurs are sceptical about this 'Thrust Area' appellation. It is difficult to see evidence of any 'thrust' given to the sector, unless granting a few subsidies are considered as a measure "to provide thrust".

The position seems to be that all sensible fish farmers, fishermen and entrepreneurs avoid falling into the trap of the verbiage and rhetoric and the frequent pronouncements by higher-ups to say that the sector is truly a "Thrust" Area. In actual fact, the Government gives the lowest priority or no priority at all to the subject of fisheries, may be unintentionally. It is possible that in some weak moment, some one in the Governmental set-up added "Fisheries" to the list of "Thrust" Areas. Beyond this, there appears to be no commitment on the part of the Government to translate the inclusion of Fisheries as one of the "Thrust" Areas into action.

This is possibly believable, as Government has never said that the inclusion meant translation into action; for, if there has been translation the state of affairs in the fisheries sector will not be so dismal as they are now. The term "Thrust" Area is at the best a catchy phrase which probably has some role to play. For example, it can be used freely at various meetings, participated by equally complacent persons who would not know or would not like to protest on the emptiness of the phrase, for the fisheries sector.

Deep sea fishing is languishing for want of a firm policy. The existing fleet is dwindling. No steps are taken to build a deep sea fishing fleet consistent with the needs of achieving maximum sustainable level of fishing. The beginning made towards diversification of fishing through introduction of tuna longliners has not been followed up, forcing the effort to set a bad example. The least the Government could have done is to consider and decide on the developmental part of the Murari Committee's recommendations, instead of just confining its attention to the recommendation relating to 'Rehabilitation Package'. This may help the existing operators but will not take the industry forward. Unless production goes up, exports cannot go up. Culture sector has no doubt progressed considerably and efforts are on to devise measures needed to contain diseases among cultured shrimps but we are far away from having an enduring solution.

Reservoir fisheries development is yet to receive the needed fillip. No concrete plans are devised as yet to provide an integrated spread effect to freshwater prawn culture. Saline water resources in Rajasthan, Haryana, Punjab and parts of U.P. remain unutilised, although possibilities of utilising these waters for fish and shrimp culture and artemia production are demonstrated. Mariculture development is not receiving much of fillip at the hands of the Government. The visible absence of 'Thrust' tends to make the entrepreneurship distrustful of the Government. Banks do not lend money any more to the sector. The participants in the sector do not receive the kind of encouragement that the agriculture sector gets, although they are made to believe in the mirage that fisheries sector is part of agriculture sector. It is difficult to see where the element of "Thrust" lies.

Fishing Chimes, Editorial 234: June(i) 1997: Vol. 17, No. 3

# Efficient Sludge Treatment: New System

Most of the shrimp farmers face the problem of sludge accumulation which gives rise to risk of disease incidence among shrimps and also environmental hazards in the water course into which the water from the farms is released. Efficient sludge removal will go a long way in the conducive maintenance of pond bottom.

Sterner Aquatech, Norway developed a new sludge treatment system for fish farms, which appears to be equally applicable to shrimp farms. A Berghein of RFRogland Research, J. Ronhovde of Sterner Aquatech, and H. Mundal of Stolt Sea farm, Norway described the system in the April 1997 issue of Fish Farming International, London. This is a combined system for both primary effluent (the main outflow from the farm) and sludge treatment.

Four drum shaped rotary sieves are used for the primary treatment at the main outlet. Another device for dewatering the backwash water from the primary treatment is used as a part of the system. The sieves rotate and backwash intermittently. This is controlled by the pressure difference between inside and outside of the drum sieves. The rotary sieves used are those manufactured by **Hydrotech,** Norway.

When one or more of the main drum sieves become blocked by the building up of waste particles and become static, the pressure difference increases to a point. At this stage dewatering drum sieve begins to rotate and backwash. When the main drum sieves are cleared, the pressure difference gets reduced and the rotation of dewatering drum sieve stops. This way of functioning greatly reduces the quantity of backwash water produced and increases waste concentration at the inner side of the outlet.

The system employs four primary flow sieves (Main drum sieves) filled with 90 µm mesh size screens, each with a hydraulic capacity of about 20 cu.m/min. The hydraulic capacity is kept on the higher side as it would allow one or two of the sieves to be inoperative while still retaining adequate treatment capacity. When 80,100 µm mesh size screens are used, they are expected to remove 50-60per cent of the total suspended solids, organic matter (as biochemical oxygen demand) and total phosphorus, but only 25 per cent of the total nitrogen.

At the farm where the system is adopted, a single 80 µm pore size screen unit is used for backwash water dewatering. The sludge water from the dewatering sieve is raised to about 34 m by using a continuously running diaphragm pump, so that solid particles are not damaged.

A conventional circular tank designed for gravity thickening of the sludge is used for the final dewatering step. The unit has a surface area of 3.3 sq. m and a volume of 5.5 cu. m. It has a steep conical bottom for sludge collection. This large capacity allows the thickening of the sludge even when the dewatering sieve is temporarily inoperative and backwash water from the primary (main drum) services is diverted directly to the thickening tank. Sludge from the thickening tank is removed frequently, often on a daily basis.

To enable the process of thickening to continue efficiently and sludge properties are maintained, the removed sludge is transferred into a 500 litre tank and lime is added to the sludge and mixed with a stirrer for 10 minutes. To achieve the required pH (pH of 12 is mentioned), 1,520g of unslaked lime (CaO) per litre of sludge, with a dry matter content of 10 percent is added. The slaked lime to be added for the stated capacity works out to 8 kg per tank. At a pH of 12 it is stated that pathogens are totally killed or atleast greatly reduced. Offensive odours and possible putrification are also eliminated.

The sludge so stabilised is stored temporarily in an eight cu.m circular sludge storage tank. Once or twice in a month the tank is emptied into silage tank trailers for transport to selected land plots where it is mixed with manure and stored prior to spreading on land. By following this system, sludge-free water is enabled to flow into the sea through the outlet. At the same time the sludge can be utilised for land crop development.

It is suggested that Indian aqua engineers associated with shrimp farms may study the system and evolve a suitable design that can be introduced for removal of sludge from shrimp farms.

Fishing Chimes, Editorial 235: June(ii) 1997: Vol. 17, No. 3

# Bhavanapadu Fishing Harbour

In the Handbook of Fisheries 1995-96 released by the Ministry of Agriculture, Bhavanapadu Fishing harbour finds place in the list of harbours operational. It is not known from what angle, Bhavanapadu is included in the list of harbours now operational. The criterion to judge the completion of the harbour is its utilisation. However, the position is that not a single fishing boat has so far entered the harbour.

The shore facilities set up such as auction hall and slipway are now reaching a dilapidated condition, without being used at all. It is ironic that the infrastructure facilities set up at this non-functional harbour are left to the action of winds which is a great injustice to the fishermen.

Mentioning these aspects, a team of office bearers of the Forum of Fisheries Professionals (Dr. D. Sudarsan, President, Capt. N. Venkateswarlu, Vice-President and Dr. K.C. Phillip, Secretary) who have visited the harbour recently said that this harbour is of crucial importance to the fishermen of the Northern Andhra. The purpose of the harbour is to enable mechanised fishing boats to conduct fishing in the adjoining northern coastal zone of A.P. without having to go all the way back to Visakhapatnam with an additional steaming time of 17 hours for landing their catches, but to unload catches at Bhavanapadu itself, availing of the shore facilities including the one for bunkering and to go back for fishing from Bhavanapadu, saving the steaming time to and fro to Visakhapatnam. Major fishing grounds off north Andhra are located off Srikakulam district where Bhavanapadu lies and also along the adjacent Orissa coast.

The real problems seem to be connected with the design of the harbour. These and drift from the south appears to have not been taken into account in designing of the training wall on the south. So much so, the sand drift from the southern side keeps piling across the harbour mouth and on the outer side of the southern wall Even in the rainy seas on mechanised boats cannot negotiate the sand barrier to reach safe anchorage points or jetties in the harbour.

It is learnt that a strong objection was raised by the State Port Department at a very early stage in the construction. The then port officer pointed out that the design must effectively take care of the sand drift from south and suggested that the matter should be referred to the Central Water and Power Research Station at Pune. It is not known whether this was done or not. The present position, however, is that the piling up of sand is so much at the wrongly designed southern wall that it may soon be fully covered with sand and spill over.

It is of utmost urgency that a study of the present situation is made by the Union Department of Agriculture in association with the State fisheries department to chalk out the measures necessary. There is, however, the heartening news that the Union Department of Agriculture has already offered to release 50 per cent of the cost involved for all the needed adjustment works. Unfortunately, the State Government, it is learnt, has not so far responded to the proposal for the release of the matching funds. This means the State Government is not interested in the welfare of the fishermen of the Northern coast of A. P. The discontent among the coastal fisher men of Srikakulam coast may explode at any time creating a law and order problem. According to indications available, fishermen are planning to adopt an agitational approach for the redressal of their grievance. It is unfortunate that even after the Government spent Rs. 3.4 crores of rupees on the harbour, they have not been able to provide the facility. Apparently all the investment made is a waste, as matters stand.

## Advice of Expert Teams

After encountering the problems during the later stages of completion of the project work, two teams of specialists inspected the site once in September, 1988 and another time in September, 1992. It appears that both these expert teams recommended more or less identical approach for the completion and utilisation of the fishing harbour.

The three main recommendations are:

1. Extension of southern wall by 60 m inclined by 5m towards the entrance channel.
2. Capital dredging of the channel so as to provide the designed dimensions after completion of extension of southern training wall; and
3. Periodical maintenance of the channel subsequent to capital dredging may be necessary.

## Observations of the Forum Team

These additional works as per the estimate made in early 1993 would cost Rs. 200 lakhs which would normally have to be shared by State and Central Governments. At a recent meeting at Visakhapainam attended by Union Ministers Mr. Yerram Naidu and Mr. B. Bulli Ramaiah, it was stated by the Ministers that the Central Government was ready to release their part of the funds but the matching funds were not forthcoming from the Government of Andhra Pradesh. Therefore, it is high time that all those interested in the development of this area in general and the fishermen community, in particular, take up the issue with the concerned authorities and press for early release of the funds by the Central and State Governments for completion of the project. Also it may be taken into account that the estimates made in 1983 may have to be revised in the light of inflation during the intervening period, and allowance may be given for the increase in prices. It is felt that this might work out ultimately to a figure which may be anywhere around Rs. 5 crores, if the entire job is to be done by Government agencies.

## Privatisation

In the context of present policy of liberalisation and privatisation, privatisation of the completion of the works required for effective commissioning of the harbour and for the effective running of the harbour could be considered. It is expected that privatisation may result in speedy completion, perhaps at a lower cost, but the agency which would be entrusted with the job would in all probability ask for a minimum of 30 years lease for running and maintaining the harbour etc.

In this connection, it may be relevant to point out that the Government of Andhra Pradesh appears to have taken a decision to privatise harbours at Kakinada, Gangavaram and Machilipatnam.

Justification for the development expenditure involved:

1. This harbour, when fully functional, will provide integrated facilities for all activities connected with fishing which would lead to economic growth, not only for the fishing community but other ancillary activities. This would also fulfil the long felt need for such a harbour in the Srikakulam district.
2. The fuel saving by fishing vessels, when the need for rushing to the Visakhapatnam is obviated, would more than offset the cost of the project.
3. A port at Bhavanapadu would reduce the manpower needs. Shorter duration voyages of fishing vessels would increase number of voyages with resultant higher production of fish.
4. The likely reduction of risk of life and damage of fishing boats during times of bad weather would be considerable and cannot be explained merely in monetary terms.

## Recommendations

1. A project on social forestry between Moolapet and Southern training wall to arrest the sand drift may have to be undertaken.
2. A ship breaking unit may be allowed to be started.
3. At least one berth may be provided for commercial cargo to be handled. The cargo items which immediately come to mind are cooking gas for distribution in the area, and salt, cashew and coconuts which are a major produce of the area, that can be sent to other places.

The observations and recommendations deserve serious consideration by the Union Department of Agriculture and the State Government.

Fishing Chimes, Editorial 236: July 1997: Vol. 17, No. 4

# American Pew Foundations' Award to Kocherry

Pew Foundation of USA has announced its triennial award for 1997-2000 to Mr. Thomas Kocherry of the Indian National Fishworkers' Forum. This is the first ever award by the Foundation to an Indian national. *Fishing Chimes* warmly congratulates Mr. Kocherry on the conferment of this prestigious award. Fishworkers and all those connected with fish and fisheries can be justly proud of this unique event. The award is a fabulous one, entailing the bestowal of US $ 1,50,000 in favour of Kocherry. In all likelihood, he would utilise the money for the uplift of Indian fish workers, through strengthening the activities of the Forum.

Fishworkers and all Indians connected with fisheries should be all the more happy that an award normally given to American researchers and scientists is awarded to Kocherry who is neither an American, nor a researcher nor a scientist. This award, thus, signifies the greatness of Kocherry in fighting against the operation of foreign fishing vessels in Indian waters, that too joint venture vessels. Apparently, the Pew Foundation considers joint ventures in fisheries sector, between Indian and foreign companies is not good for India. The Foundation must be either intensely against operation of foreign fishing vessels in Indian waters or it wants Indian waters to be free from the presence of foreign fishing vessels. Or it can be only just a manifestation of an intensively warm appreciation of the unrelenting fight put up by Kocherry against the joint ventures. However, from what the U.N. Secretary General is reported to have told the U.N. Assembly narrows down the purpose of the Award to the achievement of Kocherry in scaling down the number of joint venture fishing vessels in collaboration with the national Government. This points out that the Foundation, an American one, is against the operation of joint venture fishing vessels in Indian waters. How sound the premise on which the American Foundation is against joint ventures in the marine fisheries sector of India and to what extent the Foundation's decision for bestowing the award is rational is a matter for the Government to judge, for, it has been made a party to it for the reason that the UN SecretaryGeneral says, to repeat, the "success" of Kocherry is in collaboration with the 'national Government. To the extent we know the policy of the Indian Government continues to permit joint ventures as per the rules framed under the provisions of the Maritime Zones Act and these rules are not withdrawn. It has to be believed that the Government has 'sealed' further issue for permissions as no fresh permissions have been issued. It is for the Government to agree with the observation of the U.N. Secretary General in this regard, whether it has collaborated with, the National Fishworkers Forum. If the Government agrees, it means that deep sea fishing in Indian EEZ through joint ventures is sealed, as India has no capability to exploit the resources of the farther waters of Indian EEZ. However, while development of deep sea fishing in the Zone can materialise only with foreign help, this need not necessarily be through joint ventures. It can be brought about through bilateral arrangements. A clarifactory statement on the subject from the Government is called for to make the position clear. A point that however arises is: should the Government need an agitation by the workers to 'seal' joint ventures in national interest The stated 'sealing', particularly when it is collaborated by the Government, leads to suspicions of weakness in the policy. As it has turned out, an invidious distinction is made among entrepreneurs. The 'sealing' implies those who are in with joint ventures are given undue advantage. They do not constitute a selected lot based on merit. According to Mr. K.B. Pillai, Chairman of MPEDA, there are eleven joint ventures functioning in Indian EEZ at present. While there is no exercise to determine how many joint ventures can be there, eleven joint ventures are too inadequate to exploit the resources of Indian waters beyond territorial zone. No Indian enterprise, including Fishworkers' Forum, is equipped to exploit the resources. The attributed discontinuance, therefore means a violation of the existing policy. Government, as already stated, has to clarify the position.

For whom and for what purpose the 'collaborative sealing' has been done? Is it to benefit the few joint ventures to have the grounds to themselves? Or, is it to benefit fishworkers? If the sealing is to benefit fishworkers, has the Government considered how to equip them financially and technologically to undertake fishing on the high seas? Has the Forum or the Government done anything to bring up the workers to the needed professional level? Does the Forum itself has the capability to organise high sea fishing with its workers? Has it ever approached the Government for financial and training support to enable them to fill up the gap? Or is it the Forum itself, just to keep the EEZ free of Joint ventures, on the belief that the high seas are the source for coastal fishery wealth? Has the Forum and the Government realised that the complexion of high sea fisheries, whether demersal or pelagic is totally different from coastal fisheries and that, in case the stocks are the same the present overfishing and stock depletions in the coastal zone would not be there? Has the Government or the Forum thought of a mechanism to bring about the exploitation of the resources on our own, instead of depending on foreign collaborations? Or, is it the intention to preserve fisheries of EEZ beyond coastal zone as wild life for protection?

It is not clear in what way fishworkers are benefited or helped by the sealing. The Forum should not relax until the system of permitting joint ventures is totally scrapped, as it believes that joint ventures are against the interests of fish workers. One has to believe that the purpose behind the agitation of the National Fishworkers' Forum is solid and not hollow, and the imputation of collaboration between the Government and the Forum leading to 'sealing' of joint ventures must be in order provided there is direct confirmation by the Government, with a clarificatory statement based on an incisive analysis of the welcome development leading to the Pew Award. Now let us see what has unwittingly or wittingly come about. The EEZ not accessible to coastal fishermen is now virtually free from all foreign vessels but for the remainder of eleven joint ventures. The situation delights the foreign vessel operators as what they have been longing for has at last materialised as they wanted. The objectives of foreign vessel operators and the Forum are one and the same for the development, so far as the former is concerned it is in practical terms and for the later it is in terms of evicting foreign vessels on ideological or long range considerations, at least for the present.

The, foreign vessel owners have reasons to be happy. Firstly, so far as chartered vessels are concerned, whose operations too have been stopped, they do not have to pay 20 per cent of their gross earnings to the Indian charterers/ lessees. Secondly, they do not have the need to report at the designated port at the time of entry and departure to the Coast Guard. They do not have to declare catches and subject themselves for a check by the Fishery Survey of India. While they can avoid all miscellaneous problems while at port, they do not have to inform their position at sea to Coast Guard while in the Indian EEZ. The single problem foreign vessels have to face is the apprehension by the Indian Coast Guard or the Indian Navy. This is a risk they are accustomed to face the likelihood of such risky situations arising. They know the Indian strengths as well as weaknesses in apprehending fishing vessels, fishing far away from the coast.

It is unfortunate that the Government of India has been delaying inordinately the formulation of an updated/ revised deep sea fishing policy. We have been advocating in these columns the institution of a long term and a short term policy in this regard. The long term policy should consist of catching young selected fisherboys and others interested in fishing profession at around the age of eight and put them through an intensive course for 10-12 years at CIFNET at its adjuncts set up for the purpose at suitable centres. The training has to be free of cost. These adjuncts have to be well equipped with training gadgets and with training vessels. Bilateral assistance may be sought for the purpose. Once Academies are set up, in 10-12 years, we will have several trained men, principally from coastal fishing villages. They can be helped by the Government with the provision of high sea fishing vessels through a mechanism which can be worked out.

As the fruition of the long term policy takes time, there has to be a short term policy. One main ingredient of this policy has to be to equip vessels 10m and above for diversified fishing and impart them capability to fish in farther waters of EEZ. This reform can be brought about in two slabs. One is to equip the existing vessels of 23 m and above to undertake longlining mainly for tuna, on the high seas. Indian waters are known to have rich stocks of tuna, grossly underexploited. They can concentrate on tuna fishing during the season of availability and in the rest of the year in catching cephalopods and finfish.

As part of the second slab, mechanised boats of 10m and above can be equipped for tuna long lining on the pattern followed in Sri Lanka, Maldives and several other countries. The need is to impart the capability to the vessels by upgrading the machineries and providing longline release and hauling systems. Finance is the main constraint and in national interest, Government has to introduce an integrated system of financing linked to port-wise checking of landings, earnings and recoveries from the buyers of the catches through agreements containing a recovery system.

Taking the above suggestions into account and the stated 'sealing' collaboration of the Government with the National Fishworkers Forum, a sound policy has to be drawn up and announced by the Government, subject to the hope that Government has no intention of treating fish in waters beyond the territorial zone as 'wild life'.

Fishing Chimes, Editorial 237: August 1997: Vol. 17, No. 5

# Bycatch Escapement from Shrimp Trawls

Fitment of Turtle Excluder Devices (TED) as part of shrimp trawls has gained importance, particularly for the reason that U.S has banned import of shrimps harvested with trawl nets not fitted with TEDs. As important as lending protection to turtles is the saving of juveniles of prime and other quality fish that get into trawl nets, without means of escape. "Professional Fishermen", a reputed Australian fisheries journal, in its February 1997 issue, published a write-up entitled Progress in bycatch escapement from prawn trawls by New South Wales Fisheries", authored by Torry Gorman of Sydney, Australia, a fisheries scientist and consultant, net designer and maker. The contents throw light on the manner in which bycatches including juveniles can be excluded from trawl nets, without in anyway reducing shrimp catches. Convinced that the salient aspects of the write-up will be of interest to owners and captains of shrimp trawlers, an effort is made here to present them.

Gorman introduces the subject by saying that Malt Broadhurst and Steve Kennelly, New South Wales fishery scientists, are convinced that no single escapement device is capable of working effectively.

The scientists found that even a brief delay in hauling a trawl net produced needed stimuli that would promote escape of bycatch but the stimuli varied from species to species. This finding motivated them to work on a new design with square meshes for one of the codend panels, believing that this might stimulate bycatch species to escape during the whole of the tow and not at the end of it. They also investigated the effect of the circumference of the aftport of the codend on fish escapement without escape panel". The experiments were conducted in commercial fishing grounds with a 17 m long trawler, triple-rigged with nets having 15.8 m headline and they were towed at 2.5 knots. The middle net was different from the outer two.

The codends were 58 meshes long with a mesh size of 40 mm, made of 400/60 ply polyethylene. Four designs of codends were tested. These comprise two panels, a fore panel and an aft panel. The fore panel was 33 meshes long and the aft one was 25 meshes long. The other details are:

1. 100 meshes circumference in both panels, but of diamond mesh.
2. 100 mesh circumference fore panel; 200 mesh aft panel, both of diamond mesh.
3. 100 mesh circumference fore panel with composite square mesh escape panels of 40 mm, and 60 mm stretched mesh circumference in aft panel, 100 mesh in circumference.
4. Fore panel of 100 mesh circumference as above; Aft panel 200 mesh in circumference.

Of these, item 3 above showed good results. The 60 mm mesh panel was placed at a point in the codend where water flow was the greatest, and this allowed larger number of fishes to escape. The larger 40 mm panel increased the chances of smaller fish escaping randomly. Thus the codends with square mesh panels were found effective in allowing bycatch species to escape continuously during the tow. There was neither reduction in shrimp catch nor in the catch of incidental commercial species.

It may be noted that average catch of a trawler per day of operation along upper east coast of India is estimated at two tonnes of which the shrimp component is often at 5 to 8 per cent, table variety fish is around 10 per cent and other economical varieties are near about 40 per cent. The rest of the catch consists of juveniles and uneconomical varieties of fishes. They are not taken seriously. By this system, juveniles of economic species along with presently uneconomical ones can be saved for improving the future catch. The 100 mesh codend was seen to have increased the lateral opening of the meshes and allowed more of small fish to escape out of the codend. The weight of the bycatch in the net came down by about 40 per cent. The 200 mesh circumference of codend also allowed small fish to escape but the rate of escape was lower.

Thus, the results showed that both codends of 100 mesh or 200 mesh circumference were effective in reducing the bycatch. The experts, however, found that the most effective combination was the square mesh escape panel used in conjunction with the 100mesh circumference cod end. The notable feature of importance is that the reduction in bycatch was achieved without any reduction in weight or size composition of the shrimp catch. In addition, the mesh sizes of the escape panel were small enough to be effective at preventing the escape of commercial sized fish.

The work done has demonstrated that there is considerable scope for varying both the design of square mesh escape panels and the configuration of the codends to provide for escape of bycatch species without incurring any loss of shrimp catch. The schematic sketches of the codend, used in the composite escape panel tests as given by Torry Gorman are given at page 5 of August 1997 issue of the journal. Although the experimentation was done essentially with double rigged trawls of 15.8 m headline length on 17 m OAL vessels, the results achieved can be extrapolated for adoption on 23-25m long trawlers with 350 to 400 h.p. engines. The designing of nets for double rig (out rigger) trawling based on the work of Malt Broadhurst and Steve Kennelly can be undertaken with needed adjustments by CIFT and the nets so designed can be used for experimental fishing in the first instance. Designs for single net sten trawling for the same or of a higher size range can also be developed by CIFT following the same system. Both the designs, for double rig and single net trawling, so developed, can be tested, and once found beneficial can be provided to the commercial sector for adoption.

Fishing Chimes, Editorial 238: September(i) 1997: Vol. 17, No. 6

# Kerala takes the Lead: *People's Involvement in Fisheries Development*

Kerala Government has started an enlightened and people-based approach for fisheries development, which deserves to be emulated by other State Governments.

The Minister for Fisheries of Kerala, Mr. T.A. Ramakrishnan, assisted by his Director of Fisheries, Mr. Kamalvardhan Rao has imparted a people-based orientation for implementing a project drawn up by Kerala Agricultural University for bringing 10,000 ha of area in Kuttanad under integrated fish farming with people as the main actors. The job of the Government's functionaries will be to serve as facilitators. This approach is in refreshing contrast to the general approach for implementation of such projects by officials of the Government, through entrepreneurs.

In the marine sector, there is participation of the people, whom we call as fishermen, from a large number of villages along the coastline for fish capture. This pattern is akin to large sections of population, engaging themselves in agricultural crop production on land. Unlike the marine sector, inland populations are by and large, ignorant of the value of water to produce fish that would provide proteinaceous food to the people. Consequently, the approach all along has been to induct entrepreneurship to develop fisheries in inland waters which takes away the focus from the people. A new approach to involve people in the endeavour to develop inland fisheries is clearly called for and the Minister apparently has looked into this aspect and decided to bring in involvement of the people in the activity, with integrated farming as the starting point. There are, among people, any number of agriculturists who think that "culture" is synonymous with "Agriculture", not knowing that land with water over it is also land. The water in land depressions as also in parts of Kuttanad absorbs nutrients from the underlying land and is productive as the land itself, with the difference that the water can produce a different crop, a crop of fish. Once the people are made aware that when they integrate fish farming in water with land crops and with poultry, piggery, dairying etc., there wil be additional production, additional earnings and enhanced prosperity. This awareness will develop when people are involved in the activity of integrated farming. Some of the agricultural farmers in Punjab, Haryana and Orissa have taken to integrated farming but the activities are not people-based but entrepreneur-based. For the first time, Kerala has taken the initiative to bring a stretch of 10,000 ha under integrated farming. This reform, which is on a large scale, has far reaching beneficial dimensions. Every hectare of area would produce fishes, cereals, vegetables, milk, poultry, ducks and eggs to yield higher production and profits. The cattle, poultry and duckwastes serve as manure for fish ponds. Paddy fields too can be utilised for producing fish as additional crop. With all this potential and benefits, integrated farming would become at once popular with the people. The approach to peoplise 'integrated fish farming' would enable the utilisation of the full potential of the selected 10,000 ha of Kuttanad area in a sustainable manner. Farming Samithis are proposed to be set up by Kerala Government to formulate farming schedules.

The Chairman of the FFDA, Kottayam will be the project coordinator. A working group will be set up to advise the Management Cell, to come into being. The farmers will be the main actors in the implementation of the momentous project that seeks to bridge the gap between the plan and its effectuation in the field. With the responsibility of field operations in the hands of a section of the people functioning as the farmers, and with a dedicated management system and implementation mechanism, there can be no doubt that the desired objectives of the project will be fulfilled. This major project seeking to usher in integrated farming will shower enormous economic benefits among the people thereby bestowing prosperity mostly among the weaker sections of the society.

The process of 'peoplisation' of fisheries sector in Kerala started quite some years back, but the approach is taking a proper shape now. The process started with the setting up of 'Matsyafed' and the 'Fishermen's Welfare Fund'. Matsyafed has provided a mechanism to marine fishermen for securing equitable prices for their catches, linked to savings for utilisation in times of need. Payments go to the accounts of fishermen and this process provides a sense of security to fishermen. The setting up of the 'Fishermen's Welfare Fund' by the Government is another major reform that ensures the welfare of fishermen. The administration of the 'Fund' is linked to a scheme for the welfare of fish workers.

There are over 1.50 lakhs of fish workers in Kerala in various activities ancillary to fishing. These activities cover upkeep of fishing boats, net making, peeling of shrimps, processing of marine products, retail marketing of fish etc. The scheme seeks to provide insurance cover to the workers, besides a monthly pension and judicial assistance. The workers will contribute, as part of the scheme, a nominal amount to the 'Fund' out of their wages. Local bodies are also expected to contribute a nominal percentage of the income realised from fish markets.

These reforms apart, it is to the credit of the State Government to have rationalised the thinking among marine fishermen generally opposed to the seasonal ban, during the south-west monsoon, on sea fishing. This process of rationalisation no doubt took time but the fishermen, educated as they are, are now convinced about the long range benefits of the seasonal ban.

There is now an awakening among the people on the role they have to play in fisheries development. Once people take over the process of development in the field, there can be no looking back. Government has only to provide guidelines, organise matching funds, and a coordinating system. And the present policy of the Government is to provide these to the people for the benefit of the people.

Fishing Chimes, Editorial 239: September(ii) 1997: Vol. 17, No. 6

# Fisheries Development: Impressive Strides in Karnataka

There is a refreshing and conducive climate for fisheries development in Karnataka. This is largely due to the inimitable leadership provided by the Minister for Fisheries, Mr. Jayaprakash Hegde. This has been percolating through the departmental machinery, headed by the Director of Fisheries, Mr. Shanmukha. The strategy of the Minister, as it appears, is to identify gaps impeding the development and fill them up in order of priority.

The minister ushered in a revised policy of leasing public waters for fisheries development. The policy entrusts tanks having more than 10 ha of waterspread area but not more that 25 ha to the Zila Parishads for fisheries development. Tanks having an extent of over 25 ha will be managed by the fisheries department. Priorities in granting leases have been laid down, with the Karnataka cooperative Inland Fisheries Federation getting preference, followed by fishermen's co-operatives, SC&ST fishermen's cooperatives, unemployed graduates in fisheries science, gram panchayats, registered youth clubs and mahila mandals. While lease amounts are kept at nominal level, it will be incumbent on the lessees to stock the tanks with fish seed, take care of the crop and produce optimum quantities offish. The lease period will be three years. The inclusion of several categories of organisations will ensure development of fisheries in the waters under the umbrella of one or another public body, thereby avoiding an individual entrepreneurlessee to corner profits. The services of the fisheries department will be at the disposal of the lessees to ensure the developmental process.

Short supply of fish seed has been a major constraint in the State, mainly because of inadequacies in public participation in the endeavour to augment seed production. So much so, the minister introduced schemes to attract entrepreneurship to come forward to set up nursery farms as well as seed rearing farms. A quantum of subsidy is offered to those coming forward to participate in the programme to augment seed production. Owing to the intensification of the efforts of the fisheries department and the increasing participation of entrepreneurship in the effort, fish seed production rose from 8 crores nos. in 1995-96 to 20 crores nos. in 1996-97. A remarkable achievement indeed. This development will no doubt lead to generation of interest in the activity among people, motivating them to participate in the process.

State Fisheries Corporations all over the country, with a few exceptions, apart from not achieving their objectives have run into losses. One of the exceptions is the Karnataka Fisheries Development Corporation which, while rendering signal service to the fishing community, has also been earning sizeable profits. The profits increased from Rs. 5.98 lakhs in 1995-96 to Rs. 15.30 lakhs in 1996-97. The Corporation purchases fish from fishermen and arranges supply to the people in major cities of the state through a network of retail outlets. The healthy and profitable functioning of the Corporation is in a large measure owing to the flow of leadership from the Minister.

Another achievement of the Minister is to successfully prevail upon the Union Minister for Agriculture to set up a unit of the Central Institute of Fisheries Nautical and Engineering Training in the State for the benefit of fisher youth.

Karnataka is the first State in the country to organise fisherwomen's cooperatives. Unlike fishermen's cooperatives, surprisingly, fisherwomen's cooperatives which engage themselves in fish marketing, emerged as very efficient, successful and profit-oriented units. Tamil Nadu too has achieved similar results. Mr. Hegde has now initiated the task of expanding the role of fisher women in shore-based activities and in that context has ensured the organisation of awareness camps for fisherwomen throughout the state.

Karnataka is now well set on the path of progress in the fisheries sector.

Fishing Chimes, Editorial 240: September(iii) 1997: Vol. 17, No. 6

# Rehabilitation of Deep sea Fishing Industry

Considering the sickness of the deep sea fishing industry, as confirmed in a study conducted by an FAO expert, Mr. Guidicelli and others as well the Government of India announced a scheme for rehabilitation of deep sea fishing industry. The main cause for the introduction of the scheme is that almost all the trawlers have been either incurring losses or the earnings are just sufficient to meet the operational expenses. The scheme came into force on May 5, 1997. It offered two options. Under Option A, the entire overdue interest charged (including capitalised predelivery interest) together with 50 per cent of the principal loan as on May 5, 1997 would be frozen and the balance is to be paid in thrice yearly instalments in a duration equal to the remainder of the economic life of the vessel which is taken as 20 years. Under Option B, 30 per cent of the outstanding principal, referred to as outstanding principal, has to be paid in full as one time settlement within 3 months reckoned from August 4, 1997.

Those who opt for either option A or B would have to pay a non-refundable amount of Rs. 2 lakhs per vessel as earnest deposit. As we understand there are practically none who are interested in option A. Out of a fleet of over 180 vessels, around 80 vessels are now operating. The remaining vessels are in various stages of disuse. Some sank and others are portbound.

Owners of 66 vessels are understood to have chosen Option B, for the reason that under Option A interest has to be paid on the total loan amount and repayment problems as they are facing now would continue. Further, most of the owners are apprehensive that under this option the frozen debt would be thawed as soon as the repayment of the active outstanding principal is repaid with interest. Payment of 30 per cent of the outstanding principal would work out far less, they seem to have reasoned out. Most of the owners nevertheless now face the serious problem of mobilising the amount to be paid. Commercial banks, as is known, are reluctant to lend funds to fishing industry, more so in respect of those vessels which are II to 19 years old. Private financiers too are apprehensive to lend money to the owners.

So much so, the Association of the owners (Association of Indian Fishery Industries) pointed out to the Government the need for extending the time limit for the payment of outstanding principal amount now repayable. The Association is also understood to have sought other reliefs. All the owners who opted for 'B' paid Rs. 2 lakhs per vessel as stipulated towards non-refundable earnest money. Government has stipulated that only those who pay this earnest money would be eligible for midsea bunker facility. HSD oil taken by the vessels outside the national limits would be far cheaper than the rate at which they now get within the country. Apart from the lower price, in mid sea bunker system, there will be saving of excise duty and sales tax, which are substantial. The outgo of Rs. 2 lakhs will be more than compensated by taking midsea bunkering, within a few voyages. In case they are unable to mobilise the needed funds, the present pattern of operations could probably be continued at least till such time as the Government decides on the next course of action. The vessels are hypothecated to the Government and deciding on what should be done with the vessels, once they are taken over, will take time. Government may provide further concessions as sought by the industry, or as an extreme step, decide to auction them, whatever be the returns.

In such an event, who knows, the owners may buy the vessels through an indirect system. This will be cheaper. The past experience of auctions is that the maximum amount a vessel will fetch in an auction will be between Rs. 2 to 5 lakhs. Apparently, it would be prefer able to allow the vessels to operate and collect as much money as possible by conceding more time to pay the stipulated 30 per cent towards the loan instead of selling at a very low price. Further, this approach would atleast contribute to marine products export which now shows a tendency of sagging, because of the problems in the culture shrimp sector.

There are possibilities of the scheme ending up in a failure unless the owners find some way of securing the needed funds to pay towards 30 per cent of the outstanding principal amount. The industry is now a patient and Government is the doctor. The doctor has to put life into the patient, by administering a well formulated medicine.

Those who gave bank guarantees towards additional security over and above hypothecation of vessels and who defaulted in the payment of loan instalments faced invocation of bank guarantees. This amount was adjusted against the interest on outstanding principal amount. In the case of those whose guarantees were not yet invoked, the stand taken by the Government is to adjust this amount, and the interest thereof after invocation against outstanding principal and interest respectively. The owners feel that the guarantee amounts may be offset against the outstanding principal amount of 30 per cent and not against the total outstanding principal. The scheme, however, says that the adjustment would be made against the total principal amount due and not against 30 per cent of principal amount as repayable under the scheme.

Such of those who did not keep the guarantees alive would have now to pay that amount together with interest to avail of the benefits under the scheme. The amount will probably be offset against the interest due. The same procedure will apply to those who filed cases against the Government, financial institutions and obtained stay orders from the courts. The companies concerned would have to withdraw the cases in the first instance. The procedure as applicable to those whose bank guarantees were already invoked will be applied to all such cases.

Irrespective of the repercussions, certain provisions in the scheme dilute its purpose. Having virtually said that the outstanding principal amount repayable is 30 per cent of the total outstanding principal, it appears illogical to provide for invoking and deducting the bank guarantee amounts at source, before arriving at 30 per cent of the outstanding principal payable. Guarantees were primarily given for a different purpose and not for adjustment against loan amount due. Invocation of these guarantees for such an adjustment will further accentuate the already strained relationship between the companies and their bankers. If the Government wants the companies to survive, one graceful concession will be to return the guarantees to the owners concerned as and when 30 per cent payment is made. In order to avoid a likely complaint by those whose guarantees have not yet been invoked, a sympathetic approach of restoring the loan in such cases to the pre-guarantee invocation level and adjusting these guarantee amounts against the 30 per cent loan amount payable under the scheme will be in the spirit of the scheme.

While Mr. Guidicelli, FAO expert, came to the conclusion that the vessels now operate on no-loss-no profit basis and that they have to be equipped for combination fishing by addition of equipment for long lining, there are several who often wonder as to how the owners could continue to operate the vessels for 9-19 years on no-loss-no profit basis. The gross earnings would have gone towards the running and other expenses. When there is no surplus, the owners would normally not have any enthusiasm to continue the operations. Dedication towards the industry to that extent is difficult to visualise. In other words, there must be some kind of reservation in Governmental circles that the vessels must have been having some surplus and that the owners had been deliberately evading repayments, knowing the handicap which the Government has been experiencing in actual practice in taking over the vessels. On the other hand, the thinking can possibly be that by allowing the vessels to operate, so that atleast there will be some contribution towards exports and partial recoveries of loans and interest thereof will be possible.

On the part of the owners, the reactions appear to be different. They could get into industry with the help of the Government with great difficulty. They have been managing to get on with losses in some voyages and marginal surpluses now and then and with the help provided by the banks for quite some time. The losses by now must have however accumulated. The owners are known to owe lot of money to various suppliers and those who render services. They have also to repay the loans taken so far from private lenders to keep the vessels going. In this situation they can extricate themselves from the industry only after clearing their pending dues to the Government to whom they had already provided security, and by operating the vessels for some time.

The opening now provided by the Government in the form of rehabilitation scheme is the only ray of hope for them if they can manage to mobilise 30 per cent of the principal amount due to the Government. It might be then possible for them to operate the vessels profitably (as the investment comes down and repayment problems vanish) and utilise part of the earnings to clear outstanding loans.

It may be mentioned here that there is an inherent contradiction when paras 3.1.1 to 3.1.3 of the scheme are read together. When this is done, the meaning of the term outstanding principal alters. Para 3.1.1 gives the clear meaning that 30 per cent of the outstanding principal will be the balance payable. Para 3.1.2 confirms this. Yet the added clause in para 3.1.3 says that the outstanding principal would be reduced to the extent of guarantee amount, prior to the computation of the 'aforesaid' upfront payment (Para 3.1.2).

Once the balance 30 per cent (outstanding principal) is arrived at as mentioned in paras 3.1.1 and 3.1.2, going back to mention about the deduction of the bank guarantee amount out of 100 per cent of the outstanding principal would mean giving two contradictory definitions to outstanding principal. The real outstanding principal can only be 30 per cent of the overdue loan.

There are comments heard here and there that there are a few companies who did not provide guarantees at all. Apparently these are baseless as the scheme does not cover this category at all. In order to ensure the successful implementation of the rehabilitation scheme it is the responsibility of the Government to look into various aspects of deduction of guarantee amount from 30 per cent of the outstanding loan amount. Further, it may also be necessary to give to the companies a reasonably longer time to enable them to pool up the needed funds. As already observed, most of the companies are now encountering a serious situation under which they are not able to mobilise the needed funds for the payment of the outstanding principal amount. Government may like to reconsider these provisions of the scheme.

Indian deep sea fishing efforts are now drowning. The rehabilitation scheme is bound to relay signals that deep sea fishing in Indian waters by Indian vessels has been a failure. Further, despite this scheme and available fishery resources, there can be an eventual collapse of this industry. In order to prevent this and to ensure the position of the country as an emerging fishing nation, it is imperative that the Government evolves a foresighted policy on deep sea fishing.

Fishing Chimes, Editorial 241: October(i) 1997: Vol. 17, No. 7

# Tamil Nadu Fishery Corporation's Triumph

Among the coastal State Fisheries Corporations engaged in deep sea fishing, Tamil Nadu Fisheries Corporation emerged as a triumphant survivor. It has set an example that deserve emulation by others, by continuing deep sea fishing operations successfully since 1978. Despite this example, the other corporations never attempted to pave the path for running their deep sea fishing vessels with proficiency. The crowning glory earned by Tamil Nadu Corporation belies the notion that private sector companies function far more efficiently than public sector corporations in the deep sea fishing sector. Several private sector companies have shown such a performance that the Government of India had no alternative other than announcing a rehabilitation scheme. The other coastal State fisheries corporations closed down deep sea fishing operations in different ways. The management of the trawlers of A. P. corporation was so weak and directionless that the trawlers went out of commission and beyond repairs with unparalleled speed. The Corporation itself is now in the process of being wound up. West Bengal and Gujarat corporations sold their trawlers while winding up operations. Kerala Corporation was stoic for quite sometime but in stages, with the face saving device of entrusting the functions to Matsyafed, the closure of the Corporation and its deep sea fishing activities ensued inexorably. Tamil Nadu Corporation is the sole survivor with triumph.

What is the secret of the achievement of Tamil Nadu Corporation? It is just the magic of management. The headquarters of the Corporation is in Madras (Now Chennai). The Corporation kept a senior fisheries officer in charge of the activities for some years at Visakhapatnam until the operations stabilised. Later, imperceptibly and probably in order to reduce expenses it withdrew the officer and entrusted the management to the skippers and engineers in an inimitable way. The Corporation has four trawlers. It employs five sets of sailing officers. As and when a trawler reach the port, an officer from Chennai reaches Visakhapatnam and takes care of the sale of the catches through sealed tenders, sorts out any outstanding matters and gets back to Chennai. The spare pair of skipper and engineer on shore duty take charge of the trawler and arrange for repairs and maintenance works. Almost invariably they sail the trawlers from the port for the next voyage on the fifth day of entry into the port. The skipper and engineer who get down from the trawler take charge of shore operations and be on the job till another trawler steams in and the officers on that trawler relieve them to take command of the incoming trawler. This rotation continues.

As a measure of keen interest and involvement the state's Minister for Fisheries, and the Chairman of the Corporation who is also the Director of Fisheries make periodical visits to Visakhapatnam.

The work is so well organised that the operations take place like clock work. Repairs and maintenance works including those of main engine and other machineries, fabrication works etc., are given on annual contract so as to avoid the recurring problems of dealing with workshops on every occasion a trawler is at port. Although most of the owners feel that catches have come down, the Corporation would not allow the ratio of one tonne shrimp catch to 10 kl of fuel to be disadvantageously upset. The trawlers, as we understand, bring around 4 t of shrimp in a voyage.

The Corporation fixes a target of annual catches in a unique way. It takes the average of annual total expenses incurred in the previous years including the salaries of officers and staff connected with trawlers operations working at the Corporation's office in Chennai and fixes the target of income for the next year on that basis, keeping a suitable margin. The service rules of the Government are applied to the officers and crew. They enjoy security of service unlike the private sector companies. For the days and nights spent by the crew in a voyage, incentives come as compensation. While deckhands employed by the private sector are understood to be paid less, those on the Corporation trawlers get salaries and daily allowance at Government rates which are far higher.

The routine working of the officers is well established. The shore skipper contacts the trawlers out at sea on a daily basis and conveys information on status of catches and problems faced trawler-wise to the head office of the corporation frequently. The systematisation of the work by the Corporation in comparison to the management of private sector trawlers deserves a study by an Institute of Management or a research scholar striving to secure a doctoral degree.

Fishing Chimes, Editorial 242: October(ii) 1997: Vol. 17, No. 7

# Factors Impeding Shrimp Export Tempo

Fluctuations in capture shrimp production is a recurring feature. Once in few years a drop in the level of available stocks occurs, because of a combination of factors, one of which is that the longevity of shrimp is short, stated to be around 12 to 18 months. When there is excessive fishing of young ones and brooders in any particular year, the effect of it is seen in one of the following years. Such an effect is being felt this year on the upper east coast, as it had happened before. Over and above this, because of profuse breeding of browns and a drop in tiger and white shrimp recruitment on the upper east coast this year, the catches are mostly of juvenile browns. While reduction in white and tiger catches is experienced on this coast, some fall in brown catches may also be felt in one of the coming years.

Reports speak of a drop in capture shrimp production on the southeast and west coasts as well. The raw material stocks in the processing plants are also reported to be low now. On the culture shrimp sector, it appears that the incidence of disease among shrimp under culture is localised and several farmers are hopeful of having a normal harvest. The farms mostly in the northern districts of Srikakulam and Visakhapatnam in A.P. and a few zones in other States are stated to be affected by disease. Despite this, reduction in exports of shrimps can probably be anticipated mostly from the capture sector.

The main season for exports extends from October to December. During this period the quantum of squid and cuttle fish forms a significant part of the total exports. Most of these exports are directed at European countries. It is to be hoped that the recent ban imposed by the European Union would soon be lifted and exports will be resumed to the EU. If this does not happen, an added setback to the export earnings may take place.

The Export Inspection Agency is reported to be concentrating on checking the present conditions at various processing plants, to be followed by a special audit, with a view to assuring EU of sound hygienic and other standards of the plants. Through joint efforts, it can be hoped that the present problem will get resolved sooner than later.

Fishing Chimes, Editorial 243: October(iii) 1997: Vol. 17, No. 7

# Mid-Sea Bunkering

The Ministry of Food Processing Industries has accorded permissions to several of 41 companies operating 66 trawlers to take mid-water bunker. Around 16 trawlers were expected to take mid-sea bunker by middle of October, at a sea distance of 24 nm on the west coast, as per the permissions accorded. Government had earlier permitted a few joint venture companies to bunker their vessels outside the country with permissions to pay in foreign exchange.

For the new permissions given, the Reserve Bank of India is understood to have issued instructions to the commercial banks to open letters of credit in foreign exchange in favour of the suppliers as and when approached by the trawler owners.

The midsea supply of bunker on the west coast is reported to have been arranged from Gulf Oil Co., Dubai by an Indian entrepreneur. The midsea supplies on the east coast are being organised from Singapore based companies by US and other Indian entrepreneurs. High flash HSD oil with a flash point of 66°C will be supplied at a price around Rs. 7.20 per litre in foreign exchange terms against the local price of Rs. 11.50 per litre, it is learnt.

Each trawler would require between 300-500 kl in a year. In other words, 66 trawlers would receive a supply of 19,800 to 33,000 kl in a year. The maximum requirements may be 50,000 kl. The savings per trawler because of taking midsea bunker is estimated at Rs. 4,000per kl. This is a good incentive for toning up economics of operations.

The permissions given for mid-sea bunkering are subject to other conditions, besides the stipulation to take bunker 24. n.m away from the coast (outside customs jurisdiction). These are: a) Informing beforehand the Ministry of Food Processing Industries and coastguard on the location of bunkering and the quantity of oil to be bunkered and b) submission of voyage reports to the Fishery Survey of India soon after completion of a voyage furnishing details of area fished, species-wise catches and earnings.

The various companies concerned are reported to have worked out details of the system to be followed in regard to receiving bunker from the barge that comes with oil, by converging at the place at the appointed time.

Coupled with the Rehabilitation Scheme and the midsea bunkering facility it is to be hoped that the deep sea fishing industry, although having a modest fleet strength, will get stabilised and eventually will gain in strength. The twin reliefs provided by the Government are major steps to bale out the industry from its current crisis.

Fishing Chimes, Editorial 244: November 1997: Vol. 17, No. 8

# Deep Sea Fishing: Need for Composite System of Financing

Apex and Commercial banks have stopped lending funds for both fresh and additional capital investments and also to meet fresh or additional working, capital needs in respect of marine fisheries activities particularly deep sea fishing. Unless this situation is counteracted, Government may have soon to avail of the services of a versatile composer of obituary poetry to write the epitaph particularly of the deep sea fishing industry of the country, for installation on the frontage of Krishi Bhavan in Delhi that houses the Union Department of Animal Husbandry, Dairy and Fisheries. Surely there must be a way of formulating a strategy to evolve a sound financing system that works and that would take care of the known pitfalls in the present system of financing the fisheries sector that is now in a cadaver state. The Government has by now adequate knowledge of the pitfalls inherent in the present system.

## The Present Image

The image that the Indian deep sea fishing industry now presents to the outside world appears to convey that i) India's deep sea fishing industry is in doldrums; ii) Now is the best time to poach in Indian EEZ as the fleet strength of Indian deep sea fishing vessels has come down and likely to come down further; Hi) There is no danger of India emerging as a deep sea fishing nation, as the weakening entrepreneurship may weaken further, and iv) There is no danger of fresh investments in deep sea fishing as the banking structure is against lending money to the sector.

Smaller nations such Maldives and Sri Lanka are performing in a far more purposeful way than India in fishing for tunas (which are their main resource) and also allied fishes in their EEZs, with Maldives producing over 88,000 t/annum and Sri Lanka over 55,000 t/annum. India is stuck up in tuna sector at a little over 33,000 t/annum, that too with catches from the coastal zone. The experiment or commercial longlining with Indian vessels in our EEZ has failed.

## Golden Days are Gone

The golden days of deep sea fishing sector securing loans and repaying them, although only to some extent ended with the closure of SDFC as the financing body. While the loanees were no doubt defaulters in that era too in various degrees, the authority of the Government and the equation that the erstwhile SDFC maintained with the owners used to elicit response in respect of repayments. When the subject went into the hands of 'Designated person' the deterioration appears to have become steep; but the fault is not that of the 'Designated Person'. It was of the Government. It disrupted an ongoing system for nothing better, the straightjacketed banking system is not compatible with the characteristics of deep sea fishing industry.

## Designated Person Performed his Duty

Bankers have a set of system of recovering loans. Dutybound, the system applied by the 'Designated Person' was to auction the vessels hypothecated to the Government, when repayments are interminably delayed in spite of efforts. This approach had an irrationally negative effect. While it is talked about that one or two purchased their own vessels auctioned through proxies, others kept quiet, only to lose the vessels to ship breaking enterprises. Similar results may repeat themselves in the case of the vessels whose owners who have not opted to come under the Rehabilitation Scheme to avail of the concessional terms so generously extended for one time settlement.

## Will there be a 'Second Coming' of the DSF Scheme?

The near 'burial' of deep sea fishing in India raises the question whether there will be a Second coming' of the thrust. Despite the complacency and the lack of political will this issue will force its elf with a relentless compulsion sooner than later. The EEZ, soon to be bereft of Indian deep sea vessels in sufficient strength with be an open invitation to foreign poachers. Thai boats are reported to have been poaching in our waters with an unparalleled temerity, particularly in the shallow zones of Or is s a coast, with none to challenge them effectively. Such being the case, an indirectly open invitation to foreign vessels to poach, although unintended, in the Indian EEZ by keeping it relatively free of Indian vessels will have disastrous conse-quences. When this scenario is perceived Government may be forced to shed its present complacency towards our sea fishing and formulate a pian of action. Instead of allowing matters to drift further, it is of utmost expediency to evolve a policy and a strategy for the development of deep sea fishing in Indian EEZ and implement the same without losing further time. It has been the national experience that fishing boats were often the first to report to Government of movement of alien vessels in our waters.

## Lend and Wait for Repayments': This is not a Sound Approach

For any major scheme such as the one for financing

deep sea fishing vessels a specific organisational structure and mechanism consistent with its characteristics is of paramount importance. The present woes can be traced to their conspicuous absence. Lending funds and waiting for repayments to flow in without the support of a recovery system is too much to expect. This system would have been good during Lord Rama's period, which is now ages and ages behind. Chanting Mantras sitting under a Tamarind tree would not lead to the detachment of fruits to fall on the ground below.

Further, for meeting the financial needs of future development of deep sea fishing industry, it is just assumed by the Government that the banking structure is there to support it. Even the banking institutions thought so. Having learnt bitter lessons, now no financing body wants to touch the fishing sector even with a barge pole, thanks to the ill-considered and hasty decision of the Government to abolish SDFC.

## Where Modern System Fails, Traditional System can Hold

M/s Netsales, Cochin, fishing boat builders, have financed recently the construction of a purse seiner at their own yard in a unique way. The enterprise organised the finance and built a purse seiner and entrusted the same to a team of loanee fishermen. The main condition governing the agreement is that the returns from the catches would be shared at 30:70 or so between Netsales and the group of fishermen. Repayments based on this formula are having the effect of reducing the loan and interest payable in stages. The representative of Netsales is continuously vigilant at the harbour quay in respect of periodical arrival of vessels from fishing for claiming the share. What one can see is that where the modern system has failed, the modified traditional system has the elements of success.

## Turn a New Leaf: Seek NCDC's Help

In the present situation, in order to develop deep sea fishing, a composite approach is essential. An organisation of the owners has to be first built up. This organsiation has to work out a Project, spelling out the number, size and types of vessels to be operated and from which base and delineating other aspects and secure the approval of the Government for the same in the first instance. The organisational structure and manpower requirements and the mechanics for monitoring the movements of the vessels, among others, have to be presented in the Project. An infallible system of obtaining daily information on catches, and ensuring the invariable presence of the organisation's representative at the landing point, for taking stock of catch levels on each occasion of arrival of the vessels has to be incorporated in the Project. More than all this, the vessel owner, the organisation and processor/exporter to whom the catches will be sold would have to enter into an Agreement covering the vital aspect of repayment of loan instalments and interest. The role of each of the signatories has to be spelled out in the Agreement. The processor/exporter has to specifically agree to pay to the organisation, a specified share of the proceeds as envisaged in the Agreement to cover overheads. There will of course be several other aspects to be taken care of.

Once the contours of an effective organsational setup at each of the ports of operation is finalised, the main and most important aspect to be finalised is the source of finance. In this context, elimination of the commercial banking sector as a source of finance is inevitable. The next alternative that presents itself is the cooperative sector. The deep sea fishing industry, to prove that they will succeed in a cooperative setup, has to think of registering a Central Deep sea Fishing Cooperative Society under the Central Cooperative Act with the owners/would be owners of deep sea fishing vessels as the members. The members would have to take the highest number of shares they can, as to be eligible for financial assistance from the National Cooperative Development Corporation. On the strength of the quantum of share capital, the proposed society can apply for loans, grants, share capital contributions etc., from the National Cooperative Development Corporation with the needed supporting guarantees as stipulated by NCDC. With this assistance, the integration of the various links of the activity such as introduction of vessels, their operation, returns and repayments, through an effective management system can be effectuated. Further, post-harvest handling of catches including their processing, domestic marketing and export marketing can be achieved and integrated with the production component of the Project. Cooperative sector appears to be the only beacon light available now to guide the industry. The future depends not only on the response of NCDC but also on how the industry builds up a sound equation with this financing agency.

NCDC can have no reason not to participate as long as it is provided with a sound proposal that ties up the various links of it effectively. As long as the members feel their responsibility and avoid repetition of reported evasion of payment of instalments to Government through 'Designated Person', as at present, there should be no problem. Further, while enrolling members into the proposed cooperative, there should be a rigorous check of the past record of performance of the members and their reformed attitude. By and large, the experience has been that the skipper-owners of vessels are better pay masters. In other words, through a proforma and a Committee, members should be chosen and enrolled and they may have to be made to give a declaration of cooperation in the endeavour to participate in deep sea fishing with devotion and purpose. Frequent checks, reviews, transparency of the activities of the proposed society could make the operations clean and good.

It is desirable that the Association of Indian Fishery Industries starts a dialogue on the subject with NCDC and in that light organise a Deep sea Fishing Co-operative under its auspices so as to proceed further in the matter with the blessings and financial participation of the Union Department of Animal Husbandry, Dairy and Fisheries. It will be a tough job to come out of the past and enter into the present with fresh vigour and dedication. Yet, the AIFI should be able to accomplish the uphill task, with support from its members and the Union Department of Animal Husbandry, Dairying and Fisheries.

Fishing Chimes, Editorial 245: December(i) 1997: Vol. 17, No. 9

# Rehabilitation Scheme: Missing Reliefs

The scheme for the rehabilitation of Deep sea fishing industry introduced by the Government has given an enormous relief to Deep sea fishing companies. The scheme, as is known, is based on the recommendations of the Murari committee. The prime provision in the scheme is that the repayment requirement is reduced to 30 per cent of the loan given by the Government to the owners. This payment will be in full and final settlement of the loan. Further, the said payment frees the trawlers from their hypothecation to the Government.

There are quite a few owners who feel aggrieved in respect of the earlier deduction of the bank guarantee amounts provided by them out of interest account and not from the principal account. Under option B (one time settlement) allowed under the scheme and to which all the owners have opted on the manner of adjustment of bank guarantee amount as and when invoked would be to adjust against 30 per cent of the outstanding principal to be paid by them but this system does not cover the invocations of the guarantees made earlier to the introduction of the scheme.

There are two categories of owners *vis-a-vis* bank guarantees. In the case of some, the Government, through Designated Person (ICICI) adjusted the invoked bank guarantee amounts prior to the introduction of the rehabilitation scheme against the interest account, probably for reason of non-renewal of the guarantees in time. There are others whose bank guarantees are yet to be invoked (or already invoked) but are expected to be invoked, and adjusted against the payable balance of30 per cent of the, outstanding principal. Those belonging to the former category now feel aggrieved that just for the reason that their bank guarantees were invoked prior to the introduction of rehabilitation scheme, the relevant amounts were adjusted against the interest amount. Had those guarantees not been invoked. Then, the amounts would have been invoked now adjusted against principal amount, and not against interest account as was done.

Being a rehabilitation scheme, it would be desirable to impart a broad interpretation of the provisions under paras 3.1.3 and 3.1.4 of the scheme. The former relates to the bank guarantees yet to be invoked. Guarantees not yet invoked fall under two categories; (1) those who filed court cases and obtained stay orders to prevent invocation of the guarantees. In these cases as per par a 3.1.4, once the cases are withdrawn by the company concerned, the guarantee amounts will be adjusted against the interest account. This appears justified for the reason that the companies went to the court for no fault of the financing institution (Designated Person). However, the position has to be viewed differently in the case of such of the companies whose guarantees were invoked and the amounts adjusted against the interest account instead of principal account, earlier, although these have no connection with court cases. It would be an act of giving benefit of doubt to these companies by the Government to direct the Designated Person to give the benefit of deduction of these amounts out of the principal amount through needed adjustments (transfer of the adjustment from the interest account to the principal account). The reasonableness of this approach would have to be appreciated by the Government.

Mobilisation of funds to pay the 30 per cent of the principal amount by the companies as required under the rehabilitation scheme is an uphill task for the companies concerned, particularly at this time when the money market is so tight and constraints in securing loans well in time for making payments for meeting the obligations is an ordeal. In fact, most of the companies are scouting around in search of funds. A liberal interpretation of the provisions under rehabilitation scheme in favour of those companies whose bank guarantee amounts had been invoked prior to the announcement of the rehabilitation scheme for readjustment against the principal amount is likely to facilitate the transactions towards fruition. It is to be hoped that the Department of Banking in the Ministry of Finance (ICICI) would look into this aspect and do justice to the aggrieved.

The Government has been pragmatic and realistic in providing reliefs to the loanee owners of Deep sea fishing vessels. Apparently it has to keep in view the imperatives of enabling the Deep sea fishing industry to stabilise itself at least at this level and later to expand progressively, so as to reach the stage of sustainable utilisation of the deep sea fishing resources. However, having given several reliefs, Government has been somewhat conservative in conceding just a narrow band of time for only payments towards the one time settlement by the companies concerned after the signing of the agreements, although it is fully aware of all the problems faced by the entrepreneurs in securing credit from the banking structure or from others. The time stipulated is just a duration of 30 days from the time of signing the agreement for one time settlement. Considered from any angle, this time limitation is too short. No doubt one has to concede that whatever be the time given, the request from the industry for further extension of time may still be put forth. Notwithstanding this general tendency, considering the fact that the time of 30 days is too short, the companies deserve a reconsideration by the Government to allow a further time of at least another 60 days for the companies to fulfill the obligation.

Fishing Chimes, Editorial 246: December(ii) 1997: Vol. 17, No. 9

# Standardisation in Marine Pearl Production

Standardisation in the production of marine pearls has been achieved by the Central Marine Fisheries Research Institute under the leadership of Dr. M. Devaraj, Director of the Institute and Dr. G. Syda Rao, Fisheries Scientist of the Institute. The scientists have standardised the technology for the growth and upkeep of pearl oysters *(Pinctada vulgaris)* in landbased masonry ponds.

Under this new technology pearls of various grades ranging from perfectly round and creamy to golden yellow colour were successfully produced at Visakhapatnam last year. This development is epoch-making and has enormous commercial potential. Cultured pearls of this kind have wide market internationally.

While a few entrepreneurs, now having shrimp farms adjacent to the coast, made efforts to introduce pearl oyster culture and pearl production in their ponds, owing to lack of adequate extension support in the transfer of technology and field support from CMFRI not much of progress has so far been achieved in popularising the activity, despite its potential. The situation is anomalous and there appears to be an obvious system failure. This shortcoming has to be removed by the Indian Council of Agricultural Research by paying special attention to the problem and evolving an organisational mechanism that would lead to the spread of the technology. It is a known fact that major technologies of this kind are not meant to be kept in show case. The need is to effectively pass on the technology to the entrepreneurs and among the coastal fishermen. It is sad that the scientists developed a major technology of immense commercial import and the technology is left, by and large, to languish. Demonstrations here and there would not turn the corner. One swallow does not usher in summer.

Efforts at popularisation of the technology in a big way among the coastal fishermen and entrepreneurs have to be mounted under a major follow-up scheme. The formulation of a project for the purpose and its implementation is the need of the hour. Launching of a project for implementation at some of the farms of entrepreneurs to cover a full cycle of operations from the stage of raising adult oysters to the stage of implantation of the nuclei and harvesting of the oysters and pearls therefrom alone can yield viable results. An integrated project has to be launched to achieve results by closely involving the concerned. Special attention to popularise this programme is of utmost urgency and it is a matter of disappointment that the authorities concerned just released reports of the success in producing cultured pearls in confined waters and not much tangible has been done as follow-up measures.

It is hoped that the Indian Council of Agricultural Research will look into the subject with the attention the subject deserves.

Fishing Chimes, Editorial 247: January 1998: Vol. 17, No. 10

# Looking Back at 1997

Notwithstanding the constraints in the way of progress, there have been remarkable developments in the fisheries sector in 1997. At administrative level, the subject of deep sea fishing was shifted from the Ministry of Food Processing Industries to the Ministry of Agriculture and Cooperation in the Department of Animal Husbandry and Dairy, expected to be renamed as the Dept. of Animal Husbandry, Dairy and Fisheries. The Ministry of Environment and Forests set up the Aquaculture Authority of India with headquarters at Chennai with justice Ramanujam as its Chairman. The Authority was subsequently brought under the control of Ministry of Agriculture.

The Government of India (Ministry of Food Processing Industries) has adopted the policy of permitting midsea bunkering by deep sea fishing vessels which came into force towards the close of 1997. With the exception of Tamil Nadu Fisheries Corporation all other State Fisheries Corporations wound up deep sea fishing operations. New deep sea fishing policy has not been announced, as promised by the Government, in 1997.

National Committee on Introduction of Exotic species in India was set up by the Union Department of Agriculture. The Government of Orissa has announced a new fisheries policy in the State. The Govt. of Karnataka announced a revised inland water leasing policy. A long term policy on leasing out fisheries of rivers to small fishermen was announced by the Governments of Bihar, Manipur and Tripura.

The scheme relating to grant of subsidy on diesel oil used by deep sea fishing vessels and also by mechanised fishing vessels was operational during the year. The fishing harbour at Paradeep has reached the stage of formal commissioning.

The Central Institute of Freshwater Aquaculture was adjudged as the best among the various ICAR institutions. A new strain of Rohu, christened as "Jayanti" developed through selective breeding by CIFA was released by the Union Minister for Agriculture. The breed was developed by CIFA in association with AKVAFORSK, Norway.

The Central Institute of brackishwater Aquaculture, Chennai succeeded in inducing seabass to breed and also in raising its seed at the institute's hatchery. The institute started supplies of the seed to the farmers.

Golden mahseer was successfully induced to breed at the Lonavla farm of Tata Electric Companies. At the same farm freshwater pearls were produced using Chinese technology. This was considered as a significant additional system for the production of freshwater pearls, as a supplement to technology standards adopted by CIFA. ICAR sanctioned a research project for the setting up of mahseer hatcheries in Himachal Pradesh. In the same State sites were also allotted to farmers for the setting up of mahseer farms.

The Union Department of Agriculture launched a fishery project in Begusarai district in Bihar for the development of area specific technologies relating to fish culture and for their transfer to farmers through field training. The Union Department of Agriculture also decided to constitute a Committee for framing guidelines on freshwater fish culture.

KRIBHCO, world's premier fertiliser producing cooperative in India took up a scheme for the popularisation of fish farming in adopted villages with financial and technical assistance from Overseas Development Agency of U.K.

Aquaculture Sewage Treatment Plant was developed and successfully tested by the scientists of CIFA. Fisheries projects which were taken up with people's participation in Kerala entered their 3rd phase of implementation. A research project to step up annual per ha production of murrels was taken up at Fisheries College and Research Centre, Tuticorin, Tamil Nadu.

The Society for Rural Industrialisation, Bariatu, Ranchi, Bihar achieved significant progress in the development of fisheries of Tilaiya reservoir of DVC with the cooperation of local fishermen. The significance of this achievement is that at a time when DVC totally failed in its efforts to develop the fisheries of the reservoir, the job was entrusted to the Society which has achieved the development of the reservoir fisheries mobilising cooperation from local fishermen.

A fall in hilsa production in Ganga river continued.

FISHCOPFED, New Delhi and the Govt. of Orissa agreed to the establishment of a Freshwater Fish Production Unit in Orissa by FISHCOPFED and on export of fish by FISH FED, Orissa for sale at the outlets of FISHCOPFED in Delhi.

A new laboratory was established at Bhadra reservoir Project in Karnataka to monitor pollution levels.

A freshwater prawn hatchery and a feed mill were set up in the farm complex of CIFA under a DBT project. Demonstration of commercial culture of freshwater prawn for the benefit of farmers also commenced in the farm complex. Supply of the seed produced at the hatchery to farmers commenced and has become an ongoing process. Feed production with indigenous ingredients has gained momentum and the feed produced, having been found to be of good quality, is now under use.

An indoor water circulation system for hatcheries was developed at the College of Fisheries, Mangalore.

Successful breeding of giant freshwater prawn and rearing of the prawn larvae was achieved for the first time through an indoor water recirculation system at the Fisheries Research Station of University of Agricultural Sciences at Hasseraghatta, Bangalore.

Freshwater prawn culture has gained momentum in almost all the inland States and in all coastal States. The interest envinced in this activity by farmers in inland States such as Madhya Pradesh, inland parts of Maharashtra, Punjab, Haryana, Uttar Pradesh and Rajasthan has been a remarkable development. Integrated farming has gained strength in the States of Punjab and Haryana.

Freshwater aquaculture technology suitable for the inland waters of Andaman and Nicobar islands has been developed by the Central Agricultural Research Institute located in the islands. CMFRI developed polyculture technology for marine fishes. One outstanding feature is that exports of freshwater fishes picked up significantly during the year.

The incidence of white spot disease among cultured shrimps continued to be a challenge. The challenge was met to a large extent through various control measures thereby bringing down the level of incidence to about 70 per cent of the harvested crops. In a couple of coastal districts of West Bengal and in the two northern districts of Andhra Pradesh there was heavy incidence of the disease which has led to mass mortalities.

The challenge has led to the improvisation/addition of reservoir and settlement ponds at several brackishwater farms of the farmers. The rates of stocking of seed has been brought down by most of the farmers. They resorted to early harvesting as the main method of saving crops detected in early stages of disease attack.

Most of the large companies that set up integrated brackishwater farms suffered heavy losses. There was partial utilisation of most of the processing plants set up by these companies. Same was the case in respect of hatcheries and feed mills set up by them.

The National Fish Workers Forum appealed to the Government to ensure proper working conditions to women workers at shrimp farms.

There was a partial set back to coastal aquaculture on account of the adverse judgement of Supreme Court for demolition of shrimp farms in the CRZ which was however later stayed. Prior to the stay order 48 shrimp farms were demolished in Orissa. Supreme Court also ordered that no fresh shrimp culture cycle should be started until final judgement on the revision petitions filed by MPEDA is pronounced. The final judgement has not yet been delivered. The shrimp hatcheries being totally dependent on wild broodstock caught from the sea, there was a major increase in the price of mature and gravid females.

In order to regulate the coastal aquaculture practices, the Union Department of Agriculture formulated guidelines for the benefit of the farmers. The Marine Products Export Development Authority continued to provide advisory services to the farmers. Several private consultants too made their services available to the farmers. The MPEDA set up disease diagnostic laboratories, the latest one to be set up during the year being at Vijayawada in A.P. The technique of early detection of bacterial and viral disease among shrimps through DNA tests has been introduced through these labroratories. A few private consultants also have provided this facility.

One major development is a change in the attitude among most of the farmers, particularly those having sea water-based farms to go in for culture of species other than shrimp or to take up polyculture with shrimp. A beginning has been made by some of the farmers to diversify into culture of pearl oysters, seabass and other euryhaline species. Some concentrated focally on pearl oyster culture with a view to producing pearls based on the technology perfected by CMFRI.

A few farmers have taken up culture of freshwater mussels with a view to producing freshwater pearls.A shrimp hatchery set up under a DBT assisted scheme was commissioned at the National Institute of Oceanography, Goa. Significant progress has been achieved under a DOD scheme for raising seed of mud crab at an experimental hatchery set up at Andhra University, Visakhapatnam. Seed could be raised at this hatchery up to Megalopa stage. Several farmers particularly in A.P and T.N. went in for mud crab fattening activity by stocking water crabs. The first mud crab fattening farm was started in Karnataka.

There has been a stagnation in the marine fishing sector. Production from deep sea fishing vessels came down drastically. Three fishing companies operating tuna longliners, suspended operations. One of these companies was ordered by the Andhra Pradesh High Court to wind up the company, based on a petition by a broker.

The strength of the deep sea fishing fleet came down from around 190 numbers to less than 60 numbers of operating vessels. The owning companies are making efforts to supplement the shrimp fishing systems with diversified systems such as Tuna long lining, but without any breakthrough. A large number of vessels became defunct and some of them were sold in auction by Visakhapatnam Port Trust for the reason that the owners had not paid the port dues. Around 14 deep sea trawlers were engaged in fishing for cuttlefish, squid, lobsters, etc., successfully. They were able to export them.

Government of India announced a scheme for the rehabilitation of the deep sea fishing industry under which fishing companies have been given the option of paying 30 per cent of the outstanding principal amount as one time payment in full and final settlement of the loans given for acquiring the trawlers to the various companies. This scheme was welcomed by the Industry and most of the companies complied with the terms and conditions of the scheme.

The South East Asian Regional Cooperation Chambers and the industry suggested the taking up of joint ventures for deep sea fishing in Indian waters.

In the export sector, contrary to the fears entertained, there was a strengthening trend, particularly towards export of frozen fish. The EU imposed a ban on the export of marine products from India on the ground that the products did not conform to HACCP standards stipulated by the Union. A partial relaxation was given for 10 units, after inspecting various plants in the country but there were indications that the ban would be lifted eventually, as more and more exporters were conforming to the standards.

Sea food exports in 1996-97 registered an increase of 26.52 per cent in terms of quantity and 17.74 per cent in terms of value. 3,78,199 t of sea foods worth Rs. 4,121.31 crores were exported during the period.

The country had exported 2,92,000t of marine products from April to December, 1997 realising Rs. 3,500 crores in foreign exchange. This was against export of 2,65,500t and earning of Rs. 3,053 crores from export during the corresponding period in the previous year.

MPEDA had proposed the implementation of Rs. 200 crore pilot project in association with the Kerala State Infrastructure Development Corporation to provide a permanent solution to marine products quality. The project would include the provision of all needed processing facilities for marine products exporting units. Several new plants came up in the Bhimavaram area of A.P. which is central to the shrimp culture fishery areas of the State. The Integrated Fisheries Project set up a processing plant at Visakhapatnam with a view to popularising manufacture of value-added products. NCDC assisted Integrated Marine Fisheries Development Project with an outlay of around Rs. 23.5 crores was taken up in Karnataka.

Government decided not to extend fishing permits given to Indian companies who were operating foreign vessels under charter, lease and test fishing terms and under joint ventures subject to the expiry of the date of validity.

A new Surimi plant set up at Visakhapatnam by S.K. Bigstar Foods Ltd., in collaboration with a Korean firm succeeded in exporting a few consignments of Surimi initially but the trend could not be continued owing to problems in the procurement of raw material.

In compliance with the stipulation made by US authorities, work had been taken up for the fitment of TEDs to shrimp trawl nets. Unless there is a certificate issued by competent authority that the shrimp catches were from vessels with nets fitted with TEDs, exports to USA were not permitted.

India has sought a ruling from WTO on USA turtle stipulation that TEDs should be used on shrimp trawlers operating in Indian waters, although there was a negligible population of turtles in Indian waters. Bharat Electricals Ltd., Bangalore brought under production a fish finder in collaboration with the Central Marine Fisheries Research Institute, Cochin. CMFRI supplied these fish finders to a few fishing vessels.

The Electronic Research and Development Centre at Thiruvananthapuram developed a Fish Finder and Satellite-based Navigation Guidance System. The marketing of the integrated equipment has been taken by Aerospace Systems Ltd. The equipment is reasonably priced. It is said that mechanised fishing boats would be able to afford to instal the gadget in their boats with an element of subsidy from the Government concerned.

There were quite a few incidents involving loss of fishing boats and missing of fishermen out at sea during severe cyclonic weather conditions both on the east and west coasts of India. There were also incidents of foreign boats mostly of Myanmar and Sri Lanka reaching Indian shores owing to bad weather conditions/engine failures. Steps were taken by the authorities for sending them back to their respective countries.

Mass mortality of sea turtles off Gahirmata coast of Orissa took place towards the end of March 1997. The Gahirmata coast is proposed to be declared as a sanctuary.

With the exception of the Tamil Nadu Fisheries Development Corporation, the other corporations withdrew from deep sea fishing sector. The TNFDC emerged successfully in the operation of deep sea fishing vessels The A.P. Fisheries Corporation ran into heavy losses and consequently put its assets on sale. It is stated that there is a proposal to set up a marine products processing unit by MPEDA at Paradeep fishing harbour. COFREPECHE of France commenced working on a proposal to set up a pilot processing plant-cum-training centre at Visakhapatnam.

Ganesh Benzo Plast Ltd., Calcutta launched a product with name "Preserfish" as an additive to water before ice manufacture. It is stated that Ice so manufactured would impart greater keeping quality to the fish.

An agreement was signed between the Govt. of Orissa and the National Sea Port Company of Singapore for the construction of a port at Dhamara along Orissa coast. Central Institute of Fisheries Technology has finalised proposals to acquire a 30 m long trawler-cum-longliner for experimental purposes. Several Conferences, Seminars, Workshops were conducted during the year.

Fishing Chimes, Editorial 248: February 1998: Vol. 17, No. 11

# Fisheries Sector:
## *Focal Points for Government's Attention*

The fisheries sector of the country expects a positive developmental attitude from the new Government. The expectation is that the Government will take effective steps for the sustained and rational development of the immense fisheries resources of the country. As part of these steps, the fisheries sector also expects that the Government will set up infrastructural facilities to cover the entire country for the purpose of bringing about such levels of distribution of fish, prawn and shrimp seed as necessary using the facilities as a means to shift surpluses of seed from zones of higher production to less fortunate areas, and also to enable movement of fish from coastal areas and from reservoirs, lakes and such other larger water bodies to areas deficient in fish production for extended supplies all over the country through a well designed domestic marketing system. Further, considering the emergence of demand for freshwater prawns and major carps in certain foreign countries, the fisheries sector expects Government's initiative and support to bring the inland areas of coastal States under export endeavour.

On the freshwater sector, development of reservoir fisheries will have to receive the attention it deserves on top priority. The reservoirs constitute a major segment of the inland fishery resources of the country, now yielding, 15kg/ha on an average which is deplorably low. Tangible technologies for stepping up fish production from reservoirs have been developed and trained manpower, although in modest numbers, is available in the country. However the available manpower has to be augmented to bring about rational development of fisheries of these resources, their exploitation and utilisation. While there have been efforts in the various States to utilise the potential, the progress has been tardy, for want of adequate attention, despite availability of technology standardised by CICFRI, as already stated. The fisheries sector expects that the new Government will spare the needed attention, this sector of endeavour deserves. Reservoirs in India with an average surface of over two million ha. can produce 2 lakh t of fish annually at a modest estimate based on 100 kg of production per hectare, Development of reservoir fisheries will have the effect of upgrading the socio-economic conditions of fishermen dependent on reservoir fishing and providing employment to thousands of men and women.

What is needed is an integrated and coordinated development, just not on State-wise basis, but on a national basis in an integrated manner, through a well planned network of cold/frozen storages and processing units, ultimately leading to adequate fish supplies to the domestic markets and also contributing to exports.

The country has advanced to a considerable extent in the matter of pond and tank fishery development with the production per ha reaching a level of over 2t/ha. This is no doubt a major achievement. However, the nation cannot rest on its oars at this level of accomplishment, as much more can be done, which is possible. There are advanced technologies developed to raise the production levels through the use of two-tier culture systems, with the top-tier consisting of cage culture system and cage-free water below the cages for general composite culture or polyculture. In order to obtain a healthy growth under this intensive system, the use of aerators will have to be promoted besides improving water management systems. Schemes for augmenting freshwater production through integrated farming, running water or flow through systems etc., have to be launched.

In respect of the brackishwater farming sector there has been a major setback owing to reasons which are by now well known. These include the setback caused due to legal intervention affecting the production activity, incidence of disease, social conflicts etc. AII these problems can certainly be resolved, provided the Aquaculture Authority of India looks at the problems from various angles for the benefit of the nation. Brackishwater farming sector provides employment to several thousands of persons and contributes to exports, and it certainly deserves a down-to-earth attention of the Aquaculture A uthority and the Governments concerned.

While it is well established that brackishwater farming is ecofriendly and with practically little potential to cause pollution by itself, owing to conspiracy of circumstances and extraneous sources of pollution causing harm to brackishwater farming, there have come into the picture several factors deterring the development of this sector. Government has to pay close attention to the situation and initiate necessary measures to resurrect this harmless sector by enabling the Aquaculture Authority of India to play an active, prompt and responsive role to promote it, particularly in wastelands adjoining the coast. In U.S.A. coastal farms are classified as tourist and holiday centres, obviously because the activity is non-polluting.

Despite the fact that collection of spawn\fry from our rivers has come down, the riverine fisheries continue to be in a neglected state. The fisheries of the rivers need enrichment through specially devised stocking programmes and conservation and pollution mitigating measures. These remarks are equally applicable to the cold water riverine and stream fisheries. The major cold water fisheries such as mahseer, *Schizothorax* etc., now face the threat of gradual extinction. There is a responsibility on the part of the Government to revive the cold water fisheries and protect them, while designing means of sustained exploitation.

There is a disorganised state of affairs in respect of production of seed of fish, prawn and shrimp. Because of this situation, while some areas enjoy the availability of abundant supply of seed, the other regions have the disadvantage of having inadequate supplies, more for reason of non-availability of infrastructure facilities and slow development of the available resources. There has been inadequate attention towards creation of new resources and development of integrated farming. Integrated farming enables the utilisation of wastes from agriculture, poultry, duckery, cattle pounds, etc. It will be in national interest to promote integrated farming, more for the reason that it would not effect the utilisation of land for paddy or wheat cultivation. Running water fish culture, and two tier fish culture in farms as in Taiwan have to be promoted. The fishery industry expects that the new Government would take all these aspects into cognisance and evolve needed strategies for the further development of culture fisheries.

Giant freshwater prawn seed production and culture has now reached a takeoff stage particularly in most of the coastal States. The constraint is the meagre availability of seed of this prawn in the inland states, for the reason that for the production of freshwater prawn seed, water of certain salinity is required although for a limited number of days. This natural constraint calls for the development of freshwater prawn hatcheries at selected places along the coastline and devise a strategy consisting of developing routes for the transportation of the postlarvae seed to the inland areas of coastal States and to the inland States. The plan of action or strategy required in this context is to identify routes of supply from the coastal centres of production to carefully selected locations where seed storage farms mainly for the purpose of storing and relaying seed supplies to the waters around and for onward supplies to similar farms along well drawn routes. In other words, a national network has to be built up to provide supplies of fresh water prawn seed to the farmers in the various parts of the country. The same network can also be used for the movement of major carp seed from places of abundance to areas of low availability. The setting up of such a national grid would require utmost attention of the Government.

The task will become comprehensive if the underground saline waters in the north western belt are utilised for production of freshwater prawn seed and channelising supplies through the proposed national grid.

A similar approach has to be adopted by the Government in respect of radiating shrimp and other euryhaline fish seed supplies. There is now sufficient experimental evidence that shrimp seed and seed of euryhaline fishes can be acclimatised in gradual stages to thrive in freshwater conditions. After the needed further experimentation, the supplies of seed of these varieties also can be made available to the farmers all over the country through the proposed national grid system. This aspect deserves the attention of the Government. The proposed grid system will rationalise seed supplies, and bring about a near uniformity in price levels.

Organised fish, freshwater prawn and shrimp supplies to the domestic markets and export of freshwater prawn and fish from inland areas are the most neglected. Fish is available along the coastal belt and in inland zones where the animals are produced in reservoirs, tanks, ponds, etc. Through development of a national grid of cold storages at feasible distances and places central to the areas of consumption for the dissemination of fish supplies all over the country, one of the major bottlenecks under the present system to bring about a rational supply can be removed. Once the domestic market develops, the way will be paved for augmenting exports, not only of marine fish and shrimp as at present but also of freshwater fish and prawn which are now being exported in sparse quantities. This trend has to be reversed by the new Government in order to evolve an extensive plan of action so as to ensure not only even supplies of the products throughout the country but also to promote exports.

The nation has an EEZ extending over 2.02 sq.km of Exclusive Economic Zone. Despite this it is unfortunate the nation has no deep sea fishing policy. We did have a deep sea fishing policy until recently, but the Government has practically anulled the policy and has announced that a new policy would be introduced. This has not yet been ushered in. It is of utmost urgency that the Government introduces a new policy consistent with national interest on top priority. The present vaccum is having several repercussions, more important of which is leaving the EEZ open to owners of foreign vessels to clandestinely take advantage of the situation for their benefit.

The Indian offshore deep sea fleet which reached a strength of around 190 vessels is virtually reduced to an operating strength of around 80 vessels most of which continue to be engaged in shrimp trawling along the upper east coast of the country. Around 16 vessels out of these have taken to diversification of fishing and are now operating along the west coast of India for catching cephalopods, deep sea shrimps etc. This is no doubt a welcome development but the need is to introduce a policy inclusive of a scheme for equipping the vessels to undertake diversified fishing.

The companies have to be assisted in installing equipments such as those for undertaking monofilament tuna longlining, deep sea stern and midwater trawling and possibly gill netting. The stabilisation of the existing fleet as referred to above, minimising their interest in catching shrimps, which are already overfished, is of utmost urgency. In addition to this stabilisation of the existing fleet, it is in national interest to introduce an additional number of vessels in the OAL range of about 20-30 m equipped for multipurpose fishing, ensuring that the capital investments on each of these vessels would be consistent with the availability of resources within the EEZ and the economics of operation. In determining the number of vessels that have to be introduced, Government should keep in view the closing down of operations of a few tuna longliners introduced by Indian companies which failed to secure economic returns.

It is now known that the reason for the debacle is that the Indian enterprises introduced long liners in larger size ranges operated by Taiwanese, ignoring the fact that Taiwanese operate larger vessels for the reason that Taiwan is a distant water fishing nation and the vessels have to travel over long distances to fish in waters far away from their country and stay out for over six months at a time. This requirement led to the operation of larger long liners by them. In contrast, the vessel to be introduced by Indian enterprises have to be in smaller size ranges for operation within the EEZ of the country because of the added advantage of returning to the port within much shorter durations and as and when the holds are atleast reasonably full with catches. Accordingly, it is imperative that vessels in a size range broadly not exceeding 30m OAL have to be introduced, equipped for multipurpose fishing and with tuna longlining as one of the systems. So far as the number of vessels to be introduced and the numbers to be operated from the various ports along the coastline is concerned, the most competent agency to determine this aspect is the Fishery Survey of India which has the needed data on the available resources zone-wise, beyond the continental shelf. Any further delay in the announcement of a new deep sea fishing policy covering our EEZ would be detrimental to the interests of the nation.

The present trends in respect of development of export markets is congenial, although temporarily marred by a ban on exports from India imposed mainly by EU. Most of the varieties of fishes that are caught can be exported either as whole fish or in a processed form. This has to be kept in view by the Government in the formulation of new deep sea fishing policy which inter-all; a must also enunciate linkages with the processing infrastructure of the country. The present infrastructure needs upgradation by addition of equipments for imparting value addition to the harvested catches. Further, vessels in higher OAL range to be introduced can be equipped with onboard processing facilities and this aspect has to be kept in view in determining the strength of the fleet to be introduced under the new policy.

In general, politicians tend to accord very low priority for the development of deep sea fishing not being conversant with the importance of the subject and the fact that future fish requirements of the nation have to largely come from the seas in the next century. Unless there is a pragmatic policy for the development of fisheries of our EEZ, there will be a serious setback to augment proteinaceous food supplies so as to meet the requirements of the anticipated increase in population in the 21st Century.

Another important aspect towards which the Government has to pay attention is the narrowing down of the chasm that exists between the traditional fishing sector and the developing fisheries sector. This chasm has created a serious rift between the traditional fishermen and the industrial fishery outfits. This can be narrowed down only over a period of time and Government has to initiate remedial measures so as to gradually close the gap. What is actually necessary is to set up a few specialised fisheries adjuncts on the east and west coasts of the country, keeping them under the management of CIFNET. Fisherboys in the age group of 8-9 years have to be inducted into these adjuncts and trained for a period of 10 years. The academies have to be well equipped with the needed infrastructure facilities, if necessary, through bilateral or multilateral assistance. The fisherboys should receive training free and at Government cost. Free boarding and lodging facilities have to be provided to the candidates by the Government. The needed training vessels have to be attached to the academies, if necessary through assistance to be received from bilateral or multilateral sources. The candidates who emerge out of these academies will act as a link between the traditional sector and the developing sector and this will have the effect of unifying the two sectors in course of time and thereby extinguishing the present ill will and distance between them.

The candidates coming out of the academies will have to be assisted in securing fishing vessels and the needed working capital for operating the vessels. With these and other steps, the chasm can be expected to be closed in a few decades.

The various fisheries organisations may like to take up the various points raised in this writeup as appropriate along with others which they may have in mind and represent to the Government with a view to securing the needed reliefs.

Fishing Chimes, Editorial 249: March 1998: Vol. 17, No. 12

# Faster Fisheries Development: Consultancy, Infrastructure is the Key

There are lacunae and weaknesses in the present system of extension of fisheries technologies in India. We do not mean this system is obsolete but is certainly not working well, as is generally known.

The reason for this situation is not far to seek. There is a marked shift in the aspirations and attitudes of farmers, farming enterprises and fishing companies. These are now oriented towards faster progress, but with an inbuilt longing towards higher production per unit bereft of diseases and mortalities. The present extension system is not in consonance with this welcome trend, While sectors other than fisheries are by and large,developing faster, progress in fisheries sector is either slow or moving at snail's pace. This has brought in a distortion or an oddman out situation that prevents development in tune with overall national developmental trends. The sector cannot rest on its oars gloating over marginal annual increase in production and exports. The producers continue to be left high and dry with a gap in technology transmission. Past performances such as popularisation of induced breeding technologies etc., are no doubt there to look back with pride but we do not have much happening to hold our head higher than now in the international sphere in the context of the present attainments and the scope for the future. Seminars on Outlook of Indian Fisheries Development for the 21st Century have been able to recommend certain strategies, but the refinement and clubbing of recommendations that are akin to each other has to be done. This is not a tough job, but political will to accept them with real interest has to be forthcoming. If we go by the past record of political interest in respect of fisheries development, one will have reason to nurture doubts. Those in the sector will be fortunate if politicians recognise the importance.of the sector for national benefit. Our scientists are known for their hardwork aimed at formulating new technologies, sustainable and viable, for augmenting production. Farming and fishing communities look forward to avail of such technologies. The stumbling block, however, is that the majority of the farming and fisher classes cannot assimilate the technologies easily. Professionals have to be by their side with consistent interest to induct them to adopt new technologies, relating to mariculture, offshore and oceanic cage culture, improved brackishwater farming, freshwater fish farming, oceanic fishing, etc. The absence of an effective system to serve as a technology carrier and the needed infrastructure for the purpose, and a strategy to set the activity in motion is conspicuous. The disoriented and cadaverous extension system now in vogue is rendering more harm than good in handling the task. The attitude towards 'extension' work too, for obvious reasons, is perfunctory. This system, with a duality of functions revolves round scientists who develop the technologies and who have also to perform the secondary function of 'extension'. It may be true that they have to function through extension wings attached to the institutes but this would not mitigate their responsibility. Unless this 'extension' system is reformed, we cannot achieve faster results. Scientists can at best perform a first round introductory 'extension job'. They can be expected either to concentrate on their scientific work with some of them earmarked to do 'extension' work. It is not fair to ask them to do both. Clearly, the need is to have a wellbuilt bridge between the Institutes that develop technologies and the endusers. This bridge has to consist of professional pillars, independent but with attachments to sources of technologies and end-users. Those who construct bridges collect tolls. Same way, the professionals whom we call consultants have to receive tolls or a fees for their efforts from the end-users and possibly from the dedicated institutes whom they help by spreading the technologies developed by them to the field.

Efforts should therefore be directed at unfettered motivating of 'registered' professionals outside the Institutes' set up to perform the task of carrying technologies from institutes to the field in a fitting and dedicated fashion and help the farmers in adopting them. It is these professionals that have to bring in a new order that merges with the developmental environment around. What is meant to be conveyed in saying this has two facets. One is that while sectors other than fisheries are developing faster, the fisheries sector is at a disadvantage from various angles (lack of political and professional support, organisational weaknesses etc.) and would not fit into the general complexion of environment around. The other facet is related to conditions within the sector itself. Hypothetically, say, when there is paucity of seed production, and when grow-out systems are in readiness to receive seed, a major problem surfaces. It can be hatcheries that are slow in producing seed, or hatcheries are not there at all for species such as crabs and marine fin fishes. Seed for culturing freshwater mussels that grow to requisite sizes may not be there, or formulae for manufacture of various kinds of feeds by the farmers may not be close at hand. What would happen then? There will be no synchronisation, and links get snapped. To obviate such disastrous situations, the institution of a system of technology transfer, through professional consultants systems as bridges between scientists and end users can go a long way in establishing synchronised development. The consultants will come under pressure impelled by commercial interests, by the endusers at various stages of the activity. That will lead to impressive results, with proper coordinative efforts.

It appears that extension (a word that repels several) or squad-oriented technology transfer system must have three tiers. At the top are the research institutes or such other centres which develop and establish technologies. Their functions should be protected undiluted. The additional work that they have to be asked to handle should be restricted to evaluation of feedback information from the field and evolving solutions based on the information where needed. In no case should they be burdened with general 'extension' work. As already mentioned, they cannot be burdened with both scientific work at the experimental farms and also 'extension' work. If entrusted with both, they will have a good 'alibi' to explain away lapses, saying that they had to be busy with the other work. Individual consultants of professional calibre and consultancy firms have to constitute the second tier from the top, next to scientists at research centres. This tier is the bridge that connects the top tier with the tier below them, the field tier or tier of endusers.

Those of the first tier have to perform the function of transferring technologies to the second tier of professionals (consultants) whenever sought, on payment of an approved fees. It is these men of the second tier (consultants) that are of crucial importance in promoting the work in the field. They will have the potential to bring about a major induction of upgraded technologies into the field. These technologies will lead to increase in production.

The Three tier system as referred to above can have one weakness. Anyone having some knowledge of the subject can cheat the endusers. To eliminate this possibility, a system of compulsory registration of 'professionals' as consultants with the concerned Research Institutes should be introduced. Before according registration, a system of rigorous testing of the 'professionals' concerned at the reliant Research Institute has to be introduced. The persons who want to have recognition as consultants should apply to the Institute concerned in the form to be prescribed, appending their biodata. Thereafter, the Institute concerned should invite each of the applicants for an incisive testing of their calibre through close and informal technical discussions spread over a few days but not in a hurry. If a constituted panel of scientists of the Institute are satisfied, then registration could be granted subject to whatever conditions that are necessary. The Registrars of companies may have to be advised not to register any Fishery Consultancy Company without clearance from the Institute concerned. The end-users should be advised to avail of services of registered consultants only. Scales of fees in terms of maximum or minimum limits can be prescribed for the guidance of the endusers.

Qualified candidates who want to become professional consultants can be provided the facility of undergoing well designed courses at premier institutions such as CIFE, College of Fisheries at Mangalore, Tuticorin. Panangad etc. A minimum qualification (B.F.Sc.) or a minimum field or institutional experience can be stipulated for selection to undergo the course. This course must ofcourse, be individual oriented as Ph.D. courses, but not group-oriented. Any selected one can be taken to undergo the course at any time. A fees has to be stipulated. These candidates may be provided with opportunities to receive guidance in respect of public relationships with farmers, fishermen, industrialists, etc., in regard to tact and temper relations with others, application of technologies relating to the chosen field of consultancy at the farm of the Institute/or vessels of the Institute concerned, for a specified period of time. Once declared as competent by the constituted panel, consisting of scientists/technologists of proven track record, including a performing field professional, the candidate or the firm he represents should be awarded 'Registration'. This should be made known to the endusers through a public notification. A code of conduct for these recognised consultants should be prescribed. Their privileges *vis-a-vis* concerned Research Institutes and end-users should be made known. This will enable them to acquire knowledge of newly developed technologies at the Institutes. They can also provide feed back to the Institutes. In cases where the candidates had already acquired professional experience at the institutes or in the field inside or outside the country, and it meets with the satisfaction of the panel, due weightage could be given while granting 'Registration'.

This Registration bestows a measure of responsibility on the consultants and instils confidence among endusers. Endusers should be advised not to entertain unregistered consultants. The system will gradually eliminate 'Quacks' and arbitrary measures in farm work or in fishing vessel operations. Authorities concerned may consider the approach outlined above to usher in a new system of technology injection into the field, in the place of the present 'extension' system. The merit of the suggested system is its elasticity and it will be helpful in spreading developed technologies faster.

Consultancy firms have played and are playing a major role in spreading newly developed technologies in the west and this system deserves emulation by India. In any case, the sooner the ineffective and uncoordinated extensive system now in vogue is replaced by an effective consultancy system/network, the better it will be for the nation. The broad outlines as they occurred to us are presented in this write-up. It is hoped that the top brass in the ICAR and Government can, hopefully, give a measure of credence to what is presented here and introduce a well evolved and viable technology injection and incorporation system, to replace the present slowpaced and obsolete extension system.

Fishing Chimes, Editorial 250: April 1998: Vol. 18, No. 1

# Entry into the Eighteenth

*Fishing Chimes* is now one more volume old. It is now in the upper range of its age, having delivered 17 volumes, each of 12 monthly issues. A new volume is now borne poised to deliver another set of 12 issues, month after month, until it starts on its next annual voyage in April 1999.

We avail of this gratifying occasion to thank all the contributors, our representatives, subscribers, advertisers and all the readers for the stable support extended for the publication of the journal with confidence.

The journal has been moving over all these years at a faster pace than expected towards the realisation of its objectives. The main one among the objectives is the dissemination of information on specific approaches towards fisheries development adopted in different States to ensure faster development of diverse fishery resources of the country. With State-wise newsletters containing fisheries developmental news, coverage of salient happenings in the fisheries sector, both national and international, and the editorial notes we have been endeavouring to contribute towards the thinking process at National and State levels in shaping developmental trends and strategies.

Another objective of the journal has been to provide a medium to the younger generation of scientists and technologists to give expression to their experiences and views on application of various technologies, development of new technologies, extension problems etc. A good number of contributions we receive are from this budding category of the scientific and technological manpower. Most of their contributions, amazingly, contain fresh and novel views, indicative of the burgeoning talent among the younger generation, who would, in years to come, will be in positions of decision making in respect of scientific, technological and developmental aspects.

A further objective that we consider as very important is to pool up and publish material which is pathbreaking and introduces viable technologies.

Over these seventeen years of existence we have been adding to the pride of publishing several contributions of this category. In fact, in various libraries fisheries research centres and administrative offices, *Fishing Chimes* has gained the status of a valuable reference journal. It will be hard to find a fisheries office without *Fishing Chimes.*

Above all, one development in the past few years gives us a sense of satisfaction. A large number of students in fisheries colleges have taken subscription to the journal, which we supply to them at a discounted rate. Several students keep writing to us how useful the journal is for them to face examinations. Same is the case in respect of several who appear for ARS examinations and come out successfully.

Dear readers, you have nurtured and brought up *Fishing Chimes.* The journal is yours. We publish it for your benefit. There would no doubt be shortcomings, but these are for reasons beyond our control. It will be our earnest endeavour and longing to overcome the shortcomings, and continue shaping the journal in such a way that it would be a more useful desktop companion of the readers. Bless the journal and join us in our continuous and ceaseless endeavour to take it towards perfection, by favouring us with your valuable suggestions.

Fishing Chimes, Editorial 251: May 1998: Vol. 18, No. 2

# Deep Sea Fishing Policy

The nation has no live deep sea fishing policy at present. The complacency in introducing a well formulated deep sea fishing policy is transparent. The country cannot afford the continuance of the situation any longer. Out of 191 vessels coming under deep sea fishing class 81 vessels are in disuse, lying either abandoned or uncared for, mostly at Visakhapatnam fishing harbour. Out of the remaining, 80 vessels arc operational from Visakhapatnam base and the rest from Chennai fishing harbour and west coast ports.

Most of the vessels operate for part of the year for catching shrimp in the main, and remain idle for the rest of the year, for want of viable catch levels. This has the effect of upsetting the economics of operations of the vessels. The need is therefore to equip them for effective multipurpose operations to enable them to conduct fishing round the year.

Major tuna species constitute a known open sea resource in Indian waters, now remaining unexploited by Indian vessels. The privilege of utilisation of this resource slipped into the hands of Taiwanese, through charters, leases and joint ventures. The agitation mounted by Mr. Thomas Kochery no doubt made a dent on this policy of permitting foreign vessels to operate in Indian EEZ., but several permitted vessels continue to fish in the area. There is no transparency in respect of the number of foreign vessels still operating in Indian waters. Government in the Food Processing Ministry (which was looking after the work then) seems to have taken the view that foreign vessels permitted for operation in Indian EEZ have to be allowed to continue fishing in the area until the permitted period is over, despite provision in the conditions governing such permissions that they can be withdrawn at any time. Reintervention of Kochery is probably needed for the Government to tell the nation about the position, as to how this continuance is allowed in preference to evolving a fresh workable policy for the introduction of Indianowned vessels.

A few tuna vessels owned by Indian companies did function for some time but these operations were given up, stated to be because of losses despite good catches. The reason for the reported losses seems to be that the vessels are too large entailing heavy investments. Such vessels are meant for global operations in the same way as vessels of nations such as Taiwan. Introduction of vessels of an economical size adequate to fish in Indian EEZ alone would have cut down investments and working expenses and at the same time brought the same level of catches. Several fishing vessel owners complain that Andamans administration does not allow Indian fishing vessels of the mainland to avail of facilities at Port Blair for reasons not known. It is anachronistic that Indian vessels are denied the facilities in the territory of the nation, while vessels of Thai and of other neighbouring countries fish in Andaman waters, may be clandestinely. Indian vessels are unable to fish in the area for want of access to Andaman ports. This subject deserves immediate attention of the Government as the EEZ of Andamans is known to have rich fishery resources, particularly of tuna. In the EEZ around the mainland and around Andamam a tuna potential of around 2,00,000 t is estimated to be available but it is not utilised for want of a policy.

A planned development of Indian deep sea fishing fleet is important for stepping up fish production to meet the demand for fish in the domestic market and in the export markets. More than this, the development of a strong deep sea fishing fleet will come in handy in times of threat to India's defence from the seas around.

The small 'deep sea fishing fleet' that the nation now has is getting old. The vessels can no doubt be equipped for multipurpose operations but within less than ten years they will become defunct. This aspect has to be taken into account while working out the requirements with reference to the availability and sustainability of the resources of the EEZ. consisting of major pelagics such as tuna in the open sea and a variety of columnar fishes and demersal resources such as cephalopods, deep sea shrimps and lobsters. The policy has to revolve round introduction of multipurpose vessels.

## Financing Policy

The nation does not have a policy that works to provide financial assistance to entrepreneurs for the acquisition of deep sea fishing vessels and for recovering loans. The bulk of deep sea fishing vessels were introduced in late 1970s and middle 1980s with financial assistance extended by the erstwhile Shipping Development Fund Committee in the Ministry of Shipping and Transport. The problems of the function entrusted to Shipping Credit and Investment Company of India (SCICI), an institution set up to provide finances for the acquisition of ships and fishing vessels, but which was later merged with ICICI. SCICI/ICICI have not been able to cope up with the complexities of the fisheries part of the portfolios.

Conventional banking norms do not work in the fisheries sector. As a result SCICI/ICICI could not continue the work of erstwhile SDFC with success. For the revival of financing system for the deep sea fishing industry, it is now imperative to set up a Committee again, similar to the erstwhile Shipping Development Fund Committee, probably under the name of Fishing Development Fund Committee', in the Union Department of Animal Husbandry and Dairying. This Committee, for deciding on granting loans for the acquisition of deeps ea fishing vessels, would have to develop beforehand a conducive and workable mechanism. The Committee could approach the World Bank for start up assistance with a plan and scheme for re-orientation and for providing a re-direction to investment flows to the sector. The following suggestions would deserve consideration.

1. Identifying on both the coasts a few harbours with infrastructure, to serve as main bases for the operation of a determined number of vessels with reference to resources, from each of the harbours in the adjoining FEZ of the zone;
2. Ensuring the fulfillment of a precondition by the applicants, principally, among others, that the owner, the Committee and a chosen manufacturer exporter enter into an Agreement providing for the sale of the produce to the exporter on acceptable terms with the exporter agreeing to deduct a feasible percentage of the returns to pay to the Committee, voyagewise;
3. The chosen manufacturer exporter has to find a place in a panel drawn up for the purpose and to be approved by the Committee;
4. A senior officer of the Committee should be located at each of the nominated harbours with the function of monitoring the arrivals and departures of the vessels and also with the function of recording the details of catches before sale; He shall be present at the processing plant concerned at the time of weighment, and finalising the sale prices, based on a formula mutually agreed to with reference to prevailing export prices; and arriving at the deduction to be made for being paid to the Committee towards loan repayment before releasing payment to the owner;
5. The representatives should be provided with shore SSB Radio Sets to enable them to be in continuous touch with the vessels out at sea to get uptodate information on catch position, trends etc., on day-to-day basis;
6. Such other systems as would be necessary including hypothecation of vessels may be followed to ensure the effective functioning of the Committee;
7. A few willing commercial banks may be identified by the department to provide working capital limits to the loanee companies;
8. Preference in granting loans for acquiring deep sea fishing vessels may be given to companies proposing to have integrated operations;
9. A contingency plan for taking over the vessels of loanees found to be not subjecting themselves to the discipline of the Committee may be evolved. Under this plan a separate set up can be developed for taking over vessels of errant loanees and for having them operated under the auspices of the Committee as long as necessary.
10. Loans may be granted for the installation of additional system of fishing on existing vessels and for installation of the needed upgraded machineries. A similar system may be adopted in respect of the next generation of vessels to be multipurpose in nature, in the general OAL range of 30-40 m, subject to the above mentioned conditions.

## Construction of Deep sea Fishing Vessels

Atleast four Indian shipyards are known to have the capability to construct fishing vessels of 23 to 27 m OAL class. These yards can be enabled to acquire capability to build 30-40 m OAL multipurpose vessels, in preference to permitting import of vessels, The National Ship Design and Research Centre can be entrusted by the Government with the work of designing and preparing the shop floor drawings of these vessels. NSDRC can be authorised to supply the drawings to the shipyards at a subsidised price initially. NSDRC and the shipyards would however require initially technical collaboration from a reputed fishing vessel naval architect and an experienced captain in multipurpose operations for finalising the drawings. These professionals may have to be identified from foreign sources. Once one or two vessels based on these drawings are built and tested in actual fishing conditions under terms of collaboration with the professionals, there would no longer be need for the collaboration. This kind of collaboration is important both for the installation of mono-filament longlining system or other systems on existing vessels to improve their economics of operation and also for the construction of large sized multipurpose vessels.

The cost of Indian built fishing vessels may continue to be far higher than imported vessels. It is accordingly necessary to continue to provide subsidy to the shipyards at the present level (33.33 per cent). Foreign shipyards are known to receive subsidies of this kind from their Governments.

## Training

Facilities at Central Training Institutes may have to be strengthened to enable candidates undergoing training to acquire technological knowledge and practice in monofilament longlining and in the operations of the connected equipments, and in other diversified systems of fishing.

Fishing Chimes, Editorial 252: June 1998: Vol. 18, No. 3

# The Great Indian Saline Land Stretch

There are extensive land tracts in Rajasthan, Haryana, Punjab and western U.P. which, by and large, lie barren. The reason: soil is saline and mostly unfit to raise any land crop. That these wastelands can be brought under brackish/saline water aquaculture was experimentally demonstrated in Haryana by the Central Institute of Fisheries Education, Mumbai, when Dr. S.N. Dwivedi was its Director. Crops of shrimps, mullets, pearl spot etc., were raised from ponds excavated on the saline soils and filled with underground saline water. Further, in Rajasthan, brine shrimps were cultured in salt fields and cysts collected there from. In this background, there was an expectation that the possibilities of mounting a major inter-state project for converting these barren tracts of lands into productive, economically viable and employment-oriented fish farms would be explored through a comprehensive pre-project feasibility study. But alas, there has been no initiative taken in this direction.

The concerned State Governments and the Union Ministry of Agriculture have just left the subject in the cold storage, despite the known position that the development of the area would bestow vast benefits to the people of the zone. The benefits would of course include massive inflow of foreign exchange through exports of shrimps etc. However, as can be expected, there will be several problems to be encountered in a major effort of this kind, but the need is to get over them through appropriate solutions.

For every failure in mounting a project there will always be several seemingly convincing and equivocating reasons to show that such a project would be uneconomical and unworkable. The experimental work referred to above, although done on a small scale, did prove that the culture of marine/brackishwater aquatic animals in the area was possible. This signifies the need for a sustained effort to undertake a thorough study of the prospects for the development of the fallow land stretch for economic benefit. A close and dedicated feasibility study can establish the economic viability and the approach for the development.

Considering the importance of utilising the fallow land, the Union Department of Animal Husbandry and Dairying/ICAR would have to, in concert with the Governments of Haryana, Rajasthan, Punjab and U.P., think of setting up a Cell for undertaking a thorough investigation on the potential of the wastelands to produce shrimp, prawn and fish, keeping the socio-economic, ecological and environmental angles in view. As far as possible, such of the officials from the fisheries and other connected departments of the respective Governments can be pooled up under the Cell to be set up in the Department of Animal Husbandry and dairying/ICAR for undertaking a comprehensive study. The Cell can be headed by an experienced senior officer, known for his capabilities to plan, organise and direct a study of this nature. Luckily there are men of that calibre available in the country.

There, however, appears to be a few basic issues that need to be studied before setting up a cell as suggested. Would the pumping of ground saline water for the purpose of culture be feasible as a long term, proposition? Would the water removal cause soil subsidence? Can an adequate supply of freshwater be arranged in some or all the sections for diluting the saline water taken from the underground to the required salinity level for culture? Would it be possible to set up freshwater prawn hatcheries in the zone making use of the salt water available, instead of incurring heavy transportation charges for getting freshwater prawn seed from the coast? Can nauplii of shrimps be brought from coastal hatcheries to grow them to the post-larval stage required for stocking in the farms to beset up? Is it possible to use the underground saline water to raise brine shrimp cysts for use in rearing of tiger/white shrimp larvae? If the soils are porous, covering the bottom and sides with PVC or other synthetic sheathing be feasible? Would it be economical?

What should be the broad plan of action or contours of study to be undertaken by the proposed Cell, if the conclusions are reasonably encouraging?

It would be a national neglect to allow these vast stretches of waste lands to remain unproductive. All the possibilities of utilising the lands for aquaculture and related activities would have to be explored. Hopefully, if the results of a detailed study are positive, these would lead eventually to several benefits to the nation such as bringing the wastelands under aquatic production, generating employment, contributing to exports, foreign exchange earnings etc.

Fishing Chimes, Editorial 253: July 1998: Vol. 18, No. 4

# River and Sea Ranching

Principal fishery resources of major rivers of India now stand alarmingly depleted. Fortunately major carps have not become endangered species, thanks to well developed and widely practised culture systems. Yet the naked truth that the rivers are pitilessly denuded of the natural wealth of major carps, the heritage of the nation, persists.

Unlike major carps, several mahseer species, several species of cat fishes, such as *Wallago attu, Mystus singhala, Silonia silonia* etc., feather backs (*Notopterus* spp) and *Thynnichthys sandkhol*, the precious sandkhol carp, one of the rare living evidences of discontinuous distribution, and quite a few others are on the verge of extinction. The Central Inland Capture Fisheries Research Institute, Barrackpore, has on hand several measures to stem this trend, mostly the result of release of pollutants flowing into rivers, poisoning of fishes and over fishing. The gravity of the situation is so serious that Government has now to choose between forgetting about the vanishing national heritage of fishery wealth or taking effective measures to revive the past glory.

So far as Mahseer fishery development is concerned, in States such as Himachal Pradesh, Uttar Pradesh, Karnataka, and M.P. Some measures have been initiated but these are grossly inadequate. Unless an effective and comprehensive scheme at national level is formulated to save the various endangered species and implemented, the future of riverine fisheries will become more and more bleak and dismal. Such a scheme is needed so as to ensure implementation of the required measures in the various States through which the rivers traverse. Uncoordinated and distorted implementation of corrective measures under State schemes will be counter-productive, further, under a national scheme it will be possible to set up hatcheries in a planned way for the production of seed of required species, growing them to a ranchable size and releasing them into rivers. Coupled with this ranching programme, a project of cryopreserving male and female gametes of endangered species has to be taken up by CICFRI in association with NBFGR. Further, surveillance stations have to be set up at points central to places where poaching takes place, poisoning is done, and effluents are released, for controlling and diffusing these destructive practices. Public awareness of the evils and disastrous effects of these practices has to be stepped up through various media. Sea ranching has now assumed an inescapable importance that is more crucial than even before. It is noticed that several fish and crustacean species are fast losing their population strength, not because of anthropogenic activities alone, but also for various other reasons such as changes in hydrographic conditions, pollution etc. Alarmed at the gravity of the situation, it is laudable that the Central Marine Fisheries Research Institute has drawn up a major programme of sea ranching on the south-west coast of India, in association with the South Australian Research and Development Institute (SAUDI). This programme aims at restoring the stocks of five main depleting species *i.e.*, lobster, shrimp, grouper, pearl oyster and sea cucumber. It envisages the establishment of multispecies hatcheries at Vizhinjam, and release of genetically marked seeds in an area of about three million ha along 300 km of coastline off the southwest from Cochin to Cape Comorin, in the Gulf of Mannar and the lagoons of Lakshadweep and Minicoy Islands. Dr. Devaraj, Director of CMFRI, says that this zone is characterised by intense upwelling during southwest monsoon and the consequential high level of productivity at all trophic levels. The ranching programme is to be implemented over a period of five years by CMFRI in association with SARDI, India's Union Department of Science and technology and the State Department of Fisheries, Kerala. Funds of the order of US $880,000 for implementing the project are expected to be forthcoming by way of a grant to be released by the Australian Government under the Australia-India NewHorizons initiative. One distinct advantage of the programme will be the infusion of Australian experience in fish stock enhancement technology and establishment of project models.

Just at the time the Australian Government was to signify its formal concurrence to the contemplated ranching programme, which may also have a beneficial effect on the fishery of atleast part of the Eastern Indian Ocean that stretches from the east coast of India to the West Coast of Australia, the announcement by the Australian Government slapping sanctions on assistance to India came as a bolt from the blue. It is to be hoped that the sanctions would not cover this ranching programme which is a welfare programme of humanitarian benefit *i.e.*, a programme that will benefit several engaged in the fishing industry. It is to be hoped that the Australian Government would continue to support the programme and extend the assistance envisaged, taking into account the fact that the programme would, in final analysis, upgrade the living conditions of thousands of traditional fishermen. In the unlikely event of the Australian Government withdrawing support to the programme, the Government of India would have to explore other possibilities of implementing this crucially important programme that would bring prosperity to the nation.

Fishing Chimes, Editorial 254: August(i) 1998: Vol. 18, No. 5

# Flow of Expertise from Fisheries Colleges to Commercial Fisheries Sector

The commercial fisheries sector, over the past two decades and over, has come to rely for recruitment of candidates for management in respect of inland culture, marine capture and processing activities, mostly on those coming out of Central Institute of Fisheries Education, reputed and well established fisheries colleges and some of the conventional universities where knowledge on various aspects of fisheries is imparted. This apart, a large number of qualified candidates drawn from the various aforesaid centres of fisheries education are already in various jobs of managerial/developmental importance in several State Fisheries Departments, Central Research Institutes and several Governmental units. Quite a few have been awarded doctorate degrees. The candidates have brought credit to the august institutions that produced them.

As readers are aware, most of the fisheries colleges in the country were set up by the ICAR as part of the State Agricultural Universities. The Central Institute of Fisheries Education, Mumbai is a deemed university under ICAR's set up.

In the background of the wealth of experience gained over years in conducting the fisheries educational courses, ICAR has recently proposed a revised syllabus for B.F.Sc. course conducted at the CIFE and other colleges under its purview. The proposed syllabus lays accent on Industrial and Rural Work Experience to be gained in one semester in the final year, spread over six months. 20 credits are earmaked for this important segment of the course.

The present syllabus covers shell and fish aquaculture alone. The scope for increasing marine finfish production which lies in popularising mariculture, particularly cage culture in protected seawaters and offshore waters is not covered in the present syllabus. Recognising this, the proposed syllabus has been suitably recast to include sea farming practices of selected species.

However, marine cage culture of finfish, the activity that has to be promoted to meet the needs of the future, does not find a specific place in the proposed syllabus. The activity, however, is undoubtedly part of the term 'Sea farming practices' used in the proposed syllabus.

Commercial enterprises need men having hands-on experience, and equipped to get into stride with the job entrusted to them with ease, thereby providing relief to the employer. The general reaction of owners of fishery enterprises is that most of the candidates coming out of the colleges whom they recruit are knowledgeable. However, it has been their experience that it takes around a year for the candidates to become equal to the job they are entrusted with to handle. In other words, most of them do not have adequate hands-on experience to manage/supervise the work effectively. What has been happening now is that a candidate, who virtually picks up experience at the expense of the enterprise, tends to move away to another enterprise offering a higher salary, at the earliest opportunity. The new employer stands to gain as the candidate he has lured has experience. This lacuna is sought to be plugged in the proposed syllabus. However, allotment of around six months in the proposed syllabus to pick up the needed basic hands-on experience (industrial experience) and also rural experience is inadequate. During this short period the minimal industrial as well as rural work experience that is needed cannot be acquired. Further, on the industrial side, the candidates have to participate in field work in respect of several subjects such as hatchery operations, fish/shellfish culture cycles, operations at processing plants and so on. A candidate will get around a month at each of the places and a comment that this much of time is inadequate stands to reason.

It appears to us that the provision under the VIIth semester would need revision. It should list, as in the present syllabus, relatively more important areas in which industrial experience has to be gained by the candidates, such as hatcheries, farms, processing plants etc. Among these, each of the candidates should be asked to select one main item of his choice for acquiring hands-on training. Around 13 credits could be earmarked for this. They may be asked to opt for another up-the-sleeve item for which 4 credits could be allotted The remaining credits could go for an Arousing kind of hands-on experience in other disciplines.

The CIFE and some of the colleges have attached farms within the campus for providing hands-on training to the candidates, although not in all. Those who do not have this facility have to continue to have a tie-up with farms/hatcheries/processing plants owned by others. An ideal situation will be where a college has within its campus all needed infrastructure for hands-on training to enable the candidates to undergo the course amidst action-packed surroundings. As this is not possible in the case of most of the Colleges. The alternative for the colleges is to enter into suitable agreements with Government/private sector enterprises owning relevant units for providing the needed hands-on training. The training period will no doubt be interspersed with inspections by designated faculty members. The satisfactory completion of the training can continue to be judged as per whatever procedure is laid down. In any case, it appears in the VIII semester part of the syllabus would need a second look so as to make it more specific and helpful to the candidates in securing jobs.

Another aspect that deserves attention in the syllabus is in regard to mariculture. Sea cage culture, as already mentioned, will assume prominence and importance much sooner than we think. The reason is that marine capture fisheries in Indian EEZ have reached a plateau, in the coastal waters. The nation has to resort to mariculture to meet the production needs. Of all systems of mariculture, cage culture will prove to be the most viable one, as has happened in several countries of the west.

There has to be therefore a provision in the syllabus for the candidates to know about integrated marine cage culture systems, the designs of sophisticated cages, cage fabrication materials, stocking, feeding and harvesting systems and link-up with onshore hatcheries etc.

Fishing Chimes, Editorial 255: August(ii) 1998: Vol. 18, No. 5

# The Help Small Fishermen Need

The small fishermen/entrepreneurs invariably face the problem of securing an equitable price for their catches. Unlike exporters who have cold storages, small fishermen/entrepreneurs do not have any such facility. Setting up of cold storages by small fishermen on their own, on an individual basis is not possible at all, looked at from any angle. Firstly, their daily landings are small. Secondly they cannot invest.

In this situation, the coastal State Governments have to rally round to rescue the small fishermen from the stranglehold of those who have the advantage of buying the produce at an opportunity price from them. If the state Governments provide cold storages at important fishing harbours not having such a common facility, to be managed by Fisheries Department concerned, it would prove to be a major help to the fishermen.

In the alternative, a strong and effective cooperative body can be entrusted with the task of providing the facility. Such a step would enable the fishermen to store and regulate marketing of their fish catches and other marine produce in a way that would fetch them a good price. The facility would prove to be a major help to the fishermen. The need for this facility is keenly felt, particularly at Visakhapatnam and Paradeep, not to speak of several other fishing harbours. It is to be hoped that the concerned State Governments and the Union Ministry of Food Processing would look into this matter and take up schemes for setting up common utility cold storages. This step would improve the socio-economic conditions of fishermen and will also ensure steady supply of fish to markets lying in a good radius around the harbours.

Fishing Chimes, Editorial 256: September(i) 1998: Vol. 18, No. 6

# Crew Crisis: Severe Shortage of Fishing Vessel Skippers

India now has its own deep sea fishing fleet of 96 trawlers of which 80 are operating in the upper Bay of Bengal and the rest off the west coast This tally does not take into account the 'Joint Venture' vessels. The so called 'Joint Venture' vessels have foreign crew supported by Indian crew posted by the companies concerned. These companies are not known to have problems of manning their vessels in the Indian EEZ. In contrast, several Indianowned trawlers have been experiencing a serious shortage of skippers, although there is a massive training infrastructure by way of a well developed decades- old Training Institute (CIFNET) at Cochin coming under the Ministry of Agriculture. This Institute has two full fledged centres one at Chennai and another at Visakhapatnam. Despite a large number of candidates trained to become Mates of fishing vessels and later as Skippers of fishing vessels, coming out of the main Institute and its other centres, batch after batch, it is a matter of utter surprise that there is a severe shortage of Skippers of fishing vessels; And this shortage has surfaced a few years back and has now attained an alarming trend, notwithstanding the fast shrinking deep sea fishing fleet of India. Luckily there have been no additions to the fleet, thanks to the old policy in a state of animation and the lack of a new policy. The shortage issue deserves the urgent attention of the Union Department of Animal Husbandry and Dairying which, according to the Industrial circles and the Association of Indian Fishery Industries, is just keeping silent over several representations on the subject.

Visakhapatnam happens to be the main base of operation of deep sea fishing vessels. At least four vessels are held up at the harbour for want of skippers as at the time of this writing. The owners are making frantic efforts to recruit skippers but only by way of luring from other companies and this only means redistribution without any remedy to the problem. And this situation persists even after eight candidates who passed the relevant examination conducted by the MMD have become available.

The shortage of skippers is attributed to two main reasons. One is that several qualified candidates have left for greener pastures in foreign countries such as Nigeria, Kenya, Tanzania. Some have joined the merchant navy. Normally, even with this migration, adequate number qualified candidates ought to be available but this is not to be. It is unfortunate that CIFNET and its two centres are unable to ensure the availability of adequate number of skippers to the Industry.

Another aspect pointed out is that very few of the candidates who appear for the Skipper's examination manage to come out successful, and that too after several attempts. This state of affairs can mean two things, one of which may be that the examination conducted by the MMD is tough and not within the capabilities of the candidates concerned. The second can be that the training system at CIFNET may need a thorough review in consultation with the MMD to set right the general deficiencies in the candidates.

What have been mentioned above are longterm measures. In the short term, some reliefs without compromising manning requirements would have to be provided to the Industry. Otherwise, several vessels have to remain idle, imposing economic strain on the owners who are not at fault. The only way, to the extent we can see, is that the Department of Animal Husbandry and Dairying has to apprise the Ministry of Shipping and Transport/Director General Shipping of the situation and sack consideration on a warfooting to grant dispensation to Mates of fishing vessels with adequate experience to function as Skippers of fishing vessels for such durations as are possible and until such time as the shortage gets extinguished.

Fishing Chimes, Editorial 257: September(ii) 1998: Vol. 18, No. 6

# Cage Farming in Indian Seawaters: A Developmental Imperative

The Union Department of Animal Husbandry and Dairying is aware of the depressing scenario of stagnating exploited sea fishery output from our EEZ. Coastal waters of the EEZ are over fished. In the waters beyond, there is practically no officially permitted fishing, save fishing by two or three joint venture companies, ostensibly under Indian management but run by Chinese and Thai interests. Government did encourage the introduction of a few economically unviable tuna longliners by a few Indian companies, which had to however suspend operations after a few years of fishing having become victims of wrong advice, by whom it is difficult to say, to introduce vessels of uneconomical and unnecessarily longer size ranges on which investments were too high. There were losses, believed to be because of inadequacy of returns to repay loan instalments and interest thereof, which had led to the cessation of operations and one of the companies folding up.

In a developing activity, set backs of this kind are inevitable in the early phase. Only through setbacks, experiences are gained for modifying policies and strategies. However, unfortunately, matters have been allowed to remain where they are. This complacency, related not to tuna fishing alone but to the entire deep sea fishing sector, is not becoming of a major maritime nation like India. Island countries such as Sri Lanka and Maldives are doing very well in tuna fishing and the position is that a mighty nation like India could not perform in this field (with the exception of waters around Lakshadweep islands) even remotely anywhere near the achievements of their neighbours, despite the existence of resources. It is a sad reflection on our planning, policies and strateges in respect of the sector. The present perceptions are that the Government of India accords very low priority to the exploitation of the deep sea fishery resources of the nation and turns Nelson's eye towards poaching by foreign vessels. Such being the case, while it may be unrealistic to expect the authorities to think of promoting cage farming in the sea, it can be hoped against hope that there will be someone in the set up who recognises the importance of the activity, initiate it and impart a momentum to it, besides developing deep sea fishing in the EEZ. It will be recalled that, in the recent past a proposal by an entrepreneur to set up a cage farm near Andamans, supported by a hatchery to supply seed, was scuttled This scuttling is attributed by some to reasons other than genuine technical and developmental considerations. In any case there has been no transparency of the event.

Despite the importance of cage farming in the sea to step up fish production, no guidelines have been formulated so far for setting up such systems for the benefit of the entrepreneurs.

Several countries allow setting up of cage systems in protected bays, coves etc., and also in offshore waters, based on issue of licenses. India has several protected water areas all along the coastline and also around Andaman and Nicobar and Lakshadweep islands. At this time when the nation has reached a plateau in respect of marine capture fishing and the inability of the Government (Sadly, though) to extend fishing effort to the zones beyond the coastal one through an enduring policy and strategy, is apparent, the least the entrepreneurship can expect from the Government of India is the formulation of a policy and a strategy to set up sea cage systems and to operate them, and to encourage entrepreneurs to take to this productive activity by Imparting guidance, support and providing licenses. The activisation/reintroduction of a sound deep sea fishing policy along side cage policy too is of paramount urgency.

In fact, to restate, developmental imperatives in the face of the stagnation in seafish production and complacency in the introduction of economically viable deep sea fishing vessels demand the formulation of a viable policy for the exploitation of the known resources of the deep sea zone and for promoting sea cage culture with a supporting and an effective financing system to replace the dormant one in vogue, so ineffectively framed.

The cages can be allowed to be set up in sea plots pre-determined by the Government and allotted to selected enterprises with such terms and conditions as necessary set out in the licenses, and made valid for a viable duration taking into account the gestation period, number of culture cycles needed for earning adequate profits commensurate with entrepreneureal efforts, returns needed on investments etc. Joint ventures would have to be inevitably permitted as there is no national experience in cage farming in the sea. Unfortunately, the opportunity that presented itself through an entrepreneur, as mentioned above, was given the go by on grounds considered by several as not sound. Otherwise, by now, there would have been an accumulation of some experience.

So far as the selection of fish for cage culture is concerned, our scientists have to examine the suitability of seabass, which has been bred at the hatchery facility at CIBA last year. Once this fish is found suitable, enterprises can be allowed to culture this fish in cages by establishing the needed shore infrastructure by way of a hatchery, and for brooder and seed storage, and transportation facility mainly to take the seed to the cage systems and also for the movement of men and materials. These have also to be supported by way of training to sponsored candidates.

The management of cage systems is a totally new subject to Indian entrepreneurship. Cage sites have to be selected taking various connected criteria into account; and for this guidelines will have to be laid down. Besides this, knowledge on the manner of installing cage systems with walkways, electronic aids, feed store, general store, laboratory etc., has to be imparted to entrepreneurs. Cage dimensions, cage materials, cage cleaning systems, feed administration techniques, disease control methods and other management systems have to he made known. Considering all these aspects, there is no escape from allowing Indian enterprises to enter into joint ventures, as long as necessary, with companies engaged in similar activities preferably in the Mediterranean region where the activity is popular, particularly for sea bass and grouper, besides others. There are also several other nations such as Denmark, Norway, Australia etc., which have expertise in cage culture. A climate can be created for the entrepreneurship to negotiate joint ventures with foreign enterprises to diversify into this sector of production, for national benefit.

We cannot really hazard a guess whether the Governments at the Centre and States would take any initiative in introducing and popularising sea cage farming. It all depends on the enduring interest that the ICAR (CMFRI) and the Coastal State Fisheries Departments take in unison under the direction of the Union Department of Animal Husbandry and Dairying to introduce the activity and expand the same so as to augment marine fish production. The prerequisite is to collect basic information. For this, Groups of resource persons with a leader for each can be formed and entrusted with tasks such as (i) identification of bays, coves and other protected waters suitable for locating cage farming systems all along the coastline and areas around and between Andaman and Nicobar, and Lakshadweep and Minicoy islands ii) determining the dimensions of cage systems that can he located, the designs of the systems, identification of locations for setting up supporting hatchery systems and their integration with cage systems, communication arrangements between hatcheries and cage systems, feed needs, training support, species to be used for culture with seed produced at the hatcheries/collected from the wild, criteria for introducing exotic species, and organisational requirements iii) framing of guidelines for the benefit of entrepreneurs to set up joint ventures in collaboration with foreign companies, etc.

An organisational mechanism has to be necessarily developed to promote the activity. Dr. M. Devaraj, Director, CMFRI and his co-scientists, in a recent paper, have suggested the setting up of Sea Farmers Development Agencies for promoting mariculture. As cage farming in the sea being an important system of mariculture, their thought provoking suggestion deserves serious consideration for promoting the activity. One group can probably go into the various aspects of this suggestion and work out a project on Sea Farmers' Development Agency.

Fishing Chimes, Editorial 258: September(iii) 1998: Vol. 18, No. 6

# Prevention of Whitespot Disease among Shrimps in Grow-out Ponds

In an article entitled 'Managing Whitespot disease in shrimp', published in *Infofish International* May/June 1998 (3/98) issue, authors Y.G. Wang, M. Shariff, P.S. Srinivasa Rao, M.D. Hassan and L. T. Tan, put together a list of comprehensive measures to be taken to manage whitespot disease in shrimp. The steps cover those relating to hatcheries and grow-out ponds. So far as hatcheries are concerned, the authors have suggested their location away from shrimp ponds to prevent brooders and larvae from getting infected by any epizootics in grow-out ponds. They have also reiterated known precautions such as following good sanitary practices, treating water before use, selecting virusfree broodstock, prevention of infection during transportation and maintenance, use of immuno-stimulants to enhance resistance to diseases, use of chemicals to minimise horizontal transmission of disease among larvae [use of some commercial products such as Happy larvae (a branded product) at a dosage of 50g/m$^2$, applied from mysis stage onwards in rearing waters, has minimised horizontal transmission, and is helpful in the treatment of hatchery effluents etc.

So far as prevention of WSD in grow-out ponds is concerned, apart from known suggestions such as proper preparation of ponds, using closed systems, eradication of various carriers and practising sanitary measures, stocking virus-free postlarvae, application of probiotics etc., the authors have pointed out that low salinity reduces white spot disease incidence. It has also been advised that polyculture of shrimp with fish will reduce incidence of the disease. In the Central coastal districts of Andhra Pradesh several farmers use low salinity water for shrimp culture and also undertake polyculture with fish. The result is that there is practically no incidence of the disease, resulting in satisfactory harvest.

Considering these developments, it appears desirable that the scientists of CIBA and CIFA make a field study, conduct experiments, in this new light, at their farms and come up with a set of recommendations which would counteract the incidence of WSD. If possible and desirable, farmers can be advised to bring amenable freshwater ponds under shrimp culture.

Fishing Chimes, Editorial 259: October 1998: Vol. 18, No. 7

# Fishing Sector's Unique Features Demand a Compatibly Structured Financing System

Apex Commercial banks (particularly ICICI) and other Commercial banks happen to be the main source of finance for the fishing industry too, as in the case of the other industries. In the past over two decades, the erstwhile SCICI (now merged with ICIC1) and several Commercial Banks lent money to the fishing sector but in the process learnt bitter lessons. The chagrin of the banks over the incompatibility of the characteristics of the fishing industry with the financing norms of the banking sector began surfacing as a trickle, gradually gained in intensity over the years and eventually reached a crescendo in this decade. The disastrous result of this unfortunate upwelling has been that practically all the Commercial Banks and the apex bank ICICI have more or less closed their shutters so far as the fishing industry is concerned. NABARD too, despite the guarantees in respect of repayments that it receives from State Governments and Commercial banks, has been cautious in granting loans for the fishing sector.

The crucial role of the fishing sector as a major component in the development of the economy of the country, by way of supplying proteinaceous food to the people, providing employment to a large section of the society, and earning of foreign exchange is well known. This role is related to growth in sustainable fish production, judicious utilisation of the produce for internal supplies and for exports. Growth beyond stagnating levels is possible only when there is Governmental policy and strategy support and when investments yield sizeable incomes and profits. Whichever way we look at this subject, growth in the fishing sector as the times demand will be possible only when finance flows from the banks to the Industry and money moves back from the industry to the banks as per the terms of lending. The flow of funds to and fro having become weak, the sector is suffering and the banks feel the pinch. Their unsavoury experiences are enough to render the banks shy in lending to the fishing sector. The loser however is the sector and the developmental process.

The problem of financing the sector has now assumed an alarming proportion and it deserves a serious appraisal, with a view to evolving a solution. A sector as important as the fishing sector cannot be left in the lurch, notwithstanding the various constraints inclusive of those relating to evolving a sound policy and strategy.

Broadly stated, financiers and fishing enterprises can be seen as bound to each other mainly under two systems. One is the traditional system. Under this system, the fishing enterprise, often a group of fishermen with a leader, take money from a lender other than a bank to meet the block as well as working capital needs. This lender or his representative, conversant with the ways and vagaries of fishing, fishermen and fishing enterprises, will be present invariably at the landing point to buy the catches from his client at a concessional price and to make possible deductions towards the outstanding loan (or no deduction may be made in order to keep the party enchained) and to release the balance for the group to share. The modern banking system of the day has no organisational mechanism for emulating this tested pattern and improving upon it. Some of the fishing enterprises are alleged to come under the spell of the inherent weaknesses of the banking sector and this has relation to laxity in repayments in some way or the other. Several bankers are known to say that their enquiries reveal no fishing enterprise ever loses money. They do quote quite a few instances but their conclusions are debatable. Leaving this angle aside, it has to be conceded that, apart from the documentation, bankers trust their clients and they feel let down when repayments are not forthcoming. Their problem is : The fishing boat and her appurtenances or the fish in a farm are no doubt hypothecated to them, but they are ill-equipped either to take over and operate fishing boats or to harvest a fish farm. In fact, they say these are not their functions and that is true too. Dismayed at the loss of the money lent, they do not now look at the sector. The position, however, is that, unless they extend financial help, there can be no commercial development of the sector, which is what is happening now.

So far as the fishing enterprises are concerned, the general refrain is that no Commercial Bank is lending money to them. Let us look at the deep sea fishing sector. Several crores of rupees were given to the sector by both the Government and the commercial banking structure. Recognising the hopelessly poor repayment situation in the case of most of the enterprises (There were quite a few who were paying regularly, including the Tamil Nadu Fisheries Corporation) Government announced a rehabilitation scheme which envisaged repayment of 30 per cent of the loan given for full and final settlement of the loan given. Almost all enterprises paid within the time schedule given in the scheme and acquired unencumbered rights over the boats. It is a different question as to how they managed to pool up the money so fast. In contrast, in the culture fishery sector, there are several who do make repayments, regular or irregular, towards loans and interest. There are also several who fail to make repayments. Unfortunately bankers have very little scope to check the genuineness of non-payment. Further the fish/shrimp crop being in the water, as swimming vegetables, bankers have to depend on the integrity of the farmers and most of them do have this. There have been, however, occasional allegations of clandestine removals of the fish/shrimp crops followed by declarations that the crops were either poor or lost. Such allegations are uncharitable.

So far as capture fishing from the sea is concerned, the profits earned and repayments made to banks are obviously inter-related. From the fact that most of the enterprises have been engaged in the activity for several years, it has to be construed that they are able to reach the breakeven point atleast. Otherwise, it would have been difficult for them to continue to be in the line.

Financing of the fishing sector, in contrast to the processing sector, may have thus to take care of several inherent patches of vulnerability. These can be countered only by evolving an exclusive system of banking for the fishing sector. Withholding of finances to the sector on grounds of inability to meet the challenge only reflects the inadequacies of the banking sector to face the challenge. Banks can say: why should we face the challenge? The answer is that they have to very well face the challenge (in national interest) and find an effective solution to the problem from which they cannot run away pusillanimously. True, money is involved. The fertile brains of banking magnates have to be put together to find a way out to resolve this issue, which needs a solution. Normal banking is within the capabilities of many. But it requires men of mettle of the banking sector to find a lasting solution to the problem, which is certainly possible.

In the above background, we venture, with a measure of temerity, to outline a system that can possibly be considered for introduction with whatever changes that are needed. It can of course, be ignored too totally or partially.

1. The States having sizeable culture and capture fishery activities may be identified first. To meet the financing requirements for the commercial fishing development in each of them, in order of priority, a panel of commercial banks may be drawn up by the Department of Banking/Reserve Bank and the function of financing the fishing sector may be entrusted exclusively to these banks, one for each of the States.
2. Each of these selected banks may be advised by the Government to set up one separate Fisheries Branch (in the same way as banks set up separate branches for industries, overseas business etc.) at an appropriate centre in each of the identified States, manned by senior banking officials and with a component of technically well qualified and experienced executives, initially drawn from the state fisheries departments on deputation, and with supporting junior personnel with the needed calibre drawn from those who come out of the fisheries colleges of the country with distinction, and who are given special training in banking systems at one of the selected Colleges of Banking. These executives can, in course of time, replace those taken on deputation.
3. By using the term 'fishing sector' in this write-up, we refer to the sector that requires financial assistance of a capital nature for the acquisition of fishing vessels, their improvements, for setting up "Cage farms" in the sea, and for setting up mariculture/coastal aquaculture/inland aquaculture farms for the production of, by and large, exportable varieties of fish, shrimp, prawn, crab, pearl producing molluscs etc. Integrated projects involving the setting up of processing facilities or firm linkages with existing processing facilities for value-addition have to be treated as part of the "fishing sector'. Provision of working capital with cash credit limits has also to come within the purview of the financing system envisaged in this writeup.
4. In regard to sea fishing the provision of finance has to be with reference to the introduction of a viable number of vessels from each of the identified fishing harbours. So far as the farming sector is concerned, the financing system has to be related to a cluster of farms or other water bodies, in a compact zone with communication facilities. Such a grouping system facilitates monitoring.
5. The bank concerned must have a field-oriented monitoring system, part of the cost of which will have to be shared by the enterprises concerned. The main elements of the monitoring system, among others, to be accepted by the loanee before hand, shall be:
   (*a*) There will be a tripartite agreement between the bank, the loanee enterprise and a panel of committed buyers of the produce. It is the responsibility of the enterprise to propose such a panel of buyers who will undertake to abide by conditions of the bank in respect of guarantees, deposits, their forfeiture by the bank concerned when necessary etc. The Agreement must have a provision that the buyer of the produce agrees to maintain a no lien account with the bank into which he would be depositing a specified and mutually agreed percentage of the sale proceeds and pay the balance only to the enterprise.
   (*b*) The suggested Fisheries Branch of the bank shall set up offices at a location central to the places of commercial fishing activities. Each of these offices shall be kept in charge of a qualified and capable technocrat (with assistants) to undertake frequent visits to the farm areas and the fishing harbour concerned. So far as farms are concerned, the visits have to be both *im promptu* and also based on reports/information provided by the loanee enterprises about the calender of work and the various anticipated action dates and timings of important events such as stocking, feeding schedules, checking of growth, harvesting etc., to enable the executive

concerned to be present, if needed, at the time of the events. Likewise, in the case of sea fishing, expected dates and times of departure and arrival of vessels, times of unloading, sale prices etc., including any changes in the schedules will have to be furnished to the concerned executive of the bank's office well in time by fax/phone. The bank's office must be equipped with a radio-telephone to keep in touch with the vessels out at sea, to check catch position etc., when needed. In short, a manual containing all the guidelines as reflected in the suggested Agreement has to be given by the bank to the loanee enterprises whose acceptance of the same has to be obtained in writing beforehand.

(*c*) The executive in charge of the bank's office in the field must be made to follow a system of periodical reporting of the developments to the Manager-in- charge of the main fisheries branch atleast once a week to enable him to take corrective actions.

(*d*) The Manager of the main fisheries branch will have to make frequent inspection visits to the various centres so as to ensure alertness towards the work by the field executives and to exhort the enterprises appropriately in the execution of the work. In other words, the enterprises should feel the alertness of the Fisheries Branch but in an undisturbing way.

6. The various State Fisheries Departments may set up separate cells to maintain a liaison between the bank's offices and the enterprises concerned and to coordinate the activities. The Union Department of Animal Husbandry and Dairying would have also to set up a cell for the purpose of co-ordination, and for pooling up information on experiences in implementing the financing system in various States.

So far as funds are concerned, the Union Department of Animal Husbandry and Dairying could provide them as estimated to the identified 'Fisheries Branches' under well drawn terms and conditions. In the alternative, NABARD could be entrusted with this function at national level but not as a refinancing one.

We concede that the thoughts set about above are raw in nature but would deserve examination with a view to formulating a financing system that will work and cater to the needs of the fishing sector, which is now more or less an orphan so far as financing of commercial fishing activities are concerned.

Fishing Chimes, Editorial 260: November 1998: Vol. 18, No. 8

# Languishing Fisheries Sector: Needs Growth with Stability, Supported by Technology and Creative Financing

Survival and prosperity of any enterprise in an enduring manner is closely linked to growth with stability in a promising form. Fishery enterprises are no exception to this general axiom. Growth is part of the process of keeping an activity going on, unless resource constraints impede it.

So much so, we find marine capture fishing units tending to reach numerical expansion until a stage when the unit catches come down. When this happens the operations are rendered uneconomical and fleet strength dwindles. This stage manifests itself particularly when the Governments do not have specific regulations to monitor their fleet strengths in relation to known resources. Excessive introduction of any particular type of vessels (*e.g.* demersal trawlers) leads to differing results. One such result is that some of the enterprises will fold up. Progressive enterprises would however make efforts to take advantage of technological developments to upgrade harvests through diversification of effort to catch fishes in other water layers or from other grounds. These efforts at diversification may be by way of installation of another fishing system on board, which could be either for longlining, squid jigging, purse seining, pelagic trawling or others applicable. It is also possible that Government would introduce a system of restricting the fleet strength, type-wise with reference to assessment of stocks from time to time. In short, the trend will tend to be towards an innovative form of growth, which becomes compulsive for survival. Some may go in for on board processing systems too for optimising returns.

In the case of shore-based processing plants, growth by way of modernisation of plants for the manufacture of value-added products and reinforcing quality control and packing systems so as to withstand marketing competition and meeting the prevailing needs of importers may take place, as a natural process, which would ensure growth with stability.

Fishing harbours too pass through a growth process. Managements will be forced to expand and improve vessel berthing systems, and have systems of supplies and services upgraded, because of demand and compulsive developments.

In the freshwater, brackishwater and mariculture sectors, technological advances play a major role in the process of growth of the activities, although impeded for short spells periodically owing to threats of disease, caused by sudden changes in water quality, constraints on supplies and services and Governmental regulations. Expansion of area under culture, and intensification of culture systems so as to reap maximum sustainable harvests, are two forms of possible growth, aided by improved water management, feed management and aeration practices. Diversification of culture systems is yet another form of growth that could materialise.

So far as mariculture is concerned, the first decade of 21st Century would undoubtedly witness growth of mariculture, land based as well as sea based (cage culture), owing to force of circumstances.

Hatcheries are at present, by and large, confined to carps and crustaceans. With mariculture coming in, setting up of hatcheries for producing seed of species such as perches, groupers, breams, sea bass, crabs etc. would constitute an inevitable form of growth.

Many other growth lines may materialise. *e.g.* marine and freshwater pearl production may assume prominence and enterprises for the purpose may proliferate.

The foregoing introductory part to the theme of this editorial may be wearisome, and sound repetitive but it has to be construed as relevant to the subject of need for growth of fishery industries with stability through creative financing. Powerful opportunities in the fisheries sector for economic growth and providing employment avenues can be released through creative financing.

In the editorial presented in the October' 98 issue of *Fishing Chimes,* the need for evolving a financing system in consonance with the characteristics of fishing industry was emphasised. This writeup is in furtherance of this theme.

The fostering of a creative financing system for the development and growth of fishing industries by the Government and other connected agencies can take the industries into the 21st century, with growth and stability. Under the present climate in the country for financing fisheries industries, there can be no growth and the resultant stagnation will take the fishery industries backwards. We are witnesses to the folding up of several culture-based and capture-based industries in the recent years, mainly because of the deficiencies of the prevailing financing system *vis-a-vis* fishery industries. This state of affairs is a precursor and gives a fore-taste of the position that may prevail, to start with, in the first decade of the 21st century.

The concept of creative financing is an integrated one, involving everything that a fishery enterprise needs for its successful functioning and viable results. Under the system no aspect is expected to be taken for granted. Technological backing and its application, entrepreneurial capabilities in relation to the various components of the project, managerial inputs etc., are always adjudged by the financing body which lends money only through a portfolio administration on a continuing basis and is continuously in touch with the enterprise concerned to have a conscious awareness of the progress, so as to evaluate and monitor the same.

The system of creative financing has to be taken as differing from normal financing system in that under the normal system the financier's interest is largely limited to receiving the repayments. He would not keep a tab on the progress of the activities to know for himself when things are going awry and take remedial measures in such circumstances. A creative financier will be able to do this as his men would be knowledgeable technocrats. Considering the failure of the existing financing system, by and large, so far as fisheries sector is concerned, it has to be agreed that the present system of financing of the fishery industries would have to be reformed, if the nation has to take the sector forward with growth and stability.

We have had an opportunity of talking to a senior official of Debis Financial Services, Australia, which lends heavily to fisheries industries particularly for the acquisition of fishing vessels, new and old. Debis has 73 offices round the world with experienced staff specialised to cater to the needs of several developing sectors, one of which is fisheries. Strangely, Debis is happy to be lending to the sector, compared to the chagrin of Indian financiers in lending to Indian fishing companies. Either the Indian fishery enterprises are unreliable or the Indian banking system, so far as fisheries sector is concerned, is defective. One advice of Debis is not to borrow from an agency which has no experience and track record in lending to fisheries sector. The financing body chosen to seek a loan must have knowledge of the industry, knows how flexible it has to be in receiving repayments (in good seasons or good periods of taking increased repayments), and would accept a reasonable down payment. Debis also says that the responses of the financing body should be fast, documentation should be simple, and there has to be open communication between the lender and borrower. The financier shall take as security mortgage of the vessel only for which loan is given. Some financiers ask for security of all assets of the company but such financiers should be avoided. Once a selfcontained project report, which gives all information needed by the financier is given, the sanction of loan subject to various conditions, should not take time, according to Debis. Once satisfied, Debis grants loans but only to the extent of 80 per cent of the actual amount. The remaining 20 per cent has to be invested by the borrower.

It was surprising to learn from Debis's representative that only very rarely they come across problems relating to delayed or no repayments. The reason given for this is that their specialised staff are in continuous touch with the borrowers. This kind of follow-up makes the borrower to make repayments whenever the earnings cross the breakeven level. There will be practically no cases of willful evasion. However, when there is a continuous default for two instalments, the financier levies a penal interest of 1 per cent. Surprisingly, the representative of Debis says, they never had any occasion to take over a vessel for the acquisition of which they lent money. In fact, he says that competition exists between financiers to lend money to the fishery industries.

Under the creative financing system, the financier works as hard as the borrower to earn his interest. He does not sit back, waiting for the repayments just to flow in. He works so to say with the owners, although in a different capacity, to get the money back. There is however, one aspect. Loans for acquisition of vessels are linked to one or two brands of main engines of repute which the financiers insist should be installed in the vessels, new or old.

Caterpillar Co., has a marine financing programme (including fishing vessels) offered through its Financial Service Corporation. In addition to providing finance for acquiring new vessels, Caterpillar Financier provides finance, for marine engine power requirements. If the vessel is already mortgaged, Caterpillar Financial would, upon credit review and approval, offers financing to repay the existing 1st mortgage loan and for acquiring and installing new engines, all secured, with a new 1st priority vessel mortgage. The company offers terms of financing upto 70-75 per cent of the vessel's hard cost, upto 10 years. Construction financing is typically arranged through local banks or the contracting shipyard based on Caterpillar's commitment letter issued prior to beginning of the construction of the vessel.

The fishery industries of India, specially the marine fishing industry now faces a situation of great concern, the Upper category OAL fleet strength of 190 has shrunk to 98. Even these vessels may be phased out in the coming few years. In this situation, the Government has no alternative other than strengthening the fishing fleet under a well drawn scheme of creative financing implemented through specialised branches of selected banks, that will take care of the interests of the banks and the borrowers. The system should be so designed that the financiers will be vibrant and in close touch with the enterprises concerned and their operations, without however, interfering in any way in the normal working of the enterprises. Such a stance, with the barrower bearing part of the bank's supervisory expense, would ring in a positive mood for repayments in replacement of the present complacent trend, which is largely the result of the chasm between the financier and borrower, particularly in the intervening periods between one repayment date and the other.

There is no active financing system for fishery industries in India at present that can be availed of by the entrepreneurs. It is of utmost urgency that the Government or the Reserve Bank of India sets up a Committee to go into the question and make workable recommendations that would facilitate, when implemented, meeting the financing needs of the various fishery related industries as outlined in the earlier part of this write-up, and the points mentioned in the editorial of the October 1998 issue of *Fishing Chimes.*

The fisheries sector earns over Rs 4,500 crores in foreign exchange and the earnings would move up further impressively if a creative financing system for the acquisition and upgradation of fishing vessels and upgradation of processing plants is introduced, based on the recommendations of the Committee, proposed to be set up to go into the question.

Fishing Chimes, Editorial 261: December 1998: Vol. 18, No. 9

# Sustainable Utilisation of Transnational Fisheries Resources of Indian Subcontinent:

## Possible only through a Regional Management Mechanism

It appears that neither a basis is developed nor a direction imparted for the judicious and sustainable management of such of the marine fisheries resources of EEZs and inland capture fisheries resources of the various countries of the Indian subcontinent, which are transnational. Fishes and other aquatic animals move in the contiguous water zones which, for the respective Governments and the people, are separate political entities. Application of political boundaries to moving aquatic animals, in respect of utilisation of the overall resources in the waters which are contiguous and transnational, can have disastrous effects. The resources being common, their over-exploitation by any of the nations can lead to imbalances and severe depletion of species, which may sometimes be irreversible. There is therefore an imminent need for a common and coordinated fishery development strategy for all transnational fisheries, both marine and inland. There has to be a cognisable recognition and appreciation of this truism.

At the risk of stating the obvious, mention of the transnational marine resources between India and Pakistan, India and Bangladesh, India and Myanmar, India and Sri Lanka, India's Nicobar Island and Indonesia's Sumatra island as examples is relevant. On the inland front, there are several transnational rivers of which, those flowing through India and Pakistan, India and Bangladesh, India and Burma deserve mention. Over-exploitation of these capture fisheries resources or introduction of any new species in the transnational rivers unilaterally by any one of the countries may have inimical effects on the biotal structure, which can be harmful to the natural fauna and flora and may also upset sustainability. All these aspects are matters of concern to the nations of the subcontinent.

It is known that at one time or the other the maritime nations of the region allowed distant water fishing nations such as Taiwan, to exploit the marine fisheries resources of their respective EEZs. The extent of this exploitation, it is believed, is known to the nations concerned only through the returns taken from the captains of the foreign vessels. It can be any one's guess, however, what quantity was truly fished out and now being truly taken by such of the foreign vessels which continue to be exploiting the marine waters of the subcontinent. The number and types of foreign vessels allowed to fish by the maritime nations concerned of the subcontinent were probably decided in spurts and in an arbitrary way, and in most cases, with scant awareness of the inimical effects that will loom out of such a policy. Another aspect is that none of the countries of the region have a quantified idea of parameters such as "allowable catch".

The maritime nations of the subcontinent realised, with the more conscious ones earlier than the others, that in the matter of utilisation of the fishery resources of the respective Exclusive Economic Zones, the national companies concerned became virtually middlemen to foreign fishery enterprises, particularly Taiwanese and Thai.

An appraisal would reveal that, the belated realisation that foreign enterprises were taking way glibly a sizeable part of the marine fishery wealth in the EEZs of the nations concerned burgeoned in Myanmar, stated to be in early 1980s, and later the roots of this realisation spread gradually to Maldives, Sri Lanka, Pakistan and India in the order given. The operations by Taiwanese vessels were permitted mostly through pair trawlers and longliners, taken on charter/lease permissions by national companies or through issue of direct licenses to the foreign company concerned. Schemes aimed at providing permissions to national enterprises for operating foreign vessels on charter/leases given and to fish in the EEZs of the respective countries unleashed several by products, one of which was, as can be expected, financial advantages to both the foreign and Indian companies. A particular advantage for foreign companies was to gain access to fish in the EEZs of nations such as India and securing the permissions must have been quite a job and an achievement. One can visualise the kind of effort that must have been there. In India, the complexities of the job and the bustle of the men who emerged to handle the job ricochetted on the leadership of the small scale fishing sector, to whom, the situation provided a handle to foment agitations revolving round arguments such as encroachment of foreign vessels for fishing in Indian neritic waters, catching of fish on their migratory sojourns from offshore to inshore waters etc. There was also the factor of jealousy engulfing those unable to secure permissions or benefits out of the new opportunity of making money through charters, leases and joint ventures.

Whatever be the reasons, it so happened that Myanmar was the first nation of the sub continent to ban operations of Taiwanese vessels in its EEZ followed by

Maldives, Sri Lanka and Pakistan. India too later followed the trend, but somewhat belated, and that too because of agitations. However, so far as joint ventures are concerned, while the position in Myanmar is not known, Maldives was firm in not allowing joint ventures too. However, India and Pakistan evolved a system of joint ventures involving foreign vessels, mostly of Taiwanese and of Chinese (mainland) construction. In the case of India, vessels of Thai construction also came under Joint Venture system. Sri Lanka is understood to have allowed only one Joint Venture, involving vessels of Mainland China. This system is developed on the premise that the National enterprise that enters into joint venture with a foreign company for operating fishing vessels in its national EEZ must have vessels of the flag of its country, and the joint ventures would be on record as technical collaborations. This system was no impediment to the national enterprises. It is known that they formulated projects that provided for the acquisition of vessels on deferred payment basis from the foreign owners (who avail of joint venture opportunity).

The national companies registered the vessels in their countries, which enabled them to fly respective national flags, in the place of the earlier foreign ones. The foreign partners, to the extent known, were Taiwanese companies (as in India and Pakistan) Peoples Republic of China as in Sri Lanka and India, and Thailand and Spain as in India. The wearing of the joint venture garb is believed to have not caused any problem either to national or foreign enterprises. The shift in the policy of the countries of the subcontinent in respect of operation of foreign vessels in their respective EEZs, (withdrawal of operations under charter/lease terms) as outlined above, must have had the effect of reducing the fishing pressure.

It is quite possible that countries of the subcontinent evolved the pattern of withdrawing schemes permitting operation of chartered/leased vessels within the knowledge of each other. Having effectuated the cooperation, wittingly or unwittingly, it would be a progressive step for the countries concerned to work now towards evolving a commonly acceptable policy and system for the management of the marine fisheries resources (and inland fisheries too) of the subcontinent. Under such a policy linked with a financing system, if adopted, fishing companies of the nations of the subcontinent can acquire and operate their own vessels. The requirement of technical collaboration with foreign companies, at best, can be marginal, particularly as hundreds of national personnel must have already received training as counterpart crewmen on the chartered/leased and 'joint venture' vessels. Countries of the sub-continent, particularly India, have shipyards equipped for constructing fishing vessels of any type (any specified drawings when required, could be purchased by the Government concerned and given to the yards based on certain conditions). Training facilities too are available in the sub-continent.

A coordinated policy of this kind, *inter alia*, could delineate the optimal limits of fleet strengths of each of the nations keeping in view the sustainability of the fishery resources of the nation concerned, and the fact that fishes move freely in the entire zone unconscious of the man-made boundaries. Apart from this, there can be provision to allow a certain number of vessels of a nation to fish in the waters of an adjacent nation where they were traditionally fishing before formation of EEZs. Such an approach would reduce the problems faced in the Indo-Pakistan, Indo-Sri Lanka, Indo-Myanmar and Indo-Bangladesh border zones, presently resulting in the arrests of fishermen of one country by the other.

A regional management mechanism could be evolved for the common benefit of all the nations in the subcontinent. The mechanism of management of marine fisheries could cover the EEZs of all the maritime nations of the sub-continent. The mechanism could be such that the optimal limits of the fleet strengths of each of the nations, with reference to their own resources and allowable catches in the EEZs of adjacent nations could be determined based on mutually acceptable criteria. Likewise, the fisheries of contiguous rivers in the jurisdictions of India and Pakistan, India and Bangladesh, India and Myanmar, and India and Nepal could be managed in a sustainable way under mutually acceptable parameters and as part of a regional organisational mechanism which could be brought into being, through needed negotiations.

Fishing Chimes, Editorial 262: January 1999: Vol. 18, No. 10

# Indian Fisheries Sector in 1998: Reflections

One remarkable event that took place in the fisheries sector in 1998 was that Krishi Bhavan had again become the main temple for the fisheries sector. Government detached the subject of deep sea fishing from the Ministry of Food Processing Industries and handed it over back to the Ministry of Agriculture. In this process, Government tagged on the subject of fisheries to the Department of Animal Husbandry and Dairying and this was not to the liking of the fishing industry, whatever be the reasons. The expectation of the Industries was that the subject would be attached to the Department of Agriculture with which it was mainly there prior to creation of the Ministry of Food Processing. Many in the industry were not happy with the decision of the Government to tag on the subject to the Department of Animal Husbandry and Dairying. Some of those who are critical of the Government's action justify their unhappiness by pointing out a few areas of failures. Their finger is pointed at certain lapses which, in the main, include the non formulation of a deep sea fishing policy during the year, inability to come out with a rational policy for the development of coastal aquaculture (in the vast barren stretch along the coastline which is lying unproductive) that would have met the points raised by the supreme court on the subject in dealing with a public interest litigation, transparent action in the popularisation of developed technologies of mariculture including sea cage culture and marine and freshwater pearl culture, continued neglect in the utilisation of the vast stretch of saline lands having underground water on the northwest for fish farming, and of being unresponsive to the various recommendations made for the utilisation of the vast potential of reservoirs for augmenting fish production. The critics argue that the nation would have been on the path of producing more of aqua products, both for domestic consumption and for augmenting exports to add to foreign exchange earnings, and for providing employment opportunities, had the Government bestowed attention towards the subject. Those who are conversant with the pulsating activities in Krishi Bhavan in the earlier decades nostalgically recall them. However, the present situation may have arisen probably because of the preoccupation of the Government with subjects which are considered more important to attend.

Although Government had not announced a deep sea fishing policy, it had extended a major relief to the deep sea fishing industry by permitting high sea bunkering which helped the industry substantially.

Reservoir fishery development and mariculture continued to be subjects of inadequate interest to the Government, despite their potential to generate incomes and providing employment. Freshwater and marine pearl production for which technologies have been developed were expected to pick up during the year, but there had been little progress. It appears that there had been a system failure. The less said about commerical sea cage culture, the better would it be. There had been no positive steps to develop the activity, which is linked with granting of leases of specified areas to set up cages and with foreign collaborations. Experimental work continued to be done for rearing of tiger shrimps to a gravid condition in certain shrimp farms and also for raising mud crab seed in hatcheries but without much of a success.

Notwithstanding these observations, one encouraging aspect that can be mentioned is that during the year Government did not interfere with the ongoing activities, while at the same time not taking any new initiatives in haste, in general. It is to the credit of Government, however, that in the exports sector, Commerce Ministry/MPEDA provided a fillip to maintain and augment marine products export levels, effectively countering the onslaughts of USA through imposition of a condition for the use of TEDs, and of EU banning import of seafoods from India. Government also endeavoured to solve the problem created by China in respect of honouring the letters of credit opened by the Chinese companies towards supply of Indian marine products. Most of the letters of credit remained dishonoured and efforts continued during the year to resolve the problem. The Ministry of Commerce introduced several measures to encourage the Indian marine products exporting units. Some of these were a) providing an increase in the entitlement of duty exemption in respect of exported black tigers, b) offering subsidy on interest payable towards loans taken from commercial banks for undertaking modifications to processing plants with a view to manufacturing value-added products, c) providing soft loans upto Rs. 50 lakhs at 4 per cent interest (through the Ministry of Food Processing Industries) for the upgradation of processing plants, d) Clearance of 46 processing plants for export of marine products to EU and e) approving 6 deep sea fishing vessels for undertaking onboard processing of marine products for export to EU.

An outstanding development to the credit of ICAR/CIFA is the development of a new breed of 'Rohu', christened as 'Jayanthi', having traits of good growth, flesh firmness and improved fecundity. This breed was evolved through selective breeding of Rohu collected from various rivers of the country. The standardisation of an improved system of sewage-fed aquaculture by CIFA is another major achievement. CIFA's scientists also standardised the technology of giant freshwater prawn seed production in well designed hatcheries and also farming of the prawns in grow-out systems, both mono and polyfarming, under a project sponsored by the Department of Biotechnology.

Another positive aspect to the credit of the Government was that, (probably realising that exploited capture fisheries no longer held the key crucial for augmenting exports), there was a general non-interference in farm fishery activities, including culture of marine shrimp in freshwater. It has to be mentioned that shrimp farming would not have reached a new peak in 1998 but for the non-intrusive but effective support extended by MPEDA and the ideas for reaping healthy crops generated by MPEDA/CIBA in the minds of the farmers.

CMFRI is reported to have achieved a breakthrough in the breeding and hatchery production of grouper seed and it is hoped that a mechanism will be developed soon to transfer the technology to the farmers. CMFRI had also succeeded in breeding cuttle fish of second generation and in raising of its seed. Another breakthrough achieved by CMFRI was the hatchery production of seed of clown fish *Amphyprion chrysocaster.*

In the freshwater aquaculture scenario, all progressive farmers caught up with improved technologies of selection of broodstock of major carps and freshwater prawns for producing at the hatcheries healthy and fast growing seed, coupled with systems of effective feeding in nurseries/rearing ponds and in grow-out systems for producing marketable fish. All these would not have been possible but for the effective demonstration of technologies by CIFA and the enterprising spirit of the farmers to absorb the advanced technologies, more on their own initiative. It stands to the credit of the freshwater fish farmers of India, especially those in Andhra Pradesh, Tamil Nadu, M.P., Punjab, and Haryana that farmed fish production has been increasing and a network of domestic supplies mostly from Andhra Pradesh has spread to the various north-eastern States covering West Bengal, Assam, Tripura, Manipur etc. Karnataka is catching up by providing encouragement to entrepreneurs through subsidies for setting up seed farms, as availability of seed is the key for augmenting fish production.

The development of a result -oriented commercial-cum-technology oriented temper among the farmers is an achievement of this decade and the evolutionary process in this regard acquired a distinct shape in 1998. A parallel to this in the marine sector is the stabilisation of entrepreneurship in respect of undertaking diversified fishing activities, although on a limited scale. A few Indian trawlers are also stated to be operating in the EEZ of an adjacent nation. The strengthening of interest in the production of value-added products, development of enhanced awareness towards hygienic post-harvest handling too were part of the achievements of this decade, reaching a high point in 1998, mostly because of the initiative of the exporters themselves and counselling by MPEDA. It is due to the efforts of primary producers and the MPEDA/Ministry of Commerce that the fisheries sector could export 398,991t of marine products valued at Rs. 4661.58 crores in 1997. In 1997-98, the exports went upto 379,585t valued at Rs. 4697.48 crores (US$ 1.3 billion). There were however reports that, although the capture shrimp fishing season as well as finfish capture fishing season were encouraging, consequent to the fall in export prices, earnings dropped sharply by US $ 170 million or by13.38 per cent, and earnings plummeted to US $757 million during the first 8 months (April- November '98) from the level of US $874 million in the corresponding period of 1997. It is stated that the fall in terms of rupee earnings was not however quite as significant. The earnings fell by only Rs. 29 crores to Rs. 3,138 crores, from the level of Rs.3,167 crores in the corresponding period in 1997 (April- November'97). There may have been a decline in the export income but this is mostly attributable to the financial crisis that engulfed S.E. Asia. The fact, however, remains that the producers, exporters and the MPEDA did a magnificient job during the year by way of sustaining the level of exports, despite the various problems.

A reference has already been made about the invidious attitude of the Government towards development of fisheries, despite the fact that the sector brings in substantial foreign exchange and with potential to augment export earnings much more than as at present. It was discouraging that the Government entrusted the production segment of the activity, which forms the basis for stepping up incomes of producers as well as exporters, to a department considered not in a position to handle the subject and administratively not at par with departments such as the union department of agriculture or the department that deals with exports in the Ministry of Commerce. Had a separate Department of Fisheries been created in the Ministry of Agriculture, there would certainly have been greater and more impressive progress. It was unfortunate that Government took the view that the workload in respect of fisheries administration was not adequate to create a separate Fisheries Department. In this connection it has to be commented that the potential of the subject has to be the main criterion for forming a separate department. Once such a department is formed there will be expansion in developmental activities. The present workload cannot thus be the criterion to decide on the need for having a separate department or not for a subject like fisheries with such a vast potential to contribute to food production, exports and employment. For the reason that there is no separate department and the administration is non-professional and in non-traditional hands. The development of the sector which gives to the nation nearly Rs. 5,000 crores in foreign exchange is curbed, consciously or unconsciously.

The marine fishing fleet strength was on the decline during 1998. And there were no indications of what the Government had in the mind in leaving the fleet stagnant and declining. The expectations were that the Government would introduce integrated projects involving in this context the introduction of economically viable multipurpose vessels to facilitate exploitation of unutilised resources such as Tuna and inadequately utilised resources such as Cephalopods; but, alas, this never happened.

On the inland front, reservoirs constitute a major source for augmenting fish production. The Reservoir Project taken up with World Bank's assistance had not made progress. There were no reports in respect of setting up of an organisational structure and a mechanism to develop and exploit reservoir fisheries. Likewise the shrimp project taken up with World Bank's assistance had also not made any significant headway. The demand to bring the vast and presently unutilised saline tracts in the north-western States of the country under fish culture, had not made any dent on the minds of top brass. Professionals had been pointing out that the vast coastal lands bordering the sea could be brought under aquaculture in a beneficial way suitable for the maintenance/upgradation of the environment of the coastal waters and the coast as had been done in USA etc. However, surprisingly, Government had not applied its mind to this important segment of fisheries activity in order to utilise the potential for providing employment opportunities and socio-economic benefits to the weaker sections of the society, with an inbuilt system of protecting the environmental and ecological conditions. Riverine fisheries are in danger of excessive depletion of fish populations and there was no indication of Government doing much in 1998 to stem the trend.

Fishing Chimes, Editorial 263: February 1999: Vol. 18, No. 11

# 1999-2000: Year of Integrated Reservoir Fishery Development in Karnataka

Mr. Jayaprakash Hegde, Minister of Fisheries, Karnataka, has declared 1999-2000 as the year of 'Integrated Reservoir Fishery Development in Karnataka'.

Reservoirs happen to be the largest potential resource for augmenting fish production, not only for internal consumption, but also for stepping up exports too. While fisheries of quite a few reservoirs in various States have been developed to a modest level of annual fish production, the bulk of the reservoirs of the country have not received much of a serious attention, despite a recognition of the need, from fisheries developmental angle. Consequently, the dreams of Fisheries Departments/Corporations to bring the fisheries of reservoirs under their charge to a level of sustainable capture-cum-culture fishery balance, supported by an integrated system of development consisting of establishment of hatchery farms to meet stocking requirements, organising periodic sustainable fishing, provision of storage facilities as necessary and linkage with a marketing system including exports.

Efforts at development of reservoir fisheries have been there even prior to independence. After independence, there has been an increase in the efforts, but with sporadic successes. Considering the increase in demand for fish in the domestic market and in the fast developing export market, there is an expediency to mount organised and focal efforts for the development of reservoir fisheries.

The nation has several reservoir fishery developmental experts. Scientists at CICFRI specialise in reservoir fishery development Among the known experts, to mention a few, are Dr. Y.S. Yadava, Dr. M.R. Sinha, Dr. V. V. Sugunan, Mr. A. Sreenivasan, Dr. Desai, Dr. G.P. Dubey, Mr. R.P. Tuli and quite a few others. While the contributions of the various experts were there, the average increase in per ha reservoir fishery production continues to be very low.

India has reservoirs with a waterspread area over 20.50 lakh ha, capable of yielding over 2 lakh t of fish per annum on a conservative estimate of production of 100 kg/ha. So far as Karnataka is concerned, it has 62 reservoirs with a waterspread area of 2.25 lakh ha. When developed, they could produce 22,500 t per annum at the same modest estimate of 100 kg per ha. By adopting culture-cum-capture fishery development strategy, the per ha production could be raised even further. The augmented utilisation of reservoirs for fish production will provide employment not only to fishermen, but also to several engaged in activities ancillary to fish production. The gross income from fish that can be produced from Karnataka reservoirs per annum at 100 kg/ha would not be less than Rs. 45 crores per annum.

The situation that no cognisible breakthrough in the sector could be achieved over several decades shows that there must be several hurdles in the way. In this context, the decision of Karnataka Government to declare 1999-2000 as the Year of Integrated Reservoir Fishery Development to initiate developmental efforts is commendable and the declaration is the first step to give a formal organised recognition to the subject. The declaration does not apparently mean that during the year the task would be completed. It, however, implies that a workable course of action would be evolved and set in motion during the year. Integrated activities are specially essential for reservoir fishery development so as to bring the various components of the activity under an organisational mechanism to be built up so as to eliminate known hurdles. Reservoirs closer to each other have to be grouped, if necessary, with larger tanks, for the setting up of common facilities such as hatcheries, farms, and cold/frozen storages and marketing network. Domestic marketing system has to be developed in a manner that retail outlets can be organised under probably a system of leasing the outlets which would receive periodic supplies from the Reservoir Development Organisation to be set up. Facilities of container traffic to the nearest port of export from centralised storages can be developed and integrated with the overall system. These may be loose thoughts but during the initial year of development a sound basis, integrating the existing facilities with the organisation to be evolved could be established. In the years to follow, consistent follow-up efforts could be set in motion.

The various State Governments can consider emulating the example set by Karnataka.

Fishing Chimes, Editorial 264: March 1999: Vol. 18, No.12

# High Sea Bunkering System: Need for Extension to Mechanised Fishing Boats Sector

It was a much needed supportive step on the part of the Government of India to have accorded permissions to ocean-going fishing vessels to take bunker on the high sea at a distance of 24 n m from the coast. These permissions are a concession to the deep sea fishing industry which is stated to be reeling under losses, mostly because of the high cost of diesel. When the vessels take bunker (outside 24 n m, *i.e.*, customs limit), with the needed permission, there are savings of sales tax and excise duty, having the effect of bringing down operational costs substantially.

It is unfortunate that the concession has been limited to the ocean-going fishing vessels. The fishing vessel fleet in the OAL range below 20 m as well deserve the concession for the reason that the owners of these boats too are equally affected by the spiralling increase in operational costs (the cost of diesel oil has recently been increased by one rupee per litre) and they too fish beyond territorial waters. It is believed that the Government is aware of this but some problems may have been perceived in extending the concession. To the extent it can be visualised, the reasons, among others, may be those connected with monitoring of the bunkering operations for a massive fleet of over 40,000 boats, the scope for accidents at the time of taking bunker because of the smallness of the boats, the indirect loss in revenue by way of excise duty to the Centre and sales tax to the States etc. It should be possible for the Government to have a comprehensive work plan to get over hurdles so as to ensure that these financial benefits accrue to the small scale marine fishing sector too. The recognised Associations of mechanised fishing boat operators and mini-trawler operators can probably be advised to affiliate themselves with the Association of Indian Fishery Industries which can be entrusted with the responsibility of effective implementation of the connected operations, linked to a reporting system. One problem can be that the vessels come within the operational purview of the State Governments concerned and the permissions to take bunker on the high sea have to come from the Centre. Through an appropriate device of Centre-State cooperation, this problem can be resolved. As the need for providing the concession is justified, it should be possible for the Centre to consult the coastal State Governments and formulate a line of action. It is somewhat surprising that the Associations of Mechanised Boat Operators all along the coastline have so belatedly started raising their voice against the injustice being done to them in the matter of high sea bunkering *vis-a-vis* ocean-going fishing vessels.

## Fisheries Resource Management Society, Kerala

There has been an unbridled increase in the number of mechanised and motorised fishing boats operating all along the coastline of the country. This increase is seen to be not in consonance with the fishery resource potential of the coastal waters. Consequently, there has been an all round fall in fish catches per unit. Overfishing has manifested itself not only from the sea but also from rivers, estuaries and backwaters. Mangrove ecology is stated to have been jeopardised partly because of this. Kerala is no exception to the general trend.

This situation clearly calls for an organisational mechanism to manage marine resources. A Governmental set-up as such will not be able to produce results. The Government of Kerala, apparently having taken stock of the situation has blazed a pioneering trail by setting up a 'Fisheries Resource Management Society'. This Society was registered under the Travancore-Cochin Literacy, Scientific and Charitable Societies Act. Understandably, the Society, it is stated, would first concentrate on undertaking a census of fishing craft and gear of Kerala with reference to known resources. The objective obviously is to determine the extent of fishing effort and to what extent it is excessive compared to the resources status, both marine and inland. Carrying out this census is just one of the several proposed functions of the Society. The other steps cover taking up of a systematic cost and earning studies in the fisheries sector of the State, the scope for implementing aqua reforms, initiation of steps to improve the health standards of fishermen in the State etc. The Society will strive to understand the present standard of living of fishermen precisely in measured terms and in that light formulate policies aimed at stepping up the income levels of fishermen. The Society is presently managed by a Governing Body with the Minister for Fisheries, Kerala as the Chairman. It has the Secretaries of Fisheries and Finance of the State Government, the State Director of Fisheries, Chairman of MPEDA, the Directors of CMFRI, CIFT and School of Environmental Studies, Dean, College of Fisheries, and Chief Executives of ADAK, and MATSYAFED, besides two Government's nominees, as members. The Society is thus, a well balanced one to take care of scientific, technical, political, financial, manpower, and social aspects of fisheries resource management. The initiative taken by Kerala deserves to be emulated not only by the other coastal State Governments but also by the non-coastal State Governments to take care of inland fishery resources such as those of rivers, reservoirs and *beels*.

Fishing Chimes, Editorial 265: April(i) 1999: Vol. 19, No. 1

# CMFRI Ushers in Onshore Commercial Pearl Production

CMFRI has provided a significant sense of relief to all those who are anxiously looking for a commercial breakthrough in mariculture. It has developed onshore commercial pearl culture technology and effectuated the same at two commercial farms near Visakhapatnam on the upper east-coast of India, thereby catapulting the place and the activity into a festive focus.

Dr. G. Syda Rao, Senior Scientist at the Regional Centre of the Institute at Visakhapatnam, guided by his Director, Dr. M. Devaraj, has developed on-shore marine pearl production technology in a manner conducive to commercial application. One of the two enterprises to whom he extended the technology is now on the anvil of harvesting pearls from its onshore farm tanks; and the other is progressing to catch up.

Syda Rao's scientific and technological work is one of an integrated nature consisting of a) transporting pearl oysters from Tuticorin on the southeast coast of India to Visakhapatnam, b) growing and breeding them in on-shore cement tanks, and raising further generations of the oysters so as to have a local stock, c) implanting them with beads for producing pearls, d) harvesting pearls from the oysters experimented upon and e) validating the positive experimental results by way of successful commercial application in the field.

The technology has been adopted at the farms of two companies, both near Visakhapatnam, one located close to TASPARC's shrimp hatchery (Pearl Beach Hatcheries) and the other at Bhogapuram (Pearl Aqua). Dr. M. Devaraj, Director of CMFRI visited the farm of Pearl Beach Hatcheries recently and saw for himself the pearls in the making. Harvesting of pearls is set to be inaugurated by Dr. R.S. Paroda, Director General of ICAR in August 1999, it is learnt. It is possible that ICAR may apply for patenting this onshore technology, which is claimed to be the first of its kind in terms of accomplishing all stages of operations from raising of oysters to implanting with beads, followed by retrieval of pearls.

Sydarao's hard work, (in total disregard of reactions that he was wasting his time on something that could not be done by known stalwarts who worked on the subject before), led to the unique success which connotes contribution of prosperity to the nation through pearls.

CMFRI's Regional Centre at Visakhapatnam initiated its onshore experiments on various aspects of mariculture in its two wet laboratories equipped with tanks measuring nearly 8,000 sq ft. in all. The work taken up covered not only pearl culture and pearl production, but also snapper culture, shrimp broodstock development, crab seed production, sea weed culture, live feed culture, tissue culture of sea weeds of economic importance etc.

The first consignment of pearl oysters from Tuticorin for the experimental work was brought to Visakhapatnam in 1996. In the following years, the technology of raising the oysters and seeing them through the stages of egg laying, veliger larval formation, spat settlement, and growing the spat to adulthood, was developed. This work was followed by bead insertion with graft material (live mantle tissue) into the gonads of the oysters for pearl formation. Feeding system consisting of mixed algal species has been developed. Training in operational aspects was imparted to the owners of the aforesaid two farms and their operatives. To restate, the achievements of Sydarao were a) captive growth and broodstock development of pearl oysters, their maturation b) breeding of oysters and raising of spat and c) raising of next generation adults.

In the commercial farms, for raising of the oysters, masonry tanks of 50 to 200 t capacity are used. The tanks are 50x10x1.2m in size, and constructed in RCC. It is mentioned that filtered sea water is used for filling the tanks and for use in the hatchery tanks and in algal culture through use of specially designed slow sand filter. Neither chemicals nor antibiotics are used for treating the water. Feeding is done with live feed only. Because of these precautions no pollution is observed in the farms, each of which has a total extent of one ha, with a waterspread area of 3,500 sq.m. So far as water exchange is concerned, this is done once in every 10 days. The tanks are partitioned into compartments of 50 x 10 m each, to meet the phase-wise growth stage requirements, such as nursery, rearing and grow-out, and for holding brooder and implanted stocks. The eggs hatch out in 24 hours. Veligers are fished out and distributed in nursery units. They take 20-30 days to settle down on the substratum, studded with stones, to form spat. According to Syda Rao, very little mortality of the veliger larvae takes place. They can withstand a temperature range of 15-35°C, salinity of 16-35 ppt, and pH of 7.5-8.3 range. The stocking rate in grow-out tank units, brooder tank units and in implanted oyster stock tank units is 100 nos./sq. m. All the tanks are provided with ventilator ports for aeration all around on the upper part of the tanks. The tanks are covered with tarpaulins to cut sunlight. Each of the tanks has a feed compartment with three partitioned sections. Every day one of these sections is made ready for feed supply through a piped feed drip system. The feed consists of *Chaetoceros, Isochrysis,* and *Nannochloropsis* in different combinations. Cell concentration is maintained at 20,000 to 80,000 cells per ml. Water fertilisation in the tanks is based on a formula developed.

The oysters reach implantation size in six months. It is stated that implantation is done in the gonads with beads of 23 mm size, along with live mantle piece to serve as graft. Beads are imported from Hong Kong. It takes around six months for pearls of 35 mm to form. Within 57 months of extraction of pearls, the same oyster can be implanted again. Each oyster can yield 2 to 3 pearls in its life time.

The farms have two operatives each, one to take care of algal production and the feeding system and the other to take care of culture units, particularly of breeding tanks units, and transference of larvae/oysters in various stages of growth from one tank unit to the other. For the present, the farmers are helped by Syda Rao in the implantation work. It is learnt that persons are being trained to function as implantation and extraction technicians. According to Syda Rao, a major pearl production project in the private sector is being taken up in A.P. near Chirala. This project envisages the diversified utilisation of an existing shrimp farm for pearl production. 12 nos. of one ha ponds at the farm will be used for the work and one ha of area of the farm will be converted into RCC tanks. By end of 1999 the pearl oyster culture work is expected to be operational at this farm.

There will be the dawn of a wave of prosperity in the mariculture sector, once the coastline throbs with ecofriendly oyster farms that produce pearls and other valuable marine species.

Fishing Chimes, Editorial 266: April(ii) 1999: Vol. 19, No. 1

# Seasonal Closure for Capture Shrimp on Upper East Coast: Reflections

There has been the emergence of a movement and its strengthening in the past decade to bring about sustainable capture shrimp fishery exploitation in the Bay off the upper east coast of India. This is a welcome sign indicative of a developing awareness among those engaged in shrimp fishing activities in this zone towards ensuring sustainability of shrimp stocks. This needs to be further fostered and fortified on a sound footing with top priority and dedication, supported by the implementation of a stable, consistent and a scientifically evolved system of utilisation of the resources.

Closed seasons year after year as an arbitrary and a voluntary measure are being adopted by fishing units of upper east coast based at Visakhapatnam and Kakinada, with a view to ensuring sustenance of coastal shrimp stocks of the area. This system ought to provide distinct chances for wild shrimp stocks on the upper east coast to build up. As a recognition of the convention brought in by the fishermen, the A. P. State Government has recently banned sea fishing from April 16 to June 15. The beneficial effects of this order are to be assessed, however, from the angle that a majority of the AP-based boats fish in the waters of Orissa and further north and, consequently, the order does not really affect them, unless they are impounded during the ban period. Nevertheless the order denotes a forward stance, which eventually, ought to usher in a scientifically evolved system for the entire coast.

Notwithstanding these observations, a point for a considered decision is whether arbitrary measures of this kind should continue to be followed as at present or should be given an incisive scientific look for evolving a system that would ensure continuity of shrimp stocks on an enduring basis.

Some of the arguments in favour of arbitrary measures to protect shrimp fisheries so as to ensure sustainable level of harvests can be visualised as: (a) decline in per unit catch as seen in successive or intermittent seasons over the past several years, (b) The only way to arrest these declines is to allow shrimp to grow and give them an opportunity to breed, and (c) by the time a planned scientific study is launched, data gathered as part of the study over an adequate period are analysed, results examined and measures for implementation are formulated, the damage to the fishery could become more intensive.

Some of the arguments against arbitrary measures to protect shrimp fisheries so as to ensure a sustainable level of harvest can be (a) shrimps have a short span of life, (b) they have their early growth stages in backwaters, estuaries and such other waters and they enter the sea fishery at prejuvenile/juvenile stages. The need for conservation is more in these nursery areas and not out at sea (c) the reasons for decline in shrimp fishing levels in certain years are not clearly understood, considering that, while fishing effort is more or less the same, in some years, shrimp fishing happens to be good and not so in others, d) according to several who are conversant with shrimp fishing, the total quantity of shrimp caught remains the same year after year, with statistically acceptable variations, and e) those who are for or against closed season for shrimp have very little concern towards the effect of it on important fish species, apparently because their obsessions center round shrimps alone which have very high export value.

Neutral or objective views on the subject tend to be different: a) Increase in the number of boats fishing for shrimps takes place without any relation to decrease or increase in unit catches (b) There is no structured mechanism to gather data to study shrimp fishing effort and output trends from time to time, and to suggest measures for regulating shrimp fishing, and (c) the fishery extends over the coastal waters of more than one State and this necessitates the implementation of a closed season in the shrimp fishing waters adjacent to all the States of the east coast. Introduction of shrimp fishery seasonal closure measures with legal sanction and patrolling support, designed for the entire zone alone may serve the purpose effectively (d) Fishing goes on relentlessly for shrimp in Bangladesh waters and this defeats the purpose of a closed season on the Indian side of the Bay of Bengal to the advantage of Bangladesh boats, (e) no relationship is conclusively established between levels of fishing and shrimp stocks. Until this relationship is established, it may be difficult to decide on the various aspects of a closed season, and (f) closed season has to be declared when the shrimps are in the final stages of gonadal development or are gravid, for deriving benefits. The breeding periods of various species of shrimps differ, rendering the declaration of a common closed season somewhat difficult, although it is certainly possible when the effects of such a closed season on the subsequent fishing seasons of various species of shrimps are worked out after a detailed study.

In this situation, one point that appears to stand out clearly and probably undisputed is that no addition to shrimp fishing effort should be allowed, as any additions would tend to make unit economics unfavourable. Further, the situation that the boats pass through a prolonged lean capture shrimp season (Feb./May/June) has to be countered so as to make the overall operations economical. In other words, the low incomes during the lean period with or without closed season make the overall operations uneconomical and this needs to be tackled. One way of improving the economics is to equip mechanised boats with tuna monofilament longlining system or pair trawling system for tuna and other finfish. Boats of 9m OAL and above are amenable for installation of monofilament longlining system consisting of installation of a drum reel at the stern, for operating around 15 km of monofilament longline with around 275 branchlines with hooks. Two boats can be brought together to undertake pelagic pair trawling. The lean season for shrimps, by and large, coincides with tuna fishing season. The boats can concentrate on tuna, tuna-like fishes, sharks etc, during the lean shrimping season. The Associations concerned can approach the Government for needed assistance in this regard.

Whether a closed season for shrimp is needed or not is a major scientific topic that calls for a detailed study. Urgent steps are needed for gathering information on the following parameters by the scientists and fishery administrators with the co-operation of fishermen.

1. The total annual capture shrimp landings year-wise with 1990 as the base year and upto 1998 season and percentages of increases, or decreases of shrimp stocks, type-wise, indicating statistical validity of the variations.
2. The actual number of boats in operation and total net registered tonnage computations year to year, with indications of increases or decreases in numbers/net registered tonnage, type-wise.
3. Annual standardised unit catches, with indications of increase/decrease year to year, vessel typewise.
4. An estimate of shrimp stocks and potential annual yield in relation to the actual annual shrimping effort and shrimp landings, with observations on likely effects over future fishery and the need for reducing effort or possibility of augmenting effort in that context.
5. As already mentioned, the breeding seasons for shrimps of different varieties vary. Desirable fishing time for one variety may be undesirable fishing time for others. This calls for an assessment by the scientists and operators in respect of a breeding period that can be taken as common in general, for all the species. Such an assessment will help in the declaration of a purposeful closed season, which should not, however, have any serious socio-economic implications.
6. The need is to determine the most appropriate duration of closed season. In any case, no increase in shrimp fishing effort should be attempted until the status of stocks and potential annual yield is known. In other words, no additional boats should be allowed to be introduced even through a suitable regulation, until such time a study reveals that additional units can be introduced.
7. A system of obtaining full details of shrimp catches round the year, boat-wise, should be introduced to facilitate compilation of exploited shrimp fishery data and evaluating the same with reference to estimated stocks and potential annual yields.
8. Another major point for resolution is to arrive at a fishing strategy for the entire region. If this is not done, the exercise of observing a closed season in Indian waters will be neutralised, if some of the Indian upper east coast States and Bangladesh do not follow the same. Thus, coordination among the various upper east coast States and Regional Co-operation with Bangladesh is of utmost importance in implementing any scheme of declaring a closed season off upper east coast. As and when a scientific study establishes that a closed season during a certain period of the year would be advantageous, efforts have to be directed at building up a consensus among the various coastal States concerned and the neighbouring Bangladesh for the observation of the same, based on guidelines as agreed to among themselves and under an organisational mechanism to be established.

Studies in the middle-east appear to have proved that a closed season for shrimp is beneficial only when it is for a longer duration. For instance, in Iraq, closed season for shrimp is observed from 1st February to 30 June. In Kuwait, as a result of periodic stock assessment studies, a small area in the north of Kuwait is kept open for shrimping all the year round in certain years with a long duration closed season in the rest of the area. In Oman there is no closed season for shrimp. In Qatar the closed season for shrimp fishing is from 1st February to 30 June each year as in Iraq. In Saudi Arabia also the closed season is during the same period. Significantly, the closed season in almost all middle east countries is only for shrimp and not for fishes.

In the countries of middle-east, there is a system of licensing fishing boats. In India, the system of registration is there but there is, in general, no system of licensing. Absence of a licensing system gives freedom to the operators to fish for shrimp as they like. As voluntary closures can be subjective and convenience-oriented, it is desirable for the State Governments of the upper east coast to set up a Joint Committee to consider the subject and evolve a plan of action for the States concerned to undertake comprehensive studies as needed. The data collected during the studies undertaken State-wise, based on a uniform system can be put together by the secretariat of the Committee and conclusions drawn for implementation. The studies can probably cover the following aspects:

1. (*a*) Review of the presently available position in respect of systematic and regular collection of statistics on fishing effort, catches and size composition, species-wise and by area. (*b*) In the light of the review recommending a system to be introduced for collection of data to be supported by

stock assessment facilities, based on the data gathered, in each of the States (covering analysis of catches, fishing effort, size composition and age composition); (*c*) The analysis of data should be such that aspects such as artisanal and industrial fleet relations, stock identification and interactions with activities in Bangladesh can be understood.

2. Review of documented informations as at present on major commercial fishing and nursery grounds for shrimps and suggesting measures for charting them out.
3. Economics of operations of mechanised boats/ larger trawlers would have to be covered in the study as this aspect too is important in evolving the details of closed season and mesh regulation and for introducing any other measures to protect shrimp stocks. The other measures can include a division of the fleet and the fishing days in a season into certain blocks and relate them in such a way that the fishing pressure at any given time will not exceed a certain desired level.

In a paper on Management of Shrimp Fishery in Gulf of Mexico presented by an Expert Consultation on the Regulations of Fishing Effort (Fishing mortality) conducted by FAO in Jan. 1983 (FAO Fisheries Report on 289 Suppl. 3;n FIPP/R289 Suppl. 3 by the Fisheries Management Division, Southeast Regional Office, National Marine Fisheries Service, U.S.A), it was observed as follows:

"The biological characteristics which affect sustainable yields for penaeid shrimp are unique. They are an annual crop; very few individuals live a year and the majority harvested are less than six months old. There is no demonstrable relation between stock recruitment and recruitment overfishing and given present fishing technology this may be unlikely. That is, it may not be economically or technically feasible to take so many shrimp that the supply for the following years is affected. Because of these characteristics, fishing mortality in one year does not determine yield in the following year. The maximum sustainable yield (MSY) in number for a given year is considered to be all the shrimp available to harvest, using current fishing technology".

One recommendation made in the paper referred to above was to consider establishing shrimp management sanctuaries in important segments of nursery grounds.

According to the same paper closed seasons produced biological impacts that were not much different from open seasons. The observation was as follows:

"Simulation models, yield per recruit models and catch data were analysed for 1982 to determine the impact of the closure. The results of these studies were consistent and indicated that biological impacts were of a magnitude below detectable levels, given the precision of the modelling technique. Thus, the closure of the Florida Coastal Zone, with no consideration of the closure of the State's waters for 1982, provided no detectable benefit or loss. However, it appears that, if the closure of the Federal and State waters are considered, positive benefits could have been detected."

Commenting on the estimated benefits of the 1982 Texas closure, the same paper said, "No detectable change in landings can be attributed to the closure."

In the counculding part of the paper, it was said: "It appears that, because shrimp are an annual crop and their abundance is determined primarly by environmental conditions, the impacts of a seasonal closure can range from beneficial to non-detectable. During years of good recruitment, such as 1981, the seasonal closure provided excellent gains both in terms of number of pounds landed and value of the harvest. In years of low recruitment, such as 1982, benefits derived from the closure seem to be reduced. With several years of data, it may be possible to model the closure and predict those years in which benefits are expected. If this is possible, a more efficient closure can be established".

Keeping the observations as outlined above and other views on the subject, the issue of declaration of a justifiable closed fishing season for shrimp would probably have to be considered. Further, it may be worthwhile to consider the beneficial aspects of introducing a legal stipulation by the States concerned that each of the shrimp hatcheries should hand over to the nominated authority of the Government postlarvae of shrimp of appropriate age produced in one cycle of operations, for releasing them back into the sea for enriching the stocks.

Fishing Chimes, Editorial 267: May 1999: Vol. 19, No. 2

# Render Indian Reservoirs Sustainably Fishful

The State Fisheries Departments, the Government of India, the ICAR and the farmers, through combined efforts, have been able to achieve a substantial overall progress in respect of culture fisheries of the country. They have been able to build up the needed infrastructure by way of setting up hatcheries, farms for seed production and farming of seed to marketable fish size. There has also been a successful development of an organisational mechanism to popularise fish culture in the form of FFDAs. In the case of reservoir fisheries, however, the developmental process has been tardy, remaining, by and large, at the same level as at the time the British withdrew, with the exception of a few reservoirs. It is to the credit of the British that, at quite a few reservoirs, they set up farms and also undertook stocking of the reservoirs. The process no doubt continued after independence, but, with the production being conspicuous in a very few reservoirs, below average in some, and low to very low in others. Among the small reservoirs in which there is impressive production are Gularia, Bahra, Baghla in U.P., Chuliar and Meemkara in Kerala, Thirumurthy and Aliyar in Tamil Nadu and Markonahalli in Karnataka. Among reservoirs, other than the small ones, impressive production is noted in Yeldari and Girna in Maharashtra, Gobindsagar and Pong in H.P., Bhavanisagar and Sathanur in Tamil Nadu, Ukai in Gujarat and Gandhisagar in M.P.

There are 19,370 reservoirs in the country with an extent of 3,153,366 ha (consisting of 19,134 nos. of small ones with an extent of 1,485,557 ha; 180 nos. of medium ones with 575,541 ha and 56 nos. of large ones with an extent of 1,140, 268 ha). Sugunan (1995), giving these particulars in the FAO Fisheries Technical paper 345, estimated the present annual yield per ha from small reservoirs (less than 1,000 ha each) at an average of 49.90 kg, from medium reservoirs at 12.30 kg and from large reservoirs at 11.43 kg. So far as level of per ha production from small reservoirs is concerned, the range is given as varying from 3.91 kg/ha in Bihar to 188 kg/ha in Andhra Pradesh. In the case of medium reservoirs (1,000 to 5,000 ha each), the production is stated by Sugunan to be in the range of 7.2 kg/ha in Bihar to 624.9 kg/ha in Madhya Pradesh. In respect of large reservoirs (over 5,000 ha each), the estimated annual production is 1,143 kg/ha. The pooled production for all the three categories of reservoirs is given by him as 20.13 kg/ha. Based on these estimates, the present total annual production from Indian reservoirs has been computed at 93,650t, against the potential of 245,134 t (from 3,153,366 ha). In other words, the average per ha potential production the expert has visualised is around 80 kg. Going by this formulation, 2,45,134 t should generate gross returns of Rs. 735 crores, at an average realisation of Rs. 30/per kg as pooled value, realised out of domestic sales and exports through centralised storage plants and the connected cold chain and domestic marketing system.

There is a well argued and justified move to propagate all-male culture of Tilapia in reservoirs. As and when this takes place exports of fillets of Tilapia would have a beginning and the ongoing exports of freshwater prawn and major carps would gain further momentum. The possibilities, thus, are that around 30 per cent of the projected gross returns would be earned by way of foreign exchange. This works out to a little over 51 million US$, besides Rs. 514 crores to be earned through domestic sales. When the average annual production is raised beyond 80 kg/ha, the earnings would also get stepped up. These observations, it appears, would acquire validity only when an integrated reservoir development plan on a national scale, consisting of components such as seed production and stocking, management measures to achieve culture-capture fishery balance and sustainable exploitation, setting up of centralised ice plants and cold storages with needed processing facilities and a domestic marketing chain and an export line, is implemented. The assumption of fish production potential at 80 kg/ha by Sugunan appears to be an underestimation but it can be taken as the level that can be aimed at in the first phase.

Organisation of supplies and providing the needed service support in an effective manner is crucial for reservoir fishery development. The continued general ineffectiveness of the present haphazard developmental system in vogue in the case of most of the reservoirs is responsible for the present low level of fish production from the sector, a known major production source. This state of affairs, despite the enormous effort at development, highlights the reservoir fishery sector as a glaring gap in the overall fishery development process. An effort is made hereunder to bring out the components of an integrated plan that could be formulated to optimise fish production from reservoirs.

## Organisational Mechanism

The first step is to establish an organisation for the development of fisheries of reservoirs based on a cluster concept. This organisation can be a cooperative federation or a developmental agency to which the Government concerned could lease out the rights of development and disposal of fisheries of reservoirs concerned in the cluster. This organisation would take all needed further steps to eventually ensure the development of a sustainable fishery which will have a culture-capture balance.

Capital to be invested for establishing infrastructural facilities has to be forthcoming from the State Government concerned. They can secure funds from NCDC or any other appropriate funding organisation.

Employment of qualified and experienced personnel as required, to function under a competent chief executive who will be in charge of implementation of the project is of utmost importance. There has to be a management committee to channelise the work based on an effective policy and a strategy of development. A programme for the training of fishermen and executives including refresher training has to be part of the project. The boats and nets to be provided to fishermen and the landing points to be developed would have to be effectively covered under the programme, besides general improvements, stocking requirements and so on.

The other integral steps suggested to be taken to develop reservoir fisheries are outlined hereunder.

## Integral Step A: Grouping of Reservoirs

### Cluster Wrap

In the case of most of the reservoirs, it would not be economically viable to set up the needed infrastructural facilities such as hatchery farms and a cold storage exclusively in respect of each one of them. Considering this aspect, a compact geographical zone (which may consist of one, two or more districts) having a certain number of reservoirs with a viable extent of water area may be selected. The objective of such a selection is to have a compact reservoir cluster wrap to facilitate the development of needed viable infrastructural facilities to serve the reservoirs in the cluster.

## Integral Step B: Infrastructure

### Hatchery Farms

These may be set up with hatcheries at central locations for the development of fisheries of reservoirs. The farms should have a pre-determined spawn producing capacity, nursery and rearing ponds for raising spawn/early fry to yearlings stage, growout ponds for raising broodstock, and broodstock ponds, suitably designed.

### Reservoir-Specific Seed Storage Units

Seed that is raised at the hatchery farms, when transported to the reservoir site concerned for stocking may need storage facility for sometime before actual stocking. Further, now and then, it may become necessary to transport fingerlings to reservoir sites for raising them to yearling stage before stocking. For these reasons, separate farm space of needed dimensions has to be created close to the various reservoirs that would constitute the cluster.

### Feeder Cold Rooms at Reservoirs

Availability of ice and a cold room at each of the reservoirs for preserving fish before they are lifted to the mother storage-cum-processing centre referred to hereunder is of utmost importance. Accordingly, these would have be set up as part of the developmental endeavour. If there is justification, small ice-making units can be set up at some of the reservoirs.

### Mother Storage-cum-Processing Centre

Each of the clusters would have to be provided with a mother storage-cum-processing centre, of adequate capacity, worked out on the basis of anticipated peak arrivals from various reservoirs of the cluster for a duration of around 30 days or any other suitable criteria. Besides an ice production facility adequate to meet the daily needs of the reservoirs through a transportation system, there has to be a reception space for receiving fish that would arrive from various reservoirs of the cluster and for preparing the same prior to keeping in storage units at levels of temperatures as required and suitably kept ready in a packed form for release to domestic market and for export. A filleting machine, an IQF facility and a plate freezer may be required to be added, particularly to meet export requirements. In this connection, insulated boxes of various sizes would have to be provided for packing the fish for domestic market and cartons of various sizes for export. In addition, various other items would have to be part of the mother storage unit, to be located at a point as central as possible to the various reservoirs constituting the cluster, taking into account the nearness to a main roadhead to facilitate transportation of fish to domestic markets/to the nearest port for export.

## Integral Step C: Marketing

The Mother storage-cum-processing centre is the point of origin for providing fish for domestic marketing and for exports. So far as domestic marketing is concerned, the job to be accomplished in the first instance is to work out routes of supply, identifying the points of retail outlets all along each of the routes, with reference to buyer density. The design of the retail outlets to be set up has to be standardised and set up at each of the outlet points. The design of the retail stalls, would have to include chilled storage boxes, sale counters etc. The stalls would have to be preferably given on lease to selected persons based on wide publicity. The stall lessees would have to be assured of providing fish for sale everyday apart from various other relevant conditions.

There have to be insulated vans to carry the fish boxes, which will be delivered at each of the retail stalls by transport. Fish supplies, sales and payments etc. would be suitably regulated.

So far as exports are concerned, the system of periodic container transportation from the Mother storage-cum-processing centre to the nearest port can be organised, with all needed supporting activities.

The foregoing write-up is an endeavour to reemphasise the need to evolve a working pattern to develop reservoirs to make them fishful. Fishful reservoirs connote prosperity, with people receiving copious supplies of fish which would improve their health standards, enable fishermen to earn higher wages, and provide employment to a significant section of the society. Integrated development of fisheries would bring reservoirs under the purview of fish export basket, thereby contributing additional foreign exchange earnings.

Fishing Chimes, Editorial 268: June 1999: Vol. 19, No. 3

# Indian Marine Fishing Sector in Regression Mode

There is a gathering view that the Indian marine fishing sector has moved into the stage of stagnation and is now in the phase of regression. The maritime States and the Government of India no doubt know this for themselves. Regression is understandable in respect of coastal fishing. Coastal waters are overfished with per unit production coming down. This has given rise to a tendency towards stagnation in coastal fishing. There does not seem to be any initiatives on the part of the authorities to explore possibilities of imparting a fresh impetus to promoting fishing in the coastal zone, overcoming the over-exploitation factor and at the same time keeping sustainability aspect in view. While it has to be conceded that there is optimal utilisation of demersal resources of the coastal waters, the same cannot be said of columnar resources. The possibilities of introducing diversified fishing systems for adoption by the mechanised fishing boats for the exploitation of columnar resources can certainly be explored. There are many who believe that Indian coastal tuna stocks off the mainland and Andaman and Nicobar Islands too are sparsely exploited, in the same way as high sea tuna. This can be true because there is no fishing system adopted by coastal fishing vessels specifically aimed at catching columnar fish like tuna. It is known that coastal fishing boats of several countries have incorporated auto-monofil line systems for catching tuna and other pelagic species using small mechanised boats. Unfortunately, India has not done much in this respect.

The regression is glaring in respect of vessels of 15m OAL and above operating along the upper east coast. The population of the vessels which are all shrimpers came down to around 120 nos. from a strength of over 210 nos. There has however been some measure of diversification of fishing, from outrigger trawling to stern trawling by vessels in range of 25 -27 m OAL but the bulk of the vessels remain practically idle for over four months in a year.

The nation is unlucky in that the Government of India has no sustained interest in the utilisation of the resources beyond the coastal zone also. There had been a live deep sea fishing policy that revolved round operation of foreign fishing vessels in Indian waters under charter/lease/joint venture terms, coupled with a mandatory provision for the introduction of Indian built vessels of 20m OAL and above by the charterers/lessees concerned. However, the mandatory provision was subsequently withdrawn. A few years thereafter the charter/lease/joint venture policy too was kept under suspension. In the absence of a clear alternate policy and strategy, these developments placed the country on the path of regression in the marine fishing sector.

## Andaman Obstacle

Another surprising scenario, at least until March 1999, was the obstacle-oriented attitude of the Andaman administration towards fishing by Indian vessels owned by mainlanders, in the A and N part of Indian EEZ. Fishing vessels of the mainland were considered as unwelcome intruders and were not even extended port facilities. Only when the Association of Indian Fishery Industries brought this anomalous situation to the notice of the authorities, permission was rather grudgingly accorded for the mainland vessels to operate in the EEZ around Andamans but 24 n.m away from the coast, a regulation that is stipulated by the Government for the operation of foreign vessels on charter or lease in Indian EEZ. By implication, the Administration considers the vessels owned by the mainland companies as those equivalent to foreign vessels on lease/charter or under joint venture terms. Furthermore, the Administration has stipulated a condition of payment of a heavy royalty, for kg of fish caught. This condition tantamounts to telling not to fish in waters around Andamans, as, by any stretch of imagination, no operator can pay a royalty of that order which is stated to be far more than the market rate. The indirect effect of these two unreasonable conditions stipulated to be followed by mainland vessels is that vessels of Thai and other flags who are now known to indulge in poaching in Andaman waters could continue to enjoy the privilege, risking detection.

## Talks

A ray of hope is now seen with the Fisheries Commissioner of Andamans holding talks with the Association of Indian Fishery Industries on the subject at the initiative of MPEDA, in the last week of May 1999. The Commissioner, refreshingly enough, is reported to have adopted a positive stance and expressed eagerness to facilitate operation of mainland vessels in Andaman EEZ. However, justifiably, he is understood to have suggested to the Association to take steps for providing jobs on the vessels and for shore duties to unemployed youth of the islands and also for providing onboard training to them. He promised, as is understood, to look into the possibilities of allowing mainland vessels to fish beyond the operational zone of traditional fishermen, subject to environmental safety.

Andaman waters are known for their tuna and lobster resources and the signs of a reformed attitude towards the issue of main land vessels fishing in the waters may hopefully lead to some measure of improvement that would counteract the present regressive conditions.

It is the misfortune of the nation that the Indian Associations related to the fishing industries are generally weak. It is an interesting phenomenon that almost all those who were successive Presidents of the Association of Indian Fishery Industries opted out of marine fishing sector, with the exception of two. One of these two is Dr. C.Babu Rao who continues to be active in the sector, although he has opted out of the president ship, after a stint of a few terms. The other is the case of Dr. K. Hari Babu who runs an active integrated marine fishing enterprise and is the current President of the Association. It would be an interesting exercise for a research scholar to investigate and prepare a thesis on the subject of a good number of past presidents opting out of the industry. In fact, the Government of India themselves can assign the subject to one of the Indian Institutes of Management to go into this strange phenomenon. Those who functioned as Presidents of the Association (their companies) and who opted out of the marine fishing sector are a) Mr. Pusalkar (now no more), Messrs. N.P. Singh, N.S.H. Prasad, Dr. K.R. Prasad, Messers R.K. Varma, R.M. Shukla, and H.R. Rangarajan, the most recent of all. The name of Mr. V.S. Prasad, one of the ex-presidents and Director of Navbharat Ferro Alloys too can probably be added, as the marine products division of this company is understood to be moving towards closure.

While the reasons for the disinterest of the companies concerned are not clearly known, it appears that non-profitability of operations is not one among the possible reasons. It is possible that the companies that retracted encountered managerial problems. Further, another plausible reason can be their inability (and consequential despair) to prevail upon the Government of India to give a due and enduring importance to marine fishing sector, a vital one for the nation from the angle of providing nutritious food to the people and for adding to earnings of foreign exchange to country. It is incomprehensible as to why the successive Governments in general had failed to accord the importance that the marine fishing sector deserved. There are, of course, a few spurty exceptions of schemes for the introduction of larger vessels through imports and indigenous construction, coupled with financial support by way of loans and subsidies.

The present disinterest in marine fishing on the part of the corporate sector can be for various reasons, the chief among which may be financing support not forthcoming from financing bodies and the general disinterest on the part of the Government of India in the sector, as already stated. Marine sector yields over one billion U.S dollars in foreign exchange to the nation. These earnings can be stepped up only when fish production goes up in terms of quantity exported and quantum of earnings of foreign exchange. It is surprising that the Government continues to appear to be complacent, despite knowledge of this trend.

India, by virtue of the large size of its EEZ, the vast fishery resources of the zone, and the technology it commands, should have by now become the leader in the marine fishing field in south/south-east Asia. It is unfortunate that India does not have the kind of vibrant fishing policy particularly in relation to tuna, in the same way as Sri Lanka or Maldives has. Although a country with unrivalled tuna resources, because of absence of exploitation dominance, India is not able to acquire a dominant influence in the Indian Ocean Tuna Commission. The long range implications of this ought to be apparent to the Government of India.

It is high time for the Government of India to take an in depth look at the marine fisheries situation of the country from various angles including defence, and bring such revolutionary reforms in the approach as are necessary through a sound policy. One characteristic of developed nations is that they have an advanced and prospering fisheries status. Until India acquires this status, it will continue to be having the present level of standing among developing nations.

Fishing Chimes, Editorial 269: July 1999: Vol. 19, No. 4

# Augmenting Fish Exports: Relevance of Relying on Tuna and Tilapia

The nation has arrived at a crucial stage of inevitability of reconnoitering for avenues that would lead to stepping up of aqua exports. The items to be identified may be those of low profile in the current export basket but have the potential of contributing substantially to the export effort. Gone are the days of automatic augmentation of aqua products exports. An era has dawned for launching fresh promotional efforts revolving round newly identified exportable resources and these efforts are of paramount urgency. From dependence on coastal capture resources, whose level of production had reached a level of per unit catch decline we did move on to develop culture shrimp production, but this has also more or less reached static stage, particularly as CRZ regulations prevent further horizontal expansion.

There can be several new underutilised resources to rely upon to boost exports. Of these, what strikes as the most promising one in the capture sector is tuna, a high sea resource with awaitability estimated at 200,000 to 300,000t. As concluded at the tuna Conference held at Visakhapatnam recently, a programme to utilise 25,000t of this resource would need to be launched, as one of the measures, to boost up exports, on top priority. There is also immense scope for developing sea cage culture for groupers, perches, breams, sea bass etc. As they do in Australia, juvenile tuna can also be farmed to an exportable size. A beginning can also be made for farming squid and cuttlefish as Thais have started venturing into.

So far as inland aquaculture is concerned, one aspect that is glaringly seen is that there is very little further scope, as already stated, to augment the area under shrimp farming. This means that, there is need to explore possibilities of farming exportable species through efforts at diversification. We are now heavily dependent on farming of major carps. Unfortunately major carps have extremely limited export market. Although there can be immense scope to develop an export market for major carps, this scope can be realised only through sustained and long term efforts, based on a long term plan of action to further popularise Indian major carps, already known abroad, to the needed level. In this situation, it will be prudent, in the interests of augmentation of exports, to identify such of the exotic species which, when introduced, would not lead to any ecological imbalance, particularly, when introduced based on an effective and cautious Plan for production and dissemination of all-male seed for farming. In other words, a monitored introduction and development of brood stocks in a protected environment by approved enterprises who have the needed infrastructure or who can show proof of an arrangement to set up an infrastructure to produce all-male progeny, with a foolproof system that would not allow the brooders to escape into any water area outside, could be permitted. Without expanding the base for farming of exportable species, the scope for augmenting exports would at the best be marginal. It may be possible to produce more of freshwater prawn but the scope for this is extremely limited because of logistic problems of seed production and its supply over wider areas owing to the constraint of winter temperatures in the north.

Tilapia is one exotic fish that straddles the globe as the most eligible cultivable species from export angle. In the past decade, Dan Cohen of Israel says, tilapia had become a major source of fresh and frozen fillets in developed markets. According to him, on live weight basis, tilapia is now the third largest imported aquaculture product into USA. At present, main suppliers of frozen quality fillets are Taiwan, Indonesia and Thailand. USA consumes about 50 lakh tonnes of fish (compared to 4 lakh tonnes of shrimp) and of this, 50-60 per cent are imported in a large measure in the form of fillets, which are standardised, each of 35 *oz* weight without bones. World production of tilapia is stated to have reached 473,000 t in 1992 and it is known that the average annual growth of tilapia production over ten years from 1984 has been 9.6 per cent. Over 75 countries produce tilapia and these include 30 in Africa, 12 in Asia. China, Philippines, Thailand, Malaysia, Taiwan and Vietnam are among the Asian countries producing 390,000 t annually (85 per cent of global production). So many countries cannot be wrong in farming tilapia. Like us they too are conscious about environmental and ecological safety. If there is any reservation from these angles, the best course of action would be to devise effective measures to counteract them, instead of being unduely on the safer side by not allowing the culture of the fish at all as at present (*Fishing Chimes*, Vol. 19 No.2).

India is tilapia shy, having been disillusioned by the entry of the fish into the country, believed to be from Sri Lanka in 1950s. The fish became a pest. Later in 1980s Vorion Chemicals and distilleries in Tamil Nadu started culture of Nile Tilapia in a systematic manner by confining broodstock to brooder ponds and producing all-male seed for stocking through feed manipulation. The company marketed the fish in the domestic sector. Exports were not however tried, probably because of the official ban on culture of tilapia.

A time has now come to evolve a protected system of breeding tilapia, and to produce all-male seed. The species to be introduced for culture has to be however determined by CIFA. The seed, tested and certified; can be supplied from recognised and approved centres to the various farms for culture. Production of all-male Nile tilapia at these farms can be linked to production of processed fillets at identified processing plants for export.

CIFA can conduct experiments to evaluate, among others, the suitability of Nile tilapia, known to be a fast growing one, for composite culture, either with other fishes or for monoculture. The culture has to be linked to a protected system of producing all-male seed. Through a standardised system of culture, to be evolved by CIFA (pushing aside the traditional disrespect and apathy towards the fish), the nation would stand to gain on two fronts. One is of course the export front. The other is that the fish would yield far higher returns to the farmers. Tilapia fillets can be exported at 8-10 US $ per kg. and the procesors who buy the raw material will be able to pay for the fish at higher rates. The possible negative aspect that can be pointed out for introducing tilapia culture is in respect ecological and environmental hazards. As already stated, in many countries tilapia is cultured and benefits derived. There appears to be no reports of their indigenous ichthyofauna getting affected. And there is no reason to believe that introduction of Nile tilapia or any other viable tilapia species would be an environmental or ecological hazard, specially when a protected system of broodstock maintenance and production of all-male seed is ensured. Such a system will also be in the interests of farmers as inbreeding and consequential reduction in growth would be countered under the system. It will be uncharitable to quote the earlier unhappy experience with *Tilapia mossambica* as an example of the hazard, for the reason that this fish was allowed to proliferate and spread into the wild without any monitoring.

Considering the benefits that would flow, it is of utmost expediency for the Government to authorise ICAR/CIFA to experiment on the culture possibilities of a few selected species of tilapia, with a view to promoting culture of at least one of them, according due regard to environmental and ecological aspects as well as the economic angle. Based on the results, a well drawn national programme of culturing tilapia in tanks and ponds and in reservoirs could be developed and implemented for national benefit. A National Programme of the kind would provide phenomenally higher returns to the farmers, and would more than double the export income from aqua products. Above all, the inland farming sector would be brought under export endeavour through such a programme.

Fishing Chimes, Editorial 270: August 1999: Vol. 19, No. 5

# Checkmating Pakistani Move

According to a report in *Fishing News International* (July '99), Thailand has signed an Agreement with Pakistan providing for Thai investment of the order of US $ 140 million in Pakistan's deep sea fishing industry. The proposed investment comes to 588 crores in Indian rupees. In line with global trends, Thailand may not invest the amount for the introduction of newly built vessels. In other words, translated into assets, this investment would account for 150 vessels or more of the presently operating Thai fleet. One need not wonder if the strategy of Thailand is to gain two advantages in one move. One is to avoid fresh investment by diverting vessels out of its present fleet to fish in Pak EEZ. The other advantage is to annex new fishing grounds, thereby avoiding the problems of securing licenses and permissions now faced by individual companies. The Agreement appears to constitute a blanket permission, within the investment envisaged, probably entailing the selection of the vessels by Thai Government. The annexation of new fishing grounds by the Thai Government in this manner will undoubtedly go a long way in winning the good will of the Thai fishing industry which is always on the look out for new grounds.

This development may however have serious repercussions on the state of fisheries resources of the Indian EEZ adjacent to Pak EEZ. As at present, Pakistani vessels are known to trespass into Indian EEZ frequently. Such Pak vessels noticed by the Indian coast guard no doubt get apprehended but, being small fishing craft operated by artisanal fishermen, the dent on the resources on the Indian side may not be substantial. The problem of poaching will however become very serious once Thai vessels start fishing in Pak EEZ. The Indian and Pak waters being contiguous, Thai vessels known for fishing outside their waters, would, in all likelihood, be tempted to move into Indian EEZ. Indian coast guard may repel such intrusions but there will be generation of lot of tension in the process and sometimes leading to exchange of fire etc. The transgression of Thai vessels is possible because of the rich Tuna and other resources in Indian waters as well.

Undesirable situations can be avoided and intrusions can be stopped to a large extent by putting into operation a sizeable fleet of Indian deep sea fishing vessels with capabilities as good as Thai vessels. A strong presence of Indian deep sea fishing vessels in the adjacent area will act as a deterrent to intrusion of Thai vessels from Pak EEZ into the Indian EEZ. In this context, it is of utmost urgency to upgrade Jakhau fishing harbour, the northern most one closest to the Pak border, with the needed infrastructure facilities. The investment on this upgradation and on the introduction of deep sea fishing vessels will serve the dual purpose of matching the operations of Thai/Pak vessels and also of increasing production on the Indian side which will be largely export-oriented, as the region abounds in tuna and other fishes of export value.

Thai and Indonesian fishing vessels are known to intrude into the Indian EEZ around Andaman and Nicobar Islands. This propensity, although under check by the Indian coast Guard, can be countered effectively only by pressing into operation India's own fishing fleet. Fortunately, the Andaman Administration, which had not been allowing the operation of vessels of the Indian mainland in the waters around the islands, has now been showing signs of changing its stand. This change in attitude has to be followed up by the Indian industry. Indian vessels should operate increasingly in that part of the Andaman Sea in Indian EEZ to assert the presence of Indian fishing, which would discourage foreign vessels to operate in our water clandestinely. Such a strategy would not only go a long way in stepping up Indian exports of marine products but would also instil some measure of hesitation on the part of alien fishing vessels to trespass into Indian EEZ around Andaman and Nicobar Islands.

Fishing Chimes, Editorial 271: September 1999: Vol. 19, No. 6

# Indian Aqua Products Exports

The level of Indian aqua products exports seems to have reached the brim, looked at from the view point of the presently utilised resource base. There is, however, a vast unutilised resource base that can be inducted to push up to the understandably sagging aqua products exports. The unutilised resource base, consists of reservoirs, coastal waste lands and inland saline waste lands of the north-west, besides existing tanks and ponds. The movement for the utilisation of the unexploited resource potential for export-oriented production has to be spearheaded by the MPEDA in consort with the Union Department of Animal Husbandry, under a participatory scheme, that would envisage the association of the State Fisheries Departments concerned.

## Reservoirs

India has a large no. of reservoirs with an estimated extent of 30 lakh ha. By bringing, say, half of this area under freshwater prawn production too, at an average minimum of 5 kg/ha, 15,000 t of these crustaceans can be produced additionally per annum.. In HL form the quantity will be over 7500t, which will result in additional foreign exchange earning of around 38 million US $ or Rs. 153crores, calculated at US$ 5,000 per ton.

The projections given above are highly conservative. The production and foreign exchange earnings will be much more, when an organisational mechanism is evolved and introduced for the development of the activity, preferably in the form of Reservoir Fishery Development Agencies. Reservoirs in various zones could be grouped together with reference to their location, in proximity to each other, communication facilities etc., and provided with centralised facilities such as hatchery farms, cold storage facilities with linkages to domestic marketing network and exports.

## Coastal Waste Lands

The Coastal wastelands are estimated at 12 lakh ha. Of this, less than 1.5 lakh ha have been brought under farming. By developing atleast another 1.5 lakh ha under shrimp farming, an additional production of atleast 75,000 t (37,500 on HL basis) can be achieved, bringing in additional foreign exchange earning of US $ 188 million or Rs. 808 crores.

## Saline Lands of North West

There are vast stretches of saline wasteland in the States of Haryana, Rajasthan, Punjab and U.P.An estimated 1.5 lakh ha of these saline areas, most of which are having underground saline water can be brought under export-oriented production of shrimps/brine shrimp (*Artemia*) cysts, with the availability of some quantity of freshwater for salinity adjustment. This developmental activity would however need a detailed study from the point of view of any possibilities of soil subsidence in contiguous zones away from the area developed into farms or in farm areas themselves. An organisational mechanism is necessary to undertake the task of utilisation of the saline wastelands. Hopefully, once developed, these lands will bring in a foreign exchange of US $188 million or Rs. 808 crores per annum. Promotion of the activity needs an effective organisational framework. Suitable number of Developmental Agencies could be set up to accomplish the task on the lines of a project to be formulated in the light of the report of the expert team suggested to be set up.

The foreign exchange earnings could be stepped up further significantly by undertaking the exploitation and export of the untapped tuna resources of our EEZ with export potential and through undertaking sea cage culture, pearl oyster production etc.

Fishing Chimes, Editorial 272: October 1999: Vol. 19, No. 7

# Declining Marine Catches and Exports: Opportunities to Counter the Trend

The Indian Sea Fishing Industry is now lurching. The declining national marine fish production has been chiming alarm for the past few years.

It is needless to say that this crucial industry which fetches over Rs. 4,000 crores per annum by way of foreign exchange, deserves focal attention and support by the Government to ensure its survival and growth in national interest. The compulsions of survival have motivated the industry to approach the Government to take certain policy decisions to enable the industry to pull out of the present stagnation/decline in seafood production and to step up the same in a responsible and sustainable manner, in the background that, once there is a turnaround in the level of production, increase in exports of marine products would start registering. It is axiomatic that the level of marine products exports would go up only when there is increase in production. Diversification of products for export is only a palliative to step up foreign exchange earnings for some time. It can never be a total and a long duration solution for stepping up exports, as the same situation as at present will recur again when there is no increase in production.

Certain promotional measures taken by the Indian Government in the early decades after independence had transformed the mildly tradition-oriented and a sparsely spread artisanal marine capture fishery avocation along the coast into a booming activity through various stages of growth. Unmonitored growth took place in the later decades in the sea fishing effort in the coastal zone. Consequently, there are now in operation in Indian waters of over 40,000 mechanised and motorised boats, and a few hundreds of medium to large sized vessels with a OAL upto 27 m. The fleet operations which mostly revolve round shrimp, have gradually brought down the per unit production. The operations are thus now characterised by unfavourable economics.

The plummeting of per unit production and the consequential fall in earnings have rendered operations of several boats, small and large, uneconomical. It is a common sight to see many of them either moored indefinitely or in a sunken state, in some of the harbours. The round the year operations of the remaining vessels continue to be precarious from the angle of economics.

It has to be conceded that the growth in fishing effort, although unmonitored, is the direct result of provision of common infrastructural facilities by the Government by way of fishing harbours, boat building yards, training centres etc. Facilities such as cold storages and processing plants are examples of infrastructural facilities developed by individual enterprises engaged in the export of marine products. Apart from the infrastructural facilities, Government extended the concessional one time repayment facility of capital loans given to enterprises out of Government funds for acquiring larger fishing vessels. The concession enabled the enterprises to secure unencumbered ownership of the vessels. Another major help extended by the Government is to allow mid sea bunkering of all approved vessels of over 20 m OAL, considered to be deep sea fishing vessels, and also in supplying duty-free diesel oil to smaller vessels.

The perceptions in respect of decline in per-unit shrimp catches differ. The decline has confused the small vessel operators. They are unable to believe that the drop in unit catch levels is because of the increase in effort. The fact that the total annual catches continue to be more or less the same year after year is ignored. A theory was developed that observance of a closed season for shrimp in one season would enhance the catches during the next season. The small vessel owners of upper east coast resorted to voluntary observation of a closed season, with the owners of larger trawlers also joining. This has not however led to any tangible improvement in catch per unit.

Sensing the situation, several enterprises owning larger trawlers took tangible steps to wriggle out of the trap in which they were caught. They pulled themselves out of the satanic effects of shrimping on the overall economics of vessel operations in the long run and shifted their activities to the west coast for catching cephalopods, deep sea lobsters and deep sea prawns, considering their export potential. The more resourceful among these enterprises shifted their fishing operations to areas off Myanmar coast, with all needed permissions for catching deep sea lobsters in the main, with a good measure of success.

While some of the owners of larger vessels diversified their operations in the manner described above, some others succeeded in selling their vessels at comfortable prices to those longing to own them, mostly skippers of fishing vessels. The same trend manifested itself in the case of several mechanised boats, with persons mostly in command of the boats acquiring them from the owners. In other words, a trend towards a shift in ownership in favour of operatives has set in. The remaining larger vessel owners continue shrimping operations off upper east coast, for the reason their vessels are not either suitable to fish in depths of 150 m and over or owing to financial constraints. So far as the owners of smaller craft are concerned, they seem to wait for better times, enduring the present straits in which they are in.

Unlike countries such as China, Iceland, Maldives, Thailand, Indonesia etc., where a priority is accorded for fishery development, unfortunately, in India, fisheries sector receives a very low priority, despite the fact that the sector is one of the main foreign exchange earners of the country, and provides employment to a sizeable section of the population.

Having outlined the genesis of the declining trend of exploited fisheries of the country, the possible opportunities to counter the trend are outlined hereunder.

The economics of operation of small fishing boats in the coastal zone can be improved by weaning a substantial number of them from concentrating on catching of shrimps more or less round the year as at present, by assisting them to diversify into harvesting pelagic resources such as coastal tuna, which are believed to be under-exploited, in contrast to demersal resources which are over-exploited. The policy reform required is therefore to bring about diversification of operations of the small mechanised boats at least during certain periods of the year to undertake pelagic trawling and lining. Such a reform, coupled with a regulatory control over introduction of new boats would tend towards increasing per unit earnings. It may be mentioned here that in several countries small mechanised boats are equipped to conduct longlining and hand Iining operations, trap fishing etc.

The industry blazed a new trail for the first time by deploying quite a good number of vessels for seasonal trawling in the EEZ of a foreign nation, *viz.*, Myanmar, for harvesting deep sea lobsters and certain other demersal species. The Governmental organisations and the Reserve Bank of India issued needed permissions to enable the enterprises to conduct fishing in foreign waters and bring the catches to Indian ports for processing. The Reserve Bank and the Government are understood to have agreed in principle for the transfer of the catches on high sea as an export activity without the need to bring the vessels to an Indian port to unload the catches. This facility would tone up economics of operation further.

The achievement of the industry in this direction provides an opportunity for new policy initiatives. It is widely known that several countries enter into arrangements for fishing in the EEZs of other nations. The Government of India also could move in the same direction, in the light of the successful operations off Myanmar coast, by developing bilateral agreements with neighbouring countries of SouthAsia and South-East Asia. Studies conducted by an entrepreneur indicate the scope to extend Indian fishing vessel operations into EEZs of Indonesia and of a few other countries. Indian vessels can gain problem-free access to EEZs of foreign countries when these activities are organised under the umbrella of bilateral agreements between the Indian Government and the foreign country concerned. This is an area which appears to need the immediate attention of the Government. Simultaneously, tangible resource utilisation capabilities to harvest the huge stocks of unexploited tuna estimated to be available in the Indian EEZ, particularly around the islands have to be built up. Experts who participated in a recent round table conference on tuna have recommended that immediate steps are necessary to exploit and utilise this resource for national benefit. Government would have to set in motion the needed follow-up measures to translate the recommendations into action, as they would go a long way in stepping up the quantum of exports of marine products from India. There can be bilateral agreements for a specified duration with countries well versed in tuna longlining operations so as to ensure fruitful results. In any case, without technological help from outside, the development of the activity in Indian EEZ on our own will be very difficult and will prove to be expensive.

Another aspect is that a good part of the marine products exported from India are used as raw material for value addition and re-export thereafter by importers. Under a bilateral agreement with a selected country having expertise in the value addition technology and willing to invest, private enterprises can be encouraged to undertake the value-addition within the country itself with the foreign partner assisting in exports as well. There can also be agreements on adopting proven foreign brand names too on payment of royalty.

The resources of EEZ beyond the coastal zone are exploited intensively in some zones, moderately in some others, and not exploited at all in the remaining ones. Hitherto, the assessment has been that cuttlefishes are available in commercial quantities in the waters of the Arabian sea only. This belief has been proved wrong during 1999 season. A few companies deployed their trawlers in this season to catch cuttlefish in the EEZ of the upper east coast with encouraging results, thereby demolishing the myth that west coast waters alone are the store house of cuttlefish with a substantial quantity.

In view of the new experience in commercial operations for cuttlefish, Government has to initiate measures to encourage the industry to diversify the fishing activities to catch cuttlefish and possibly squid from the EEZ beyond the coastal zone of the east coast.

There are several who are convinced that deep sea lobsters and prawns are not just restricted to the known grounds off west coast of India but they are also there in all likelihood as a garland off the entire Indian coastline in the 150-200 m depth zone and beyond. This conviction suggests that there have to be efforts at checking whether they are really available as believed. The job of commercially-oriented exploration to check their availability will have to be undertaken by the FSI, as the industry is not equipped to undertake the activity on its own. However, under a suitable arrangement between FSI and the Association of Indian Fishery Industries, a scheme for commercially-oriented surveys could be drawn up and implemented for national benefit. It has to be however kept in view that all the vessels owned by the members of the Association are not suitable for such operations.

Another aspect is the denial of access to the Indian vessels of the mainland to fish in waters around Andaman and Nicobar Islands and also around Lakshadweep Islands. A ray of hope that may however improve the situation is that, unable to withstand repeated representations from the industry, the Andaman and Nicobar Administration has conceded that the vessels could fish in waters beyond 24 nm from the coastline of the islands. This relaxation has opened up an opportunity for examining from various angles the justification for the 24 nm limit. One aspect is that the concession seems to have no scientific, technical or social basis. The islands are oceanic. The deeper zones are not far away from the coast. There is no way of conducting fishing in the deeper zone using the traditional boats. In other words, by allowing the trawlers to operate within 24 nm, but beyond 12 nm (as is the practice off mainland) of the coast, the economic condition of traditional fishermen of the islands who are far less in numbers will not be affected. On the other hand, by securing employment on these trawlers, fishermen of the islands can improve earnings and upgrade their living conditions. Andaman and Nicobar Administration has therefore to consider allowing the vessels of the mainland to fish in the EEZ of Andaman and Nicobar, outside the zone of operation of the traditional boats or beyond 12 nm from the coast. It may be noted that, while the vessels of the mainland are not allowed to operate in A and N waters, vessels of Thailand and other countries are known to fish in the same waters unauthorisedly.

Governments of Philippines, Maldives and Sri Lanka have set up FADs (Fish Aggregating Devices) in their respective EEZs to help their fishing industries. FADs attract the gathering of fish beneath these and around. Congregations of this nature facilitate capture using tackles such as purse seines and lines. It is worthwhile for the Government to examine the possibilities of setting up FADs around Andaman and Nicobar and Lakshadweep Islands and at some selected points off mainland, on the lines developed by the countries referred to above. CMFRI is known to have attempted some experiments on fishing with the help of FADs, but nothing further is known.

Fishing Chimes, Editorial 273: November(i) 1999: Vol. 19, No. 8

# Fishing Industry Victimised: Orissa Cyclone's Disastrous Affront

The sinewy super cyclone that lashed the Orissa coast left behind it a crippled fishing industry as it crossed the coast at Paradeep. Several in the industry including a skipper lost their lives. A larger number of vessels including two larger ones sank. The fate of several fishermen who are still to be accounted for is not yet known. Several shrimp farms and hatcheries along Orissa coast have been destroyed.

The aftermath of the havoc is frightening. The economics of the industry are thrown into a disarray. The families of those who succumbed to the cyclone are in the lurch. Those who lost their boats, those whose vessels ran aground, those whose vessels were battered, and those who lost their nets are now in a state of economic distress, with little relief forthcoming.

The major victims of the calamity are the fishing boats and the crew, mostly those belonging to AP. The boats took shelter at Paradeep Port to escape from the onslaught of the cyclone but it pursued them mercilessly at a velocity of over 250 km per hour, resulting in loss of boats and men.

Several of the boats are understood to be without insurance, probably for the reason that the owners could not have adequate earnings to meet the cost of insurance.

The saving grace was that, when the entire Paradeep port area and the town were in total darkness and disarray with no communication facilities, captains of larger fishing vessels had been able to speak with the vessels at Visakhapatnam fishing harbour over radiotelephone and relate to them the calamity that overtook them. These vessels were the only ones having lights in the entire area because of the generators they have. Even the Coastguard had to resort to using radio-telephone of a fishing vessel to contact their own patrol vessels. Fortunately the larger fishing vessels had adequate food stocks which they could share with others.

Ports are supposed to be places of safe anchorage, particularly in times of cyclonic weather. It may not be in order for the authorities to defend themselves by saying that because of the extraordinary vigour of the cyclone, some of the vessels sank, some ran aground and some suffered damages. There can be other reasons such as those related to the design of the harbour, overcrowding of vessels etc. It is desirable to identify the specific causes for the losses suffered by the fishing industry in the port, which is supposed to provide protection to the vessels and institute required remedial measures to enhance safety for them while at port, in the future.

For no fault of theirs several vessels berthed in the port suffered losses. Insured or not insured, the losses suffered by the boats would have to be compensated by the Government concerned for the enterprises to survive. While the calamity took place in Orissa State, most of the boats affected are those of enterprises in A.P.

Cyclones are an annual recurring feature striking some part of the coastline or the other. While crossing the coast, cyclones invariably cause destruction to properties and several human deaths take place, mostly in the coastal villages affected, as these receive the first battering impact. Now the disaster that struck Paradeep shows that fishing ports are no exception to the onslaught of cyclones, despite the fact that they are designed to be safe anchorages.

The small mechanised boats come within the purview of State Governments. Larger trawlers, being deep sea fishing vessels, come under the control of the central Government. These features may give rise to quite a few technical issues, as to who should provide relief. Further, the loss of boats and damages to farms would have a retarding effect on exports of marine products. It is of paramount importance for the Government of India to set up a task force consisting of experts drawn from the central and State Governments, Non-Government Organisations (NGO's), and fisher leaders to make recommendation and to produce an 'Action Manual' that would provide guidelines to set in motion the needed measures of relief with expedience in cyclone affected fishing villages and fishing harbours.

It appears that, if and when a task force is set up, an aspect it has to look into is the need to have port-wise and village-wise permanent Rapid Action Forces with members drawn from Revenue, Police, Fisheries, Non-Governmental Organisations and fisher leadership, to swing into action as a cyclone strikes. Where necessary, representatives from port and coast guard may have to be associated.

Another aspect is that each of the villages should be provided with an efficient and a working system of conveying weather warnings, particularly those relating to severe ones, by Radio telephone from transreceivers or by VHF Units set up at centralised locations, to the nominated local leaders in each of the village. Each of these villages should be provided with Radio telephone or VHF or both. It is heartening that this requirement is already covered by a scheme to provide RT facility to coastal fishing villages in A. P., but the scheme has to be extended to all other coastal States too, if not done already.

The leaders and others on the task force should be given intensive training on the manner in which they have to deal with pre and postcyclone compulsions, apart from being provided with an 'Action Manual' which would clearly tell, step by step, the tasks to be undertaken, such as safe mooring or anchorage with fenders and with adequate space between boats to prevent battering, for safe storage of nets and implements, evacuation of men, women and children to safe locations, calling boats out at sea from shore based transreceivers or VHF Units to come back forth with to the base, procuring and keeping stocks of rations etc. for suppliying boats. All boats should invariably be provided either with radiotelephones or VHF sets, with mother units at the villages concerned. Well before the onset of cyclone season refresher courses should be conducted in coastal villages and ports by the Rapid Action Force (RAF) concerned. Their nominated leaders should be trained before they take over full fledged leadership.

It is possible to counter cyclone disasters with a careful long term planning and by evolving and implementing a near foolproof system. With an organisational mechanism built up in a manner that it can be activised in the shortest possible time, perils now being unleashed by cyclones on the fishing industry could be minimised. In short, the aim of the plan should be to equip each of the fishing villages with all needed infrastructural facilities to keep the boats, nets and other requisites safe and to provide means of safety for the lives of fisher families during cyclones. A system of having a village level relief fund could be explored inclusive of operational guidelines.

Fishing Chimes, Editorial 273(a): November(ii) 1999: Vol. 19, No. 8

# Farm-reared Tiger Shrimp Broodstock Production

It is familiar knowledge that shrimp hatcheries in the country face the problem of short supply of disease-free tiger shrimp broodstock. Dependence is totally on wild exploited broodstocks which are known to harbour dormant white spot virus which becomes active when a deterioration in the state of health of the shrimp takes place and is congenial for the virus to proliferate. In order to get over the problem, the solution put forth has been to culture farm-bred broodstock. There has been no breakthrough in this respect in India, although scientists are very much aware of the problem and are probably working on it, but what is being done is not known. What all could be gathered is that tiger shrimps do attain maturity in farm ponds but very few eggs, mostly unviable, are released. A senior scientist of ICAR said quite a few years back that the problem could be solved. Dr. S.N. Dwivedi once observed that one way of ensuring gravidity in farm conditions will be to undertake cage farming of juvenile shrimps of advanced growth in certain zones of the sea having the needed salinity. The present position in respect of the subject, however, appears to be in a standstill state. The situation can be likened to dependence on wild major carps for induced breeding until early 1960s, when the scientists could succeed in growing major carps to a stage of gravidity for induced breeding work in farm conditions. At the International Workshop on Development of Sustainable Management Practices in Shrimp Farming held in Bhubaneswar on 30-31 July, 1999, there was a startling revelation from Dr. Denis Gasnier of COFREPECHE of France that IFREMER had developed at its Tahiti Research Centre in French Polynesia broodstock of one strain of tiger shrimp which was now in its 12th generation of production and that broodstock production was well under control at the Centre. Full monitoring of environmental conditions for breeding and usage of specific feed was being done for the purpose. He said that it was now a common practice to supply seed produced out of domesticated brooders. It is of utmost priority for our scientists concerned to find a way of producing broodstock of tiger shrimp at our farms for use in shrimp hatcheries, same way as IFREMER is stated to have accomplished.

Fishing Chimes, Editorial 274: December 1999: Vol. 19, No. 9

# Indian Tuna Tangle: Unravelling Possible Only Thro' External Inputs

The Tuna Round Table Conference held in Visakhapatnam in July 1999 made certain pathbreaking recommendations to achieve the objective of effective utilisation of the tuna resources of the Indian EEZ, estimated at 230,000 t. The recommendations envisage that one vessel each of the Integrated Fisheries Project, Fishery Survey of India and of the private sector need to be modified for undertaking monofilament tuna longlining in the first instance, to be followed eventually by conversion of 30 per cent of the deep sea trawler fleet in the OAL range of 23 to 30m for tuna long lining. The Director, MPEDA, Fisheries Development Commissioner, Director General, Fishery Survey of India and the Directors of Integrated Fisheries Project and Central Institute of Fisheries Technology, among others, gave shape to these recommendations.

The fishing industry is now waiting with bated breath to know about the follow-up action. So far as the fishing industry is concerned, one company has offered to modify its vessel for tuna longlining too but at the same time sought financial assistance from MPEDA. There has been a response to this approach in a meeting but it has fallen short of the expectations of the company. This apart, it is learnt that the overall guidelines to translate the recommendation for field application have not yet been finalised by the Fisheries Division of the Union Department of Animal Husbandry and Dairying, which is the nodal department. The major constraint to be reckoned with in this respect seems to be largely related to the financing aspect, confidence creating technical inputs including preparation of drawings for the modifications, selection and installation of equipments, imparting training related to commercial operations on the one hand and avoidance of the past pit falls on the other in the related developmental endeavour by three Indian tuna companies whose performance was rated good initially but later found flagging down. Several plausible reasons were put forth for this denoument. The main reason, among others, identified was that investments on the vessels were unduly high. The overall length and other specifications of the vessels were disproportionately excessive, compared to the dimensions of high sea zone of the Indian EEZ and its proximity to the Indian major fishing ports. Another highlighted point was that the vessels could not secure permissions for fishing in the EEZs of other nations for the reason that they were already under fishing stress. Without covering other grounds, vessels of the type the Indian companies acquired which were meant for global fishing endeavour, would understandably become uneconomical when their operations are restricted to the national EEZ. In this light, the conference recommended introduction of medium sized vessels suited for long duration voyages within Indian EEZ, in order to make a fresh beginning as mentioned above, so as to bring down the investments and at the same time optimising returns.

That the recommendations made have to be implemented in a manner that there will be no recurrence of the mistakes of the past is well understood. Further, it was unfortunate that foreign men brought in by two other Indian companies to install monofil longlining system on their vessels failed in fulfilling the job entrusted.

From the foregoing account, it can be seen that the need is to find avenues of financing and to entrust the job of modification to a proven professional enterprise, which would also undertake demonstration of monofil long lining with the equipment installed by them and also train crew members in the fishing system on commercial lines. No more risks can be taken. Wasteful investment, disappointments and setbacks in the utilisation of the resource profitably would have to be countered. Tuna is a known foreign exchange earner through exports. The resource position is well documented and it is well vindicated by Taiwanese chartered tuna vessel operations.

South Africa, a recent entrant into monofil longlining for tuna through medium-sized vessels, is reported to be doing very well in its EEZ.

Monofilament longlining for tuna is an achievement of the later part of the 20th century, pioneered by USA. While monofilament longlining has come to stay in several countries. USA has the prominence among them. In the southern hemisphere, Australia and New Zealand have taken to the system in a big way. Seychelles has emerged as a major base for French tuna vessels.

The approach to be adopted in ensuring an infallible entry into high sea tuna exploitation is to team up either with USA or Australia for the purpose, may be for around three to five years in the first instance. In this context, it is desirable that the Government of India enters into an enabling bilateral agreement either with US or Australian Government to participate and assist Indian enterprises in the endeavour. The suggested agreement could cover assistance by way of supply of monofil tuna longlining equipment and providing technical expertise for its installation on vessels of 12m to 30m OAL as an additional system by the nominated U.S./Australian Agency, with the Indian side taking care of accommodation and local hospitality of the experts and the needed workshop technician and other support *i.e.*, supply of locally available items of equipment for undertaking the installation work. In addition, the collaborating agency could provide the needed operational training to the nominated crew members in association with CIFNET in some of the longlining installed vessels of the first batch.

Since the tuna resources position in the Indian EEZ is well known, assistance might be forthcoming from the foreign Government concerned, or a lending agency as nominated, with mutual acceptance. The enabling collaboration agreement could cover joint ventures between Indian and foreign enterprises, supported by a financing arrangement by the World Bank or Asian Development Bank, covering, *inter alia*, provision for the sale of the catches to the nominated joint venture partner. There can be a welldrawn mechanism for providing finance, technical inputs and for monitoring the operations including marketing.

An alternative to the above mentioned suggestion, to be within the frame work of the bilateral agreement, could be taking to a project with World Bank's assistance, providing for expertise from FAO. The components of assistance could be more or less the same. The companies could be given loans under the project through a financing institution like NABARD or ICICI, who would recover the loan amounts in instalments. The project could be implemented by a suitable organisational mechanism set up for the purpose.

There could be several other alternatives for the utilisation of the resource. The need now is for the Government to decide on the most feasible strategy that would lead to the fastest utilisation of the resource in a sustainable manner for national benefit. The development of tuna longlining activity will provide employment opportunities and contribute substantially to export earnings which are now on a downward trend. The introduction of tuna longlining in a big way by equipping vessels of12-30 m OAL for the purpose is expected to vastly upgrade the marine products export scenario and may well take the present level to a peak crossing Rs. 5,000 crore level per annum. The main requirement is a mindset on the part of the Government devoid of complacency towards the development of fisheries sector, particularly relating to tuna, and to initiate measures on top priority for promoting the system, considering its advantages.

No National Government in the world that neglects its fisheries development is known to flourish. With the exception of Mr. Kironmoy Nanda, West Bengal's Fisheries Minister, the nation has not seen any minister in charge of fisheries either at State or Central level, finding a continuing berth. And reasons for this, other than political stability and such as those related to degree and quality of commitment towards fisheries development, may need an analysis. The moral is that the Central Government should pay much more focussed attention for the development of nation's fisheries, since fish has been identified as the main additional source of food of the future to keep the multiplying population fed, besides contributing to foreign exchange earnings.

Fishing Chimes, Editorial 275: Jan/Feb 2000: Vol. 19, Nos. 10&11

# Editor's Foreward...

Aquahatcheries have come to occupy a crucial place in fisheries development. As is known, the reason for this is that they have become the main source of seed supply for aquaculture, for raising brooders for induced breeding and for ranching/revival of depleting wild stocks. Readers are aware that there has been an unprecedented but somewhat unregulated growth of aquahatcheries. The somewhat sudden upsurge of demand for fish and crustacean seed in the past two decades and the unpreparedness of the authorities concerned to devise ways and means of monitoring and co-ordinating the seed demand and supply factors have led to a somewhat odd situation of groping in the dark in respect of future planning for aquaseed production. Readers will agree that for any further increase in the number and capacities of hatcheries, the need for which may soon arise because of the vast scope for bringing new areas under aquaculture, there is need for updated information on the hatchery situation from time to time.

An initial effort made by *Fishing Chimes* to gather information on the present status of aquahatcheries in the country had not been encouraging. Barring information on State-owned hatcheries, it was seen that there was a near absence of a comprehensive inventory of hatcheries in the private sector, (which are more preponderant than those in Govt. sector) in several States. It was felt that this lacuna would soon prove to be a major handicap not only for the planners and administrators but also to the farmers, equipment suppliers and several others in the line. The production of this special issue on hatcheries is an attempt towards neutralising this lacuna. It has to be admitted, that while the information presented in the issue is substantial and indicates trends, it can be said that it is closer to being a complete inventory. *Fishing Chimes* would continue its efforts towards presentation of supplemental information in the later numbers of the journal; and it is hoped that this would become possible soon. Nevertheless, the Editor sincerely hopes that the contents of this issue would by themselves be of tangible benefit to the readers.

The history of aquahatcheries in India goes back to the early part of the 20th century when the British set up a few trout hatcheries, the most conspicuous of which are the one's located in J& K, in U.P. and at Avalanche in Nilgiris of Tamil Nadu, but now in the Biosphere zone and probably non-functional. After this exercise, there had been a gap of several decades in the development and practice of aquahatchery technology in the country, not taking into account the bundh breeding system that has been in vogue in Bengal from times immemorial, but not adequately known in all parts of the country until a few decades back.

It has to be reckoned that the present day major carp hatchery system had evolved from bundhs (Bungla bundh being the most advanced). The process of evolution of aquahatcheries culminated in the development of circular, and modified circular Chinese hatcheries and ecohatchery systems. The designs of these systems are either wholly of the Chinese style or with several modifications to the Chinese prototype to suit Indian conditions. In the transition phase from bundh to the circular hatchery systems, hapa and jar systems came into vogue and these too continue to be in use. All these developments took place in the second half of the century. On the other side of the carp hatchery development canvas is the emergence and proliferation of shrimp hatcheries along the coastline in the last two decades of the 20th century based on Galveston system of USA and Taiwanese (Japanese) extensive system and also quite a few freshwater prawn hatcheries based on green and clearwater systems.

One distinct event of the century in the fisheries sector was the introduction and spread of induced fish breeding system adopting hypophysation technology. It imparted a conspicuous economic uplift to the culture fisheries sector of India. In this context, it will be pertinent to mention that the nation is indebted in full measure to Hiralal Choudhury and K.H.Alikunhi and of course to several other scientists for the revolution that took place in the culture fishery sector because of this technology. The record of events related to the aforesaid revolution will be incomplete without a mention of HCG which came into use in the heels of the practice of administering pituitary hormone extract, followed by the later introduction of Ovaprim, a synthetic hormone product by Glaxo Company which obtained the product from *M/s. Syndel Laboratories*, Canada. This introduction was followed in recent years by the release of an indigenously manufactured similar synthetic hormone product known in the market as Ovatide. Ovatide is manufactured and sold by Hemmo Pharma and this product too has come to be accepted and used widely by fish breeders.

An equally momentous breakthrough in aquahatchery sector came with a ricocheting sequel, when TASPARC, an organisation set up by MPEDA established a shrimp hatchery near Visakhapatnam in A.P. The hatchery was designed based on Galveston technology and the resounding success of the system induced several in the following years to set up shrimp hatcheries along the coastline based on the same system. The Editor is grateful to TASPARC and its Project Director, Dr. K. Joshua for having given to the Editor a supporting hand in the designing of this special issue.

Another prototype shrimp hatchery too was established by OSSPARC, an organisation on the same lines of TASPARC created by MPEDA. The hatchery of OSSPARC is located at Gopalpur-on-Sea and this hatchery adopted the French technology. Some of the entrepreneurs adopted the Taiwanese extensive system but its popularity was marginal. As a result, as a follow-up of the successes of the first batch of hatcheries, over 190 shrimp hatcheries in all, have sprung up all along the coastline.

There are quite a few new hatchery technologies developed by the Central Institutes but they continue to await extension to the commercial sector. The constraints that detain the extension process are not however known. One new technology reported to have been developed by the CIBA is in respect of hatchery seed production of the Sea Bass *(Lates calcarifer). Fishing Chimes* published a note on the subject (Vol. 17 No. 6). Thereafter there is no information to show that the technology has been extended to the commercial sector, except that some numbers of the seed produced were supplied to a farmer. CIFA has developed the technology of production of freshwater pearls. Likewise, CMFRI has developed the technology for the production of marine pearls. The hatchery production of the spat of the two species acquires value only when the technologies of production of the spat of both the categories is extended to the farmers, coupled with the pearl production technologies. Unfortunately, it appears that this has not been possible for the two institutes, leading to reservations in the minds of entrepreneurs in respect of perfection of the technologies. As the technologies are well standardised, there do not seem to be any viable reason for not extending the technologies. Atleast one entrepreneur who ventured into the production of marine pearls with extension support from CMFRI now feels chagrined that he was left in the midstream. The fact that, sometime back, a commercial firm under the name "Tamil Nadu Pearls" in collaboration with a T.N. State Government enterprise could not succeed in running the enterprise strengthens the reservations about the constraints. Regarding freshwater pearls also, it is possible that there is some constraint that is impeding the extension of the technology, the development of which is by now widely known. It is desirable that ICAR reviews the position and identifies effective steps for the extension of the technologies to the commercial sector. For the reasons narrated above material in respect of hatchery technologies of the species referred to have not been included in this issue.

It is also known that CMFRI could successfully produce sea cucumber seed. Very little is known beyond this. Considering the demand for sea cucumbers in the foreign markets, entrepreneurs would be interested in the production of sea cucumber seed for farming purposes. The problem is one of transmission of the technology. In view of this position, no notes on sea cucumber hatchery seed production could be included. There are quite a few other species, the hatchery seed production of which is understood to have been achieved on an experimental basis, but these could not be covered in this issue for obvious reasons.

The three layered significance of the year 2000, combining the onset of the third millennium, the 21st century and the year 2000, has bestowed on the Editor the pleasant duty of devoting a few pages of this special issue to summarise the history of fisheries sector of India.

This special hatchery issue provides not only the state-wise particulars of aqua hatcheries in the country, to the extent they could be gathered, but also several papers authored by reputed scientists and specialists on various aspects of aqua hatcheries. The Editor expresses his gratitude to all of them for having enriched the contents of this special number.

In conclusion, the Editor hopes that this issue would prove to be of immense value to the readers.

Fishing Chimes, Editorial 276: March 2000: Vol. 19, No. 12

# Exim Bank (With new US Creditline) Can Rescue Collapsing Indian Deep sea Fishing Industry

There is an alarming stagnation in the Indian deep sea fishing industry. The number of units in operation has been coming down over the years with no replacements. Several of the vessels constituting the fleet have crossed their estimated life span and the others may soon be crossing the barrier. In this situation, the survival of this industry would depend on the introduction of a revival project aimed at restructuring/strengthening the existing fleet, which is stagnating and is now face to face with an imminent collapse. This critical situation is widely known but is being allowed to languish further.

It is true that the owners manage to keep the vessels operational through timely renovation measures, such as changing of worn-out hull plates and machinery parts, and in some cases replacement of machineries too. These upkeep measures are laudable but the owners find the exercise disproportionately expensive, compared to the order of earnings. The periodical maintenance costs, which have virtually turned out to be in the nature of investments, are met out of earnings, with the entailing economic disadvantages that are now being increasingly felt. This process, while tending to add to investments, has not been bringing in any compensating additional incomes. These upkeep investments will make sense only when they proportionately augment incomes and serve the purpose of yielding atleast marginal profits after balancing the expenditure and income. This being not so, it would be prudent to couple the upkeep investments as at present with productive investment related to installation of an additional fishing system (besides the trawling system) that would neutralise their profit reducing effect. This additional system can be one of installing systems for monofil longlining, squid jigging, gill netting etc.

There can be no escape from restructuring of the existing fleet on the above lines for counteracting the adverse effects of upkeep investments of a capital nature in an economically viable manner, because of the age factor of the vessels. In other words, the long term survival of the Indian deep sea fishing industry is demandingly linked to the fleet restructuring/expansion programme, which however entails marginal capital induction.

As at present, there are no avenues for securing funds from the commercial banking structure to bring about the needed reform. As is known, capital, technology and manpower are the real constraints in the way of restructuring. Restructuring entails installation of an additional fishing system (as already stated) besides periodic renovation, provided the industry is confident of the position in respect of non-shrimp resources that are commercially viable. If the non-shrimp resources are considered to be not viable for installation of an additional fishing system on the vessels, the position will be that the balance of economically viable life of the vessels even with renovation would prove to be tapering down. In fact, several owners give vent to their genuine distress that with the upkeep expenses going up, the operations are becoming increasingly burdensome with the incomes remaining either static or coming down.

Fortunately the Indian EEZ has a known, sustainable and sparsely exploited tuna resource base. This situation, and the dismal state of economics of the ageing vessels clearly indicate that there is need to plan for sustainable and viable fishing with additional investments as needed that would yield adequate returns to profitably take care of the same. The additional investment would consist of not only periodical upkeep expenses, (change of plates, machinery parts, machinery replacement etc.,) but also the cost of purchase and installation of an additional fishing system on the vessels.

## Capital

The present hostile situation regarding capital induction is known to all concerned. The banking structure as well as the Government are non-responsive to the capital investment requirements of the Industry. This state of affairs has pushed the industry into *a cul de sac*. There are no efforts at solving the problem which is left to languish, totally unattended to.

Reports speak of a new credit limit of US $ 500 million from US Export Import Bank to Exim Bank of India for the acquisition of machinery, technology and other goods and services that are needed by Indian entrepreneurs for business growth and export activities, extended under a Memorandum of Understanding signed on March 24, 2000 between US Commerce Secretary and Managing Director of Exim Bank of India. This provision opens up an opportunity to develop and implement a project, for instance for installing monofilament tuna longlining equipment in the present fleet of around 90 vessels of 23-27m OAL, to be imported from USA, the main known source for such equipment, availing of funds under the above mentioned credit line. USA is a pioneer in monofilament tuna longlining and this country happens to be in a position not only to supply the equipment but also the technology for the designing, installation and operation of monofil tuna longlining system as an additionality on the vessels. There are quite a few reputed US tuna fishing companies with global operations who will be interested to have commercial participation in an integrated endeavour to utilise the known extensive tuna resources of the Indian EEZ that would give them the benefit of importing Indian tuna into USA.

The capital investment needs for the purchase and installation of equipment (at an average of $ 150,000 per vessel for 100 vessels), are estimated to be of the order of $ 13.5 million. In addition, around $2.5 millions may be required for upgrading the facilities at one or two existing processing plants in India for the manufacture of valueadded tuna products.

It may be recalled that one recommendation made at the recently held Round Table Conference on Tuna of Indian EEZ was that the trawlers now in operation (23-27 m OAL) might be equipped with an arrangement for conducting monofilament tuna longlining, under a compact, space-friendly and economically viable system. It was unanimously agreed at the conference, that implementation of this strategy would be the most economical means to upgrade the present fleet with the lowest possible financial investment and at the same time contributing to the stepping up of exports and economic returns inclusive of export earnings.

While there can be a few other ways too of improving the capabilities of the present fleet of vessels now getting on in age, the immediate need seems to be to translate the recommendation of the Round Table Conference into action through collaboration with a sound enterprise of USA.

## Technology

The technology of designing additional fishing installations and also in respect of the operations has to come from fishing vessel naval architects having experience in the line, either in USA or from any of the other countries having the needed expertise. The import of technology, preferably as part of a package, is however very crucial and unavoidable, atleast in the first round. Further, taking a risk in entrusting the job to indigenous talent, uninitiated into this kind of job may prove to be costly, unless an indigenous shipyard teams up with an experienced foreign enterprise to accomplish the additionality. Once the system takes roots, the manufacture of the items concerned and their installation would eventually become an indigenous activity. As it is, Covema Filaments in Cochin is known to be manufacturing and exporting monofilament longline.

## Man Power

We have practically no commercially trained and capable manpower for the operation of monofilament longlines. It is accordingly imperative that professional foreign crew have to be initially inducted under the suggested collaboration programme for such duration as is necessary for Indian crew to pick up the skills. As already mentioned, there are quite a few reputed and globally known tuna fishery companies in USA such as Casamar, Star Kist and Bumble Bee, engaged in tuna fishing, processing and marketing. It is possible that representatives of these firms would participate in the forthcoming International Tuna Conference (sponsored by INFOFISH) to be held in the last week of May 2000 in Bangkok. Participation by representatives of MPEDA, AIFI and the Union Department of Animal Husbandry and Dairying of the Ministry of Agriculture at this Conference would provide an opportunity to hold preliminary discussions with the representatives of the participating foreign companies on the question of formation of joint ventures with Indian enterprises for promoting integrated tuna fishing operations from catching to marketing and with an export line-up.

In order to explore the possibilities of collaboration and to determine the modalities of it, it appears necessary for the AIFI and MPEDA to hold discussions on the subject with the representatives of the Government of India beforehand so as to determine the criteria for negotiations with foreign companies at the forthcoming Tuna conference in Bangkok in the first instance. Thereafter, an Indian team with an appropriate composition can participate in the said Conference, and with a prior arrangement conduct negotiations with the representatives of foreign enterprises either at the conference or elsewhere with a view to establishing joint ventures acceptable to them and eventually to the Governments of both the sides. As part of the process there can also be pre-project exchange visits by the representatives of the two sides for a close understanding of the points of view of each other and in that light pave the way for signing MoUs in the first instance to be followed by Agreements.

In this background, guidelines for preparing a project, with an organisational mechanism for implementing the project as part of it may have to be worked out. In this connection, variables such as diffused ownership by small companies and manner of control have to be finalised and followed. There may be scope of inducting capital, technology and manpower by ways other than those mentioned above. These need to be explored.

The crisis in the deep sea fishing industry can surface any time. The warning signals as observed now are ominous. Overburdened with commitments consequent to mounting expenses and static or declining incomes, the signals may collide with the weakening situation, creating the problem of facesaving in respect of the international standing of India in the deep sea fishing sector. The nation has a status to keep and we cannot be complacent, particularly when Pakistan is known to have entered into a contractual arrangement with a foreign company for fishing in its EEZ and with the possibilities of these foreign vessels intruding into the Indian EEZ through Gujarat waters. The impending crisis should not be allowed to manifest itself, for want of capital, technology and manpower. There is a responsibility on the Government to devise and implement a plan of action to obviate the crisis. It is to be hoped that the Ministries of Commerce (MPEDA), Agriculture and AIFI would certainly be able to meet this challenge.

Fishing Chimes, Editorial 277: April 2000: Vol. 20, No. 1

# Options for Registering Cognizable Aqua Products Export Growth

Indian aquaproduct exports, now practically static, can go up only when a wider production base is established by bringing all conducive inland lentic waters of the heartland of the nation under exportable aqua production. This can be achieved through the preparation and implementation of a major master plan on a national basis, with the participation of the State Governments and MPEDA. The plan should be such that it integrates the various components from rights over resources to export marketing, with coverage of infrastructure facilities and other essential components. Continuation of neglect of bringing the vast inland lentic resources other than shrimp/prawn under export basket will perpetuate the present stagnation in inland aqua products exports. An enduring increase in exports of inland aquaproducts is possible only when means of augmenting production of exportable species from the vast non-coastal inland lentic resources are effectively taken care of. Value addition to existing exports may bring in extra foreign exchange earnings for sometime but sustainable increase in earnings through this route would not be possible in the long run.

Capture shrimp production graph no longer displays a rising trend. Farmed shrimp production sector is riddled with WSSV disease problem on one hand and is prevented by a Supreme Court order from any further expansion in the CRZ on the other. These factors cannot be expected to be conducive to expansion of farmed shrimp production any more. Mud crab farming will take a long time to gain in strength and cannot therefore be expected to tilt the balance much in favour of substantially augmenting exports, as this activity also suffers from more or less the same CRZ constraints as shrimp farming. So far as mariculture is concerned, the nation has not yet reached the take off stage. In any case, the on-going stay order of the Supreme Court will come in the way of taking up mariculture in the CRZ. The option left for augmenting exports of marine aquaproducts from farming sector have thus narrowed down relentlessly to a point of concentration on sustainable utilisation of untapped potential zones other than in CRZ.

Means of expansion of marine capture fishing effort has now dwindled down virtually to a few options. One is to exploit the tuna fishery resources of the Indian EEZ. The other is to intensify surveys for locating new deep sea lobster and deep sea prawn grounds in the EEZ which exist beyond 150m depth, so as to expand exploitation base for the species eventually. The third is to promote sea cage farming in enclosed waters such as coves and offshore waters for producing exportable species such as perches, groupers, breams and even tuna as is being done in Australian waters. In this context, it has been a refreshing development that MPEDA's Executive body has recommended recently the introduction of these promotional programmes. Surveys in respect of deep sea lobsters and deep sea prawn grounds are on-going programmes of FSI and it is to be hoped that there will be emphasis on these. It is good to see that the AIFI has been maintaining liaison with the MPEDA and the Fisheries Division of Union Department of Animal Husbandry.

The freshwater farming sector, with the exception of farming of freshwater prawns in a few pockets, stands neglected from the export angle. The urgent need is to promote freshwater prawn farming in all possible lentic - waters of the heartland of India too, supported by a network of hatcheries for freshwater prawn seed production to be set up along the coast with a view to organising channels of supply of seed so produced for stocking conducive inland lentic waters.

**There has been a refreshing development amidst the bleak picture in respect of exportable aqua production in non-coastal lentic waters at a workshop conducted in Bhopal on 26 and 27 February, 2000 by the Madhya Pradesh Council of Science and Technology. At this workshop, a path-breaking recommendation was made to promote farming of tiger shrimp in the cultivable lentic waters of Madhya Pradesh, keeping export earnings in view. Another momentous recommendation made at the workshop was that monosex farming of Tilapia in an environmentally safe manner should be taken up to step up exportable farmed fish production and to bring in higher returns through export of fillets of the fish. These two recommendations are indeed timely. When implemented, these would lead to expansion of export-oriented production base in Inland States. The Deputy Director General (Fisheries) ICAR who participated in the workshop is understood to have supported the two recommendations.**

It may be added here that shrimp is now extensively farmed in freshwater at several east coast centres and the same system can well be extended to the lentic water areas of States such as M.P. The higher cost of shrimp and prawn seed, because of transportation from hatcheries located along the coastline, may add to the costs but the farming activity would emerge as profitable once the supporting infrastructure and the needed facilities are developed.

So far as Tilapia is concerned, the opposition to its farming, although voiced by well-intentioned circles, needs a thorough reappraisal. The aspect for consideration is that the fish is already there in the country (introduced in late 1950s of 20th century) and now it is sought that a faster growing Tilapia species may be introduced for farming based on mono-farming system involving males only, and with the seed production to take place at a few approved hatcheries and seed farms for supply to the farmers under a well monitored system, and with the needed precautions. The non-acceptance of this option by some, on the premise that a few pairs may escape into the wild, only betrays the lack of resolve to accept the challenge and highlights the pusillanimity and overcautiousness of those at decision-making level. The sooner a revised policy is formulated and a strategy is evolved on this issue, the better it will be for the nation. Government could consider setting up a task force to work out details in this respect.

**The Indian major carps are eminently suited for export as they have all the features that merit their export. MPEDA has to strengthen promotional efforts to popularise the fishes in various foreign markets. The exports of major carps are now limited to a few middle-east countries. Foreigners would certainly like the general appearance, firm texture of flesh, fewer bones and other good qualities of these fishes.**

Another opening for augmenting exportable aqua production is to formulate a policy and strategy to utilise the saline underground water and saline tracts now lying waste in north-west India. The Fisheries Division of the Union Department of Animal Husbandry and Dairying could certainly initiate serious and purposeful measures to bring these assets under exportable aqua production.

Those in authority are anxious to augment aquaproducts exports and are exploring ways of doing it. The aspects to be considered in this context as brought out above, among others, are known to the authorities concerned. What is now needed is to expedite the formulation of a policy, a strategy and a plan of action to put these available farmable resources to produce exportable aqua species and to create an infrastructure for their processing and export. The sooner the complexion of the situation is upgraded the better it would be for the nation.

The reforms reiterated above, when implemented, would not only augment exports and foreign exchange earnings, but also would rein in substantial employment opportunities and unleash an enormous economic activity.

Fishing Chimes, Editorial 278: May 2000: Vol. 20, No. 2

# National Project on Export-Oriented Inland Aquafood Production Essential

India's overall aqua production base has been reasonably reliable in respect of growth until recently. The wilting of the base set in when the marine capture sector reached a no-growth level, with overfishing manifesting itself in coastal waters. Aggravating this, the Deep sea fishing fleet strength of the country has also registered a steep decline.

The known major and sparsely exploited resource of the EEZ, besides deep sea prawns (red rings), is Tuna. There have been initiatives to equip the remaining deep sea fish fleet for tuna longlining too and for developing professionally trained crew for manning the longliners but with little progress. In the meanwhile, some of the trawlers diversified into fishing for deep sea lobsters, cephalopods, and deep sea prawns (red rings). The bulk of them have however continued to confine their operations to catching of coastal shrimps.

The declining trend in the elasticity of the production base was stemmed to a significant extent by promoting farming of shrimps in coastal farms in a big way. Several farmers also diversified into farming of freshwater prawns. All these efforts are laudable but a lot more can be done, as the authorities are aware, by reforming, widening and strengthening the exportable aquafood production policies with export-orientation in a big way.

There is a keen awareness that bulk of inland lentic water of the country is well suited for augmenting export-oriented aquafood production and the potential can be availed of through implementing a major project conceived as an unitary one but implemented in phases as the area to be covered is very wide.

## Deep sea Fishing Policy

The nation has no cogent deep sea fishing policy at present. When one looks at these policy parameters critically, they would be seen as chillingly negative and the nation does not deserve them. A glaring need is staring at us in respect of exploitation of known stocks of tuna in the EEZ and in regard to survey of resources of deep sea lobsters and prawns off the entire coastline with the concomitant and inescapable imperative of bringing in technology and equipments for the exploitation of the resources, particularly tuna, from developed nations such as USA and Australia which have expertise in tuna fishing, particularly longlining. These aspects do not find a place in the policy.

In this background, it is refreshing to know that the Government of India has set up, although long delayed, an Expert Group for the formulation of a comprehensive policy for the development of marine fisheries resources of the Indian EEZ. The terms of reference given to the Group are understood, to cover a) ascertaining the present status of exploitation of marine fishery resources by the traditional, motorised traditional, mechanised and deep sea fishing vessels, b) formulating programmes for the upgradation of the capabilities of the small-scale sector to fish in deeper waters, c) working out the needed strength of area-wise resource-specific deep sea fishing fleet comprising tuna longliners, purse seiners, squid jiggers, bait (pole and line) boats etc., d) evaluating the capacity of the present deep sea fishing fleet and suggesting modifications and redeployment, if necessary, e) estimating and identifying sources for meeting the investment requirements of the marine fisheries sector, f) assessing the need for joint ventures and taking vessels on lease from foreign fishing companies, g) identifying the human resource development needs of the marine fisheries sector and formulating programmes for meeting such requirements, and h) suggesting conservation measures taking into account the code of conduct for responsible fishing and other global initiatives for sustainable development of marine fisheries.

MPEDA has moved ahead in the right direction even prior to the setting up of the Group in respect of the material points of reference to it that have a bearing on augmenting marine products exports. The measures taken have led to an increase in the value of sea food exports for the year 1999-2000 by 10 per cent over the achievements in 1998-99. The exports touched a level of Rs. 5,096 crores ending 1999-2000, crossing the Rs. 5000 crore barrier for the first time. In dollar terms also exports increased by 7 per cent to a level of US$ 1.184 billion (from US$ 1.107 billion in 1998-99). In terms of quantity, exports grew by 12.2 per cent from 3.03 lakh t to 3.40 lakh t.

The export endeavour has been adroitly piloted in an extremely critical situation that displayed symptoms of stagnancy. As a result, the growth of10 per cent exports in terms of quantity and 7 per cent in terms of value could be achieved in 1999-2000, despite the white spot virus menace that continued to afflict the shrimp farming sector. The achievement is attributable to crucial measures taken by MPEDA for stepping up the fishing capabilities of small mechanised boats and for needed value addition to the catches.

The boat owners have been assisted through subsidies to equip their boats with GPS, among others, to facilitate reaching the same grounds of tested potential unerringly. While this development played a major part, it is the effective effort of the fishermen in conducting fishing for deep sea prawns (red rings) at depths upto 400 meters, with an amazingly innovative and daring net dragging and hauling prowess at which few fishing technicians could conceive of that, enabled a measure of addition to the exportable production.

Another aspect that has made a contribution to the augmentation of export earnings is value-addition. Earnings from value-added exports grew by a significant 75 per cent during 1999-2000. The bulk of the contribution in this direction came from surumi exports made by four companies, three on the west coast, one each run by Amar Cold Storage and Hindustan Lever Ltd., in Gujarat, Gadre Marine Exports, Ratnagiri in Maharashtra, and Big Star Marine Exports at Visakhapatnam now under lease with Hindustan Lever Ltd.

Value addition is undoubtedly one way of augmenting exports. But value addition has its own limitations, such as technologies and the constraints on raw material availability. Therefore, while paying attention towards value addition, there has to be planning for increased efforts directed at expanding the raw material production base in order to establish an enduring system of export, and bringing in technologies.

R.A.M. Varma observes in a write-up on value addition in "Sea Food International" (March 2000 issue) that India's seafood industry must be realistic about the concept of adding value. He says that much of the emphasis on adding value to each and every product is not a worthy campaign. He adds that developing countries like India have to realise and accept the limitations on value addition.

The measures referred to in the preceding paragraphs represent steps for the beginning of the widening of the marine production base of exportable species. In this context, resolutions passed by the MPEDA's Board at its meeting held on 28 April, 2000 would gain a significant place in the history of marine fisheries development of the country. These resolutions have some measure of relevance to the terms of reference given to the Expert Group referred to above.

The Board passed two important resolutions. One of this was to equip two deep sea trawlers with monofilament longlining equipment for undertaking a pilot effort to catch the unexplored tuna stocks, mostly of Yellowfin, in the Indian EEZ. The other resolution was to initiate preparatory work for setting up cages in open sea for the farming of exportable species such as groupers, perches, seabass, breams etc., with supporting shore-based infrastructure by way of hatcheries for the production of needed seed for the aforesaid farming activity.

These two momentous resolutions need supporting action right from now so that by the time the pilot operations are completed, the authorities concerned can be in readiness to organise facilities for extending technical as well as financial support for follow-up expansion work in that direction.

The Indian deep sea fishing industry is now moving towards collapse. A fleet of strength 190 vessels that was there until recently has now shrunk to 90 and very soon the number will undoubtedly plummet further. One strong reason for this is that no enterprise is in a position to invest on new vessels for obvious reasons. In this situation, the main opening available for the enterprises is to prevail upon the Association of Indian Fishery Industries to prepare an integrated project with support from the Ministries of Agriculture (FSI) and Commerce (MPEDA) for renovating the existing trawlers by changing of worn-out plates, replacing or repairing machineries as required and for the installation of additional equipment for conducting, say, tuna longlining. Such a project could include effort at equipping two trawlers with longlining equipment to test feasibility, as now proposed by MPEDA. Once found feasible, the next phase of the project to equip the present fleet like-wise, with technical and financial assistance drawn from acceptable foreign enterprises, can be taken up. Through this approach the need for massive investments on new vessels can be obviated to some extent.

A project of this nature could be promoted by the Association for the benefit of its members with support from the Indian Government and Export-Import Bank of India. The Export-Import Bank of India could tap funds out of the one billion US $ credit line stated to have been earmarked by the US Exim Bank for being provided to Indian enterprises, linked to import of technology and equipment from USA. Tuna longlining system is an achievement of the USA's fishing enterprises and the import of related technology and equipment from USA fits in eminently well to develop tuna longlining in Indian EEZ, known for its untapped tuna resources.

Similar projects could be worked out for setting up sea cage culture system in Indian EEZ.

It is common knowledge that India has opted for globalisation of its economy for national benefit. Globalisation is of particular relevance to the fisheries sector for the reason that new technologies have to be injected and investments have to be attracted to flow in, for establishing new export-oriented production units. It is true that, some years back Government had decided not to accord permissions for establishing joint ventures in the fishing sector. The conditions of existence at that time must have been responsible for the decision. The present situation is such that there is an imperative need to expand the production base to augment aqua products exports, taking into account any indicative willingness of foreign fishing business enterprises to enter into joint ventures with the objective of taking part in fishing in Indian EEZ for species that have market abroad in a processed form, to conform to the requirements of consumers in markets of their interest. In this context, there is no alternative to introducing a cautious policy of bringing in technologies and also investments. It will be an impracticable approach to perceive that technologies could be developed on our own and the needed investments could also be generated within the country. Past history proves this. For instance, efforts were made by Indian entrepreneurs to undertake tuna longlining on their own but they had to face failure for reasons that are related to technical constraints and lack of experience in the operations. In regard to investmental needs, irrespective.of feasibility parameters, Indian banks are not in a mood to extend financial assistance to the deep sea fishing sector. Investments have to therefore necessarily come from outside, from those who have interest in developing the aqua product trade in association with Indian enterprises.

It now stands proven that upgradation of capabilities of mechanised boats is feasible in several respects. CIFT has to now develop formalised designs based on the innovations made by owners of mechanised boats to catch deep sea prawns and probably deep sea lobsters. Mustad Co., has developed an autoline system well suited for installation on small boats. This also deserves study.

Several estimates had been made in the past by various workers on the area-wise requirements of resource-specific vessels for exploitation of fisheries of the EEZ. These could be good guidelines to refine the estimates and recommend the requirements. Vessels equipped for mid-water trawling, deep sea demersal trawling and longlining, among others, would be needed for operation all along the coastline.

The syllabi, of marine fisheries courses conducted at the fisheries colleges and other institutes must have a major component in respect of management. Senior captains and engineers with experience may have to be put on the job of preparing the syllabus, taking into account the syllabus adopted in countries such as Taiwan and South Korea.

CMFRI has to be further geared up to continue to play a major role in the sustainable utilisation of the resources. Several countries follow catch-quota system for national vessels governed by licenses, daily reports, and periodic inspections. The Australian system is hailed by several as the best system that has not only stabilised the fisheries resources of their EEZ, but has also stepped up incomes to stakeholders over a period of time.

**Freshwater prawns and tiger shrimps would constitute the two main species for augmenting export-oriented farm production throughout the length and breadth of the country, excluding the temperate zone. It may not, however, be desirable to farm these two species together, but separately as part of polyfarming with fish species. It is now established that tiger shrimps could be farmed in freshwater and at a workshop held in Bhopal recently it was recommended that the farming of these shrimps be taken up in the waters of Madhya Pradesh, probably to start with. The farming activity of this species and the freshwater prawn may have to be developed in such a way that, over a period of time, the seed relating to the two species produced in the coastal zone gains a spread effect into the inland zone through a process of advancement, zone by zone. In other words, the infrastructure could be developed progressively to provide the supplies of seed from the hatcheries on the coast to around 100 km inland in the first phase. In the second phase a further spread effect could be achieved.**

Integrated with this process, systems that involve the setting up of processing and frozen storage plants at convenient centres for pooling up of the produce and processing and storage of the same thereat for their transportation to the nearest ports, can be promoted in the private sector by providing land, electricity and other facilities and the needed incentives. It will be agreed that aquaproducts exports can go up substantially only through a major and well planned effort and all those concerned will have to apply their mind to work out ways and means for accomplishing this stupendous job.

The vast inland lentic waters of the Indian heartland are eminently suited for export-oriented aquaculture system. In other words, these waters are a dollar mine awaiting to be developed and exploited for augmenting aquaproducts exports. Once these waters are brought under export-oriented aquafarming system, the foreign exchange earnings from fisheries sector will move up several fold to new heights. Task forces are no doubt required to be set up for working out details and preparing projects that would have nation-wide application.

An area of at least 10 lakh ha of tanks and ponds, and large, medium and small reservoirs can be brought under a system that would involve farming of prawns and shrimps, and male Tilapia and this could hopefully be achieved in a period of 5 to 10 years. Presuming a minimum average exportable production of 200 kg/ha per annum can be promoted, there can be an additional production of 2 lakh t per annum. When exported, this produce can bring an additional one billion US dollars per annum, which will have the effect of doubling the present level of foreign exchange earnings out of export of aquafood products. Value addition or other devices would certainly augment value of exports but these measures can only be supplementary but not a substitute to launching a massive project to suck in the vast export production potential of inland lentic waters into the basket of the on-going export endeavour, besides stepping up marine capture fishing effort for exploiting tuna and other resources.

Fishing Chimes, Editorial 279: June(i) 2000: Vol. 20, No. 3

# National Fish Farmers' Day

It was on 10th July, 1957 the first breakthrough in induced breeding of Indian major carps was achieved at Angul, in Orissa. This was achieved by Dr. Hiralal Chaudhuri, who carved out a name for himself in international arena as an eminent culture fishery scientist and a gifted biotechnologist in fish breeding. This success achieved by him, under the guidance of Mr. K.H. Alikunhi, totally transformed the complexion of Indian freshwater fish culture. The impact of the achievement was so powerful that within a few years after this, a major shift from the collection of riverine carp spawn in vogue till then to production of the spawn through application of hormone induced breeding technology as introduced by Dr. Hiralal Chaudhuri took place. The result has been, as we experience now, a total transformation of the technology of raising major carp spawn and spawn of other fishes through hatchery systems and also of the related fry and fingerling raising technology as an integral part of freshwater fish culture practices.

We now see for ourselves the qualitative and quantitative change in inland fish production over the years, which has ushered in the socio-economic upgradation in the lives of those engaged in carp fish seed production and carp fish farming. The introduction of the technology has led to a new wave of employment potential in the inland fisheries sector. Inland fish production which was around 2.50 lakh tonnes in 1957 moved up to the level of over 7 lakh t annually by 1970s, 14 lakh t in 1980s and at the end of the 20th century to an annual level of around 25 lakh tonnes. Above all, India's present prestigious status in the global inland fish production, next only to China, is essentially due to the breakthrough that was achieved on 10th July 1957, which is indeed a red letter day for the nation and for its fish farmers, inland fishery scientists, professionals and administrators.

It is very unfortunate that such an important day in the history of the inland fishery development of the country is just taken for granted and is practically forgotten. The readers would agree that this day has to be enshrined with a unique place in the history of fisheries of India.

In this context, in order to commemorate the historic and momentous event, we appeal to the Government of India and the State Governments to declare 10th July as the National Fish Farmers' Day. Highlighting in this manner will enable the memory of the event to be carried from generation to generation. It is suggested that the observation of this Day may start from in the year 2001 *i.e.*, the celebration of the first Fish Farmers Day on 10th July 2001, will be fittingly appropriate. The time available in the meanwhile should be adequate for the Governments and the farmers' organisations to initiate preparatory work on the celebration of the first National Fish Farmers' Day as suggested. This can be celebrated in a telling manner through organisation of meetings, seminars etc., and by honouring Dr. Hiralal Chaudhuri and Mr. K.H. Alikunhi.

We appeal to Mr. Nitish Kumar, Hon. Minister for Agriculture, Government of India, the Secretary, Joint Secretary and Fisheries Development Commissioner in the Union Department of Animal Husbandry and Dairying to favourably consider the suggestion and set in motion the steps needed to ensure the celebration of the first National Fish Farmers Day on 10.7.2001 in Delhi, and in all State Capitals, by Central and State Fisheries Departments, Fish Farmers Associations, Fish Farmers Co-operatives etc.

Fishing Chimes, Editorial 280: June (ii) 2000: Vol. 20, No. 3

# Tuna Mystery of Indian EEZ

A mystery seems to pervade tuna exploitation scene of Indian EEZ. Readers are aware that, after allowing Taiwanese longliners to fish in Indian EEZ through charters, lease and joint ventures for long, the Indian Government decided to discontinue the policy thereof with a clarification that these activities would however be allowed to continue to last until their permitted validity expired. This policy has the effect of eventually clearing the Indian EEZ of all foreign longliners with the exception of those permitted to be owned (with registration in India and with Indian flag) by atleast one Indian company but believed to be virtually operated by foreign owners under an arrangement that is an inversion of charter/lease conditions that prevailed. In other words, but for the 'paper' acquisition of ownership (whether the change in ownership is on record in the country of origin as well is not known) the operations and sharing of earnings is believed to remain virtually the same as under charter/ lease systems. Similar inverted operations are believed to be there in Pakistan's EEZ too, with the vessels acquiring Pakistani registry and flag but with the fishing operations and management believed to be in foreign hands too as in India. The information is that 10 such foreign longliners of Chinese origin are being operated by an Indian company in the Indian EEZ, and around 14 longliners of foreign origin by Pakistani interests in Pakistani EEZ and one longliner probably of Chinese origin in Sri Lankan water.

In the past over 15 years six companies imported line fishing vessels and one company a purse seiner into India.

Fishing Chimes, Editorial 281: July 2000: Vol. 20, No. 4

# Tiger Shrimp Domestication

The inescapability of developing the technology in India to domesticate tiger shrimps has been coming into sharper and sharper focus during the past one decade. The present status of domestication of tiger shrimp in India for the past over ten years is generally similar to the problem faced in the case of Indian major carps in middle 1950s of the last century in respect of securing brooders.

The present inevitable dependence on wild brooders of tiger shrimp harvested from the sea for the production of its postlarvae at the hatcheries has exposed the Rs. multicrore shrimp export industry to serious vulnerability. The relentless onslaught of white spot virus that comes with the wild tiger shrimp broodstocks and gets transmitted to their progeny has been reducing and downgrading farmed output of the animal. The quantum of fall in farmed production of tigers in 1999-2000 was put at over 20,000 t with an average value of Rs. 500 crores. With the general trends indicating a continuing fall in farmed production of tiger shrimps, the urgency for the intensification of efforts to domesticate tiger shrimp so as to build up farm-raised broodstocks for their postlarval production does not need emphasis.

Indian Inland fishery scientists have global recognition. They have several achievements to their credit. Yet, in the background of quite a few countries achieving success in raising domesticated tiger shrimps for several generations, the Indian tiger shrimp hatchery owners wait with bated breath to hear the news that tiger shrimp has been domesticated by our scientists. It is possible that work is going on in various Indian scientific institutions on the subject but there is no transparency.

Now let us look at the global position. At the International Workshop on Development of Sustainable Management Practices in Shrimp Farming held in Bhubaneswar on 30 and 31 July, 1999, Dr. Denis Gasnier of *COFREPECHE,* France said that IFREMER had developed at its Tahiti Centre in French Polynesia one strain of *P. monodon* (12th generation) in captivity. Broodstock production of the species there, was well controlled. It was now a common practice at Tahiti Centre to supply seed produced out of domesticated brooders. It was also mentioned by him that a domesticated project would cost US $ 800,000 and working expenses would come to US $ 750,000 with consultancy fees extra. He also spoke on genetic improvements under experimentation to have disease resistant strains. At the same conference, Dr. K. Joshua, Project Director, TASPARC, Visakhapatnam, suggested that a study on domestication of tiger shrimps should be taken up simultaneous to efforts at collection of pathogen free brooders of the species.

In Thailand, a private company named 'Shrimp Culture Research and Development Company Limited (SCRD) was established under the aegis of the National Center for Genetic Engineering and Biotechnology. The company has several partners including commercial enterprises, the Thai Frozen Foods Association and some shrimp farmers' associations. The progress achieved by this company is stated to be promising. Utilising the facilities provided by C.P. Aquaculture, one of the partners in the SCRD, the company is reported to have successfully demonstrated the feasibility of domestication and has so far produced the third generation of stocks.

The Central Scientific and Industrial Research Organisation of Australia, in association with a private sector partner, Seafarm Private Limited, is reported to have succeeded in breeding domesticated tiger shrimp. It was demonstrated that the reproductive performance of domesticated tiger shrimp was akin to the performance of wild broodstock. A programme is stated to be now on the anvil to undertake selective breeding so as to have progeny with more beneficial traits.

M/s. High Health Aquaculture Inc of USA advertises for the sale of domesticated tiger shrimp stock.

In this background, it is pathetic that India is far behind in producing domesticated shrimp brooderstocks. It is to be hoped that ICAR would accord top priority to the development of tiger shrimp brooder stock so as to eliminate the huge economic losses the farmed shrimp industry is incurring because of disease incidence. It is understood that the Department of Biotechnology awarded projects on the subject to Annamalai and Andhra Universities but there is no transparency of what is being done and at what stage the work is. We do hope that the Union Department of Animal Husbandry and Dairying and the Union Department of Agricultural Research and Education would intervene and ensure that the nation would not lag behind but would strive to be atleast at par with the work going on in the other countries on this subject of vital economic importance for the nation.

Fishing Chimes, Editorial 282: August 2000: Vol. 20, No. 5

# Future of Fisheries Development Rests on Technocrats of Quality and Competence

Graduates and Post-graduates in fisheries coming out of fisheries colleges of India and also from related departments of certain universities, batch after batch, get absorbed in various positions not only in Central and State Governmental Fisheries Departments but also in commercial enterprises running hatcheries, fish farms, marine fishing boats, processing plants, etc., in positions such as managers and technical hands. Some of the qualified candidates set up their own enterprises.

The basic features of the Indian fisheries educational system are impressive. Despite this, murmurs are often heard, both from Government Departments and managements of private sector enterprises that the quality and competence of some of the candidates coming out of the institutions are below par. One reason for this mentioned uncharitably is that those that fail in securing entry into Medical colleges and others who have no hopes of gaining entry into them, join fisheries colleges. In this background they express the view that the future of fisheries development will be shaky, if the capabilities of the bulk of emerging generation of fisheries technocrats originating from the institutions is not toned up.

It is a well known tendency on the part of successive older generations to highlight or impute inefficiency, perfunctoriness at work, inadequate technical knowledge, lack of devotion to duty and quality consciousness etc., among some of the new technocrat entrants into the sector. Imputations of fall in standards from generation to generation are not new. In fact, observations of this nature are stated to be on record, even from 18th century. It appears that, because of the expansion of fisheries development activities, and the consequential increase in the fisheries educational and training facilities, the number of qualified candidates has been on the increase in the past couple of decades. While a certain percentage of these qualified candidates seeking employment emerge as competent technocrats of quality, many believe that the rest of them would be mostly of average capabilities. Probably, as the bulk of those employed fall under the later category, the employers keep expressing their disenchantment and chagrin, when they notice that the targets of their expectation are not realised. Apart from this, those who have attained seniority tend to comment generally on the fall in the standards of the juniors working with them.

While there can be considerable exaggeration in these perceptions, there can also be some substance. The enterprises that have come up based on new technologies, now and then point out the problem of inefficiencies among the technical personnel employed by them.

In this situation, a way out has to be found to tone up the standards of quality and competence, particularly among candidates of average capabilities who come out of the various institutions qualified as fisheries technocrats. More than those who impart education to them, the candidates themselves would have to adopt a well planned system of acquiring a deep insight into the subject matter. It should also be possible for the faculty members to intensify efforts to inculcate an awareness in this regard among the students. This is mentioned for the reason that it is not easy for the candidates of average standing to know about the system to be followed in organising their studies outside the class room in order to have an indelible and an indepth understanding of the subject matter. This is mentioned only to emphasise the responsibility on the part of the faculty members to infuse among the students the need for undertaking indepth studies of the various facets of the subject outside the class room. Such an infusion would certainly enable a large section of the students to come out of the institutions as technocrats of quality and competence far beyond the average level so as to play a dedicated role in the development of fisheries of the country. In other words, the faculty members and the students have to further strengthen their cooperative approach for a common purpose. When this is done, the parameters of quality of learning and competence of the candidates would get honed up. These parameters, of honed quality, would be eminently helpful in the later period of the actual involvement of the candidates in the industry. Thus, the degree of success achieved by the candidates in the post-college career would mostly depend on their hold on developed/developing technologies.

In the culture fishery development, as is commonly known, we have three broad phases consisting of growth promotion, harvest and postharvest activities. Drawing an analogy, in their journey through fisheries education, the candidates, as is known, pass through the study phase (comparable to growth promotion activities), the harvest phase (appearing for examinations and securing the degree), and the postharvest phase (career or marketing of the talent of quality and competence in the subject acquired during the study phase).

Let us examine the study phase briefly. This phase, being of a learning process must impart an infusion of total awareness of its crucial importance among the candidates from various angles. Several students, particularly the average ones, are often seen to be not having the needed sharpness or awareness for retaining the specific details of various important technologies in a lasting memory. Instead, as is generally observed, they just follow the routine of a superficial understanding of the subject in a perfunctory way. The inculcation of a proper study culture leading to the retention of the specific details of various technologies would vastly help the candidates in their future career. The adoption of a proper study culture would also provide a sound orientation to the average students to fare well at examinations in order to secure an outstanding rank. It is known that many students fail to answer the questions succinctly and well at the examinations, despite having the knowledge. This indicates that the candidates have to acquire the faculty of faring well at examinations and in a way to impress the examiners by practising an improved study culture and an improved way of performing at examinations.

In the course of their studies, students often come across several sentences in the books or class notes which cannot be readily grasped. Experts say that whenever such a situation arises such difficult sentences should be broken into their various components and reassembled again. In this process, it is said that the meaning would emerge clearly and register on the mind of the candidates.

Generally, authors build up a paragraph revolving around a single point and with a connectivity to the point made in the next paragraph. This being the case, the approach to evolving a study culture of ones own has to be to extract the central points of each of the paragraphs and noting them down. Many students keep note books by their side for jotting down notes. This system can be improved upon. A general suggestion is that atleast three readings of each of the chapters should be given with a reasonable timebreak in between. On each of these occasions notes can be taken in three different note books. The idea behind this suggestion is to ensure a progressive perfection in the presentation of the points. By the time the entry in the third round is done, the candidate would have made his perceptions crisp and clear with an incisive insight.

In the harvest (examination) phase, because of a more enlightened study phase as referred to above, a superior way of presentation of answers in the examinations would materialise. This becomes possible because of the continuous and persistent preparation during the study or culture phase. The harvest phase can be expected to lead to the scoring of an outstanding level of marks which would pave the way for a better post-harvest phase.

Success in the post-harvest (career phase) operations is mostly dependent on depth of knowledge, talent and competence. Once an employer is impressed over these parameters of an employee, he will not leave him and the development of this kind is the first stepping stone for building up a lucrative career with a good reputation.

Considerable progress in the development of fisheries of the country has been achieved. While this achievement is the result of a combined effort on the part of technocrats, scientists, farmers and fishermen, the main contributory factor behind is however the inputs of a band of motivated technocrats with capability to generate an array of activities to augment fish production and upgrade its utilisation, culminating in its domestic and export marketing. When well motivated candidates with competence, coming out of the fisheries educational institutions provide reinforcement to the continuity of the ongoing endeavour, there will emerge such a change in the scenario that would provide a miraculous boost to fish production. For this to happen, an enlightened approach towards the quality of knowledge acquisition may have to be fostered among the students who are the scientists technocrats of the future. Stated differently, the approach to be adopted may have to be such that it would lead to the emergence of a far superior category of technocrats of quality and competence, for employment both in the Governmental and private sectors. To sustain and be ahead in studies, the plus factors have to be taken care of. The candidates have to strive to have as part of their buildup, the gist and steps of technologies, particularly those related to operation of various types of hatcheries, culture practices, capture fishing systems and processing and marketing aspects.

The depth of knowledge that the fisheries students acquire no doubt dictates the course of their career. In addition, it would also contribute considerably to the overall development of fisheries of the nation. In this context, what is of paramount importance is the need for the emergence of students out of the fisheries educational institutions with a deep sense of confidence in respect of their readiness to effectively take part and contribute to the development of fisheries of the country.

It is possible that the few thoughts put down here may generate reactions of various hues, but the objective of the exercise would be served if it provokes a critical appraisal of the present pattern of studies and leads to the evolution of an improved and more effective system of study culture among the students, to enable them to achieve a distinctive rank and a bright career thereafter. The fisheries graduates and postgraduates, who come out of the fishery educational institutions with an awareness of their crucial role, would be rendering a signal service to the nation as the pillars of future fisheries development.

Fishing Chimes, Editorial 283: September(i) 2000: Vol. 20, No. 6

# Tiger Shrimp Domestication

## *CMFRI's Progress in the Taming Game*

The latest we have come to know is that CMFRI has made a remarkable and a reassuring beginning in respect of domestication of tiger shrimp. The experiments in this regard, in progress at its Mandapam centre, have progressed to the stage of raising $F_2$ generation of the shrimp. It would probably take another three years to reach the stage of raising F6 generation, at which stage the Indian shrimp hatcheries can expect to receive the first ever supplies of pond-grown tiger shrimp brooders for raising disease-free postlarvae of the species.

The crucial importance of raising domesticated broodstock is well known. The availability of such brooders will diminish the present dependence on capture tiger shrimp brooders, which are known to carry and transmit white spot virus to the progeny. The surfacing of the disease in culture ponds stocked with such infected postlarvae has the disastrous effect of causing mortality and of diminishing returns to the farmers. In this context, CMFRI deserves to be congratulated on the promising beginning it has made towards production of domesticated tiger shrimp. The domesticated ones in a mature condition can be induced to breed through eyestalk ablation as is now being done with capture tigers. It is heartening that CMFRI is now engrossed in this work without let or hindrance and is pursuing the same with redoubled vigour so as to achieve tangible results in the shortest possible time. In order to generate an awareness of the crucial importance of the work among shrimp hatchery owners and managers, CMFRI may have to think of shifting the venue of the experiments to a centre along upper or middle east coast where there is a relatively large concentration of shrimp hatcheries. Such a step will be conducive for a proximal interface and exchange of views between the hatchery owners and the scientists at various stages of the work and also for the inculcation of a feeling of participation in the endeavour among the managements of hatcheries.

It has to be admitted that India is far behind certain countries such as French Polynesia, New Calodonia and USA in the matter of domestication of tiger shrimp. In USA, at least one company advertises the availability of farm-raised tiger shrimp brooders. Australia has achieved a breakthrough recently in raising viable broodstocks of tiger shrimp.

Once the technology of shrimp domestication is perfected in India, it would be in nature of things that CMFRI would initiate follow-up measures to extend the technology through a training programmes among integrated farming groups, so as to enable them to produce their own stocks of disease-free brooders eventually. In any case, it is to be hoped that the technology transfer would be smooth and benevolent, free from hurdles that many in the fisheries industrial sector are understood to be experiencing in respect of transfer of technologies in respect of activities such as marine and freshwater pearl culture etc., to them from the ICAR institutions concerned.

Fishing Chimes, Editorial 284: September(ii) 2000: Vol. 20, No. 6

# A Sequel on the Anvil

The over 20 year old Bay of Bengal Programme that has ushered in a quiet revolution in the small scale fisheries sectors of countries of the Bay of Bengal Region is now poised to transform itself into another incarnation. Sponsored by the Governments of Denmark, Japan and the Member countries of the region and with FAO as the main executing agency, BOBP in its three phases of existence during the period (Third phase will come to a close by December 2000), brought about significant changes in the professional life of artisanal fishermen, highlighted by an upgradation of the quality of their life.

Many believe that the achievements of BOBP have not been publicised in a manner that they deserve. They are probably even underplayed, in tune with FAO's culture of doing its best and just leaving matters at that. What has been done by BOBP is to bring out periodical reports that have limited circulation, and not within the reach of the common man. Popularisation of beach landing crafts, motorisation of crafts, introduction of new net designs, postharvest technologies such as hygienic curing of fish including smoking etc., introduction of insulated fish boxes for transportation from landing centres to markets, and accentuation of the role of fisherwomen in coastal fishery development along Indian coast, building up of awareness in respect of problems involved in the use of gear such as push nets and bag nets in Bangladesh waters, working out migratory routes of tuna in Sri Lankan and Maldives waters and in respect of ornamental fish trade of Sri Lanka, striving to launch community-based fisheries management in Thailand and so on, are to the credit of the BOBP. It will not be an exaggeration to say that BOBP showed the way for the development of small scale fishermen, with the merit of the benefits not being smuggled away by capitalist interests.

It is axiomatic that so much of spade work that has been done would have to be fostered by member nations of the region themselves so as to ensure a homogenous and coordinated follow-up for continuity of the process of socioe-conomic development of fisher community that BOBP assiduously set in motion. This process would have to be the responsibility of the nations of the region themselves from now on. It would be in order, for these have an organisation of their own in tune with the commonalities of their aspirations.

The 24th meeting of BOBP Advisory Committee with representation of the various Governments of the countries of the Bay of Bengal Region held at Phuket in Thailand in October 1999 too visualised the need for a successor Inter-Governmental Organisation (IGO) of the region, considering that BOBP's final phase will come to a close by this year end. The Committee desired that an IGO may emerge as a sequel to continue the mission of BOBP. The Committee has apparently had in view an evaluation of the good foundation for small scale fisheries development and management that BOBP laid in the past.

The IGO which is now on the anvil, with four countries of the region *i.e.*, India, Bangladesh, Sri Lanka, and Maldives having already given their clearance to the concept, would no doubt activate, coordinate and fortify the management efforts of the constituent nations for the benefit of the small scale fishery sector of the region.

Further, while BOBP laid the foundation and promoted the activities to a considerable extent, many of the activities still remain to be further developed. Broadly staled, these are : Vessel operations monitoring system to ensure sustainable capture fishery exploitation in tandem with the introduction and enforcement of well conceived regulations, development of systems of realistic estimation of exploitable potential from time to time, ensuring quality of fish and fish products, bringing together various fishing interests to a common line of thinking and action, coupled with a system of dissemination of resource status information among their members etc. It has to be conceded that these tasks can be performed inculcating greater public awareness and participation through an IGO, particularly at this stage of development, in preference to an organisation in the nature of BOBP, with sponsorship of international hues.

At the present juncture, the most expedient task the proposed IGO has to concentrate upon is to promote a sharp awareness among the fishermen to follow the FAO's Code of Conduct for responsible fisheries, highlighting the long term benefits of observing the code.

Let it be ardently hoped that the IGO, expected to be in position in early 2001, would build a superstructure, on the sound foundation laid by its parent, imbued with viable values that would endure and last for ever for the benefit of the fishermen and women of the region, from generation to generation. Let us all extend a warm welcome and support to the IGO and wish god speed to it in the implementation of the programmes to be soon taken up.

Fishing Chimes, Editorial 285: October 2000: Vol. 20, No. 7

# Inland Culture Fishery Sector of India: Future Scenario

The present global picture of inland shrimp culture fishery is bogged down with environmental issues. Arguments about the farfetchedness of the views put forth by environmentalists highlighting the inimical effects of effluents released from aquaculture farms continue to have a marginal impact on the general belief that the inland shrimp culture activity is harming the aquatic environment. India is no exception to the trend. For this and other reasons not fully known, the environmental lobby has been gaining in strength. So much so, in preference to making investments on excavated static aquaculture systems prone to various risks, several countries, particularly USA, are moving towards setting up units based on closed recirculation farming systems for fish production. The investments for setting up these systems are no doubt high but, presently, particularly in developed countries, the returns are also high. Developing countries like India can probably follow the system restricted to production of shrimps and prime fishes such as Tilapia, having export demand. Such an approach may render the investments viable and at the same time eliminate the environmental hazards presently commented upon.

Closed recirculation systems based on Fluidised Sand Bed Biofiltration technique are reported to have been successful, apparently in raising viable crops, with a water exchange rate of less than 2 per cent per day. Under this technique, it is stated that water is let in at the bottom of a sand bed. As the water travels, it lifts the sand, making the static bed viscous and fluidised. The system, as designed, will enable, it is pointed out, 100 per cent removal of ammonia in each spasm of water passage through the bed, thereby keeping the inflowing water in an excellent condition all through. A reference to literature on the system will certainly facilitate our research institutes to work out designs of such systems.

Another closed recirculation design is stated to be based on 'Zero Water Exchange, Aerobic, Heterotrophic' principle involving a stocking rate of 500 shrimp pl/cu.m. without any disease problem. Yet another closed design mentioned is based on linkage with 'Algae Raceways for Bioconversion of Dissolved Wastes' aided by solar energy for the production of what is described as 'Organic Detrital Algae Soup' to serve as feed for the stocked animals. There are also other closed systems that are stated to be coming into vogue. These are a) those involving floating bio-filters as bio-clarifiers, and b) those with devices that provide more of bottom surface area through installation of multifold structures for providing more of grazing area for the animals under farming. In other words, there is a general trend in developed countries such as USA to usher in the next generation aquaculture systems by way of closed recirculating systems which would facilitate a) Quick removal of solid wastes, b) Cost-effective removal of suspended solids, c) Effective removal of dissolved organic matter d) Removal of dissolved inorganic wastes such as ammonia, e) Removal of nitrates, f) Supply of Oxygen and removal of carbon dioxide g) High rate of water turn over, h) pH adjustment, i) proper feeding practices, and j) effective management to ensure safety and to avoid risks in culture operations and to ensure ecological balance devoid of disease incidence. It is stated that open recirculating systems, taken up in Thailand etc., have been found to be ineffective over a period of time.

The day is not far of for Indian entrepreneurs to compulsively opt for closed recycling systems for farming of shrimp and exportable varieties of fish. Continuance of the static farming system (particularly for shrimp) in excavated ponds as at present will sooner or later become uneconomical because of disease incidence, increasing costs of disease control, problems of risk management, threats from environmentalists who seem to enjoy tacit support from various quarters and so on. It is thus high time for the Central Institutes of Freshwater Aquaculture and Brackishwater Aquaculture and Central Marine Fisheries Research Institute to work on closed recirculation systems so as to catch up with the progress in this respect, now in full swing in developed countries such as USA. In this context, a reference to the presentations by experts in the journal 'Global Aquaculture Advocate', June 2000 issue, would be a revealing exercise.

Fishing Chimes, Editorial 286: November 2000: Vol. 20, No. 8

# Recommendations of Meets on Fisheries Topics: Destined Dormancy

As is known, Fisheries sector has diverse focal areas of developmental endeavour, the aim of all which is to ultimately ensure sustainable fish production, and augmented incomes to fishermen and other stakeholders. Scientists, technologists, techno-administrators and several others who devote their time and attention to secure data and knowledge on these focal areas with a view to finding solutions to problems that impede progress in the developmental work from scientific, technical and socio-economic angles. In this background, Fisheries Conferences/Symposia/Seminars/Workshops/Colloquia/Brainstorming sessions etc., are organised from time to time to pool up, discuss and evaluate the presentations of the various participants made thereat. In the light of these exercises, the organisers will have a set of recommendations drawn up by the sessional chairmen concerned at the events. With the approval given at the plenary sessions, the said recommendations are sent for the consideration of the authorities concerned for their approval to implement them, the premise being that, such an implementation will facilitate the achievement of positive progress, ultimately in terms of increasing fish production.

Sometimes Committees/Working Groups are also set up mostly by the Central or State Governments to make recommendations on specified subjects with certain terms of reference. The recommendations made by such sponsored Committees/Groups receive prior attention by the Government concerned for processing, for the particular reason that they must have been set up for a purpose. In contrast, the recommendation made by the other Meets such as seminars are dealt with on a different footing. It will be a miracle if any of the recommendations made by Conferences, Symposia, Seminars/Workshops etc., receive follow-up attention.

The organisers of various Events, other than Governmental Committees/Working Groups, who too work hard for months together not only to pool up funds to conduct the Events, but also to identify and invite experts and specialists to present papers relevant to the topic of the Meet concerned. They prepare for the benefit of the participants working documents such as abstracts of papers, the session-wise details of the programme etc. They ensure through chairmen of the various sessions that the papers are tellingly presented by the contributors.

While there are several instances of approvals and follow-up action on recommendations of Committees and Groups set up by Governments, it is very hard to find instances of the recommendations made by Meets of other categories such as Conferences, Symposia, Seminars, Workshops etc., that would need approval not at the level of organisers but by some one higher-up. The recommendations will no doubt be there for reference when needed.

The routine practice of being complacent towards recommendations made at Meets other than Governmental Committees and Working Groups deserves a reform. If not reformed and the prevalent system as aforesaid is allowed to continue, a good number of useful recommendations aimed at rearrangement of research priorities, strategies for reorienting production preferences etc., would continue to remain shelved, thereby blocking progress. There have been recommendations made for implementing certain strategies of development of reservoir fisheries, a grossly neglected sector, at several gatherings of experts in the past. There have been several recommendations too made in respect of improvements to culture fisheries, initiating a programme of developing the saline zones of the northwest for fish farming, and in respect of many others, which, when implemented, would augment fish production as well as exports.

These and other recommendations made by various conferences, seminars, workshops etc., continue to remain dormant for no plausible reason.

In order to remedy the situation, a centralised cell in the fisheries division of the Union Department of Animal Husbandry and Dairying can be set up. The mechanism governing the functioning of the cell can revolve round covergence of all recommendations made by conferences, seminars etc., along with background material at this focal point. Upon receipt of the recommendations, the officer kept in charge of the cell could take all needed steps either for the approval of the recommendations in the form as recommended or with modifications or for the rejection of the same. The organisers concerned should have the right to know about the fate of the recommendations within a reasonable time. The development of a mechanism of this kind would prove to be a logical step for the fulfillment of the expectations of the organisers of the various Events. The season for Conferences, Seminars, Workshops etc., is now on and the Central and State Governments would have to do their best to translate into action such of the recommendations that are good out of those received from time to time. It is suggested that such of the recommendations not considered worthwhile and do not deserve approval can be rejected under due intimation to the organisers without much of a time lag.

Fishing Chimes, Editorial 287: December 2000: Vol. 20, No. 9

# Reflections on Certain Aspects of Technology Transfer

Old timers in the fisheries sector, particularly in the categories of farmers, technocrats and scientists often recall with nostalgy the development and application of important technologies that upgraded the complexion of commercial fisheries sector in the few decades immediately after independence. The impact of these had led to a marked progress in the socio-economic prosperity of fishermen, fish farmers, entrepreneurs and others who took to the fisheries sector.

Outstanding examples of developed technologies that put the fisheries sector in focus as an important source for augmenting protein-rich food production are not far to seek. In fact, they are enshrined in the history of fisheries development of India. Mechanisation/motorisation of fishing boats, trawl net operations, processing technologies and so on brought in a distinctive change in the marine fisheries front. So far as the inland fisheries sector is concerned, we have seen an equally historic fast-forward spread of technologies all over the country in respect of induced fish breeding, fish hatchery technologies, composite fish farming, integrated fish farming, penaeid and non-penaeid prawn hatcheries, the seed production of these species and their farming, etc. In respect of industrial production, technologies related to manufacture of quite an impressive array of items such as chitin, chitosan, surgical sutures, fish/prawn pickles have been developed by the Central Institute of Fisheries Technology and several of them have been passed on to entrepreneurs.

The technologies of the first phase such as induced fish breeding, fish hatchery technologies and composite fish farming were passed on to the fish farmers by the scientists with a deep sense of pride and achievement. The transfers of technologies evolved in the subsequent phases became somewhat business-like, involving elaborate procedures, payments etc. Some of these transfer efforts, particularly those relating to marine pearl oyster culture and pearl production, freshwater mussel culture and pearl production from such cultured mussels are still on the drift.

The basic function of a scientist is to develop viable technologies that will generate profits and in a form suitable for application in the field. While fisheries scientists perform this function diligently, persons not in the know of the problems of scientists, express certain views which are often uncharitable.

One can however agree with a point of view, with some reservations, that only some of the scientists succeed in standardising sound technologies that are eminently suited for field application and augmenting aqua production.

The function of scientists mainly revolves around research inputs. Over and above this, if they are burdened with the task of extending the technologies as well, and that too on a wider scale and in various States, it would result in the dilution of their scientific responsibilities. Considered from this light, the utmost they can probably be called upon to perform is to conduct training classes and demonstrate the relevant technology at the institute concerned.

The dependence for extension seems to be mostly on the scientists for taking the developed technologies to the field. The effectiveness of this route, so far as major technologies in respect of fisheries are concerned, is perceived as considerably weak by several, in any case where hand-to-hand transfer of technologies is considered essential. The situation probably calls for the introduction of a mechanism that ensures an effective transfer of technologies, taking practical aspects, mostly related to the limitations and constraints scientists and entrepreneurs face in the process. For example, the inflexible terms and conditions governing transfer of technologies act as a bottleneck and lead 'to several problems in the technology transfer process'.

Scientists would have to be looked upon mostly as a source to develop and standardise viable technologies for release and to critically examine their application in the field from time to time so as to facilitate corrective measures.

They should not however be entrusted with the function of extension too, as that would divert them from scientific work. The users of the technologies are those mostly coming within the jurisdictions of the State Directorates of Fisheries. These Directorates have a wide network of offices manned by technical personnel, percolating down to block level. Thus, there can be no better extension agencies to propagate, popularise and ensure implementation of new technologies, than the State Fisheries set up, which is an organised one. In fact, the Departments, through their district staff and Fish Farmers Development Agencies, extend fish farming and other technologies to the target groups. As it is, the ICAR institute concerned also conducts training courses for the inservice candidates sponsored by the departments with the objective of equipping them to extend the technologies.

The need therefore is to strengthen the mechanism of transferring technologies standardised by scientists through the medium of Extension Scientists to the field officers of the State Fisheries Departments. As soon as a technology is perfected, the institute concerned should make it possible for extension scientists to activate their interface with the State Fisheries Departments in the direction of imparting new technologies. It is true that candidates sponsored by the State Fisheries Departments are given training at the institute concerned as it is, but this has to be provided for one round of operations for effectivness and for this captive infrastructural facilities as at present need expansion. After going back to the State concerned, the respective candidates, in turn, could be put on the job of providing training in the subject to other field officers/functionaries at one of the farms/units, in the concerned states, identified and prepared for the purpose.

There can be certain technologies which may not be amenable for transfer through fisheries departments. The technologies relating to these activities can be transferred direct to the technocrat-entrepreneurs capable of picking up the relevent technologies, mobilising the needed investments and managing the enterprises to be set up by them. The selection of candidates can be through advertisements instead of waiting for the entrepreneurship to come to know of the technologies and approach the institute concerned. A modest course fee can no doubt be prescribed. A system of this kind, in the place of the present practice of asking for payments or deposits from entrepreneurs who approach, towards technology transfer charges, often seen to be prohibitive, may be more conducive to ensure the adoption of the technologies.

Another system can be to make the technologies available to professionally oriented commercial organisations in the first instance. These professional bodies, in their turn, can pass on the technologies to the needy persons/enterprises subject to such terms and conditions as approved by the authority concerned.

Fishing Chimes, Editorial 288: Jan/Feb 2001: Vol. 20, Nos. 10&11

# On Nation Wide Application of Aquaculture Technologies Conducive for Export-Oriented Production

The growth of shrimp and prawn culture over the past one decade is entwined with the application of advanced technologies, mostly to cater to the export production needs of the species. The export-orientation of the culture activity has motivated the farmers to invest heavily, as they continue to do, on inputs such as feed, chemicals, probiotics etc.

The application of export-oriented aquaculture technologies for the production of shrimp now takes place along the coastline of India. The same can be said in respect of prawns too, with the difference that this activity has made marginal inroads into the inland zones of Coastal States and Inland States too.

There is a vast scope for producing shrimp and prawn in inland lentic freshwater bodies located on the mainland, away from CRZ. In the States such as Rajasthan, Haryana, Punjab and adjoining areas of Uttar Pradesh, the extensive underground saline water resources available could also be well utilised for export-oriented aquaproduction.

There is an inescapable need to popularise export-oriented aqua production all over the country for the reason that marine capture shrimp production, which has all along been the mainstay of Indian exports, and continues to be so, is now stagnant. In order to augment value-oriented export level, efforts are now being made by MPEDA and the exporters to push up the status of exports upwards from year to year mostly through augmentation of production of exportable species, mostly shrimp and freshwater prawn from the culture sector in the Coastal Zone. There are however inexorable limitations in this direction.

This situation demands a strategy to progressively bring the inland freshwater lentic resources over the spread of the length including breadth of the country and the saline zone of the north west, with probably the partial exception of the temperate zone, under export-oriented production.

There is an amply demonstrated evidence that, the drawing up and implementation of a well-designed project for phased application of export-oriented aquaculture technology, (from the outer limits of CRZ, aimed at utilising the production potential of the mainland lentic water resources) for the production of exportable species such as the tiger shrimp and giant freshwater prawn, deserves and requires immediate attention.

The implementation of such a project, taking care of the training component, will pave the way for ushering in prosperity in the inland rural areas through generation of more of entrepreneureal avenues, employment, exportable aquaproduction and higher incomes.

Exported species, in the main, are The Giant Freshwater Prawn, Tiger Shrimp, Sea Bass and Tilapia.

## Giant Fresh Water Prawn

As is known, the main constraint in extending the farming of this prawn in mainland areas distant from the coast is the problem of production of its seed in non-coastal areas because of the general non-availability of saline water there at. As is also known, most of the hatcheries are located close to the coast. For the production of prawn seed the availability of salt water of around 12 ppt, although for a short duration, is essential. Efforts have been made at a few inland centres distant from the coast to produce postlarvae of the prawn using both constituted salt water and also transported sea water from the coast. There are also instances of prawn PLs transported from the coast and stocked in tanks and ponds of the mainland far away from the coast but there have been practically no economically viable successes.

## Tiger Shrimp

Postlarvae of Tiger Shrimp obtained from hatcheries along the coastline are now being farmed in freshwater tanks and ponds within CRZ and areas adjacent to it as a successful commercial activity in the central and southern coastal districts of A.P and probably in other Coastal States as well. Being euryhaline, the post larvae and juveniles of Tiger Shrimp are seen to get acclimatised to freshwater conditions very fast. In fact, in certain places far distant from the coast, the stocking of tiger shrimp postlarvae in tanks and ponds is going on, leading to good harvests. One advantage of farming of shrimp in freshwater is seen to be that the white spot virus is generally unable to survive in freshwater conditions. At a workshop conducted in Bhopal on 26 and27 Feb 2000 by the M.P. Council of Science and Technology, in which the Dy. D.G. (Fisheries) ICAR participated, a recommendation was made to promote culture of Tiger Shrimp in cultivable lentic freshwaters of

MP. This recommendation holds good for other States too. Farmed Tiger Shrimps raised in freshwater ponds in coastal districts of A. P. now constitute a significant part of Indian aqua exports.

## Sea Bass

Hatchery seed production technology of this euryhaline fish has been developed at CIBA. The technology is eminently suited for commercialisation. Sea bass fillets have a good export market.

## Tilapia

There is a good export market for fillets of this fish, mostly in USA and Europe. Through adoption of monosex culture in closed systems, coupled with seed production in farms reserved exclusively for Tilapia monosex seed production, and governed by stringent regulations and their effective enforcement to prevent escape of the fishes into the wild, it should be possible to farm Tilapia, in tanks and ponds. The most widely accepted species in USA and Europe is the Nile Tilapia. At the Bhopal Workshop mentioned above Tilapia culture on the lines indicated was recommended. It is heartening to learn that Government has permitted two companies to take up culture of Tilapia and this may well prove to be a good augury.

Regarding feasibility aspect, the suggestion to bring inland freshwater lentic resources such as tanks and ponds under marine shrimp and freshwater prawn farming, sea bass and Tilapia farming (and also in the saline zone of the northwest under marine shrimp and Tilapia farming) may seem weird and impracticable on first thoughts but a closer examination will reveal that a project of the kind suggested is eminently feasible and the constraints that may come in the way are amenable for removal through a planned strategy.

A Team of experts from Governmental organisations or a professional consultancy organisation can be entrusted with the job of preparing a master plan and a project report for the progressive coverage of tanks and ponds of the mainland, away from CRZ, for export-oriented production of the species. It is believed that the entire mainland with or without the temperate zone can be covered by the export-oriented farming activity over a period of 5 to10 years.

There will be need to have certain guidelines for working out the master plan/project report on the subject. The suggested team has to be entrusted with the responsibility of working out an organisational mechanism for implementing the suggested master plan/project, taking into account the inter-state implications and Centre's role. The manner in which the public and private sector participation has to be promoted also deserves particular attention.

Taking into account the number of ponds and tanks and their extent, convenient land stretches from west to east on the one hand and east to west on the other, beyond CRZ and in sizeable but convenient Blocks may be demarcated in the peninsular zone and also in the coastal northeast and north-west parts of the country and beyond. Likewise, the northern part of the country can also be demarcated appropriately, from south to north, excluding the temperate zone where Shrimps, Prawns and Tilapia would be difficult or impossible to grow.

In each of the Blocks, (to be divided into SubBlocks) places where PL/Seed Storage and Dissemination Units (SDU) with all needed facilities can be set up may be identified, taking into account road connections, power supply, and other facilities. The purpose of these SDUs will be to serve as centres for storage of PL/Seed (brought from the hatcheries) for acclimatisation, storage and dissemination for stocking in tanks and ponds within the Block concerned. The Blocks, divided into SubBlocks, can be numbered. Storage and Dissemination Centres as required may be set up at convenient locations in a way conducive to receiving and disseminating PL/Seed within the Block area or for transfer to adjacent Blocks. Thus, these storage units would serve as transit centres too for the movement of PL/Seed to the other units within the same Block or to the adjacent Blocks.

Routes may be formed from the hatcheries to the nearest Blocks and to SDUs therein and also from Block to Block. To start with, PL/Seed may be transported to the SDUs in the first set of Blocks adjacent to the CRZ and from there to the adjacent series of Blocks and so on as per a working plan to be drawn up. PL/Seed may also be transported to the farther Blocks as well, if the distances in between are conducive for such a transport with acceptable levels of mortality.

A Coordinating mechanism between hatchery managements and the enterprises who will own or run SDUs may be promoted and standardised through a suitable mechanism for achieving the objectives of the project.

Some of the existing private seed farms too can be upgraded and recognised as SDUs. In the alternative, a network of such Units can be promoted in the private sector by the State Fisheries Departments by providing such incentives as needed for setting them up in the private sector. Another way is that the State Fisheries Departments may organise this infrastructural facility and activate the same as SDUs either directly or through competent and qualified persons or private companies to run them as per terms and conditions to be stipulated by the Government.

Facilities may be organised based on a Plan for pooling up the harvested catches of exportable species from tanks and ponds at various points and transferring them to centrally located processing plants for processing and for later transportation of the packed products in containers to the nearest port for export.

A pilot project, keeping the above mentioned guidelines in view, among others, can be drawn up and implemented in the first instance through joint or tripartite participation of i) the concerned coastal State, ii) the receiving State and iii) the private or public sector enterprise concerned. The solution to the problems faced as worked out, while implementing the pilot project would facilitate the formulation of a nationwide project to be implemented by a nominated Authority to promote export-oriented production from lentic waters of the mainland beyond the CRZ.

The vast inland lentic waters of the Indian mainland are eminently suited for establishing an export-oriented aquacultiire system. In other words, these waters are a dollar mine awaiting to be developed and exploited for augmenting aqua products exports. Once these waters are brought under export-oriented aquaculture system, the foreign exchange earnings from fisheries sector will reach newer heights. Task forces are no doubt required to be set up for working out details which may vary from region to region.

It is believed that an area of at least five lakh ha of tanks and ponds can be brought under an export-oriented farming system involving farming of prawn and shrimp, and monosex Tilapia and it might be possible to achieve a coverage of this level in a period of 5 to 10 years. On the presumption that a minimum average exportable production of 300 kg shrimps/prawns ha/annum can be promoted, it can be estimated that an additional production of 1.5 lakh t per annum can be achieved in 5 10 years. When exported, this produce can bring an additional 1,200 million US dollars or Rs. 5,400 crores per annum.

Looked at from any angle, coastal aquaculture is a non-hazardous, environment friendly activity. It is a boon, laden with economic benefits in a substantive measure to the coastal poor. Its importance stems from the fact that it brings into use and renders saline coastal lands that lie barren and unproductive now into assets that yield aquacrops, thereby providing employment and income to those interested and in the line.

The ushering in of coastal aquaculture has turned out to be of considerable unexpected benefit over the years mostly to small farmers. No doubt, larger houses plumped in, to participate in the activity in a big way, but had to wriggle out of it, particularly the farming part as fast as they entered, the main reason being that the subject is not in tune and not compatible with their style and culture of functioning.

## Coastal Aquaculture: Environment – Friendly and Socio-economically Benign

The opposition to coastal aquaculture seems to have arisen because the larger houses and enterprises moved in. Those in the opposing groups who were, as it seems, keenly on the look out for situations of the kind, appear to have misjudged and geared up for a mammoth movement to rectify the situation, unwittingly little realising that coastal aquaculture is, by and large, non-polluting, environment-friendly and well-suited for the economic uplift of coastal poor. By the time they had probably realised the truth, it was too late for retracting. They had to probably justify and strengthen their standby mustering additional points. Among these are points such as salinisation of adjacent wells and paddy fields because of coastal aquafarms.

The fact is, responsible and well managed coastal aquafarms, be they for shrimps or for other aqua animals, are eco-friendly. It is somewhat farfetched to dub coastal aqua farming as harmful. Coastal aquafarms would not ordinarily salinise the adjacent lands or wells. Nor would they pollute the environment as alleged. In fact, they act as a benign, useful and transitional live buffer between sea and land, as long as biodegradable inputs are applied, and this is being done, by and large. Where this is not done it should be insisted upon. Just because of the possibilities of certain things happening, coastal aquaculture cannot be dubbed as an enemy of environment. In fact, criticism of the kind to which the sector is subjected to can be levelled against any activity. The criticism does not stand to reason but only calls for remedial measures where needed. Studies in many areas on these aspects were conducted and these had clearly shown no adverse impact on the environment as evidenced by I) the report of NEERI submitted to MPEDA in Feb. 1995, 2) CMFRI's study on EIA in 1997 and 3) scientific results presented in the First Indian Fisheries Science Congress 2000 by Sumanta Dey, B.K. Das and B.P. Gupta *et al.* (21-23, Sept. 2000, Chandigarh). Salinisation has not occurred beyond a distance of 25m from a shrimp pond according to MPEDA's report of Feb. 1995. Where called for, the remotest possibilities of salinisation can be offset by creating a buffer zone of sufficient width filled with freshwater adjacent to coastal aquafarms. Considering the socio-economic importance of the activity, institution of remedial measures is important here.

There can be no doubt that aqua farming can co-exist with rice cultivation. The examples are *Pokkali* fields of Kerala and *Bheris* of West Bengal. Abroad, Kung Krabean Bay ecosystem in Thailand can be a fine example of environmental compatibility in which belts of mangrove forest, shrimp farms and rice fields co-exist on the landward side, and lagoon or sea on the other side. The shrimp farm discharge finds its way into the sea through the mangroves, which play the role of treating any of the wastes present in the effluent before its joining the sea. In any human activity relating to farming, there is no way of avoiding outflow of used water, as in the case of agriculture too. Further, the rich nutrients present in the released from coastal aqua farms waters also contribute to the fertility of the sea for increasing the natural fish production. Quite recently, the Scottish parliament has rebutted claims of WWF that marine aquafarms are sources of environmental danger with scientific data (API Newsletter Vol. 4(1), Nov. Dec. 2000;

Shrimp farming at present is practised, by and large, only by small scale farmers who own or take on lease small water bodies adjacent to the coastal waters. These farmers undertake mostly traditional type of farming with improved methods and management practices. In this connection, it may be mentioned that the farmers have learnt to overcome the dreaded white spot viral disease that affects shrimps under farming, by resorting to polyfarming with finfishes or seaweeds or through reduction of salinity.

The farmers do not now think at all of stocking their farms with wild shrimp seed. This is because of over 200 shrimp hatcheries set up along coastline with Government's support, obviously because of the eco-friendly nature of shrimp farming. Government would not have supported the setting up of the hatcheries, if it was a harmful proposition.

Generation of employment for rural poor is taken care of by coastal aquaculture as in the case of other forms of culture systems. As employment opportunities at all levels of education are available, rural youth now find employment at the farms near their own villages. This is the only rural based modern activity having global dimension with wide social impact.

Coastal aquaculture, as already stated, generates employment opportunities. More than 4 persons/ha directly and 4 persons/ha indirectly are provided with employment and nearly 8 lakhs of people are benefited. The annual income from the sector is estimated to be Rs. 34 lakhs/ha.

Let us consider examples of other countries who have forged ahead in aquaculture. China is now number one in coastal aquaculture, relegating India to a position next to it. Further, Smaller countries like Thailand and Vietnam are overtaking us in coastal aquaculture with their respective Governmental support. Increased employment opportunities in South East Asian, European, Scandinavian and American countries due to aquaculture, particularly coastal, are well documented. In USA coastal aquafarms are tourist spots too. Surely, if they are pollution-ridden tourists would not go there.

Many wild stocks such as shrimps and spiny lobsters in our coastal waters are heavily exploited. Most of them are becoming endangered. To replenish such stocks sea ranching of hatchery raised seeds is the only way. Such initiatives will maintain the level of wild stocks and fishers will not be deprived of their legitimate income due to dwindling stocks. Since hatchery technology for shrimps is already available, we can easily augment the natural stock in the sea by sea ranching of shrimp seeds.

Coastal aquaculture is a boon to the nation. It can feed millions of our undernourished poverty stricken population with cheap nutritious protein-rich food, besides earning valuable foreign exchange which can contribute to the building of our national economy on a firm footing. And all this in an ecofriendly manner, in line with other food production-oriented sectors. Being practised only in low saline lands, which have remained fallow/unproductive or barren for ages, it raises national wealth, particularly through exports, a national priority.

As Aquaculture Foundation of India reminds us, classification of land as agricultural or otherwise was done several decades ago by the British when conditions were different. Now a reclassification, based on a scientific survey is absolutely essential to delineate areas suitable for coastal aquaculture but unfit for agriculture. It may not be surprising if such a survey reveals that lands which were classified as agricultural ones earlier are no more so. Construction of dams/barrages across the rivers and reclamation of land in the backwaters/estuaries for urbanisation have taken place extensively leading to salinisation of the lands, identified originally as agricultural lands. The potential of brackishwater area is estimated to be 1.4 million ha as against 0.1 million ha under cultivation at present.

The sea is already a dumping ground for pollutants, due to discharge of untreated toxic wastes from large scale industries such as of chemicals, pharmaceuticals, fertilisers, textiles, plastics, atomic plants, heavy metals, oil explorations in the sea and oil refineries, ports, pesticidal residues from agricultural fields and domestic sewage which are let into the sea directly or through rivers and creeks. Shrimps cannot afford to pollute the environment and survive at the same time. There is no evidence to show that shrimps, for that matter, any aquatic animal has died because of pollution of farms or its effluents. To avoid pollutionary effects, aquafarmers have already resorted to the well known methods of water treatment such as settlement ponds, bio-ponds etc., so as to ensure pollutant free water supply to farms. Similar methods are adopted before discharging the farm effluents into the creek or sea. Therefore, ascribing environmental pollution to the shrimp farming and condemning it on that score is unjustified and looks invidious.

Wherever coastal aquaculture has been developed, well connected roads, electricity, drinking water facilities and even health care centres have come up. Allied activities have also strengthened the rural economy leading to the overall development of the area concerned. Thus, coastal aquaculture provides an unparalleled opportunity for social cohesion and balanced spatial distribution of development to the disadvantaged coastal poor.

In fact, those interested in environmental safety should start a movement for the spread of aquaculture for the socio-economic well-being of coastal poor, for bringing vast stretches of coastal lands under farming, as the barrenness of these lands are proving harmful to the society from various angles.

Coastal Zone Regulation Act of 1986 and subsequent notifications in Feb. 1991 appear to have not taken into account the future social and economic needs of the people of the coastal zone of the nation. It has to be recalled that, while hatcheries, fish curing yards and other activities such as of fishing harbours which require water front facilities have been declared as permitted activities in the CRZ notification of Feb. 1991, and the reason for not including aquaculture i.e, raising and rearing of live organisms in ponds/tanks is somewhat odd. How can any one be against culture of harmless, dumb and water-friendly animals ? Apparently, this may have been either due to oversight or inadvertance, since aquafarming also is a waterfront activity just like a hatchery which is literally akin to aquafarming involving the rearing of the organisms in improvised tanks.

In conclusion, we have to observe that those who oppose coastal aquaculture may evaluate the position in an objective perspective and work towards a regulated and already eco-friendly proven coastal aquaculture, that is so beneficial and in the nation's interest.

Fishing Chimes, Editorial 289: March 2001: Vol. 20, No. 12

# Prognostic Reflections on Fishery Resources on Indian EEZ

Going by what scientists say, Indian EEZ now holds an unexploited fishery potential of 1.1 million t. (Total estimated potential is 3.9 million t of which 2.8 mill t were exploited in 1999-2000). Most of the balance of potential is stated to be existing in the oceanic zone. There are doubts about the capability of the nation for exploiting the available balance of resources as estimated. Regarding renewal scenario of exploited resources, It is known that coastal shrimp stocks are the main annually renewable resource. Other resources such as lobsters and several species of finfishes take more than a year, in some cases several years, for renewal and this may have the effect of reducing the present annual output in the future.

This situation projects a confusing picture in respect of future annual growth of marine fish landings. This picture also confuses those in the industry about their future. It also tends to discourage those who are inclined to take to marine fishing line. Taking into account possible limitations over the exploitation of available fishery resources coupled with the general unwillingness of the banking structure to extend financial help (not their fault) to the industry, particularly for the introduction of new units, what is now being thought of is to upgrade the capabilities of medium sized vessels of 15-27 m LOA for diversification of fishing effort, with possible subsidies from Government and investments to be raised by the enterprises concerned. To illustrate, there has been some progress on the west coast in respect of upgradation of vessels of 50 ft range for demersal fishing in deep sea zones and there are some new initiatives to upgrade shrimp trawlers of 23-27 m OAL on the east coast for undertaking tuna fishing by installing monofil longlining equipment. It may be mentioned here that there are several who ardently hope that these initiatives would be successful, and would lead to the exploitation of the presently unexploited pelagic resources such as tuna, etc. The quantity of catches that can be obtained through these innovations may result in a marginal addition to the catches.

In this challenging background, it would be a great feat if the Government succeeds in framing a tangible deep sea fishing policy covering the pelagic and deep sea zones of the oceanic area where most of the balance of the estimated resources are believed to occur but with a long way to go for exploiting them. Even with optimism that these could be exploited in short term through certain management measures, it is difficult to say how sustaining the additional production would be, if we go by the past experience with deep sea lobsters, squids, cuttle fishes etc.

A Group of Scientists that went deep into the revalidation of the fisheries resources of EEZ had apparently no new evidence that would have enabled them to take the resource estimate of 3.9 mill t upward. So much so, the Group is understood to have confirmed the earlier estimate of 3.9 million mt with an insignificantly marginal increase. The Group must have been conscious that, over and above the present average annual production of 2.8 mill t, some part of the balance that is available goes into the holds of intruding foreign vessels, which must be having the effect of reducing the estimated available balance of potential of 1.1 million m t further.

Until recently there were reservations in respect of the quantity of fisheries resources of the EEZ which were estimated at 3.9 million t in 1980s. Several optimistic fishery professionals hold the view that there must be much more than 3.9 million t of resources considering the farflung dimensions of the Indian EEZ and the overall static nature of Indian fishing effort in general. This optimism is now in jeopardy with the Group set up for revalidation of resources more or less confirming the earlier estimate of 3.9 million t.

Let us presume that 3/4 of the EEZ computed at 1.50 million sq. km is intensely exploited to yield 2.8 million mt per annum (nearly 1.8 mill t per million sq. km or 1.8t per sq. km) on an average. It is difficult to say whether the Indian EEZ is fertile or barren from the angle of presently exploited yield per sq. km of the Coastal Zone from which almost the entire quantity of presently exploited 2.8 million t come from. The assessment of the resource at 3.9 million mt must have been quite objective and not swayed by the inelasticity of annual production, the result of the present static level of fishing effort at pelagic/columnar and in certain demersal layers, as it is seen now. The resources are no doubt renewable but how much of time it takes for renewal of various species constituting the fishery other than coastal shrimp is not clearly known.

With an estimate of around 1,83,000 t of tuna resource known to be available in the pelagic zone of the shelf area for exploitation and an unestimated quantity of 200,000t in the pelagic layer of oceanic zone, there is optimism that these could be exploited to augment production but there is no means of achieving this as at present. It is to be hoped that the pilot project now afoot for equipping trawlers of 23-27 m OAL for monofil longlining would eventually become a full fledged project in the tenth plan, with provision for setting up the needed infrastructure and for providing the needed training to prospective candidates. The additional hope is the operation of used long liners which some of the entrepreneurs are learnt to be importing under a deferred payment system. Their operations in the Indian EEZ would have the effect of stepping up exploitation of pelagic resources, particularly tuna and tuna like fishes and pelagic sharks. The criticism that imports of this kind tantamount to operation under charters or leases do not seem to have much substance, as exploitation under Indian banner and the initial accrual of the total sale proceeds in foreign exchange to the nation are important. It has to be appreciated that acquisition of ownership by means of this kind of device may appear to be the best option to several, particularly, when there is no other means of securing finance to acquire them. The main point in doubt can however be the *de facto* ownership versus legal ownership, particularly when dual registration is possible.

It is evident that the industry has no faith in the availability of economically viable resources in the offshore, deep sea or oceanic zone. If the industry has that kind of faith they would not employ most of the trawlers owned by it primarily for exploiting known resources in the coastal zone, mostly shrimps. Not having this faith and experiencing that shrimp fishing can be done only for a few months in a year, several companies have managed to move their vessels into EEZs of Myanmar and one company into the EEZ of Indonesia. Surely, they would not embark on such a risky diversion, if they are convinced that Indian waters too have adequate resources. Further, they would not subject themselves to the problems they now face particularly with the customs when they bring their vessels from Myanmar/Indonesia laden with catches harvested from the EEZs of the above mentioned countries.

The confusing scenario, confounded by confirmation of the earlier resource status, must have been responsible for the ongoing timelag in the formulation of high sea deep sea fishing and pelagic fishing policy. Not knowing the problem that is probably riddling the minds of the planners in finalising these policies, the entrepreneurship tends to uncharitably blame them for the delay.

The present situation can probably be summed up as an enigmatic one, with an evidence of inadequacy of resources to go for fishing beyond the coastal and its adjacent zone and lack of means to move into the oceanic zone for exploiting its available estimated demersal and pelagic resources. This baffling situation is exemplified by the virtual absence of truely Indian deep sea (pelage/demersal) fishing fleet, and the needed infrastructural facilities at places central to possible areas having balance of the estimated potential.

It appears that we have a three-fold problem before us for exploiting the estimated balance of unexploited resources available in the oceanic and adjacent zone. One is the lack of vessels that can exploit these resources. The other is lack of the needed infrastructure. The third is the bleak picture in respect of availability of funds for investments. There can be three alternatives to resolve this impasse. One is to find money for investments to implement a project to introduce vessels, develop infrastructure at key centres such as Andamans once there is confidence about the resources there at. The other way is to have bilateral agreements with countries such as USA/Japan/Korea/Australia for exploiting on a sustainable basis relatively better known pelagic resources such as tuna or demersal resources such as deep sea lobsters, prawn etc., with provision for individual companies from both sides to enter into joint ventures to undertake integrated operations from production to export marketing. The third option can be to allow Indian companies to acquire foreign vessels on hire purchase or deferred payment basis providing for cooperation in all aspects from production to export marketing and for repaying the cost of vessels out of earnings within a specified period. It will be prudent to adopt one of these alternatives, in the place of giving scope for poaching vessels to take advantage of our inaction and make money to the detriment of the nation's right to have that money. By adopting one of the alternatives atleast part of the money would accrue to the nation.

Fishing Chimes, Editorial 290: April 2001: Vol. 21, No. 1

# Forecast on Fisheries Schemes of Tenth Plan

We are now on the threshold of Tenth Plan. This threshold status tempts those interested but who are not within hearing distance of the sound and fury that signifies all that is going on in respect of finalisation of fisheries schemes of the plan, to guess what could be the likely contours and content of the plan that would be hammered out by the specialists. Having listened to a few enthusiasts who are good at the guessing game based on the trends, a formulation is attempted here to be checked later for its nearness to the actual plan when it comes out of the cod end.

It will be auspicious to begin the guessing game in respect of inland fisheries schemes likely to find a place. Starting from northern altitudes. It is possible that development of cold water fisheries would receive prior attention, may be through a scheme aimed at providing encouragement and support for the setting up of commercially-oriented integrated culture fishery units including hatcheries related to cold water fishes, principally mahseer and trout. Probably a programme of ranching of cold water streams and promotion of game fishing would also find a place in the plan with the needed emphasis on sustainability, responsible fishing, environmental safety, eco-friendly approach etc.

As they climb down from the extreme north to Punjab plains or Rajasthan desert zone, or western U.P. or Haryana, the planners may think of the great saline stretch on the north-west and ways of utilising the potential of the area for shrimp and prawn culture and for Artemia production. They may however get bogged down when they are told about the pitfalls and past experiences in respect of these lines of endeavour. Further, there can be several logistic problems that may confront them. How could the whole thing be planned, where would the money come from, how to train men, how to organise flow of seed and other inputs for culture, and how to organise marketing, exports or domestic. How to produce and harvest Artemia cysts, how to pack and market them and other issues also may come up. Development-oriented men the planners are, they can be expected to visualise the contours of the schemes and arrive at well conceived and implementable schemes that will yield viable results and shower prosperity on the farmers eventually. Pessimists, even among planners, who, alas, do exist, may pile up hundred and one arguments against the proposals. It would be interesting to know whether these projects will come up for consideration, and if so, what the result would be.

As the scanning is continued further down towards the rest of the plains and the plateau zone, the most obvious area that would strike a planner would be the need for reservoir fishery development. The developmental planners, scientists and technocrats know that without a cluster approach and an effective organisational mechanism and net-working in respect of supplies and services related to seed and other inputs, consistent operations to achieve culture - capture balance in these waters will not be possible. Further, the vital need for integration of phased harvesting, linkages with centralised cold storages/value-addition units and transportation channels and marketing linkages with domestic as well as export channels in the developmental effort is well known to the planners. Even with all these components, the task of an effective or viable reservoir fishery development is a daunting excercise. This ought to encourage and embolden intrepid planners to take on the challenge. Reservoir fishery development is a major area that has been eluding planned action. One can therefore expect reservoir fisheries development would be one of the main planks of the fisheries plan document and it would be interesting to know in what manner it would be transposed in the plan.

It is common knowledge that our wetland fisheries are the most neglected. Many are aware of the amazingly single handed achievement by G.N. Mitra a few decades back in respect of reclaiming and bringing under fish production a large number of swampy wetlands in Orissa when he was Director of Fisheries, Orissa, besides introducing measures for developing Chilka lake fisheries. What Mitra did was totally eco-friendly but his sterling work doesnot seem to have been emulated by other States for the development of their wetlands. West Bengal and Assam have tried and they continue to try development of fisheries in their *beels* but not with much success. Collair and Pulicat lakes are wetlands getting increasingly eutrophic. But for converting marginal areas into fish/ shrimp ponds, not much has been done for their development. In West Bengal, the Minister for Fisheries takes personal interest in wetland development but cognisible development is still at a distance. The Chief Minister of A. P. has recently exhorted villagers to desilt all tanks and ponds in disuse for storage of water. This concept can be converted into an integrated activity covering fish culture too. Agricultural fields, particularly those located in upland areas, can be desilted to the extent necessary for gathering water during rains and utilising the same for integrated paddy farming with scampi: It would be an admirable achievement if a tangible scheme for development of fisheries of wetlands finds a place in the Tenth Plan Fisheries Chapter.

One aspect that would obsess an inland fishery maniac is the manner in which the economics of operation of the present pond fisheries would be sought to be improved in the Tenth Plan. Unless an alternative or an addition to major carps is planned, taking export slant into account, not much in terms of improvement in value earnings from inland pond fishery sector, can be expected. Supplies to the North-Eastern States including West Bengal from various other States, particularly Andhra Pradesh have reached a near saturation point and a trend towards fall in prices may soon materialise. This critical situation would present a problem to be tackled. One can expect that the planners would, in all likelihood, rivet their attention towards giant freshwater prawn culture and may be also towards tiger shrimp farming which is now being cultured in freshwater, as value-imparting stocking additionality that deserves consideration. Logistics of supply of seed of Scampi and tiger shrimp, which can be produced economically along the coastline only, may receive attention for incorporating an economically viable plan involving dissemination of seed supplies from the coast through a relay or networking system from production centres and also for linking the post-harvest activities with a cold chain system having centralised plants at convenient places to facilitate processing, preservation and storage for feeding domestic market, with provision for organising export marketing too. Who knows, the planners may even think of recommending all this to be done in the private sector through a well-planned organisational mechanism. In any case, it would be interesting to know the manner in which they look at the subject, if it comes to their attention. One thing seems to be clear: Without bringing in the export angle, present economics of pond fishery development would not improve.

The planners are likely to be bewitched by the strides made in Punjab and Haryana in respect of integrated fish farming. What has been achieved by Punjab and Haryana farmers would deserve to be emulated by other States. Would the planning group recommend a scheme on integrated farming for application on a national basis? We will of course know once the plan document is out.

Now let us turn our attention to the marine sector. One can probably expect that a scheme for upgradation of the capabilities of the presently operational trawlers up to 15-16 m OAL for fishing in farther waters, along the coast line emulating Kerala/Karnataka model, would find a place in the fisheries plan. With all the clamour about tuna as the main exploitable resource available in Indian EEZ, another well thought-out scheme for promoting tuna fishing with larger vessels, in all likehood, would find a place. Problems of financing and technology transfer in respect of tuna fishing would no doubt come up for the closest consideration and a strategy to counter a possible existence of regional or global intrigues or manoeuvres to corner the bulk of the benefits out of fishing in the Indian EEZ, by known foreign interests to the detriment of Indian industry would in all likelihood be hammered out. The intentions of foreign tuna vessels to prevent Indian tuna fleets from coming up, met with success so far, but Tenth Plan strategies may counteract these overtures. It is also possible that the planning group may think of a project for providing the additionality of tuna monofilament fishing equipment on the presently operational small craft too, for the reason that monofil system is very compact and can be incorporated on them. May be the group would think of improving infrastructural facilities appropriately for post-harvest activities related to tuna exports in logistically central locations in Andamans and Lakshadweep zones and probably on the north-west too. A well thought out project that may have provision for bilateral agreements with countries such as Australia, New Zealand, USA, etc., in respect of inputs, transfer of technology, export marketing etc., would probably be identified, as India is almost totally deficient in expertise related to monofil tuna long lining system, keeping in view the past aborted efforts for placing India as one of global tuna players, which was probably because of short comings in planning.

Another aspect that may engage the attention of the planners is the imperative of introducing coastal/high sea cage culture system, supported by shore-based hatcheries. They would, in all likelihood, recommend the inclusion of a scheme on this, considering that the marine catches have reached a plateau. Several aspects of Fisheries Education and Training would almost certainly receive attention under the plan, for, the implementation of the plan would depend mostly on availability of well trained personnel with hands-on experience and with a sound background of fisheries technology and fisheries personnel management, both in regard to culture and capture aspects and in regard to value-added processing and marketing in both domestic and export fronts.

Government would have to rely on private sector in a large measure for implementing the schemes that directly aim at increasing production, either from inland or marine sector. To translate such schemes interaction, private sector enterprises would have to be helped with financing facility provided by a Fisheries Development Fund to be set up in the Fisheries Division of the Union Department of Animal Husbandry and also through permitting the enterprises to enter into joint ventures with genuine foreign companies, under provisions of bilateral umbrella agreements to be entered into with the Governments concerned. We have to see how planners would tackle the financing problem. Let us all look forward to the release of a unique Tenth Plan fisheries document, the first one of the new millennium.

Fishing Chimes, Editorial 291: May 2001: Vol. 21, No. 2

# Small Mechanised Boats: Diverification for Tuna Needed

Assessments made indicating substantial exploitable tuna fishery potential spread over coastal, shelf as well as high sea segments of Indian EEZ are well known.

Fishing in Indian EEZ by small mechanised boats, while confined mostly to the Coastal Zone, extends in some measure, towards the shelf zone. The fishing effort is, however, predominantly shrimp-oriented. Pelagic fishing is confined, by and large, to catching of sardines and mackerels with purse seines, boat seines and ring seines.

Aimed fishing for tuna, in Indian waters is confined to the waters around Lakshadweep islands. Stray catches of tunas from coastal waters are known to take place along the coastline of mainland and in Andaman waters. There was an organised effort in middle 1990s at tuna longlining (multfilament) in Indian EEZ, but it was short lived.

In this background, suggestions have been made for exercising initiatives aimed at installation of monofilament longlining equipment not only on 23-27 m OAL trawlers but also on small mechanised fishing boats, as a diversified system for catching tuna. The motivation for the suggestions is to upgrade the economics of operations of boats on the one hand and utilisation of available tuna resources on the other. It was felt that such initiatives, particularly in respect of small mechanised fishing boats, would result in the augmentation of exportable marine fish production significantly. For example, when, say, 2,000 mechanised fishing boats are equipped for monofilament longlining for tuna with 10 km line with 150 hooks, each for operation from each for 200 days in a year, there will be a tuna production of around 22 t in a year per boat, computed at minimal hooking rate of 1.5 per cent. This would result in additional tuna production of the order of 44,000 t per annum, from the coastal zone and the zone adjacent to it. When exported for canning, the income from this much quantity will be, at 60 cents US $/kg, 26.40 million US $ or Rs. 121.44 crores. When the number of boats for such conversion is stepped up keeping sustainability in view, the earnings would of course zoom up. As and when infrastructural facilities for pooling up the catches from the boats in fresh chilled condition and their air lifting are organised, the stage will be set for exporting tuna fish in sashimi form to Japan for securing returns, several fold high.

India is not alone in respect of the need for equipping of small mechanised fishing craft for diversified fishing for tuna. Brazil has a similar need and is reported to have moved forward in fulfilling the same. This is reflected in a paper authored by Dr. Matt Broadhurst and Dr.Fabio Hazin at the Universidade Federal Rural dePernambuco, in Recife, Brazil, and published in April 2001 issue of an Australian journal, *Fishing Boat World* (p. 14-16). According to the paper, some of the operators of small mechanised craft of 8- 12m OAL, reacting to the problem of decreasing returns, experimented with sub-surface longlines to target stocks of tuna, sharks and bill fishes, with encouraging results, as early as mid 1980s.

The authors say "Sub-surface longlines are considered passive fishing gears, relying on the chemical and visual stimuli of bait and other attractants to direct targeted species to the hooks. In contrast to several other fishing methods, longlines are considered relatively efficient at selecting those individuals that are targeted, whilst avoiding large quantities of incidental catch". The authors have listed out several key factors such as vertical distribution of hooks in relation to maximum abundance of target species, type and size of hooks and their spacing along the main line, setting methods, etc. Integration of these various factors was earlier considered not possible for smaller vessels of 8-12 m OAL, the most commonly used artisanal vessels, but not so now.

New technological developments during 1997 have enabled the coastal fishing enterprises to incorporate sub-surface tuna longlining system in artisanal mechanised fishing vessels of 8 to 12m OAL. These technological developments, such as monofilament main lines, chemical light sticks, (designed to increase attraction to hooks), improved hook design and different types of bait, were incorporated. These gear changes dramatically improved the CPUE of many species and particularly of the sword fish. The operators found that the effectiveness of these relatively new configurations of sub-surface longlines, introduced in 1997, meant that profitable catches can be maintained using fewer hooks set over shorter main lines. Such reductions in gear have facilitated the adoption of long lining system by small artisanal vessels.

Analysis of operation of 10 large longliners including several leased foreign vessels, operated off the coast of North-eastern Brazil had shown a total catch with unit effort which remained fairly stable at an average of about 2.4 (100 hooks). The longline catches by smaller boats are also seen to be close to this figure. Encouraged by the development, the local Governments of North East Brazil provided a fillip to the installation and operation of longlining system from the smaller boats, as an effective alternative to existing subsistence-based small scale fisheries.

The authors say that they have recently completed a preliminary study to assess the feasibility of transfer of the technology of operation of tuna monofil longlining system to small mechanised boats. This was done by comparing the relative catch rates of a converted artisanal vessel of "11m length using approximately 300 hooks set along a 20 km main line", against those from a large leased vessel of 24 m in length using 1200 hooks set along a 40 km longline".

Both the longliners (24 m and 11 m OAL) fished for a similar number of days during the same period (9 months between 1997 and 1998) but in different areas. The operation of the artisanal vessels was limited to 100 km from shore while the leased vessel fished upto 500 km from shore. The results had shown that the relative abundance of commercially important species within the operational range of these smaller vessels was more than sufficient for economically viable fishing. The net financial return per hook set was similar to the larger, leased vessel and the total income was almost 10 times greater than that derived from traditional fishing methods during the same period.

The authors have pointed out: "The work described above provided evidence to suggest smaller longliners can achieve commerically viable catch rates and we subsequently adapted several vessels (8 to 11m in length) to accommodate the equipment required for sub-surface long lining. While the catch rates of these vessels have been comparable to those mentioned above, there is a lack of any information about the effects of different gear configurations on their performance. Such information is required to maximise efficiency, minimise the catch and facilitate regulation of effort".

The authors also conducted elaborate experiments on the kind of bait to be used. Squid and mackerel were the baits experimented upon. They had found that hooks, baited with vertically-oriented mackerel, had significatly more contact by fish and less bait remaining at the end of each set. Although they retained significantly less total fish by weight and fewer sword fish, the results also indicated that "visual stimulus is an important factor influencing efficiency of sub-surface longlines. It may be feasible, therefore, to examine the utility of simple methods for enhancing bait appearance. This might include artificial baits used in conjunction with small portions of natural baits. The successful development of cheap baits would reduce many of the costs of artisanal longlines and promote their involvement in this fishery".

Another aspect pointed out by the authors was that the installation of tuna longlining system would result in the movement of boats away from fisheries close to the shore, thereby reducing fishing effort on stocks of coastal species.

Now that our 10th Plan Fisheries Schemes are in the final stage of preparation, it may be desirable for the authorities concerned to seriously consider inclusion of a scheme for equipping smaller coastal fishing craft with sub-surface tuna long lining equipment.

Fishing Chimes, Editorial 292: June 2001: Vol. 21, No. 3

# India's Sil/Fishing Vessel Import Scheme: Implications and Impact

The Government of India issued an Extraordinary Notification on imports in Gazette of India, Part II - Section 3 - Sub-section (ii). This embodied the addition of a new chapter, namely – "1A : General Notes regarding import policy" to form part of schedule I of ITC (HS) Classification of Export and Import items, 1997-2002. As per Annexure A to the notification, Item 10 of the Table in Annexure A permits import of trawlers and other fishing vessels against Special Import Licences (SIL).

The marine fisheries resources of Indian EEZ, particularly of the demersal layer of the coastal zone and onwards to mid-distance towards the shelf stand over-exploited. This has resulted over years in the annual catches plummeting inexorably. Despite this alarming situation that is causing concern, it is incomprehensible that the Government of India included a blanket provision for the import of fishing vessels in the notification aforesaid. There being no conditions governing the imports except surrender of SILs, several entrepreneurs availed of the facility and are understood to have imported 19 longliners and 13 trawlers of Taiwanese and Thai origin, after fulfilling formalities such as registration as Indian vessels. They are also understood to have obtained RBI's permission for paying the cost of vessels imported on deferred payment system.

The notification has to be construed as not only entrepreneur -friendly but also as nation-friendly but partly. It would have been totally nation-friendly too had the permissions for imports of the vessels are allowed selectively and made subject to such regulations as are imperative to ensure rational and sustainable utilisation of our marine fisheries resources. This selective approach has to be considered to be all the more necessary in respect of Taiwanese and Thai-built vessels. The need for this caution stems from the harmful experiences of the Nation, during the period of operations of Taiwanese and Thai vessels under charter terms etc., which led to the withdrawal of charter, lease and joint venture schemes.

MMD, Customs and Coastguard go by the contents of the said notification, for clearance of the imports. The absence of provision for clearance of the imports by the subject-matter department *i.e.*, Union Department of Animal Husbandry and Dairying beforehand, has made the imports wholly entrepreneur-friendly without any relation to the possible adverse impact on the current fishery resource position. After the damage has been done by fast-track action by some of the professionals who earlier held charter/lease permissions for Taiwanese vessels, (who could now smoothly import the vessels), the flaw was noticed and a provision is stated to have been added as an amendment to the notification that prior clearance of the subject-matter department is necessary before permitting imports. It is however a moot point whether Government could reverse the imports already made as these took place as per the notification and before the addition of the proviso.

Taiwan and Thailand, in the main, have traditional fishing interest in Indian EEZ. Taiwan is a global fishing nation with a very wide reach covering EEZs of a large no of countries, developed, developing and under-developed. So far as Thailand is concerned, it has more of regional interest covering Indian, Myanmar, Vietnamese and such other EEZs. The traditional interest of Thailand is far older than that of Taiwan. Taiwanese interest in the resources of Indian EEZ came to light in early 1970s. In comparison, Thai vessels are known to be fishing in Andaman waters and waters off West Bengal and Orissa even prior to independence.

Korea too is interested in fishing in Indian EEZ but this interest is believed to be not of the same magnitude as that of Taiwan and Thailand.

All these three countries indulge in lawful fishing, (by taking permissions) to the extent possible, in the Indian EEZ, in the same way as they probably do in the EEZs of other nations such as Pakistan. In doing so, they have a compulsive motivation, strengthened by the support and encouragement they receive from their respective Governments. Being nations with an inordinate appetite for fish and fish products, a characterestic which cannot easily be appreciated by an average Indian, their Governments have no option other than supporting their fishing companies to spread out into waters of other nations with what all they can do to ensure the entry of their vessels into alien waters, for the reason that fisheries of their own EEZs already stand over- exploited. For example, if deregistration of a vessel is necessary to enable her to get registered in another country, it is stated that, for tactical reasons, the Government concerned, may do so.

At a time when these countries are understood to be worried about the withdrawal of chartering/leasing schemes and the embargo placed on any further joint ventures by the Indian Government, the present Notification of Indian Government announcing SIL scheme involving provision for import of fishing vessels by Indian enterprises must have come as a boon and balm to Taiwanese and Thai interest and the Indian professionals in charter business in the past, to resume their activities again in a lawful way. In fact, the SIL notification provided an excellent opportunity to transform the banned chartering and similar systems into a regular 'Indian-owned' fishing activity, while at the same time preserving all the features of the charter system, including provision for permissions to their crew members for vessel operations. One difference between the past and the present with no disadvantage is: under charter system 20 per cent of gross earnings were going to the Indian side and the rest to the owners. Now as '0wners', the Indian side receives first the entire amount earned out of 'their exports', meeting expenses out of these returns as required by the 'foreign' owners on bunkering, wages to crew and under other heads while also ensuring that a minimum of 20 per cent of gross earnings remain as their earnings.

Quite a few issues which are related to the policy aspects, and the manner in which the old charter conditions are camaflouged deliberately, leading to exposure of national inadequacies and weaknesses to other nations, arise from this situation. In other words, what appears to have happened is that, Government banned chartering of foreign fishing vessels and related systems, only for bringing back virtually the same systems in another form. Under the SIL/Import system, the vessels imported are registered as Indian vessels, may be with proofs of deregistration of earlier registrations. Payments towards the cost of imported vessels is allowed ostensibly through deferred payments, but this is only another form of allowing the 'sellers' (who obviously continue to have real ownership) with their crew to fish and to take the catches stored in the vessels as an 'export', but in actual terms taking back their own catches, thanks to the access secured by having their vessels technically imported by Indian enterprises. This inverted system too no doubt yields foreign exchange but it also displays a transparent deception that is quite good for record as India's own operations although they are actually otherwise. This SIL/import policy is, in actual fact, a continuation of the earlier charter and other systems by way of another *avataar*. Further, joint venture system involving Taiwanese/Thai/Chinese vessels permitted earlier, is now confined to a few for the reason the joint ventures could not be cancelled, because of the provisions governing them: The withdrawal of the scheme later put a stop to such joint ventures anew as a further development. The present import system has removed all inconveniences such as frequent checks and clearances involved in the earlier charter/lease/joint venture systems. The camaflouging of the continuation of the same policy of charter/leases in the guise of imports whether meant or not, in an improved manner, is thus apparent. Looked at from any angle this is a national shame. Old wine is accomodated in a new bottle.

It can no doubt be conceded that imports of fishing vessels are necessary. However, the imports can only be allowed on case to case basis, keeping in view the resources position, need for upgrading capabilities of small enterprises to exploit unutilised pelagic resources such as tuna. One wonders: whatever has happened to the Union Fisheries Division to remain quiescent when such a disastrous formulation was being made. The post-notification stipulation providing for imports of vessels only after the needed clearance is given by the Fisheries Division of Union Department of Animal Husbandry and Dairying can be of use, if any, for applications received from now onwards only, as the damage has already been done. Very few countries outside the 'elite' nations of 'Professional Calibre' would be interested in accepting exports based on deferred payments without guarantees. As imports of a sizeable number of vessels have already been permitted by the Government, if the Indian enterprises concerned happen to go to a court of law to contest the seizure of some of these vessels reported to have been made by the coastguard not only the notification of the Government itself could come in the way, making the imports ineffective, but possibly there can be other justifiable reasons for the seizures and it is difficult to visualise what the Government has up its sleeve.

As already indicated, there appears to be no influence whatsoever of the present status of our fishery resources on the introduction of SIL/Fishing vessel import scheme. Further, the motivation for extending the SIL scheme to cover import of fishing vessels too is not clear to the industry. So much so, the Association of Indian Fishery Industries has expressed itself against the import provision in the present form, which was apparently made without consulting it, or for that matter, may be even without consulting the Union Department of Animal Husbandry and Dairying. If this scheme is allowed to be continued, as it is, the Nation would be owning its EEZ, from fisheries angle, mainly for the benefit of foreign fishing vessels of the 'elite' group whose aim can only be to keep the main fisheries of our EEZ within their grip and confine the Indian Industry to its present pathetic status.

Having said this, we have to concede that there is no escape from dependence on foreign expertise for the introduction of vessels, particularly longliners for fishing for tuna, whose stocks in the EEZ remain mostly unexploited (excluding the stocks taken away by poachers), and of trawlers for deep sea fishing in the oceanic depths. It has also to be conceded that the provision for imports in the notification is justified. What is not justified is the blanket nature of the provision that is totally oblivious of the resource dynamics of Indian EEZ, non-availability of trained men, needed infrastructure to the fullest extent etc.

The fishing industry and the Government are the best judges of the ongoing joint ventures with the Taiwanese (organised through Singapore or Hongkong companies, for the reason that Taiwan has no diplomatic relations with us) and Chinese and Thais. The role of the companies of these countries in the joint ventures is one-sided and dominating. For them, the joint ventures are just a mechanism to gain access to our EEZ. Accordingly, the operations are not designed either for transfer of technology or for eventual real transfer of the vessels registered ostensibly under Indian flag. The Indian companies can be totally at the mercy of their foreign partners, who can just take back their vessels, either to come back at their convenience or not to come back at all. It looks that for them the main role played by the Indian counterpart is to secure needed clearances or facilitating needed clearances in India. This situation is not honourable. However, there is one merit and that is the activity yields some foreign exchange without much of investment and without any effort at fishing management. The vessels can never be amenable for operations by Indian companies by themselves and with Indian crew. Efforts were made by Indian enterprises in the past to operate one or two seized Taiwanese vessels but these were unsuccessful.

The system of payment of the cost of the vessels to the 'foreign owners' in instalments deserves support, if the vessels could later be operated by Indian crew alone. At the time of permitting import itself this has to be established and the authorities should be convinced of this. If this is considered not possible, there can be no point in permitting such imports, as the vessels (Taiwanese, Chinese, Thai, Korean) remain as foreign vessels for all practical purposes, although labelled as Indian. Taiwanese vessels that operated in Indian EEZ under charter/lease terms are known to shift their fishing venue to other EEZs when fishing in Indian EEZ is dull for them. The days covered under such fishing activity used to go unaccounted for and used to form part of travel days and stay at the place where catches were sold. This being the pattern, the same can be expected to be applied to the 'Indian owned' vessels of the 'elite group' (known to have registration in another country besides Taiwanese registry, prior to registration in India).

In the case of exports of catches of these vessels, the Indian side has only one option, *i.e.*, selling the catches (which are actually harvested by the foreign owner from whom the Indian owner could secure technical ownership) to the same 'foreign owner'. The invoice preparation would follow the same pattern as under charters etc., which is generally known, but in an inverted manner.

In a long term perspective, it is desirable not to allow import of fishing vessels under deferred payment terms from countries of the 'elite group' such as Taiwan and Thailand who have traditional fishing interest in Indian EEZ, unless it is established beyond doubt that the imports from these countries are not akin to charter/lease systems in a different form.

Efforts at marine fisheries development through imports of vessels without linkages with associated matters such as promotion of fishing capabilities, upgradation of shore infrastructure etc., and without linkages and integration with related activities would not yield tangible and purposeful results. It is therefore desirable to link the provision to permit import of fishing vessels to integrated fishery projects to be taken up under bilateral agreements with developed friendly countries whom the Government can trust. In the Indo-Pacific region we have two developed countries in the fisheries field. These are Australia and New Zealand. Mexico has expertise and it is stated that Indonesia has taken to longlining in a big way. Government can negotiate, develop and sign a bilateral fisheries agreement with one of them. The main aspects of the Agreement can revolve round the promotion of integrated longlining in Indian EEZ and probably elsewhere too for mutual benefit, with identified points of responsibilities applicable to each of the sides, one of which will be according permissions to individual companies of the respective countries to enter into joint ventures and also on other aspects that such joint ventures could cover. Provision for providing financial assistance by the foreign country concerned to facilitate imports of vessels and other equipments by Indian enterprises, guaranteed by Indian Government, transfer of technologies including those related to post-harvest operations, and training and education aspects etc., could be part of the suggested bilateral agreement.

Camouflaged imports that would have all ingredients of erstwhile charter/lease system would no doubt be entrepreneur-friendly and would also bring in foreign exchange. Such imports would not however be wholly nation-friendly in that the operations will be totally controlled by the foreign fishing companies with the role of the nation getting virtually reduced to providing access to the vessels to fish in Indian waters, and with the Indian entrepreneurs playing the role of dummy owners and receiving some returns as foreign exchange at their mercy. This would certainly be a national shame and would reflect our inability to develop on our own pelagic resources and deep sea fishing and leave an interpretation of playing into the hands of global players in the fishing game. The nation has to avoid this shame and this is possible only through taking up projects under bilateral agreements with friendly countries as suggested in the preceding paragraph.

Fishing Chimes, Editorial 293: July(i) 2001: Vol. 21, No. 4

# 10th July 2001 Marks the Birth of Special Fish Farmer's Day

Major carps were successfully induced to breed for the first time at Angul in Orissa on 10th July 1957 through administration of pituitary harmone, by Prof (Dr.) Hiralal Chaudhuri. This pioneering work, over years, has led to aquaplosion in the country principally through quality major carp seed production and supplies to fish farmers for farming, from hundreds of hatcheries (based on induced breeding) that sprang up in the country, in the wake of the technology developed by the Professor.

In order to commemorate this day on which such a momentous breakthrough that dramatically transformed the fish culture sector in India, the Government of India recently declared 10 July as Special Fish Farmers' Day. The first Special Fish Farmers' day after the declaration was celebrated at various centres all over the country by fish farmers and fishers on 10 July, 2001.

The declaration is directly related to a grateful recognition by the Government, of the unique and solid impact of the breakthrough, which has altered in near totality the predominantly capture-oriented inland fishery of India into a culture-based one. The event elevated fish culture, over years, to the status of a recognised industrial activity which has since been playing a growing role in strengthening rural economy and in the national aqua product export endeavour. It highlights the signal importance of aquafarmers in stepping up aquaproduction.

It was left to Ms. Nita Chowdhury, a Senior Officer of the Indian Administrative Service (one of the few officers of the Service, known to have a dedicatory slant towards fisheries development) and also the Union Joint Secretary (Fisheries), to take cognisance of the unique significance of 10 July in the history of culture fishery development of India. The momentous import of the event dawned on her at the time when she conferred the first Hiralal Chaudhuri Fish Farmer Gold Medal, instituted by Jayashree Charitable Trust, at a well attended function held in December 2000 at CICFRI, Barrackpore. She apparently perceived the catalytic role the celebration of 10th July as Fish Farmers' Day can play in infusing and fortifying the determination of fish farmers to achieve sustainable higher levels of culture fish production and in augmenting the incomes and prosperity of fish farmers, backed by the opportunities the annual celebration would provide for them to reflect upon their past performance, and to evolve measures for further improvements in their farming profession. So much so, she initiated the landmark proposal, which, by virtue of its importance, could earn the well-deserved approval of the Prime Minister to the proposal. This approval has empowered farmers and all others concerned to observe 10th July every year as Special Fish Farmers 'Day.

The lead celebration of the Day took place at the Central Institute of Fisheries Education, Mumbai on 10 July, 2001. Mr. Nitish Kumar, Union Minister for Agriculture inaugurated the celebration, in which Mr. Ananda Rao Devakote and Mrs. Meenakshitai Patel, Cabinet Minister and State Minister respectively of the Government of Maharashtra, Ms. Nita Chowdhury, Union Joint Secretary (Fisheries), Dr. K. Gopakumar, Deputy Director General (Fisheries), ICAR, Dr. S. Ayyappan, Director and Vice-chancellor, Central Institute of Fisheries Education, Mumbai among others, participated. They spoke in laudatory terms about the significant contributions made by fish farmers in raising aqua production of the country and thereby adding substantially to aqua exports. A large number of farmers and fishers participated in the function, which was celebrated with gaity, fervour and enthusiasm. The second annual Hiralal Chaudhuri Best Fish Farmer Award of Jayashree Charitable Trust was conferred on this occasion on Mr. Ch. Srikanth, a progressive aqua-farmer specialising and contributing with distinction to scampi seed production and scampi culture in Nellore district of Andhra Pradesh, by Mr. Nitish Kumar, Union Minister of Agriculture and Railways.

The Special Fish Farmers' Day was celebrated on 10th July in almost all the States. In Kerala this was accompanied by a Workshop on 'Fish Farming Problems and Solutions', inaugurated by the Union Fisheries Minister, Prof. K. V. Thomas. Mr. Thomas said the Government was planning to hold a discussion on the proposed changes to be brought about in aquaculture and fish marketing sector, involving all agencies in this field, in the presence of the Chief Minister, Mr. A.K. Antony on August 27 in Kerala. He said that there were similar discussions in on West Bengal, Bihar and other States too. There were similar celebrations, with participation by the fisheries ministers concerned, farmers and officials in all the States.

Fishing Chimes, Editorial 294: July(ii) 2001: Vol. 21, No. 4

# Tenth Plan Fisheries Schemes Refreshingly Forward-looking

The successive fisheries plans were all along dominated by certain repetitive schemes, despite the fact that quite a few of them such as those related to established seed production and culture activities, mechanisation of fishing vessels, statistics etc. Considering that the need for renewing such schemes, plan after plan, atleast in the later five year plans, has been tapering out, a new orientation seems to have been given to the tenth plan fisheries schemes.

In this background, it is refreshing to know that the Fisheries Division of the Union Department of Animal Husbandry and Dairying ushered in certain path-breaking schemes, refreshingly different and aimed at achieving a new set of objectives in tune with the emerging priorities and needs of the present and the future.

One major line of action envisaged in the 10th plan is the induction of the vast inland culture fishery sector of the country into aqua-product export basket, while also simultaneously envisaging the establishment of a domestic fish marketing network. This single initiative alone is capable of adequately transferring the fisheries developmental sector into a throbbing activity in the inland zones that can result in a higher level of fish production, higher level of exports, employment etc. When culture of species such as giant freshwater prawn and tiger shrimp is propelled into the numerous tanks and ponds and reservoirs located in the guts of the country, and into the vast saline lands now lying waste in the North- Western part of the country with the needed infrastructural development, a major change in the socio-economic life of the weaker sections of the society would manifest itself in the zones. The export endeavour will add to the national foreign exchange earnings in a substantial manner. The creation of the domestic fish marketing network that would spread far and wide will take fish within the reach of the common man. At the same time, this network would enhance opportunities to exporters to bargain for higher rates from importers, and the fish farmers would also be able to realise higher incomes, because of export demand.

Another major plank envisaged is the mounting of a scheme for the exploitation of high sea pelagic fisheries resources such as Tuna and also the deep sea resources such as lobster in the Indian EEZ around the mainland and around Andamans. The introduction of this scheme fills a demand in that direction by the fisheries industries. The implementation of such a scheme would counteract the present overtures of Taiwan, a distant water fishing nation to exploit to their advantage the fishery resources of Indian EEZ unhindered. The presence of the Indian Tuna and other fishing fleets introduced under the scheme will result in a near total discouragement to foreign vessels to fish in our waters. At the same time, the nation would stand to earn considerable foreign exchange through exports.

A noticeable flaw in the past planning has been the lack of focus on mariculture including sea cage culture. This lapse can be expected to be overcome by implementing well drawn schemes proposed under the plan, aimed at promoting mariculture and sea cage culture. The development of this activity would be a boon to the coastal fishers for augmenting their incomes and to upgrade their standards of living, and would also promote export-oriented production.

The development of reservoir fisheries has all along been undertaken in a perfunctory manner. A cardinal proposition in the 10th Plan is to take up in a systematic way the development of capture-culture balance in the various reservoirs of the country in an integrated manner and apparently with a cluster approach to provide the needed infrastructure facilities for production and supply of seed for stocking, for harvesting the produce, for its transport to and storage at centralised units and channelising the same to feed the proposed domestic marketing network and for exports.

The introduction of suitable organisational mechanisms for the development of all major types of resources such as reservoirs, coastal fisheries etc., on the lines of FFDAs and BFDAs is envisaged under the 10th plan. This reform will enable the various projects to gain momentum and run on a sound-footing.

In the past two decades, we have been observing a major depletion of fisheries resources of rivers and of coastal sea waters. Not much attention has been paid all along to restore the depleting fisheries wealth of these resources. In the 10th plan, it is heartening to note that a programme for ranching of rivers and coastal waters is to be enshrined. This programme would eventually restore the national fisheries heritage. Though late, it is gratifying that such a thoughtful and useful scheme has been included.

The upgradation of professional skills of traditional fishermen is much talked about, but the talk is not matched by specific schemes or projects aimed at achieving this upgradation in a purposeful manner. The 10th Plan is expected to launch organised programmes for the benefit of traditional fishermen in various fishery activities, linked to the generation/creation of new employment opportunities.

The implementation of the various plans envisaged will be possible only when there is financial support. Accordingly, it is envisaged in the plan to introduce a conducive and structured financing policy for the development of the various projects in the plan, taking into account the unique characteristics of the fishery sector.

The Union Fisheries Division and the Planning Commission deserve to be congratulated for identifying the future lines of fishery development so thoughtfully and purposefully.

Fishing Chimes, Editorial 295: July(iii) 2001: Vol. 21, No. 4

# Tiger Shrimp Hatcheries Need Supplies of Disease-Free Brooders

India has the problem of raising desease-free shrimp post-larvae for the reason that the brooders they have access to are from the wild, and they are often infected with white spot virus. Of late, considering the problems of producing tiger shrimp nauplii at the hatcheries has come in as boon. Almost all the hatcheries now obtain nauplii stocks from separate nauplii production centres that sprang up at various points. These units supply nauplii to hatcheries, where they are reared into post-larvae.

Efforts at producing 100 per cent WSSV-negative post-larvae at hatcheries continue with mixed results. Some of the hatcheries have set up their own PCR labs. Several PCR laboratories central to shrimp culture ponds, have also been set up by MPEDA. Despite this innovation, however, the occurence of disease in nauplii and upto post-larvae stage, and in culture ponds continues and the disease could not be controlled to a satisfactory level as yet.

One solution identified to eliminate this menace, is to stop using domesticated brood shrimp stocks. In several countries such as New Caledonia, Hawai and a few other States of U.S.A., Australia etc., technology for the production of domesticated shrimp stocks has been developed. With the availability or raised brooder stocks it has now become a common practice in these countries for broodstock producers to supply domesticated virus-free tiger shrimp brooders to hatchery owners. One advantage enjoyed by these countries is that tiger shrimp is not indigenous to them. The initial stocks were imported by them and the shrimp has established itself in these countries. One precaution they take is to ensure that the tiger shrimps do not escape into wild waters. Emulating the pioneering work done in these countries, in India too the Central Marine Fisheries Research Institute has started a project for production of domesticated shrimp brooders. It is learnt that the Institute has succeeded in raising broodstock upto the third generation and work is on for raising the fourth generation. It is stated that the brooders have to be developed atleast upto the 7th generation for using them for hatchery seed production.

The present status, as mentioned above, gives an indication that it may take at Ieast another 2 or 3 years for CMFRI to supply domesticated brood tiger shrimp stocks to the hatcheries. This means that, during this period of anxiety, there is the phase of economic losses, particularly because of the falling international prices of exported tiger shrimp. Farmers are not in a position to buy PLs of tiger shrimp from hatcheries at a price not more than 0.12p per piece, because their earnings have come down. Against this, the PL production costs at the hatcheries per piece are stated to be around 0.25p. This kind of financial disadvantage is likely to continue until such time as the international shrimp export prices move up, but this is not expected in the foreseeable future.

The sale of domesticated brood shrimp in U.S.A and other countries, where domesticated brood shrimps are produced, has become a routine feature. Advertisements captioned 'Domesticated Shrimp Broodstock available for sale' are noticed at least in U.S.A. The advertisements also say that the broodstock supplied is disease-free and is used in commercial hatcheries all over the world.

It will take time for the Indian enterprises to reach such a stage. In the meanwhile it will be a very helpful and a purposeful act on the part of the Government to permit import of Tiger Shrimp brooders by the hatchery owners. For example, these could be imported from Australia where tiger shrimp brooders are produced on a regular basis and sold.

The problem of the Government can be that granting permission for such imports is not feasible as we have no quarantine system to check imported live fishes, shrimps etc. It will be good for the Government to quickly introduce this quarantine system at specified airports and permit import of brood tiger shrimp, through them. Only such of those who have the needed infrastructure for keeping the brooders in healthy live condition and for radiating supplies therefrom to various hatcheries can be considered for granting import permits.

According permission for such imports could be continued until such time as the CMFRI succeeds in raising brood shrimp stocks atleast upto 7th generation. The repercussions of not permitting the imports as suggested could be the continuance of disease regime among cultured shrimp and strengthening adverse reputation among importers of tiger shrimps. Permissions for imports of domesticated shrimp stocks are far more desirable than continuing with the disease menace which has serious implications.

Fishing Chimes, Editorial 296: July(iv) 2001: Vol. 21, No. 4

# Tuna Tangle

India could not as yet solve the problem of utilising the tuna resources of its EEZ. Consequently, known foreign tuna longlining interests in Indian EEZ are believed to continue to adopt their own methods of exploiting the resources. While the Government approved of a pilot scheme as a prelude to equip two privately owned shrimp trawlers with tuna longlining equipment by providing 50 per cent subsidy thereof and also conceding to take care of all expenses in respect of training of crew, no progress could be made so far. The reason for this is believed to be the overtures of traditional foreign interests to keep to themselves the monopoly over utilisation of tuna resources of Indian EEZ they seem to have. Their intention is believed by several as one aimed at helping the owners in various ways so that they can resume or continue to be in the same status as under the erstwhile Indian charter and joint venture schemes, keeping the operations and marketing under their thumb. An adverse point is that there is evidence to show that, in the past, Taiwanese operators deliberately avoided training of Indian personnel under charter and joint venture schemes.

The fact that India is unable to utilise its tuna resources (with the exception of skipjack around Lakshadweep) despite various efforts over years leads one to perceive that there can be some invisible external force, to match Indian's own inaction, which is preventing the utilisation of the resources by Indian enterprises. This theory may be construed as based on a wild guess, but the dramatic exit of three Indian companies that operated five longliners from the activity, while saying all along that they were landing good catches; and the recent import of longliners by Indian enterprises involving inverted charter operations, cast a glimmer of doubt that, may be there is some strategy behind. These observations may be in the nature of a flight of imagination, but would certainly deserve examination. Foreign interests cannot be blamed as what they do is to take care of themselves and it is for India to counter the overtures. The availability of tuna resources in the Indian EEZ is proven. Taiwanese proved this for India. Their known interest to secure a footing in the Indian EEZ to enable them to continue the operations reconfirms this. FSI and CMFRI have been telling time and again about the tuna wealth Indian EEZ holds. Probably India is to be blamed for not utilising the resources, despite availability. India also appears to others as not effectively taking care of its rights.

India plays a prominent role in the Indian Ocean Tuna Commission. This role can be diversified for the development of integrated tuna fishing operations in the Indian EEZ on a sound and enduring basis, in collaboration with a reliable foreign tuna fishing nation that is friendly towards India and having expertise in tuna longlining. On hindsight, it looks as if the bringing in of the pilot scheme as referred to above has dislocated the line of action aimed at conversion of existing trawlers to undertake tuna longlining too and made the industry to lose time as is now continuing to happen. The scuttling of the pilot scheme has given time for other interests to ensure that India remains where it was in the tuna sector.

Indian tuna stocks, as in other EEZs are exposed to predation by whales. It is estimated that fishes etc., 4 to 5 times of global fish production, are consumed by whales. They eat 1-4 per cent of their body weight per day. In other words, compared to global fish production of 90 million t, whales consume 360 to 450 million t, of which tunas are a good part. Because of India not catching tuna of its own EEZ, two things are happening; one is that foreign interests exploit the resource and Indian enterprises have practically no chance of catching them, as at present. The other is that whales get them.

In the light of the foregoing appraisal, it is imperative that the Government of India tackles the situation with all the attention it deserves and not allow the situation to drift any further, as it would be advantageous to others, chiefly Taiwanese interests. The need now is not for a pilot scheme but for launching a major integrated project to exploit and utilise tuna resources of Indian EEZ for national benefit. The scheme may have to be drawn up in such a way that we do not get trapped into undesirable situations once again. There are countries such as Australia, New Zealand, USA and several others, whose expertise in integrated monofil tuna longlining operations can be tapped in a manner that the country acquires all round expertise without being dominated but conceding such benefits as justified to the collaborating country. One aspect is certain. We are not in a position to build up expertise in tuna sector on our own. We have to depend on others and it will be in national interest to work towards achieving this, pledging some of our interests for the best possible use until we acquire the needed expertise, as was done in 1950s in respect of shrimp trawling.

Fishing Chimes, Editorial 297: August 2001: Vol. 21, No. 5

# Shrimp Famine Stalks Upper East Coast

A paradoxical combination of two factors poised against the economics of operation of shrimp boats are presently crippling the shrimp sector of the upper east coast. One of these is the unprecedented drop in capture shrimp landings at the various centres along the coast this season. The other factor that aggravated the situation was the slide in global shrimp prices, particularly in Japan and USA. The net effect of these two factors is that shrimping vessels, particularly those operating from harbours of A. P. are forced to confine themselves to harbours and other anchorages in the zone, with all consequential repercussions such as mounting costs, job insecurity etc. This development has brought all associated capture shrimp fishery activities to a grinding halt.

The vessels that operate from various centres of upper east coast, as elsewhere along the coastline, are dependent on the shrimp component of their catches for economic viability. This component has come down to such a low level that they are unable to recover even the cost of diesel oil. So much so, all the fishing ports and landing points are jam-packed with anchored boats.

In States like Maharashtra and Karnataka, it is learnt that there is exemption from payment of sales tax and also excise duty. In the States of the east coast no such exemption is given. The granting of similar exemption in the coastal States of the east coast too would provide some relief to the boat operators. Adding to woes of the operators of the industry is the abnormal increase in berthing charges at Visakhapatnam harbour from Rs.230/- to Rs.2069/- per boat per month. The long standing demand of the Mechanised boat and other Associations to prevent vessels of foreign origin from operating in sand heads and other areas of upper east coast and to subsidise diesel oil prices, in view of the mounting operational costs, went unheaded by the Government. Another grievance is that wild shrimp seed is exploited unchecked by several for supply to farmers. Boat operators feel that this has a depleting effect on shrimp fishery. For this reason, they want this activity to be banned. Under the weight of these problems, which remain unheeded by the authorities, the members of the A.P. mechanised boat operators association have taken to staging an indefinite dharna at Visakhapatnam to draw the attention of the Government for the redressel of their grievances.

While boats could stop fishing to cut down further losses, the culture fishery sector had no such easy option. Postponement of harvesting, hoping for an improvement in shrimp price structure, ignoring the readiness of the crop for harvesting, would give rise to added expenditure on feed and other items without commensurating benefits. The growth of shrimps, if so postponed, will be just marginal and not adequate to offset the additional costs. Consequently, the farmers had no alternative other than harvesting. Whereas in the last season they could secure a return of around Rs. 500 per kg of shrimp harvested, now the exporters could offer only less than half of this, because of the declining global market for shrimp. This is so because, unfortunately, exporters are not in a position to hold on to the stocks for long, for want of adequate storage capacity at shore plants for the purpose.

In general, marine fish do not command a lucrative price in the States of West Bengal, Orissa and in Andhra Pradesh. For this reason, the quantum of shrimp catches in the landings of the boats alone influence the economics of operations. At the best, returns from sale of fish catches form marginal earnings to the operators and these are not of much consequence to them. For this reason, the per unit earnings of boats in these States have tumbled down to a hopeless level, forcing most of the owners to close down operations. The accumulated dues on cost of diesel oil and port dues are the burden that the operators have to discharge now. In contrast, in the State of Tamil Nadu, prime fish component of the catches fetch an attractive price in Chennai city and this appears to make a crucial difference between the level of earnings of boats that operate in the upper and lower east coasts. So much so, despite poor shrimp catches, the boats of Tamil nadu and southern A.P. coastal zone are believed to have been able to reach break-even level.

Various contributory causes are attributed to the present depressing situation. One is that Thai vessels are stated to be dominating the grounds of the West Bengal and Orissa coasts which the Indian boats are unable to counter. It is alleged by Vessel Owners Associations that coast guard does not effectively intervene despite complaints and they could not do anything by themselves as the foreign vessels are stated to be well equipped with grenades and arms to protect themselves. Ironically, these vessels fly Indian flag as they are technically Indian owned but are stated to be totally controlled by Thai owners and Thai crew and operated with Paradeep as the base. Another aspect mentioned is that the number of bottom-set gill net units that had been put into operation off West Bengal coast has led to poor trawl net operational results. It is difficult to say how far this can be true. In any case, there is a general consensus that the closed season as experimented upon recently in all the four coastal States on the east coast is no solution for conserving annual crops like those of shrimps. A further view expressed is that, for complex biological reasons in a three or five year cycle shrimp fishery suffers a setback. This had happened before, although not so severe as in this season. It is felt that the situation became so bad not so much for other reasons but because of operations of imported vessels with Thai crew, which are stated to be conducted in relays by batches of vessels round the year. The allegation is that different sets of vessels but with same marking and documents manage to be in the grounds on a relay system, thereby operating relentlessly and without break, with no let or hindrance.

Export prices crashed from US $18/kg for 16/20 count in June last year for head-on black tiger to less than US $10 per kg by July end this year, according to the President of the Association of Indian Fishery Industries. This is because of flagging consumption of shrimps in Japan and a trend towards lower prices in U.S.A, it is stated. However, the latest position is that there has been a marginal increase in export prices now.

Charles Woodhouse, Principal in the Philadelphia Law firm of Woodhouse Shanahan Hart PC and who is on the Board of Directors of Choice Canning Co, Inc, one of India's largest shrimp exporters, in his regular column entitled' "Farm Shrimp News" in Fish Farming International, wrote as follows in July 2001 issue of the tabloid, under the caption "Market bottoms at prices not seen in five years" :

"Long-time readers will remember my oft-quoted maxim from The Old Man of the Sea – 'You'll run out of money before they run out of shrimp'. This year should teach anyone in the trade that there are plenty of shrimp in the world.

Yes, there is all this silly talk about diseases and viruses - but the truth is that the world's shrimp farms had a record year in 2000 and there are still lots of shrimp left to eat. Enough shrimp to destroy traders who began the year with major positions.

The bell weather Ecuador 41/50 shell-on tail now sells for $3.50 a pound in New York. Those of us who did not commit suicide in 1995 after the market dropped from $5.00 to $3.50 have now seen the market drop from $6.00 to $3.50.

There is also some strange stuff going on in the market. South American 26/30 tails are defying gravity at $7.40, while I can buy a perfectly sound Thailand 26/30 tail for $5.00-difficult to explain!

This year promises to be another record one for US shrimp imports. Through March, they were 18 per cent above comparable 2000 levels at 155 million pounds (70,308 tonnes).

"As usual, Thailand is the number one supplier to the US, with India now in second place instead of Ecuador. India has also moved into number two position as a supplier to Japan, replacing, Indonesia - with Thailand, of course, top as always".

"In Japan, unlike the US, major importers have been cautiously looking for the bottom before making fall (autumn) season commitments with suppliers. Japanese imports were only 66,000 tonnes through April, which is even with the comparable 2000 level, despite much lower prices".

## Future Contracts

An interesting story comes out of Japan, with the KANEX (Kansai Commodities Exchange) considering and rejecting the possibility of introducing a Frozen Shrimp Futures Contract.

As many of our readers will already know, a commodities Futures Contract is a financial 'derivative' instrument which serves as a proxy for a physical transaction which is called a Forward Contract.

In the 'real world' commodities, producers and users enter into Forward Contracts for their physical sale and purchase of actual products. They, as well as speculators, then 'hedge' physical positions (or gamble) by trading Futures Contracts.

These contracts are finally settled at maturity without physical delivery. Yes, you could theoretically take delivery, but this seldom happens. Anyone foolish enough to do this with shrimp would get a nasty surprise when 'generic' shrimp was delivered.

Apparently the KANEX executives and traders studied shrimp and determined that, as we all know, shrimp are not a fungible commodity.

For shrimp, value is determined by species, freshness, processing, grading, brand identity, and it takes a lifetime of experience to understand why one case of shrimp can be sold any time at a profit and another will sit in the warehouse until you have to dump it.

The Minneapolis Grain Exchange (MGEX) still continues its listing of White Shrimp contracts, but it has never been able to develop significant volume. A trip to the MGEX website reveals that "there is no open interest in any contract months" for either listing.

Always remember, the Old Man of the Sea says: 'An expert in the shrimp business is someone with more than 20 years' or less than one year's experience. 'It seems that, before repeating this mistake, the KANEX executives were smart enough to find one of the old experts rather than a young one.

The present alarming shrimp fishing situation along upper east coast has rendered capture shrimp business uneconomical to the exporters too, because of relentlessly falling export prices, and with pressures from producers on the exporters to buy their produce, however low it is, at higher prices. This pressure is compounded with the availability of cultured tiger shrimp in abundance with farmers, who have no option other than selling their produce to the exporters.

It is believed that export prices would gradually pick up from now on and this throws a ray of hope that the situation may improve. With cultured shrimp harvesting on the east coast having almost come to a close, the exporters, if they have to buy, have to depend on capture shrimp only from now onwards, it is felt. The outlook, however, is vague.

Two long term solutions to guard against such situations are put forth. One is to take immediate steps though an appropriate project to equip the boats for diversified fishing in farther waters. The other is to introduce a system of declaring fish famines officially, as they do occur now and then, to enable the Government concerned to assist fishermen to tide over such famines. A permanent Fish Famine Fund can be set up with a corpus provided by the Government and fishing vessel owners making contributions, as stipulated, to the Fund on a continuing basis during fishing seasons. The level of contributions by fishing units, and the procedure for collections, have to be determined by the Government and a recognised Association can be entrusted with the job. Government may make matching contributions on an annual basis based on audited accounts to enable the Association concerned to distribute these funds among the registered beneficieries to tide over financial crises as per the guidelines laid down, as and when fish famines are declared. In Maharashtra and probably in one or two other States a system of declaring fish famines is understood to be prevalent. Keeping all aspects in view, the Government of India can recommend to the coastal Governments to introduce the system of declaring fish famines. In this context the States can be advised on the criteria to be kept in view for declaring a fish famine. In the same way as the fishermen's accident insurance schemes, the Government of India could also consider assisting coastal State Governments with a measure of matching contributions. The other ancillary lines of relief to cover day-to-day economic problems of fishermen are : 1) Confine operations of recently imported fishing vessels to the zone beyond territorial waters and that too only with all-Indian crew and not permitting any foreign crew to work on the vessels, 2) Organising training programmes to Indian crew for working on imported liners, 3) To stop import of trawlers, 4) Bringing berthing charges at Visakhapatnam harbour at par with other harbours, and 5) Regarding the demand for ban on wild shrimp seed exploitation, it is stated that the ban is already there and steps are understood to have been taken for implementation but the problem is one of effective enforcement. Here it has to be mentioned that the operators, who want wild shrimp seed capture banned may have to take special interest in catching gravids for sale to hatcheries. Unwilling to give up this remunerative practice probably for the reason that they will be exploited in any case for export, they seek a ban on wild seed capture. These observed stands are contradictory to a large extent. They observe a closed season to protect stocks and the same principle should be applicable to avoid catching of gravids too. It appears that what the operators should ask the Government is to undertake development of strains of domesticated tiger shrimp brooders for supply to hatcheries and for a stipulation that hatcheries should release one round of post larvae produced by them into the sea, until such time as domesticated brooders are developed for supply to hatcheries.

Fishing Chimes, Editorial 298: September 2001: Vol. 21, No. 6

# ...on Generation of National Fisheries Policies

The shape of emerging needs and the positive compulsions that they generate generally exercise a profound influence on the Government concerned in the formulation of policies in respect of developing sectors such as fisheries. Traditionally, the primary function of any democratically elected Government is to take care of the welfare of the people on one hand and maintain law and order on the other. In respect of discharging these responsibilities Governments have the support of a well established mechanism that provides information on conditions of existence for instituting such measures as are necessary to discharge the related responsibilities.

Fisheries developmental activities in India are relatively less traditional in nature, with just around 50 years of accumulated experience in the country. While the various Governments have departments to promote and monitor these developmental activities, the source that generally provides motivation for initiating a developmental work is the people. Politicians are of course the chiefs that run Governments and in this process they reach out to be in touch with the pulse of the people, engaged in various commercially-oriented developmental sectors for giving shape to policies thereof. They know that this approach is basic for the purpose. Those engaged in the various commercially oriented developmental sectors are aware that their prosperity depends on the understanding of their requirements by the Government concerned and the policies adopted by them. In this context, those engaged in the sector form into Associations/Federations to represent their needs to the Governments concerned. There is thus a mutual dependence, those in industry (commercially oriented development sector) telling the Government concerned through their Associations what they want, and industry telling about their problems to the Government concerned for formulating policies on that basis. Any disbalance in the equilibrium will result in an ill-fated developmental trend.

In the same way as in any developing industry, in the fisheries industries too, upgradation or expansion is a continuing process. Solution to one problem often leads to the emergence of another problem. In other words, the developmental process unleashes a string of situations that call for total alertness on the part of the Associations to project the problems faced and the policy measures needed to get over them for the attention of the Government concerned.

In this background, it is axiomatic that all fisheries associations must maintain a continuing and dedicated interface with the authorities concerned regarding the issues facing the industry and the policy options to be excercised. Such continuous contacts will promote a healthy relationship with a perspective between the Government and the Association concerned. Unfortunately most of the Fisheries Associations are deficient in adopting this approach. They ignore the need for timely representations to the Government concerned on vital aspects.

As mentioned already, it is the politician who has his finger on the pulse of those engaged in the fisheries industries (in the same way as in the case of other industries), and the ministers in charge of fisheries happen to be politicians. Once policy matters based on representations made by the fishery industries are brought to their notice, they respond far more incisively than others. They evaluate the ground situation and pronounce policy and strategy directives.

Thus, the process of policy or strategy formulation generally takes off from the level of Associations. The need is therefore for the various Associations to strengthen themselves in the right direction, to function confidently and effectively in their relationship with the senior functionaries and the ministers in the Government concerned. They must identify and nominate such of the members who are knowledgeable and talented as their emissaries and spokesmen and lend all needed support to them so as to enable them to articulate the points of view of the Association concerned before the authorities effectively. The general pattern to emerge has to be such that they should be having meetings with the authorities as frequently as possible and be a known sight in the corridors of Krishi Bhavan or in other power-centered Bhavans, carrying the message from the industries to the authorities concerned. Unless a culture of this kind, purposeful and devoted and imbued with an equation is developed, there will be bleak chances of emergence of policy and strategy decisions from the Government concerned as the related fishery industry needs. At present, there seems to be a ridiculous situation of some of the Associations not being quite sure of what kind of policies they want and what policies to be continued. For example, the Association of Indian Fishery Industries wants a deep sea fishing policy to be announced by the Government but one hopes that it would be able tell the Government what kind of policy it wants beyond reference to what Murari Committee recommended. The impulse is to suggest banning of import of Thai and Taiwanese vessels, on the premise that they continue to be really owned by enterprises of those countries with the total crew comprising their nationals and using the Indian flag as flag of convenience just for gaining entry into Indian EEZ. It is to be hoped that it will propose a well argued alternate policy option. In fact, the Association appears to need to develop a system of meeting the concerned in the Government and make representations on vital issues confronting the industry from time to time. True, it sends letters to Government on the needs of the industry from time to time but these would not have the impact of personal representation of policy needs comprehensively.

Governments cannot normally be expected to take the initiative and responsibility for launching important fisheries policies unilaterally without taking the industry into confidence. In fact, Governments long to receive representations from the industry/its associations from time to time oriented towards augmenting fish production and on various post-harvest aspects in a sustainable and eco-friendly manner through removal of hurdles in the way.

There are however some organisations that are active. The National Fish Workers' Forum and a few state-oriented Associations are seen to be active in demanding certain policy interventions by the Government fairly effectively. So is the case with the Seafood Exporters' Association of India. It has a network of State-wise centres that project their problems to enable the main association to take them up with the Centre for policy decisions. Thus, so far as seafood industry is concerned, Government gets the needed feedback from it and the industry receives policy support as justified, from the Government. In the case of culture and capture fishery sectors the position is however not however that bright. Marine capture fishery stocks, by and large, are now at a critical point of exploitation and the problem faced is largely one of regulation and conservation of the resources of coastal zone on one hand and exploitation of coastal pelagic and high sea pelagic resources and demersal resources of the deep sea of the Indian EEZ. In the present situation of stagnation in marine fish landings from Indian EEZ by Indian fleet, and lack of fishing fleet equipped to catch tuna, the main resource that is available for stepping up exploitation in the zone, the potential future developmental prospects of the Indian fishery sector are predominantly intertwined with the growth of the three-pronged culture fishery segment, consisting of freshwater, brackishwater and seawater lentic units.

The farmers do have their Associations mostly at district levels. The Aquaculture Foundation of India continues in its endeavours to have effective linkages at district and state levels but this situation has not enabled it to emerge as a strong force to counteract the strategies and activities of the environmental lobbies (believed by many to be misplaced) effectively for the benefit of the farmers. Several innovative technologies that would totally make aquaculture environmentally safe are now available but not much could be done to prevail upon the Government to evolve policies for the popularisation of these technologies (closed recirculation systems, zero-water exchange, pond lining, non-polluting feeding systems etc.). The lack of strong leadership among farmers has become a major handicap in that no effective representations related to policy matters reach the state Government concerned and the central Government. Shrimp hatchery owners have Associations, may be in a couple of States, and they have the good luck of having the positive support of MPEDA, mainly for the reason that, shrimp exports can go up only when the hatcheries function well and supply seed to farmers. Despite this advantage, the Hatchery Associations have not taken up issues such as development of domesticated brood shrimp stocks for production of PLs at the hatcheries. The hatchery owners seem to be reconciled to courting the wrath of the Government and others for being responsible for overfishing of wild broodstocks. If these Associations are strong they could have certainly taken up the issue strongly with the Government concerned and obtained helpful policy decisions in respect of domesticated broodstock development. As the present aquaculturists lobby is mostly district-oriented and weak, no policy formulations are taking place in respect of viable and sustainable activities such as sea cage culture and other types of mariculture which involve legal support in respect of leases of seawater areas etc. The recent initiatives of the Government on its own in respect of utilisation of waste saline lands for aquaculture, development of reservoir fisheries etc., should put the private enterprises of fishery sector to shame, as these initiatives should have come first from the farmers and entrepreneurs.

In sum, what is sought to be highlighted is that strong, well informed and widely representative National level fisheries apex bodies with supporting organisations at State and District levels should emerge to strengthen the hands of the Central and State Governments to take sound and development-oriented policy decisions for the effective and sustainable utilisation of fisheries of the country. As long as this does not happen, the state of fishery development of India will continue to vacillate between stagnation and slow growth. It also appears necessary for the Central Institute of Fisheries Education to conduct well designed courses periodically based on a well drawn syllabus, for the benefit of the nominated office-bearers/members of the Associations to enable them to develop a wider perspective of the subject and to educate them on policy aspects and as to the form in which to represent their problems to the Government concerned to enable them to evolve suitable policies and strategies for their implementation.

Fishing Chimes, Editorial 299: October 2001: Vol. 21, No. 7

# On Aggregated National Fisheries Administration

Mr. Nitish Kumar, the then Union Minister for Agriculture, while speaking at the Special Fish Farmers' Day function held at the Central Institute of Fisheries Education, Mumbai on 10th July, 2001, voiced the determination of his ministry to take the needed steps for upgradation of the Fisheries Division in the Department of Animal Husbandry and Dairying into a separate Fisheries Department. He had explained that this upgradation which was long overdue, was essential to bring about faster development of inland and marine fisheries resources of India for national benefit. In this context, he referred to the long standing demand from various fisheries organisations of the country for this reform. It was pointed out that the process of fisheries development was getting retarded due to lack of linkages in respect of decisions for promotion of the developmental effort among the concerned union departments that handle various aspects of fisheries.

Fisheries Research is the domain of ICAR, which functions under the control of Union Department of Agricultural Research and Education. The results of fisheries research, in general, fail to reach the field for application, for want of needed co-operation and co-ordination from other agencies concerned. Past experiences reveal that, at the time the central fisheries research institutes were under the Fisheries Division of Union Department of Agriculture, speedier transfer of technologies was taking place. Composite fish culture and carp hatchery technologies could be disseminated in the field without time lag. Compared to this, it is now seen that, as examples, freshwater and marine pearl culture and other technologies, developed by central research institutes after the separation of fisheries research establishments from the Union Department of Agriculture could not be extended effectively to the field. So is the case more or less with seabass culture. The work being done at Central Marine Fisheries Research Institute (CMFRI), and Fishery Survey of India are complementary to each other. With the former being under ICAR and later under Union Department of Animal Husbandry and Dairying, there is inadequate co-ordination which has the effect of delaying developmental process. The resource estimates made by CMFRI and the survey work of Fishery Survey of India do not move in unison, which was not the case when CMFRI was under Union Department of Agriculture. There are several instances of such lack of co-ordination, leading to developmental setbacks.

Government have recently expanded the list of items that can be imported with licenses issued by Director General of Foreign Trade who functions under Ministry of Commerce. As part of this, several licenses for imports of fishing vessels have been issued initially by the aforesaid Directorate under the Ministry of Commerce, without ascertaining the impact of issue of such permissions on marine fishery resources. Only when the Union Fisheries Division pointed this out, although after some damage to resources was done, the need for taking prior clearances from the Union Fisheries Division was realised. The Mercantile Marine Department under Ministry of Shipping and Transport is understood to have registered the imported vessels as Indian vessels. In doing so, it is observed by several that the non-conformity of some of the specifications of these imported vessels with those of India has been overlooked. Several other lacunae in the agreements between buyers and sellers and in operational aspects were commented upon. For example, Thai built trawlers, which ostensibly came to be owned by Indian enterprises, were seen to be recently fishing in coastal waters along upper east coast to the utter disadvantage of Indian fishermen. All these would not have taken place, had the process of issuing import licenses and registrations was initiated at the level of subject matter ministry.

One can see the relevance of observation of Mr. Nitish Kumar, the then Minister for Agriculture to aggregate all matters related to fisheries and place them under the control of a separate Union Department of Fisheries to be formed under the Ministry of Agriculture. Such a set-up would bestow on this new department greater onus and responsibility in handling all matters related to fisheries. For instance, when an enterprise wants to import a vessel the application would then first go to this Department, from where, with the needed clearance will go to Director General of Foreign Trade. Such a channelised system would provide the needed safety to the resources and protection to national interests. The same logic would apply for registration of fishing vessels by MMD, and on apprehensions of foreign fishing vessels by coastguard. Several exemptions, subsidies and so on presently given by various departments to the fisheries sector would also get duly co-ordinated, if proposals reach the Fisheries Department first for scrutiny and clearance.

The nation deserves to have a separate fisheries department. It has a coastline of over 8000 km, one million fishermen, a couple of lakhs of traditional boats, over 40,000 mechanised boats and 70 nos. of surviving larger trawlers out of a strength of 200 nos. a decade back. Had there been a separate fisheries department, the decline in fleet strength could have been arrested, and multipurpose vessels for exploitation of demersal and pelagic fisheries would have been introduced. Above all, Indian fisheries development would have been in harmony with those of other nations, atleast of smaller ones such as Srilanka and Maldives which have made great strides in tuna fishing. In culture sector too, while there was progress, India lags behind in various respects even in comparison with smaller nations in Central America. The main reason for this situation is the absence of a separate Union Fisheries Department.

The earlier view held by the Government that the work load in respect of fisheries is not adequate to warrant the formation of a separate department, to say the least, sounds hollow. One cannot expect workload to go up when determined and devoted steps are not taken to utilise the available resources effectively for national benefit. When the present divisional set up could result in an annual production of 5.4 million t per annum (and 7th position among world nations in fish production until recently, which has now dropped down to 8th position, for obvious reasons, earning 1.4 billion US$ in foreign exchange, one can imagine the great strides the sector can make, if only it is elevated to the status of a department. Here it has to be mentioned that, among all States of India, West Bengal has been consistently occupying top position in production, mainly because it has a separate department for fisheries at Government level.

In the above background, although Mr. Nitish Kumar is presently not the Agriculture Minister, his public official pronouncement needs to be brought to the attention of the present Minister for Agriculture, Mr. Ajit Singh by the Union Fisheries Division to enable him to followup the proposal made by his predecessor, to its logical end.

Fishing Chimes, Editorial 300: November 2001: Vol. 21, No. 8

# Trout Farming Expansion in Himachal

Pursuant to a proposal submitted by the Government of Himachal Pradesh, it is learnt that the Government of India has provided special financial assistance of Rs. one crore to that State for implementing a scheme on – "Promotion and Development of Coldwater Aquaculture in Himachal Pradesh", as a 100 per cent Centrally Sponsored one. The scheme envisages (i) Substantial financial assistance to trout and carp fish farmers ranging from Rs.20,000 to 25,000 per unit and first year inputs @ Rs. 10,000 per unit; (ii) Expansion of existing trout farms of the State; (iii) Setting up of an additional feed-mill; and (iv) Programme on habitat restoration of selective trout and mahseer stretches of the State.

The sanctioning of this Assistance Scheme by the Central Government is, in fact, a recognition of the excellent work under the Indo-Norwegian Trout Project, Patlikuhl (Kullu Valley) accomplished by the Fisheries Department of Himachal Pradesh under the leadership of its Director, Dr. Kuldip Kumar. The project succeeded in evolving the much-needed viable technology of production of trout seed on a large scale as well as compounded artificial feed for rearing table-sized trout. The project could also generate keen interest among trout farmers of the upland areas of the State. Under the new trout farming scheme, each farmer would be given financial assistance of the order of 50 per cent of investment needed to the extent of Rs.25,000 per raceway. The assistance for purchase of first-year inputs *viz.*, trout seed and trout feed has been limited to Rs.10,000 per farmer. Similarly, in the case of running water carp farming unit the financial assistance has been enhanced to Rs.20,000 against Rs.4000 currently being given including the cost of inputs. It is a great advance that the department proposes to set up 25 trout farming units and 100 carp farming units during the current year alone.

The promotion of fish farming in the State would go a long way under the Centrally Sponsored Project on the anvil for exploiting the fishery potential of the fast-flowing high altitude streams/tributaries on the one hand and for generating employment among the hill inhabitants on the other.

Concomitant to this expansion of the farming programme in the State, the project also includes other related components *viz.*, (i) expansion and remodelling of existing trout farms in the State so as to increase their capacity of trout seed production; and (ii) establishment of an additional feed-mill so as to meet the soaring fish feed requirements of the category of new fish culturists now coming up in the State. Another component of the scheme relates to 'Habitat restoration' which envisages the utilisation of angling potential of selected stretches of the State's streams for trout and mahseer fishing. It is learnt that under the 'Habitat restoration' scheme, work would be undertaken for deepening of certain stream beds through removal of silt, and widening of stretches and stocking them with trout yearlings. Based on the success of this excercise, it is good that other streams/tributaries would be taken up for bringing about such habitat improvement, as being planned.

There is good news that, in another significant decision, the Norwegian Government has agreed in principle to provide financial assistance of Rs.50.00 lakhs for the setting up of a disease investigation pathological laboratory at Patlikuhl trout farm. A formal proposal for providing the grant is understood to have already been sent to Norwegian Embassy through Government of India. The establishment of such a diagnoistic laboratory, the need for which is keenly felt, would remove problems of disease cure and control now faced by trout culturists, due to the lack of this facility. The laboratory would be equipped with the latest equipments for histological studies and identification of bacterial, viral, fungal and protoplasmic pathogens. It is understood that another proposal has also been taken up by with Norwegian agencies for training of in-service personnel in the prophylactic and curative aspects of trout diseases and this is in the right direction.

As a follow-up to the initiation of number of a Central and State sponsored subsidy-oriented schemes in Himachal for the promotion of carp and trout farming, it is learnt that the Himachal fisheries department has initiated measures to intensify the connected "Extension" activities. In this context, it is learnt that it has been decided to utilise the services of All India Radio, Shimla by the department, in broadcasting regularly, selected talks on the subject by Fisheries Experts. An agreement to this effect has been signed with AIR, Shimla, to start 'Fisheries School' which envisages direct interaction between the departmental officials and registered fishermen/fish farmers. It is learnt that an extension capsule has been designed by the department covering 13 serials to be broadcast within a period of three months. Under the said scheme the provision seems to be that all the registered fishermen would have the option to write letters to AIR, Shimla, incorporating their queries regarding any problems being confronted in the fish farming programmes, it is stated. These would be suitably answered by the experts of the department and broadcast by A.I.R., Shimla.

It will be good if similar programmes can be taken up in other States too with trout farming potential, such as Arunachal Pradesh, Sikkim, Tamil Nadu and Uttarakhand.

Fishing Chimes, Editorial 301: December 2001: Vol. 21, No. 9

# Lined Ponds

The system of covering the bottom and sides of fish ponds with high density polyethylene sheets is spreading fast in several countries. The reasons for this are simple. It becomes often difficult to completely remove the accumulated sludge from pond bottom. This drawback often turns out to be congenial to disease incidence during the culture phase. Further, the removal of sludge from the bottom as part of pond preparation for the next crop takes not only considerable time but often turns out to be incomplete. The transposition of a barrier (High density polyethylene sheet) between the bottom soil and the superjacent water column prevents mixing of the wastes with the bottom soil. Further, it facilitates the removal of the wastes such as faecal matter and left over feeds fast with a measure of thoroughness. This enables the commencement of the next farming cycle in the shortest possible time after a harvest. The other stated advantages are that a farmer need not worry about possible inimical bottom soil composition any more,and the number of aerators for use can be reduced (two aerators are enough for a 2000 sqm pond. The later one helps in saving energy costs. HDPE sheathing used for pond lining is estimated to last for around 15 years. The additional capital expense involved for lining a pond with HDPE will prove to be economically beneficial, as the augmented returns through higher production will pay back the investment on the lining in a few years. Higher production comes from reduction of risks of disease incidence and improved growth because of cleaner environment, and possible savings in feed costs.

While lined ponds are advantageous in several respects, it has to be borne in mind that the advantages will accrue only when there is an alert management. For example, in lined ponds water transparency will be high, because of the separation of water from the bottom. For this reason, growth of phytoplankton tends to be deficient. This naturally affects zooplanktonic growth too. In order to solve this problem, experienced farmers having lined ponds resort to fertilisation of pond waters so as to encourage growth of plankton.

It is desirable that the Central Institute of Freshwater Aquaculture conducts experimental farming of shrimps/ prawn/fish in a lined pond in order to formulate a standard technology and to bring out the economics of farming of various commercially important species. The results of such an experiment can be popularised among farmers, once these are seen to be beneficial to them, through training and field extension programmes. Work of this nature can also be undertaken by the Central Institute of Brackishwater Aquaculture, Rajiv Gandhi Centre for Aquaculture of MPEDA at Mayiladuthurai, in Tamil Nadu and at the sub-centres of Central Institute of Fisheries Education, Mumbai.

Colleges of Fisheries and University Departments imparting fisheries education and having attached farms can also conduct farming experiments in lined ponds for demonstration to their students on one hand and to assess results on the other.

A recent survey conducted by Global Aquaculture Alliance revealed that several farmers in Americas have taken to pond lining technology and most of them reported positive results. In Asian countries(Malaysia and others), all farmers who have adopted this technology and responded to the survey of GAA reported positive results from lined ponds. One conclusion of the survey is that, as confidence increases, use of lined ponds could gain greater acceptance.

Fishing Chimes, Editorial 302: Jan/Feb 2002: Vol. 21, Nos. 10/11

# 2001 in Restrospect: Development, Lessons and Events

The year 2001 rolled out, leaving in its wake considerable disenchantment among various fishery industries, be they related to capture or culture fishing effort, processing, or exports. All these sectors sustained a setback during the year. On the marine capture side, the coastal fishing as well as the offshore/deep sea fishing sectors suffered serious losses, partly due to failure of prime shrimp fishing season (mostly because of late onset of breeding and consequent reduction in sizes at capture) and partly because of a major decline in shrimp export prices (because of heavy stock holding, particularly in Japan, and also because of lower count problems) for a good part of the season. The farmed shrimp fishery sector also had its alarming share of disappointment and suffering, as the export prices of farmed shrimps too declined because of buyer resistance. Added to this, the shrimp farming sector had the misfortune of the white spot virus disease (WSSV) attack, mostly because of heavy rains and sudden fall in water temperature during the two main farming seasons. Another debacle was that flood waters engulfed shrimp farms, particularly along the east coast, taking the farmers unawares and unprepared, resulting in loss of crops. This setback had also burdened the farmers with additional investments for the following crop. As the policy of insurance companies was not to insure shrimp/fish crops, farmers had to mobilise needed funds etc., on their own, mostly depending heavily on processors/exporters, as the banking sector, by and large, was averse to extend financial assistance to farmers.

The freshwater culture sector, which is mostly related to major carps, also had to endure severe problems in raising crops, for the same reason, as in the case of shrimp crops. So much so, fish exports from A. P. to North-eastern States suffered a setback.

There were agitations by the mechanised boat operators at several places along the coastline, seeking subsidy towards spiralling fuel costs and relief from relentless operations of a large number of foreign fishing vessels in Indian EEZ and that too often close to the coast. Their grievance was that these foreign fishing vessels had been cornering the resources that rightfully belonged to the Indian stakeholders and this had led to fall in their catches and earnings thereof, which, as it turned out, were not even adequate to meet their vessel running and maintenance expenses. The same problem was faced by the larger fishing vessels too.

Unable to withstand the fall in catches and earnings thereof, a company owning six trawlers managed to shift them for fishing in the EEZ of Indonesia. Around another ten trawlers also shifted operations to fishing grounds in the Indian EEZ around Andaman and Nicobar Islands. The unfavourable situation has also led to a selling spree of quite a few trawlers, but, fortunately, all of them changed hands at the same base only and continued operations. The swapping activity was confined mostly by way of sales to captains of fishing vessels. Another noticeable development was that the strength of larger fishing vessel fleet had dwindled down to around 70 nos. over the past few years from a fleet strength of over 190 vessels. Reduction in fishing effort however took place as some of the companies sold their trawlers to foreign enterprises, and some became defunct.

The Government of India relaxed regulations in respect of importing fishing vessels by Indian enterprises. As a result, several Indian companies, most of those who were earlier in the business of 'operating' foreign vessels on charter, purchased a good number of trawlers as well as tuna longliners and registered them as Indian vessels. In a way, this has been a good development in that these initiatives increased the strength of Indian fishing fleet, despite a belief nurtured by several that these vessels continue to be owned by the foreign enterprises concerned, and the device of Indian ownership has enabled them to legally gain access to Indian EEZ. This interpretation would not however have any force, as long as our Government has no policy (in respect of financing etc.) to enable Indian entrepreneurs to acquire vessels under their real ownership. The consoling point is that something is better than nothing, although this consolation is shameful to India, a South Asian Power. It is stated that these vessels are being operated by foreign crew under an arrangement for export of the catches to the erstwhile foreign owners. A good number of these trawlers chose to operate them in the area of operation of other Indian owned trawlers, thereby creating fishing competition and sharing of stocks. This situation has created a problem to the Government as well as the fishing industry. There was a litigation initiated by one or two or more out of those who acquired vessels under the new regulations, when the operations were interfered with by the coast guard but the situation appears to be returning to normalcy.

Indian Marine Products Exports are reported to have received some setback in the sense that the anticipated increase in farmed shrimp exports is believed to be of an order adequate just to compensate the visualised drop in exports because of expected cumulative fall in capture shrimp catches. In any case, the outlook is that the marine products exports during 2001-2002 may either reach the level achieved in the previous year, or marginally exceed it. It may be mentioned here that during the year 2000-2001, marine products exports grew by 23.3 per cent, touching a level of Rs. 6,300 crores.

Most of the processing plants in the country could secure the approval of EU for exports to EU countries. Another development is that there have been initiatives to process and export value-added products, encouraged by several subsidies instituted by MPEDA to promote value addition to exported products.

The Coastal Aquaculture Authority of India, supported by MPEDA and Coastal State/UT Governments, could motivate farmers engaged in shrimp aquaculture in Coastal Regulation Zone (CRZ) to register themselves with the Authority.

This has given fillip to a trend directed at regulating and accounting for aquaculture units in operation in the Coastal Regulation Zone as per the norms laid down by the Supreme Court in its September 2001 judgment. Another development is that, at the instance of the Supreme Court of India, the Coastal Aquaculture Authority has submitted to the Court a comprehensive report on the present status of aquaculture in the CRZ, highlighting the fact that culture activities in the zone are being in the nature of traditional and modified traditional operations and also stating that there has been practically no adverse impact on the environment. In order to tackle the problems of incidence of white spot virus and other diseases, particularly in the coastal aquaculture sector, MPEDA has set up several PCR laboratories in all the Coastal States where there is concentration of the activity. This facility has come as a boon to farmers. Over and above this, MPEDA announced several subsidy schemes, one of which was to help private sector hatcheries, by way of handsome subsidies, in the setting up of dedicated PCR laboratories at their hatcheries.

The Central Institute of Fisheries Education has set up a research laboratory in Udaipur, Rajasthan to undertake investigations on fish production using underground saline water available in the arid agro-climatic region of the State.

CMFRI has made considerable progress in the production of domesticated brooder/gravid tiger shrimp. The work is stated to be now in the stage of production of fourth generation ($F_4$) of progeny. Efforts are understood to be on to secure the needed funds to step up the pace of the ongoing work. The development of domesticated brood shrimps will go a long way in eliminating WSSV among farmed shrimps.

The marine fishery potential of the Indian EEZ has been revalidated by a Group of Scientists set up by the Government. The report submitted by the Group is understood to have indicated that there is practically no change in resources position.

Ms.Nirja Rajkumar, Joint Secretary, Fisheries, in the Union Department of Animal Husbandry and Dairying, has been re-elected as Co-chairman of the Indian Ocean Tuna Commission. Shortly after this election, however, Mr. P.K.Pattanaik, has taken over as Joint Secretary, (Fisheries). Similar undesirable development struck the industry earlier, when Ms. Nita Chowdhury, the then Joint Secretary, (Fisheries), was the Co-chairman of IOTC. She was transferred, unfortunately though, at a time when she was revitalising the developmental activities in a focal manner, particularly in respect of tuna fishery development of Indian EEZ. Nevertheless, the choosing of an Indian for the Co-chairmanship has highlighted the important position India holds in the tuna arena. Some work has been done by MPEDA during the year for the introduction of a Pilot Tuna Project involving the installation of tuna longlining equipment on two shrimp trawlers in the commercial sector, to undertake tuna operations in the Indian EEZ, the work to be entrusted to a Japanese Company. While it may take some more time for the maturation of the project, in the meanwhile, it has so happened that, as the year was drawing to a close, a good number of trawlers, having come across sizeable shoals of big eye and yellowfin tuna in the Upper Bay of Bengal, operated handlines and, in a few days, each of the vessels could catch around 3 to 5 tonnes of these fishes, each weighing around 4-5 kg. This incident lends proof to the tuna potential that Bay of Bengal holds. In fact, it is observed by several that one reason for the convergence of several foreign tuna vessels into the eastern belt of Indian EEZ is the tuna potential it commands.

The Government of India declared 10th July as Special Fish Farmers' Day. The importance of this day leading to such a declaration is that on 10th July 1957, Dr. Hiralal Chaudhuri achieved a breakthrough in inducing major carps to breed, through administration of pituitary hormone injections. The event took place at Angul in Orissa. Hiralal Chaudhuri Best Farmer's Gold Medal, instituted by Fishing Chimes Jayasree Charitable Trust, 2001 was awarded, on the occasion of the celebration of the Day on 10th July 2001 at the Central Institute of Fisheries Education, Mumbai, to Mr.C.Srikanth, a progressive fish farmer, for his remarkable achievement in hatchery production of freshwater prawn seed and also a high level of farmed giant freshwater prawn production.

Tenth plan fisheries schemes have been finalised during the year. The accent of these schemes is stated to be on augmenting and extending inland culture fish production, particularly in the coldwater belt of the country. The reason for this slant is that the development of coldwater fisheries of the country could not progress all along at the same pace as in the warm water States of the country. An exception to this probably is that trout culture fishery development has received a remarkable fillip in the State of Himachal Pradesh, in the recent years. The Norwegian Government extended technical and financial assistance in this regard to the State, besides assistance for the setting up of a trout hatchery. Well designed steps have been taken in the State to popularise trout culture activity among the emerging trout farmers.

Another achievement in the year 2001 was the consolidation of the technology of shrimp farming in freshwater ponds. The related technology, innovated by the farmers themselves, has become fairly well established in the inland areas adjacent to the east coast, but away from it, particularly in the central zone. The shrimps grown in freshwater now stand accepted in Japan and other importing countries.

There have been cognisable efforts on the part of several coastal fish farmers to diversify efforts towards farming of species other than shrimp, particularly for the reason that shrimp is vulnerable to disease. The main alternative species favoured by most of the farmers is Seabass, farming technology of which has been standardised by the Central Institute of Brakishwater Aquaculture.

There has been focal patronage extended by the farmers in respect of production of the giant freshwater prawn seed and also farming of the prawn. This activity has spread not only in the inland cultivable waters adjacent to the coastal zone but has also moved farther into inland areas in Madhya Pradesh, Maharashtra and so on. Polyfarming of major carps with prawns/shrimps has started gaining favour among aquafarmers. Further, there are several who are contemplating to secure Government's approval for the farming of monosex Nile Tilapia/GIFT for the reason that the fish has a wide market in the form of fillets in USA and in several European countries. In several south east Asian countries, farming of this fish has become popular, for the reason that it is eco-firendly with practically no negative effects on the environment. There is a global view that asserts its positive qualities, particularly in respect of eco-friendliness. Monosex farming is undertaken in these countries only as a measure of abundant caution. A trend towards setting up of aquaculture clubs and clinics, particularly in the State Andhra Pradesh, has set in.

The Asian catfish, *Clarias gariepinus,* which strayed into Indian inland waters from Bangladesh, has spread into almost all the States of the country, despite the ban imposed by the authorities on its introduction.

Fish production from Govind Sagar reservoir crossed a record level of production of 900 tonnes by mid Jan. 2002 and this may cross 1200 tonnes level by March 2002.

Tawa Resevoir in Madhya Pradesh yielded for the first time 327 t of fish in 2000-01 under the management of Tawa Matsya Sangh.

Santha Marine Biotechnologies Pvt. Ltd., Hyderabad added during the year a Rs. ten crore marine algae culture facility for the production of Vitamin A. The plant is located at Kalla Mozhi village near Tiruchandur in Tamil Nadu.A multifilament fishing net plant, set up by Safa Marines Ltd., Bhubaneswar, was inaugurated during the year by Norwegian Ambassador in India, at Balasore, Orissa. This company has also introduced a new 23 m LoA, trawler 'Sai Sadoba' with Haldia in West Bengal as the base.

A breakthrough in the larval and post-larval production of the Ganga river prawn, *Macrobrachium birmanicum choprai,* has been achieved at the Central Institute of Freshwater Aquaculture.

The Central Institute of Brackishwater Aquaculture, Chennai, shifted into its own building during the year. The regional centres of Central Marine Fisheries Research Institute and the Central Institute of Fisheries Technology came into being in Visakhapatnam during the year, through upgradation of their earlier sub-centres at the place. The Fishery Survey of India has taken up the construction of its research centre in Mumbai. CICFRI's North-eastern Regional Centre was inaugurated at Dispur, Guwahati on 6 November, 2001.

The Department of Fisheries, Kerala, has initiated a housing and colonisation scheme for Kerala fishermen. The State has also pioneered in entrusting fishery development work to panchayats under a local self Government set-up, through a project known as Janakeeya Matsya Krishi. The Government of Madhya Pradesh has decentralised the administrative powers related to fisheries development to the District level officers. CIBA developed shrimp feed technology for processing and production of different grades of pelleted feeds for use in improved extensive and semi-intensive culture of tiger as well as Indian white shrimps. Private sector feed manufacturing companies such as CP Aquaculture, Higashimaru, Gold Mohur, Avanti Feeds etc., have stepped up their role in supplying quality feeds to the farmers. Godrej Agrovet Ltd., set up a sophisticated Aquafeed Plant in Association with Tai Aqua Company, Taiwan, at Chennai. Santir Aquatic Pvt. Ltd. set up a captive PCR Laboratory at its hatchery located near Kakinada in A. P. Several other hatcheries also set up such PCR Laboratories at their hatcheries. A PCR lab was also set up by the A.P. State Fisheries Department in collaboration with the State Institute of Fisheries Technology, Kakinada.

The Ministry of Health and Family Welfare issued the final notification dated 2 May, 2001 laying down doses of irradiation in respect of fresh sea foods, frozen seafoods and dried sea foods.

Indian Farmed Shrimp Production crossed 100,000t mark per annum. Among all the States engaged in this activity, A. P. emerged as the top most with 74,000 ha brought under shrimp farming, producing nearly 45,000t of shrimp.

The endangered turtle, *Kuchuga sylhetensis* has been located for the first time in Cooch Bihar District of West Bengal.

CMFRI could secure an order for setting up a Marine Pearl Nuclear Implantation Unit at Sharjah, UAE.

A Model Marine Plywood Fishing Boat, specially designed for A. P. fishermen and constructed by South Indian Federation of Fishermen's Societies, Trivandrum was launched by Prof. K. V. Thomas, Kerala Fisheries Minister during the year.

Ornamental fish production, mostly for export, has made new strides. High levels of this production have been achieved in Tamil Nadu and West Bengal. The activity is being pursued as a cottage industry, particularly in South 24 Paraganas District of West Bengal.

There are indications that the Government as well as the industry may have a few lessons to take, as reflected by the experiences gained in the process of implementation of fisheries development programmes in the country during the year. An effort is made hereunder to put them together: There has been a persistant demand from various fisheries associations for the past several years for the creation of separate Union Fisheries Ministry or atleast a separate Union Department of Fisheries. There have been also any number of recommendations in this regard made at several conferences, seminars and symposia. The subject of fisheries, with enormous potential to contribute to nutritional security of the country and to augment export earnings from the sector substantially (from the present annual level of Rs. 6,300 crores), certainly deserves serious consideration of the Government. A general grievance is that, ignoring level of export earnings from fisheries sector which are more compared to Animal Husbandry and Dairying Sectors, that the fisheries sector has also the potential to contribute to the nutritional security of the nation, and it deserves and needs a fillip to play its due role and also highlighting the same, the benign Government has continued to be complacent about this vital need. This neglect is on a flimsy and untenable ground that the fisheries sector has not generated sufficient work for the Government. One expects that Government would have the general perception that, when only it gives fillip to a potential activity such as fisheries development, there will be generation of work that will be productive too and contribute to national economy in a greater measure. We have to expect the Union Agriculture Minister to bestow his attention on this aspect from this perspective.

The main marine resource that now stands virtually unexploited is tuna. The Government cannot really have any excuse for dabbling with this issue any more, particularly when India holds the Co-chairmanship of Indian Ocean Tuna Commission. There is a strong view that Government should launch a major project as a joint venture with a country having good reputation and without undue selfish motives in respect of the integrated development of tuna fisheries of Indian EEZ. The moves of certain countries trying to gain access into Indian EEZ through devious means to achieve their commercial objectives need to be check-mated through formation of joint ventures direct with friendly countries having known expertise in monofilament longlining, which is required to be developed in India on top priority to stem the incursion of poaching foreign vessels through irregular routes.

Several countries who export sashimi tuna to Japan have developed not only landing and berthing facilities for their tuna vessels, but also facilities for airlifting the catches to Japanese market by small planes within a short time of arrival of vessels at port. One reason for the checkered tuna fishing effort in Indian EEZ by genuine Indian Vessels is the absence of facilities and funds. Entrepreneurs have been for long voicing the cruciality of setting up such facilities, at a location in Andamans (Andamans happens to be a crucially central location for tuna fishing, as evidenced by not only the locational aspect, but also the convergence of several foreign vessels to fish in the area, braving risks of apprehension), and at selected ports on the mainland, atleast one on the east coast and one on the west coast. Learning a lesson from the present state of affairs, Government should take steps to introduce an integrated project as a joint venture from the stage of infrastructure development suited for airlifting of chilled sashimi tuna to Japan for marketing, with linkages for the introduction of an integrated tuna fishery programme in collaboration with a country such as USA or Australia or New Zealand, providing for transfer of technical expertise. It is widely known that a Tuna Project of this kind has special significance of being a substantial foreign exchange earner.

The Indian exports of shrimp take place largely in block-frozen form. It is common knowledge that importers utilise block-frozen shrimp for re-export in value-added form. A lesson has already been learnt that value-added shrimp products have to be exported in the place of block-frozen product for higher earnings, but progress in regard to the swiching over to value-added system is very slow, despite introduction of schemes by MPEDA for providing incentives by way of subsidies for installation of the needed equipment at the processing plants and for adopting improved packing systems conforming to international standards. Greater attention is needed to hasten the switching over process.

There has been a long standing proposition for encouraging diversified culture for the production of species suitable for both export and domestic consumption through development of coastal wastelands, and fallow saline stretches in Haryana, Rajasthan, Punjab and Western Uttar Pradesh. The initiative, stated to have been taken by ICAR for a survey in this regard with Australian co-operation is timely and it deserves to be pursued to its logical end.

The fall in export prices of certain marine products has left the industry high and dry with no alternative in sight. This emphasises the need for development of a strong nation-wide domestic marketing network to act as a deterrent to reduction in export prices by the importing countries, and as a mechanism for providing fish supplies to the Indian population in an enduring manner at reasonable and stable prices. An efficient domestic marketing system will act as a signal to importers that regulation of export prices as they now do cannot be the same when India has a domestic fish marketing network supported by a cold chain.

It is good that efforts are on for production of tiger shrimp brooders through a domestication programme. This process has to be stepped up for achieving results in the shortest possible time. While India may not need foreign collaboration in this regard, the present need, however, is for funds and facilities. ICAR should not have much of a problem in this regard, considering the crucial importance of development of domesticated tiger shrimp brooders. Until such time a breakthrough in. this regard is achieved, shrimp hatcheries would have no alternative other than dependence on wild brooders. The continuance of this dependence would only mean the prolongation of the present menace of outbreak of white spot virus disease attacks among farmed shrimp, with all the attendent problems of economic losses to farmers and fall in export earnings.

Fishing Chimes, Editorial 303: March 2002: Vol. 21, No. 12

# Tuna Fishing in Indian EEZ: Training Needs

It is learnt that there are 20 used foreign tuna longliners, whose ownership has been acquired by Indian enterprises, and one more under a joint venture which are presently in operation in the Indian EEZ. Information on the zones of Indian EEZ in which these liners are operating, the species of tuna they are harvesting, their average weight range and quantities harvested is hard to come by. This information, provided tuna caught by these vessels is being exported, should be there with Customs/ MPEDA and with Fishery Survey of India, in case the consignments are inspected by its scientists. 'Prime' of MPEDA of recent months does not include particulars of tuna exports. Apparently no tuna exports have been taking place.

There is a general impression that Indian - owned tuna and other tuna foreign vessels imported recently and added to the Indian fleet tend to operate in total isolation from the mainstream of the Indian fleet. This trend is probably attributable to the fact that most of the crew on these vessels are of foreign origin and the operations are directly export-oriented, the exports taking place probably on their own bottom, unlike the other Indian vessels which do not undertake tuna fishing. The general system of Indian vessels is to sell their catches to Indian processors or their agents.

Information on the types of used foreign longliners acquired by certain Indian companies, *i.e.*, whether of multi or monofilament style, and of what length of line, number of hooks operated, areas of operation and other relevant details would be of great value to entrepreneurship, now longing to equip their shrimp trawlers with tuna longlining equipment. MPEDA has already initiated a pilot scheme, in association with the Association of Indian Fishery Industries, for equipping two privately owned shrimp trawlers with monofilament long lining equipment, to operate 72 km of monofilament long line (60 km effectively) with around 1,500 branch lines. These vessels, so upgraded, are expected to be operational by May 2002 or even earlier. It is possible that the needed number of foreign crew members, having knowledge of the grounds and the operations will be employed to work on the vessels, particularly because Indian crew have little experience in tuna fishing and knowledge of tuna grounds. The operations can be expected to yield encouraging results. The dissemination of information on the results and the operational aspect of these upgraded vessels and in respect of the operation of the newly acquired used foreign tuna longliners by Indian enterprises would enable the owners of other trawlers to add longlining equipment to their vessels too.

CIFNET imparts to its students practical training in trawling and other fishing methods including longlining. Now that the future opportunities lie in the direction of monofilament longlining, it is desirable to add to CIFNET's fleet monofilamant longliners too. It would also be a sound training proposition to add multi-purpose vessels equipped for stern trawling and also for monofilament longlining. Each of CIFNET's Centres can be provided with one such training vessel for providing onboard training to the candidates undergoing the course. Such a step will facilitate imparting of training in monofil longlining, in careful retrieval of fish and their transfer into the hold, proper preservation of tuna, and removal from the hold, and unloading in a correct manner.

There are very few captains in the country who have experience in monofilamant longlining. Once CIFNET equips itself to impart training in monofilamant longlining, interested skippers of fishing vessels can be given an opportunity to receive refresher or regular training in monofil longlining to meet the demand that may soon arise. In the same way, instructors at CIFNET may be deputed to countries such as Australia/USA to receive onboard training in monofil longlining at the concerned training centres in those countries. Such a step would provide a sound orientation to operations under tuna monofil longline operational training programme. Quick action by the Government in this direction would enable CIFNET to provide trained candidates to the industry, by the time the present fleet of trawlers get equipped with tuna longlining system and new tuna longliners are added to the fleet.

Fishing Chimes, Editorial 304: April 2002: Vol. 22, No. 1

# Time for FSI to Share its Mandate

The Fishery Survey of India (FSI). first set up in 1947 as Exploratory Fisheries Project in Mumbai, has emerged over the past 50years as an unique fishery survey organisation in the south-east Asian region. Spread over all along the Indian coastline in a manner that provides access to conduct fisheries surveys in the Indian EEZ from a number of Bases, FSI's headquarters in Mumbai has developed a well structured system of monitoring the survey work conducted from its various bases. It has been extending a steadfast support to the Indian fishing industry and stood by it in several ways all these years. At the time when CIFNET was not yet established, FSI provided on-board training to deck hands during survey cruises of its vessels and later to a number of CIFNET's trainees to enable them to become Skippers. Mates. Engineers etc. The trained candidates, mostly those trained before CIFNET was set up, became the nucleus to meet the manpower requirements of poineering companies such as Union Carbide and New India Fisheries. The commercially attractive fishing results achieved by these companies blazed a new trail for the Indian fishing industry to blossom in their fishing operations conducted in areas beyond the traditional coastal zone.

The primary credit for locating shrimp grounds off Orissa and West Bengal coasts goes to FSI. This location eventually led to the growth of the industry. Later, FSI located fishing grounds of several commercially important fishes and crustaceans in the Indian EEZ. Tuna grounds, particularly around Lakshadweep and Andaman and Nicobar Islands and in the high sea zone off the nation's mainland were located by FSI but the credit for commercially exploiting this resource went mostly to the chartered longliners, mostly of Taiwanese ownership. Revalidation of deep sea lobster and deep sea prawn grounds off certain zones of Kerala coast, and location of spear lobster grounds around Andaman and Nicobar Islands, Squid and Cuttlefish grounds along west coast, besides several other viable resources have been accomplished by FSI.

The estimated fisheries resources of Indian EEZ have been revalidated at 3.92 million t recently, but this estimation is at the same level as before. Against this potential, the level of annual exploitation, reached as in 2001-02, was 2.8 million t of the resource, thereby leaving a balance of I. 2 million t. which works out to about 28 per cent of the total estimated resource. Most of this balance is reckoned to be in the deep sea zone in the form of both pelagic or demersal resources, and apparently not accessible to the Indian fleet.

Many in the industry believe that the estimate of fisheries resources of Indian EEZ is a conservative one. Their reasoning is based on the premise that, apart from 2.8 million t exploited by the Indian Industry; an additional quantity is exploited by foreign vessels too clandestinely. There is no way to know the quantity that gets exploited in this manner. Once poaching is prevented, and the Indian enterprises are helped either in expanding the fleet strength with suitable types of vessels or in upgrading the present vessels as needed, a line of action for utilising the remaining resources not accesible at present to our fleet would take shape. And this approach would help in locating the grounds containing this estimated balance of resources not being exploited at present by Indian vessels.

Another reaction of the industry is that there can be an underestimate of the resources, deliberately made as an abundant measure of caution. An overestimate means invitation of focal criticism and accusation of inaction in organising the utilisation of the available resources. The critics reinforce their observation with a seemingly logical argument which runs a follows: Around 40,000 mechanised/motorised fishing boats and around 500 nos. of vessels of15m LOA operate in the territorial waters and a little beyond. Around 65 larger vessels of 23-27 m OAL operate in the waters believed to be beyond the coastal strip. Unable to withstand the economic strain because of poor catches, the managements of a few larger vessels shifted their operators to a zone away 'from the Indian EEZ'.

The combined effort of all mechanised/motorised boats and the larger vessels now yield an annual production of 2.8 million t. the bulk of which is taken from the shelf zone. Stated differently, a substantial area beyond the coastal zone and also the shelf zone is left unfished by the Indian fleet, despite the fact that the resources of this area are sizeable. In this connection, there is a belief in the industrial circles that there are uncharted grounds in the Indian EEZ for deep sea lobsters, deep sea shrimp and cephalopods, mostly of squids and cuttle fishes, which may be lying in irregular patches at undisturbed depths at around 200 m and beyond. It is believed that in the 'pelagic waters of high seas there are unexploited tuna resources, over and above similar stocks followed and captured by a couple of distant water nations.

According to the critics, it is hard to believe that in an area of 2.02 million sq. km of its EEZ. India has such a low level of resources that can yield only 2.8 mill t/a year. In other words, the view is that survey work needs expansion, although there are constraints in the way. One major constraint is that FSI, despite efforts, has not been able to undertake surveys from commercial angle, because of working problems. Night fishing goes against working norms of crew employed to function on Government-owned vessels. At present, surey work is planned, based on a scientifically designed grid system, which is not that conducive for inparting a commercial orientation, particularly because of the working norms which the crew insist upon, as is their right.

FSI plans its survey work utilising the available fleet taking into account the recommendations made by the main and base-wise adivsory committees it has constituted.Further. FSI has the system of disseminating the results of surveys for the benefit of the industrial enterprises. If these devices have promoted an interface between FSI and the industry, these need to be strengthened further with a commercial orientation, and this is possible through stimulation of interest among the concerned by way of location of specific and viable fish stocks of commercial interest among those in the industry. Knowledge in this regard will motivate the industry to acquire new fishing vessels for exploiting such stocks. For want of this knowledge and quite a few other reasons such as advancing age of vessels arid constraints in securing financing facility, the strength of industrial fishing fleet of the country has plummeted from around 190 nos. to 65 nos.

In this background, there is an imperative need for FSI to share its mandate with the industry in a such a manner that the fishery enterprises can experience for themselves the various aspects of survey work and at the same time contribute to the diversification of survey effort to such newer zones where they can expect to locate hitherto unknown grounds of deep sea shrimp, deep sea lobster, squids and cuttlefish and of other prime fisheries. The Fishery Survey of India can take the initiative of drawing up a scheme that provides for sharing of survey responsibilities between itself and the Industry, taking into account the views to be offered by the Association of Indian Fishery industries. Broadly visualised, FSI may lease out to selected fishing companies the survey vessels presently in operation, for an adequately long duration governed by the needed terms and conditions. One of these conditions may provide for atleast one representative of FSI to be on board each of the survey vessels to ensure the recording of voyage-wise data in the required format and to ensure that the voyages are purposeful, taking into account the commercial needs. Based on the daily reports from the vessels received at the various bases of FSI. guidance can be imparted to the fishing companies in respect of new viable grounds located, for exploitation. A formula for the transfer of services of the crew members willing to work under commercial terms and conditions with the lessees may be arrived at through mutual consultations between FSI and the leasing company concerned. The lease amount may be fixed, taking into account the depreciated cost of the vessels, the likely operating costs, and the objectives. While the areas to be surveyed could be left to the lessee concerned, based on specific proposals put forth. FSI would have to extend the needed guidance to ensure that the surveys progress in a way to deliver results as would be useful. The agreement to be entered into between FSI and the leasee has to be of course finalised through mutual consultations between the Association of Indian Fishery Industries and FSI. The excercise of sharing responsibilities would no doubt bring about a change in the present working style of the various bases of FSI and of the Director General and his scientists at the Main Centre. All the daily data received from the leased vessels, their analysis and the conclusions drawn would provide a new complexion to the work. Further, the reactions of the industry in respect of the operational worthiness of the survey vessels which have become old would have the effect of inducing the Government to strengthen the fleet. The results of the survey conducted under the proposed new scheme will bring out sharply the commercial fishing fleet requirements for exploiting the resources as estimated from time to time Once this aspect is clearly known, the need to help the industry for securing financial support for acquiring new fishing vessels would come out sharply for the Government to institute such measures as are necessary for strengthening the survey fleet.

Fishing Chimes, Editorial 305: May 2002: Vol. 22, No. 2

# ....Towards Dawn of Nation-wide Export-Oriented Aquaculture Era

## *(of Giant Prawn, Black Tiger and Nile Tilapia)*

A shift in the aquaproduction scenario of India towards shrimp that crept in some years back has now manifested itself. The reason: Marine capture fisheries output that was the sheetanchor of the national fish production plummeted to its lowest level over years. As part of this trend, capture shrimp production has faithfully followed in sympathy. In this situation, fortunately, culture fish production came as the saviour and restored the overall national trend. However, this remedial adjustment has left in its wake two shrotcomings. One is that the increase in culture fish production has been mostly in respect of major carps which have no export value and the increase in their production has led to diminishing returns to the farmers. Another is that Black Tiger (BT) and giant freshwater prawn (GFP), while having marginally added to overall national ouput have contributed to augmenting exports and earnings thereof. Farmers and exporters could not however earn much of profits because of the depressed global aquafood market situation. The outlook is fortunately improving now.

This situation clearly points out an inescapably unitary line of action to increase export-oriented production, *i.e.*, working towards the dawn of nation-wide increase in GFP and BT culture production. Farmers have to be enabled to diversify into culture of Nile/GIFT tilapia too which has a strong export market, mainly to USA, in whole as well as filleted form. In this background, there is an indisputably emergent responsibility on the part of the authorities concerned to draw up and set in motion an aggressively bold policy to promote the spread of the culture of these species over the length and breadth of the country, of course barring the temparate zone where they do not survive.

An appraisal of the present scenario and the indications that it throws for the authorities to act upon in respect of policy formulation, increasing production, augmenting incomes of farmers and stepping up exports is attempted hereunder.

### Giant Freshwater Prawn (GFP)

An estimated 30,000 ha of water area supported by over 70 hatcheries has been developed in the coastal States. In inland States such as Madhya Pradesh, Punjab, and Haryana also, there has been some measure of GFP farming coverage. A begining has been made by the farmers of Punjab and Haryana who brought 10 and 3 acres of water area respectively under GFP farming. The credit for this initiation goes to the Central Institute of Freshwater Aquaculture, whose unit in Ludhiana has made this possible. In fact, very recently, 80,000 pls of the species from CIFA, Bhubaneswar, and 60,000 pls from Nellore in A.P were airlifted to these States. The cost of this transport is stated to have come to a reasonable 85p per piece. Earlier 35,000 airlifted pls were stocked in 3 acres of area in the zone yielding a production of 1.311 ha in one farm and 211 ha in another farm. This demonstration has led the Government of Haryana to allot Rs 2. 75 crores for the promotion of GFP farming in the State, mostly because of the initiative excercised by Ms. Asha Sharma, Secretary (Finance) and Commissioner (Fisheries), Haryana. The Director of Fisheries, Haryana, has the distinction of setting in motion measures for promoting the activity. It is learnt that Amalgam Group with head quarters in Kerala and an office in Delhi has offered to extend marketing assistance to Haryana and Punjab farmers both within the country and for export as well.

In Madhya Pradesh a promising beginning in GPF seed production has been made by Prof. T.A. Qureshi of Barkatullah University using constituted sea water. A modest start has also been made by the farmers of Chattisgarh State by getting GFP seed from A.P. and stocking their tanks with them. In fact, after its formation in November 2000, the new State has shown keen interest in the promotion of prawn farming. It is learnt that the State Cabinet has taken a decision in prinicple to go in for a joint venture with an Indian entrepreneur to demonstrate commercially viable and export potential of GFP farming. Trials in this regard have already begun at Demar Fish Farm (Dhamtari) in an area of 4 ha, by an A ndhra Pradesh-based entrepreneur. Apart from this, polyfarming with GFP has been also initiated at six places. This consists of polyfarming of *Macrobrachium malcolmsonii* (River Prawn) with fish and this is being done in Ambikapur and Raigarh. Monofarming of *M.rosenbergii* alone has been taken up in Jagadalpur, Dantewara, Bilaspur and Durg areas. At Dantewara and Jagadalpur, prawn seed is supplied on 100 per cent subsidy basis to tribals. The prevailing interest in prawn farming is no doubt a welcome feature but there are a few aspects which need to be looked into. In Tripura, a GFP hatchery has been set up by the State Fisheries Department.

In Andhra Pradesh itself 22,084 ha in the coastal zone and 259 ha in the interior (Telangana area), far away from the sea coast, have been developed for GFP farming, yielding an annual production of 20,910 t (2000-2001). In Kerala, Karnataka, Maharashtra, Orissa and West Bengal too a considerable extent of area has been brought under GFP farming.

Considering the stage set for making a headway in promoting GFP farming so as to achieve an addition to aqua product exports, MPEDA may have to think of drawing up a massive plan on a priority basis, for extending GFP farming to all amenable water resources in the country. This plan can include measures for a quick estimate of lentic resources that can be developed for GFP farming and the kind of organisational mechanism needed for the seed of GFP to be taken from coastal centres to the States concerned (by air, van. relay shifting), their acclimatisation, storage, and supply to farmers, monitoring of culture endeavours, creation of channels for pooling up catches, their presentation, storage and transport to exporting outlets (harbours). The task is no doubt arduous but it is feasible and deserves to be undertaken for the beneficial utilisation of the resources and for augmenting exports.

MPEDA, the Union Fisheries Division, ICAR's Fisheries Research Wing, and State Fisheries Departments need to openly express their views on the subject in order to provide strength to the movement already started in Haryana. Chattisgarh etc.. so as to contribute to the national wealth. With its expertise MPEDA can certainly pave the way for significantly contributing to the preparation of a project (after receiving a mandate from the various States). The project can provide for quick surveys of water resources, and for institution of pilot schemes involving training programmes too in each of the States and later embark upon a follow-up expansion phase in the light of the pilot phase results. The existing organisational set up in the form of FFDA can well be utilised for this development programme. MPEDA can also come up with a scheme for providing handsome subsidies, to be shared by the State Fisheries Departments and Union Fisheries Division. The training costs can also be subsidised. The suggested project could include provision for infrastructure needs such as centralised seed storage units, centralised ice plants, product storage and processing units and transportation and marketing facilities. It is encouraging that Mr. Tharakan, Chairman of Amalgam Group, has already inducted his Delhi unit into the activity.

Regarding BT farming, there are clear cut possibilities of farming these in the saline zone of Haryana, Punjab and Western U.P. The main need that can arise in this venture is to bring down the salinity of water to 15-20 ppt by taking in freshwater from nearby Indira Gandhi canal or other sources. The other constraint is related to seed supply. By setting up centralised seed storage farms, PLs brought in by air-cum-road transport method can be stored in them and utilised for stocking farms to be set up in the zone. Through a devoted and consistent effort, the required infrastructure facilities can be developed by the State Governments concerned. In the alternative, a commercial organisation can be encouraged to set up the infrastructure by providing land and other facilities. There can be no problem of salinisation of adjacent agriculture fields on account of promoting the farms for the reason that the entire zone is saline. In Thailand, salinity problem in rice fields is experienced because sea water is taken inland through canals for feeding shrimp ponds and there was the incidence of salinity from canal waters seeping into adjacent fields. No such contingency can arise in the case of saline lands converted into shrimp farms in the States mentioned.

So far as the extension of BT farming in freshwater ponds is concerned, some have reservations about the feasibility of this practice. They believe that farming of BT in freshwater can be possible only where there is some measure of bottom soil salinity. This cannot be correct as BT farming is being conducted in several tanks which have no bottom soil salinity any more than in the areas farthest from the sea. Presuming that there can be such a problem in the freshwater tanks/ponds in the States far away from the coast, the issue can be resolved by taking up a pilot project covering selected tanks and ponds in the States such as Madhya Pradesh, Bihar, U.P (eastern) etc. The results of the pilot programme will resolve the issue once and for all. In fact, Dr. K. Gopakumar, former Deputy Director General (Fisheries), ICAR exhorted the farmers of MP to take up BT farming in their ponds, at a seminar held a couple of years back in Bhopal.

Turning to Tilapia, it is good to know that at last, the Ministry of Agriculture is convinced that the exotic Nile Tilapia (*Tilapia nilotica*) needs to be allowed to be imported into the country. Apparently, the ministry is swayed by the consideration that the fish has an enormous potential as an exportable commodity. Two companies, one of which is Waterbase Ltd, have now the permission to import Nile Tilapia, it is learnt. They can now set up dedicated farms at centralised locations, where monosex seed of the species can be produced and supplied to the farmers. In this context, many would not agree with the view of Dr. P.K. Ramachnadran, Vice-President, Waterbase Ltd, Chennai, published as part of an interview he gave to Aqua feed (Vol 5, issue 1, 2002) that Tilapia has no local market in India. While his perception may be in order, the permission given for introducing the fish in India is apparently not for meeting the domestic demand, but is meant mainly to cater to the surging demand principally in USA for whole dressed Tilapia and also its fillets in IQF or block-frozen form. He has not however mentioned this aspect in his interview. In any case, there must have been a motivation for seeking and securing the relaxation from the Government for farming this fish.

In conclusion, we appeal to MPEDA to formulate a project as indicated above, and seek approval of the Government and co-ordinate its implementation, in order to mobilise the efforts of lakhs of inland fish farmers for producing aqua crops that have export orientation too. These efforts, over a period of time, would go a long way to counteract the declining trend of exports. As Robet S.Hooey of Aquastar says, Indian GFP and BT have gained good acceptance in US market and this is a positive signal to produce more of them mainly for export.

Fishing Chimes, Editorial 306: June 2002: Vol. 22, No. 3

# Utilisation of Fisheries of Indian EEZ: Reflections on Policy Options

The Indian marine fishing industry has been eagerly awaiting, for the past few years, the announcement of a comprehensive national policy for the development of fisheries of Indian EEZ by the Government of India. The latest known on the subject is that a policy formulation group set up by the Government for the purpose has submitted its report quite sometime back. Even allowing for the generally known complacency of the Government towards development of fisheries of the nation, the ongoing delay in the announcement of the policy may soon prove to be against national interest.

India, possessing 2.017 million sq km of EEZ and situated strategically in the Indian ocean region, possesses all the positional advantages of being a leader (Alas! not-excercised) among the nations of the region. Considering the underexp/oited open sea and deep sea marine fisheries resources and over-exploited coastal resources, the nation deserves to be endowed with a purposeful, forward-looking and dynamic policy for their sustainable utilisation. And such a policy has to replace the present one, which is mainly corrective in nature, and is aimed at ridding the country of undesirable operations of chartered, leased and joint venture foreign vessels. This corrective job having been more or less accomplished, no doubt Government is working on replacing these undesirable operations with a more nationally-oriented system, but the time lag is far too inordinate. So far as coastal fishing is concerned, there have been no initiatives at diversification of fishing effort in the Zone, so as to reduce fishing concentration on conventional resources. There is thus a policy vacuum.

This policy vacuum has led to an interesting development that has unwittingly served to test the possible swing in the mood of the industry in respect of strengthening of the depleting larger vessel fishing fleet. The test took place through a notification of ministry of commerce amending the rule related to import of vessels to provide for acquisition of used vessels by Indian enterprises. This notification, as can be understood and appreciated, primarily invoked the interest of those with experience and contacts with owners of foreign fishing vessels who gave their vessels earlier on charter/lease basis to them for operation in Indian EEZ. The result is that several of them have availed of the opportunity that the amendment provided. It is learnt that 32 vessels have been imported by them, registered under Indian flag and put in operation, with an arrangement for the payment of the cost of the vessels on 'deferred' basis. As it turned out, for the foreign fishing vessel owners, this was a good alternative opening to gain access once again to fish in Indian EEZ and, for the Indian enterprises it provided an opportunity to revive the earlier practice of vessel-oriented exports without unloading and through an export-link up, although in a different capacity. Technically the imported vessels are now their own. despite the belief in certain circles that the *de facto* ownership continues with the foreign owners. For the Ministry of Commerce, it seems that the imports are apparently a means to augment value of marine products exports, which have been sagging, of late.

As can be seen, this vessel import route, based on 'deferred' payments, solves the problem of investments and at the same contributes to exports. Despite these advantages, several in the sector feel that this route may give rise to certain diadvantages. Firstly, this route is not amenable for utilisation by the general run of fishery enterprises. Another aspect is that, for the operation of the vessels, reliance has to be totally on foreign crew. The induction of Indian counterpart crew will be in the nature of a formal presence, as the foreigners, for obvious reasons, do not have interest in providing training to Indian crew, as had been the past experience under the erstwhile charter/lease scheme. And, as may have been noted, the present used vessel import activity happens to be an inverted system of the erstwhile charter scheme. In a way, there can be no sense in having this system with the present contours, which tantamounts to a virtual revival of the erstwhile scheme (old wine in new bottle). Further, the Indian 'owners' may not practically have any control over the movements of the vessels once they leave the port. The past practice had been that they move to areas in or outside the region depending on the fishery complexion. In other words, the return of the vessels for the next voyage can be beyond the control of the 'Indian owners'. Whatever be the conclusion, considering that the main resource available for exploitation is pelagic tuna, and there being practically no trained Indian hands in tuna longlining, there seems to exist an urgent need for expanding training infrastructure, either on the imported tuna vessels or elsewhere, to build up a cadre of trained men in tuna long lining for utilising their services on a truely Indian tuna fleet which will undoubtedly emerge soon.

Only a small section of foreign vessel owners, that too mostly those from Taiwan and Thailand, have interest in fishing in Indian EEZ for traditional reasons and because of their close knowledge of Indian situation. Because of this, only those with earlier contacts with foreign vessel owners could succeed in the import of used vessels. This situation appears to have unleashed professional conflicts in the industry. Several other issues could have also surfaced. Consideration of these aspects, in all likelihood, may have delayed the announcement of the new policy by the Government.

It is not that all those who imported used fishing vessels are happy. Reports speak of registration of 32 used foreign vessels under Indian flag. Half of these are trawlers and the remaining are tuna long liners, as the information goes. Most of these have been idle, despite registration as Indian vessels, because of an apprehension by coastguard for some reason or the other and some are stated to have fled because of fear of apprehension (Despite the fact that they are registered as Indian Vessels). The latest development, however, is understood to be that, Government have decided to allow the tuna long liners to operate probably because longlining is a passive fishing system and the trawlers are not allowed apparently for the reason that trawling is a dynamic fishing system. This approach, if true, indicates that it can be there in the policy to come, which may not, in any case, be delayed for long.

'Arranged' import of used vessels with export linkage is certainly an easier option but many consider this as a solution that compromises national honour, because of its inherent manipulative features, as attrubuted. It is not known whether the Government is considering other options that would be much more direct for effectively utilising the known unexploited resources of Indian EEZ such as high sea tuna, oceanic squid and cuttlefish, deep sea shrimp and lobsters. Whatever be the options Government may be considering, it appears that a sustainable option may be there in developing an integrated institutional mechanism that would channelise the energies of such of those fishery enterprises that have a record of experience in fishing with larger vessels to participate in the operations for catching tuna, squid, cuttle fish, deep sea shrimp, lobsters etc.. with an integrated approach.

The Ministry of Agriculture set up an integrated Fisheries Project in 1970s with the objective of demonstrating and promoting integrated projects in the private sector, as a follow -up to the effective and devoted work done by its predecessor, the Indo-Norwegian project. Keeping this historic promotional work done by this Project in view, the Integrated Fisheries Project could now be geared up to perform the function of promoting projects for the integrated utilisation of the aforesaid resources, possibly by taking the help of NORAD, which has vast experience in the line. In this connection, it will be recalled that NORAD had earlier helped India so much in a consistently dedicated manner in the introduction and popularisation of fishing with mechanised boats in India's coastal waters and also in the survey of deep sea fishery resources. Their assistance could be sought again to take the industry further forward, at this juncture of being at cross roads for want of finance and expertise in certain respects. In the alternative, assistance from Australia which too has experience in long lining for tuna and in the exploitation of deep sea resources, could be availed of.

Apart from taking help in the promotion of high sea fishing, more specifically for longlining for tuna, deep sea fishing for lobster, shrimp, squid etc.. there is another angle for obtaining Norwegian or Australian help.

As is widely known, there have been no tangible efforts so far at introducing and promoting sea cage farming in India. It is significant that this aspect had also been highlilghted by no less a person than Dr. K. Gopakumar who was Deputy Director General (Fy) ICAR at the CIFT-Industry Meet held in Visakhapatnam recently. This can be interpreted as an admission of the lapse.

Both Norway and Australia have enormous experience in cage farming in protected as well as open sea waters. While cage farming has become a routine activity in Norway and several other countries as a sustainable method of augmenting fish production, India remains uninfluenced by the idea. However, some years back, one Indian entrepreneur made strenuous efforts at great expense to install a cage farming system in Andaman waters but either shortsightedness or some other factor prevented the authorities concerned from encouraging the endeavour. Thus a great opportunity of introducing sea cage farming was lost.

In this background, marine fisheries policy could include the promotion of cage culture of fishes such as groupers, snappers, sea bass etc., in protected as well as offshore waters of Indian EEZ, taking due care of aspects such as providing sites on long term lease to entrepreneurs supported by needed survey and other help from NORAD or from Australia for setting up hatcheries on shore, for providing seed supplies to cages and for providing training and for other inputs. Sea cage farming can add to national fish production and exports significantly and compensate for the declining marine capture fishery output.

There are examples of several countries augmenting their sagging fish production through cage farming. Two recent ones out of these are mentioned as follows.

Cage culture of Yellow fin tuna has been taken up in Mexican waters, yielding encouraging results. India is now not in a position to visualise taking up any such project and this is unfortunate. Greece provides another example of the operation of tuna cage farms on a commercial scale.

A report in *Fish Farming International* (May 2002 issue) says that a privately owned Mexican company has just begun growing yellowfin tuna in cages off Mexico 's Pacific coast. Cage culture of tuna in waters around 360 m distant from an island and at a depth of about 40 ft has been taken up. There is a promising future for warm water mariculture on the continental shelves, near islands where the presence of strong currents, lack of pollution and low nutrient waters offer excellent sites offshore for cage farming, it is mentioned. The new habitat created by the cages and the small quantity of faecal discharges into the open sea are seen to significantly increase the concentration of marine life around cages, besides the fish that are grown in the cages.

Barges equipped with onboard freezer for holding frozen feed consisting of sardines and mackerel for feeding cage fish, accomodation for crew, power generating system, and with facility for cage observations are stated to have been pressed into use in that country.

The present day offshore grow -out cages are well advanced in design with lower maintenance requirements and greater endurance in rough weather. Polyethylene cages as big as those having 42m dia are used. 48 ha of area is allotted for setting up a cage farm to each of the enterprises by the Mexican Government.

Tow-cages, different in design to grow-out cages are towed to centres where tuna juveniles are available and which are caught by tuna vessels. Such cages are used for collecting tuna juveniles for transfer to grow-out cages. For harvesting, a net is placed inside the main cage to concentrate the fish on it and a gaffing platform from which to pull fishes individually is used. They are killed on site, and carefully cleaned before rapidly chilling them for quick transport to the market.

It is also reported in *Fish Farming International* (May 2002 issue) that in Greece, a company by name Nireus SA operates three marine fish hatcheries, and 18 cage farms, four EU certified packing plants and one processing plant.Each of the cage farms has a grow out capacity to produce 1,187 t/cage. 19 m dia HDPE circular cages are used, besides barge feed transport system. The possibilities of securing expertise from NORAD or Australia or Nireus Group of companies (1st Km Koropiore-varis Ave, 19400 Koropi,Greece: Fax +30 10 6626803, Website: www.ttirtus.com) can be explored for the introduction of commercial sea cage farming in Indian seas. CMFRI can probably play a significant role in this respect, through taking up experimental programmes of sea cage farming and eventually taking up schemes for providing training to entrepreneurs in cage farming technologies, as part of a collaborative project, thereby paving the way for the popularisation of sea cage farming in Indian EEZ.

Fishing Chimes, Editorial 307: July 2002: Vol. 22, No. 4

# On Loktak Lake Fishery Development

Loktak is a shallow and somewhat acidic freshwater lake, with an extent of around 20,000 ha, situated at Latitude 24.50" N and Longitude 93.80" E in the north-east of India in the State of Manipur, close to the city of Bishnupur. Although relatively shallow, it has a few points where the water depth goes upto 25 feet. It is known for its unusual infestation with weeds, both dead and live, and also heavy siltation. This infestation, piled up over years into various floating shapes, oblong, circular, rectangular etc, with a depth of 5 to 6 feet. These are known as 'Phums'. They cover about half of the surface of the lake. *Phums* vary in size and can be as large as one ha. In a way, they have turned out to be a sort of fish aggregation devices, that facilitate fish harvesting. Over years, the fishermen dependent on the lake fisheries improvised a certain alteration in the structure of several of the *Phums*. Leaving a rim all round, they have removed the rest of the *Phum*.

The truncated area is used as a feeding place for fish which are offered fresh weeds, rice bran and ground nut oil cake powder as feed to attract their aggregation.

The lake is being stocked with carp fingerlings since 1966, and October to March is the season for *Phum* fishing. Fishermen encircle a *Phum* with a net to collect the fish. It is stated that *phums* yield between 15 - 40 t of fish annually, besides catches taken through use of hook and line etc. *Phums* get dispersed sometimes moving over long distances within the lake. When this happens, fishermen improvise fresh *phums* close to their habitations. Another feature one notices is that thatched huts are built on *phums* here and there to serve mostly as lodges for visitors or for fishermen's use.

In order to utilise the potential of the lake for beneficial purposes, the state Government has set up Loktak Development Authority. While the main purpose of setting up the Authority was to build a hydro-electric project, the Authority is also charged with the responsibility of development of fisheries of the lake. As part of this assignment, it has taken up the task of clearing the lake from *phums* and scattered weeds. Further, to facilitate stocking, the Authority has set up two major carp hatcheries and is understood to have been advised by the fisheries department to raise grass carp seed at the hatcheries for stocking the lake.

There is another lake, about 5,000 ha in extent, close to Loktak lake. This lake is known as Pumlen lake. *Phums* occur in this lake also. Pen culture has been introduced in this lake and the few pens that are there are owned by individuals.

The estimated total annual fish production of Manipur State is stated to be 16,000 t, against the requirement of 40,000t. A strategy to step up fish production, mainly through the utilisation of the potential of Loktak lake is understood to be in tne formative stage at the Directorate, to fill up the gap.

In formulating a strategy for augmenting fish production from the lake, the problems presently faced in this regard would no doubt be kept in view. One such problem, as already mentioned, is weed infestation which is now being tackled by the Loktak Development Authority. Once this problem is got over, the next step, which can be taken up simultaneously, as the weed clearance progresses, is to move towards a capture -culture fishery balance. Taking into account the configuration of the lake and the rivers/streams that flow into the lake, water that flows out through a canal for running the turbines of the hydro-electric system, and the Imphal river that flows alongside, a total fishery-oriented survey needs to be undertaken, aimed at demarcating plots for giving on long lease to farmers for forming marginal ponds of appropriate sizes and also for setting up cages of suitable dimensions in amenable parts of the lake. Marginal ponds and cage systems are preferable to setting up of pens for the reason that marginal ponds can protect the highest lake contour and cages do not lead to silation as pens are prone to. Keeping the ecological conditions in view, the design of the marginal ponds and the disposition of the cage batteries would have to be planned to be set up in such a way that they do not impede water flow into and out of the lake. In other words, sufficient water space would have to be left in between cages and between the cage batteries, and also underneath the cages. The farming of giant freshwater prawn in the cages can be undertaken for six months, covering the rainy season in between. Under cage farming system, with the benefit of removal of metabolites because of water flow, the prawn PL stocking rate can be pretty high. This implies that the production per ha can go up, say to the extent of 20 t/ha or more. Assuming that 2,000 ha of water area of the lake, out of its 20,000 ha are brought under cage farming, the production from the lake can go up from the present maximum level of 40 t to 40,000 t. Visualising another 1000 ha of marginal area are brought under production, there can be a further addition of 2000t of production. In the marginal ponds either monofarming of prawns or polyfarming of major carps/grass carp with prawn can be take up. In the capture area grass carp finger/ings/yearlings can be stocked along with common carp and rohu or as found practicable at 100 fingerlings per ha. Freshwater prawn seed can be produced within the State using constituted salt water at 14 ppt. If this is found to be not practicable, freshwater prawn postlarvae can be imported from States like West Bengal, Orissa and A. P. and used for stocking after conditioning the seed suitably in rearing farms to be set up. The farmers can be helped through subsidising suitably the cost of air transport of seed which can be shared by the State and Central Governments. Needed training programmes can be mounted taking the help of CIFE or other competent organisations.

In order to test the practicability of the suggested programme, a pilot scheme may be taken up in the first instance to set up and operate marginal ponds, one or two batteries of cages, and a rearing farm. The cage dimensions, the manner of fixing them, the system of safeguarding stocks in the cages during flood season, the feeding system, testing growth, watch and ward arrangements etc., may have to be well thought of and finalised before adoption.

Loktak lake is a great asset in the context that Manipur Slate in which the lake is situated, needs to augment its fish production. Further, once it is proved that giant freshwater prawn can be farmed in the marginal ponds and cages setup in the lake the prospects of their exports come to the fore, necessitating the setting up of ice plants and cold storages and probably processing plants having freezers and value adding equipments, preferably in the private sector. There is obviously an enormous scope in the direction of augmenting fish production from the lake for domestic consumption as well as for exports, and it would be worthwhile for the Government of Manipur to direct focal attention on the suggested approach which is pregnant with great opportunities of enhancing the economic conditions of fishermen and other stakeholders in the fisheries sector. This line of action has to be, however, strengthened through prior consultations with experts from Union Department of Animal Husabandry and Dairying, ICAR and Ministry of Commerce (MPEDA) as they can impart a desirable direction to the endeavour considering the export angle. Steps taken in a proper direction will undoubtedly contribute to the prosperity of the State. History tells us that any geographical entity replete with fish production thrives.

Fishing Chimes, Editorial 308: August 2002: Vol. 22, No. 5

# Dilemma of Indian Deep Sea Fishing

Indian EEZ's coastal zone and its adjacent waters are known to be overfished and this has been causing concern. The cause for this does not end here but extends to the Indian marine fisheries scenario beyond the coastal zone too, not because of overfishing but for other reasons. Capture marine fish production from the EEZ has reached a stagnant level of about 2.8 million t per annum against an estimated fishery resource availability of 3.9 million t. The strength of larger fishing vessels that operate beyond the coastal zone, some of which occasionally undertake deep sea fishing too, has plummeted from 190 to 65 nos. For want of avenues of employment because of this, a large number of candidates trained by CIFNETfor operating larger vessels are without employment. They can secure employment only when the fleet expands, and fleet expansion can take place only based on extent of unexploited resources and location of new resources. The ongoing deep sea fishery survey work to locate new fishing grounds is handicapped as most of vessels of the fleet of survey vessels of Fishery Survey of India (FSI) which undertakes viable survey activities have become old and also fleet must have shrunk. The estimated fishery resource status of Indian EEZ, as now taken into account, is considered by several fisheries professionals as subjective, and is believed by them to have been worked out on a low key only to prevent criticism of Governmental inaction in respect of augmenting commercial fleet. This, it is believed, has kept the exploited output stagnating. The feeling is that the approach in estimating the fisheries resources of EEZ is biased, and this bias is to explain away and justify inaction. The situation has been allowed to drift by the Union Fisheries Division to such an alarming stage that the financing bodies are reluctant even to accept an application for loan to acquire a fishing vessel, it is often commented. The responsibility for the emergence of this state of affairs is squarely on the shoulders of the Government, as it could not develop an enduring system of promoting and monitoring the activity, apparently because they could not know the true resource position. While the Indian estimate of the resources indicate that expansion of fishing effort in the zone beyond coastal waters through direct Indian investment is not worth the while, the opposite is the perception of foreign interests, who not only poach but are also interested in gaining official access to fish in the EEZ of India, some way or the other. Thus second hand foreign fishing vessels, acquired by Indian owners but totally operated by foreign interests, are doubted to be fruits of a plan, partly to compensate for fall in exports consequent to the shrinkage of Indian fishing fleet of larger vessels and withdrawal of charter/ joint venture schemes some years back. The official revival of the ingress of foreign vessels, through the second hand vessel import scheme, to operate in the Indian EEZ can only be an indirect expression of of repentence of withdrawal of charter and joint venture scheme. The replacement of the old scheme with an equally inappropriate new one tends to admit the Government's difficulty to develop the nation's own deep sea fishing fleet, despite its shrinkage to 65 nos. There can be one conjectural explanation that the resources are conservatively estimated bordering on under-estimation because of the situation that Indian industry is not in a position to invest, and Government has not announced any plan to provide financing support in this regard. Apparently for this reason, Government decided to permit import of used foreign vessels, ironically by those not in the main stream of the industry. Thus, the foreign interests regained access to fish in Indian EEZ in the guise of Indian owned vessels. These vessels are known to be totally under operational control of foreign interests who take away 100 per cent of the catches, in the form of exports made by the Indian owners, of course only after throwing back the low priced material.

The root cause of this situation is related to policy weaknesses, which are allowed to continue. In fact, Government has no deep sea fishing policy for the past several years, a really deplorable situation. All would agree that a major nation like India should have a policy to sustainably exploit its deep sea fishery resources. Unless the weaknesses are identified and steps taken to extinguish them, there can be no way to bring about sustainable marine fishery development of Indian EEZ to an optimal level.

The FSI has a hoary past. In the evolution to its present form, no doubt FSI grew in respect of surveys consistent with its strength. It is axiomatic that results of survey work should be the main basis for the estimation of fisheries resource potential. Unfortunately, FSI has handicaps in effectively extending its survey work to the entire deep sea zone. For this reason, whatever support it provides in respect of resource estimation to CMFRI, which by itself may be having its own limitations to work out a well found resource estimate, becomes a marginal contribution to the effort. In other words, FSI which ought to be the leader in assessing the fishery resources of the EEZ, suffers from the weakness of having old and inadequate survey fleet and consequential inability to be not in a commanding position to estimate the fisheries resources of EEZ, based on its survey results. So much so, CMFRI, which has the initiative of resource estimation, probably performs its function, withstanding whatever limitations it may have.

One Survey vessels Sagar Sampada, and a fleet of small survey/research boats that can operate only in the coastal zone do not provide the gamut of data needed to develop a well found resource estimate. In case CMFRI over-estimates, there can be an unleashing of criticism regarding non-utilisation of resources. Out of 2.8 million t of annual production, 2.2 million t come from the coastal zone, which is stated to be far in excess of sustainable level. Considering this, it is conceded by all concerned that the zone is over-exploited. The area beyond contributes just 0.5 million t per annum, against the situation that there is not much of conclusive evidence of the extent of resources beyond the coastal zone. Most of the direct evidence from the zone beyond the coastal zone had come from the operations of chartered foreign vessels that operated till recently in the Indian EEZ. Many consider that the declarations in respect of the catches given by the chartered vessels did not represent the true state of affairs and that they were adjusted from voyage to voyage as per the understanding between the owners and charterers. In this scenario, in the next round of resource assessment, the resource assessing authority can have two options. One is to frankly admit the constraints in the way. Second is to follow the unadmitted tradition of placing the resource estimate at around 25 per cent above the level of exploited resource status, as is being, by and large, followed right from the dawn of independence or probably earlier. It can however, be admitted that there is an inherent safety in this system. It does not embarass any one and will not highlight inaction. It provides justification for the declining deep sea fishing fleet of the country, and allows non-investment oriented introduction of foreign vessels through the kind of import system that has been ushered in *i.e.*, long liners for tuna, and trawlers for bottom and mid-water fishing in Indian EEZ to exploit the estimated marginal resources beyond the present level of exploited resources. It also justifies the absence of any major initiative to expand India's own fleet, beyond permitting import of used fishing vessels, and allowing their operation by the foreign owners from whom the Indian enterprises imported them. The other reason is that the resource position beyond coastal zone as estimated may not merit this. It also helps in the present situation of having no avenues to organise investments to encourage new vessel introduction. The unexploited resources being marginal at 25 per cent of the estimated resources, there can also be an argument on need for expanding fishing effort. The non-coastal fisheries resource scenario and the follow-up action to utilise the resources have all the ingredients of convincing ourselves that we are doing our best in the circumstances. Tuna is the main marginal resource highlighted to be available, and around 16 used foreign-operated longliners under Indian ownership with export tie-up are believed to have been introduced for operations in Indian EEZ. The number of imported foreign fishing vessels introduced and registered under Indian ownership are not known to the mainstream industry and the importers have apparently emerged as a new category, with no intention of joining the mainstream of the industry. So much so, information on the number of such vessels introduced and their operations are, by and large, restricted to the Union Fisheries Division, the new class of owners, those foreigners who operate them, the customs, probably coastguard, and the MPEDA who would have been consulted as export angle is involved. When the introductions are above board, there can be no reason for the Government to be so secretive, as is generally perceived.

One silver lining is that a pilot scheme conceived in 1996 to equip two Indian trawlers with longlining equipment has materialised in 2002. But for AIFI and MPEDA this initiative would not have been there and it is hoped that the initiative would open a new chapter in India's high sea fishing effort.

Having delineated the trend of views on the subject as above, and with the objective of provoking a discussion on this crucial topic, a point that is sought to be made here with some temerity is that the existing co-ordination between FSI and CMFRI (the later also conducts resource surveys), needs to be strengthened. This is to be done by way of ushering in a well structured and comprehensive surveys, that would facilitate coverage of survey work upto the deepest zone of the EEZ in a such a manner that meaningful, if not viable, results flow in, particularly in respect of deep sea squids and lobsters that are believed to be there as a garland in the deep sea zone all around the peninsula. The ultimate objective of the Government is one of estimation of the resources of EEZ as realistically as possible and in this context, it is essential to plan to utilise the resources through introduction of additional vessels as required, along with concommitant financing for acquisition of vessels, and also for crew training.

In the light of the foregoing observations, it is felt that it will be in national interest for the Government to set up a committee of experts, consisting of those within and outside the Government, to incisively examine the alarming situation and make recommendations on an organisational restructuring of the resource-related functions to be performed by FSI and by CMFRI, preferably by amalgamating them under one leadership so as to achieve well found estimation of the resources and plan for optimally sustainable utilisation of them over a reasonable period of time, with the needed linkages with the commercial deep sea fisheries sector from the angle of domestic fish supplies as well as exports.

Fishing Chimes, Editorial 309: September 2002: Vol. 22, No. 6

# Regulation of Shrimp Aquaculture

The Aquaculture Authority was notified by the Union Ministry of Environment and Forests on 6 Feb., 1997 in compliance with a directive given by the Supreme Court in its judgement of 11 December 1996. The purpose of the directive was to have a competent authority to regulate shrimp aquaculture in India, (which is mostly confined to the Coastal Regulation Zone which extends to 500 m from the coastline), in a manner that would ensure development of the activity in a sustainable and eco-friendly manner. The operative part of the direction was that only traditional shrimp farms as in existence at the time of the aforesaid judgement in the CRZ should be allowed to continue farming activities. However, there was a relaxation that they may also adopt improved traditional farming system, thereby noting that semi-intensive and intensive farming operations should be totally given up. Comprehensive guidelines in regard to traditional and improved traditional farming have been recommended by a Committee set up later by the Authority. According to this Committee, the adoption of these guidelines shall enable farmers to achieve production ranging between 1 to 1.5 t/ha/crop. Two crops per annum can be taken leading to an annual production of 2 to 3 t/ha, according to the Committee.

A Sub-Committee was also set up by the Authority to formulate guidelines for incorporating effluent treatment systems as part of shrimp farms. Accordingly, this Sub-Committee, which took into account the existing legal provisions in respect of effluent treatment and the norms for sustainable development and management of brackishwater aquaculture prescribed by the Ministry of Agriculture and the mandate of the Authority that all shrimp farms of 5 ha waterspread and above within the CRZ and 10 ha and above of such area outside CRZ should have effluent treatment system, recommended a set of guidelines for being followed by the farmers. The Sub-Committee also recommended that farms having a water extent upto 2 ha could be grouped and an effluent treatment system for each of such groups might be established.

These recommendations incorporate a design and lay out of effluent treatment system, and water exchange schedule too, and it is supported by a cost estimate for the construction of a typical effluent treatment system in an area of 0.5 ha.

The guidelines recommended by the aforesaid Committee and Sub-Committee, constitute the spring board for the field application of the directives of the Supreme Court. As a measure of further follow-up action, the Authority ensured the setting up of State and District level Committees to receive, examine, and submit applications from farmers for registration of their farms in the CRZ, in a prescribed form, to the Authority for consideration and registration on merits.

Although around two years have elapsed since the time farmers have been told to apply for registration of their farms with the Authority, it is stated that only a small number of farmers have submitted their applications so far. Appeals, exhortations and articulation of consequences of not applying for registration, such as likely demolition of unregistered farms have not changed the situation perceptably.

As long as the farmers keep their operations traditional and improved traditional, many believe that the position of such farmers would be in conformity with the orders of Supreme Court. It has not probably registered on their minds that registration of farms is a mechanism for regulating the activity, particularly in respect of effluent treatment and it would be in their own interest to have their farms registered. This registration requirement has become a long drawn process, and uncharitably believed by many that it should not come in the way of expediting the process of setting up a network of effluent treatment system all along the belts having shrimp farms. They feel that an alternative way of achieving the installation of effluent treatment systems at the various farms with an extent of 5 ha and above could be found, pending registration of farms, so as not to delay the critically needed effluent treatment systems. The farmers are no doubt inexcusably tardy and probably evasive in respect of registration but, the fact that they contribute to over Rs. 2,000 crores of foreign exchange through shrimp production they achieve, presumably well within the norms of traditional and improved traditional farming and within the limitations of their rural background, may, as a consolation, justify the adoption of another approach to hasten the setting up of effluent treatment systems which would not transgress the orders of the Supreme Court. Any alternative approach, several say, need not be one of diluting the requirement of registration, but one of according greater importance to effluent treatment of waste farm waters which is of considerable urgency from the point of view of pollution control.

There are observations overheard that the Authority can direct the Coastal State Governments for undertaking a survey of the shrimp farms in CRZ along their respective coastlines through their fisheries departments and prepare an inventory of the farmers and their farms with whatever essential or other details that can be ascertained. On the basis of such surveys, a state-wise plan of setting up effluent treatment systems, farm-wise, or on the basis of groups of farms, as the case may be, can be worked out. Such plans can incorporate centralised treatment systems involving settlement/ sedimentation ponds/biological treatment ponds for feasible stretches as may be determined.

The formulation of such a programme can probably be taken up simultaneous to the on-going process of registration. In fact, when work starts on the survey and planning of effluent treatment system, the farmers would be motivated to register themselves, with the hope that the Government may share the cost of waste water collection and treatment before release into open waters. In other words, in the interests of pollution control, farmers may appreciate the work to commence at the distal end of the objective and converge ultimately at the registration point backwards.

Fishing Chimes, Editorial 310: October 2002: Vol. 22, No. 7

# Welcome Leads

The fisheries sector now has two conspicuous welcome openings to take the sector forward. One of these is related to culture fisheries front and the other to marine capture fisheries activity. There are indications that these leads are being followed up by the authorities and it has to be hoped that in the tenth plan there will be encouraging progress in widening and spreading the related follow-up work with the transparency that it deserves to enthuse the industry and to accrue benefits to the nation.

The culture fishery lead relates to bringing the tanks and ponds of the heartland of the country, under giant freshwater prawn culture. An initiative in this direction, to bring the lentic waters of the northern coastal plains of the country under freshwater prawn farming is now in progress. An approach of this kind was once considered as dreamy and impractical. Nevertheless, this innovative idea has been now taken by some in authority and several enterprising farmers as a challenging proposition, considering its positive and beneficial aspects. Consistent efforts over past several years have now provided a measure of momentum to field activities related to the endeavour, conspicuously in Haryana and Punjab, and these have to be pursued further for being translated into the realm of total reality.

In fact, with the financial support of the Centre, a project with substantive outlay has been taken up in Haryana. In Punjab too, Jawans of a CRPF camp near Jalandhar and some farmers in the State are now engaged in the venture. A commercial activity of taking prawn seed from Orissa and A.P and feed from Kerala and Tamil Nadu has come into being. In Tripura, Madhya Pradesh and Chattisgarh too the beneficial aspects of this diversified activity have come to be recognised. There has been an initiative to set up a freshwater prawn hatchery using constituted salinewater in Madhya Pradesh, in fact the first one outside the coastal belt. In Tripura, PLs of the prawn are understood to have been produced by Mr. Bhattacharya, an enterprising scientist of the State Fisheries Department, using constituted salinewater and PLs are also understood to have been supplied to the Fisheries Department of Manipur, another State very keen to introduce the giant prawn into the State for culture. In Rajasthan, prawn under culture have reached maturity and the stage is set for the breeding phase. The work is monitored by Dr. Jain of CIFE. In Chattisgarh, farmers have made some headway in farming prawn in collaboration with Ananda Foods of Bhimavaram, A.P., securing supplies of seed and feed from the same State. Mr. Jamil Ahmed, a retired Director of Fisheries, Orissa set up a prawn hatchery near Raipur which is expected to become operational soon. At the time of writing this editorial, encouraging information has trickled through from Punjab that the Research Centre of CIFA at Ludhiana has succeeded in raising PLs of the prawn at a hatchery set up with DBT assistance at the place.

The foregoing narration clearly indicates the need for consolidation of the efforts and providing a focal direction to the same for expansion. This is particularly necessary, considering the export potential of the prawn.

It is quite possible that the authorities concerned are on the job in silence, but there will be generation of enthusiasm and vibrancy when there is transparency and when there is a nation-wide integrated project covering the hatchery, nursery and farming components of the activity taken up in inland States by the Centre. Such a project, with provision to extend technological inputs and also financial help to the States, besides having a significant component of training to selected field staff/ farmers in respect of hatchery design, its setting up and operation, and nursery and farming operations and management, would prove to be a great step forward. CIFA could be the main force behind the hatchery design, hatchery operations and training programmes. An infrastructural network for storage, processing and export covering identified points to constitute the network, would have to be planned to be set up, preferably in the private sector by providing the needed incentives by way of land and other facilities with provision of subsidies. A list of technologists experienced in the line can be drawn up and circulated among the States to avail of their services in an appropriate manner. The State Governments concerned and the farmers have to prevail upon the Fisheries Division of Union Department of Agriculture, CIFA of ICAR, and State Departments of Fisheries and MPEDA to participate in this major project for benefitting the farmers, the industry and the nation.

It is laudable that, as a beginning in this direction, Haryana has taken up with central assistance the implementation of a major project on promotion of freshwater prawn culture in the State with a total outlay of 3.20 crores. It is to be hoped that the other States such as Rajasthan, U.P, Madhya Pradesh, besides others, would emulate Haryana's initiative. The promotion of the farming system in existing tanks and ponds and in farms that can be developed in the potential wastelands, vast in extent, particularly in the north-west and in coastal areas, will lead to all round prosperity.

The other lead we have is in the high sea capture fishing sector. The results of exploratory surveys carried out by Fishery Survey of India, have been providing clear estimates of existence of sizeable tuna stocks in the EEZ. These results could not be translated into an enduring commercial activity all these years. As a prelude to promoting utilisation of tuna resource potential, although belated, MPEDA, in association with the Association of Indian Fishery Industries took up, under a Pilot Project, the installation of monofilament long lining system on two shrimp trawlers of 23.50 m LOA. Trial longlining with these two vessels for a few days in the high sea zone off Visakhapatnam yielded encouraging results. These results have generated instantaneous interest among mini-trawler (15 m OAL) operators too. Their association is now engaged in working out a project in this regard to be submitted to the Government for help. Small mechanised boats can also be equipped with tuna longlining equipment.

As MPEDA is aware, the real challenge lies in a) having a sizeable number of the trawlers of the existing fleet equipped or monofil longlining (as introduction of new longliners is prohibitively expensive), and b) organising post harvest facilities. The two Pilot Project vessels are now exposed to the disadvantage of constraints in marketing of the catches in a sashimi form when only the returns will be remunerative. The reason is that, for earning viable returns, the exports have to be in sashimi style and for this the fishing duration of tuna vessels would have to be limited to around seven days and in any case not more than ten days, for quality reasons and for earning a good price. The export has to be in chilled condition packed in ice. The catches can be sold in frozen condition too to canneries also but the returns will be far lower. For export, packed in ice, the catches have to reach the air port close to the exporting centre for immediate shipment to Japan, the main importing country. The export has be done in ice-filled coffins. Air transport of small consignments pose the problem of shipments from an Indian airport to a Japanese airport (Tokyo) as this involves a couple of transhipments, to be well taken care of. Even under a marketing collaboration arrangement, this practical problem will have to be solved, only by inducting a tangible number of existing vessels, equipped in the same way as the pilot project vessels, for tuna longlining. Once there is a sizeable fleet, group fishing of vessels will be facilitated, and the periodic landings in seven-day intervals will facilitate direct air lifting by small planes from a chosen Indian airport to Tokyo. The need of the hour, therefore, is to find the quickest way of inducting atleast around 30 vessels, for undertaking tuna longlining, as recommended by the Round Table conference of 1999, to realise the objective of the exercise and to enthuse the two companies who offered their vessels for the installation of tuna equipment under the pilot scheme and following it up until the job is done. The main objectives of the pilot project, as is known, are to induct diversification of fishing effort, which is now predominantly shrimp-oriented, so as to exploit the available tuna stocks, for improving the sagging economics of operation of the vessels, and to add to export earnings. There is confidence in the industry that, under the leadership of MPEDA and AIFI, the purpose of the pilot project would be taken to its logical end, instituting all needed measures in this regard on a war footing.

Fishing Chimes, Editorial 311: November 2002: Vol. 22, No. 8

# Alarming Signals

The fisheries sector is at cross roads, causing concern. Let us first take a look at the marine fisheries sector which has now come to be characterised by certain depressing features. The economics of operations of small mechanised boats are in total disarray. Of late, even during prime fishing seasons, the operators have been experiencing a major drop in per unit catches, particularly of exportable species (crustaceans and cephalopods), on which the economics of operations depend. As if this is not enough, as a major contradiction, the export prices, particularly of shrimp have been sliding, coupled with a pronounced negative aspect of raising diesel prices and other operational expenses. This trend that has set in portends bad times. Finding a way to reduce the number of mechanised boats in order to enhance per unit catches can be one of the options to correct the situation but this is very difficult to excercise, considering the socio-economic implications. Another option can be to introduce a system of controlling additions to the fleet and banning replacement of decommissioned boats. In this situation, an additional but a positive initiative that deserves preferential application is to explore the possibilities of upgrading the capabilities of the boats for midwater trawling/longlining. In this respect, the Central Institute of Fisheries Technology has to be mandated to take up R&D work in close association with Coastal State Fisheries Departments. The State Fishermen's Training Centres can play a participatory role in providing training to fishermen in the operation of these systems of diversified fishing. In fact, the Union Department of Animal Husbandry and Dairying, in association with the Coastal State Fisheries Departments could sponsor a pilot project, to start with, covering all Coastal States, for implementation by CIFT. The recommendations forthcoming as a result of the implementation of the pilot project could be followed up suitably.

Another alarming signal that has emanated loud and clear is the declining fleet strength of 23-27 m LOA range vessels. Unable to withstand the sharply declining economics of operation, several companies sold their vessels to foreign companies. Further, many vessels have fallen into disuse. These developments reduced the fleet strength to around 65 from about 190 nos. Of the remaining vessels, the ownership of a good number of vessels changed hands, those selling them recouping losses to the extent possible, and those buying them, mostly skippers, embarking on new fishing initiatives.

In this background, a pilot project has been taken up by MPEDA for equipping two trawlers for undertaking monofil tuna longlining as a measure of diversification and this initiative has emerged as a hope to resurrect the languishing economics of operation of larger trawlers. The two trawlers equipped for longlining under the pilot project have now completed their first commercial voyage and exported part of the catches, in sashimi grade although modest, because of several factors. In this first trial voyage, however, considerable experience is understood to have been gained in respect of tuna fishing grounds, although they could not as yet locate thermocline zone close to which tuna hovers. It is learnt that the owners of the pilot vessels mainly face three problems. One is the location of grounds. The vessels are understood to be not equipped with sonar or other equipments for locating shoals of tuna. The other is the cost of imported bait. When import of bait becomes inevitable, there is however an expectation that the MPEDA may subsidise the cost for some time. The third problem is the high air freight for export of sashimi tuna from India to Japan. There are indications that MPEDA may subsidise this cost also until the activity picks up. Another aspect is that, there seems to be a conspicuous absence of co-ordination between the owners and the Fishery Survey of India. The later has considerable experience in tuna exploratory surveys. In this context, it appears that there will be an improvement in the operations, if FSI too is made part of the pilot project, particularly as the Indonesian crew now operating the vessels do not seem to be having adequate knowledge of tuna grounds of Indian EEZ. A point to be noted in this context is that, among those having close knowledge of Tuna grounds of Indian EEZ, Taiwanese master fishermen appear to be in the forefront, with over ten years of successfully recorded exploitation of tuna resources of the Indian EEZ. The desirability of securing the services of a Taiwanese professional master fisherman for locating the grounds and to help in post-harvest operations therefore needs to be looked into with the urgency it deserves, particularly as the availability of resources is confirmed by CMFRI, and as we have no expertise to reach the correct grounds in the EEZ. It is, however, encouraging that the Japanese contracting company which has provided technical services for conducting tuna fishing operations by the two vessels is striving to the extent of being equal to the challenge, which includes the problems of post harvest activities, particularly in relation to export marketing. With tuna catches from just two vessels, that too from unfamiliar grounds, organising preservation, packing and exporting them is a tough job.

Considering the nation 's inexperience in monofilment longlining for tuna and absence of a financing channel to equip more number of trawlers same way as the pilot ones for tuna longlining, it may be worthwhile to explore possibilities of finding a friendly and knowledgeable foreign country having experience in the field and interested in investing in the installation of tuna longlining equipment on existing shrimp trawlers, and also on the needed shore infrastructure. Such a foreign country can extend technical assistance in respect of operations including training to Indian crew and also help by undertaking export marketing. The recovery of investments could probably be governed by acceptable provisions in the agreement to be entered into between the Government of India and the foreign Government concerned and by entering into back-to-back agreements between the Indian owners and the Government of India.

The next coginisible alarming signal is that columnar fishes of Indian EEZ continue to be the least exploited. A project for exploiting these resources, encompassing all categories of vessels including mechanised fishing boats would need to be taken up, coupled with the setting up of needed post-harvest facilities.

So far as deep sea bottom fishing from the Indian side in the EEZ is concerned, it is known that at depths of 400 m and over, there has been practically no commercial fishing effort. There is some estimate of demersal resources available at these depths, the reliability of which is however not certain. In any case, it would be difficult to believe that the deep sea demersal zone is barren. In order to utilise these resources, a commercially-oriented exploratory survey by FSI in collaboration with a private sector company would deserve to be taken up, by obtaining a used vessel equipped for deep sea trawling from a developed country, linked to training of Indian personnel and processing of catches in value-added form and marketing.

In the brackishwater culture sector, the major problem faced, as is known, is the widespread incidence of white spot virus disease among tiger shrimps under farming. Farm management, ensuring hygienic and fertile water conditions without allowing waste matter to gather and accumlate in farm ponds, is fast catching up among most of the farmers. With inproved water management system gaining ground, the other problem to be tackled will be one of stocking the farm ponds with disease-free shrimp post-larvae. The present system is one of utilising wild shrimp brooders for production of post-larvae at the hatcheries. Testing of brooders before subjecting them to breeding, and of postlarvae before stocking, is being done for positive or negative disease bearing virus signals but the results are said to be not that reliable. The ultimate solution identified is to develop domesticated shrimp broodstock. CMFRI could make some progress in this direction. While the dimensions of shrimp domestication problem are known, and the work thereon is presently limited to CMFRI, the depth and the extent of effort needed to expand the work and achieve results do not seem to have registered on the minds of the authorities concerned. Thailand has progressed to a stage of $F_3$ (or so) generation of domesticated tiger shrimps. The enormous effort that must have been put in to reach this stage can be visualised, taking into account the vast efforts put in so far to reach F1 (or so) stage in India. Probably realising and estimating the extent of further effort needed for establishing an enduring and sustainable source of disease-free tiger shrimp brooders in Thailand, M/s. Inve, a Dutch company of international repute has now been involved by authorities in that country for strengthening the programme to take it to its logical finale. Australia too, which has progressed upto $F_8$ (or so) stage of domesticated brooder tiger production, is hesitant to provide hatcheries with supplies of domesticated shrimp brooders developed, apparently waiting until establishing a firm base that would ensure sustained supplies of brooders to the hatcheries.

Such being the position, the obvious conclusion one can draw about the work on the subject now going on in India in the hands of CMFRI can be termed, in comparative terms, as too unequal to the dimensions of the effort and the experimental infrastructure needed. A project of this kind requires a major direct support from ICAR, commensurate with the enormity of the problem. For a crucial purpose of this kind, so intertwined with the socio-economic life of the farmers, and export earnings, the project on this subject should have been accorded top most priority by ICAR, and CMFRI should have been allowed to take up a collaborative project under its direction with the work extending into the arena of the shrimp hatcheries in the private sector with financial help flowing in for expanding the working base of the brooder development work. It is difficult to conceive of a project more deserving than this for the closest attention of ICAR/Union Department of Animal Husbandry and Dairying. Seeking help from Department of Biotechnology, as is being done now, indicates the lack of interest on the part of ICAR in supporting this important project that deserves direct and active support financially and otherwise by it.

Turning to the freshwater sector, it has to be mentioned that a serious signal that is gaining in strength is the declining incomes of the farmers of major carp culture sector. This trend can be stemmed only by providing adequate preservation and storage centres for enabling sustained marketing based on demand situation, coupled with transport facilities and creating a demand for Indian major carps in foreign markets. The external market for these excellent fishes, well comparable with prime fishes of other countries is now unfortunately extremely limited but through a publicity strategy and organisation of exports of these fishes with Central support, a remarkable reform that would lead to enhanced national benefit could be ushered in.

Another aspect that is gaining ground is giant freshwater prawn culture in the northern States. The growth of the activity has to be fostered by the State Governments concerned, taking cue from the initiative excercised by the Government of Haryana in this direction. The promotion of this activity (from production to export) will be an uncommonly major excercise, calling for production promotion coupled with seed transport from centres of availability to farm sites, setting up hatcheries in the northern belt wherever suitable saline water sources that are available can be utilised for running shrimp hatcheries and, by provision of infrastructural facilities, etc.

In respect of all the measures needed for taking positive measures to meet the challenges posed by the warning signals, the initiatives have also to spring from the fisheries associations, related to both culture and capture fisheries, in order to strengthen the Governmental approaches that could emerge, so as to induce the authorities concerned to concede the attention that the signals deserve for immediate action.

Fishing Chimes, Editorial 312: December 2002: Vol. 22, No. 9

# Fishing in Indian EEZ: Guidelines

The Government of India (Ministry of Agriculture, Department of Animal Husbandry and Dairying) has recently issued an order, along with guidelines for conducting fishing operations in the Indian Exclusive Economic Zone, as an enclosure. The guidelines have been issued by the Union Department in the capacity of being the nodal Department for developing fisheries in the Indian Exclusive Economic Zone (EEZ), under provisions of entry 57 in list 1 of the Schedule of the Constitution. This order, bearing No. 21005/1/2001-FY (Ind) dated 1" November, 2002 says that the guidelines will be binding on all deep sea fishing vessels operating in the Indian EEZ from 1st November, 2002. It is also mentioned that any violation of guidelines by the vessels would be viewed seriously and penalty/punishment as deemed fit would be imposed on the defaulter. There are 21 guidelines and 2 annexures (Annexure A-l, A-2; and Annexure B) to the guidelines.

These guidelines, as ordered, could be expected hopefully to serve the purpose of regulating/monitoring deep sea fishing operations in the Indian EEZ. They can be considered as opening up a new chapter in the history of Indian deep sea fishing industry and this important development is apparently visualised by the authorities as a measure to promote deep sea fishing in Indian EEZ, which is, unfortunately, now in a bad shape.

The measure would have been more reassuring if it is more clearly related to the legal aspect too, but this may probably be unimportant. At present, the nation has only two legal provisions governing sea fishing. State-wise Marine Fishing Regulation Acts are one of them. The other is the Maritime Zones of India (Regulation of Fishing by Foreign Vessels) Act and the Rules notified hereunder.

These notified Rules are applicable to foreign vessels only and not to Indian owned vessels. In other words, no legal basis appears to exist at present to regulate operations of Indian vessels beyond the territorial waters. Had the provision in respect of deep sea fishing in the constitution was adequate to regulate fishing in the Exclusive Economic Zone, Government would not have enacted Maritime Zones of India Act (Regulation of Fishing by Foreign Vessels), but would have issued guidelines as has been done now.

The definition of a deep sea fishing vessel is given in the enclosure to the order as fishing vessel of 20 m overall length and above. The purpose of this definition, as seen from the guidelines is to prevent stern demersal trawling by them in the zone upto 12 nm from the shoreline and also upto a distance of 24 nm from the shoreline on both the coasts in specified zones. On the east coast, this zone extends from Nizampatnam to Paradeep. The purpose of banning stern demersal trawling in the entire EEZ seems to stem from the new system of according permissions for import of used deep sea fishing vessels, with flexibility in respect of payment of the cost involved on deferred basis where applied for, and also to have joint ventures for operations of the vessels as well as for exports, apparently after discards. If these special category of vessels are allowed to undertake demersal. trawling as in the case of other Indian vessels for export of catches (no doubt after throwing out uneconomical species), fishing pressure on bottom resources can be high. Therefore, the solution thought of seems to have been to ban all bottom trawling by deep sea fishing vessels. This provision in the guidelines has however left, an exception that arises, uncovered under this concept. Around 18m OAL trawlers based on DANIDA 's Tadri design now conduct deep sea stern trawling upto a depth of 150m on the west coast. The present definition enables them to continue demersal trawling as at present beyond territorial waters, which is an obvious advantage to them and a disadvantage to vessels of 20 m OAL and above. In fact, vessels of 20 OAL and above now conduct stern trawling within territorial waters also, but outside the zone of operation of small mechanised boats. Under the present guidelines they cannot do so.

The guidelines refer to 200 shrimpers of 20 m OAL and above operating in the Indian EEZ. There are, however, only 65 such vessels now operating in the zone. The owners of these vessels are now in the throes of a serious economic crisis and, if they have to stop bottom trawling for which the above vessels are equipped, their economic plight can be terrible. The owners of these vessels hope that the Association of Indian Fisheries Industries would take up the legal and other aspects of the issue emerging out of the related guidelines with the Union Department of Animal Husbandry and Dairying for a solution, without the need for a legal approach, in case the guidelines are not considered legally valid.

The first sentence under the first guideline starts with the expression "permission in writing (LoP) is required from the nodal ministry".... What the abbreviation "LoP" means is not clear, as it does not tally with the preceding expression "permission in writing ". LoP may mean Letter of Permission but it cannot be presumed that this abbreviation gives that meaning, as the Order is not supposed to be based on conjectures.

It is stated in the covering order that "the guidelines will be binding on all deep sea fishing vessels operating in the Indian EEZ from the date of issue of the order". The guidelines have to be binding on the owners of the deep sea fishing vessels and not on deep sea fishing vessels, being inanimate entities. The deep sea fishing vessels do not violate the guidelines (as mentioned in the order) but those who operate the deep sea fishing vessels can violate the guidelines.

Another point that has to be mentioned is that Entry 57 in list 1 of 7th Schedule of the constitution does not make any reference to EEZ. It covers only that part of the sea which is beyond 12 nm from the shoreline, which now happens to be part of the recently declared EEZ.

As per the provisions in the guidelines, five types of fishing by vessels of 20 m OAL and beyond are permitted. These do not include bottom trawling. As is generally known, there is no bottom stern trawling conducted by Indian vessels now at a depth beyond 200m and in any case beyond 500 m depth. In other words, the demersal resources from this depth onwards are not at all exploited by the Indian deep sea fishing industry. In this context, it appears that there should be provision to allow deep sea demersal stern trawling at least beyond the depth of 500 m, if not beyond 200 m depth. Another aspect to be noted is that vessels permitted to conduct mid-water/pelagic trawling can switch over, without being noticed, to bottom stern trawling. Firstly, the rationale for preventing the exploitation of un-utilised resources beyond around 200 to 500 m depth is not clear. Secondly, even presuming that this seemingly unreasonable stipulation is correct, it may not be practicable to prevent bottom trawling at these depths.

There are two sets of Annexures to the guidelines, numbered A-l and A-2, which can cause confusion.

The guidelines, whether valid under law or not, appear to need a further look by the Union Department of Animal Husbandry and Dairying.

Fishing Chimes, Editorial 313: Jan/Feb(i) 2003: Vol. 22, No. 10&11

# Shrimp Export Scene

## Brewing Battle between Black Tiger and Vannamei Shrimps

### A Portending Development that may Impact Indian Shrimp Exports

The adage that says big fish eat the small may well get reversed as the brewing battle between Tiger and Vannamei shrimps becomes decisive. *Litopenaeus vannamei* (also known as *Penaeus vannamei*, and popularly called as Pacific White Shrimp and as 'Chinese White' in U.S. market), in terms of its relative merits as an export commodity, has unleashed and kept alive a debate on the captioned issue over the last two years.

It is common knowledge that shrimp has a prominent place the U.S market. Species like *Penaeus setiferus* and *P. vannamei* are most extensively farmed in almost all the Central and South American countries like Mexico, Brazil, Ecuador etc. and their market destination is mostly USA. C.P. Bahari (CPB) shrimp project in Indonesia has been producing 100 mill PL of *P. vannamei*/month for stocking its farms. Indonesia produced 5000 mt of cultured *P. vannamei* in 2002 which is expected to go up to 20,000 mt in 2003. In this country, while tiger production is 4t/ha, vannamei production is 10 t/ha. This shrimp is now the leading farmed shrimp in Taiwan with 15,000 mt of annual production compared to 10,000 mt of *P. monodon* per annum. A trend towards vannamei farming is distinctly visible in Thailand, Philippines, Vietnam etc., and of course in China. China has made commendable progress in Vannamei farming. For the year 2002 it was expected to produce around 3,50,000 mt of Vannamei shrimp.

The American importers and consumers, wooed by the lower prices, appear to be now increasingly veering towards this wonder shrimp *P. vannamei* which they now tend to rank on par with black tiger in terms of quality, taste and appearance. The import of black tiger in quantity terms into USA has been however on the rise continually despite the down turn in the economy and invasion of Vannamei shrimp imports.

This scenario apart, Vannamei has now surfaced as a potential competitor, if not a Black Tiger exterminating one and is posing a threat to black tiger's survival as a major shrimp export item. In the aqua export front, most of the Asian countries like India, Thailand, Vietnam and Indonesia are solely dependent upon black tiger farming as a major source of earning foreign exchange. This sole dependence on a single species is now inflicting damage and has become a risky factor for these countries, in view of high capitalisation of the black tiger farming as part of export industry.

Vannamei is seen as having distinctive advantages both to the farmers and consumers alike, as compared to the black tiger. Low production cost and high average farm output (around 4 to 5 mt per hectare) are the clinching factors that have been attracting the farmers towards Vannamei, while low product price and its physical resemblance to wild caught shrimp, *Penaeus indicus*, lures the consumers at retail level. The Chinese experience, in general terms, with Vannamei has been described as so good that the shrimp farming in the country is becoming synonymous with Vannamei farming. In view of its inherent merits now noticed, market for Vannamei, especially in U.S. is expanding, and it appears that no alternative is left for the shrimp farming industry of India too to respond in conformity.

Coming out of the initial reluctance, countries like Thailand and Vietnam, after taking cognisance and fully appreciating the writing on the wall, have recently given final nod for Vannamei farming in their respective countries. Needless to say, Vannamei production and export obvisously will be at the cost of black tiger. This is already proven and in fact felt by other shrimp exporters to U.S. As Vannamei is farmed in an intensive mode, medium and smaller count shrimps from 31/40 through 71/90 are harvested more in quantity and in these count segments black tiger is bound to suffer as its prices are destined to be uncompetitive. Trends in this direction are already visible even at the existing demand-supply levels and a much clearer picture will emerge once Vannamei supply increases.

So far as India is concerned, the situation seems to be not much different. The shrimp farming and export industry has already been in troubled waters since February 2001 due to factors like economic slow down in importing countries like Japan and USA and also failure of black tiger crop in the wake of outbreak of viral diseases. Shrimp farmers and exporters in large numbers are increasingly vanishing out of the industry with huge losses which indicate that the Indian industry is again heading towards a major threat to its survival, with Vannamei farming spreading at a fast pace in other countries, and being severely outside this trend, with all the accompanying disadvantages. The Vannamei invasion appears to be relentless. It is not an exaggaration to say that, till date, few of the Indian shrimp farmers know or have knowledge about Vannamei farming. It has to be confessed that Indian farmers are less organised and knowledgeable than their counterparts in South-East Asian countries. They need to be properly guided and trained with respect to new trends emerging in the global markets and the necessity to adapt themselves to the changing times.

Commencing from the ensuing summer crop, reports indicate that all the shrimp producing countries in South East Asia are planning to shift over, to a major extent, to Vannamei farming. It is high time for India too to explore the possibility of taking up Vannmei farming on a large and expanding scale, taking into account the ecological aspects, which, hopefully, may not be different to those related to our own native white shrimp stocks. In the same way as *Clarias gariepinus* spread all over the country, there are possibilities of Vannamei too spreading all over the shrimp farming zones of the country as two enterprises already have permissions to bring them in. Looked at from any angle, this topic requires policy inputs, organisatinal effort and deep insight into all relevant aspects to monitor both the spread and or control of the spread of the species.

The following areas are specially very critical for Vannamei farming in India: i) Identification of Shrimp farming regions suitable for Vannamei culture; ii) Importation and development of proper broodstock; and iii) Provision of hatchery facilities for producing high quality PLs of Vannamei; iv) Regulation and monitoring of Vannamei farming by providing technical assistance to the shrimp farmers, and Product development and marketing assistance to conform with global standards.

The endeavour encompassing the above areas would be very expensive and unaffordable in the hands of individual farmers. Governmental agencies/bodies like MPEDA would have to necessarily play a promotional role as was done in the case of black tiger seed production and farming in late 1980s.

To start with, it is advisable for MPEDA to consider persuading its own hatcheries, OSSPARC and TASPARC, to go in for importation of healthy Vannamei broodstock from countries where the species is disease free and take up the production of Vannamei PLs for supply to the farmers. Unlike black tiger, disease-resistant broodstock of Vannamei has been developed in Hawaii and in several other places. Hawaii is in the forefront in supplying SPF (specific pathogen free) and SPR (specific pathogen resistant) broodstock of Vannamei. For a healthy and sustainable Vannamei farming the quality of broodstock is of great importance. Its supply and quality requires to be regulated to avoid failures at a later date. Entrusting initially this important job to the private hatcheries may not be a step in the right direction. Instead, it is desirable that organisations like OSSPARC and TASPARC, who have pioneering experience in introducing shrimp seed production technology of this kind into the country, can be actively involved in the subject in the wider interest of the industry. Once the species is domesticated at these hatcheries, the private hatcheries can be encouraged to go in for production of Vannamei PLs. OSSPARC and TASPARC today do not have much of a role to play in the matter of black tiger seed production and as such they can focus on this important diversification effort in order to transform the future of the Indian shrimp farming industry on the right lines, as the conditions now demand.

Fishing Chimes, Editorial 314: Jan/Feb(ii) 2003: Vol. 22, Nos. 10/11

# Indian Fisheries Sector: 2002

It can be said that the year 2002, by and large, ended disappointingly for the Indian Fisheries Sector. Yet, an incisive appraisal of certain events indicates leads towards progress.

India is not alone in its quota of disappointments in 2002. Most of the nations world over experienced bad fishing time during the year. In India, failure of monsoons, changes in the hydrobiological conditions of water resources, dwindling fishing fleet, and, disease incidence in shrimp culture sector can be perceived as some of the reasons for the debacle, among others.

Looking at the brighter side, a highlight that can be perceived is the launching of a valiant pilot effort to introduce monofilament long lining system for catching tuna, the available stocks of which in the Indian EEZ were estimated by CMFRI at 640,000 t. The stocks were well exploited until recently by Taiwanese longliners. In this light, preliminary results of the pilot effort, although not encouraging, lead to the conclusion that the dimensions of the effort would have to be on a scale adequate to establish an integrated operational system from the stage of locating tuna stock to the logical end of exporting the harvested ones in sashimi form, taking timely care of the crucial linkages in between, with professional inputs. Results of longlining with just two vessels that too not having sonars to locate grounds seem to have left the problem of location of tuna shoals that move very fast, by and large, unsolved. Further, the small number of tunas caught in any of the few voyages of the vessels emphasises the need for group fishing, linked to effective location of shoals, in order to generate economically adequate quantities of tuna for airlifting them to the Japanese market in sashimi form. It is difficult to reconcile ourselves by equivocating that the estimated tuna stocks in Indian EEZ, made by CMFRI were not proven, in the face of the situation, to say again, that Taiwanese longliners that operated earlier in the Indian EEZ were swinging into their vessels good quantities of tuna, and the recently acquired used longliners by Indian enterprises too continue to do so with foreign crew. The present situation, to say the least, is anomalous and there has to be something wrong somewhere. In this context, it would appear that our own inadequacies in respect of the dimension of the endeavour have to be blamed. It will be, however, too far fetched to blame interested global tuna fishing nations of some kind of conspiratorial manipulation to confine India to its traditional fishing effort around Lakshadweep islands and to block fishing beyond so that tuna of Indian EEZ will continue to be in the non-Indian domain. This and other problems related to production, preservation and export management are no doubt receiving the closest attention of the authorities.

A progressive measure that was introduced, as already mentioned, was to allow import of used tuna vessels by Indian enterprises, on deferred payment basis, despite certain disadvantages. There was, however, one advantage in that, without any immediate investment, it had become possible to augment production and foreign exchange earnings. This development was no doubt a blessing in the present grave situation of a steep decline in the Indian fishing fleet of vessels of 20m OAL and above to a level of 65 nos. from around 195 of them, in recent years. The problem of investments apparently came in the way of restoring the number of vessels in the field towards the earlier strength, and commensurate with the resource position. The investments problem being very intricate, the decision to permit import of used vessels must have become inevitable, with probably no other non-compromising alternative to adopt.

Another event of significance during the year as it drew it a close was the notification of guidelines for operation of deep sea fishing vessels in Indian EEZ.

An unique continuing development was the strengthening of the successful initiative of marine capture fishery enterprises on the west coast and on Kanyakumari coast for the operation of vessels of around 18 m OAL for conducting deep sea demersal trawling. Some vessels had also taken to gillnetting, and operations for catching tuna on the west coast with encouraging results.

An outstanding feature of the year was the initiation of steps by the authorities to upgrade the hygienic conditions of several fishing harbours. Fishing Harbour Management Committees were proposed to be set up. There had been significant scaling up of facilities at processing plants, with a good number of them equipped for value addition to the exportable products. In the matter of Indian marine products export destinations, USA had emerged as the topmost one relegating Japan to the next position.

The decline in capture shrimp catches had been, to a large extent, made up by way of stepping up farmed shrimp as well as farmed freshwater prawn production. This has substantially made up for the downward trend in marine products export value. A cumulative extent of about 1,52,400 ha had been brought under shrimp farming in the CRZ, which had led to an annual farmed shrimp production of about 1,00,000 t. So far as freshwater prawn production is concerned 36,640 ha had been brought under this activity, giving an annual production of over 24,000 t.

The number of shrimp hatcheries along the coastline registered a level of around 240 nos. with a capacity to produce seed over 11,000 million nos./annum. Likewise, the number of freshwater prawn hatcheries in the country went up to nearly 60 nos. with a production capacity of 1500 million nos./annum. Use of immunostimulants and probiotics to control disease among farmed shrimps gained momentum. MPEDA set up several PCR testing labs. Several shrimp hatcheries also added to their set up dedicated PCR labs to facilitate detection of WSSV. MPEDA provided subsidies for the setting of PCR labs by hatcheries, for installing value-adding equipments at processing plants, and for several other improvements.

MPEDA implemented several subsidy schemes some of which are for providing subsidies for automatic flake/chip ice making machines (25 per cent) subject to a maximum of Rs. 2 lakhs, for generator sets at 25 per cent of the cost subject to a maximum of Rs. 2.50 lakhs, for upgrading deficient cold storages at 25 per cent for insulation and 25 per cent for upgrading the existing diffusers subject to a maximum of Rs. 3.50 lakhs, for acquisition of processing machinery and equipment for production of value-added marine products at 25 per cent of cost subject to a maximum of Rs. 15 lakhs, for distribution of insulated fish boxes at 50 per cent subsidy and also in respect of several other items. MPEDA provided subsidy assistance to develop scientific shrimp/prawn farming at 25 per cent of the cost subject to a maximum of Rs. 33,000/ha, restricted to 1.50 lakhs for developing up to 10 ha of water area per beneficiary.

Considering that the tempo of shrimp exports was hampered because of excessive use of antibiotics, Government banned the use of 20 harmful antibiotics, particularly in shrimp aquaculture (hatcheries and farms).

The Supreme Court passed a judgement against collection of contributions towards Kerala Fishermen Welfare Fund. A scheme for extending financial assistance for the setting up of ornamental fish breeding units in Madhya Pradesh was taken up by MPEDA. MPEDA took up a proposal for opening an office in the North-East for promoting ornamental fish trade in this zone.

*Peneaus vannamei* is reported to have been introduced into the country. Sizeable quantities of this shrimp are believed to have been produced in farm ponds at a few coastal centres. A view had gained strength that encouragement to farming of *Penaeus vannamei* will greatly help in augmenting marine products exports, as had happened in the case of a few countries of the world. A breakthrough is reported to have been achieved at Andhra University by D.E. Babu and his team in the breeding of mud crab and raising its seed (upto Megalopa stage) at the crab hatchery set up at Visakhapatnam with DBT's assistance. These were reported to have been grown into crablets in a nursery pond. Rearing of rock lobster larvae to phyllsoma stage was also reported to have been achieved at the same facility.

Government reduced duty on imported Artemia from 40 per cent to 15 per cent. 'Bio-remid Aqua' was launched by NEOSPARK, stated to be effective in controlling diseases among shrimps under farming.

Cultured shrimps/freshwater prawns contributed to nearly 60 per cent of exports in quantity and 86 per cent of the value of exported Indian aqua products. A.P emerged as the main contributor of farmed shrimps and prawns to the export endeavour. In port-wise exports in 2001-2002, Chennai topped the list with a quantity of 4,15,161t valued at Rs. 1570.17 crores followed by Kochi (72,035t/Rs. 930.87 cr), Visakhapatnam (22,154 t/Rs. 771.81 cr), Kolkata (17,692t/Rs.523.94 cr), followed by lesser export values from other ports.

MPEDA conducted a number of training programmes, workshops/seminars for the benefit of stakeholders in respect of HACCP audit, eco-friendly and sustainable shrimp farming, freshwater prawn farming, and use of chemical and antibiotics in aquaculture.

Five training programmes on 'Seafood Processing/Packing and Inspection for export to Japan were conducted at Kolkata, Visakhapatnam, Bhimavaram, Chennai and Cochin by Japan International Co-operation Agency, as arranged by MPEDA.

Two 2-week refresher training courses on Microbiology were organised by MPEDA at CIFT, Cochin.

Closed water recirculation system was demonstrated by MPEDA in private ponds in the States of Gujarat and Karnataka with subsidy assistance to the tune to Rs. 10.96 lakhs.

CMFRI submitted a project on the development of domesticated broodstock of tiger shrimp to ICAR while also continuing the work already taken up on a modest scale. Freshwater prawn culture made unique inroads into non-coastal States, such as Chattisgarh, Tripura, Haryana, Punjab, Himachal Pradesh, Rajasthan, Madhya Pradesh etc. Freshwater prawn introduced into Rajasthan was observed to have attained maturity in the saline waters of the State. In Punjab, CRPF jawans took to major carp and freshwater prawn farming with great success, helped by CIFA. There was a breakthrough in the breeding of freshwater prawn in Tripura and Chattisgarh. Hatcheries had also been set up and are under operation in these two States. In Punjab also there had been progress in this direction. Government of India sanctioned a major scheme for the development of freshwater prawn farming in Haryana. Anil Lamba, Balaji Singh and several other farmers of Haryana raised freshwater prawn crops successfully in their farms. Freshwater prawn farm effluent treatment system was developed at MPEDA's Tamil Nadu Freshwater Prawn Production and Research Centre under the leadership of A.L. Muthu Raman. Rajiv Gandhi Centre for Aquaculture, Myladuthurai, set up at Karukkala Cherry village, Karaikal, UT of Pondicherry a farm and demonstrated thereat pond farming and also cage farming of seabass. The Centre also set up a seabass commercial hatchery at Thoduvai, Thirumullai Vasal, Sirkazhi taluk, Nagapattinam district, Tamil Nadu. The units were set up under the leadership of Y.C Thampi Samraj, Project Director of the Centre. Cage culture experiments were successfully conducted in M.P. by CIFE, and in Tamil Nadu by Water Technology Centre of Tamil Nadu Agricultural University. Advanced major carp fry was produced in a running water cage culture experiment conducted by Water Technology Centre, Tamil Nadu Agricultural University, Coimbatore. CIFRI demonstrated pen culture in Samagiri *beel* in Nagaon Dt, Assam. Aquaculture as part of participatory integrated watershed management was introduced by Central Soil and Water Conservation Research and Training Institute, Dehra Dun.

Fish farming as an integral part of irrigation system was initiated by the Water Technology Centre, Tamil Nadu Agricultural University. Fisheries-based integrated farming system was taken up successfully by Dr. C.S. Singh, former Dean (Fisheries), College of Fisheries, G.B. Pant University of Agriculture & Technology, Uttaranchal, at Singharia, Kunraghat, Gorakhpur, U.P. Sustainable development of fish culture and also freshwater prawn culture in Jharkhand picked up momentum. Breakthrough was achieved in grass carp breeding in Sikkim.

In Himachal Pradesh, trout farming in the private sector had been further developed successfully with assistance from NORAD, under an Indo-Norwegian Project. The programme had yielded good results. There had been further development of Trout Fisheries in J and K too with Norwegian Aid. Around 1,000 t of annual fish production from Gobind sagar was recorded by HP Fisheries Department. Mass mortality of fish took place in Sarsa river, Solan Dt, Himachal Pradesh.

The culture of Asian catfish, *Clarias gariepinus* had become popular in several States. This exotic fish had entered into Yamuna, Sutlej and Godavari rivers, causing concern. Skate species *Himantura indica* was reported from Sarayu river that flows through north-east plains of Bihar, by Prof. Ashok Kumar Jha.

Vietnam model of aquaculture development and also setting up of freshwater prawn hatcheries on Vietnam pattern of green water technology, as developed by B. Madhusoodan Kurup of the School of Industrial Fisheries of Cochin University of Science and Technology, Kerala, were stated to be on the anvil in Kerala. People's campaign for promoting aquaculture was also launched in Kerala. Research projects with Dutch collaboration for improving sustainability and productivity of *Pokkali*-based shrimp farming in Kerala was taken up by the School of Industrial Fisheries of Cochin, University of Science and Technology.

Preparation of pelletised fish feed using hand-operated palletisers was demonstrated to farmers by College of Fisheries, Ratnagiri.

A Fisheries Development Mission was launched in Tamil Nadu. Similar Mission was also planned by the Government of Orissa. The proposal for shifting of Directorate of Fisheries, Karnataka from Bangalore to Mangalore was dropped.

M. Visweswariah Aquarium at Brindavan Gardens, Bangalore, was inaugurated. Integrated development of fisheries of Hesaragatta and Markonahalli reservoirs was taken up. Bombay Aquarium Society celebrated its golden jubilee.

Attainment of record size of *Gudusia chapra* in water bodies of Samspur bird sanctuary was reported. Aquaculture is reported to be becoming popular among north Rajasthan Agriculturists. Freshwater pearl culture experiments are reported successful in Maharashtra too.

Tiger shrimp farming in South Gujarat is reported to be fast picking up Pacific white shrimp (*Penaeus vannamei*) farming was assessed to be a challenge to tiger shrimp farming in India.Use of immunostimulants in fish culture highlighted by' several professionals for producing healthy shrimp crops.

Indian river prawn was successfully bred by D.R. Kanaujia, A.N. Mohanty and Shalini Soni, Scientists at CIFA. They also developed String Shells Technology for effective harvesting of the post-larvae of Indian river prawn under hatchery conditions (It was reported that this prawn was also successfully bred in China). Successful breeding of common clownfish *Amphiprion percula* under captive conditions was achieved in A and N Islands by K. Madhu and Rema Madhu of the Unit of Central Agricultural Research Institute, Port Blair.

Eco-restoration of Chilika lagoon through opening of a new mouth near Magarmukh was achieved. Chilika lagoon was taken out of Montreux record.

6th Convocation of CIFE was held on 14 November, 2002. East Coast Regional Chapter of Indian Association of Aquatic Biologists was set up at Visakhapatnam. AIR Ranchi pioneered in broadcasting episodes on fish culture. Fishermen oj Madhubani District of Bihar were trained in improved methods of fishing by scientists of Burla Centre of CIFT located in Orissa. CIFT demonstrated fish handling and processing in the NEH region. CIFT constructed two canoes successfully using rubber wood at an economic price and these were well received.

India was elected as co-chairperson of Indian Ocean Tuna Commission. It was also represented on an Indian Ocean Tuna Working Party. IOTC working party meetings on tropical tunas were held in Seychelles and China. Midwater trawling along upper east coast by Fishery Survey of India yielded encouraging results. Import of used tuna longliner has picked up. New Fishing harbours planned to be taken up in A.P.

A new record of a stranded whale, *Balaenoptera indica*, at Shivrajpur on Dwaraka Coast in Gujarat was reported.

A new rule that requires skippers must have Radar Observer's (Fishing) certificate was introduced. Kerala Government reduced the ban period for monsoon trawling to 45 days. There were reports of capture of turtles off Orissa and A. P coasts, during their migration towards the coast for laying eggs. Turtle awareness Centre opened in Orissa. Intrusion of nine foreign trawlers detected near Kendo island on Indo-Bangla Desh border.

EU introduced regulations for labeling of seafoods marketed in E U. The Surimi plant at Visakhapatnam was shifted to Brahmwar in Karnataka by Hindustan Lever. Technique of quality improvement of sundried *Acetes indicus* (Jawala prawn) through radiation processing was developed by Food Technology Division, Bhaba Atomic Research Centre, Trombay, Mumbai. This Centre also standardised technique of radiation processing of aquafarmed fish products.

FDA lowered its minimum acceptable level of chloramphenicol in marine products imported into USA from 5 parts per billion to 1 ppb.

Dried fish maws export has been brought under export (Quality Control and Inspection) Act, 1963.

Shrimp export prices declined. US emerged as a major importer of marine products from India. Kerala Fish Export Zone was reported to be on the anvil. Allanasons, a pioneer seafood processing and export company, had started production of seafood items in consumer packs at facilities certified as in accordance of EU regulations. E U scrapped ban on import of shrimps from Vietnam and Pakistan.

New team of members have been inducted into MPEDA's top governing body. Jose Cyriac, Chairman, MPEDA has been renominated as member of Export Inspection Agency.

Fishing Chimes, Editorial 315: March2003: Vol. 22, No.12

# The Challenge of India's Declining Coastal Fishing Output

Exploited fisheries of India's coastal waters have remained static at around 25 lakh t per annum since 1992-93. This has resulted in the decline of coastal marine catch per unit to such an uneconomic level that most of the stakeholders are forced to lead a life of misery and desperation. In contrast, inland fish production moved up from 17.86 lakh t in 1992-93 to over 30 lakh t/annum in 2001-02, mostly because of a significant rise in culture fishery output. The policy initiatives such as promulgation of closed seasons in the coastal waters have not been taking the marine production upwards. Because of this and other reasons scope for increase in coastal fish production continues to look bleak; The coastal States and the Central Government are aware of the situation and are believed to be exploring ways to extricate the affected fishermen from the precarious and appalling situation.

Until around middle of the nineteenth century, agricultural land in India was an occupation-based property, available for utilisation by the occupants for raising agri-crops or for agriculture. As a variation, but following another pattern, marine waters, historically, had been a common property utilised by fishers for capture fishing as they wanted. In course of time, land was surveyed and ownership of land stretches under occupation was conferred on the individuals concerned by the rulers. Similar step could not, however, be taken in respect of marine waters for obvious reasons. Consequently, the resource continued to remain a common property with the only reform introduced being one of promulgation of State-wise Marine Fishery Regulation Acts which brought in common rights of fishing for those belonging to the State concerned in the territorial or coastal waters as a significant reformatory step. Fishers or fishery enterprises not belonging to the State concerned and not eligible for registration as per the rules are required to be permitted by the authorities of the concerned States to fish in their territorial/coastal waters. This reform could not, however, increase production but only helped in protecting the interests of fishers of the State concerned.

This state of affairs highlights the need to consider adoption of the kind of approach that is being followed in advanced countries for introducing mariculture on long term lease basis in amenable zones of coastal waters for augmenting fish production. There was no motivation in India for thinking of such an approach for adoption in Indian Coastal Zone for the reason that no overfishing/ depletion of fishery was discernible until recently. This approach, however, now deserves to be considered for adoption because of sheer necessity, to salvage the detereorating coastal capture fish production. This approach, while it is a very difficult one, can be adopted now, for the reason that, at the time land survey was taken up also, there were reservations that this would be difficult. Accepting the challenge, however, it was ushered in and strengthened into a system over a period of time and this has now come to be established. The glaring difference between surveying of land mass and coastal waters that was faced was that, in the case of the former there was *de facto* ownership by virtue of occupation at the time when the survey started, whereas in the case of the coastal waters there was no such occupation for initiating a similar excercise. However, there is the prevalence of traditional common *de facto* rights, while the general ownership is vested in the State.

It may be a long drawn and snare-ridden process, to follow the same pattern followed as on land to create private ownership rights over segments or plots of coastal waters,but there can be no escape from conducting an allotment/license-oriented survey of coastal waters, in order to identify stretches of water sheets that can be brought under mariculture by way of setting up cages or pens through granting of long term leases to eligible stake holders, followed by granting of ownership rights/leases, based on a system to be evolved. All well protected water stretches can come under 'A ' category of waters, followed by others under other designated categories. The Grid system followed by Fishery Survey of India for conducting surveys can probably be followed for identifying cage-friendly zones, within each of the Grids.

It is axiomatic that there has to be a legal basis for the granting of leases of identified water stretches in the first instance, prior to conferment of ownership rights, if feasible, as was done in the case of land ownership. This basis can be established by incorporating needed amendments to the Marine Fishery Regulation Acts.

Granting leases for setting up mariculture units in the coastal waters has become a well established practice in some countries, particularly USA..

Under the provisions of the prevailing law in respect of each of the Coastal States of USA, the concerned State Department of Marine Resources conducts a public hearing on each of the applications received for the exclusive allotment of a marine water site to cultivate fish using net pen techniques or for other compatible uses. Public notices are advertised by the Department calling upon any interested persons to attend the public hearing to ask questions and give testimony. The notices incorporate several details. Two such public notices, out of several so published,and taken out from April 2002 and January 2003 issues of *Commercial Fisheries News* published from Stonington, Maine State, USA are reproduced on the left part of this page. As may be seen, one of them relates to granting of a lease to a foreigner and another to a national.

Similar notices are issued for renewal of leases and for cancellation of leases. One set of these notices as published in April 2002 issue of *Commercial Fishing News* is also reproduced on the left as part of this write-up.

It will be in national interest of India to utilise its Coastal waters for farmed fish production, partcularly at this juncture when the plummeting coastal capture fishery per capita output is causing serious concern from various angles, *i.e.*, production, incomes, socio-economics, exports and so on.

There are however certain problems like thefts, damages to nets, patrolling, obstruction to navigation etc., to be taken care of by the licensee concerned. Those who take up farming in capture waters in other nations also face such problems but they are able to counter them and, in the same way, Indian enterprises have to also face and solve them consistent with the local conditions and regulations.

Taking cognisance of the advantages of promoting mariculture in coastal waters, the Union Department of Animal Husbandry and Dairying may have to consider setting up a Group chaired by its representative and consisting of representatives of state fisheries departments and legal experts to study the legal frame work and the systems in the various Coastal States of USA and in European Countries such as Norway and to recommend measures for according permissions for setting up farming systems in coastal waters, to propose the kind of back-up or infrastructural facilities needed for assisting in the setting up of farming facilities, to suggest measures for establishment of shore hatcheries for the production of seed of farmable species, for provision of shore to farm and farm to shore transportation facility for carrying seed, feed, personnel etc, and above all, for the designing of farming systems and for recommending operational guidelines.

Progressive and reformatory measures to introduce farming systems in coastal waters, with foreign technical assistance, which may certainly be needed to start with, is of utmost urgency to a) make the best use of Indian coastal waters for augmenting aqua production, b) to restore and strengthen the socio-economic conditions of dependent fishermen and other stakeholders and c) thereby to strengthen fish supplies for domestic market and for exports. Let us not be behind other progressive nations in respect of mariculture any longer.

Fishing Chimes, Editorial 316: April 2003: Vol. 23, No. 1

# Status of Indian Marine Fisheries Development: An Appraisal

India has been facing a consistently alarming trend of deterioration in the performance of its marine fisheries sector, in terms of fleet strength and consequential exploited output for the past several years. This trend is ascribed by several to static and narrow objectives of development behind ongoing policies. This perception could be wrong, but whatever be the reasons for this, looked at from any angle, the fact remains that the scenario is depressing. It is hard to find support from the marine fisheries sector, to affirm that the Government have introduced long term and enduring policies to upgrade present fishing effort to cover farther areas to harvest known tuna and other resources in coastal waters and beyond and to exploit deep sea and oceanic fishery resources. For over a long time past, we have been subjected to listening to repetitive statements that the marine fishing/ fisheries policy has been with the Government for approval. There must be valid reasons for the inordinate delay on the part of the Government in according approval to the policy as proposed.

So far as exploitation of fisheries resources beyond coastal zone is concerned, nations generally plan to develop their own capability. In the case of India too, such an effort had been made in the past to develop its own fleet. A scheme was taken up in late 1970s to import vessels for outrigger shrimp trawling and bottom stern trawling. Under this scheme, 28 shrimp trawlers with outriggers and also fitted for stern trawling were imported from Mexico. An additional number of such trawlers were permitted to be imported later. This and other initiatives led to the expansion of the fleet by way of adding 23 to 27 m OAL vessels mainly for shrimping, taking the fleet strength to 195 nos., by 1990. A further effort directed at the introduction of upgraded vessels would have ensued as the next stage of development, but the promulgation of the Maritime Zones of India (Regulation of fishing by foreign vessels) Act in 1981 must have changed the direction of thinking. In the wake of this Act, the issue of specific notifications was seen, aimed at according permissions for chartering/leasing of foreign fishing vessels by Indian enterprises on the condition that the Indian enterprise concerned would introduce its own vessels, equivalent to the number of vessels chartered but not necessarily the type of vessels chartered. Joint ventures, boiling down to charter/leasing practices in actual practice, were also allowed. There have however been no focal initiatives directed at exploiting the waters beyond the Coastal Zone of the EEZ including the deep sea zone through introduction of Indian owned vessels with the requisite capability. The provisions of the Maritime Zones Act and the rules framed thereunder are believed to have motivated Taiwanese built foreign vessel owners to create/ organise a situation of offering their vessels on charter/ lease to Indian enterprises and this line of short-term or transient fleet development went on upto around 1994 when the charter/lease policy was withdrawn. Under this scheme, in all 189 vessels (mostly of Taiwanese ownership but registered in Panama, Honduras etc.,) operated in Indian EEZ, for certain durations as convenient to owners, from 1985 to 1993. Thereafter, in 2002, the old wine came in a new bottle by way of introduction of a policy of permitting import of second hand fishing vessels, with the legal ownership (as distinct from *de facto* ownership) in the name of the Indian enterprise.

These stages of development as set forth above tend to indicate dependence on measures, interim or palliative in nature, characterised by the absence of a long term developmental perspective, to target the unexploited fishery resources beyond the coastal zone. Alongside these developments, there has been a sway of intiatives by foreign interests (mostly Taiwanese) to gain access over fishing in Indian EEZ for their benefit all along, taking advantage of the Indian policy 'avataar' in vogue at the time, but having the same thread forming the core.

In overall terms, one finds that marine fish production from Indian EEZ has been relentlessly plummeting over years for the reason that the main thrust of fishing is confined to coastal strip, neglecting the resources of farther waters of EEZ. So far as coastal fishing is concerned, the mechanised boats operating all along the coastline have now reached a stage of having either static or falling per unit catches. They are now in severe distress. The owners of a large number of these boats are forced to moor their boats at the landing points. The adverse economics that have engulfed them have stemmed from the inexorably declining resources, shared compulsively by an astonishingly large number of coastal fishing boats, with some measure of competition also from the few larger vessels which are known to transgress into the coastal zone. The enormous increase in the cost of fuel and other operational expenses and the declining market prices are the other adverse factors. Can we perceive any well-directed and enduring policy measures to counteract this situation of distress that faces the small mechanised boats? Subsidies conceded by some of the coastal States have been found to be not of real help in ultimate terms. While larger vessels in the range of 15m and those 20m OAL and above are allowed to have access to mid sea bunkering, the smaller boats which need the facility, as much as or even more than the larger vessels, are denied this facility, on the ground that it will be unmanageable to organise supply of duty-free oil to thousands of boats.

It is known that all mechanised boats are virtually engaged in bottom trawling, mostly for catching shrimps, and that there is a drastic depletion in not only shrimp stocks but also of finfish and cephalopods in the coastal zone. The situation apparently calls for policy initiatives to counter the situation. There is however no known indication of any such initiatives exercised by the authorities. Experts say that the pelagic and columnar resources of the coastal zone, some of which have export potential and others can be imparted export potential, are inadequately exploited. In the light of this undeniable assessment, one policy option can be to find ways and means of diversifying operations of these boats to exploit the non-demersal resources according to seasons of availability in a sustainable manner, supported by training programmes in these diversified operations.

The nation is more or less stuck with bottom trawling operations by small mechanised boats for the past four decades and over. The coastal fishing operations have to be extricated out of this involvement in bottom resource exploitation and in this context, there is a responsibility cast on the Government to institute measures to upgrade the skills of fishermen in the direction of diversified fishing to enable them to operate vessels that can fish not only in farther waters, but also to exploit pelagic and columnar resources of coastal zone. The neglect in exercising this option is largely responsible for the overconcentration on exploitation of demersal resources.

There are known coastal tuna resources in Indian EEZ. In this context, a programme of equipping small mechanised boats with monofilament tuna longlining system deserves to be taken up. Such a programme has been successfully introduced in Brazil. Further, under a DANIDA-assisted project implemented in Karnataka, vessels of 15 m OAL range, equipped for operation upto around a depth of 300 m were successful introduced. A few of such vessels now operate on the west coast (Karnataka, Kerala and Goa zones) and off Kanyakumari Coast of Tamil Nadu. This initiative deserves to be emulated by the other coastal States, but this is possible only with the policy and technical support of Government of India.

There are around 400 mini-trawlers of around 15 m OAL engaged in trawling for shrimp along east coast, a commodity with declining per boat catches. These trawlers can be helped by a policy intervention for equipping them in the same way as the aforesaid vessels of DANIDA design. Further, there are 18 trawlers of 23.5 m OAL of unfinished construction (under the control of the Government) languishing at the Kakinada yard of East Coast Boat Builders. At this time when there is an urgency to organise efforts at oceanic and deep sea fishing in the totally unexploited zone of the Indian EEZ, it would be a good and wise policy initiative to find ways and means of reviving the construction work of these trawlers by installing engines of around 800 hp and winches of adequate capacity to fish at depths upto 500 fathoms and beyond.

At most of the fishing centres, as is known, boat-owning stakeholders continue to be indebted to money lenders. Efforts at bringing them under a co-operative fold have largely failed. Because of this, there is not much of change in their relationship with the financiers at village level. In order to impart a direction to this traditional activity, there has to be a policy thrust aimed at ensuring the emergence of an effective organisational mechanism that ensures provision of an integrated financial, technical and managerial support at major landing centres. In other words, this mechanism should enable the organisation to be set up to mobilise investments including those needed for setting up its own processing plant and also for setting up its own post-harvest facilities with Government support, and administrative, financial and technical help coming from dedicated professional Non-Governmental Organisations (NGO's). The integration and linkages of the system have to be governed by regulations as framed by the Government. Once a system of this kind is introduced, the financial needs of the stake holders can be possibly met substantially. Through such a system, the socio-economic strength of the stakeholders would have opportunities of upgradition.

So far as the larger vessels of 23 m OAL and above are concerned, their operational status is pitiable. It is unfortunate that this segment of the industry, which contributed so much in the past, is now left in the lurch with an undeserved neglect directed at them. The fleet which was around 195 vessels strong until a few years back, has gradually dwindled to around 65 numbers as at present. Most of the owners of these vessels are on the look out for buyers. This slimming process is related to a) the age of the vessels, most of which are 15 years old and some even over 20 years old and b) there have been no replacements. All of them were introduced for shrimping, although a few of them could diversify into catching lobsters, deep sea shrimps and cephalopods for some time. Despite awareness that the vessels have become old and need replacements with vessels capable of exploiting deep sea and oceanic resources at pelagic and demersal levels in consonance with the present resource disposition, there has been an inexplicable delay on the part of the Government in introducing needed measures to grapple with the situation, many in the industry feel. They feel that there has been a neglect of the paramount need to have a deep sea fishing fleet at par with other developing nations, not just to protect the Nation's honour and status as a major maritime force, but mainly to utilise the high sea pelagic, columnar and above all, totally unexploited oceanic deep sea resources. An offshoot of this dilation is that trawler owners, several of whom have now become bankrupt, are now drifting towards other means of survival.

One initiative taken to upgrade the activities of these languishing larger vessels is of equipping two of them on a pilot scale with monofilament tuna longlining system. It is to be hoped that this scheme, now stated to be under reorientation so as to extend the pilot effort to cover various links, with a fleet of 20 upgraded trawlers, from production of tuna to their export marketing, will soon become re-operational with an outlay of Rs. 7.5 crores. This move is no doubt a significant one but it does not neutralise the vast neglect to which the sector has been subjected to over years, despite its recognised role in providing proteinacious food to the people and contributing heavily to the export endeavour.

Those in the industry have been able to perceive so far, in the main, only two significant developments in the history of Indian marine fishing after independence, besides exploratory surveys. One is the introduction of small-mechanised fishing boats with Norwegian assistance and a little later nylon twine and monofilament for making fishing nets followed by introduction of outboard motors for upgrading capabilities of traditional craft. Introduction of FRP boats also took place in the wake of these efforts. The other effort was the scheme for introduction of 28 nos. of Mexican built 23.15 m OAL trawlers, equipped for outrigger and stern trawling, which later gained momentum to swell the fleet strength to 195 nos. in the range 23 to 28 m OAL, (as already mentioned) through imports of vessels from countries like Australia, Holland, Korea and construction at Indian yards. What all was done later, as also already mentioned, was by way of treading the path of permitting vessel chartering/leases and joint ventures bereft of any real stakes, but based on operations by foreign interests linked virtually to payment of a percentage of gross incomes to the Indian side by the owners of vessels who played the role of buyers of catches also. The chartering/lease systems were later abolished, only to give way to the new system of importing 'used vessels', based on legal registration in India and an arrangement for the payment of the cost of vessels to the *de facto* owners in instalments for record, 'out of earnings', and the vessel operations remaining virtually in the hands of the *de facto* owners.

There are many in the industry who are disappointed that the Indian Government could not as yet evolve a policy of enabling the Indian fishery enterprises to have *de facto* ownership too of these vessels, befitting the status of India, as a major maritime nation.

Development of a national policy for the utilisation of our marine fisheries would have to revolve round Indian investment and total management by Indian enterprises for introducing new or old vessels and for strengthening other infrastructure facilities under a healthy system, that provides for training of crew and managerial personnel and in marketing. This is what the industry looks for, with willingness to enter into joint ventures.

The Indian shipyards have to acquire expertise for building vessels for diversified fishing methods one of which including tuna longlining. It is known that financing institutions within the country and international organisations such as the World Bank too will come forward to invest, once the projects are seen to be sound and have the approval of the Government of India. It is unfortunate that instead of looking at options of this kind, Government's encouragement went all out for the import of second hand vessels with depleted life span and with the *de facto* owners of the vessels probably having their own fishing strategies within Indian EEZ and also after leaving Indian EEZ. The objective of the foreign vessel owners in conceding legal ownership and retaining *'de facto'* ownership is believed by many as a strategy just to gain an unharassed access into Indian EEZ, followed by operational freedom. Let us hope that this belief is misplaced. By permitting total employment of foreigners on these second hand vessels, the candidates trained at our own institutions are denied opportunities of employment.

In this background, it is important for the Government to have a focal reappraisal of the situation with a view to evolving a policy that the nation can be proud of.

Right from the time India became independent, its fisheries sector has not received the attention it deserves, despite the fact that it has vast fishery resources, provides employment, provides proteinacious food to the people, and contributes substantially to our foreign exchange earnings. Almost all nations have separate fisheries ministries but India does not have one. This is somewhat abberant and inexplicable. There is a general surmise that traditionally Government of India does not consider its fisheries important. This surmise is based on the logic that right from the time of independence no prime minister thought of having one Minister exclusively for fisheries development. The same bent of mind continues despite having an EEZ of over 2 million sq. km and an enormous fisheries potential for development beyond the coastal zone. Other nations such as USA, UK, Thailand, Taiwan and nearer to us Sri Lanka and Maldives are prospering despite other problems, because they do not neglect their respective fisheries sectors. There have been any number of appeals from the industry, fisheries professionals etc., to the Government to create a separate union ministry or department for fisheries but the resolve of the Government not to concede to the proposal is unshakeable. This disregard puts out wrong signals that there must be a strong reason for the Government to adopt the present policy, which unfortunately benefits certain foreign nations having interest in the fisheries of EEZ of India. It is possible that several countries nurture the surprise that a major nation of dimensions of India does not have a separate fisheries ministry. Of course, some of them will be happy that India does not have one. The Government of India probably thinks there is not enough workload to have either a separate ministry or a department for fisheries ignoring the fact that, only when there is a separate set up, there will be expansion and intensification of work aimed at utilising the fisheries potential of the EEZ for domestic consumption as well as exports. The fact that the fisheries sector earns over Rs. 6000 crores in foreign exchange annually for the country has not tilted the resolve of the Government on the subject.

A major lacuna in Indian fisheries administrative system is that the Government has no specific field set up to promote fisheries development to move towards sustainable utilisation of the nation's marine fisheries resources. Countries like UK, USA, Canada, and Australia have separate Sea Fisheries Authorities to implement the policies of their respective Governments. Keeping this in view, in the same way as the Ministry of Commerce set up Marine Products Export Development Authority to promote maritime products exports, the Union Ministry of Agriculture too should set up a Sea Fishery Development Authority of India to promote the activity. A set up in the form of a Fisheries Division as at present is designed to develop policies and review their implementation but is not equipped for promoting and monitoring field activities. FSI, CIFNET, IFP and CICEF under the Fisheries Division are equipped for surveys, training, demonstration of integrated operations and for setting up harbours respectively but not for directly promoting fishing effort and monitoring the same.

Another aspect to be mentioned here is that the Prime Minister of UK has recently taken charge of a Committee set up to develop a strategy for further promoting Britain's fishing industry, although there is a separate fisheries minister to look after fisheries development. So much is the importance given by the UK Government for the development of their fishing industry. This highlights the need, for India, which has far higher marine fisheries resources than UK, to have a separate Ministry for Fisheries, as has been done in most of the developed and developing countries, besides setting up a Sea Fishing Authority to promote the utilisation of India's marine fisheries resources beyond its territorial waters.

Fishing Chimes, Editorial 317: May 2003: Vol. 23, No. 2

# Giant Freshwater Prawn Broodstock: Quality Improvement

Giant Freshwater Prawn (GFP) hatchery managers, by and large, are now inclined to opt for brooders of the crustaceans taken from the wild in preference to farm raised ones. The reason is the same as articulated in the case of major carps *i.e.,* the inevitability of inbreeding among the brooders raised from hatchery seed results in stunted growth of progeny. For the same reasons as CIFA resorted to in the selective breeding of Rohu, there is need now to consider launching a programme that would ultimately arrest stunted growth of hatchery-raised post larvae of GFP. In other words, the view that a research programme deserves to be taken up by CIFA at this stage of the phenomenon of stunted growth among farmed GFP raised from hatchery seed is gaining strength. Mr. M. Sudarsan Swamy, President, All India Shrimp Hatcheries Association has highlighted this aspect by pointing out that GFP hatchery owners now prefer to have wild brooders for hatchery work to avoid production of PLs prone to stunted growth by subjecting broodstock raised from hatchery produced seed.

The urgency of undertaking a research programme to find a solution to the problem has been already noticed in Thailand. Further, Sudarsan Swamy says that our GFP hatcheries need 60,000 brooders by 2005. We cannot certainly depend on wild stocks to meet a requirement of this order. The need is therefore to develop a genetically improved stock of GFP for generating supplies to the hatcheries. As and when the progeny of the genetically improved stocks too may start showing signs of yielding PLs that may exhibit stunted growth in course of time, needed precautionary/remedial measures at appropriate stages can be instituted.

According to Apisit Buranakanoda (Asian Aquaculture Magazine, May/June 2002 issue) a project for the genetic improvement of GFP was taken up as early as 1998. The study team collected samples of GFP from all rivers of Thailand. The best out of them with the most encouraging head-trunk ratio, were selected and utilised for the breeding programme to evolve a strain that gave a head-trunk ratio of 1:3, found to be the meatiest strain. An ingredient of the programme is to have a system of exchanging farmed males and females by the farm owners before supply to hatcheries. Further, all male grow- out production is a pre-condition to develop GFP farming as the males grow faster than females. In order to fortify all-male stocking programme, research is stated to be underway at the Genetic Engineering and Agricultural Biotechnology Centre of Research and Development Institute of Kasetsart University in Thailand to standardise a system of sex reversal from females to males before stocking.

Fishing Chimes, Editorial 318: June 2003: Vol. 23, No. 3

# Associations Related to Indian Fisheries Sector

Marine and Inland Fisheries Sectors of India have grown several fold in the past over five decades. Annual fish production from Inland Fishery Sector rose from 2.18 lakh t in 1950-51 to 3.0 million t in 2001-02. So far as annual marine fish production is concerned, it went up from 5.34 lakh t in 1950-51 to 2.3 million t in 2001-02. Exports moved up from 57,819 t valued at Rs. 463.31 crores in 1989-90 to 400,000 t valued at over Rs. 6,050 crores in 2001-02.

This cascade in production is the result of the dedicated team work of innumerable persons, both in Government and private sectors. The scientists in the research institutions, technologists and technical men in Central and State Fisheries Departments, coupled with the entrepreneurs, technologists, scientists, managerial personnel and millions of fishermen and other workers were the heros who contributed to the achievement.

Based at the various research institutions, several Associations and societies function. Examples are the Indian Fisheries Association C/o Central Institute of Fisheries Education, Mumbai, Inland Fisheries Society of India C/o Central Inland Fisheries Research Institute, Barrackpore, West Bengal, Society of Fisheries Technologists (India), C/o Cental Institute of Fisheries Technology, Matsyapuri, Cochin, and Marine Biological Association of India C/o Central Marine Fisheries Research Institute, Cochin. The Asian Fisheries Society, Indian Branch, located at Mangalore, promotes effective interaction and co-operation among fisheries scientists and technocrats. In respect of the working of these societies and associations with particular reference to the journals they publish, several suggestions came out for strengthening their working at the Sixth Indian Fisheries Forum held recently in Mumbai. These covered the need for certain uniform and co-ordinated norms for the benefit of the fisheries scientists engaged in research work on various topics of importance for strengthening the scientific base of fisheries developmental work in the field. A view was also expressed that these bodies need to revv up their thinking process so as to provide a positive impact on applied aspects of policy related to fisheries science and technology.

There are three inportant national level associations *i.e.*, Seafood exporters Association of India, Association of Indian Fishery Industries and All India Shrimp Hatcheries Assocation in the fisheries industrial sector which play a significant role, the first one in augmenting marine products exports and safeguarding the interests of processors and exporters, the second in the promotion and looking after the well being of fishery industries, particularly of the larger fishing vessel operations, and the third one to take care of the technical and commercial activities of shrimp hatchery owners. There is also a national fish workers union to protect the interests of fishery workers and mechanised boat operators. State-oriented associations in this regard also exist. On the culture fishery front there is a Confederation of Fish Farmer's Welfare Association and also several State-based associations to take care of the interests mostly of shrimp farmers.

Several day-to-day problems confront the members of the Associations. So much so, the office-bearers have to divert their attention primarily to resolving these problems. Most of these problems arise because of gaps in the intent of the policies of the Government and the follow-up action on them at field level that arise for some reason or the other.

One main function of the various Associations, as is known, is to identify the problems confronting the stakeholders, finalise policy/strategy solutions, bring them to the notice of the Government and prevail upon the Government to approve of them. This function is, by and large, taken care of by the Associations. They keep the Government informed of the important developments in the industry and seek such policy and other support as they need. A couple of results noticed because of the efforts were a) Setting up shrimp nauplii production centre in Andaman and Nicobar Islands, and Disease diagnostic labs by MPEDA, b) Export of value-added products, c) Promotion of culture of giant freshwater prawn etc. The associations however continue to fail in prevailing upon the authorities to take up certain programmes which should have been taken up. These include programmes of development of domesticated shrimp broodstocks, introduction of ranching and cage culture programmes, introduction of diversified culture systems, disseminating technologies of marine and freshwater culture, stemming inbreeding problems posed by brooders used for hatchery operations etc.

On the marine capture fisheries sector, there is need for the mechanised boat operators and other associations to upgrade their leadership capabilities in the direction of formulating measures for suggesting to the Governments concerned for restoring coastal fishery wealth, other than those which Coastal State Governments traditionally think of. For example, they can demand promotion of marine cage culture of economically viable species supported by a chain of their hatcheries, as they have in several countries. They can also demand sea ranching of important fishes. No doubt these measures call for lot of planning, investment, induction of expertise and training but Governments are there to think of such measures to help fishermen.

Regarding riverine capture fisheries, it was generally thought that riverine fish populations would improve for the reason spawn collections from rivers have been almost given up, because of the popularity that induced breeding technology has gained. However, the expected increase in the riverine fish populations had not materialised. Fishermen's organisations have to prevail upon the Government to introduce certain measures like ranching supported by a chain of hatcheries and seed farms along the banks for producing the needed seed for the purpose concerned. Unfortunately there are no fishermen 's organisations to represent this aspect to the authorities and probably an NGO can step in to fill the void.

On the culture fisheries side, there are quite a few associations, mainly connected with shrimp culture activities. These are mostly district or area based and their objectives vary. It is generally seen that they function in an isolated manner. Shrimp culture sector has a plethora of problems, highly sensitive, calling for technical skills. MPEDA extends enormous support, technical and otherwise to solve them. It is learnt that MPEDA has now taken the initiative to promote shrimp farmers associations all along the coastline and integrate them to form into an apex body to serve the two-way purpose of having a mechanism for gathering field level information with representations/suggestions for improvements for introducing remedial measures.

The farmers of non-coastal inland culture fishery sector are the most disorganised. There are practically no associations to take care of their interests. Most of them are not aware of what they are missing. The sector is in an imperative need of having farmers' associations to represent to the Governments concerned on what they need. A compaign to bring about awareness and on that basis to encourage formation of associations at district levels to be linked to State level associations by the enlightened among the farmers is essential. Only when such Associations are there Governments can be prevailed upon to upgrade the activities to bring them to the stage of making the fish culture activities in the interior areas export-oriented for the well being and prosperity of the farmers. The leadership being generally weak, field support mostly comes from consultants and from dealers of feed, medicines, and other inputs. The progressive farmers from each of the Associations could take the initiative of fortifying the organisational structure of these Associations and thereafter of forming State-wise organisations to tackle the problems of farmers with needed confidence.

Fisheries Departments, both at State and Central levels deserve to have the backing of strong, purposeful and representative associations. Such associations will have the capability to provide realistic field picture with helpful recommendations for policy/strategy formulations. The development of a correct equation between fisheries departments of the Governments and the associations is of utmost importance in this context.

Fishing Chimes, Editorial 319: July 2003: Vol. 23, No. 4

# Fish Farmers Need Inland Culture Fish-based Expansion of Export Efforts

It is common knowledge that Indian Culture fish production is predominantly major carp oriented and that the culture production of major carps has been swelling up over the past four decades, influenced by successive technology-oriented breakthroughs in major carp seed production and their farming. The annual major carp production of India has now reached a level close to 3 million tonnes. The major carp production would have gone up further but for a forced slow-down of the growth trend caused by supply glut and uneconomical offtake causing drop in earnings of farmers. The inadequate domestic fish marketing network and the near absence of export market for major carps have led to the present economic plight of the farmers. This deplorable situation is unfortunately gaining strength, despite the fact that the Indian major carps *i.e.,* Catla, Rohu and Mrigal, are traditionally known fish for their excellence, with an inviting appreance, firm flesh texture and a tongue tingling taste. Yet, the fishes failed to fetch an external market, with the exception of an insignificant movement to a couple of middle-east countries and a few consignments to the west, to cater to the needs of migrated Indians.

There will be a worthwhile external market for major carps only when there are exports to USA. Unfortunately, no dent could be made so far to sell the fish in that country. We export shrimps, prawns and several other items to USA supported by excellent business relations with the importers. Despite the contacts Indian exporters have with the USA market and the need to promote exports of major carps to augment incomes of fish farmers, it is inexplicable that no market could be established for Indian major carps in USA etc.

There was an occasion to discuss this issue with a businessman of USA in the line. It arose when he was in a mood of appreciation of the taste of a fried fillet of Rohu. Picking up his appreciation as the starting point, an effort was made to elicit his viewpoint as to why US importers of Indian fish items are generally averse towards placing orders for purchase of Indian major carps. His spontaneous observation was that Indian businessmen did not probably know that American people generally consider carps as fishes of poor quality. So much so, no US fish importer, barring a few who are in the business of meeting demand for major carps from those hailing from Indian sub-continent, would take the risk of importing them into USA. While so commenting, and reiterating the sterling quality of Indian major carps, he hazarded the suggestion that major carps could be referred to by another name that will be appealing to the American perceptions. For example, he said that, in commercial parlance, marine prawns are referred to as shrimp and freshwater prawns as freshwater shrimp only to mobilise consumer support. Apparently, the above mentioned changes were developed to enlist acceptance of these in US market.

In this background, Indian major carps can probably be rechristened and popularised as Indian Bass or Indian Freshwater Bass or may be as Indian Bream or Indian Freshwater Bream. There is no scientific logic in promoting these names, but commercial logic is important. It is true that carps are neither Basses nor Breams but in business the names can have the potential to catch up. The giant freshwater prawn is not a shrimp but it sells in USA only when sold as freshwater shrimp. In Europe they sell when they are sold in the name of scampi, although an illogical name.

It is desirable that, MPEDA, Union Fisheries Division, State Fisheries Departments and Farmers' Associations, look into the problem of finding an export market for Indian major carps, according due importance to it and evaluating all conceivable ideas on the subject for finding a solution. The serious part, as is known, is that the inland fish farmers, whose mainstay is major carp culture, are now in great distress, facing a situation of diminishing returns from major carp culture and with no schemes introduced to help them to get normal returns as before.

Shrimp farmers are vocal and have gained experience in securing the attention of authorities to solve their problems. In contrast, inland fish farmers are mute, with practically no leadership and scanty Governmental attention. Their present plight deserves serious attention of the authorities to find an export market for major carps in whole form, or in processed form as loins or as fillets.

Fishing Chimes, Editorial 320: August 2003: Vol. 23, No. 5

# Recreational Fishing: Commercial Potential

Commercially lucrative and eco-friendly recreational culture fishery opportunities are burgeoning in the country. These are scattered for the time being but there are encouraging reports of their success. At the Sixth Indian Fisheries Forum held at the Central Institute of Fisheries Education, Mumbai in December2002, Mr.Chandrasekhar Bhadsavale made a presentation on the need for integrating fish farming scenario with tourism. Citing his own efforts, he said that he had achieved initial success that indicated bright prospects in this direction. Tourists who visited his farm displayed close interest in watching induced breeding operation of major carps. It was mentioned that he could also provide the attraction of rides on water buffaloes to tourists. He set up bamboo-based cottages by pond sides for the stay of the tourists at US$ 25/- per night. He provided well equipped kitchens in the cottages with cookery needs such as gas stove, utensils including frying pan and the other needs besides dining facilities. Those who want to prepare fried fish on the pond side could take the needed items to the place with ease. He demonstrates angling to newcomers and also provides to them rods, lines, baits and other needs,. With word going round, tourists now seek reservations in advance for visiting and staying at the farm.

There has been a similar development in West Bengal. A tank, located in a hamlet near Miriktown, known as Mirik lake, has been taken up for development as a recreational fishing spot as part of Mirik tourist project and efforts are on to improve the facilities further. A large number of tourists now come to the lake for angling.

The latest we hear is about development of recreational fishing in a Meghalaya community pond, located in Mawthung village. Perched at a height of 950m, the tank has been developed by the village elders into a recreational spot for angling. The revenue realised through sport fishing in the pond is utilised for village welfare activities which include the running of the Mawthung upper primary school. The Umium reservoir in Meghalaya also attracts a large number of tourists these days for angling.

These developments, as outlined above, indicate the benefits of promoting angling-oriented recreational activities on a commercial scale at conducive centres with potential in this direction. In fact, according to available information, the owners of several coastal fish farms in USA throw open their farms for recreational angling, providing all needed facilities, particularly on holidays.

The initiative to promote recreational angling can come from the Union Fisheries Division in association with the Union Department of Tourism. A model project integrated in nature can be drawn up and circulated among the State Governments for being taken up through their fisheries departments. There can be provision for providing extension services in respect of recreational tourism, coupled with an element of subsidy so as to popularise the activity. The beneficial effects of such a scheme would make the selected centres lively and vibrant and would generate enthusiasm and income as well among the enterprises concerned and State Departments of Tourism. CIFE and the Fisheries Colleges can conduct short-term training courses for the benefit of entrepreneurship. Candidates can be sponsored by the State Fisheries Departments and other organisations, based on nominations received from village panchayats, municipalities and from enterprises. The course can be designed in a demonstrative way and conducted at selected coastal farms, reservoirs, tanks and ponds. Initiatives taken on these lines will have the effect of promoting a new line of recreational fishery with commercial potential and the trainees will be grateful for equipping them in this direction.

Fishing Chimes, Editorial 321: September 2003: Vol. 23, No.6

# India's Marine Fishing:

## Scenario Before and After Declaration of EEZ

The Indian marine fisheries scenario was glorious and vibrant until mid-1980s. Until this period, the fishing activities were mostly confined to the Coastal Zone. Later these were marginally extended beyond the coastal zone for catching mostly deep sea lobsters, deep sea shrimps and also cephalopods. The Progenitor of this great era, cherished by the fishermen, was the Indo-Norwegian Project. This Project, ably executed by Union Department of Agriculture, under the leadership of Late Devidas Menon, and by Naval Architects sent by NORAD, two of whom were Paul Zeiner and Peter Gurtner, lighted lamps in the lives of thousands of coastal fishers of the country. Under the Project, not only traditional boats were mechanised by the installation of inboard engines, but mechanised boats of convenient new designs were introduced and later new designs were.also introduced by CIFT. In recent years, fibreglass boats and outboard motors for propulsion were introduced. The project ushered in a sea change in the socio-economic conditions of fishermen. This achievement, however, lost its sustainability in recent years, because of overfishing in coastal waters and because of upsurge in the running expenses of boats. The phenomenon of overfishing materialised due to the benevolence of the coastal State Governments by way of aiding and encouraging the multiplication of the fleet to fish in Coastal Zone. The result, as we see now, is a dangerous depletion of coastal fish stocks.

In late 1970s, the location of shrimp grounds along the upper east coast led to introduction of medium sized outrigger trawlers by Union Carbide first, followed by imports of 23-25 m LOA trawlers essentially for shrimping but also equipped for stern trawling, and introduction of similar trawlers constructed indigenously, from late 1970s to late 1980s. The operations of these vessels paved the way for some of stakeholders to diversify during non-shrimping months for catching deep sea lobsters and shrimps and also cephalopods but this diversification could not be sustained because of finance problems of the stakeholders and drop in operational efficiency of the vessels. Indigenously constructed mini-trawlers in 15m LOA range were also introduced in the later part of this period by private fishing entrepreneurship for coastal shrimping and purse seining in the coastal zone.

As this process was on, unwittingly though, the subsequent policies of the Government propelled the Indian fishing industry, which was raring to go beyond the Coastal Zone, mostly into the lap of Taiwanese/Thai fishing interests. This was an iniviting opening to these interests and also to Indian enterprises (to get over the situation of non-availability of investments required particularly to acquire larger vessels to undertake fishing in farther waters). Another aspect is that the development-enthused Indian enterprises as it enabled them to obtain permission for taking fishing vessels on charter/lease/joint venture from these foreign sources. Under these systems it was the responsibility of the owners to make investments in respect of operational expenses, marketing of the catches, and recovering the costs out of sale proceeds, all according to an approved pattern. The great advantage for the Taiwanese/Thai interests under this system was gaining fishing access into Indian EEZ, through the device of charters etc. By and by, it was mostly the Taiwanese who came forward to avail of the opportunity.

Probably for the reason that their country has no diplomatic relations with India, the resourceful Taiwanese were seen to have taken the help of certain companies in Singapore/Hongkong which probably sprung up for this purpose. These companies were providing to Indian enterprises vessels of Taiwanese origin (but with dual registration in another country such as Panama, Honduras, etc.) ostensibly on charter/lease/joint venture to comply with Indian regulations, but, as could be deduced, for the purpose of gaining fishing access into Indian EEZ. Interestingly, there were exactly similar developments in Pakistan, and a few other countries of South Asia. The force behind this pattern was perhaps not clear at that time, but the objective of this force was, as can be seen now, was to gain access into EEZs of South Asian countries like India. So far as India is concerned, 149 Taiwanese chartered/leased joint venture vessels operated in Indian EEZ from 1985 to 1993. In 1994 Government abolished the charter/lease system, but retained the provision for joint ventures to last as long as the validity of the approvals was there.

An interesting development around this period was, however, the breakthrough achieved by DANIDA in equipping an (about) 18m LOA trawler for fishing at depths upto 200m and little beyond along north Karnataka coast. This contribution of DANIDA led to the development of a sizeable fleet in 18 LOA range along the south-west coast and Kanyakumari coast with good results.

At this point of narration, a digression to attempt an explanation for the seemingly obsessive interest on the part of Taiwanese distant water fleet to gain access into Indian Exclusive Economic Zone, which was notified in 1976, is called for. The regulations to control fishing by foreign vessels in the EEZ were later notified in 1981 by the Government of India. These two notifications seem to have acted as alarm signals to Taiwanese fleet owners, habituated to sending their distant water fishing vessels for fishing in Indian waters till then. Their concern over this development that was unfavourable to them was however neutralised to a large extent with the introduction of foreign fishing vessel charter/lease joint venture schemes by the Indian Government. Their concerns however came back to square one with the Government of India winding up the schemes in 1994. This winding up, although based on the recommendations of Murari Committee, took place without any arrangement to fill the vacuum simultaneously.

In this background, the recently introduced system to permit the import of used vessels has gained momentum. The owners of vessels of Taiwanese construction are understood to have availed of the facilities under the scheme. To start with, under the scheme, used trawlers and longliners were permitted 'to be imported and registered in India with provision for payment of the cost in instalments as the voyages progressed. The instalments system was inevitable as no commercial bank was coming forward to 'lend money' for the acquisition of fishing vessels, small or large, by Indian enterprises. This suited the 'sellers' also as it brought in the system of export tie-up between the 'sellers' and 'buyers' with the operational function, understood to be in the 'sellers' hands, until such time as the total cost is paid and the longer it is delayed the better it would be for the 'sellers' to have access for a longer period. This situation, it is felt by many, has virtually brought in a demarcation between legal ownership and *de facto* ownership. The other development, as can be expected is that, although these imported vessels are legally Indian, they became a distinct category, separate from the other Indian fleet. The used vessel import scheme was initially detested by those in the main stream of the industry. However, later, perceiving the financial benefits with practically no investment and constrained by the losses being incurred in the operation of shrimp trawlers, they have started showing interest to come under the scheme.

While the number of imported used vessels now in operation in Indian EEZ is not clearly known, one can say that the activity has spread happiness among all concerned. The Taiwanese 'owners' are happy for having gained access into Indian EEZ, known for its quality Yellowfin tuna catches, very much relished in Japan, the country where they sell the fish. The Indian companies are happy as they derive financial benefits without the need for capital investments, and the Government probably is pleased because of the utilisation of the resource in a legal way and because of foreign exchange inflow.

These developments should have made the Union Department of Animal Husbandry and Dairying to announce the policy incorporating the used vessel import feature too into it, in fulfillment of the longstanding promise of announcing the policy soon. But this has not happened until the time of this writing. Apparently there must have been some valid reasons for the delay. These reasons can be only be conjectural in nature. One reason can be that FAO is now working on a programme of undertaking a study of development of Bay of Bengal as a large ecosystem. Once ingredients of this programme are announced, may be some of these could be included in the policy. The delay in announcing the policy may be for this reason. Further, the interest evinced by Taiwanese alone and no other country, in ensnaring the Indian enterprises to the system of'selling'vessels on deferred payment, an easy option, and cornering the benefits thereof may be causing some concern to the Government, particularly for the reason that a separate category of businessmen in the line have sprouted, to the neglect of the main industry which is now very weak. The proposition is good, but how come Taiwanese alone, through Singapore based or other companies, come forward and others such as Australian and U. S. companies do not evince interest? We do not know. Further, introduction of used vessels of this kind that provide the primary benefit of access to' Taiwanese vessels into Indian EEZ, a valuable concession, is probably considered as a demeaning one for a major nation like India and this aspect is probably generating second thoughts. In other words, the long range implications of the used vessel import system, that too with a predominant tilt towards Taiwanese vessels in real terms are also possibly weighing on the minds of the planners. If this conjecture is wrong, then there must be some other reason for the long delay in announcing the policy.

Another aspect is that many perceive a deliberate international conspiracy to prevent India from becoming a global fishing power. So far, India has not been able to introduce commercial fishing with any deep sea vessel of its own construction, capable of fishing enduringly in the deep sea and oceanic zone. The reason for this can be our own complacency or international conspiracy or both. As evidence of this line of thinking, the failure of a few past efforts to introduce tuna vessels really of Indian ownership is cited. Keeping these aspects in view, it is possible that an enduring way of introducing Indian owned deep sea and oceanic vessels (with an integrated system of financing, crew training, processing, export marketing systems) is being thought of, but this view can only be again a conjecture.

Another matter of concern that may be confronting the Government may be the dwindling of the real Indian (medium length) fleet of 23-27 m OAL from about 195 to 50 nos. now. This decline is something that can be understood, as the vessels have become old and it is a miracle that as many as 50 are still operative, although mostly for coastal shrimping as before. After all, something has to be done to restore normalcy, not for coastal shrimping, but for catching cephalopods, deep sea lobsters, deep sea shrimps, finfishes etc. from the deep sea zone and in the totally unexploited oceanic zone.

There is no known reaction from the Government on the sharp decline in the fleet strength of the medium sized Indian fishing fleet. Looked at from the resources angle, the decline in the fleet strength probably does not cause concern. These vessels compete with small mechanised craft in the exploitation of coastal shrimp and may be, from this angle, it is good that the fleet has shrunk. The virginity of oceanic and deep sea resources is of course another matter that is probably being looked into. The resources of Indian EEZ are estimated at 3.9 million tonnes. This estimate is related to exploitable resources. Of this quantity, 2.8 million t are now being exploited annually leaving barely 1.1 million t behind. Do we need to invest heavily on exploitation of these ageing stocks? This issue can have several angles. One is that out of 2.8 million t now being exploited, 2.2 million t are estimated to come from the coastal zone, equal to its estimated exploitable resources. The rest of the harvested quantity also probably comes from this zone, as we are inadequately equipped to fish beyond the coastal zone. This is presumably construed as having led to overfishing of the coastal zone. It is possible that an analysis of this complex situation and formulating measures to cope up with the situation is delaying policy formulation. We do not know.

Are we one of those developing countries continuously enchained, in some dominant form or the other, to foreign fishing interests such as those of Taiwan? Is this good in the long run ? If we are helpless to develop deep sea and oceanic fishing on our own, equipped as we are as at present, is there no way of teaming up with a really supportive and friendly developed fishing nation, if necessary, with World Bank's finance, to develop a sustainable and integrated deep sea and ocean fishing industry that we can confidently call as our own, in the same way as we teamed up with Norway in 1960s to introduce small mechanised boats and called the activaity our own ? and by activating our shipyards known to be capable of building deep sea/ocean going fishing vessels, but needing some technical and technological inputs as may have to be acquired from outside? These and other associated aspects certainly deserve consideration. Fortunately, an Association that represents deep sea/ oceanic fishing industry is taking care of the situation, but it is difficult to say where the inadequacies lie. Does ihe industry also think that it is not yet time for the Government to notice and act ? We do not know.

While all over the world cage culture systems are coming into vogue to counteract the trend of declining marine capture fisheries, the complacency on the part of the Government in not telling the industry of their policy in this direction is inexplicable. In short, our inability, as a major nation, in developing our own deep sea fishing fleet and cage culture systems is a matter of great concern. We cannot equivocate and console ourselves for long that imported used vessels are there to fill the void. Further, diversification of coastal fishing effort so as to reduce fishing pressure on coastal demersal fishing stocks, which is no doubt engaging the attention of the Government, has to be brought about as soon as possible, by way of adding longlining system mainly for tuna and also facility for mid water trawling on the coastal fishign boats. This is of utmost expediency and probably the Government is looking into this aspect too.

Fishing Chimes, Editorial 322: October 2003: Vol. 23, No. 7

# ....On Utilisation of Tuna Resources of Indian EEZ

It Is axiomatic that the tuna resources of the Indian EEZ have to be utilised for optimal national benefit, preferably through operation of vessels of total legal and *de facto* ownership of Indian enterprises. To start with, the presently owned Indian vessels can be equipped with inevitable external inputs, besides availing of such inputs for export marketing. For translating this aspect into practice, as a beginning, the initiative has to come primarily from the Indian marine fishery industry.

The present status of tuna exploitation in Indian waters is in a tangle, and the Association concerned, the Association of Indian Fishery Industries (AIFI), has to work out a solution to come out of the tangle and place it before the Government for clearance. This solution has to be an integrated, by taking into account the tuna resource estimates of Indian EEZ and in neighbouring international waters, the efforts made so far to utilise the resources, and all the failures of the past in this regard and of the present position. The stake the industry has in the tuna sector is of crucial importance and the solution suggested would have to generate motivation in the Government to lend support. In fact, support of this kind did sprout in abundance in the past decades but had not yielded the anticipated results and there must be a reason or reasons for this.

Let us recapitulate the features of the present tuna situation of Indian EEZ. On the positive side, among others, these are: a) There are adequate tuna resources as evidenced by the pole and line catches around Lakshadweep, catches from handlines operated by fishermen along the upper east coast and by Taiwanese-operated longliners registered under Indian ownership; b) The keenness of Indian enterprises to diversify into tuna fishing; c) The encouraging fishing results from the pilot project under which two trawlers equipped for longlining were operated, and reports of good catches by foreign longliners operating in and outside Indian EEZ, and also in Indonesian waters, and d) there is an indication of a foreign enterprise interested in joining hands with Indian companies to set up a tuna cannery in India, e) The *de facto* owners of Indian owned longliners are understood to have succeeded in initiating an air transport service from Chennai to carry chilled tuna to Japanese market.

On the negative side, among others, the following can be listed: a) The interest of Taiwanese longliners is restricted, by and large, to gaining access into EEZ for catching tuna without probably parting with expertise; b) They are understood to be cornering 95 per cent of gross earnings from sale of tuna caught from their vessels under Indian ownership; c) In general, no other nation, with the exception of one case of Mainland China, is interested in joining hands with Indian enterprises in Tuna fishing endeavour; d) No lessons are learnt by Indian enterprises from nearby Srilanka in respect of tuna fishing and no initiatives are exercised in picking up tuna fishing skills from that country; e) Our training Institutes have no motivation for providing training with focus on tuna fishing; f) The experimental project for conversion of two Indian shrimp trawlers for longlining yielded encouraging results so far as fishing is concerned but had failed in respect of post-harvest/export activities as there appears to be no clear and feasible provision for an effective integration of the activity with sashimi grade chilled tuna export; g) The scheme that permits the import of used tuna longliners has the affect of adding to exports but is deficient in respect of transfer of fishing and post harvest skills to the Indian side; h) The operational aspects of the scheme exposed our shortcomings in the way of emerging as a tuna nation, despite having resources, a point that would remain in the fishing history of the nation as a failure, placing India at par with Pakistan in South Asia, and i) There is no financial support either from the Government or from commercial banks for the acquisition of tuna vessels by the industry. Consequently the nation had to take a compromising stance of permitting import of used tuna vessels, acquiring legal ownership but leaving *de facto* ownership indirectly to the foreign enterprises.

There is no expertise with us in respect of equipping vessels for tuna fishing and in regard to tuna fishing operations (with the exception of pole and line fishing in Lakshadweep waters) and also in export marketing. Dependence on others to acquire expertise on all these aspects has become inescapable owing to financial, manpower and marketing constraints. The result is that the alternative of acquiring used longliners of foreign origin had to be per force preferred but this had led to some measure of criticism that articulated a compromise of national interests through a devious system that presents an explainable front but the fact that remains is that, while the gross foreign exchange earnings from exports of tuna through this activity which have no doubt added to their total inflow, the net earnings are stated to be as low as 5 per cent of the gross earnings.

To be on our own, a Japanese company was chosen to equip two old Indian trawlers for longlining, mainly for tuna. The Japanese company not only accomplished the installation job exceedingly well but also made the vessels operational with the crew provided by them. The company however failed in integrating the operations with export of sashimi grade chilled tuna. This is understandable, as such a linkage is possible only when there are a good number of vessels returning in batches to the shore after fishing with a viable number of tuna, to be unloaded for export to Japanese market by air availing of the facility of small carrier planes, to be organised by the Japanese partner, shouldering the responsibility of export marketing. In the absence of this linkage, the project failed. The operations however have shown that there was no resource constraint.

The failure in organising chilled tuna export system seems to have been covered in a recent effort by Taiwanese interests, as already mentioned, by initiating such a service which it is hoped will be enduring and beneficial to Indian legal owners.

If there is a consensus that the present system of permitting import of used longliners is in order, although this provides access indirectly to the Taiwanese owned vessels to fish in Indian EEZ, there has to be reconciliation that it ignores the *de facto* foreign ownership. In this context, it may be noted that the Indian legal owners have to clear the debt related to the total cost but it is doubtful whether this can ever happen with the *de facto* ownership resting on foreign enterprises. Even when the dues are cleared, the Indian side will face the problem of operating the vessels as the skills are not there. There can however be a way out to plug the lacuna of foreign *de facto* ownership.

A workable solution can be to develop a project for the conversion of the majority of the shrimp trawlers of Indian ownership for tuna longlining under a project to be developed by the Association of Indian Fishery Industries (AIFI) in collaboration with the Taiwan Deep Sea Tuna Boat owners Association, with prior clearance of the Indian Government. Such a project can be submitted by AIFI, soon after signing, to the authorities for clearance. It is to be believed that the AIFI would be able to evolve a realistic project that would bring about the installation of longlining equipment on around 40 Indian shrimp trawlers inclusive of mini trawlers. The project suggested being a commercially-oriented one and non-political, it can be expected that the Government of India would give its prior clearance for the negotiations, which would have to cover the various aspects such as investments, technical inputs related to drawings, equipments and their installation, crew commitments, fishing areas and operations, and export marketing aspects involving air transport. So far as the Taiwanese Association is concerned, as long as the arrangement gives them a legal access into Indian EEZ, and the catches are entrusted to them for marketing in Japan, they are likely to agree. The additional advantage is that Taiwanese captains know not only tuna grounds in the Indian EEZ well, but also in international waters and in several other EEZs too. An agreement of the kind suggested will have the main advantage of inducting our own vessels for catching tuna, in contrast to the indirect device of securing legal ownership of the vessels for record and depend upon Taiwanese or others totally for the operations. In this process, the Indian vessels which are now passing through a phase of adverse economics of operations, will have the opportunity of strengthening themselves for securing viable returns and for also picking up capabilities of operating the vessels on their own.

Fishing Chimes, Editorial 323: November 2003: Vol. 23, No. 8

# Microfinance in Support of Coastal Fisherwomen

NABARD has been playing a significant role in the improvement of the socio-economic status of fisherwomen of coastal fisher communities by way of refinacing commercial banks for extending financial support to the target stakeholders. In this endeavour, the apex body has the support of FAO, World Bank and the Central and Coastal State Governments. The refinance support extended by NABARD, so far as socio-economic uplift of coastal fishers is concerned, is mostly by way of providing finance to self-help groups, and is related, by and large, to survival activities of fishers who deal in dry and fresh fish marketing. While continuing the policy, NABARD is now shifting the focus gradually towards entrepreneurship development. It now aims at providing loans towards capital investment, technology infusion and capacity building. So far as FAO is concerned, it accords an exclusive focus on fisherwomen's programmes in relation to Self-Help Groups and income generating activities within the system, among others. FAO is involved in credit initiatives for fisherwomen and the organisation sees microfinanace programmes as a means for fishing communities to gain access to much needed credit services that are appropriate to their needs. The Union Department of Animal Husbandry and Dairying is keen on exploring possibilities of setting up an institution to meet the financial needs of groups of women interested in fish production, processing and marketing.

In this background, NABARD, in association with FAO, conducted a national workshop on the subject in July 2003 in Goa. A follow-up development in this context was that, in the light of a recommendation made at the workshop, the Union Department of Animal Husbandry and Dairying set up a Core Group to identify the needs of Coastal fisher communities such as credit, infrastructure, technical and other related needs, with reference to the steps taken by the various Departments/Agencies/ Organisations to cater to them and to work out a Perspective Plan of Action for their development.

The setting up of the Core Group is a welcome development. The perspective plan to be recommended by the Group, after taking into account the ongoing status of the activity, would undoubtedly lead to a reform of the present system, now lacking, to some extent, in direction and co-ordination. While quite a few agencies do concentrate on implementing development -oriented activities, many others just believe in dry/fresh fish marketing as the main means for members of Self-Help Groups to earn their livelihood. There are, however, quite a few cases of entrepreneurial initiatives. In Kanyakumari district of Tamil Nadu, the Life Care Trust has motivated fisherwomen to take to lobster fattening. Women Groups at Keelamanikudi near Chidambaram in Tamil Nadu have taken to backyard ornamental fish breeding. Mariculture of mussels and oysters has been taken up by women groups in Kerala. M.S. Swaminathan Foundation has taken up promotion of pearl culture in the Gulf of Mannar area. Sea weed collection and sale have been also taken up by Women's groups in Tamil Nadu. In Maharashtra women self-help groups are encouraged to take up crab culture. While all these are positive aspects of movement towards entrepreneurship, most of the groups continue to deal in dry and fresh fish marketing routine which does not lead to any visible professional upgradation, although it is a viable and consumer service-oriented activity.

The Core Group would no doubt take cognisance of these and other features while formulating its recommendations, which will cover not only fisherwomen but also fishermen, as per the terms of reference.

Another aspect that needs the attention of the Core Group is the total absence of co-ordination of the work of the women's groups under various NGOs. Each NGO co-ordinates the work of the groups under its purview but there is a conspicuous absence of a co-ordinating system to streamline the work of each of the NGOs and bring into being a networking system among them. There has to be a networking of the activities of the various NGOs at least in each of the States. In otherwords, looked at from the angle of the recommendations of the workshop that emphasised the implementation of various developmental activities that have been suggested, the State Fisheries Departments, in association with the State Women's Welfare Officers, have to think of evolving an organisational mechanism, to consist of promotional agencies to be set up at major fish landing centres, and regional centres to coordinate the work of primary promotional agencies at the landing centres; and a State level agency at the Directorate, to promote, monitor and co-ordinate the work and to help the department to formulate policy measures. The suggested state level agency can take the initiative of introducing a proper networking system. The attention of the Core Group in this direction, besides other aspects, would be not only in the interests of the stakeholders but also of the development of the coastal zone economy.

Fishing Chimes, Editorial 324: December 2003: Vol. 23, No. 9

# ...On production of All-male Giant Prawn Progeny

Amir Sagi of the Institute of Applied Biosciences of Ben Gurion University, Negev Beersheva, Israel has been working on the topic "The Androgenic Gland and Monosex Culture of Prawns: Biological Perspectives". Experimenting with immature specimens, he succeeded in achieving complete sex reversal of males into females, and females into males. The reversal from male to female was achieved by removing androgenic gland (Androctomy) and females to males by androgenic gland implantation.

The revelation of this revolutionary achievement was a major highlight of the International Symposium on Freshwater Prawn held in Cochin and organised by the College of Fisheries, Panangad, Cochin in August 2003.

While saying that the technology of sex reversal as developed required further research inputs for perfecting the technique, Sagi had indicated that the standardisation of the same was round the corner and that this was linked to the identification of androgenic hormone in decapod crustaceans. Once this is achieved, it would be a short step to the developmentof the technique of microsurgical androctomy in juveniles, amenable for adoption with ease at technician's level, for utilising the androgen gland so separated for achieving sex reversal, adopting the most feasible method that is considered as the best for the purpose.

It is widely known that males of giant freshwater prawn (GFP) grow faster and reach a significantly larger size compared to females. It is also known that several farmers have been able to segregate males out of the hatchery-raised seed before stocking by observing the Appendix Masculina on the second pair of walking legs of males and probably by locating the genital opening on the fifth pair of walking legs. It appears that C. P's technicians were the first to popularise this method of segregation among the Indian GFP farmers. The disadvantage of the process however is the neglect of female juveniles. The way to prevent this neglect is to achieve an all-male progeny and Amir Sagi highlighted in this context the need for the perfection of the technique of microsurgical androctomy for the removal of the gland from male specimens.

The androgen gland is located at the distal part of the vas deferens of male reproductive system. As could be understood from Sagi's presentation, this gland is to be removed and planted in the body of juvenile female. Such a juvenile will grow into a functional neo-female, explained Sagi. When such neo-females are mated with normal males, all male progeny would result. Emphasis was also laid by him on the development of biosynthetic products of androgenic gland to facilitate production of neo-females by farrners/hatchery technicians.

In all likelihood the Central Institute of Freshwater Aquaculture (CIFA) is either currently conducting research on the aspects mentioned above or is on the anvil of taking up the work, considering its crucial importance, to eventually step up production of giant freshwater prawn.Over 36,000 t of the prawn are now being produced against the production target of 50,000 mt per annum to be achieved by the end of this decade. This target was considered as too low by Michael New who was the chief guest at the symposium. In this background, it is of utmost expediency for CIFA to intensify efforts in the direction of develping the technology of all male GFP seed production. In fact, CIFA could undertake a project on the subject probably involving selected Indian universities for broadbasing the research effort for quicker results.

One point sharply made was that all male GFP prawn seed production technology as and when finally developed, inclusive of the one for microsurgery of androgenic gland, should be accessible to the farmers through the extension services of the Government concerned and not allowed to slip into the hands of multinationals. A farmer-based total technology needs to be developed to keep the cost to be incurred by the farmers on the lower side. However, the need for development of bio-synthetic products and their production may manifest itself at a particular stage. When this happens, multinationals and/or large companies may step in. It is very difficult to say now, what the future holds, but one aspect is clear. Androgen gland holds the key for the production of all-male progeny of GFP.

Fishing Chimes, Editorial 325: Jan/Feb 2004: Vol. 23, Nos. 10&11

# Balance Sheet 2003

The year 2003 has witnessed certain progressive events in the fisheries sector.

India had achieved fourth position in fish production among all fish producing nations of the world and the credit for this goes to the various State Fisheries Departments, the Union Fisheries Division, the farmers and fishermen.

MPEDA, by any reckoning, had strengthened itself as the most outstanding fisheries developmental organisation of the country. Marine products exports in 2002-2003 set a record, earnings US $ 1424.90 million (Rs. 6881.31 crores), the exported quantity being 467,297 t. This achievement was apparently the result of team work and a determined leadership in the face of a critical situation. On one side were the falling capture shrimp landings; on another side was the menace of disease afflicting farmed shrimps. Efforts were mounted to counter these adverse factors with a good measure of success. The setting up of PCR labs to detect viral diseases among shrimps and quality control labs at several centres all along the Coast to detect antibiotic residues in shrimps is a major development of the year in the shrimp hatchery and marine products export sector. Apart from ushering in diversification of products and upgraded quality standards at processing plants and at shrimp hatcheries, MPEDA took initiatives at introduction of giant freshwater prawn culture in inland States such as Bihar. It had the distinction of providing fillip to ornamental fish culture and exports of ornamental fish.

The Government of J and K is reported to have mounted initiatives for stepping up trout production from the present 100t/annum in order to promote their export to Europe. The programme includes production of trout fingerlings for supply to farmers and for developing infrastructure for facilitating trout exports.

Nine units of Liquid Chromatograph-Mass Spectrometer, a monitoring equipment to detect residues, particularly in products for export, have been set up at various Centres along the Coastline, the latest having been established at Veraval. The unit at Veraval is being upgraded to analyse hormone residues and dioxanes too.

An ICAR's project for Tiger shrimp SPF broodstock development at a cost of Rs. 135 lakhs had been positioned for being implemented soon by Principal Scientists at three Centres of ICAR at Visakhapatnam Centre of CMFRI by Dr. Maheswarudu, at CIBA's Muthukadu Centre near Chennai by Dr. Ravichandran, and at Port Blair Centre of CARI by Dr. Soundararajan. Another development had been that MPEDA had initiated planning for a Project to be taken on an island of Andaman and Nicobar UT with a satellite unit along Orissa Coast under a consultancy arrangement with Mr. Andrews Kuljis of USA, who was earlier engaged for the establishment of Tasparc.

The Central Marine Fisheries Research Institute had succeeded in the development of tissue culture for the production of pearls. It had also succeeded in producing image pearls in the body of marine bivalve molluscs. It had taken up a project in Andamans for the production of cultured black pearls from the pearl oyster *Pinctada margaritifera.*

A major farmer-friendly event of the year with immense potential had been the welcome and firm entry of giant freshwater prawn culture into the lentic waters of non-traditional and distant States from the coast such as Haryana, Punjab, U.P., Madhya Pradesh, Bihar, Chhattisgarh, Tripura, Manipur etc. The development of seed and feed supply lines to these States by air and road had been a major development in which CIFA, BMR hatcheries, Waterbase, Higashimaru, and Godrej Agrovet too had played a prominent role. Haryana Government stood foremost in this endeavour, having taken the initiative of launching a major giant freshwater pra wn culture project for the benefit of its farmers. The State Fisheries Department of Tripura, an Inland State, had earned the distinction of setting up the first ever functional Giant Prawn Hatchery, based on use of constituted saline water, a couple of years back and the operations became well stabilised during 2003. A private sector GFP hatchery was set up in Chhattisgarh. The Cochin University of Science and Technology, Cochin, with the initiative of Prof. (Dr.) B. Madhusoodan Kurup, had the credit for bringing in Greenwater hatchery technology of producing giant freshwater prawn seed. The College of Fisheries, Cochin, under the leadership of Prof (Dr) Mohan Kumaran Nair, extended to the field giant prawn hatchery design and operating technology. Monosex (male) giant freshwater prawn culture had gained momentum during the year, in which C.P. Aquaculture played an important part. In Tripura and Manipur, CIFE had played a major role under the leadership of Dr. S.C. Mukherjee and Mr. S. Krishna Reddy in the transfer of giant prawn seed production and culture technology. Prawn culture technologies were introduced in Haryana, Punjab and U.P. by Dr. Hardayal Singh of CIFA, with Mr. B. Saharan, Director of Fisheries, Haryana playing a major role. Dr. A.K. Upadhyay of MPEDA introduced the giant prawn culture technology in Bihar.

The release of Jayanthi Rohu developed by CIFA with Norwegian collaboration in A. P. and other States had been a major development of the year. Experimental pen culture in reservoirs for seed as well as fish production had been successfully conducted in Tungabhadra reservoir and in a water body in Madhya Pradesh. This had opened up new opportunities for inland fishery development. Assam Government introduced a new fish seed policy and Orissa Government a new reservoir fishery policy.

In the wake of the introduction of the system of entrustment of fishery development functions to the people themselves through Panchayathi Raj Institutions in Kerala, the Governments of Karnataka, Madhya Pradesh, and Rajasthan too took initiatives for the transfer of functions of fisheries development to Panchayat Raj bodies.

The introduction of Indian owned secondhand tuna vessels of Taiwanese and Thai construction is believed to have gained a certain momentum during the year. Further, the Association of mini-trawler owners under the leadership of its President, Dr. Y.G.K. Murthy had mounted initiatives for installation of monofil longlining equipment on their trawlers, mainly for the reason that they were deep in economically adverse operational straits, because of a major fall in shrimp catches. Rajiv Gandhi Centre of MPEDA and Pancham Aquaculture commenced commercial production of seabass seed and its supply to farmers. Pancham Aquaculture succeeded in the breeding of *Marsupenaeus japonicus* and raising its seed for supply to farmers on a commercial scale.

Through permissions given by the Government to two Indian companies, the Pacific White shrimp (*Penaeus vannamei*) gained entry into the country, the seed of which is understood to be under production at certain hatcheries and also this shrimp is farmed and exported as a measure of withstanding shrimp export competition from a few countries.

Steps for the formulation of FAO's Bay of Bengal large marine ecosystem project had been initiated during the year. The studies and the recommendations thereof are expected to improve the fishery complexion of Bay of Bengal.

Several initiatives for the upliftment of professional capabilities of fisherwomen and for lending financial support to them had been taken during the year by NABARD and others.

A major revival of fisheries of Chilika lake had taken place after the opening of a new mouth.

The industry continues to await the announcement of the Marine Fisheries Policy of the Government of India. There are rumours that, for unknown reasons, the Government is hesitant to announce the policy. However, there are encouraging indications that tenth plan marine fishery development schemes have been formulated keeping the unannounced policy in view and that these schemes were in an integrated form. Let us hope that the policy itself would be soon announced.

Adverse economics have haunted the mechanized fishing boat sector and continue to torment it. The reasons for this had been the high cost of diesel and fall in shrimp prices. While larger vessels continued to enjoy offshore bunkering facility, this facility was not extended to the small boats. Unfortunately, even with the offshore bunkering facility and reduction in fleet strength from 190 vessels to around 50 nos. larger vessels too were under the spell of unfavourable economics.

In USA, an antidumping case had been filed by the shrimp producing interests in eight States of USA against India and a few other countries such as China, Vietnam etc. This had caused enormous anxiety and the Seafood Exporters Association had engaged a legal firm in USA to fight the case.

Shrimp producers and exporters continued to face serious problems of reduced and unfavourable market prices. Culture shrimp sector was particularly affected as Japan reduced the quantum of their imports on grounds of muddy-mouldy flavour attributed to farmed shrimps of Bhimavaram zone of A.P.. Freshwater fish culture sector suffered severe losses during the year because of overproduction and limitations of domestic marketing system and absence of export market.

Fishing Chimes, Editorial 326: March 2004: Vol. 23, No. 12

# Aquaculture and Aquafarming

An effort is made in this column to examine the scope of the terms 'Aquaculture'and 'Aqua farming'. These two expressions, apparently, owe their origin to terminology in respect of land-based crop production system in vogue, known from times immemorial as 'Agriculture' and Agrifarming' or 'farming'. The reason for the derivation of captioned terms, related to lentic water sector from those of land-based crop production, is obviously the close relationship between crop production from land and water. As is known, in land production, the cultivated plants derive sustenance direct from soil, aided by rain water or other water sources. Similarly, in ponds, tanks and such other confined waters (and capture water resources too) the water column first absorbs nutrients from the soil and these generate the production of phytoplankton to start the food chain. This culminates in the production of fish or other aquatic animals and plants. This broad similarity must have led to the derivation of the agri-based terms 'Aquaculture' and Aquafarming'.

These two terms are recognised and accepted round the world with no exception. Thus we have the World Aquaculture Society and several such other global bodies which deal with aquaculture adopting the terms aforsaid. In India we have Institutes of freshwater aquaculture and brackishwater aquaculture. Further, recently a new Society of Aquaculturists has been started in India with Chennai as its headquarters, besides the Aquaculture Foundation of India at Chennai. Yet, no standard dictionary lists the term 'Aquaculture'. A few dictionaries list the term as 'Aquiculture', obviously for the reason that it owes its origin to 'Agriculture'. This aspect,however, is of academic interest. The derivation of the term 'Aquiculture' in dictionaries obviously follows the style of the term 'Agriculture'.

The terms 'Aquaculture' (or simply 'culture') and 'Aquafarming' (or simply 'farming') are often used to signify the same activity. It appears that, in the same way 'Agriculture' covers the whole range of integrated operations from raising of seedlings in nurseries to farming of the seedlings to raise crops, 'Aquaculture'has to cover not just the grow-out systems but the seed-raising activity too. Thus, a person who raises fish seed, stocks them in his pond (and also may supply to others) and obtains a fish crop therefrom, has to be called as an 'aquaculturist' or fish culturist' or a person engaged in aquaculture (of fish, crab, shrimp etc), or only fish culture as the case may be. On this premise, a person who obtains fish seed or other aqua seed supplies from outside for farming in his pond, can be termed not as a 'aqua or fish culturist' but only as a 'fish-farmer' or 'aqua-farmer'. Extending this premise further, those who are engaged in specific types of fish culture or fish farming can be referrred to as 'Major carp culturist' or 'Major Carp farmer', and so on, as applicable. Those engaged in composite farming or polyfarming (as may be distinguished from composite culture/polyculture) would have to be categorised as just 'Fish Farmers' or 'Aqua farmers'. In contrast, those engaged in the production of various kinds of fish or other aqua seed and also engaged in their farming would deserve to be referred to as 'Fish culturists' or Aqua culturists', as the case may be. These are integral terms. Thus, for clarity of expression, it is desirable to distinguish a 'fish culturist' or 'aquaculturist' as a higher category of professionals compared to a fish farmer' or an aqua farmer.

The proposition set out above is also applicable to mariculture, say of holothurians, mussels, clams etc. so as to distinguish from those engaged in their seed production alone and those engaged in integrated operations. Here two points deserve to be noted. One is that the activity that integrates fish raising in association with pig, cattle, duck and poultry production is rightly referred to as integrated farming and not integrated culture. The other one is the inexplicable evolution of the term cage culture (although the term cage farming is also rightly used). The dimension of breeding and of seed production either in cages or at land based hatcheries for culture in cages has to also come under the term cage culture, and grow-out system alone in cages has to be referred to as cage farming.

The frequent interchangeable use of the terms 'culture'and 'farming' also deserves an examination. According to dictionaries the word 'culture' has the dimension of multiplication, whereas the term 'farming' denotes only the grow-out activities. Thus, in agriculture, a person who raises seedlings and links the activity to grow-out production is called as an agriculturist. An agriculturist can be a farmer too, but a farmer cannot be termed as 'Agriculturist'. Extending the same logic to aqua sector it has to be construed that a person who raises seed and also farms them has to be referred to as an Aquaculturist. He can also be referred to as an aquafarmer as farming is part of the term aquaculture. A person who does not raise seed on his own but procures them from outside to meet his stocking needs is an aquafarmer (major carp farmer, mussel farmer, etc., composite aqua-farmer, integrated aqua farmer, and so on) but not as aquaculturist.

Readers are invited to make available to the Editor their valuable views on this presentation, that seeks to establish the difference between aquaculture and aquafarming.

Fishing Chimes, Editorial 327: April 2004: Vol. 24, No. 1

# ...Towards Shrimp Culture Sans Disease

Nothing short of a major revolution in shrimp farming system can bring back the glorious pre-1994 happiness to the shrimp farmers. In the post-1994 era, several well-intentioned initiatives to restore normalcy in shrimp farming were taken by the authorities which were followed by the farmers to the best of their ability. Invited by the Indian authorities, NACA intervened to stem the onslaught of disease scourge and consequential deterioration in quality of the production output, with some measure of impact. Several well intentioned enterprises devoted to supplies and services and consultants stepped in to help farmers to wriggle out of bad water quality, disease and growth problems, and achieved some remedial contribution. Despite all these efforts, the problem of the disease affliction, particularly white spot, was more or less defiant,refusing to relax its hold. Innovative pond management techniques introduced by NACA's team did play some curative part in the study area. The net position however is that despite the benefit of being surrounded by well wishers, several of the farmers continue to be in the mire, running deep into debts. Among the three main players, the authorities, suppliers of antidotes and the farmers, it is difficult to say who has a feeling of achievement but probably not the farmers.

It is not that India is alone in this plight. There are many other countries too in the same kind of straits and the nearest to us among them are Bangladesh and Sri Lanka, but, according to reports, these countries are a few steps better off than India. Ecuador, Brazil and several other countries have had their share of woes but it is said that they are emerging out of the disease and production problems. Taiwan, Thailand, Vietnam, China and others also are stated to be gradually wriggling out of the stranglehold of viral and other diseases that have been crippling their shrimp farming sector too. In India, as elsewhere, the problems start with the shrimp hatcheries along the coastline where wild shrimp brooders are used for PL production. We have a battery of Polymerase Chain Reaction (PCR) laboratories along the coastline, but despite their active role in detecting PLs that are positive to the white spot virus for rejection, the mortality inflicted by the virus on shrimps under farming continues. There are, no doubt, steps under way to promote utilisation of disease - free monodon stocks of Andaman sea for PL production, and there is hope in the air that these steps will improve the situation. Some years back nauplii produced using Andaman tiger shrimp brooders were reported to have been brought to the Indian mainland and raised into PLs, only to be dismayed by the surfacing of white spots on them. It is possible that this may have happened for reasons other than the quality of nauplii and in all likelihood, the result of the present effort would be different.

There are also reports of efforts in the direction of raising domesticated brooder shrimp stocks at three centres in the country and as a result, it is certainly possible that in about six or eight years viable and disease-free brooder stocks would be produced for supply to hatcheries.

There are other efforts to improve the growout systems in a manner that they would produce disease free shrimps. These have been set in motion in the country with outside expertise too. Experts from NACA, and those hailing from elsewhere tried their best to experiment on production of disease free shrimp crops but only with partial success. The present status of the shrimp culture situation, however, seems to be that, whatever be the efforts, the white spot disease refuses to quit in a substantive measure. There are no doubt stray exceptions which do not, however, change the general picture much.

Now there appears to be a consensus in all shrimp farming countries that the disease problem can be extinguished only when the pond bottom conditions could be totally made free from waste feed and faecal droppings on a continuing basis and the quality of water is maintained free from any contamination, and the stocked PLs are disease free. Having tried pond lining and also aeration system, the general approach that seems to have now emerged is to switch over to a water recirculation system that provides for sedimentation of suspended solids in the water in a sedimentation pond before it is taken into a grow-out pond, in which it is kept flowing towards the outlet in such a manner that the waste feed and faecal matter get filtered and the resultant water returns back into the pond because of recirculation system. It is also recognised that the setting up of the recirculation system is expensive but it is felt that a way of investment to meet the cost has to be found, as the higher/healthy crops that are expected to be produced will be totally disease-free and would bring in higher returns to cover the periodical repayments with interest and at the same time leaving a comfortable margin.

## Initiative in Brazil

The disease problem being perplexing, many believe that there can be a solution only in theory, notwithstanding application of several antidotes and other measures. The antidotes and other measures, it is felt by some, do not totally solve the problems of ineffective removal of accumulated bottom contamination and its intertwining with bottom soil. Many advocate that the bottom can be made soil-free for ensuring effective removal of wastes, preferably through recirculation. While there are no reports of any lasting breakthrough by way of any method other than recirculation from any part of the world, there is a refreshingly encouraging news of legislative measures in Brazil, aimed at ushering in a new order of shrimp farming practice, based on recirculation system supported by a sedimentation pond. The Government of Brazil took a bold decision to eliminate imperfect practices through a legislation, which was introduced in October 2002. The legislation has made it mandatory for all shrimp farms of Brazil to use sedimentation basins or lagoons regardless of farm size. In addition, the use of water recirculation systems by farms is made compulsory, with an exception that these systems are to be conditioned for use, giving due regard to locational, technical and economic viability.

In a new shrimp farming region in the Pedencias and Porto do Mangue countries of Rio Grande do Norte, of Brazil, about 80 per cent of 6,000 ha being developed will be using recirculation systems and 100 per cent will have sedimentation basins according to Brazilian Farmers' Association. This development is reported in Global Aquaculture Advocate (Vol 6 No 3). In the same country, in the State of Ceara, in an area of high salinity, top priority is also being given for the setting up of recirculation systems and sedimentation basins. Further, in regard to effluent quality, it is stated that there was an ample agreement between the Brazilian Government and the progressive private farm sector that mandatory use of sedimentation basins or lagoons would represent an advanced and viable water treatment that would contribute to the minimisation of the possible environmental impacts from effluents especially Total Suspended Solids (TSS) and Biological Oxygen Demand (BOD).

An encouraging aspect reported is that the use of sedimentation basins and recirculation systems have already been validated in Brazil. Technical viability and environmental benefits, in detail, were presented during the Special Shrimp Industry Session held at the recent World Aquaculture Meet 2003 in Salvador, Brazil.

India too has a plan to introduce a legislation but it is different from the Brazilian legislation. It is meant for setting up an Aquaculture Authority. In order to get over the present disturbed shrimp farming conditions, there is a general feeling that a set of reforms that would touch the basic aspects and not symptoms alone is essential. These basic aspects have become increasingly intertwined over years with the bottom soil, water conditions and feed dispensation. There appears to be no way of countering these, without resorting to sedimentation and filtration-oriented recirculation systems and effluents treatment system, as realised in several countries. MPEDA, having realised the need to introduce closed water recirculation system, is stated to be on the anvil of introducing a regular scheme for encouraging farmers to set up recirculation system in their farm ponds. In its Annual report (2002 - 03), MPEDA says "In order to prevent the viral diseases in shrimp farms, one of the measures to be adopted is the closed water re-circulation system, where there is no water exchange, which brings the carrier of viruses into the farm. Low stocking density and feeding rate in closed system will reduce water pollution. In this closed water re-circulation system, the required quantity of water is initially stored in reservoir-cum-treatment pond, where it is treated and is used for the culture purpose. The drained water from the culture ponds is brought into the ETS and then to the Reservoir-cum-Treatment pond from where the treated water flows in to the culture pond. The financial assistance provided by MPEDA is @ 75 per cent of the capital cost, subject ot a maximum of Rs. 6.0 lakhs per beneficiary for demonstrating this technology in farmer's pond. After assessing the efficiency of these demonstration programmes, a regular scheme is being introduced for encouraging the farmers to set-up re-circulation system, for which a modification is suggested to Ministry in the existing ETS subsidy scheme".

It will be a herculean task in India to bring about such a reform as in Brazil, as it involves not only the standardisation of designs to meet the needs of various categories of situations that may involve not only different design and economic compulsions, but also the problems of investments. However, there appears to be no escape from initiating the reform.

Recirculation systems, linked to effluent collection and treatment before release, that can be considered for adoption (notwithstanding the investment aspect) for phased replacement of the traditional system for shrimp farming (extensive, modified extensive) are of three categories : i) sedimentation pond-associated pond recirculation system; ii) Raceway system and iii) system of two parallel raceways connected at the ends to enable circular flow.

A Study Group inclusive of a representative from NABARD and headed by an experienced professional in recirculation systems may have to be set up by the Union Department of Animal Husbandry and Dairying to examine various aspects of the subject and recommend measures for a phased introduction of one or the other recirculation systems, regional or state-wise, taking into account the severity and scope for extinguishing the disease menace, and the trail blazed by Brazil, the technology induction problems, the investment problems, and trained manpower implications etc. The technology induction problems relate to components of recirculation system such as biofilters and mechanical filters and several other aspects, and designing of systems and also identifying feeds that would be compatible with their use. In this context, it has to be borne in mind that shrimp farmers in Thailand have been successfully using closed pond systems to reduce risks of attacks by shrimp pathogens to the farming system. Active suspension ponds, which reduce the requirements for water exchange have also been demonstrated in shrimp cultured in Belize.

The heavy investments the farmers have to make for installing the recirculation system (with the anticipated higher returns) have to be taken into account, in the background of present depressing state of the Indian shrimp farming industry. The study team could consider, if and when set up, recommending the need to prepare and send a Project to the World Bank for financial assistance for appraisal and sanction. The project could propose an organisational mechanism too, apart from the components of financial inputs from the Government by way of subsidy, training etc. The Group could also consider proposing NABARD as the Agency for channelising the proposed financial assistance of the World Bank.

The problem being a massive one involving lakhs of ha, the suggested Group could also be asked to examine and recommend the need for setting up an Apex National level Co-operative or Corporation to promote the spread of a reformed system of shrimp farming all along the coastline on the lines to be worked out, linked to the anticipated improved operations of the existing hatcheries with the usage of broodstock harvested from Andaman sea and the stock to be developed through domestication programme.

Only a part of the CRZ is under shrimp farming now. Once it is established beyond doubt through experimentation or pilot effort that the composite recirculation systems will keep the coastal shrimp farms free from disease incidence and environmental pollution and will lead to a higher level of healthy crops, Government could bring the details to the notice of the Supreme Court to reconsider its earlier decision of restricting the activity to traditional or improved traditional system of farming in the CRZ. An initiative on these lines may bring back the earlier glory of shrimp farming, endowed with sustainability, environmental compatability and higher productivity and an improvement in the earnings to the farmers.

Fishing Chimes, Editorial 328: May 2004: Vol. 24, No. 2

# Marine Fishing: Perspectives of Diversification

There has been a relentless upsurge of bottom trawling effort all along Indian coastline during the past three decades and over. As part of this, so far as vessels of 20m OAL and above are concerned, their strength crossed 190 nos. by middle 1980s because of the shrimp trawling lure. Most of them having been decommissioned, year to year, the operating fleet strength has now come down to around 40 nos. Shrimp has been the main target species for all these trawlers. The reduction in the fleet strength of trawlers should have normally increased the catch per trawler. However, there has been either no change or there has been a gradual decline in shrimp catches per trawler. This can be due to biological reasons arising out of overfishing and the constraints in the export of capture shrimp to USA because of the compulsion of using turtle excluder device in shrimp trawls and arising out of impact of antidumping duty threat. These factors have eventually rendered the economics of operation of these trawlers unviable. Recognising this adverse development and the proclivity on the part of the owners towards diversification of fishing effort from shrimp trawling to other fishing systems,principally longlining for tuna, the Union Department of Animal Husbandry and Dairying introduced a scheme in the tenth plan for the conversion of existing trawlers of 20 m OAL and above for resource-specific fishing for which a back-ended subsidy of Rs. 15 lakh per vessel was provided. The provision is that this programme would be implemented through the Fishery Survey of India (FSI) by suitably modifying the imported technology through the ICAR's Central Institute of Fisheries Technology (CIFT). While the source for import of the technology is not mentioned (apparently this is left for the initiative of the owners), it is stated that necessary consultation would be available from CIFT.It is further stipulated that the proposals for modification of trawlers for alternative resource-specific fishing would be processed through FSI and that the conversion should be carried out by the owners by adopting the imported technology, with the needed modifications.

In view of the general position that the tuna resources of the Indian EEZ are the least exploited in comparison to the EEZs of several other countries, there is a general interest among the Indian owners of vessels of 20 m OAL and above to seek assistance under the aforesaid scheme for the installation of tuna monofil longlining system on their trawlers. In fact, with the assistance extended by MPEDA a couple of years back, two Visakhapatnam - based shrimp trawlers of 23 m OAL were equipped for conducting longlining for tuna. A Japanese company was entrusted with the job which included not only the installation of the needed additional equipments but also the demonstration of operational aspects, provision of crew etc. While the fishing results of these trawlers (which conducted a few tuna voyages but later gave up) are stated to have shown improvement in catches from voyage to voyage, the owners, according to reports, found the marketing part a stumbling block. The lesson learnt was that an activity of this kind would have to be an integrated one, effectively covering all aspects, from fishing through preservation of catches on-board to unloading them at port and their transference to a small plane for carrying to Japanese market. Further, there was a realisation that the involvement of the chosen importer or buyer was essential to ensure acceptability of the quality of the exported fresh chilled sashimi tuna packed in ice, in Japan. As is known, Japan is the main destination for sale of tuna in sashimi form for securing remunerative returns.

The identified lacunae as part of the lesson learnt were thus in respect of export-oriented post-harvest operations. The catches have to be in a fresh chilled condition and would have to reach the Japanese market in around ten days of their capture so as to protect quality in that state of preservation. The transportation has to be necessarily by air and the needed number of tuna fish would have to be assembled and packed in 'coffins' for being loaded into a plane chartered for the purpose either by the exporter or the buyer as per the prior arrangement. The requirement in this context is that an adequate number of vessels, in turns, would have to be returning from the fishing grounds to the port, within seven to ten days of departure, to unload the tuna for packing in ice in 'coffins' and loading into the plane. This will become possible only when a certain number of vessels operate under a co-ordinated system conforming to a schedule. It is thus axiomatic that an organisational pattern has to be built into the system.

The experiment as earlier referred to failed because there were only two modified vessels that were made operational for tuna longlining. As it happened, their landings were grossly inadequate to undertake an organised marketing effort. A co-ordinating agency could not be put in position because of this. One of the owners did manage to sell the catches to an exporter of frozen tuna loins but the arrangement did not last. Further, the two modified vessels functioned more or less independently. The lesson was that there had to be at least 15 to 20 vessels brought under the fold of a coordinating agency, that would implement a schedule of departure and arrival in turns and also take care of other needed operational aspects.

In this background, the requirement of co-ordination of operations of the vessels becomes important, to be made part of the present scheme introduced by the Union Department of Animal Husbandry and Dairying.

Under this scheme, operators of the vessels planning to diversify fishing effort would have to apply to Fishery Survey of India. This implies that each of the enterprises would have to look after its own interests. The number of vessels that can be subjected to diversification being limited, their operations have to be necessarily co-ordinated so as to ensure quality of the exported material and its timely delivery. This material, for each round of export, has to consist of not less than a certain number of fishes for ensuring economically viable operations. Therefore, the need will be to bring together all the vessel owning enterprises concerned together to decide on having a co-ordinating agency. As it happens, all these enterprises being members of the Association of Indian Fishery Industries (AIFI), those intending to avail of the subsidy facility offered under the scheme can authorise AIFI to take over the function of coordination. When this is done, a separate field-oriented set up, as a wing of AIFI, would be required to be established and suitably equipped with ship-to-shore and shore-to-ship communication and other required facilities. AIFI would have to also take upon itself the task of obtaining the needed quotations, preferably on a global basis, for equipping all the vessels concerned, for monofil longlining for tuna with needed on-board facilities, besides providing other shore facilities needed for effectively playing the role of export co-ordination. Obviously, a company with wide experience in tuna longlining and close contacts with Japanese importers would be the best suited for performing this job that entails integration of several functions.

The financial assistance of 50 per cent of the cost of modifications as subsidy, limited to Rs. 15 lakhs, offered by the Government may look generous but this alone would not make the project successful. This provision, looked at from any angle, would need reconsideration for expansion. Investment is needed not only to meet the cost of modifications but also towards the cost of longline, sonar, and manpower training requirements, besides working capital needs. Another aspect is that, with the ongoing generally negative attitude of bankers towards extending financial support to fishing companies, there is need to have provision for financial assistance by way of loans, principally against hypothecation of vessels and providing a realistic and feasible margin money. So as to ensure recoveries there can be a built-in system of pro-rata deduction of instalment amounts towards repayments out of export returns, under an arrangement for transferring the same to the lender, who can possibly be the same bank as would handle the export documents. This subject merits a discussion in all its aspects at a meeting of all concerned to be convened by the Government. These days, most of the commercial banks have adequate funds to extend finance for developmental projects and they may not really need refinance. Notwithstanding this, NABARD could be motivated to extend refinance to the identified banks, under a guarantee to be extended by the Government.

The role assigned under the project to FSI and CIFT in the implementation of the project goes a long way towards its success. One gap in the scheme is, however, noticeable. This is that the MPEDA appears to be outside the frame of it. MPEDA, which implemented the earlier scheme must be having a clearer idea of the reasons that led to the failure of the same and their experiences and views in respect of the first experiment that they handled would be a great help in the implementation of the present scheme. Many would agree that MPEDA would have to be in, for the scheme to be totally successful.

In the finalisation of the cost of the modifications to the trawlers, as approved by Fishery Survey of India and the Central Institute of Fisheries Technology, and possibly MPEDA, the owners of the vessels/their organisation (AIFI), must have a major part in the negotiation with the short-listed tenderers concerned in respect of cost of modifications, before the offer is finally cleared by the Government and the financing institution concerned for extending the needed assistance. The yards/centres proposed for carrying out modifications would have to be an important aspect of the final decision on the tender concerned. All aspects related to this important project would no doubt be thoroughly considered so as to avoid later setbacks. There are conflicting views on the relative efficacy of the original tuna monofil longlining system developed in USA and the later one developed by the Taiwanese. The system installed by the Japanese company on the two Indian modified trawlers a couple of years back is stated to be based on the Taiwanese system, which is considered by many as having some deficiencies. This aspect may have to be looked into while dealing with the project.

In conclusion, it has to be mentioned that the owners should, at this stage itself, take up the issue of developing the scheme as a Co-ordinated Project so as to ensure well regulated exports, which alone can yield viable returns. Individual vessel-based applications for subsidy assistance without a co-ordinating mechanism may result in the repetition of the past unsavoury performance of the two vessels modified for tuna longlining.

Tuna longlining is a passive method of fishing that is environment friendly, Vessels of 15 to 20m OAL are presently operated in the offshore waters of quite a few countries for this kind of fishing. Considering that the strength of deep sea fishing fleet of India has dwindled down to around 40 nos. from a level of around 190 nos., and heavy investments are needed to restore the fleet strength to the earlier level, authorities would have to evolve a scheme for equipping all presently operating vessels in the range of 15 m OAL for tuna longlining too. Over 600 nos. of such vessels are reported to be engaged in trawling at present all along the Indian coastline. All these can be modified to conduct longlining for tuna in the open sea zone for short duration voyages of around seven days. In fact, it is learnt that, a Co-operative based at

Visakhapatnam with members owning such vessels is engaged in working out a scheme to diversify into longlining for tuna, but they can succeed only when there is active support from the Government and the financing bodies. It is encouraging to know that NCDC has come forward to finance a pilot project of the Co-operative to start with. The initiatives on this line of action of modifying 15 m OAL trawlers for open sea longlining have to be shared between the authorities and the owners, (who have formed into a co-operative) with a co-ordinating effort forthcoming from the authorities concerned.

Fishing Chimes, Editorial 329: June 2004: Vol. 24, No. 3

# CMFRI'S Appraisal of Status of Exploited Marine Fisheries of India

India, in the same way as in many other countries, is passing through a critical phase in respect of its exploited marine fisheries resources. They now hover around 2.6 million t per annum, threatening to decline further. The per unit catches have been inexorably coming down in varying degrees all along the coast line. Operators of all categories of vessels face the problem of declining incomes, with some of them having no alternative other than selling their vessels. The strength of vessel fleet of over 23mOAL has declined to around 40 nos.

The situation is alarming. The stakeholders confine their operations to known grounds. There are however some who have extended operations to farther waters to the extent the capabilities of their vessels can take them. One major constraint they face is the lack of well appraised knowledge of the unexploited resources and the grounds of their occurrence and lack of vessels capable of exploiting these resources.

Gone are the days of emphasis on exploitation of known resources. In the present day situation, coverage of unexploited fishing grounds has become inescapable. Apart from knowledge of grounds, training in new methods of fishing such as longlining, upgradation of existing vessels or acquisition of vessels equipped to cover fresh or relatively unexploited grounds, availability of shore facilities for imparting value addition to processed products and stregthening linkages for effective domestic as well as export marketing, have now emerged as aspects of paramount importance for the earnings of stakeholders to move up.

In this background, one basic formidable handicap is the lack of analysed information on the exploited marine fisheries resources. The contours of action to plan and extend fishing activities can be determined only when the position of remainder of fish stocks is known.

In this puzzling situation, we now have, like a rare and refreshing development, a book on the Status of Exploited Marine Fisheries Resources of India, edited by Mohan Joseph Modayil and A.A.Jayaprakash of Central Marine Fisheries Research Institute, Kochi. The contents reflect that, while stocks of some species are overfished, there are other stocks like deep sea demersal species and open sea pelagic species that are underexploited. Coastal marine fisheries face severe depletion. These are indicative of serious deficiencies in marine fisheries management.

The aforesaid book is a compendium of papers authored by reputed biologists on status of 22 commercially important exploited fish groups, four marine crustacean groups (Penaeid shrimps, non-penaeid prawns, crabs and lobsters), marine molluscs (gastropods, bivalves and cephalopods), marine turtles and mammals, and sea weeds. These papers are apexed by an appraisal paper that eminently summarises and arrives at conclusions in relation to the various presentations. This apart, the book includes a well presented appraisal of the exploited Marine Fishery Resources of India, besides chapters on economics of fishing and fish marketing, trend in landings, potential yield from Indian EEZ and on forecasting of fisheries.

This book presents, by any reckoning, a valuable account of the output of CMFRI on exploited fishery resources in recent years. The contents are a practical guide, primarily on the status of the exploited marine fisheries of the Indian EEZ for the benefit of the industrial sector and also to the new generation of entrepreneurs. As could be seen, the contents would infuse enormous interest, particularly among the new generation of qualified young men looking for opportunities to come up and also among those in the industry whose morale is often seen as sagging, because of drop in the exploited fisheries, more pronouncedly of the crustaceans. The contents of this book are also a great source of knowledge enrichment, a means of widening the vision and perspectives of fisheries students and also of those who look for opportunities to undertake productive marine fisheries research. Some of the presentations such as those related to tuna and live bait for tuna, carangids, ribbon fishes, perches, catfishes, threadfin bream, bulls eye and sea weeds are very revealing in that there is a substantially unexploited chunk of these resources that could be exploited in a sustainable manner.

This book deserves to be in the libraries of various universities and institutions dealing with fisheries develpment, and fisheries economics.

*Fishing Chimes* congratulates all the talented scientists who contributed valuable and well prepared papers on the topics dealt with by them. M.Srinath deserves a special mention here for the excellent appraisal of the exploited Marine Fishery Resources of India.

Mohan Joseph Modayil and A.A.Jayaprakash accomplished a splendid editing job. The preface of the Editors so well reflects the rich contents of the book. As Dr. Mangla Rai, the Director General, ICAR, says, in his telling Foreword, this publication will be welcome by all those concerned with the development of marine fisheries.

Fishing Chimes, Editorial 330: July 2004: Vol. 24, No. 4

# Thoughts that Gripped on Fish Farmers Day

The Fish Farmers Day, celebrated on 10 July, 2004, has a history. At a meeting held in Dec 1999 at the Central Inland Fisheries Research Institute, Barrackpore, West Bengal, the first Hiralal Chaudhuri Best Fish Farmer Award was given to late Mr. N. Nilratan Ghosh of West Bengal by *Fishing Chimes* Jaysree Charitable Trust. It was presented by Dr. Nita Choudhury, the then Union Joint Secretary (Fisheries). On that occasion, an appeal addressed to the Government of India to declare 10th July as Fish Farmers Day was voiced. The reason behind this was that, over four decades back, on 10th July 1956, for the first time, Dr. Hiralal Chaudhuri, renowned Indian Fishery Scientist, succeeded in inducing Indian major carps to breed through administration of fish pituitary hormone extract. This breakthrough has transformed the inland culture fishery sector to the present glorious status of major carp seed production all over the country. This upsurge has led to an unprecedented increase in annual major carp production to around 2 million t annually. The technology of carp seed production having become popular, the system of carp spawn collection from rivers had a natural death, resulting in the much needed restoration of riverine carp fisheries. These developments eventually resulted in the declaration of 10 July as Fish Farmers Day. The first Fish Farmers Day was celebrated on 10th July, 2001.

Fish Farmers Day is now an important occasion to pay tribute to fish farmers, fishery scientists, technocrats and fishery workers for their achievements in the culture fishery sector. All these categories of persons, working together, gave to the nation technologies of composite culture, polyculture, integrated farming, pond fishery management systems etc. The fisheries educational institutions such as Central Institute of Fisheries Education (CIFE), Mumbai and various Colleges of Fisheries have been imparting education and training in fish farming and the related central fisheries research institutes have been playing an important role in conducting periodical training programmes in updated technologies. The State Fisheries Departments, particularly the Fish Farmers Development Agencies, have been playing a major promotional role in this line. Let us acknowledge with gratitude the dedicated role of all these organisations in the promotion of fish farming in the country.

On the Fourth Fish Farmers Day, several functionaries all over the country are believed to have taken a pledge to mount intensive efforts towards reaching the goal of achieving sustainable farm fish production, to make our tanks and ponds fishful, and to upgrade the socio-economic lot of fish farmers.

The Supreme Court has recently pronounced that Pisciculture (a commonly used technical term to denote fish farming or fish culture) is not part of Agriculture. This is a historic decision that endows pisciculture (or fish culture or fish farming) with a new status. The decision has unleashed a new situation. Firstly the longing on the part of fish farmers (or pisciculturists or fish culturists), to be treated at par with agrifarmers (or agriculturists), mostly for securing subsidies and other reliefs, has lost steam. While reconciling themselves with the changed situation, they may probably begin to feel proud of having a separate status, away from agriculture.

The emerging trend among Agrifarmers in certain States to diversify into aquafarming (fish or shrimp or prawn or crab) indicates that atleast some of them consider aquafarming or aquaculture as comparatively much more profitable than agrifarming. This trend has been a precursor to the system of integrated fish farming that covers horticulture, piggery, poultry, duck farming, and cattle raising, besides agriculture too. This is a welcome trend. At the same time, in the present situation of inadequate post fish harvest facilities that hinder efforts at value addition, the emergence of situations of fall in prices of fish and other aqua produce because of the diversification on the part of some of the agrifarmers towards aquafarming (which has added to fish production and thereby causing fall in prices) deserves a close examination for evolving needed remedial measures such as setting up of storage facilities for organising regulated marketing, so as to prevent drop in income levels of aquafarmers.

A digression at this stage of the narration will be in order to understand the definitions of the terms fish farmer, fish culturist and pisciculturist. A fish farmer can be defined as a person who stocks his farm with fish seed for growth to produce table fish for the market. What he does is actually farming and not culture. The word 'Culture' has a different meaning. A fish culturist, as may be defined, is one who produces not only fish but also seed, to utilise the same for growing them in his farm ponds. He may of course also supply the seed to others for farming or culture as they want. Differently stated, the term culture involves multiplication (*e.g.*, culture of microorganisms such as bacteria, virus etc. or fishes like Tilapia, murrels, certain catfishes etc.). If one says that he is culturing fish the expression will be technically correct, provided the purpose is to achieve their multiplication too besides growth. However, the exclusively growth-oriented activity to raise a crop of major carps or others can be termed as farming.

Returning back to the main theme, the Indian farmers, by and large, are mostly stuck up with composite farming of major carps, whose production has increased enormously. This increase has led to fall in fish prices in the market. In order to balance the economics of operations, farmers, particularly of Andhra Pradesh, have taken to export of farmed fish to various States of North-east including Tripura and Bihar. A major problem now is to counter the general drop in fish prices and this is possible only with Government's support. The support has to come by way of promoting infrastructure as needed for domestic marketing and export, particularly in the non-coastal States. Setting up markets is part of the effort towards development of infrastructure. The other integrated requirements are centralised cold/frozen storage facilities besides ice production, processing facilities and transportation network linked to markets.

In coastal States several farmers diversified into production of giant freshwater prawn because of availability of seed and feed. Production of seed of the prawn is difficult in non-coastal States. However, States like Tripura and Chattisgarh have succeeded in raising the seed through pilot hatcheries set up and operated using constituted saline water. Such efforts are on in Punjab, Haryana and Rajasthan too. Haryana is ahead of others in one respect. The State has launched a major scheme with assistance from the Centre for popularising giant prawn culture. Other non-coastal States can emulate this initiative. The non-coastal States however meet their present requirements of freshwater prawn seed from coastal States. Feed is also obtained from outside, although this is expensive. Further, the economics of giant prawn production depend on export marketing. The channels for importing seed and feed into these States and infrastructure for promoting exports are now taking shape, with the exporters playing a role. Farmers and entrepreneurs in various States such as Punjab, Haryana, UP, Bihar, Chattisgarh, Tripura, Manipur etc. are now involved in the endeavour and it is to be hoped that in this decade there will be encouraging developments.

We have the problem of securing a remunerative price for major carps. The problem needs an indepth study to formulate a plan of action which, it appears, would have to be related to increasing production in a sustainable way at reduced costs, provision of infrastructure for storage and regulated release of production to domestic and export markets, coupled with all strategic measures needed for developing the same. The domestic fish and fish product marketing system needs enormous improvement. Further, Indian major carps deserve to have a good export market. No serious effort has been made so far to achieve this. Therefore, efforts are needed in this direction, not only to augment foreign exchange earnings but also to help the farmers to realise improved returns.

Rajasthan, Haryana, Punjab, Western UP have vast tracts of saline land with underground saline water reserves. Under the renewed leadership of CIFE considerable research work is in progress to achieve utilisation of these areas for aquafarming. There has been some measure of success in farming shrimps and prawns and a variety of fishes in this Zone in ponds with water of varying salinities as required. Once the technologies are standardised for wider application, there will be ushering in of prosperity among the various interested sections of the society of the region.

The latest sustainable and eco-friendly systems of intensive culture that are economically viable because of integration with a well planned marketing system have taken roots in quite a few countries but these have not yet been introduced in India. Two of these are a) Pond cage farming system, and b) Recirculation system.

**Pond cage farming system** : Under this system, cages are set up to cover half of the pond area from surface to column leaving about 2 ft of water at the bottom. In between the cages, aerators are set up. The cages are stocked with fish seed. In the cage-free pond area prawns are stocked. The waste feed from cages that drops down increases fertility of pond waters. Any stagnating material is removed during water exchange and periodic cleaning. The production from pond cage system is considerably more than intensive farming systems in vogue.

## Recirculation System

Under this system, there is a reservoir pond and a set of connected farm ponds. The ponds are either lined with polyethylene sheets or built in masonry. Water from the reservoir pond is led through filtering screens into the ponds from where it is taken back to the reservoir pond. Waste material settles down at the bottom of the reservoir pond. The water, treated with lime etc. as needed, is again taken into the ponds. The recirculation process is adjusted in a way to maintain the needed minimal depth. The pond stocking density is at a level far higher than the normal system because of the recirculation of water which removes waste matter. Feeding and maintenance schedules are followed as determined.

Farmed shrimp production has gained focus because it brings in foreign exchange through exports. The white spot disease incidence among farmed shrimp has however created problem of sustainability of the effort. In this situation, it is good news that a project in collaboration with NORAD has been taken up to develop genetically upgraded domesticated broodstock free from white spot and other diseases for supply to shrimp hatcheries for disease free seed production and supply to shrimp farmers. CIFE and CIBA are major collaborating partners in the project. It is auspicious that this project has been approved on the eve of the Fish Farmers Day, 2004.

In the light of the present situation of falling returns from major carp farming, farmers and other fishery functionaries should strongly recommend to their respective Governments and MPEDA (a) to mount a special project to establish an export market for Indian major carps, and processed major carp products, (b) to institute special steps to establish giant freshwater prawn (GFP) farming in the non-coastal States too supported by hatchery produced seed supplies and also feed supplies, (c) to create needed supporting infrastructure for preservation, processing, transportation of major carps and their products and GFP and its products to ports for export and for benefit of non-coastal States, (d) The non-coastal States may follow the lead provided by Haryana Government by way of securing sanction for financial assistance for a GFP project from the Centre, and (e) Authorities may be prevailed upon to strengthen Fish Farmers Development Agencies to enable them to continue to be a potent media for transfer to farmers commercially viable technologies of seed production and farming.

Fishing Chimes, Editorial 331: August 2004: Vol. 24, No. 5

# Separate Fisheries Ministry is Urgent, in National Interest

The significant progress the nation achieved by 2002-03 in the form of augmenting the annual fish production to a level of nearly 3 million t in the inland sector and to about 2.6 million t per annum in the marine sector, and the marine products exports per annum to a level of Rs. 6800 crores, has now, discouragingly though, acquired a distinctly downward trend following a phase of stagnation.

The reason for this trend is that the tempo of progress has lost the power of endurance. This is because there is no policy in position in tune with the emerging developmental needs. Further, unfortunately, the fisheries sector does not have all the support that it desperately needs from the Government. Many believe that there is no national fisheries policy in position. This can be only partly true, for, there is actually a decade-old policy in position, although it is outdated. In the culture fishery sector it continues to be propped up mainly by technologies of induced breeding, composite farming and of late integrated farming, without any declared policy direction in respect of spread of freshwater prawn and other systems of culture across the country, particularly for utilisation of saline water areas of north west and introduction of upgraded systems of farming such as recirculation systems, closed or open. On the marine capture side there are no serious policy initiatives visible to divert the cognisibly counter productive coastal fishing effort that now concentrates heavily on shrimp fishing. The exploitation of tuna resources of Indian EEZ, by direct Indian effort in areas other than those around Lakshadweep islands, is virtually neglected. It is now in the hands of vessels of flag of convenience. Marine cage culture systems, very much needed now to rehabilitate fishers who have lost their livelihood because of depletion of resources, increase in fishing costs, fall in fish and shrimp prices etc, have not yet been introduced. On the export side, luckily value addition came into the picture in recent years although on a marginal scale. In fact, a good part of the increase in production in the past few decades took place, as many believe, mainly because of the initiatives of stakeholders, but this trend too has now slowed down. T.M. Choudhary, Secretary, Association of Indian Fisheries Industries has described the marine fishing industry as' a shrinking and sunset industry. Clearly the top fisheries set up at politico-administrative level at the centre would need to have strength, time and identity to perform, in co-ordination with others concerned. In other words, the fisheries developmental policy can acquire a solid shape and it can gain a real and a lively fillip only when there is a separate Union Ministry for Fisheries.

For example, owners of Taiwanese longliners who operate in Indian EEZ with the flag of convenience would do their best to ensure that we do not develop capabilities at tuna longlining and we have some experience already in that direction. This strategy has to be smashed. An administrative Head of a separate fisheries ministry alone, as many captains of fisheries industries assert, would be able to devote undivided attention to resolve the crucial national as well as international fisheries problems of this severity that now confront the nation. This comment acquires weight when looked at from the angle of the recent decision of the Supreme Court that clarified that pisciculture is not part of agriculture and a fish farmer is not the same as agri farmer. Riverine, reservoir and lake fisheries and marine fisheries are in any case not part of agriculture. Further, for the same reason, forests are not part of agri group. So much so they were brought under a separate Ministry by the Government. For fisheries to remain recognised as part of domestic animal groups, Supreme Court has clearly said that fishes are not domesticated animals. Thus, looked at from any angle, there can be no justification for fisheries to be under the Ministry of Agriculture, unless the subject has to continue to be so for other reasons.

Norway helped India in the mechanisation of fishing boats which upgraded India's coastal fishing. Over 53,000 mechanised boats and over 44,000 motorised boats now operate along India's coastline, but this situation has led to overfishing, mostly of bottom fish stocks, for no fault of Norwegians. The problem that has arisen as a result of overfishing is one that can be solved through diversification of fishing effort and of upgradation of a good number of vessels to fish in waters beyond the coastal zone. Larger vessels have to be introduced to exploit fisheries of waters beyond the Coastal Zone, as the existing fleet of such vessels, which by itself is small, has dwindled from a level of over 190 nos. to around 40 nos., only because they were concentrating on shrimping, an activity that is claimed to have become unviable. The rise in operational costs of fishing vessels, accentuated by increase in price of diesel oil and coupled with fall in market price of shrimps and fish, is a problem threatening livelihoods of fishers.

India has not been able to exploit its tuna resources and had to therfore subject itself to a humiliating situation of allowing Taiwanese tuna vessels with Indian flag of convenience to fish in our waters, for want of a way of utilising the resource as we have no tuna fishing vessels. And this is taking place when neighbouring Srilanka and Maldives have made remarkable progress in tuna fishing and export. These and many other International fisheries problems of the nation need undivided and exclusive attention.

India's position in the international fishery canvas has been fading of late, apart from its status having become hollow in the Indian Ocean Tuna Commission. This observation is often heard in fisheries industrial circles.

On the inland culture front there are any number of problems that need exclusive attention. Our reservoir fisheries are poorly developed. Increase in major carp production and fall in major carp market prices owing to absence of regulated domestic market have placed the farmers at a disadvantage. Further, there is the challenging problem of bringing the vast neglected saline tracts of Rajasthan, Haryana, Punjab and Western UP under fish and prawn/shrimp production. The growing giant prawn production initiatives in the North and Central India lack support of a supplies and services network. These and many other problems need focal attention at top policy and administrative level and in this context there can be no alternative other than formation of a separate Union Fisheries Ministry, to induct this much needed reform.

It is not fair to say, as has been said before by those in authority, that (presently) a 'small' subject like fisheries does not deserve a separate Ministry. Only the ignorant can say this. With over 9,000 km of coastline, 2 million sq miles of EEZ, 3.9 million t of exploitable marine resources, millions of ha of reservoirs and tanks and ponds, and several rivers and other flowing waters, one cannot say that fisheries of India do not deserve a ministry. Only those who have no vision can revel in such an observation. Presently fetching over Rs. 6,400 crores in foreign exchange, an earning of this order very few sectors can boast of, experts say that fisheries sector has the potential to fetch over Rs. 10,000 crores in foreign exchange annually.

There have been any number of recommendations made at several conferences, seminars, symposia, workshops etc. highlighting the need for a separate Ministry for Fisheries at the Centre. There is no evidence so far that these appeals have had any impact at the top (Although there is a whisper in the air now that the matter may be lined up for attention). In this situation the various fisheries associations and other such organisations in the country, headed by one of the premier bodies such as Seafood Exporters Association, Association of Indian Fisheries Industries,Forum of Fisheries Professionals of India, Apex Co-operative Federations etc. should spearhead a move to convene a top level conference of delegates of all fisheries non-Governmental bodies to highlight the Issue, to deliberate upon it threadbare and come up with a resolution that appeals for the formation of a separate Union Ministry of Fisheries in the national interest of saving our fisheries wealth and rendering its exploitation sustainable, improving the socio-economics of our fishers and for providing employment to thousands of fishery workers, to augment foreign exchange earnings and follow it up vigorously through an Action Committee until such time as the objective is realised.

Fishing Chimes, Editorial 332: September 2004: Vol. 24, No. 6

# Fishing Fleet Matters

Having enjoyed the pride of possessing over 190 vessel strong fishing fleet of larger vessels of 23-27 m OAL until recently, India has now descended down to the level of holding a shrunken fleet of around 40 nos. of these vessels, most of which are 18 to 20 years old. This count does not, however, include the foreign longliners, ostensibly of Indian ownership and flying flag of convenience (Indian flag), which authorises fishing in Indian EEZ.

The surviving vessels can be said to be in the queue to reach the ultimate stage of contributing to further reduction of the fleet soon. The drop in the fleet strength, from about 190 nos. to the present around 40 nos. has been taking placing in stages, because of disposal of vessels by sale, seizure for non-payment of dues to port trust, or abandonment due to heavy repairs and operational losses, stranding, sinking, capsizing and so on. In this situation although already late, those in the industry feel that steps need to be mounted the soonest to revive the larger fishing vessel fleet of the nation, through implementation of purposeful and well-conceived measures, aimed at sustainable utilisation of the resources beyond the coastal zone. Only when such a plan is implemented, India will be able to have a viable fleet of larger vessels of her own, they feel. Otherwise the nation may continue to have addition of foreign vessels with flag of convenience. This manner of proliferation may be a good option to achieve and report increase in marine products exports and earnings thereof, but the outflow of 95 per cent of the foreign exchange earnings, as understood, to reach the real owners of vessels of flag of convenience in some form of the other, causes concern to many. However, it is this outflow that sustains the operation of these foreign vessels in Indian EEZ and the '*de facto*' owners bear the burden of the 'arranged' Indian ownership and of the Indian flag as long as they are in Indian waters because of this outflow, some observe. And, according to what is heard, the Indian owners are aware of and unfamiliar with the fishing method of these vessels, not knowing much about the operations. However, they may be knowing adequately about the owners of foreign vessels operating in Indian waters under Indian flag of convenience, because it is they who take the catches.

The introduction of larger vessels consisting of 23-27 m LoA vessels equipped for outrigger trawling (for shrimps) and also for stern trawling for fish etc. took place a little over two decades back. The owners engrossed themselves in lot of shrimping from these vessels, but as years went by, they faced disillusionment at incurring progressively unbearable operational losses. So much so, several of the vessels, at one stage, diversified into fishing for deep sea lobsters, deep sea prawns, and cephalopods. These strategies worked but they could not be sustained, because of long time taken for renewal of stocks, limitations of grounds etc. Some of the owners deployed their vessels for fishing in Myanmar and in Indonesian waters and probably elsewhere. Some sold away their vessels. The remaining vessels (around 40 nos. now as already stated) continue to be wedded to shrimping, reconciling themselves with the economics of fluctuations in catches of shrimps and declining incomes thereof. They are aware of the rising market for finfish but they are unable to avail of the trend for various reasons. In order to help the owners to diversify their fishing effort, Government have announced certain subsidies for the installation of tuna longlining or other equipment for diversified fishing from these trawlers and the owners do come forward to avail of the assistance, with the hope that no calamity of the kind that took place recently would befall their vessels either before or after the installation, and also probably believing that the past calamities were exceptions.

India has a small-sized fishing vessel fleet of remarkable strength consisting of around 50,000 nos. of mechanised boats, over 30,000 nos. of motorised boats and probably over 1,50,000 non-mechanised boats. Most of these concentrate on coastal fishing. (The exceptions are a few 15m LOA range vessels venturing out for fishing beyond the coastal zone). Observations are often made to say that the owners of larger vessels should desist from operating their vessels in the coastal zone competing with the smaller vessels for the same resource, particularly shrimps.

While the fishing companies owning larger vessels have the disadvantage of old age of their vessels and the associated hazards, the seafood exporting companies face a serious shortage of capture seafood raw-material and also of mounting losses. So much so, the seafood exporters have taken the initiative of forming a consortium of around 45 of them to gain the needed combined strength to deal with the alarming situation and to find a way of securing a package of reliefs, primarily a one time settlement from the commercial banks with whom their indebtedness has grown because of the sickness that has engulfed seafood export business, for no serious fault of theirs. The Association of Indian Fishery Industries is also probably doing its best to help its members, all of whom are stated to be in distress, waiting for a way to come out of the present bad situation. In this predicament, the need seems to be for the Association to organise a consortium of all the willing companies in distress and help them to get a package of reliefs as needed, in the same way of securing of reliefs as contemplated by the seafood exporters. The main reliefs the fishing companies need are i) tested information on resources position beyond the Coastal Zone or beyond the area of operation of mechanised boats, ii) The kind of vessels required to exploit these resources, pelagic, columnar and benthic, beyond the coastal zone., iii) From where and how to secure the vessels needed for exploiting these resources, iv) technology inputs needed and their source, v) trained manpower requirements and source, vi) additional shore infrastructure needs and vii) marketing support etc.

Government could certainly lend all needed support to the industry in respect of all the aforesaid points, but the details are to be articulated by the industry in clear terms and in a unitary voice.

Production, processing and exports are interdependent activities, although those who deal with them in most cases function separately. There are, however, quite a few enterprises who deal with all the three components, (production, processing and exports) by themselves. So far as producers are concerned, the main drawbacks they suffer from are the old age of their vessels and the high costs of maintenance, low returns from catches and dilemma in respect of upgradation for diversification of fishing effort with reference to investments, and detereorating economics because of the age of vessels.

The production work on one hand and processing and export operations on the other, which now virtually have independent runs, can gain strength and support, financial and otherwise, only when they are well integrated. And this integration, which is now somewhat lacking, becomes possible once those handling these components acquire strength to achieve the needed linkages. The suggestions in this context are:

i) Let all interested producing companies, on the same lines as is being followed by exporting enterprises in setting up the Forum for Revival and Reconstruction of Seafood Export Industries, form into a Forum or consortium for the purpose of working towards receiving financial and other support from the Government and others concerned for its members to acquire larger vessels. The forum or consortium has to project their declining assets mostly consisting of overaged trawlers, which need upgraded replacement.

ii) Apart from the present enterprises engaged in the operation of larger vessels, there can be many small boat operators interested in upgrading their activities and there can be others interested in entering the line. In this background, Government may consider issuing a notification seeking expression of interest by companies that are viable in terms of one or more criteria related to sea fishing beyond the territorial waters of the EEZ, some of the criteria being a) past fishing experience with small or larger vessels in the zone, b) ability to mobilise margin money and to secure financial support by way of a loan for acquiring larger vessels for open sea fishing (pelagic/ columnar and demersal), or readiness to participate in a project that aims at developing a larger fishing vessel fleet for exploitation of fishery resources beyond territorial waters and integrating the production with processing and exports/internal marketing and providing needed infrastructure.

As the responses trickle in, motivated by the notification, a concurrent initiative that may be needed on the part of the Government is to set up a high level Empowered Committee to deliberate on various vital interrelated issues so as to formulate an integrated project for ushering in a fleet of larger vessels of needed types in numbers as determined. This Committee headed by the Chief of the Fisheries Division of the Union Department of Animal Husbandry and Dairying, and having members representing ministries of finance, commerce, shipping etc., and planning commission and others, can examine the various issues involved for developing a deep sea fishing fleet. The issues, as are known, are related to a) entrepreneurship, b) Tested information on resources and determination of the types and nos. of vessels that have to be introduced, c) development of needed designs and shopfloor drawings in India preferably in collaboration with one or more of reputed yards of Japan, China, USA, Australia as identified through an effective publicity, d) selection of one or more shipyards of India for the construction of vessels based on the developed designs and drawings in India in collaboration with the selected foreign shipyard/yards, e) organising a financing package with the funds as could possibly be arranged by the foreign collaborating shipyards or by an Indian Bank like the State Bank of India by securing funds from NABARD and guaranteed by Government of India, for being lent to the Indian fishing companies for release to the constructing shipyards concerned and f) selection of the fishing companies to participate in the project out of those who would respond to the notification calling for applications and who would agree to conform to the requirements as stipulated.

An alternative way of ushering in India's own deep sea fishing effort, which is not there, is to work on a project to be taken up for the purpose with bilateral or World Bank's Assistance.

There can be many other routes to introduce deep sea fishing effort in Indian EEZ, the basic criteria of them being the Will and Determination on the part of the authorities and the industry in this respect. It is imperative that a policy, strategy and a plan of action to vest India with the status of a deep sea fishing nation is developed the soonest; to prevent the spread of 'flag of convenience' culture, which has already taken roots, equating India with several other derisively viewed nations because of this undesirable system.

Fishing Chimes, Editorial 333: October 2004: Vol. 24, No. 7

# Historic Commercial Tuna LL Success with Total India Effort–In Visakha-Kakinada Stretch of Bay of Bengal

It was on the 8th day after new moon in September 2004 (22 Sept., 2004) the 16.5m LOA shrimp trawler, MFV Gamini, improvised for tuna longlining (12.5 km long main line with 250 branch lines), left Visakhapatnam fishing harbour on her first ever tuna longlining commercial voyage southwards, although an experimental one. The voyage was conceived jointly by the Managing Directors of Gees Marine Products Pvt Ltd (Mr. Gordian Kagoo) and of Hemo Seafood Pvt Ltd (Capt A.B. Raj). The vessel had a crew of eight, commanded by Capt A.B. Raj and with Capt Ignatius as co-skipper. One innovative aspect is that the crew included three fishermen of Thoothoor fishing village near Kanyakumari, having experience in hook and line fishing for sharks and skipjack. The names of these three as could be gathered are 1) Cyrus 2) Saji and 3) Jinny.

At pages 10, 11, 12 of this issue a detailed report on this voyage authored by Premchand and Paul Pandian of Fishery Survey of India is presented.

The voyage lasted for nine days. The vessel came back to the port on 30 Sept with a remarkable catch of 28 nos. of yellowfin tuna. Together they weighed 1512 kg, each weighing between 40-68 kg. There were other fishes too in the catch, two swordfish, one sail fish and eight sharks. The tunas were lined up for export marketing immediately after the landing. The success of the two companies in longlining for tuna and also marketing was largely due to the timely advice and support extended by Brig S.K. Aggarwal, President, Association of Indian Fishery Industries. About a year back two trawlers of 23m OAL, one of his company and another of B. Srinivasa Rao were equipped for tuna longlining by a Japanese company and longlining for tuna was stated to have been conducted from these vessels by Indonesian crew. The results of fishing as well as marketing were however disappointing.The experiences gained in that effort enabled S.K. Aggarwal to advise Gees and Hemo in a tangible way. Five pieces (200 kg) of A+ and A grade Sashimi grade yellowfin tuna were sent by air to Canada where they were well accepted and were stated to have fetched prices as high as US $9 and 5 per kg respectively for A+ and A grades. The other yellowfin tuna, which were of B grade were sent to Srilanka and these are understood to have fetched around US $ 3 per kg.

The maiden tuna longlining operation by Gamini has demolished the myth surrounding tuna longlining prevalent in Indian fisheries circles that the activity is not wholly within the capabilities of Indian crew and that it entails superior skills that are presently the preserve of Taiwanese, Chinese, Korean, Indonesian, Thais, Japanese, Australian, American, Mexican etc. In any case, the perception that tuna longlining skills are not the cup of tea of Indian crew now stands disproved.

Gordian Kagoo, Raj and Ignatius rediscovered the talent of Thoothoor fishermen in the operation of longline. After an incisive investigation to locate talent they zeroed in on Thoothoor fishing village, about 40 km from Kanyakumari. Several fishermen of this village possess extraordinary fishing skills which include traditional prowess for operating hook and line for catching sharks and skipjack on high sea. Gordian, Raj and Ignatius found them as the men they need for operating tuna longline. Their serendipity in this regard at once became precious and refreshingly encouraging and this changed the scenario at once. Having talented men with the needed skills in their own area (The owners hail from Kanyakumari area and Thoothoor is a fishing village near Kanyakumari), it struck them that yearning ignorantly for skilled tuna longlining men from far away places was something absurd. With the sudden rediscovery of Thoothoor, it was a short step for the entrepreneurs to get three skilled fishermen from there to Visakhapatnam to prepare the longline on the vessel itself and undertake fishing for tuna with it. With surprising deftness and ease, the three Thoothoor fishermen prepared and operated the line. The result of the operation was the hooking of 28 yellowfin with an average hooking rate of 2.8 per cent per set and a overall hooking rate of 6.8 per cent in four longlining sets during the total voyage period of nine days, each round lasting a little over one and a half days.

In the background of the encouraging and exciting development as put forth above, it is to be hoped that the authorities and the industry would formulate a wholly new line of action for the introduction of a tuna longlining fleet, in the place of the ongoing flag of convenience system. Fortunately we now have the availability of needed inputs. These are: a) vessels (23-27m, 15-18 m OAL) which, although old, can be utilised for installing longlining system, obviating the need for heavy investments for the time being for building new ones; b) Monofil line is presently manufactured in India (and exported); c) hooks are made in India (and exported); d) Imported line reel drums and line haulers installed in a couple of vessels earlier can serve as models for manufacture of the parts for assembling them in India; e) skilled crew (those from Thoothoor) to serve as starting point for producing more of them particularly at training centres; and, f) of course knowledgeable skippers and second hands. With a modest investment of around Rs. 50,000, Gamini is stated to have been equipped for tuna longlining with about 12.5 km of longline, holding 250 branchlines. This indicates that organising finance for equipping the present fleet of large and mini vessels by the owners should not be an unsurmountable problem. They could form into a co-operative and approach NCDC for financial support.

Many of those connected with the operation of tuna vessels of flag of convenience are of the view that operation of such vessels is for national benefit. In the present situation, if such vessels are not operated (in the absence of *pucca* Indian flag vessels), foreign poaching vessels will exploit the resource cornering the benefits. To get over the problem, all possible Indian vessels can now be equipped for long lining with indigenously produced monofil line and other equipment. So far as manual operation is concerned, Mr. Gordian Kagoo says that a study of this system of operation of long line as followed in Gamini has clearly indicated that it causes crew fatigue and it would not be a proposition conducive to operational efficiency of longline. Reel drums and line haulers can be made in India following available designs. Items like the hydraulic motor, as would be required for operating the system could be imported, if not available locally.

With the authorities offering subsidy for equipping presently operated mini and larger fishing vessels for diversification, and in the light of the encouraging results of the tuna longlining voyage of MFV Gamini in the Upper Bay of Bengal in September 2004, as reported in an article in this issue and as explained in this editorial, the important task to be taken care of now is the organisational aspect. As S.K. Aggarwal and B. Srinivasa Rao say the main aspect to be taken care of is to have an enduring and reliable tie up with an importer/importers of Sashimi Tuna and other possible tuna products and work backwards on the kind of packing/transportation arrangement needed for export of harvested tuna in fresh chilled Sashimi form mainly to Japan on one hand and as packed loins or in cans to EU,USA and others on the other. Total involvement of the importer in the integrated operations from production to final phase of export would be reassuring and mutually beneficial. It is advisable to start work on adding longlining equipment on the existing vessels, with the symbiotic involvement of the chosen importer.

The long term step needed is of course to take up a project for the introduction of a fleet of newly built vessels for exploitation of tuna and other resources of Indian EEZ aimed at domestic supplies as well as for export with the needed integration of the various components of the project including the financing and training part. The tuna resources of Indian EEZ are estimated at around 4,80,000t.

A way is now seen for the development of India's own integrated tuna fishing industry. The manner in which this way will be strengthened and utilised in a planned and successful manner depends on the authorities and the industry and the wise selection of importing enterprises, capable of extending all needed support. Finance could be raised by appproaching NCDC with a well drawn project report. The support of the prospective importers could be enlisted to meet part of the margin money requirements.

There is now a clear transformation in the tuna related climate and for this the three Thoothoor fishermen, Gordian Kagoo, A.B. Raj and Ignatius on one hand and Fishery Survey of India on the other deserve to be warmly congratulated for their purposeful and target-oriented work accomplished with silent and firm efficiency. The men of Gamini in particular did a job, that could not be accomplished in the past, by mobilising all inputs of Indian origin.

Fishing Chimes, Editorial 334: November 2004: Vol. 24, No. 8

# On Extricating Indian Shrimp Culture Sector from its Present Travails

MPEDA/NACA on one hand and ICAR/ AKVAFORSK of Norway on the other have kindled an elastic hope among shrimp farmers of India that they could wriggle out of the persisting disease and other problems haunting culture shrimp sector.

A post-harvest problem that farmers face is that average shrimp production level per hectare in Indian CRZ, placed at 200 kg, is stated to fetch gross incomes that are just adequate to meet operational expenses. There is no way of increasing production by either extending the area of farming or by way of adopting upgraded methods of achieving higher production, owing to the orders of the Supreme Court against both, to protect environment.

While under the project launched by MPEDA/ NACA, a well conceived programme is being implemented on an experimental basis in a selected zone that is affected, to extinguish the shrimp disease scourge and to introduce a sound pond management system in another vital direction, CIFE and CIBA of ICAR have teamed up with AKVAFORSK of Norway to implement a project on Genetic improvement of Black Tiger Shrimp, with the objective of developing a strain that would be free from white spot syndrome virus or of any other disease - causing micro-organisms and which could be grown to gravid stage for being utilised through ablation or otherwise for raising post-larvae in hatcheries for supply to farmers.

Under the MPEDA/NACA project there has been an encouraging progress by way of introduction of some field level reforms, aimed at ushering in co-operative action among farmers through formation of aquaclubs to take decisions on the manner of pond management, flow of supplies and services, and on developing linkages with harvesting, storage and marketing. So far as ICAR/ AKVAFORSK project is concerned, it was launched on 4 July 2004 and considerable spade work was understood to have been accomplished for commencing the task of raising the first generation progeny.

Apart from the two lines of action as aforesaid, taken to normalise the shrimp PL production work in a manner that the PLs would be totally healthy and to introduce a well-designed crop management system so as to reap good harvests, MPEDA has also taken up a scheme to catch and utilise tiger shrimp brooders in Andaman waters which are found to be generally disease-free, unlike those in the coastal waters of the Indian mainland.

Digressing a little at this stage of narration, it would be relevant to recall the manner in which the farmed shrimp sector took roots in the country until the time of the striking of the disease wave. By any reckoning, as the knowledgeable ones would agree, the spread of the activity was haphazard and without much of a direction. So much so, the alarming conditions that took over led to a public interest litigation which resulted in the orders of the supreme court restricting shrimp farming activities to traditional/extensive or modified extensive system of farming and also to the freezing of the activity as at the time of date of delivery of the judgement. So much so, around 10 lakh ha of potential area of CRZ considered suitable for shrimp farming continues to be fallow. The level of shrimp production is stated to be around 200 kg/ ha as already mentioned, and it is not clear how far it tallies with the ground realities. The assessed low rate of production is attributed to the non-adoption of either upgraded or intensive production methods. The farmed production is presumed to have come from a harvest of around 100,000t in the zone and this much production in understood to work out to an average of one t/ha. Actual production figures are not available but the above mentioned assessment appears to have been made from the cultured shrimp exports. While the emerging trend is an increase in quantity of farmed shrimps in the composition of exported shrimps, there has been no matching increase in the share of farmed shrimp production in overall shrimp exports, stated to be for the main reason that in the CRZ not only the production area is frozen but also the technology too is blocked at traditional/extensive/modified extensive level for protecting the environment.

Another aspect widely known is that, a good part of harvested shrimps is heavily infected, mostly with white spot, because of which a steep drop in farm gate returns is taking place.

The reasons generally identified for the disease incidence are: 1) Stocked PLs are mostly infected ones, and 2) the formed sediments at the bottom turn lethal to shrimps over a period of time because of the mixing up of left over feeds and faecal matter with bottom soil and this helps the white spot virus or other harmful micro-organisms to proliferate. Removal of sediment and exchange of water was seen to be of partial help as part of the wastes get inextricably mixed with bottom soil rendering it harmful and these wastes may or may not be part of sediment removed from time to time.

Having experienced these and other problems, farmers in USA and probably in certain other countries too have been switching over to closed masonry recirculation systems. The high investments involved for setting up these systems are justified by pointing out the tension-free operation, the higher and disease-free shrimp production, and the augmented incomes generated by the sale of shrimps harvested from tanks equipped with close recirculation system, and also reduction in water requirements, as frequent exchange of water can be avoided under this system. As the production from a closed recirculation system is mostly semi-intensive/ intesive in nature, far better harvests are reaped.

So far as the present Indian situation is concerned, no farmer can go beyond traditional/extensive or modified extensive system of shrimp farming as most of the land having sea water supply facility is located in the CRZ. Further, according to orders of the Supreme Court no new shrimp ponds can be set up in the CRZ. Apart from this, the problem of salinisation of agricultural lands adjoining shrimp farms has to be taken into account. When closed masonry system is adopted, this problem will not be there. As it has turned out, stepping up of disease-free and healthy farmed shrimp production has now assumed a measure of importance, because of the steep decline in capture shrimp production. As and when it may become possible to convert the presently fallow lands of CRZ into masonry shrimp ponds with recirculation system by approaching the supreme court and obtaining reconsidered orders, the chances of disease incidence becomes minimal enabling shrimp ponds to be brought under semi-intensive/intensive system. One aspect to be mentioned here is that many hold the view that sticking to traditional or extensive or modified extensive culture system admits of our technological inadequacies to face the challenge posed by disease problem.

One way out can be for the MPEDA to have a design incorporating all good features of standard US design for setting up a closed or open shrimp farming system with a recirculating arrangement, totally free from the disadvantages of ponds with earthen bottoms which in course of time become centres of harmful microbe proliferation and most of these microbes are believed to be of disease causing potential. No doubt masonry systems with recirculation facility will be expensive compared to earthen ponds but the aspects for and against the proposition of introducing closed or open recirculation system could be incisively studied. Such a study may reveal that the advantages would be more. Firstly, in a recirculation system, intensive farming can be undertaken and this generates adequate funds to pay back the cost of construction in a certain time frame. The disease problem would get eliminated because of recirculation and removal of waste matter from time to time. The chances of salinisation of adjacent agricultural lands can also be eliminated. So far as the problems that can be visualised, the main one in this regard is related to the orders of the Supreme Court that prohibit the setting up of any new ponds in CRZ. The other problem concerns sizable investments that the activity would attract.

CRZ regulations exempt the setting up of hatcheries in the zone. These hatcheries are of masonry construction. The suggested pond recirculation system being also of masonry construction, MPEDA could approach the Supreme Court to permit the setting up of one pilot recirculation system in the CRZ in the first instance and for its operation. The court would naturally have an independent evaluation of the impact of the system on the environement, based on the operational report furnished. If the evaluation meets with its approval, exemption could be sought for the setting up of such units with recirculation system in the CRZ by the entrepreneurship, based on the cleared design and as approved by MPEDA.

As is known, the extent of area under shrimp farming in the CRZ, developed till the date of orders of the Supreme Court, stands frozen. This phase of development therefore has to be taken as Phase I (or the only phase so far) of shrimp farming in CRZ. There has to be now a Phase II that would take us beyond phase I, if the Supreme Court approves, so that a fillip can be given to the utilisation of the fallow area of CRZ, through allotment to trained farmers who can take good care of the environmental conditions as prescribed. Once the setting up of recirculation systems in the CRZ receives the approval of the Supreme Court, based on the possible proposals of MPEDA, all measures needed for a regulated development of such units would no doubt be taken, with a rational method of effluent treatment and disposal through an interconnected system covering all the ponds, zone wise. It can then be said that a golden period would dawn, showering prosperity among farming enterprises, farmers and farm workers. Inflow of foreign exchange would go up, because of increased exports of shrimp and shrimp products. There will be sustainable growth of shrimp farming with stability, viably compensating for the drop in capture shrimp fisheries and contributing to the opportunities of employment.

Fishing Chimes, Editorial 335: December 2004: Vol. 24, No. 9

# Marine Fishing Policy has Some New Features

The best part of the recently released marine fishing policy by the Union Department of Animal Husbandry and Dairying is the provision for encouraging Sea cage farming and the setting up of fish aggregating devices out at sea. This apart, another main aspect is that the policy seeks to bring fishing in coastal waters both traditional and the on going upgraded systems into focus to match with the activities of the stakeholders in the deep sea sector so as to achieve harmonised development of marine fisheries, both in the territorial waters and extra territorial waters. It also highlights, as part of the policy, the objective of ensuring socio-economic security of the artisanal fishermen whose livelihood solely depends on this vocation. Further, fishing by boats upto 12m LOA would be treated at par with agriculture (and not as part of agriculture) and operations by vessels of over 12m and upto 20m LOA would be treated at par with small scale industries.

As a support to achieve these objectives, the policy envisages a departure from open fishing access concept in the territorial waters, by way of introduction of stringent management regimes. In order to mitigate the hardship these measures may bring in, particularly at this time when the artisanal fishermen are in the throes of an economic crisis, the policy aims at promoting fishery exploitation in deep sea and oceanic waters particularly to reduce fishing pressure in territorial waters. As part of the proposed policy to divert traditional fishing effort towards exploitation of deep sea and oceanic waters, it is envisaged not only to organise technology transfer to small scale fishermen and encourage subsistence level fishermen to upgrade their activities not only in this direction but also to provide infrastructure support to the fisheries industrial sector. Further, the policy envisages motorisation of 50 per cent of traditional craft only as a measure to continue subsistance fishing in a better way in nearshore waters. This is obviously a measure to equip traditional fishermen who now operate in territorial/coastal waters with improved craft. This would of course step up working expenses, with disproportionate increase in incomes and tending to deplete the fishery of the zone further.

Considering that the coastal/territorial waters of India too have good stocks of tuna, it would have been a good policy to encourage and lend support for equipping 50 per cent of the present fleet of mechanised boats to conduct tuna long lining (with 10-15 km of long line on each of the boats) as has been done in Brazil. Tuna long lining being an eco-friendly system, a policy measure of this kind would automatically fulfill the objective of augmenting the sagging incomes of stakeholders and would also enable exploitation of the hitherto unutilised resources of the coastal waters, mostly tuna and tuna-like fishes, and add to export earnings. Further, a policy measure of this kind would reduce the present pressure on other fishing resources in the territorial water.

Another policy parameter envisaged is to encourage small mechanised boat owners through incentives for the acquisition of multiday fishing units. This encouragement probably aims at enabling such of the boat owners, other than those who now conduct multiday operations, to carry considerable quantities of ice, mostly for preserving shrimps catches on board, to acquire vessels that can fish beyond their present area of operations.

While most of the policy indicators as mentioned above are more in the nature of old wine in a new bottle, refreshingly enough, some of them break new ground to cover open sea cage farming and installation of FADs. Cage farming would of course call for selection of sites, their allotment to those who can set up supporting hatcheries and also for organising demonstrations and training by the authorities. Technology of cage fabrication and their installation has to be brought in.

The provision for promoting cage farming is the best part in the policy, for, it has the potential of increasing production that can be mostly export-oriented, without depleting the natural resources. It has also the scope to open up a major economic activjty that would alter the complexion of the industry. The mention of 'open sea'apparently implies the coverage of the sea beyond territorial waters with the activity and this provision is to be welcomed.

As already mentioned, the policy also includes the setting up of fish aggregation devices and this is laudable. CMFRI experimented on this system some years back with good results, but it was not followed up. Considerable planning would of course has to be there to identify spots for installing the devices and on identifying the vessels which would exploit the aggregated fishes.

So far as fishing in deep sea and oceanic waters is concerned, the policy renews the earlier objective of encouraging acquisition of resource-specific vessels of over 20 m LoA, designed for catching tuna and squid. These will be permitted now for import by wholly owned Indian enterprises, after screening their applications and the policy provides for specific incentives for wholly Indian owned vessels for venturing into international waters and for concluding fishing arrangements (Joint Vertures) with other nations under license. This provision can also be extended to the oceanic zone of EEZ whose resources are now stated to be under-exploited.

A major lacuna perceived in the ongoing bifurcation of developmental functions in the Union ministries is that marine fishery production part is mainly handled by the Coastal State Governments and the Union Department of Animal Husbandry and Dairying, while export promotion work is with MPEDA (Ministry of Commerce). This bifurcation needs to be removed by forming a single agency that would be able to integrate and co-ordinate both these activities for improved and sustainable results and to stem fall in exports as has happened recently.

The policy envisages greater participation of co-operatives, NGOs and Local Self Governments in the developmental process. This will go a long way in taking the process closer to the people. Human resources development for promoting fisheries sector was mentioned as a policy measure in the case of Andaman and Nicobar Islands but not for Lakshadweep Islands. This initiative could also be made applicable to Lakshdweep Islands. Further, as tuna fishing grounds are located around these Islands and these are central for airlifting of the catches in value-added sashimi or in other forms, mostly to Japan, there is need to develop one centre in each of these Island groups for the purpose. The pattern followed in this regard in one of Indonesian islands close to Andaman and Nicobar Islands can be emulated, by way of developing the needed infrastructure.

One conspicuous lacuna in the policy seems to be the absence of any provision for providing focused training in tuna longlining and pole and line fishing to fishing second hand trainees at CIFNET. Highlighting of this aspect in the policy appears necessary. In the foreign tuna vessels now in operation in Indian EEZ, there are reports that the Indian counterpart crew are treated as dummies, which indicates that the sea training part related to tuna fishing needs greater attention.

The policy document does not make any mention of the desirability or otherwise of continuance of the recent system of allowing operation of foreign vessels with the Indian flag of convenience. The Government's policy in this regard with the rationale behind it needs to be included in the policy, as this is a recent development which has not been sanctified by any of the earlier announced policies, but only by the recent letters of permissions issued, which too are not covered by any policy announcement.

Fishing Chimes, Editorial 336: January 2005: Vol. 24, No. 10

# Open Sea Fishing in Indian EEZ Dominated by 'Indianised' Flag of Convenience(FOC) Vessels

It is believed by many that there has been a policy misdirection in the development of India's open sea fishing fleet. The policy in vogue now enables adventurous Taiwanese enterprises to gain entry for their vessels into the Indian EEZ, through an 'Indianised' 'Flag of convenience' route, dressed up as Indian registered vessels.

The Law of the sea has a provision that, the nations traditionally fishing in the waters of another nation prior to declaration of EEZs would gain a right to fish in such waters, in case the nations concerned fail to have at the soonest a fleet of its own to fish in its EEZ.

Even after three decades of declaring its EEZ, India has not been able to develop its own deep sea fishing fleet of a size consistent with its resources. So much so, countries like Thailand, Japan, Taiwan and Korea which used to fish in Indian EEZ prior to its declaration, may now be able to secure the right to fish in Indian EEZ under provisions in the Law of sea. There are indications that the Indian Ocean Tuna Commission may soon raise the point that fishing vessels, particularly tuna vessels of nations as aforesaid have a right to fish in Indian EEZ, as there are no Indian vessels, particularly tuna vessels, operating in the Indian EEZ, but for the few Indianised ones of Taiwanese ownership recently introduced. More of these vessels however may probably gain permission to fish in the EEZ with reference to resource position although India's marine fishing policy that has been announced recently does not say anything specific in respect of introducing any particular number of open sea fishing vessels within a time frame.

Taiwan has emerged in the past few decades as a great distant water fishing nation. Being a small island with an EEZ that is not consistent with the size of its vast fishing fleet and its capabilities, Taiwan had no alternative other than moving into its past grounds (which now form EEZs of other nations round the world), for survival. It could carry on in that direction as there was no resistance. With the concerned nations developing awareness of the resources in their EEZs, problems of global fishing have become severe for the Taiwanese, calling for evolving a lasting solution to face them.

This situation must have set in motion a search by Taiwanese for ways of gaining fearless entry into the EEZs of certain nations (which are familiar to them) to test their luck. This process of intrusion in some form or the other into the EEZs of other nations seems to have become inevitable for them. As time went by, because of countersteps taken by some of the transgressed nations, the intrusion narrowed down to certain underdeveloped and developing countries of which those of the Indian sub-continent happened to be some.

Capabilities of Taiwanese at open sea fishing are stated to be remarkable. Owing to various compulsions, owners of Taiwanese vessels seem to have chosen to fan out their vessels into various possible areas, so as to keep their surplus fleet strength profitably operational. In adopting this strategy, they seem to have faced flag problem. In order to get over this, two steps were taken by them, so far as tuna vessels are concerned. One was to team up with the Japanese interests to set up in Tokyo the Organization for Promotion of Responsible Tuna Fisheries (OPRT), with the support of the two Governments (Japan and Taiwan), in December 2000. The aim of the OPRT, as declared, is to eliminate flag of convenience tuna longlining vessels operating throughout the world. Having formally declared this objective, the members seem to have resorted to having their vessels 'sold' to companies in certain other countries with tuna resources in their EEZs, with an arrangement for the retention of operations in their hands and on conceding part of the income to the new "owners" conforming to 'FOC system in relation to the concerned new owners'. So far as Indian EEZ is concerned, while the vessels will continue to operate as 'Indianised' Taiwanese ones flying Indian flag, for international or national scrutiny they will be nationally registered vessels. An ingenious system indeed, which serves Taiwanese interests and safeguards the new owner's national interests too.

**Unlike other countries such as Thailand, most of the Indian populace have no obsessive attachment to fish as a food item. This national characteristic seems to have become the hallmark of the Government of India too. So much so, many say that in India, fisheries aspects do not get from its Government the same attention and priority as the other major food related subjects get. This basic characteristic has now left the country with a marginal number of around 40 nos. of authentically Indian larger vessels of 23 to 27 m OAL, mostly engaged in coastal shrimping but which are referred to as deep sea fishing vessels. There has been an addition of around 35 tuna longliners (exact number could not be ascertained), which are stated to be actually of Taiwanese ownership but are registered in India too and fly the India flag which is, for the Taiwanese, a flag of convenience for gaining access into Indian EEZ, in the**

**name of the Indian company concerned. This implies compliance with local regulations in respect of manning etc, although probably under the virtual operational control of the Taiwanese. It is alleged by some that Taiwanese register their vessels in the same way in several other countries such as Pakistan and in certain countries of South America and Africa in order to gain access into the EEZs of those countries.**

**The conquest of Taiwanese distant water fishing vessels to gain coverage of all the oceans has been progressive in nature. Before Taiwanese adventure gained strength, Japan, Korea, Thailand and probably Indonesia were also engaged in a similar conquest, but unable to withstand Taiwanese forays, their intrusionary activities seem to have dwindled.**

The area of operations of Taiwanese vessels at one time seem to have included a few developed countries and that was before the declaration of Exclusive Economic Zones (EEZs). After EEZs were declared they seem to have continued to do so, absorbing the risks of seizure. Seizure of such vessels by the nations concerned was sporadic for sometime but soon it became widespread, acting as a deterrant.,

Such seizures mean loss of vessels, arrests of crew, economic losses and court cases to the Taiwanese. These happenings seem to have motivated the Taiwanese to encourage the other Governments concerned, mostly developing and underdeveloped, to take their vessels on charter or lease. They seem to have succeeded in this strategy enabling their vessels to be operational in the EEZs of many countries (Including India and Pakistan) for several years as chartered or leased vessels.

There are reports that the going was good in this direction for a long time. The fishing activity mainly consisted of bull or pair trawling. This kind of fishing which was known to have been done on a massive scale led to overfishing. Over fishing or non-sustainable fishing became conspicuously noticeable later. Probably for the reason that their operational losses were heavy because of fall in catches of trawling because of overfishing, the activity seems to have been, by and large, discontinued by the Taiwanese in several zones. However, they did not withdraw from the chartering/leasing system (because of losses) on their own, but they seem to have tactfully encouraged the chartering or leasing nations concerned to withdraw the systems, so as to deliberately give the upper hand to them and concede the credit of the withdrawal to them.

Tuna longliners of Taiwan had been concentrating all along mostly in the Atlantic and Pacific with some measure of presence in the Indian and southern ocean. With depletion of tuna stocks in Atlantic and Pacific and the simultaneous highlighting of the availability of unexploited tuna stocks in the Indian Ocean, and particularly in the Bay of Bengal, Arabian Sea and Andaman waters so far as India is concerned, the Taiwanese seem to have extended their strategy to gain entry into the EEZs of India and other South Asian nations, and as part of it to revive their activities in the Zone.

Not being in a position to find a way of suggesting to the national Governments concerned such as India to resume the system of chartering or leasing, they seem to have thought of making efforts to extend the concept of 'nationionalised flag of convenience system in respect of their long liners *vis-a-vis* certain countries of which India is apparently one. This was the time when the Government of India was thinking in terms of developing resource-specific fishing.

The Taiwanese fishing vessel owners association is understood to be a well managed one having a reputation of developing close relations with other nations having good fisheries potential. Whatever be the reason, there seems to have been the successful introduction of the of 'flag of convenience' vessels of theirs in several coastal nations. The concept has gained acceptance primarily for the reason that it eliminates heavy investments on outright purchase of these vessels, which are very expensive. Under this system, the Taiwanese vessel concerned gets registration in the foreign country concerned as its national vessel thereby acquiring the right to fly the flag of that country. An advantage of this system to the Taiwanese company is that the physical control over the vessel would continue to rest with it for various technical, investmental and operational reasons and, it can retain the strength of its crew at numbers as needed some way or the other, giving the cognisance to the prescribed limitations. As the 'purchase' of Taiwanese vessels will be on deferred payment terms linked to voyages performed, the full transfer of ownership would take place only after all the instalments are paid. Till then, the Taiwanese company would manage to retain physical possession. They know that, full payment of the cost is something that would not normally happen and by the time it would ever take place, the vessel would have crossed its working age. Further, it can send these 'FOC vessels to fish in the EEZ of the registered country according to its convenience from time to time with reference to catch positions. In the event of fishing season in the EEZ concerned is not encouraging but is good in the EEZ of another country where also the Taiwanese vessel concerned has 'FOC' facility, the efforts would be to move the vessel to those grounds. This strategy has the effect of reducing the number of low profit Taiwanese vessel voyages in certain EEZs. One disadvantage for the Taiwanese, however, is that the sale proceeds would have to first go to the account of the vessel 'owning' company concerned which would release the amount under various accounts such as operational expenses from time to time.

**India was on the right track when it started developing a 'deep sea fishing fleet' in middle 1970s. The activity started with the introduction of 28 nos. of trawlers of 23 m OAL fitted for stern trawling and outrigger trawling. Later, several such vessels were added, taking the fleet strength to over 190 nos. The vessels having became aged, a conspicuous decline in the fleet strength started in 1990s. Consequently the presently operational vessels.mainly devoted to shrimping, now stand at around 40 nos. To this has to be added around 35 nos. of Indianised 'Taiwanese' longliners that have recently joined the fleet.**

The process of formulating a purposeful policy to build up the open sea fishing fleet with reference to availability of resources in Indian EEZ and adjoining international waters should have been initiated by the Government in late 1980s but this was not to be as could be understood from the known developments.

Many in the Indian industry feel that, instead of succumbing to the easier option of permitting the acquisition of the Taiwanese longliners (and leaving them to be under the control of the Taiwanese owners for all purposes, particularly in respect of coming back into the Indian EEZ for the following voyage as their convenience would allow) by Indian enterprises, a project could have been developed on the tested model of the GOI's scheme of middle 1970s (which was successfully implemented at least to the extent of introducing 28 out of 30 vessels targeted as already stated), with such modifications as are needed for recoveries of instalments and in regard to the guarantee system to be finalised in respect of the foreign lending bank or Indian lending bank. Such a project, fortified with an integrated training programme, and a collaboration brought about between a reputed Indian shipyard, with good track record (such, as Alcock Ashdown/Bharati Shipyard) of building fishing vessels and a reputed longliner building foreign shipyard, and providing for well negotiated linkages with importers of sashimi grade tuna and tuna products and with interested bankers, would have established an enduring basis for the development of Indian tuna fishing industry. Many believe that even now it is not late to take up a major project involving the construction and introduction of around 50 or more longliners- cum-trawlers of 20-25m OAL for operation in the Indian EEZ. The foreign collaborators could also be probably identified taking the help of FAO and/or our Missions abroad. Financing bodies such as the World Bank or Asian Development Bank could lend the needed support. The beneficiaries could be identified through an open advertisement based on select criteria.

Having said this much on the topic, it may also be mentioned here that, sooner or later, the present system of permitting purchase of Taiwanese longliners, which is an 'indianised' 'Flag of Convenience' system, would have to be replaced by a more plausible and enduring system and it will be good if the Government could consider looking at the subject again, keeping the various related aspects in view, as the Nation cannot be enchained to Taiwan in the matter of open sea fishing in the Indian EEZ for long. Another aspect to be considered is that Taiwanese may either abandon or take back their old vessels on the ground of not having received the full payment towards their cost. Difficult situations of this kind could be avoided by developing India's own real fleet not encumbered by the interests of other nations such as Taiwan.

Fishing Chimes, Editorial 337: February 2005: Vol. 24, No. 11

# Cage-based Pond Farming

Majority of Indian fish farmers are stuck up either with major carp farming or with major carp-based composite farming. These systems, once profitable, have now, over years, become disadvantageous to the farmers. Overall increase in major carp production has had the effect of bringing down market prices of major carps to an unviable level. In this situation, it is now imperative to identify and introduce an alternative system that would fetch viable returns to the farmers. A development in this context is that farming of giant freshwater prawn, shrimp and catfishes have emerged as alternative systems of farming, particularly in the Coastal Zones. These alternative systems have also made inroads into certain inland States of the country with encouraging results.

In this background, the Central Institute of Freshwater Aquaculture has been highlighting the results of its success in the diversification of farming systems towards catfish farming and freshwater prawn farming, the adoption of which could extricate the farmers from their present plight. The scope for the spread of catfish/giant freshwater prawn farming however has certain limitations. Mass scale production of seed of these species has natural constraints and their seed cannot be produced in the same quantities as in the case of major carps at one time. The impact of this on aqua production can be corrected by introducing the system of cage-based pond farming for augmenting the level of production. Under this system, while finfishes like major carps can be produced in good quantities in cages set up in ponds, the drop in returns because of this increase in production can be made up by farming giant freshwater prawns/shrimps too in the open pond area that is free from cages, as these crustaceans command higher prices because of their export potential and demand. Considering these aspects, the Central Institute of Freshwater Aquaculture could consider initiating experimentation and subsequent follow-up work on popularising cage-based pond farming that would help farmers to counteract the losses that are presently sustained by them because of main concentration on major carp production. While major carp production could be continued or even enhanced for higher incomes under this system, although at a reduced price per kg as at present, once this production is combined with exportable items such as giant freshwater prawn/shrimps, the overall returns could be enhanced to a large extent through optimisation of the component of export-oriented production.

## Open Pond Farming Combined with Cage Farming in Pond

One way of optimising viable production from a pond is to combine cage farming system with open pond farming, in the place of the present system of totally open pond farming based mainly either on major carp farming or composite farming. This combined system envisages the adoption of polyfarming practices, with cages set up to cover half of the area of pond surface and covering columnar water to about 3/4th of the pond depth by the cages, leaving the remaining 1/4'h of cage-free zone water underneath for prawn/shrimp farming. This kind of combination and integration of pond cage and open pond farming system has been successfully tested and commercialised by Team Aqua Corporation (TAC) of Taiwan as early as 1997.

TAC has specified a minimum depth of 2.5 m (8 feet) for a farm pond to undertake cage-cum-open water aquafarming. The surface area of each of the cages is 27.56 sq.m (5.25 x 5.25m). The depth of each cage has to be 2 m. The cubic volume of each cage thus works out to 55.12 cubic meters. According to TAC, the production capacity of each cage is 1.2 to 1.4 tonne (This can be rounded up to an average of one tonne). On this basis, a cage, in terms of each sq m can produce around 36 kg in a production cycle. This level of production per sq m, gives an indication that around 36 nos. of advanced fry can be stocked per sq m of a cage or 18 nos. per cubic meter. (In Taiwan Tilapia is used for stocking the cages. In India major carps may have to be substituted). As the volume of each cage is 55 cu m, its production capacity works out to nearly 1000 kg. Stated differently, in order to achieve 1,000 kg of production from a cage, each cu m of its water space has to be stocked with 18 nos. of advanced fry, on the assumption that each advanced fry could be grown upto a weight of 1 kg. There is no indication about the percentage of mortality and this has to be taken into account approximately, in estimating the net production. It is possible that there will be a significant percentage of mortality.

The cages set up in half of the area of a one ha pond would cover 5000 sq m or 10,000 cu m, as the cages will be 2m deep. On the basis the advanced fry to be stocked in the zone covered by several cages comes to nearly 1,80,000 nos. per ha, it can be reckoned that nearly 180 tonnes can be produced from cages in half of the area of one ha pond (with the other half of the surface and columnar area cage-free). According to TAC, this level of production could be achieved from fishes farmed in cages alone. The FCR has not however been mentioned by TAC. However, it can probably be taken as 1.5:1 for the achievement of the high level of production projected at an average of one tonne (carrying capacity) per cage. To repeat, the cages will be set up to cover half of the surface and columnar area of pond upto a depth of 2 m.

The remaining water area, *i.e.*, around 0.5 m of water below the cages and the remaining open area not covered by the cages can be utilised for farming freshwater prawns/tiger or other shrimps after acclimatisation, as decided. Microorganisms that spring up utlising the waste food dropping down from the cages and which move and settle down at the bottom would serve as feed for the stocked advanced crustacean post-larvae moving in the cage-free zone, besides microplankton that will be available in the zone. In TAC system there is no indication of any application of supplementary feed to the crustaceans stocked for growth in the open farming section.

One aspect to be mentioned here is that according to TAC, fish alone can be grown in the cages leaving the rest of the pond zone for the production of prawns/shrimps. The rationale of this system is explained by TAC as follows: The kind of bacteria that proliferate in a fish pond (in this case in fish cages) are different from those occurring in a shrimp/GFP pond (in this case in the open pond zone). Gram positive bacteria dominate in the fish cages. In the fish-free or cage-free area (open zone) of the pond concerned, there will be dominance of Gram negative bacteria. While these are prone to cause diseases to fishes, it is stated that these Gram negative bacteria do not cause any harm to crustaceans under farming. According to TAC, for this reason, the cage-free zone is utilised for farming crustaceans (freshwater prawns/shrimps). Thus, the pursuit of polyfarming of fish and crustaceans in a pond (fish in cages set up to cover half of the pond area from surface into columnar layer, and crustaceans in the cage-free zone) not only enables effective utilisation of planktonic blooms but also ensures that the fishes being farmed in the cages are less susceptiable to disease outbreaks, ultimately leading to a more productive harvest.

Cages improve the quality of the harvested fish. While they have to be obviously fed with an appropriate feed, it is stated that, prior to the harvesting, normally, fishes should not be fed. One stated advantage in cage farming is that the fishes grown in them would not develop muddy flavour. There will be also more efficient FCR.resulting in reduction of feed costs.

Cage farming saves time and labour. Two persons can harvest the fish in a cage without difficulty. Sick or diseased fishes in cages can be segregated conveniently for observation or treatment. According to TAC, by placing a tarpaulin (pre cut and approximating the dimensions of net used for harvesting) inside the net, one can isolate pond water inside the cage and apply chemicals to it for treatment of disease. Volume of chemicals used gets drastically reduced due to reduction in volume of water for treatment.

Recording of the various parameters, such as rate of feeding, growth rate and observations in respect of diseases would be easier in the cage farming system.

The distance between the base of a cage and the pond bottom should be atleast 50 cm so as to ensure a good pond water circulation. Since the stocking density as well as production is high in inland cage farming practices, it becomes essential to increase aeration and also provide for a standby generator in case of power failure or interruption. Upto 8 sets of cages (5.25 x 5.25x 2.0 m each), there may not be need for the installation of an aerator. However, where more than 8 sets are installed, it would become necessary to install an aerator central to the batteries of cages. In general, a set of two batteries of four cages each that converge with space in between for installing an aerator is the pattern of installation of cages with needed supports.

## Gram-Negative and Gram-Positive Bacteria

Gram-negative bacteria are those which fail to stain with Gram's reaction. The reaction depends on the complexity of cell wall and has for long determined a major division between bacterial species.

Gram-positive bacteria are those with a cell wall of comparative simplicity which allows it to be stained according to Gram's method.

Besides the cost of stocking of the cages with major carp fry, the cost of feed would be the main recurring expenditure. At an FCR of probably 1.5:1, the cost of feed may have to be worked out. The feed, which would have to be nutritive, need not however be a very expensive one. Powedered rice bran and groundnut oil cake at 1:1 ratio or as is generally used, can serve as a good feed. However, experimentation will be necessary to arrive at the nature of feed to be given and the quantities to be fed. Probably, rate of stocking in the cage can also be suitably adjusted to prevent any likely mortalities or the requirements of production, consistent with the determined stocking schedule and environmental aspects. Phased harvesting can also be adopted for realising optimal output and returns.

So far as prawn/shrimp farming in the bottom and the rest of the open zone is concerned, the rate of stocking of PLs can be as per the prevailing norms, to produce on an average 0.8 to 1.5 t per ha, until such time as improved environmentally conducive systems are developed. TAC specifically says that fish alone should be farmed in the cages and shrimps/prawns have to be farmed in the bottom waters and in the open pond area remaining uncovered by the cages. Some experimental work appears necessary to check whether shrimp/prawn can be farmed in the cages and bottom living fishes can be grown in the cage-free area. The work done by TAC of Taiwan has apparently led to the system of farming fishes alone in the cages and shrimps/prawns in the cage-free area because of the Gram-positive and Gram-negative bacteria factors as already mentioned. There appears to be a need for having a clear understanding of the influences of these two categories of bacteria in polyfarming and monofarming systems for possible realigning of these systems as now practised.

While the system of polyfarming using pond cages has been found to be environment-friendly in Taiwan and probably in a few other countries, it may be desirable to conduct further experiments, taking Indian conditions into consideration, in order to establish the viability of the system, in relation to impact on the environment and economics.

Fishing Chimes, Editorial 338: March 2005: Vol. 24, No. 12

# Utilisation of Tuna of Indian EEZ– Focal Look at Fishing Industry's Needs

The stakeholders engaged in wild sea shrimp fishing have been in the throes of an economic crisis for the past some years. The reason for the crisis is that they indulged in overfishing of shrimps, taking advantage of their fast growth and the habit of breeding once or twice in a year, which helps, as is believed, in restoring the stock level.

The phenonomenon of overfishing is not confined to India alone but it is widespread. An FAO Report says that poor sector governance has enabled the creeping practice of overfishing to continue and negatively affect fisheries. Management of shrimp fisheries has become a political and economic process, requiring changes in institutional, legal, and regulatory frameworks, and greater participatory role of the private sector.

The motivation to overfish is related to the export demand. So much so, the overfishing, continued relentlessly over years to such an extent that the progeny resulting from the left-over brooders, apparently had not been contributing adequately to sustain the shrimp stocks to the level needed. Consequently, there has been a progressive drop in shrimp landings at a fast pace to result in the uneconomic status of the shrimp fishery. Closed season to prevent fishing particularly for shrimps was no doubt declared all along the coastline as a remedial measure but improvement in the situation has not been significant.

A search for alternatives ensued as the situation was moving mercilessly towards the expected denouement. Some tried fishing for squids and others for lobsters etc., but the results were not enduringly encouraging. The search continued until it focused on tuna, mostly the yellowfin kind, which is in greater demand, particularly in sashimi style of preservation (first grade) and also in non-sashimi form as loins, steaks and paste and for canning (2nd grade).

In 1980s, a few companies imported used tuna vessels (a purse seiner, a long liner and a pole and line vessel). Their operations however failed. Later, there were stray initiatives by a couple of companies to equip their vessels for longlining utilising the services of foreign technicians but there was no success. Then there was a fresh initiative in 2004 for equipping existing 23m OAL vessels for tuna longlining. Two trawlers MFV Sameera and MFV Madhavi were equipped for longlining supported by financial subsidy from MPEDA. The installation of longlining equipment and successful longline trial fishing was done by Sanko Busan Co. of Japan. Despite this success, the owners continued to face problems. These related to organising export of mainly sashimi tuna and this could be tackled only marginally. Efforts were also mounted by owners of mini-trawlers of around 16m OAL through organising themselves into a cooperative and approaching NCDC for loan assistance. The word going round now is that NCDC is favourably inclined and is processing the application.

In this scenario, as a separate effort, a 16m long trawler Gamini was equipped for longlining by G M N Gordian Kagoo and A.B.Raj. The vessel was equipped for operating 15 km of longline with 250 branch lines and hooks. The operations have been successful but the problem of linkage with export marketing continues to elude solution.

The developments thereafter are: The owners of eight trawlers of 23 m OAL have taken up modifications to their vessels for longlining (Mayura I and Mayura II of Samro Food Processors, Michael and Superna of Gees Marine Products, Jordon of ARM Marine Enterprises, River Krishna of Hemo Sea Foods, Napoleon of SAF Marines, and Gloria of Kagoo Marines. Four out of them have left for fishing on 14 March, 2005, and the others are expected to follow soon. The results of fishing by the four vessels already out at sea are reported to be good.

Another development that lent proof of existence of good tuna resources in the coastal waters of the Bay of Bengal too was the heavy landing of tuna and tuna-like fishes by manual longlining by non-mechanised boats for the past few seasons, particularly off north A P coast.

In this setting, a seminar on Tuna was held at Hotel Taj Residency, Visakhapatnam on 9 March 2005 by MPEDA, in association with the Association of Indian Fishery Industries and Seafood Exporters Association of India.

The remarkable aspect of the event was that Mr. Sahideen of M/s Sanko Busan Co., Japan, Mr. P. Vasudevan of Lakshadweep Fisheries Development Corporation Ltd. and Mr. G S Rajeev, Plant Manager, Matysafed, Kerala made focal presentations mainly on post-harvest logistics upto the stage of exports, that highlighted the important links and details of this phase of work, not yet clearly known to most of the Indian operators.

The seminar, predominantly participated by the representatives of the marine fish producing and exporting companies, came out with crucial recommendations that would take the industry forward.

The seminar was inaugurated by Mr. J. Dange, Principal Secretary (Agriculture, Animal Husbandry and Fisheries) Government of Maharashtra. Mr. Dange and Mr. Gorakh Megh, Commissioner of Fisheries, Maharashtra, gauged the depth of interest in tuna fishing among the various participating enterprises. They seem to have made an assessment of the possibilities of introducing the activity by Indian enterprises to waters off west coast, particularly off Maharashtra coast.

Mr. Mohan Kumar, Chairman, MPEDA, besides highlighting the importance of tuna as a viable resource in Indian EEZ for sustainable exploitation and export, explained the various incentives offered by the Authority in equipping trawlers for longlining and also the support extended by it for strengthening post-harvest operations of longlining, in order to achieve growth in remunerative exports of sashimi as well as non-sashimi quality tuna. Referring to the various operational problems, he exhorted the participants to suggest measures to counter and solve them. Reacting to this appeal, the participants experienced the excitement of sharing views among themselves, mostly on logistics of post harvest tuna longlining operations. Noting that the past failures in respect of effectively undertaking export of tuna have taken place because of inadequate attention and because of absence of an organisational mechanism to firmly take care of aspects such as transfer of tuna from vessels to plants, proper storage and upkeep at the plants, ensuring supply of quality packing material, timely and proper transportation of tuna packs to the airport, effective arrangements in respect of their freight and forwarding, linkage for arranging chartered flights to take the consignments mostly to Japan in a good condition, etc. The representative of Atlas Logistics Pvt. Ltd. who participated in the seminar offered to provide all needed logistic support, for post-harvest operations and this was well received.

One important outcome of the discussions at the seminar was a recommendation that viable groups of long liner owning companies may be formed and such groups may be registered. These groups may organize common facilities, group-wise, to take care of the provision of needed supplies and services on time to the members of the groups concerned. There was a general agreement that this lacuna that is now holding up tuna export should be removed in the shortest possible time, by instituting the needed measures in the light of the above recommendation, under the leadership of MPEDA and in the light of the practical aspects presented by the experts in post-harvest tuna operations, these being Sahideen, Vasudevan and Rajeev.

A point was raised by some of the participants that, while the industry was happy at the provision of subsidy, the position was that most of the owning enterprises were not in a position to raise all the balance of investment needed. On this premise, they requested that a financing facility would need to be identified and activated. The participatory response to the seminar was overwhelming. The interest evinced by the participants reflected their desperate longing to find a solution to diversify into tuna longlining and the anxiety to receive help by way of a clear-cut line of action with needed support.

There was a suggestion put forth by Dr. K.Haribabu, Vice-Chairman of MPEDA, that owners of small mechanised boats also should be assisted in equipping their boats for tuna longlining, adding that such a step would prove to be a viable measure to rehabilitate them from the present economic crisis they are in now. It may be noted here that, at a workshop on correct use and upkeep of fishing aids held by CIFNET in Visakhapatnam in Nov. 2004, a similar suggestion was made, while also proposing that imparting of training in tuna longlining to small scale fishermen should be the responsibility of CIFNET. It was also suggested that equipping the small boats for tuna fishing should be the responsibility of CIFNET/IFP.

To sum up, the seminar on tuna held on 9 March 2005 in Visakhapatnam has shown the way to forge forward in respect of effective utilisation of tuna resources of Indian EEZ in sashimi/non-sashimi form, from production to exports. MPEDA, the Indian Fishery enterprises and the foreign delegates played a key role in the deliberations. The follow-up action is now in the hands of Indian Enterprises, AIFI, SEAI and MPEDA.

Fishing Chimes, Editorial 339: April 2005: Vol. 25, No. 1

# The Beginning: 'Fisheries' Secured a Place in a Union Dept's Name

The decades-long struggle by the fisheries fraternity of the country for achieving a separate Union Ministry for Fisheries has generated its first outcome recently. The Union Government has at last conceded to tag on the word 'Fisheries' as an adjunct to the present name of the Department, which is Department of Animal Husbandry and Dairying, under the Ministry of Agriculture. The benevolent addition of the word 'Fisheries' to the name of the aforesaid Union Department may not however really alter the dimensions of its purview over the subject, for the reason that the multifarious activities of fisheries development in the country may continue to remain distributed in several ministries at the Centre. And an entrepreneur would have to continue to struggle running around from one Ministry to another for help, directions and clearances.

The best the addition of the word 'Fisheries' in the name of the Union Department of Animal Husbandry and Dairying can serve is to set all concerned to become more conscious of the reasons for the non-inclusion of Fisheries as one of the subjects of the Department so far and what the present inclusion portends. The subject of Fisheries is a multifaceted one, and some of its various integral parts are looked after by quite a few other Ministries. This basic position, seen now in the context of the extension of the name of the Department, may hopefully motivate the top men to work towards gradually shifting the various components of the subject of fisheries to the Union Department in the direction of strengthening the Fisheries Division under the Union Department of Animal Husbandry, Dairying and Fisheries (DAHDF). This strengthening, again hopefully, would centralise all fisheries functions at this point (Union Fisheries Division) so as to quicken decisions for speeding up developmental progress.

At least examples of a few problems that have arisen out of the scattered nature of the integral aspects of the subject of fisheries deserve attention. MPEDA, under Ministry of Commerce, also now promotes Culture fisheries, such as of shrimps, prawns and crabs. However, this function is basically, in fact, in the hands of the Union Department of Animal Husbandry, Dairying and Fisheries and also in the purview of State Fisheries Departments. The reason for the aforesaid approach on the part of MPEDA is that the imperatives of the situation force it to transgress into this area of work, because of its responsibility to promote exports of marine products.

It has to depend on a strengthened production base to achieve augmented exports, and on the related aspects like training, information dissemination, etc., so as to step up the production of needed raw-material for processing and export. In exercising this needed initiative, to promote export-oriented production it probably has no alternative other than to commit itself to an overlapping of functions, which are of course justifiable. This overlapping is reflective of a struggle to reach the export target MPEDA's export activities which are now considerably dependent on farmed fishery output, as capture fishery output is on the decline.

In this context, if the Aquaculture Authority continues to be under the Union Department of Animal Husbandry, Dairying and Fisheries, its efforts to achieve farmed production as needed for exports may face blocks, as the norms followed by the Aquaculture Authority are not designed to achieve optimal sustainable production, but to keep it down to traditional level. An anachronistic approach indeed, as many feel, but it cannot be found fault with, as it is only following the directive of the supreme court. The need for the DAHDF in this situation is to formulate a development plan, in consullation with MPEDA, that would facilitate achieving optimal sustainable aqua production in an eco-friendly manner and seek approval to the plan from the Supreme Court because it has earlier prescribed a set of norms. They may have to be reviewed now by the court, considering the developments in the ground situation and technology of culture production.

Though the Ministry of Food processing in its new Avataar has returned the deep sea fishing portfolio to the Union Department of Animal Husbandry, Dairying and Fisheries, post harvest processing of fish is still in its mandate. Yet, it wouldnot take over the Integrated Fisheries Prqject from the Union Department of Animal Husbandry, Dairying and Fisheries.

The Department of Ocean Development, which was originally mandated to focus on non-living resources and Antarctic studies, has, of late, been happily venturing into various aspects of fisheries, including coastal fisheries, sea safety and even Aquaculture, it is learnt. Likewise, some say that the Union Department of Biotechnology has become another full-fledged fishery activist with very little biotechnology content in its activity or the projects related to fisheries it sanctions, but they may be wrong. Looking at ICAR, its mammoth research set up seems to be continuing its association largely with the Union Department of Agriculture and Co-operation only, to the neglect of theAnimal Husbandry, Dairying and Fisheries subjects, which have been carved out to form a new department in the same ministry. The net result seems to be that there appears to be practically very little linkage between Animal Sciences Research of ICAR and developmental efforts of the Department of Animal Husbandry, Dairying and Fisheries (DAHDF). The overgrown ICAR is bursting at its seams. Even so, it is learnt that it has been adjudged as non-performing. It is being mentioned that it has been strongly recommended that its non-performing activities should be pruned. So much so, many believe that, ICAR has already succeeded in downgrading fisheries in the mighty currents of agriculture research. Assuming that all the fisheries related subjects are brought under the nodal DAHDF, it is any body's guess whether the Department would be able to do justice to the Sector. This may be the only Union Department, which can boast of not having been able to create a single post in fisheries in the past one decade, despite the need in the context of remarkable developments in the sector, worldwide. There is more to the credit of DADF in this direction. It has the distinction of also lapsing over forty per cent of its technical posts under its purview due to nonfilling up of the vacancies in time. Many senior posts suffered downgrading, as their analogous posts in other fisheries establishments got upgraded, the knowledgeable observers say.

Fisheries are a thrust area with immense potential for poverty alleviation, contributing to food security and for foreign exchange earning. A strong ministry for fisheries or atleast a dedicaied Department of Fisheries with support from ICAR for promoting Fisheries Research would be the minimum configuration that can work. Parallelly, the Union Department should maintain and the best technical team in its Fisheries Division to address the diverse challenges that the sector faces.

Fishing Chimes, Editorial 340: May 2005: Vol. 25, No. 2

# On ways of Stemming Decline of Coastal Fisheries Resources

References are frequently made to an alarming trend of decline in Coastal fishery stocks. Simultaneously, efforts are also on to stem this decline. One such effort is to declare closure of fishing during main fish breeding season. However, it will take several years to experience any improvement in the fishery complexion because of this measure. Another suggestion to improve the situation is to reduce the strength of coastal fishing fleet. But this is not well received as a feasible option to exercise as such a step would create unemployment among a large number of fishermen. Further, while decommissioning of part of the present fleet may improve per unit production, it may not lead to an improvement in the total production. In this background, a way out has to be found for inducting the affected fishermen, who will be hard pressed to eke out a living because of decline in per unit landings, into an alternate avocation. In this context two options, one related to sea farming with cages, and the other to equip the coastal boats to undertake non-demersal (pelagic) fishing in and beyond coastal water need consideration.

## Cage Farming

No determined efforts have been made in India so far to promote sea cage farming, in contrast to the emulatively successful and professionally established cage farming practices prevalent in several countries, the closest of them to us being in Thailand, Vietnam etc. None of these countries have been experiencing any serious problems in the pursuit of cage farming; Nor any serious ecological problems are faced. In April 2005 issue of *Fishing Chimes* a detailed account of the cage farming system authored by Dr. Mohan Joseph Modayil, Director, Central Marine Fisheries Research Institute, has been published.

Cage farming can shower several benefits, provided it is developed in a well conceived and in an integrated manner. Acquisition of technology for fabrication and installation of cages, technology of shore-based hatchery operations for the production of yearlings of cultivable species for stocking cages, management of cages under farming, system of storage and use of feeds as adjuncts to cage systems, running of ferries from shore to hatcheries and vice versa, manpower development to meet the operational requirements of cages, manner of harvesting the produce and marketing of the same and providing training to selected candidates, preferably to those sponsored by organisations of small scale fishermen, at centres set up for the purpose, preferably by CIFNET, are the main requirements.

The activity calls for State-wise coastal promotional set up as separate wings of the State Fisheries Departments. Once infrastructure requirements covering all aspects are taken care of, the next step would be for the Governments concerned to organise identification of areas for setting up cages, taking into account all related aspects. Once these sites are identified, as is done in USA and other countries, applications should be invited (as also done in USA etc) from prospective farmers/farming enterprises to allot the areas concerned on lease, subject to the terms and conditions, as stipulated. Nowadays, cages are being erected not only in Coastal Zones but also in open sea areas. These cage operations would boost up fish production, in an eco-friendly manner substantively, by and large, largely compensating for the fall in capture fishery output in a progressive manner. Cage farming operations have been developed in countries such as Australia, Mexico, and in certain Mediterranean countries to such an advanced stage that tuna juveniles are stocked in cages, fattened and live tuna-laden cages are towed to harbours adjacent to marketing places for unloading the tuna so raised.

In short, cage farming is a viable alternative to compensate substantively for the fall in capture sea fish landings and this deserves to be given the needed attention by the authorities concerned. Further,the activity will generate employment to a large number of fishermen, trained to function as cage farmers.

Another avenue to provide a productive and upgraded avocation to the presently trawling-oriented fishermen will be to develop projects under which a predetermined number of mechanised boats can be equipped for long lining for tuna with about 10 km of line each. This diversification can certainly be achieved and when achieved it will go a long way in reducing demersal overfishing. At the same time, not only the coastal tuna resources can be exploited and utilised, but also these resources beyond the coastal zone can be availed of, thereby helping to replace the flag of convenience foreign tuna vessels now operating under LoP system.

Fishing Chimes, Editorial 341: June 2005: Vol. 25, No. 3

# Endangered Fish Species of India: Their Restoration to Normalcy A National Project Needed

Several Indian fish species are moving towards an endangered status and some out of them are stated to be in a vanishing stage. While estimates of the number of endangered fish species of India made by various experts are understood to differ, all are however stated to be one in pointing out the precariously endangered status of a considerable number of fish species.

According to Chandramohan *et al* (1999), 47 out of 327 fish species of India are critically endangered, 92 endangered and 82 vulnerable. P. Das and A.K. Pandey (1999) have referred to 21 vulnerable species as compiled by Menon (1989), of which four are endangered and 17 threatened, according to them. NBFGR has tentatively identified four endangered, 21 vulnerable and two fish species that have become rare. In addition, according to them, there are also 52 species, of indeterminate status.

Generally stated, a gloomy and alarming picture of the present status of endangered species confronts us. In this background, efforts are stated to be on for the revival of stocks of the endangered species concerned, although these may be in an unco-ordinated and diffused manner. All the major river systems of the country now have this problem of endangered species.

In this context, it may be stated that, so far as mahseers are concerned, there is a general awareness that the Golden Mahseer (*Tor putitora*) of North Indian rivers, the Deccan Mahseers (*Tor khudree* and *Tor tor*), and also the Chocolate Mahseer (*Acrossochilus hexagonolepis*) of the Kaveri and Bhavani river systems, and a few other species such as *Tor mahanadicus* etc., require urgent attention in respect of revival of their stocks to keep them out of the endangered zone.

In the present alarming situation, the apparent task that needs immediate attention is to institute tangible measures towards the goal of restoring the endangered fish populations to their past glory. The restoration has to be such that, while increasing the endangered fish population, a capture balance of the populations is maintained in the river systems/reservoirs through well planned yearling production and carefully managed stocking programmes. In this context, the setting up of specialised hatcheries at selected points close to the rivers and reservoirs, to be well managed by trained and dedicated personnel, is crucial for achieving the goal, besides intensifying efforts at conservation of the present stocks, with adjustments as are necessary from time to time. The seed production has to be supported by cage farming systems to be set up in the rivers/reservoirs concerned. The objective of this support, as will be obvious, is for the purpose of raising yearlings of the endangered species for fulfilling judiciously planned stocking programmes.

In order to prepare the ground for developing a National Project, one suggestion can be that a Committee of Experts/a Task Force with experts drawn from various State Fisheries Departments, the Central Fisheries Institute concerned and Union Department of Animal Husbandry, Dairying and Fisheries may be constituted by the Government of India. The following indicative terms of reference may deserve consideration.

## Indicative Terms of Reference

1. To assemble and review available information on endangered fish species of the country and in that light finalise a list of all such fishes, resource-wise (rivers, reservoirs, streams etc.,) with such details as are necessary.
2. To assess the annual yearling requirements of endangered species for stocking the waters that now stand depleted of them, so as to restore the populations, resource-wise, over a determined period of time, and in order to establish capture-culture balance of each of the resource.
3. To identify locations with reference to availability of communication and power supply facilities for setting up hatchery farms to produce yearlings of the needed species for stocking, taking into account the existing infrastructure (hatchery farms available in the zone).
4. To determine the capacities of each of the hatchery farms suggested for establishment and to recommend model designs, indicating the investment needs, man power and seed raising requirements, etc.
5. To suggest measures to introduce supporting cage farming systems to raise yearlings for stocking rivers/reservoirs concerned under a well designed system of accountability.
6. To recommend a stocking procedure linked to the setting up of intermediate short-term seed retention units so as to facilitate storage and for conditioning.

7. To recommend an organisational mechanism to be adopted to implement the programme; and
8. To suggest guidelines in respect of monitoring production at hatchery farms and at cage farming systems and also stocking programmes with reference to adjustment of stocking levels.

## An Attempted List of Endangered Fish Species is given hereunder

- ☆ **Himalayan Rivers:** *Labeo dero, L. dyocheilus, Oreinus sinuatus, Schizothorax esocinus, S. nigar, S. richardsonii, S. curviformes.*
- ☆ **Mahanadi:** *Puntius sarana, Labeo fimbriatus, L. calbasu. Rhinomugil corsula.*
- ☆ **Krishna-Godavari System:** *Puntius kolus, P. dubius, P. sarana, P. porcellus, Labeo fimbriatus, L. calbasu, L. pungusia, L. gonius, L. boggat, Thynnichthys sandkhol, Cirrhina cirrhosa, C. horai* (Extinct).
- ☆ **Tungabhadra (Karnataka):** *Osteochielus thomassi, Silonia childrenii.*
- ☆ **Kabini river (Karnataka):** *Cirrhina reba.*
- ☆ **Krishnaraja Sagar (Karnataka):** *Puntius carnaticus.*
- ☆ **Deccan Rivers** : *Puntius pulchellus.*
- ☆ **Rajasthan (Ghagar River)** : *Labeo calbasu.*
- ☆ **N.E.States:** *Barilius bola.*
- ☆ **U.P.**: *Barilius bola*, and *Glythis gangeticus. Labeo fimbriatus* is a threatened species in Rihand reservoir.
- ☆ **Ladakh:** *Gymnocytris biswasi* (extinct).
- ☆ **Tamil Nadu (Kaveri - Mettur Reservoir):** *Barbus dubius, Barbus carnaticus, B. chrysopoma* and *Cirrhinus cirrhosa.*
- ☆ **Tamil Nadu (Kodaikanal Lake and Palani Hill Streams):** *Barilius gatensis.*
- ☆ **Tamil Nadu (Kaveri - Bhavani System):** *Silonia silonia and Pangasius pangasius* (threatened.).
- ☆ **Tamil Nadu (Amaravati, Tirumurthy and Aliyar Reservoirs)** : *Labeo fimbriatus.*
- ☆ **All South-Indian Rivers:** *Labeo kontius. Puntius pulchellus.*
- ☆ **Kerala:** *Channa microlepis* and *Clarias dussumeri* in Pampa river.
- ☆ **All Rivers** : *Channa microlepis* and *Notopterus notopterus*
- ☆ **Exotic fishes:** Gourami is not seen now-a-days, many say.

  There can be many other fishes in an endangered state in the inland sector. The particulars relating to them could not be listed here for want of information. It is possible that some of the particulars given above would require readjustments/corrections. Further, the river dolphin (*Platinesta gangetica*) is considered as an endangered species.
- ☆ **Euryhaline Fishes:** *Lates calcarifer* and *Hilsa hilsa (Tanualosa ilisha)* are considered to be endangered species but this may need confirmation by our scientists. It is possible that there may be few more endangered euryhaline fishes too.
- ☆ **Marine Fishes:** In the marine sector, it is mentioned that a few cat fishes have become rare and are on their way towards an endangered status. A close study is needed to identify if there are any other species that are to be considered as endangered in the marine sector.

Fishing Chimes, Editorial 342: July 2005: Vol. 25, No. 4

# 'Catches' in Fisheries Work

As the saying goes, 'In every fishing round there will be a 'catch'. While this can be basically true, looked at from a wider perspective, it can connote a deeper meaning.

Since the past some years, there has been a significant slide in fish landings from marine capture waters of India. This phenomenon is a 'Catch' in the fishing situation. This 'Catch' has been having an adverse impact on the economics of fishing operations and the socio- economic life of the stakeholders.

The fishing season that has started now has posed a kind of 'Catch' in marine capture fishing, atleast along the upper east coast. Shrimp fishing in the zone has a start with good 'Catches' at least for the larger vessels in the first voyage, but this welcome development, unfortunately, is riddled with very low returns. This is an unexpected blow to the stakeholders. Added to this is another adverse 'Catch'. The diesel prices zoomed up rendering the situation doubly critical and dilemma prone. As the returns are not adequate even to meet the diesel oil bill the operators have now come under the spell of this 'Catch', and are drawn into a dilemma whether to continue operations or discontinue them. Both ways they have to confront adverse 'Catches'.

The authorities are also helpless. The problem is intricate. There can be three routes to tackle it, one is to extend subsidies and the other is one of advising the stakeholders to diversify into passive methods of fishing. The third option can be to suspend fishing activity, awaiting better times. The spurt in the oil price is so high that it would entail the granting of a higher rate of subsidy, which could stretch the outlay beyond a reasonable limit. As an alternative, the authorities can also look for a way of tackling the problem by further encouraging stakeholders to switch over to passive systems of fishing like tuna longlining, squid jigging, and operation of lobster traps, which consume less of oil. Tuna, exported in sashimi style, particularly to Japan,would fetch a good rate. Oil consumption can also be less in the case of squid jigging and trap operations for lobsters. The trawl net operations can however be resorted to as a supplementary effort only when passive methods also do not yield economically viable catches during certain periods.

Of late, there has been the dawn of a refreshingly new outlook among a section of the fishing stakeholders to move away from shrimp trawling. Authorities too are anxious to achieve a reform that would shift the focus from shrimping, but several 'Catches' prevent progress in this direction. These 'Catches' relate to financial support and organisational reforms. Once a reformist outlook fully takes over and a fillip with all needed precautions is given to it, an enduring solution involving the emergence of a diversified system can be expected to emerge, with trawling, particularly for shrimps and other demersal species, serving as a buffer and a standby system.

Diversifications in marine fishing system can be expected to work well, even in times of no slump in the prices of the main harvested species such as shrimps and cephalopods.

There are several 'Catches' in Indian fisheries sector. The nation's active Indian fishing vessel fieet of 23 m oAL and above which was at a level of over 190 nos. until recently has now dwindled down to around 40 nos. At the same time the LOP/FOC tuna vessels dominate in our EEZ, particularly during the tuna fishing season. This is an undesirable situation and there has to be a bolder way of coming out of this 'Catch 22' situation.

Another aspect is that the returns from some of the Indian marine products exports, particularly shrimps, have been sliding relentlessly. In the case of a diversified item like tuna, it is possible to obtain a good export price, particularly for those of sashimi quality, provided the export marketing system is well fortified with the participation of an importer of a progressive bent of mind from business angle. For such a fortification there have to be group-oriented organizations to handle the export process, in association with the co-opted importer.

The high level of farmed major carp catches of the country is an achievement that has gained in strength in the last four decades but with a 'Catch'. Market prices of major carps have plummeted in the various parts of the country with the exception of north-east where there is an unfulfilled gap in demand for various reasons. For this reason major carp catches from various States of the country, particularly from Andhra Pradesh make a beeline to the north-eastern States of the country, with the hope of returns of an order that would cover the costs of production and hopefully leaving a margin.There has been no significant export of these fishes for want of demand. There were suggestions for marketing major carps in the west under attractive commercial names to be popularised in that part of the world. There has been evidence that the taste of flesh of major carps is far superior to several other favoured fishes in the West. This feature can be encashed by introducing major carps in USA and Europe under attractive names and through an aggressive publicity campaign by marketing them either as whole fish or as fillets in an appropriately processed form. This campaign has to be very effective to counteract the prevailing prejudice against carps. The 'Catch' in the way to proceed in this direction is pessimism, lack of initiative and assumption of doubtful results. A way can certainly be found to solve the problem, once a sustained movement is fostered in the direction of finding an export market to Indian major carps in USA and Europe, with the determination to crown it with success, for the benefit of the farmers and the Nation.

Fishing Chimes, Editorial 343: August 2005: Vol. 25, No.5

# .....On Introducing Super Tilapia (GIFT) Farming in India

Super Tilapia of ICLARM (now known as WorldFish Centre) is farmed at present in various countries such as Israel, China, Taiwan, Vietnam, Philippines, Thailand, Indonesia, Malaysia and also in Bangladesh. Super Tilapia, otherwise known as GIFT (Genetic Improvement of Farmed Tilapia) was developed by the scientists at WorldFish Centre (formerly known as ICLARM) through selective breeding of several strains of Nile Tilapia. This grows faster and survives better than the original fish.

According to available information farmers across Asia are increasingly culturing Super Tilapia (GIFT). This fish lowers the farmers' cost of production, thereby optimising their income. It also allows them to practice low cost, environmentally friendly aquaculture. According to available information, GIFT survives well even in polluted waters. It can be farmed well under extensive systems too, avoiding even the application of commercial feeds, it is said. At the same time, it is stated that the fish also responds well when extra feed is given. GIFT benefits both the rich and poor farmers. An example given is that a small farmer owning 1.56 ha of pond in the Philippines could earn as much as US $ 3,100 per year by farming GIFT, without the use of commercial feed and with no additional expense.

GIFT was developed by former ICLARM from several strains of Nile Tilapia, a popular farmed fish in Asia, adopting traditional selective breeding techniques. It is neither genetically modified nor transgenic. It is stated to be now in its ninth generation of selection and has yet to reach its maximum potential yield. Selective breeding of Nile Tilapia is continuing in a number of countries seeking to develop a GIFT variety tailored to local growing conditions, it is mentioned. Tilapia is rapidly expanding its market share, helped by the drop in capture fisheries. The American Tilapia Association reported that retail sales of Tilapia had surpassed that of trout since 1995. Indonesia, Thailand, China and Taiwan are leading Asian exporters of Tilapia and its products to the growing US and European markets. The US is the world's largest importer of Tilapia and Tilapia products. Bangladesh and Srilanka also export Tilapia to US and other countries.

Tilapia has come to be the most studied tropical food fish. Only the temperate Atlantic salmon has seen more research performed on it. Sometimes known as "Everybody's fish", Tilapia farming is popular because it is vigorous and tolerates crowding incredibly well in fish ponds, it is pointed out.

GIFT consumes rice bran to weeds and even sewage but it is mainly plant-eating. It can reach marketable size in four months when reared in cages, and in six months when grown in ponds rich in natural food and without the use of commercial feed, it is stated.

It is stated that demand for GIFT fry exceeds supply at the moment. The reason for this is stated to be poor management and technical inefficiency at many hatcheries. However, it is stated that efforts are underway to improve hatchery operations and, in some countries, Governments are training workers to maintain GIFT broodstocks properly.

Farm trials in a number of countries have shown strong gains in GIFT yields, compared with local strains, between 25 per cent (China) and 78 per cent (Bangladesh). The cost of production of GIFT is also stated to be markedly lower, compared to other cultivated species, over 30 per cent in the Philippines and Bangladesh and about 20 per cent in China, Thailand, and Vietnam.

## Philippines

So far, acceptance of GIFT is most striking in the Philippines. GIFT fry now accounts for over 30 percent of Tilapia fry supplied to Philippines farmers. Tilapia was unknown in the Philippines 25 years ago.

## Thailand

GIFT is also spreading fast in Thailand. It is possible that what is good for Thailand will probably be good for India too, but a study of the present status of GIFT farming in Thailand in relation to its effects on ecological conditions may be helpful to have an answer.

## China

In China, the world's biggest producer of Tilapia, GIFT is spurring Tilapia farming and consumption in the warmer central and southern regions. Commercial breeders now have access to several State hatcheries. Chinese experience is stated to be that Tilapia is easy to grow and is more resistant to disease than carps. The diversification in China towards GIFT farming deserves to be considered for emulation in India. A confirmatory study covering various angles of the subject can be undertaken to start with. While Chinese farmers continue to farm their carps, they have diversified into GIFT farming, apparently for augmenting production and incomes.

## Indonesia

In Indonesia, GIFT is usually farmed in freshwater lakes. However, it has been found to grow well on natural food in tanks and ponds and this is sparking enthusiasm among farmers to take to GIFT farming. Farmers growing carp in freshwater lakes in West Java have switched over to GIFT, an indicator for Indian farmers to consider following similar practice. As GIFT meat is thicker and firmer than that of local Tilapia strains, it is stated to be the leading farming species in Indonesia.

Big seafood exporter in Indonesia, the Swiss-owned Aquafarm Nusantara, is understood to have taken up farming of GIFT in lake Toba, a famous crater lake in Sumatra, for export to the US, having discovered that GIFT yields are 35 per cent higher than normal strains.

## Malaysia

Red and black Tilapia is farmed in Malaysia. The Malaysian Fisheries Research Institute and erstwhile ICLARM hoped to develop an improved GIFT strain that can adopt itself well to local water bodies and which will contribute to Governmment's target of raising aquaculture output substantially by 2010.

## Bangladesh and Vietnam

As GIFT is fast growing and high yielding, it is considered ideal for farming in seasonal ponds that dot Bangladeshi countryside. The widely cultivated Indian major carps do poorly in these seasonal water bodies. GIFT is also making inroads into Vietnam, where carps and riverine catfish are currently the freshwater species farmed. Half the animal protein intake comes from fish in these two countries.

Indian Government is understood to be against the introduction of GIFT, a widely accepted fish for farming in several countries, in India. It is to be noted here that there are no reports of any harm done by GIFT (Super Tilapia) to the fish fauna of the countries into which it was introduced. This could be probably confirmed by the Indian authorities concerned by sponsoring an indepth study of the ongoing GIFT farming activities and their effects on local fish fauna in countries such as Thailand, China, Indonesia, Philippines and near home in Bangladesh. Further, Vorion Chemicals, Chennai, undertook monosex culture of Nile Tilapia until recently for a quite a few years, with no known reports of either its escape into the wild waters or of ecologically adverse effects of this fish on native fish fauna. Yet, probably on account of the reason that compulsions that forced Vorion to take up Nile Tilapia monosex culture, although it was not part of their profession, were no longer there, the company gave up Nile Tilapia farming.

Nila Tilapia/GIFT (Super Tilapia) has to be considered as an eminently viable species. GIFT has become popular in neighbouring Bangladesh and Sri Lanka with no reported adverse repercussions. Fillets of this fish are being exported from these countries, mostly to USA. So far as India is concerned, there is no material to conclude that, if introduced in India, GIFT (Super Tilapia) would harm the native fish fauna. In the light of the fact that Nile Tilapia, cultured by Vorion Chemicals, had not been reported to have entered the wild, it could be considered that introduction of GIFT (Super Tilapia), which is a selectively bred version of several types of Nile Tilapia, would not be an ecological hazard if introduced. Further, GIFT (Super Tilapia), introduced by the respective Governments in South-east Asian Countries, as referred to above, has brought prosperity to the farmers of those countries, besides augmenting production for domestic consumption as well as exports. There are no reports of any adverse effects on account of the introduction of GIFT. Qualities of GIFT are different from *Tilapia mossambica*, which was introduced into Indian waters long back in 1960s and acquired a bad reputation of having become a pest. This experience should not be a deterrent to the introduction of GIFT, a far superior strain with sterling farming qualities and export potential.

In the light of the developments beneficial to the farmers taking place in neighbouring countries because of introduction of GIFT for farming, without causing any adverse effect on indigenous species and on the environment, it is desirable that Indian Government takes steps to formulate a pilot project in this respect that would provide setting up of a few GIFT hatcheries at selected points and formulate guidelines for their operation and supply of certified seed, monosex or otherwise, to the farmers, for its farming. It may be noted here that in USA too Tilapia farming has been introduced and American Tilapia Association is understood to be promoting this. In the present scenario that displays symptoms of stagnation in aqua products exports and fall in the earnings of farmers, the crucial importance of introducing GIFT deserves the consideration of the authorities concerned. At a technical session held in Visdkhapatnam as part of Fish Farmers Day celebration on 10 July, 2005, the following recommendation was made.

"Considering that 'Genetic Improvement of Farmed Tilapia' (GIFT) fish is being farmed and exported in fillet form by several countries such as Taiwan, Israel, Thailand etc. to USA, Europe and Japan in large quantities and that there are no serious environmental hazards reported from those countries, the Session recommended that Government may reconsider the issue with a view to allowing introduction of GIFT fish in India, its seed production and farming under a regulated system and its export."

In case there are confirmed and compelling reasons and justifications for not allowing introduction of GIFT (Super Tilapia) for farming in India, the Union Department of Animal Husbandry, Dairying and Fisheries may consider coming out with an explanation to the farmers pointing out the reasons for not permitting its introduction and tell them about the unfortunate reasons for depriving them from augmenting their incomes through this route as is being done by countries such as Thailand, Bangladesh and Sri Lanka. The Indian farmers have a cause to prevail upon the Union Department to review their present decision, which is not in consonance with the policy adopted by the other Governments of the region, whose concerns are no less important than ours, and either favour the introduction of GIFT (Super Tilapia) in India, subject to whatever conditions that may have to be stipulated, or recommended an alternative which will enable the farmers to augment their incomes. It is also unfortunate that the Associations of Fish Farmers in the country are so complacent that they have neither prevailed upon the Government to allow introduction of GIFT into the country, nor have they explained their stand. Apparently the Associations are weak and unless they strengthen themselves, there can be no way for the farmers to secure relief from the silent agony they have been enduring because of the present culture fishery situation revolving predominantly round Indian major carps, with declining or static incomes and rising expenses.

Fishing Chimes, Editorial 344: September 2005: Vol. 25, No. 6

# ...On Restructuring of Open Sea Indian Marine Fishing Fleet

The operators of open sea marine fishing fleet of India acquired world class capabilities at bottom trawling during the past over three decades. Behind this acquisition is an enormous supporting effort, specially by the Government of India, and providers of related supplies and services for ensuring uninturrupted fishing operations. Besides entrusting to certain Indian shipyards a few decades back the construction of a large number of open sea fishing vessels for developing an Indian built fishing fleet, to operate beyond the coastal zone of the country, the Government of India set up infrastructure for training of fishing operatives to run the open sea fishing fleet. Fishing harbours for providing safe berthing, maintenance of vessels and unloading and marketing facilities for the catches landed have also been established by the Government. The entire strength of the fleet of larger vessels of 22m OAL and over consists of bottom trawling vessels. This strength went up to a level of around 190 nos. by 1980s, which had however dwindled down in the past over two decades and now it is at a level of 40 nos. of operating vessels. An encouraging aspect in this scenario, however, is the emergence of a fleet of over 300 mini-trawlers in the size range of 15m OAL operating in the coastal waters and the zone adjacent to coastal zone all along most of the nation's coastline.

The dwindling of open sea fishing fleet strength should normally connote an increase in per unit catches, particularly of shrimp, because the fishing concentration by the reduced no of vessels too was mostly on catching of shrimps from the open sea closer to the coastal zone. However, this has not happened. The reason for this disappointment is attributable to the position that the fishable stocks too came down, much faster than the phasing out of the vessels. As a consequence, economic problems engulfed the operatons of the vessels. These problems include, among others, depletion in stocks, specially of shrimps, increase in fuel price, and general decline in returns, mostly because of a major slide in export prices of most of the marine products. The general efforts to counter the trend have been by way of imparting value addition to the products for export on one hand and through diversification of fishing effort towards harvesting of more remunerative species, on the other. In respect of efforts aimed at the later category, by way of either acquiring vessels to meet the need or through equipping the existing vessels for diversified fishing, the quest to find a source of finance for the purpose became pronounced. There are no doubt several financing bodies willing to provide funds but the problem has been that these bodies need guarantees in respect of repayments, and no agency other than the Government could be identified to provide them. Here it has to be mentioned that Government did provide such guarantees'(although indirectly) for the acquisition of new vessels by enterprises in the past, but had to face disillusionment because of non-payment of loan instalments by some of the owners to the financing agencies, leading either to efforts at invocation of guarantees by them or their resorting to one time settlements.

Government and the financing bodies had to subject themselves to the aforesaid embarassing and difficult situation because of the lapses on the part of the borrowers, deliberate, or otherwise. In this respect, however, Government has to now look back for an evaluation of the past events. Before standing guarantee indirectly for the of loans extended to the owners by financing bodies, the Government should have put in place a field-based mechanism to monitor the operations of these vessels, to keep track of quantum of landings, to whom the catches would be sold and the anticipated returns thereof. They should have made sure that the system of disposal of the catches was as per their approval and under their surveillance. The system could have been to have a certain percentage of the returns earmarked towards repayments at source. The estimated cost (of having an in-built monitoring mechanism of this kind) to ensure repayments could have been recovered as an additional prorata levy along with the loan repayment amounts from the borrowers.

In this context, it can be conjectured that the Government is now apparently unwilling to provide even indirect guarantees because of past experience, and also because of an adverse change in the fisheries resource status in the EEZ, the result of over-exploitation of bottom resources, (particularly of shrimps), which has led to the crisis arising out of a steep drop in per unit catches and, unfortunately, also a general decline in export prices too. This has led to the inescapable alternative of organising diversification of fishing effort.

## From Guarantee to Subsidy

A two pronged strategy seems to have been evolved now by the Government to face the problem of investments for diversification. This strategy bypasses the need for the Government to provide any kind of guarantees to financing bodies that may be willing to extend loans for acquisition of new vessels for diversified fishing or for conversion of existing vessels for diversified fishing. One of the prongs aims at encouraging the owners of vessels to instal monofil tuna longlining equipment on their vessels and for the Government to extend 50 per cent subsidy on the cost of such installation. This prong has in a way obviated the need for providing guarantees for such conversions, although finding money to meet the balance of capital investment is seen as a continuing hurdle by the owners.

The other prong of the strategy, as is known, is related to introduction of LoP (Letter of Permission), FoC (Flags of Convenience) tuna longlining vessels having needed facilities for storage and preservation of catches on board, and with needed endurance to be out at open sea and conduct economically viable fishing for a long duration in a single voyage. Although at the risk of compromising the status of India as one of the top fishing nations, Indian fishing firms have been given the opportunity to buy used tuna long liners from owning foreign enterprises for operation in the Indian EEZ under a system of deferred payment of the cost over a period of time subject to Letters of Permissions (LoP) issued by the Government of India. This LoP system has provided an alternative to the problems of investments and guarantees. The inspiration for the adoption of this system is understood to have come from the fishing industry of Taiwan, a well known global fishing nation.

Taiwanese longline vessels owners, who have been traditionally fishing in Indian EEZ (before and after its formation) and who have sound knowledge of tuna grounds of the zone, seem to have thus turned out to be a source for the generation of a solution to the funding problem. This Taiwanese solution, it is said, is on the same lines as presented by them to a few other nations too. It is simple and could be summarised in the following words: Buy our vessels and register them under your flag. Payment towards the cost? Pay this in instalments as convenient and with no compulsions, voyage by voyage. Technology? We will operate the vessels for you. Exports ? We will take care of them and credit the returns to the Indian owner in whose name the vessels will be registered in India. The Indian owner will just pay atleast our operational and other expenses in foreign exchange out of the earnings from voyage to voyage. A further amount will be adjusted against loan instalments payable. The vessels will be of real and total Indian ownership as and when the full cost is repaid.

This system has solved two problems, one is of that of Taiwanese vessels, and the other of the Indian industry to have access to funds to buy vessels, at one stroke. In other words, the Taiwanese vessels can gain fearless access to fish in Indian EEZ. The other is that the exports will be in the name of the Indian enterprise concerned, and no immediate investments are needed. This system has led the way to the introduction of what we now know as Indian owned 'LoP (Letter of Permission), FoC (Flag of Convenience) vessels, which follow the pattern in vogue in several other victim countries. Saying again for emphasis, this system has solved for the Taiwanese the major problem of their vessels fishing in Indian waters as foreign vessels. These formerly poaching Taiwanese owned vessels now fly Indian flag, because of holding Indian registry too, although full transfer would take place only when the entire cost as per agreement is paid; and when this will take place is a matter for the future. The operations of these vessels in Indian EEZ are seen as legal but these are also seen as illegal by Indian fishermen and their Unions. Further, they also see several disadvantages in the system adopted. Firstly, an access is given for the Taiwanese vessels to operate in Indian EEZ with all advantages going to them, although they sportively agree for the depiction of the exportable quantity harvested as exported by India, and for recording of the foreign exchange earnings as of India to start with, for the reason that the vessels are registered as Indian vessels too, besides probably other registrations. The outflow of substantial foreign exchange by way of operational expenses, crew wages etc., from the foreign exchange earned, leaving a meagre part of it behind is a different matter. Further, what is important is that, in factual terms, the operations are by foreign vessels, and for their financial benefit. Operations of this style classify India as one among those fishing nations which succumb to strategies of this nature.

Keeping these and other relevant aspects in view, Government may therefore now consider a project approach for the introduction of a pre-determined number of open sea fishing vessels with capabilities of diversified fishing for operation in Indian EEZ beyond territorial waters or the coastal zone. The latest status of the fisheries resources of Indian EEZ beyond the coastal zone or territorial waters may have to be assessed once again by the Central Marine Fisheries Research Institute (CMFRI) in association with the Fishery Survey of India (PSI). Thereafter, it is desirable that FSI determines the types and number of vessels to be introduced, taking into account the present number of around 40 vessels of 22 to 27m OAL vessels operating in the EEZ, beyond coastal zone. This worksheet could form the basis for the preparation of a project report for the introduction of a planned number of vessels of the types needed. In order to attract a financing agency, and to protect its interests, a well designed guarantee system as suggested in a preceding paragraph, could be considered for formulation, the guarantee coming from the Government of India.

It is to be believed that the Government of India does not hold the all-time view that the only way of restructuring the Indian fishing fleet to operate beyond the coastal zone or territorial waters is through introduction of LoP/Flag of Convenience Vessels, just for the reason that this system would not involve immediate investments.

Once the guarantee system is streamlined in a manner acceptable to all, particularly the financing bodies, an upsurge of economic activity by way of open sea fishing vessel building at various Indian shipyards in collaboration with one or more friendly fishing nations like Norway, Australia, Denmark, Holland etc. can be visualised, besides improvements to landing and berthing facilities and in respect of processing for value addition etc.

So far. as the coastal zone is concerned, the problem is not one of addition to the existing fleet strength but one of either reducing its strength or deploying a predetermined number of vessels of the fleet, which are now equipped for stern bottom trawling and engaged exclusively in harvesting of shrimps. This deployment has to be for undertaking diversified fishing beyond coastal zone. Over years, it is stated that the catches from these boats per unit have come down, resulting in a steep drop in incomes and the consequential repercussions. In this background, there is a widespread view that Government should take up new projects, one for upgrading of a certain number of these boats for fishing beyond the zone, and also another for decommissioning some of them. In regard to the later category, there have to be schemes for providing an alternate employment avenue to those who would have lost their means of living, because of decommissioning. One such scheme can be in the form of introduction of sea cage farming, taking care of allotment of sites on lease for setting up cages, providing training in cage making, their installation, and stocking them with yearlings generated by supporting hatchery systems and other avocations, like harvesting etc., mainly to the displaced stakeholders concerned.

Fishing Chimes, Editorial 345: October 2005: Vol. 25, No. 7

# Diversified Aquaculture Technology Transfer

## *MPEDA's Unique Initiative*

The setting up of Rajiv Gandhi Centre for Aquaculture (RGCA) for Aquaculture Technology Transfer by MPEDA as early as 1995 for the purpose of achieving the much needed emergence of diversified systems of aquaculture (to extricate the related farmers from the fluctuating risks being generated by shrimp culture) goes down as one of the few major land (aqua) marks in the history of culture fisheries development in India. This focal initiative has strengthened the ongoing efforts directed at diversification of brackishwater aquaculture. The inspiration for this initiative can be traced to the progress achieved in the line by the Central Institute of Brackishwater Aquaculture. As a follow-up, RGCA now concentrates mainly on transfer of technology related to farming of diversified brackishwater species with export potential such as Asian Sea bass and Mud Crab. More of such cultivable species such as lobsters may get added to the list sooner than later.

The first major land (aqua) mark in the inland aquaculture arena constituting a breakthrough was in the freshwater fish culture sector, in the form of induced breeding of major carps which revolutionised inland freshwater fish production in the country. In the wake of this major event ushered in by CIFRI four decades back, we now see that MPEDA, having earlier introduced shrimp hatchery technology into the country and promoted shrimp farming technology all along the coastline of the country, later concentrated on infusing corrective steps not only aimed at production of viable shrimp brooders, but also disease-free shrimp seed and healthy shrimp crops. In this background, MPEDA took the unique initiative of ushering in export-oriented diversification of aquaculture to cover species such as Asian Seabass, Mud crab etc.

The aim of this write-up is to place on record the historic contributions of MPEDA in respect of ushering in the era of export-oriented diversified brackishwater aquaculture and to set up the needed institutional and field infrastructure for training of executives selected to serve as agents to promote the application of the technology for the development of diversified culture systems of aqua production.

In a short span of ten years of its inception RGCA has made remarkable progress. Having developed a close relationship with the Central Institute of Brackishwater Aquaculture, (which has developed technologies mainly for the culture of Asian Seabass and mud crab), RGCA has eventually acquired its own personality.

The main objective of RGCA being diversified aquaculture technology transfer, the foundation stone for the Building Complex for the purpose has been laid on 19 August, 2005 at Karaimedu village in Nagapattinam Distict of Tamil Nadu, by Mr.Kamal Nath,Union Commerce Minister.

Apart from promotion of diversification of brackishwater aquaculture (seed production and farming), RGCA also has the responsibility to develop sustainable mud crab capture fisheries. As a step towards this, crablets were released on the same day into Pazhayar estuary, and this important event took place from the hands Mr. Mani Shankar Aiyar, Union Minister for Petroleum and Natural Gas.

So far as the main aspect of diversification of aquaculture is concerned, RGC has already logged impressive progress. In respect of Asian Seabass it set up a hatchery at Toduvai, a coastal village of Nagapattinam district of Tamil Nadu. This hatchery is already operational and Asian seabass seed produced there at is already being supplied to farmers, with needed extension support.

In order to popularise the farming technology of Sea bass, a demonstration farm has been established at Karaikkal, Union territory of Pondicherry by RGCA. At this farm, farming of Seabass in cages, a novel technology to raise finfishes in controlled conditions, was demonstrated, besides farming of the fish in open ponds by using farm made formulated diet.

A technical collaboration programme between RGCA and the Department of Primary Industries, Queensland State, Australia is also being processed by MPEDA, the purpose being to upgrade and modernise the hatchery and farming technology of Seabass at par with international standards.

The development of Technology of Sea bass seed production and farming is the stepping stone for streamlining the technology for other valuable marine and brackishwater finfishes such as Grouper, Cobia, Snapper, Tilapia etc, which are having enormous export potential. RGCA has drawn up clear plans to develop these technologies soonest, it is learnt.

In order to conserve the natural population and sustain the export of live mud crabs, hatchery technology to produce mud crab seeds under controlled conditions, and farming technology for producing mud crabs have been developed. In this context, RGCA established a pilot scale Mud Crab Hatchery too at Thoduvai village and successfully started producing mud crab seeds on a commercial scale. RGCA has also established a farm with nursery and grow out ponds for mud crab at Karaikkal. At this farm demonstration of nursery rearing of mud crab seed and grow-out system is in progress.

The demonstration of commercial mud crab seed production by CIBA and at RGCA has made a significant impact. Mr. E. G. Eraniappan, a rural aqua farmer, picked up the technology from CIBA/RGCA and set up a mud crab hatchery (Periyar Mud Crab Hatchery) at Vandalur near Chennai and succeeded in producing mud crab seed. He thus emerged with glory and distinction as the only mud crab seed producer in the private sector of India, as at present.

These developments constitute a major breakthrough in the Indian aquaculture scene. And these have grouped India with the four known countries in the world *viz.*, Philippines, Vietnam, Australia and Indonesia who alone could so far suceed in this line. RGCA has established a facility at Neelankarai, Chennai to demonstrate to the farmers the manner of collection of the juvenile lobsters and growing them in controlled conditions by providing formulated and natural diets until they grow to a marketable size.

RGCA has also now initiated a project for the demonstration of production of specific pathogen free broodstock of tiger shrimp with a view to producing healthy shrimp seeds for their farming. This is an unique project which requires total biosecurity in the operations. For this reason, it has been established in Andaman and Nicobar Islands by availing of technical consultancy from M/s Aquatic Farms Ltd., USA.

RGCA has a proposal on hand to develop technology for improving the quality of broodstock of freshwater prawn through selective breeding. It is also proposed to develop a technology for producing all male population of the prawn for increasing the productivity of these farms. In this connection technological collaboration is being sought from Ben Gurion University, Isreal.

Artemia cyst biomass is an inevitable live feed item required for hatchery operations related to marine and freshwater organisms. So much so, as a compulsion, it is being regularly imported into India mainly from USA for meeting the feeding requirements of larval stages at the hatcheries. In order to achieve self-sustainability in the production of Artemia, RGCA now proposes to establish a project to develop technology for the production of Artemia cysts in solar salt pans at Tuticorin, Tamil Nadu by availing of technical consultancy from Cantho University, Vietnam.

## Training Complex

A Training Complex is proposed to be set up at RGCA. It will have the following functions; a) Transferring Aquaculture Technology to farmers; b) Providing hands-on training in commercial aquaculture; c) Serving as a venue for Seminars, Symposia and Workshop on various significant issues on Aquaculture; d) Publishing a Journal of international repute on applied Aquaculture Technologies; e) Facilitating training in seafood pre-processing, quality control, product development and marketing, by national and international institutes like MPEDA, ICAR's Institutes, EIA,xINFOFISH, ICLARM, NACA, BOBP, etc., and f) Leasing of space and facilities to Government organisations and approved NGOs for conducting short term courses, Seminars and Workshops.

Hands- on training courses with a duration of two years would be taken up at the Centre, in respect of i) Production of Sea bass seeds; ii) Production of sea bass fingerlings; iii) Culture of sea bass in cages set up in ponds; iv) Production of mud crab seed; v) Nursery rearing of mud crab seed. vi) Grow-out farming of mud crab; vii) Advances in scampi aquaculture; viii) Advances in shrimp aquaculture; ix) Hatchery production of genetically viable male Tilapia seeds; x) Farming of Tilapia; xi) Lobster fattening by using cost-effective feeds; and xii) Artemia cyst production technology.

The emergence of Rajiv Gandhi Aquaculture Technology and Transfer Centre for Aquaculture (RGCA) constitutes a major advance to promote diversified technologies of brackishwater aquaculture. The outlook for the success of this endeavour, initiated by MPEDA, being very promising, the fisheries sector of India can look forward to the strengthening of the ongoing efforts at culture fishery diversification all along Indian coastline and the dawning of a prosperous era that would flourish with an advanced class of socio- economically developed farmers engaged in the production of export-oriented aqua species, and a contented and a well placed network of processors and exporters, all of whom working towards augmentation of aqua products exports supported by sustainable development of diversified aquaculture production.

Fishing Chimes, Editorial 346: November 2005: Vol. 25, No. 8

# On Past and On-going Utilization of Fishery Resources of Indian EEZ

Let us take a hindlook on the major policy initiatives of the Government of India for the utilisation of the fisheries resources of the Indian EEZ, since its declaration in 1976.

Around this year the Exploratory Fisheries Survey Project located rich shrimp grounds off Orissa Coast. This location mainly motivated Union Carbide to introduce two imported shrimp trawlers of 28 m OAL range in the first instance to be followed by the introduction of few more indigenously constructed shrimp trawlers in the same range. Earlier New India Fisheries Co., introduced a couple of Japanese built trawlers.

By 1978 under a scheme for introduction of multipurpose trawlers of 23 m OAL, equipped with outriggers was implemented. 28 of them were imported from Mexico. In 1980s there were further imports of similar multipurpose trawlers from Holland, Australia etc. Simultaneously, multipurpose trawlers constructed indigenously at various Indian yards were introduced taking the fleet strength to over 190nos.

There were efforts in the sixth plan period to import used vessels for catching tuna under the 100 per cent export-oriented scheme of Union Ministry of Industries. A few of them were imported but operations could not take off.

It can be said that the introduction of larger vessels took place in a climate of inexperience. The effort constituted a beginning, for there had to be one. There were no doubt some estimates of fishery resources of the EEZ to go by at that time, in addition to the results of the fishing operations in Indian EEZ by a few Indian and several fishing vessels of quite a few countries, more prominently of Taiwanese and Thai origin. There was some presence of Japanese and Korean vessels too.

In middle 1980s came the chartered vessels operations phase in the Indian EEZ with a sprinkling of joint ventures dominated by Taiwanese bull trawlers, single stern trawlers and long liners. In this phase, a good number of them were permitted to operate in the zone.

The operations of chartered/joint venture vessels in Indian EEZ continued up to 1994, a long duration of around 10 years. This record breaking continuity is attributed to mutual interests. Taiwanese interest was obviously related to gaining access for fishing in Indian EEZ. The interest of the Indian charterers was apparently related to 15 per cent returns out of gross earnings out of the operations with practically no investment. Besides utilisation of the resources, the Government's interest was aligned to the gross export earnings that would accrue because of exports of the catches without much of post-harvest effort. There was also the fear that, if there was no national fishing effort, IOTC may take steps to induct foreign fishing vessels to operate in the EEZ.

When the going of the Taiwanese vessel operations was good, a problem arose. Taiwanese, it was stated, had to reconcile with heavily depleted trawling grounds in the Indian EEZ which was of course their own making. They overfished the demersal resources. Unable to bear mounting losses sustained by them in Indian and also in Pakistan's EEZ, there were indications that they were preparing themselves to pull out, Murari Committee recommended the winding up of the charter scheme, and this was accepted by the Government. This was a blessing to Taiwanese interests.

Taiwanese were aware of the sizeable tuna stocks, in Indian EEZ. This was related to their own fishing experience in the operation of tuna longliners in the zone. This knowledge seems to have prompted them to take an indirect initiative to motivate the Indian entrepreneurs to buy their tuna longliners on hire-purchase basis. This bait has led to the incarnation of the earlier charter scheme in the form of the new system of hire-purchase or the flag of convenience system, which is a familiar one to the Taiwanese but new to India. This system that came into being gave two benefits. One is that the Indian fleet gained in strength without any significant investment Secondly there was addition of foreign exchange earnings from exports of capture fish.

This is no doubt a good development but with undesirable side lights too. The situation is reflective of the inability of the Nation to develop a grip over the enduring utilisation of fish resources of Indian EEZ with vessels fully owned by Indian Enterprises themselves, instead of depending on flag of convenience vessels under a system of allowing foreign-owned vessels to be registered as Indian vessels on record, while being fully aware that the vessels are under the *de facto* ownership of Taiwanese enterprises and run by them as such for all practical purposes. Both the sides are known to be aware that full payment of the cost of vessels is not likely to be made through hire-purchase system for various reasons, and both the sides also do not like this to happen. The vessels being under the full operational control of the Taiwanese owners, what they would do when the full payment towards the cost is not made, can be guessed.

This subject of focal national importance is dealt with by the Union Department of Animal Husbandry, Dairying and Fisheries directly. The Department has two field level organisations dealing with fishing vessel operations and these are the Fishery Survey of India and the Integrated Fisheries Project. Their present role in respect of introduction of Indian owned fishing vessels for commercial operation in the Indian EEZ beyond territorial waters is very weak. At the same time, it is known that several of the Indian shipyards are capable of building fishing vessels of the required standards, as indicated by their past record. Several of these are now in operation in the Indian EEZ.

Consistent with the fisheries resources position in the Indian EEZ, it should certainly be possible to organise introduction of indigenously built fishing vessels to enduringly exploit these resources. The main constraint in this regard is the absence of an authorised and an organised field mechanism for the purpose. This organisational mechanism can probably be created by merging the Fishery Survey of India (FSI) and the Integrated Fishery Project (IFP). When this is done, the resultant integrated unit will have the ready means of working out the numbers, types and other particulars of vessels to be introduced in the Indian EEZ, based on resources estimates and results of surveys. Once the no of vessels, type wise, to be introduced as worked out, receive Government's approval, the suggested component (the present IFP) for amalgamation with FSI can undertake further action for the introduction of vessels, following a formulated strategy that would cover financial, technological and operational inputs. Strategies for selection of fishing companies, the number of vessels to be supplied to them, type wise, and for entrusting the work to approved Indian shipyards, for evolving a financing mechanism, ensuring follow-up documentation work such as tripartite agreement between a) boat owners, b) financiers as selected and c) the processors/exporters to be supplied with catches, and their cost recovery system etc. The shipyards concerned could be encouraged to form into consortia for entering into agreements with proven foreign shipyards for supply of designs and drawings and to extend technical support on payment of mutually agreed fees. The suggested IFP section of FSI can work out all the details and undertake the co-coordinating work between the prospective vessels owners, shipyards (both Indian and foreign), and financing bodies. The needed strengthening of the IFP component of the suggested set-up with specialists to take care of the co-ordination work of various linkages, particularly related to financial inputs, vessel building etc., would need to be taken care of. In other words, the need is to have an organisational set-up and mechanism outside the Union Department but under its control, to achieve introduction of Indian-owned larger fishing vessels to sustainably exploit the open sea resources of Indian EEZ, both pelagic and demersal.

The task is to face the challenge of introduction of Indian-built vessels for the benefit of the Nation, by the Nation and for the Nation. Government could consider setting up a Committee of experts drawn from various specialised fields concerned to advise the Government on a strategy to be followed to come out of the stranglehold created by a foreign global fishing interests to keep the nation enchained to them and to continue to derive benefits to realise its interest.

Fishing Chimes, Editorial 347: December 2005: Vol. 25, No. 9

# Fisheries Scenario of Kolleru Lake in A.P

Kolleru lake is located in West Godavari and Krishna districts of A. P. It has an extent of around 90,000 ha. It is basically a freshwater lake but has a connection to the sea through a channel that is not wide enough to facilitate free flow of water during heavy rains. A major resource of commercially important fish and crustacean species, the lake provides livelihood to a large number of fishers and others,related mostly to fish farming and fish marketing.

Over years and in stages, Kolleru lake emerged as an important source of supply of fish to the States in the north-east part of India. In the initial years, over 50 years back, the supplies, mainly of climbing perch, used to be directed at Hooghly market of Kolkata. Later the supply basket expanded to cover air breathing fishes such as magur and singhi for several years, to be followed by the addition of major carps later. The successful major carp supplies to the north-eastern States from the Kolleru zone are the result of the enormous progress in major carp seed production and farming in the Zone.

The emergence of a focal awareness of the potential market for major carps in the north-eastern States (besides West Bengal), such as Assam, Manipur, Tripura and Meghalaya, and also in Orissa and Bihar gave a major thrust to major carp production in Kolleru Zone. The market, mainly for major carps in the north-east, is so inviting that, content with even meagre margins, several entrepreneurs, as a complement to their normal avocations, plumped into the fish farming arena. It is said that these entrepreneurs included, among others, prominent politicians too.

Entrepreneurs of socio-political status, stated be mostly hailing from upper strata of society, are given the credit of forming ponds within the lake, mostly in the marginal areas. Initiatives in this direction, mentioned as having been overdone to such an extent that the routes of water flow out of the lake have been narrowed down. This has led to a condition of water overflowing from the lake into adjacent paddy fields and other areas. During recent heavy rains there was submergence of agricultural lands and village lands. This occurrence, which is not uncommon, however, has led to an embarrassing situation for several politicians who are believed to have set up tanks mostly in the perpheral areas of the lake. These tanks are stated to be of varying dimensions with atleast a couple of larger ones of around 200 ha each or even more. In all, according to reports, there are around 500 tanks set up in the lake. The recent experience has been that these tanks, stated to be with an estimated total extent of around 30,000 ha, have been impeding free outflow of water from the lake, causing submergence of paddy fields and other lands around, there being only one outlet debouching into the sea. This outlet, being narrow, impedes quick outflow of lake waters during heavy rains. There have been recommendations in the past by experts to widen the channel and to construct a regulator to prevent ingress of sea water but this has not been acted upon so far.

Assessing that the capture fishery of the lake was not adequate to provide significant economic benefits to the people whose livelihood is intertwined with the lake, in 1960s, the State Government allowed fishermen's co-operatives to excavate ponds in the marginal areas of the lake for fish farming. This approach adopted by the State Government stood the test of time, in any case for quite a few decades thereafter, without upsetting ihe environmental conditions of the lake in any significant way. This policy had also upgraded the socio-economic conditions of the persons of the community hailing from weaker sections of the society dependent on lake fisheries who formed into fishermen's co-operatives with the support of the State Fisheries Department. This initiative, besides leading to positive benefits by way of increasing fish production and incomes of stakeholders, has motivated several entrepreneurs to set up a number of ponds in the lake area, besides taking on lease several tanks set up by fishermen's co-operatives. By virtue of their abilities and status, they could achieve higher levels of fish production, the surplus out of which found its way to States in the north-east which have been experiencing a shortage of fish supplies. This development brought name and fame to the fisheries sector of A. P. The fish supplying enterprises of A. P established enduring and sustainable routes of fish supply to markets in Orissa, West Bengal, Assam, Meghalaya, Tripura, Manipur, Orissa, Bihar etc., getting over several problems faced initially in establishing routes of supplies. There are reports that Andhra fish, although occasional and in small quantities, reach markets as far as Jammu. The entrepreneurship managed to have the fish boxes designed well to meet the rigours of transport, and produce them in fiberglass and thermocole for packing fish in ice hygienically. The pioneering work done by them in respect of supply of fish to north-east Indian States would live in history, in verse and song. Their accomplishments are revolutionary, taking fish in a wholesome condition over several hundreds of kilometers from A.P. to meet the requirements of people who need fish at those destinations. They not only built up a sizeable business running into hundreds ofcrores of rupees, but also achieved close bonds between people, so far apart. They deserve praise. Instead, they are now receiving brick bats.

The Government was silent when the tanks were formed apparently as they found the activity in order. Now it is suddenly felt that the tanks are impeding water outflow from the lake. May be for reasons of jealousy or genuine apprehensions of ecological and other damages such as flooding of agricultural fields in times of heavy rains etc., the entrepreneurs who brought prosperity to the area are being harassed undeservedly. This year the flooding diverted greater focus on the tanks set up in the lake which is however neither a sudden nor a new development as has been projected. The result of this projection was the intervention of the State Government to breach and demolish most of the tanks set up in the lake area. One impact of this was on the on-going fish production from the tanks in the marginal areas of the lake. The good relations built up over years by A.P. fish exporting enterprises with the people of north-eastern and other nearby States have suddenly become weak. There has also been a setback to commercial relations infish trade between A.P and north-eastern States of India.

The reports on environmental damages to Kollair lake that have received wide publicity are considered by many as subjective. An objective study would however reveal the exact position. It is not that flooding of paddy fields has taken place for the first time in the area. This is a recurring phenomenon once in few years. There are comments that this flooding was taking place even before the marginal tanks were set up. This time, however, it is stated that several new factors including those of a political nature came into picture. Kolleru lake has a dominating impact on socio-economics of almost every section of the society of the area and whichever section that is in command would have an upper hand in the affairs of the lake.

A trend of this nature is understandable. What is not understandable however is the failure of the successive Governments to plan and usher in a fish production system, which would have to be there on a enduring basis, to ensure a capture-culture fish population balance of the lake, without impairing the ecology, the water dynamics of the lake and the eco-tourism potential. It is not clear how fishery development of the lake would come in the way of seasonal migration of birds to the lake. The point made that fish farming caused pollution of the lake is grossly exaggarated. The pond owners, to the extent known, being enlightened, follow good management practices. In fact, the presence of fish would tone up the migration of birds. Had an assignment to ensure the aforesaid balance been entrusted to the Central Inland Fisheries Research Institute, it would have offered a viable plan of action, that would have ensured fish production from the lake at a sustainable level, for yielding all round benefits that would be socially, economically and environmentally viable. All said and done, the main blessing the lake bestowed on the people is its fish.

Healthy development of fisheries of the lake, through promotion of mutually dependent fish capture and culture systems, is certainly possible. Promotional efforts of this nature would certainly upgrade the socio-economic conditions of the people dependent on the lake. Further, besides exports of fish to north-eastern States of the country, exports of fish, prawn and shrimp to other States and also to other countries could be promoted, besides supplies for local consumption. Eco-tourism can also be developed as has already been done in certain parts of the country. When fishes are there in plenty in the lake everything else would also prevail in abundance. Abundance of fish in the lake will attract birds to the lake.

The damages have been done. In order to prevent recurrence of these damages, the State Government needs to seriously consider referring the issue of development of the fisheries of this important basically freshwater lake, one of the largest of the Asian region, for the benefit of the people, to a reputed National Fisheries Research Organisation like the Central Inland Fisheries Research Institute to undertake a detailed study and formulate an eco-friendly plan for promoting sustainably optimal aqua production from the lake, to consist of fish shrimp, prawn, crab etc., and recommend appropriate measures there of. The State Government would have to consider widening the channel that connects the lake with the sea, instead of blaming the development of fish farming in certain lake areas for the retarded outward water flow from the lake.

Fishing Chimes, Editorial 348: January 2006: Vol. 25, No. 10

# Hind Look at 2005

2005 has to be construed as the year in which several progressive measures on fisheries development were initiated. One of such measures, although not specifically seen to be directed at flag of convenience long liners, was seen to be having the potential to discourage the sway of such vessels in Indian EEZ. This measure concerns the initiative of equipping a batch of Indian trawlers by their owners for conducting tuna longlining too. The results of the operations of this step towards diversification have been very encouraging. Several tuna consignments in sashimi style have been exported to Japan as a result of this. Apart from this venture with some initiative, the owners owe a lot to the Union Department of Animal Husbandry, Dairying and Fisheries for the support given by way of a substantial subsidy of 50 per cent of the cost of modifications. One aspect noticed is that there are indications of some measure of discomfiture on the part of the Taiwanese flag of convenience vessel owners at this development. Another significant development is a joint venture agreement signed in December 2005 between the Government of Andhra Pradesh and an American tuna organisation bearing a name with a global touch, for the exploitation of tuna, particularly yellowfin of Indian EEZ. The A.P. Government may have by now applied to the Centre for the clearance of the joint venture proposal.

Another significant development in 2005 relates to sea cage farming. This is the result of the initiative launched by CMFRI under the leadership of its Director, Dr. Mohan Joseph Modayil. The Union Department ofA.H, Dairying and Fisheries has approved of a project on cage farming and in pursuance of this the CMFRI has called upon interested enterprises for expression of interest to participate in the project.

There have been other interesting developments during the year. Giant Freshwater prawn farming has made inroads into several inland States anew, other than Punjab, Haryana and Madhya Pradesh, in which the activity was taken up much earlier. These States are Uttar Pradesh, Bihar, Chhattisgarh, Rajasthan, and possibly a few others. Of greater significance than the spread of the farming of the giant prawn (which involves transport of seed and feed from coastal centres) is the coming up of a giant prawn hatchery in Chhattisgarh, as the next one to the one set up in Tripura earlier.

Successful rearing of larval stages of freshwater prawn in inland saline waters was achieved by CIFE's Centre at Udaipur in Rajasthan. This Centre also achieved a breakthrough in giant freshwater prawn and artemia seed production in these waters. CIFA's unit at Ludhiana developed a system of enabling over-wintering of prawns under farming.

Rajiv Gandhi Centre for Aquaculture successfully completed a trial run in the production of mud crab seed in April 2005. A more important and a significantly breath-taking development is the setting up of a mud crab hatchery in the private sector by one Mr. V. G Eraniappan in a coastal village near Chennai. The hatchery is now operational, supplying crablets to farmers, the first significant supply having gone to a farmer in a village near Nellore in A.P.

Fishery Survey of India has added two longliners to its fishery survey fleet.

Mr. Chintalapati V. Narasimha Raju of West Godavari District in A.P. has succeeded in commercialised hatchery seed production and farming of Singhi.

The Seafood Exporters Association of India signed MoU on fisheries development in Mauritius with the Mauritius Chamber of Commerce.

There were several disappointing developments too. Fish, shrimp and prawn farmers experienced in various parts of the country problems of flooding, loss of crops and drop in farm gate prices. Exporters of marine products to USA, particularly of shrimps and prawns, became victims of anti-dumping duty levied by the U.S. Government.

(The inclusion of developments as listed above are based on those recorded in *Fishing Chimes*).

## General Events

### (Based on those Recorded in Fishing Chimes)

- ☆ Union Dept. of A.H. and Dairing renamed as Department of Animal Husbandry, Dairying and Fisheries.
- ☆ A National Marine Fisheries Census was conducted by CMFRI.
- ☆ An LCMSMS lab was set up at State Institute of Fisheries Technology, Kakinada, A.P. It was inaugurated by the Chief Minister of A.P.
- ☆ Floating cage culture was launched in Meghalaya on 4 May 2005 by Mr. J.D. Rynmba, Minister for Fisheries, Meghalaya.
- ☆ Rapid Immunodot test kit for WSV was introduced by College of Fisheries, Mangalore.
- ☆ CIFNET started B.F.N.Sc. course, affiliated to Cochin University of Science and Technology.

- ☆ West Bengal State Co-operative Federation Ltd. (Benfish) entered into its silver jubilee year.
- ☆ Foundation stone was laid for a Building Complex of Rajiv Gandhi Centre for Aquaculture Technology Transfer at Karaimedu Village, in Nagapattinam Dist of Tamil Nadu by Mr. Kamal Nath, Union Minister for Commerce on 19 August, 2005.
- ☆ Dr. T.V.R. Pillay, a renowned International culture fishery expert passed away on 9 Feb., 2005 in Bangalore.
- ☆ Dr. T.V.R. Pillay Aquaculture Foundation was set up in Nov 2005.
- ☆ Mr. Abraham J Tharakan was re-elected as the President of Seafood Exporters Assocaition of India.
- ☆ Mr. T.M.Chowdary was elected as the President of Association of Indian Fishery Industries.

## Awards
## (As Recorded in *Fishing Chimes*)

- ☆ World Food Prize was awarded to Dr. Modadugu Gupta.
- ☆ Meen Mitra Awards were conferred by the Govt, of West Bengal on M/s. Niranjan Sihi, J V H Dixitulu, and Sotyendranath Boral on 10 July, 2005.
- ☆ Bhoomi Nirman's Best Shrimp Farmer Award was given to Mr. Manoj Sharma, a shrimp farmer of Gujarat.
- ☆ Zoological Society of India's Gold Medal was awarded to Dr. D E Babu.
- ☆ Dr. Saly N. Thomas, senior scientist of CIFT was chosen for the Jawaharlal Nehru Award.
- ☆ Dr. I. Karunasagar received the International Award on sea safety called 'Academic Contributor of Biennium' from the International Association of Fish Inspectors.
- ☆ Dr. I. Karunasagar received Rafi Ahmed Kidwai Award.
- ☆ Dr. Vijayakumaran, scientist of CMFRI received SOFTI's best scientific paper award
- ☆ Matsya Maha Sangh, Bhopal received a Cheque for Rs.40 lakhs from GOI for performance in fish marketing and providing houses to fishermen.
- ☆ M/s A.B. Patil, Anil Kumar Tarad, Kamlesh Gupta, Ashis Kumar Sarkar, Bulusu Satyanarayana, Deepak Roy and Manok Sharma received Awards given by CIFE on Fish Farmers' Day.
- ☆ Hiralal Choudhury Best Fish Farmers' Award was given to Mr. V.G. Eraniappan by *Fishing Chimes* Jaysree Charitable Trust, Visakhapatnam.
- ☆ EnRA fellowship was awarded to Dr. A.K. Pandey of Central Institute of Freshwater Aquaculture, Bhubaneswar.
- ☆ Bilasbhai Ketin Matsya Vikas Puraskar 2003-04 was awarded to Shaikh Rahimullah Khan by the Director of Fisheries, Chhattisgarh.
- ☆ CIFT's Research Team won International Smart Gear Award.

Fishing Chimes, Editorial 349: February 2006: Vol. 25, No. 11

# Indian Freshwater Fish Farming Problems of Major Carp Monopoly

Marine fishes have until recently been the main component of Indian fish production. This complexion has now changed. While the share of annual marine fish production in the total Indian fish output has become more or less static, the annual share of Inland fish production in it, mostly dominated by farmed fish, is currently almost at par with marine fish production (At about 3 million tonnes per annum).

The credit for the rise in inland farmed fish production goes to the fishery scientists, technocrats and fish farmers and the nation is grateful to them. Their achievements have been great, production going up from 600 kg/ha in the decade after independence to over 2200 kg/ha now, mostly from ponds brought under Fish Farmers Development Agencies and also by the progressive and enlightened farmers outside FFDAs. Looked at from the angle of composition of the produce, it has to be admitted that the bulk of the production consists of Indian major carps, supplemented by common carp, a few other exotic carps, catfishes, magur, singhi, etc. A recent development is that giant freshwater prawn has gained a place among farmed species.

Anamolously, the impact of increase in the major carp production on overall Indian freshwater fish farming sector has become the bane of the stakeholders concerned. Farmgate prices of major carps have dwindled in a manner that is negatively disproportionate to the investments, particularly in regard to cost of feed, which has been moving up. Farmers are the main victims of this unfortunate trend. Fishes other than major or minor carps (Air breathing fishes etc) secure a better price for the farmers but farm production of these is marginal. Several farmers have no doubt diversified into farming of some of them, but this is just marginal.

So far as aquafarming policies are concerned, with the exception of fishery leasing policies, it is difficult to come across any policy consciously introduced by the State Governments in respect of aquafarming. The only exception seems to be the developmental/regulatory measures in respect of wetlands, introduced by the Government of West Bengal. The developments in fish culture that have been taking place from time to time are because of force of circumstances and farmer/ entrepreneureal interest in technologies evolved by our Research Institutes. Another significant development, based on technologies seen in recent years in inland States is the policy to popularise giant freshwater prawn farming introduced by the State Governments of Punjab, Haryana, Rajasthan and U.P. The introduction of this policy is, however, in isolation, without post harvest linkages to processing, preservation and marketing.

Production of fish will be purposeful to a farmer when it is supported by an infrastructure, a network of supplies and services and a developed or developing market.

Fish farming in India is undertaken by a farmer or by a farmer- entrepreneur, taking for granted the existence of a viable market for his produce. Looked at from the past perspective, fishfarming in India grew from a traditional background to the present stage, not based on a specific policy to bring all cultivable water areas under production, but based on a leasing system of confined water areas owned either by the Government or local bodies to whom they are entrusted by the Government concerned. The motivation behind the leasing system is not to specifically increase fish production but to earn revenue for the State to the extent possible. The emergence of privately owned fish ponds is of recent origin, more conspicuously in Punjab and Haryana, developed in agricultural lands. In some of the States, there are scattered initiatives in the private sector at converting fallow lands of various categories into fish ponds, but this development is of recent origin. In Rajasthan, Haryana, Punjab and Western U.P there are vast extents of saline lands with underground saltwater reserves but there are no conspicuous policy initiatives to bring them under aqua production, although suitability of these lands for setting up fish farms has been amply demonstrated by the Central Institute of Fisheries Education. The Department of Fisheries, Rajasthan has produced Artemia and their eggs in saltpans of Rajasthan. These are valuable feed items in shrimp and prawn hatcheries, but there has been no conspicuous policy follow-up.

It has to be agreed that the Nation does not have an overall plan designed to spread fish farming across the length and breadth of the Country, particularly in such of the areas where it will be justified, taking into account the environmental, socio-economic, market-orientation and sustainability angles into account. Generally stated, the development of fish farming in the country continues unabated in a haphazard and undirected fashion, despite the existence of well organised State Fisheries Departments.

There is an urgent need for demand - based/demand creating farmed fish production in India. There can be no two opinions about undertaking State-wise studies of the ultimate possibilities of utilisation the existing lentic resources for market-oriented development of fish production in them. It is true that all information is there but no graphic picture in its totality is available. Once a graphic picture of this kind is developed as Part One of an exercise in this direction, the second part would have to follow, to consist of the actual development process.

An effort is made hereunder to highlight some aspects of the fishery development system in respect of our tanks and ponds. What is being done now is to follow a loose or a vague system to link fish production work with the ultimate goal of profitable marketing of fish for viable returns. An unlimited market for fish, particularly for major carps, is presumed. We rarely come across a proposal that talks about the demand for fish. It is taken for granted that it is boundless. Most of the time, we come across material that highlights targeted fish production levels with no reference to markets. The economics of operations given in any project proposal give calculations of 'annual income' and 'annual expenditure'. The elasticity of demand and its effect on fish prices is rarely referred to. Fluctuations in domestic market prices of farmed fish, mostly on the declining side, are now come across very often. This disturbs the viability of incomes of farmers which are of paramount importance and many farmers feel that, while much is being done to protect their interests by the authorities, they do not feel the impact. The general approach is mainly to look into feasibility of pond fish production from considerations of area of pond, preparation aspects, stocking and feeding, but with very little attention towards marketing aspects of the produce.

There is an imperative need to develop a State-wise domestic marketing system that spreads over all important centres so as to avoid unremunerative sales. And these domestic systems would have to be interlinked through a national policy. The domestic marketing system should have to involve centrally located fish storages, ice-making and supplying units and fish transport facility from these centres. Like-wise there has to be a transport facility to supply fish and ice from the centralised storage centres to various fish supply outlets set up along well drawn routes, with the outlets entrusted to selected enterprises capable of marketing of fish. Once an organised domestic marketing system is established and activated, price regulation/stabilisation will ensue and there will be the effect of increase in earnings of stakeholders at various levels backwards from the marketing end upto the point of producers (Fish farmers) and a firm trend of price fluctuations based on actual field factors will establish itself. The domestic market price levels will have to then form the main criterion for farmers or related entrepreneurs for regulating their farming activities as needed in the place of the present faith-oriented system of stocking, mainly with major carps/exotic carps.

While an organised and regulated domestic marketing system is important for the farmers to improve their declining earnings, there can be a real uplift in their earnings towards viability only when there is an export dimension to their produce. Presently, their produce essentially consists of major carps, and these do not have export value. Unfortunately, in the main importing countries such as USA and those in Europe, any fish called as a carp is considered as a low class fish. This attitude goes against the interests of Indian farmers. The silver lining in this gloomy situation however is that, served without revealing the name, the preparations made out of Indian major carp flesh are well accepted by visiting foreigners to India. Once the consumers are told about the source of the dishes, as it happened quite a few times, the foreigners underestimate their hosts.

India, as is known and without doubt, is a carp country. All Indians can be proud of this. At the same time, it is essential to counter the problem of low returns that carps fetch to farmers. In order to extricate farmers from the plight into which major carps have pushed them, the authorities would have to consider, as an additional alternative, the introduction of GIFT (Genetically Improved farmed Tilapia) developed by World Fish Centre or Nile Tilapia, into India, taking all needed precautions in respect of ecological safety. GIFT/Nile Tilapia have enormous demand in USA and Europe and several countries have already taken advantage of this. Government of India has to consider this issue for a policy decision in favour of GIFT/Nile Tilapia introduction for the benefit of the farmers, supported by centralised hatcheries with attached farms for producing all-male GIFT/Nile Tilapia fish and a system that prevents mother stocks from escaping into wild waters.

It is felt that there is a way, although a hard one, to solve the present problem faced in exporting Indian major carps. This route is to invest the Indian major carps with names that would be attractive to the common fish consumers of the west, and by reconciling ourselves to the fact that unless this is done, exports of major carps to Europe and USA will continue to be elusive; and this export dimension is of crucial importance to augment the incomes of the primary producers. After christening the major carps for the purpose of export with new names such as Indian Freshwater Bream, Indian Freshwater Roach, Indian Freshwater Wrasse/Indian Freshwater Snapper, or other selective ones, a sustained promotional campaign has to be launched in a systematic manner without getting discouraged in between, keeping in view the objective and its long term dimensions. MPEDA with all its experience would certainly be able to bring about the reform, through a strategy that will have its stamp. Apart from efforts at exporting rechristened ones as whole fish, value-added exports in the form of frozen fillets or in other forms could also be promoted.

Fishing Chimes, Editorial 350: March 2006: Vol. 25, No. 12

# Extension Anxieties

India has made remarkable progress in the development of its fisheries in the past 50 years. In 1950-51, India's fish production was 7.5 lakh t (5.34 lakh t marine and 2.16 lakh t inland). By 2005-06 the Indian fish production zoomed up to a level of around 6.3 million t (Consisting of 3 million t of marine and 3.3 million t of inland). This works out to an increase of around 800 per cent in over 56 years or on an average an increase of 14 per cent per annum. It is stated that India is presently the world's fourth highest fish producer and the second highest inland fish producer after China, although the gap in production between the two countries is enormous, with the inland production of China stated to be over 53 million t.

Indian Fisheries Scientists established in the past an enviable record in the development of improved fish production technologies and their transfer to end users. In contrast to this past position, the present situation is perceived to be different. While some professionals hold the general view that prevailing extension system is weak and perfunctory, there are others who feel that the spread of developed technologies is related to a large extent to the economic viability of the end products in relation to the results of application of extended technologies.

Past experience shows that technologies that provide a sustainably higher production of economic viability and yield higher returns get quickly extended to the end users, neutralising various problems in the way. It can be said that technologies under this category that have already taken roots in the country include, 1) induced breeding of fishes, 2) composite fish culture, 3) poly culture, 4) marine shrimp farming, 5) freshwater prawn farming etc. In contrast to the quick progress achieved in the transfer of the above mentioned technologies to the end users, extension efforts directed at developed technologies such as farming of sea cucumbers for export, and pearl production from marine pearl oysters and freshwater mussels, proved to be, by and large, discouraging. This anti-climax was mainly attributed to unfavourable economics of the related commercial operations. Encouraging scenario that is now emerging in the above mentioned background is that technologies for seed production and farming of mud crabs, spiny lobsters, seabass and certain air breathing catfishes developed at ICAR's research institutes are attracting the attention of aquaculturists.

It is a welcome development that an impetus is being given by CIBA/MPEDA for promoting mud crab seed crablets production. A few enterprising aquaculturists diversified into the mud crab arena (production of its seed and its grow-out/fattening operations). A larger number of the farmers have to be now induced to diversify into this line through extension effort and also provision of grants/subsidies. A measure of awareness of the benefits that the adoption of these technologies would bestow, being already there among some of the farmers, the present need is to foster this diversification trend through a well conceived strategy.

Technologies for pearl production from marine oysters developed by CMFRI evoked initial entrepreneurial interest. Tamil Nadu Industrial Development Corporation and SPIC in Tamil Nadu embarked on marine pearl production under the banner of a joint venture company called "Tamil Nadu Pearls" but it was short lived. Some constraints hampered the commercial success of this enterprise. There were also disappointments in the extension of technology developed by CIFA in respect of pearl production from freshwater mussels. Both these setbacks have been attributed to some missing dimensions in the technology transfer which must have been identified but not known to have been revealed. In all likelihood these may be related to the coordination aspects of production work with innovative marketing strategies to secure viable returns.

Sea cucumbers are a much relished food item in several countries of the far east. CMFRI developed the technology of seed production and farming of sea cucumbers over a decade back. The technology however, still awaits transfer to entrepreneurs. All initiatives at popularising the technology seems to have been suspended for the reason that wild stocks of sea cucumbers happen to be prohibited items for exports in dried or other forms. The fact that farmed sea cucumbers in a processed or other forms can well be exported under certification from the concerned authorities is probably overlooked.

The way in which the exotic white shrimp, *Litopenaeus vannamei* was introduced into India and subsequently allowed to be cultured indicates the need for an effective regulatory mechanism for monitoring entry of alien species into the country. The demand for White shrimp in the global markets could have been tapped by promoting the culture of the native white shrimp, *Fenneropenaeus indicus,* a much hardier and disease resistant species than its exotic counterpart.

An improved version of the tilapia known as GIFT (Genetically Improved Farmed Tilapia) developed by the World Fish Center is widely farmed in Thailand, Israel, Taiwan and several other countries in an eco-friendly and commercially viable manner and it is exported in whole and value added form to various global markets, particularly of USA and Europe. In this background, time has now come for India to reconsider its stand on farming of Tilapia, involving its geneticially improved version. The country needs to study the prospects of introducing and culturing the GIFT strain for its export from India through a well conceived technology extension mechanism, which has to involve protected breeding centres and setting up of all- male seed production units, to supply the seed so produced, duly certified, to the farmers. It has been said that the hormone content in the sterile feed given to the fry for sex reversal into males would disappear from the flesh of the young ones after reversal into males in about two months in grow out ponds. The grown-up fish would thus be totally free from any steroids.

On the marine front, after having experienced a glorious phase of successful extension of commercially viable technologies in the past, the nation is now passing through a stagnation period, with the only encouraging initial progress by way of adoption of tuna longlining technologies, mostly the result of self-help extension effort by the enterprises concerned.

The aforesaid past phase of marine fisheries extension effort consisted of introduction of mechanised boats under the erstwhile Indo-Norwegian Fisheries Project, the introduction of fibreglass beach landing crafts with OBM by the Bay of Bengal Programme and also improved versions of mechanised boats (45ft OAL) for deep sea trawling as developed by Indo-Danish Tadri Project in Karnataka. Subsequently a few versions of the stern trawlers in the range of 15m OAL referred to as mini trawlers, sona boats and so on have been introduced for bottom trawling in coastal waters.

There was also considerable extension effort in the area of fish preservation and processing which provides the needed linkages between production and marketing. Extension efforts to impart value addition to the processed products have also been on the increase in the recent years.

A recent development is the focal attention directed at the introduction of marine cage farming. The result of this was that the Union Department of Animal Husbandry, Dairying and Fisheries sanctioned a project on marine cage farming to be implemented in Indian waters by the Central Marine Fisheries Research Institute. This project involves enormous extension work in respect of fabrication and erection of cages, setting up of hatcheries, introduction of feeding systems and so on.

ICAR's Fisheries Research Institutes are the main sources of new technologies for the spread of which extension work has to be carried out. Besides these, a few organisations set up by MPEDA and a few research institutes under the CSIR and State Fisheries Departments, Universities and Fisheries Colleges are engaged in the development of new technologies. Although the possibility of duplication in research efforts is quite possible, this could be avoided through improved coordination and regional networking systems.

There is a general perception that technology extension as a part of Indian fisheries set up does not receive the attention it deserves. The style of fisheries extension work now in vogue is believed to be in the nature of a routine excercise, with inadequate involvement of the State fisheries developmental machinery which has a lot to do with the extension of technologies. There is a view that a well designed system of extending technologies to farmers and other stakeholders through formation of viable groups of them headed by trained technology transmission leaders and the total involvement of the members as part of self help groups is needed, with the system supported by the needed infrastructure. In this context, the extension model followed by MPEDA/NACA for extending hygienic shrimp pond management practices in Bhimavaram zone of West Godavari District of A.P., deserves to be emulated with such modifications as are necessary. In this zone, aqua clubs consisting of shrimp farmers were formed and trained to function as self help groups to spread among themselves the message of hygienic shrimp pond management.

The Krishi Vigyan Kendras (KVK) could play a vital role in achieving the much needed extension support to Indian Fisheries scenario by disseminating technologies to the target groups. Agencies like FFDAs and the related NGOs can also be involved and their services can be utilised for effective transfer of technology from the research centres to stakeholders.

Considering the imperativeness of toning up the fisheries extension machinery, the Union Department of Animal Husbandry, Dairying and Fisheries would have to look into the desirability of setting up of a Committee of Experts to review the present status of Fisheries Technologies Extension System and its allied aspects and recommend measures to augment and tone up the effectiveness of the system in a form that the impact of the various technologies extended are widely adopted in a manner that would be useful in stepping up fish production in India on a substantial scale.

Fishing Chimes, Editorial 351: April 2006: Vol. 26, No. 1

# Enter Silver 'Jubilee' of *Fishing Chimes*

The maiden editorial of *Fishing Chimes* that appeared in its Inaugural Issue (April 1981) carried the Caption '*Enter Fishing Chimes*'. Sentimental over this, this editorial is captioned "Enter 'Silver Jubilee' of *Fishing Chimes*".

Fishing Chimes is now in its Silver Jubilee year. The journal swam through 25 years to celebrate its tryst with this cherishable event. This performance is the result of the sustained growth in the readership interest and the encouraging opportunities given by the readers to keep the journal victoriously marching forward.

*Fishing Chimes* was inaugurated In April 1981 by Mr. Goka Ramaswamy, the then Minister for Fisheries, Govt of Andhra Pradesh. As a follow-up generated by this, the subscribers became its Sultans who have been keeping the journal afloat with their horse power. The contributors of sterling papers to the journal turned out to be its kings, and they showered their contributions over the journal to act as organic fertilisers for it to blossom. The advertisers played the role of Archidukes. They kept the magazine chiming. And, finally, those that ruled over the journal happened to be its readers. The gratitude of the editor and his supporting staff goes to all the aforesaid patrons.

The journal recorded in its successive issues, among others, events related to several culture fishery technologies. These related to seed production of giant prawn, shrimp, major carps, exotic trout and also the Innovative efforts at seed production of singhl, magur etc, on the one hand and farm production and marketing aspects, both domestic and export, of most of the aforesaid species, on the other.

The distinguished representatives of the journal, most of whom are fisheries experts, played a significant role In the enrichment of the contents of the journal. They strove to keep the readership abreast of developments in their respective States. H.N. Chandrasekhariah, P.K. Samanta, Armaan U. Muzaddadi, K.R. Narayanan, K. Slmhachalam and Basheer Ahmed are the representatives. D.B. James and H.S Badapanda have been Resident Editors in Tamil Nadu and Orissa respectively. Fishing Chimes records with gratitude their dedicated support. Santhi Jammi is the representative of the Journal in USA and she has been lending considerable support. One important contribution of her is to provide information on the system followed by US authorities for according permission for the setting up of cages in demarcated sea plots for fish farming.

M/s. Raamakrishna Printers Private Ltd., headed by Mr. G. Ramakrishna, its Managing Director, have been the printers of the Journal for the past over 20 years, supporting us all through and extending all co-operation. The State Bank of Hyderabad has been the Bankers of the journal all through, never giving us a chance to complain.

The spread of innovative lines of endeavour such as setting up of hatcheries for giant freshwater prawn seed production in non-coastal States such as Trlpura, Chhattisgarh and Punjab and their farming in these and other States such as U.P., Rajasthan, Bihar etc., were among the emerging developments chronicled.

Mohan Joseph Modayil, S.D. Tripathl, I. Karunasagar, M.A. Upare, N.K. Agarwal and Venkatesh Salagrama are the honourary and honoured advisers of Fishing Chimes. They have been playing a benovelent role in providing timely advice on piloting the journal through a path that would keep the readership abreast and enlightened on various technological developments in the sector.

Several major achievements in the fisheries sector were covered In the span of the silver jubilee period of Fishing Chimes. One such achievement of the nation was a remarkable increase in the level of annual fish production. It was 2 million t in 1981 and it rose to 6.2 million t by 2005. Indian marine products exports rose from 78,542 t valued at Rs. 21.89 lakhs In 1981 to over 4 lakh t valued at over Rs. 6,500 crores (US S 1.48 billion) by 2005.

On the fisheries educatlon front, the journal covered several important events. One of these was the declaration of the Central Institute of Fisheries Education as a Deemed University and the others were in respect of setting up of Animal Husbandry, Veterinary and Fisheries Universities in Tamil Nadu, West Bengal, Karnataka and A.P., besides a Fisheries University set up in Maharashtra, recently. Several Fisheries Colleges affiliated to Agricultural Universities have come up in various other States such as Tripura, Bihar, Assam, Orissa, Kerala, Gujarat, Rajasthan and Uttarakhand. All these colleges started fisheries graduate courses. Some of them now offer post-graduate fisheries courses too. Further, a number of Universities provide facilities to conduct research leading to doctorate degree.

The growth phase of Fishing Chimes has been the witness to a plethora of remarkable developments most of which go to the credit of related central fisheries research institutes, to the Union Ministry of Agriculture and State Fisheries Departments. The revolution ushered in by the introduction of induced breeding of major carps in 1958 acquired strength in the following years, mostly after 1981. They had a spread effect with the entry of modified versions of Chinese type of circular eco-hatcheries. Improved rearing systems for seed production of major and other carps followed. The systems of seed production of several cat fishes also came into vogue. Polyfarming and integrated farming systems gained a significant place in the overall farming scenario. CIFA's units popularised improved systems of paddy-cum-fish farming, particularly in north-eastern States. The freshwater cage culture system, first started in a reservoir in Karnataka has had a spread effect, as promoted by CIFRI, in various States. But the system has not yet taken roots to the extent needed. Fish Farmers Development Agencies have emerged as a viable mechanism for spreading farmed fish production technologies. Training programmes conducted by CIFRI/CIFA and others for the spread of improved farming technologies have become a potent vehicle for the purpose.

While the journal has witnessed all these and several other beneficial developments, there are certain outstanding events recorded which have a historic flavour. Chief among these are the specially designed hatcheries that came up in good numbers along the coastline for shrimp and prawn seed production. The prawn hatchery system has gained entry into the inland States, operating now mostly on constituted salt water. More important of the other innovations are the system of transport of prawn seed from coastal centres to interior points in a good condition and the spread of prawn farming into interior States like Punjab, Haryana, U.P., Bihar, Madhya Pradesh, Chhattisgarh, Tripura, Manipur, and so on.

A professor challenged some years back that, if any one can prove that the tiger shrimp can be farmed in freshwater, he would deserve a nobel prize. That professor is no longer there in this mortal world, but Fishing Chimes had the opportunity to record the beginning and the spread of tiger shrimp farming in over 4,000 ha of freshwater ponds mainly in the Krishna and West Godavari districts of Andhra Pradesh. Besides this, the system of 'refarming' of yearlings of major carps has come up in certain coastal districts of A.P. and at a few other centres, with excellent levels of growth and recovery. This refarming system is a unique one. What is done is to impart a stunted growth to the major carp seed, through higher levels of stocking, mostly for 10-12 months, and later transfer them in viable numbers into grow-out farm ponds. The experiences are that such stunted yearlings grow astoundingly fast with practically little mortality, yielding lucrative returns.

Another achievement chronicled in *Fishing Chimes* is the emergence of integrated fish farming system., developed by the scientists at CIFRI/CIFA. A further recorded development is that farm grown Major Carps and Catfishes, mostly from A.P., are no longer sent to the market as before in a haphazard manner. They are packed well in ice in plastic crates and exported in a highly hygienic way, to States such as Orissa, West Bengal, Assam, Tripura, Manipur, Meghalaya etc. This activity, while generating an enduring economy, had bridged the gap in the fish availability in the aforesaid States to a large extent.

There are a few other encouraging developments during the later part of 25 year period, as recorded in Fishing Chimes. These are i) the development of a genetically upgraded Rohu species by CIFA, christened as Jayanthi Rohu, in association with AKVAFORSK of Norway which has now spread among the Indian fish farmers to a significant extent. Another development is the production of seabass seed by CIBA initially and later by Rajiv Gandhi Centre. It is another matter the technology could not as yet have a cognisible spread effect among farmers. Yet another achievement is the production of mud crab seed in a specially designed hatchery, not only at the one set up by CIBA, and at the hatchery of Department of Zoology of Andhra University by Prof. D.E. Babu, but also by a professional aquaculturist of Tamil Nadu, Mr. V.G. Eraniappan, the owner of Periyar Mud Crab Hatchery, located near Chennai.

*Fishing Chimes* has recorded that the first freshwater prawn hatchery in India was set up by the Central Institute of Fisheries Education, followed by the second one by Central Institute of Freshwater Aquaculture. The later spread of giant prawn hatcheries along certain parts of the Indian coastline and in a couple of interior States such as Tripura and Chhattisgarh derived inspiration from these achievements. Closed recirculatory hatchery system for giant prawn was successfully experimented upon at the Hassarghatta Research Station of Karnataka Agricultural University and recently the larval cycle has been reportedly closed at the Ludhiana Centre of CIFA.

*Fishing Chimes* recorded the abrupt winding up of the fishing vessel financing system successively by SDFC, SCICI and ICICI and the emergence of the ongoing vacuum in the system of extending financial assistance for acquisition of larger fishing vessels. *Fishing Chimes* also witnessed the glorious period of Indian shipyards playing a significantly encouraging role in the construction and supply of fishing vessels and also the collapse of the system for which the responsibility is squarely on the Government, whatever be the lapses of others. *Fishing Chimes* also recorded the steep decline of the population of Indian-owned larger fishing vessels from around 190 nos. to around 40 nos. as at present. The development is also the result of the short-sighted and what can be termed as an improperly conceived alternate policies of the Government for augmenting the fleet strength. Fishing Chimes dealt with these aspects in its editorials.

The role of Banking structure in extending financial assistance to fishing industry has turned out to be on a low key during the 25 year period. The assistance extended by NCDC and other co-operative financing bodies also appeared to be on a low key.

An unique role was played by MPEDA by introducing the technology of shrimp seed production by setting up two shrimp hatcheries one in AP and another in Orissa in 1980s, as recorded in *Fishing Chimes*. These two hatcheries, known as Taasparc and Ossparc, having achieved the purpose for which they were set up, are reported to be now concentrating on demonstrating diversified systems of seed production for the benefit of the farmers who now face problems of raising disease-free shrimp crops. These two hatcheries made history by serving as models for the setting up of a large number of shrimp hatcheries all along the coastline. Another major contribution of MPEDA has been the promotion of shrimp farming. This was an additional initiative on a different footing to the promotional work by brackishwater fish farmers Agencies promoted by the Union Ministry of Agriculture and the Coastal State Fisheries Departments. MPEDA, in association with NACA brought in the idea of aqua clubs for the production of disease-free shrimp crops, as a boon to farmers. Another historic contribution of MPEDA, among others, was to set up PCR Labs for testing shrimp postlarvae, mainly to pre-test and facilitate supply of disease-free shrimp seed by the hatcheries. Mobile PCR Labs were also introduced by MPEDA. CMFRI took up a project for developing farm-raised broodstock of shrimp and progressed upto raising of the third or fourth generation of these shrimps but according to the chronicled information in *Fishing Chimes*, it wound up the work. Central Institute of Brackishwater Aquaculture now handles another project in association with CIFE and AKVAFORSK, Norway, as per notes in *Fishing Chimes*.

The rise and fall of fishing vessels chartering system between middle 1980s and 1994 has been documented in *Fishing Chimes*. The development of the retrograde system of allowing operation of flag of convenience vessels in the Indian EEZ and problems that sprouted therefrom have been covered in the journal.

The multinational company Union Carbide entered into the integrated deep sea fishing sector in a big way but bowed out later. Same was the case with several others like EID Parry, Britannia, and Greaves Cotton (New India Fisheries). Waterbase, a large house, preferred to enter into integrated farming system, withstood several constraints and is now seen to be stabilising itself. It has developed a collaboration with Inve, an international giant in aqua field. Manufacture and supply of aquafeeds of international quality and export of soft crabs are some of the strong points of Waterbase. C.P, Alltech, Novazymes and a few other such companies of international repute have also gained a foothold is aqua field supplies and services. Avanti Feeds developed collaboration with Thai Union in quality aqua feed manufacture and supplies to farmers and has emerged as a prime producer and supplier of aquafeed. Quite a few large companies like Hindustan Lever are well settled in processing and export of marine products.

In one of its editorials, *Fishing Chimes* told its readers about the charters scenario and its later replacement by flag of convenience vessel system. This system, practised by Taiwanese vessel owners is not new to them. All of them are aware of the strategy to register their vessels in another country under the ownership of an identified owner in that country, following the system of enabling the adopted owners to conform to the system of meeting all expenses and sharing the net incomes, earned voyage-wise, as per agreement, while retaining physical control over the vessels.

In respect of the Indian charter system that lasted upto 1994, stories in circulation suggested that, by 1994, Taiwanese trawlers totally depleted accessible bottom stocks of Indian EEZ beyond Coastal Zone. This led them to adopt an alternative system so that they can continue to be in the grounds to undertake pelagic fishing and this they did adroitly, as believed. They dealt with the problem in such a way, according to the stories they left behind, that Government of India would have to ask them to leave Indian EEZ. This was done by the Indian Government, stated to be based on a recommendation of a Committee called Murari Committee, to close down the charter scheme. But, according to the concerned circles, Taiwanese owners soon diverted attention towards the alternative of exploiting pelagic resources (Tuna) of the Indian BEZ. They however took a measured time to drive through this alternative idea for the consideration of the Indian authorities. This idea was to have a situation of transfer of the ownership of their vessels to Indian companies on hire purchase basis, which, in other words, meant flag of convenience system so called by others, as Taiwanese would not surrender their actual ownership but only on record to satisfy the Foreign Government concerned.

The exemplary inputs by the erstwhile Bay of Bengal Programme (now known as BOBP-IGO) for the uplift of professional skills and socio-economic conditions of small scale fishermen were documented in Fishing Chimes. Due to the efforts of BOBP, fibreglass beach landing craft with outboard motors became popular among small scale fishermen. BOBP also introduced high opening trawl nets and several other innovations too.

CIFA succeeded in producing freshwater pearls. CMFRI has made considerable headway in introducing technology of production of marine pearls. 'The transfer of these technologies for commercial application would no doubt reach a satisfactory level soon. These institutes have also to their credit the development of several other innovative technologies. CIFA, CIFRI, CIFT and NBFGR have rendered yoemen service, particularly to the fisheries sector of north-eastern States, by introducing several technologies for the benefit of the entrepreneurs and farmers of the zone. These include farming of prime fishes, particularly in paddy fields, and breeding and production of ornamental fishes.

CIFT developed several beneficial technologies and transferred them to the field. Some of these are in respect of value addition to crustacean, molluscan and finfish products, and production of sutures from fish guts. Innovative technologies of producing chitin and chiosan were developed, by the institute and these were transferred to the private sector in India as well as to enterprises abroad.

Several beneficial products, other than those for disease control entered the scene. These include Ovatide of Hemmo Pharma, and Ovaprim of Glaxo for induced breeding of fishes and Telenet for inducing growth in fishes through generation of sound waves electronically, introduced by Telenet company, Kolkata. Oxyflow is another well tested and established product introduced by Guybro Chemical, Mumbai for the benefit of farmers. It is designed to ensure needed oxygen supply for fish in ponds.

The fisheries educational Institutes, during the past 25 years, have rendered signal service by generating trained manpower for the sector. On the training side, CIFNET has been the main provider of manpower for the operation of larger fishing vessels. Unfortunately, with the stagnation in the growth of large-sized commercial fishing fleet, many of the hands trained at CIFNET left the country to take up jobs abroad.

There have been several efforts in the past 25 years for the development of reservoir fisheries. *Fishing Chimes* saw the emergence of a few outstanding Indian experts in Indian reservoir fishery development. One is Dr. V. V. Sugunan and the others are Mr. A. Sreenivasan, Dr. G.P. Dubey, Dr. Kuldip Kumar and in recent years, Dr. V.B. Sakhare. There are many more and the name of these are on record in *Fishing Chimes*. The reservoir cluster concept supported by the needed infrastructural facilities for each of the cluster, has been highlighted in the journal. In the past 25 years, wetlands fisheries development has gained importance, with quite a few of them taken up for development with world Bank's or other outside help. One carp hatchery project with Hungarian help taken up during the period was recorded as one of partial success.

In its sojourn towards silver jubilee year, *Fishing Chimes* had the unique opportunity of recording about the top position that West Bengal has eked out for itself in respect of annual fish production. This State has the unique distinction of having Mr. Kironmoy Nanda as the Minister for Fisheries for over 23 years and continuing in the position. He has established a record of rendering unique service to various sectors of fishing industry of the State.

Andhra Pradesh gained the distinction of being the highest producer of farmed shrimp, among all the States of India.

The 25 years period has witnessed several initiatives at fisheries development in *beels, mauns* and other wetlands resources. Cage and pen culture in these resources as well as in reservoirs has gained momentum.

The major achievement that deserves attention is the rise in per ha freshwater fish production in the country to over two tonnes. Like wise, shrimp production in several brackishwater farms has crossed a production level of one t/per ha. Fresh strides have been made in mud crab production and in live crab exports.

The Indo-Danish project in Karnataka developed a design for the upgradation of around 15 m long mechanised boats for fishing in the EEZ beyond coastal waters and the design became very popular with the entrepreneurs. Hundreds of boats conforming to this design have been introduced, particularly on the west coast and are operating triumphantly.

The realistic and field-oriented approaches at augmenting export-oriented production adopted by MPEDA and the Union Department of Animal Husbandry, Dairying and Fisheries have resulted in the successful addition tuna monofilament longlining system on six shrimp trawlers in 23 m OAL range to start with. The credit for the achievement predominantly goes to the owners of the vessels. Many more trawlers are in the line for similar conversion to emulate the successes, utilising the subsidy schemes introduced by the Union Dept of AH, Dairying and Fisheries and also by MPEDA.

High sea bunkering for larger vessels came to be introduced in 1998. Supply of diesel oil at subsidised rates to small fishing crafts, around this period, has been introduced by several Coastal State Governments, with the States concerned and the Centre sharing the costs.

Although there has been an enormous delay in promoting sea cage culture, the clearance of a project on the subject, prepared by CMFRI, by the Union Department of Animal Husbandry, Dairying and Fisheries took place in 2005. It is to be hoped that this important project would start functioning the soonest and will yield expected results to stabilise marine fish production levels.

An outstanding achievement during the period was the success in the induced breeding of Mahseer by Tata Electric Companies. This Company, under the leadership of Dr. S.N. Ogale succeeded in breeding golden mahseer and supplying its seed to several State Fisheries Departments.

The development of cold water fisheries received particular attention in the past several years. Trout farms have come up in J and K with EEC support and in Himachal Pradesh with Norwegian help. Raising of seed of trout and using the same for enriching fisheries of streams and rivers and also for stocking ponds have gained momentum in these States.

Garware Wall Ropes, a pioneering Indian company, has gained the distinction of supplying fishing net webbing to several Western countries. Monofilament longline export to various western countries has been achieved by Covema filaments, Cochin and Safa Marine Industries Ltd., Bhubaneswar in Orissa set up by Capt. Haren Mahapatra. The later is an Indo-Norwegian JV project.

On the negative side, *Fishing Chimes* recorded some discouraging developments. There were opinions critical of policies in respect of development of deep sea fishing, domestic fish marketing, and reservoir fisheries development. So far as promotion of deep sea fishing is concerned, the criticism has focus on the vessel financing system which has been allowed to disintegrate. Larger fishing vessel fleet strength had dwindled down with no efforts at reviving the same on the lines as needed, despite having well equipped Indian shipyards interested to take part in this programme. Marine fish production became stagnant with a trend towards decline and with little initiatives at exploiting the deep sea resources lying untouched by Indian effort. There has been no development of a co-ordinating mechanism between the estimated resources, survey results and vessel introduction efforts to sustainably avail of stocks beyond 200m depth zone. Several marine fishing companies and shrimp farming companies that sprang up in 1980s unceremoniously closed down soon after, drowning a large number of share holders into severe financial plight.

During its Silver Jubilee period, *Fishing Chimes* recorded initiatives and efforts at utilisation of the saline tracts in Rajasthan and other north-western States of the country, now lying waste, for aqua production. Unfortunately there has been no breakthrough, despite the fact that feasibility of shrimp and giant prawn production has been demonstrated by CIFE. In this background, recently, a Unit of CIFE which was engaged in promoting farming activities in the saline waters of Rajasthan is stated to have been ordered to be closed down, instead of strengthening it to tone up its performance.

A very useful diploma course in fisheries that was being conducted by CIFE had been closed down during the silver jubilee period of *Fishing Chimes*, not taking into account the import of the excellent services the diploma holders had rendered and continue to render.

The upgradation of the skills of small fishermen to fish in farther waters has been a long cherished dream. The proposition has been to catch them young and equip them with an enlightened vision so as to enable them to improve their socio-economic conditions. This continues to be a distant dream, despite the fact that the Prime Minister of India, as early as Nov 1985, as recorded in Fishing Chimes, observed that there was need for catching fisherchildren young to orient them to adopt better methods of fishing, while also emphasising that there was need for demarcating fishing rights at sea for them as on land.

There had been a demand since 1980 for the setting up a separate ministry for fisheries at the Centre. This could not make any impact on the successive Governments, except that, just a year back, the term 'and fisheries' was added at the end of the name of the Union Department of 'Animal Husbandry and Dairying'.

In the 25 year of journey of *Fishing Chimes* one misplaced readjustment of Fisheries Administration which was happily corrected later on, although late, took place. This was by way of creating a separate Ministry of Food Processing Industries, into the orbit of which the subject of deep sea fishing was sucked. The Fishery Survey of India wasrtagged on to this ministry while the related activities such as fisheries education, training, marine fisheries research, Integrated Fisheries Project remained under the Ministry of Agriculture. One outcome of this was that the Fisheries Development Commissioner, who is under the Agriculture Ministry, was kept aloof from deep sea fishing work. Happily enough the subject of deep sea fishing came back to Agriculture Ministry in 1998, as recorded in Fishing Chimes. The post of Fisheries Development Commissioner has been downgraded to the rank of Joint Commissioner a few years back, indicative of a revised assessment of the importance of the subject of fisheries in the national set up.

**CMFRI developed in 1988 a new multicrop system of cultivating pearl oysters, mussels, other shellfishes and finfishes, at the bottom of the sea using specially designed cages but there had been the bottleneck of technology extension, according to a surmise.**

**CIFRI succeeded in breeding hilsa and raising its seed but follow up measures to standardise and extend the technology to the field had not gained momentum. The technology of raising seed of major carps round the year was also developed by the Institute and it has had a measures of spread effect.**

Depletion of riverine fisheries had been a glaring feature in all the 25 years. This had been an inexplicable development, despite the fact that collection of carp spawn from rivers had been significantly given up because of the introduction of induced breeding technology.

India became a prominent member of Indian Ocean Tuna Commission. A problem that has been generated in this connection was the operation of Taiwanese vessels as Indian registered vessels under dual registry. As Taiwanese are known to register their vessels under other national flags because of their expediencies, while retaining Taiwanese flag (Dual registration is allowed by Taiwanese Government, it is learnt), the problem of real ownership is understood to be coming in the way of recognised operation of the vessels as those of Indian vessels, so far as IOTC is concerned.

Some of the other major developments that found a place in the successive issues of *Fishing Chimes* are summarised here under.

1. An Indian Station called 'Dakshin Gangotri' has been set up on Antarctica, with the Indian flag. Several expeditions took place. A task force on krill project development has been set by the Union Department of Ocean Development.
2. Marine Fisheries Regulation Acts were enacted by several Coastal States.
3. West Bengal and A.P. emerged as the foremost centres of aquafarming. Bhimavaram zone in A.P emerged as a study centre in respect of major carp and shrimp farming for several States. CIBA's farm, the farm of Rajiv Gandhi Centre for aquaculture have come to be centres of study of sea bass and mud crab farming. The Periyar mud crab hatchery set up by V.G. Eraniappan and the one set up by the Dept. of Zoology, Andhra University have also became places of visit by prospective entrepreneurs.
4. Amalgam group announced some years back a plan for setting up nation-wide domestic fish marketing system. No further information on the initiative was available.
5. Vorion chemicals took the initiative of farming Nile tilapia in a farm fed by treated effluents of its brewery. There was considerable progress and the farmed fishes were sold at specially set up retail stalls in Chennai. Despite the reported success, the farm was closed down.
6. Fish Finder was developed by the R&D centre of Bharat Electronics at Trivandrum.
7. On shore commercial pearl production was standardised by CMFRI.
8. MPEDA set up a SPF shrimp Broodstock and SPF shrimp nauplii project centre in Andamans.
9. Coastal Aquaculture Authority Bill was passed by the parliament and it became an Act. The Coastal Aquaculture Authority has been constituted.

10. The proposal for setting up a National Marine Fisheries Development Board was mooted as early as 1993. It was expected to be set up in 1993 itself. According to an editorial of Jan. 1993 in Fishing Chimes, the position at that time was as follows: "The proposal to set up a National Marine Fishery Development board had been receiving the attention of the Government for quite sometime and it is expected that in 1993 the Board would be set up with wide ranging powers. The likelihood of the Board being entrusted with the function of financing fishing vessels is also visualised. A proposal for the setting up of a National Fishing Harbour Authority has also been under consideration. The present stage is stated to be that the proposal to set up National Fishery Development Board has been cleared by the EFC and has been sent to the Cabinet for clearance.
11. Fish Farmers Day was approved by the Government in 2000 and is now celebrated every year on 10 July.
12. Hiralal Chaudhuri Best Fish Farmer Award was instituted by *Fishing Chimes* Jaysree Charitable Trust. This Award has been given so far to farmers from West Bengal, Andhra Pradesh, Manipur, Tripura and T.N.
13. There had been a rise in the setting up large brackishwater and coastal marine farms by several newly floated companies but they had to face closure because of court orders that only traditional farming can be undertaken in CRZ.
14. Several Indian fishing companies entered into joint ventures but not much is known about their present status.
15. There have been no tangible efforts at finding ways of exporting Indian major carps.
16. Sea weed farming has gained momentum through SHGs in Tamil Nadu.
17. CMFRI standardised technology for clam production.
18. No tangible progress could be made in respect of mariculture,commercial pearl production, sea-cucumber farming, *Azolla* culture, snail culture etc.
19. Tamil Nadu Fisheries Corporation was the last public sector Corporation to operate larger fishing vessels. It continues to be active in other respects also, as is the case with other corporations in Karnataka, West Bengal and a few others.
20. A Plan for utilising treated sewage for fish farming was developed by CIFA.
21. Murrel culture Project, taken up by the College of Fisheries, Tuticorin, was stated to be successful but there were no reported follow up measures.
22. Shrimp exports had to face antidumping embargo from USA, antibiotic problem, mainly from EU, and muddy-mouldy flavour problem mainly from Japan.
23. Stipulation of use of TEDs by USA retarded shrimp fishing efforts at sea.
24. White shrimps were bred in 1984 at Vallarpodam hatchery of MPEDA in Kerala. Instead of asking for the reactivation of the success for promoting its exports, a slant towards Pacific Shrimp has been shown by entrepreneurship.
25. The era of deploying Indian trawlers for fishing in the EEZs of other countries such as Myanmar, Indonesia etc started in 1990s, with considerable success.
26. MPEDA/Union Dept of Animal Husbandry, Dairying and Fisheries introduced schemes for extending 50 per cent subsidy for equipping trawlers for diversified fishing methods such as longlining.
27. MPEDA extended subsidies for setting up shrimp hatcheries, PCR Labs, installation of certain processing equipments for value addition etc.
28. The introduction of the first few batches of trawlers of 23-27 m OAL turned out to be an end by itself and it did not serve as a stepping stone for further introduction of larger vessels to cover farther areas beyond shrimping zone and for harvesting stocks of deeper sea zone, both pelagic and demersal.
29. Mid-sea bunkering scheme for fishing vessels was launched in 1997.

Note: The above mentioned developments (as recorded in *Fishing Chimes*) could not be listed in Chronological order. It is possible that a few of them may have been also not listed inadvertently).

Fishing Chimes, Editorial 352: May 2006: Vol. 26, No. 2

# Nation-wide Fish Marketing Network

There have been suggestions for the past over 30 years to develop a Domestic Fish Marketing Network in the country but these have not yet moved towards the desired result. The main reason for the failure, as assessed, has to be attributed to the weak initiatives in this regard.

The suggestions are that the network of the domestic marketing system has to progressively cover all the States of the country, linking centres of production to consuming points. For establishing the network, State Governments have to play a major role. For them to do this, guidelines for setting up a State-wide network and linking it to national system would have to be circulated by the Centre among the State Governments for follow-up, in relation to an overall plan of action envisaging the development of a national network of the system, to be monitored by the Centre eventually. In other words, this national programme would have to entail the preparation of a master plan for its implementation in a co-ordinated manner. Considering the massive work involved, a separate organisatonal set up both at Central and State levels with a functional integration mechanism at the Centre, to link the efforts at State levels with the national grid, becomes imperative. The needed funds have to be earmarked and a well briefed set of officials under a dedicated leadership has to be placed in position for the project to gain the needed shape on the ground and acquire strength to make the system a vibrant force to keep the people supplied with fish on an enduring basis in a hygienic condition. An organised supply of whole some fish to the consumers, which is the long desired and cherished final link of the ultimate objective of fisheries development, has to be achieved and the network is needed in this context.

There is an overwhelming consensus among those concerned at various levels dealing with fisheries that a domestic fish marketing network on a national scale is of utmost importance and is a necessity considered from any angle. Such a network will streamline and strengthen the marketing activity which presently functions in the country in patches and in a haphazard manner, with resultant in effective coverage and economic disparities.

Sometime back the Amalgam Group in Cochin, under the leadership of Mr. Abraham Tharakan, announced initiatives in the direction of setting up of a domestic fish marketing system, but the present status of the efforts is not known.

A few decades back, there had been a route formed in Karnataka for the supply of fish to the people, operated under the co-operative sector, from coastal production centres in South Canara District to consumers in Bangalore city and in Coorg district. There are also certain known routes of fish supplies that developed from heavy production centres both along the coast and from inland areas, mainly to metropolitan cities like Mumbai, Kolkata and Chennai. Another development, of late, is the flow of farmed fish from States like Andhra Pradesh to Orissa, West Bengal, Bihar, Assam and several other north-eastern States. A study of the dimensions of this marketing activity would provide a wealth of information to serve as a guiding basis for the development of a nation-wide fish marketing system. In fact, it should be possible to absorb the present coverage by ongoing routes into the overall planning, to be made by organising additional facilities as needed. Other routes on the same lines but with improvements to the ongoing ones could be promoted, incorporating solutions to the problems that the fish transporting vehicles now face while crossing State borders, for repacking fish with fresh ice at identified points, etc. Provision of infrastructural facilities at determined points for production and supply of ice, for repacking of fish along the long distance routes of fish transportation can well be achieved with a measure of effort.

The way to proceed further on this vital subject seems to consist of first appointing a National Level Committee to prepare and present to the Union Department of Animal Husbandry, Dairying and Fisheries a detailed plan of action for establishing a nation-wide domestic marketing system. This Committee, supported by a secretariat, could be entrusted with the task of first undertaking a detailed study of the existing domestic fish marketing system in the various States and in the Inter-State arena. After a critical appraisal of the structure and working of this system, the Committee could recommend intra State and inter-State marketing systems for adoption, covering infrastructural, management and other needs. Matters related to Inter-State trade, the facilities that are required to be provided to ensure quick passage of vehicles through State borders, and for refilling of fish crates with ice and repacking etc would have to be covered. State-wise requirements have to be formulated with estimates of investment on infrastructure, manpower and other financial needs. The Committee may also have to be asked to recommend locations for setting up of new ice plants, cold storages and processing plants in the various States, besides places for setting up retail stalls and on measures needed to upgrade and integrate the present domestic marketing activities with the system recommended. The Committee could also consider and suggest measures for creating an export market particularly for major carps, keeping in view the need for securing higher returns for this predominently higher component of the pond harvests, for the benefit of the farmers.

It will also be necessary for the suggested National level Committee to prevail upon the State Governments to set up State-level Committees to work in tandem with the Central Committee to recommend the blue print of network of marketing system in the State concerned, based on guidelines provided by the Central Committee. Taking these recommendations into account the National Committee could finalise its recommendations.

It is desirable that Chairmen of the National and State Level Committees are chosen from a category of experts who have knowledge and experience not only in the marketing system of fish but also of other consumables such as eggs, fruits etc.

In conclusion, it has to be mentioned that, in ultimate terms, the efforts directed at increasing fish production will not be rewarding, if the final link of the chain, which is marketing, is not well taken care of. The Union Department of Animal Husbandry, Dairying and Fisheries has to therefore pay focal attention towards creation of the national fish marketing system. This has to be done so as to ensure that both the producers and consumers would not continue to be subjected to the ongoing but avoidable hardships.

Fishing Chimes, Editorial 353: June 2006: Vol. 26, No. 3

# Deficient Political Support: Bane of Fisheries Sector

Politicians keep watching the broad needs of the people of different categories in their areas of operation to assess and take care of them as needed. With a close knowledge of needs of their constituents, they raise issues related to developmental aspects for the attention of the Governments concerned and prevail upon them to institute needed measures and implement them. All these they do to be of service to the people. This common line of promotional effort in a democracy is fortified by the initiatives as well as other measures covered by the developmental mechanisms of the Governments concerned. In other words, the developmental process, by and large, is the result of what the politicians do in their quest to take care of the needs of the people by way of finding answers to the extent possible in respect of the various sectors of activities under their purview.

Let us look at tne sectors of agriculture, animal husbandry, forests, commerce and industry, for example. In respect of these, the politician knows the needs of the related stakeholders, and works with interest towards fulfilling them. It is a different story, however, in the case of fisheries sector, which functions in isolation, almost totally so, in respect of the bulk of its activities. Only a few conflict-oriented issues of the sector come to the notice of politicians, most of whom tend to judge the situations against the interests of fisheries stakeholders.

Unlike the agriculture sector, the basic issues related to fisheries arena continue to remain unresolved. In respect of agriculture, its basic issues are well taken care of in contrast to the fisheries sector, for the reason the politician has an interface with the leadership in agriculture sector. The leadership in fisheries arena, besides being weak all over, has not yet been able to develop the required level of interface with the political mainstream to inject into it the needed focus for the kind of development the sector needs.

It is perplexing that politicians are generally apathetic towards problems of fisheries sector. This cannot be for the reason that fishermen, who hail from a weaker section of the society do not have monetary and other supportive strength to attract their tangible attention.

The fisheries sector is starved of financial support, particularly for acquisition of vessels for fishing, or at least to stem the decline in the fleet strength noticed of late. Further, the farm gate prices of fish have come down. Farmers have no avenues of securing viable prices. No politician is known to take up these issues either with the Government or with the bankers. There are also no political initiatives at bringing the vast saline tracts of north-west under fish production.

Fortunately, in this background and as a ray of hope, there is an announcement that the Union Department of Animal Husbandry, Dairying and Fisheries has decided to set up the National Fisheries Development Board to deal with the problems of fisheries development. This Board, which will be hopefully set up soon, may be able to find answers to the current problems of the industry. This Board can, however, function well only when there is political support.

It can be said that politicians prefer to see the bureaucrats determine the fate of the fisheries sector of the country. This trend can be changed only by the professional leadership in the fisheries sector. It only can eradicate the trend of complacency and reverse the present mood of the industry, helplessly reconciling itself to its fate. The sector should work against those who provide access to the fisheries resources of the Indian EEZ to foreign interests under legally tenable garbs. One such garb is, allowing the operation of foreign vessels in the flag of convenience mode, instead of introducing bold measures for developing India's own real fleet, with the needed financing and training support and enhanced facilities for post-harvest operations. There are no indications of any such moves as at present. Let the fishing industry be not an onlooker of the situation which can lead to its own destruction.

Until such time the fisheries are recognised by politicians as an important sector with immense potential to produce and provide proteinacious food for the people, and to earn foreign exchange for the nation, standing of the sector in their eyes will continue to be nominal, perfunctory and without depth. Unless politicians are educated on the subject through Seminars, Conferences and field visits to fisheries centres that can be eye openers, their outlook would not develop to the stage of representing the sector in the perspective, it rightfully needs.

For example, politicians are not, in general, conscious of the fact that the Indian exclusive economic zone of the marine realm is 2.02 million sq. miles, an area about equivalent to the Indian land mass, and capable of yielding annually 3.9 million t of fish and other marine food products, more valuable than the land produce of the country from nutritional angle too. The need is therefore to educate the politicians on this aspect and enable them to know that water is as productive and as valuable as land.

Unlike several neighbouring countries like Bangladesh, Myanmar, Sri Lanka, Thailand, Vietnam, Indonesia, Philippines and others like China, Taiwan and South Korea, the status of fisheries sector of India is glaringly different in various respects. Fisheries are not considered by many in the country as a respectable profession, at least as respectable as agriculture. Government of India probably thinks it is not a subject that deserves to be given the status of even a department in its set up, although countries much smaller and with less of fisheries resources compared to India have separate fisheries ministers, for obvious reasons. One such reason is that fishes are a great source of protein and also oil that lends protection to human heart. These can be available in abundance once there is a commensurating effort which can be generated only when there is adequate political intervention. The reasons for not having a separate Fisheries Ministry/Department at the Union level in India are not known, despite the voicing of a over two decade old well justified demand in this direction. One reason for this situation seems to be that the union Government thinks that there is inadequate workload to elevate the subject even to the status of a department. This thinking grossly overlooks the fact that only when there is a separate Department/Ministry for Fisheries, work load consistent with its potential will develop to lead to higher fish production, for promoting internal fish and other aqua products consumption and for augmenting aqua products exports. Only when there is a separate Ministry/ Department the needed attention can be directed at achieving several reforms.

Fisheries in other countries are as important as agriculture and animal husbandry. However, for historic reasons, it is not so in India. It has to be however conceded that, despite the apathy of the politician and the Governments in the past fifty years, fisheries sector has gained some attention, but not to the extent it deserves. The reason for the aforesaid neglect is that fisheries are somewhat away from the nation's mainstream of food production activities, such as agriculture, animal husbandry and dairying. In most of the countries vegetarians are uncommon. This is not so India. Moreover, traditionally, many in India consider that vegetarianism is an upgraded eating habit. The impact of the sum of the situation is that the bulk of the politicians have very little exposure to the fisheries sector and its importance. In their interface with the people, in general, the subject of fisheries rarely comes up. So much so, in the absence of representations and demands from the people, the politicians have little awareness of the subject, thus only incidentally raising the needs of the sector in parliament/ State legislatures.

The need is therefore to educate the politicians first. And this crucial role has to be played by the fishermen's associations and other fisheries professional organisations. Very few of the politicians know that, as already stated, the Indian EEZ has almost the same extent as the Indian land mass and there is a vast coastal fisher population of which over 53 lakhs of them are full time fishermen. There are over 2.2 million ha of tanks and ponds for augmenting farmed fish production. It is unfortunate that there is a very marginal interface between those in the fisheries sector and the politicians. Had there been a sound communication channel between the leadership in the fisheries sector (although weak) and the other politicians, the problems would have been different. Issues like domestic fish marketing system, securing a viable price for the fishes harvested by our farmers, bringing the vast inland saline stretch of the North-West India under fish production, development of sea cage farming, coverage of farther areas of the Indian EEZ by sustainable fishing, stemming of ongoing decline in the strength of larger fishing fleet of India which has already fallen from a three digit number to a two digit figure, upgradation of the professional skills of coastal fishermen to enable them to cover waters beyond their traditional zone to improve their catches sustainably and augment their incomes and living standards etc., would not have been dragging on for over 40 years without any viable solution.

The present need is for the various fishery related Associations to wake up, educate and motivate the politicians to pay attention to solve the aforesaid and other problems of the fisheries sector of India. Let it be realised that without political support and intervention, development of fisheries of the country will continue to be at the present stagnant State, and leaving all of us to reminisce over the few past beneficial developments that took place such as mechanisation of fishing vessels, introduction of synthetic twine, setting up infrastructure for stepping of marine products exports (which are also now stagnating), introduction of induced breeding technology, circular eco-hatcheries, composite farming, polyfarming and integrated farming technologies.

Fishing Chimes, Editorial 354: July 2006: Vol. 26, No. 4

# Need for Tuna Fishery Development Agency

There is an urgent need for the recently approved National Fisheries Development Board to pave the way for the setting up of Tuna Fishery Development Agency with nation-wide jurisdiction, as soon it starts functioning.

Because of its export potential, shrimp could impart momentum to marine capture fishery development revolving round it all these years. This momentum resulted in overexploitation of shrimp stocks. The consequential depletion in stocks has led to the search for alternate exportable species. Squid and cuttlefish claimed attention but stocks were not adequate to fill the general demand gap. Tuna too was in the focus. So far as tuna in concerned, for the past several decades, there have been rhetorical discussions but not much could be done because tuna demanded a totally different set up and facilities for harvesting them. The obvious avenue to be explored has to be therefore to develop an organisational mechanism to achieve an integrated development of the vast tuna resources so as to achieve the ultimate aim of exporting them in sashimi form, by and large. Overlooking this, the nation plumped for an easier option that holds the potential of harming national interest. This option was to place the activity of harvesting and exporting tuna of the Indian waters under a system of engaging foreign vessels and recording their ownership as of Indian enterprises, but with the physical ownership remaining undisturbed with the foreign owners. This system is known in international parlance as one of flag of convenience operations. One benefit enjoyed because of this system has been that this yielded demonstrative results, although entailing a substantial measure of economic disadvantage to the Indian side and also to Indian pride. There has been probably no way of avoiding this flag of convenience course of action, for, unless tuna fishing which is new to India is demonstrated, Indian initiatives do not take shape. The Indian owned foreign tuna vessels, all longliners, did give shape to Indian initiatives and clearly demonstrated the tuna stock availability at commercially viable levels and also indicated the patterns of post-harvest operations.

This irrefutable and encouraging development should have led to the introduction of truly Indian owned tuna longliners but this was not to be, mainly because an organisational mechanism is lacking and also because of investment constraints. Now that the National Fisheries Development Board, as cleared by the Cabinet, will soon be set up, one of its priority tasks would have to be to take urgent steps for setting up a National Tuna Fishery Development Agency, keeping in view the initiatives exercised and the random and checkered measures taken so far to develop tuna fishing in Indian EEZ.

In this directionless situation, one hopeful development was the taking shape of the idea of equipping Indian shrimp trawlers of 23m OAL range for tuna longlining. There was the availability of a G.A. drawing for modified deck arrangement to facilitate monofil longlining, developed by Conrad Birkhoff, a German naval architect, in 1980s. It, however, remained dormant for nearly two decades, until two of the owners of shrimp trawlers triumphantly installed the longlining equipment on the decks of the vessels, inspired by this design and taking advantage of the subsidy given by the Union Department of Animal Husbandry, Dairying and Fisheries. They conducted commercial operations with the modified vessels. They, however, experienced problems in integrating post-harvest tuna export marketing work with the fishing phase. To say briefly, the situation at this stage was: Resource was there, technology of equipping trawlers for longlining too was there. There was demonstration that Indian crew (with traditional lining skills) could operate and harvest tuna through monofil longlining system. In any case, the nation has the infrastructure for imparting training in longlining at CIFNET. The subsidy-aided financing route for installation of longlining equipment on the deck was put in practice. What remained weak was the export marketing linkage, which could not be accomplished.

A leader emerged at this stalemate stage. He is Capt.J.M.N.G.Kagoo, Chief of SAF Enterprises and Gees Marine Products. He took the initiative of forming a group of six trawlers, including two of his own companies mentioned above, for the installation of longlining equipment on their decks. The other vessels of the Group were two of Samro Food Processors (Pvt) Ltd. (headed by Mr. M. Sivaram) and one each of A.R.M.Fisheries (Pvt) Ltd.(Capt.A.B.Rqj is the Managing Director), and Hemo Seafoods (Pvt) Ltd. whose managing director is Capt. Melkiyas. This group of four, with six vessels, led by Capt. Kagoo modified the deck arrangements in a manner suitable for tuna longlining. They integrated the production effort with export marketing imperatives, thereby ensuring the integration of the links of their sterling work consisting of the fishing phase, sashimi quality maintenance after removal from the hold, pre-export requirements such as quality grading by a visiting representative of the importing company etc., transportation to the airport, loading the consigment into the aircraft, etc. are fulfilled. In 2005-06, the group made 70 shipments consisting of 250 t of sashimi grade tuna (exported to Japan), and 45 t of headless, tailless and gutted tuna to USA. 75 t of loins were sold to an exporter in

Tuticorin. They sold locally another 75 t, these being not of exportable quality. All A+ and A grade tuna were exported at FOB rates of 3.5 to 4 US $/kg and B+ and B grade at 3.0 to 3.5 US$/kg to Japan. Exports to US were of B+ grade, at US $ 4.5/kg.

The four pioneers who achieved the breakthrough in sashimi tuna export from Visakhapatnam are now referred to as those belonging to Kagoo Group in knowledgeable marine fisheries circles. This group acknowledges that their success is attributable in a large measure to the support, advice, and encouragement from the MPEDA on the one hand and from the Fisheries Division of the Union Department of Animal Husbandry, Dairying and Fisheries on the other.

The Kagoo group, having now gained operational experience in several tuna longlining voyages they conducted and in marketing of tuna in sashimi and loin forms are now in a position to authentically articulate the kind of gaps they have come across in accomplishing integrated tuna fishing operations. In this context, there is a responsibility on the Association of Indian Fishery Industries to bring the gaps to the notice of the authorities for their appraisal and in that light to initiate measures to close them.

As could be ascertained, one of the major problems the enterprises face is in respect of berthing facilities. Tuna being a roving fish, its catching points in the EEZ are unpredictable. Therefore the captains have to berth their vessels for unloading catches at a port near which around 7 days of fishing are over and also for taking supplies and for various essential purposes. As problems are faced by the members of the group in this regard, the Union Dept. of A.H., Dairying and Fisheries would have to take up the matter with the Ministry of Shipping and Transport and ensure that tuna vessels are allowed to be berthed at any of the harbours along the Indian coastline. This they can do only when they are told about the problem. The vessels cannot be out at sea engaged in fishing for more than seven to eight days. Otherwise the tuna will not be in sashimi condition. There is no infrastructure available by way of sheds for handling chilled tuna at any of the harbours for export in that condition. They have to be well packed before export and this has to be done after packing in the shortest possible time. The group is ready to build its own sheds for the purpose but the harbour authorities, whom the operators are stated to have contacted in this regard, are apathetic. It is a miracle that the group managed to export the few consignments mentioned above, while withstanding the ordeals admirably well. The AIFI can surely take up the matter with the concerned authorities, for this is one of its functions. What is learnt is that Mumbai and Cochin port authorities in particular are very adamant to concede the facility. They refused to allow the tuna vessels to berth in their respective harbours, it is mentioned.

Another problem the tuna vessels operators face is related to refrigerated transport. Chilled tuna has to be exported by air from the international airport nearest to the harbour where tuna vessels concerned, are berthed. Refrigerated vans are needed to carry the chilled tuna cargo from the harbour. Unfortunately, Kagoo group does not own any refrigerated van. They have to hire one. The group has been making efforts in vain in this direction but with little success. AIFI has to do something tangible in this respect by way of prevailing upon MPEDA to find a way to overcome this problem. The members of the group are not in a position to acquire a refrigerated van, but they are in readiness to take one on hire. The logistics of providing this facilities to the group can certainly be worked out by the authorities concerned.

The removal of the above mentioned bottlenecks would provide a fillip to integrated tuna longlining operations, linked to exports, in the Indian EEZ.

Tuna of Indian EEZ, particularly the shoals that frequent the Bay of Bengal, offer an enviably attractive opportunity for sustainably stepping up their catches and exporting them. The task involved in this regard is a mammoth one. Unfortunately, the dimensions of the task seem to have been underestimated. It involves efforts at harvesting of over 2 lakh t of tuna. At present, there are eight nos. of modified shrimp trawlers of 23 m OAL range and of true Indian ownership for harvesting tuna. Outweighing this number, there are a good number of flag of convenience vessels of Indian registry operating in the Indian EEZ. However, how many of them are there will be known precisely only when the Government adopts a transparent approach, which is not the case at present.

Whatever be the reason, India may have to be judged as having been on the wrong path, so far as the strategy for the utilisation of tuna resources of the Indian EEZ is concerned. After having allowed flag of convenience foreign tuna vessels to operate in the Indian EEZ, wisdom seems to have dawned recently to give credence to the alternative that the existing shrimp trawlers which are in distress because of depletion of shrimp stocks, could also be equipped with monofil longlining system for harvesting tuna. Subsidy has been announced by both Union Dept. of Animal Husbandry, Dairying sind Fisheries, and also the Ministry of Commerce (MPEDA) to help the owners to take to this route.

As already mentioned, six shrimp trawlers, modified for tuna longlining, operated in 2005-06 and the operations were successful. This is the first breakthrough of a historic nature. It has the potential to pave the way of conversion of around 40 such trawlers for tuna longlining and to form the basis for developing a truly Indian owned tuna fleet. This development in fact deserves a. celebration for encouraging the pioneering group and also for motivating others to avail of the subsidy assistance being given to strengthen the trend. Unfortunately the follow up action is however on a very low key. This attitude is, in a way, symptomatic of the apathy towards Indianisation of the tuna fishing activity in the Indian EEZ. The sooner an initiative for transforming the complexion of the Indian marine fishery industry into a prosperous phase is taken, the better it will be.

Having seen the prosperity that ought to have rightly accrued to India, but has gravitated towards Taiwan (the flag of convenience vessels are stated to be of Taiwanese origin), an American interest has come forward to enter the sector. This interest has made some progress towards forming a venture with the A.P. Government, which has a dismal past record of operating deep sea fishing vessels. The A.P.Fisheries Corporation, which managed the vessels, having run into heavy losses, was wound up.

If one looks back at the past of India tuna fishing endeavours, one aspect will be clear. The efforts are disjointed, lacking in direction and without any organisational, integrational or projectisation focus. This approach cannot be the way of handling an opportunity of utilising an estimated tuna resource of 213,000 t of the Indian EEZ. If the Government had developed a project for the utilisation of atleast 100,000 t of this resource through true Indian effort, its implementation would have led to an inflow of foreign exchange of at least US $ 300 million or Rs. 13,000 million on full development. With no indication of an approach of this nature, the attention was focused on introducing Indian registered flag of convenience foreign fishing vessels to exploit the tuna resources of the Indian EEZ. The result of this was an ostensible inflow of foreign exchange, but only to be released back substantially to the foreign vessel owners, under various accounts (crew wages, cost of fuel, other running expenses, repairs, maintenance etc.). There are observations that the catches do not get inspected before the vessels leave Indian EEZ.

Keeping in view the successful demonstration of the way of effectively utilising the tuna resource of the Indian EEZ as done by the Kagoo group, for the benefit of small to middle level enterprises, Indian Government has to now think big and refuse to settle for less, in terms of setting up an organisation for utilising the resource for national benefit. There has to be the development of an integrated project to be implemented nation-wide by a Tuna Fishery Development Agency to be set up and headed by competent professional in the line. It is to be hoped one such professional would be available. The recently created National Fisheries Development Board can take up the job of working out an organisational mechanism for implementing the project through the Agency. A few major fishing harbours, having the facility of international airports nearby, can be chosen to set up the administrative and operational bases. The axiomatic need to cater to the international sashimi tuna market can be fulfilled by the introduction of the system of group fishing, so that, in turns, they can engage themselves in voyage rounds of 7 to 8 days, and take the ice-chilled tuna to the nearest harbour having an international airport closeby or within a reasonable distance, for taking the tuna consigments packed in ice in coffins for air shipment. Kagoo group, following the Indonesian example, adopted the system victoriously. This achievement deserves to be taken as a model by the suggested Agency for prescribing the guidelines of handling catches resulting from group fishing.

The pattern will have to be to develop viable groups of vessels for operation from selected harbours chosen as the bases of operations. The Agency will have to have its Managerial Units at each of these harbours, with needed supporting facilities and staff.

One special feature of the suggested Agency will have to be the possession of its own financing wing (with funds provided by the determined source), to be managed by a set of persons with financing background and earlier exposed to commercial fishing aspects. This wing should have the support of a financial guarantee, to be given by the Government for granting loans for the acquisition of vessels etc. as per norms prescribed. The Agency should adopt a well designed mechanism for keeping control over the operations of the financed vessels. The scope and means of hypothecation of the loaned vessels to it has to be its strength. The Agency should have an attached Fishing Vessels Rehabilitation/Health Restoration Unit. Its function would have to be to take over the vessels, the owners of which have not paid the loan instalments and to run them as the financier's responsibility, entrusting the job to a spare set of crew to be maintained, while also taking needed legal action on the loanees as per loan agreement terms. At least one Australian foreign fishing vessel financing company follows the system, which is stated to be effective. Further, the Agency's financing unit should be totally vigilant over the movements and operations of loanee vessels, by being continuously in touch with them, particularly in respect of production and the export activities, which would have to be shared by the members of the Group, in accordance with a condition to be included in the loan agreement.

To say briefly, an organisational pattern and mechanism on the above lines, giving due consideration to the points made above and any others too would have to be developed, with group fishing as the core of the developmental work.

To start with, each of the existing 23-27 m OAL trawlers can be equipped with longline with 1,500 branch lines, each with a hook, with provision of subsidy. This line of action deserves consideration. There are around 40 such vessels and by equipping them all for monofil longlining, there will be a production of 160 t of tuna per vessel per annum with gross sale proceeds, estimated at an average of US $ 3 per kg. When 40 such vessels are inducted, the foreign exchange earnings will be of the order of 19.2 million US $.

It is estimated that 20.5m OAL Vessels, each equipped for longlining with 120km of longline (1200hooks)can land, on an average 120t of tuna each annually. A fleet of such vessels of a determined strength can be developed in the next phase under another integrated project to be introduced.

For utilising 100,000 t of the resource, around 800 vessels of 20.5 m OAL may be required. This introduction could be planned in phases through appropriately conceived integrated projects, based on an evaluation of past results and linked to trained manpower requirements too.

In Brazil, small mechanised boats of 36 ft range are reported to have been equipped with 10-15 km of longline each. The vessels are stated to be conducting tuna longlining operations with results that have toned up the economics of operation of the vessels and upgraded the socio-economics of the owners. The suggested Agency can take up a scheme for the installation of 10-15 km of longlining equipment on board after developing a suitable design. Although small mechanised boats come under the State sector, a way for the Agency to take up this work can be identified.

According to a report in a recent issue of Bay of Bengal News, Maldives have introduced a tuna longliner in the small scale sector, based on a new generation design. The suggested Agency can also study the possibilities of introducing this design with the support of the National Fishery Development Board.

A few thoughts on the utilisation of the tuna resources of the Indian EEZ are chronicled above. It is for the readers to draw conclusions from their views and give expression to them in word and deed.

Fishing Chimes, Editorial 355: August 2006: Vol. 26, No. 5

# Some of the Tasks Ahead for National Fisheries Development Board

The approval accorded by the Indian Cabinet for the setting up of National Fisheries Development Board constitutes an unique and a much needed step to take the Indian fisheries sector further forward. The. Board, when constituted, will facilitate filling up of the major gaps dogging the Indian fisheries sector. Let it be hoped that, what the Union Department of Animal Husbandly, Dairying and Fisheries could not achieve directly so far, would be taken care of by the National Fisheries Development Board, which will no doubt be constituted soon.

## Saline Waters of North-West: Fisheries Development

Vast extents of saline lands of north-west, mainly in Rajasthan and in parts of adjoining States of Punjab, Haryana and U.P, known to be unfit for agriculture, are lying fallow. These lands hold underneath them enormous quantities of saline water, unfit for agricultural use. Under the leadership of Dr. S.N. Dwivedi, when he was Director of Central Institute of Fisheries Education, Mumbai there were initiatives at farming a few salinity tolerant species in a saline zone of Haryana. The results have been encouraging but the logistics of providing a spread effect to them continue to be in the process of exploration from various angles such as land utilisation, linkages with domestic and export-oriented production and marketing and the socio-economic impacts of the activity. Taking into account the encouraging aspects of the positive outcome of the experimental work of CIFE, time is now ripe to mount a major integrated project for the development of fisheries in the saline zones of the States mentioned. The opportunities available for the development of fisheries of the saline zones hold enormous potential for upgrading the livelihoods of the weaker sections of the society in the areas in several respects.

Rajasthan alone has over one million ha of land, salt-laden and supported by a saline water resource. Punjab, Haryana and Western UP also have vast extents of saline land and saline water resources.

Out of all the States in the saline belt of the north west of the country, Punjab is in the forefront in regard to the utilisation of this precious resource, with the State Government sanctioning a project for saline water fisheries development in the districts of Faridkot, Ferozepore, Muktsar, Moga, Bhatinda and Mansa, with an initial annual outlay of Rs. 66 lakhs. NFDB can take steps for extending the approach adopted by the Government of Punjab to cover all the north-western States endowed by nature with saline lands and supporting salt water resources. It will be to the credit of NFDB to usher in a new era of economically viable livelihood to the weaker sections of the society through fisheries developement of the area, linked to domestic marketing as well as exports of the produce.

Development of Fisheries of the saline water zone is only one of the several opportunities to knock at the door of the National Fisheries Development Board. The general absence of a domestic fish marketing system coupled with a cold-storage oriented infrastructure for distribution of fish, to be pooled up at centralised locations is felt, but with no solution in sight. The lack of this infrastructure has been having the disastrous effect of a serious drop in farm gate prices, particularly of major carps, the mainstay of inland farmed fish of India. As if this problem is not enough, Indian major carps, excellent fish, have practically no export market. Schemes are needed to fill up the above mentioned gaps, besides others, and these can be ushered in with the needed intervention by the NFDB.

The inland States of Haryana, Punjab, Chattisgarh, Bihar and U.P. have achieved some progress in the farming of giant freshwater prawn. These initiatives have provided adequate evidence to establish that the nation-wide spread of giant freshwater prawn farming, which has enormous potential for augmenting production for internal consumption, and also for exports to augment foreign exchange earnings, is possible. What is needed to bring about this socially and economically potent reform is a nation-wide integrated project that spreads from the coastline to the guts of the nation in its tropical belt. The project would have to cover its various links such as seed production, movement of seed over the length and breadth of the interior of the country, extension of culture technology, harvesting system, pooling up and storage of the produce and their domestic and export marketing. Being an activity of nation-wide dimensions, NFDB will be the most suitable agency to conceive of a project and organise its implementation, from seed production to marketing.

## Open Sea Tuna Stocks Utilisation

In the marine sector, the NFDB has to intervene through well planned projects, for equipping the small scale fishermen to come out of the shackles that prevent them now from moving beyond the coastal waters, to sustainably exploit the offshore resources. One suggestion has been to equip them to exploit the pelagic tuna resource, which is only marginally exploited now from the coastal water and the waters beyond. There can also be a project for the exploitation of open sea tuna by medium and large sized vessels under a well conceived integrated programme. NFDB has to consider these opportunities, among others, in the marine sector, and prevail upon the Government to set up a National Tuna Fishery Development Agency, as suggested by *Fishing Chimes* in its July 2006 issue. An organisational mechanism of this kind alone can facilitate the effective utilisation of the tuna resources of the Indian EEZ for national benefit and for preventing foreign interests from exploitation of the resource for their benefit.

Fishing Chimes, Editorial 356: September 2006: Vol. 26,No. 6

# Marine Fisheries Census, 2005

Sponsored by the Union Department of Animal Husbandry, Dairying and Fisheries, the Central Marine Fisheries Research Institute (CMFRI) conducted the Marine Fisheries Census of India, 2005. It accomplished the challenging task with enviable competence. The Report on the Survey, taken up in 2005, was released on 25 July 2006 in New Delhi by Mr. P. M.A. Hakeem, Secretary, of the aforesaid Union Department.

The earlier Marine Fisheries Survey was conducted in 1980, also by CMFRI. The present survey of 2005 updates the Census data of 1980 admirably well. This updating, as is known, is essential to provide a direction to the planning and development process of the marine fisheries sector, which is a continuing process. The Central Marine Fisheries Research Institute richly deserves appreciation by the fisheries sector, the Union Department of Animal Husbandry, Dairying and Fisheries, and the Planning Commission for providing a comprehensive and authentically updated basic data related to Indian marine fisheries.

The status of development of marine fisheries sector of the country, good or not so good to be seen as assessed, by the survey, goes to the credit of the stakeholders and the Government. Annual marine fish production increased from 15.55 lakh t in 1980-81 to three million t in 2004-05, because of increase in effort. By 2005 the number of mechanised boats increased to a level of 58,911 nos. from around 34,000 nos. earlier. The no. of mechanised boats engaged in trawling (out of the total of 58,911) touched a level of 29,241 by 2005. The no. of motorised boats rose from around 26,000 nos. earlier (1980) to 75,591 nos. now (2005). The present census however indicates that the fisherfolk as such own 35,806 mechanised boats, 52,971 nos. of motorised boats and 96,661 nos. of non-mechanised boats. So far as the total strength of non-mechanised boats are concerned, while it was around 1,80,000 nos. in 1980, the present strength as per 2005 Census is 1,04,370 nos., a decline of 40 per cent.

Thus, in the past 25 years there has been an increase of mechanised boats by around 25,000 nos. (80 per cent) and in the case of motorised boats there has been an increase by nearly 200 per cent, and there has been an understandable fall of 40 per cent in the population of non-mechanised boats. These developments are encouraging.

There is however one aspect that causes concern. This is related to the indications of a decline in the sense of attachment of the fishermen to their traditional profession. At the time of the earlier census the number of marine fishing villages along the coastline was 3600. According to the present census this number has come down to 3206. Further, the fisher population has also dwindled from 5.35 million nos. at the last census to 3.52 million nos. now (2005 Census). The fall in the no of villages is partly attributable to the destruction of some of the fisher villages along Tamil Nadu and Pondicherry coasts during the recent tsunami but the surviving population must have been rehabilitated. In any case, the no. of villages devastated may not be as high as 384. The fall in population indicated in the census could therefore be for other reasons too. It is possible that the evolving economic and social compulsions have led to the decline in the population. One among these can be the declining fish catches and returns and the lack of adequate infrastructure facilities for handling fish for value addition to augment returns. The other reason can be the social pressures. Fisher villages on the coast used to be isolated, but no longer so now because of road communication facilities to nearby towns. With the limited educational facilities in the villages, and influenced by other factors, heads of fisher families must have found job opportunities, other than fishing, in the neighbouring towns. A possible trend towards decline in fishing population must have been dealt with in the census report.

After the 1980 census, towards the end of the next decade around 192 larger trawlers of 23-25 m OAL were added to the fishing fleet. Of these, only around 40 vessels are active now. This detail must have found a place in the census report.Sponsored by the Union Department of Animal Husbandry, Dairying and Fisheries, the Central Marine Fisheries Research Institute (CMFRI) conducted the Marine Fisheries Census of India, 2005. It accomplished the challenging task with enviable competence. The Report on the Survey, taken up in 2005, was released on 25 July 2006 in New Delhi by Mr. P. M A. Hakeem, Secretary, of the aforesaid Union Department.

The earlier Marine Fisheries Survey was conducted in 1980, also by CMFRI. The present survey of 2005 updates the census data of 1980 admirably well. This updating, as is known, is essential to provide a direction to the planning and development process of the marine fisheries sector, which is a continuing process. The Central Marine Fisheries Research Institute richly deserves appreciation by the fisheries sector, the Union Department of Animal Husbandry, Dairying of Fisheries and the Planning Commission for providing a comprehensive and authentically updated basic data related to Indian marine fisheries.

The status of development of marine fisheries sector of the country, good or not so good to be seen as assessed, goes to the credit of the stakeholders and the Government. Annual marine fish production increased from 15.55 lakh t in 1980-81 to three million t in 2004-05 because of increase in effort. By 2005, the number of mechanised boats increased to a level of 58,911 nos. from around 34,000 nos. earlier. The no. of mechanised boats engaged in trawling (out of the total of 58,911) touched a level of 29,241 by 2005. The no. of motorised boats rose from around 26,000 nos. earlier (1980) to 75,591 nos. now (2005). The present census however indicates that the fisherfolk as such own 35,806 mechanised boats, 52,971 nos. of motorised boats and 96,661 non-mechanised boats. So far as the total strength of non-mechanised boats are concerned, while it was around 1,80,000 nos. in 1980, the present strength as per 2005 Census is 1,04,370 nos., a decline of 40 per cent.

Thus, in the past 25 years there has been an increase of mechanised boats by around 25,000 nos. (80 per cent) and in the case of motorised boats there has been an increase by nearly 200 per cent, and there has been an understandable fall of 40 per cent in the population of non-mechanised boats. These developments are encouraging.

There is however one aspect that causes concern. This is related to the indications of a decline in the sense of attachment of the fishermen to their traditional

profession. At the time of the earlier census the number of marine fishing villages along the coastline was 3600. According to the present census this number has come down to 3206. Further, the fisher population has also dwindled from 5.35 million nos. at the last census to 3.52 million nos. now (2005 census). The fall in the no of villages is partly attributable to the destruction of some of fisher villages along Tamil Nadu and Pondicherry coasts during the recent tsunami but the surviving population must have been rehabilitated. In any case, the no. of villages devastated may not be as high as 384. The fall in population indicated in the census could therefore be for other reasons too. It is possible that the evolving economic and social compulsions have led to the decline in the population. One among these can be the declining fish catches and returns and the lack of adequate infrastructure facilities for handling fish for value addition to augment returns. The other reason can be the social pressures. Fisher villages on the coast used to be isolated, but no longer so now because of road communication facilities to rearby towns. With the limited educational facilities in the villages, and influenced by other factors, heads of fisher families must have found job opportunities, other than fishing, in the neighbouring towns. A possible trend towards decline in fishing population must have been dealt with in the census report.

After the 1980 census, towards the end of the next decade, around 192 larger trawlers of 23-25 m OAL were added to the fishing fleet. Of these, only around 40 vessels are active now. This detail must have found a place in the census report.

Analysis of the data provided by the census and the conclusions drawn therefrom are of immense importance.

These conclusions form the springboard for initiating measures for further development of the marine fisheries of the country, avoiding pitfalls. These include village centered fisheries education and training, provision of needed further infrastructural facilities for landing and berthing of fishing vessels, for post-harvest handling of catches, for preservation of harvested produce and their storage,provision of transportation facilities wherever needed and all these as a routine system for securing improved returns, and development of closer beneficial interactions among fishery stakeholders in the villages and the participating outsiders and for fortifying the common endeavour of upgrading standards of living and improving the socio-economic conditions.

Analysis of the data provided by the census and the conclusions drawn therefrom are of immense importance. These conclusions form the springboard for initiating measures for further development of the marine fisheries of the country, avoiding pitfalls. These include village-centered fisheries education and training, provision of needed further infrastructural facilities for landing and berthing of fishing vessels, for post-harvest handling of catches, for preservation of harvested produce and their storage,provision of transportation facilities wherever needed, and all these as a routine system for securing improved returns, and development of closer beneficial interactions between fishery stakeholders in the villages and the participating outsiders and for fortifying the common endeavour of upgrading standards of living and improving the socio-economic conditions of fishes, in the main.

Fishing Chimes, Editorial 357: October 2006: Vol. 26, No. 7

# Aquaculture in Indian Coastal Zone

## *Eco and Farmer-friendly Legislation Introduced*

The Coastal Aquaculture Authority Act 2005 (CAA Act) is a virtual boon for the development of aquaculture in the Coastal Zone of India. The provisions enshrined in the Act reflect a project-oriented framework for the aforesaid development on a firm and steady footing. The guidelines of development notified under the provisions of the Act, which would obviously be followed, would usher in a focally significant phase in the coastal aquaculture part of blue revolution, that would concede to the stakeholders economically viable benefits far higher than what inland freshwater fish farmers are able to secure. Fortunately, the Act excludes the anachronistic condition that prevailed until recently that no new farms should be set up in the coastal zone. The Act relates the setting up of new farms to a survey of the coastal zone. Further, what gains emphasis is that, while the conferment of ownership rights over coastal land plots vests in the State Government concerned, the regulatory functions stand vested in the Coastal Aquaculture Authority (CAA), who would register the names of the allottees in their record as recommended by the District/State Committees and approved by it, already in vogue. The ownership registrations by the State Revenue Departments can be expected to follow this as a matter of formality.

The incomes of farmers in the case of shrimp/mud crab farming in brackishwater farms of the coastal zone per ha will be about 10 times more compared to freshwater farmed produce obtained mostly beyond the coastal zone, and which mostly consists of major carps, marketed for domestic consumption.

The coastal land stretch of India is sparsely developed for brackishwater farming, as at present. Taking this into account, the Government of India has now launched a reformed framework to bring the coastal zone, with the exception of specified areas such as mangroves etc (detailed in a following paragraph), under commercially viable farmed aqua production, subject to the regulatory measures of Coastal Aquaculture Authority (CAA). This is a major leap forward over the pre - CAA situation, which was characterised by the norm that no new aquafarms should be set up in the coastal zone. There was then no appreciation of the concept that the zone could be utilised for aqua production in an environmentally safe and sustainable manner, for the benefit of the nation.

The reformed framework contained in the Act is one that presents expanding export-oriented production opportunities. Out of 2 million ha of the coastal zone, presently 1.80 lakh ha are now under aqua production. Conceding that, out of the balance of the area (1.820 million ha) 0.820 million ha are covered by Agricultural lands, salt pan lands, mangroves, wetlands, forest lands and lands for village common purpose, the balance of one million ha, when developed for export-oriented aquafarming, would increase the present farmed products exports from the present 93,000 t with export value of around Rs. 2570 crores to 4,65,000 t with an estimated export value of Rs. 12,850 crores.

In order to achieve this expansion, it is expected that CAA, under the provisions of the CAA Act, would conduct a total survey of the Coastal Zone, which would be followed by demarcation of farm plots, interlinked with farm effluent collection system into centralised channels and from there to pass through an effluent treatment plant into the sea in a pollution-free status. Obviously, the allotment of these plots, other than those which stand already alloted with ownership rights with registration by CAA, will be followed by bestowal of ownership rights by the State Government concerned to meet regulatory needs.

Another provision under the CAA Act is that the CAA should extend support to Coastal State Governments to construct common infrastructure - common effluent system. Besides fixing standards for all coastal aquaculture inputs, it is stipulated under the Act that the CAA shall carry out and sponsor investigations/studies/ schemes relating to environment protection and demonstration. It has also to authorise research/ developmental organisations to carry out research/ investigation work on farming aspects of the Coastal Zone.

As per the provisions of the Act, the CAA has several other functions. The important of these are; i) To collect and disseminate data and other scientific and socio-economic information in respect of matters related to coastal aquaculture; ii) To direct the owners of the farms to carry out such modifications as are necessary to minimise the impacts on coastal environment by various aspects including stocking density, residual levels/use of antibiotics, chemicals or other pharmacologically active compounds; iii) To order seasonal closure of farms for ensuring sustainability of the coastal aquaculture practices; and iv) To order closure of coastal aquaculture farms in the interest of maintaining environmental stability and protection of livelihoods and for any reasons considered necessary in the interests of coastal environment.

The Act also embodies elaborate provisions for registration of coastal aquaculture farms with the Coastal

Aquaculture Authority and its renewal. The aspects of consideration of applications for registration, such as their passage through District Level Committees and State Level Committees before reaching the Coastal Aquaculture Authority for final clearance; and for renewal of registrations, including particulars of fees payable, are covered, apparently for the benefit of the farmers. The specimens of the forms to be used are also given.

The Coastal Aquaculture Authority has also published guidelines for regulating coastal aquaculture that cover site selection, construction and preparation of shrimp farms, water quality management, seed production, seed selection and stocking, feed and feed management, health management of shrimps, use of chemicals and drugs, harvest and post harvest management, farm hygiene and management and so on, that are equally pertinent.

The Coastal Aquaculture Authority (CAA) has formulated certain guidelines for a sustainable and environment-friendly utilisation of the Coastal Zone, utilising the powers vested in it under the Coastal Aquaculture Authority Act, 2005. Apparently, keeping in view the past haphazard growth of shrimp farms in the CRZ and the problems of regulating the working of the ongoing farms, the CAA has been given powers to regulate brackishwater aqua farming to a distance of 2 km from the HTL of seas, rivers, creeks and backwaters of coastal zone. The provision applies to a distance upto which the tidal effects are experienced and where salinity concentration is not less than 5 parts per thousand (ppt). For this purpose, the salinity measurements shall be made during the driest period of the year, as per the stipulation.

## Surveys

The axiomotic role of survey of the Coastal Zone for the regulated development of brackishwater farming in it is enshrined in the guidelines notified under the Act. This survey and its results are of utmost importance for the fisheries development of the zone. The results of this survey would lay the foundation for designing the layout of the farms in the zone and for conferring CAA - registered rights on the present farmers and to those to whom allotments would be made, based on consideration of the applications to be received from farmer entrepreneurs by the District Committees, to start with, to be followed by State Committees and CAA.

It is difficult to visualise the final strategy which the Coastal Aquaculture Authority would follow to accomplish the task of bringing the two km wide coastal zone under brackishwater farming (the CAA has no jurisdiction over freshwater bodies with less than 5 per cent salinity even within the zone). While the coastal aquafarming sector has to await developments, it will be pertinent to reflect on the scenario that may emerge, for what it is worth. This is attempted hereunder.

The ownership of Coastal lands, barring such plots of land already under the registered ownership of some, vests in the State Government. Once the CAA of India has with it a broad layout of farms to be set up in the coastal zone including inlets, outlets and location of effluent treatment plants etc., for centralised management, the stage will be set for:

1. Allotment of farm pond plots in the sections of lay-outs designed for single/group management;
2. Integration of presently owned farm ponds into the independent/group management system of farm lay-outs to be evolved;
3. Dealing with the issues of allotment of plots, conferring ownership rights by the State Government to the allottees and on related issues like technological and manpower support, organising supplies and services, particularly of those related to inputs such as lime, manures, seed, feed, probiotics etc and in respect of harvesting of produce, its storage, marketing etc.
4. Organising training related to farm management, recording data, conducting refresher courses, periodical workshops, seminars etc to review the progress with reference to production levels, eco-friendly conditions, economics of operation etc.

The task ahead would be looked at by the authorities from specific angles. These, to the extent they can be comprehended, are:

1. In the case of the existing tanks and ponds with clear ownership rights, the owners of these water bodies have to apply to the District level Committee concerned for recommending their applications to the State Committee/Coastal Aquaculture Authority (CAA) for the regularisation of their holdings. In the total survey of the coastal zone to be conducted by the CAA, these may have to be integrated with clear identity and in a way that the water inflow, outflow and effluent treatment systems get merged in the overall design.
2. In all likelihood, the survey and farm plots earmarking work would be undertaken by CAA with reference to the district boundaries, considering that the District Committees would have to first consider applications for farm plot ownership, (which would no doubt be invited through public notification as is done in USA), followed by the State Committee concerned.
3. The water supply into farms and water outflow from them are expected to be part of an inter-connected overall system. It is to be expected that the CAA would entrust the work to a regulating service subordinate to it, which would also extend support to State Governments to construct common infrastructure-common effluent treatment system. There may be a need on the part of the State Governments to set up cold storages at needed points in the coastal zone for the facility of the farmers to store their produce for some time, until they are able to transport it to the market. For this, the needed road connections would probably have to be provided.
4. The CAA, in all likelihood, has already taken steps to have a cadre of experienced personnel to

undertake surveys of the coastal land, formation of district-wise plots and for further follow-up action.

5. It is to be visualised, as already mentioned, that the conferment of ownership rights of farm plots would vest in the State Revenue Departments concerned, as finally cleared by the CAA; and the regulatory functions would be vested in the CAA under CAA Act; and
6. The CAA would, in all likelihood/prescribe the format in which each of the owning farmers have to maintain periodically records of pond water conditions, from pond preparation to harvesting, and also in regard to the periodic status of fish growth etc.

In conclusion, it may be said that a shrimp/crab/fishful era of Coastal aquafarming will soon take shape under the leadership of CAA, and let it be hoped that it will be one of showering prosperity on the aquafarmers, upgrading their socio-economic conditions and enabling them to contribute significantly to national well being.

Fishing Chimes, Editorial 358: November 2006: Vol. 26, No. 8

# Promotion of Organised Domestic Value-Added Fish Marketing

## *Linked to a Network of Value-Adding Coastal Fish Yards*

Fish curing yards were set up by the British along the coastline of erstwhile Madras Presidency, probably in middle 1930s or even earlier. Their existence continued after Independence too. However, they have now become a vanishing tribe. A few of them may still be functioning along the coastline of the reorganised States of A.P., Tamil Nadu, Kerala and Karnataka.

A review of the present situation, characterised by the absence or low key presence of Fish Curing Yards will reveal the relevance and imperative need for such yards although with a difference, in all major coastal fishing villages even today. This difference that can be visualised is that these yards, if brought into being again, have to be made structurally compatible with the developing scenario of opportunities for value addition to fish in varied ways.

Looking back, the Britishers have to be admired for visualising and introducing the concept of fish curing yards so long back A common feature then and now too is the problem of utilisation of miscellaneous and small sized shoaling fishes like anchovies, sardines etc. The solution to this had been one of sun drying and salting, the later for larger ones without a ready lucrative market in fresh condition. This solution, by and large, continues to be relevant except that a market has now developed for certain small-sized fishes mainly for conversion into surimi and fish meal and for larger ones for marketing in fresh/iced condition, or after value-addition. This has become possible because of the progressive development of road connectivities to coastal fishing villages. These have facilitated transportation of such of the readily available but unmarketable heavy quantities of fishes from either the landing points or from other centres to serve as raw material for surimi, fish meal and fish feed manufacturing plants.

There was the burden of excise duty on salt in the period prior to independence. So far as fish curing with salt is concerned, in order to ensure hygienic curing of fish in healthy surroundings in coastal fishing villages, the then British Government applied the device of exempting salt used for fish curing from levy of excise duty, linked to the stipulation that all catches to be sun dried/salted should be brought compulsorily to the yard where facilities were provided by the Fisheries Department for the purpose. Further, the yards served as a vital link between the departments and the fishers. This system worked well until the abolition of excise duty after independence. This abolition, coupled with the increasing demand for fish, development of road connections between consuming centres and fishing villages, and the accessibility so established to nearby centres of demand for fish, led to dilution of focus towards fish curing yards. This also happened because at that time the potential of these yards that dotted a good length of the Indian coastline, was not visualised for imparting different kinds of value addition to prepare products such as fish pickles, fish meat balls, fish wafers, fish papads etc, for domestic marketing/export.

Coastal fishing villages and fishing harbours are the basic units that throb with operations of fishing boats and fish landings. This feature entitles them to be considered as centres of prioritisation for imparting value addition, particularly to low value catches. This value addition, in the form of sun drying and salting is being done at present in vacant places adjacent to landing centres, and in available spaces within the premises of some of the fishing harbours too. Looked at from the angle of hygiene, however, there have to be concerted efforts at discouraging these practices. As a replacement, value adding fish yards may have to emerge as a reincarnation of the earlier fish curing yards. These yards deserve to be set up at all important fishing villages and as part of the infrastructure at fishing harbours, equipped with facilities for needed value addition to fish in the form of value-adding coastal fish yards with attached cold storage units and fish drying platforms, salt curing infrastructure for fish and also facilities for making fish pickles, fish papads, fish wafers, fish meat balls, retort pouches of shrimp, fish fillets, etc. These value adding yards would have to be manned by experienced and dedicated technical hands headed by a chief yard officer. In fact, the setting up of such yards can be under a centrally sponsored scheme in the Xlth plan. The setting up of yards of this kind all along the coastline at identified centres will provide a pulsating scenario of integrated marine fishing activities. The scheme would have to provide for training programmes to selected fishermen and women who would function as the media for integration of low priced fish output with value added processing and packing for domestic as well as export marketing, employing the trained hands. These trained hands, apart from employment at the yards, would be in a position to take up preparing fish pickles etc as a cottage industry. For this to happen, the trained fishers can be organised into self help groups with provisions for financial help, supplies of needed inputs and provision of linkages for marketing of the products, loose or packed.

Besides sun drying and salt curing in the open areas of the suggested value adding fish yards, the output, as already mentioned, may also consist of various products,besides dry/salted fish packets and various other products amenable for being made at cottage level. The techniques of preparing several types of value added products have been standardised by CIFT and are now available for adoption.

Domestic fish marketing has two angles. One of this is related to fresh fish and dried/salted fish distribution and marketing. So far as fresh fish distribution and marketing is concerned, while some channels of supply to urban centres from coastal fishing villages have come into being, lot more remains to be done and efforts are on for establishing a nation-wide/region-wise distribution network and these would bear fruit over a period of time. Regarding dried/salted fish traditional systems continue to operate successfully with the commodities moving from production centres, mostly to hill areas and other zones where fish availability is scarce. Fresh fish for surim manufacture and dry fish for fish meal production are transported by vans from centres of production to the manufacturing points.

The other angle concerns distribution and marketing of value-added products. The spread of department stores in towns and cities has already provided an avenue for sale of value-added fish products. Several department stores now receive mostly supplies of canned fish and shrimp, fish pickles etc from the processing units concerned for marketing. The activity however has certain limitations. The suppliers are few and the supplies are also limited owing to price constraints etc. Once a wide network of value-adding fish yards are set up at potential fishing villages all along the coastline, the cost of supplies of value-added products to department stores will come down. This will help the stores to supply the value-added items to their customers at a reasonable price.

The benefits of a scheme for setting up value-adding fish yards in identified fishing villages all along the coastline will be many. These, among others, are: 1) They will upgrade the activities in fishing villages, augment incomes of the fishers, and bring about an enormous improvement in their socio-economic conditions; 2) Each of the potential coastal fishing villages will be the proud possessors of infrastructure for the profitable utilisation of miscellaneous and low value fishes at the value-adding fish yards; 3) the value-adding fish yards will serve as centres for providing training in value addition to fishermen and women and will provide employment opportunities to them; 4) The activities at the yard, to be developed directly by the authority concerned or through an agency with a franchise, will invest an identity and shower a measure of importance to the village concerned, besides bringing in prosperity to the inhabitants; 5) The operations at the yard being commercial in nature, as they pick up vitality with the strengthening of qualitative production in quantity and consequential effective distribution and marketing linkages, the yards concerned would eventually become self-contained, reducing the dimensions of financial inputs needed from the Government gradually; and 6) The yards could also be brought under a registered body for effective functioning.

## Need for Intervention of NFDB

The suggested scheme for setting up value adding fish yards to dot the country's coastline would entail the need for intervention by the National Fisheries Development Board, which could introduce the scheme and support State Fisheries Departments for positioning of separate Cells at the District Fisheries Offices in each of the Coastal States and UTs, for monitoring the work at the said yards. The Integrated Fisheries Project (IFP) has wealth of experience for providing training to fishers in adding value to fishes and also in packing such products. The project has also skills at marketing these products. These attributes of IFP could be well utilised by the NFDB in gearing up the integrated operations at the value-adding fish yards, the setting up of which deserves to be considered favourably by NFDB. Another aspect is that the activities of the stakeholders who dry their catches in the sun and of those who salt their catches would need to be channelised towards developing a distribution and marketing system to cover both domestic and export activities to countries such as Myanmar, Nepal etc. These have to be co-ordinated by the State Fisheries Directorates and by the Board. Like-wise, a system for marketing of the value-added products through department stores and other channels has to be promoted with the needed positioning of sales staff.

Fishing Chimes, Editorial 359: December 2006: Vol. 26, No. 9

# Promotion of Tuna Fishing in Upper Bay of Bengal of India

Over 30 years back the well utilised exploited fisheries of Indian seas was mostly finfish based. This complexion later gave way to diversification of fishing effort mainly towards shrimp fishing in the Indian waters. Of late, however, owing to the depletion of capture shrimp stocks, mostly for reasons of overfishing, there has been a quest at identification of a further resource with export potential, other than shrimp, occurring in open sea zones of EEZ as well as in the coastal zones. This includes the zone off the upper east coast of India. The first successful translation of this line of endeavour off this coast was the installation of tuna long lining equipment on six commercial trawlers of 23 m OA L in 2005. Over 400 t of fresh chilled tuna caught in the Bay of Bengal off North A.P. by these vessels was exported in 2005 from Chennai by air, mostly to Japan. This became possible because of the granting of subsidy of 50 per cent (subject it to a maximum of Rs. 15 lakhs) towards the cost of the equipment and its installation aboard these six trawlers by the Union Department of Animal Husbandry, dairying and Fisheries. The six vessels, equipped for long lining as mentioned above, have been continuing tuna fishing operations successfully. This encouraging development has motivated not only the owners of several other 23m OAL vessels to apply for subsidy for similar installation of long lining equipment on their vessels but also those who own 15 m OAL vessels. These vessels can also be equipped with longlining equipment for open sea tuna fishing.

Vessels of 20 m OAL and above alone were considered as ocean going vessels until the time the Union Department of Animal Husbandry, Dairying and Fisheries (Dept. of A.H.D and F), as part of its order No. 21001/3/2006 Fy (Ind) dated 10 May 2006, issued a set of guidelines to govern the operations of fishing vessels of 15-20m OAL as well, equipped for longlining, as these are also brought under the category of'deep sea fishing vessels. In other words, under these guidelines, Deep Sea Fishing Vessels are defined as any fishing vessel registered under IMS Act, 1954, capable of engaging in deep sea fishing, with 15 meter overall length and above. This definition paved the way for initiatives to equip 15m OAL vessels with longlining equipment for catching tuna in the Indian EEZ, beyond territorial waters.

In its order of 10 May, 2006 mentioned above, the Union Dept of A.H, D and F stated that, with its approval, monofilament longliners of 15-20 m OAL could also operate in the EEZ as per the guidelines stipulated, one of which was that the vessels must hold a Letter of Registration (LOR) from the Union Dept. of A.H.D and F. The prescribed form to apply for the same had been given as part of the guidelines. This form, duly filled in has to be submitted with a non-refundable fee of Rs. 5,000 in favour of Pay and Accounts Officer, Union Dept of A.H. Dairying and Fisheries, subject to conforming to the several other guidelines stipulated by the said Department for compliance by the owning enterprises to undertake longlining, in the EEZ beyond the territorial waters. In addition, a notorised Affidevit on Rs. 10 Non-Judicial stamp paper, taken in the name of individual owners/ companies, as an undertaking in the format prescribed, has to be furnished. There are a few other stipulated documents too to be furnished to the Union Dept of A.H,D&F. All these documents have to be sent with a covering letter on letter heads of individual companies concerned, addressed to Joint Secretary (Fisheries), Union Department of Animal Husbandry, Dairying and Fisheries, Krishi Bhavan, New Delhi- 1. The permitted enterprises have to also submit voyage data to the Fishery Survey of India in the format prescribed.

The facility aforesaid, that enables operation of 15-20m OAL vessels equipped for tuna longlining in the EEZ beyond territorial waters, constitutes a major step forward to bring about augmentation of the tuna fleet strength for harvesting the estimated catchable annual potential of2,13,000 t of tuna resources of Indian EEZ. It is learnt that several enterprises have already applied for the issue of needed Letters of Registration in respect of their 15 m OAL vessels. Another development that strengthens the facility provided as mentioned above in the utilisation of tuna resources of India EEZ is the introduction of another scheme for the purpose by MPEDA. The quantum of subsidy to be conceded under this scheme towards the cost of installation of tuna equipment on vessels (presently equipped for trawling only) will be based on assessments and the recommendations made by the Committee set up by MPEDA, which consists of technical experts from CIFT, FSI, CIFNET and MPEDA. The Committee will recommend to the MPEDA the quantum of subsidy to be given (to be determined based on the assessment made for the improvements needed and the balance of economic life of vessels in the case of vessels that have crossed more than 12 years of age) to vessels of 15m OAL and above. This Committee is understood to have inspected 27 vessels of 15 m OAL at Visakhapatnam and on that basis determined the improvements needed and assessed the quantum of subsidy that each of these vessels deserved (50 per cent of the cost/Rs. 15 lakhs per vessel whichever is lower). The assessed vessels would operate under the fold of a Co-operative Body registered under the name Integrated Tuna Fisheries Mutually Aided Co-operative Society Ltd.

Estimating that around 30 t of tuna per vessel could be harvested by a 15-20m OAL vessel in a year (in 200 fishing days) and based on the estimated potential of the resource, it is possible that there will be scope for introduction of a good number of 15-20m long vessels, upgraded from their present design to function as tuna vessels for a good part of an year.

As the name of the Society to be formed implies, an integration of the activities of the vessels coming under the fold of the Soceity has to be achieved and this will be possible only when there is shore infrastructure for the needed post-harvest operations and for exporting fresh chilled tuna, mainly to Japan, USA and Europe. Presently, the system followed by the six vessels now in operation in the upper Bay of Bengal zone bordering A.P. State is to conduct longlining for around 7 days at a time, and transfer the catches into one of the six vessels. This vessel carries the produce to Chennai for air transport of the consignment from there to the destination. One constraint in this work that the operators face is the problem of inadequate post-harvest facilities there, as at other ports.

A large number of FRP non-mechanised boats too now conduct long lining for tuna in the Upper Bay of Bengal. Small mechanised boats have not yet entered the tuna fishing field in any significant manner. It would be desirable for the Union Department of Animal Husbandry, Dairying and Fisheries and NFDB to advise the Coastal State Governments to introduce a scheme to equip small mechanised/non-mechanised boats (specially FRP boats) for tuna long lining with around 10-15 km of head rope to hold with clamps around 150 vertical lines each 15m long, with hooks at the lower end, to hang down into the water from the headrope. The needed financial and technical assistance can be extended by the Government concerned under a centrally sponsored scheme. These boats can operate in the territorial waters, if not beyond, to exploit the tuna resources in them. It has to be mentioned here that, according to reports, Brazil has succeeded in implementing a scheme aimed at upgrading the capabilities of small mechanised boats for tuna longlining with rewarding results. Tuna longlining being a passive method entailing operations mostly in the pelagic zone, mechanised boats can operate, unencumbered by problems that are faced in trawling which has the dimension of depth, calling for installation of higher horse power engines in the boats in relation to depths.

There is also need to introduce a scheme for the introduction of larger vessels designed for fishing for tuna too, supported by needed improvements/additions to facilities at fishing harbours to deal with tuna landings. A scheme for providing finance for introducing these vessels of appropriate size range could be successfully implemented, provided the final phase of the operations of these vessels is well taken care of for recovering instalments of loan repayments out of returns realised from sale of catches, the buyer of which has to be identified well in advance and made party to a tripartite agreement among the financier, operator and the buyer identified.

To sum up, there is need to equip the existing fleets of vessels for operating tuna longlines on the one hand and to work out and implement schemes for introducing additional vessels for the purpose, providing needed post-harvest facilities, and taking into account financing arrangements needed and also gearing up Indian shipyards to build new tuna vessels as required, if necessary with foreign collaboration and taking into account the economics of operations in countries like Australia, Japan, Taiwan, South Korea, China and Thailand. These countries have found integrated tuna fishing operations viable, although on a global scale. Many of these vessels operate in the water of countries like India, Pakistan and several others which include quite a few South American countries too.

Fishing Chimes, Editorial 360: January 2007: Vol. 26, No. 10

# On Promotion of Export Market for Indian Carps

There is an unfulfilled longing among stakeholders concerned to have an export market for Indian carps, specially the Indian major carps. This has, by and large, remained unfilled. It is true that small quantities of major carps are exported to a few countries having migrant or settled Indian population who still retain their aptitude towards these fishes. There has however been no cognisible breakthrough in the export of Indian carps to countries like USA and those in Europe who traditionally import Indian aqua products other than carps. Cursory studies have indicated that in USA and Europe, for instance, carps are considered as fishes of inferior quality. They fetch somewhat lower returns, but in the present Indian scenario of discouraging farm gate prices, export prices can be far better than them, providing improved returns that will enable the farmers to meet the high expenses they have to incur, particularly on feed and have a margin. As is known, whatever they produce has to be sold now within the country, withstanding several constraints. One of these is that farmers have to depend heavily on major carp farming, unlike other countries who have several viable choices. Indian farmers too no doubt can have choices but these are mainly of exotic fishes, some of which are illegally brought into the country and the others have problems of seed availability and farming and marketing limitations. The exclusive dependence, by and large, on major carp farming has generated problems of inconvenient surpluses in fish production inconsistent with the present situation of not having a nation-wide network of a cold-chain and retail outlets as needed, as part of the domestic fish marketing system, which is inadequately organised. The locations of fish ponds being scattered, there is no developed system for pooling up catches at various centralised points and facilities of transportation to the marketing outlets.

The shortcomings of the domestic fish marketing system coupled with the absence of an export market for major carps have pushed the major carp farming sector of the country into a disadvantageous position. In other words, the farmers are face to face with the transparent risk of securing uneconomical returns.

As a ray of light in this seemingly inescapable despondency on the part of stakeholders due to lack of opportunities for exporting Indian major/minor carps, *Fishing Chimes* has now come across a write up on carps, under the feature 'Supplies and Markets/News' at P19 of October 2006 issue of Seafood International.

According to this feature, imports of carps at global level are marginal at around 50,000 t annually; the main importers are Hong Kong, now part of China. Germany imports over 6,000 t of carps. Austria, Belgium, Malaysia and Canada also import carps, although in small quantities. The write up in Seafood International says that in late 1980s HongKong entered the market as an importer of live carps. HongKong has remained the largest importer of carps since then, followed by South Korea, Germany and Poland. The report further says that in Asia, trade in carps is dominated by China and Malaysia as exporters, while Hong Kong and South Korea are the largest importers. In Europe, according to the write-up, the largest exporters are Belgium, Bosnia/Herzegovina and Lithuania, while the main importers are Germany, Poland and Slovakia.

It is stated that the introduction of live carps in international trade was one of the reasons for the dramatic increase in global trade of carps in 1998. According to the write :up, the highest prices appear to be paid by Poland, Serbia and Montenegro, followed by Germany. The import prices are mentioned as U$ 2.27 per kg. In these countries, strangely, live carps command the lowest of prices, while fresh or chilled carps fetch the highest prices.

Contrary to the general impression in India, carp is the most popular freshwater fish after trout in Europe, according to *Seafood International*, Oct 2006 issue. Hungary, the Czech Republic, Poland and Germany are the main consumers. The journal says that the Central European Countries are the most promising markets for the fish. Saying that in some areas like the Mississippi delta, carps have become a menace, it was observed in the feature that the solution to get over the problem would be to develop the market for carps so that the increased catches can be disposed of economically. This approach applies equally well to the present problem of increasing catches of major carps or other carps faced in India. Pointing out the negative image westerners have about carps (although they do have a market but with limitations), the journal suggests that carps be called as 'silver cod' so as to generate greater marke interest. Considering the foregoing account, it is desirable that measures are initiated for exporting Indian major and other carps to countries of Central Europe in the first instance. The measure to be taken can include a) Popularising Indian Carps under appropriate names that can capture the interest of the people of that part of Europe; b) Sending a team of specialists to Central European countries to discuss the subject of exporting Indian carps with the concerned in those countries and prepare ground for sending trial consignments for testing market reaction and price aspects; and c) sending trial consignments to identified agencies in those countries for marketing and for further follow-up action.Establishment of a market for Indian major and other carps in Central Europe can well prove to be a major step in the direction of helping our fish farmers in upgrading their earnings and thereby their economic status.

Fishing Chimes, Editorial 361: February 2007: Vol. 26, No. 11

# Allowing Exotic Cultivable Aqua Species into India: Need for a Realistic Policy

The post-independence history of introduction of exotic aqua species into India is characterised by case by case decisions that probably became compulsive because of certain situations. They do not give the impression of being based on a long term policy that spells out specific criteria in respect of differing situations of introduction that can be visualilsed. Prior to independence, Trout, Tench, and Crucian carp and its versions were introduced in a section of the temperate belt of the country. In the tropical zone Gourami was introduced. In the post-independence era, authorised/unauthorised introductions of exotic fishes took place. These included Common Carp, Silver Carp, Grass Carp, Tilapia, and the Catfishes, *Clarias gariepinus* (Thai magur), *Pangasius sutchi* (Pangus), and probably a few others. There has also been the introduction of *Litopenaeus vannamei* (Vannamei shrimp or Pacific white shrimp).

Introduction of Tilapia (*Tilapia mossambicus*) into India took place in late 1950s. It entered into India from Sri Lanka, as is believed, during the days of the erstwhile undivided composite State of Madras. The fish, being a prolific breeder, as is known, became a pest, particularly in South India to start with. Its latest invasion has been into Jaisamund lake of Rajasthan, where also it continues to be a pest. Some years back, M/s. Vorion Chemicals of Tamil Nadu was allowed to introduce Nile tilapia for monosex farming in ponds containing the effluent of the factory of the company before its release into the sea. The activity continued for a few years but it was later discontinued for some reason. There were successful initiatives at organising domestic marketing of the fish in and around Chennai city. Yet, the system was left to languish. Another aspect is that the fish also has considerable potential for export in value-added form. Taking advantage of this, it was expected that the company would diversify into value-added processing of the fish, but it did not happen.

*Tilapia mossambicus* continues to relentlessly exercise its presence in most parts of the country. Besides this fish, which proved to be a pest, the exotic carps (Common Carp, Grass carp and silver carp) were stated to have been later introduced. They have now established themselves in India, with the silver carp securing a firm foothold in Gobind Sagar reservoir.

*Clarias gariepinus* and *Pangasius sutchi* are believed to have been brought in or strayed into India from Bangladesh and utilised for farming. They are however now seen to have strayed into some of the Indian rivers. Information in circulation is that these species, despite the ban imposed on their introduction in Indian waters, are understood to be clandestinely farmed in several States of the country.

So far as shrimps are concerned, it is common knowledge that, Government permitted two Indian enterprises to import Vannamei shrimp (Pacific white shrimp). These enterprises started producing its seed. The present position, however, is that while farming of Vannamei shrimp stands banned, there is unconfirmed news that the two permitted companies are producing seed of Vannamei probably for supply to other farmers too.

The introduction of exotic fishes, mainly trout and tench, in the temperate zone, has justifiably facilitated the development of stocks of these fishes in temperate waters, which were having sparse fish population prior to the introduction. The introduction of Gourami which was also done prior to independence in the tropical zone of the south did not cause any harm to the local fish population. In fact, it so happened that its population dwindled.

The introduction of exotic carps, both prior to independence and thereafter, does not appear to have caused any serious harm to the Indian fish fauna. One reason for this is that India is basically a carp country. If silver carp has created a problem in Gobind Sagar, it is attributable to reasons other than its 'invasion'. Unlike Tilapia, it is not a prolific breeder to become a pest. Tilapia has become a pest in Jaisamund Lake in Rajasthan and in several other water bodies in the country. In contrast, the problem with silver carp is the abundance of its catches, mostly from Gobind Sagar and general lack of post-harvest handling facilities including cold storage facilities and distribution net work.

The attributed cause of inadequate demand for silver carp, as stated, is only an excuse to cover up the absence of needed post-harvest handling facilities. The standpoint in respect of dealing with the issue of exotic fish introduction in the country has all long been one of ostensible inflexibility on one hand and of surrendering to seemingly helpless situations when exotics intrude into the country some way or the other. Thai Magur and Pangus entered into our farming system without any approval from the Government, in contrast to Nile Tilapia and Vannamei which were introduced with permission from the Government. Monosex seed of Nile Tilapia was used for farming by Vorion Chemicals in Tamil Nadu as long as it was undertaken by them. Vorion Chemicals, for some reason phased out from farming of Nile Tilapia. Having introduced the fish and its mono-sex farming, probably it would have been a prudent policy on the part of the Government to promote at other selected centres a protected system of having monosex seed production centres for supply of the seed to farmers. The introduction of an alien fish like Nile Tilapia and its monosex farming was not probably considered unjustified at the time it was introduced as it would keep the population of the fish under control. It appears that the same point of view continues to be valid as monosex seed production and farming of Nile Tilapia or other suitable exotic fishes, when carefully done, would not affect the native wild fish fauna.

GIFT Tilapia was also not introduced into India, while several other countries introduced this fish. This decision by some countries must have been based on the same considerations as are applied for the introduction of Nile Tilapia. So far as India is concerned, the earlier unpleasant experience with *Tilapia mossambicus* may have also influenced the decision not to introduce GIFT, and also of not doing anything further in respect of Nile Tilapia. Probably, on a similar thinking Vannamei farming is not allowed any further, overlooking the fact that its stock developed by the two permitted companies has the potential to spread in some way or the other.

The foregoing account, presented on the basis of the existing situation, is indicative of the absence of a well directed policy in respect of introduction of exotic aqua species in the country, after independence.

The present status of thinking on the subject of introduction of exotic cultivable aqua species is such that it indicates the need for incisive deliberations to evolve an enduring and purposeful policy on the subject. This policy may have to aim at replacing the present system of allowing private enterprises to introduce exotic species first on an experimental basis and later imposing a ban on such activity on some consideration or the other as happened with Vannamei and probably with Nile Tilapia. The policy for consideration can be to entrust the experimental work related to the introduction of exotic aqua species to the Central Fisheries Research Institute concerned. If the results justify introduction of the species concerned, further steps could then be taken to provide a spread effect to the species, incorporating such precautions as may be needed. A policy that discourages the clandestine or other way of entry of aqua species through neighbouring countries can also be developed. Keeping these aspects in view, it is desirable that the National Bureau of Fish Genetic Resources (NBFGR) takes the initiative to convene a brain storming session with the participation of experts in the line to discuss the problem from various angles and to evolve guidelines to facilitate the development of a realistic and enduring policy on the subject of introduction of exotic aquatic species into India.

Fishing Chimes, Editorial 362: March 2007: Vol. 26, No. 12

# Desert Fishery Development

## of Aquafarming in the Saline Zone of North West India

### *National Fishery Development Board's Intervention Imperative*

There is a major gap in the development of fisheries of the North-West part of India. This gap is related to the continued fallowness of its saline water stretch, located in Rajasthan and Western parts of Haryana, Punjab and Uttar Pradesh, so far as fisheries are concerned. The Central Institute of Fisheries Education (CIFE), Mumbai exercised an initiative some years back at promoting farming of mullets, milkfish and shrimps in the saline waters of Haryana and Rajasthan. The initiative however languished subsequently despite encouraging results. This is attributed to the inadequacy of efforts directed at establishing linkages among various supplies and services related to inputs and technological support. It is unfortunate that the activity more or less petered out eventually.

This is a baffling situation that defies a convincing explanation for its presence. The scenario is that, on one hand there are assets in the form of vast stretches of saline water resources both underground and above ground, awaiting utilisation for fish/shrimp/prawn production, and, on the other hand, there are appropriate technologies available for application in such waters with modifications as are needed, for fish/shrimp/prawn production. It is unfortunate that there is only a perfunctory interest with shades of illogical complacency in the pursuit of this activity for augmenting the aforesaid production from the barren saline zone of north-west, despite the potential. The waters, in part, are now used only for salt production.

Now let us recall the status of saline water resources, say of Rajasthan, which is having a substantially extensive saline zone, when compared to the other such States having similar potential. Rajasthan is having surface saline waters in the form of five natural saline lakes of which 'Sambhar' is the largest one with an average waterspread of about 182 sq. km. It has also extensive tracts of ground saline water. 60 per cent of the area of the State has also saline aquifers. The salinity of the waters of these aquifers varies from 2 to 300 ppt. It is estimated that the ouput of these aquifers ranges from 50 to 300 Cum/day. The availability of saline ground water is the highest in Barmer district. The other principal districts with ground saline water are Jaisalmer, Pali, Jalore, Jodhpur, Churu, Bikaner, Hanumangarh. Sriganganagar and Nagour. Relatively lesser amounts of saline ground water is also available in Ajmer, Jaipur, Tonk, Sirohi and Bhartapur districts. There are also smaller water bodies in the arid region which are limnologically distinct and are called 'Playas'. Some of these are lowlying and are of low salinity. The extent of such waters is estimated by ISRO at 1,18,519 ha. Another source of saline water in Rajasthan is the result of secondary salinasation of water-logged areas that took place along the extensive canal network of the Indira Gandhi Canal Project. These areas have an extent of over 45,000 ha as at present.

The species that could be successfully raised in inland saline water, besides brine shrimp (*Artemia salina*), are the giant freshwater prawn, milkfish, mullet, sea bass, and tiger shrimp. Experience indicates that milkfish (*Chanos chanos*) and mullet (*Mugil cephalus*) could be grown in saline water upto 40 ppt salinity. It has also been noted that milkfish can tolerate salinity even upto 70 ppt. Giant freshwater prawn (*Macrobrachium rosenbergii)* could be farmed in the water having salinity upto 15 ppt. It is stated that polyfarming of tiger shrimp (*Penaeus monodon*) with certain compatible prime fish species has been found to be more profitable compared to its monofarming. Tiger shrimp could also be farmed in Rajasthan's saline waters having salinity upto 15 ppt and having concentration of potassium of 200 ppm and with addition of potash or potassium chloride.

There is an estimation that 3.5 t of fish/ha/year or prawns/shrimp at 1.5 t/ha/year can be produced in a sizeable extent of saline waters of the State.

While the estimate of extent of inland saline waters in Punjab, Haryana, Western U.P. and in Tamil Nadu could not be accessed, it is believed that these are extensive, allthough far lower than the extent in Rajasthan.

In the light of the foregoing account and other particulars that can be pooled up, it is desirable for the National Fisheries Development Board to intervene to promote the development of fisheries in inland saline waters of Rajasthan and in the other States mentioned, in association with the Stale Fisheries Departments concerned. As would no doubt be visualised, a major project for fisheries development of these waters has to be mounted, preceded by a pre-project survey, for achieving sustainable utilisation of the resources for fish/shrimp/prawn production in an integrated manner. Following the primary stage of surveys, the project has to consist of designing of farms, their construction and other aspects such as, organisation of seed supplies and other inputs, development of farming system and farm management system inclusive of the final linkages in respect of pooling up and storage of produce at various centrally located places, and providing transportation facilities, to culminate in effective connections with domestic and export marketing systems.

Fish farming in the saline water stretches of desert areas may sound anomalous to some. Such persons may have to note that in Israel, over the last decade, a scientific hunch on use of underground brackishwater has turned into a bustling business. Israel's scientists are stated to have realised that they were on to something when they found that brackishwater drilled from underground aquifers hundreds of feet deep could be used to raise warm water fish. Prof. Appelbaum of Jacob Blaustem Institute for Desert Research at Sede Boqer Campus of Ben Gurion University, pioneered the concept of desert aquacullure in Israel in the late 1980s. Fish are presently grown using underground saline water in selected desert zones of Israel with success. Urial Safriel, an Ecology Professor at the Hebrew University of Jerusalem is reported to have observed that 'most development in Israel, still driven by the Zionist ethos that the desert was some mistake of God, and that 'we have to correct and make the desert bloom'. In other words, he exhorted that desert saline waters could be utilised for fish production. He showed to the farmers that they could use the water in which the fish were reared, for irrigating their desert land crops. The organic waste produced by the farmed fish acts as a fertiliser, he is reported to have said.

In conclusion, it is suggested that the National Fisheries Development Board (NFDB) may take the initiative of bringing the desert waters of the north-west under fish production.

Ways can be found through an enduring effort to bring the six saline water lakes of Rajasthan, having a large extent, under capture fish/shrimp production and also under cage farming, as possible. By taking this step a substantial addition in fish production can be achieved. 45,000 ha of water logged saline areas of the State could also be brought under farming to yield an annual production of over 45,000t, Underground saline waters located by ISRO. which have an extent of over 1,00,000 ha could be similarly utilised through planned development, based on application of appropriate technology for the purpose. By bringing this potential area under beneficial aqua production, it might be possible to produce atleast 1,00,000 t of exportable aqua species that can be valued atleast at US$ 300 million. In addition, large quantities of brackishwater species so produced but with no avenues of export can be channelised for domestic marketing. Taking into account the enormous potential of desert aquafarming and the avenues of employment and socio-economic development of weaker section of the society, can be opened up. It is to be hoped that the National Fisheries Development Board would mount efforts in the direction of utilising the potential for national benefit. This suggestion deserves projectisation and purposeful follow -up efforts thereof with determination.

Fishing Chimes, Editorial 363: April 2007: Vol. 27, No. 1

# Scenario Before, and After–Letters of (Fishing Vessel) Permission (LoP) System

In the past thirty years and over, the Open Sea Fishing Policy related to Indian EEZ beyond the territorial waters underwent a couple of distinctive developments. The basic policy, initiated in middle 1970s was one of introducing totally Indian owned vessels. This policy had a glorious and dignified start. Apart from four trawlers permitted for import by Union Carbide from USA in early 1970s, 28 nos. of 23.14m OAL Mexican built trawlers, equipped for both stern trawling as well as lateral trawling were imported by the permitted companies under a scheme launched by the Union Department of Agriculture. This was followed by further imports of a few vessels as permitted, mostly from Australia and Holland, similarly equipped as the Mexican trawlers but in the OAL range of 25-27 m. These imports were financed by the erstwhile Shipping Development and Finance Committee (SDFC). Later, the Shipping Credit and Investment Company of India financed the acquisition of several more fishing vessels. In all, the fleet strength went up to over 190 vessels by late 1980s.

This phase of trawler introduction signified a golden beginning that also generated indigenous larger fishing vessel building capabilities at certain Indian shipyards like Mazagoan Dock, Alcock Ashdown, Goa Shipyard, Bharti Shipyard, Hooghly Docking etc. Most of these vessels, both imported ones and the Indian-built ones mentioned above operated for over 20 years and some of them continue to operate.

A merciless twist to this excellent beginning was later imparted by the authorities. The main feature of this twist was the conspicuous absence of a long range view on fully owned Indian open sea fishing vessel introduction programme. The need for the sustainable utilisation of the fisheries of EEZ beyond territorial waters, an axiomatic approach to be supported by a plan of introduction of the required number of the resource-oriented Indian vessels over a feasible period, was overlooked. Not only this, the authorities at that time seem to have taken a ruthlessly anti-developmental view, based just on the single premise that repayments of loans given by the then SDFC and the SCICI were not forthcoming from the loanees as per the agreement. Instead of introducing an effective system of recoveries of loans by way of the introduction of the needed mechanism which was necessary for an activity of this kind, matters were left unmonitored and this led to the sprouting of understandable weaknesses of evasion or delayed repayments by the loanees. Instead of countering this trend, and introducing an effective system of recoveries in the interests of development, the authorities chose the easier option of winding up this common financing facility (extended by the then SDFC and the SCICI). This short sighted policy led to the subsequent events of increased foreign fishing vessel aggressions into our EEZ, besides maneuvres by distant water fishing nations, chiefly Taiwan, to gain a regularised access for fishing in the Indian EEZ. There was the view that these developments had been clearly against national interests.

Despite the abundance of fisheries in the EEZ beyond territorial waters of the country, estimated at nearly two million t, which included various species of tunas and tuna-like fishes, occurring mostly in the pelagic zone, and of crustaceans and others in the demersal zone, the policy deficiency, which was one of lack of provision for maintaining and increasing Indian owned fleet strength, led to shrinkage in the numbers of Indian vessels that constituted the fleet, from about 190 numbers to around 40 numbers, at around the turn of the century. These vessels have been diversified by their owners to function as carrier vessels for ONGC's operations in the Bay of Bengal. They had to resort to this step, as their fishing operations, mostly revolving round catching of depleted stocks of shrimps in the shore-ward zone or the coastal waters of the EEZ of the Upper Bay of Bengal, became uneconomical.

This scenario led to the stage of the opportunistic commercial interests in the country gaining relationship with one or more foreign nations having distant water fishing fleets, to work towards exploitation of fisheries of open sea zone of Indian EEZ, at pelagic and demersal levels. What all was needed was for the two interests to gain a nexus to create a situation in which the Indian authorities would avail of their services for exploiting the fisheries resources of the Indian EEZ. Not apparently having adequate vessel strength for exploiting the fishery resources of the national EEZ, the Indian authorities had to humiliatingly plunge first into a scheme of permitting Indian enterprises to charter foreign-owned vessels to fish in Indian EEZ, with an understandable stipulation that for every vessel chartered, one indigenously built vessel owned by the enterprise concerned should be introduced. Over 200 or more such vessels were permitted to be chartered in stages for quite some years subject to the above conditions, by the Indian authorities. The benefits of these operations however substantially went to the Kitty of foreign owners. What were shown as foreign exchange earnings through exports of the catches were gross figures, a substantial part of which was siphoned out as operational expenses, charter hire etc., by the foreign owners.

After some years of operation of these chartered vessels, the foreign owners, who were virtually operating the vessels with their crew in the Indian EEZ (The charter system was only a device for their vessels to gain entry into the Indian EEZ), with the Indian enterprise functioning as the charterer but with practically no operational involvement, suddenly found that they had heavily depleted the bottom fish stocks, thus rendering the operations uneconomical for them. This adverse situation that the foreign owners faced at that stage induced them to find a way of getting out of the charter agreements. Unwilling to initiate remedial steps from their side, probably so as to maintain the confidence the Indian authorities have in them, the foreign owners seem to have managed in such a way that the situation of depleted stocks had an impact on the Union Ministry of Agriculture too. This eventually took place through a recommendation of Murari Committee that advised the Government to withdraw the charter scheme.

This recommendation was accepted by the Government. The foreign vessels under charter, mostly of Taiwanese construction, happily obeyed the orders and withdrew, obviously because the decision was totally in tune with their business interests. But then, the next problem that confronted the foreign vessel owners was how to come back to exploit the rich pelagic tuna fisheries resources of Indian EEZ, estimated at 213,000 tonnes. To come back by way of charters, there was no chance, as the Indian Government withdrew the charter scheme, as also wanted by them. The only and the best alternative for them in this situation, as also apparently seen by them, was to find another way of re-entry into Indian EEZ; and this had to be as Indian owned vessels (instead of chartered vessels), while at the same trme maintaining their real ownership in tact. Information in circulation is that they adopted a similar strategy in quite a few other countries. The strategy, in specific terms, was to have the vessels registered in the countries in whose EEZs the foreign vessels want to fish in the name of a pseudo-owning enterprise in and of the victim country, on the conceded condition that they would be willing to receive the payment towards the cost of the LoP vessel concerned in a certain number of voyage-wise instalments, payable by way of surrendering the catches in value terms as reflected by export figures. They would facilitate the transport of the catches to the importing destination by the same vessels and the export earnings would be surrendered in the name of the 'new' owners of registry (LoP owners). The vessels would of course be operated by the foreign crew, supplemented by the counterpart national crew. This approach, as extended by the real owners with success in the case of India too, has been noted by certain Indian fisheries industrial circles also as a facade to gain fishing access into the EEZ of the country.

While it will be superfluous to dwell more on this aspect, the point to be noted now is the arrangement between real owners and LoP owners that is known to have emerged. This, as is known, is a camouflaged flag of convenience system. The registration of the foreign fishing vessels as Indian fishing vessels took place as per the provision for import of vessels allowed by a notification that was. issued by the Ministry of Commerce some years back. However, the Union Department of Animal Husbandry, Dairying and Fisheries is stated to have expressed disagreement for the operation of these vessels registered as Indian vessels by foreign crew in Indian EEZ. A compromise had to be therefore struck later, as could be understood. This compromise was that the Union Department of Animal Husbandry, Dairying and Fisheries would issue 'Letters of Permission'for all such foreign vessels registered as Indian vessels, for fishing in Indian EEZ, based on applications received from the Indian enterprise concerned in the prescribed form for the issue of the Letters of Permission.

The past decisions taken to withdraw financing support by SDFC/SCICI for the acquisition of open sea fishing vessels of 20 m OAL and above had thus led the nation over the past few decades into the present miry situation of compromise, loss of pride and loss of fishery wealth for the benefit of others. Not having a fishing fleet of needed strength, the nation is now in a situation of inability to avail of the fisheries resources of its EEZ on its own and to helplessly induct foreign interests to do this for the nation, undergoing all attendant humiliations.

India is an important member of IOTC. This membership and the right to fish for tuna in the Indian Ocean/Indian EEZ is inextricably related to India's tuna fishing presence in the Indian Ocean. As at present, India has a marginal tuna fishing activity in the Indian EEZ/ Indian Ocean. IOTC allots to its member countries quotas of tuna to be fished from the Indian Ocean (including the national EEZs) to its member countries. While the nation concerned has full rights over the tuna resources in its EEZ, IOTC has the right to allot the resources left unutilised by this nation, beyond a stipulated duration, to other eligible nations. In other words, if these resources are not utilised by the nation concerned on its own within the specified period, stated to be 10 years under the Law of the Sea, this nation will lose the quota allotted. It can then be allotted to other nations who apply for allocations. In this situation, it has been felt that LoP vessels flying Indian flag and inducted for tuna fishing in Indian ocean will help in retaining Indian rights over the tuna stocks of Indian EEZ and the open ocean beyond. It is however a moot point whether this would really become possible, considered from the angle that all the LoP vessels (particularly of Taiwanese ownership and of other registries), as understood, must have been already registered with IOTC, as they are known to have been fishing in the Indian ocean for tuna all along under one flag or the other (flag of convenience). It will be good if the foreign vessels concerned (LoP vessels) give up their earlier registration and plump for a single registration as an Indian vessel. But such a decision on their part is believed to be extremely doubtful. The reason for this is that the LoP vessels are mostly distant water fishing vessels with arrangements of various categories entered into with different countries. Their fishing plans are of a global nature *i.e.*, they tend to shift their fishing operations from EEZ to EEZ, in relation to the information they get on the potential fishing grounds where stocks happen to be available at the time. This contingency explains the reason for their sudden disappeance from a particular EEZ and surfacing in another EEZ. In the process of these adjustments, several lapses seem to get committed in respect of declaration of catches etc. As mentioned earlier, the purpose of certain foreign fishing vessels in securing registration in another country is just to gain access into that country's EEZ; and all other developments that take place thereof are subsidiary and are not of any material concern to them. What is of interest to them is their physical ownership of the vessel concerned and its catch and they are the least bothered in what manner the other concerned country accounts for the catches.

So far as India is concerned, there are news snippets under circulation in fisheries industrial circles which indicate the ordeals through which Indian LoP owners pass through to achieve their objective of earning money, while also ensuring that the real owners have access into Indian EEZ as and when they want. One point in circulation in Indian fishing circles, which can be true or otherwise, is that a) some of the Indian LoP owners had their vessels inspected by IRS in foreign ports (Sri Lanka is quoted), and b) tlie real owners commence fishing in the Indian EEZ because of this inspection without touching any Indian port. Normally, they are expected to touch an Indian port atleast to record their entry and also exit from the Indian EEZ. It is to be believed that the LoP owners intervene and ensure that the LoP vessels touch an Indian port for the declaration of the catches and for the assessment of the exact quantity exported. It is also to be believed that, in all certainty, declarations of catches are verified by the officers of FSI who inspect the catches on board. Another aspect mentioned in fisheries industrial circles is that, the placement of counterpart Indian crew does not always take place as per the regulations and the news in circulation is that there are only few LoP vessels with the placement of Indian crew on them as per the related stipulation. It is said that LoP vessels which have been operating for the past 6-7 years continue to operate with foreign crew alone, whereas the guidelines stipulate that such vessels should have 100 per cent Indian crew at this stage of operations, and the cost of such vessels should have been fully paid by them by the time of this much of duration. Many in the industry however say that this condition is not generally followed. It is also said that there can be variations between the quantity of tuna actually exported and the quantity declared. While this may or may not be correct, it is felt that declarations take place just as a formality to be fulfilled, in relation to the actual payment received from the concerned importer of tuna, as on record. Some in the industry allege that the Indian LoP vessels are those which are already registered with IOTC under a different flag other than the Indian flag. Consequently tuna catches by these Indian LoP vessels in Indian EEZ are not being recorded by IOTC as of Indian vessels, it is commented. Now IOTC wants to have tuna fleet expansion plans of its member countries so as to allot country-wise tuna quotas in the EEZs or outside them for exploitation based on historic data. It is not clear whether India has any significant historic data of tuna exports. It has, of course, in the main, recent data of exported tuna catches of the six trawlers of 23 m OAL modified for tuna long lining and also of around ten mini trawlers similarly modified for tuna long lining related to past one or two years. Thanks to the initiative of the Union Department of Animal Husbandry, Dairying and Fisheries and of MPEDA, there would soon be the induction of a few more Indian-owned 23 m long fishing vessels and also mini-trawlers equipped for tuna long lining.

An aspect that can be of concern is the information in circulation in fisheries industrial circles that some of the permitted LoP vessels get substituted by the real owners. Another alleged lapse pointed out by some in the industry (which may not withstand verification) is that Coast Guard's inspection of some of the LoP vessels is being carried out at outer anchorages of certain Indian ports without any relation to the completion of customs formalities. This can facilitate departure of the LoP vessels from Indian EEZ soon after this inspection.

In the foregoing background, it has to be agreed that the systems of chartering of vessels as followed earlier or of encouraging flag of convenience/LoP systems as at present, can at best be undesirable palliatives, that turn out to be devices for enterprising persons/companies of India to be benefitted out of these systems as long as they can. Authorities, in all likelihood, continue the present system of LoP for want of an alternative.

The one and the only alternative to be adopted as could be seen, in the light of the foregoing background, is to have straightforward Indian operations by totally Indian-owned vessels and not by chartered or LoP vessels. The drawback in this proposition, however, is the investment component. There is a vacuum in this respect after SDFC/SCICI were closed down. Commercial Banks are stated to be not interested in lending funds for acquiring fishing vessels. This means that there is no way of attracting refinancing facility from NABARD. In any case, the present situation is that most of the entrepreneurship is not in a position to invest 10 per cent of the cost of vessels to secure a loan, even if it is available.

In this situation, it is learnt that the Association of Indian Fisheries Industries has proposed that the National Fisheries Development Board (NFDB) should take the initiative of launching an integrated scheme in this regard. The substance of the scheme, as could be ascertained from the Association is that the NFDB should take up the construction of around 100 tuna longliners of about 20m OAL in the first instance, based on a tested design to be acquired from a reliable source or as designed at CIFT, including shop floor drawings, for construction at selected Indian shipyards having past track record of building quality fishing vessels (Alcock Ashdown, Bharti Shipyard, Goa shipyard, Hooghly Docking etc) or at a few selected foreign shipyards. NFDB could invest funds for the construction and acquisition of the vessels. The vessels so constructed could be leased out by NFDB under a set of well drawn conditions, to professionally competent entrepreneurs (future lessees/owners), to be selected based on particulars given in prescribed application form as received, in response to an advertisement and carefully verified. The Lease Agreements to be entered into would have to be quadripartite in nature, signed by authorised representatives of NFDB, the applying company, the predetermined buyer of the landings having good connections with importers, and who will be one having sound credentials, and the Bank concerned. The Buyer and the Bank as well as the lessees will unconditionally undertake to agree for the deduction of the stipulated proportionate amount towards the lease money at source out of voyage to voyage earnings, to be paid to NFDB by the Banker concerned. When these payments reach the level of the total cost of the vessel concerned, NFDB would have to agree to transfer the ownership of the related vessel to the enterprise concerned. There have to be of course many other conditions but they have to revolve round the aforesaid aspects. There can be a reasonable subsidy to be conceded by the Government but only at the time of payment of the last few instalments of the lease amount to NFDB and linked to good operational performance and payment of lease amount in regular instalments as stipulated in the Agreement.

Another aspect of importance is that NFDB should permit the operations only from selected ports. An organisational field structure and mechanism has to be developed to monitor the operations of the leased vessels. As the ownership, as suggested, vests with NFDB until the full lease amount is paid, the lessees would not have the problem of making any capital investments initially. Payment towards the cost will be by way of payment of instalments of lease amount. They have to mainly concentrate on securing funds towards working capital but this aspect has to be done by the enterprise concerned in a manner that convinces NFDB.

NFDB has also to take care of shore infrastructure for post-harvest handling of tuna and for monitoring export of tuna in sashimi or other forms by the lessees. The export part has to be promoted in the field by the buyer and the lessee by developing an enduring arrangement with a reliable importer and by arranging exports of viably adequate quantities. In fact, the lessees have to form into viable groups that could develop linkages with importers. A programme for the training of the Indian crew for operating tuna longliners has also to be integrated with the vessel introduction programme.

To sum up, the immediate need is to replace the LoP system by the introduction of wholly and truly Indian owned tuna fishing vessels in a period of five years or so, supported by crew training programmes, taking into consideration the various aspects discussed in this write-up. The sooner the present system of providing entry to foreign vessels in a camouflaged way to fish in Indian EEZ and carry the fishery wealth of the zone, to the utter disadvantage of the nation is replaced, the better it would be.

Fishing Chimes, Editorial 364: May 2007: Vol. 27, No. 2

# Tuna Exploitation in Indian EEZ: Low Key Indian Effort

## *(As highlighted at Tuna 2006, Bangkok)*

Anwar Hussain, Managing Director, Abad Fisheries, Kochi, India, presented an informative paper on the topic ' Tuna Industry Situation and Outlook in India' at the 9th INFOFISH WORLD TUNA TRADE CONFERENCE (Tuna 2006) held in Bangkok from 25-27 May, 2006. This presentation threw light on several important aspects of the topic. While the highlight is the present status of the utilisation of tuna fishery resources of the Indian EEZ by Indian effort, it excludes details of the support extended by the Union Department of Animal Husbandry, Dairying and Fisheries to owners of six trawlers of 23 m OAL range for equipping their vessels for tuna longlining too. The paper contributed by Anwar Hussain does not also touch upon the significant phase of past operations by a good number of tuna vessels chartered by Indian enterprises in Indian EEZ (now no longer in vogue) and also on the ongoing LoP (Letter of Permission) system (that came in the place of aforesaid chartered vessel operations). The LoP system has enabled owners of several foreign tuna longliners to use it as a device to gain entry into Indian EEZ in the guise of vessels of Indian Registry. This aspect has not been touched upon in the paper.

The contents of Anwar Hussain's paper, nevertheless, present not only a succinct picture of tuna fishing situation by Indian effort in the Indian EEZ but also the status of Indian export of tuna from 2003-04 to 2005-06, which is mostly related to the operations of foreign longliners that functioned during the period as LoP vessels. Some of them continue to do so. LoP system, as is known to many, has facilitated certain foreign tuna longlining vessels to gain a regularised entry into Indian EEZ for fishing.

Anwar Hussain perhaps deserves all praise for not making any direct reference at Tuna 2006, to LoP (Letter of Permission) foreign tuna vessels now operating in Indian EEZ, as such a mention would have placed India in a poor light before the international gathering. The absence of a reference to the Indian LoP foreign vessels operating in the Indian EEZ has however lent focus to his point in respect of essentiality of foreign participation in some form or the other for tuna exploitation in Indian EEZ. At the same time, according to another observation of his in the same presentation, India, despite its location with a linkage to the international tuna route that passes through waters of EEZ of Maldives and Srilanka, the level of exploitation of the tuna resources of Indian EEZ by Indian effort and their utilisation continues to be somewhat behind the output of its neighbours Ie., Srilanka and Maldives. While it is to the credit of these two countries that they could do well on the tuna front and that too without any foreign participation, Anwar Hussain shields the failure of the Government of India and the Indian industry from their lapses by putting across 'the view' that foreign participation' is essential in some form or the other for tuna exploitation in Indian EEZ. In other words, according to him, what Sri Lanka and Maldives do not need, India requires. It may however be noted here that Sri Lanka and Maldives do avail of foreign help but this they do only in regard to the exports of tuna but not in respect of their tuna production. Further, India may not also need foreign assistance for equipping their trawlers for tuna longlining, as demonstrated by six Indian entrepreneurs who not only successfully accomplished the addition of longlining system on their trawlers, but also demonstrated commercial longline tuna fishing in Indian EEZ by such additionally equipped vessels on a continuing basis.

Thus, the developing Indian situation is, in fact, in essential terms, no different from that of Sri Lanka and Maldives in respect of foreign participation so far as tuna harvesting is concerned. To say once again, brushing aside 'the view' that foreign participation' is essential for exploitation of tuna in Indian EEZ, Indian owners of six nos. of trawlers of 23m OAL range installed tuna longlining equipment on their vessels, with the help of the Union Department of AH, Dairying and Fisheries, and are doing well without any foreign participation. MPEDA also helped owners of vessels of less than 20m OAL to accomplish similar installations on them. The owners of all these vessels had only to import longline monofil drums, reels branch lines, line haulers and hooks in the main, and this cannot be construed as foreign participation. It will be pertinent to mention here that, earlier a couple of Indian entrepreneurs took the help of foreign experts to achieve the same result, but without success.

One aspect is now clear to many. This is that, while the Indian Union Department of AH, Dairying and Fisheries has taken the first step to extend financial assistance by way of providing subsidy to owners of six trawlers of 23 m OAL for the installation of tuna longlining equipment on them as mentioned above, MPEDA is also now playing a major role in initiating a project for extending subsidies for installation of longlining equipment not only on 23m OAL vessels but also on much smaller vessels of around 15m OAL for the installation of tuna longlining equipment. MPEDA has also come up with a scheme for providing subsidy for the introduction of new longliners upto 20m OAL by Indian fishing enterprises. These initiatives are apparently related to augmenting Indian tuna exports.

An aspect to be reconciled with is that the scope for extending subsidies by the authorities in respect of old vessels for installing tuna equipment will have limitations. Most of the 23-25 m vessels have become very old and will have to be phased out sooner or later. The looming problem is thus one of investments for the introduction of new vessels (both for replacement and also for strengthening tuna fleet),for which the commercial banks or other financing organisations have to come forward to extend financial assistance. Anwar Hussain has not touched on this problem but it is to be hoped that the authorities are aware of the problem and would come up with a solution for attracting investments. It is understood that AIFI is of the view that NFDB should organise construction of around 100 nos. of tuna longliners of around 20 m OAL based on an operationally sound tuna longlining vessel design and lease them out to selected tuna fishing enterprises. The leasing, no doubt, has to be organised based on well structured terms and conditions, one of which has to be related to deduction of proportionate/ stipulated lease amounts, voyage wise, at source, by the Bank concerned for transfer to NFDB.The leasing system as visualised will solve the problem of investmental funds faced by the enterprises concerned.

The author gave particulars of annual tuna exports from India. These rose from 6,137 t in 2003-04, to 12,658 t in 2005-06. It is believed that the quantities exported were mostly harvested by LoP vessels (which cannot be construed as truly Indian vessels). A quantity of around 880 t are understood to have been exported in 2005-06, by the owners of six or some more truly Indian owned vessels.

Another revealing point made by Anwar Hussain is that, against an estimated tuna resource of 278,000 t in Indian EEZ, the annual level of tuna landings in India plummeted from 45,167 t in 2000 to 35,721 t in 2002. The position of tuna landings in 2005-06 can be assessed from the quantity of processed tuna exported from India, which was 12,658 t, although mostly from LoP vessels.

Maldives, although a small island nation in the Indian ocean, has a strategic location. It lies adjacent to the ocean path of migration of tuna. M.Adil Saleem, Managing Director of Maldives Industrial Fisheries Co., Ltd, in his paper presented on the topic 'An Overview of Maldives Tuna Industry' at Tuna 2006, estimated that landings of tuna in Maldives in 2007 would reach a level of 80,000 to 90,000 t. The Corporation is also striving hard to increase its tuna processing capacity.

So far as Sri Lanka is concerned, according to Roshan Fernando, President, Seafood Exporters Association of Srilanka, who presented a paper on Tuna Industry in Srilanka' at Tuna 2006, tuna landings in 2006 at Columbo harbour were close to 5,000 t. While the tuna landings in Sri Lanka at its 12 fishing harbours, 35 minor anchorages and about 600 minor fish landing centres, are not mentioned, it can be presumed that the annual tuna catches must have been sizeable, considered from the angle of the total annual export level of nearly 15,000 t, as in the year 2006.

While India follows the strategy of allowing foreign tuna vessels to operate in Indian EEZ as LoP vessels, the system adopted by Sri Lanka is different. The Board of Investment of Sri Lanka extends concessions to private sector companies to apply for Fish Landing Licenses to induce foreign deep sea fishing vessels to unload their catches at a Sri Lankan Port (mostly Columbo). The licenses are issued under Sri Lanka's Fish Landing Regulations of 1996. The companies use the licenses to sign up with foreign tuna vessels to land their catches at Sri Lanka Harbours. It can be said that the difference between the Indian LoP system and the Sri Lankan Fish Landing system is that while India claims ownership of the catches, Sri lanka is interested only in providing handling and exporting facilities of the catches without claiming any ownership rights over the catches. It is enough for Sri Lanka to have a certain number of foreign vessels unloading their catches of a particular quantity at its port/ports for export. That country's satisfaction is apparently related to the maximal utilisation of the harbour facilities that the country has, and the income generated by the activity.

As already stated, the tuna resources of Indian EEZ are estimated at 278,000 t This estimate would probably have been at a higher level, had there been no interception of this restlessly moving fish on its migratory movement from the West towards the East through the Indian ocean in the Zone close to the south of the Indian peninsula. The fishes are intercepted by Maldivean and Sri Lankan fishing effort as they pass through the respective EEZs, well before the fish stray into the Indian EEZ. Impeded in their movements northwards both in to the Arabia Sea and the Bay of Bengal because of gradually increasing shallowness of the seas in that direction, the tunas, stuck up in the diversion pockets, tend to get back to the principal route of easterly migration, which is towards the south of Sri Lanka. Despite the fact that there are no traditional tuna fishing systems in the Indian EEZ (with the exception of Lakshadweep), a successful and cognitive beginning towards tuna fishing has been made recently in the Bay of Bengal off the east coast of India by converting some of the trawlers for tuna longlining. While this success is now being pursued for further expansion of the activity, the effort could certainly be broad-based and directed through three identified directions,aimed at stakeholders and linked to training programmes. One such direction, as followed in Srilanka and Maldives now is to develop a fleet of small mechanised boats for undertaking pole and line tuna fishing with fibreglass poles and/or longlining. This step will help the bulk of the stakeholders, who now mostly undertake stern trawling with small mechanised boats (earning either marginal profits or sometimes sustaining losses), to augment their incomes through tuna catches. Another step, as has already been taken by the authorities, is to convert the overaged trawlers for longlining, taking all the attendent risks of incurring heavy maintenance costs. However, the third one, an enduring and really desirable step, is to promote a major project for introducing, say, around 100 nos. of longliners in the 20 m OAL range, preferably built at the Indian shipyards, who already have experience in the construction of larger fishing vessels and who can easily pick up and adopt the technology of constructing tuna longlining vessels.

Fishing Chimes, Editorial 365: June 2007: Vol. 27, No. 3

# Indian Freshwater Prawn Trinity: Farmed Production

## *National Level Projectisation Imperative*

The International Symposium on Freshwater Prawn, held in Cochin, India in August 2003 under the auspices of the College of Fisheries, Cochin, was a 'prawnmark' event. Apart from presenting a global picture of the status of giant prawn culture, the proceedings of the symposium provided potential leads for an enduringly progressive development of freshwater prawn culture. Despite the potential value of these leads, it appears that not much of follow-up action has been taken on them.

Three prime freshwater prawn species occur in India. In order of their economic importance, as visualised by many professionals, these are : a) The Giant freshwater prawn *(Macrobrachium rosenbergii)*: b) The River prawn *(Macrobrachium malcolmsonii)*, and c) The Gangetic prawn *(Macrobrachium gangeticus)*.

The initiatives taken in the various coastal States, particularly in Andhra Pradesh and Kerala in the seed production and farming of giant freshwater prawn (GFP) are remarkable. While the efforts in this direction in the States of Tripura, Manipur, Chhathisgarh, Bihar, Uttar Pradesh and probably in one or two other States are unique, one discouraging aspect is however glaringly clear and this is related to the scanty national level attention towards this trinity of prawns bestowed by the authorities at national level. The only exception in this respect is the understandable interest being evinced by the Marine Products Export Development Authority in respect of the GFP, one out of the three.

Considering the economic importance of the Indian freshwater prawn Trinity *i.e.* giant prawn, river prawn, and gangetic prawn, the tested technical feasibility of farming of one of them le., GFP, in inland lentic water, as far away from the coast as in Uttar Pradesh, has established a clear need for formulating a national project on sustainable production of the aforesaid trinity of prime Indian freshwater prawns, if not for GFP alone. This need is not just for strengthening Indian aqua exports, but it is also relevant for the revival of the fast declining riverine stocks of all the three prime species.

So far as the culture of giant freshwater prawn (GFP) is concerned, India has achieved considerable progress. In 2005-06, there was a production of 43,395 mt which was a little over 20 per cent of its global production. While this is a remarkable achievement, there are also discouraging reports indicative of deteriorating sizes (both in length and weight) of hatchery-produced progeny of GFP due to inbreeding. All male farming of the species has been resorted to by some of the farmers as one of the means of counteracting this trend. The other strategy adopted in this context has been to utilise brooders collected from distant riverine sources for producing the seed.of GFP in the place of farm-raised brooders.

While further work on these lines would certainly be useful, what is pertinent at this stage is to review what follow-up action has been taken by CIFA, College of Fisheries, Panangad, Kerala and others concerned on the production of all-male progeny of GFP, as a follow-up to a revelation at the symposium held in August 2003. This revelation was: Amir Sagi of the Institute of Applied Biosciences of Ben Gurion University, Israel, who spoke on the topic of monosex culture of prawns, announced the success achieved in the complete sex reversal among prawns. The reversal from male to female was achieved by him by removing androgenic gland (Androctomy) and from females to males by androgenic gland implantation. He pointed out that the technology of sex reversal was linked to the identification of androgenic hormone in decapod crustaceans. The androgenic gland (as stated) is located at the distal part of the vas deferens of male reproductive system. According to Sagi, this gland is to be removed and planted in the body of juvenile female. Such a juvenile, according to him, would grow into a functional neo-female. When such neo-females are mated with normal males, all-male progeny would result, Negi explained: Development of biosynthetic products of androgenic gland to facilitate production of neo-females by farmers/hatchery technicians also requires attention.

In this background, Indian GFP farmers have been eagerly expecting that soon CIFA would come out with a standardised technology of production of all male GFP progeny for adoption by the concerned enterprises in the line. The availability of all male progeny through induction of neo-female production system, possibly in collaboration with the Institute of Applied Bio-sciences of Ben Gurion University in Israel, would usher in a new Indian era of quality GFP production and exports. In this context, in the same way as a project on production of healthy stocks of tiger shrimp under the leadership of CIFE (with CIBA as its principal associate) in collaboration with AKVAFORSK of Norway has been promoted by ICAR, there could be efforts mounted to bring about a project between CIFA and the Institute of Applied Biosciences of Ben Gurion University of Israel, to achieve the creation of neo-females of GFP as the prime source for production of all-male GFP seed.

Research efforts on the above lines would have to be directed not only in respect of the giant freshwater prawn, but also to protect the potential for promoting farming of the Gangetic prawn *(Macrobrachium gangeticus)* and the river prawn *(Macrobrachium malcolmsonii)*. These two prawns too grow to a good size in terms of length and weight (although not to the extent GFP grows). These prawns are eminently suited for the hatcheries raising their own seed in hatcheries.

The Gangetic prawn (GP) has received virtually scanty attention until now towards its culture fishery development, in glaring contrast to the level of attention GFP has received. This prawn (GP) stands overexploited now from the various principal Indian rivers such as Ganga, Padma, Hooghly and Brahmaputra. According to Tiwari, males of Gangetic prawn (GP) attain a size that ranges from 16 to 19 cm and females 13 cm. Further, Singh and Srivastava reported a size range of 6 to 22 cm in males and 6 to 17 cm among females of Gangetic prawn (GP). D.Roy, saying that the GP now faces capture over-exploitation and gross negligence in respect of its stock improvement (pushing the prawn to the verge of extinction), has appealed for measures towards evolving and perfecting methods of its seed production and farming. He says that needed initiatives are now of paramount urgency to ensure production of seed of this prawn at commercially viable level, while reducing larval mortality to the minimum extent possible. Underscoring the need to promote the seed production and farming of the Gangetic prawn (GP), he suggests that there have to be efforts, to start with, to transport the brooders in a biosecure manner to coastal prawn hatcheries for raising their seed. Once this experiment proves successful, the seed so raised or by other means can be used both for farming and also for stocking in the rivers like Ganga at an appropriately viable size. Pointing out that work should also be done to induce gonadal maturity of females for easy availability of brooders, he feels that efforts have to be also made to examine the possibility of construction of some bypass channels at the site of Farakka dam on Ganga to provide passage to GP as well as other migratory fish species to save biodiversity.

Some of us may have reservations in respect of successful breeding of the Gangetic prawn in a hatchery. These are allayed by D R Kanaujia and A.N.Mohanty, scientists at CIFA by explaining that the technology of seed production of this prawn has been developed at CIFA, although recently. It is now time to extend and popularise this technology by setting up hatcheries at selected points adjacent to rivers such as Ganga, Padma and Brahmaputra. In Tripura, Manipur and U.P., the giant freshwater prawn has been bred and its seed raised successfully.

In 1960s, lot of successful experimental work was done in the farming of the river prawn *(Macrobrachium malcolmsonii)* by the scientists of CIFRI. This work was done at Kadium farm in A.P. located near the Godavari anicut at Dowaliswaram. The early juveniles of the species used to ascend upstream in swarms over the anicut in their quest to reach the main river stretch above the anicut. This upstream migratory instinct was taken advantage of for collecting the early juveniles for transfer to the farm ponds at Kadium. The activity was however short-lived, despite its success, as the market conditions at that time were not encouraging for freshwater prawn culture. Further, owing to the construction of a barrage upstream of the anicut, the ascent of the juvenile river prawns had virtually stopped. As already stated, however, according to D.R. Kanaijjia and A.N. Mohanty, the seed production and *grow-out* technologies in respect of river prawn *M.malcolmsonii* have already been developed and are now being practised in a small way. What is however needed now is the popularisation of the technologies.

The Indian Trinity of Freshwater prawns now receives varying levels of attention from the Government, researchers and farmers. These are not in consonance, not however only with the potentialities of the Trinity to sustainably contribute to national aqua production and exports, but also with the needs for the upgradation of the socio-economic well being of the core stakeholders. As outlined in the preceding paragraph, GFP is now having its sway over several inland States although in a haphazard way, without a direction and a system.

Nation-wide promotion of integrated freshwater prawn farming (leaving the uncongenial cold water zone) has to be promoted by a nominated Central Agency. The first phase of this promotional work has to be in the form of Project survey. This survey should aim at identifying locations for setting up hatcheries along the coastline and at other centres of saline water availability inland in addition to freshwater. An investment and management system for these hatcheries can be evolved, linked to a training plan. Visualising these hatcheries as if they are in position, an integrated system for supply of the seed to the farmers in near and farther pond zones, through centralised storage and supplying seed farms to be set up, either attached to hatcheries or otherwise, has to be evolved. The execution of the Project sections that fall into the various States can be entrusted/handed over to the State fisheries departments concerned, with the overall co-ordination handled by the Central Agency concerned. The broadly indicated Project approach, as outlined above, needs to be linked to the farming phase by the stakeholders, fortified by training, infrastructure facilities and marketing linkages.

In the coastal States where prawn farming has been picking up, the problem of inbreeding and consequential stunted growth in progeny is faced. The experiences in the farming of GFP gained so far (in the coastal as well in the few inland States concerned) could no doubt be used for spreading the activity in a full measure and in a refined way in the coastal as well as inland States concerned. Some initiatives have already been taken by MPEDA for promoting sustainable GFP culture in various Coastal States. Considering MPEDA's interest in the promotion of export of processed giant freshwater prawn, it is to be hoped that it would extend.its attention towards the promotion of the other two prawns too of the Trinity, *i.e.* Gangetic prawn and river prawn too in the States concerned by taking the help of CIFA in the preparation of a Project report, that would have a national dimension, and through it promote the activity with the co-operation of the States concerned, by way of efforts directed at export promotion of the Gangetic and river prawns. It is to be hoped that efforts in the direction would be mounted, by the Centralised Agnecy to be nominated. It is hoped that there would be a general consensus that MPEDA would prove to be the best Agency to function as the Centralised outfit; and that MPEDA would definitely be convinced about the potential for the export- oriented production of all the three freshwater prawns, nationwide.

Fishing Chimes, Editorial 366: July 2007: Vol. 27, No. 4

# Ignored Facet of Value-Addition

In India, the application of the technology of value addition to fishery products is largely associated with exports. While the share of value-added items out of the exported Indian marine products has been gaining in strength in the past few years, so far as India's domestic aqua products' market is concerned, despite the vast scope, the efforts at introduction of value-added products for marketing within the country are conspiciously deficient. In an economically fast developing nation like India, fish and shellfish pickles, ready-to-serve fish curry in flexible pouches/retort pouch packaging, fish wafers, fish soup powder and several other such products will have an extensive domestic market spreading not only over cities but also over towns and even over larger villages. While it has to be conceded that a few progressive Indian enterprises have ventured, although marginally, into the production and supply of value-added products in the domestic sector, the potential remains virtually untouched.

One plausible reason for this situation can be policy inadequacies. The present policy of the Government in respect of value-addition, as already mentioned, has its focus on exportable marine and brackishwater products. The only exception is the inclusion of the giant freshwater prawn. This is understandable as MPEDA essentially deals with exports of marine products. However, there are others like Union Department of Animal Husbandry, Dairying and Fisheries and the ICAR who would have to look beyond marine products exports but there appears to be no clear indications of this being done.

The concept of value addition to marine products needs to be looked at from a holistic perspective and not in relation to exports alone. It has to be looked at also from the point of view of augmenting the stagnant earnings of fishers. The extension of the system of application of the technology of value addition to fishes for domestic marketing has a socio-economic dimension too. Once it is established that income from a particular fish species can be stepped up through value addition, its price in raw form would tend to increase, thereby contributing to the strengthening of the value status of the other fishes and the increase in the incomes of fishers.

During any fishing season, there will be catches of at least one or more value addable fish species at several landing points along the coastline. Such landing centres, in all likelihood, already stand identified. In this scenario, what has to be done is to determine confirmatively such landing locations through a survey, taking into account their distance to the nearby cities, towns and larger villages, the estimated consumer fish demand at these places, the available wholesale/retail marketing infrastructure thereat and other related aspects such as the types and numbers of craft and tackle, number of fishermen engaged in their operation, present system of disposal of fish landings etc.

Armed with the basic information as outlined above, the coastal State Fisheries Departments can constitute task forces to work out integrated plans of action to set up a network of Value Adding Fish Centres (VAFC) along the coastline of the State concerned with needed intra-State and inter-state linkages.

The basic design of these value adding fish centres (VAFC), to consist of i) material reception and storage area, ii) pre-processing/raw material preparation area iii) processing section, and iv) packing room and (iv) final product storage room, may have to be drawn by the Central Institute of Fisheries Technology (CIFT) which has the distinction of developing technologies for the manufacture of several value-added fish products whose domestic marketing, in particular, would add to the earnings of fishers who toil hard to harvest fishes from the sea. These fishes form the source of their livelihood. They serve as the raw material for value addition at the VAFCs. As is known, some of the value-added fish products standardised by CIFT are: canned shrimp and fishes such as sardines, clam meat, crab meat, oyster meat, tuna fillets etc., fish wafers, fish soup powder, fish pickles, ready-to-serve fish curry in flexible pouches/retort pouches, solar dried fish etc.

The design of the value adding fish centres (VAFC) would have to provide for the installation of all or most of the equipments for the production of the above mentioned value added items.

The present situation continues to be one of a noticeable quantity of fish bycatches and miscellaneous fishes fetching very low returns. These impose an adverse impact on the earnings of fishers. The obvious basic need is therefore to instil an attitude of fish value addition among coastal fisher populations. The mechanism to bring about this has to revolve round setting up of fish value additing centres (FVAC) at selected points along the coastline. As already mentioned, these centres will have to be set up based on a design to be developed by CIFT. Persons trained by CIFT will have to be inducted to run the FVACs as long as needed, *i.e.*, until trained coastal fishers can be deployed to take over the work. CIFT can develop a manual in the local language concerned for the benefit of the functionaries, on the lines Britishers developed a manual for fish curing at the yards set up by them, during their regime, along a good length of the Indian coastline.

The Integrated Fisheries Project was set up in 1970s for the purpose of bringing about integration of fish production, its processing (mainly at coastal centres) and organising marketing at various consumer centres. The Project can now be directed to fulfill this role which is its responsibility, as it was set up for the purpose of bringing about integration of fish production, its processing at coastal centres as for as possible and organising marketing of the products at various prospective consumer centres, through upgraded or improved marketing system or by organising new outlets as needed.

Under a central project, to consist of State-wise sub-projects to be taken up for spreading the practices of value-addition along the coastline, value-adding fish centres (VAFC) can be set up, initially on a pilot scale at selected coastal points in each of the coastal States and later increased to cover the entire constituent coastline of the State concerned. The investments to be made for setting up the VAFC, to come from an identified source, will certainly turn out to be beneficial to the fishers. These centres will have to emerge as the nuclei for spreading economic prosperity among fishers, earning name and fame as centres of supply of value-added fish products at economically viable prices to the people of the country. They have to be linked to the marketing infrastructure in cities, towns and larger villages around. The Centres will then make an enduring contribution for the socio-economic uplift of the coastal fishers on the one hand and provide protein-rich fish and fish products to the national population on the other.

Fishing Chimes, Editorial 367: August 2007: Vol. 27, No. 5

# Through issue of LoPs to Indian Enterprises to Usher in Foreign Fishing Vessels into Indian EEZ–India Opts for a Debatable System

An appraisal of the Policy initiatives of the Union Department of Animal Husbandry, Dairying and Fisheries in respect of the sustainable utilisation of the sizeable tuna resources of the Indian EEZ would reveal that they are consistent with the national needs. Not having taken any viable steps for building up a national tuna fleet for the recurrent harvesting of the migratory tuna available for most of the year, it has contented itself by helping the present owners of over two decade old six Indian shrimp trawlers in the 23 m OAL size range with a subsidy, to enable them to instal tuna longlining equipment on them. The apparent essentiality of follow-up measures for building up a national fleet of tuna fishing vessels in the needed OAL range and of needed strength seems to have been either put aside or ignored. This is an unfortunate way of dealing with this requirement, particularly when there has been a visible intrusion of Taiwanese-owned tuna longliners into the Indian EEZ either illegally or under a regularised facade like LoP system. This situation has given India the compulsion of undermining the objective of protecting, and regulating tuna fishery of Indian Ocean Zone and bracketing Indian EEZ as an area where IUU fishing takes place. Earlier, such intrusions into the EEZ of Pakistan were also learnt to have been regularised in that country and also in several other countries through the device of Letters of Permission or Licensing. Unwittingly. India has also been sucked into a similar system. The Union Department of Animal Husbandry, Dairying and Fisheries apparently found the LoP system in order, and taken steps for introducing this system of fishing operations in the Indian EEZ, with a predominant component of foreign crew.

These aforesaid Letters of Permission (LoPs) provide the much needed protection to the Taiwanese vessels, which probably have registrations of Panama or of such other countries. With similar Letters of Permission issued'by few other countries. Taiwanese vessels are known to move from one EEZ to another, depending on the information they get on tuna shoals directly from satellite data or through global information monitoring services that provide the details. The information in circulation is that, despite the so-called ownership of vessels bestowed on the possessors of Letter of Permission which provide for their registration with Indian MMD, the vessels in reality are stated to continue to be under the physical ownership of the Taiwanese owners. For having facilitated hassle-free access into the Indian EEZ, the real owners are believed to concede a mutually acceptable financial benefit to the LoP possessor on voyage to voyage basis. It is stated that LoP vessels have to be also registered with MPEDA as per the conditions stipulated in the LoP. The vessels are also to be surveyed as per the prescribed schedule by MMD.

According to the information in circulation, whether it deserves to be believed or not, very few LoP vessels are registered with MPEDA and none of the LoP vessels are registered with it as export - oriented ones. There are some who go to the extent of saying that several of them are also not even registered with MMD. The Indian Coast Guard also has problems with the LoP vessels, some say. There are news snippets in circulation that some of the LoP vessels leave the Indian EEZ without declaring the catches at any of the Indian ports. It is possible that this is a conjecture but deserves verification. In any case, according to the information in circulation, the catches of most of the LoP vessels do not go on record with MPEDA as part of Indian marine products exports. Let it be hoped that this information too is not correct. Further, the LoP vessels, supposed to be under Indian registry when they commence operations in Indian EEZ. are operated mainly by foreign crew (75 per cent of the total strength). The remaining are Indian crew (provided they are there) and the word that goes round is that they play a dummy role on board. The captain and chief engineer of LoP vessels are stated to be always Taiwanese: and whether this kind of placement is in total contradiction of the Indian ownership of the vessels under the letter of permission is not known.

The LoP vessels, although they are *de facto* foreign vessels with the operations essentially under Taiwanese control, are focused as Indian vessels, while the need for the development of India's own non-LoP tuna fleet which is of crucial importance seems to have been ignored for all practical purposes. There are some who express the view that Taiwanese interests always take care, through some device or the other, to ensure that an effective and viable truly nation-owned tuna fishing fleet does not come up. They obviously feel that, if such a fleet develops, they will have no place in the nation's EEZ, either under LoP system or otherwise.

The Ministry of Commerce and the MPEDA (that functions under it, deserve to be congratulated for their stance, as noticed by several, against LoP vessels. As already indicated, for some reason or the other, very few LoP vessels are understood to have been registered under MPEDA. Very few LoP vessels thus report their catches to MPEDA, as per the information that makes the rounds. There are also doubts expressed by many about their reporting the quantum of their catches from Indian EEZ to Fishery Survey of India.

The Association of Indian Fisheries Industries, an over 30 years old organisation, is totally against the LoP system. They support the initiatives of MPEDA to encourage introduction of truly Indian owned tuna vessels. Groups of LoP vessel 'owners', who could not countenance the opposition to LoP system.are stated to have formed into separate Associations, stated to be around five and mainly consisting of a few motivated members in each and who of course have a pro-LoP stance. Now there is no one to listen to the stand of the AIFI (except hopefully its own members). So much so, at a meeting convened recently (16 July, 2007) by the Union Department of Animal Husbandry, Dairying and Fisheries to consider the subject of development of fisheries in the EEZ, the problems of operators of LoP vessels were mainly taken up for discussion.

Let us look into the implications of LoP foreign tuna vessels operating in Indian EEZ. from the angle of the Indian Ocean Tuna Commission (IOTC). It is learnt that IOTC has not so far recognised any of the LoP Vessels operating in Indian EEZ as Indian vessels. The reason for this seems to be that the names of the Taiwanese LoP vessels are already on its record as registered vessels of other nations. Further, it is learnt that India has not yet furnished the schedule of Indian Tuna vessels to be introduced in the coming years for operation in the Indian EEZ and the adjacent open Indian ocean. Apparently, the Union Department of A.H. Dairying and Fisheries also does not believe that LoP vessels can be construed as true Indian vessels. If this belief is there, they would have furnished to IOTC the list of LoP vessels as Indian-owned vessels.

One characteristic of the on going offshore Indian marine fishing industry (not taking into account LoP vessels) is that it lacks the ability to impart a sustainable interest in the activity among entrepreneurship. At one stage the strength of offshore Indian fish fleet touched a level of over 190 nos. The strength of the fleet has gradually come down to around 40 nos. now. Several of these vessels are stated to have diversified to function as carrier vessels to ONGC. Despite fisheries resource potential of Indian EEZ, the weakening of the sector has been thus taking place. This situation places a responsibility on the Government to reverse the trend through measures that would generate confidence among entrepreneurs to participate at least in the lucrative tuna fishing industry. This would become possible only when the authorities encourage the Indian fishery entrepreneurship to move in this direction, instead of chilling such proclivities further by enabling foreign interests like those of Taiwan who are known world over about their ability to conquer the fishing opportunities in the EEZs of other possible countries, for the reason that its EEZ is small and it has a large tuna fleet of over 700 vessels, developed for global spread. The LoP approach, had to be probably adopted by India for some reason or the other and this is presently enabling Taiwanese vessels to have access to tuna resources of Indian EEZ. without bestowing any direct benefit either to the National exchequer or to the National aquaproducts export performance. No wonder, MPEDA is stated to be disillusioned at this, as the word going round indicates.

The situation that has been compellingly inviting introduction of measures to utilise the tuna resources of Indian EEZ, is known for the past two decades. Yet. the Union Department of Animal Husbandry, Dairying and Fisheries has not taken any direct and purposeful steps to meet the challenge, commensurate with the dimensions of the task, for national benefit. What seems to have been mainly done was one of regularising the overtures of Taiwanese tuna vessels to exploit the resources of Indian EEZ for their benefit, besides helping owners of six age-old Indian shrinip trawlers of 23m OAL range for the installation of tuna longlining equipment on their decks. Because of this facility, these vessels have diversified into tuna longlinirig in the Indian EEZ and of course these have been performing well. There have however been further marginal steps taken in that direction and in a low key. From the point of view of the vastness of tuna resources in the Indian EEZ, to say again, these steps so far taken for developing India's own tuna fishing fleet would have to be considered as marginal in nature. The real need is to take up a major scheme for tlie introduction of an estimated number of new Indian tuna vessels of the requisite size range aimed at asserting India's interests in the sector and to eliminate the intrusion of foreign tuna vessels into Indian EEZ, in some way or the other.

There can certainly be several alternatives to promote an Indian Tuna fishing fleet for operation in Indian EEZ. by ways other than providing access to Taiwanese fishing vessels, and that too in the quickest and shortest possible way. There can be a simultaneous programme for the introduction of tuna vessels under a long term plan. The National Fisheries Development Board (NFDB) can take care of this aspect in a manner that will benefit the nation and the industry. MPEDA is understood to have already taken up a scheme for building up a fleet of tuna vessels of intermediate range and this initiative can be strengthened further in collaboration with NFDB. Another alternative can be for the NFDB to take up a massive programme of upgrading Indian small mechanised fishing vessel fleet for undertaking tuna longlining, coupled with a training programme for their operation in the farther waters of Indian EEZ, as Brazil is stated to have done. Simultaneously, a long term programme of building intermediate range tuna vessels at selected Indian shipyards which could be brought under the fold of a consortium, can be taken up. NFDB can buy the shop floor drawings of suitable vessels from a country like Australia, which is understood to be doing well in the operation of intermediate range tuna longliners. The drawings so acquired can be made available to the suggested consortium of shipyards, linked to the requirement of obtaining applications from fishery enterprises to take the tuna vessels constructed based on the acquired designs and shopfloor drawings supplied to selected fishing vessel construction yards. The vessels so constructed can be given to the eligible applicants on terms of lease-cum-transfer of ownership basis by NFDB, once the agreed cost is fully paid by the lessees by way of periodical lease amounts.

It has to be conceded that Taiwan is one of the few tuna fishing nations, with a large fleet of tuna fishing vessels most of which are designed and equipped for long lining in distant waters, away from Taiwan. As it happens, these distant waters also include the EEZs of other countries, where Taiwanese vessels are to undertake fishing, but it is not apparently illegal so far as they are concerned. However, this kind of fishing being too risky, it is believed by many that Taiwanese enterprises have had to inevitably find ways of accessing the EEZs of other chosen countries of known tuna potentiality in their EEZs. for deploying their vessels for fishing for tuna in them. One has to appreciate and concede that they have probably no other alternative other than doing this for their survival in the fishing industry, but the affected nations, probably India being one among them too, face problems of either adjusting themselves with the situation or of extricating themselves out of it.

It has to be admitted that, owing to national inadequacies and prevailing foreign commercial interests in tuna resources of Indian EEZ, there are several conflicting aspects to be taken into account in arriving at long term as well as short term policy decisions that would lead to the accrual of socio-economic benefits to the nation in the largest possible measure. These decisions would have to be taken to replace the present situation of all the advantages (but for an insignificantly marginal financial benefits accruing to a few Indian enterprises at US $ 35,000 per large vessel, per voyage of six months and US $ 2,000 per small vessel gravitating, towards Indian LoP owners from foreign vessel owners who gain access to fish in Indian EEZ because of the LoP system. In order to consider these conflicting aspects and recommend well considered policy measures or the utilisation of the tuna resources of the Indian EEZ for national benefit, the Govenment may have to consider setting up a high level committee headed by the Secretary. Union Department of Animal Husbandry, Dairying and Fisheries/the Union Department in the Ministry of External Affairs which deals with the Law of the Sea, and with members drawn from the Central Marine Fisheries Research Institute, The Central Institute of Fisheries Technology, the Fishery Survey of India, the Association of Indian Fishery Industries and such other organisations and also of knowledgeable experts on the subject to represent the points of view of the fishery industries.

Fishing Chimes, Editorial 368: September 2007: Vol. 27, No.6

# Community-Based Minor Irrigation Tank Fisheries Development in A.P.– As Part of World Bank-Aided Irrigation Project

Minor irrigation tanks in India are one among several prime sources of fish production. The State Fisheries Departments concerned and/or those who take the fishery rights over the minor irrigation tanks on lease undertake the development and exploitation of the fisheries in these tanks.

The main claim over the waters of irrigation tanks being with the agri-farmers, the fishery developmental interests suffer sometimes. This is because water is often released from the tanks for irrigation by the authorities without taking into account the water level needs for the safety of fishes and for their survival. Problems are also faced by those in charge of tank fisheries development, in respect of escape of fishes from unprotected surplus weirs/ irrigation water outlets. Further, often there will be need to set up pens or cages in the marginal areas of the tanks for seed raising or for storage of fish. Setting up of these in an irrigation tank is sometimes not allowed by the irrigation authorities. Tree stumps, boulders and such other structures tend to cause problems in fishing operations.

In this background, the basic approach of the Tank Management Project of the World Bank is seen to be one of providing for co-ordination of the various developmental aspects such as irrigation, fisheries, horticulture, shore vegetation etc. This approach deserves to be welcomed, as it constitutes a much needed reform.

The Memorandum of Understanding on the Andhra Pradesh Community-based Tank Management Project (that covers fisheries development too), was signed between the Government of Andhra Pradesh and the World Bank on 20 April, 2007.

Andhra Pradesh has around 74,000 nos. of minor irrigation tanks. It may be mentioned here that the MOU signed between the A.P. Government and the World Bank is in fact a follow-up to the enactment of Andhra Pradesh Farmers Management of Irrigation System Act of 1997 which has a provision for participatory management of irrigation tanks. Andhra Pradesh appears to be the first State in India to opt for participatory irrigation management.

One of the components of the World Bank Project mentioned above provides for tank system management and for improvement of tank-based livelihoods. As part of minor irrigation system improvement, fisheries of about 3,000 tanks spread across 21 districts of the State are proposed to be developed in three batches of 500, 1,000 and 1,500 tanks respectively. The average waterspread area of all the tanks is estimated at over 2,00.000 ha with an average of 67 ha each per unit. Once all the 3,000 tanks (with an area of 2,00,000 ha) are brought under fisheries development, they can be expected to yield around 500 kg/ha/annum on an average, yielding a fish production of about 1,00,000 t per annum. Fish yield of this order from irrigation tanks, which would connote a significant increase in fish production from them, can be expected not only to substantially increase the incomes of fishers but, dependent on sustainable fish output from the tanks, would also reduce economic vulnerability in the fisher communities that depends on fishery output of the tanks for their livelihood.

As already mentioned, the Project has a focus on improving fish production and fish productivity of the tanks. For this purpose, the project includes financial provision for upgrading fish production practices from selected minor irrigation tanks through improved stocking, feeding, management and harvesting techniques for intensive fish/prawn cultivation and also for training, capacity building and expert visits of representatives of Fishermen's Co-operative Societies and departmental staff concerned. The approach will also provide a spread effect of the fisheries developmental initiatives over the other minor irrigation tanks, not presently brought under the Project.

So far as implementation is concerned, there is provision in the Project for the needed support, primarily at the district level. This would be provided by the State Department of Fisheries. There will be district level management teams dedicated to tank fisheries development which will be part of the Project Management Unit. This Unit will work under a State level Commissioner of Command Area Development Authority.

The Project says that, at the tank level, the focal point of organisation and implementation will be the Water Users Association. The important stakeholders of the fishery community will be members of this Association. The Association would receive 50 per cent of the rental income from lease/auction of fishing rights to tanks. Non-Government Support Organisations recruited by the Project will facilitate community organisation for the developmental process.

It would be desirable for the other States to emulate the precedence set by the State of Andhra Pradesh and take up similar co-ordinated projects for the development of fisheries of irrigation tanks along with the other components such as irrigation, horticulture, shore area development etc.

Fishing Chimes, Editorial 369: October 2007: Vol. 27, No. 7

## *To Expand its Open Sea Fishing Fleet–India Needs a Related Financing Policy*

# Starved of Investments, Open Sea Fishing by Indian Vessels in the Indian EEZ Suffers

The Government of India would have to take a policy decision that would facilitate flow of investment capital to the Indian Open Sea Fishing Industry to add new vessels to the already shrunk national open sea fishing fleet, as required. In the present situation of stagnation in the fleet strength, the Government has preferred to permit foreign fishing vessels to fish in EEZ under an LoP (Letter of Permission) system with no benefit that the nation can be proud of. The LoP vessels (which are of Taiwanese ownership) are those that have also not been recognised as Indian vessels as yet by IOTC (Indian Ocean Tuna Commission), as the information goes.

Several issues of fish production from the Indian EEZ rattle the Indian marine fishing industry. One among these issues is the harvesting of Tuna from its EEZ, for the exploitation of which the industry is now cornered to depend mostly on Taiwanese vessels taken under the LoP system. In the general absence of a mechanism to extend financial support to the industry for the acquisition of tuna longliners, the stakeholders now function under the shadow of the LoP programme that provides access mostly to Taiwanese vessels to fish for tuna in the Indian EEZ, as per the policy of the Indian Government. Here, it has to be, however, conceded that there have been efforts at adding tuna long lining equipment on existing shrimp trawlers, (which are already over-aged) through a scheme of subsidy/grant introduced by MPEDA/Union Department of Animal Husbandry, Dairying and Fisheries. Under the scheme of MPEDA, 42 nos. of medium sized shrimpers of around 50 ft OAL have also been equipped for tuna longlining and soon another 27 vessels of OAL from 50 to 72 ft are expected to be launched soon by Mr. Jairam Ramesh, Union Minister of State for Commerce.

As the tuna resource of Indian EEZ to be exploited is of the order of 0.28 million t, the initiative stands glaringly countered by the present policy of the Government itself that allows operation of foreign vessels in Indian EEZ under LoP system. There are now quite a few enterprises that hold these LoPs. Surprisingly, as understood, the Government of India gets no share of the returns from the catches harvested by LoP vessels. The bulk of the earnings go back to the real owners towards operational and other expenses, leaving a marginal fixed amount for LoP holders, as per the information in circulation. The GOI is a mute spectator watching the tuna resources of Indian EEZ being taken away in a regular and approved way by the foreign vessel owners, because of LoPs. The foreign owners operate the vessels with their captains commanding them, although these are ostensibly supposed to be operated by the LoP holders.

Australia, Iceland, and probably quite a few other developed countries have a class of commercial banks devoted to financing of integrated marine fisheries projects. Fishing vessels, whose acquisition requires heavy financial support, happen to be the main component of these integrated projects. These banks specialise in providing funds towards capital investments for the acquisition of fishing vessels and for the installation of related shore-based infrastructure on the one hand and towards the working capital requirements on the other. They have an appropriate follow-up field mechanism to be in close touch with the operations of the vessels so as to be in a fortified position of receiving back in instalments, as programmed, the loan amount, within the stipulated repayment period.

The Indian commercial banking sector, with the exception of ICICI for a short duration, has not so far been able to extend financial support for larger fishing vessels of 20 m OAL and above, although a few banks lent support for the acquisition of small to medium sized mechanised fishing boats in the overall length upto around 15m. So far as working capital is concerned, several banks have extended this facility for larger (ocean going vessels) vessels too.

While several developed nations now have specialised fisheries banks, engaged in providing investmental as well as working capital loans for the marine fisheries industries, so far as financing of ocean-going fishing vessels is concerned, India was at par with the developed nations by 1970s itself, but not so now. The Union Ministry of Shipping and Transport set up the Shipping Development Fund Committee (SDFC) which extended loans for the acquisition of ocean going fishing vessels. SDFC was however later wound up but without a substitute arrangement. A company known as Shipping Credit and Investment Company of India came into being. It lent funds to several enterprises for the acquisition of ocean going fishing vessels. The function of financing of ocean going fishing vessel acquisition later devolved on a specialised branch of Industrial Credit and Investment Company of India Ltd. However, this company also later withdrew from the function. All the three bodies mentioned above extended finance for over 190 vessels, of which around 40 are stated to be now operational.

There must be some factors that may have led to the failure of all the three successive financing bodies (SDFC/SCICI/ICICI) in recovering the loans extended, atleast to a reasonable level. Surprisingly, compared to SCICI and ICICI, the results generated by SDFC,a Governmental organisation, were understood to be somewhat better, although the basic features of the performance of all the three bodies were comparable.

There have been no known efforts made by the Government of India to analyse and identify the causes for the debacle in the financing system and on measures to be taken to set right the situation as it had emerged. The borrowers were of course to be blamed for the failure in making timely repayments of the loan amounts taken, although the lending bodies must have adopted an effective system of follow-up action integrated with a close appraisal of the fishing performance of the financed fishing vessels from time to time. There is however no information available on this aspect.

While all the three bodies mentioned above must have been as purposeful as the fisheries banks that have come up in developed nations, they could not survive, apparently for the reason that they failed to recover the loans extended. The ICICI, the last ocean-going fishing vessel financing relic, managed to wriggle out by declaring and accepting one time settlements.

SDFC/SCICI/ICICI, at the time they were dealing with the subject, seem to have contented themselves with the provision for hypothecation of loanee vessels in their favour in the loan agreements. In the history of the aforesaid financing bodies there were only stray instances of taking over of hypothecated vessels for non-payment of loan instalments, although this lapse of non-payment was not occasional. The hesitation in taking over the hypothecated vessels could have been for the reason that the financing bodies were not equipped to take care of the operational aspects of the vessels, after their confiscation. Had there been a mutually agreed provision in the loan agreement that provided for the purchaser of the catches to deposit on behalf of the loanees a mutually agreed percentage of the sale proceeds with the financing agency concerned towards the loan taken, coupled with a supervisory arrangement at the landing point, and a provision for RT contact during the voyage period with the vessels out at sea to know the catch position from time to time, the situation would have been different. Further, as the granting of loans by the aforesaid financing agencies had to be restricted to a few vessels for each of the applicants, that too to those other than large houses and without a specific provision for integration of the project with shore processing infrastructure and/or exports, the landings had to be necessarily sold to outsiders, including owners of processing plants. This weak marketing linkage must have had the potential of being porous in respect of sale proceeds. Further, the loan agreements could have been tripartite with the pre-identified acceptable buyers of catches also as a party.

Considering the rich fisheries resource potential of Indian EEZ, the imperativeness of augmenting aqua production to meet the nutritional needs of the people and for stepping up aqua products exports, the essentiality of augmenting foreign exchange earnings, the need to prevent aggression of foreign fishing vessels into Indian EEZ, and for the sustenance of India's standing as one of the major global fishing nations, there can be only one alternative of further action for the Government at this stage of the situation; and that is one of introducing an enduring system for financing acquisition of ocean going fishing vessels, keeping in view the reasons for the failures of SDFC, SCICI and ICICI in the endeavour, and also the remedial measures to be infused into a new financing system to be hopefully introduced. In this context, a study of the working systems followed by foreign banks specially the ones in Australia, Iceland (Glitnir Bank) and in other advanced countries like USA and in Europe would be useful. A delegation from Glitner Bank which was in India recently, evinced interest in supplying the knowledge and expertise they have on the subject to the Indian seafood industry through its global network. The delegation was also keen to explore the possibilities of investments in India in the fisheries sector.

The borrowers, as will be appreciated, will justifiably have a few reasons to explain delays in the timely repayment of the loan instalments. These can be a) drop in quantity and quality of catches, b) declining returns, c) repairs and increase in running expenses of vessels, d) experienced manpower shortage, and e) probably a few others. These aforesaid possible problems being part of the industry, the borrowers cannot always be blamed totally for their unintended lapses in repayments. In fact, the financing body is expected to have a system that would counteract these setbacks and at the same time take care of loan recoveries. As several knowledgeable professionals say, the loan agreement has to be fortified by making the identified purchaser of the catches of the vessel concerned a party to it, thereby making the loan agreement a tripartite one. There can be a recognised or an approved list of purchasers valid for the recognised centres for landing of catches. In a tripartite agreement, as mentioned above, there can be inclusion of a condition therein that says that the purchaser shall agree for the deduction of a certain percentage (out of the sale proceeds payable to the vessel owner) at source, to be paid to the financing body directly, for the adjustment of the same against the outstanding loan amount in respect of the vessel concerned. There can be problems in making a provision of this kind in the loan agreement but these can certainly be sorted out.

The characteristics of the fisheries sector are such that no scheme for extending financial assistance for acquisition of fishing vessels can be successful without the broad involvement of the financing agency in the critical operational aspect of the project (marketing of catches) for which financial assistance is extended. Unfortunately, unlike the allied land based activities (including these of the fisheries sector such as processing plants, hatcheries) such as those of agriculture, cattle and other domestic animal raising activities including poultry, it can be said that assessment and evaluation of results of fishery projects, being aqua-related, become arduous and call for a different strategy. The approach to be adopted may not be akin to the normal monitoring approach of bankers in respect of the other projects, referred to above. In fact, this aspect mainly led to the establishment of fisheries banks in various countries. It is understood that these fisheries banks have field wings which are continuously in touch with the loanee vessels in operation in a manner that prevents the loanee vessel owners from thinking of finding a way of postponing repayments. Further, they remain in touch with the vessels in the fishing grounds to follow the catch position etc.

The Indian fisheries industries are unhappy that the Indian Commercial Banks are reluctant to lend funds for the acquisition of larger fishing vessels. There are however a few exceptions. As mentioned earlier, this hesitancy is for the reason that they are not equipped to deal with financing of this kind. They do not have separate fisheries wings and they do not also see the need for setting up such wings. Doubtless there are many enterprises which need loans for acquiring fishing vessels, but this aspect has no cognisible impact on the working canvas of the Commercial banks in general.

Looked at from the angle of the vast exclusive economic zone that India has and the rich fishery resources it contains, particularly of the valuable tuna stocks, there can be no two views on the imperative urgency to build up a viable tuna fishing fleet of India's own, instead of allowing foreign nations like Taiwan to exploit the tuna fishery resources of Indian EEZ under some strategy or the other. For building up of a tuna fishing fleet in the correct way, in consonance with national prestige, India should have a good financing system that can take care of all conceivable resultant aspects. The Indian fisheries sector brings in a foreign exchange of over Rs.8000 crores annually to the nation. The sector also has the distinction of providing proteinacious food to the people, besides being employment-oriented. It plays a major role in the upgradation of the socio- economic conditions of fishers. Despite the importance of the sector, the Indian commercial banking system does not accord the needed priority that the fisheries sector deserves. In this background, and taking into account the failures of SDFC/SCICI/ICICI in extending investmental finance in an enduring manner for the acquisition of ocean going fishing vessels, the Government of India would have to favourably consider setting up a high Level Committee preferably under the Chairmanship of the Secretary, Union Department of Banking, with members, one each, drawn from the Union Department of Animal Husbandry, Dairying and Fisheries, the Union Department of Commerce/MPEDA, the Association of Indian Fisheries Industries and from two identified Commercial banks, one a nationalised one and the other a private one. The terms of reference to this Committee can revolve round the issue of extending financial assistance by the Commercial banks for the acquisition of ocean going fishing vessels in all its ramifications.

The suggested Committee can also probably be asked to consider the desirability of entrusting the function of financing of fishing vessels to the National Fisheries Development Board (NFDB), while taking into account the fact that NFDB by itself cannot handle the work all alone. It is desirable to consider recommending that it may enter into a collaboration with a financing body that has previous experience in the line, such as ICICI. While NFDB can be in overall charge taking particular care of the field activities, (inclusive of shore and voyage contacts by RT), through establishment of regional centres at identified places along the coast with qualified field staff, whose main job would have to be to have an appraisal of the landings by each of the financed vessels, the returns realised through their sales to one of the approved buyers, who would have to be made a party to the loan agreement, mainly for the purpose of recoveries at a certain percentage of the sale proceeds to the account of the financing institution and the rest going to the vessel owner concerned. The lending part of the work could be taken care of by the collaborating bank. The funds for lending can be partly contributed by the collaborating bank or as mutually agreed to.

Until such time the Government is able to usher in a financing system for the acquisition of ocean going fishing vessels for operation in the Indian EEZ and beyond by Indian enterprises and a fleet of such vessels of adequate strength is developed, Government may have to consider encouraging joint ventures between Indian enterprises and reputed enterprises in identified countries known for sound operations, to facilitate exploitation of fisheries resources of Indian EEZ and beyond in the open sea particularly for harvesting tuna, in the place of the present LOP system.

Fishing Chimes, Editorial 370: November 2007: Vol. 27, No. 8

# Reservoir Fisheries Development Through Partnerships

A National Workshop on the captioned subject was held in Bhopal in July 2007, organised by CIFRI and NFDB. This event kindled hopes of emergence of a tangible solution for the development of reservoir fisheries of the country. One reason for this is that, since its creation, NFDB has been endeavouring to fortify such of the fisheries sub-sectors that have not been able to forge ahead owing to inherent organisational or other deficiencies. As part of this approach, the gaps in reservoir fisheries development that have been impeding an integrated effort towards stepping up the average national per ha fish production from reservoirs and the ongoing inadequacies to fill them up must have motivated NFDB to intervene in this aspect of inland fisheries development. CIFRI being the focal organisation having a national level grasp over the reservoir fisheries developmental situation, it is understandable that NFDB has teamed up with it to optimise overall national reservoir fisheries production, banking on the State Fisheries Departments to reorganise and play their part in an upgraded manner following a partnership strategy. It is to be expected that the proposed partnership would work towards promoting an organisational structure and mechanism, supported by infrastructural facilities to be set up at individual reservoirs, at levels of reservoir clusters to be formed for organising common facilities of supplies and services, and at regional levels, so as to ensure sustainable production with needed linkages to the stage of marketing.

The aforesaid Workshop, participated by a galaxy of fisheries experts, was inaugurated by H.E. Dr. Balram Jhakar, the Governor of Madhya Pradesh.

The initiatives for the development of reservoir fisheries date back to pre-independence days. These have been mostly in the form of stocking of seed, aimed at reaching the stage of culture-capture balance, and eventual harvesting of both the natural fish stocks and those that are the result of culture, through certain patterns of leasing systems. In a few reservoirs, the setting up of cages for seed/fish production has been taken up in recent years. The result is the spread of efforts at reservoir fishery development. In fact, from quite a few reservoirs, cognisible results, predominantly attributable to stocking, have been on record. The earliest of these to be recalled was from Mettur Dam (Stanley reservoir) in Tamil Nadu, with the addition of a few more from the same State in recent years. There were also encouraging results in that direction in recent years from Gandhisagar in Madhya Pradesh, Nagarjunasagar in A.P., Tungabhadra reservoir, Krishnarajasagar and others in Karnataka, Gobindsagar in Himachal Pradesh and quite a few others. While the various State Fisheries Departments continue their efforts at developing the fisheries of reservoirs in their respective States, and at leasing them out as per rules prevalent, the CIFRI pursues its mission to evolve and introduce reservoir-specific and region-oriented technologies for the development of reservoir fisheries. The concept of partnerships in the development and management of reservoir fisheries development has apparently emerged in this process.

It was expected that the presentations made at the Workshop would bring out the various features of the partnerships concept with the required depth and clarity. While this did not seem to have come about, one aspect that emerged was that the NFDB would provide funds for reservoir fisheries development to the States concerned who may join hands with it to achieve the targeted goal. The details of the goal would no doubt be known soon.

So far as the presentations at the Workshop are concerned, while the Director of CIFRI spoke on technical issues related to reservoir fisheries development referring to management options, there could have been more of discussional focus at the Workshop on management and organisational aspects of reservoir fisheries.

Dr. S.N. Dwivedi spoke on scope of fisheries development in Madhya Pradesh. Mr. Moti Kashyap, Minister for Fisheries, Madhya Pradesh hoped that the workshop would provide a road map in respect of fisheries development in the State. H.E. Dr. Balram Jhakhar, the Governor of Madhya Pradesh unlike the above mentioned participants, however, emphasised on fisheries development of reservoirs, stressing on the need for pin-pointed action thereof. Dr. V.V. Sugunan laid emphasis on community-based fisheries management of fisheries. He had not however placed before the Workshop several other points that he highlighted in his valuable report on Indian reservoir fisheries published by FAO as Fisheries Technical Paper No.345 in 1995. Dr. S. Ayyappan spoke on the general role of NFDB with particular reference to development of reservoir fisheries and funding for the purpose. Mr.Ajit Bhattacharya, Union Joint Secretary (Fisheries) spoke on reservoir fishery leasing policies particularly in respect of inter-state reservoirs. Dr. P.V. Dehadrai said that the Government of Bihar had declared aquaculture as part of agriculture, but he had not explained in what way this declaration would help in upgrading reservoir fisheries development. It was also mentioned by him that the concerned should directly interact with the Reservoir Management Authority, without giving any indication of which authority he was referring to. He, however, desired that NCDC (not NFDB) should focus on infrastructure development rather than on loans and subsidies. The representatives of several State Fisheries Departments spoke on fisheries development in their respective States with marginal or no reference to reservoir fisheries development. Only Mr. Rajeev Kumar, Director of Fisheries, Uttarakhand spoke on reservoir fisheries development in his State, emphasising on the training aspect. Dr. V.R. Desai, former Director of CIFRI, felt that inadequate fish seed availability for stocking reservoirs was a major constraint in the development of reservoir fisheries. He also pinpointed the need for reassessment of reservoir fisheries resources and fish seed requirements for reservoir fisheries enhancements. Dr. G.P. Dubey spoke on reservoir fisheries development in Gandhisagar and Dr. Dwivedi felt that the Gandhisagar model of development could be emulated by the other States for reservoir fisheries development in their respective States. One solid suggestion pertinent to the topic of the Workshop came from Mr. Udai Varma of Narmada Valley Development Authority. He proposed a three tier co-operative model, based on suggestions in this respect by Mr. Samar Dutta of IIMA.

A laudable outcome of the Workshop, as could be noted, was the main recommendation that laid emphasis on the partnership aspect between CIFRI (as the technological provider nation-wide) and the NFDB as the fund provider to the various State Governments for the execution of the action plan (to be drawn up) to improve fish production (of reservoirs). This recommendation is, however, silent on the organisational mechanism and management strategy to be adopted at State, regional and central levels.

As Dr. Kohli of CIFE observed at the Workshop, funding alone would not bring about reservoir fisheries development. As is known, the basic initiative for reservoir fisheries development happens to be in the hands of the State Fisheries Departments. For exercising this initiative, they need a workable integrated plan of action that covers all the reservoirs in the States concerned. It can be said that the main factor that has been impeding the development of reservoir fisheries is the general absence of integrated initiatives in that direction. To start with, these have to be related to the preparation of projects in each of the States at a) individual reservoir level; b) reservoir cluster level; c) regional level and at d) State level. There has to be a co-ordinating set up at national level (NFDB level). Adequate attention has to be bestowed on the preparation of the aforesaid plans before setting the process of reservoir fisheries development in motion. NFDB, in association with CIFRI and the State Governments can have these projects prepared in the first instance. For the preparation of these projects right from individual reservoir level, the main components of integration to be taken care of are; i) location of or identification of sources of seed supply; ii) setting up of seed storage units near reservoirs for keeping the seed for sometime (when needed) before releasing in the reservoir concerned;iii) provision of facilities for periodic sampling and for final harvesting; iv) organising centralised cold storage facilities for the harvested fish either from the reservoir concerned or a cluster of reservoirs; and v) provision of transport facilities to the nearest fish market from the cold storage concerned.

Reservoir-wise projects and also projects for clusters of reservoirs with provision for infrastructure facilities and needed trained manpower may have to be also drawn up. Providing for the needed linkages among these clusters, regional projects may have to be also prepared. Such regional projects could be put together to form the State projects. Inter-state projects and national level set up can be developed at NFDB level. These projects can aim at interstate supplies of yearlings for stocking and also for inter-state supplies of harvested fish as per requirements.

It is felt that the completion of the basic ground work as mentioned above will have to be the starting point to undertake purposeful reservoir fishery development. This can probably be accomplished through the setting up of Reservoir Fishery Development Agencies headed by Chief Executive Officers. These can be set up in respect of each of the reservoirs as decided, or for a cluster of reservoirs. Regional and State level agencies may have to be also set up for coordinating the activities. Further, there would be need to have a reservoir fisheries developmental set up at national level (NFDB).

As already mentioned, the lacuna in respect of an organisational mechanism (and management strategy too) is responsible for the present underdeveloped state of Indian reservoir fisheries. The continued neglect of the situation would only imply its perpetuation. The NFDB has to consider advising the State Governments to set up Committees to make appropriate recommendations for the evolution of an organisational mechanism for the development of fisheries of reservoirs in their respective States by providing them with a set of guidelines for the purpose. Once the State level organisational set up comes into being, a national level system at NFDB for the co-ordination of the State-wise activities could be introduced. The infusion of an organisational set up to handle the massive task of development of fisheries of reservoirs in India having a total area of around three million ha with a potential to produce sustainably at least 150,000 t of fish annually at a minimal average fish production rate of 50 kg/ha, is of paramount urgency. To believe that this massive task could be achieved without the backing of a full fledged organisational mechanism and a management strategy will continue to be a mirage.

Fishing Chimes, Editorial 371: December 2007: Vol. 27, No. 9

# Diamond Jubilee of CIFRI: Vintage CIFRIANS Reminisce

2007 is a year of pride and reflection to the fisheries fraternity of India. The pride stems from the fact that two of its prime fisheries research institutes *i.e.*, Central Inland Fisheries Research Institute and Central Marine Fisheries Research Institute, are now in the Diamond Jubilee year of rendering yeoman service to the fisheries sector of the Nation. In the momentously active growth period of 60 years, the aforesaid institutes had to bid farewell to several of its scientists who retired from service. The retired scientists who were once in vantage positions at the institutes have proved themselves to be the best to tell about the glorious ascendency of CIFRI to the present status. In order to place on record their impressions on the rise of CIFRI for the benefit of the posterity, the Director of CIFRI has brought out a commemorative publication entitled 'Reminiscence ~ CIFRI (1947-2007), containing the reminiscences mostly of several retired scientists of the Institute. This publication will no doubt be followed soon by a similar one in respect of CMFRI.

The aforesaid publication was released on 10 July, 2007. It is a document that would qualify to be a historic one. The reminiscences of a galaxy of scientists ranging from Dr. Hiralal Chaudhury to Dr. V.R.P. Sinha and a few others reflect their calibre and maturity to succinctly chronicle the past landmark events of the Institute, enshrined in their memory. They also vividly tell us about their remarkable inputs that propelled CIFRI to its present magnificent status.

Dr. Hiralal Chaudhury, in his write-up, apprises us of the beginning of CIFRI by saying 'The Station that time existed with primitive facility, without electricity, instruments, even drinking water was difficult to get". Dr. (Mrs.) Sarojini Pillai said that in those days, the Chief Research Officer had his living quarters on the campus premises, the only one thus privileged'. Dr. V.R. Pantulu said that he went to the Institute to take charge of his post expecting to enter a grand research laboratory, but, instead, what confronted him was a pathetic conglomeration of non-descripit, dilapidated military hutments of world war two vintage. Paying a tribute to late Dr. H Srinivasa Rao, a former Director of CIFRI, Dr. Pantulu said 'If indeed anybody deserved the title of 'FATHER OF THE INSTITUTE' it was Dr. Rao'. Dr. Pantulu said that it was Dr. Rao who took the initiative for having the military hutments at Barrackpore, in which the Institute was located, declared as unfit for human occupation and initiated steps for the construction of the present magnificent building in which CIFRI is now housed. Dr. Pantulu further said, 'Admittedly, the prominence of the Institute in the National arena was due to the success in induced breeding of major carps. This breakthrough was achieved by Dr. Hiralal Chaudhury. Dr. Chaudhury, then as now, is a very unassuming, amiable gentleman. A good songster with a sonorous voice, he used to entertain his friends, of whom I had the privilege of being one, with East Bengal songs', He Said. Presenting a nostalgic account of his historic experiences abroad from 1965 commencing from his stint at Mekong Secretariat, Pantulu concluded his reminiscences by saying that 'I have never looked back at the Institute after I left it not because I did not want to but because I could not bear to'.

Dr. R.D. Chakravarty, reminiscing on the history of CIFRI said "The CIFRI started in 1947 at Maoirampore, Barrackpore, in hutments in use reportedly during the World War II as bakery and has developed into a premier centre for inland fisheries research in Asia". Saying that the science of applied fisheries was new when it was started in India, he observed that researchers at the CIFRI were then required to do and learn. The lack of reference material during the early years was common, he said, adding that fisheries science as is known now developed much later, based on work done in research institutes such as CIFRI. "As such, it may not be improper to say that it is the CIFRI that laid the foundation and developed modern freshwater aquaculture in our part of the world" he pointed out. He highlighted the tough field work of early years of CIFRI.

Prof. HPC Shetty recollected with gratitude and fondness his long association with CIFRI, pointing out that the Institute mothered a few related national institutes, namely, CIFA, CIBA, NBFGR and NRC on Coldwater Fisheries. Saying that in CIFRI, in those days, it was a case of survival of the fittest for the scientists, he observed that the present scenario was different with periodical assessments for promotion in ICAR institutes. Earlier, only one out of a dozen or more meritorious candidates for a post would get promotion. In the present set up, every eligible meritorious candidate got the promotion, it was mentioned. Prof. Shetty told the participants about historic services as the project co-ordinator of spawn prospecting investigations and as the major player in the work of the Fish Seed Committee, earlier in 1960s. He recalled with nostalgy his association with stalwarts like Dr. Bhimachar, Dr. V.G. Jhingran, Dr. K.H. Alikunhi, Dr. V.R. Pantulu and others.

Dr. Apurba Ghosh, recollecting his experiences in CIFRI, highlighted with pride his association with Dr. T.V.R. Pillay, who played later a pivotal role in propagating global aquaculture development programmes in underdeveloped and developing countries of Asia and Africa through FAO.

Dr. V. Gopalakrishnan, with a conspicuous clarity and a focus on details related aspects of his association with the past major events at CIFRI, articulated his thoughts on current status of Inland Fisheries of India. Now in USA, it is amazng how well the events that took place so long back are etched in his memory.

Dr. T. Rajyalakshmi said that, during the early days of CIFRI, with no infrastructure in respect of transportation, reference books and laboratory equipment, the scientists worked with dedication, adopting innovative ideas for developing research, establishing survey and sampling systems and gathering data. She paid rich tributes to the stalwarts. Dr. B.S. Bhimachar, Dr. T.V.R. Pillay, Dr. V.R. Pantulu and Dr. V.G. Jhingran, who guided all those at the Institute in proper direction. Highlighting some of the aspects of fisheries developmental work handled in those days that covered Hooghly-Matla-Roopnarayan estuarine systems, the Farakka problem, and many others such as reservoir fisheries development, studies on mangroves and brackishwater aquaculture, she said that she learned to 'walk' at CIFRI.

Dr. S.P. Ayyar presented a vivid account of his impressions of CIFRI that accumulated during his long sojourn in CIFRI. He said: 'The Institute, founded by the eminent Fishery Scientist Dr. S. L Hora and nurtured by doyens like Drs. H.S. Rao, T.J.Job, B.S. Bhimachar, V.G. Jhingran, K.H. Alikunhi, Dr. H. Choudhuri and M.T. Philipose, to mention a few, has grown in the last 60 years into one of the most prestigious Fisheries Research Institutes in the world. The trail blazing R&D achievements and the technology packages developed by CIFRI in inland fisheries and aquaculture had been singularly instrumental for the great strides made in these fields by our country'. It is gratifying to note that a large number of positions in various Fisheries Research Institutes, Fisheries Colleges, State Fisheries Departments and Private Fisheries Industries, as also at different other places are manned by CIFRI trained personnel, he said.

Dr. S.D. Tripathi chronicled his memories of CIFRI, laying focus on his experiences as the scientist in charge of the Training Section of CIFRI, his experiences during his association with NGOs and farmers and in handling coordinated projects.

Dr. P. Das journeyed down the memory lane telling about his rich experiences in spawn prospecting in West Bengal. He observed that the glorious period of CIFRI began from 1971 onwards, when Dr. V.G. Jhingran initiated four All-India Co-ordinated Research Projects, *viz.*, operational Research Projects, Krishi Vigyan Kendras, Training Centres and the most required Fisheries Extension Services.

Dr. Maniranjan Sinha, saying that his memories of CIFRI went back to 1961 when he joined the Institute at an age of 19 years, and relating his rich experiences at the Institute, observed that the 'road ahead for CIFRI is much longer and difficult. The challenges faced by CIFRI are very many. The onus of providing solution to the current research problems is on CIFRI and the whole nation is looking towards it for the same. With its glorious past, I am confident that CIFRI would be able to meet these challenges in years to come'.

Dr. Modadugu V. Gupta, World Food Laureate said: 'with the fish stocks from open waters declining due to various anthropogenic causes, CIFRI has a daunting test of coming up with management solutions and technologies for sustaining and increasing inland fish production. CIFRI has come a long way in 60 years of its existence and I wish CIFRI success with efforts to contribute to the welfare of the nation through sustained productivity of inland waters and management of ecosystem health'.

Dr. Raman presented a very lucid account of his experiences as a Principal Scientist at CIFRI and a gist of his achievements in the culture of freshwater prawn and studies of fisheries of Pulicat lake.

Dr. C.S. Singh gave an account of his journey through CIFRI and his present work at his own fish farm in his village in Uttar Pradesh.

Dr. P.V. Dehadrai alleged that, while four important air-breathing fishes were induced bred at Khanpur near Delhi, the support of the co-ordinated project of CIFRI could not be secured for the spread of the technology, probably because of apprehensions about diluting the credits the Institute was gaining through composite fish culture. He referred to one embrassing episode of a weird claim to culture of the so-called vegetarian frog *(Rana hexadactyla)* which really hurt the scientific integrity of the Institute. He was, however, happy that the sad story ended without any severe repercussions on the Institute. Observing that after Dr. Jhingran, the leadership in the CIFRI had been waxing and waning, he was happy to note that Dr. Maniranjan Sinha and Dr. K.K. Vass after him had shown pragmatism and commitment towards the Institute, by adding new infrastructure, reorganising divisions and inculcating team spirit to promote second line of leadership.

Dr. V.R.P. Sinha dealt with the remarkable developments that took place under the All India Co-ordinated Research Project on composite Fish Culture and how the farmed fish production rose under the project to 2,000 kg per ha from a level of 200 kg/ha. He related that this increase was because of the work done by 429 FFDAs which came into existence, and saying that the technology followed by FFDA was that evolved by CIFRI. He also gave an account of the present status of development of major carp farming in different States of India.

Dr. K.K. Vass, the present Director of CIFRI, relating his experiences at CIFRI, particularly under the Directorships of Dr. Arun Jhingran, Dr. S.P. Ayyar, Dr. V.R.P. Sinha, and Dr. M. Sinha, said that he was proud to have been associated with CIFRI through trial and tribulation from its silver to the diamond jubilee period.

He said that the scientists at the Institute shifted their focus from production system research to ecosystem based management, a big challenge.

While Dr. Sugunan was nostalgic about the Howrah Station which was the connecting link in his journey to Barrackpore, Dr. Dilip Kumar was proud to say that CIFRI gave him the depth in fisheries science and educated him on its purpose.

Dr. S. Ayyappan paid a fishful tribute to CIFRI. Saying that the dream of every student of Fisheries was to visit the CIFRI, he observed that the Silver Jubilee Souvenir of CIFRI had become a reference collection. He said that the challenges that the inland fisheries faced were many, stakes were high, and the potentials were also huge. He added: "CIFRI has the right mix of experience and expertise, credentials as also the new tools, achievements as well as the vision. There are great expectations from stakeholders and the Institute surely would provide everything required for sustainable and profitable fisheries. At this juncture of Diamond Jubilee of CIFRI, I cannot pay a greater tribute to the Institute than to say that, "if I were to start my career all over again, I would join CIFRI, Barrackpore".

Dr. N.C. Dutta, while highlighting the achievements of CIFRI in a historical perspective put in focus its main contributions such as induced breeding of IMCs, the technologies of composite fish culture, polyculture and integrated farming.

The publication 'Reminiscence 1947-2007' of CIFRI can be described as a historic document. The various contributing authors to the document have related the development - oriented scientific and technological personality of the CIFRI in a focal perspective.

Fishing Chimes, Editorial 372: Jan/Feb 2008: Vol. 28, Nos. 10&11

# Flash Back 2007: Year of Tuna in India

Quite a few fisheries events took place in India in 2007. One prime event out of these is the cognisible emergence of the tuna sector, outshining all others. It is common knowledge that initiatives to promote the integrated development of the Indian tuna sector have been there for long, although on a perfunctory scale. The first serious initiative in this regard was taken in October 2002. In this month, a historic beginning was made by the industry, with the support of the authorities, by inducting two modified shrimp vessels of 23 m OAL range (one each of Sagarika Seafoods and of Lalitha Seafoods) for diversification into tuna longlining. The tuna fishing results from these vessels were considered as encouraging but the marketing part of the catches, more or less, are stated to have turned out to be disappointing, whatever be the reasons. As time went by, in late 2005, there was, however, a revival of the aforesaid initiative by the Fisheries Division of the Union Department of Animal Husbandry, Dairying and Fisheries. Encouraged by the subsidy offered by the Union Department mentioned above, four Visakhapatnam-based fishing enterprises added tuna longlining equipment to six of their trawlers of 23m OAL range. So equipped, the entrepreneurs started tuna fishing operations in mid-2006 from these vessels, with an integrated operational strategy, which had provided for an effective export marketing tie up. Consequently the operations have been gaining in strength since then and the results have been very encouraging. There was, however, a time lag in respect of giving this fillip to the activity to reach this encouraging situation. One concluded reason attributed to this time lag was that the Union Department of A.H, Dairying and Fisheries took the aforesaid remarkable success in its stride without imparting expeditiously the needed follow-up drive to motivate the other trawler owners to emulate the example set by the group of owners of the six successful vessels. One reported observation made in knowledgeable circles to explain the delay was the bestowal of overconcentration by the said department in authorising operations of LoP tuna vessels mostly of Taiwanese ownership in the Indian FEZ. Another unreconcilable aspect often mentioned by some is that till November 2007, the top officials of the Union Department of Animal Husbandry, Dairying and Fisheries had not thought of even expressing their appreciation of the remarkable achievement of the group of owners of the six vessels, who gave the first ever performance of achieving the harvest and export of around 700 t of chilled tuna besides exporting a few thousand tonnes of frozen tuna loins etc in 2006-07. On 12 October 2007, however, although belated, at a meeting convened by MPEDA, the entrepreneurs concerned were first appropriately honoured by the Association of Indian Fisheries Industries. This was of course later followed by another meeting convened by the Director General, Fishery Survey of India at Chennai on 19 November, 2007 at which the entrepreneurs were deservedly honoured for their performance.

Thus, the efforts directed at tuna fishery development of Indian EEZ by Indian owned tuna vessels, which started in October 2002, took around 48 months to yield enduring results not only in respect of tuna production but also of its exports as an integrated activity. The credit for this achievement of course goes to all the owners who pioneered in the line. A good share of the credit for the achievement in a substantial measure also goes to the fishermen of Kanyakumari zone who mostly conducted the deck operations and who, with their experience in longlining, made the operations superbly successful.

The formal recognition of the success of tuna fishing operations in the Indian EEZ having finally come forth from Union Department of Animal Husbandry, Dairying and Fisheries and also from the Union Department of Commerce (MPEDA) in 2007, this year (2007) is entitled to be enshrined in the history of development of Indian marine fisheries, as the Year of Tuna in India.

In this background, the emerging outlook is that (encouraged by the subsidies offered by the Government) the owners of the remaining fleet of shrimp vessels are likely to opt for conversion of their vessels for tuna lining. Further, it is possible that Government may consider a scheme for the introduction of Indian-owned newly built/ second hand 18 m OAL tuna longliners for operation in the EEZ, as has been advised by the Australian tuna fishing expert George K. Skoutarides.

MPEDA, apart from its contribution and achievement in the tuna sector, has other laurels to its credit in 2007. It set up the National Centre for Sustainable Aquaculture (NAcSA) at Kakinada in A.P. This centre initiated measures for the promotion of sustainable aquaculture. MPEDA also has the distinction of establishing in Cochin in 2007 an organisation called NETWORK (Network for fish quality management and sustainable fishing). This organisation is mainly concentrating on promotion of sustainable capture fishing and quality management of captured fish.

RGCA of MPEDA achieved a breakthrough in seabass culture. This achievement has opened up a new opportunity of diversification for the aqua farmers from their present practices, particularly of those in the shrimp farming line, to diversify from their present concentration on this activity. RGCA is now equipped to produce seabass seed at its hatchery and supply them to the farmers, and also to extend to the farmers the technologies of hatchery seed production of seabass and of farming them

Another major stride made by MPEDA has been to develop infrastructure in Andamans for the production of SPF tiger shrimp brooders and seed. This progress will have a far reaching reformatory effect on the disease-prone shrimp farming operations currently taking place at most of the farms located along the Indian coastline.

**What would appear to be a significant correlation to the experimental sea cage farming taken up by CMFRI during the year, MPEDA has entered into an MOU with a professionally reputed Norwegian enterprise, Innovation Norway, for the promotion of sea cage farming in Indian marine waters. This is a major advance in the Indian marine fisheries sector.**

The other fisheries events that have lent pride to the nation are the entry of CMFRI, CIFRI and FSI into the Diamond Jubilee year and the CIFT into the Golden Jubilee year. There have been other events of significance that took place in the year 2007, such as Indaqua 2007, introduction of crab culture project by Chilika Development Authority, experimenting with mechanised pond fish harvesting system by CIFA and many more.

Fishing Chimes, Editorial 373: March 2008: Vol. 28, No. 12

# Upgradation of Onboard Storage Facilities on Small Trawlers

## *MPEDA's Subsidy Scheme bestows multiple benefits*

There have been focal efforts mounted by the Union Fisheries Division as well as by MPEDA for promoting the addition of equipment on larger fishing vessels, mostly engaged in shrimp trawling now, for undertaking tuna longlinig too. The stakeholders have been one with the authorities in respect of this approach as it would enable them to extricate themselves from the problem of spiralling operational costs in respect of shrimp trawling on the one hand and declining unit prices of shrimps on the other. However, only some of the owners who could secure bank or other kind of finance could avail of the incentive offered as subsidy by the aforesaid authorities for the addition. Among the vessels so far so equipped are those in the category of OAL of 20 m and above, mostly of 23 m OAL range.

The fleet strength of vessels of the above mentioned range in the country is, however, such that even when all of these vessels which have facilities for storage of tuna caught, are equipped for longlining, the impact in terms of stepping up of national tuna exports would not be adequately on the higher side and not so impressive in relation to the available tuna resources of the Indian EEZ.

It is logical to presume that MPEDA took into account this aspect as well as quite a few other important factors while introducing a subsidy scheme recently for addition of refrigerated/insulated fish storage facilities on small fishing boats of less than 20 m OAL. The main one out of these aspects can be the problem of organising finance for the construction of new tuna vessels of, say, 18m OAL, as has been recently recommended by the Australian tuna expert, Mr. George K. Scoutarides. Another aspect that must have obviously come up for consideration was the issue of organising the conversion of smaller vessels now engaged in trawling/gill netting for undertaking, for instance, tuna longlining too in the EEZ. The precedence set by Brazil, in the matter of assisting small vessel owners for having augmented facilities for tuna longlining can be recalled in this context.

The Union Fisheries Division does not seem to have taken any initiative for promoting the upgradation of onboard facilities on small trawlers. The reason for this may be the constraint in organising conversions of this nature for the reason that the smaller boats conduct fishing mostly in territorial waters which come within the purview of the coastal State Governments, under the Constitution. Of course, the Centre could always extend assistance to Coastal State Governments for bringing about the reform.

It is laudable that MPEDA has announced the subsidy scheme for the installation of insulated/ refrigerated fish hold/RSW system and ice making machinery on board mechanised fishing vessels of less than 20m OAL. This is a highly encouraging initiative not only for the benefit of the stakeholders but also for augmenting tuna production from the EEZ. All the additions mentioned above are needed for post-harvest handling of harvested tuna. The scheme does not, however, cover assistance for the installation of longlining equipment on the boats, probably for the reason that longlining from most of these vessels can be done manually. The main investments may have to probably cover those for setting up insulated/refrigerated fish holds, which are needed, but the owners would not be able to have them without assistance. So far as investment on longlines with hooks of around, say, 20 to 25 km length on each boat is understood to have been considered but MPEDA must have felt that this will be within the means of the boat owners. It, however, appears that financial assistance by way of subsidy for the installation of longlining system too is well justified.

The small boat operators would no doubt be grateful to the MPEDA for introducing the subsidy scheme. One benefit to be noted in this context is that, while the additional installations enhance the capabilities of the small fishing boats, they would also acquire the flexibility of undertaking multipurpose fishing systems that would cover trawling as well as longlining. The prime advantage is however related to the upgradation of economics of operations.

The subsidy scheme would undoubtedly receive widespread positive response from the boat owners. Besides guidance in respect of submitting applications for subsidy, MPEDA would no doubt take the needed initiative now for providing at the various fishing harbours the needed additional infrastructure facilities and linkages for the pre-export processing and packing of the tuna catches and also for export marketing.

It can be expected that, as a result of the scheme, nearly 30,000 small fishing boats (equipped ultimately with additional onboard facilities) would eventually be able to land at least at the rate of around 3 t of tuna (besides others) per boat on an average in a year as additional tuna landings, the total probably coming to around 90,000 t in a year, fetching around US $ 220 million more per annum.

Fishing Chimes, Editorial 374: April 2008: Vol. 28, No. 1

# Active Political Support Needed for– Faster Fisheries Development of India

Technologies in respect of developmental subjects like fisheries by and large, are evolved by scientists, for field application. Once a technology is developed, there will be initiation of follow-up action by the concerned to the stage of its successful adoption by the related stakeholders. As an essential part of this endeavour, a flow of help to the stakeholders with financial support by way of subsidies and/or infrastructure support, wherever necessary, would have to be organised by the authorities. In order to secure support of this kind, which often becomes essential, initiatives of public bodies such as associations/societies of fisheries stakeholders have to come into play, to prevail upon the authorities, when needed, to extend the needed help. This kind of support from associations/societies is marginal, as at present, and it is also with very little impact. Politicians, except the very few who are ministers dealing with fisheries in the Government concerned, are known to be generally complacent towards fisheries development. In order to secure the involvement of the politicians concerned, the office bearers of fisheries associations and fishermen's co-operative societies should appraise the politicians, preferably elected representatives of the area concerned, of what the stakeholders need and prevail upon them to highlight the problems of stakeholders in a sustainable manner as a support to their representations to the authorities concerned, so as to ensure their intervention.

Several in the Fisheries Industries and also those in the Fisheries Departments generally happen to be against political interference in fisheries developmental matters. This stance cannot be considered as in order. We live in a democracy and all developmental activities have a direct or indirect relationship with people, and politicians represent the people. They know the pulse of the people. A scientist or an extension worker will have his own limitations in popularising a technology among stakeholders. These limitations can be neutralised and the extension work can be taken forward with support provided by the Associations/Societies and with the positive intervention preferably of elected people's representatives concerned, who will have the approac to the authorities in the Governmental set up, to secure what the stakeholders need, such as subsidies, related infrastructural facilities etc. The concerned politician, once he knows what his constituents need, can liaise fast, convince the authorities concerned and ensure that the stakeholders get what they need to implement a new technology. For example, a stakeholder needs certain facilities to take up cage culture or to install certain essential equipment in his boat. When a politician intervenes in a positive way, there will be a sharp result that would help the stakeholders. After all, the politician concerned has to prove himself.

From time to time, as new technologies become essential, authorities work towards developing them for introduction. In this process, in the light of the commercial potential of the new technologies concerned that they come to know, the stakeholders too get interested in their introduction. For achieving this, they also need a viably strong support (public/political) to achieve a quicker mode of their introduction. It may be mentioned here that there are quite a few potential technologies, which, when implemented, can upgrade the living standards of fishers, and these are awaiting introduction both in the marine and inland fisheries sectors. So far as the marine fisheries sector is concerned, it is now widely known that marine capture fishery output has been dwindling and that this trend has an adverse impact on the socio-economic conditions of fishers. One alternative mooted to get over the problem is to introduce marine cage culture. For the introduction of this activity, there has to be a major scheme that would cover location of sites for the setting up of cages, selection of persons for allotment of these sites and providing training to them, organising infrastructura support for fabrication and supply of cages, and thei installation. The activity also needs support by way f setting up hatcheries and attached farms for producti n and supply of seed of appropriate size for stocking c ges. The stakeholders have also to be trained in taking ca e of the crop, harvesting of the crop and in marketing of he produce. As India does not have the technology of cag culture, CMFRI has recently taken the initiative of starting experimental cage culture. As a further step in this direction, MPEDA has entered into a joint venture with Norway for the introduction and promotion of cage culture. In order to impart an integrated fillip to a massive activity of the nature of marine cage culture, the stakeholders need support from local leadership. The activity also calls for planning from various angles and an organisational structure for implementing the plan has to be built. Once there is good support to the authorities on the promotion of cage culture, their hands will be strengthened to go ahead with the implementation of the programme with a sense of expediency for national benefit and for the benefit of stakeholders.

Besides promotion of cage culture, there are other lines of endeavour that would need public/political support. One among these is the taking up of a project for equipping small mechanised boats for tuna longlining, as has been done in Brazil. This is what the owners of the boats now need, to wriggle out of the present socio-economic plight they are in, because of the steep fall in capture shrimp and fish output. Another aspect is that the Indian EEZ has now become a virtual monopoly of foreign tuna fishing vessels, particularly of Taiwanese ownership. There is public/political opinion and support to prevail upon the Government to have the policy of introducing really Indian owned tuna vessels to replace the foreign vessels now exploiting the tuna resources under the system of LoP.

The LoP system may be a shield to protect India from its lapse in not developing its own capabilities for exploiting tuna in its EEZ as required under the regulations of Indian Ocean Tuna Commission, but, unfortunately, it has had the effect of showering benefits over the owners of foreign tuna vessels (Taiwanese-owned tuna longliners) on the one hand and on the LoP holders on the other, with marginal benefits to the nation. In national interest, politicians would have to be educated to oppose the LoP system and to herald a movement for the introduction of fully Indian owned vessels.

On the inland aquaculture side, a major issue that has to be pointed out is that public/political support is needed for the utilisation of the vast saline lands, now lying fallow, in the north west of the country (Rajasthan, Haryana, Western U.P. and Punjab). The development of fisheries in these lands through their conversion into farms for production of salt water fish would contribute not only to the production of commercially viable and export-oriented species such as shrimp, sea bass and others but also to the upgradation of the incomes and socio-economic conditions of the people of the area. Public workers and politicians have to recognise this challenging situation and prevail upon the State Governments concerned to take up projects to achieve the transformation of these lands from fallowness to a productive status.

Fishing Chimes, Editorial 375:May 2008: Vol. 28, No. 2

# The Beginning

## *Sea Cage Culture: CMFRI's Successful Experiment*

It is anachronistic that sea cage culture has not so far made its entry into Indian seas. This is in glaring contrast to the enormous progress in this activity made is countries like China, Vietnam, Thailand, Malaysia, Philippines, Taiwan etc., so far as Asia is concerned. Sea cage cu ture aspect continues to be a grey area of Indian fisherie development. However, now there is a ray of light tha can expand and dispel this shortcoming. The CMFRI has conducted successfully a sea cage culture experiment in the upper Bay of Bengal (Visakhapatnam sea). A trial harvest of the cage, although conducted partially, yielded an encouraging result, indicative of a good crop. A heartening success with potential for progress indeed.

The prevailing impression in fisheries circles within and outside the country, that India could not so far venture into sea cage culture, may assume an encouraging complexion from now on, because of the experimental success. We can visualise the future scenario of cage culture in Indian seas now, taking into account the determined stance of the authorties to introduce measures to counteract the declining marine capture fish output and to upgrade the economic status of the stake holders concerned and one of these measures, a focal one, will have to be to promote sea cage culture. An additional reason for this optimism is that, on one hand the Central Institute of Brackishwater Aquaculture and Rajiv Gandhi Centre for Aquaculture have developed the technology for hatchery production of sea bass seed and also the pond - based grow - out technology for this fish; and on the other CMFRI has now demonstrated, as mentioned above, farming of sea bass in a sea cage. Diversification in cage culture from sea bass to other fishes like groupers, snappers, cobia etc., will be a short step. CMFRI played a pioneering role in establishing the operational aspects of the activity, although on an experimental basis, with support from the Union Department of Animal Husbandry, Dairying and Fisheries. Thus, the CMFRI presented to the nation the success of the first experimental result in sea cage farming in India. This success has been achieved without availing of any external technical help. It set up the cage in the coastal waters off Visakhapatnam which is visible to onlookers from Ramakrishna Beach. The cage was moored by CMFRI at a depth of 11m and at a distance of about 300 m from the coastline bordering the city. The cage was first set up on 30 April, 2007. There were, however, the initial problems. These were resolved and an improved version of the same cage was set up on 23 Dec., 2007. This date has to be considered as a historic one. 1,400 nos. of Asian seabass fingerlings obtained from RGCA with an average weight of 14.5g each were stocked in this cage. There was partial harvesting of the produce on 26 April, 2008. The cage harvesting programme was inaugurated in a befitting manner by Prof. Mohan Joseph Modayil, then Member of the Agricultural Scientists Recruitment Board. A point to be noted is that it was also this professor who started this cage culture experimental work at the time when he was Director, CMFRI. Dr. G. Syda Rao, Scientist-in-charge of the Regional Centre of CMFRI, Visakhapatnam, Dr. L. Krishnan, Principal Investigator of the Project, Dr. N.G.K. Pillai, the present Director, CMFRI are the other scientists behind the outstanding success of this first ever successful effort at sea cage farming in Indian waters. During the inaugural harvest, a token 550 kg of the fish were harvested and the rest were retained in the cage for further growth. IFP unit at Visakhapatnam took over the harvested sea bass for value addition and marketing.

The initial success in the cage farming experimental effort, as mentioned above, is an event with an enormous significance. It paves the way for India to eventually emerge as the first nation in South Asia to promote sea cage culture on a large scale. It is of course known to all concerned that China and South-east Asian nations such as Thailand, Malaysia, Vietnam and a few other countries such as Taiwan are far ahead of India in cage culture. In this situation, the encouraging development is that a beginning has been made, with a supportive initiative by MPEDA by way of signing of an MoU on the subject with Innovation Norway. Under the provisions of this MoU, the main activities to be promoted would be: Detailed survey of the sites, environmenental impact assessment, demonstration of technology, commercialisation, capacity building and downstream activities like processing, value addition, marketing etc. Private participation for early commercialisation of the cage farming is also being planned.

As is known, sea cage culture has two components. One of these is the component of seed (advanced fingerlings) production for stocking cages. In this regard, CIBA, RGCA have developed the related hatchery technology which has already been extended to Pancham Aquaculture in the private sector. There being no further necessity of experimental effort to be mounted for seed production to stock cages, what is to be taken care of now is a follow-up action to identify suitable sites, along the coastline for setting up hatcheries for seed production and supply for stocking cages as the activity develops, taking into account the planned zones of concentration of cages.

The expected first phase of setting up of sea cages would no doubt be confined to the territorial waters, for a long time to come. As is known, territorial waters come under the jurisdiction of the coastal State Governments, as per the Constitution of India. Considering this, the State Governments have to be involved in the process of promoting the erection and operation of cages in the territorial waters, from the planning phase itself. Identification and allotment of sites to selected enterprises with training support has to be taken up by the Coastal State Governments concerned with the needed support from the Union Department of Animal Husbandry, Dairying and Fisheries and ICAR.

It may be mentioned here that the U.S. Government has a system of identifying the sites for the location of cages and notifying such sites as part of the exercise of calling for applications for allotments. The applicants who apply for allotment of sites notified are interviewed by the committee or officials concerned and allotments are made on that basis. The Governments of Australia and Norway and several others also have a system in regard to allotment of cage sites. It is desirable that the Department of Animal Husbandry. Dairying and Fisheries study these systems, and in that light, develop India's own system for being followed by the coastal State Governments, at this stage itself. Simultaneously, the needed set up for identification of sites for erection of cages, the manner in which cage erecting has to be promoted, and the system of mounting training programmes for erection of cages and for cage farming, linked to development of hatcheries along the coast and promotion of marketing links could be worked out and placed in position. Considering the enormity of the preparatory work to be done, it may be essential to set up coastal State - wise cage culture promotion cells and a co-ordinating cell at the centre to impart a fillip to the promotional activity of a sea cage culture system in India.

Fishing Chimes, Editorial 376: June 2008: Vol. 28, No. 3

# ...On Pangs of Vannamei Introduction in India

The general policy of the Government of India is against entry of exotic aqua animals and plants into the country. This policy is in tune with the approach on this aspect by other countries too, such as Australia. Despite the aforesaid policy, there has been the entry of several exotic fishes into the country, tolerated by the authorities. Exotic *Tilapia mossambicus, Pangasius sutchi, Clarias gariepinus* and a few others like silver carp, bighead carp, grass and common carp have been brought into the country in the recent decades, after independence. In the preceding British regime, exotic gourami and cold water fishes like trout, tench and crucian carp were introduced. There have been, however, no focal evidences to identify and establish the harm done or Compatability noted with national fish population by the exotics to the native populations. There were fears that Tilapia would have an adverse effect on indigenous fish but this had not happened in a conspicuous and enduring way in the country, silver carp dominated Govind Sagar reservoir for years but the population of this fish in the reservoir is reported to have dwindled considerably as at present. Broadly looked at, however, it appears that the exotic fishes that entered Indian waters have not caused much of a serious damage to the indigenous population. This observation may sound equivocal but the fact remains that for whatever that had happened in respect of entry of exotic fishes into India, there is no way of wriggling out of the damage, that may have been done. At the same time, there is no reliable evidence of any adverse impact of exotics on Indian fish fauna. In any case, the nation has to reconcile itself, in case it is established that has there been any damage done, instead of getting obsessed over the past follies of ours, it is high time to learn lessons from the repercussions of the past introduction of exotic fishes and formulate a line of action to set right any unlikely adverse effects that may have emerged, because of the past productions of exotics.

The issue that is of concern now is the inescapable and compelling situation in respect of fulfledged introduction of Vannamei. Two years back, rightly or wrongly, two Indian commercial enterprises had been permitted to import brooders of *P. vannamei* for undertaking pilot operation of its seed production and farming of the seed, instead of entrusting the trials to the concerned Central Research Institute for recommending a line of follow-up action (positive or negative). Because of what appears to be an insulation of the results of the pilot work on Vannamei shrimp production and export as being done by the two enterprises concerned, the farmed shrimp producers and exporters now feel the adverse impact of the result of the invidious approach and the seeming inaction of the authorities in respect of further follow-up action. The two companies, according to the information available, seem to be doing well. Vannamei seed is produced, grown to marketable size. These are processed and exported by them. It is learnt that, compared to the problems encountered in the seed production and farming of black tiger shrimp such as disease incidence, by and large, these are marginal in the case of Vannamei shrimp. The other advantages experienced in Vannamei farming are stated to be higher stocking rates, and relatively low production cost compared to black tiger farming. While Vannamei's growth period upto the harvesting of marketably stage is stated to be 60-90 days, in the case of tiger shrimp the farming period to reach marketable size is 90-120 days. The two companies concerned must have been reporting the progress and results of their pilo work to the authorities concerned, on the results of the pilot farming of Vannamei and the export performance in the past two years, as they could assess. The authorities must have conducted an appraisal by now and taken a decision on the introduction of Vannamei for farming in the country. This has not been apparently taken until now, forcing the Seafood Exporters Association to approach the authorities concerned for remedial action. Undoubtedly, there has been a long delay in this regard and this must have been a boon to the two enterprises now in an enviable position and a penalty for the others.

*Penaeus indicus*, the Indian white shrimp, does not have the advantageous features of higher stocking rates and faster growth as in the case of Vannamei. As the growth performance of *Penaeus monodon* (Black tiger shrimp) is far better than that of the Indian white shrimp, most of the farmers have opted for the farming of black tiger. Now that Vannamei has come in, which has a far better growth in a shorter duration, the farmers long to take up its production and rightly so, as has been done in several countries. In the present situation, they entertain hopes of clearance from the authorities for diversifying into vannamei farming. The exporters too expected that the Union Dept of Animal Husbandry and dairying would accord permission for the introduction of Vannamei, but this has not happened so far. Fish farmers do not have the organisational strength in the same way as agri-farmers have, to prevail upon the authorities to equip them to undertake Vannamei farming. To compensate for this shortcoming, MPEDA is understood to be rightly supportive of the need to lift the present restriction that confines Vannamei farming only to two enterprises and to introduce in its place a nation-wide permission fo farming of vannamei. It is learnt that the Union Commerce ministry is in favour of promoting Vannamei farming, along the coastline and into the interior to a possible distance. The present situation, however, is understood to be causing concern to the authorities, inluding the Union Department of Animal Husbandry, Dairying and Fisheries, but this Department, for some reason or the other, seems to be unable to bring in the reform apparently for reasons of their own. MPEDA must be the most concerned at the present situation, particularly in the background of China, Thailand, Vietnam, Taiwan and many other countries having gone far ahead of India in their exports of shrimps, mainly because of Vannamei dimension, which has not apparently caused any environmental problem to them. Now, the ray of light is that there seems to be an awareness of the international picture *vis-a-vis* India's performance at the top level of the Government. So much so, many in the industry now believe that the Government would soon permit the farming of Vannamei in India without restrictions and with an arrangement linked to import of brooders by a centralised organisation and dissemination of the same to various farmers to start with. In such a case, the imports would no doubt be from reputed sources known for supply of Vannamei SPF broodstock of quality.

In case the authorities are unable to accord permission to all Indian aqua farmers to undertake Vannamei farming, the farmers have the right to be told through a public statement that the farming of Vannamei is not allowed in India. At the same time, a decision has to be taken on the withdrawal of the permission earlier given for farming this shrimp to two companies.

## Larval Stages and Growth

This species, (Vannamei) as could be learnt, has six nauplii stages, three protozoeal stages, and three mysis stages in its life history. The carapace length (CL) of *L. vannamei* postlarvae ranges from 0.88 to 3.00 mm. The larval stages (1.95- 2.73 mm CL) can be recognised by the lack of a thoracic spine on the 7th sternite, and relative rostral length against the length of eye plus eye stalk ranging from 2/5- 3/5, rarely 4/5. The most distinguishable morphological character is the development of supra-orbital spines in the second and third protozoea. Colouration is translucent white. The shrimp is most commonly known as the "Pacific white shrimp". The body of the species often has a bluish colour that is due to a predominance of blue chromatophores which are concentrated near the margins of the telson and uropods (Eldred and Hutton, 1960). It grows to about 230 mm [9 inches], it is stated.

***Foot Note:*** *Penaeus vannamei*, the Pacific White Shrimp, is now known taxonomically as *Litopenaeus vannamei.* The popular name is white leg shrimp. This shrimp is the native of the Eastern Pacific (from Mexico to Peru). The main sources of this shrimp are Equador, Mexico and Brazil but SPF brooders of Vannamei are being produced in Hawai and probably at other places.

Fishing Chimes, Editorial 377: July 2008: Vol. 28, No. 4

# Indian Fisheries Sector Needs Fortifying Measures

Symptoms of a catastrophe are gathering strength over years around several facets of Indian fisheries sector. The authorities would no doubt have an assessment of the situation so as to set in motion measures to enable normalcy to stage a come back. Progress can be there only when there is a return to normalcy.

Large and small mechanised fishing vessels now face a serious problem of increase in operating costs. Because of the steep rise in diesel oil price, vessels have cut short the duration of their voyages. The domestic marine fish prices are stagnant, despite increase in vessel running costs. This situation has a serious adverse effect on marine fish production levels, which have been on the decline in recent years. By encouraging small mechanised trawling boat operators to diversify into longlining operations, which is a passive method, there can be an economic relief to them because of possible increase in earnings from sale of fish which are likely to consist of a good quantity of tuna, now enjoying export demand. There are no doubt some initiatives by the authorities in this regard in respect of medium to large vessels but the smaller vessels are left untouched. There can be the induction of a major project to cover this gap, as has been done in Brazil. By entrusting a project of this kind to NFDB, with provision for a field set-up to keep track of the landings, there can be a major progress in the earnings of stakeholders and also in export level.

So far as the development of a fleet of larger tuna vessels in concerned, as has been achieved in several countries, it is imperative to take purposeful steps in the place of the present LoP system, for introducing truly Indian-owned vessels under a major project, following the successful introduction of a good number of large fishing vessels for operation in Indian waters as was done the middle 1970s, promoted by the Union Department of Agriculture when the subject was with it then. It is difficult to comprehend the reasons for preferring the LoP route, although the Nation had the past successful experience in operating her own larger fishing.

There is recognition that marine cage culture system to be introduced in Indian waters not only to counteract the trend of stagnancy in marine fish landings, but also to increase fish production, and thereby providing an avenue for the fishers to augment their present income level. MPEDA entered into an Agreement with Norway in this regard on the one hand and CMFRI conducted a pilot scale experiment in cage farming on the East Coast on the other successfully. Further follow-up action is felt by many as very tardy which is the least expected.

On the shrimp farming front, the situation needs considerable corrective action. While there have been several interventions to achieve disease-free farmed black tiger production with a remarkable measure of success, Indian white shrimp, however, continues to be neglected. Added to this complacency is the questionable permission given to two private sector enterprises to introduce *Peneaus vannamei*, the Pacific white shrimp. While these two companies have the monopoly, all other shrimp farming enterprises are at a disadvantage. Many entertain the view that the authorities should have entrusted the pilot job of experimenting with the plus/minus aspects of introducing *P. vannamei* to a Central Research Institute like CIBA. This institute would have tested the desirability or otherwise of introducing the species in Indian waters. In case the result of its experimentation is in favour of introducing the species in the country, it would have recommended this along with a set of guidelines. The present feedback is that all the farmers and the concerned associations are now agitating for permission to introduce *P. vannamei* and to farm them with a view to promoting export of this shrimp, as has been successfully done in China, Thailand and several other countries. In this background, it is imperative that Government should come out with a clear plan of action, so as to put a stop to the confusion that now prevails.

Sea cucumbers are a protected marine species, under the Wild Life Act. Like any other wild animal, land or water-based, they require total protection and the Act rightly ensures this. However, in the same way as protected wild animals are kept in Zoos, their progeny allowed to grow, and later some released into the wild environment, there is a need to grow and breed sea cucumber too in a protected environment supported by a hatchery for the production and release of progeny back into the sea. At a stage of development of this activity to a reassuring level of systemising the sea cucumber seed production and operations for releasing young ones back into the sea, a well-formulated item of providing an export orientation to the farmed sea cucumbers can be introduced linked to such legal provisions as may be needed and to be provided. Under the system all the operations would have to be checked, monitored, and certified to have been cleared from all angles.

There are several other fisheries developmental aspects that need the intervention of authorities. These include surveys, and allotment of sites to selected beneficieries, supported by training in the respective fields, for augmenting production and for socio-economic development of fishers and promotion of fishery enterprises. These include mariculture and cage culture, promotion of small scale enterprises for manufacture of cages and for their installation, and for setting up shore-based hatcheries for induction of seed for stocking cages. Training programmes for the development of man power to man the aforesaid activities is essential.

There is a vast extent of fallow saline land of over one lakh ha in Rajasthan, Haryana, Punjab and western U.P. While some of these saline lands have water both over them and underneath, some have water only underground. When developed for saline water farming/brackishwater farming (with freshwater taken from nearby resources), there will be a major upgradation of the scenario. There will be increased aqua production that will usher in prosperity through domestic as well as export marketing of the produce in fresh or value-added form. A fishery-oriented industry will emerge, providing opportunities of employment and socio-economic development of the zone. The National Fisheries Development Board needs to excercise an initiative in the aforesaid direction, in association with the State Governments concerned, preferably by promoting a mega-project for the purpose.

It is to be hoped that the authorities concerned would conduct a review of the present status of the aforesaid activities and draw up strategies for greater attention towards development.

Fishing Chimes, Editorial 378: August 2008: Vol. 28, No. 5

# Fisheries Development in Inland Saline Waters of India

## *NFDB has to Intervene*

There is a vast extent of saline waters, both underground and overground, in the north - west region of India. In this region, Rajasthan envelops the largest extent of such waters, followed by western U.P., Haryana, Punjab and, probably marginally, in Gujarat. There are different estimates of the area available, which have a range from 50,000 ha to one million ha. Assuming that a minimum extent of 50,000 ha, in all, can be brought under crustacean production, to yield an average production of 2t/ha of exportable species (shrimps and fish varieties for value addition) with an average value of US$ 3000 per t, it can be estimated that there would be an additional annual foreign exchange inflow of the order of US$ 450 million gross or Rs.18,000 million on an average.

At the rate of five beneficieries per ha of the aforesaid area suggested to be brought under aqua production, there would be 2,50,000 beneficieries, with each having an average annual individual gross earning of Rs. 72,000 and Rs. 36,000 net (Rs. 3,000 per month on an average). The level of income may improve when the production levels move up.

The main initiative for this developmental activity has to come from the State Governments concerned, who would have to be appraised of the potential and the results.

**As an encouraging step forward in this direction, there were some initiatives, particularly from the Central Institute of Fisheries Education, when Dr. S.N. Dwivedi was its Director. He succeeded in the farming of Tiger Shrimp *(Peneaus monodon)* at Sultanpur in Haryana. Owing to problems of logistics of organising supplies etc, the follow-up action seems to have come down later on, eventually bringing it to a virtual standstill. A little later, there was another significant development in Haryana as a remarkable success in raising giant freshwater prawn seed and also in its utilisation for raising freshwater prawn crops, under the leadership of Dr. Hardayal Singh of CIFA at that time. It is unfortunate that there was no tangible follow-up on this success too. In Rajasthan, Dr. Jain of CIFE is reported to have achieved a remarkable success in this direction but not much was heard thereafter.**

**Out of all the States of the North-West of the country having saline belts, Punjab was in the forefront in regard to providing funds for utilisation of saline waters of the State for export-oriented aqua production. The Government of Punjab sanctioned, in 2006, a project for saline water fisheries development in the State. Looked at in a historic prospective, this was a courageously progressive step but it needed support from the Centre, particularly for the reason that, for the first time, the State of Punjab bestowed its focus on the fisheries development of its saline water resources. The subject was new to them and they had to look for expertise from other sources like CMFRI/CIBA and also from others to organise the logistics of supplies of imputs, and for providing training and extension support back-up. They would have been able to mobilise all this support, had the centre lent the needed support.**

**In the August 2006 (Vol. 26 No. 5) issue of *Fishing Chimes,* in the editorial, an appeal was made to the National Fisheries Development Board, which was set up by then, to convert the fallow saline stretch into a productive zone. Extracts from this editorial are reproduced here under:**

**"Out of all the States in the saline belt of the north west of the country, Punjab is in the forefront in regard to the utilisation of this precious resource, with the State Government sanctioning a project for saline water fisheries development in the districts of Faridkot, Ferozepore, Muktsar, Moga, Bhatinda and Mansa with an initial annual outlay of Rs. 66 lakhs. NFDB can take steps for extending the approach adopted by the Government of Punjab to cover all the north -western States endowed by nature with saline lands and supporting salt water resources. It will be to the credit of NFDB to usher in a new era of economically viable livelihood to the weaker sections of the society through fisheries developement of the area, linked to domestic marketing as well as exports of the produce".**

**"The inland States of Haryana, Punjab, Chattisgarh, Bihar and U.P. have achieved some progress in the farming of giant freshwater prawn. These initiatives have provided adequate evidence to establish that the nation-wide spread of giant freshwater prawn farming, which has enormous potential for augmenting giant-prawn production for internal consumption, and also for exports to augment foreign exchange earnings, is possible. What is needed to bring about this socially and economically potent reform is a nation-wide integrated project that spreads from the coastline to the guts of the nation in its tropical belt. The project would have to cover its various links such as seed production, movement of**

**seed over the length and breadth of the interior of the country, extension of culture technology, harvesting system, pooling up and storage of the produce and their domestic and export marketing. Being an activity of nation-wide dimensions, NFDB will be the most suitable agency to conceive of a project and organise its implementation from seed production to marketing."**

The saline lands, and saline water stocks of the North-west constitute a major economically important resource with aqua production potential. The probable lack of grasp over the plan of action needed to bring these assets into production cannot be an excuse to prolong the present complacency for ever. NFDB has to intervene, mobilise scientific, technological and developmental facets of the subject, in association with the related State Governments/State Fisheries Departments through brain-storming and co-ordinating efforts, while refusing to unreasonably succumb to views adverse to the realisation of the objective. CMFRI, CIBA, based on a detailed indepth study, should be able to suggest a plan of action for evolving an organisational mechanism with the participation of the various States concerned and leading to an export-oriented and commercially viable field activities supported by investments, infrastructure, training and other operational aspects that would provide linkage to the various stages of the work, finally leading to effective export/domestic marketing. The firale to be aimed at is to bring about an upgradation of the socio-economic status of the depressed categories of persons to be inducted as a new class of aquafarmers.

Fishing Chimes, Editorial 379: September 2008: Vol. 28, No. 6

# Indian EEZ: Tuna Fishing Scenario

## *LoP System's Benefits are One Sided*

The tuna fishing scene of the Indian EEZ beyond the Coastal Zone is dominated by large Taiwanese tuna longliners of over 40 m OAL. Around 60 nos. of these longliners are understood to be now in operation under Letters of Permission (LoP) issued by the Indian Government (Union Dept. of AH, Dairying and Fisheries). These LoPs, issued in favour of Indian fishing companies by the Government to have 'Indian owned operations in the Indian EEZ,' are believed by many in the Indian marine fishing circles as having lent a protective facade to the Taiwanese owners, not only to exploit the tuna stocks of Indian EEZ, concerning which they have enormous knowledge, but also to take them away, allegedly with no declaration or declaration of only part of the catches, mentioning very low export rates thereof. So much so, a fishery entrepreneur (G. Prithviraj) has given a different expansion form to the abbreviation LoP as 'License of Poaching', mentioned to be so talked about in marine fisheries circles. In any case, the strengthening opinion in the Indian marine fishing industry is that. the operations of LoP tuna longliners are emerging as a constraint to the viability of operations of non-LoP tuna vessels. The operators of these (non-LoP) vessels and their associations were vehemently protesting against LoP vessel operations until recently, but not much of opposing sentiments are now heard from these associations. The reasons for this are not known. Further, one aspect which is ventilated by many is that LoP system, which is believed to have been internationalised by Taiwanese, concedes access to their longliners to fish not only in the Indian EEZ but also in the EEZs of several other countries. The introduction of this system is attributed mainly to the initiatives of Taiwanese vessel owners or their Association. Apparently they need the access to the EEZs of other nations, as theirs is probably already overfished. So far as the 'victim' countries (such as India and others such as Pakistan and those in Middle East, and along African and South American coasts, as is understood) are concerned, not only the resources of these EEZs are exploited through various strategies but also by ensuring that the 'victims' do not equip themselves to utilise their tuna resources to the disadvantage of the Taiwanese interests. Taiwanese owners now gain, as they are the actual operators in the EEZs of others, and who take away the catches in their vessels. What the Indian LoP holders get is not, however, transparent. Simultaneous to the LoP scheme, had the Government conceived of an integrated scheme for import/indigenous construction of tuna vessels in a suitably structured form, by now a totally Indian owned and a suitably sized tuna fishing fleet of the required strength would have been built up for the sustainable exploitation of our estimated 247,000 t of tuna stocks, supported by a training programme under a joint venture or otherwise as needed, with linkages from production to processing and marketing. There are no doubt some marginal initiatives in this direction by MPEDA and by Union DAHD and F but these are very weak.

The simultaneous development of an integrated project as mentioned above would have at least facilitated the phasing out of the LoP system by now. The latest order that Indian crew on LoP vessels has to be a minimum of 25 per cent (for all time) clearly shows that LoP vessels would operate virtually with foreign crew, a concept that reflects the position that the vessels are not only not truly Indian but we also do not want them to be truly Indian. Small countries like Vietnam, Sri Lanka, Indonesia, Philippines etc are far ahead of India in respect of tuna fisheries development. All these countries could develop tuna fishing without a system like LoP. This approach can probably be emulated by us.

The LoP system in India has unleashed several undesirable overtones. These can be stemmed only when LoP system is withdrawn. The latest 'reform', as already mentioned, is the 25 per cent Indian crew component stipulation governing the LoP system. According to this condition, the Indian crew on LoP vessels would always be a minimum of 25 per cent, against the earlier norm hat this strength would be gradually increased to 100 per ent on 'Indian' vessels and that the foreign crew would be reduced in stages, as the Indian crew picks up prowess in onboard operations. It is difficult to imagine the con iderations that led to the decision of the fixation of 25 per cent minimum of Indian crew for all time. This helps the foreign owners and of course also the 'Indian owners' who do not want the Indian crew component to go up for obvious reasons.

Indian EEZ is a much coveted fish poaching avenue not only for Taiwanese fishing vessels, (with the operation of some of them probably regularised through Letters of Permission) but also of other countries. We have Thai vessels poaching in Indian EEZ and several Indian vessel owners have been referring to their presence as noticed in the Bay of Bengal and Andaman Sea. The latest addition to the poacher's list is Sri Lanka. The tuna longlining vessels of this country have been seen fishing for tuna in Indian waters as north as Visakhapatnam.

A report on this that appeared in Deccan Chronicle of 10th September 2008 is reproduced at page 47.

It is learnt that a Committee headed by Dr. S Ayyappan, Dy. DG (Fisheries), ICAR met on first September 2008 to formulate guidelines for LoP vessels. The LoP Associations were invited to participate in the meeting but not the Association of Indian fishery Industries which is an over 30 year old Association that represents Indian marine fisheries industries and is well recognised by the Government.

It is widely known that the owners of Indian LoP vessels are 'Indian'. It is also widely known that the actual ownership (and control) over the vessels continues with the Taiwanese. One pertinent point here is that these vessels are not registered either with IOTC as Indian vessels or with the MPEDA in respect of Indian marine products export aspect. For the later reason, the possibly undeclared catches that are being taken away by the LoP vessels (which apparently constitute exports) are understood to be not reported to MPEDA. Further, the Vessel Monitoring System (VMS) for the LoP vessels has not so far been introduced. So much so, these vessels ave the facility of entering the Indian EEZ and moving out without any system-oriented watch. Irrespective of the position whether they are deemed as Indian vessels or Taiwanese vessels (probably under a third country registration), their movements and the exported catches they take with them would have to be at least on MPEDA's record. The present position seems to be that no Indian authority knows how much of tuna is actually being exploited by LoP vessels and taken out to another country. Presently, it is very convenient for the owners concerned to operate LoP vessels in Indian EEZ and take way the catches without touching any Indian port. Before registration as LoP vessels ('Indian owned vessels'), what all is done is that they are produced for inspection by the authorities of Indian Registry of Shipping and necessary certificates are obtained, which are accepted by the Indian coastguard. It is learnt that the LoP vessels on their first entry for fishing for tuna in the Indian EEZ, anchor outside the chosen Indian port limits. On intimation, the Coastguard officials inspect the LoP vessels to accord the needed clearance. Thereafter, the Indian counterpart crew board the vessels. So far as the MS Act is concerned, those conversant with provisions in it say that the above mentioned procedure violates the provision in the related Section of the Act which stipulates that the vessels have to be inspected by MMD beforehand.

Besides 60 LoP tuna vessels now in operation, ten more letters of intent (covering around 40 vessels) are understood have been cleared by the empowered Committee which met on 1st September 2008. The strength of LoP vessels thus will go up now to 100 nos., aggravating the disadvantages that the truly Indian-owned vessels are now subjected to.

G. Prithvi Raj, Managing Director of Priyansh Fisheries (Pvt) Ltd. recently sent a couple of representations on the subject to the Hon. Prime Minister of India. These are reproduced on pages 9 and 10 of this issue. The contents of these representations throw light on the harmful impacts of the present LoP system. Considering the various damaging aspects of the LoP system brought out in the representation, it is of utmost importance for the Government to consider measures for the removal of the factors seen to be causing national harm.

- ✰ G. Prithvi Raj says that the status of the positions in respect of the following has to be clear in order to understand well the Indian Tuna Scenario.
- ✰ The cost of acquiring LoP secondhand 45 - 50m Tuna Long liners the cost of each of which is in the range of Rs. 6-8 crores. The source of funds for making 10 per cent initial payment towards this cost is to be known.
- ✰ The Maintenance costs of LoP vessels such as those of Dry Docking, Engine Spares etc., have to be reflected in the Books of Accounts. The vessels are dry docked every two years. Who is paying for maintainance such as dry-docking has to be made known.
- ✰ It is known that the employment of the Foreign crew members at the present strength is sufficient to conduct the entire fishing operations. Indian crew have to be taken only to fulfill the condition stipulated, although without the need. The justification and purpose of fixing Indian crew component at 25 per cent of the total foreign crew is to be made known.
- ✰ The Foreign crew salaries are being paid by the Indian Company, as stated. TDS has to be therefore deducted therefrom and paid to the Government by the Indian owners.
- ✰ The Indian owners do not have any control over the vessels at present. There must be a way of berthing the vessels, at any Indian Port for a joint Inspection by the Govt officials concerned and representatives of the industry.
- ✰ Income tax has to be paid by individual companies to the Govt of India, particularly for such of the vessels which are now in the sixth year of operations. As per the Govt letter dated 11/8/2006, the cost of these vessels should have been fully paid up by now.
- ✰ After the vessels complete fishing operations in the Indian EEZ (from Oct - April each year), the owning companies must be informing the Govt about the catches and the waters in which they fished. The income thereof has to be reflected in their Books of Accounts.
- ✰ The necessity for the 'Indian' vessels to leave the Indian EEZ has to be clarified as they are no longer foreign vessels. They can declare the catches, export them and resume fishing from an AP Port like Vizag.
- ✰ The LoP companies have to clarify whether they are registered with Marine Products Export Development Authority as an Exporter.
- ✰ The LoP companies have to abide by the IOTC resolutions. Otherwise, the countries of their registry may be declared as encouraging IUU fishing and thus undermining the resolution of IOTC.

Fishing Chimes, Editorial 380: October 2008: Vol. 28, No. 7

*In this Era of Upsurge of Improved Culture Fisheries Technologies*

# Lentic Water Fishery Rights: Leasing Policies Need to be made Matchingly Production-Oriented

The inland lentic water fishery rights leasing Policies in vogue in the various States of India, by and large, have their roots in the pre-independence scenario of Inland Fisheries development of the country. Later, there have been upward adjustments in the parameters of the aforesaid leasing policies parameters from time to time in most of the States. These revisions, by and large, had not, however, taken into account the possible optimal sustainable production that can be achieved through application of the chosen advanced culture fishery technology. A good number of candidates educated and trained in the advanced fish culture technologies and coming out of the several fisheries colleges in the country now can be attracted to take on lease one or more of these water bodies for fisheries development but this is possible only through needed policy readjustments. Initiatives in this direction are of utmost importance. As mentioned, competent manpower in inland fisheries sector is now available mainly for the purpose of augmenting inland culture fish production. This can be achieved largely by inducting this available manpower to promote fish production in lentic resources of the country such as tanks and ponds, which have an extent of over 2 million ha, under an optimally sustainable level of production. The possible increase in farmed fish production, assumed at a minimal annual average of 500 kg/ha, can well be of the order of one million tonnes per annum and which can be valued at over Rs. 5,000 crores. Looked at from this angle, there can be no escape from the conclusion that there has to be an effective integration between competent culture fishery developmental manpower and the policy of leasing out fishery rights over tanks and ponds, besides reservoirs and other capture – culture fishery resources, such as *beels, chaurs, mauns* etc.

For long, right from British days, the focus, in respect of selection of fishing/fishery rights lessees, had been on men with experience in organising exploitation of fisheries, both in confined waters like tanks and ponds and in culture – capture resources such as reservoirs, flowing water stretches of rivers and canals and in lakes, *beels, mauns, chaurs* etc. The structure of the leasing policy continued for long to be on the aforesaid pattern. As time went by, however, reforms were introduced to improve the system. One of these reforms entailed the preferential awarding of fishery leases to panchayats for some years and later, when the rights were transferred to panchayats, totally on a preferenetial basis in favour of fishermen's co-operative societies, initially for a duration of one year, but in most of the States this duration has been enhanced mostly to three years for the reason that this would encourage the societies to nurture a long term interest in the fishery promotional work. In case there is no society, the provision was that the lease would be decided in favour of the highest bidder. The lessee concerned, depending on lease terms, has the freedom/obligation to develop the fishery and undertake harvesting periodically within the lease period.

Over years the on-going general pattern of the inland fishery leasing policy, as presented above, has been gaining in strength, mostly as a marginal means of revenue earning by the Government concerned, although some of the lessees are also known to be stocking fish seed for augmenting production and incomes thereof. The reasons for this are obvious. CIFRI (and CIFA, its successor in respect of culture fishery development) developed several technologies of sustainable fishery development in tanks and ponds. These technologies range from composite fish farming, mixed farming, integrated fish farming, phased stocking and harvesting systems, paddy-cum-fish farming, cage farming etc. In the course of time, the uncertainties of securing quality fish seed for stocking cultivable waters have almost disappeared, mostly because of the introduction of induced fish breeding technology. The technology of raising of fish seed to a stockable size has emerged as a standard practice. Along side the development of culture (seed production on one and and grow-out farming on the other) technologies and conditioning of the farmers for their adoption, as a ready mentioned, there has been the conspicuous emergence of fisheries education and training infrastructure in the various parts of the country, which has generated qualified fisheries manpower nation-wide. This manpower has the capability of organising and undertaking fish production work in tanks, ponds, reservoirs and other lentic water bodies in a sustainable and socio-economically viable manner. There are also indications of self help groups taking initiatives towards lentic water fishery development through leases or otherwise.

Thus there has been a major transformation in the trained manpower availability in the country, particularly for the development of inland lentic fisheries resources. There is no longer the need to continue to depend solely on the class of lessees who would pay the lease amounts as decided (mostly based on the average of incomes of past three years and occasionally with some increase in the highest bid amounts) by the Government concerned. As at present, generally stated, the State Governments, in dealing with fishery leases, do not relate fish production and incomes from leased fishery resources. With the availability of means of application of advanced production technologies through the educated and related trained manpower that comes out of the several fisheries colleges/institutes spread over the various parts of the country the situation has changed. The concept which has been generally prevalent is that the educated and trained manpower has to look for employment mostly in fisheries departments. This concept needs revision in a manner that would attract those coming out of fisheries educational/training centres to participate as entrepreneurs in the culture fisheries developmental activities too to justify the expertise they have and to have the pride of earning a lucrative livelihood.

Conducive conditions, for the aforesaid transformation to be ushered in, have to be brought about mostly by the State Fisheries Departments. Firstly, there has to be a change in the ongoing fishery leasing policies. The present system of conferring fisheries leases on fishermen's co – operatives on priority basis no doubt has justification all these years, but this system needs to be replaced in stages and in the shortest possible time to cover educated and trained manpower too. The reason for the introduction of this reform is that, in the present state of development, the fish seed production and farmed fishery development work has to be the domain of professionally ompetent fish culturists, either as individual entrepre eurs or in groups or in the form of co-op societies or companies. The main function of fishermen, whom they will engage, besides other skilled hands, is to undertake other jobs such as those related to fish seed production and fish farming, once they gain experience in these lin s of endeavour. In other words, as mentioned earlier, the ob of producing fish seed and undertaking fish farming shou d vest in the hands of trained manpower having expe tise in all aspects of culture, now available in ab ndance. They would certainly perform in a far better anner than the present category of fishermen/traditiona contractors. The trained persons, unlike fishermen and t e traditional contractors would be knowing all about f sh farming, from the stage of fish breeding and seed p oduction, pond preparation, stocking of ponds or other confined bodies with seed, and about fishery management and the harvesting and marketing aspects of the fishery bodies concerned. It is thus desirable to introduce and develop the system of giving confined water fisheries rights on lease to qualified persons for reasonably long durations, with such supports as they may need. For example, NFDB can be motivated to arrange flow of working capital as loan from NABARD to enterprises of trained hands, in case they are unable to invest on their own. There can be also support from MPEDA for the export of such of the produce (other than shrimps, prawns and crabs) that can have an outlet in that direction.

A system of registration of qualified persons (Technocrats), individually or in groups/associations or registered companies of technocrats with the State fisheries Departments needs to be introduced. All such registered candidates/bodies would have to be considered as eligible to apply for fishery developmental leases as notified, providing the details thereof. A Committee may be set up by the State Government concerned to examine all such applications and select the best out of them for allocating priority numbers. A system on the suggested lines and as appropriate can be introduced in phases, starting with an identified number of water bodies to be first brought under the reformed system. These can be increased in phases so as to cover all the waterbodies. The Union Department of Animal Husbandry, Dairying and Fisheries can consider setting up an Inter – State Committee to consider the subject and make recommendations for approval and adoption by the various State Governments.

So far as reservoir fisheries development is concerned, it is desirable to form convenient clusters of them to set the production process in motion. This can be organised through development of viable infrastructure for supplies and services in order to set the production process in motion. Such clusters can be brought under a separate organisational mechanism. Reservoir Fishery Development Agencies can be set up for the purpose. These agencies may be entrusted with the organisation of the needed supplies and services to the lessees. The suggested Agencies can also set up hatchery farms. The advanced fingerling/yearlings produced at these farms can be made available to the lessees for stocking. The suggested Reservoir Fishery Development Agencies can also employ trained farm hands for the production of advanced fingerlings/yearlings for supply to the lessees concerned for stocking. Harvesting could be entrusted to professional fishers and their societies.

A policy initiative on the lines indicated above as a basis is of utmost expediency to bring about a change in the fishery leasing systems in the country for the effective utilisation of the confined water resources for realising sustainable fish production, and for ensuring optimal incomes from the produce to the stakeholders and to the State concerned and at the same time providing quality fish to the consumers. Guidelines in respect of monitoring, supervising and checking of progress of work could be drawn up for being followed, by the committee suggested above, including the prescription of periodical returns to be submitted to the next authority concerned.

Fishing Chimes, Editorial 381:November 2008: Vol. 28, No. 8

# ...On Regeneration of Sea Cucumber Stocks in Indian EEZ

Overexploitation of sea cucumbers is a global phenomenon. The alarming decline of stocks of these holothurians worldwide should have normally motivated atleast the developed maritime countries to impose a ban on fishing of these animals. However, this had not happened. In contrast, in India, the Union Ministry of Forests and Environment banned catching of all species of sea cucumbers from Indian waters by including these under Schedule I of Wild Life Protection Act, 1972, which includes the names of the wild animals that are totally banned from capture. The banning was done with the objective of protecting their stocks. The inclusion of sea cucumbers under Appendix II in the list of the Convention on International Trade in Endangered Species of Wild Life Fauna and Flora (CITES) was also recommended by India so as to conserve their populations.

The quantity of processed sea cucumbers exported from India, compared to the other exporting countries of these animals is low. During the period of 10 years from 1978-1987, an average of 36 t of sea cucumbers valued at Rs 26.7 lakhs were exported annually from India. While, during 1996-97, 70 tonnes of processed sea cucumbers (Beche-de-mer) were exported from India, in 2001 only 3.8 tonnes of the product were exported, indicating the effectiveness of the ban imposed. Because of the low level of exports of the commodity, the processed sea cucumber exports from India did not find a place in the Book entitled 'Sea cucumber – A compendium of fishing statistics' published by FAO in 2003.

The quantity of sea cucumbers exported from India may have become meagre because of the ban on their exploitation, compared to the export levels of the other countries. At the same time, it is heartening that there was an unanimous appreciation of the timely imposition of the ban, which was extolled as a measure that is totally in order, at a recent workshop on the subject held by CMFRI at Chennai on 25 Aug 2008.

A galaxy of experts participated in the aforesaid workshop. Several of them offered suggestions in the nature of supportive measures to the ongoing ban and also aimed at revival of the sea cucumber population to their pre-depletion level of stocks. One of the suggestions entailed a requirement that is not possible under the existing provisions of Wild Life Protection Act. This was one of collection by fishing a judiciously calculated number of brooders of sea cucumbers from the sea, fishing of which is, however, a banned activity. This suggestion was in the nature of an appeal voiced by the experts seeking that the present provisions in the Act may be amended appropriately to facilitate collection of brooders of sea cucumbers from the wild by CMFRI, exclusively for the purpose of breeding and producing juveniles for ranching.

The devising of strategies to enhance the natural stocks of sea cucumbers, based on "standardised technology' readily available with CMFRI is immediately called for, as observed by Dr. S. Ayyappan, who expressed the veiw that it is an essentially supplemental requirement to strengthen the purpose of the ban. This line of action was supported by the participating scientists, pointing out that the efforts at revival of sea cucumber stocks by banning of their fishing, should be strengthened by this innovative intervention in the form of sea ranching, utilising the seed produced in hatcheries to be set up and raising them into juveniles for ranching, provided that the needed inclusion of a provision to permit this approach is facilitated by a suitable amendment to the Act, as suggested by Dr. P.S.B.R. James, the eminent marine fisheries scientist who was behind the enormous and purposeful research inputs on the subject of the restoration of sea cucumber stocks in Indian waters to the old glory. It may be noted in this context that the internationally practised hatchery technologies of the prime sea cucumbers *(H. scabra* and *H. spinifera)* were first developed by the scientists at CMFRI and later adopted by quite a few other countries. As pointed out by Dr. P.S.B.R. James, the immediate need is to develop a small technology model of a sea cucumber hatchery not only to demonstrate the viability of sea cucumber seed production and raising them to juvenile stage but also the sea ranching aspects of the juveniles so produced. Another aspect that should receive immediate attention, subject to suitable amendment of the Act, is to take steps to covert a few selected shrimp hatcheries for sea cucumber seed production, as suggested by Dr. S. Ayyappan. As the purpose of sea ranching of sea cucumbers is one of enriching the natural stock of the animals, it is required that this activity should be undertaken in a well planned manner taking care of the scientific and technological aspects. The programme of ranching, in terms of numbers and periodicity as adjudged to be appropriate, coupled with an effective monitoring efforts so as to ensure sound post ranching results, is needed, as observed by Dr. D.B. James, a pioneer marine fisheries scientist of CMFRI in sea cucumber research and who played a major role in the success achieved in the hatchery production of sea cucumber seed, and known to have been emulated later by certain other countries, as earlier mentioned.

The present need is thus the institution of measures on a priority basis for rebuilding of the natural stocks of sea cucumbers. This could be achieved through well designed stock enhancement programmes, as articulated by Dr. P.S.Asha, Principal Scientist of CMFRI, who is currently working devotedly on the subject of restoring the stock level of sea cucumbers in the Indian EEZ to their past level, taking care of related management techniques. These include the production of formulated feed for broodstock, development of advanced nursery management techniques, effective ways of mass releasing of hatchery produced sea cucumber juveniles and several related aspects.

The senior forest officers of the Government of India who participated in the workshop conveyed their full support to CMFRI in its endeavours to restore the stock level of sea cucumbers in the Indian EEZ, thereby conveying in principle their acceptance of proposal for amending the Wild Life Protection Act to pave the way for the production of needed quantities of juveniles of sea cucumbers for sea ranching and releasing them into the sea and monitoring the developments thereafter. The workshop of August 2008 has thus bridged a strategy gap between the need to set up hatcheries for sea cucumber seed production, raising the seed to juvenile stage and using them for ranching on one hand and the present situation of legal hurdles to procure brooders of sea cucumbers from the sea and using them for their seed production at the hatcheries and the attached farms to be set up, for raising their juveniles for sea ranching, more particularly in the over-exploited zones such as Gulf of Mannar and Palk Bay, on the other.

Fishing Chimes, Editorial 382: December 2008: Vol. 28, No. 9

# ...On Promotion of Cage Aquafarming in India

Major strides have been made in freshwater and marine cage farming in the past over four decades in several countries. The main motivation behind this development has been the relentless decline in capture fish stocks, which is the result, mainly of overfishing. The top freshwater cage aquafarming countries are China, Vietnam, Indonesia, Philippines, Russian Federation, Turkey, LaoPDR, Thailand, Malaysia and Japan. India is far behind, with only marginally noticeable cage farming experiments conducted mainly in Madhya Pradesh and Karnataka. They seem to have yielded some results, but with perfunctory follow-up or no follow-up action to give a spread effect to the activity. Regarding Marine and brackishwater cage farming, the position is more or less, one of a non-starter, but for one experiment in marine cage culture by CMFRI. India does not figure in the list of top ten countries that have made a mark in this line [These countries are Norway, Chile, China, Japan, U.K., Canada, Greece, Turkey, Republic of Korea and Denmark, (Including Faroe Islands)]. Another aspect is that, in India the live fish trade in mainly based on capture and holding them in cages in the waters of Andaman and Nicobar islands, as recorded by FAO (Tech paper 498).

It is a matter of great concern that there has been an undesirable and undefendable neglect of development of the cage farming sector in India. Conceding that the development of cage farming falls in the State sector, it stands to reason that initiative in this regard rests with the Union Department of Animal Husbandry, Dairying and Fisheries (UD of AHD&F), as the co-ordination of the activity spread over several States and UTs is needed.

Thus, an organisational structure headed by the Union Department of AHD&F and ramifying into the various States/UTs needs to be set up to promote cage farming. While the functions of formulating and circulating guidelines for the setting up of cages in inland confined waters (freshwater and brackshwater lakes, *beels, mauns* and other wetlands, reservoirs, large irrigation or other tanks and/or such other waters), and in marine coastal waters would have to vest in the UD of AHD&F, and each of the States and UT Governments would have to set up their own centres with linkages to district-wise or region-wise units to promote cage farming.

As is known, cage farming has an enormous potential to augment fish production and to counteract the declining trend in capture fisheries. It is possible that there is either an inadequate comprehension or appreciation of this potential or there is a deliberate neglect of this farming aspect in Indian top fisheries hierarchy. There can be no other plausible reason for not bestowing any cognisible attention to this potential sector of fisheries development. The outcome of this situation is that, India, which has received global recognition as No.2 in inland fish production next only to China, is no where in the picture in respect of cage farming, compared particularly with China, Vietnam, Indonesia, Philippines and several other countries in respect of freshwater cage farming, and to Norway, Chile, China, Japan, UK and several other countries in regard to marine cage farming. In this background, the imperative initiative needed is to mount a major project commensurate with the potential, to promote cage farming in India. Such a promotion will not only increase fish production but would also generate employment opportunities both in public and private sectors.

Cage farming is virtually a new fish production activity for India. Overlooking the past general neglect in respect of promoting cage farming either under a scheme or a project (with the exception of the recent experimental sea cage farming effort of CMFRI) and taking advantage of the contours of the developmental aspects of cage farming in other countries, particularly in China, Vietnam, Indonesia, Thailand and Malaysia in respect of freshwater cage farming, and in countries like Norway, Chile and China in regard to sea cage farming. India has to now forge ahead in this sector.

As can be visualised, the aforesaid line of development can be accomplished only through a planned effort that has to be mounted in stages in a progressive manner. In stage I, selected faculty members from the various fisheries colleges in the country can be deputed for durations as determined, through a prior arrangement, to selected cage farming nations for receiving training in 1) selection of sites for setting up cages; 2) selection of cage netting material and utilising it for fabrication of cages in relation to sizes there of; 3) installation of cages in the sites; 4) integration of cage farming with hatcheries and associated farms for securing fingerling supplies in relation to stocking density, feed and feeding schedules etc; 5) cage management, with reference to testing of growth periodically, protection from thefts etc; 6) harvesting systems and 7) transportation and marketing.

In Stage II, a system of earmarking the selected cage site allottees for training in the activity, in the various fisheries colleges, the nominees of whom would have been trained by then in selected cage farming countries. Oncs the allottees are trained, the respective sites can be handed over to them to go ahead with the follow-up work such as cage fabrication, their installation and the associated integrated cage farming operations. The allotment of cage sites would no doubt be undertaken by the State Governments concerned, based on applications received on prescribed forms and approved with reference to prioritised categories of applications. It is desirable that in these, a category related to those with fisheries educational qualifications can be considered for inclusion. Further, authorities concerned may think of introducing a diploma or certificate course in cage farming at one or more fisheries educational institutions for considering such trained candidates as part of a category of those with fisheries educational qualifications for the allotment of cage sites.

The selected candidates would need financial support for the purpose of buying cage material, organising the fabrication of cages, their installation and all other follow up measures until the stage of marketing of cage – grown fishes. This support, as part of an integrated approach in this regard until marketing of the cage – raised fish may have to be provided by the State Governments concerned in association with National Fisheries Development Board. A set of returns can be prescribed to the allottees for their periodical submission to the authorities concerned to enable them to assess the progress. Apart from providing the needed shore infrastructure facilities to ensure integrated operations, there has to be an organisational mechanism to take care of the various stages of field operations and this can be in the form of, say Cage Fishery Development Agencies, to function under the control of appropriate suprivisory State level fisheries organisations.

Cage farming is a serious activity with enormous potential for augmenting quality fish production which is needed for raising nutritional and health standards of people, to promote exports and also to create new avenues of employment. Its introduction in India will open up a new chapter in fisheries development of India that would also contribute to countering of the ongoing declining trend in capture fisheries. The development of this sector has to be planned in the perspective of the realistic dimensions of the scope and potential of the subject and implemented in this light. In this context the initiatives excercised and the inputs put forth in this sector by countries such as China, Norway, Thailand, Malaysia, Indonesia, Taiwan, etc., in cage farming would have to be borne in mind.

Fishing Chimes, Editorial 383: Jan/Feb 2009: Vol. 28, Nos. 10/11

# Fisheries Events, 2008

## *(Including those that took place in the last quarter of 2007)*

The developmental highlights of the Indian Fisheries Sector during the year 2008, (leaving out the consequential developments that took place because of international factors/financial/economic depressions and their impact on aqua production and on domestic and export supplies), enable us to identify certain positive trends of promise for augmenting fish production sustainably and to strengthen positive and beneficial inpacts that are now being generated for the socio-economic development of the stakeholders. On the marine front, despite the policy of introducing LoP vessels, considered by many as not in national interest in the long run, there have been positive developments in respect of promoting the conversion of small and medium sized mechanised boats for tuna long lining. There has been a silent revolution launched by several boat owners who have taken to tuna long lining after converting their boats for the purpose. Several stakeholders have taken to exporting their tuna catches and also tuna products such as loins. These developments have become possible because of financial and technological help extended by the authorities. Another promising development is that the Government has permitted the introduction Vannamei shrimp into the country to compensate for the problems being faced by the stakeholders in raising healthy and well grown tiger shrimp. Inparting value addition to aqua products has resulted in new heights of endeavour in the sector. The North-Eastern States of the country have received special attention from some of the Central Institutes *i.e.*, CIFRI, CIFT, and CIFE and this has ushered in a new era of fisheries development in these States. Another development is that in States such as U.P, Bihar, Tripura and several other States there has been progress in respect of production and utilisation of seed of Giant Freshwater prawn, with a special outlook focused on the seed production and farming of Ganga river prawn too. The programme of value addition to fish through their various products has gained in strength. Commercial sea cage farming has been poised to make a remarkable entry, because of the successful demonstration of sea cage farming by CMFRI and the MoU signed between a Norwegian enterprise and MPEDA on the subject. The infrastructure development for strengthening and expanding the fisheries developmental activities in the country have come to be poised for expansion. The fisheries education and training institutes have contributed a lot for the development of fisheries man power. There have been several other initiatives in most of the States and UTs of the country, taken afresh to strengthen the ongoing fisheries developmental status. Portable hatcheries have been introduced by CIFA which have been gaining in popularity.

The various events that took place in the fisheries sector in 2008 (and also in the last quarter of 2007) are chronicled hereunder based on the particulars recorded in *Fishing Chimes.*

- ☆ 4 Oct., 2007: Brackishwater Aquafarmers Meet, Ratnagiri, Maharashtra.
- ☆ 30 Oct. to 3 Nov., 2007: Training on ornamental fish trade and aquarium maintenance held at College of Fisheries, Raha, Assam, sponsored by NABARD.
- ☆ 16 Nov., 2007: National level consultation workshop on Environmental Management Reform for Sustainable Farming, School of Industrial Fisheries, CUSAT, Cochin, Kerala.
- ☆ Nov., 2007: Training-cum-demonstration on skill upgradation in net fabrication and mending, conducted in Orissa.
- ☆ 19 Nov., 2007: Conference on monofilament tuna longlining technology, held in Chennai.
- ☆ 20-23 Nov., 2007: 8TH Asian Fisheries Forum, Kochi, Kerala, conducted symposia on gender and fisheries, shrimp aquaculture, fish health, living aquatic resources, productivity enhancement, harvest and post harvest technology, value addition, fishing technology, economics and marketing, exports, aquatic biodiversity, environment, pollution impacts, biotechnology, genetics of molecular biology, and human resources.
- ☆ 29 Nov., 2007 to 3 Dec., 2007 : A.P. Farmers visited Assam on an invitation from that Govt.
- ☆ 29 Nov., 2007: National Seminar on Emerging Trends in Fisheries Technology in Maharashtra, held in Mumbai.
- ☆ 15-16 Dec., 2007: Symposium on Ecosystem Health and Fish for Tomorrow, held at CIFRI, Barrackpore, West Bengal.
- ☆ 23 Dec., 2007: Freshwater Cage Culture Experiment's Success, CIFE, Mumbai.
- ☆ 27-29 Dec., 2007: Eminent Scientist Award 2007 presented to Dr. A K Pandey, Scientist, CIFA.
- ☆ 29 Dec., 2007: RGCA's Demonstration of Seabass Cage Culture Technology, Seabass Hatchery, Technology at Nagapattinam, Tamil Nadu.

- ☆ 3 Jan., 2008: 95th Indian Science Congress. Presentation of Excellence Award to Dr. V R P Sinha by the Prime Minister.
- ☆ 4 Jan., 2008: Golden Jubilee Celebrations of CIFT concluded.
- ☆ 4 Jan., 2008: 'Matsyakumari' - Designed and built by CIFT launched.
- ☆ 27 Jan., 2008: Freshwater Prawn Farmers Meet on Prawn Farming in Saline affected Agricultural Wastelands of Sangli Dist, Maharashtra organised by MPEDA and NABARD.
- ☆ Feb., 2008: MPEDA's Subsidy Scheme for Installation of Insulation/Refrigerated Fish hold/ RSW System and ice making machinery, on board small fishing vessels introduced.
- ☆ 7-8 Feb., 2008: National Workshop on Development of Fisheries for Domestic Marketing of Fish and Fishery products, held at College of Fisheries, Muthukur, Nellore, A.P.
- ☆ 8 Feb., 2008: 92nd Birth Anniversary of Late Dr. G N Mitra, eminent fisheries developmental scientist celebrated in Bhubaneswar.
- ☆ 8-10 Feb., 2008: India International Seafood Show, held in Kochi.
- ☆ 12 Feb., 2008: National Science Day and National Productivity Week were celebrated together at CIBA, Chennai.
- ☆ 17 Feb., 2008: CIFE's Molecular Biology Lab Inaugurated at Kakinada by Dr. S. Ayyappan, Dy.D.G.(Fisheries), ICAR.
- ☆ 18-19 Feb., 2008: Workshop on New Directions and Dimensions in Fisheries and Aquaculture, held at Rajahmundry, Andhra Pradesh by Nannaya University.
- ☆ 18-19 Feb., 2008: 'Stakeholders' Meeting on Tuna resources held in Goa at which modification of fishing vessels for tuna longlining was emphasised.
- ☆ 25 Feb. to 7 March, 2008: Indian Fisheries Professional Team visited Vietnam, headed by Dr. M.Sakthivel.
- ☆ 26-29 Feb., 2008: Training Workshop on Taxonomy of Echinodermata.
- ☆ 28 Feb., 2008: Avanti Feeds launched Mermaid feed, a pellet feed.
- ☆ 29 Feb., 2008: M K R Nair took over as Fisheries Development Commissioner, Govt. of India.
- ☆ 13 March, 2008: Seafood Exporters met the Prime Minister and submitted a Memorandum.
- ☆ 14 and 15 March, 2008: National Workshop on Aquaculture Nutrition and Feed was organised by RGCA at Chennai.
- ☆ 23 March, 2008: Meeting of Marine Products Exporters and other stakeholders conducted at Visakhapatnam. It was noted at the meeting that 25 vessels were converted for longlining in Visakhapatnam region by that time. Several constraints in the tuna vessel operations were pointed out.
- ☆ March, 2008 : New Officer bearers of Society of Aquaculture Professionals, Chennai elected with Mr. D. Ramraj as President of SAP for the period April, 2008 to March, 2010.
- ☆ March, 2008: Dr. S. Ayyappan inaugurated ornamental fish breeding and rearing unit at College of Fisheries, Udaipur.
- ☆ March, 2008: Meeting on offshore cage farming of Norwegian and Indian representatives of MPEDA and of four Indian companies was held. It was decided to start three demonstration projects, one on East Coast, one on West Coast and another in Andaman with Norwegian technology and assistance.
- ☆ March, 2008: Dr. S A. H. Abidi appointed as Member of the Azad University, Planning and Development Board, Lucknow.
- ☆ March,2008: Southern Seafoods Ltd. Visakhapatnam and its associated company in Port Blair commenced processing and export of Tuna Loins from Visakhapatnam directly to Japan. It has plans to process and export tuna Loins from Andamans too.
- ☆ 26 March,2008: A.P. Govt. tied up with an US Firm for exploitation of Tuna in Indian EEZ. It formed a new company under the name Andhra Pradesh Marine Fisheries Development Co. Ltd The name of the U.S. company is World Tuna Development International.
- ☆ 31 March,2008: Union Govt. has brought shrimp production under Videsh Krishi Gnana Upaj Yojana Scheme which provides DEPB additional 3.5 per cent benefits for shrimp exports.
- ☆ April, 2008: CIFT formulated Fish ice cream.
- ☆ April, 2008: Dr. N.G.K. Pillai took charge as Director, CMFRI.
- ☆ 12-14 April, 2008: Juvenile Fish Excluder-cum-Shrimp Sorting Device: Awareness cum-Demonstration Campaign held in Maharashtra.
- ☆ 16 April, 2008: G B Pant University bestowed honorary Doctorate degree on Dr.Modadugu Vijaya Gupta, World Food Prize Laureate and Aquaculture Specialist.
- ☆ 26-27 April, 2008: NBFGR celebrated its Silver Jubilee.
- ☆ 1-10 May, 2008: Technology transfer programme of Marine Ornamental Fish breeding sponsored by the Union Department of Biotechnology was conducted at Annamalai University, Parangipettai, TN.
- ☆ 20 May, 2008: R. Paul Raj took over Member Secretary, Coastal Aquaculture Authority of India.
- ☆ 22 May, 2008: Seminar on Biodiversity Regime : Emerging Challenges and Opportunities', held at Mangalore on 22 May, 2008.

- ☆ 3 May, 2008: Mr. Om Prakash took over as Joint Commissioner (Fisheries) at the Centre.
- ☆ 3 May, 2008: Integrated Fisheries Project, Cochin renamed as National Institute of Fisheries Post-harvest Technology and Training.
- ☆ 28-30 May, 2008: Tuna 2008 held in Bangkok, Thailand.
- ☆ 30 May, 2008: C.T.Betgeri took over as Director, Central Institute of Coastal Engineering for Fishery, Bangalore.
- ☆ May, 2008: Portable FRP Carp Hatchery, developed at CIFA in 2004 and installed at several places, has been installed at College of Fisheries, Dholi, in Bihar.
- ☆ 26-30 June, 2008: National Association of Fishermen conducted NFDB sponsored training programme at Udaipur.
- ☆ 28 June, 2008: National Seminar on WTO held at CIFT, Cochin.
- ☆ June, 2008: MPEDA has closed down its shrimp hatchery in Andhra Pradesh.
- ☆ 1 July, 2008: Ambedkar E Eknath has taken over as Director, Central Institute of Freshwater Aquaculture, Bhubaneswar.
- ☆ 3 July, 2008: Brackishwater Aquafarmers Meet 2008 was held at Kakdwip Research Centre of CIBA.
- ☆ July, 2008: P Krishnaiah took over as Chief Executive of National Fisheries Development Board, Hyderabad.
- ☆ July, 2008: G Syda Rao took over as Director, Central Marine Fisheries Research Institute, Cochin.
- ☆ 10 July, 2008: 'Fish Farmers' Day celebrated at CIFA, Bhubaneswar, in Arunachal Pradesh, at various centres in Kolkata, at CIFE's Kakinada Centre and at many other centres.
- ☆ 10 July, 2008: Fishing Chimes Jayshree Charitable Trust, Visakhapatnam, as part of fish farmers' day celebrations, awarded Mementoes with Citations to M Sakthivel, President, Aquaculture Foundation of India and to S Nagireddy, Expert in Aquaculture.
- ☆ 10 July, 2008: National Fish Farmers Day Celebrated in Udaipur, Rajasthan
- ☆ 10 July, 2008: Fish Farmers Day Celebrated in Karnataka (by UAS, Fisheries Dept and KFDC)
- ☆ July, 2008: M C Nandeesha received Sahameitri Award from the Cambodian Dept. of Fisheries.
- ☆ 24 July, 2008: Seminar on Harvest and Post-harvest Technology for Tuna was held in Cochin.
- ☆ July, 2008: Jawaharlal Nehru Award for outstanding post-graduate Agricultural Research was presented to Dr. (Mrs.) P S Asha, Scientist of Tuticorin Centre of CMFRI by Mr. Sharad Pawar, Hon. Minister for Agriculture in Delhi
- ☆ July, 2008: Sardar Patel Best ICAR Award for 2007 was presented to CMFRI. In Delhi. Dr. G Syda Rao, Director of the Institute received the prize money of Rs.5 lakhs and a plaque from Hon. Union Minister for Agriculture, Mr. Sharad Pawar.
- ☆ 31 July, 2008: B K Behara, Joint Director, MPEDA has retired from service on superannuation.
- ☆ 6&7 August, 2008: Scroll of honour was presented to Bikas Chandra Mohapatra by Mr. Peter Reid, Team Leader, N R International on behalf of Orissa Watershed Development Mission of Agriculture, Department of Government of Orissa.
- ☆ 18 August, 2008: Rashtriya Vidya Saraswati Puraskar bestowed on Prakash Shingare, Associate Professor, Kharland Research Station, Panvel, Maharashtra
- ☆ 22-24 August, 2008: ILDEX India Exhibition was held at Pragati Maidan, New Delhi.
- ☆ 23 August, 2008: CIFT-designed fiberglass boats constructed for operation in Jaisamand lake and other water bodies of Rajasthan by fishermen dependent on the lake fisheries was launched at a function conducted by the State Fisheries Department, Rajasthan College of Fisheries, MPUAT, and CIFT at Udaipur.
- ☆ 25 August, 2008: Workshop on strategies for conservation and resource enhancement of sea cucumbers of India was held in Chennai.
- ☆ August, 2008: RGCA firmed up continuous operations at the Broodstock husbandry centre and at the hatchery at Thoduvai. Installations of recirculation systems were also completed during the year. It also organised a harvest mela in which 12 tonnes of seabass with an average weight of 600-g each were harvested.
- ☆ August, 2008: Moana Technologies, Hongkong set up shrimp seed production centre in Srikakulam District, Andhra Pradesh.
- ☆ August, 2008: Organic Aquaculture project of Rosen Fisheries, Kerala took off, based on international standards.
- ☆ Sept., 2008: Anand Vasanth Asnotikar took over as Minister for Fisheries, Karnataka.
- ☆ 8-10 Sept., 2008: INFOFISH/MPEDA Regional Workshop on Production and Marketing of Coldwater Fish Species was held at Mahila in Himachal Pradesh.
- ☆ 22-24 Sept., 2008: A training programme on compliance with quality standards was held in Kochi by the School of Industrial Fisheries, Cochin University.
- ☆ 28 Sept., 2008: New Office Bearers of Association of Indian Fisheries Industries elected Dr. Y G K Murthy as the the Association's President.
- ☆ 1 Oct., 2008: N S H Prasad, former Vice-Chairman of MPEDA left mortal world.
- ☆ 11 Oct., 2008: Bismi Feeds (P) Ltd. Perumthottam Village, Sirkali Taluk, Nagapattinam district in Tamil Nadu launched a new shrimp feed.

- ☆ 14 Oct., 2008: Japanese Tuna Importers' team visited Visakhapatnam.
- ☆ 14-18 Oct 2008: International workshop on Green Certification of Ornamental Fish, was held in Kochi.
- ☆ Nov 2008: Vannamei import scheme notified.
- ☆ 13 Nov 2008: Workshop on Reservoir Fisheries, was held at CIFT, Cochin.
- ☆ 20 Nov 2008: FVO Mission of EU visited Goa.
- ☆ Nov 2008: Fresh registration of new mechanised boats taken up in Kerala through a Committee.

Fishing Chimes, Editorial 384: March 2009: Vol. 28, No. 12

# SPF Shrimp Broodstock Development

## *India Far Behind*

India is one among the top shrimp producing and exporting nations. India has a well established research and shrimp production management base, capable of promoting a sustainable and disease-free farmed shrimp production. Yet, the internationally renowned Indian fishery scientists have conspicuously failed in the production of SPF broodstock of any of the commercially important shrimps. As a result, they have downgraded India's standing in this line of endeavour by necessitating the compulsive import not only of SPF broodstock of tiger shrimp but also of Pacific White (*Penaeus vannamei),* now in the pipeline. So far as Indian white shrimp (*Penaeus indicus*) production is concerned, unable to produce SPF broodstock of this shrimp, an easier way of importing SPF broodstock of Vannamei shrimp has been resorted to, thereby subjecting the nation to the ignominy of relegating Indian white shrimp (*Penaeus indicus*) to the background, in contrast to the successful production of this shrimp of SPF quality by countries like Saudi Arabia, Equador and a few others. The net Indian position is that the nation not having expertise, now imports SPF broodstock of tiger shrimp and also vannamei shrimp, while virtually neglecting Indian white shrimp. This glaring situation depicts the alarmingly downgraded status of India in respect of technological abilities at producing SPF shrimp broodstock, to a position far lower than countries such as Saudi Arabia. Could any of the fisheries scientific organisations in India, relate its achievements in the SPF broodstock production of any commercially important shrimp with pride? CMFRI can probably cite its success in this production of SPF tiger shrimp seed upto F3 stage but no further. It had to stop at the F3 stage of SPF tiger shrimp seed production. What is the reason for this? It was irrationally asked to stop the work by the ICAR. This happened a few years back. Had it continued the work, probably by now the nation would have had SPF tiger brooders. Probably reconciling itself to the neglect of production of SPF tiger shrimp in the country, NFDB has done the next best that is possible by importing SPF broodstock of tiger shrimp for their seed production at a hatchery set up in North A.P. at a place known as Sompeta, for providing SPF shrimp seed to Indian farmers.

Disease invasion among shrimp under farming in India has brought down its production level, thereby imparting a negative impact on marine products exports. This development has alerted the authorities and the stakeholders including the exporters to find a solution to the debacle. The easiest of this, as finalised was to import SPF shrimp broodstock.

As is known, the bulk of shrimps exported from India consists of black tiger which is one of its indigenous species. Far lower quantities of Indian white shrimps too are exported but these are mostly of capture origin. The imported Pacific white shrimp (*Penaeus vannamei*) farmed by a couple of Indian enterprises (who were permitted to import their broodstock/seed) also became part of to export basket, although in a very small quantity.

In the South Asian region, but not in the south-east Asain region, India is in the forefront in respect of development and application of technologies of shrimp production in brackishwater/freshwater ponds. Scientists of CMFRI, some years back, succeeded in inducing the Indian white shrimp to release eggs through eyestalk ablation. They were later raised to the stage of seed for stocking. However, shortly after this, for some reason or the other, CMFRI lost its interest in promoting the activity of inducing Indian white to breed and in raising its seed. One reason for this could be that the work was no longer in the purview of CMFRI, as CIBA came into existence by then.

On the same logic as mentioned above, the work connected with the development of SPF broodstock of tiger shrimp should have come into the hands of CIBA after it came into existence. However, as it happened, CMFRI continued to deal with the work which progressed upto the development of $F_3$ SPF larval stage. At this point the work was stopped, only to be further carried out by CIFE and CIBA, in association with a Norwegian enterprise. While there is no information on the present status of the work, there are observations that not much of progress has been made. One guess is that, had CMFRI been allowed to continue the work after $F_3$ larval stage too, probably there would have been some cognizable progress by now.

Another focal initiative for raising SPF black tiger broodstock was taken by MPEDA. It had started a project in Andamans for the production of domesticated Shrimp for raising SPF shrimp seed, a few years back. There is however no information on the outcome of the project.

The immediate need is to make available SPF tiger shrimp seed to the farmers. In the above mentioned situation, the authorities, obviously, had no alternative other than depending on an outside source for organising the supplies of the seed. At this juncture, the National Fisheries Development Board took the initiative of entering into an understanding with a foreign company of repute for supply of SPF tiger shrimp broodstock to be utilised for producing the seed at a hatchery in the country and for supplying the same to the farmers. This chosen hatchery is located, as already mentioned, in north Andhra Pradesh at Sompeta from where the supplies of seed produced would be channelised to reach the farmers. While there is no information on how long the imports of SPF broodstock would continue, in all likelihood, there will be a plan of action to raise brooders out of the adults produced in grow-out ponds (utilising SPF seed) and utilise them as the next generation brooders, while taking steps to prevent inbreeding. Now that Vannamei SPF broodstock is being permitted for import by identified enterprises, it is to be believed that an arrangement for the production of successive batches of SPF brooders of these shrimps will also be an integral part of the plan of action, either at a centralised hatchery farm or at several identified hatchery farms.

The present situation in respect of the Indian white shrimp (*Penaeus Indicus*) is anamolous. While this shrimp is widely recognised as an exportable shrimp in various countries including India too, the difference in the Indian context is that it is relegated to a lower status, with little interest evinced in producing SPF broodstock of the shrimp. In contrast, in countries like Saudi Arabia/ Equador, and others, SPF seed production of the species is part of their commercial shrimp production and export activities. In this background, it is considered by many that the promotion of interest in the development of SPF broodstock of Indian white shrimp is of utmost importance and this thinking deserves to be given the needed attention.

Let it be hoped that the authorities concerned would work towards the utilisation of the scientific talent in the country for the development of expertise in the development of SPF shrimp broodstocks supported by needed facilities, and trained hands, for the production of recurring batches of broodstock of the three shrimps *i.e.*, Tiger, Pacific White and Indian White. It is possible that, after stabilising the seed production and supply system from the hatchery set by it in North A.P., NFDB, would replicate the model at other chosen centres along the Indian coast.

Fishing Chimes, Editorial 385: April 2009: Vol. 29, No. 1

## Empowering Marine Fishers Through Value-Addition Strategy

# Marine Fishers' Earnings Can Go Up Only when *Identified Coastal Centres-Based Value Adding Infrastructure is Set Up*

Indian marine fishers are in a lurch. Although the earnings from sale of fishes landed by them now are proportionally higher (compared to past earnings because of general increase in prices), these continue to be far lower and inadequate for most of the fishers to have a fair living. The Reason: The prime types of fishes landed, as sorted out, are auctioned at the landing point itself and the rest are diverted either for drying or salt curing. The on–the–spot auctioning of landed fishes (other than those excluded for drying/salting), while it fetches immediate returns that the fishers naturally need for their daily expenses, the disadvantage of this system, however, is that equitable returns from their landings are elusive to the fishers. As it happens, the fishes, after the first sale at the landing point, pass through a few intermediaries subsequently, may be two or more, before the final sale to the consumers at the urban or other markets. At the final sale point, the price can be three or more times higher than the first price of procurement at the landing point, generally during the first (auction) sale. The present system thus denies to the fishers a share of this increase in final sale price. The exception to this routine, however, occurs when one of the family members of the concerned fisher takes the fish for direct sale mostly in the retail market, but such instances are few and far between. The net loss in the earnings of a fisher because of the aforesaid traditional system is estimated in knowledgeable circles as around at least three times of what he actually receives at the landing point under the present system. Intermediaries are of course entitled to have their earnings because of their initiative, but this cannot be allowed to be a hurting disadvantage to the fishers. Unless the authorities subscribe to the view that intermediaries too have a right to make money, the present system that denies the accrual of the possible extent of the final sale proceeds of their hard earned catches, to the fishes would have to be neutralised.

The situation is so alarming that the National Fisheries Development Board has to recognise the prevailing situation and initiate measures to do justice to marine fishers. These measures should aim at imparting value to the fish catches at the landing centres themselves, at identified centrally located points. At these points the needed infrastructural facilities would have to be provided to facilitate integrated operations, from the stage of receiving of raw fish to the point of outflow of processed value-added products, for the purpose of radiating them therefrom among sales outlets. The arrangement should be such that all fishers who register themselves with the authorities in charge of the value-adding plants would be entitled to hand over their catches at the plant, at which, after value-addition, the final products would be sold to registered buyers (sales outlets which can be department stores or other such outlets) at determined rates, out of which an entitled share, as agreed to, will be conceded to the raw fish suppliers to the value-adding plant concerned. A part of this, as agreed to, would be paid in advance to the fishers at the time of supply and the balance due would be paid at the agreed point of time.

The deeply entrenched problem of marine fishers being exploited by middlemen, by way of cornering of the final market sale proceeds,a good part of which should have accrued to them (marine fishers), can be solved only through the ushering in of a major National Project of an integrated nature for rendering financial justice to them. The highlight of this suggested project would have to be to add value to fish at the related coast-based infrastructure facilities (value-adding plants) as mentioned above and sharing a commensurating and viable part of the returns with the stakeholding fishers.

The investment for setting up the said coast-based value adding infrastructure would have to be borne by NFDB, who would have to also organise its implementation.The pre-project work would have to naturally consist of survey of landing centres along the length of the coastline with reference to the number of fishing units, their annual landings, number of stake holding fishers, possible centralised places of locations of value adding plants along the coastline, identifying the chain of fish marketing outlets in cities, towns and larger villages at feasible distances to the landing points around,besides taking care of all other related aspects. Armed with this data, the project can be planned to consist of various links from the point of fish landings and their segregation in relation to value-adding potential and value addition at the chain of plants that would include packing and transportation of finished products to the various marketing outlets. Besides the value-adding plant's design, specifications and other particulars of the machineries and equipments to be installed, the aspects to be taken care of would have to be brought out in a working manual of the plant along with the operational details from the points of fish landing to the stage of marketing of value-added products. The selection and registration of marine fishers, Fishermen's Co-operative Societies and others in the line for providing raw material supplies and the regulations to govern their functions would have to be drawn up by the officer designate or the Project Director, for being followed by the operatives. The regulations concerning payments to the stakeholding fishers concerned have to be also worked out by the concerned in the project.

In regard to the Organisation to be set up to run the suggested national -wide Project, the choice would have to obviously converge on the former Integrated Fisheries Project, now aptly renamed as the National Institute of Fisheries Post-Harvest Technology and Training. NFDB can consider commissioning this organisation to prepare the Project Report that would of course have to cover all the integrated components from fish landing points to final stage of marketing of the produce in value added form. A Project of this kind would not only provide enhanced incomes to fishers but would also ensure supply of hygienically processed value added fish products to the people.

Another important part of the suggested venture, to be considered by NFDB, would be to set up, to star with, one model Coast based Pilot Fish Value-Addition Plant, representing those to be set up eventually all along the coastline as part of the project. Such pilot plants, one to be located one in each of the Coastal States at selected points, can be set up first and made operational. At these pilot plants the working systems can be tested by the National Institute of Fisheries Post-Harvest Technology and Training, so as to facilitate the extension of similar plants to other districts along the coastline to serve as a basis for replication all along the coastline.

It will be recalled that, in the pre-independence period, the British rulers set up Fish Curing Yards along the coastline of erstwhile Madras State, that also covered a small part of the coast of the present State of Orissa, present Coastal AP and Tamil Nadu,also a part of the present Kerala State (Malabar district), and also a small part of present Karnataka State in the southern part. These yards were set up, with the provision for conceding exemption of excise duty on salt, for adding value to fishes through salt curing/sun drying. Fishers were not allowed to do this value addition anywhere outside the precincts of the fish curing yards. Since then, there have been major developments in fish value addition technology, in any case in the past couple of decades, and these deserve to be popularised as a follow-up to the past value-adding tradition of production of hygienic salt cured and dried fish. The hygienically salt cured fishes at fish curing yards in the British days were fetching relatively higher returns. In the same way, the present day value-added fish products too are securing higher returns, but the benefits of these are not going to the fishers. The need now is therefore to ensure that, through the setting up of a chain of fish value adding plants along the coastline under an organised system as indicated in the foregoing write-up.It should then be possible to enable coastal marine fishers to earn higher incomes that would be commensurate with their hard work.

Fishing Chimes, Editorial 386: May 2009: Vol. 29, No. 2

# Impact of NFDB on Pace of Indian Fisheries Development

The National Fisheries Development Board (NFDB) of India was inaugurated by the then Union Minister for Agriculture Mr. Sharad Pawar on 9th September, 2006. In his inaugural address, while promising to ensure proper input availability for the promotion of fisheries activities from production to marketing and for integrating these activities effectively, he laid focus on the vast potential for development of mariculture and also cage culture, particularly in the coastal waters of the country. He also emphasised the need for improving post – harvest handling and marketing facilities of fishes for the benefit of stakeholders. Pointing out the major gaps in the cold chain planning at different levels in the country, he gave an indication that NFDB would address the issue of linking production centres with marketing outlets in a big way. He referred to the immediate need for diversification of farmed species and farming systems and for their integration with other farming systems. Pointing out that overconcentration on the production of major carps was causing economic distress among the farmers, he brought to the fore the need for popularising aqua shops as single one stop facilities for the supply of seed, feed and other inputs for the benefit of the farmers. One potential aspect specially mentioned by him was the development of fisheries of reservoirs and also ox-bow lakes. Another important point highlighted by him was in respect of fisheries and aquaculture development of the coldwater zone of the country, which, he observed, was presently in an undeserved state of neglect in respect of development.

Dr. S. Ayyappan, the then chief executive of the NFDB, explaining that Rs. 2100 crores were allocated to the Board for utiilsation towards the various activities to be undertaken from 2000 to 2012, detailed the scheme-wise allotments in respect of subjects such as intensive aquaculture, reservoir fisheries, coastal aquaculture, deep sea fishing and tuna processing, mariculture, sea ranching, sea weed cultivation,infrastructure development for post-harvest processing, fish dressing and for solar drying of fish, domestic fish marketing, and so on.

Follow-up action in regard to the implementation of the various schemes listed above must obviously be in progress and the stakeholders concerned must be in the know of the developments. The others interested in these aspects may have to probably look for opportunities to know about the progress through means of other plans that are now available. The sterling work now in progress, understood to have been initiated by NFDB is in respect of reservoir fisheries development in some of the States, in respect of the assistance extended for conducting seminars etc. and in regard to several other developmental activities which would come to be known to those interested only through word of mouth or through occasional information published in newspapers, but which happen to be not seen often. However, considering the purpose and importance of the items of work entrusted to NFDB by the Union Department of Animal Husbandry, Dairying and Fisheries, it will be agreed by all concerned that release of periodical special bulletins depicting the progress and status by NFDB in regard to the fisheries development under various schemes/subjects/topics is imperative, as these will be of benefit of all those interested in fisheries development.Such bulletins are released by FSI, CMFRI and others.

The Union Department of Animal Husbandry, Dairying and Fisheries specifically entrusted certain functions to NFDB as publicly mentioned by the Minister of Agriculture; and the Chief Executive of NFDB at the time of the inauguration of the organisation.In this context, those in the Industry would naturally be anxiously looking forward to know about the progress thereof at reasonably periodic intervals.

Promotion of seaweed cultivation is on the agenda of NFDB. However, beyond what was accomplished by Aquaculture Foundation of India and fishers' self help groups in south Tamil Nadu, which was widely publicised, information on the interventions of NFDB in the various coastal States on this developmental activity awaits to be told. So is the case in respect of sea ranching and the setting up of specialised hatcheries to produce eventually advanced fingerlings of identified species for ranching, and the proposed set up for undertaking the work. The same situation is applicable in respect of cage farming too. The role NFDB intends to play for the development of cage farming along the coastline in co-operation with the coastal State Fisheries Departments (identification of sites, selection of beneficiaries for their allotment, Training Programmes and Centres, and for providing a set up for the purpose, and information there of and on such other organisational matters, would probably be announced by and by).

There are many who are anxious to know about the strategies and technologies adopted for the development of fisheries of reservoirs, *beels* and other wetlands; the promotion of India's own technologies for the preduction of SPF broodstock of shrimp/prawns as has been successfully achieved by countries like Saudi Arabia, in relation to the upgradation of the design and working methods in the respective hatcheries, farm ponds and diversification of farming systems in relation to improved seed production packages and so on.

There are vast stretches of saline lands on the north - west of India which are lying fallow. An isolated begining was made to utilise these lands in Haryana by Dr.Dwivedi over a decade back with some measure of success but for some reason, there was no follow up action.

There can be no doubt that remarkably encouraging results of several advanced systems of fisheries development have been either achieved are in progress in the country, but these alone are not enough. The stakeholders have to be told about them in an effective manner to generate further progress.

Fishing Chimes, Editorial 387: June 2009: Vol. 29, No. 3

# Management of Fisheries of India

A major achievement in the fisheries sector of India in the past few decades is one of managing Indian Fisheries wealth in a remarkable manner monitered by the Central and State Governments, with sustained production results, despite the fact of not having an organised Indian Fisheries Service. Many have felt the need for a separate Fisheries Ministry at the Centre although no need has been felt for having a Fisheries Service.This is probably because of existence of a net work of qualified fisheries executives, most of whom are post graduate/graduate degree holders in fisheries, degrees taken mostly from fisheries universities, around 16 of which are there in the country.These candidates mostly occupy the fisheries executive positions from sub – divisional to district, regional levels and the top position of Director of Fisheries. The exception to this ongoing system is the induction, mostly of Indian Administrative Service Officers either as Directors or Commissioners of fisheries of the State concerned, but this deviation is attributed by several to two reasons; one is that IAS officers returning from other positions back to the main line in the State concerned needing a good placement; and vacancies of positions like Director of Fisheries can provide a stop-gap arrangement ; The other is that an officer requests for placement at a particular place. In any case, it has to be believed that the State Governments would like to have a fisheries professional primarily as the Head of their Fisheries Department.

Another aspect that the Centre should have to consider favourably is to have a separate fisheries service, considering the extent of fisheries resources of the country and their potential to augment food production and to upgrade the socio – economic conditions of lakhs of fishers and other stakeholders dependent on fisheries. In the same way as the forest sector has a separate Union department, fisheries too deserve to have a separate department, exclusive for the subject as its wealth is not less important for the nation than forest wealth.

The quality of fisheries management policies in force, in terms of area coverage and depth of their application, determines the contours of the impact of the policies on the status of sustainable fisheries development in the national zone concerned. Several of the countries, big or small, in the frontline in the development of fisheries such as China, Australia, Canada,USA and Taiwan, Israel, Vietnam, Thailand etc., have positioned experienced fisheries professionals at the top to propel their fisheries developmental process forward. In contrast, in India, both at the Centre and in the States, it is believed that, at the helm of affairs at the top, there are politicians with very little experience in fisheries.So far as the complexion of the political leadership in the fisheries sector is concerned, there are no encouraging signals generated as a result of the recent election. The only exception seems to be that one Mr. Vatti Vasant Kumar,a newly appointed Minister in A.P. State, has vast experience both in marine capture and Inland culture fisheries, but he has been placed in charge of rural development in the State and not of fisheries. At the Centre, we have a well reputed politician as the cabinet minister in charge of fisheries, but the general comment is that the policy decisions have been in favour of introducing Taiwanese vessels functioning as Indian vessels for fishing in Indian EEZ, totally neglecting the introduction of fully owned Indian deep sea fishing vessels. So much so, Indian EEZ is now virtually allocated for fishing by Taiwanese high sea vessels in the garb of Indian LoP vessels. Had India introduced a scheme for the introduction of deep sea fishing vessels of its own construction at Indian shipyards (which have the experience in the line), by now, Indian owned deep sea fishing fleet would have come up, fecilitating the phasing out of the Taiwanese-owned vessels introduced in the garb of Indian vessels.Another aspect is that,while several nations went ahead in marine cage culture decades back, India is striving now to enter the line. Had the top leadership taken a tangible initiative much earlier, India would have by now been in the list of countries in the commercial marine cage culture activity.

In regard to the State sector,fisheries professional management at the top level can be discerned only in some of the States, and at various levels of application in the other States. So far as State Fisheries Corporations are concerned, while the one in the State of A.P. was closed down, alleged to be because of apparent mismanagement, in other States one would not come across any indication of impressive performance.The reason for this state of affairs is apparently the unsuitability of the top persons in charge who are probably not from fisheries background to perform well and register good results. So far as State Fisheries Departments are concerned, several of the top positions are in the hands of non-fisheries hands, mostly those belonging to Indian administrative or a state service.

There have been many recommendations made at several conferences and other gatherings of experts for the setting up of a separate ministry of fisheries at the Centre to promote fisheries development, but this has not been accepted so far. In any case, it is doubtful whether the setting up of a separate Ministry/Department would make much of a difference as long as the administrative serivce officers without fisheries background manage the affairs of the newly created Department, if at all it is created. There can be a reformed situation only when experienced fisheries professionals are given the opportunity to serve as Directors of Fisheries in the State sector.

India has the distinction of being the second in respect of Inland fish production in a global perspective, the first position going to China. In terms of per unit fish production (marine/Inland) or in other respects such as fisheries management, India may have to be rated as behind, compared to even small countries like Vietnam and to others such as Taiwan, South Korea etc. This situation stares at us, despite the fact that India has a well developed fisheries infrastructure that covers all phases of fish production linked to the marketing phase. So far as fisheries management personnel is concerned, it is known that India produces fisheries graduates as well as post-graduates. They come out from several fisheries colleges affiliated to fisheries universities, besides post-graduates in fisheries coming out of the well reputed Central Institute of Fisheries Education, Mumbai (which is a deemed university) and other fisheries universities. Several of these qualified persons are now in important positions in several fisheries organisations,by virtue of having acquired vast experience in the fisheries managerial aspects. There are also a number of fisheries experts who happen to be part of the fisheries branch of the Indian Agricultural Research Service. In fact, experts that hail from the fisheries branch of this service are in charge of a few fisheries organisations in the North-eastern and certain other States of India.

Looked at from any angle, fisheries experts of Indian Agriculture Research Service, although they are perceived presently from a certain angle only in relation to fisheries,are eminently suitable to effectively manage the fisheries affairs of the State Fisheries Departments, once looked at from a development – oriented angle. As at present, of course, in most of the States there are several qualified and experienced fisheries officers (promoted from senior positions) heading the fisheries departments concerned. However, information is that there are nine States (Andhra Pradesh, Gujarat, Kerala, Maharashtra, Orissa, Rajasthan, Tamil Nadu, U.P. and West Bengal) in which the fisheries departments are headed by the officers of the Indian Administrative Service.This is understandable,as in the past too, some of the Officers of ICS and IAS rendered yeoman service to the fisheries sector (and some of them continue to do so). Late K.N. Anantharaman of the ICS, Late G.V.S. Mani of IAS, and quite a few of others from these services rendered unique services for the development of fisheries in the erstwhile composite State of Madras and later in Andhra.

It has to be clarified here that the non-fisheries professional Directors of fisheries in various States too render welt-directed services for the development of fisheries inclusive of the fisheries stakeholders.Their handicap in this context however is the limitations in respect of their technical and technological background. No doubt they endeavour to pick-up the needed background knowledge for doing justice to the position, but before this is done they often get transferred. In fact, according to what is heard, the administrative service officers are posted as Commissioners or Directors of Fisheries for a short duration, until the officer concerned could be given a place to serve in another job where his services could probably be better utilised. Considering the availability of professionally capable fisheries experts with needed qualifications for appointment as Commissioners of Fisheries/Directors of Fisheries/Managing Directors of State Fisheries Corporations or such other organisations it is felt by many that State Governments should not trouble officers of Indian Administrative or other central services to take up the aforesaid positions. Instead, the system of inducting senior fisheries professionals in the Indian Agriculture Research Service to the positions of heads of fisheries departments may preferably be adopted,in case a suitable departmental officer for the post could not be identified.

Management aspects at micro level also deserve focal attention. The intra technological aspects such as those at **seed production centres** *i.e.* hatcheries (say, in respect of brooder maintenance, inducing them to breed, taking care of spawn, raising it up to the stages as required such as fry and later to subsequent stages),and many similar others of micro stature, would constitute important and crucial parts of fisheries management. While the function of effectively implementing technological aspects of this kind vests with managers at the appropriate level, the effectiveness of this manoeuvre will depend on the proven technological awareness of the subject by the manager and his associates concerned. And here comes the importance of having a competent man conversant with the related technologies for imparting the needed alertness and effectiveness among the technologically positioned working personnel.

The Indian Fisheries Universities/Colleges have the distinction of producing well accomplished fisheries graduates (and post-graduates) for taking part in fisheries developmental process as managers, both in public and private sectors. Some of them establish their own ventures. While these candidates are a great asset, an aspect to be specially looked into in depth in this context by the fisheries educational institutions is the current level of fisheries management aspects covered in the education and training parts of the related syllabus and set right deficiencies, if any, by updating the syllabus.

Frequent technological developments in the fisheries sector have been taking place in several countries and India too is naturally one among them. The implementation of the emerging technologies will be faster once the top management of the fisheries departments is in the hands of professionally well qualified Directors, so as to stimulate the required trend of alertness downwards to the needed level. While fisheries administrative aspects can be well looked after by those in the line, field managerial aspects concerning fisheries, both marine capture and culture, inland capture and culture fisheries, reservoir fisheries (capture and capture-based culture), and of tank/pond/farm fishery would need the attention of professional fisheries managerial hands from top to bottom, so as to take care of all related aspects. The distinction between general administrators and professionally oriented fisheries experts becomes crucially important in this context.

Fishing Chimes, Editorial 388: July 2009: Vol. 29, No. 4

# Future of Fisheries Development Rests on Technocrats of Quality and Competenance

Graduates and Post-graduates in Fisheries coming out of fisheries colleges of India and also from related departments of certain universities, batch after batch, get absorbed in various positions not only in Central and State Governmental Fisheries Departments but also in fisheries commercial enterprises, that include aqua hatcheries, fish farms, enterprises connected with marine fishing boats, processing plants, export enterprises, etc., in positions such as managers and technical hands. Some of the qualified candidates set up their own enterprises.

The basic features of the Indian fisheries educational system are impressive. Despite this, murmurs are often heard, both from Government departments and managements of private sector enterprises that the quality and competence of some of the candidates coming out of the institutions are below par. One reason for this, mentioned uncharitably, is that those that fail in securing entry into Medical Colleges and others who have no hopes of gaining entry into them, join fisheries colleges. In this background, they express the view that the future of fisheries development will be shaky, if the capabilities of the bulk of emerging generation of fisheries technocrats originating from the institutions are not toned up.

It is a well known tendency on the part of successive older generations of fisheries executives to highlight or impute inefficiency, perfunctoriness at work, inadequate technical knowledge, lack of devotion to duty and quality consciousness etc., among some of the new technocrat entrants into the sector. Imputations of fall in standards from generation to generation are not new. In fact, observations of this nature are stated to be on record, even from 18th century. It appears that, because of the expansion of fisheries development activities, and the consequential increase in the fisheries educational and training facilities, the number of qualified candidates has been on the increase in the past couple of decades.While a certain percentage of these qualified candidates seeking employment emerge as competent technocrats of quality, many believe that the rest of them would be mostly of average capabilities. Probably, as the bulk of those employed fall under the later category, the employers keep expressing their disenchantment and chagrin, when they notice that the targets of their expectation are not realised. Apart from this, those who have attained seniority tend to comment generally on the fall in the standards of the juniors working with them.

While there can be considerable exaggeration in these perceptions, there can also be some substance. The enterprises that have come up based on new technologies, now and then point out the problem of inefficiencies among the technical personnel employed by them.

In this situation, a way out has to be found to tone up the standards of quality and competence, particularly among candidates of average capabilities who come out of the various institutions qualified as fisheries technocrats. More than those who impart education to them, the concerned candidates themselves would have to adopt a well planned system of acquiring a deep insight into the subject matter. It should also be possible for the faculty members to intensify efforts to inculcate an awareness in this regard among the students. This is mentioned for the reason that it is not easy for the candidates of average standing to know about the system to be followed in organising their studies outside the class room in order to have an indelible and an indepth understanding of the subject matter. This is mentioned only to emphasise the responsibility on the part of the faculty members to infuse among the students the need for undertaking in-depth studies of the various facets of the subject outside the class room. Such an infusion would certainly enable a large section of the students to come out of the institutions as technocrats of quality and competence far beyond the average level so as to play a dedicated role in the development of fisheries of the country. In other words, the faculty members and the students have to further strengthen their co-operative approach for a common purpose. When this is done, the parameters of quality of learning and competence of the candidates would get honed up. These parameters, of honed quality, would be eminently helpful in the later period of the actual involvement of the candidates in the industry. Thus, the degree of success achieved by the candidates in the post-college career would mostly depend on their hold on developed/developing technologies.

In the culture fishery development, as is commonly known, we have three broad phases consisting of culture (seed production, farming), harvest and post-harvest activities. Drawing an analogy, in their journey through fisheries education, the candidates, as is known, pass through the study phase (comparable to culture activities), the harvest phase (appearing for examinations and securing the degree), and the post-harvest phase (career or marketing of the talent of quality and competence in the subject acquired during the study phase).

Let us examine the study phase briefly. This phase, being of a learning process must impart an infusion of total awareness of its crucial importance among the candidates from various angles. Several students, particularly the average ones, are often seen to be not having the needed sharpness or awareness for retaining the specific details of various important technologies in a lasting memory. Instead, as is generally observed, they just follow the routine of a superficial understanding of the subject in a perfunctory way. The inculcation of a proper study culture leading to the retention of the specific details of various technologies would vastly help the candidates in their future career. The adoption of a proper study culture would also provide a sound orientation to the average students to fare well at examinations in order to secure an outstanding rank. It is known that many students fail to answer the questions succinctly and well at the examinations, despite having the knowledge. This indicates that the candidates have to acquire the faculty of faring well at examinations and in a way to impress the examiners by practising an improved study culture and an improved way of performing at examinations.

In the course of their studies, students often come across several sentences in the books or class notes which cannot be readily grasped. Experts say that whenever such a situation arises such difficult sentences should be broken into their various components and reassembled again. In this process, it is said that the meaning would emerge clearly and register on the minds of the candidates.

Generally, authors build up a paragraph revolving around a single point and with a connectivity to the points made in the subsequent paragraphs. This being the case, the approach to evolving a study culture of ones own has to be to extract the central points of each of the paragraphs and noting them down. Many students keep note books by their side for jotting down notes. This system can be improved upon. A general suggestion is that atleast three readings of each of the chapters should be given with a reasonable time-break in between. On each of these occasions, notes can be taken in three different note books. The idea behind this suggestion is to ensure a progressive perfection in the presentation of the points. By the time the entry in the third round is done, the candidate would have made his perceptions crisp and clear with an incisive insight.

In the harvest (examination) phase, because of a more enlightened study phase as referred to above, a superior way of presentation of answers in the examinations would materialise. This becomes possible because of the continuous and persistent preparation during the study or culture phase. The harvest phase can be expected to lead to the scoring of an outstanding level of marks which would pave the way for a better post-harvest phase.

Success in the post-harvest (career phase) operations is mostly dependent on depth of knowledge, talent and competence. Once an employer is impressed over these parameters of an employee, he will not leave him and the development of this kind is the first stepping stone for building up a lucrative career with a good reputation.

Considerable progress in the development of fisheries of the country has been achieved. While this achievement is the result of a combined effort on the part of technocrats, scientists, farmers and fishermen, the main contributory factor behind is, however, the inputs of a band of motivated technocrats with capability to generate an array of activities to augment fish production and upgrade its utilisation culminating in its domestic and export marketing. When well motivated candidates with competence, coming out of the fisheries educational institutions provide reinforcement to the continuity of the on-going endeavour, there will emerge such a change in the scenario that would provide a miraculous boost to fish production. For this to happen, an enlightened approach towards the quality of knowledge acquisition may have to be fostered among the students who are the technocrats of the future. Stated differently, the approach to be adopted may have to be such that it would lead to the emergence of a far superior category of technocrats of quality and competence, for employment both in the Governmental and private sectors. To sustain and be ahead in studies, the plus factors have to be taken care of. The candidates have to strive to have as part of their build-up, the gist and steps of technologies, particularly those related to operation of various types of hatcheries, culture practices, capture fishing systems and processing and marketing aspects.

The depth of knowledge that the fisheries students acquire no doubt dictates the course of their career. In addition, it would also contribute considerably to the overall development of fisheries of the nation. In this context, what is of paramount importance is the need for the emergence of students out of the fisheries educational institutions with a deep sense of confidence in respect of their readiness to effectively take part and contribute to the development of fisheries of the country.

It is possible that the few thoughts put down here may generate reactions of various hues, some of which can be disapproval, but the objective of the exercise would be served if it provokes a critical appraisal of the present pattern of studies and leads to the evolution of an improved and more effective system of study culture among the students, to enable them to achieve a distinctive rank and a bright career thereafter. The fisheries graduates and post-graduates, who come out of the fishery educational institutions, with an awareness of their crucial role, would be rendering a signal service to the nation as the pillars of future fisheries development.

*(This editorial, earlier published in August 2000 Issue of Fishing Chimes, is reproduced here, particularly for the benefit of the fresh batch of students who would have joined or would be joining the fisheries colleges this academic year).*

Fishing Chimes, Editorial 389: August 2009: Vol. 29, No. 5

# Giant Freshwater Prawn : Monosex (Male) Farming

Of late, there has been the emergence of a consensus among fisheries scientists and fish farmers that monofarming of male giant freshwater prawn seed deserves focal initiatives. The obvious reason for the identified approach is based on tested results that showed that males of GFP grow far faster than females, providing distinctively higher yields that provide increased returns to GFP farmers on one hand and to the exporters on the other. At present, in India, there are over 40,000 ha of water area brought under GFP farming, supported by around 70 hatcheries leading to supply of 1.83 billion GFP seeds to farmers annually. The GFP farming activity has now spread from the Coastal Zone to the inland States too, supported mainly by certain channels of seed supplies that have come up to serve the interior ponds and tanks away from coastal belt. Further, a few of the inland States have set up their own hatcheries utilising either locally available underground or overland saline waters or constituted saltwater.

The slow spread of giant freshwater farming in the non-coastal States of India is traceable to logistics of taking the seed from the hatcheries, most of which are located along the coastline producing seed in a healthy condition for stocking in distant tanks and ponds. One problem in this context however is that seed of GFP has the disadvantage of stunted growth because of inbreeding at the points of their breeding and in the course of raising of their seed. There were efforts at getting over the problems by organising cross - breeding between distant stocks but with limitedly encouraging results.

In this background, taking cognisance of the fact that males of GFP grow faster, efforts have been mounted to segregate males out of the juvenile stocks for stocking. At one stage, the field workers of C.P. Aquaculture undertook the male segregation work in a coastal district of A.P. based on the location of male genetic pores, despite the fact that this kind of segregation has the disadvantage of rejection of a large number of female specimens.

In this situation, there has been a new development,as part of the results of research at Ben Gurian University in Israel under the leadership of Prof Amar Sagi. The results of this research work on giant freshwater prawn were presented at a Symposium on Freshwater Prawn held in Aug 2003 under the auspices of the Fisheries College of Kerala Agricultural University, Kochi, India. These results are stated to have opened up an opportunity for ICAR to advise the Central Institute of Freshwater Aquaculture, Bhubaneswar to take up experimental work on producing all male progeny of GFP and extend the encouraging, results among the concerned stakeholders. A summary of presentation on the subject made by Amin Sagi at the said symposium is given hereunder:

Prefacing that males of the freshwater prawn, *Macrobrachium rosenbergii* grew faster and reached a significantly larger size at harvest compared to females, Amin Sagi said that, because of this feature, culture of monosex all-male population was advantageous compared to the normal mixed sex population. Pointing out that sexual differentiation in crustaceans was regulated by the androgenic gland, he explained that the androgenic gland had been described in a variety of crustacean species as exerting morphological, anatomical, physiological and behavioural effects. In these respects, the androgenic gland had been shown to play a pivotal role in the regulation of male differentiation and in the inhibition of female differentiation. In *M. rosenbergii*, complete sex reversal was achieved by androgenic gland removal (andrectomy) from males at an immature stage, resulting in female differentiation, including the development of ovaries, oviducts and female gonopores. Similarly, androgenic gland implantations into early stage immature females led to the development of testes, sperm ducts and male gonopores. *M. rosenbergii* specimens that had thus undergone sex reversal proved to be capable of mating with normal male specimens and producing progeny. Based on the above, production of monosex (male) prawn populations via intervention in the androgenic gland was a needed step. Pointing out that a feasible technology was still to be developed and an androgenic hormone had not yet been identified in decapods, two lines of future biotechnological research and development pathways were proposed by Amar Sagi: 1) Micro-surgical andrectomy in juveniles, leading to the development of functional neo-females to be mated with normal males to produce all-male progeny and 2) Elucidation of androgenic gland biosynthetic products to enable future biochemical or molecular interventions.

Included in this issue of *Fishing Chimes* is a comprehensive paper on the subject of GFP male progeny production, authored by Professors C. Mohan Kumaran Nair and Salian. Readers of *Fishing Chimes* particulary those at the top in the fisheries hierarchy may like to go through the contents of this contribution and consider the emergent need to produce all male progeny of giant freshwater prawn stocks for the particular reason that this step would not only lead to augmentation of value-oriented aqua production but would also add to the national export endeavour. There will also be the addition of value-adding efforts towards the promotional work related to the expansion of GFP male progeny production work all along the length and breadth the country.

CIFRI has the distinction of gifting Induced fish breeding technology to the Nation. CIFA is a chip of CIFRI. It carries the blood of CIFRI and therefore shares the distinction of introducing the technology and of distinctively transforming the traditional fish seed production work all over the country into an advanced technology, while at the same time leaving the riverine seed stocks to have freedom of their own growth and movement.

In the fisheries sector, after the introduction of induced breeding technology, there has been no epoch-making development of the same distinction. Let it be hoped that the introduction of all-male GFP seed production technology would be a second revolutionary development, after the introduction of induced breeding technology, having the impact of a major socio-economic development, among the stakeholders and for augmenting export-oriented production.

Fishing Chimes, Editorial 390: September 2009: Vol 29, No. 6

# Role of FFDAs: Promotion of Pond-based Cage Farming

The Fish Farmers Development Agencies in the country have emerged in the past three decades as a powerful mechanism for organising supplies and services (including post-harvest linkages) to fish farmers, in the pursuit of their professional activities. The highlights of supplies and services extended by FFDAs are related to the extension of financial support to farmers on the one hand, and in respect of organising for their benefit, supplies of various inputs that include supplies of seed and other needed requisites for pond preparation and for promotion of crops, taking care of the needs in respect of disease prevention and control, judicious feeding, post harvest management and marketing of the produce at remunerative rates, on the other. FFDAs also help farmers in adopting applicable systems of farming such as composite farming and integrated farming, besides the diversified systems that cover farming of fishes such as Asian sea bass, *(Lates calcarifer)*, the catfish, *(Pangasius* spp*)*, the Climbing perch *(Anabas testudineus)* etc. Farmers are also helped in the production of major carp fish seed through induced breeding. The adoption of the developed farming technologies by the farmers, as extended to them by over 440 FFDAs in the country, have upgraded the average annual Indian farmed fish production per ha from 600kg before their formation, to an average of 2,226 kg per ha as at present.

FFDAs are achievers. Being so, their continuing professional longing would be to achieve higher levels of fish production per ha. This implies that they await to be told by the State Fisheries Department concerned on the latest advanced sustainably higher yielding farmed fish production technologies that could be extended by FFDAs to the stakeholders concerned. An encouraging answer to this question, although it may sound presumptive, is the popularisation of pond cage farming system among the stakeholders.

Taiwanese farmers have pioneered in pond cage farming system. This system combines pond cage farming with open pond farming in the same pond. This system, as practised in Taiwan, integrates cage fish production, mainly of Tilapia with acclimatised black tiger shrimp seed, either to freshwater or to ponds of low salinity waters. This system deserves emulation. However, since Tilapia is a banned species in India, other value-oriented species such as seabass (or even major carps) can be raised in pond cages. In the open pond water farming part and in the water underneath the cages of the system, black tiger shrimps, acclimatised to low salinity water or freshwater conditions as required can be raised. Vannamei shrimp also can be raised in the cage-free zone.

The system, as mentioned above has the following features:

1. The adoption of polyfarming practices in ponds with cages set up to cover half of the area of pond surface and related columnar water, to about ¾ th of the pond depth by the cages, leaving the remaining 1/4th of water of cage-free zone underneath for prawn/shrimp farming. This kind of combination and integration of pond cage and open pond farming system has been successfully tested and commercialised by Taiwanese farmers. An example is Team Aqua Corporation (TAC) of Taiwan who standardised this system as early as 1997, and which was adopted by others.
2. TAC specifies a minimum depth of 2.5 m (8 feet) for a farm pond to undertake cage-cum-open water aquafarming in it. However, this depth could probably be readjusted to prevailing Indian conditions. In the TAC system, the surface area of each of the cages is 27.56 sq.m (5.25 x 5.25m). The depth of each cage is suggested to be 2 m by TAC. This stipulation brings the cubic volume of each cage to a level of 55.12 cubic meters. According to TAC, the production capacity of each of the cages of the above mentioned dimension per annum is 1.2 to 1.4 tonne (This can probably be rounded up to an average of one tonne). On this basis, a cage of the aforesaid dimension, in terms of each sq.m, can produce around 36 kg in an annual production cycle. This level of production per sq m gives an indication that around 36 nos. of advanced fry can be stocked per sq m of surface area of a cage or 18 nos. per cu.m. As the volume of each cage is 55 cu m, its production capacity works out to nearly 1,000 kg. Stated differently, in order to achieve 1,000 kg of production from a cage, each sq m has to be stocked with 36 nos. of fingerlings or each cu m of its water space has to be stocked with 18 nos. of fingerlings, on the assumption that each of the fingerlings could be grown upto a weight of 1 kg or marginally more. In order to counter any possibility of mortality, precautionary measures as needed and applicable need to be taken. The measures suggested by TAC on this aspect are not available.

The cages set up in half of the area of a one ha pond would cover 5000 sq m or 10,000 cu m, as the cages will be

2m deep. On the basis the fingerlings to be stocked in the zone covered by several cages comes to nearly 1,80,000 nos. per ha, or 90,000 nos. per 0.5ha, it can be reckoned that nearly 90 tonnes can be produced from cages in half of the pond area of one ha pond (with the other half of the surface and columnar area being cage-free). According to TAC, this level of production could be achieved from fishes farmed in cages alone. The FCR not having been mentioned by TAC, it can probably be taken as 1.5:1 taking into account the achievement of the high level of production projected at an average of one tonne (carrying capacity) per cage.

The remaining water area, *i.e.*, around 0.5 m of water below the cages and the remaining open area not covered by the cages can be utilised for farming freshwater prawns/tiger/vannamei or other shrimps (if needed, after acclimatisation as decided). Micro-organisms that spring up utilising the waste food droppings into the zone below the cages, and which move and settle down at the bottom would serve as feed for the stocked advanced crustacean post-larvae moving in the cage-free zone, besides microplankton that will be available in the zone as indicated under the TAC system. In this system, there is no indication of any application of supplementary feed to the crustaceans stocked for growth in the open farming section(water zone underneath the cage zone). Probably the faecal droppings from the cages and the bacteria and other microorganisms generated therefrom will constitute adequate food material for the crustaceans in the bottom water zone.

One aspect to be mentioned here is that, according to TAC, fish alone can be grown in the cages leaving the rest of the zone for the production of prawns/shrimps. The rationale of this system is explained by TAC as follows: The kind of bacteria that proliferate in a fish pond (in this case in fish cages) are different from those occurring in shrimp/GFP holding pond area (in this case in the cage-free or open pond zone). Gram positive bacteria dominate in the fish cages. In the fish-free or cage-free area (open zone) of the pond concerned, there will be dominance of Gram negative bacteria. While these are prone to cause diseases to fishes, it is stated that these Gram negative bacteria in the water below in cage zone do not cause any harm to crustaceans under farming. According to TAC, for this reason, the cage-free zone is utilised for farming crustaceans only (freshwater prawns/shrimps). Thus, the pursuit of polyfarming of fish and crustaceans in a pond (fish in cages set up to cover half of the pond area from surface into columnar layer, and crustaceans in the cage-free zone) not only enables effective utilisation of planktonic blooms but also ensures that the fishes being farmed in the cages are less susceptible to disease outbreaks, ultimately leading to a more productive harvest.

Cages improve the quality of the harvested fish. While they are obviously fed with an appropriate feed, it is stated that, prior to the harvesting, normally, fishes should not be fed. One stated advantage in cage farming is that the fishes grown in them would not develop muddy flavour. There will be also more efficient FCR, resulting in reduction of feed costs.

Cage farming saves time and labour. Two persons can harvest the fish in a cage without difficulty,it is stated. Sick or diseased fishes in cages can be segregated conveniently for observation or treatment. According to TAC, by placing a tarpaulin (pre-cut and approximating the dimensions of net used for harvesting) inside the cage, one can isolate pond water inside it and apply chemicals to it for treatment of infected fishes. Volume of chemicals used gets drastically reduced due to reduction in volume of water for treatment.

Recording of the various parameters, such as rate of feeding, growth rate and observations in respect of diseases would be easier in the cage farming system.The distance between the base of a cage and the pond bottom has to be as maximal as possible, so as to ensure a good pond water circulation. Since the stocking density as well as the resultant production is high in inland cage farming practices, it becomes essential to increase aeration and also provide for a standby generator to guard against power failure or interruption. Upto eight cages (5.25 x 5.25x 1.5 m each), there may not be need for the installation of an aerator, it is stated. However, where more than 8 cages are installed, it would become necessary to install an aerator central to the related cage batteries. In general, a set of two batteries of four cages each that converge leaving space in between for installing an aerator is the pattern of installation of cages followed by TAC, with needed supports.

## Gram-Negative and Gram-Positive Bacteria

Gram-negative bacteria are those which fail to stain with Gram's reaction. The reaction depends on the complexity of cell wall and has for long determined a major division between bacterial species.

Gram-positive bacteria are those with a cell wall of comparative simplicity which allows it to be stained according to Gram's method.

Besides the cost of stocking of the cages with major carp fry, the cost of feed would be the main recurring expenditure. At an FCR of probably 1.5:1, the cost of feed may have to be worked out. The feed, which would have to be nutritive, need not however be a very expensive one. Powdered rice bran and ground nut oil cake at 1:1 ratio or as is generally used, can serve as a good feed. However, experimentation may be necessary to arrive at the nature of feed to be given and the quantities to be fed. Probably, the depth of cages and rate of stocking in the cage can also be suitably adjusted to prevent any likely mortalities or to meet the other requirements of production, consistent with the determined stocking schedule and environmental aspects. Phased harvesting can also be adopted for realising optimal output and returns.

So far as prawn/shrimp farming in the cage-free open zone of the bottom of the pond and rest of the cage-free open zone is concerned, the rate of stocking of PLs can be as per the prevailing norms, to produce on an average 0.8 to 1.5 t per ha, until such time as improved environmentally conducive systems are developed. TAC specifically says that fish alone should be cultured in the cages and shrimps/prawns have to be farmed in the bottom water and in the open pond area remaining cage-free. The work done by TAC of Taiwan has apparently led to the system of farming fishes alone in the cages and shrimps/prawns in the cage-free area because of the Gram-positive and Gram-negative bacteria factors, as already mentioned. There appears to be a need for having a clear understanding of the influences of these two categories of bacteria in polyfarming and monofarming systems for possible realigning of these systems as now practised.

While the system of polyfarming using pond cages has been found to be environment-friendly in Taiwan and probably in a few other countries, it may be desirable for the chief executives of FFDAs to look into the desirable parameters of areas of cages/open areas and the stocking schedules, taking into account the presently available features in relation to the Taiwanese stipulation of depth of cages at 1.5 m, which may not be feasible for pond depth as in Indian farms, which are normally far within 2 m length.

Fishing Chimes, Editorial 391: October 2009: Vol. 29, No. 7

# Development of India's Own Genuine Distant Water (Deep Sea) Fishing Fleet

A peep into the past brings back to our memory the efforts mounted by the Government of India for the development of distant water (Deep sea) fishing in the Indian EEZ, soon after the enactment of the Territorial Waters, Continental Shelf, Exclusive Economic Zone and other Maritime Zones Act, 1976 (80 of 1976). These efforts constituted a historic beginning, soon followed up by initiatives by the Government at introducing larger fishing vessels of 23-27 m OAL, designed for conducting distant water (Deep sea) fishing in the Indian EEZ, both by import as well as by indigenous construction. So far as Imports were concerned, these were permitted by the Government on the condition that an equal number of vessels as allowed for import would be constructed at Indian shipyards by the permitted importers. This unique initiative led to the progressive development of an Indian distant water (Deep sea) fishing fleet with a strength of 191 vessels by the year 1990. This level of fleet strength has, however, dwindled down to around 48 nos. by middle of the year 2009, a very discouraging development. It was more so because there were no well directed efforts at restoring the Indian Distant water (Deep sea) fishing fleet strength to the past level, under the genuine ownership of Indian enterprises.

Owners of fishing vessels of a few advanced fishing nations (particularly Taiwan), mostly under the leadership of their associations, are habituated to exploiting the fisheries resources available in the farther zones of EEZs of several other nations, of which India happens to be one. In this scenario, the Indian situation of a declining distant water (Deep sea) fishing fleet became an encouraging feature for certain foreign interests, mostly those of Taiwan. The owners/crew of distant water (Deep sea) fishing fleet of Taiwan,who have intimate knowledge of the fisheries of the Indian EEZ, are seen and also known to be well entrenched in IUU fishing in the farther waters of Indian EEZ. The Taiwanese point of view in this context, as assessed, seems to be that, in the event of development of Indian distant water (Deep sea) fishing fleet, the economic interests of distant water fishing fleet of that country operating in the Indian EEZ would suffer a serious setback. In this context, a hind look at the events that have taken place particularly since 1980s would indicate the kind of remedial measures launched by Taiwanese fishing interests from their point of view, so as to enable them to continue their fishing activities in the Indian EEZ on an enduring basis.

The remedial measures launched by Taiwanese fishing vessel owners in the above mentioned situation to protect their stake in the Indian EEZ indicate a possible loophole in the policy of the Indian Government in respect of foreign fishing in the Indian EEZ, particularly in relation to Taiwanese interests in the Indian EEZ. The strategy adopted by the Taiwanese to continue their fishing operations in the Indian EEZ in the past are assessed as follows:

a) Working towards a system of entrusting their fishing vessels nominally but officially to Indian enterprises on charter and at the same time having full control over fishing operations and in the dealings related to the catches; b). Operating vessels in the Indian EEZ by giving them on lease to Indian enterprises but retaining full control over fishing operations and in dealing with the catches; and c). Managing to secure permissions to enter into joint ventures with Indian fishing enterprises. These three systems are, however, not in vogue at present, but are replaced by an LoP (Letter of Permission) system, now in force, which is explained in later paragraphs.

Under all the three systems mentioned above, in actual practice, the vessels were known to operate under the control of the foreign (Taiwanese) crew, the reason being that the set up on the vessels was such that it would be impractical for the Indian crew to conduct and manage fishing operations on these foreign vessels. A counterpart Indian crew as approved would of course be on board but only by way of their formal presence and with no role in actual fishing, as had come to be known.

The Taiwanese measures outlined in the preceding paragraphs, and which were formalised by the Indian Government as reflected in its policies at that time, seem to have served as a face saving façade to both the sides. The catches in respect of all the three aforesaid categories of operations came to be deemed as Indian catches but taken away by the foreign vessel owners, with the Indian side deeming them as its exports. The Taiwanese vessels continued their fishing operations in the Indian EEZ under the three categories referred to, particularly the chartering system, with practically no problems until a few years back. At this stage, however, the Taiwanese vessels faced a serious problem of depletion of fish stocks in the Indian EEZ, primarily because of their own relentless operations. With several undreds of trawlers, mostly paired and single, equipped for demersal and pelagic fishing, pressed into operation by them under the systems mentioned above, of course with Indian clearances, Taiwanese vessels swept the waters of the EEZ, rendering the resultant operations at certain stage uneconomical even for the Taiwanese owners themselves.

This development forced them to find a way of abandoning their trawling operations, probably for a few years to come, until the stock position revived. However, unable to withdraw the fleet on their own unilaterally, obviously for tactical reasons, they seemed to have looked for a way out and which appears to have ultimately taken the shape of a Committee set up by the Government of India, for the purpose of suggesting measures for countering the depletion, of which one was to be withdrawal of the Taiwanese trawling fleet,which was favourable for the Taiwanese operators too. Eventually, the Committee, headed by Mr.Murari, recommended the abolition of the charter and leasing schemes which was of course accepted by the Government of India, and apparently welcomed by the Taiwanese.

At this stage, the tuna fishing activities in the Indian EEZ, particularly by Taiwanese longliners,came into focus As long as the Taiwanese trawler fleet was operating, their longlining fleet, stated to be already there in Indian EEZ, was not well noticed because of their operational merging with their trawler fleet as seen from a distance. With the voluntary exit of their trawlers, the Taiwanese fishing interests concerned have come face to face with the problem of securing formal permission of the Government of India for the operation of their longliners in the Indian EEZ. Based on their experiences in such situations in the EEZs of a few other nations, it is believed that Taiwanese interests worked towards the introduction of the LoP (Letter of Permission) system. Under this system, with the LoP issued by the Government of India, the vessels would be registered as Indian Vessels and the formal ownership would get transferred to the Indian side,once the cost of the vessels is totally paid in instalments out of earnings from sale of the produce. As all the crew would consist of Taiwanese, issue of permits, based on the applications given, would be as per eligibility. Now a large number of longliners of Taiwanese origin operate in the Indian EEZ under the LoP system, but, so far none of the vessels could become truly Indian, firstly because there were deficiencies in the payments towards cost of vessels(which are believed to be to the liking of Taiwanese owners) and secondly because the Indian side is not in a position to take over and operate the vessels on its own.

As already mentioned, Indian Government had introduced in the past 191 deep sea fishing vessels, some imported and others indigenously constructed. There were no major problems in the indigenous construction of the vessels. The reason for this is that India has a well developed infrastructure for the designing and construction of larger fishing vessels. Apart from this, India also has a well developed Training institute in Cochin, for providing training to candidates in fishing vessel operations. This institute (CIFNET) also has two regional centres, one in Chennai and the other in Visakhapatnam. While a good number of trained candidates come out of the main Institute and its two regional centres, many of them take the recourse of joining the merchant navy for the reason that the strength of the Indian distant water (Deep sea) fishing fleet has come down and also because there have been no additions to the fleet. As an offshoot of this situation, the activity of upgradation of the existing smaller boats and construction of medium sized boats has come about. These vessels do not require skippers, fishing second hands and engineers of fishing vessels.

The considerations of the Government that led to the abandonment of the proven system of introducing Indian-owned newly built distant water (Deep sea) fishing vessels to fish in the farther waters of Indian EEZ are obscure. Until recently, there were financing avenues available for the construction of distant water (Deep sea) fishing vessels through erstwhile SDFC and later through SCICI/ICICI. These avenues were however closed down, probably because of the problems of recovery of loans given, although hurdles of this kind, common to most of the lending agencies, can certainly be crossed over. The loan recovery efforts need planning and a close interface with the borrowers but this was not there. In this background of the past, the revival of the system of introduction of distant water (Deep sea) fishing vessels, both indigenously constructed and imported with the needed organisational safeguards, would have been a better option in the place of the present LoP system, so as to have a fleet that is genuinely Indian. An option of this kind, would certainly have enabled the utilisation of the existing fishing vessel building infrastructure in the country. In any case, it should have been possible to promote, simultaneous to the LoP system, indigenous construction of distant water (Deep sea) fishing vessels.

In the past, the vessels built at Indian Shipyards, such as Alcock Ashdown, Bharti, Hooghly Docking, Cochin Shipyard, Mazagoan Docks etc., proved to be of an excellent quality, comparable to those built at foreign shipyards.

It is a national shame that, despite having the needed infrastructure for constructing distant water (Deep sea) fishing vessels at several well developed shipyards, the Government has resorted exclusively to the system of LoP vessels, instead of striving to build a truly Indian-owned ocean going fishing fleet. Looked at from any angle, what emerges in the background outlined in the preceding paragraphs, is the paramount need on the part of the Government to review the position and evolve sound policy guidelines to promote indigenous construction/import of the needed versions of deep sea fishing vessels of genuine Indian ownership, in the place of the ongoing LoP system which, more or less, wholly benefits the foreign fishing interests.

In this situation, the Federation of Indian Fishery Industries that has been formed recently may prevail upon the Union Department of Animal Husbandry, Dairying and Fisheries to entrust the function of developing India's own distant water (Deep sea) fishing fleet to the National Fisheries Development Board (NFDB) or any other such organisation in India.This Body can assess the number of distant water (Deep sea) fishing vessels to be introduced and work on a programme of their introduction by the enterprises concerned. This assessment has to be integrated with a financing programme, to be launched by it, if necessary, in association with a major Indian Bank with a progressive outlook. The main weakness in the earlier distant water (Deep sea) fishing vessels introduction programme, was the absence of a system to take care of the post harvest activities at the landing centres so as to ensure recovery of the instalments due from the sale proceeds. This shortcoming would have to be set right. The vessel introduction programme can cover the ntroduction of vessels built at Indian shipyards of course having past experience, coupled with the possible imports from certain countries like Australia, Holland, etc., in this respect.

Fishing Chimes, Editorial 392: November 2009: Vol. 29, No. 8

# Foreign Fishing Vessels, Simple Permit-Based Access into Indian EEZ Mooted

## *A Historic Challenge to be Faced– By the Indian Marine Fisheries Enterprises and the Federation of Indian Fishery Industries*

Recently, on 9 September, 2009, there has been a historic development. On this memorable day, the authorised representatives of Indian marine fisheries Industrial Associations from various States/UTs met at Visakhapatnam under the chairmanship of Dr. Y.G.K. Murthy and formed the Federation of Indian Fisheries Industries. This formation has been of course overdue for long. In the absence of a national body to represent to the Government the problems of marine fisheries industries,several problems faced by the Indian marine fisheries sector continue to be unresolved. The formation of the Federation may extinguish this lacuna.

The present trend of national marine fisheries developmental policies has to be construed as virtually one sided, dominated by the views of those in the Government. The main reason for this is that there is no national level representative body of marine fisheries stake holders to take care of their overall interests. The only national body that counts is the Association of Indian Fisheries Industries formed in late 1970s with Delhi as its headquarters, under the leadership of the representatives of larger companies such as Union Carbide, Greaves Cotton (New India Fisheries), EID Parry, Britannia and others who were playing a major role in the deep sea fishing industry at that time. In the later years, all the aforesaid companies exited from the fisheries sector leaving the burden of management of the Association to smaller deep sea fishing companies, most of which are located in A.P and Tamil Nadu. The exit of the larger companies from the marine fisheries sector can be indicative of several depressing conclusions that can be drawn. Eventually, the headquarters of the Association was shifted from Delhi to Visakhapatnam, where it continues. The outcome of these developments, however, is that the national character of the Association became diluted, although it is now and then consulted on certain issues by the Union Department of Animal Husbandry, Dairying and Fisheries, and earlier by the undecided Union Department of Agriculture. However, considering that the stakeholders, most of whom have their vessels operating from Visakhapatnam in A.P., and as they are the only members of the Association, it has not been able to function on a national basis. In this situation, the formation of the National Federation on 09.09.2009 with the participation of representatives from the marine fisheries associations in all the coastal States/UTs of India has paved the way for making needed representations to the Union Department of Animal Husbandry, Dairying and Fisheries and also to the Coastal State Governments for the integrated marine fisheries development of the Indian EEZ, from the stage of introduction of distant water fishing vessels to the final link of domestic and export marketing of fish, crustaceans, molluscs and their products,in the main.

There are several aspects of concern to the marine fisheries industries on which the Federation would have to represent to the Government of India for redressal from time to time. One of this is in respect of the proposed Marine Fisheries Regulation and Management Bill. The provisions in the draft of this Bill are causing concern to those in the industry, because they are more inviting to foreign marine fishing companies interested in their operations in the Indian EEZ under the provisions of the proposed Bill that provide unfettered access to their vessels into the Indian EEZ just with a permit given by the Government. However,at the same time, the provisions would impede the growth of the fleet owned by Indian deep sea fishing enterprises. It may be recalled in this context that from times immemorial fishing by Indian vessels in Indian marine waters happens to be undertaken with freedom not requiring any permits or licenses from the Government. This tradition has been there right from the pre-British and also in the following British regime and no subsequent Indian Government thought of altering this system that stood the test of time. Registration of Indian fishing vessels with the Government is the only stipulation that has been introduced in the recent decades and which is already extended to LoP vessels.Under the proposed Bill mentioned above, the owners of Indian fishing vessels have not only to obtain a permit for fishing in the marine waters of their own nation, but they are also treated at par with owners of foreign owned vessels, who also need to have permits to fish in Indian EEZ. Further, an unfortunate and demeaning aspect is that Indian owned as well as foreign-owned fishing vessels, according to the provisions in the proposed bill, would be treated alike. In this background, there can be a situation in which the Indian-owned vessels may be subjected to step-motherly treatment in one form or the other because of the new and upgraded status of foreign - owned vessels. This can further impede the growth of genuinely Indian owned fishing fleet.

The provisions in the Maritime Zones of India Act presently in force clearly lay down that the control over the fisheries of Indian territorial waters vests in the State Governments/UTs concerned. The proposed legislation, however, dilutes this provision, bringing in an ambiguity about the vesting of fishing rights over the territorial waters in the Coastal State Governments.

## Kerala Seeks Review of Marine Fisheries Bill

Kerala Fisheries Minister demanded recently that the Government of India should review the proposed Marine Fisheries (Regulation and Management) Bill in consultation with the States on its provisions.

The Minister told the media that the proposed Bill would pave the way for plundering of the fisheries of the marine waters of Kerala (and other States) by foreign fishing vessels. The State Government opposed the Bill strongly, he said.

Those in the industry say that, while giving a free hand to permit-holding foreign fishing vessels, the proposed Bill would generate constraints in the operations of Indian - owned vessels interested in distant water fishing in their own EEZ. Further, the new Bill stipulates severe penalties to Indian fishing vessels for fishing in areas in the Indian EEZ outside the zone mentioned in their permits. As at present, the LoP vessels, actually operated by the crew of foreign owners, are stated to be fishing in the Indian EEZ as they want without let or hindrance. Such being the experience, one can visualise the ill-effects of fishing by foreign fishing vessels permitted to fish in the EEZ, under the proposed legislation.

Provision for a plan for the management of fisheries of Indian EEZ has been given a place in the Bill. The related provisions under this inclusion can, unwittingly though, be applied in a subjective manner causing harm to the fishing interests of Indian fishing vessel owners in the Indian EEZ. This apart, the provisions in the proposed Bill would need to be brought closer to the Indian stakeholders as per the provisions in the Maritime Zones of India (Regulation of Fishing by Foreign Fishing Vessels) Act, 1981. It may be mentioned here that the main purpose of the new legislation as now contemplated has to rightly spring from the urgency of building up of a truly Indian owned distant water fishing fleet to the needed level, as this has been declining fast. It has already declined from a strength of around 191 vessels about a decade back to the present level of around 40 vessels. The new legestation has to also stem the upsurge of the number of foreign owned vessels, LoPs or otherwise, operating in the Indian EEZ, whose strength has been going up fast.

An incisive appraisal of the provisions in the proposed bill would indicate the possibility of a reverse impact by them on national needs. The provisions tend to encourage introduction of foreign fishing vessels instead of working towards introduction of Indian owned vessels. It appears that the present need is to constitute a Committee consisting of identified top Indian marine fisheries professional experts and of those having relevant specialised legal expertise and of an objective bent of mind to examine the subject incisively and to propose the needed revision to the present draft Bill, for the consideration of the Government (If it cannot be dropped), before placing the same in the Parliament for its consideration. In fact, it appears that legal provisions as at present in respect of management of fisheries of EEZ and the rules framed thereunder are adequate to take care of the present problem of stemming the aggression of foreign fishing vessels into the Indian EEZ and to encourage the building up of Indian - owned fleet of deep sea fishing vessels. It needs to be ensured that the proposed legislation does not impart the reverse effect of encouraging the entry of foreign fishing vessels into Indian EEZ and making matters difficult for increasing the strength of an Indian owned distant water (deep sea) fishing fleet. What all is needed is to set up a supporting financing and related monitoring mechanism for the building up of an Indian - owned deep sea/distant water fishing fleet. Once such a fleet is developed, it would discourage inflow of foreign fishing vessels into Indian EEZ. With foreign fishing vessels operating in Indian EEZ, it would be very difficult for the Indian distant water fleet to grow and flourish.

Another reform the Federation has to work on is in respect of imparting fisheries professional education to fisherchildren needed at that level all along the coastline by catching them young, in order to inculcate in them a deep sense of dedication for protecting our waters from the unauthorised entry of foreign vessels into our EEZ and to develop them as sentinels of the coast. In this context, it would be pertinent to recall that during their rule over India Britishers set up fisheries schools in several coastal fishing villages. After the Britishers left India, instead of multiplying the number of such schools for the benefit of the fisherchildren, either they have been closed down or their purpose has been diversified. Taking this background into consideration, the Federation has to prevail upon the Centre to team up with the Coastal State Governments concerned to set up fisheries schools in coastal fisher villages so as to impart effective fisheries education to the fisherchildren, not only in respect of upgraded fishing technologies inclusive of crafts and gear, but also in regard to instilling a deep sense of safeguarding the coastal waters from the entry of foreign vessels and other intruders, and be of support to the coast guard. Further, the fisherboys could be given training at sea in association with the coastguard in respect of dealing with the intruding foreign vessels and crew.Of course, if the proposed Bill becomes an Act, there may not be any intruders. All presently poaching foreign fishing vessels are likely to apply for permits and may get them, as the provisions in the providing Bill are conducive for this.

Another aspect to be looked into by the Federation is to help the Indian fisheries stakeholders in earning improved returns from their catches at the landing point itself. This can be achieved by prevailing upon the authorities to set up value-adding processing plants at selected centres along the coastline. The provision of this infrastructure will lead to an increase in the prices of fishes at the landing points themselves for the benefit of the Indian fishers.

It is to be hoped that the Federation would submit needed proposals to the authorities on the above mentioned and several other relevant reforms needed and follow them up effectively so as to ensure that 1) the provisions in the proposed bill to be placed before the Parliament would take care of the interests of marine fisheries development of the country, as the industry requires, 2) will spread professionally - oriented and also coastal protection - oriented fisheries education among fisherchildren and 3) value-adding infrastructure is provided in centrally located coastal villages so as to ensure higher returns to fishers at landing points, linked to a domestic marketing network covering towns and large villages around for the marketing of the value - added products at retail outlets.

Fishing Chimes, Editorial 393: December 2009: Vol. 29, No.9

# India–Asean Free Trade Agreement and the Indian Seafood Sector

The most important contribution of fisheries sector to the Indian economy is one of providing livelihood to many poor fishery households, especially those located in the coastal areas. The sector provides employment to over 11 million people, engaged fully or partly in fishing, fish processing, fish marketing and in fishery related subsidiary activities. While the exact number of persons engaged in secondary fisheries activities would be difficult to quantify, it is estimated that so far as women are concerned some half-a-million of them work in the pre- and post-harvest fishing operations. This sector thus provides livelihood support to a large number of households through income generated from domestic and export fish trade.

The nation has made enormous investments in the fisheries sector. There are about 280,000 fishing crafts in the marine sector and the gross investment in them and in fishing equipments is estimated to be Rs. 80,000 million. Most of these are in the private sector. Currently, there are some 399 processing plants having a daily freezing capacity of 7,283.36 tonnes of fish products. Besides, there are 471 cold storages in the country. The total estimated capacity of these is 89,274 tonnes. The number of fishing vessels which mainly contribute to the export market is calculated as 12,660. From a subsistence-based livelihood activity pursued by a group of largely poor and rural artisans, marine fisheries sector has acquired the hues of an urban-based, capital intensive commercial sector, earning sizeable sums of foreign exchange for the country.

India has entered into trade agreements with several nations, which allow duty free exports and imports, inclusive of fisheries items. Under the agreement signed between Government of India and Sri Lanka in 1999, the two countries decided to progressively reduce and eliminate obstacles in the way of bilateral trade through a free trade agreement. A number of fisheries items are allowed to be exported into India by Sri Lanka, duty free, under this agreement. India also signed an agreement with SAFTA (South Asian Free Trade Area) in 2004, which comprises the SAARC nations. Preferential trade agreements too are signed between India and other South Asian Nations such as Bhutan under SAPTA (South Asian Preferential Trade Area). In the case of fisheries items, Bangladesh and Maldives get preferential treatment, the former under SAPTA, and the latter in respect of certain items under SAPTA. Under the initiative of the Economic and Social Commission for Asia and the Pacific (ESCAP) an agreement was signed in 1975 for trade expansion through exchange of tariff concessions among seven developing member countries of the ESCAP region. These countries, namely, Bangladesh, India, Laos PDR, Republic of Korea, Sri Lanka, the Philippines and Thailand, agreed to a list of products for mutual tariff reduction. Certain fisheries items get concessions in terms of import tariffs under this Agreement. A preferential trade agreement was also signed by India with the Mercosur countries (Argentina, Brazil, Paraguay, Uruguay) in 2004. While ornamental fish appear in the preferential list offered by India, no fisheries item appears in the preferential offer list of the other countries.

The latest is the ASEAN Free Trade Agreement between India and the Ten Member Association of South East Asian Nations which will come into effect from 1st January, 2010. ASEAN (Brunei, Singapore, Cambodia, Indonesia, Malaysia, Laos, Myanmar, the Philippines, Vietnam and Thailand) and India signed the long-awaited Free Trade Agreement (FTA) in Bangkok, ending more than six years of intensive negotiations. Under this agreement ASEAN and India will lift import tariffs on more than 80 percent of traded products that include fishery products, between 2013 and 2016.

In the shrimp export markets, China, Thailand, Vietnam, Indonesia, Mexico, Greenland and Ecuador are the major competitors to India.The relative compound growth rate of shrimp export from India indicates that the country is lagging behind other shrimp exporting countries both in terms of volume and value. Moreover, some of these countries import Indian shrimp for reprocessing. More than 60 per cent of India's exported marine products to south-east Asian countries are re-exported by them after processing.

Fish imports have not been very important to the Indian economy. There was a small surge in fish imports into India in the mid-1990s which accounted for less than one percent of the net exports. These imports were made mainly to counteract the prevailing under-utilisation of the capacities of processing factories in some of the States (notably in Kerala). When this strategy did not work out to be viable, the share of imports slid back once again. India imported fish worth Rs. 890 million during the financial year 2006–07. Out of this, Hilsa alone, worth Rs. 516.90 million, was imported from Bangladesh for the consumption of the people of West Bengal and for export for the use of Bengali community abroad. The Seafood Exporters Association of India feels that the import of fish for processing and export will not affect the livelihood of fishers. In this context, the Association's point of view is that India has a very well developed processing industry for taking care of all exportable catches of Indian fishers, and the processing units follow strict quality control norms that are EU certified. Further, entrepreneurs of this sector have upgraded their processing facilities by investing heavily. Despite these developments, ironically, however, 80 per cent of the processing capacity continues to remain unutilised. This is due to the fact that there is a substantial short supply of raw fish. It is now proved that the Indian strategy of imports will not solve the problem. What is needed is increase in indigenous production to utilise the unutilised processing capacity. About 10 lakh workers are engaged in the sea food processing industry in India and their livelihood is now at stake because of raw material shortage which it has not been possible to cover up in an economically viable manner through imports of raw fish.

The Free Trade Agreement with ASEAN has met with stiff resistance from some of the State Governments and also the agricultural sector. Much of the opposition has come from Kerala. Its concerns centre on seafood sector too, besides plantation sector. It is feared that import of fishery products into the country would cause downfall of prices. Some of the common varieties such as cuttle fish, pomfret, anchovy, and ribbon fish are caught in China and Thailand as well. Importing of these items will dampen their prices in India. This will also adversely affect fishermen for whom the catches of the above mentioned fishes are their sole source of livelihood. The outlook is that it will be unprofitable for Indian fishermen to catch these fishes, once they are also imported, as this will cause an undesirable impact on the prices of these fishes caught from Indian waters. Further, the imports will create a problem of survival for the small fishermen. It is not only the fishermen, whose livelihood depends on this industry, that are affected, but also other stakeholders dependent on the industry. These include boat workers, loading and unloading persons and many others. This employment and livelihood angle has to be taken care of, in case imports of fish into India are allowed.

Several non-Governmental bodies have raised concern about India- ASEAN Free Trade Agreement. The Kerala Swathanthra Matsyathozhilali Federation and Focus on the Global South, a regional research and campaign group, are prominent among these groups. They are of the view that the Central Government's decision to sign a Free Trade Agreement (FTA) with ASEAN would have an adverse impact on the livelihoods of thousands of people engaged in the fisheries and agriculture sectors, especially in the Kerala State. FTA would hit the State of Kerala more severely because of the large number of people engaged in fishing, fish vending and processing. In recent years, the fish stocks have been depleted in Kerala waters due to overfishing by Indian vessels and also because of introducing foreign fishing vessels. Fish prices have also crashed due to imports of cheaper varieties of fish, forcing many to give up fishing. Further liberalisation of fish imports to utilise the idle fish processing capacity will aggravate the ongoing problems of the fishing community. The biggest threat would come from imports from Thailand, the world's largest exporter of farmed shrimps and also from Vietnam, the world's eighth largest seafood exporter.

Despite the aforementioned position, officials of the Ministry of Commerce and Industry are of the view that the India-ASEAN Free Trade Agreement (FTA) safeguards the interests of domestic fishers and farmers. Government organisations support import of fish to utilise the surplus processing capacity and to contribute to marine products export by way of re-exporting the products produced from imported fish. About 10 lakh workers are engaged in the sea food processing industry in India. Studies conducted in this area show that an additional 5 million job opportunities can be created by importing fish for processing in India utilising the spare capacity. Looked at from a narrow angle, when fish is imported into India from other countries and processed for re-export, it is possible that this may benefit the Indian economy. However, it has to be seen that, at the same time, the imports would hamper the livelihood of poor fishermen, whose fishing activities will suffer, once fish imports are allowed, even though for re-exports in a processed form.

The plight of fisher community in the country is in bad shape. A great majority of fishermen of Kerala State are indebted. The total debt of about 12 lakh fishermen of Kerala State is estimated to be Rs. 1,500 crores. According to a report by the State Fisheries Department, at least 70 per cent of marine fishermen of the State are in debt.14 per cent of these fishermen have not repaid the loans taken by them. Those who have taken loans from money lenders have been able to repay only the interest. Global recession has decreased the export prospects of fishery products. Besides this, the fall in the availability of fishes along the Kerala coast has made life of fishermen miserable. Many have taken loans from money lenders at higher interest rates. As a result of global recession, prices of several species of fish have come down. Impact of ASEAN agreement will further worsen the condition of fishermen in the country. Traditional fishermen of Kerala fishes own fishing units of low capital investment. The reason is that they are unable to get loans from banks and other financial institutions as they have no valuable properties to hypothecate. Hence they are forced to be at the mercy of commission agents and middlemen. Although several measures have been initiated to mitigate the difficulties of fishermen, such measures are yet to produce results. An effective co-operative marketing movement alone can save the fishermen from many of their difficulties. In this background, the work of Fishermen Debt Relief Commission announced by the Kerala State Government has to be speeded up to save the fishermen from their present socio-economic and security problems. The continuance of the problems without redress may lead to unrest and clashes in the coastal belt. Several such factors are to be taken into consideration before implementing any long term trade agreement on fisheries with other countries.

Fisheries sector of the country has to adopt several measures to cope up with the demands of changing trade conditions and take advantage of the new opportunities. For successfully facing the competitions from abroad, domestic seafood producers and processors have to invest in the further development of infrastructure, capacity building, and in management and in institutional frameworks for technical support and monitoring. It is necessary to identify products for trade, which is not going to have detrimental effect on the livelihood of poor fishermen. Further, technical skills of fishermen and other stakeholders in the sector have to be upgraded. To achieve this, there is a need to develop a uniform country-wide training framework and implementing it at the field level for the benefit of different stakeholders. Hygiene and quality control need to be ensured right from the moment of capture of fish. Improving hygiene standards and preservation facilities onboard will reduce most of the problems related to contamination further down the chain towards processing and marketing. There is a need to provide basic training to fishers on implementing good management practices (GMP) at all stages from harvest till the fish reach the processing factory.

The most important need is one of reducing the cost of production to make the products more competitive in domestic and international markets. Part of the high costs associated with exports is considered to be due to the requirements of compliance with trade measures in respect of sanitary and phyto-sanitary aspects and non-tariff trade barriers. A major part of the costs also originates from factors like over-capitalisation, long supply chains involving many intermediaries, process inefficiencies and shortcomings in monitoring.These have to be controlled. Value addition is one area for improving export performance of the seafood industry in India. Thus it can be seen that a combined effort is required to make the Indian seafood industry competitive internationally and to take advantage of such trade agreements.

*(Inputs in respect of this editorial are from Prof. (Dr) D.D. Nambudiri).*

Fishing Chimes, Editorial 394: January 2010: Vol. 29, No. 10

# Why is the Giant Sleeping Still?

Reservoirs are often referred to as a 'sleeping giant of Indian fisheries'. Although a cliché, this figurative expression is worth pondering over. Are the reservoirs, as a fishery resource, really a sleeping giant? and if yes, why is it sleeping still? And how can one kick this giant awake? Trying to answer these questions will throw light on the opportunities the reservoirs offer and the challenges we face to translate these opportunities into production and thereby bridging the gap between the actual and potential production from these water bodies. During 1995, the country had more than 3 million ha of reservoirs and today it could be well above 4 million ha. The fish production and yield from Indian reservoirs have always been abysmally low. During 1995, fish yield from small, medium and large reservoirs were 50, 12 and 11 kg/ha/year respectively (average 20 kg/ha) and the total fish production was around 94,000 tonnes. Although this estimate may not be very accurate and it is possible that the fish yield could have increased since 1995. It is well accepted that the Indian reservoirs are producing fish way below their potential. If the consistently high yield rates achieved in the reservoirs of some other developing countries and the performance of some well managed/stocked reservoirs and beels are any guide, substantial hike in yield is possible from Indian reservoirs.

Going by the unused opportunities that the reservoirs offer for both economic and social development, one is justified to call reservoirs a sleeping giant. China, the world leader in reservoir fisheries, records a mean yield of > 800 kg/ha from its reservoirs and those from Sri Lanka (>300 kg/ha) and Cuba (>100 kg/ha) are also high yielding. Similarly, some well managed *beels* of West Bengal, where fisheries are managed on the basis of stocking and recapture, have recorded yields up to 2000 kg/ha/year. These indicate the high untapped fish production potential that our reservoirs carry. To begin with, even at a very modest rate of 500, 250 and 100 kg/ha/year respectively from the small, medium and large reservoirs of India, nearly one million tonnes of fish can be produced from Indian reservoirs every year by adopting fisheries management norms, based on scientific advice.

Reservoir fisheries development has some interesting economic and social dimensions which deserve a mention. Producing 1 million tonnes of fish from aquaculture entails high investment in the form of pond preparation, feed, seed, medicines, labour and other inputs, apart from heavy demands on water, a resource which is becoming increasingly scarce day by day. Developing fish production in 1 ha of fish pond will cost Rs. 2 to 3 lakh, in addition to other inputs. At an average yield of 2 tonnes/ha (as obtained under FFDA ponds), producing 1 million tonnes of fish from aquaculture ponds will cost a formidable Rs. 10,000 crore only for pond digging, leave alone other inputs. In contrast, increasing fish yield from reservoirs cost virtually nothing. The only expenditure involved is stocking, which is very negligible, as opposed to the heavy capital investment required in aquaculture. There is also a social dimension of reservoir fisheries development. The profit obtained in aquaculture ventures accrues to an entrepreneur, investor or a group of individuals as 'return on investment'. In contrast, the benefits due to increased fish production obtained from a reservoir (under a good governance regime), are shared by a large number of fishers- the key stakeholders. That means this is a large cake and each stakeholder gets a slice, albeit small. Thus, the reservoirs provide opportunities for inclusive growth, which is economically sound and socially justifiable. Yes, reservoir sector is the sleeping giant of Indian fisheries.

## Why is the Giant Sleeping Still ?

The challenges that come in the way of awakening this giant can be summarised under two broad categories: i) Technology challenges, the poor application of scientific management tools and ii) Governance; and management challenges, the lack of enabling socio-economic environment. Of these, technologies for developing reservoir fisheries are relatively simple and do not demand very high technical skill. These can be applied by anybody with some basic management skill and normal intelligence. Still, the rate of adoption of scientific advice for reservoir fisheries is very low, and most of the man-made lakes in the country are still being managed on a very arbitrary manner, leading to low productivity. Many things need to be done to correct the situation. For starters, one has to decide on the appropriate fishery management system to be followed in a particular reservoir. Very few persons understand the fish production processes as applied to reservoirs. Some consider reservoir as a capture fishery resource and some others believe this is a culture fisheries unit, but in fact it is neither. Fisheries management of reservoirs is the best explained as 'sustainable fishery enhancement', which is defined as a 'range of practices/processes by which qualitative and quantitative improvement is achieved from water bodies through exercising specific management options'. This is something intermediate between culture and capture fisheries. Enhancement *inter alia* includes 'culture-based fisheries (stock and recapture)', 'stock enhancement (enhanced capture fisheries)', 'species enhancement (introduction of species)', 'environmental enhancement (fertilising to increase plankton)', 'management enhancement (introducing new management options)' and 'enhancement through new culture systems (cage culture, pen culture, FADs, etc)'. Reservoirs offer scope for one or more forms of enhancement. The most suitable enhancement strategy for a particular reservoir is chosen, based on its morphometric, edaphic and biological characteristics. The two most common forms of enhancement suitable for and followed in Indian reservoirs are culture-based capture fisheries and stock enhancement.

It is not the complexity of technology that comes in the way of achieving higher production from Indian reservoirs, but the lack of appropriate governance arrangement and management regimes is the real challenge. The reservoirs in India are common property resources, generally managed based on community activity. Thus, organisation of the community that manages the system plays a key role. Quiet often, on account of inadequate awareness, empowerment and motivation, the community remains incoherent and disorganised whose members at times act at cross purposes. This not only weakens their ability to negotiate with the other sections of the society, but also make them easy prey to the unscrupulous elements like money lenders and middlemen. This is the bane of reservoir fisheries throughout the country. In isolated pockets, where the community is well organised and works under good institutional support in the form of effective cooperative societies, Self Help Groups (SHGs) etc., very high yield and equitable distribution of income are reported. The reservoir fisheries will be successful only when the community that fishes in the water body is under a sound governance set up and the community owns and manages the fish stock. All stakeholders should take part in the decision making process and the benefits accrued by implementing improved scientific norms should be equitably shared by all. Co-management (where the representatives of the community and Government take part in decision making process) is the most ideal for the reservoir fisheries department. The State (the State Government, local Self Governments, NFDB, etc) can play a pivotal role in improving the governance systems of reservoirs by providing an enabling policy environment for this purpose.

Inadequate marketing channels and marketing infrastructure often act as disincentives for the community to produce more fish by managing the resource in an appropriate manner. The emerging marketing opportunities and consumption pattern in the country, especially the supermarket culture, can be suitably utilised to the advantage of fishers. Proper arrangements including post-harvest processing and value addition will go a long way in improving production and these aspects need to be integrated into the management/development process.

Ownership of reservoirs does not always vest with the Fisheries Department and in many cases it has no access and authority to manage the fisheries in reservoirs. In an ideal situation, even if a reservoir is owned by other Departments, at least the fishing activities should be within the purview of the Fisheries Department of the State Government concerned. The DAHDF and NFDB can take a lead in persuading the States to follow a common policy on this issue. However, the State Fisheries Departments need to shift from a 'revenue generation' approach to a 'development approach' and similarly, the enforcement (command and control) approach should give way to a participatory (co-management) approach. Stocking should be the responsibility of the community that manages a reservoir and the focus of the State agencies should be to encourage/enable/empower them to undertake this responsibility, rather than doing the stocking themselves. The role of Government agencies/ NFDB shall be to facilitate/promote/demonstrate the process of stocking and provide incentives in the form of initial seed money/revolving fund or soft loan. The need is to create a demand for fish seed from the reservoir sector, which the fish seed industry can meet. Outsourcing and contract seed rearing on 'buy back arrangements' are also worth trying.

In short, the major governance challenge in reservoir fisheries is the lack of appropriate community organisations, institutional arrangements, and an enabling policy environment. Reservoirs, for sure, are a major fishery resource, but they remain highly dispersed under a plethora of control regimes with very weak governance and policy support. It is an uphill task to bring all the reservoirs under a proper fisheries management framework due to a multitude of challenges as explained above. Strong policy support at the National and State levels is essential for realising higher productivity in reservoir fisheries. Fisheries being a State subject, action for creating such policy environment is eventually the responsibility of State Governments. Nevertheless, DAHDF and NFDB can take a predominant part in building- up capacity in the States and act as facilitators for creating policy guidelines. A national action plan embracing the following aspects is essential to plan and execute such a mammoth task:

- ☆ To begin with, all reservoirs in the country need to be brought under a GIS-based inventory database. For this purpose, some existing anomalies (in some States) on definition of the reservoirs need to be resolved.
- ☆ The reservoirs need to be identified based on their morphometric, edaphic and biological characteristics in order to choose the best management strategy including the suitable target species.
- ☆ Depending on the above characteristics, management decisions such as stocking requirements, seed production options, development of enhanced capture fisheries norms, need for HRD and empowerment etc., can be determined.
- ☆ All these are fed into a computer-based management system that would generate advice for the policy makers at macro-level as also for the reservoir managers at a micro-level.

☆ A cluster approach will be ideal for developing seed production infrastructure, market links, marketing infrastructure and centres for processing and value-addition.

The guiding philosophy of reservoir fisheries should be to ensure that the fruits of increased production should percolate down the communities that include traditional fishers, local communities and sometimes the rehabilitated oustees. Private Public Partnership (PPP) is welcome, but while engaging private players, the interest of fisher communities should be protected. Government also has the responsibility to ensure that national policies on environment and exotic species are not violated. What needs to be done includes, utilizing all available scientific advice on the subject, building of reservoir development capacity in the States, guiding the State players to bring reservoir fisheries under a common policy framework for development, and creating a vehicle that facilitates and co-ordinates action plans and ensures production, processing, value addition and marketing of fish. It is an uphill task, but well within the realm of possibility through concerted action.

Yes, the giant can be awakened through a fairly hard kick, but who could possibly give that kick? Can NFDB do this, or do we need a Reservoir Fisheries Development Agency (RFDA) ?

*(Inputs in respect of this editorial have been provided by Dr. V.V. Sugunan)*

Fishing Chimes, Editorial 395: February 2010: Vol. 29, No.11

# Pilot Project on Pond Cage Farming in a Selected State Needed

There has been a major revolution in the complexion of fish production from Indian waters. A decade back the total annual Indian aqua production was of the order 3.77 million mt consisting of (2.23 million mt from marine capture and 1.54 million mt from inland waters, mostly culture sources). In contrast, while the annual Indian marine capture production in the recent years touched 3.3 million mt and is stagnant at that level, in the inland sector there has been a remarkable rise in the annual aqua production, mostly from tanks and ponds. This has boosted the inland annual aqua production from 1.54 million mt to over 3.7 million mt as at present. This achievement has led to the conclusion that the scope for augmenting Indian aqua production now lies mostly in the Inland aqua farming sector and this can be further widened by popularising inland pond cage system among the farmers. The results of the pioneering experiments on pond cage farming system conducted at Rajiv Gandhi Centre for Aquaculture at Karaikal in the UT of Puducherry, confirms this. Cage farming experiments were also conducted by CMFRI in selected coastal waters of the country with encouraging results. However, the results showed that between the two (Pond cage and Sea cage systems), the development of cage farming in ponds and tanks (or even reservoirs) taken up on priority to start with, or simultaneously with the development of cage farming in the selected coastal waters stretches all along the coastline, will provide a major fillip for augmenting quality aqua production that would yield higher economic returns to the farmers and other stakeholders in the line.

So far as promotion of pond cage farming is concerned, the authorities have to take into special consideration the long time lag in promoting this activity in the country, compared to the widely known progress achieved in this activity in countries like Taiwan, Vietnam, Thailand etc. There have been also experimental successes recently in this regard at the farm of Rajiv Gandhi Centre for Aquaculture in Karaikal in the UT of Puducherry. The pond cage farming work accomplished at this Centre has established that from a pond cage, (reckoned in terms of production from one ha of water area) fish production of the order of 12 t can be obtained with a distinct scope of raising it to 25t annually. Impelled by this achievement, it can be expected that RGCA would initiate a programme on a national scale to extend the technology of pond cage farming all over the country, possibly through the organisational mechanism of FFDAs. A programme of this kind, as a project, can probably be entrusted by the Union Department of Animal Husbandry, Dairying and Fisheries for implementation to Directorates/Commissionarates of Fisheries of the various States, with the technological support of RGCA.

The dimensions of the water area for being covered by the suggested project are extensive. For this reason and the potential of the project and taking into account, the ingredients of the project which are multifaceted, the ICAR/Union Department of Animal Husbandry, Dairying and Fisheries may consider the need to constitute a Working Group to go into the various aspects and formulate guidelines for the preparation of a pilot project on the subject to be taken up in one of the States for follow-up application in other States progressively. In the background that there are 28 States and 7 Union Territories in the country having 626 districts and over 440 Fish Farmers Development Agencies, (with each of them headed by a chief executive officer and each of these also having the District Collector/Deputy Commissioner concerned as the Chairman of the FFDA concerned), the following aspects as outlined hereunder, among others, may have to be suggested to the proposed Working Group for its consideration and suggesting the various aspects of taking up a pilot project in the suggested States.

As already mentioned, the technology of pond cage farming has been developed and it is mainly available in the country with the Rajiv Gandhi Centre for aquaculture at Karaikal in the UT of Puducherry. What is the strategy to be followed for the transmission of the technology among the pond owners, and other stakeholders in the pilot project area? In this context, the following aspects among others, may have to be kept in view.

Drawing up of a training programme for application in the State to the pilot project covering all essential aspects of pond cage farming is suggested, to be conducted at RGCA in such numbers of batches as may be required for a duration of 4 days per batch and the training of around 10 Chief executive officers of FFDAs, per batch at a time. The main objective of RGCA in conducting the training programme being one of developing and demonstrating the technology of cage farming to the trainees, it would have to undertake and accomplish this challenging job, based on the outcome of the discussions on the subject with the Departments of Fisheries of the State concerned, Union Fisheries Division and also the National Fisheries Development Board which may have to play a major role in financing and monitoring this promotional activity through a grant. The training would equip the CEOs to take the farmed pond fish production in the country from the present annual average of 2.5 t/ha to atleast 12 t/ha to start with.

1. The suggestions outlined above in respect of transfer of technology of pond cage farming through training among the CEOs of FFDAs, when implemented,may contribution to increase in fish production and Tri level of earnings of farmers.

   As the trained CEOs of FFDAs return back to their bases after completing the training on varying dates,the commencement of the transfer of technology to the farmers could be undertaken by them on appropriately staggered dates, until all the farmers coming under the FFDA concerned are trained.

2. Simultaneous with the promotion of the training programme, the farmers have to be supported in respect of the needed capital investments for the acquisition of pond cages and their installation in their ponds and also equipment for water quality testing etc. There may be need for providing funds by way of subsidy to the FFDAs in this regard and also towards recurring costs for the acquisition of seed, feed, transportation costs etc. for some time.

3. Under an Agreement, with a financing body, a system of extending financial help to the farmers can be introduced. Under this, the nominated Bank or any other Agency, as identified, may provide the needed funds to the farmers towards the cost of cages and for their setting up in the ponds as needed, under a guarantee that may be given by the National Fisheries Development Board or any other organisation as agreed to, on mutually accepted conditions that would cover the no. of installments of repayments, rate of interest etc. A suitable provision in respect of margin money can also be included.

4. Another alternative in respect of the financial help can be the formation of a consortium of the farmers or groups of FFDAs to identify and negotiate with a reliable cage supplying enterprise of long standing and enter into an Agreement with it for the supply and installation of the cages in the ponds concerned with the undertaking/guarantee as needed and as mutually agreed to. There has to be a supporting system of stocking of the cages with mutually acceptable types of seed of the needed sizes coupled into a system of supply of feed as agreed to; and on marketing of the produce, inclusion of an inbuilt arrangement with the cage supplying enterprise for the recovery of the cost of cages etc., based on a mutually agreed no. of installments annually/half-yearly along with interest. The development of an integrated system of this nature is necessary for the success of the project.

5. The cage supplying enterprise would channelise the cage supplies to the farmers through the FFDAs concerned. This enterprise would install the cages and take part in the stocking of the cages concerned with seed, and in organising feed supplies, so as to ensure aqua production and viable returns therefrom.

6. There can be an arrangement under which, out of the total proceeds, the cage and other inputs supplier would get the proportionate part out of the sale proceeds of fish or shrimp or both towards the cost of the cages and any other supplies towards the instalments of the repayment of loan amount and the interest thereof. All these aspects would however be governed by the finalised provisions in the Agreement.

7. The Agreement can be tripartite in nature, the parties being i) Fish Farmer's Development Agency concerned, ii) Cage Supplier-cum-Financier, of the integrated programme and iii) The guarantor, as mutually agreed to.

A co-ordinated initiative, promotional in nature as outlined above will further develop farmed fish production under the aegis of Fish Farmer's Development Agencies. The average per ha production per annum from fish ponds under FFDAs is now estimated at 2.5 t/ha. Compared to this, under pond cage farming system followed, particularly in Taiwan, the fish production level per ha is placed at 40 t/ha/annum. In comparison, the level of production in pond cage farming achieved at Rajiv Gandhi Centre for Aquaculture is 12 t/ha/annum. In this background, it would be a right move to introduce a pilot project for strengthening FFDAs to take up pond cage farming, probably following the pattern demonstrated by RGCA with such readjustments as needed, and also taking into account those followed in Taiwan. The pattern followed by Team Aqua in Taiwan is to set up cages to cover about half of the surface area of the ponds leaving around 2 ft of water area under the cages too. In the cages, fishes are farmed and in the rest of the area either shrimps or prawns are farmed, yielding a production of 40 t/ha/annum, fetching very high returns. Shrimps/prawns should not be grown along fishes, in cages set up in ponds. The reason is that the gram-negative bacteria released by shrimps/prawns will harm fishes. Fishes release gram-positive bacteria which would not harm either fish or prawns/shrimps.

The adoption of the system would go a long way not only in the strengthening the socio-economic status of farmers and the other stakeholders but would also contribute to the increase in foreign exchange earnings through exports.

Fishing Chimes, Editorial 396: March 2010: Vol. 29, No. 2

# Separate Ministry for Fisheries: Widespread Demand

The Federation of Indian Fisheries Industries and several other Indian fisheries organisations have come out with a relentless demand for the carving out of a separate Union Ministry for Fisheries. The demand aims at imparting the needed drive to bring about the introduction of the much needed schemes for the all round and well covered development of fisheries of the country particularly for the building up of the required level of India's own deep sea fishing fleet. This demand is made to facilitate a sustainably optimal level of utilisation of the fisheries resources of Indian EEZ, for the benefit of the stakeholders and of the people.

The demand stems mainly from the ongoing situation of foreign fishing vessels exploiting the fisheries of the Indian EEZ, in the garb of Indian registry fishing vessels, also known as LoP (Letter of Permission) vessels, a new abbreviation to represent the foreign vessels brought into Indian EEZ, believed to be attributable more to the ingenuity of interested foreign fishing vessels owners, mostly Taiwan, taking advantage of the steep decline in the strength of the truly owned deep sea fishing fleet, from 191 nos. in 1990 to around 40 nos. now. This trend came about mainly because the Government of India has ignored the subject of marine fishing in the Indian EEZ beyond territorial waters to such an extent of bringing it down to a low priority status, despite its importance and the role of fish as i) proteinous food provider to the people, ii) a substantial foreign exchange earner, iii) part of a sector of industrial potential, and iv) a means of upgrading the living conditions of fishers and their socio-economic development.

Most of the nations, even those far smaller than India and having less of fisheries potential than of India have separate fisheries departments/separate fisheries ministries for the purpose of promoting and utilising their fisheries resources beneficially. Vietnam, Malaysia, Indonesia, Taiwan, Myanmar, Sri Lanka etc., are among such countries. So far as India is concerned, a few years back, responding to an agitation, what the Government did was to tag on the subject of 'fisheries' as one of the subjects under the Department of Animal Husbandry and Dairying, in the third position of importance. This addition has elevated fisheries from the status of Division in the Union Department of Agriculture to a third place in the aforesaid Department. This step has not brought about any change in the situation.

The present developmental set-up of Indian fisheries at national level presents a pathetic picture, as an insignificant corner in a Union Department that looks after Animal Husbandry and Dairying too. Considering the vast fisheries resources, both marine and inland, the present set-up of the Fisheries part of the Union Department is such that the technical hands in it, being very low in strength, are unable to cope up with the needs. Currently, there is no Fisheries Development Commissioner in the Department. The administrators thus happen to be in charge are also not fisheries technical experts. The result is that fishing in the EEZ of the country is virtually dominated by foreign fishing vessels.

As early as in its April 1995 issue, *Fishing Chimes* came out with an Editorial pleading for forming a separate Union Department of Fisheries because of the potential and importance of the subject. In the period thereafter, the role of fisheries in human nutrition, socio-economic development of stakeholders concerned, and in various other prime respects has heightened so much, that it is now imperative to form a separate Union Ministry for Fisheries at the Centre to fulfill and assert its legitimate position in the national set up, not only as a nutritious food provider to the people, but also to work towards strengthening its status in the international scenario. Apart from being listed as a country of top fisheries rank in FAO, India is also a prime member in the Indian Ocean Tuna Commission. These aspects entitle it to have a separate Fisheries Ministry/Department.

There has been focus on the utilisation of India's potential for marine fisheries development, which is presently confined largely to the coastal zone and the generation of employment opportunities in it. In contrast, the fisheries set-up and the marine fisheries developmental policies in relation to the EEZ launched by the centre are so weak that India now occupies an insignificant position in respect of its own deep sea fishing fleet in the International scenario. The nation, which is having an EEZ of over 200 n.m all along the coastline of 8,000 km, now has just around 40 deep sea fishing vessels of its own. Its policy in respect of utilisation of the resources beyond territorial waters revolves around permitting foreign fishing vessels to fish in the Indian EEZ, by issuing Letters of Permissions to Indian citizens to bring in such vessels for exploiting the deep sea resources of Indian EEZ mostly for their benefit. India has neither a system of providing financial help to Indian enterprises for introducing their own deep sea fishing vessels nor a plan of action for organising construction of deep sea fishing vessels for Indian enterprises at the various Indian shipyards which are well equipped to undertake construction of such vessels. The present dependence on issuing Letters of Permissions to Indian fishing enterprises to bring in foreign vessels with crew for exploiting the resources mostly for their benefit is unfortunate. It is incomprehensible and also inexplicable that the Government of a major nation like India cannot successfully work towards enabling its enterprises to have their own truly owned vessels.

In this background, it is of utmost expediency for the Government of India to carve out a separate Ministry for Fisheries or atleast a separate unclubbed Union Department of Fisheries instead of continuing to perpetrate the downgraded status of the Indian deep sea fishing sector with all its disadvantages of being at the tail end of the Union Department of Animal Husbandry and Dairying, despite its importance. As a separate Department/ Ministry on the subject of Fisheries, it must be put in charge of a separate Minister so as to invest it with the needed authority to strengthen itself, consistent with its legitimate status, to expand and contribute to the development of the Nation.

At the time of irrationally taking out the subject of deep sea fishing out of the Ministry of Agriculture in middle 1990s, and merging it with the Ministry of Food Processing Industries, an editorial on the need for the creation of a separate Union Fisheries Department in the Union Ministry of Agriculture was published in *Fishing Chimes*, as early as in April 1995, explaining the repercussions of such a merger. The results of the subsequent decisions taken as an aftermath of merging the subject of deep sea fishing into the Ministry of Food Processing Industries and realigning it back with the Union Department of Agriculture, and later with the Union Department of Animal Husbandry and Dairying, have proved to be an even worse development, as is seen now, as a virtual entrustment of deep sea fishing in the Indian EEZ to foreign fishing vessels, in a form that has led to a situation that gives most of the benefits of fisheries of Indian EEZ to foreign vessel owners.

Fishing Chimes, Editorial 397: April 2010: Vol. 30, No. 1

# Gaps

Fisheries Development of India has made rapid strides in the recent years. However, as the dimensions of the developmental coverage increase, any gaps that remain would show up. An effort is made here to list some of these gaps, as could be perceived.

## Gap 1: Utilisation of Distant Water/Deep Sea Fisheries of Indian EEZ by Indian-Owned Vessels

Against the estimated extent of marine fisheries resources of 3.9 million tonnes of Indian EEZ, 3.163 million t are now being exploited annually (2009), the catch level being 88 per cent of the estimated resource. This level of achievement is extraordinarily impressive, considered from the angle of the limitation in the strength of the fleet for the coverage of distant and deep sea waters of the EEZ. The Indian fishing fleet, overall, consists of over 1,80,000 nos. of traditional boats, 44,000 nos. of motorised traditional boats, and 53,684 nos. of small mechanised boats and around 3,000 vessels of 16 m OAL range, all of which operate within a depth range of 200 m. In addition, there are around 40 nos. of larger vessels of Indian ownership of 23-28 m OAL,that can cover deep sea and distant water zone and, over 100 nos. of LoP vessels, mostly of Taiwanese (55 m OAL) ownership, but with some of them of Thai ownership (but of less OAL compared to Taiwanese vessels). In other words, the estimated exploited fisheries of Indian EEZ by truly Indian owned vessels is mostly within a depth of around 200 m and, of this, 3.163 mill t (2009) consistuting 88 per cent of the fishery available is being harvested. In addition, the LoP vessels must be harvesting a good part of the remaining 12 per cent (0.74 million t) beyond 200m depth (may be 300m) and within also, which may include several economically important species such as squids, cuttle fishes, deep sea lobsters, deep sea prawns and various kinds of fish such as perches, breams, tunas etc., Apparently, the possible illegal entry of foreign fishing vessels to fish in the Indian EEZ is related to the exploitation of mainly these and apparently other presently un estimated resources. The quantity which they exploit and take away in their vessels or on the related mother vessels is not known to us.

From the aforesaid indicative background, a few focal thoughts can arise. Of the estimated fisheries resources of Indian EEZ [3.9 million t (or 1.95 t per sqkm)], 3.163 million t (1.6 t per sqkm) are now being exploited by Indian effort (not taking into account the possible quantities exploited by foreign vessels (LoP and others)on an average, and most part of this resource is exploited by small crafts but within territorial waters, the area of which is 1.78 lakh sqkm, constituting about 9 per cent of the total extent of EEZ. Thus, as per the present estimate of the fishery resources of the EEZ, a predominant part of the estimated resources (88 per cent) is being exploited within about an area of 9 per cent of the EEZ (Upto 200-300 km depth). Conceding that the rest of the area of the EEZ (91 per cent)is almost barren (Containing only 0.74 million t of the resources or 0.40 t (400 kg) per sq.km, there can be two conflicting conclusions. One is that the foreign vessels such as of Taiwanese ownership would not be making efforts to gain entry into the Indian EEZ for exploiting these resources unless they are sizeable, probably far more than 0.74 mill t. The other conclusion can be(probably a far fetched one) that the estimate of fishery resource of Indian EEZ may happen to be much more than 3.9 million t. One can also possibly arrive at another uncharitable surmise that the estimate of the resources of the EEZ is put at 3.9 million t, only to have the satisfaction that we are now almost fully exploiting the resources of our EEZ. (with the inclusion of catches additionally exploited by LoP vessels) and therefore there is no need to introduce schemes for promoting the strength of Indian deep sea fishing/distant water fishing fleet, as that would involve purposeless investments for introducing a further number of these vessels. Thus, the prevailing resource estimate of the Indian EEZ, in other words, justifies the Centre's present approach of being distant to the subject of reviving the facility of providing loans to fishery enterprises for acquiring distant water/deep sea fishing vessels. This conclusion however, may not be also in order, looked at from the angle that the alternative chosen by the Government was to enable foreign vessels to fish in Indian EEZ through LoP system by Indian parties, thereby enabling foreign vessels to operate in Indian EEZ enduringly. While this approach of the Government coupled with its move to introduce a Marine Fishing Regulation Act, may have its plus points, one aspect is very clear. There is presently a major gap in the manner of promotion of sustainable exploitation of the distant water/deep sea fishery resources of Indian EEZ. The Federation of Indian Fishery Industries and Coastal State Governments are conscious of this gap and they would no doubt work towards ensuring that the gap is rationally filled up by the Centre for the benefit of the Industry and the Nation.

## Gap 2: Fisheries Development in Inland Saline Lands

There are vast extents of saline lands in several States such as Haryana, Punjab, U.P and Rajasthan, with saline waters over and also underneath them. They are estimated at over one lakh hectares. By organising the development of these areas for export-oriented aquaculture, atleast 500 kg of exportable production per ha can be achieved. Such a development would augment exportable production (by atleast 50,000 t per annum) and contribute to additional foreign exchange earnings of around Rs.1,000 million annually. It would not only bring waste lands (although partly used for salt production) under productive economic use but would also bring about employment opportunities. This suggestion deserves the attention of the Union Department of Animal Husbandry, Dairying and Fisheries, the ICAR (Central Institute of Brackishwater Aquaculture), and of the Governments (Fisheries Departments) of the respective States. Authorities may have to consider filling up this gap.

## Gap 3: Rice-fish Farming

In India, rice-fish farming, despite its feasibility and profitability, is marginally followed. One exception is that the farmers of Arunachal Pradesh are known to follow this system. Probably it is undertaken in a few other States too. The system, however, deserves to be promoted in all the States as it would be beneficial to the farmers. The system is predominantly followed in China and also in several other south-east Asian countries. The Union Department of Animal Husbandry, Dairying and Fisheries may consider promoting this technology in association with the Union Department of Agriculture and exhort the State Fisheries Departments to promote the system in collaboration with the State Agriculture Departments.

## Gap 4: Pond Cage Farming

The nation has achieved remarkable progress in pond fish farming, taking the average per ha production to 2.5 t/ha and to a peak of over 12 t/ha on an average in the case of the output of several progressive farmers. While this is a laudable achievement to the credit of CIFRI, followed up by CIFA later, it was the farmers who picked up and adopted the technology in its integrated form linking seed production with grow-out farming, under the systems of composite fish farming, diversified farming and integrated farming. These efforts have led to a startling increase in inland fish production of the country, taking it to No. 2 position the global scenario, only next to China. The annual inland fish production of India now stands at nearly 3 million tonnes (most of which is farmed fish production), almost at par with annual marine fish production which is at the level of 3.167 million t/annum as in 2009. While the production level of farmed fish is at a level that makes us proud, the global trends are such that Indian farmers would have to be enabled to adopt a more advanced technology. Countries like Taiwan have made such breathtaking advances in the area of pond cage farming, producing under the system 40 t of fish per ha on an average. At an experimental level, India too has no doubt made considerable progress in pond cage farming. At the Rajiv Gandhi Centre for Aquaculture, 12 t/ha in pond cage farming has been achieved in the experimental work done so far. The production will eventually reach higher levels at par with reported Taiwanese achievement. The initial experimental achievement of RGCA is a distinctive contribution for the upgradition of pond farming technology of India. Now that the technology has been established by RGCA and cage material and cage fabrication, and cage installation and maintenance technologies are now available in the country, the stage is now set for the Centre and the State fisheries departments to take the needed follow-up action to empower all pond-owners/farmers to set up cages in their ponds/tanks and take up fish farming in them. Probably there are initiatives already taken in this regard but there is no information there of. We have over 440 Fish Farmers Development Agencies in the country. RGCA can organise training programmes in pond cage farming for the benefit of the Chief Executive Officers and others of the FFDAs. The State authorities concerned can strengthen FFDAs, technologically and financially so as to equip them to organise setting up of cages in the ponds in their area of operation, stocking them with juveniles and growing them to marketable size. It is now essential to initiate follow-up action to disseminate the achievement of RGCA in cage farming all over the country, by introducing a project with nation-wide application.

## Gap 5: Sea Cage Farming

CMFRI has rendered a unique service to the marine fisheries sector by successfully conducting sea cage fish farming on an experimental basis at selected locations along the coastline of the States like Gujarat, Kerala, Tamil Nadu, A.P. and probably others. Sea cage fish farming is a technology of great potential for augmenting fish production. One aspect of this is that it has the potential of progressively compensating the drop in capture marine fish catches. The other one is that it will contribute to the socio-economic development of the coastal fisheries communities. As is known, there has to be a well thought out plan of action so as to impart an enduringly effective spread effect to the activity in the coastal waters. To start with. CMFRI is now poised to extend the technology of cage farming all along the coastline. It may be mentioned here that the ownership of territorial waters is vested in the State Government concerned. Therefore, cages can be set up either by the State itself or by those to whom the State grants lease rights. In this respect, the State Governments can probably follow the US system. In USA, the authorities concerned conduct a survey and identify plots in the sea for setting up cages. After this exercise is done and the plots are demarcated with identity particulars, a public notification calling for applications for allotment of the sites is issued. The application received in response are screened, and the applicants are interviewed for selection. Once the allotment order is issued further action is taken. The coastal State Governments can also probably follow the system of granting leases of identified cages sites to coastal village fishermen's co-operative societies on bare-site basis or on installed cage lease system.

ICAR (CMFRI) has to consider advising the Coastal State Governments concerned to take the needed steps in the direction mentioned above, coupled with initiatives for advising the applicants concerned to take measures to have cages and install them for stocking with juveniles, for providing feed for the stocked fishes, and for organising surveillance to prevent poaching etc., On the part of the State department concerned, it has to take steps to organise the setting up of hatcheries with attached nursery/rearing ponds for producing juveniles of the type of fishes needed, for stocking in cages, and set them up at selected points along the coast as a chain to facilitate supply of juveniles to the cages that would come up eventually. While this gap has now been narrowed down with the experimental success in sea cage farming achieved by CMFRI, follow-up measures have to be taken as indicated in the proceeding paragraph.

## Gap 6: Coastal Marine Fisheries Development Agencies

Many marine fisheries developmental bodies have come up in the country. The process started from the time of British rule itself. One focal aspect that came out clearly all through right from the British days was that fishers were central to marine fisheries development. So much so, fisheries schools and fish curing yards were set up by the British rulers as early as the beginning of the 20th century. There has been over six decades of fisheries development in the country after independence which reflects part upgradation of fishing system from country boats to motorised/mechanised boats. Over all these years, out of 2,80,491 nos. of boats, 1,81,284 boats remained traditional (41 per cent), 44,578 became traditional motorised (15.8 per cent) and 54, 629 (19.13 per cent) have been mechanised. These details indicate that even in over six decades the successive maritime State Governments and the Central Government failed to bring about total transformation. One reason for this is the low average educational background of fishers, both men and women. Alongside the developmental efforts, the State Governments have to promote awareness of developmental needs among the fishers coupled with an effective organisational system for the further promotion of education among coastal fishers coupled with steps for upgradation of professional fishery-related standards, as required at that level. For this purpose, in the same way as Fish Farmers Developmental Agencies and Brackishwater Fish Farmers Development Agencies have been set up, there is an imperative need to set up Coastal Fisheries Development Agencies, district wise or otherwise as needed, for bringing about the needed level of development. Such Agencies, when set up, can play a pivotal role in the developmental process, not only for the development of craft and gear and training of fishermen in undertaking advanced categories of fishing and handling of harvested fish, but also in respect of securing a better price for the fish catches. This will, however, be possible only when the suggested Coastal Fishery Development Agencies would be enabled to set up value addition plants, one or two, in viable pre-determined zones. Instead of selling the catches at the landing point, fishers can take the catches to one of the value addition plants where the management concerned will buy the catches at prices as agreed to earlier for the concerned category of catches. A system of this kind will automatically upgrade the price of the catches at the landing point. The managements of the plants can build up a marketing network for the value-added products covering retail outlets in nearby towns and larger villages. The suggested reform would lead to the further socio-economic upgradation the economic standards of fisher families.

The advantages of setting up Coastal Fishery Development Agencies deserve to be evaluated by the authorities, for follow-up action.

## Gap 7: Fishing Harbours; Infrastructure Facilities

Tuna landings at several fishing harbours along the Indian Coastline are on the increase. Tunas are exported in whole and also in the form of loins and fillets. Prior to export, the products are to be packed. Tuna and tuna products require pre-processing facilities prior to packing. These are not there at most of the fishing harbours. At one or two centres, exporters have set up their own tuna pre-processing facilities. In order to facilitate and speed up the tuna export process, it is important to provide the facility for pre-processing work at the harbours themselves. The authorities concerned have to look into this aspect and see that this gap is covered.

Fishing Chimes, Editorial 398: May 2010: Vol. 30, No. 2

# Professional and Socio-economic Uplift of Marine Fishers

## *Need for Interactive Dialogue System*

Marine fishers, because of their dependence mostly on sea fishing and the related activities, have come to be reckoned as a distinctive category of population. This is obviously because most of the marine fishers live in villages located along the coastline far from the main stream of the society. The Coastal State Governments extend several monetary benefits to the coastal fishermen such as off-season allowance, contribution towards insurance premia, distress relief, subsidy for motorisation of fishing boats etc. In some of the Coastal States, financial assistance to fisherwomen in respect of fish marketing is also provided.

Appraising the marine fishers in respect of the various supportive measures extended by the respective Governments to them is the responsibility of the State Fisheries Departments. While the fisheries departmental officers visit the fishing villages often to appraise the fishers about these, the process has its own limitations for the messages to register on their minds as needed.

This gap has to be filled up and this can be done by adopting one of the folklore systems followed in rural areas, to effectively convey the reformative developmental messages. The systems to be adopted are dramas, *kathas*, dances etc., The spread of the developmental measures and new technologies to be introduced can be achieved effectively through following one or more of the systems of the kind mentioned above.

A unique initiative in this direction has been taken in Karnataka recently with a remarkable impact. Mr. Vasantha Hegde, Assistant Director of Fisheries, Kumta in that State organised a Dance Drama (Yakshagana) aimed at educating fishermen in responsible fisheries activities. This proved to be very effective in transmitting the message across to register on the minds of fishers for them to understand the objectives of responsible fisheries utilisation and also to keep them informed of assistance provided by the State Department of Fisheries to them under different schemes.

The fisheries extension wings in each of the Divisions/Districts of each of the State Fisheries Departments, as a first step, can prepare a script through which they can depict and describe the various developments, past and present, and those in the formative stage. This can be given to a village dance/drama troupe for adoption, while also explaining to them the background, the purpose and the manner in which the theme has to be presented to the fishers that assemble The performance of the trained troupe concerned has to be so modulated that it would have a telling impact on the audience. It would have to be so organised that the message — conveying groups move from one fishing village to another. After an extension/information spreading show of the kind mentioned, it can be so planned that the concerned field level staff of the Department would undertake an action-oriented follow-up work that will lead to the anticipated results. The aforesaid troupes can move from village to village, based on a schedule given to them.

The other aspect to be taken care of by the State Fisheries Departments is in respect of imparting fisheries education among fisherchildren. In most of the fisher villages there are primary schools. What the children need, however, is education and training in respect of fisheries resources, kinds of fishes that occur, their relative abundance, crafts and tackles used for fish harvesting, post harvesting activities, value addition, marketing etc. These aspects, besides the details of the facilities and programmes under which financial and other support is extended by the authorities being part of the on-going fisheries education, can be covered under the Dance and Drama programme mentioned above.

Dance and Drama programmes are not new to rural areas. The system has to be given the needed fisheries orientation for the benefit of fishers.

Fishing Chimes, Editorial 399: June 2010: Vol. 30, No.3

# CIFRI's 64th Foundation Day

## *A deserving occasion to vow to raise Hilsa to the Status of a Farm Fish*

The 64th Foundation Day of CIFRI was celebrated on 17 March 2010 at its Headquarters in Barrackpore, Kolkata.

Being an event of historic significance, it will be in fitness of things to reminisce over the decades-old dedicated attachment of CIFRI to the cause of Hilsa fisheries development, among others. The scientists of the Institute bred Hilsa and raised its seed. As a follow-up of this remarkable achievement, CIFRI has to now take a vow to devote a part of its main attention towards raising the status of Hilsa as a prime farming category of fish at par with others like major carps. This could be done in association with CIFA, which is its own offspring. The tested technology of raising hilsa hatchlings in a hatchery is with CIFRI, although it may have to be probably further refined. The technology of raising hilsa seed to the needed size for stocking in tanks and ponds may have to be also firmly developed. There is field evidence that hilsa seed survives and grows well in freshwater tanks and ponds. The editorial in January 1985 issue of *Fishing Chimes* says that Hilsa culture had been carried out successfully in West Bengal. Some measure of success in breeding of hilsa in tanks was stated to have been achieved during that year in this State by CIFRI. Further inputs in this direction would lead to the development of an integrated technology that would enable the induction of hilsa into the category of a farm fish that can be raised to the size of a brooder, bred and seed raised and grown in ponds and tanks to marketable size. This can certainly be achieved. When this happens, it would be a boon not only to the fish farming class but also to the consumers who prefer to have hilsa dishes often as part of their diet. Hilsa farming can certainly be achieved through the implementation of a planned strategy in that direction.

The Central Inland Fisheries Research Institute of India stands apart as a unique fisheries research body with a rich historic background that has global hues. In the course of its glorious growth, it has given birth to several distinctive achievements, besides those related to hilsa. Over a period of time, these achievements gained in strength and led to a spread effect with encouraging results thereof. One outstanding and unparalleled achievement of CIFRI, as we all know, is the introduction of technology of breeding of fish through injection of pituitary hormone extract into their body. This has revoluntionised and replaced the system of dependence on collection of fish spawn/fry from rivers during flood season for rearing them to a stockable size before releasing them into tanks and ponds. The induced breeding technology has, as is known, ensured the convenient, easy and assured availability of fish seed to the farmers for stocking in their ponds and tanks. The technologies of composite fish farming, polyfarming and integrated farming have been evolved and introduced by the institute and these have generated additional benefits to fish farmers. One of CIFRI's offsprings, the Central Institute of Freshwater Aquaculture continues the work on upgrading the systems. It has improvised a portable hatchery unit and also a mechanised pond fish harvesting system, among others. It has demonstrated effectively in various States of the country the technology of achieving higher fish/prawn production per ha because of which several farmers in various States are now producing fish/prawns at far higher levels per ha compared to the earlier decades. The Institute also demonstrated multi-cropping systems that have facilitated reduction in the expenditure on feeds. Another feather in the cap of CIFA, as the offspring of CIFRI, is that it gifted to the Indian fish farmers Jayanti Rohu a selectively bred quality Rohu that grows longer and weightier by 17 per cent comparative to the original rohu.It also opened up new avenues of breeding and seed production of catfishes. It also gave to the farmers an effective specific called "CIFAX" for controlling/treatment of epizootic ulcerative syndrome among fishes, besides an ELISA-based immunodiagnostic kit to diagnose fish bacterial diseases, and also CIFACURE to treat microbial infection in ornamental fish. The Directorate of Coldwater Fisheries can also be considered as an offspring of CIFRI.

It will be interesting to note that Bangladesh is on the job of pond farming of Hilsa. the country's most delectable foreign exchange earner. Hilsa in ponds is a saviour as its very existence in river conditions is threatened by pollution, climate change and even over- consumption.

Researchers at Chandpur Fisheries Research Institute have taken to farming of hilsa in ponds, seeking to belie a common perception that hilsa fish can survive only in rivers. People in Bangladesh as well as West Bengal in India swear by the taste of the full -of- bones yet melt-in-the mouth hilsa, with its preparation considered a must during festivities. Hilsa eggs, fried and spiced up, are also eaten with relish in the both countries, though some say this has led to over consumption of the fish. Work has been on since 1988 in Bangladesh to breed the Hilsa in ponds, but there is no breakthrough until now. Chief Scientific Officer of the institute, Anisur Rahman hopes to succeed this time, it is reported.

A 1.5-inch long newly born hilsa fish, also called fry, was released into a pond under the supervision of the institute. Three ponds inside the institute have been readied for hilsa farming. A full -scale experimental rearing of hilsa in the ponds began in May 2010. The researchers say that fishermen net 10,000 to 19.000 tonnes of fry a year from these ponds.

CIFRI now concentrates on weight development of riverine and estuarine fisheries, on reservoir fisheries and on wetland fisheries such as those of *beels, mauns* etc. So far as riverine fisheries are concerned, there is some relief now. The inevitable pressure that was there earlier on spawn and fry collection from rivers has almost disappeared because of the introduction of induced breeding technology. CIFRI's attention now has a shift towards evolving measures for a sustained development of capture fisheries of rivers and capture-culture fisheries of reservoirs It has also formulated steps to counteract proliferation of exotic fishes in rivers, one example of which is that of *Clarias gariepinus. Pangasionodon hypophthalmus.*, an exotic riverine catfish have also made its entry, mostly into the rivers of the eastern States of the country. Farmers have brought the seed of these fishes into the country from Bangladesh and they are now being farmed in West Bengal, Andhra Pradesh, and several other States, mostly on the north-east. In fact, it is stated that the Government of India has now accorded clearance for the farming of Pangus (*Pangasionodon hypopthalmus*) in the country.

The dedicated work done by CIFRI in promoting fish production from wetlands is widely known. However, there is a lot more to be done by way of further demonstrating in an enduring manner the setting up of the cages in wetlands and taking steps for developing the needed supportive infrastructure facilities for promoting cage farming in an integrated manner, under an organisational set up to be suggested to the State Governments concerned. Cage fish farming in some wetlands and rivers has already been introduced in a few States like Assam, West Bengal, Karnataka and others but, considering its potential for augmenting fish production and upgrading the incomes of the farmers. CIFRI has to prevail upon the various State Governments to promote the activity under well formulated projects. The State Governments can introduce a system of issuing permits to the farmers for setting up the cages in demarcated plots, supported by a financing system, on a subsidy basis.

Developmental fisheries research work on reservoirs and lakes, as is known, comes under the purview of CIFRI. With the exception of a few stray cases, the status of level of fish production from these resources continues to be low and disorderly. NFDB has taken up a project for lending support to the State Fisheries Departments for stocking of several reservoirs in those States with advanced fingerlings. While this is a good step forward, what needs to be considered is the grouping of the reservoirs into viable clusters. Each of such clusters would need to be provided with the required organisational and infrastructural support, which has to be one of centralised yearling production, (cluster-wise), and supplying them for stocking in the reservoirs within the cluster concerned, keeping the capture-culture balance in view and linked to the system of regulated harvesting of the fish, their storage and marketing, to be organised by the State Fisheries Departments concerned, in association with CIFRI.

Fishing Chimes, Editorial 400: July 2010: Vol. 30, No.4

# Utilisation of Fisheries of Indian EEZ: Policy

## *–Investment Constraints Led to Foreign Vessel (LoP) Induction?*

It is axiomatic that the fisheries resources of the Indian EEZ, estimated at 3.9 million tonnes, will have to be utilised in a sustainable manner for the benefit of the nation. Out of the aforesaid estimated resource potential, the latest known annual status of its exploitation level is close to 3.20 million t (around 88 per cent). The strength of India's present fishing fleet owned by Indian enterprises and capable of fishing in deeper waters beyond territorial zone of EEZ to a certain distance offshore is of two categories, those of over 20 m OAL and above, and of 15-16 m OAL. Those of former category consist of around 40 nos. There are 70 vessels of the other category (of 15-16 m OAL). In other words, the total fleet strength of Indian owned vessels as mentioned above can be reckoned as 100 nos., as some of them (around 10 nos) are non-operational. So far as foreign vessels being operated in the Indian EEZ under the Indian LoP (Letter of Permission) system by Indian enterprises concerned, the latest known strength of these vessels is 92 nos., of 35 m OAL and above, each of which can be deemed to be equivalent to the average capacities of about five presently operated Indian owned larger vessels. The present annual output (of around 3.2 million t) is believed to be taken mostly from the territorial waters by the mechanised and non-mechanised fishing boats, besides the 100 nos. of larger vessels. Here, it has to be mentioned that there can be reservations to concur with this surmise for the reason that there are frequent reports of illegal fishing in the Indian EEZ and the present output of around 3.2 million t mentioned may not include illegal catches. The quantities that are covered by the illegal fishing that may be taking place can be substantial. While the estimate of the fisheries resources of the Indian EEZ (3.9 million t) has a factual basis, many express reservations saying that it is hard to believe that out of 3.9 million t of the estimated resource of Indian EEZ, as much as nearly 3.20 million t are being exploited by the Indian owned fishing fleet, considering that very few of the EEZ's farther sea fishing vessels of real Indian ownership and also of LoP authorisation are now operating beyond the territorial waters of the Indian EEZ. The share of exploited fishery of Indian EEZ by these vessels, big or small, can be estimated only from their fishery exploitation levels. Information concerning this could not be ascertained.

The general position as can be discerned seems to be that the resources of the EEZ are being almost fully exploited. This is a great supporting point for the present approach of the Government of withdrawing investmental support to Indian enterprises for acquiring their own larger vessels. This is highlighted by the winding up of vessel financing facilities (closure of SDFC/SCICI) along with the non-promotion of the needed field set up and the climate in which the enterprises could respond to the personnel of the financing bodies on the job in the field for timely repayments of loans taken by the enterprises concerned. In other words, the setup of the erstwhile financing bodies (SDFC/SCICI) at the catch landing points of the vessels was so weak that it gave opportunities for the loanees to postpone or delay repayments. This weakness helped in proving that repayments of loan instalments were not being made by them regularly. One adverse impact of the withdrawal of the vessel financing system is that the growth of Indian deep sea fishing fleet has been stemmed, paving the way for the ushering in of foreign-owned vessels, with attribution of Indian ownership, under the Letters of Permission (LoP) given by the Indian Government to Indian enterprises, for operating them in Indian EEZ. The vessels would of course continue to be under the possession of foreign owners until such time as the cost of vessels is fully repaid to them and this is not likely to happen (has not happened so far) as long as the vessels continue to be operated by foreign crew and there can be several strategies of the foreign vessel owners for ensuring that the Indian side would not be able to repay the dues fully at any stage. The main aim of the foreign vessel owners, while opting to give their vessels to the Indian LoP holders, can be believed to be one of gaining entry for their vessels into Indian EEZ for fishing. Once this is achieved, their main problem is solved. The foreign vessel owners, particularly Taiwanese, are believed to be familiar with the fishing grounds in the Indian EEZ. In contrast, the Indian enterprises/Indian crew have rarely had an opportunity to gain working knowledge of the grounds beyond territorial waters of their own EEZ. Thus, the Indian enterprises have the handicap of not having the needed knowledge of the grounds and operational aspects on one hand, and they seem to have no source of funding for acquiring vessels of their own other than what the foreign vessel owners concede from voyage to voyage, on the other. There is of course the availability of Indian crew trained at CIFNET and also having fishing experience. However, as there is no addition of truly owned Indian vessels to the Indian fishing fleet, there is no scope for them to secure jobs on larger vessels to undertake fishing in the farther waters of the Indian EEZ. So much so, several of the trained Indian fishing vessel officers leave the country to take up jobs on fishing vessels of other countries.

The various Indian fisheries associations, particularly the Association of Indian Fisheries Industries, the Federation of Indian Fisheries Industries and the South

Indian Federation of Fisheries Societies appealed to the Government to withdraw the policy of allowing the operation of foreign fishing vessels in the Indian EEZ through the device of LoP, but this appeal has not elicited response. In this situation, the Association of Indian Fishery Industries has also filed a writ petition in the High Court of Andhra Pradesh seeking a direction to authorities concerned for the stoppage of the operation of LoP vessels in the Indian EEZ. Acting on this, the High Court of Andhra Pradesh gave a direction to the authorities to consider the related representation of the AIFI dated 1/4/2010 and pass an appropriate order for the stoppage of operation of LoP vessels in accordance with law on or before 20/7/2010 and communicate the same to the petitioners (AIFI).

What is now needed is a dedicated approach on the part of the Government/authorities concerned to introduce a well formulated policy to promote fishing in the farther waters of Indian EEZ by truly Indian owned vessels, for the exploitation of the hitherto sparsely (hopefully) utilised resources of the zone, such as squids and cuttle fishes, which have high export value, and with the needed linkages upto their marketing stage. The columnar zone of Indian EEZ is known to be having rich resources of squids and cuttle fishes but India does not have vessels that can exploit these resources and it is not known whether LoP vessels report the catch particulars of these species. The authorities have also to reverse their present perceptions in respect of exploitation of the fisheries resources of the Indian EEZ beyond territorial water through LoP vessel operation system and in their place entertain progressive ideas to promote India's own larger fishing vessel fleet, keeping in view the vessel building infrastructure that India already has and the opportunities of utilising this for building up the fleet. Once the Government demonstrates that it will promote utilisation of the fisheries resources of the farther waters of Indian EEZ also by Indian enterprises for national benefit, there will be the emergence of follow-up activity leading to the dawning of prosperity upon Indian stakeholders dependent on the optimally sustainable utilisation of the nation's marine fisheries sector beyond territorial zone.

Fishing Chimes, Editorial 401: August 2010: Vol. 30, No.5

# First Time Ever: A Woman Fisheries Scientist Generates History

## *Dr. Meena Kumari becomes Chief of ICAR's Fisheries Set-up*

An unprecedented event of a historic hue took place in ICAR in the last week of June 2010. This was : Dr. B. Meena Kumari, Director, Central Institute of Fisheries Technology, had been elevated by the Council to the Top Fisheries Position as Deputy Director General (Fisheries) in its set-up. This outstanding aspect highlights the meritorious and development - oriented fisheries research contributions of Meena Kumari, primarily in respect of post-harvest fisheries operations, marine as well as inland. Further, her name now stands enshrined in the history of fisheries research and development of India, as the first woman fisheries scientist to have reached the pinnacle of recognition as the top-most fisheries research scientist of the country. One can also construe that her achievements tally with the import of her name, which connotes fish. *Fishing Chimes* believes that Meena Kumari was christened with that name by her parents, guided by a premonition, that one day she would be at the top of the fisheries research set-up of India. What is sought to be conveyed is that the first part of her name 'Meena' means fish, the first incarnation of Lord Vishnu in Hindu mythology.

Traditional marine post - harvest sector has been the main domain of Indian Coastal fisherwomen. Supply of fish to the people for consumption is a noble task, as fish keeps them healthy and coastal fisherwomen perform this as their traditional role. In the pre-harvest stage, fisherwomen engage themselves in the fabrication of fishing nets, and in the post-harvest stage they undertake marketing of fish in fresh as well as processed forms (sun-dried/salt cured). Yet, in the top tiers of the national fisheries set up, we do not come across an extension of the related or allied roles at a higher level by any women fisheries experts. Until as at present, we come across only a few names of women fisheries scientists who could reach an applaudable status in the Indian fisheries developmental scenario. Of these, we have with us as top seniors, to the extent known, Dr. Sarojini Pillai and Dr. T. Rajya Lakshmi. They held top fisheries positions and are now leading a retired life. So far as those in the on-going top fisheries positions are concerned, one is Dr. B. Meena Kumari, who has now reached, as already highlighted, the topmost fisheries research position in ICAR. There are also quite a few other distinguished women fisheries scientists in high positions, one of whom is Dr. S. Girija, Director, NIFPHATT. It may be mentioned here that, so far as the State sector is concerned, the status of women scientists in top fisheries developmental positions can be considered to be in an improving mode. Dr. Madhumita Mukherjee is the Director of Fisheries, West Bengal, and Ms. Ramalakshmi is the Director of Fisheries, Puducherry.

Dr. Meena Kumari reached the top leadership status in fisheries research line of ICAR, following the route of hard work and achievements. Having entered ARS in 1977, she secured thereafter a Ph.D. degree from the University of Kerala, conferred in 1989 for her thesis on the topic 'Ecology of Fouling', supported alongside by the bestowal of Young Scientist Award, instituted by the Academy of Environmental Biology, Lucknow. In the same year, she also bagged Punjabrao Deshmukh Women Agricultural Scientist Award, 2002 of ICAR, for her outstanding contributions to the fisheries field. Another laudable aspect is that, besides several fellowships, she became the proud possessor of six other prestigious Awards. The result of all these achievements is that, from the position of a scientist in 1977 at CIFT, she rose to the status of the Head of Division (Fisheries Technology) at the same Institute by the year 2007; and, in Nov 2008 she earned the Directorship of CIFT. To the admiration of well-wishers and others, as earlier mentioned, she rose to the status of Deputy Director General (Fisheries), the top fisheries research position in ICAR in June 2010, in the place of Dr.S. Ayyappan who was elevated to the position of the Director- General of ICAR.

All through her career, besides undergoing several professional training courses on various field - oriented fisheries topics, she accumulated experiences in the implementation of as many as 20 fisheries research projects that covered not only marine coastal and deep sea fishing gear, but also 20 projects that included those on reservoir fishery exploitation. She worked on the upgradation of traditional fishing gears, lobsters traps and also on crafts such as those made of rubber wood, on aluminium canoes and on several others with good results. She implemented around 18 externally funded projects too. Further, it was to her credit that she made significant research contributions in respect of remote sensing of marine fisheries resources and also on time - series analysis of research data on Indian oil sardine, in relation to physical, chemical and oceanographic parameters of Kerala waters, which have the longest record of variations in these.

She developed and popularised a collapsible lobster trap made of ms rod and welded mesh, coated with plastic.

This had helped fishermen to carry with them a good number of Traps in one and the same trip for augmenting lobster catches. She also developed and popularised collapsible fish and crab traps for operation in Inland waters, including estuarine waters. Introductions of a large meshed gill net 'or operation in Lakshadweep waters and also a large meshed purse seine and ring seine for operation in the waters off the south - west coast are to her credit. On the Inland fisheries side, her accomplishments cover introduction of framed gill nets, vertical line gill nets, and monolines for catching demersal predatory fishes, besides vertical drop lines, and a trammel net. She also has to her credit aspects such as identification of small or compact aquatic systems by remote sensing for efficient management, and geo-informatics aspects for locating fish shoals.

Her achievements are many more, and under her leadership, those in the fisheries research line in ICAR will have the opportunities of upgrading their skills and achievements. There is a well entertained hope among marine fisheries stake holders that under her leadership, the fisheries research set-up under ICAR will take special initiatives to upgrade the traditional role of coastal fisherwomen in respect of value - added fish processing by way of organising the setting up of well designed fish value-addition units at identified coastal centres along India's coastal line. The expectations are that the technologies of value addition to fishes, would be extended, by way of organising the setting up of one pilot value - addition plant at a selected coastal centre, to start with, to be followed up by their extension, all along the coastline, supported by well designed training programmes in the line, for the benefit of fisherwomen The system may have to be that the plants would be operated by the trained fisherwomen, with needed linkages to cover purchase of fresh fish from landing centres by the fisherwomen and plant staff for processing and packing them into value-added produce and for marketing them in various towns, cities and other marketing points through a chain of outlets to be identified.

Fishing Chimes, Editorial 402: September 2010: Vol. 30, No.6

# Fisheries of Indian EEZ Beyond 12 n.m from Coastline

## *Needs and Utilisation*

The fisheries resources of India EEZ are estimated at 3.9 million mt. Of these, it is stated that 3.3 mill mt, are now being exploited. Bulk of to exploited fisheries from Indian EEZ comes from the territorial zone of the EEZ, which extends upto a distance of 12 nm from the coastline. In this Zone, fishing is conducted by Indian-owned vessels of small and medium lengths and to some extent additionally by around 40 nos. of Indian-owned vessels of over 20 m OAL, which also exploit the waters, although marginally, beyond 12 nm width from the Coast. It is generally believed that, while the fisheries of the territorial waters are fully exploited by Indian - owned fishing fleet, the utilisation of fisheries beyond the territorial waters is mostly done by the LoP vessels, (Indian registered but mostly of Taiwanese/Thai ownership/possession), supplemented by the operations of the Indian owned larger fishing vessels, as already mentioned.

The fisheries resources position of the Indian EEZ is that only 0.6 million t (6 lakh t) are now available for exploitation, that too in the farther waters of the EEZ, that are of course, known to be under exploitation by foreign - owned but 'Indian registered' vessels.

Traditionally, until the recent times, Indian fishers are not known to be venturing into the sea to conduct fishing beyond coastal waters. In this background, the first ever initiative was taken by the Indian Government to introduce larger fishing vessels to exploit the fisheries resources of Indian seas beyond the territorial waters in around late 1960s and early 1970s by permitting certain larger houses (Union Carbide, EID Parry, Britannia, ITC, New India, Spencers, Tatas etc.,) and later Konkan, Universal Sea Foods and Fisheries Corporations (AP, Tamil Nadu, Kerala, Gujarat). After operating their vessels for some years, all these companies and corporations wound up operations of their fishing vessels. At the time of doing so, most of them gave the impression that they were winding up the operation for the reason that the SDFC was not extending loan facility to them for expanding their fleets. Thereafter, in stages, nearly 70 other companies, wound up their larger fishing vessel operations, in some way or the other (by selling their vessels to foreign companies or other Indian enterprises). In some cases the vessels were seized and auctioned by the financing bodies concerned. In the case of those who sold away the vessels on their own, it is believed by many that, utilising these amounts and probably the saved surplus amount earned out of the earlier operations, in the main, some of them are believed to have diversified into other sectors, mainly real estate. There was no intervention from the authorities concerned to study the problem and solve it to stem the trend of phasing out of the marine fishing sector. The situation of stated non-recovery of loan instalments from the owners was believed to be the result of the financing bodies not having a well planned and an effective set up at the landing points (harbours) so as to have a close watch over the quantum of catches, unloaded after each of the voyages and sold to exporters/processors/others, so as to check the actual level of returns and assess the reasonableness of the returns. This lacuna may have had the effect of poor recoveries towards the loans given. Because of this and probably other problems not taken care of, it was believed that there were problems of shortfalls in repayments and the effect of this was that the larger vessel fleet strength dwindled from around 190 nos. in 1991 to the present level of around 40 nos. At the same time, there have been additions to the fleet at the initiative of the entrepreneurs who raised funds to introduce trawlers (mini-trawlers and others) in the size range of 12m -15m.

Notwithstanding the reason conjectured above, the plummeting of the fleet strength of larger vessels from 190nos to around 40nos is mainly attributable to the entrepreneurs/industry. There were alleged diversions of earnings from fishing vessels towards other investments like real estate by several companies as mentioned above (other than large houses, who wanted to have loans for strengthening their fleets, but could not raise them and invest for this purpose). Because of problems of this kind, the Indian marine fishing industry has now come to the present deplorable stage of the authorities allowing foreign vessels to operate in Indian EEZ under the LoP system. The Government must have come to the conclusion that there is no escape from this in the scenario that came to be evolved because of the dwindling of the strength of Indian-owned fishing fleet on the one hand and the lack of means for providing funds for investing on new vessel acquisition. While the reason for the dwindling of the truly Indian-owned fishing fleet is attributable to the observed follies of fishing enterprises and of their associations, it would have been an appropriate step on the part of the Government to evolve measures that would set right the situation. Instead of introducing a deliberate system of providing access to foreign fishing vessels to gain a fishing sway over the Indian EEZ. The LoP system brought in by the Government can be a well-reasoned justification, put forth deliberately to cover up, if one can say, the indiscrete decision to allow foreign vessels to virtually gain fishing access into our EEZ.

The introduction of the LoP system by the Government would tantamount to a deliberate and a planned one, (A revival of Special Import License system to bring in foreign fishing vessels that was there a decade ago), although it may not actually be. Motivated or otherwise, the erstwhile SDFC, which was providing loans for the acquisition of fishing vessels of over 20m OAL. was closed down, although there was evidence that the level of recoveries of loans given by SDFC was considered by many as satisfactory, in comparison to several other sectors. In any case, considering the nature of the subject and its potential to increase fish production, provide employment to the people and to bring in foreign exchange, it can be said that there was no justification for closing down SDFC. Instead of closing it. Government could have taken alternative steps of a positive nature, instead of entrusting the job to a banker who has no background of the subject. Their inexperience in the line became clear later, from the way they dealt with the loanee fishing companies. Compared to professional fisheries bankers in certain other countries, particularly in Australia, the manner in which the concerned Indian Bank functioned was non - professional so far as the fisheries sector was concerned. Many expressed the view that the erstwhile SDFC performed fairly well and most of their loanees were rated as good in respect of repayments. A remark that was heard in respect of closing down of SDFC at that time was that it was done to eventually introduce the LoP system, which may or may not have political motivation. Several Indian ship yards (Mazagaon Dock, Goa Shipyard, Bharati Shipyard, and several others) constructed a good number of larger fishing vessels whose performance was highly rated even by foreign shipyards and foreign fishing companies. Instead of further promoting and encouraging the Indian Shipyards to strengthen themselves in the line and serve the fishing companies by constructing and supplying fishing vessels to them, the authorities had totally discouraged the yards, thereby blocking the avenue of Indian fishing entrepreneurs to have Indian built larger fishing vessels.

Taking the aforesaid aspects into account from all angles, at least now, the Indian Government should take purposeful steps to a) revive larger fishing vessel financing system and b) organise the induction and re entry of Indian shipyards into the line of construction of fishing vessels, based on shop floor drawings of the kind of vessels required for operation in the Indian EEZ and/or in open waters beyond. In this context, it has to be agreed that the National Fisheries Development Board is the most appropriate and a naturally befitting body to be entrusted with the job of providing financial assistance to fishing enterprises for acquiring fishing vessels, mostly on the lines of erstwhile SDFC, which was extending loans for the acquisition of fishing vessels by fishing companies, supported by linkages in respect of monitoring of fishing operations and having an effective field level control at landing points so as to ensure effective recovery of loans given, voyage wise. NFDB can set up a separate full-fledged wing having the structural features of a fishing vessel financing body, such as those most successfully run in Australia or in USA. The weakest point in SDFC/SCICI/ICICI system, as already explained, is the ineffective appraisal of the harvested quantity of fishes/crustaceans etc in terms of quantity and quality, and ensuring maximum possible returns, out of which a predetermined amount towards repayment of loan can be adjusted from the buyer concerned, who would of course be one out of the buyers approved by the financing body. So far as funds are concerned, it should certainly be possible for securing approval for a project prepared for the purpose either from Asian Development Bank, or World Bank or another reputed financing body, supported by a guarantee from the Government of India. One of the financing bodies can certainly be expected to provide funds to the proposed financing wing of NFDB to deal with the work. The scope of extending financial assistance by the suggested financing wing of NFDB could no doubt cover other developmental subjects such as cage farming, setting up of hatcheries, etc.

Not providing specific financing facilities to the Indian entrepreneurship for acquiring larger fishing vessels to fish in Indian EEZ would appear to be an inappropriate policy, whatever be the faults on the part of related Indian entrepreneurship. The procedure to be followed in respect of submitting applications for loan, their scrutiny and criteria for sanctioning of loans can no doubt be in the nature of revival of the system adopted by the erstwhile SDFC, with the needed readjustments, taking into account the procedure in USA and/or Australia or in such other countries. Aspects such as the margin money, rate of interest, no. of instalments etc could certainly be arrived at, taking in account various related aspects.

There can be a plan to introduce around 200 larger fishing vessels to be built at various approved Indian shipyards, by providing the needed incentives to them and by formation of consortia of shipyards or by other means. In case of necessity, imports from countries like Australia, Holland etc. could also be considered on a contract basis for a supply of good number of vessels so as to get a price discount by linking the supply with the making available of designs and shop floor drawings, coupled with a simultaneous programme of building such vessels at Indian shipyards. In any case, the ignominy or permitting foreign vessels to operate in Indian EEZ under cover of devices like LoP would have to be erased by the Government in the shortest possible time.

Fishing Chimes, Editorial 403: October 2010: Vol. 30, No.7

# Pond Fishery Development Scenario in FFDAs

## *Need for Review*

Tanks and ponds in the country, having a total extent of over 22 lakh hectares constitute a major resource for the production of fish, prawn and shrimp both for domestic consumption as well as for exports. By 1950s tanks and ponds emerged as recognised potential sources, mainly for fish production to start with, and later for diversification into production of prawn and shrimp. By late 1970s, the strengthening of this recognition has led to the formation of an organisational mechanism for the promotion of tank/pond farming, that is linked at one end for seed and feed supplies, and at the other end for the supply of various other inputs that the farmers need. Pond fishery management, harvesting and marketing of the produce too are also covered by the mechanism. A project on the setting up of this organisational mechanism, christened as 'Fish Farmers Development Agency (FFDA)' was formulated [by your Chief Editor when he was Deputy Commissioner (Fisheries) in the then Union Department of Agriculture], for the clearance of the Government. The said project, having been cleared by the said Union Department, was placed before the Central Board of Fisheries. Approved by the Board, it was later cleared by the Planning Commission. The process of setting up FFDAs in the various States and UTs of India gained momentum thereafter. The first batch of FFDAs were set up in several States and UTs of the country in the year 1974-75, followed by successive batches of them in the other States/UTs in the subsequent years. The number of FFDAs increased to 50 by 1977-78. In this year, the annual average per ha production level was estimated at 50 kg/ha. As years went by, the no. of FFDAs went up to a level of 379 by 1991-92 with the annual per ha production moving up to 1970 kg. By middle 1990s, while the average national annual fish production from FFDAs moved up to 2500 kg, in Punjab it crossed 4 t/ha and in Haryana, West Bengal and A. P. 3 t/ha. As at present, several farmers (in A.P. and in quite a few other States) are known to be producing, on an average over 7 tonnes per ha, although some of them are outside FFDA.

The remarkable performance of the farmers under FFDAs is attributable to the technical, financial, training and coordinating support extended by the FFDAs, under the leadership of their Chief Executive Officers concerned. The owners of the ponds and tanks, coming under the purview of the FFDA concerned receive support in getting fisheries leases of Government/local body tanks and ponds, training, technological advice, financial assistance by way of loans from commercial banks and subsidies from the Government. The farmers also get support in securing the inputs needed in the fishery management of ponds and also for their application in pond preparation prior to stocking. Support is also extended for securing supply of seed for stocking as needed and for obtaining feed supplies and organising daily feeding of fishes, periodic water management and testing of growth of stocked seed etc., leading finally to harvesting of the crop and its marketing. The FFDAs strive to update the farmers in respect of various farming systems followed by them, such as mono-farming, composite farming, and integrated farming.

The aqua farming developmental pattern through FFDAa has acquired an established status over the past 30 years. While some of the innovations in farming technologies that took place during this period have been absorbed from time to time, a clear picture of the present status of the application of the technologies and the impact thereof awaits to be assessed. The range of production level, be they of fish or shrimps or giant prawns, or of crabs, that could be achieved by FFDAs (through their constituent farmers) have to be made known to all stakeholders who are widely dispersed, to serve as a guideline for them to strengthen their efforts at augmenting aqua-production from their tanks and ponds. These efforts can be diverse, covering various phases of farming (pond preparation, stocking with fingerlings/yearlings, phased harvesting and restocking with yearlings, and integration with marketing). In respect of all these, there have been progressive developments in relation to FFDA monitored tanks and ponds and also among those under the purview of fisheries co-operatives and Associations for the promotion of integrated aquafarming activities of an upgraded nature. The outcome of these efforts has been an increase in aqua-production, consisting of not only fishes, but also crustaceans and the resultant stepping up of supplies to domestic markets and also for exports.

The present status of pond/tank fishery developmental activities from an organisational view point, as outlined above, indicates the imperative need to conduct a state-wise review of the status of the impact of the work being done by FFDAs and other organisations such as farmers associations, Farmers Co-operatives and others; and in that light, the contours of the organisational mechanism with upgraded objectives, aimed at achieving maximally sustainable results have to be formulated. This will be possible only for the Union Department of Animal Husbandry, Dairying and Fisheries, by way of constituting a high level Committee of identified experts in the line, with a well-ingrained scientific and technological bent of mind. This Committee could take into account the achievements by bodies other than FFDAs too who have a record of experience in integrated commercially-oriented aqua farming on sustainable lines, and these consisting of collective farming systems that take care of integration of the various components of the activity ranging from arranging pond fishery leases/arranging finance for setting up new ponds/renovating existing ponds, providing training, organising an ensuring supply of various inputs and their application, promotion and management of crops through application of the developed technologies that are conducive for sustainable crop production and its marketing, by adopting a well formulated system, linked to the needed infrastructural facilities and market conditions. Now that the introduction of Pungas (*Pangasianodon hypopthalamus*) into India has been cleared by the authorities, the Committee could also extend advice on the manner in which the incorporation of this species in the various FFDAs could be undertaken.

Systems such as contract fish farming and collective fish farming have already come into vogue outside FFDAs. These are believed to have stood the test of time. The Committee suggested for being set up in the foregoing write-up, could perhaps look into aspects of this kind in an appropriate form for being incorporated in the activities of FFDAs. Persons with qualifications/experiences in fish culture could be utilised for promoting pond/tank fishery system based on a cluster concept that takes care of supplies and services linked to domestic/export marketing. M/s. Oceanaa of Chennai are understood to have achieved a breakthrough in this respect. So is the case in respect of a few farmers' groups in the Central districts of Coastal A.P., through mutual agreement. NFDB, in association with the State Fisheries Departments concerned can introduce a system of registration of enterprises of farmers interested in setting up such clusters and lend support to such enterprises to undertake the job of bringing together a certain number of stake holders in a compact area to undertake pond/tank aquaculture (fish, shrimp, prawn etc, mono/composite/integrated) and lend support to them by way of providing needed supplies and services that would include linkages with marketing avenues, both domestic and export.

Once the State Fisheries Departments, in association with NFDB develop a restructured system of promotion of integrated aquaculture system, there will ensue an upgraded style of functioning of FFDAs under which cluster aqua-farming of viable dimensions could be promoted for enabling the stake holding farmers to generate a sustainable level of aqua produce that would significantly take forward the farmed aqua production levels and incomes thereof, through organised domestic marketing and exports of fish, shrimp, prawn and crab produced in the country.

Fishing Chimes, Editorial 404: November 2010: Vol. 30, No.8

# SPF Shrimp Broodstock Development in India

## *No Take-off as yet, despite having Expertise and Facilities*
## *Union AHD&F Dept's Silence, as could be perceived, is baffling*

India is one among the top shrimp producing and exporting nations. India has a well established research and shrimp production management base, capable of promoting a sustainable and disease-free farmed shrimp production. Yet, the internationally renowned Indian fishery scientists have conspicuously failed in the production of SPF broodstock of any of the commercially important shrimps. As a result, they have downgraded India's standing in this line of endeavour by necessitating the compulsive import not only of SPF broodstock of tiger shrimp but also of Pacific White (*Penaeus vannamei*), now in the pripeline. So far as Indian white shrimp (*Penaeus indicus*) production is concerned, unable to produce SPF broodstock of this shrimp, an easier way of importing SPF broodstock of vannamei shrimp has been resorted to, thereby subjecting the nation to the ignominy of relegating Indian white shrimp (*Penaeus indicus*) to the background, in contrast to the successful production of this shrimp of SPF quality by countries like Saudi Arabia, Equador and a few others. The net Indian position is that the nation, not having expertise, now imports SPF broodstock of tiger shrimp and also Vannamei shrimp, while virtually neglecting Indian white shrimp. This glaring situation depicts the alarmingly downgraded status of India in respect of technological abilities at producing SPF shrimp broodstock, to a position far lower than countries such as Saudi Arabia. Could any of the fisheries scientific organisations in India, relate its achievements in the SPF broodstock production of any commercially important shrimp with pride? CMFRI can probably cite its success in the production of SPF tiger shrimp seed up to $F_3$ stage but no further. It had to stop at the $F_3$ stage of SPF tiger shrimp seed production. What is the reason for this? It was irrationally asked to stop the work by the ICAR. This happened a few years back. Had it continued the work, probably by now the nation would have had SPF tiger shrimp brooders. Probably reconciling itself to the neglect of production of SPF tiger shrimp in the country, NFDB has done the next best that is possible, by importing SPF broodstock of tiger shrimp for their seed production at a hatchery set up in North A.P. at a place known as Sompeta, for providing SPF shrimp seed to Indian farmers. Although two years have passed, there is no report of success in raising of seed from the SPF brooders imported from Hawaii. It is learnt that Indian representatives from NFDB and CIBA visited Hawaii in USA where the company which supplied the SPF tiger shrimp brooders stocked at Sompeta farm is located but no other information thereof is available.

Disease invasion among shrimp under farming in India has brought down its production level, thereby imparting a negative impact on marine products exports. This development has alerted the authorities and the stakeholders including the exporters to find a solution to the debacle. The easiest of this, as finalised, was to import SPF shrimp broodstock.

As is known, the bulk of shrimps exported from India consist of black tiger which is one of its indigenous species. Far lower quantities of Indian white shrimps too are exported but these are mostly of capture origin. The progeny of imported Pacific white shrimp (*Penaeus vannamei*) farmed by a couple of Indian enterprises (who were permitted to import their broodstock/seed) also became part of the export basket, although in a very small quantity.

In the South Asian region, but not in the south-east Asian region, India is in the forefront in respect of development and application of technologies of shrimp production in brackishwater/freshwater ponds. Scientists of CMFRI. some years back, succeeded in inducing the Indian white shrimp to release eggs through eyestalk ablation. They were later raised to the stage of seed suitable for stocking. However, shortly after this, for some reason or the other, CMFRI lost its interest in promoting the activity of inducing Indian white shrimp to breed and in raising its seed. One reason for this could be that the work was no longer in the purview of CMFRI, as CIBA came into existence by then.

On the same logic as mentioned above, the work connected with the development of SPF broodstock of tiger shrimp should have come into the hands of CIBA after it came into existence. However, as it happened, CMFRI continued to deal with the work which progressed up to the development of $F_3$ SPF larval stage. At this point, the work was stopped, only to be further carried out by CIFE and CIBA, in association with a Norwegian enterprise. While there is no information on the present status of the work, there were observations that not much of progress had been made. One guess is that, had CMFRI been allowed to continue the work after $F_3$ larval stage too, probably there would have been some cognisable progress by now.

Another focal initiative for raising SPF black tiger broodstock was taken by MPEDA. It had started a project in Andamans for the production of domesticated shrimp for raising SPF shrimp seed, a few years back There is. however, no information on the outcome of the project.

In the background as outlined in the preceding paragraphs, the immediate need is to make available SPF tiger shrimp seed to the farmers. In the present situation, the authorities, obviously, had no alternative other than depending on an outside source for securing shrimp brooders and organising the supplies of the seed raised therefrom. The National Fisheries Development Board took the initiative of entering into an understanding with a foreign company of repute for supply of SPF tiger shrimp broodstock to be utilised for producing the seed at a hatchery in the country and for supplying the same to the farmers. This chosen hatchery is located, as already mentioned, in north Andhra Pradesh at Sompeta from where the supplies of seed produced would be channelised to reach the farmers. While there is no information on the results of subjecting the imported tiger shrimp for breeding and raising seed therefrom, in all likelihood, there will be a plan of action to raise brooders out of the adults produced in grow-out ponds (utilising SPF seed) and utilise them as the next generation brooders, while taking steps to prevent inbreeding. Now that Vannamei SPF broodstock is being permitted for import by identified enterprises, it is to be believed that an arrangement for the production of successive batches of SPF brooders of these shrimps will also be an integral part of the plan of action, either at a centralised hatchery farm or at several identified hatchery farms.

The present situation in respect of the Indian white shrimp (*Penaeus indicus*) is anamolous. While this shrimp is widely recognised as an exportable shrimp in various countries including India too, the difference in the Indian context is that it is relegated to a lower status, with little interest evinced in producing SPF broodstock of this shrimp. In contrast, in countries like Saudi Arabia/ Equador and a few others, SPF seed production of the species is part of their commercial shrimp production and export activities. In this background, it is considered by many that the promotion of interest in the development of SPF broodstock of Indian white shrimp is of utmost importance and this thinking deserves to be given the needed attention.

Let it be hoped that the authorities concerned would work towards the utilisation of the scientific talent in the country for the development of expertise in the production of SPF shrimp broodstocks supported by needed facilities, and trained hands, for the production of recurring batches of broodstock of the three shrimps *i.e.* Tiger, Pacific White and Indian White. It is possible that, after stabilising the seed production and supply system from the hatchery set by it in North A.P., NFDB. would replicate the model at other chosen centres along the Indian coast. Looked at from an overall angle, the perceived silence of the Union Department of Animal Husbandry, Dairying and Fisheries on the alarming Indian failure in the production of SPF shrimp brooders/seed is baffling, if not intriguing.

## A New Dimension to Tiger Shrimp Culture in Australia

There have been interesting reports from Australia in the middle of 2010 on monodon production of 17.5 t, on an average, per hectare in one crop. A press release from Australia stated that the maximum production of monodon, observed at certain Australian shrimp farms was 24.2 tonnes/hectare. However, with the generally declining yields of *P.monodon* at several Australian shrimp farms, the trend on the part of farmers and entrepreneurs of Australia too is to look at *P.vannamei* for augmenting production. This has been countered to a significant extent, because of higher monodon yields, of late, at certain farms particularly of Food Futures Flagship.

The high production was because of a broodstock of tiger shrimp which was developed as a result of a ten year research by the Food Futures Flagship of the CSIRO. The report had further stated. 'CSIRO' is an abbereviation of the Commonwealth Scientific and Industrial Research Organisation. The CSIRO started the Food Futures Flagship programme. This is headed by Dr. Nigel Preston whose doctoral studies were based on the survival of larvae of Penaeid shrimp, mostly of monodon.

The above mentioned press release is reproduced hereunder, with explanatory clarifications here and there. "An Australian company says its specially bred prawn (shrimp) has the potential to "revolutionise" the local and international shrimp aquaculture industry".

After, eight generations of selective breeding Australias Commonwealth Scientific and Industrial Research Organizations (CSIRO) scientists and the country's aquaculture industry have bred a black tiger prawn (shrimp) that is producing record yields.

One of CSIRO's partners. Gold Coast Marine Aquaculture, this year produced average yields of 17.5 metric tons per hectare, more than double of the industry average. Several ponds produced 20 metric tons per hectare and one of them produced a record yield of 24.2 metric tons per hectare.

In the past two years, the shrimp has also won five gold medals at the Sydney Royal Easter Show, including the highest award possible, "Champion of the Show." With about 50 percent of all prawns (shrimps) sold in Australia currently imported, developing an Australian prawn (shrimp) that breeds in captivity and is completely sustainable is a major gain for both the local prawn (shrimp) industry and consumers seeking Australian seafood, said CSIRO.

If the rest of the Australian black tiger prawn (shrimp) industry adopted the new breeding technology, Australia's annual shrimp production could increase from 5,000 metric tons to 12,500 metric tons, adding AUD 120 million (USD 105 million or EUR 85 million) annually to the industry's value by 2020.

"The new prawn's (shrimp's) yield has exceeded all our expectations. The average industry productivity for farmed prawns (shrimps) is only five tons per hectare, so this year's average yield of 17.5 tonnes per hectare is a major leap forward," said Dr. Nigel Preston, leader of the CSIRO Food Futures Flagship prawn research project. "These huge yields can be replicated year after year, which means consistent supply of a reliable and high quality product — all vital factors for the long-term growth and prosperity of the Australian prawn (shrimp) farming industry".

According to a clarification given, the project has sourced broodstock locally from the Gulf of Carpentaria and its scientists have also utilised the now common biotechnological techniques of genetic markers to prevent inbreeding among close relatives. *Fishing Chimes* got in touch with Dr. Nigel for further information on the average growth of the shrimp and on factors like salinity tolerance, F.C.R. and others. Although Dr. Nigel declined to give specific information thereof citing commercial compulsions, some information could be gathered from CSIRO's paper on the topic published in the "The Rising Tide, Proceedings of the Special Session on Sustainable Shrimp Farming, World Aquaculture 2009".

The publication states that CSIRO had achieved an average individual farmed tiger shrimp growth rate of 63 g in 12 months on an average with a first generation-raised tiger shrimp females in 1997. Later on, there has been a steady growth and they observed an average individual farmed tiger shrimp growth of 128.6 g in 2002. Further, in 2006, they had achieved a farmed individual tiger shrimp growth of 134.1 g on an average, in 12 months.

Regarding salinity tolerance and the stocking densities, Dr. Nigel stated through an email that the normal stocking densities in Australia were at 35 - 40 nos./sq.m. and the salinity tolerance was in the range of 15- 38 ppt. However, Nigel was confident that the salinity tolerance could be extended further.

## Comparison

An inevitable comparison with the Indian programme being developed by NFDB in association with MOANA TECHNOLOGIES will be in order here. Although not much information is available regarding the programme of NFDB under implementation at Sompeta in Srikakulam Dist. of Andhra Pradesh, there is one chief difference noted between the two. The programme in Australia is based on locally sourced broodstock, while that at Sompeta farm in Srikakulam district in Andhra Pradesh is based on broodstock sourced from outside India. Further, the details on when the programme will become mainstream supplying the seed required, reported to be 7 billion, to satisfy the vast indigenous requirement, is still not known. The desirability of a team from NFDB to visit the CSIRO programme to understand its feasibility and for its application in India and also to assess the possibility of a partnership to put the indigenous programme on the fast track deserves to be studied for follow up action, as considered necessary.

## The Future

The future of the shrimp sector of India depends on a lot of variables that need to be addressed at. Of late, the choice has been between Vannamei and Monodon. Monodon has lost its market share to Vannamei world over, owing to the later having the advantage of availability of SPF broodstock and a ready source of supply of Vannamei seeds produced therefrom to the market. However, there are certain aspects to be taken care of in respect of Vannamei. One is that the SPF seed will stay SPF only when the related bio-security protocols are adhered to, and the needed extent of publicity is given thereof. Secondly, competing against China in the production and supply of Vannamei to the export market needs a considerable concentrated effort. Countries like Philippines have found it more profitable to stay with the domestic Vannamei market than exporting them. However, there are no big players, as at present, in the line.

One major advantage of tiger shrimp is that it has a traditional export market and many of the Indian farmers are into its farming. The main problem that the owners of hatcheries have been ventilating all along in this context is the non-availability of disease-free broodstock. Therefore, programme of Monodon broodstock development is accelerated for the farmers to benefit a lot, while also taking up the promotion of Vannamei's SPF broodstock development along side. The funding initiative for a partnership of such a course of action mainly rests with NFDB. The enterprises concerned/their associations have to take up the issue with the NFDB in a persuasive and focal manner so as to achieve commercially viable field operations.

The imperative need and importance of SPF shrimp broodstock production in the country and utilising the same 'producing quality shrimp seed for supply to the farmers is known to all concerned. As a follow-up to this axiomatic necessity there have been efforts mounted by the related developmental organisations in the country including research institutes and the Deemed Fisheries University (CIFE). Several years have passed but there is no breakthrough as yet. In contrast to this situation, several countries such as USA., Equador, Saudi Arabia and Australia have succeeded in raising SPF shrimp brooders and in raising their seed.

India has global reputation of having scientific and technological talent in the field of Fisheries, and this certainly is inclusive of abilities to produce SPF broodstock shrimp and raising their seed therefrom. It is believed that this observation will not be denied. In this background, it is baffling that Indian fisheries scientists and technologists could not accomplish a breakthrough in this aspect until now, as is known.

Taking into account the contents of the attempted foregoing narration, readers would no doubt reflect on the subject. (The clarification shrimp within brackets after 'Prawn' in added at *Fishing Chimes*).

Fishing Chimes, Editorial 405: December 2010: Vol. 30, No.9

# On Taking the Fisheries Industry Onboard

## ....*for Ratification of ILO C188*

The International Labour Organisation (ILO), in association with the Ministries of Labour (MoL), and Agriculture (Union DAHD&F) of the Government of India conducted a Consultative Workshop at Kochi. It was held on 31st August and 1st September, 2010 for the purpose of deciding on the ratification of ILO-Convention 188. Representatives of various trade unions, international (ICSF) and national NGOs and those of State and Central Departments of Fisheries, fisheries research organisations, and DG shipping, (apart from the Ministry of Labour, Ministry of Agriculture and ILO) participated in the event.

The virtual absence of the stakeholders/ representatives, from the fisheries Industry, despite invitation, was conspicuous at the Workshop. The fact that the organisers had invited the representatives of the fisheries stakeholders but they failed to attend the Workshop was an indication that something was wrong somewhere. In their absence, as the Workshop was in progress, a feeling of an orchestrated one-sided attempt to support the ratification of C-188 was in the air. Dr. K. Vijayakumaran. Director-General, Fishery Survey of India articulated a note of caution that it was not appropriate to push forward C-188 for ratification, without eliciting the views of stakeholders by holding consultations with industry, preferably at an important fishing hub. (fortunately it is stated that the Joint Secretary, Ministry of Labour had agreed with this suggestion). The absence of press and media representatives in the meeting or lack of any communication to the print and visual media created an impression that the organisers of the meeting were trying to push something important through a recommendation *i.e.*, notification of ILO C188 without giving scope for a wider consultation.

It may be noted here that, at an earlier meeting, an attempt to push forward a shipping regulation on fishing vessels did not succeed because of a total disagreement from the fishing industry.

In general, the philosophy of providing better working conditions to the fish workers of the world conveyed by the C-188 deserves to be welcomed. No doubt, India would be one with the idea, but it has to, at the same time, work towards improving the living and working standards of its fish workers to upgrade them to the world standards progressively. In this respect, however, India needs to address reform in its own way, taking due consideration of the characteristic features of the Nation's fishing industry. The mode of translating the philosophy into a pragmatic action plan has to be evolved in the national context, with the participation of the major stakeholders (the vessel owners) and the State and Central Departments of Fisheries.

Overall, the C-188 depicts an idealised situation. Looked at from this perspective, so far as India is concerned, this document gives the impression of having been drafted based on a vague idea about the production systems and production relations prevailing in its marine fishing Industry. (On this aspect, a suggestion was made to the officials of the Ministry of Labour and those of ILO to undertake a field study at important fishing centres in the country. On a positive note, the ILO consultant at the meeting expressed his desire to visit the local fishing harbours, it is learnt).

The implications of the various clauses of C-188 have to be discussed in the socio-cultural scenario of the Indian fisheries sector. For example: If the age of entry of a person into fishing will be 14 years as proposed and approved, it can lead to the collapse of many family enterprises in traditional sector. If we go by a decent working time of eight hours for fishing, as provided, even a sustainable utilisation of the resource may not be possible, thereby causing socio-economic hardship to the fishers. Another converse aspect is that, whatever be the fishing duration stipulated, the catches, particularly the trash fishes stacked on board may get spoiled and stink, impacting on the health of fishers and on the economics of operations. The terminology of describing fishing as a hazardous job in the document and the inherent ambiguity of the provision with respect to subsistence and commercial fishing therein needs clarification.

The organisers reiterated the total flexibility of the various clauses and stipulations in the ILO C188. At the same time there was an air of urgency in pushing the document towards ratification. The major concern is about the mechanism of implementing the provisions therein and of their compliance. One of the members present at the workshop mentioned a case where the initial provision of voluntary reporting of catches had come to be mandatory in the course of time, forcing the signatories for strict compliance and reporting.

The need for improving the living standards of our fishers is, no doubt, very important and the Government would be concerned about it. The process of achieving the transition to a higher standard, however, warrants the cooperation of all concerned. The country can upgrade its fishing effort in a phased and sustainable manner without dependence on dictations from an external agency by evolving a well planned policy strategy. India can evolve its own mechanisms and instruments under existing legal framework with a bottom-up approach, while seeking support from friendly countries and organisations.

The idea of top-down approach in law making is being widely criticised at various levels and the need for bottom-up approach is being advocated in its place all over the world. However, the process of pushing C-188 from an international body with the involvement of some vested interests, by totally ignoring the aspirations, concerns and capabilities of the primary stakeholders is quite undesirable. ILO has apparently come out with the document probably not with many buyers from the third world. It is likely that India stands approached now for signing the document, the provisions in which will be useful for various agencies, some of which are probably acting as sales agents.

A concern about the increasing complexity of fishery environment, aptly expressed by the NFF chairperson, was quite valid. The international agencies have made the system more complex by their intervention. Several dimensions have been added to fisheries sector to make it more and more complex. The import of jargons of occupational safety, child labour, human rights etc in fisheries context is likely to add further dime sions. India may be one among the few countries where policy discussions are held in a language alien to its people. While the sophisticated expressions of the expert consultants create a make-believe climate of heavenly developments, the livelihood issues of fisherfolks get lost in the maze.

It can be said that there is an incredible national tendency of a general nature to ratify conventions sometimes without properly understanding the long term implications and often failing to create the necessary infrastructure and institutions to secure compliance. The C-188, in its present form, may be quite flexible and appear harmless. However, in the long-run, if we are accused of non-compliance, then we will be inviting trouble. The need for creating better environment in the fishing industry must certainly be our concern. However, it need not necessarily be dictated by any external agency.

Ministry of Labour and the Trade Union leaders (who have no real stake in the fishing Industry) will be happy to see that the C-188 is ratified at the earliest. The onus of meeting the obligations will, however, squarely fall on the shoulders of the Union Department of Animal Husbandry, Dairying and Fisheries and the State Departments of Fisheries. The aforesaid Union Department is already facing a lot of problems in implementing the regulatory instruments in the fishing Industry. State Fisheries Departments are struggling with the implementation of MFRAs. While the Union Department of Animal Husbandry, Dairying and Fisheries and the State Departments have to focus attention on resolving various current issues, the ratification of C-188 will further add to the complexity of the regulatory environment and increase their burden of implementing the provisions in the instruments.

The Industry cannot shy away from their responsibility of adjusting itself and of evolving in consonance with the emerging global scenario. Instead of avoiding the process of consultation and blaming the Ministry of Agriculture or Ministry of Labour at a later date, the Fisheries Industries should actively engage themselves in the discussions and dialogues and in that light agree to adopt the possible measures in course of time. They must ensure that what is not feasible for them is not imposed on them either by the Indian authorities or by the International agencies. They should oppose the moves that would make their own existence miserable.

One outcome of the concerns expressed by the Director-General, Fishery Survey of India at the aforesaid workshop, was that the Union Ministry of Agriculture had suggested to the Union Minister of Law to constitute a task force with Dr. K. Vijayakumaran, Director - General, Fishery Survey of India, as its Chairman to express its views. It is understood that MoL had agreed to the constitution of the said Task Force but excluding the Director-General, Fishery Survey of India. The proceedings of the Workshop communicated by the MoL did not take into account the concerns and views expressed by Fisheries Survey of India at the Workshop.

Fishing Chimes, Editorial 406: Jan and Feb 2011: Vol. 31, No.10 &11

# Coastal Regulation Zone Notification

The Coastal Regulation Zone Notification 2010, issued by the Ministry of Environment and Forests (MoEF), Government of India, attracted critical responses from various sections of the society. This notification describes the few provisions that are proposed to be incorporated in the existing Coastal Regulation Zone Notification, 1991. The current process to revise the CRZ 1991 notification is the result of the report of the Expert Committee appointed by the Ministry of Environment and Forests under the Chairmanship of Dr.M.S.Swaminathan, to go into the objections against the Coastal Management Zone (CMZ). The Expert Committee had recommended amendments to the CRZ notification 1991 with a view to (i) strengthening coastal protection and (ii) to fortifying provisions for housing of fishermen and for their livelihood. MoEF organised public consultations in all Coastal States and the Union Territory of Pondicherry. The draft, however, has reportedly ignored most of the views expressed by the fishing community and environmentalists. It has failed to stick to the mandate given by the Swaminathan Committee and has used it simply as an opportunity to further tamper with the CRZ Notification 1991 and to accommodate other interests.

The objective of the Coastal Regulation Zone Notification 2010, as declared by the Ministry, was to ensure livelihood security to the fisher communities and to other local communities living in the coastal areas, for the conservation and protection of coastal stretches, its unique environment, and its marine area and to promote development in a sustainable manner based on scientific principles, taking into account the dangers of natural hazards in the coastal areas and in sea level rise due to global warming.

The Coastal Regulation Zone Notification 2010 affected about 4.0 million sea-going fishermen in the country living in 3,202 marine fishing villages, who depend directly on the sea for their livelihood. The Ministry had earlier issued the Coastal Regulation Zone (CRZ) Notification 1991 under the Environment (Protection) Act, 1986. Basically this CRZ Notification 1991, which is still in force, seeks to protect and regulate the use of the land within 500mts of the coast and 100mts along the tide influenced water bodies. All developmental activities, proposed to be located in this zone, are regulated under this Notification. It classifies the coastal stretch of the country into CRZ-I (ecologically sensitive areas), CRZ-II (built up of municipal areas), and CRZ-III (rural areas). In the draft CRZ Notification 2010, in addition to the classification? CRZ-IV (Islands of Lakshadweep and Andaman and Nicobar) includes the water areas upto the territorial waters and the tide influenced water bodies. A separate draft Island Protection Zone Notification has been issued for protection of the islands of A&N and Lakshadweep under EPA, 1986.

As per the Notification, between 0-200mts there is a No Development Zone where repairs and reconstruction of housing of local communities are provided. Between 200-500mts, construction of houses of local communities and tourism projects, including green field airport at Navi Mumbai, is permissible.

Ministry of Environment and Forests had invited public comments on the pre-draft CRZ 2010. Subsequently, after due processing of comments and further legal procedure, the draft CRZ Notification. 2010 was issued under Environment (Protection) Act, 1986. inviting suggestions and objections from the people within sixty days from the date of issue of the draft notification. Thereafter, the draft CRZ Notification, 2010 shall be finalised and notified in Official Gazette. Public comments to the draft legislation, once finalised, would be further taken on record before the new legislation is notified.

The draft CMZ Notification, 2008, attracted a large number of representations especially from the fisherfolk and the local communities. Only activities that require the waterfront and foreshore facilities should be permitted in the CRZ Unfortunately, the draft continues to legitimise all the activities which do not have such a justification and have crept into the notification over the years. These include "non-polluting" industries in Special Economic Zones. Nuclear power plants, power generation by non-conventional energy sources, "green field airport" in Navi Mumbai. storage of petroleum products, fertilisers and chemicals, large-scale housing projects, generous exemptions to tourism projects, etc. It is worth noting that some of these activities, including Nuclear Power Plants and the Navi Mumbai airport, are permitted in CRZ-I areas that supposedly enjoy the highest level of protection in the CRZ notification.

A cumulative impact study of ports has been imended by the Swaminathan Committee, and till me a moratorium has to be implemented on new ports. No such study has however been conducted and raft has a peculiar formulation to permit ports in "stable coasts" that are not subject to erosion. Given that ports themselves have contributed to the erosion of many stable coasts, this measure will only mean the eventual spoiling of the coastline. The cumulative impacts of many other activities including foreshore facilities for thermal power plants (which pump in enormous quantities of sea water for their water requirements) also need to be studied.

Draft CRZ Notification 2010 contains provision of certain "special considerations" to Greater Mumbai, Kerala and Goa. This threatens the very foundation of the CRZ regime. Instead of the CRZ being an All-India regulation, based on common rules, this idea opens the door for special favours to individual States or Areas. A serious look into the special considerations shows that, with the exception of the favour shown to the builder lobby in Greater Mumbai, there is nothing very special about the so-called considerations. Another danger posed by the section on special considerations is that concessions in housing are provided for "local residents" or "coastal communities". This may mean the eventual ouster of fishing communities from the coast by others with better means.

The provision to revise Coastal Zone Management plans every five years gives the bureaucracy the scope to tamper with the zones and accommodate new interests from time to time. The provision for a fresh classification of the coast into zones that may provide an opportunity conveniently reclassify CRZ-I areas also as CRZ-II or III and so on is objectionable. The island territories (Andaman and Nicobar, Lakshadweep) are excluded from the purview of the CRZ with the intention of creating a separate Island Protection Zone (IPZ) notification for them. The IPZ is nothing but the CMZ for the islands and it will signal the de-regulation of the islands from the point of view coastal regulations.

Fishermen are provided no relief in the CRZ 2010 Notification with regard to housing. Its many fishermen are still treated as illegal occupants of their traditional homeland. Draft notification has completely failed the fishing community. Despite talk of recognition of fishing community rights, the notification has not done justice to fishing community housing, social and cultural needs All it does is to allow additional housing in the 200-500 m zone CRZ-III, ignoring the fact that many fishing villages are entirely within the 200 m no-development zone in some of states.

Fish workers across the country have called for a mass protest against the draft Coastal Regulation Zone (CRZ) 2010, demanding withdrawal of the draft notification West Bengal, Orissa and Andhra Pradesh wings of the National Fish workers Forum (NFF) have warned of a nationwide struggle, if the Ministry of Environment and Forests went ahead with the current Coastal Regulation Zone pre-draft notification 2010.

Spearheading the protest, the National Fish Workers Forum has also issued a list of demands, including a substantial representation from the fishing communities in the Coastal Zone Management Authority, at both the State and National levels. They have also demanded that the recognition of the rights of the traditional fishing communities be theirs, calling for the rights of fishing communities to repair, reconstruct and develop their housing in tune with the natural population growth. They spoke on their keen interest in the CRZ areas, including in urban zones.

In the past 20 years, the Ministry of Environment and Forests has made several attempts to widen the arm pit of the CRZ rules for the protection of coastal belt. There was an attempt to form Ocean Regulation Zone (ORZ) similar to CRZ. Later, CRZ and ORZ were combined to incorporate coastal pachayats and municipal areas under the rule and it was called Coastal Management Zone (CMZ) Notification, 2008. When objections were strengthened against this, the Ministry had dropped this and decided to strengthen CRZ.

The Coastal Regulation Zone Notification, 1991, has been amended about 25 times after taking into consideration the requests made by the various State Governments, Central Ministries, NGOs etc. In addition, there are several office orders issued by the Ministry of Environment and Forests, clarifying certain provisions. Such an act of Subordinate Legislation by executives is going unhindered for the past two decades. That is happening by ignoring the Indian Parliament and the Legislative Assemblies of the Coastal States It is high time that a halt has to be made to such irregular acts by beaurocrats and a law has to be made by an Act of the Parliament. For that purpose, a bill has to be moved in the Parliament and considered by the elected representatives of the people.

CRZ Notification of 2010 has generated several apprehensions. This has to be discussed in the Parliament and in the State Legislative Assemblies. The biodiversity of the Indian Sea Coast and associated innumerable water bodies and livelihood concerns of the fisher community cannot be allowed to be tampered by vested interests either in the form of beaurocrats or a business lobby in India Elected representatives of the people have to rise to the occasion and act with will and conviction, holding high the long term interests of the nation.

***Source of Inputs : Prof. D.D. Nambudiri***

Fishing Chimes, Editorial 407: March 2011: Vol. 31, No.12

# Paddy-cum-Fish Farming

## *Pond Cage Farming and Sea Cage Farming*

The relative quantum of contribution of annual fish production from Indian fisheries resources until a few years back had been on the higher side from the marine sector at a level of around 3.6 million tonnes. Subsequently, there has however been a trend of decline in the annual Indian marine fish production because of drop in the strength of India's own fishing fleet capable of operating beyond its territorial waters. In comparison, there has been an increase in India's fish production from inland water resources, mostly in the farming sector, besides reservoirs etc. So much so, the present position is that, atleast for some time to come, the increase in Indian fish production will largely depend on the progress in fish farming sector, inland and marine. (The marine farming sector is now poised to come up in the form of sea cage farming). As days go by, there can be an increase in the capture fish production in the marine sector, although the present trend of capture fishing in Indian waters beyond territarial water by truly Indian owned vessels is not encouraging.

So far as increase in fish production in the Indian inland fish farming sector is concerned, the outlook is distinctively indicative of new avenues of development. These relate mainly to cage and other forms of farming in wetlands and in respect of development of pond cage farming, as is being remarkably pursued in Taiwan, Vietnam, Thailand etc. Another line of Indian fish roduction in India is to promote fish farming in agriculture/irrigation water resources. There are vast areas that are available for the purpose and promotional work in them in the direction of augmenting fish production can be developed by inland fisheries research centres, in association with agricultural production centres by way of inducting such persons who come up with interest and aptitude to take to this line. A promising beginning has already been achieved in this direction, one in Arunachal Pradesh for paddy-cum fish farming and another for the promotion of Paddy-fish farming in Ballia district of U.P. in canal - irrigated areas. Taking these as examples of success, interested State Fisheries Departments can promote development of paddy-fish farming on similar lines, under a co-ordinated project to be launched by the Union Department of Animal Husbandry, Dairying and Fisheries to promote the above said categories of farm fishery development, which will have an enormous potential of augmenting the incomes of fish farmers coming under this line of promotional endeavour.

Reproduced hereunder are extracts of reports on development of integrated Paddy- Fish Farming in inland –Irrigated Area of Ballia district in U.P, authored by P.N. Singh (Head), S.K.Singh, P.K.Singh and A.K.Gupta of Department of Agriculture Chemistry and Soil Sciences and A.K. Gupta of Department of Zoology. S.M.M. Town P.G College, Ballia- 277001. (Extract taken from Vol.29 No. 2 of *Fishing Chimes*, May 2009).

Experts say that integrated paddy - fish farming is one of the most rational methods of utilisation of waterlogged areas of agricultural lands for fish production too. In areas where paddy fields retain water for 3-8 months during the cropping year, the strategy of dual cropping can provide an additional production of fish too. In this context, an extensive survey of Doharighat Sahayak Pariyojana (DSP) canal command areas in Ballia district in U.P. was conducted by the authors on the second line of development mentioned above from 01.04.2002 to 31.12.2003. Experiments were also conducted in the aforesaid duration to explore the possibility paddyland-based fish farming, in areas where farmers face the problem of water logging, seepage, salinisation and rise of water table. Indian major carp species, Rohu, (*Labeo rohita*) Catla, (*Catla catla*) and common carp, (*Cyprinus carpio*) and silver carp (*Hypothalmicthys molitrix*) were farmed along with salt tolerant paddy varieties, USAR-1 and Mahsuri. The selected paddy varieties were grown adjacent to canal sides at all amenable points. After 120 days of farming, weights of rohu, catla, silver carp and grass carp, present in the ratio of 8:15:18:19, were noted. The cultivation of USAR-1 paddy variety along with fish farming was assessed to be encouraging in this region because of the promising paddy yield at 3.69t/ha-1 in comparison to Mahsuri variety at 3.251/ha-1. The district of Ballia is situated between 250 to 26° N latitude and 830–12 E to 840- 40 E longitude in the eastern- most corner of Uttar Pradesh. It is located in middle Gangetic plain region (zone IV) of the Agro - climatic region of India. In order to meet the irrigation requirement of the district, Doharighat canal in 1955-56 and Doharighat Sahaik Pariyojana (DSP) canal in 1981 were excavated in Ballia district. For the last forty five years, continuous use of canal waters for irrigation has greatly affected the physical and chemical properties of the arable land of command area. Soils of canal command areas have been severely affected with salinity, alkalinity and sodicity, due to prolonged seepage, submergence and waterlogging condition. Obviously paddy is the main choice crop for this region due to its unique adaptability under varied unfavourable environments. Keeping this in view, an experiment was conducted to find out the suitability of DSP command area for paddy - fish farming. Integrated paddy and fish farming is recognised as a sustained form of aquaculture with agriculture and significant quantity of the world's farm- produced fish comes through this kind of farming (Mathias *et al.,* 1994). Ghosh (1992) reported that more than 137 countries in the world had been practising paddy-fish farming since late 1970s. In India, paddy-cum-fish farming was advocated by Hora (1951), Iyengar (1953 and1962), Tripathi (1963), Sinhababu *et al.* (1983, 1991, 1992), Sinhababu (1993), Sinhababu and Sarkar (1998) and Prabhudeva *et al.* (2004). Paddy-fish farming is practised on a commercial basis in a few countries, *viz.,* China, Bangladesh, Indonesia, Korea, Malaysia, Philippines. Thailand, Vietnam etc., where fish farming is reported to have been integrated with almost all important plant crops *i.e.,* paddy, wheat, oil seed, horticulture and animal components like dairy, poultry, duckery and piggery (Ayyappan *et al.,* 2004). A lot of research has been carried out on different aspects of paddy - fish integration system (Khoo and Tan, 1980: Likangmin, 1988, Lightfoot *et al.*, 1992: Huazhu. 1994 and Sinhababu, 2001). However, no information is available on paddy-based fish farming in canal command areas whose soils have become now problematic The yield of crops (paddy and fish) depends on the soil type and its quality, climatic conditions and irrigation facility. Earlier experiments on fish farming in paddy fields of West Bengal had shown that the fishes showed rapid growth in paddy fields than in ponds (Jhingran. 1982) and the survival rate was high.

Tage Moda, Director of Fisheries, Arunachal Pradesh gave an account of promoting paddy- cum-fish culture in Arunachal Pradesh: An extract of profile (related to Paddy- cum - Fish culture) of this published in Vol.29 No. 10 of January 2010 issue of *Fishing Chimes* is reproduced here under:

## Paddy-cum-Fish Culture

This Scheme has taken roots and has gained popularity, particularly in the Apatani plateau of Lower Subansiri district of Arunachal Pradesh. Generally, under this scheme, Common carps are raised in paddy fields during a period of three months in a year, thereby achieving fish production of 300 kg/ha/3 months with little effort and without application of supplementary feed. Realising the dual benefits derived from this scheme, people of middle belts as well as those in some high altitude areas are gradually opting for paddy-cum-fish farming system so as to raise a dual crop *viz.,* paddy and table fish simultaneously by application of the related technology imparted by the department. As of now, the programme is being popularised throughout the State to achieve maximal production levels of fish and paddy.

Taking into account the problems in augmenting fish production in the marine sector of the country, the authorities should consider concentrating on increasing fish production from reservoirs and wetlands in inland sector through introduction of cage farming in them on one side and also by promoting pond cage farming system on the other, following the technology adopted in Taiwan, Vietnam, Thailand etc. So far as pond cage farming is concerned, the promotion of this activity can probably be entrusted to Fish Farmers Development Agencies by the State Government concerned, based on a set of guidelines to be formulated by the Union Department of Animal Husbandry, Dairying and Fisheries.

## Sea-Cage Farming in Marine Coastal Waters

CMFRI has succeeded in the production of farmed fish in sea cages at several points along of the Coastal Waters. Lot of further work has to be probably done by the Union Department of Animal Husbandry, Dairying and Fisheries and ICAR in association with Coastal State Fisheries Departments to identify sites for setting up Cages, to work out a programme of cage acquisition from an identified source, evolve a mechanism of pooling up needed financial source for providing Cages to be allotted to farmers by equipping them with financial support, organising a system of advanced seed production and for supplying the same to the farmers etc. There can be many issues to be looked at and sorted out, particularly in respect of determining site dimension and allotting them to the prospective farmers. Once the system of sea-cage farming is standardised, there can be a major contribution to fish production from sea cages.